Microbiology

a clinical approach

[redacted]

Microbiology
a clinical approach

Anthony Strelkauskas
Jennifer Strelkauskas • Danielle Moszyk-Strelkauskas

Garland Science
Taylor & Francis Group

Garland Science

Vice President: Denise Schanck
Editor: Elizabeth Owen
Assistant Editor: Sarah Holland
Senior Media Editor: Michael Morales
Assistant Media Editor: Monica Toledo
Production Editors: Georgina Lucas and Simon Hill
Design and Typesetting: Matthew McClements, Blink Studio Ltd.
Illustrator: Matthew McClements, Blink Studio Ltd.
Copyeditor: Bruce Goatly
Proofreader: Susan Wood
Indexer: Liza Furnival

Front cover image shows the spirochete *Treponema pallidum* corkscrewing into tissue. Image from Science Source/ Science Photo Library.

ISBN 978-0-815-36514-3

Library of Congress Cataloging-in-Publication Data

Strelkauskas, Anthony J., 1944-

Microbiology : a clinical approach / Anthony Strelkauskas, Jennifer Strelkauskas, Danielle Moszyk-Strelkauskas.

p. ; cm.

Includes index.

ISBN 978-0-8153-6514-3

1. Medical microbiology--Textbooks. I. Strelkauskas, Jennifer. II. Moszyk-Strelkauskas, Danielle. III. Title.

[DNLM: 1. Communicable Diseases--microbiology. 2. Infection--microbiology. 3. Microbiological Phenomena. QW 700 S915m 2010]

QR46.S87 2010

616.9'041--dc22

2009041852

Published by Garland Science, Taylor & Francis Group, LLC, an informa business,

270 Madison Avenue, New York NY 10016, USA, and
2 Park Square, Milton Park, Abingdon, OX14 4RN, UK.

Printed in the United States of America

15 14 13 12 11 10 9 8 7 6 5 4 3 2 1

Visit our web site at **http://www.garlandscience.com**

Mixed Sources
Product group from well-managed forests, controlled sources and recycled wood or fibre
www.fsc.org Cert no. SW-COC-002985
© 1996 Forest Stewardship Council
FSC

Preface

Microbiology affects our lives every day. We are surrounded by microorganisms—they live in our bodies, on our bodies, in the air we breathe, in the ground we walk on, in the water we drink, and in the food we eat. Health care professionals see evidence of this every day. Sometimes, the presence of microorganisms is beneficial or even essential to good heath; but in many cases microorganisms can be harmful or even deadly. By understanding the fundamental principles of microbiology, health care professionals can help manage the problems caused by these organisms.

The study of microbiology began in the late nineteenth century with the investigation of the causes of disease. Since then there has been a vast increase in knowledge about the structure, physiology, and genetics of microbes, and microbiology now encompasses environmental microbiology, insect control, and biotechnology. To do justice to the whole of microbiology would take a book much longer than this one. Therefore, our focus is on the relationships between microbes and infectious disease, since this is the area most relevant to human health. This book is the result of many years of teaching microbiology to students interested in the health care professions, in particular the nursing field, and it is specifically designed for this audience. Thus, from start to finish, the book provides a clinical approach to microbiology through its emphasis on the roles microbes play in disease and infection.

The book contains several unique features that make it an effective tool for teaching microbiology to a clinically oriented audience. The text begins with a special Learning Skills section, written by Dr. Peter Susan, which introduces students to practical strategies for improving reading comprehension and retention of information. Clinical themes are interwoven throughout the individual chapters, and the book as a whole, by emphasis and repetition of key clinical concepts. In particular, we emphasize what we call the big five, which are the five requirements for infection: namely, "get in," "stay in," "defeat the host defenses," "damage the host," and "be transmissible." These requirements are introduced in the first chapter and revisited throughout the text. The first chapter also presents a series of case studies designed to grab the readers' attention and demonstrate the vital importance of studying microbiology.

All chapters start by addressing the two questions most often asked by students: "Why do I need to know this?" and "What do I need to know?" Short summaries of individual sections of each chapter, called Keep in Mind, are interspersed throughout, in order to help students digest important facts and concepts before moving on to the next section.

There are three types of end-of-chapter questions. Self Evaluation and Chapter Confidence questions test basic comprehension. Depth of Understanding questions require the student to integrate important concepts in a more challenging way.

And Clinical Corner questions ask students to apply what they have just learned to a specific clinical setting or problem. Answers and explanations for why they are correct are available on the Student Resource Website.

In addition to these pedagogical features, chapters are filled with carefully selected images to facilitate understanding. Photographs and micrographs have been selected for both their beauty and their ability to illustrate important concepts. Glimpses of the wonder of the microscopic and molecular worlds are also found in the MicroMovies which were developed to make some of the challenging topics more accessible to students. Icons in the margin of the book indicate when a movie that corresponds to a particular section or figure is available.

This book is a communal effort; many people have helped us and they are credited in the Acknowledgments that follow. We would like to thank you for choosing our book. It is important to us that the book is accurate, authoritative, and interesting. Despite our best efforts, it is inevitable that there will be errors. We encourage readers who find errors to email us at science@garland.com in order that we may continue to refine the text in subsequent printings and editions. Your input is very important to us and we thank you in advance for your comments and questions.

Tony Strelkauskas

Jennifer Strelkauskas

Danielle Moszyk-Strelkauskas

Acknowledgments

The inspiration for this book came from the hundreds of students we have known, studied with, and taught. We owe them a great deal, and hope this book will benefit future students who want to enter the health care professions. The inspiration provided by the students came to life through the efforts of many people, and we would like to acknowledge their contributions. We thank Garland Science, our publisher, for believing in the project. Special thanks go to Matt McClements, whose incredible artwork and design made this book easy to read and understand. Bruce Goatly copy-edited the manuscript and we are grateful for his editing skills and command of the English language. Mike Morales and Monica Toledo edited and produced the animations and multimedia. Thanks also go to Sarah Holland for her patience with authors and her universal good cheer, and to Georgina Lucas and Simon Hill for all their help in production of the book. We are grateful to Sherry Fuller-Espie (Cabrini College, Pennsylvania) for writing the Question Bank, and Joanne Dobbins (Bellarmine University, Kentucky) for reviewing all the questions. We have saved the best for last: the person who held us all together, who encouraged, cajoled, pushed, and pulled, but always kept things running smoothly. Though she could be a psychiatrist, she is the best editor there could ever be. We have no way to thank Liz Owen enough for making our work so enjoyable. Thanks, Liz.

The authors also want to thank special people involved in the project:

AJS — I want to thank my beloved wife, Jennifer, for so many years of love and encouragement, and my daughter (and co-author) JES and my son Daniel for being the best children a father could have. The successful completion of this book is because of them.

JES — I'd like to thank my dad (and coauthor), AJS, for being everything anyone could ask for in a dad. You are my inspiration. You made being smart synonymous with being cool and I can never thank you enough. Thanks also to my mom for sharing with me her unshakable faith that we can "make it all work" because she could and did.

DM-S — I want to thank my husband, Danny, for being a constant through all the changes. Thanks also to my parents and my sister for all the support and words of encouragement.

In writing this book the authors and publishers have benefited greatly from the advice of many microbiologists, immunologists, and instructors. We would like to thank the following for their suggestions in preparing this edition.

Waseem Ahmed (Community College of Allegheny County); Ralph Alcendor (City Technology College, CUNY); Mary Allen (Hartwick College); Mordechay Anafi (York University); Josef Anné (University of Leuven); Silvio Arango-Jaramillo (Palm Beach State College); Janice Barney (Mount Wachusett Community College); Preena Bhalla (Maulana Azad Medical College); S. A. Bhatt (North Gujarat University); Christine Bezotte (Elmira College); Laurie Bradley (Hudson Valley Community College); Kathleen Boyle (Southeast Arkansas College); Carol E. Carr (John Tyler Community College); Yasemin Congleton (Bluegrass Community and Technical College); Paul Cos (University of Antwerp); Chester R. Cooper (Youngstown State University); Christina Costa (Mercy College); Don C. Dailey (Austin Peay State University); Anjana J. Desai (University of Baroda); Beverly Ann Dixon (California State University, East Bay); Kristiann Dougherty (Valencia Community College); Khrys Duddleston (University of Alaska); Hussein El Ebiary (South Puget Sound Community College); Hanan El-Mayas (Georgia State University); Pamela Fouche (Walters State Community College); Wim Gaastra (University of Utrecht); Eric T. Gillock (Fort Hays State University); Louise M. Hafner (Queensland University of Technology); Jenny Hardison Clark (Saddleback College); Janelle Hare (Morehead State University); Diane Hartman (Baylor University); Pamela K. Hathorn (Oklahoma State University); George A. Jacob (Xavier University); Amy Jessen-Marshall (Otterbein College); Jeanne Kagle (Mansfield University); Judy Kaufman (Monroe Community College); Dubear Kroenig (University of Wisconsin); John M. Lammert (Gustavus Adolphus College); Sue Lang (Glasgow Caledonian University); Carol R. Lauzon (California State University, East Bay); Steven Leidich (Cuyahoga Community College); Jared Q. LeMaster (Cuyahoga Community College); Roger Lightner (Nashville State Community College); Holger Hill (Free University, Amsterdam); Anne Mason (Mesa Community College); Ethel M. Matthews (Midland College); Janie Milner (Santa Fe Community College); Richard Myers (Missouri State University); Edwin Noboa (GateWay Community College); Lourdes Norman-McKay (Florida Community College); Gordon Plague (Fordham University); Madhura Pradhan (Ohio State University); Nadia Rajsz (Orange County Community College); Clifford M. Renk (Florida Gulf Coast University); Jackie Reynolds (Richland College); Beverley Roe (Erie Community College); Sarah Richart (Asuza Pacific University); ValJean Rossman (Community College of Allegheny County); David Jesse Sanchez (Los Angeles City College); Susan Salter (University of Tasmania); Lisa Sedger (University of Technology, Sydney); Heather Seitz (Johnson County Community College); Prafull C. Shah (Valdosta State University); S. P. Singh (Saurashtra University); Theresa Stanley (Gordon College); Terry A. Tattar (Edison State College); Lewis Linton Tomalty (Queen's University); Olga E. Vazquez (Valencia Community College); Ernesto Lasso de la Vega (Edison State College); Helen Walter (Mills College); Wan Wei (Texas A&M University); Janice Yoder Smith (Tarrant County College); Malcolm Zellars (Georgia State University).

About the Authors

Anthony J. Strelkauskas, PhD

After earning a PhD from the University of Illinois Medical Center, Chicago, Tony Strelkauskas completed postdoctoral research at the Sidney Farber Cancer Institute, Harvard Medical School, before becoming a professor at the Medical University of South Carolina, where he taught immunology and microbiology to medical students. He is now the lead instructor for the microbiology course at Trident Technical College, Charleston, South Carolina, where his students have repeatedly nominated him for Who's Who Among American Teachers.

Jennifer E. Strelkauskas, DVM

Jennifer Strelkauskas is pictured here with a 6-day old baby alpaca (called a cria) after giving the newborn its first examination. She earned her DVM from Auburn University, Alabama, and is currently practicing veterinary medicine in Hood River, Oregon.

Danielle Moszyk-Strelkauskas, MD

Danielle Strelkauskas earned her MD from the Uniform Services University of the Health Sciences and then completed an internal medicine internship at the Naval Medical Center, San Diego, California. She is currently the Chief Academic Resident in Emergency Medicine at the Naval Medical Center, San Diego, California.

A Letter to Our Students

Welcome to microbiology. This is one of the most important courses you will take as you prepare for a career as a health care professional. It can be a complex course and you may be anxious about it, but you're not on your own.

We have spent our careers in health care as a clinical researcher, a veterinarian, and a medical doctor. So we know the difficulties and rewards of your studies, since each of us also traveled a similar road. We have written this book to help you succeed in your studies and your career. We hope you enjoy learning microbiology with this book, and put the knowledge to good use. There are two important things you should know:

1. This book was specifically designed and written for you: that is, students preparing for careers in health care.
2. Drawing from our professional experience, we focus on the things most important to know, and have tried to make them as accessible as possible.

In the next section, we explain how the book is geared specifically for students preparing for careers in health care. We also review the variety of interactive activities designed for this book that are available on the Student Resource Website, located at **www.garlandscience.com/micro.**

We hope you take advantage of the help this book and the Student Resource Website provide. We know if you use these resources, you will be better prepared for both this course and your career. We wish you the best of luck and all success.

Book Features for Students

The following features of this book were designed to help you succeed in both your studies and the practical application of that knowledge to your profession.

Clinical Focus

Because you are planning a career as a health care professional, we purposely limited the scope of this book to discussions of infection and disease. These include topics you will encounter throughout your career. We have tried to weave a continuous clinical thread through the scientific topics we explore and where appropriate highlight the role in infection and clinical significance of basic science with a special icon ⊕.

Special Topics

The book begins with a special Learning Skills section, written by Dr. Peter Susan. It overviews practical strategies for improving reading comprehension and retention of information. It should help you study more efficiently, and ultimately learn more from the book and the course. We also include entire chapters on specialized topics such as emerging and re-emerging infections, bioterrorism, and resistance to antibiotics. As future health care professionals, you will need to understand the basic science that underlies clinical interventions in these areas.

Accessibility

Although there are some difficult concepts in microbiology, we have tried to minimize these difficulties by using everyday language to explain things as simply as possible. At the end of each section, we include a brief review of the most important topics and concepts just covered, called **Keep in Mind**. These should help you keep the preceding information in mind as you move forward.

Just for fun, we also include Fast Facts sections throughout the book. These are designed to highlight a particularly interesting microbiology topic or fact, while providing a little break from your studies.

Clinical Terminology

In many cases, discussions of infection require the use of clinical terms that may be hard to understand. Definitions for these terms can be found in the Glossary.

Illustrations, Photographs, and Tables

The book figures are bold, bright, and user-friendly, and we believe they will make difficult topics easy to understand. The numerous photographs and micrographs were chosen to illustrate the organisms and infections we discuss. The tables present a great deal of information in a digestible format that's easy to understand.

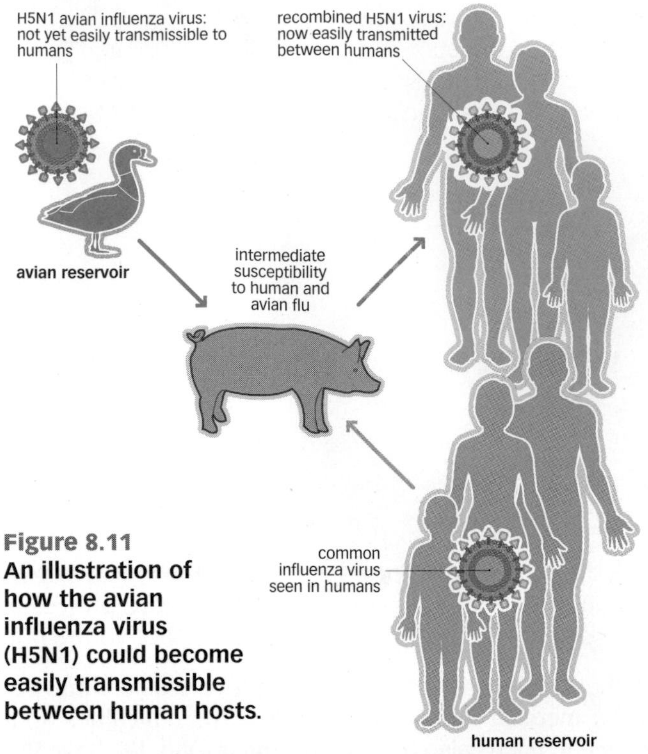

Figure 8.11
An illustration of how the avian influenza virus (H5N1) could become easily transmissible between human hosts.

End-of-Chapter Questions

There are three types of end-of-chapter questions. The Self Evaluation and Chapter Confidence questions test retention of facts and concepts. The Depth of Understanding questions require you to integrate important concepts in a more challenging way. And the Clinical Corner questions ask you to apply what you have learned to a specific clinical setting or problem. The questions, the correct answers to the questions, and explanations about why the answers are correct, are available through the E-Tutor feature on the Student Resource Website (see opposite).

Pathogen List

The Pathogen List, located near the end of the book, is a quick and easy reference to the pathogens mentioned in the book, their characteristics, and the diseases they cause. This will be useful not only for this class but also throughout your career.

Viruses				
Latin name	**Nucleic Acid Type**	**Morphology**	**Disease**	**Chapter**
adenovirus	dsDNA		acute respiratory disease, diarrhea, gastrointestinal infections	12, 13, 22
alphaviruses	+ ssRNA		encephalitis	28
arboviruses	+ ssRNA + ssRNA − ssRNA		hemorrhagic fever, hepatic necrosis, West Nile fever (fever and encephalitis)	8, 24, 25
arena virus	− ssRNA		zoonotic central nervous system infections	13

Student Resource Website (www.garlandscience.com/micro)

The Website contains multimedia designed to help you master the concepts and terminology presented in the book. The main features are the E-Tutor, MicroMovies, Bug Parade, Flashcards, Glossary, and Lecture Notes.

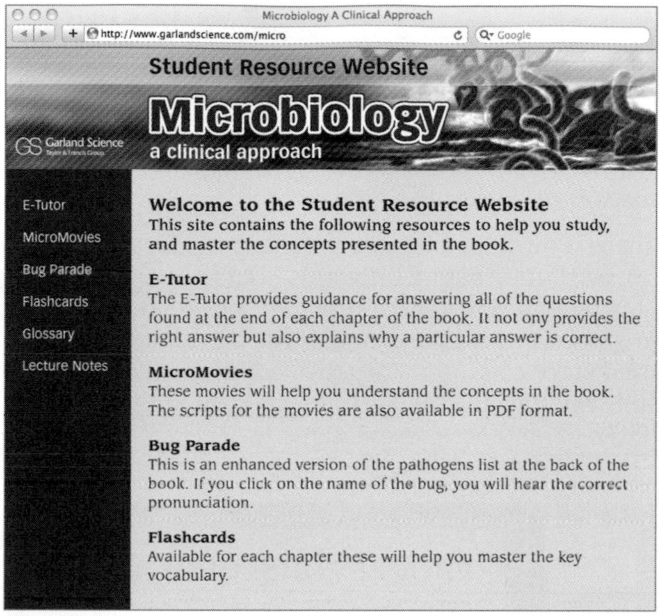

E-Tutor

The E-Tutor provides guidance for answering all of the questions found at the end of each chapter. The E-Tutor not only provides the right answer, but also explains in detail exactly why a particular answer is correct. The E-Tutor covers the Self Evaluation and Chapter Confidence questions, the Depth of Understanding questions, and the Clinical Corner questions.

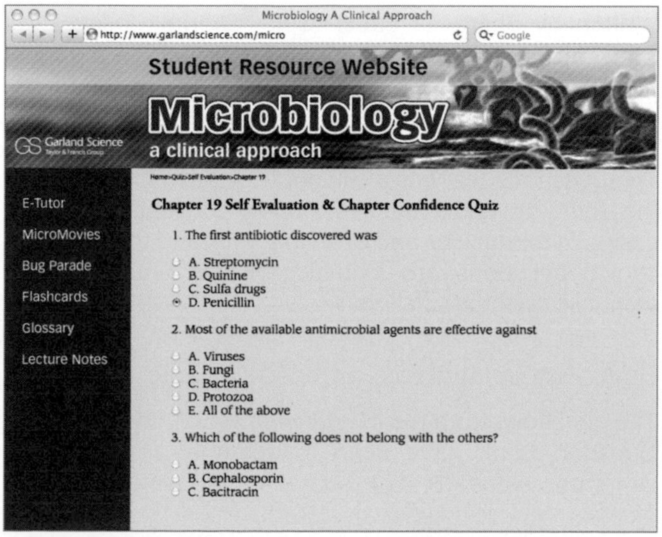

MicroMovies

We created a number of short movies (animations and videos) that will help you understand the concepts presented in the book. The movies are referenced in the textbook with an icon that indicates when a corresponding movie is available on the Website. The transcripts of the voice-over narration for each movie are also available on the Website in the Movie Scripts PDF.

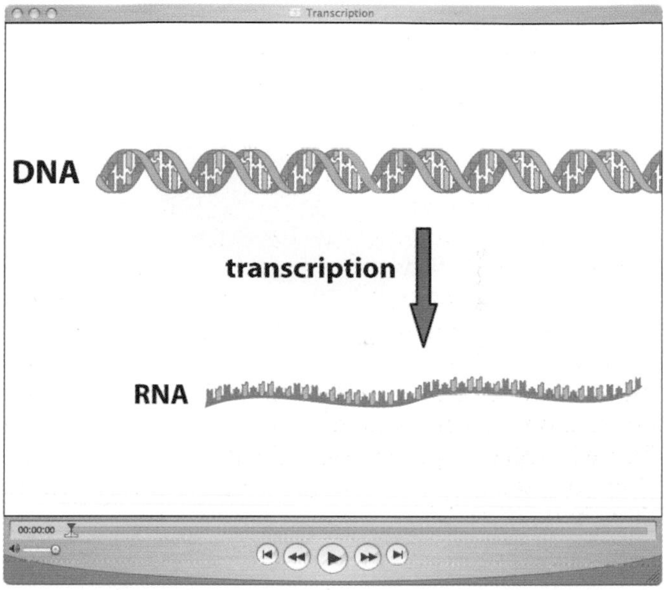

Bug Parade

The Bug Parade is an enhanced version of the Pathogen List found at the end of the book. If you click on the name of the bug, you will hear the correct pronunciation. This can be very useful since many of the Latin names are difficult to pronounce.

Flashcards and Glossary

Interactive flashcards and a searchable Glossary are available on the student resource site, and will help you master unfamiliar terms.

Lecture Notes

These topical outlines are designed to help you follow a lecture on a particular chapter. You may want to print these out, bring them to class, and annotate them during the lecture.

Resources for Instructors

All instructors who adopt the book will be entitled to the full complement of resources described below. These resources will be supplied on the Instructor's CDs and the Classwire™ course management system. Please contact **science@garland.com** for further information.

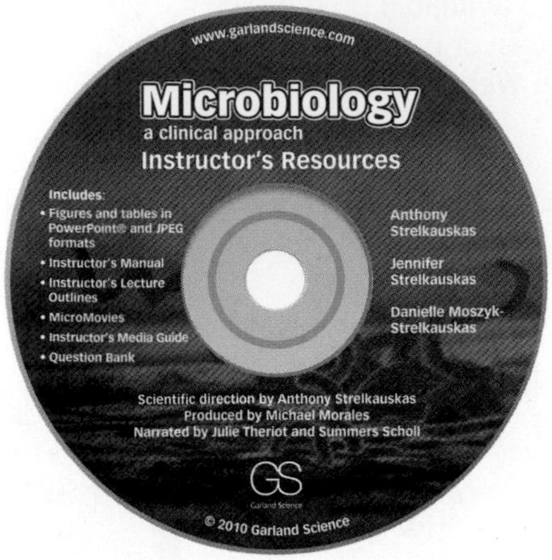

Students are provided with a variety of supplemental study aids, which are described in the previous section, and available on the Student Resource Website located at **www.garlandscience.com/micro.**

The Art of Microbiology: A Clinical Approach

The images from the book are available in two convenient formats: PowerPoint® and JPEG. They are located in folders on the Instructor's CD or can be downloaded from the Web via Classwire™. On Classwire, the individual JPEGs are searchable by figure number, figure name, or by keywords used in the figure legend.

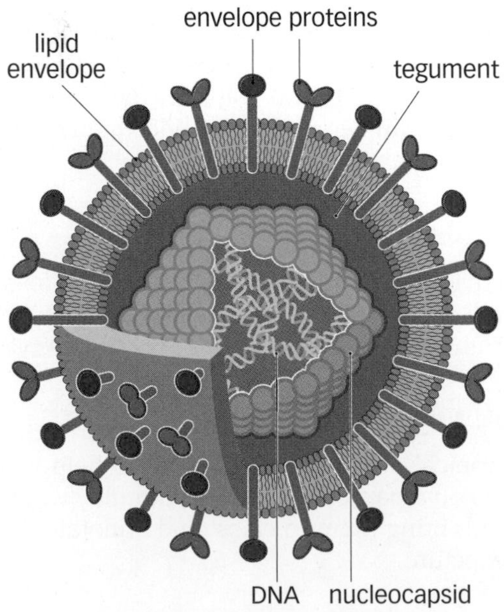

Figure 12.4 An illustration of the herpes virus, which has an enveloped complex icosahedral structure.

Instructor's Manual

To facilitate the design of a course around this book, the authors provide: a sample syllabus, based on their course; detailed guidance on each chapter; presentation strategies; clinical connections; instructional goals; and discussion of potential problem areas for students.

Instructor's Lecture Outlines

These PowerPoint presentations provide a complete set of lecture outlines for this course integrated with illustrations and tables from the book. There is one presentation for each chapter. The presentations can be used in the classroom "as is," or can be easily adapted to suit your course. The Lecture Notes on the Website are based upon these presentations.

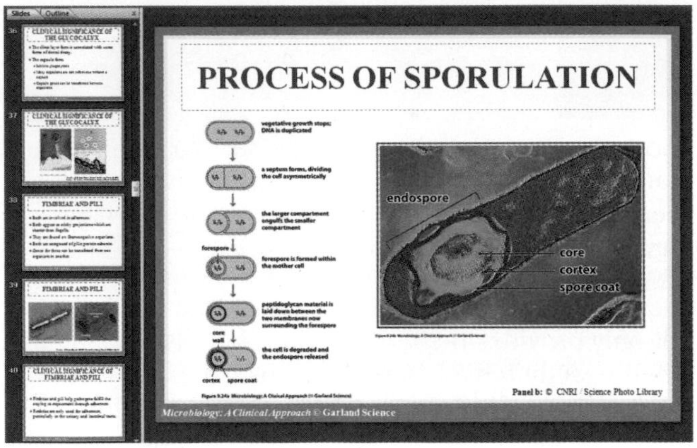

Question Bank

Written by Sherry L. Fuller-Espie, Cabrini College, the Question Bank contains over 800 questions in a variety of formats: multiple-choice, true-false, fill-in-the-blank, matching, and depth-of-understanding questions organized by book chapter. Answers are also given. The questions test basic retention of scientific facts and the ability to understand and apply concepts. They are designed for quizzes and examinations, and the multiple-choice questions are suitable for use with personal response systems (clickers).

Diploma® Computerized Question Bank

The questions from the *Microbiology: A Clinical Approach* Question Bank have been loaded into Diploma test generator software. The software is easy to use and

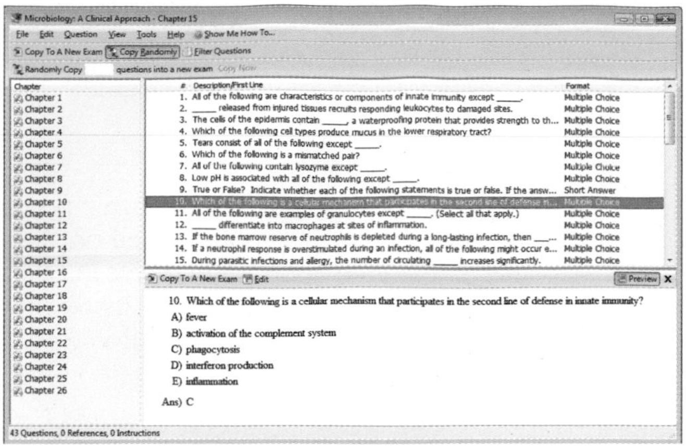

can scramble questions to create multiple tests. Questions are organized by chapter and type, and can be additionally categorized by the instructor according to difficulty or subject. Existing questions can be edited and new ones added. It is fully compatible with several course management systems, including Blackboard®. The Diploma computerized Question Bank is available on a separate CD to qualified instructors.

MicroMovies

Short movies have been developed to complement material in a select number of chapters, with a special emphasis on molecular genetics, virology, and immunology. The movies are identified in the text with a special icon.

The movies are located in folders on the Instructor's CD. Each movie has a voice-over narration, and the text for this voice-over is located in the Instructor's Media Guide.

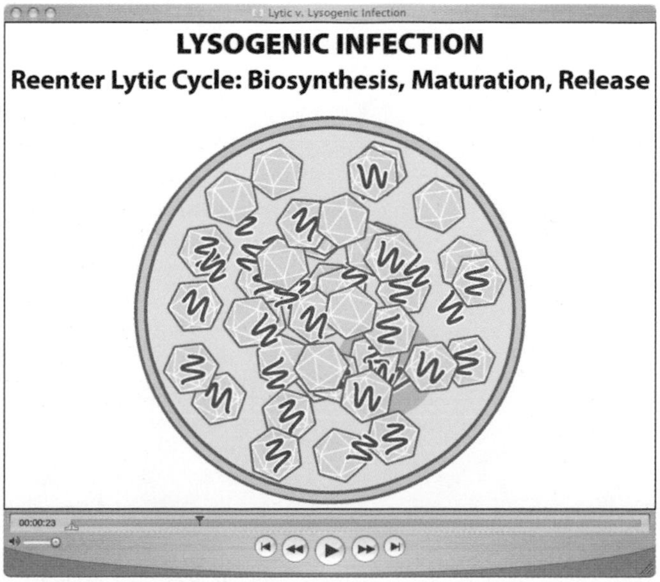

The movies can easily be imported into PowerPoint presentations, and are available in two handy formats: WMV and QuickTime®. The WMV versions are suitable for importing movies into PowerPoint for Windows®. The Quick Time versions are suitable for importing the movies into PowerPoint for Macintosh®. Students are provided with the movies and movie scripts on the Student Resource Website.

Instructor's Media Guide

This PDF overviews the multimedia package available for students and instructors. It also contains the text of the voice-over narration for all of the MicroMovies.

Classwire™

Nearly all of the instructor's supplements described above are available online via the Classwire course management system.

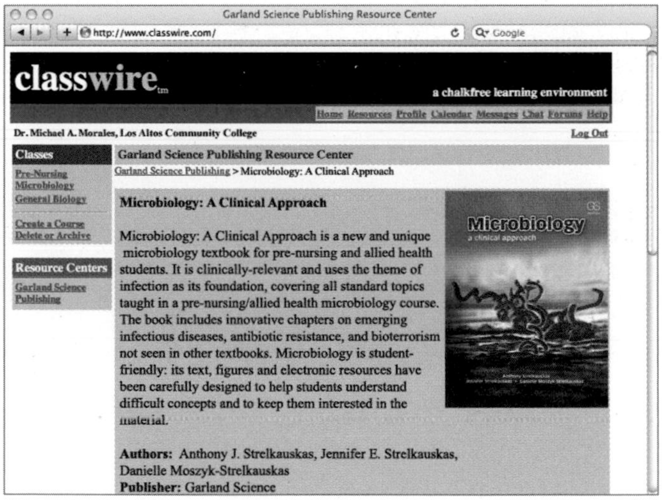

The system also provides access to instructor's resources for other Garland Science books. In addition to serving as an online archive of electronic teaching resources, Classwire allows instructors to build customized websites for their classes. For additional information, please visit **www.classwire.com/garlandscience** or email **science@garland.com**.(Classwire™ is a trademark of Chalkfree, Inc.)

CONTENTS

Overview

CONTENTS

Section View

PART III Characteristics of Disease-Causing Microorganisms

Selecting an Antimicrobial Agent	420	EVOLUTION OF ANTIBIOTIC RESISTANCE	465
Types of Chemical Agent	421	MECHANISMS FOR ACQUIRING RESISTANCE	467
Phenol and Phenolic Compounds	422	Inactivation of Antibiotic	467
Alcohols	423	Regulation of β-Lactamase Activity	468
Halogens	423	Efflux Pumping of Antibiotic	469
Oxidizing Agents	423	Classification of Efflux Pumps	469
Surfactants	424	Regulation of Efflux Pumps	470
Heavy Metals	424	Modification of Antibiotic Target	470
Aldehydes	425	Modification of Target Ribosomes	471
Gaseous Agents	425	Alteration of a Pathway	472
PHYSICAL METHODS FOR CONTROLLING		•	
MICROBIAL GROWTH	426	MRSA, VRSA, VRE, AND OTHER PATHOGENS	472
Heat	426	CONTRIBUTING FACTORS AND POSSIBLE	
Refrigeration, Freezing, and Freeze-Drying	428	SOLUTIONS	474
Filtration	428		
Osmotic Pressure	428	PART VI Microbial Infections	
Radiation			
	429	Chapter 21: Infections of the	
A Word About Hand Washing	430	Respiratory System	481
Chapter 19: Antibiotics	425	ANATOMY OF THE RESPIRATORY SYSTEM	481
	435		
HISTORICAL PERSPECTIVES	435	PATHOGENS OF THE RESPIRATORY SYSTEM	482
Antibiotics Are Part of Bacterial Self Protection	436	Bacteria that Infect the Respiratory System	483
ANTIBIOTIC SPECTRA	437	BACTERIAL INFECTIONS OF THE UPPER	404
ANTIBIOTIC TARGETS	441	RESPIRATORY TRACT	484
The Bacterial Cell Wall	442	Otitis Media, Mastoiditis, and Sinusitis	484
Penicillin	443	Pharyngitis	484
Cephalosporins	443	Scarlet Fever	485
Carbapenems	444	Diphtheria	485
Monobactams	444	VIRAL INFECTIONS OF THE UPPER RESPIRATORY	
Glycopeptide Antibiotics	444	TRACT	487
Isoniazid and Ethambutol	444	Rhinovirus Infection (the Common Cold)	487
Polypeptide Antibiotics	445	Parainfluenza	487
The Bacterial Plasma Membrane	445	BACTERIAL INFECTIONS OF THE LOWER	
	445	RESPIRATORY TRACT	488
Synthesis of Bacterial Proteins Bacterial Nucleic Acids		Bacterial Pneumonia	488
	447	Chlamydial Pneumonia	489
Bacterial Metabolism	448	Mycoplasma Pneumonia	490
ANTIVIRAL DRUGS	449	Tuberculosis	490
Acyclovir	449	Pertussis (Whooping Cough)	492
Ganciclovir	450	Inhalation Anthrax	494
Foscarnet	450	Legionella Pneumonia (Legionnaires' Disease)	494
Ribavirin	450	Q Fever	495
Amantadine	451	Psittacosis (Ornithosis)	495
ANTIFUNGAL DRUGS	451		473
Polyenes	451	VIRAL INFECTIONS OF THE LOWER	400
Azoles	451	RESPIRATORY TRACT Influenza	496
Griseofulvin	451		496
Other Antifungal Antibiotics		Respiratory Syncytial Virus (RSV)	498
	451	Hantavirus Pulmonary Syndrome (HPS)	499
DRUGS FOR PROTOZOA AND HELMINTHS	452	FUNGAL INFECTIONS OF THE RESPIRATORY	
DEVELOPMENT OF NEW ANTIBIOTICS	453	SYSTEM	499
New Targets for Bacteria	455	Pneumocystis Pneumonia (PCP)	499
Peptide Fragments	455	Blastomycosis	501
New Targets for Viruses	456	Histoplasmosis	501
The Cost of Research and Development	456	Coccidioidomycosis	502
TESTING OF ANTIBIOTICS	457	Aspergillosis	503
Chapter 20: Antibiotic Desistance		Chapter 22: Infections of the Digestive	
Chapter 20: Antibiotic Resistance	463	System	507
BACKGROUND	463	를 통해 하는 것이 있는 것이 없는 것	
		BACKGROUND	507

PARASITIC INFECTIONS OF THE BLOOD	604	Tools	638
Chagas' Disease	604	CELL CULTURE	638
Filariasis	605	Mammalian Cell Cultures	639
Chantar 26: Infoations of the Chin		Viral Cell Cultures	639
Chapter 26: Infections of the Skin		Adult Stem Cells	639
and Eyes	611	Embryonic Stem Cells	641
ANATOMY OF THE SKIN	611	GENETIC ENGINEERING	642
BACTERIAL INFECTIONS OF THE SKIN	612	Monoclonal Antibodies	642
Infections of the Hair Follicles, Sebaceous Glands,		Recombinant DNA Technology	643
and Sweat Glands	613	PROTEOMICS, GENOMICS, AND BIOINFORMATICS	646
Folliculitis, Acne, Impetigo, and Erysipelas	614	Microarray Analysis	649
Scalded Skin Syndrome	615	Forensic Science	649
Gas Gangrene (Clostridium perfringens)	616	Biosensors and Nanotechnology	650
Cutaneous Anthrax	617	그 이 경험 가지 않는 사람들에 가지 않는 사람들이 되었다. 그런 사람들은 사람들이 되었다. 이 그 나는 사람들이 되었다.	
VIRAL INFECTIONS OF THE SKIN	617	INDUSTRY AND APPLICATIONS	650
Measles	617	WE KNOW THAT WE CAN, BUT SHOULD WE?	652
Rubella (German Measles)	618	Chapter 28: Bioterrorism	(55
Smallpox (Variola)	618		655
Chickenpox and Shingles	619	WHAT IS BIOTERRORISM?	656
Herpes Simplex Type 1	620	HISTORY OF BIOTERRORISM	656
Warts	621	BIOLOGICAL WEAPONS	658
FUNGAL INFECTIONS OF THE SKIN	622	Anthrax	659
Cutaneous Candidiasis	622	Botulism	660
Dermatophytosis	623	Plague	661
PARASITIC INFECTIONS OF THE SKIN	623	Smallpox	662
Cutaneous Leishmaniasis	623	Tularemia	664
Lice	624	Viral Hemorrhagic Fevers	665
INFECTIONS OF THE EYES	624	PROBABILITY AND EFFECTS OF A BIOLOGICAL	
Conjunctivitis and Other Eye Infections	625	ATTACK	666
Neonatal Eye Infections	626	Warning Signs	666
Loaiasis	626	It Takes a Village	667
		Multiple Chaice Answers	
PART VII The Best and the Worst:		Multiple Choice Answers	669
		Glossary	671
Important Issues in Microbiology		Datha was 1 int	
Chapter 27: Biotechnology	633	Pathogen List	693
BIOTECHNOLOGY DEFINED	634	Figure Acknowledgments	703
THE INDUSTRY	634	Index	707
History of Biotechnology	636	IIIUGA	707

Learning Skills: Using Your Brain Effectively

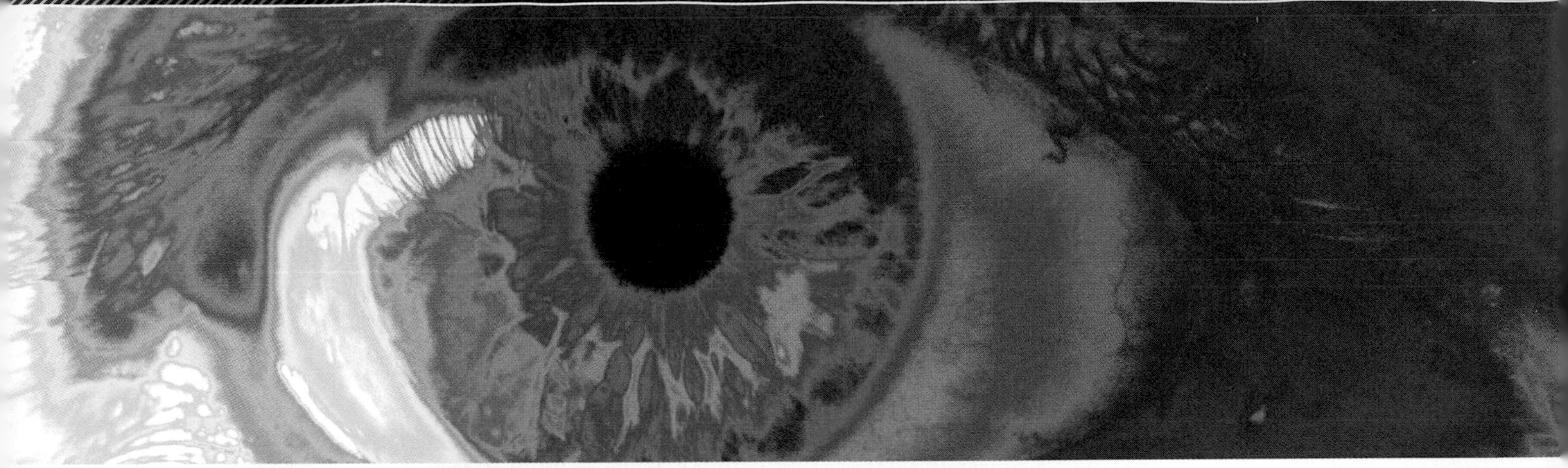

YOUR BRAIN'S ON/OFF SWITCH

You are equipped with one of the most powerful learning tools known to exist: a human brain. The "ON" switch is not as simple as other tools. With many tools, if you flip the "ON" switch, the tool will begin to operate as it was designed. Unlike other tools, your brain has its own will — your will. If you do not desire to learn, your brain simply stays turned "OFF" and learning cannot happen. To learn, you must want to learn (see below, "Motivation to keep the brain 'ON'"). In addition, your brain tries to be efficient. If you believe that you cannot learn something, your brain will not learn. When this happens, your brain simply avoids what it believes is wasted effort. Your brain can learn efficiently only when you believe that you can master the new knowledge. If you think you cannot learn a subject, then learning that subject will be impossible or, at best, extremely difficult. The "ON" switch for your brain first requires you to believe not only "Yes, I want!" but also "Yes, I can!"

Once your brain turns "ON," it operates by connecting new ideas to ideas that you have already mastered. If you fail to relate new ideas to your past experience (fail to make sense of it), then learning slows. This can cause a capable person to falsely think: "I cannot do this," and your brain switches "OFF." You must connect new ideas to what is familiar to keep your brain turned "ON" to mastering the new knowledge or skills. As you connect ideas together, they form a growing network of knowledge. We experience this as greater understanding and skills in a subject. As the network of ideas for a subject grows, related topics become easier, faster, more interesting, and more fun to learn. When you learn enough, you take interest in a subject and become more powerful at turning "ON" new learning in that subject. You can foster interest in anything: desire to learn it, believe in yourself, and actively connect with the new ideas.

If a student lacks background in a subject, it is more difficult to connect to the new unfamiliar ideas. Again, the brain will tend to turn "OFF." Imagine attending a lecture or reading a book for which you lack enough background: most people get frustrated, lost, or bored. To reduce this, educational systems use "prerequisites," which include knowledge and skills that you are expected to master before attempting a course of study. Teachers also tell students to read ahead before class, which helps students avoid getting lost during a lesson. If instructional materials are beyond a learner's background, the learner can get frustrated or, worse, will think, "I can't." In this case, a learner can achieve success by seeking the prerequisite knowledge from easier resources. A student who needs more fundamental background must resourcefully seek and use lower-level books, films, websites, classmates, tutors, and so on. Using such resources takes time, but the student will more rapidly develop the necessary network of fundamental ideas. Once the necessary "stepping stones" are mastered, the student can stay turned "ON" to more unfamiliar and complex ideas in a subject. Persistence in connecting new ideas to your familiar ideas will make the new ideas more familiar. The newly achieved familiarity then makes future learning in the subject even easier, because now there are more familiar ideas available for making connections. You should think how new ideas help make sense of things. As understanding increases, your background and skills increase, your interest increases, and you begin to enjoy learning about the subject.

You can direct your brain to turn "ON" to any subject. Obviously, more unfamiliar subjects take more time. Having a poor background in a subject can be overcome, but you must commit to the necessary time and effort. To start learning, you should honestly evaluate yourself and identify weaknesses in your background for a subject. Then you should develop a learning strategy that includes mastering the more fundamental prerequisite skills first. The time you will need for studying relates inversely to your background skills. That is, if you have less background for a subject, learning must require more time. The past is gone. Obviously, you cannot instantly increase your background to reduce the time needed for study. However, proper planning can help. If your background in a subject is low, you must plan enough study time into your learning strategy. Choose your commitments carefully to avoid being overwhelmed. For example, a college student could choose a "preparatory" course, to master fundamental skills, before enrolling in a more difficult course. In the workplace, an employee can master skills for a higher position before applying for a promotion. In summary, to switch your brain "ON" to learning, you must:

- **Believe you can learn**
- **Have enough background or time to learn**
- **Want to learn**
- **Spend your study time personally connecting to the new ideas**

Only you can make yourself believe that you can learn a subject. If you lack an essential belief that you can do it, you should recruit a support network of people to encourage you and foster positive expectations. This will most probably require face-to-face support from friends, family, advisors, counselors, therapists, and so on — a human touch. This chapter does not address time management and planning skills. Again, advisors and counselors can help, and there are many self-help books and courses that can help you develop these essential skills. The next three sections of this chapter address motivation: why you want to learn. The remaining sections of this chapter address learning methods for understanding concepts and developing communication skills.

Because your textbook is designed for a microbiology course to prepare students going into health care, you are most probably preparing for a career in health care. If not, then you are at least a current or future consumer of health care. Most diseases that health care must address are caused by infectious particles (viruses and prions), microorganisms, and other parasites. Microbiology focuses on the study of these pathogens (disease-causing things) and how they interact with their surroundings, including the human body. This book focuses on the microbiology of human health care; however, many concepts you will master also apply to the health of other organisms that people use for food, for materials, or as loved pets. Your ability to turn your brain "ON" to microbiology concepts will help you reach toward becoming the best possible worker in or consumer of health care that you can be.

INTERNAL MOTIVATIONS TO KEEP YOUR BRAIN "ON"

During learning, your brain wants to know "Why am I learning this stuff?" If you lack interest or sufficient reason to learn something, your efficient brain turns "OFF" to the new ideas. Again, learning is slowed or stopped. If you repetitively memorize something without knowing how it fits with your personal knowledge, you will quickly forget it. Your brain will not want to waste memory on information that it cannot or will not use. You must have motivation to keep your brain on a learning task. If motivated, you will keep your brain turned "ON" and ideally develop new connections that build your knowledge and skills. When properly motivated, you will learn. When you ask, "Why am I learning this stuff?" all possible reasons are from two fundamental sources:

- **Internal motivations driven from within yourself**
- **External motivations driven from outside yourself**

Of these, internal motivations are the most powerful. Humans are born with internal motivation for learning. As an infant, you learned instinctively. You were a self-motivated learning machine. No one had to say, "You have to learn." When the infant you was faced with something unfamiliar (most things, at that time), you automatically learned as much as you could. People incorrectly limit this curiosity to childhood: sometimes referring to it as "that child-like curiosity." However, adults can have it, too! This was, still is, and always will be your natural curiosity. Probably too often, people forget to use it. This natural curiosity is your personal motivation for self-improvement. If you tap into your powerful natural curiosity, new learning becomes faster, more satisfying, and more fun. New ideas change from being something you "have to" learn to something you "want to" learn. For example, you already experience this "want to" learn with your favorite subjects and hobbies. Rhetorically speaking, "Can you accept a new subject, as though it were a new favorite hobby?"

Are you wondering, "What happened to that powerful unfettered curiosity of my youth?" As you accumulated knowledge over your past, you remembered things by connecting them together and using the knowledge. Inside your brain, you created stories of how things are organized and how things work. You created stories and explanations that we can call "models" of all possibilities that you ever experienced or imagined (truth and fantasy). In childhood, you created new models to make sense of the world. Necessarily, learning was driven by a broad curiosity for the unknown. After all, in your infancy, it was all unknown. As you gained experience, your models of the world (real and imaginary) grew and became interconnected. Now, new ideas that more easily interconnect with your existing models are learned more easily. The models that you created in your past have developed into your present interests and skills in certain subjects. Interestingly (excuse the pun), interests tend to limit your naturally broad curiosity. You tend to seek ideas that connect more easily to what is already familiar. Thus, as you gain experience, you tend to narrow your broad curiosity and focus on interests that are more developed. Interests help us to focus more on particular sets of ideas for a deeper understanding of those subjects. Internal motivation is driven by your naturally broad curiosity and your more focused interests that continue to develop over time.

Difficult or boring subjects are less familiar ones, in which you have yet to develop interest. In short, you need more time and background to develop interest in these "foreign" subjects. If you exercise your power to engage your naturally broad curiosity, you can overcome a lack of interest. If a subject is unfamiliar, your natural curiosity can still drive the creation of new stories and explanations in your mind. As you gain abilities to understand the new subject, learning becomes easier. As you gain more background in a subject, you make it a part of you. Obviously, more personal background means more opportunities to make more personal connections to the subject. Interest starts to grow. Learning new ideas is an extremely creative process that requires imagination driven first by natural curiosity and then by growing interest as the new subject makes more and more sense. Of course, none of this happens unless you turn your brain "ON": want to learn, believe you can learn, and find enough time to connect to the new ideas, such as creating stories, explanations, and uses for the ideas.

Internal motivation for learning is a drive for self-improvement. What makes it so powerful is that the learning is its own reward. Internal motivation makes you feel good whenever you learn something new. This then makes you want to learn more. When you learn even more, you feel even better and want to learn even more. Broad curiosity and interests can drive you to want to understand anything and everything. Obviously, there is too much possible knowledge for one person to master within his or her lifetime. Because time is limited, each of us must focus our learning. External motivations help us focus on subjects that help us to fit better in the world around us.

If you have chosen a career in health care, you will need to understand concepts of microbiology, because you will encounter diseases in your work. If not, you still cannot avoid your own health care needs, not to mention those of your family, friends, and other people who will inevitably get infectious diseases. These needs provide powerful external motivations to help you focus on concepts presented in this textbook and course. However, to get the most out of your study time, you should develop internal motivations, too. Internal motivations will give you the satisfaction and rewards of self-improvement. This will help you tie more ideas together, faster, and with greater understanding than by the external motivations alone. In the same amount of study time, internal motivations will help you become an even more powerful worker in or consumer of health care.

EXTERNAL MOTIVATIONS TO KEEP YOUR BRAIN "ON"

External motivations drive learning, based on influences outside you. Undoubtedly at times in your past, you felt that you had to learn. For example, you had to learn things required by parents, teachers, or perhaps employers. Currently, your personal interests are internal motivations, but many developed from outside influences that required you to learn. For these, initial learning was driven by external motivations. When you persisted long enough in a subject, you made enough sense of it (connections to yourself) to eventually acquire a self-motivated interest in the subject. Thus, starting from external motivation, you gained internal motivation for a subject that otherwise might never have developed.

External motivation helps you to learn subjects that will help you fit better into the world around you. Educational systems, such as a microbiology course, are essentially systems of external motivation that attempt to promote your abilities to

- **Understand new concepts**
- **Develop skills using concepts**
- **Communicate concepts with other people**
- **Apply concepts to solve problems**

If successful, education maximizes your power to help "make the world a better place." Ultimately, external motivations help you survive and thrive in your environments, including "becoming productive members of society." External motivation is driven by seeking external benefits (gains, pleasures, or rewards) or avoiding external costs (losses, pains, or punishments). This can help you choose different, more productive, directions for your life.

For learning, external motivation is vulnerable. If apparent benefits decrease or costs increase, external motivation decreases. Thus, we could lose external motivation and reduce effort even when continued effort would bring success. In jobs and educational systems, external motivation is vulnerable in another way. For ease and logistics, social systems focus on correcting what a learner does incorrectly. Much less time, if any, is spent reinforcing what the learner already does correctly. Humans make mistakes. If you are only externally motivated, then your mistakes tend to make you dwell on the resulting losses, pains, or punishments. Obviously, this leads to discomfort, promotes negative attitudes, and slows the development of curiosity and interest. External motivation alone cannot maximize your learning.

Do you limit your learning potential by relying mostly on external motivations? If you primarily learn for purposes such as just getting that job, just getting that paycheck, or just getting through with it, then you probably have limited yourself to external motivation. Again, this makes you vulnerable to negative attitudes. If you truly learn because you want to understand more, to have greater abilities, or to be a better person, then you gain internal motivation. Internal motivation "sees" errors and mistakes as opportunities for greater learning. Internal motivations maximize learning and make you resistant to discouragement when external motivations try to correct mistakes.

The needs for health care are external motivations. If your interest in this book is as a consumer, then you might focus on general concepts and only details that apply to your health or the health of your loved ones. However, if you have chosen a career in health care, then the more microbiology concepts you can master, the better worker you will be. In professional health care, you will see many patients. The more patients you see, the more likely it is that you will run into the variety of different phenomena that are studied in microbiology. All health care workers should be externally motivated to learn as much as possible. Internal motivations will make you want to know it all. However, so much is known about microbiology that your limited study time will never allow you to master it all. Use external motivations to focus on fundamental concepts first and to choose details that you need most. For example, the objectives of a formal course or instructions from a teacher or employer are externally assigned to guide you in particular directions. Thus, these external motivations will help you to make the best progress for your given circumstance.

MOTIVATIONS OF "GOOD LEARNERS"

For best learning, you must seek to add internal motivations to external motivations. Good learners always develop and use internal motivations. Are you a good learner? If you are internally motivated, then each new achievement, no matter how small, yields internal benefits, such as feelings of personal growth and joy of understanding. Internal motivation is your best protection against the vulnerabilities of external motivation. **Poor learners stay reliant on external authority or circumstance to force them to learn. Good learners develop a personal drive to understand a subject.** When poor learners finish a course of study, they do not want to return to concepts through which they "suffered." When good learners finish a course of study, they continue to think spontaneously about things they learned. Good learners enjoy returning to the concepts they mastered. The best learners continue to imagine new and different ways to make connections between all the different ideas that they ever achieved. Thus, after a learning requirement ends, learning continues in good learners but ends in poor learners. Once the threat of punishment or promise of reward ends, external motivation will not promote continued learning. When a course of study ends, a weak learner is "glad it's over," but a strong learner is "glad to have begun!" Feelings you use to motivate your learning are your choice. For better motivation, an internal "wanting" to learn works much better than an external "having" to learn. Good learners use both external and internal motivations for their different strengths:

- **Internal motivations, to make learning about anything enjoyable and continuous**
- **External motivations, to focus learning on the needs of your circumstances**

As a health care consumer, there is still a chance that you or someone you care about could get even a rare disease. As a health care worker, you will meet many patients, which increases the odds of encountering various diseases. Furthermore, there is a 100% chance that you or someone you know will get disease from infectious organisms. All of us have had and will get infections. This is strong external motivation to learn about microbiology, especially as it applies to health care. If you have chosen a career in health care, then you must already have some personal interest in understanding, preventing, and treating diseases. Furthermore, as a human being, you should strive to be the best possible at your job. Your interest and desire for personal growth in health care provide strong internal motivations for learning new and unfamiliar concepts. Allow your motivations to help you become a good learner of microbiology.

QUESTION NEW CONCEPTS TO HELP YOUR BRAIN

If you want to learn, believe you can learn, and have planned enough time for a subject, you are ready to do the actual learning. Education includes three main learning objectives:

- **Ability to understand concepts**
- **Ability to communicate concepts correctly**
- **Ability to apply concepts to solve problems or achieve goals**

First, you must understand the concept. That is, you must imagine answers to who, what, why, when, where, and how-type questions about the concept. You must imagine how this concept relates to other concepts — how it connects to them. If you have trouble understanding a difficult concept, you probably need to identify and master more basic concepts first. As you piece together enough basic concepts, you will eventually understand the more difficult concept. Society depends on precise and clear communication. Unless you can efficiently communicate a concept, you may thoroughly understand it but its use is limited and you cannot prove that you know it. If you cannot communicate clearly and correctly, then from society's view you do not have the knowledge or understanding that is in your mind. Obviously, you must already have sufficient understanding if you are to communicate a concept. If you understand a concept and can communicate it, then you can go one step farther: apply the concept to solve problems or achieve goals.

A concept is a general idea about something, often a set of related things. Concepts can be about anything: physical objects, properties of objects, processes of change, abstract ideas, and even things that are completely imaginary. Even the idea of "concepts" is itself a concept. Concepts are units of knowledge that your brain uses to deal with information and interrelate different ideas. Your brain learns by making connections between your various ideas and experiences. A concept is how an idea can connect to other ideas. As you imagine how a new idea relates to familiar ideas and experiences, you build connections to that new idea. Eventually, the new idea becomes more familiar, it makes more sense, and you begin to understand its concept. If you learn enough to recognize a new, never before experienced, presentation of an idea, then you have gained an understanding of the concept. Now the idea is familiar, and you can more easily understand additional new ideas that relate to this concept. Concepts help us organize relationships. They allow us to answer fundamental questions of understanding: who, what, when, where, why, and how? Actively answering questions about an idea will help you master its concept. As you increase your repertoire of concepts, you also want to connect different concepts together. More complex concepts are just ideas that tie together multiple simpler concepts. Thus, as you understand and tie more concepts together, you begin to master even more complex concepts. Your learning becomes easier, because you gain concepts in your brain, which you can use to connect to even more new ideas and experiences. You get smarter, faster.

If you memorize a separate new fact or idea, then you still have not mastered a concept. A new idea or fact that does not connect to other ideas is useless to your brain. Because your brain operates efficiently, it tends to forget separate useless ideas or facts. A common error by students is to repeat the same idea, in the same way, over and over. Repetition is good for motor skills (muscle coordination) but fails to develop concepts. Without connections and uses, the brain might never recall an idea — it's forgotten forever. Furthermore, your brain cannot recognize different versions of the same idea that are too different from a memorized one.

Concepts allow us to generalize: recognize and group together all examples of the same idea. However, to develop a concept, we must make connections. You will make faster progress by actively considering relationships with a given concept, such as:

- **How is the concept defined?**

- *Collect different definitions for the same concept.*
  - *Create a definition in your own words.*

- **How many different examples of the same concept can you list?**

- *List some examples.*
  - *Explain why these examples belong with this concept.*

- **How is this concept distinguished from different closely related concepts?**
  - *List closely related concepts that might be confused with this one.*
    - *Explain why this concept is different from the closely related concepts.*
- **Does this concept have an opposite? If so, what is its opposite?**
  - *Describe the opposing relationship.*
- **How does this concept interact with other concepts to form more complex concepts?**
  - *Describe how it is part of something bigger.*
- **How do smaller concepts interconnect to form this concept?**
  - *Describe how it is composed of smaller things.*

A much better strategy than repetitive memorization is to learn something new about an idea each time you study it. Make new connections, like the ones listed above. New connections will develop your understanding of the concept. Thus, learning more makes a concept easier to understand, which actually makes it easier to learn even more!

Hopefully, as someone interested in health care, you have already become motivated to learn microbiology. A good learner will want to use productive learning techniques, such as asking questions about the ideas you are trying to master. As you study microbiology, remember to ask questions related to infectious organisms, how they reproduce, how they get through the body's defenses, and how they produce disease. How does each concept relate to the other concepts? Is it an organism? Is it a part of an organism? How do the structures of an infectious organism interact with structures of the human body? How are smaller structures combined to form larger structures? How do they work? Is the concept a step-by-step process of change? What are the necessary steps of the process? Does this process contribute to larger, more complex processes? As you proceed to study, constantly ask and find answers to such questions. This will allow you to forge new connections between ideas and develop your understanding of microbiology concepts.

ALWAYS USE THE POWER OF IMAGINATION TO HELP YOUR BRAIN

Hopefully, this has never happened to you, but perhaps you have seen a teacher get frustrated with a student who says, "What do I have to know?" or "What's the least I've got to study?" or "What's the minimum to pass the course?" Such statements indicate a lack of motivation and a predisposition to memorize, not understand. If so, then the problem of poor motivation must be addressed by the student. Otherwise it will continue to block efficient learning. A teacher might tell you, "You must know it all." Of course this is impossible, but it is meant to aim the student in the right direction: make as many new connections as possible, which will require starting with the most fundamental first. Unfortunately, a student lacking sufficient motivation will not get the implication and will just get frustrated. Once sufficiently motivated, you can choose creative learning experiences and use your powerful imagination to master concepts. Though motivated, you might still block your learning by a lack of confidence. **Figure 1** shows a flow chart to start your learning. Before anything else, you must first believe that you can learn and you must want to learn the subject. This will help unlock your imagination, which you need for efficient learning.

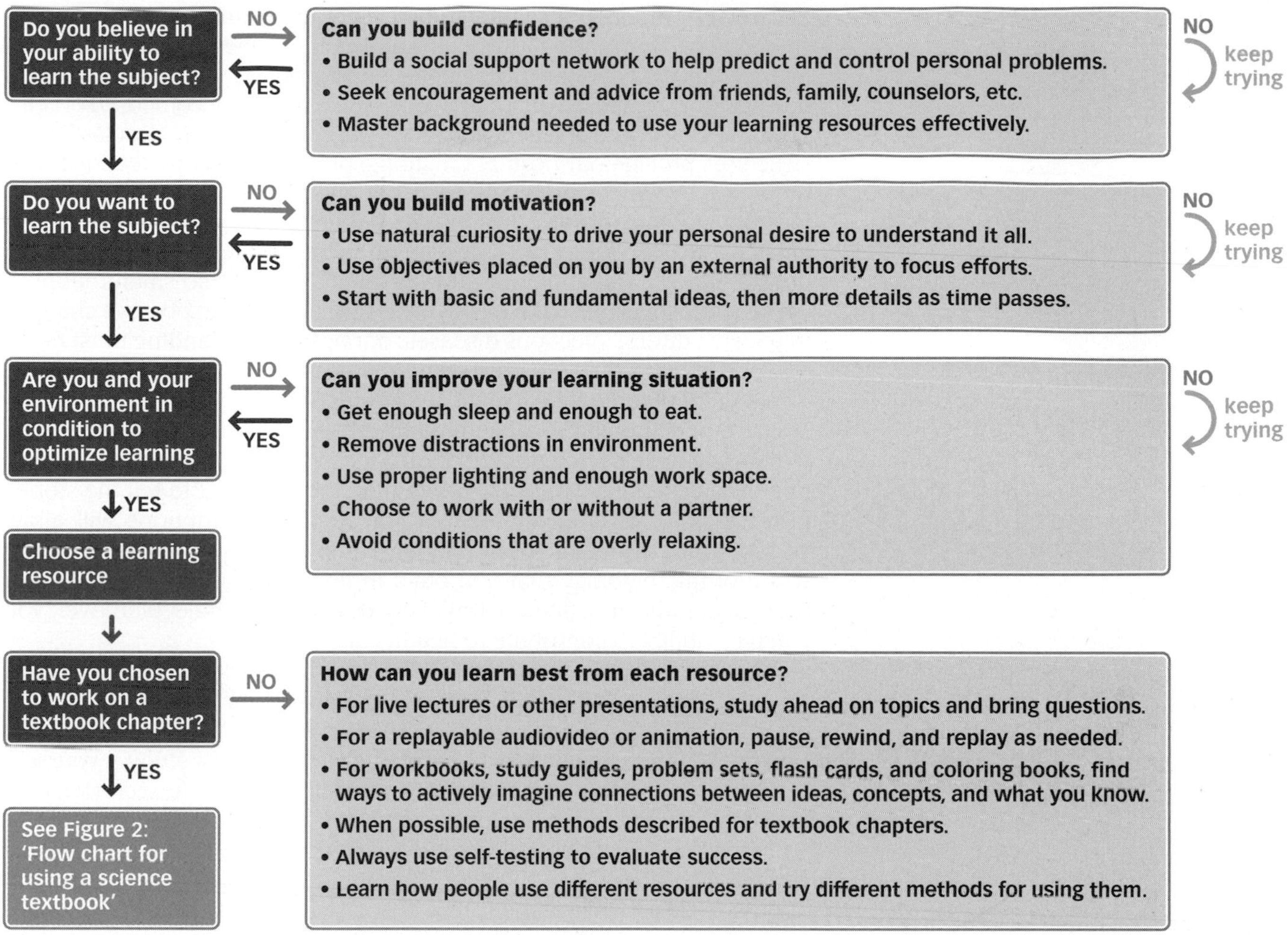

Figure 1 Flow chart for getting started on learning. This chart lists some issues that you face each time you start studying (whether you think about them or not). This takes you up to choosing a learning resource. Note that there are many different learning resources that are used in a wide variety of ways. If you want to develop strategies for using learning resources, then seek books or other resources that focus on learning skills. This chart (and this book) cannot detail the many possibilities. However, this might help get you started.

Only after you master a concept can you use it to learn faster, enrich your knowledge, gain skills, solve problems, and make life better. To master a concept, you must develop a sufficient understanding. Understanding concepts is a creative process that can be done either by direct experience or by imagination. Some concepts can be learned by going out and physically experiencing related circumstances. However, your brain has a much more powerful ability: **imagination**. Imagination frees you from the constraints of direct experience. You learn most concepts by imagination, because many concepts are too inefficient or are impossible to experience physically. For example, we can imagine that the sun is 93 million miles away, without traveling those 93 million miles. Imagination also allows us to learn concepts communicated from other people.

Many concepts are developed over centuries of thinking and learning, through the experiences of many people. A complex concept can require millions of person-hours to figure out. This is much more than one human lifetime, for just a single concept! However, our ancestors recorded ideas about concepts and developed explanations or stories to help us understand the concept. Now we can read and rapidly imagine what took long times and many people to figure out. Within seconds, we can imagine things that no one has ever physically experienced: incredibly long distances, incredibly large objects, extremely small objects, or very abstract ideas. Without imagination, many concepts are impossible to learn. With imagination, you can learn from other people's past work and master a concept that took many lifetimes to build — but you can do it in just minutes or hours, instead of years, decades, or centuries!

The power of modern health care derives from an understanding of concepts down to the molecular level: the molecules of chemicals and how they interact. Microbiology involves understanding organs of the human body, how they are composed of tissues, which themselves are composed of cells, which are composed of organelles, which are composed of molecules. These structures provide our normal functions, protection from infections, and the routes by which organisms infect and move through our bodies. Most infectious organisms are cells or particles much smaller than cells. Like all stable matter, they are composed of the molecules of chemicals. The prefix micro- in microbiology literally means "small," so small that phenomena are invisible to the naked eye. Although we can observe changes in patients during infectious diseases, our best understanding must reach down to microscopic structures and the chemicals that compose them. This requires imagination to build the worlds of microbiology in your mind. Micrographs, drawings, or animations of these worlds will be examples to help you develop your imagination. Once you imagine enough of these microscopic worlds, you will explore how they connect to tell the stories of microbiology. Ultimately, reading or hearing descriptions will allow you to form pictures in your mind, and then to change and improve the world of microbiology that you build in your own imagination. As you more accurately imagine microbiology down to the molecular level, you will gain ability to contribute to health care.

USE LEARNING RESOURCES EFFECTIVELY TO HELP YOUR BRAIN

When ready to learn, the most important thing you must study is yourself. In **Figure 1**, after building confidence and motivation, you next choose your learning circumstances, including your learning resources. Of course, you are your most important tool for learning. You must be in as good a condition as possible to optimize learning (for example fed and rested). Then choose an environment that best supports your ability to concentrate and imagine. This includes choosing the learning resources, such as this textbook, that present the concepts that you intend to master. You cannot make necessary connections to understand a concept until you can relate it to yourself and your existing knowledge. Complex or detailed concepts require an understanding of more basic concepts on which they are built. As you approach a new concept, you must consider whether your background and experience are sufficient. Resources that explain concepts come in many forms: texts, diagrams, figures, lectures, videos, experts, and other learning resources. Each learning resource assumes that you have reached a particular level of knowledge and skills, and also assumes that you have certain communication skills. As you face a new learning resource (lecturer, book, video, and so on), ask yourself:

- **Do I have the necessary communication skills to interpret this resource?**
- **Do I have enough background concepts to comprehend this resource?**

Communication skills include not only the ability to send information to other people but also your ability to receive and understand information. The latter is essential for you to use a learning resource. For example, if you cannot understand the French language, then you will never understand an explanation in French, no matter how simple the explanation. This is obvious, but less obvious is when you attempt to use a resource that assumes a higher level of background knowledge and skills. You might be able to read the words, but you might not be able to interpret the intended message. Can you imagine what the speaker or writer is trying to say to you? Different disciplines often have different specialized ways of communicating — their own languages. For example, a high-level resource for a science will assume that you have mastered some basic scientific terminology used in that discipline. Most post-grammar-school science assumes some ability to use tables, graphs, charts, figures, and standard symbols. For your sake, you should quickly identify which assumed skills you still need to use a resource. Then, before proceeding, you should do whatever is necessary to master those skills. Otherwise, you could get lost, become frustrated, and fail to learn, even though you are making an effort. Successfully identifying and mastering the necessary communication skills and background concepts can save large amounts of wasted effort.

Resources come in a wide range of difficulties from low-level to high-level. Higher-level resources make more assumptions about what you know (and can do) than lower-level ones. If you find explanations or presentations difficult or impossible to understand, they probably assume a knowledge of basic concepts or communication skills that you must master first. Rather than continuing with a resource that is incompatible with you, your best strategy is to identify what the presenter (author or lecturer) assumes that you (the audience) should know. If anything in an explanation is unclear, then seek and use other learning resources to master those concepts or skills. Seek resources that present concepts at a lower more fundamental level — ones that assume less background ability. Children's books and videos can be very useful. Be careful, because simpler resources sometimes use over-simplifications that are incorrect at higher levels of understanding. Over-simplifications are chosen to speed up a mastery of more basic concepts at the expense of producing a little misconception. As you work your way up to higher-level resources, over-simplifications and misconceptions will appear as contradictions. Do not ignore the contradictions. By resolving contradictions, you will further increase your higher-level understanding. As you master enough background, you will find greater success with using the higher-level resources. For example, if you do not understand how to read a graph, it will be impossible to learn about a concept presented on a graph. However, you certainly can find a resource that teaches about graphs. Once you have learned how to use graphs, you can go back to the graph of a concept and learn about the concept. A common mistake made by students is repeatedly using a resource again and again, without identifying and mastering the background skills that are necessary for successful learning from that resource. Formal courses of study often include "required" resources. Keep in mind that you might need additional resources to master basic concepts and skills to use the course's "required" resources effectively.

At this very moment you are using a learning resource: this text. By now it should have helped you believe that you can learn and want to learn microbiology. If you got the message, then you learned how motivations keep your brain turned on. You have begun to question your own learning, and you are ready to seek ways to improve learning. You know to ask questions of yourself and find the answers. You know to always take advantage of your powerful imagination. You should start picturing yourself studying better and enjoying it more. If any of this paragraph is not true, then this learning resource has not been entirely successful. If the previous text of this chapter has not been successful, what is the problem? Like any learning resource, this text makes assumptions about you, the reader. First, it assumes that you have a certain level of reading comprehension skill and vocabulary. If you cannot comprehend this text, then you might need reading and English skills that are not covered in this book. Otherwise, as you proceed, you will have even greater difficulty, because in addition to this assumption, the textbook will make additional assumptions. You should have an understanding of basic biology and chemistry concepts, and you should know the correct way to read a science textbook.

If you read a chapter of a science textbook like you would read a novel (straight through, uninterrupted, from beginning to end), you are approaching the science text incorrectly. Your study of this book will be more productive after mastering science textbook reading skills, because it assumes you have these skills. You might seek help from a reading specialist or a science teacher. If an explanation uses unfamiliar vocabulary to explain a new concept, you will not be able to imagine the new concept. You must master the unfamiliar vocabulary used in the explanation. You might need to use an earlier part of the book, look terms up in a dictionary, or seek simpler presentations in lower-level biology or chemistry books. Once you have mastered the assumed knowledge and skills in your book, you will more effectively use the text to continue building your understanding of microbiology. **Figure 2** summarizes a commonly used approach to using a science textbook.

ALWAYS USE SELF-TESTING TO HELP YOUR BRAIN

Remember: the most important thing you must study is yourself. A common cause of unproductive studying is a failure to self-test. If you self-test, you can discover how much learning happens and confirm what you achieve. If you fail to self-test, you could waste large amounts of time, which happens when you continue to use a study method that is not working. For example, self-testing could help you identify how well a particular learning resource is working. If learning is not fast enough, you can focus on identifying and fixing the problem. Self-testing is essential to make studying as efficient as possible. By definition, the more efficiently you study, the more you can learn in the same amount of time. Who wouldn't want to do that?

Here is a general approach to self-testing. First, select a concept or group of related concepts that you think you can make sense of within 15–20 minutes. If you have many concepts to master, you should start with the more fundamental or basic ones. Then decide on a strategy for mastering the concepts. There are many ways to learn. Which ways will you try during the next 20-minute interval? Will you read part of a textbook? Will you watch an animation or short video? Will you discuss concepts with someone who already knows them? Will you create a diagram, a table, or some other sketch of the ideas? Will you create a story, an explanation, a joke, a song, or a dance? Will you get up and try to "act out" the concept? Studying is a creative process, because you must create your own understanding of the concept in your mind. This will require new experiences, real or imagined. Whenever possible, choose a method that gets you actively involved with the learning. Once you select target concepts and a learning strategy, it's time to begin actual learning. Limit yourself to about 20 minutes. Try setting a timer with an alarm, then study the concepts as you planned. When the alarm sounds "time's up,"

Figure 2 (opposite) **Flow chart for using a science textbook.** Textbooks are still a standard resource for most formal coursework. This chart lists a strategy for approaching chapters of a textbook. This chart incorporates a commonly recommended strategy for using textbooks, called SQ3R, which refers to its major steps: survey, question, read, recite, and review.

Have you chosen to work on a textbook chapter?

NO → See Figure 1: 'Flow chart for getting started on learning'

YES ↓

If available, have you examined resources that could help focus efforts?

NO → **Do you have an idea of concepts and skills that your instructor or employer expects you to master from this?** (YES ←)

- Imagine and list possible expectations.
- Examine any objectives, lists, or outlines given to you.
- Examine any notes or summaries for lectures or presentations on topics covered in this chapter.

YES ↓

Have you surveyed or scanned the chapter?

NO → **Do you have an idea of how concepts and skills are organized and presented in the chapter?** (YES ←; NO: keep trying)

- Read the chapter title and the section subtitles.
- Look at any boldfaced and italicized terms.
- Examine graphs, diagrams, tables, and other figures.
- Read any summaries or conclusions.
- Read questions at ends of sections and at the end of the chapter (when present).
- Start to match information with learning expectations.

YES ↓

Have you questioned what you will be reading?

NO → **Have you formulated and written questions that you expect to answer from reading the chapter?** (YES ←; NO: keep trying)

- Rewrite titles and subtitles into the form of questions.
- Write questions for concepts presented in graphs, diagrams, tables, and other figures.
- Write questions that your instructor or employer might expect you to answer from this chapter.
- Write questions that you want to be able to answer.

YES ↓

Have you actively read the chapter?

NO → **Are you imagining what is written and making notes of what seems important?** (YES ←; NO: keep trying)

- As you read, answer questions that you formulated.
- Read to understand meanings, not just to answer the questions.
- Write down important ideas, including those that help meet learning expectations of instructor, employer or self.
- Master any unfamiliar terms used as though you know them; try the glossary, index and other resources (dictionaries, lower-level textbooks, etc.)

YES ↓

Have you recited the terms and concepts without looking at the textbook?

NO → **Can you explain the concepts to someone else?** (YES ←; NO: keep trying)

- Write a letter to a friend describing what you learned.
- Read your questions, answers, and notes out loud.
- Do written self-testing and self-evaluation to identify what you do and do not know.
- Research to make explanations more complete and answers to questions more concise.

YES ↓

Have you reviewed what you have learned?

NO → **Do you revisit your new knowledge and continue to connect new ideas together?** (YES ←)

- Re-read the chapter to see what makes more sense.
- Re-examine your notes often and imagine new connections to your past experiences and learning.
- Imagine and invent ways to use the knowledge to present to others or to solve problems.
- Create stories, songs, jokes, skits, etc., to review your new concepts by using them in different ways.

YES ↓

Use your new understanding to help you better connect to future learning

stop the learning, and start self-testing. Put all your books, notes, and practice away. Take out a blank sheet of paper and write an explanation of what you just tried to figure out. For example, write a letter to a friend explaining the way you connected ideas together to make sense of the concept. Your written answers record what you have learned, providing data about yourself that you can evaluate.

The next step is to evaluate what you learned in that 20-minute interval. Far too often, students limit self-testing to thinking answers, instead of writing answers. If you do this, you will check your resources to decide whether the ideas have gotten "into your head." When you only have thoughts to check, you face a fundamental flaw in self-evaluation. Your brain automatically "fills in" gaps in stories and explanations. In addition, your brain likes to feel progress. As you look at resources to check your knowledge, you can easily think you know things that in fact you still do not know well enough. Furthermore, this "in your head" evaluation will not catch things that you know but still cannot recall. Thus, as you check your books and notes, your brain will recognize the presentations and can make you believe, "Oh, I thought that — I know that," when in fact you still cannot explain the concept correctly and completely. In this case, you mistakenly think that you mastered the concept, and move on, without making sufficient progress.

A written answer for self-testing is a permanent record of what you accomplished during your study time. Your brain cannot magically make words or drawings appear that you did not write. Now you can evaluate your written answer objectively. Did you leave any important ideas out? Did you make sense? How can you improve and expand on your recorded understanding? Furthermore, you can take your answer to a friend, classmate, tutor, or teacher to get help in evaluating your answer. Your written answer is data about what you could do at that time, and you can use it to make better plans for future learning. Now you can spend more time making connections to things that you forgot to include in your answer. Learn something new about the forgotten ideas to make them more a part of you and easier to recall. Finally, you can save your results and compare them with future efforts, allowing you to gauge your progress. In addition, you can record the date, time, and study method to help improve your choices for future learning experiences.

What if you study for 20 minutes and the timer goes off, but you can write only little or nothing about what you tried to learn? Was that 20 minutes a waste of time? Not yet. However, if you repeat the same study method, you should expect a similar result: little or nothing. If you do this, you will waste 40 minutes: the first 20 minutes and another 20 minutes in repeating a learning method that already proved unproductive. Do not ignore your self data. Your self-test is only useful if you evaluate and act on the results. At the end of an unproductive 20-minute study interval, instead of repeating a mistake, try to figure out what went wrong. Ask yourself, "What must I change to improve studying?" It might not be the study method itself. Do you understand all the background concepts and communications in the resources you used? Are there too many distractions in your learning environment? Have you eaten enough? Have you slept enough? Identify the problems and fix them. If the study method is the problem, then try something different. Develop a variety of ways that you can use to study. Different types of concepts and skills can require different learning methods to optimize efficiency. In addition, people have personal learning styles, so what works for someone else might not be most efficient for you. Try various study techniques, study their effectiveness for you, and learn which work best for different learning objectives.

In all endeavors, including health care, people refer to ideas that are parts of larger stories and explanations. People attempt to communicate more efficiently by assuming that you already know the stories and explanations. Of course, to understand what your co-workers mean, you must already have those stories and explanations in your mind. When problems arise, you often will not have time to seek resources and learn what you need to know. As a health care professional, you are expected to have that understanding in your head already. Self-testing is the only way to make sure that you get the knowledge and skills that you must master, before a demand is placed on you. Otherwise, when a need arises, you might not be able to understand what is happening, and you will not be able to solve the health care problems that you face.

USE ACTIVE LEARNING TO REPLACE INEFFECTIVE METHODS THAT TURN YOUR BRAIN "OFF"

Ineffective studying might not just waste time and effort: it can convince your brain that you cannot be successful. Remember: if you believe that you cannot learn, then your brain turns "OFF". Failure to make necessary changes to improve learning can lead to frustration, and frustration can lead to anger or sadness. If anger is aimed inward at your own unproductive behaviors, then you might attack them, eliminate them, and turn yourself around. If your anger is aimed outward at your learning resources, then your attacks might eliminate valuable resources (such as your teachers or books). If you get angry at the subject matter, then your brain will dislike it, maybe even hate it, and refuse to learn it. Remember: if you do not want to learn something, then you will not. Sadness is your body's natural response to deal with uncontrollable circumstances. Sadness helps shut down unproductive efforts. If frustration leads to sadness, then of course, you will dislike and avoid studying. Even worse, anger and sadness can make you believe that you cannot learn the subject. Unproductive studying will cause you to lose the "wanting" and "believing" that are essential to keep your brain turned "ON" because your brain does not want to waste effort,.

Study time must be as active, creative, and effective as possible. Studying cannot be done by just "putting in the time." If you are passive about learning, then you are just looking at stuff or listening to stuff. If you fail to really imagine what ideas mean and connect them to other ideas, you will not master the concepts. Proper self-testing can reveal a lack of learning. If you find that you are not learning, you must take immediate action. As stated above, it is essential to stop ineffective studying as soon as possible. First, you must believe you can learn, and plan enough time to learn based on your personal background. These ideas were addressed in the first section of this chapter, "Your brain's ON/OFF switch." In addition, you must want to learn, which we detailed in the next three sections of this chapter, dealing with motivation. Finally, you must develop and use methods to make a personal connection to new ideas. In the four sections just before this one, we have covered questioning, imagination, learning resources, and self-testing.

The importance of studying and evaluating yourself cannot be overemphasized. This has been known since ancient times: in ancient Greece they called it "*gnōthi seauton*" — know thyself. To get the most out of your learning, you must constantly figure out what you do and do not know, and prioritize your efforts. By actively piecing together explanations and imagining ways in which concepts can be used, you will give your brain reasons to remember the concepts. For this course of study, as your abilities grow, new microbiology concepts that you encounter will more quickly become familiar, and you will be able to solve more problems that you face as a health care worker or consumer.

COMMUNICATION: IDEAS MOVING BETWEEN DIFFERENT PEOPLE'S BRAINS

Learning concepts requires you to know yourself, gain your own experience, and connect to ideas that are in your own mind. Learning to communicate requires you to know other people, share experiences with them, and connect to ideas that are in their minds. If you are brilliant at understanding a concept, then you must also communicate it correctly. Without proper communication skills, society will conclude that you still lack the knowledge and skills. To show what you know, you must correctly interpret messages from other people and concisely communicate your ideas to other people, using the standard language established for each discipline of study. Errors in interpreting messages or constructing messages will indicate a lack of knowledge and skills, even ones that you actually have!

Human beings are social animals with advanced abilities to communicate. Our powerful communication skills allow us to coordinate activities, create societies, and reach achievements greater than one person could ever do alone. If you cannot communicate, then as far as other people are concerned, your knowledge and skills do not exist. Lack of communication skills will make your good ideas of limited or no use to others. Education is how we learn to operate effectively with other people. This course provides education for the discipline of microbiology. Communication is an essential goal of education. A person can learn many concepts but still be poorly educated. Your approach to education should develop not only understanding but also your communication skills. Communication allows you to integrate with society effectively. First, you must understand concepts, and then you can develop skills at communicating those concepts. Avoid reversing the process. If you do not understand a concept and attempt to communicate it, then you could send someone the wrong idea. This also puts you at risk of sounding ignorant, incompetent, or foolish. Such appearances undermine confidence in you, and reduce or eliminate your ability to work with other people. Start by understanding concepts, to communicate them well. Learn to communicate effectively, to work well with other people. When you work well with other people, your efforts can be added to a society, to reach your greatest potential. Communication skills that you develop in this course will allow you to become a more effective worker or consumer in our society's health care system.

Communication uses language. Human language uses words, symbols, and a variety of specialized communication forms, which are standard "placeholders" for concepts and their relationships. You cannot effectively send or receive messages unless you understand how the language is received. Communication skills fall into two major categories: (1) receiving messages and (2) sending messages. Of these, sending messages is more difficult, because to be able to send a message you must already know how the message would get received. Receiving messages involves all skills for interpreting language in its various forms. To understand messages, you must know enough background concepts and standard terminology. You must also be able to interpret specialized communication forms that are used to express relationships, such as diagrams, figures, tables, graphs, and equations. Messages that you receive can teach you new concepts. The learning resources, described previously, contain messages specialized to help you master new concepts. However, you must have enough background skills to use them. Learning resources attempt to send other people's understanding to your mind. Obviously, if you cannot understand the language of a resource, then you cannot learn from it. A good receiver of messages must have enough background language skills and must know how to listen carefully, how to read well, and how to comprehend messages (that is, imagine meanings).

Receiving messages can be challenging, but sending messages is even more challenging. For receiving, you need only to recognize standard language and interpret a message. You imagine the ideas of the message in your own mind. You can understand the message but still might not be able to send the message. The goal of any message that you send is to create ideas in someone else's mind. For sending, you must create a message that can be received and interpreted by someone else's brain. You must already understand how messages get received and predict how the targets of your message would comprehend them. To send an efficient message, you must:

- **Understand all the concepts to be sent in the message**
- **Figure out your target audience**
- **Choose language that the target audience uses**
- **Create a complete and concise message focused on achieving its purpose**

First, use learning skills to master the content that you intend to send (concepts, language, and communication methods). Then, figure out whom you want to communicate to; in other words, the audience. Is the message to your boss, to an instructor, to a friend, to a colleague, or to a more general audience? What can you assume about the language skills of your audience? Next, decide what language method and terminology would work best for sending your message. Choose these on the basis of the language ability that you expect from your target audience. This is where you, the sender, place a demand on your audience to have sufficient background. If the message is for a less educated audience, it must use simpler language. However, simpler language tends to be less concise and requires more words than for a more educated audience. If the message is for a more educated audience, then it should use more advanced terminology and be as concise as possible. For an educated audience, failure to be concise carries a risk: if your message uses terms that are too simple, too general, or unnecessarily wordy, your message might communicate that you lack an understanding of the concepts. Every time you communicate, endeavor to choose the most precise terms that still include everything that must be addressed in the message.

We have already seen that learning involves evaluating yourself. To learn better communication, you must learn more terminology and language skills, practice communicating, and evaluate the "practice" messages that you create. As mentioned before, recorded messages are best for effective self-evaluation. For verbal messages, try audio recordings that can be played back for self-evaluation. If you understand *and* can communicate with microbiology concepts, then you will more readily understand what health care personnel say to you about infectious disease. Furthermore, you will express knowledge that will gain respect and increase confidence of your patients, co-workers, and other health care professionals. An important part of communicating at higher levels is to achieve sufficient vocabulary. Vocabulary allows you to communicate more concisely and avoid appearing ignorant to educated people.

VOCABULARY: A TOOL FOR YOUR BRAIN'S COMMUNICATION SKILLS

Vocabulary refers to the use of terminology to communicate a subject. Terminology refers to a set of abstract terms, words, symbols, and phrases that are used to communicate ideas and their concepts. Every discipline of study has specialized terminology used in the vocabulary for that discipline. The ability to define terms is necessary but it does not mean that you have mastered vocabulary. If you cannot use a term, you have not mastered its vocabulary. To have a good vocabulary, you must understand the terms, speak the terms, write the terms, and use the terms correctly in messages you create. Your vocabulary is only as good as terms that you can use to interpret ideas from messages and send ideas in your own created messages. As your vocabulary increases, you will understand higher-level messages and can use terminology to create more concise and precise messages. Learning the terminology (symbols, words, and phrases) generally begins as you master the concepts. When you learn how to use a concept, your brain naturally wants to learn how to communicate it. Mastering vocabulary can provide even more conceptual connections in your brain.

For vocabulary, the first skill you learn is to recognize terms and imagine their meanings (recall definitions). You should also begin to distinguish new terms from other terms that look or sound similar. A useful way to make connections to new terminology is to break a term into its parts. What is the meaning for each word in a phrase? For larger words, what are the meanings for each word part: prefix, suffix, and word root? Try to learn meanings for the parts of each term. Better dictionaries show origins of words and define word parts. If the word you are learning is not in your dictionary, look for words that use the same or similar word parts. Can you identify shared meanings between new terms and terms you already know? If you find words that you already know, these can help you connect a new word to your personal background, making the new term become familiar faster. Keep in mind that words and word parts occasionally have more than one meaning in different contexts. However, for a new term, each time you recognize a word or word part, your curiosity can help you wonder whether it is used in a way that is similar to your background knowledge. This active thinking helps you work with the new concept. Often the same or similar word parts in different terms will have a shared meaning. When this happens, it makes conceptual connections that help you remember all the related terms better, and you more rapidly become familiar with the new terminology.

Your vocabulary of prefixes, suffixes, and word roots will give you an additional ability. If you encounter new terminology that you have never seen before, you still might be able to predict meaning from examining the word parts. When you master a large enough vocabulary of prefixes, suffixes, and word roots, you will be able to begin identifying possible meanings of new terminology. Your vocabulary plus the context in which the new terminology is used will sometimes be sufficient for you to figure out the meaning or a close approximation. Thus, a growing knowledge of word parts will give you a greater ability to understand terms that you have never encountered before. How amazing is that?

Language terms (phrases, words, word parts, and symbols) are very abstract. For example, the word "head" is not a head. This word is just a series of four inked letters on this page. Being abstract, the word "head" can have different meanings, depending on its context. When you see the word "head" you might imagine an expanded structure on the front of an animal containing sensory organs, such as eyes, and openings into the body, such as a mouth and nose. "Head" also might make you imagine the expanded end of a nail or bolt. It might make you think of a person in charge of a group of people, such as a department head. From a sailor's perspective, "head" refers to a toilet. The point is that the same formation of ink on a page is so abstract that it can have very different meanings. If you have never used the term before, then it would have no meaning at all — it would be in an unknown language. You must master the concepts that words represent, so that the context of communication (sent or received) allows understanding to occur.

The abstract nature of new terminology poses a challenge: how do you learn a foreign-looking new term? The short answer is: practice. For terminology that you receive in messages, the more messages that you read or hear, the better you will imagine the meanings. Obviously, you first must thoroughly understand the idea to which the term refers. Then each time you hear or see the term, imagine its meaning. If the word sounds or looks like something familiar to you, then create a fantasy story in your mind in which the familiar idea interacts with the new concept. Now the unfamiliar term can make you think of a familiar idea, and when you recall your fantasy story it will help you remember the meaning of the unfamiliar term. Note that the fantasy story does not have to be realistic, because you are just trying to recall a meaning in your mind, not use it to send a message. However, your fantasy story can help you in reverse, too. That is, when you want to use the unfamiliar term (but cannot remember it), your fantasy story can help you recall your familiar term that sounds or looks like the unfamiliar one. This might then be enough to trigger recall of the unfamiliar term. Now you can use the unfamiliar term to send a message. As you use a term more, it will become more familiar, and even easier to recall. Eventually, you will not need memory tricks, because when you imagine the idea or concept, the vocabulary term automatically will come to mind.

VOCABULARY: EXPRESSING CORRECT TERMS

The easier part of vocabulary is recognizing the meaning of a term that you are given. However, you have not mastered a term's vocabulary until you can recall it and use it correctly to send messages. Sending messages to other people requires highly coordinated muscle movements to write or pronounce the terms. Series of coordinated muscle movements are called motor skills. Our brains learn motor skills by repetition. For example, walking, dancing, singing, riding a bike, driving a car, and tying shoes all require repetitive practice to master the motor skills. The same is true for producing language. In fact, your brain has a specialized language center dedicated to motor skills for producing terms. You must practice language motor skills to get good at them. After learning a new concept, each time you imagine its meaning also think of the term. Imagine how the term is pronounced and say it correctly. Imagine how the term is written, and write it correctly. The more you practice the term, the better you will produce it.

Of course, to send good messages, you must be able to speak or write terms correctly. If you use incorrect pronunciations or spellings, then you can send incorrect information or make your audience believe that you are ignorant and do not understand the concepts behind the terms. Be careful: language centers in human brains have good memories. If you allow yourself to pronounce or write a word incorrectly, the incorrect motor skill will be programmed into your memory. Avoid misspelling and mispronunciation whenever possible. Unfortunately, some terms are not pronounced the same way that they are spelled. For such terms, try learning two pronunciations: one for speaking and another for spelling. The pronunciation that matches the spelling is called its phonetic pronunciation. Interestingly, people who are "poor spellers" often have great spelling memories. Unfortunately, they taught themselves incorrect motor skills that still get recalled when they attempt to use the terms. Each time you misspell or mispronounce a term, you must overwhelm the incorrect motor memory with the correct skill. Achieve this by repeating the correct spelling or pronunciation multiple times. For newly acquired terms, research has shown that each incorrect attempt must be accompanied by at least seven correct repeats. If you have a history of expressing a term incorrectly, then many more correct repeats might be necessary.

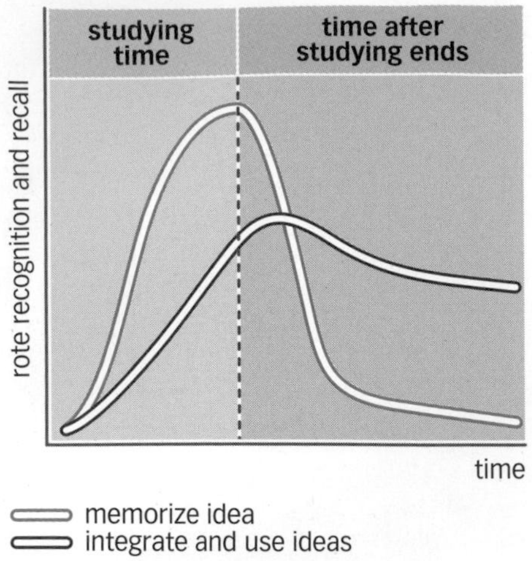

Figure 3 Effect of study method on rote recognition and recall. Study methods include repetitive memorization of concepts (memorize ideas; blue curve) and conceptual integration and using ideas (integrate and use Ideas; red curve). Increases in the graphs indicate learning curves, whereas decreases indicate forgetting curves. The vertical dashed line indicates when studying stops. Note that the learning curve for conceptual integration and using ideas continues past the dashed line, because these activities often keep the mind thinking about concepts even after studying ends.

Most poor spelling and poor pronunciation can eventually be overcome by repetitive practice. However, if you commit to writing and speaking correctly the first time and every time, then your language motor skills will improve more rapidly with the least amount of work.

Repetitive memorization works well for the spelling and pronunciation of terminology, because these are mostly motor skills. However, speaking and writing about a concept cannot be done unless you connect to and recall those motor skills. Again, these connections can be made by practice, but now the practice needs to show your brain the usefulness of the concept and its corresponding term. You must imagine using a concept in a variety of different ways, and then create different messages using the correct terminology. Each different correct use will improve your understanding of the concept, and each different message will improve your ability to recall the proper term. As you practice more uses, your brain will value the concept more and will recall the correct vocabulary more easily. As your understanding grows, your skill at recognizing new, different, novel presentations of the concept will grow. This skill is called general recognition or generalization.

REACH BEYOND MEMORIZATION TO INTEGRATION OF CONCEPTS

Integration of concepts involves actively relating concepts to each other, allowing us to improve generalization skills. Memorization does not improve generalization. Repetitive memorization or rote memorization only increases rote recognition skills. Rote recognition is your ability to recall a concept, but only when a new experience matches the way you learned the concept. Often, when people refer to memorization, they mean rote memorization, which only gives them an ability to do simple recognition. Rote memorization is seductive, because you can rapidly learn to recognize and recall many terms in a short time. This gives an impression of rapid learning, although limited to rote recognition. Rote memorization does not help you recognize broad uses for a concept. If usefulness of a concept is limited, then your efficient brain rapidly forgets it (see the blue graph in **Figure 3**). When memorization stops, forgetting happens quickly and most memorized concepts eventually get lost. Rote memorization is only useful for concepts and terms that you must recognize in the same way that you learned them. Furthermore, you can only solve simple recognition problems, which must be solved within the limited time before forgetting occurs.

Conceptual integration is much more powerful than memorization. Conceptual integration means adding ideas together, relating different concepts to each other, and making more connections between different ideas in your brain. This increases your understanding of a concept, which increases the usefulness of a concept to you. As understanding and usefulness increases, then your efficient brain values the concept and keeps it for future use. Making connections for each concept takes time, so recognition and recall of concepts increase more slowly than for rote memorization. In the same study time using integration of concepts, you learn to recognize less than with memorization. However, forgetting is significantly inhibited by integration of concepts. After studying stops, integration allows you to remember many times more concepts for much longer periods than rote memorization (see the red graph in **Figure 3**). Furthermore, conceptual integration allows you to solve more complex problems, which involve more than just rote recognition.

Rote memorization produces very limited recognition. In essence, you have not mastered the concepts and are unable to solve problems that require more understanding. General recognition or generalization is an ability to identify concepts even when you experience them in ways that you have never seen before. This requires a firm understanding of the concepts, which does not occur much during memorization (see the blue graph in **Figure 4**). If you repeat the same thoughts about the same concept, without making new connections to other ideas, you limit yourself to success in rote recognition and will not develop your ability to generalize. Using memorization alone, it takes more time to do general recognition and problem solving, because you need much more time to think through connections that you failed to make during studying (not shown in **Figure 4**). How can you tell whether you rely too heavily on memorization? General recognition and problem solving must often be done quickly within a limited time. If you rely too much on memorization during learning, then when problem-solving time is limited, you will:

- **Feel rushed for time (or run out of time)**
- **Feel like you do not know what you are doing**
- **Have little or no confidence in your answers**

Whereas rote memorization helps little with generalization and problem-solving skills, conceptual integration and practice in using ideas significantly develops these difficult skills. In essence, as you make more connections to concepts, you increase your understanding, which is essential for these more complex skills. Thus, as you integrate ideas and use them in different ways, you will more easily be able to do general recognition and problem solving (see the red graph in **Figure 4**.) Again, because your brain is learning uses for the concepts, it will value them more and you will retain more ability to generalize and solve problems with these concepts. Conceptual integration and practice using concepts is the only way to achieve your greatest ability to understand and solve the most difficult problems. If you use study time to gain deeper understanding, you will be able to do more general recognition, solve more problems, and perform skills in less time and with more confidence than someone who studied too much by rote memorization.

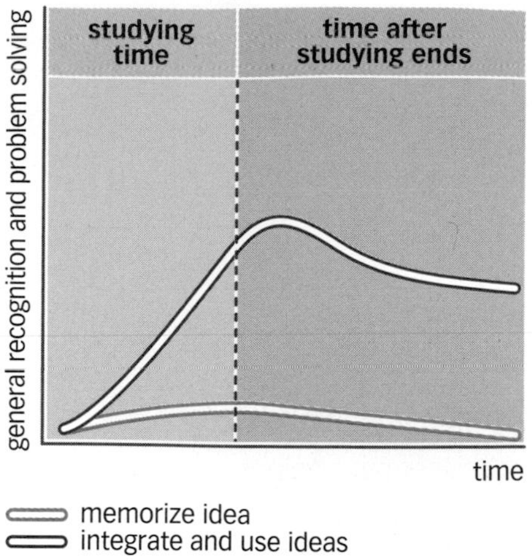

Figure 4 Effect of study method on general recognition and problem solving. Study methods include repetitive memorization of concepts (memorize ideas; blue curve) and conceptual integration and using ideas (integrate and use ideas; red curve). Increases in the graphs indicate learning curves, whereas decreases indicate forgetting curves. The vertical dashed line indicates when studying stops. Note that the learning curve for conceptual integration and using ideas continues past the dashed line, because these activities often keep the mind thinking about concepts even after studying ends.

As a health care consumer or worker, you will face problems addressed by microbiology concepts. The problems will rarely, if ever, present themselves in the same way that you learned the concepts. Thus, it is incredibly important to reach for the highest levels of understanding, so you can generalize and recognize concepts in the novel ways in which they will show up in health care. Greater breadth and depth of understanding of how the concepts relate to each other will provide greater ability for you to solve health care problems.

SUMMARY: USING YOUR BRAIN EFFECTIVELY

As needed, re-read this chapter or seek other resources to develop your learning skills. To optimize your brain's work for learning, review these reminders:

- **Remember always to think about studying yourself.**
- **Want to learn, believe you can learn, and strategize your learning, to get the most progress out of the study time that you spend.**
- **Foster your curiosity and interest in new concepts. Learn them as though you will always remember them. Enjoy learning and reward yourself with feelings of self-accomplishment each time you master another concept.**

- Plan ahead, to give yourself the time you need to reach your learning objectives.
- Actively learn new concepts and test yourself often. Use recorded answers to evaluate your learning, what you master, and what you still need to master.
- Starting with convenient or assigned resources, strive to understand concepts. If these resources expect skills that you lack, then seek simpler resources to master more basic concepts and skills. Afterwards, return to higher-level resources for greater success.
- There are many ways to study. Do not limit yourself to one method, and move forward from memorization to the integration and use of concepts.
- Examine different presentations of a concept and different uses. Imagine common examples. Consider closely related concepts that might lead to confusion. Study opposites for concepts that have them.
- Remember: mastering concepts requires much more than just memorizing words, definitions, and examples. To master a concept, you must relate it to other concepts.
- Analyze smaller parts of a new concept that combine to make the idea complete. Figure out how the concept integrates with other concepts to form more complex ideas.
- Be creative; imagine concepts in your mind and practice using them. Invent questions and answers about who, what, where, when, why, and how. The questions can also be used for self-testing.
- Create stories, jokes, songs, drawings, tables, graphs, classification schemes, and so on, that relate concepts to each other and to your personal background. For more difficult concepts, try a greater variety of study approaches.
- Learn to communicate well with each concept. Develop your vocabulary. Recognize and correctly interpret terminology sent in messages to you. Recognize novel presentations of concepts in your experiences and recall proper terminology that you would use to send messages.
- Practice sending messages using your new vocabulary, and correct any mispronunciations and misspellings with enough repetition.
- Whenever possible, strive to understand concepts beyond mere rote recognition: understand enough to generalize concepts, solve complex problems, and reach goals using the concepts.

PART I

Foundations

The study of microbiology may be the most important and fascinating course a health care professional ever takes. To understand the information we will be discussing as we travel through the chapters of this book, we have to prepare a sound foundation. **PART I** of this book contains chapters that will work toward establishing that foundation.

The opening chapter gives us some good examples of the involvement of microbiology in our everyday lives. It also gives us our initial look at infectious diseases and the concepts of virulence and pathogenicity, which we will see repeatedly as we proceed.

In **Chapter 2** we review and re-establish the principles of chemistry that are required for understanding many of the topics to be discussed in later sections of the book. Basic concepts of bonding, pH, and types of biological molecules will be very important as we discuss molecular structures and infectious disease mechanisms.

Chapter 3 gives us a working knowledge of the basic mechanisms associated with metabolism. This is important because it is metabolism that controls the growth of microorganisms. When we use the term "growth" in microbiology, we mean an increase in the number of organisms. For pathogens, which are our main interest, growth is an integral part of the infectious process.

The final chapter in **PART I** (**Chapter 4**) builds on what was introduced in **Chapter 1** by reinforcing our fundamental understanding of pathogenicity. We accomplish this by focusing on the relationship between the host and the pathogen. This chapter also provides a preliminary look at the differences between prokaryotes (bacteria) and eukaryotes. Finally, the chapter describes the structure of the eukaryotic host cell, which is the target of infectious disease. Here, in addition to refreshing your memory about the structures found in the eukaryotic cell, we examine the role of many of these host cell structures during the infectious process.

When you have finished **PART I** you will:

- Have a foundation in the chemistry required for the study of microbiology.
- Be able to relate how bacteria need and acquire the energy needed for growth.
- Have developed a foundation for understanding pathogenicity and virulence and the relationship between the pathogen and the host.
- Have reviewed the structures of the host cell that are involved in the process of infection.

All of these things will make the chapters in **Part II** and the rest of the book easier to understand and will establish a thread of continuity that will continue throughout the book. Enjoy the journey!

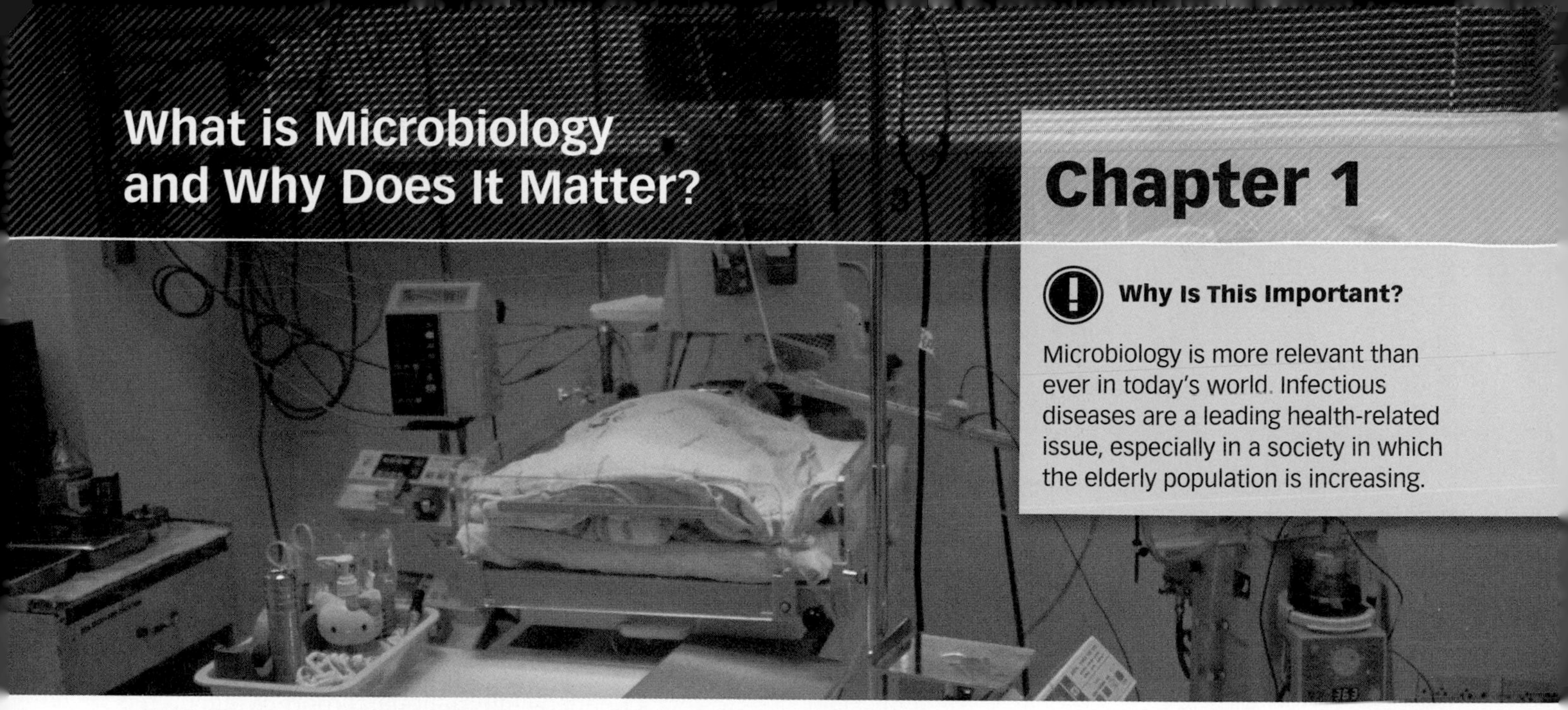

What is Microbiology and Why Does It Matter?

Chapter 1

Why Is This Important?

Microbiology is more relevant than ever in today's world. Infectious diseases are a leading health-related issue, especially in a society in which the elderly population is increasing.

OVERVIEW

In this chapter we will look at the reasons why as a health care professional it will be important for you to understand microbiology. Rather than take the traditional approach and talk about the historical facts associated with this field we will look at microbiology from the perspective of infectious disease and how it affects us every day.

We will divide our discussions into the following topics:

What Is Microbiology and Why Does it Matter?

- CASE STUDIES
- THE RELEVANCE OF MICROBIOLOGY TO HEALTH CARE

"Infectious disease is one of the few genuine adventures left in the world. The dragons are all dead and the lance grows rusty in the chimney corner. ... About the only sporting proposition that remains unimpaired by the relentless domestication of a once free-living human species is the war against those ferocious fellow creatures, which lurk in dark corners and stalk us in the bodies of rats, mice and all kinds of domestic animals; which fly and crawl with insects, and waylay us in our food and drink and even in our love."

Hans Zinsser

The above quote fits very well with today's concept of microbiology and infectious disease. However, Hans Zinsser, who was one of the early pioneers in microbiology and infectious disease, actually made this remark in 1934. This concept of the importance of microbiology still holds true today. If you look at the newspaper or listen to the news it is surprising how many times microbiology and infectious diseases are mentioned.

In September 2005, hurricane Katrina devastated the gulf coast and destroyed much of New Orleans. In the aftermath of the hurricane, one of the greatest dangers became infectious diseases that could result from the lack of sanitary conditions. These conditions can be seen after many natural disasters, and infectious diseases can be many times more difficult to deal with than the physical elements of the disaster. Let's look at a few other examples of how microbiology affects everyday life.

What Do I Need to Know?

To get the most out of this chapter, please review the following terms from your previous courses in biology, anatomy, physiology, or chemistry: antibiotics, antiseptics, bioterrorism, bioremediation, biotechnology, carrier, disinfectants, gall bladder, multi-drug resistant tuberculosis, normal bacterial flora, peptidoglycan.

Figure 1.1 A crowded hospital waiting room is a perfect environment for the spread of infectious diseases such as tuberculosis. Many of the people here are already debilitated from illness, which increases the chances that they could become infected.

CASE STUDIES

Special Delivery...

You go to the mail box and take out the mail. While looking through it you notice that one of the letters has a fine white powder on it that blows off as you walk back to your house. As you open the letter, more of the powdery substance fills the air and you can't help breathing some of it in. Without knowing it, you have inhaled spores of the bacteria *Bacillus anthracis* and contracted **inhalation anthrax**. You will be dead in a matter of days, and there is little anyone can do to prevent it. Does this sound like a sci-fi movie? It's not. In fact, it happened in 2001 and could happen again because **anthrax** is a potential weapon for bioterrorists.

Ivan Goes to Chicago...

Ivan has been a prisoner in Russia for 5 years. During this time he has contracted multi-drug resistant **tuberculosis**. He has served his time and after being released, he decides to visit relatives in Chicago. He travels on a full flight to New York's Kennedy International airport. While on the plane he experiences several severe coughing spasms, which he thinks little of because he has active disease and has experienced them for several years. In New York he changes to another full flight and lands safely in Chicago. After several days with relatives, he feels poorly and goes to the county hospital for treatment. Because Ivan is not bleeding or experiencing chest pain he is told to take a seat in the waiting room, which is full (**Figure 1.1**). While waiting, Ivan once again begins to cough. In point of fact, Ivan is a one-man epidemic.

He is contagious, and with each cough he is expelling infectious *Mycobacterium tuberculosis* organisms in an aerosol. If the droplets are small enough, they can be expelled a considerable distance and hang in the air for a relatively long period. During Ivan's travels, he has exposed all 300 people on the flight from Russia to New York, many of whom have changed planes in New York and exposed their fellow passengers to the pathogens they got from Ivan. In addition, while waiting in the hospital, he has exposed everyone else in the waiting room. More importantly, he has managed to bring a deadly pathogen into an environment (the hospital) where people are debilitated from being ill or **immunocompromised**, a condition in which their immune system is not functioning properly. If these people become infected, they will have difficulty dealing with the tuberculosis pathogen that Ivan is carrying. Because of today's rapid methods of travel, this kind of scenario is not that far-fetched. Experts estimate that some respiratory diseases could be moved around the world in less than 48 hours.

Hamburger Havoc...

Your 6-year-old daughter loves hamburgers, so you decide to take her to her favorite hamburger restaurant. She delightedly gobbles up a huge burger but that night she experiences a bout of **diarrhea** and the next day begins to vomit. The diarrhea and vomiting continue for 2 days and are accompanied by severe abdominal cramping. You suspect food poisoning and continue giving her fluids, but you notice that her stool now has blood in it. By the time you take her to the doctor, she is anemic and in renal failure and is immediately put on dialysis. The doctor finally tells you that she has been infected by enterohemorrhagic bacteria called *Escherichia coli* (**Figure 1.2**), which are found in improperly processed ground beef, and had she not been seen in time she would have died. This kind of **bacterial hemorrhagic disease** is a problem that is caused by the high volume of food production and has resulted in significant changes in the beef industry (**Figure 1.3**).

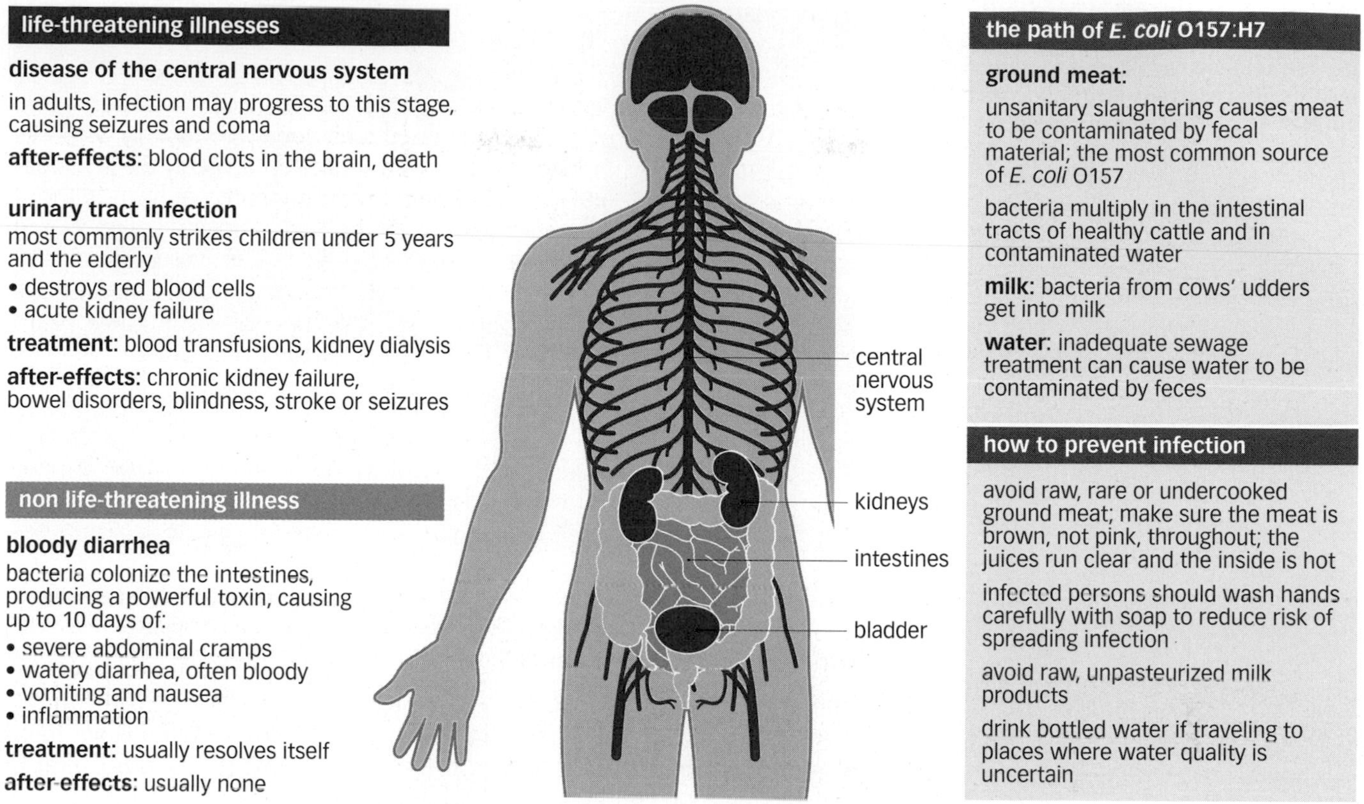

life-threatening illnesses
disease of the central nervous system
in adults, infection may progress to this stage, causing seizures and coma
after-effects: blood clots in the brain, death
urinary tract infection
most commonly strikes children under 5 years and the elderly
• destroys red blood cells
• acute kidney failure
treatment: blood transfusions, kidney dialysis
after-effects: chronic kidney failure, bowel disorders, blindness, stroke or seizures
non life-threatening illness
bloody diarrhea
bacteria colonize the intestines, producing a powerful toxin, causing up to 10 days of:
• severe abdominal cramps
• watery diarrhea, often bloody
• vomiting and nausea
• inflammation
treatment: usually resolves itself
after-effects: usually none
central nervous system
kidneys
intestines
bladder
the path of E. coli O157:H7
ground meat:
unsanitary slaughtering causes meat to be contaminated by fecal material; the most common source of E. coli O157
bacteria multiply in the intestinal tracts of healthy cattle and in contaminated water
milk: bacteria from cows' udders get into milk
water: inadequate sewage treatment can cause water to be contaminated by feces
how to prevent infection
avoid raw, rare or undercooked ground meat; make sure the meat is brown, not pink, throughout; the juices run clear and the inside is hot
infected persons should wash hands carefully with soap to reduce risk of spreading infection
avoid raw, unpasteurized milk products
drink bottled water if traveling to places where water quality is uncertain

Figure 1.2 The enterohemorrhagic bacterium *E. coli* O157:H7 can cause life-threatening illness.

The Hospital Can Be Dangerous...

Uncle Harry went to the hospital for a hip operation. The surgery went well and the family expects Harry home in a matter of days. You have volunteered to get Uncle Harry and bring him home from the hospital. When you arrive, Harry is not in his room, and you find that he has been transferred to intensive care (**Figure 1.4**). Upon arriving at the intensive care unit you see that Uncle Harry is in a room with a sign outside that reads **MRSA** Authorized Personnel Only. The doctor tells you that Uncle Harry has developed a **nosocomial** (hospital-acquired) infection with a bacterium called *Staphylococcus aureus* that is resistant to the **antibiotic** methicillin. Harry stays in intensive care for three more days and dies at 8 a.m. on the fourth day. You and the rest of the family are in shock because Harry was not sick and was going into the hospital for a simple elective operation. In fact, your Uncle Harry is one of over 20,000 people who die each year in the United States from nosocomial infections.

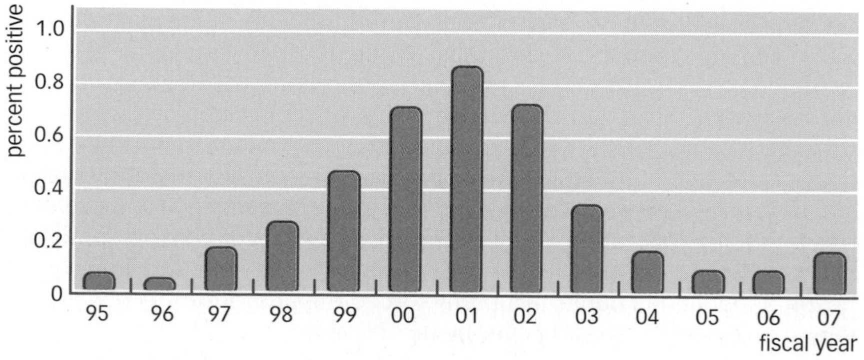

Figure 1.3 Results of the examination of raw ground beef for the presence of *E. coli* O157:H7. Notice the increases in positive samples between 1999 and 2003. The meat-processing industry has incorporated new safeguards to protect the public.

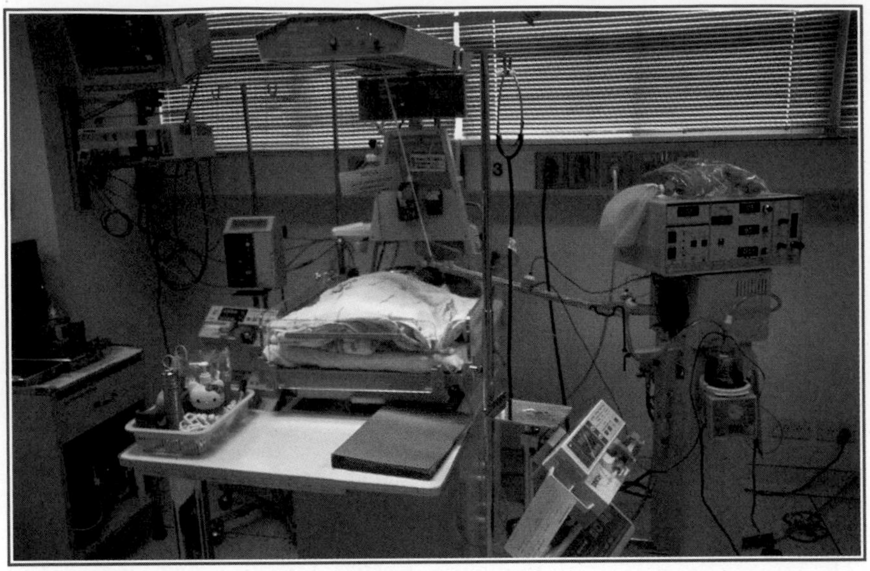

Figure 1.4 An intensive care unit that is closed to all except authorized personnel as a result of infection with methicillin-resistant *Staphylococcus aureus*. The same kinds of conditions are used for VRSA (vancomycin-resistant *Staphylococcus aureus*). These kinds of hospital infection are on the rise because of the increase in antibiotic resistance among bacteria.

Did You Wash Your Hands?...

"Employees Must Wash Hands Before Returning To Work." You can see this sign in many places throughout the world. Why? What does hand washing accomplish and why is it important? There was a hospital in which over 7000 children were born each year. However, an epidemic of fever in this hospital, caused by Group A streptococcal pathogens normally found on your skin, killed 600–800 mothers each year (a death rate of 10%). When the records of these events were carefully examined, they showed that in the ward where these deaths were most prevalent, only doctors and medical students delivered the babies. In the other delivery ward, the babies were delivered by midwives and there was a very low incidence of these infections. On looking more closely at the facts, it was found that many of the doctors and medical students had moved directly from performing autopsies to the delivery room without properly washing their hands. When they began to wash their hands thoroughly, the number of deaths caused by this infection dropped to less than 1%. Incidentally, these events occurred over 150 years ago, and the work of Dr Ignaz Semmelweis, who figured out this problem, was instrumental in developing the standards for hospital hygiene today.

Mary, Mary, Quite Contrary...

Mary was a cook. She loved cooking and worked for a very wealthy family in New York in 1906. The family had visitors one day and one of them infected Mary with **typhoid fever**. Interestingly, Mary did not get sick but she became a carrier. Within a short time several people in the house where she worked developed typhoid fever, so Mary moved to another home. The same thing happened there and at two other places where Mary worked, and she was finally forced to seek employment under assumed names. After 6 years the New York Board of Health caught up with Mary and, although she was unwilling to cooperate, forced her to be examined in a hospital. The examination showed that Mary's gall bladder was infected with typhoid-causing bacteria and that the only way to stop the infections was to have the gall bladder removed. Mary refused to agree to the operation and was therefore sentenced to isolation in the hospital. For 3 years, Mary was "imprisoned" in the isolation ward of the hospital, but finally public sympathy forced the authorities to release her with a promise she would never be a cook again.

Five years later, 25 cases of typhoid fever occurred in a woman's hospital, and 8 of the infected people died. Coincidentally, the cook for the hospital suddenly disappeared. It did not take long for the authorities to discover that the cook was Mary. When she was arrested, Mary defended her actions by saying she had a right to be a cook and that she would continue to cook. She was sent back to the quarantine ward of the hospital and was forced to remain there for the rest of her life (23 years). Remarkably, Mary died of a stroke, not typhoid fever. We will see that there are many infections that are latent and carried by individuals who do not get sick themselves but can pass the infection to others.

It's for the Birds....

Many people get a flu shot each year to protect them from the flu. One of the most compelling problems in microbiology and infectious disease today is **influenza**, which is a viral disease. As we will see, there are numerous strains of this virus and some have been very destructive. In fact, there have been three severe epidemics of influenza, the worst of which occurred in 1918 (the Spanish flu) and killed 30–50 million people in 1 year (**Figure 1.5**). Fortunately, it has not been seen since, but there is the potential for genes from the virus that infects humans to combine with genes of a type of influenza virus that is currently seen in birds (avian influenza). This could result in the appearance of a pandemic similar to that seen in 1918. If this occurs, will a flu vaccine help and will there be enough of the vaccine to go around, or will decisions have to be made about who gets it? Even more importantly, how many people will die?

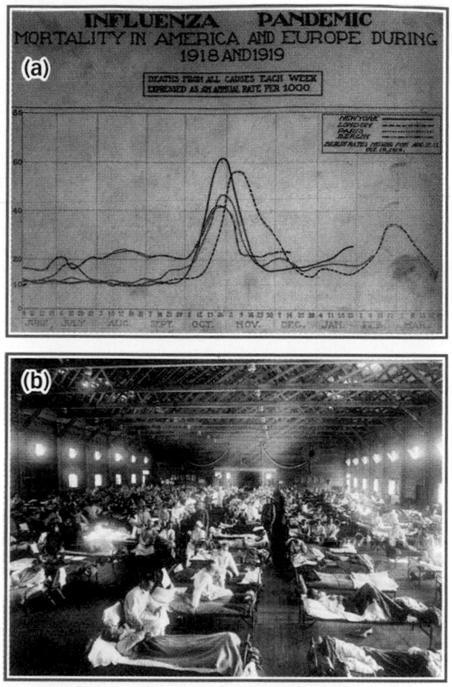

Figure 1.5 The 1918 influenza pandemic. Panel **a**: The Spanish Flu pandemic of 1918 was a worldwide infection that killed an estimated 50 million people. Panel **b**: In hospitals of that era it was difficult to limit the spread of infection, especially when so many became so sick, so quickly. Could we handle this type of epidemic today?

Keep in Mind

- Microbiology has relevance to everyday life.
- We do not live in a sterile environment; therefore we interact with microbes all the time.
- Travel allows the movement of infectious diseases around the world in a relatively short time.

THE RELEVANCE OF MICROBIOLOGY TO HEALTH CARE

The above examples are just a small sample of the type of microbiological problems that we have to face. The better that health care professionals understand how a pathogen causes a disease, the more they can help patients. As a health care professional you may be involved in cases like these. How will you react? The chapters that follow are designed to help you if and when that time comes.

Infectious disease accounts for a large percentage of health care. If we look back through history, there has always been disease. In fact, spores of bacteria have been found that are over 25 million years old. In the nineteenth century, tuberculosis and other pulmonary infections were the scourges of the world and were the leading cause of premature death. Although the causes of some of these infections were being discovered, there was little that could be done to stop or prevent them.

The science of microbiology dates back to pioneers who isolated specific microbes and proved that they could cause disease. This was the first golden age of microbiology (1875–1910), when microbes were directly associated with the diseases they caused and was followed by advances in public health that lessened the number of deaths resulting from infection.

In the first half of the twentieth century, scientists studied the structure, physiology, and genetics of microbes and found links between microbial properties and disease. During this period, the discovery of penicillin in 1929 and sulfonamide drugs in 1936 opened the door for effective treatments of infectious diseases. By the end of the twentieth century, advances in the fields of molecular biology and genetics had increased our knowledge at the molecular level. Furthermore, development of vaccines and antibiotics as well as better sanitation methods changed the face of infectious disease (**Figure 1.6**).

Now in the twenty-first century, after most infectious diseases were thought to be under control, uneasiness has begun to grow around the world. Diseases we thought were conquered are reappearing, and resistance to treatments that were previously completely effective is beginning to grow. New diseases are emerging and organisms that were thought to be harmless are now being found to cause disease in certain circumstances. The potential for bioterrorism has gone from fiction to fact, with anthrax being used to kill innocent people. For students in the health sciences, the fundamental understanding of microbiology has never been more relevant.

Infectious Disease

It is important to remember that, of the thousands of species of bacteria, viruses, fungi, and parasites, only a tiny fraction are involved in disease of any kind. In fact, most of these organisms live their lives, essentially ignoring humans. Furthermore, of the tiny number involved in infections, only a fraction cause disease in humans; the rest infect other animals or plants. Among the human pathogens, the potential to cause disease is referred to as **virulence**, and we can categorize pathogenic microorganisms based on their degree of virulence.

Many bacteria and some fungi are part of what we call the normal microbial flora of the body, where they naturally colonize the skin and mucosal surfaces. Most of the time these organisms are harmless and in some cases they actually provide us with important products such as vitamins. In certain circumstances the organisms can become pathogenic, but when they do they are only mildly virulent. In contrast, there are

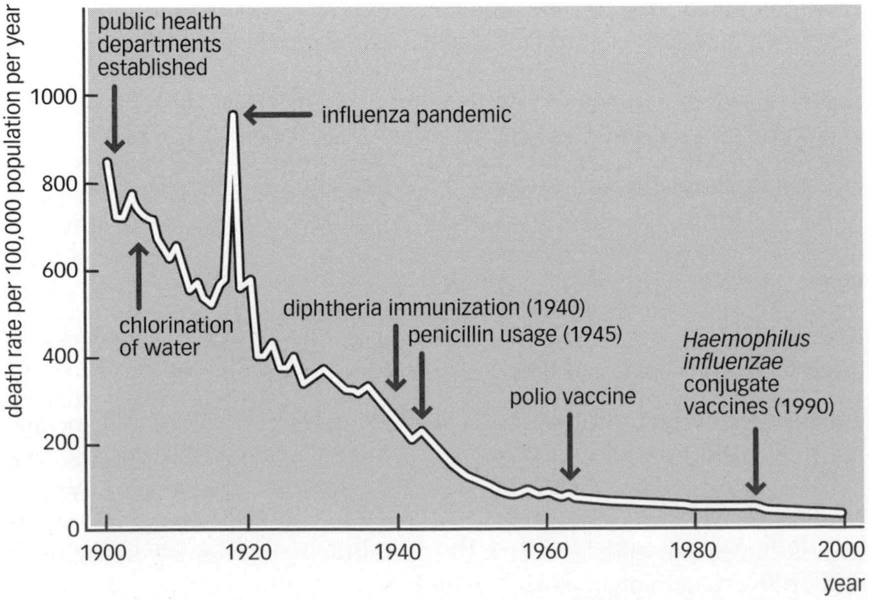

Figure 1.6 Death rates for infectious disease in the United States during the twentieth century. Notice the decline after the introduction of public health measures, immunization and antibiotics.

pathogens, such as *Yersinia pestis*, that are always highly virulent and always associated with disease. *Y. pestis* causes **bubonic plague**, which involves disease that comes on suddenly with great severity, and death in one-third to one-half of those infected. For most pathogens the basic aspects of how they cause disease can be looked at from three perspectives: epidemiology (the study of the factors determining the frequency and distribution of disease), pathogenesis (the study of how disease develops), and host defense.

Epidemiology — As we will see in our discussions of the disease process (Chapters 5, 6, and 7), pathogens must accomplish five tasks to successfully cause disease. First they must get into the body or the cell. Then they must be able to stay in, defeat the host defenses and cause damage to the host. Last, they must be able to be transmitted to a new host so that the infection will continue. In epidemiology, each pathogen is looked at from this perspective and classified with regard to transmission. Some are transmitted by air, some by food and water, some by insect vectors, and still others through person-to-person contact. In addition, geography has a role in infectious disease: some pathogens are found worldwide, whereas others are restricted to certain geographic areas. Knowing how an organism gains entry and how it spreads are vital to our efforts to care for infected individuals and to protect those around them. It also helps in our understanding of the epidemic spread of disease.

Epidemic outbreaks of disease are fostered by factors such as poor socioeconomic conditions, ignorance of the cause of infection, natural disasters, and poor hygiene. The possibility of recurrence of old pandemics such as the Spanish flu of 1918 remains a problem today and we are already seeing an extended pandemic with HIV infection and **AIDS**. Moreover, bioterrorism presents us with even more menacing possibilities as in the case depicted above, in which the inhalation of human-produced aerosols of anthrax spores could produce a lethal **pneumonia** on a massive scale.

Pathogenesis — It is not easy for pathogens to produce disease in humans. This is the primary reason that so few organisms are successful pathogens. Multiple factors known as **virulence factors** are required for the organism to persist in the host, cause disease, and escape the host defenses so that the infection can continue.

Pathogens can use a variety of methods to damage host cells and tissues. Bacterial pathogens can produce digestive enzymes and toxins, while the intracellular multiplication of viruses seen as part of viral infections produces massive numbers of viral particles, which cause the death of the infected cell. In fact, some of the damage associated with infection is due to an overcompensating host defense. For example, fever, pain, and swelling are universal symptoms of inflammation that result from host defense reactions designed to fight infection. If these reactions are too severe, they can damage the host. Symptoms can also be associated with the particular organs that are involved in the infection. For example, coughing, diarrhea, and nervous system dysfunction are symptoms seen in the respiratory, digestive, and central nervous systems.

Host defense — We will see throughout our discussions that infection is essentially a complex competition. The pathogen uses multiple methods to survive and thrive, while the host defense (immune system) becomes involved immediately in an attempt to kill or expel the pathogens. Much of the outcome depends on the success or failure of the host defense. The best example of host defense failure is the disease called acquired immunodeficiency syndrome (AIDS) in which the defense mechanism of the host is compromised and infections that would have been easily defeated end up overwhelming and killing the host.

There are two basic types of defense against infection. The first is the nonspecific defense referred to as the **innate immune response**, which represents the first line of defense by the host. Nonspecific responses are truly dynamic, elegant, and lethal. As we will see in Chapter 15, these responses involve a variety of cellular mechanisms coupled with chemical and mechanical factors. In many cases these are all that are needed to defeat the infection. The second line of defense is called the **adaptive immune response** and is also very lethal and specific. In addition, the adaptive immune response has the benefit of memory, which means that whether you see that same infection 4 or 40 years later, the response to it will be immediate and powerful. It is the adaptive defenses that keep us safe in a hostile atmosphere. The critical role of these systems can be seen in burn patients who have lost part of an important innate barrier (the skin), and in patients with AIDS, who lose the adaptive immune defense.

Consequently, for a pathogen to be successful it must find a way to defeat, evade, or hide from these potent host defenses, and many pathogens have developed intriguing ways of doing so. Some will attack the defending cells, while others will change their looks (a form of camouflage) such that the adaptive defense response is fooled. Some pathogens will use the defending cells as a place to hide from the host defense.

Treatment of Infectious Diseases

During the course of our discussions we will focus on some of the treatments that are used for specific infections. Many potent and successful therapeutic tools to treat us and protect us from infection have been developed over the past 75 years. These include antibiotics as well as **disinfectants** and **antiseptics**. As we will see later, antibiotics can have adverse side effects and they therefore have to be selective. They must kill the disease-causing microorganisms but not harm the patient. The first antibiotic ever discovered was penicillin, which also happens to be naturally selective. It prevents the formation of one of the components of the bacterial cell wall called **peptidoglycan**, which is not found in human cells. Unfortunately, there are very few drugs that are selective for bacteria and even fewer for fungi and parasites, which have close structural and metabolic similarities to our cells.

Treatment becomes even more difficult for viral infections because viruses are, by definition, intracellular parasites that reside in our cells. If treatment does not negate the virus before it invades the host cell, the only defense is to kill the infected cell. Fortunately, our defense mechanisms can accomplish this task, but there is still the collateral damage of losing the cell.

The best treatment for any disease is prevention, and for infectious disease this involves public health measures and immunization. To take effective public health measures, we must understand the transmission mechanisms of infection so that we can interfere with those mechanisms. Public health measures involve activities such as disinfection of water supplies, increased care in the preparation of food supplies, insect control, proper hygiene and sanitation, proper care in waste removal and treatment, and many other measures that prevent contact with infectious agents. Immunization requires that we understand the immune mechanisms and that we design vaccines that will successfully stimulate protection.

Microbiology Isn't Just About Infectious Disease

Before we proceed in our examination of microbes that cause infection and disease, it is important to understand that some organisms are very beneficial. They are responsible for recycling vital elements such as nitrogen in the soil, and they also interact in many other ways with animals, plants, and the environment where they convert elements such as carbon, nitrogen, oxygen, and sulfur into usable forms. Bacteria and fungi also have a role in returning CO_2 to the atmosphere when they decompose organic waste and dead plants and animals.

Fast Fact

The bacterium *Caulobacter cruentus* uses a stalk to adhere to rocks. The tip of the stalk secretes an adhesive made of sugars and proteins. Scientists have measured the strength of the adhesive, which they call holdfast. It is estimated that a 10 cm^2 surface slathered with holdfast could potentially hold about 70 tons!

Bioremediation and recycling — Microorganisms are used to recycle water during sewage treatment (**Figure 1.7**), converting the waste into useful byproducts such as CO_2, nitrates, phosphates, sulfates, ammonia, hydrogen sulfide, and methane. Microbes have been routinely used for bioremediation since 1988, cleaning up toxic waste generated in a variety of industrial processes. In these cases, the organisms use the toxic waste as a source of energy, and in the process they decontaminate it. They can also clean up underground wells, chemical spills, and oil spills as well as producing useful products such as enzymes that are widely used in cleaning solutions.

Insect control — Bacteria and their products have been used extensively to control pests such as caterpillars, bollworms, corn borers, and fruit leaf rollers, all of which can damage crops used for food. In some cases they are applied as part of crop dusting and in others bacterial genes are engineered directly into the plant. These types of agricultural pest control are considerably safer for the environment than the old methods which employed chemicals that polluted soil and water and indiscriminately killed pest and non-pest, some of which are beneficial to humans.

Biotechnology — In the burgeoning world of biotechnology, bacteria and viruses are used in a variety of amazing ways. For example, many of the drugs we take for granted such as insulin are mass-produced in genetically engineered bacteria. We will look at the dynamic field of biotechnology in Chapter 27.

Figure 1.7 Waste water treatment is one of the areas in which bacteria provide beneficial and required service to humans.

Keep in Mind

- When we understand how microbial pathogens cause disease, we are better able to treat and defeat these infections.
- There are thousands of species of bacteria, viruses, fungi, and parasites, but only a tiny fraction are involved in disease.
- The potential to cause disease is referred to as virulence, and pathogens can be categorized on the basis of their degree of virulence.
- Part of the disease process is based on the host's ability to defend itself against infection by using the innate and adaptive immune responses.
- Microbiology is not just about infectious disease: it is also part of the recycling industry, biological remediation projects, and in particular the biotechnology industry.

SUMMARY

- Microbiology is very relevant to our everyday lives.
- We are exposed to potentially dangerous pathogens on a daily basis.
- These pathogens possess virulence factors which allow them to persist in the host, evade host defenses and cause disease.
- Pathogens must accomplish five tasks to be successful in causing disease. They must get in, stay in, defeat the host defenses, damage the host, and be transmissible.
- Microbiology is not just about infection and disease. In many cases microbes can be very beneficial to humans.

We live in a cloud of infectious organisms that could be potentially lethal to us. Consequently, our understanding of how microbes cause disease and how we defend ourselves is basic to our survival on this planet. The following chapters are intended to increase that understanding.

SELF EVALUATION AND CHAPTER CONFIDENCE

MULTIPLE CHOICE

Answers are given in the back of the book and help can be found in the student resources at www.garlandscience.com/micro

1. Infectious diseases are closely associated with
   - A. Wars
   - B. Natural disasters
   - C. The development of antibiotics
   - D. Famine
   - E. All of the above

2. Virulence can be defined as
   - A. Opportunistic infection
   - B. Adequate nutrition
   - C. Degree of pathogenicity
   - D. Limited rates of growth

3. Epidemic outbreaks of disease are fostered by all of the following except
   - A. Poor hygiene
   - B. Decreased birth rates
   - C. Poor nutrition
   - D. Poor socioeconomic conditions

4. The infectious process involves the following two steps:
   - A. Virulence factors and epidemiology
   - B. Transmissibility and virulence factors
   - C. Virulence factors and adaptation
   - D. Adaptation and transmissibility

5. Damage to the host can result from

   A. The host response
   B. Virulence
   C. Multiplication of the pathogen
   D. All of the above
   E. None of the above

6. Which of the following are universal signs of inflammation? (Use the code 1=coughing, 2=pain, 3=redness, 4=heat, 5=swelling.)

   A. 1, 2, 3, 4 and 5
   B. 2, 4 and 5
   C. 2, 3 and 5
   D. 1, 3 and 5
   E. 2, 3, 4 and 5

7. The first host response is

   A. Specific
   B. Minimal
   C. Nonspecific
   D. Partly specific
   E. None of the above

8. The second line of host defense is

   A. The innate response
   B. The accessory response
   C. The chemical response
   D. The adaptive response
   E. All of the above

9. Treatment of disease relies on which of the following?

   A. Antiseptics
   B. Disinfectants
   C. Antibiotics
   D. None of the above

10. Which of the following uses of microbes is not beneficial to man?

   A. Insect control
   B. Crop fertilization
   C. Mineral conversion
   D. Organic decomposition
   E. All of the above are useful

DEPTH OF UNDERSTANDING

Questions listed here require you to bring together the concepts you have learned in this chapter into a discussion format. This helps you to increase your depth of understanding of the material you have learned. Help can be found in the student resources at www.garlandscience.com/micro

1. Seven thousand years ago the measles virus was found only in cattle. Now it is only seen in humans and never in cattle. Explain the reason for this change.

2. Some experts recommend using microbes instead of chemicals for the increased production of agricultural products. What are the benefits and drawbacks of such a change?

Fundamental Chemistry for Microbiology

Chapter 2

Why Is This Important?

An understanding of chemistry is essential for the topics we will be covering and will help you better understand cellular structure and function, which are paramount for your study of microbiology.

OVERVIEW

This chapter is not intended to be a thorough examination of chemistry. Instead, it is designed to provide a basic understanding of chemical bonding, and the structure of biological molecules. It is important to understand these concepts, because the structures and functions of microorganisms are molecular and many of the pathogenic effects of infection occur at the molecular level. Therefore, to understand the infectious process we need to understand basic chemistry.

We will divide our discussions into the following topics:

Fundamental Chemistry for Microbiology

- CHEMICAL BONDING
- WATER
- ACIDS, BASES, AND pH
- BIOLOGICAL MOLECULES

Chemistry is the underlying foundation of many of the things we will study in later chapters of this book. In fact, life on Earth is based on a series of chemical reactions. We will see that many microbial pathogens infect and damage cells and tissues. Therefore, we need to understand how these cells and tissues are constructed and the information presented here will give us a foundation for this construction. Tissues are organized in the following sequence:

atoms
↓
molecules
↓
cells
↓
tissues

What Do I Need To Know?

To get the most out of this chapter, please review the following terms from your previous courses in biology, anatomy, physiology, or chemistry: anion, ATP, ADP, cation, dehydration synthesis, electrons, glycerol, fatty acid, molecule, peptide bond, protein, solution, solvent, and solute.

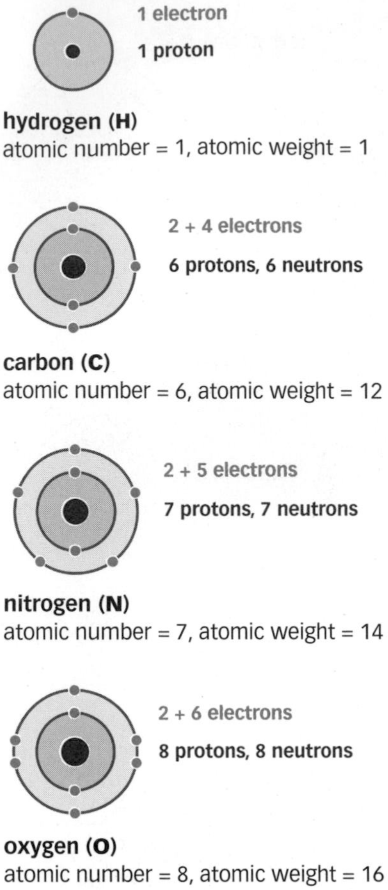

Figure 2.1 The atomic structures of four atoms of biological importance. Hydrogen is the simplest of the four, with only one proton and one electron. In this case, the outer shell is incomplete because it needs two electrons. This causes hydrogen to bond easily to other atoms. For carbon, oxygen, and nitrogen the first shell is filled with electrons but the outer shells are not. It is the outer orbital where chemical bonding takes place.

CHEMICAL BONDING

To understand this sequence, we can look at it as a series of building steps. Atoms bond together to make molecules, and molecules are then joined together to make cells. Finally, cells associate to make tissues. It should be pointed out that these tissues are the basis of organ structure, but the infectious processes that we will be looking at are usually found at the level of cells and tissues, so we will focus on these.

We will begin by looking at how atoms bond together. Atoms are composed of three types of particles: protons, neutrons, and electrons (**Figure 2.1**).

Protons — These are located in the core (also known as the nucleus) of the atom and have a positive charge. The **atomic number** of any atom is equal to the number of protons in the core of that atom.

Neutrons — These are also located in the core of the atom but they have no charge. In certain circumstances, the number of neutrons is equal to the number of protons, but in some cases the number of neutrons can be different from the number of protons. The atomic weight of an element is equal to the total number of protons plus neutrons.

Electrons — These atomic particles have a negative charge and are located in shells (or clouds) that surround the core of the atom. In a neutral or uncharged atom the number of electrons equals the number of protons.

In the atom, electrons are found around the core in shells. The shells occupied by the electrons have limitations as to the number of electrons they hold. In the case of the first shell, only 2 electrons can "fit", while in the second shell there is room for 8 electrons. Although the third shell can actually hold 18, it is stable when 8 electrons are present. The last thing we need to understand is that the shells are more stable if they are full of electrons (that is, 2 in the first and 8 in any subsequent orbital). If an electron shell is not full, that shell is unstable and is capable of bonding so as to make its orbital full and stable. The giving and receiving of electrons between atoms is the basis for ionic bonding.

Ionic Bonds

As a way of illustrating ionic bonding, let's look at two elements. The first is the element sodium, which has an atomic number of 11. We therefore know that the element sodium has 11 protons in the core. We also know that there are 11 electrons around the core in shells. If we apply our numerical restrictions, there are 2 electrons in the first shell, which makes it full, and there are 8 in the second shell, which makes it full too — but there is one electron left, which occupies the third shell. Because this shell is not full, the atom is unstable and this instability makes it possible for the sodium atom to donate the electron to another atom.

Now let's look at the element chlorine, which has an atomic number of 17 and therefore has 17 protons. If there are 17 protons, there will be 17 electrons, and if we fit them into their shells we will have 2 in the first, 8 in the second, and 7 in the third. As we saw in our example of sodium, the outer shell is not filled with electrons and is therefore unstable.

If sodium and chlorine were to meet, they could solve each other's problem because sodium has a single electron and wants to donate it. Chlorine has only seven electrons in its outer shell and needs eight to be stable, so it is ready to accept an electron. That is precisely what happens (**Figure 2.2**): the lone electron from sodium moves to the outer shell of chlorine, thereby filling the outer shell and making the chlorine atom stable. When this giving and receiving takes place, the overall electrical charges of both sodium and chlorine also change and they are classified as **ions**. Because sodium has lost an electron it becomes more positive and is referred to as a **cation**. In contrast, chlorine has gained an electron, making it more negative, and it is now classified as a chloride **anion**. The opposing charges of these cations and anions are attracted, and the two ions form an **ionic bond**. The construction of biological molecules involves a second type of bond called a **covalent bond**, in which electrons are not given or received but instead are *shared*.

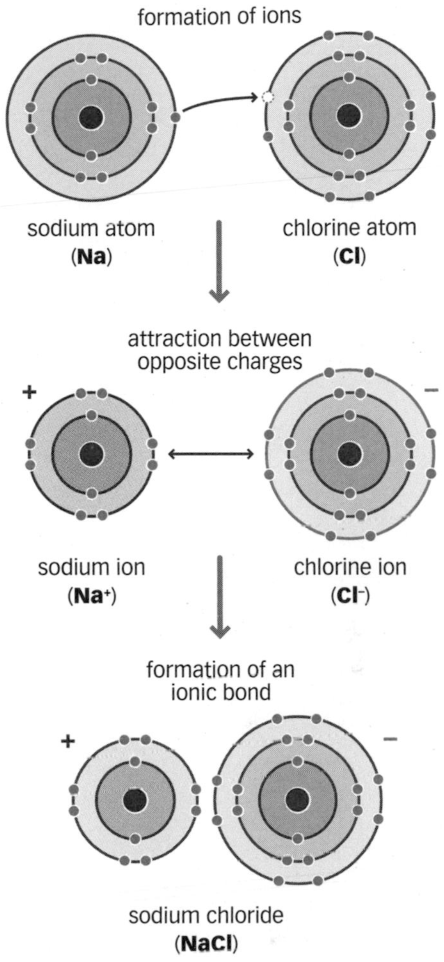

Figure 2.2 The formation of an ionic bond. In this sequence of events sodium and chloride form an ionic bond. The first step is the donation of an electron from the sodium atom to the chlorine atom. This ionizes the atoms, allowing the ionic bond to form.

Covalent Bonds

The second type of chemical bond is the covalent bond. Covalent bonds are the type of bonds seen in biological molecules. In contrast to the ionic bond, the covalent bond is formed when electrons in the outermost shells of the atoms are shared. If one pair of electrons is shared, a single covalent bond will form; if two pairs of electrons are shared, there will be a double covalent bond. In addition, covalent bonds can have polarity, which can be viewed as the equality of sharing that occurs. For example, in a **nonpolar covalent bond**, the sharing of the electrons between two atoms is equal and in this case, the bonds are electrically neutral.

However, if the sharing is unequal (one side is pulling more than the other) there can be a polarity, which results in a weak electrical charge. The best example of this **polar covalent bond** is seen with water, which consists of two hydrogen atoms that are bound covalently to one oxygen (**Figure 2.3**). In this sharing, the oxygen pulls on the hydrogen's bonding electrons more than the hydrogen pulls on the oxygen's bonding electrons. This results in a slightly positive charge on the two hydrogens and a slightly negative charge on the oxygen.

When we look at the mechanism of covalent bonding, atomic structure also has a role. We can use carbon to demonstrate this bonding. In fact, biological life on this planet is based on carbon, which has an atomic number of 6. Therefore, we know that it has six protons in the core, and six electrons in shells around the core. Because it takes two electrons to fill the first shell, there will be four left in the second. Carbon therefore needs four more electrons to fill the shell. This can be accomplished by bonding to other atoms.

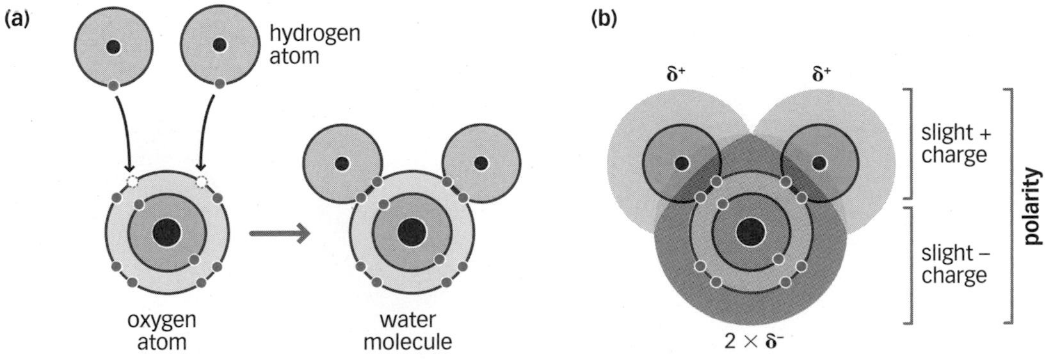

Figure 2.3 The formation of a covalent bond. Panel **a**: The water molecule is formed through covalent bonds in which each hydrogen shares its electrons with oxygen. Panel **b**: This is a polar covalent bond in which the water molecule has a slight negative charge (δ^-) on the oxygen side of the molecule and a slightly positive charge (δ^+) on the hydrogen side of the molecule.

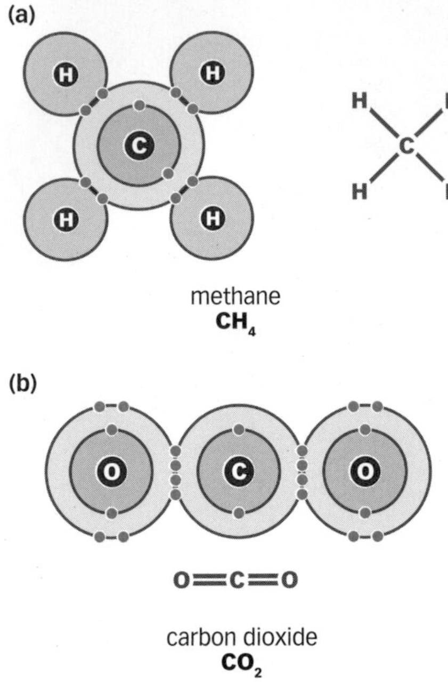

Figure 2.4 The binding properties of carbon. Panel **a** shows the ability of carbon, which has only four electrons in the outer orbital, to bind covalently to other atoms (in this case hydrogens). Panel **b**: The formation of CO_2 takes advantage of the same binding properties of carbon. However, in this case two pairs of electrons are shared with each oxygen, causing the formation of double covalent bonds.

Because covalent bonding involves sharing, it is not necessary to move electrons from one atom to another. All that needs to be done to create a covalent bond with a carbon atom is to bring other atoms near enough so that their outer shells are close to its outer shell. We can illustrate this with hydrogen atoms, which have only one electron in the outer shell and need two electrons to be stable (**Figure 2.4**). Because carbon has only four electrons in its outer shell, four hydrogens can bond covalently to carbon by sharing their electrons. When the four hydrogens are covalently bonded to it, the carbon will have eight electrons in its outer shell, which makes it stable. Each of the hydrogens will have its own electron plus one shared from carbon, giving the hydrogens two in their outer shell and making them stable as well.

Carbon's ability to bond covalently with other atoms makes it useful for building biological molecules such as carbohydrates, lipids, proteins, and nucleic acids. In addition, each of these biological molecules will have a chemical shape that will be directly related to the function of the molecule. Part of achieving and maintaining that shape depends on the last type of bond, the **hydrogen bond**.

Hydrogen Bonds

These are chemical bonds that are found between and within molecular structures, and they help to give molecules their shape. Hydrogen bonds form because of the polarity of molecules seen in covalent bonding. If we think about this in simple terms, we can use the analogy of an atom sharing with another atom but being still somewhat attracted to a third atom.

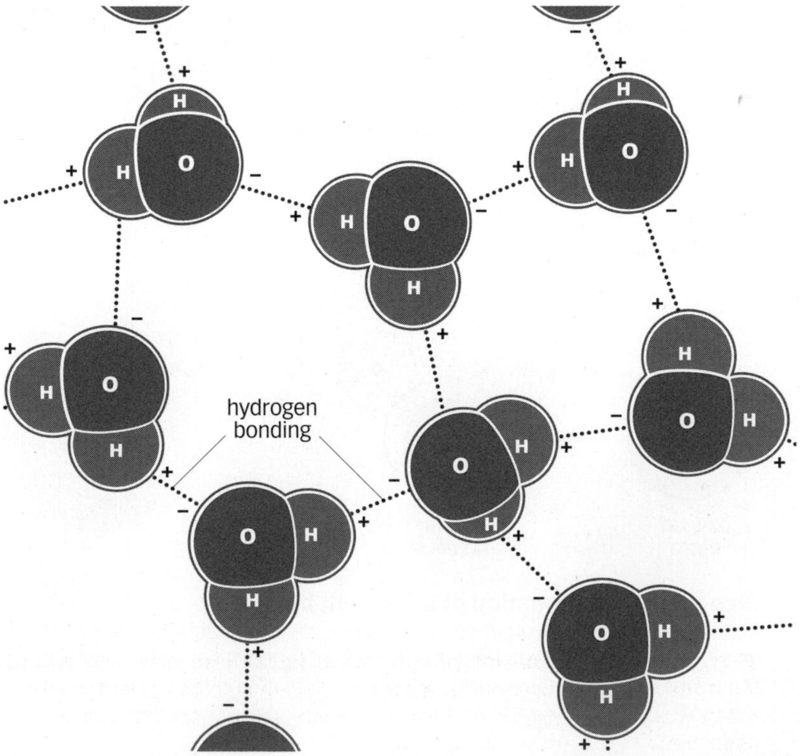

Figure 2.5 Hydrogen bonding between water molecules occurs because of the slight positive and negative sides of the water molecules. These hydrogen bonds are the weakest of the chemical bonds and can rapidly break and reform.

If we consider water, this becomes easy to understand. In **Figure 2.5** we see that the polarity of water allows one molecule to be attracted to another through the slightly positive side of one water molecule and a slightly negative side of another.

Hydrogen bonding gives water a fluid property when these weak bonds between water molecules are breaking and reforming quickly. If we lower the temperature, the breaking and rejoining slows down and the water becomes ice. If we heat the water, the bonds break so fast that there is little reforming and the water molecules dissipate as steam.

Keep in Mind

- Atoms are composed of protons, neutrons and electrons.
- Atoms can bind together to make molecules, which can join together to make cells.
- There are three types of chemical bond: ionic, covalent, and hydrogen bonding.
- Ionic bonds form when electrons are donated to or received by atoms.
- In covalent bonds, electrons are shared by atoms.

Fast Fact

Water makes up a large percentage of the cell. Proteins, which are the key structural and enzymatic components of the cell, and nucleic acids, which control structure and function, work only in solution.

WATER

Water may be the most important component for life because it has properties that are very important for physiological functions. We will look at three of them: solubility, reactivity, and heat capacity.

1. **Solubility** — There are a remarkable number of molecules that can dissolve in water. In any solution there is the solvent (which in this case is water) and the solute (which is the material that is dissolved). If we consider ionic bonds and the polarity of water, it is easy to see how water can dissolve other compounds. For example, let's use sodium chloride as the solute. We already know that it is a compound made up of positive sodium (Na^{+}) ions and negatively charged chloride (Cl^{-}) ions. Water molecules, which are polar, can come between these elements and essentially surround them, forming **spheres of hydration** (**Figure 2.6**). This dissociates (separates) them and keeps them separated. Consequently, when you put salt in water it dissolves. However, if you continue to add salt, it stops dissolving. This is because there are no longer enough water molecules to surround the ions, and some of the sodium will stay bound to the chloride.

2. **Reactivity** — Chemical reactions normally occur in water, and water can also participate in reactions. When dehydration (the removal of water) is used to build molecules, it is referred to as **dehydration synthesis**, and when water is used to break down molecules, it is called **hydrolysis**.

3. **Heat capacity** — Heat capacity is the ability to absorb and retain heat. Many chemical reactions give off heat as a by-product; water, which has a high heat capacity, can absorb this heat.

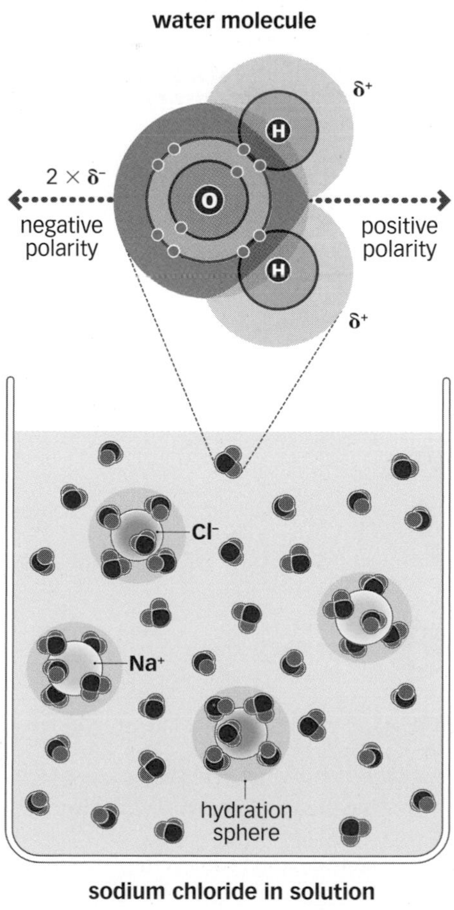

Figure 2.6 The formation of spheres of hydration, which surround compounds that dissolve in water. In this example, the polarity of the water molecules causes the dissociation of the sodium and chloride ions and the formation of spheres of hydration.

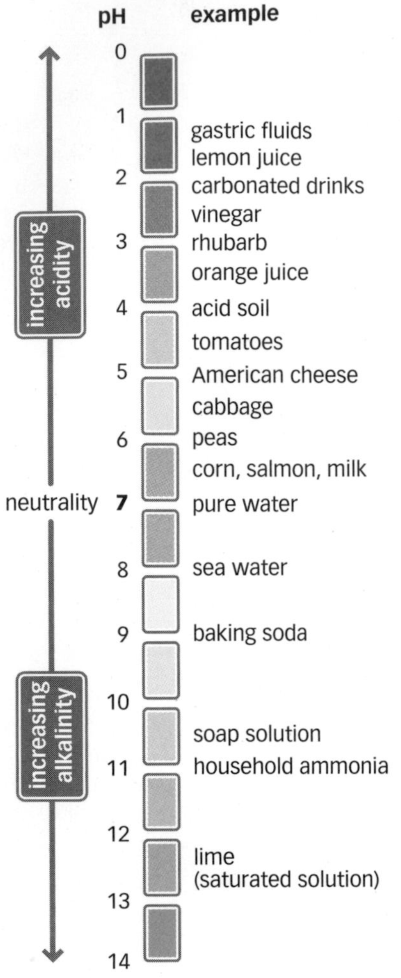

Figure 2.7 The pH scale shows the acidity or alkalinity of common substances.

ACIDS, BASES, AND pH

From a chemical point of view, most of life on Earth lives in a neutral environment. However, there are microorganisms that can live in highly acidic or alkaline environments. When we look at an acid, we can view it as a donor that donates hydrogen ions (H+) to a solution. These hydrogen ions can also be donated as part of a chemical reaction. In contrast, the hydroxyl ion, OH–, which is referred to as a base, can neutralize H^+ ions. A solution that contains an excess of hydroxyl ions is more alkaline, or basic. The concept of pH is basically a definition of the acidity or alkalinity of a solution. On the pH scale (**Figure 2.7**), high H^+ ion concentration is related to low pH, and vice versa. The pH scale goes from 0, which is the most acidic, to 14, which is the most alkaline, with 7.0 being a neutral pH. Pure water has a pH of 7.0 and is completely neutral.

Keep in Mind

- Water has several properties that are important for physiological functions, including solubility, reactivity, and heat capacity.
- pH measures the acidity or alkalinity of a solution.
- Acidity can be considered as the amount of free hydrogen ions (H^+) in a solution.

BIOLOGICAL MOLECULES

Now that we have examined bonding, water, and pH, we can look at biological molecules, which are also referred to as organic molecules. There are four major categories of biological molecule: carbohydrates, lipids, proteins, and nucleic acids. All of these molecules are naturally occurring and all use carbon as the primary building block for their structure. Recall from our discussions above that carbon has the ability to form covalent bonds and link up in long chains that can form a wide variety of structural types of molecule. In addition, carbon-based biological molecules occur in all living things as well as in the remains of living things. Carbons can form long chains and these chains can have hydrogens and other atoms such as oxygen and nitrogen attached to them. We can also subdivide these molecules into parts such as the functional group, which is that part of the molecule that participates in chemical reactions and gives the molecule its particular chemical properties.

All living organisms require energy, and large organic molecules can provide this energy. Organisms harvest the energy in organic molecules by breaking the chemical bonds that hold the molecules together. In fact, the major biological molecules can be viewed within the context of the energy they contain and the ease with which that energy can be obtained. This energy is used to synthesize ATP, which is the form of energy that cells use to do "cellular work" such as active transport or protein synthesis. We discuss these principles in the next chapter.

Fast Fact

Some carbohydrates are found in the cell walls of bacteria, where they have a significant role in infection.

Carbohydrates

Carbohydrates can be viewed as the most easily used source of energy. Plants produce carbohydrates such as cellulose and starch, which animals eat for energy and store as **glycogen**. Microorganisms break down carbohydrates from their environment and can also synthesize carbohydrates.

All carbohydrates contain carbon, oxygen, and hydrogen, with hydrogen usually being found in a 2:1 ratio to the other atoms. We can group carbohydrates into three major categories based on how many building blocks are

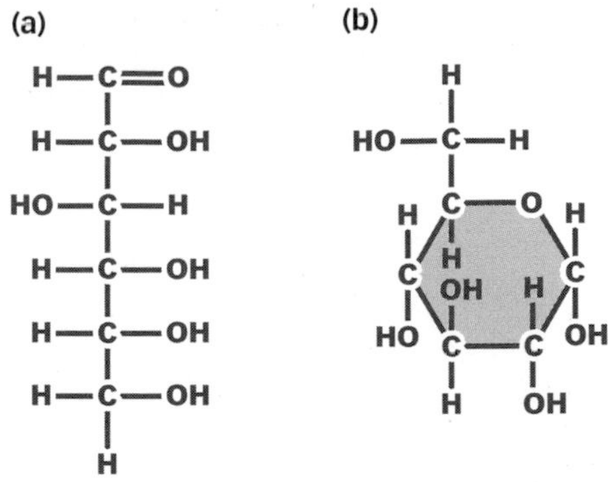

Figure 2.8 The structure of glucose. Panel **a** shows the structural formula in the straight-chain form. Panel **b** shows the structure in the ring form, which is the most common form in nature.

involved. A monosaccharide is the smallest type of carbohydrate and is the building block used to make larger carbohydrates. A disaccharide has two monosaccharide building blocks joined together. In some cases, we can join many of these monosaccharides together to form polysaccharides.

A monosaccharide consists of a carbon chain with several functional groups attached to it; a good example is the glucose molecule (**Figure 2.8**). It has the formula $C_6H_{12}O_6$, meaning that there are 6 atoms of carbon, 12 hydrogens, and 6 oxygens all bonded into the structure. This molecule can also be represented in a ring form, which is the form it would fold into in a solution. Disaccharides are formed when two monosaccharides are linked together. This occurs by dehydration synthesis in which the removal of water results in the joining of two monomers into a dimer (**Figure 2.9**). Polysaccharides (**Figure 2.10**) are formed in the same way except that the molecule can become very large with monomers repeatedly added on to the end of the growing polysaccharide molecule. The polysaccharide form is the storage molecule for carbohydrates and is a tremendous source of energy because single monomers can be broken off and metabolized individually.

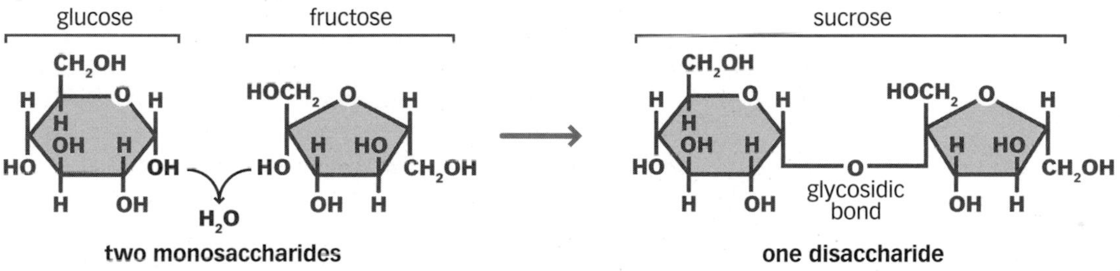

Figure 2.9 The formation of a disaccharide is accomplished through a dehydration synthesis reaction in which water is removed and two monosaccharides are joined together.

Lipids

Lipids are a chemically diverse group of substances that includes fats, phospholipids, and steroids. These substances are relatively insoluble in water, which makes them very useful as elements of cellular structure, in particular the plasma membrane. Lipids can also be used as energy sources, and some lipids actually contain more energy than carbohydrates. In contrast with carbohydrates, lipids contain more hydrogen and less oxygen.

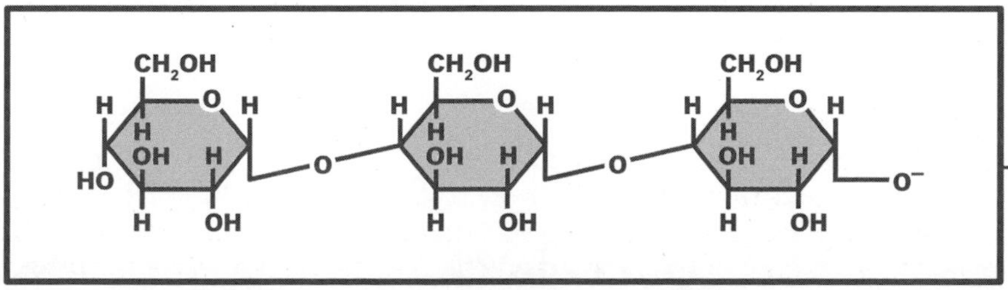

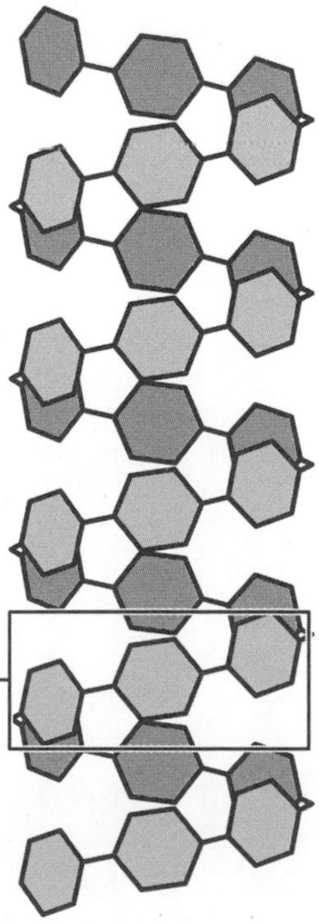

Figure 2.10 Polysaccharides such as starch are formed by a dehydration synthesis reaction in which individual monosaccharide building blocks are added to a growing chain.

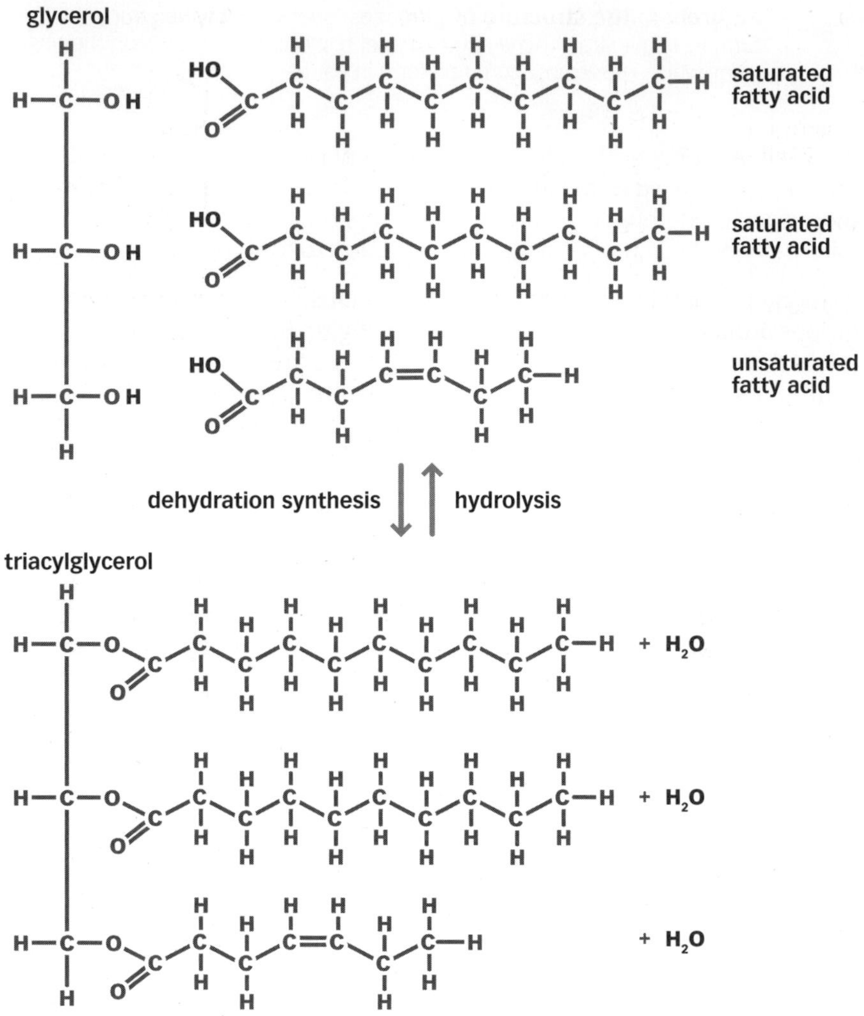

Figure 2.11 The formation of a triacylglycerol which involves the combination of one glycerol molecule and three fatty acids through dehydration synthesis.

Fats

Fats are lipids that contain the three-carbon molecule glycerol and one or more **fatty acids**. These fatty acids are long chains of carbons with bound hydrogens. As with other organic molecules, dehydration synthesis is used to synthesize fat. During this process, bonds form between the glycerol and one end of each of the fatty acids (**Figure 2.11**). These can be monoacylglycerols, diacylglycerols, or triacylglycerols, depending on how many tails are attached.

Fatty acids can be saturated or unsaturated, depending on how many hydrogen atoms are attached to pairs of carbons in the tails. Saturated fatty acids contain all of the hydrogens that can possibly be bound. In contrast, the unsaturated forms have lost hydrogens and formed double bonds at the locations of the missing hydrogens (**Figure 2.12**).

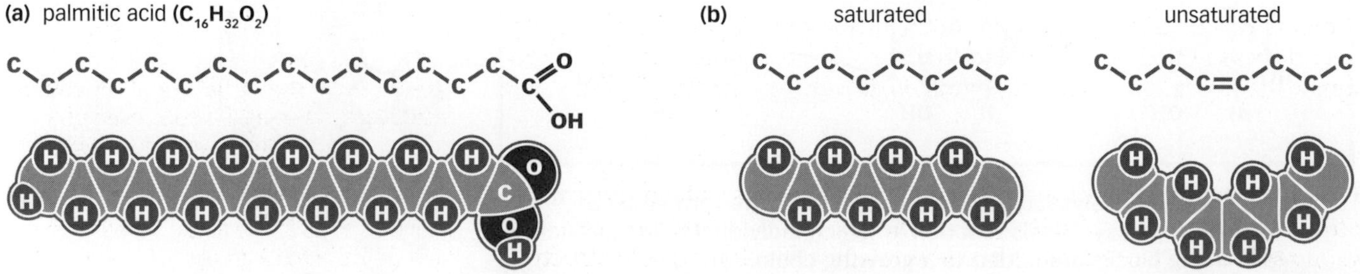

Figure 2.12 The common characteristics of fatty acids. Panel **a**: Palmitic acid demonstrates the two common characteristics of fatty acid molecules: the long chain of carbons and the reactive carboxyl group (COOH). Panel **b**: A fatty acid is either saturated (contains the maximum number of hydrogen atoms) or unsaturated (has one or more double bonds).

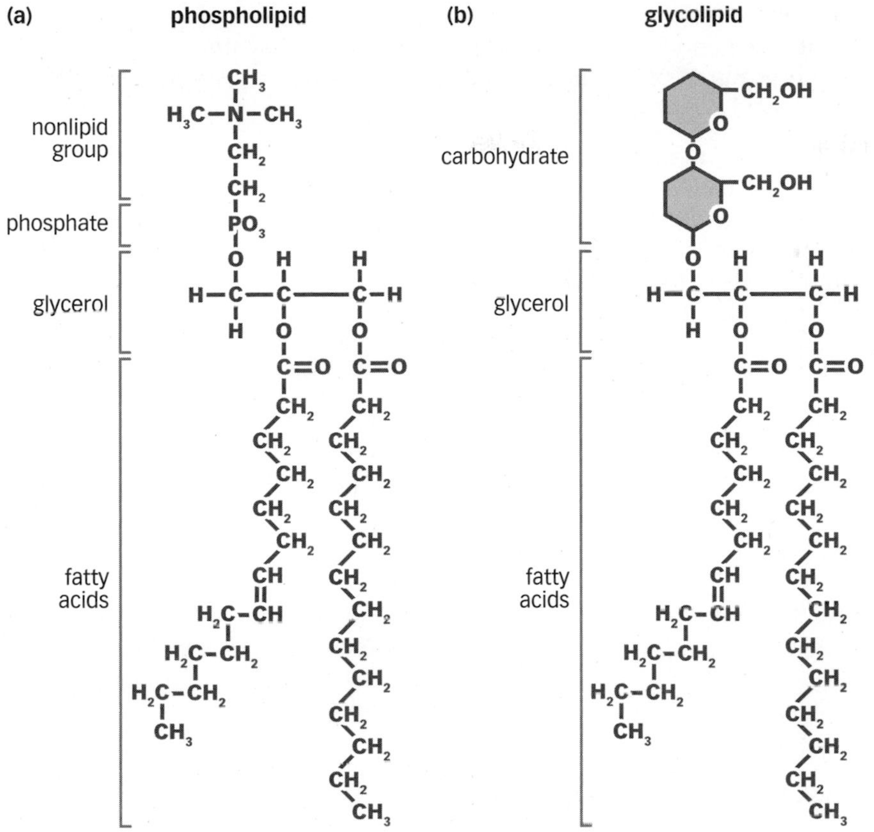

Figure 2.13 Phospholipids and glycolipids are two important biological molecules. Phospholipids have a phosphate group attached to the molecule, whereas in glycolipids a carbohydrate molecule is attached.

Phospholipids and Glycolipids

Some lipids have other molecules attached to them, such as carbohydrate (to form a **glycolipid**) or a phosphate (to form a **phospholipid**) (**Figure 2.13**). The phospholipid molecule is the foundation of a cell's plasma membrane structure, giving it unique properties. The phosphate end of the molecule is hydrophilic, which means it can interface with water, but the opposing end of the molecule is hydrophobic and cannot interface with water. These phospholipids are used to form a barrier between the water outside the cell and the water inside the cell.

Steroids

Steroids are four-ring carbon structures that are different from other types of lipid. One of the most important steroids is cholesterol, which is insoluble in water and is found in the cell membrane of some eukaryotic cells and also in the bacterium *Mycoplasma*.

Proteins

Proteins are one of the most important of the biological molecules. They are very diverse in both structure and function, and each has a specific three-dimensional configuration (shape) that is directly related to function.

Properties of Proteins

Proteins are made up of amino acid building blocks, which are carbon structures that have at least one amino group (NH_2) and one carboxyl group (COOH) (**Figure 2.14**). These two parts of the amino acid structure

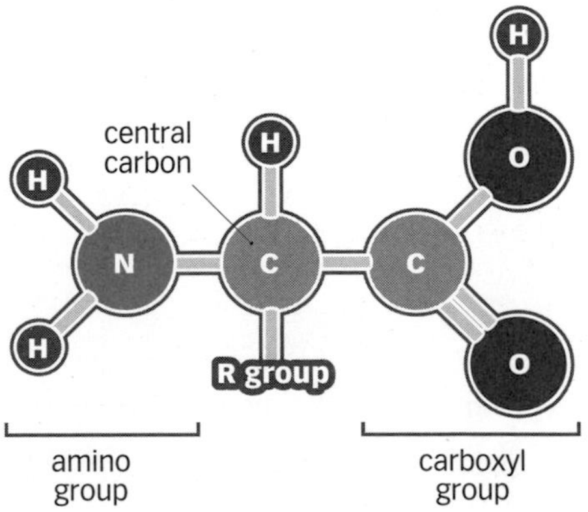

Figure 2.14 The structure of an amino acid. Each amino acid is composed of a central carbon atom to which four different groups are attached. The amino group, the carboxyl group, and a hydrogen atom are found on all 20 amino acids, but each amino acid differs at the R group.

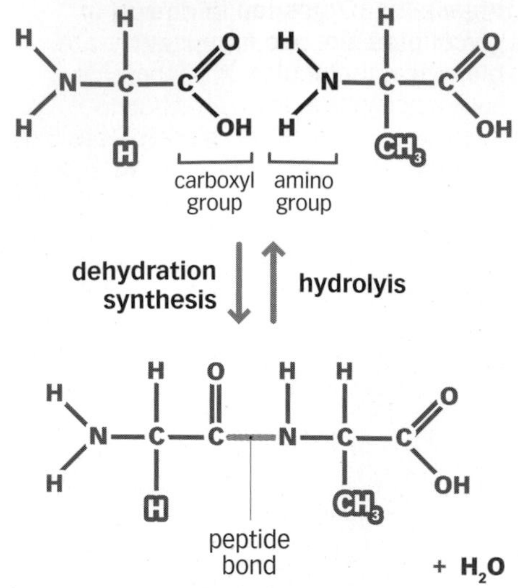

Figure 2.15 The formation of a peptide bond, which links amino acids together. In this example, alanine is bonded to glycine. Notice that the bond is caused by dehydration synthesis and always involves the amino group of one amino acid and the carboxyl group of the other. This linking of amino acids can continue and form a polypeptide chain.

are important because they are involved in the formation of a **peptide bond**. The peptide bond is formed by dehydration synthesis in which the amino group of one amino acid bonds to the carboxyl group of another amino acid (**Figure 2.15**).

Proteins are made up of long sequences of these linked amino acids, which are called peptides. These peptides can vary in length from two amino acids (dipeptide) to many amino acids (polypeptide). Some amino acids contain sulfur atoms, and these can bond to sulfur atoms in other amino acids to form **disulfide bridges**, which are important in the folding of the protein (**Figure 2.16**).

Protein Structure

As mentioned above, proteins are three-dimensional molecules, and the structure of the molecule is critical to the function of the molecule. This three-dimensional structure can be broken down into four levels (**Figure 2.17**).

1. The primary level is the sequence of the amino acids in the polypeptide chain.

2. The secondary level consists of a folding or coiling of the polypeptide that is brought about by the sequence of the amino acids and hydrogen bonds that form between amino acids. This secondary folding is usually seen as either a helix or a pleated sheet form.

3. The tertiary level involves the folding of the chain upon itself. This folding confers the major three-dimensional structure of the polypeptide and is held in place in part by hydrogen bonding and the formation of disulfide bridges. This folding can result in several shapes, including globular and fibrous (threadlike) structures.

4. Quaternary structure occurs in very large proteins for which more than one polypeptide is joined together. This joining together of individual polypeptide chains also occurs through the formation of bonds and between amino acids in the individual polypeptide chains.

It is important to understand that the three-dimensional structure of a protein is susceptible to disruption by chemical changes that affect the hydrogen bonds and other weak bonds that hold the structure in place. Factors such as pH and temperature can change the structure of the protein and thereby the function of the protein. This is referred to as protein denaturation, which can be a lethal event for microorganisms.

Types of Protein

There are a variety of proteins, but structural proteins and enzymes are among the most important. Structural proteins contribute to cellular structures such as the cell membranes and the cytoskeleton that maintains the

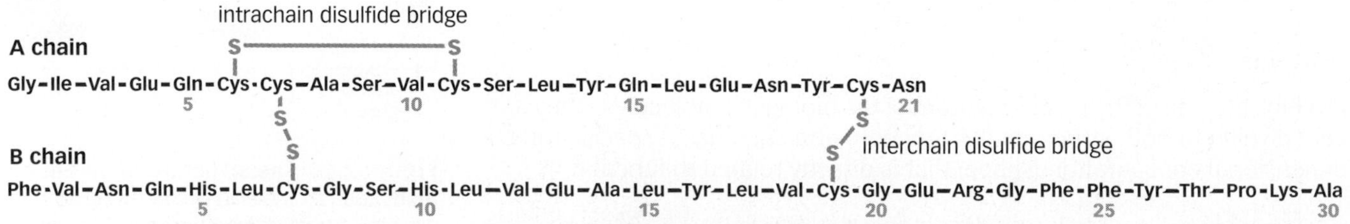

Figure 2.16 The formation of disulfide bridges occurs between amino acids that contain sulfur atoms. These bridges contribute to the three-dimensional shape of polypeptides, and they can be intrachain (within a polypeptide chain) or interchain (linking two chains together).

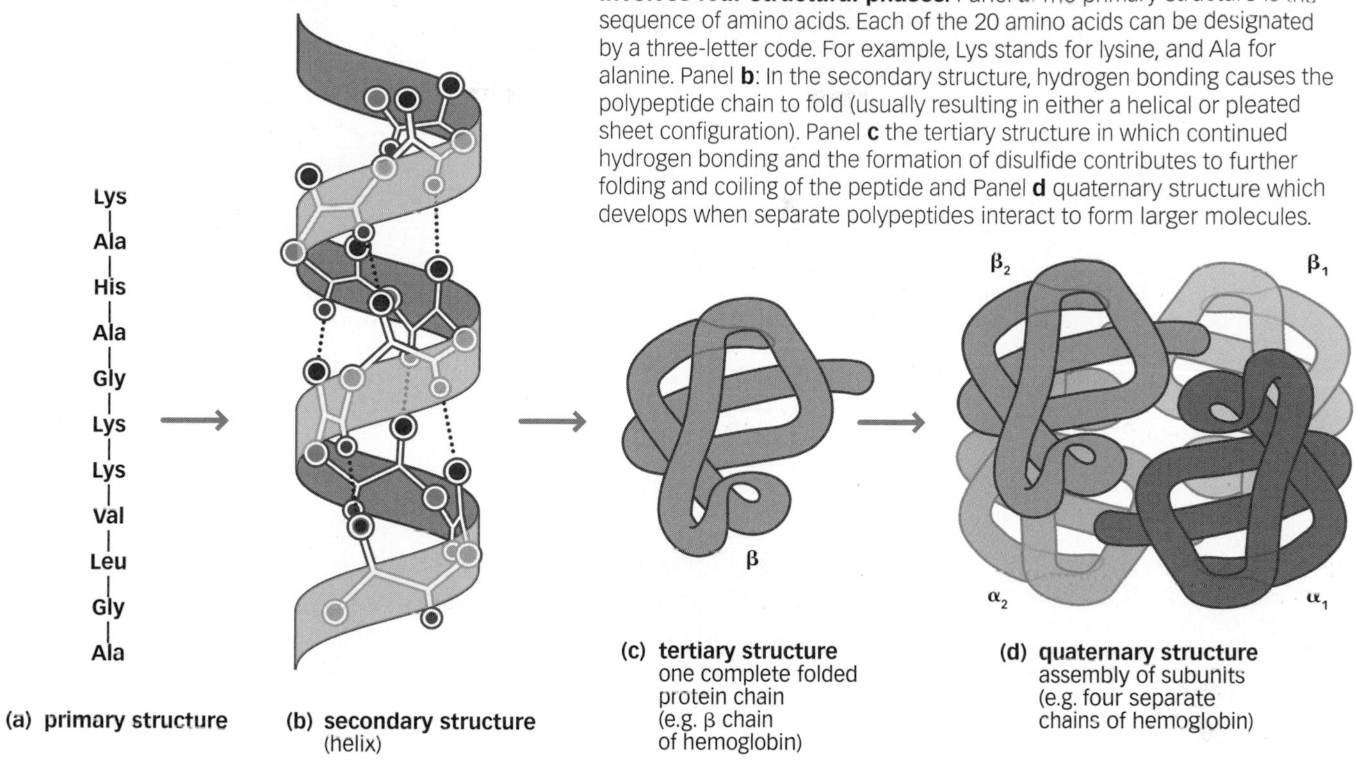

Figure 2.17 The three-dimensional configuration of a protein involves four structural phases. Panel **a**: The primary structure is the sequence of amino acids. Each of the 20 amino acids can be designated by a three-letter code. For example, Lys stands for lysine, and Ala for alanine. Panel **b**: In the secondary structure, hydrogen bonding causes the polypeptide chain to fold (usually resulting in either a helical or pleated sheet configuration). Panel **c** the tertiary structure in which continued hydrogen bonding and the formation of disulfide contributes to further folding and coiling of the peptide and Panel **d** quaternary structure which develops when separate polypeptides interact to form larger molecules.

shape of the cell. Structural proteins are also involved in the motility seen with flagella (structures that allow an organism to move). In contrast, the enzymatic proteins are involved in many cellular functions such as metabolism. In metabolism, there are a series of chemical reactions that allow for the build-up or breakdown of organic molecules. This series of enzymatic reactions depends on the energy of activation, which is the energy necessary to make the reaction happen. Enzymes lower the energy of activation, thus making the reaction go faster (**Figure 2.18**), by

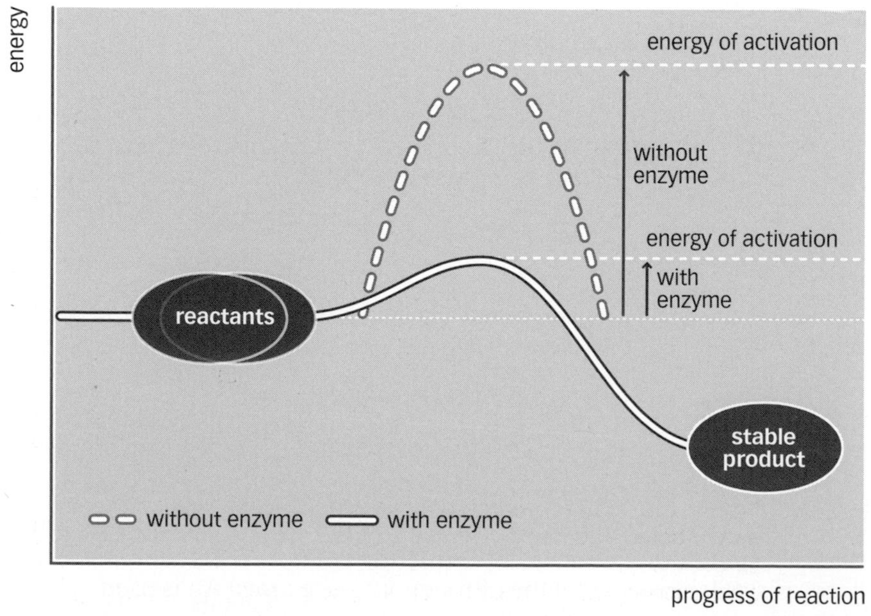

Figure 2.18 The effect of enzymes on the energy of activation. Chemical reactions require energy to proceed, and this energy is referred to as the energy of activation. Enzymes lower this energy and allow the reactions to proceed faster.

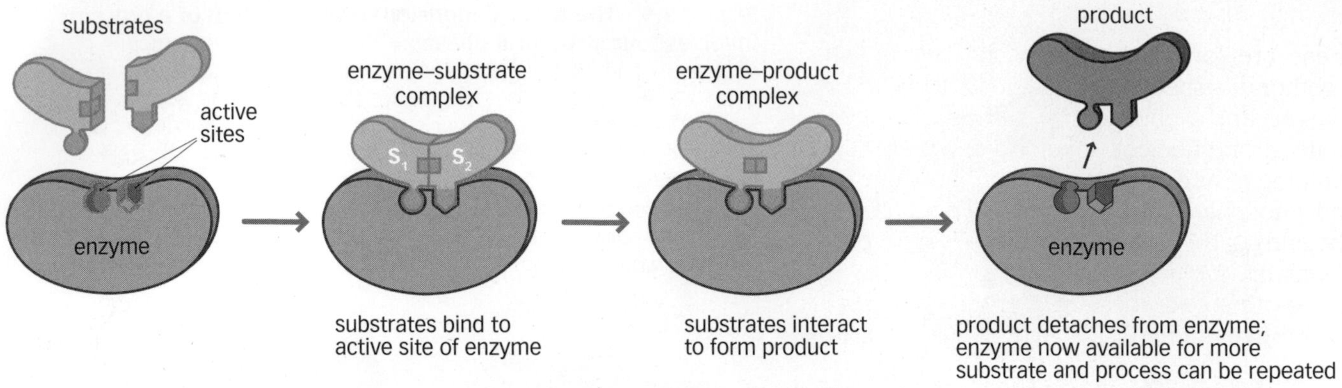

Figure 2.19 A simplified illustration of enzyme structure and function. Each enzyme has a specific shape that contains one or more active sites. Substrates "fit" into these sites and undergo rapid chemical processing into products. Notice that if the shape changes, the enzymatic function can change.

holding the two reactants close together. In these reactions the reactants are placed on the protein in active sites, which result from the shape of the protein folding (**Figure 2.19**). With this in mind, it is easy to see that if the shape of the protein is changed, the function of the enzyme can be compromised. We discuss enzyme activity in further detail when we look at metabolism in the next chapter.

Nucleic Acids

These organic molecules are involved in cellular information and also function as energy molecules. There are two types of informational molecule: deoxyribonucleic acid (DNA) and ribonucleic acid (RNA). Each building block of DNA consists of a nitrogenous base (adenine (A), thymine (T), guanine (G), or cytosine (C)), the sugar deoxyribose, and a phosphate (**Figure 2.20**). Both adenine and guanine are purines, which are double-ring structures. In contrast, thymine and cytosine are pyrimidines, which are single-ring structures (**Figure 2.21**). As we will see

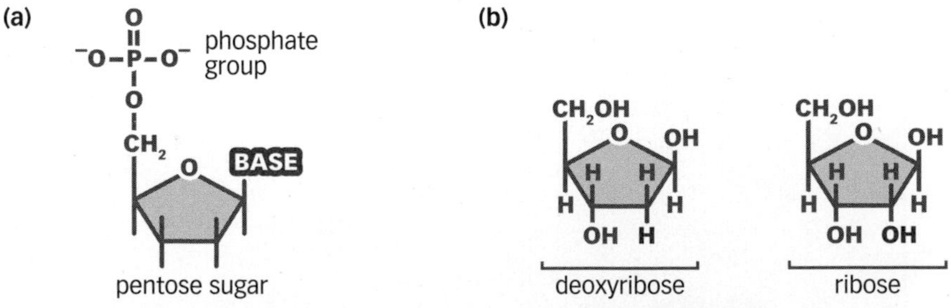

Figure 2.20 The structure of nucleotides. Panel **a**: The basic structure of nucleotides, which are composed of a phosphate, a pentose sugar, and a nitrogenous base. Panel **b**: The pentose sugar deoxyribose is found in DNA, whereas the sugar ribose is part of the RNA molecule.

in Chapter 11 the differences between the purine and pyrimidine structures are important for the overall structure of DNA. The structure of RNA bases is similar except that the nitrogenous base uracil (U) is used instead of thymine.

Structure of Nucleic Acids

These are long polymeric structures in which nucleotides are linked together in a special way. Each nucleotide contains the nitrogenous base, a sugar, and a phosphate. There are covalent bonds between the phosphate of one nucleotide and the sugar of the next, and we can view this chaining of nucleotides together as a spine made up of alternating sugars and phosphates with the nitrogenous bases extending inward from it. The ends of the spine are chemically different: one end is called the 5' end, in which the 5' carbon of the sugar molecule is attached to the phosphate. At the other end, the 3' carbon of the sugar is unattached (**Figure 2.22**). It is here that additional nucleotides can be linked.

For cells and most viruses that use DNA as the genetic material, the molecule is double-stranded, meaning that there are two strands that go in opposite directions (antiparallel, as shown in **Figure 2.23**). The antiparallel nature of the strands has a significant role in the replication of DNA, as we will see in Chapter 11. In DNA, the bases face each other, and because of the chemical nature of these bases they can hydrogen bond to each other. This occurs in a specific way, in that A always bonds to T and G always bonds to C. In addition, the bonding of A to T involves two hydrogen bonds, whereas the bonding of C to G involves three hydrogen bonds. It is important to remember that these hydrogen bonds are weak and that they are easily broken during the replication process, during which the chains separate. The bonding of A to T and G to C is referred to

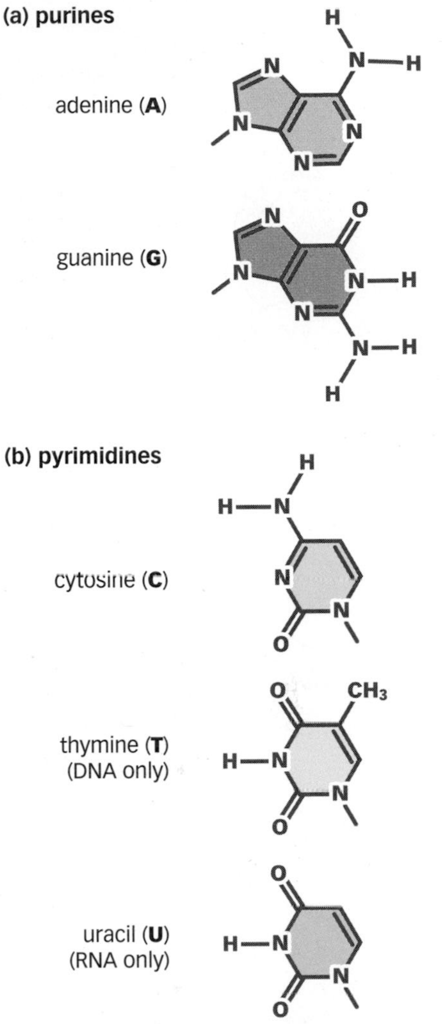

Figure 2.21 The structure of purines (adenine and guanine) and pyrimidines (cytosine, thymine, and uracil).

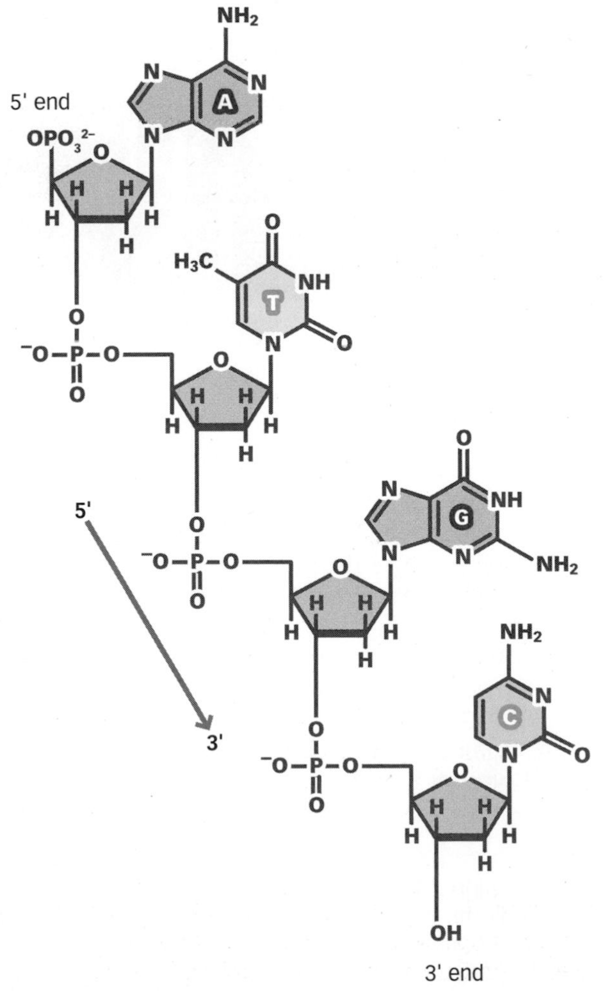

Figure 2.22 The structure of the DNA strand, which is composed of a phosphate and sugar backbone linked together in a 5' to 3' direction. Notice that the nitrogenous bases project inward from the backbone.

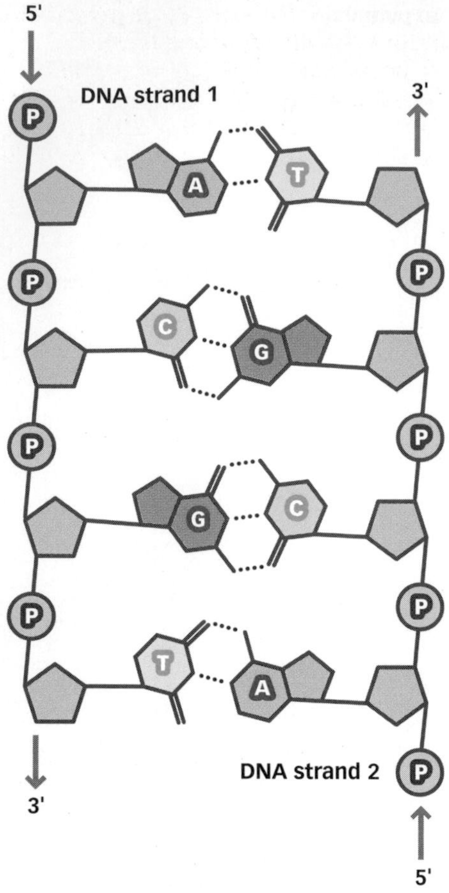

Figure 2.23 The DNA strands are antiparallel, with the strands running in opposite directions. Notice the bonding between the bases, in which there are two hydrogen bonds between the bases A and T and three hydrogen bonds between the bases G and C.

as *complementary base pairing* and is very important for both replication and gene expression. Because of the hydrogen bonding that occurs within other parts of the structure, the DNA molecule is twisted into a double helix (**Figure 2.24**).

DNA is the genetic material for all life on this planet — animals, plants, and bacteria. It carries the instructions for producing all of the structures found in the cell and also for the genetic regulation that is required for successful growth of the organism. DNA is faithfully replicated and carefully regulated so that there are no mistakes and the information that is handed down from generation to generation is correct. However, we will see in Chapter 11 that mutations which can change genetic information can occur.

Ribonucleic acid is a single-chain molecule that incorporates ribose sugar and is made by the same bonding mechanisms that link DNA nucleotides together, with uracil used instead of thymine. It is important to note that complementary base pairing occurs between DNA and RNA and is the basis for the transcription of DNA into messenger RNA, which is used to encode a protein sequence.

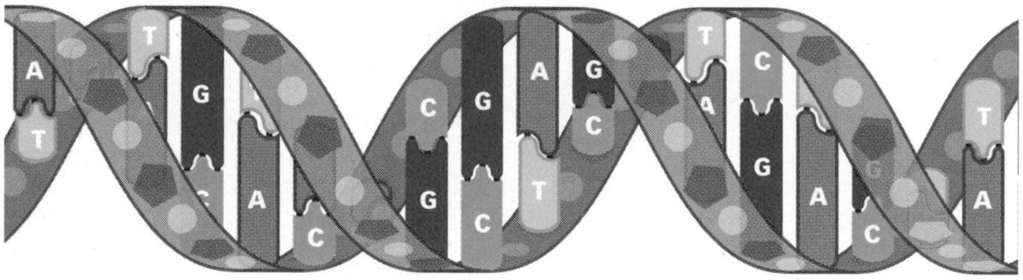

Figure 2.24 An illustration of the double-helical structure of DNA.

ATP

As we mentioned above, energy is required for all the processes in the cell, and some of these processes, such as protein synthesis, can be very energy-consuming. The major energy molecule in biological systems is ATP, which contains the nitrogenous base adenosine, a ribose sugar, and a chain of three phosphates bonded to the sugar (**Figure 2.25**). The bonds between these phosphates are high-energy bonds that when broken yield energy. When the outermost phosphate is removed from ATP it yields ADP (adenosine diphosphate). In most cases, this ADP can be recycled back into ATP if a phosphate is added, and the ATP can then be used for energy again. The recycling between ATP and ADP is important to the overall physiology of cells, but making ATP from ADP requires energy.

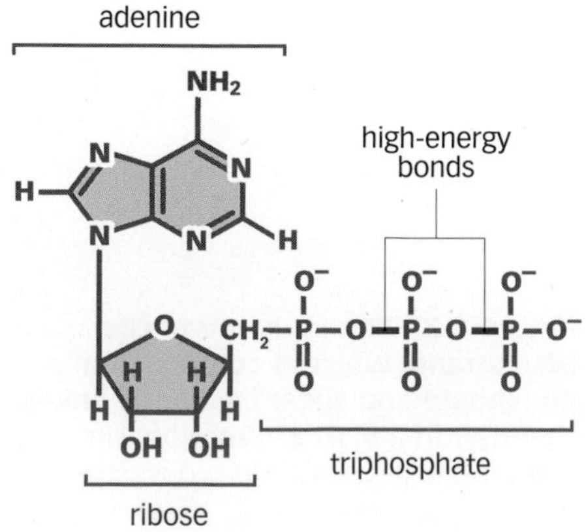

Figure 2.25 The nucleotide adenosine triphosphate (ATP), which is the immediate source of energy for most active living cells. Breaking the high-energy bonds between the phosphates releases energy that can be used for cellular functions.

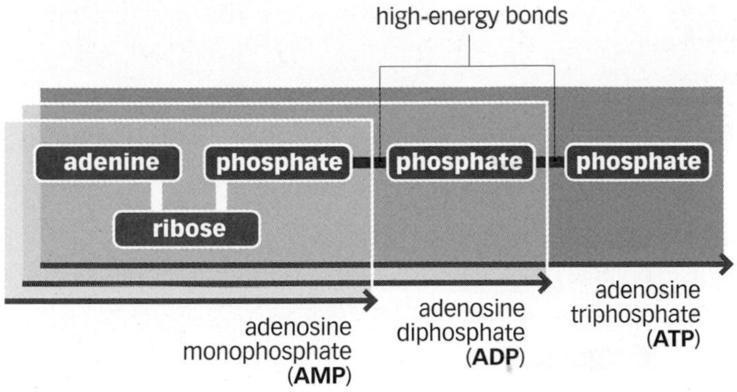

Figure 2.26 The removal of phosphate from ATP yields ADP and further removal yields AMP. It is important to note that the molecule can be recycled. For example, the addition of phosphate to ADP can form ATP.

In some cases, the phosphate from ADP can be removed to yield AMP (**Figure 2.26**). The breaking of this bond also yields energy, but less than when ATP is converted to ADP. Energy derived from breaking the bonds of ATP can be used for synthesis reactions, for motility, and also for the active transport of materials across the plasma membranes of cells. ATP must be continually replenished; this is carried out through metabolism, which we will cover in the next chapter.

Keep in Mind

- Biological molecules use carbon as the primary building block of their structures.
- There are four biological molecules: carbohydrates, lipids, proteins, and nucleic acids.
- The three-dimensional structure of a protein is directly related to the function of the protein.
- ATP is the major energy molecule in biological systems.

SUMMARY

- Atoms are composed of protons, neutrons and electrons.
- Atoms join together to make molecules, which can join together to make cells.
- There are three types of chemical bond: ionic, covalent, and hydrogen bond.
- Ionic bonds form when electrons are donated or accepted.
- Covalent bonds form when electrons are shared.
- Hydrogen bonds form between hydrogen atoms and are the weakest form of chemical bond.
- Biological molecules use carbon atoms as their building blocks.
- There are four types of biological molecule: carbohydrates, lipids, proteins, and nucleic acids.
- ATP is the major energy molecule of the cell.

In this chapter we have explored some of the chemical principles and structures that will help us understand the information to come. We saw how bonds are formed and how molecules are constructed. These molecules are the foundation of cells and tissues, which are the targets for infectious disease. In the next chapter, we will apply this information as we begin to examine metabolism.

SELF EVALUATION AND CHAPTER CONFIDENCE

MULTIPLE CHOICE

Answers are given in the back of the book and help can be found in the student resources at www.garlandscience.com/micro

1. Which of the following is the type of bond holding Na^+ and Cl^- ions together in NaCl?
   - **A.** Covalent
   - **B.** Ionic
   - **C.** Polar covalent
   - **D.** None of the above
2. Which of the following types of bond is found between the atoms of water molecules?
   - **A.** Hydrogen
   - **B.** Covalent
   - **C.** Ionic
   - **D.** Polar covalent
3. Which type of molecule contains glycerol?
   - **A.** Nucleic acid
   - **B.** Protein
   - **C.** Carbohydrate
   - **D.** Lipid
   - **E.** All of the above
4. Which type of molecule contains an amino group as part of each building block?
   - **A.** Carbohydrate
   - **B.** Protein
   - **C.** Lipid
   - **D.** Nucleic acid
   - **E.** None of the above
5. Which of the following statements is not true?
   - **A.** Enzymes are used in a chemical reaction.
   - **B.** Enzymes increase the probability that a reaction will occur.
   - **C.** Enzymes are proteins.
   - **D.** Enzymes lower the energy of activation of a reaction.
   - **E.** None of the above.
6. What type of bond is found between carbons in an organic molecule?
   - **A.** Ionic
   - **B.** Covalent
   - **C.** Hydrogen
   - **D.** Nonpolar covalent
   - **E.** None of the above
7. Structurally, ATP is most like which of these molecules?
   - **A.** Carbohydrate
   - **B.** Nucleic acid
   - **C.** Lipid
   - **D.** Protein
8. Which molecule is composed of a chain of amino acids?
   - **A.** Lipid
   - **B.** Carbohydrate
   - **C.** Fat
   - **D.** Nucleic acid
   - **E.** None of the above
9. Which of these molecules make up the plasma membranes of cells?
   - **A.** Nucleic acids
   - **B.** Carbohydrates
   - **C.** Lipids
   - **D.** Water
   - **E.** None of the above
10. Starch and glycogen are made up of
    - **A.** Amino acids
    - **B.** Lipids
    - **C.** Fatty acids
    - **D.** Carbohydrates
    - **E.** Glycerol

DEPTH OF UNDERSTANDING

Questions listed here require you to bring together the concepts you have learned in this chapter into a discussion format. This helps you to increase your depth of understanding of the material you have learned. Help can be found in the student resources at www.garlandscience.com/micro

1. Describe the binding that can occur between element X, which has an atomic number of 11, and element Y, which has an atomic number of 9.

2. Describe how protein building blocks are bound together to make polypeptides.

3. Discuss the formation of spheres of hydration and why they are so important in biological systems.

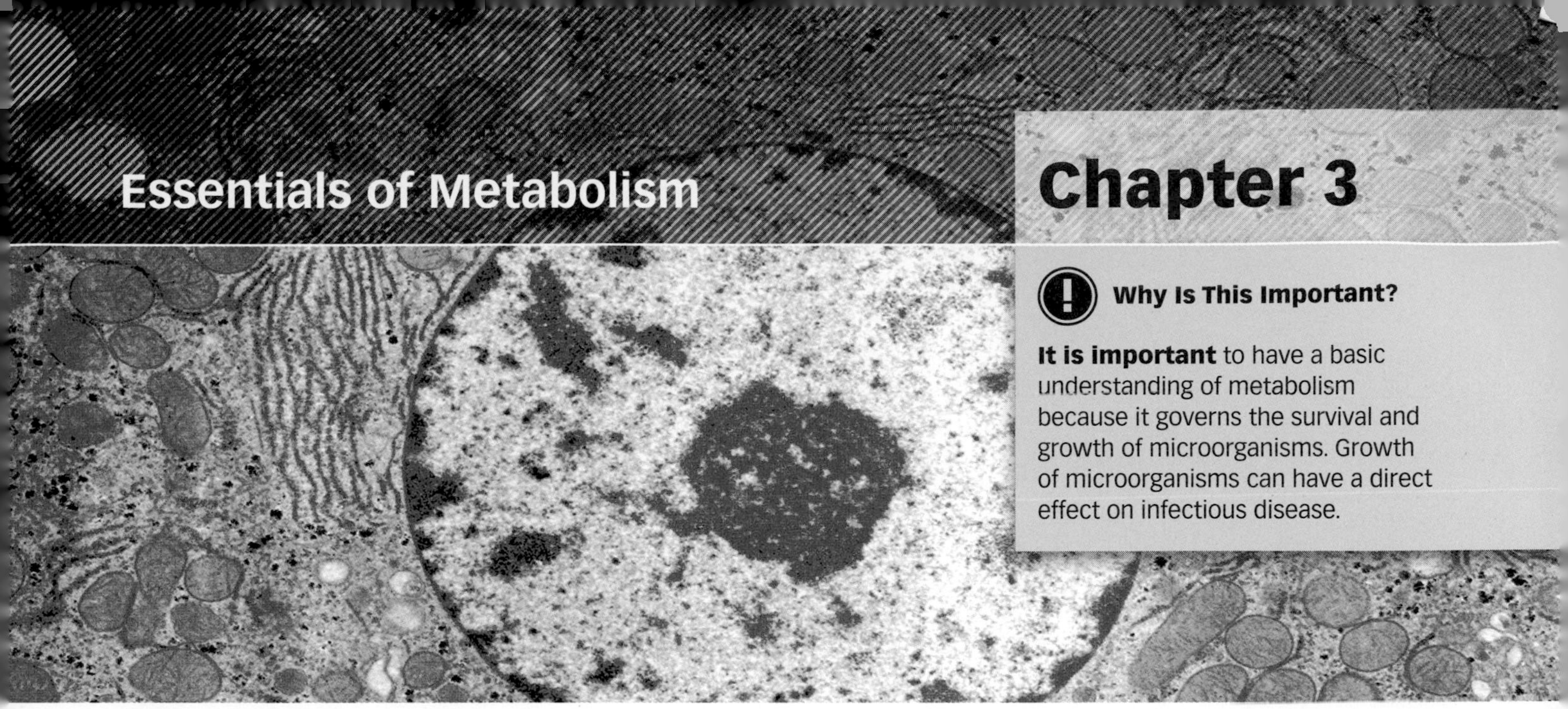

Essentials of Metabolism

Chapter 3

Why Is This Important?

It is important to have a basic understanding of metabolism because it governs the survival and growth of microorganisms. Growth of microorganisms can have a direct effect on infectious disease.

OVERVIEW

Before we look at the effects that infectious diseases can have on humans, it is important to get a basic idea of how metabolism works. In many cases, infectious agents must be present in the body in great numbers if they are to do harm, and attaining these high numbers requires an orderly progression of metabolic functions followed by cell division. Metabolism involves catabolism, in which molecules are broken down and energy is released, and anabolism, in which energy is used to build molecules. These mechanisms are subject to regulation, which coordinates the hundreds of independent biochemical processes involved in metabolism.

We will divide our discussions into the following topics:

Essentials of Metabolism

- BASIC CONCEPTS OF METABOLISM
- ENZYMES
- CATABOLISM
- ANABOLISM

BASIC CONCEPTS OF METABOLISM

Because we emphasize pathogens and infectious disease in this book, you might ask yourself why it is important to understand **metabolism**, which is defined as the chemical processes that go on inside any living organism. Simply put, pathogenic organisms (like all other organisms) require energy to live: no energy, no life. Energy obtained through catabolism is one of the products of metabolism. Therefore, if there is no metabolism, there is no life. As we will see in later chapters, the severity of an infection can be linked to increases in the numbers of pathogens in the infected host. Metabolism makes energy available to an organism, and energy is linked to growth. Consequently, the better an organism can metabolize, the better it grows. For pathogens, good metabolic function makes the pathogens more successful at causing disease. In this chapter, we look at the basics of metabolism, but before we examine these mechanisms we need to learn a few new terms.

In biological systems, carbon and energy are required for growth. Microorganisms have two ways of obtaining the carbon they need to live: autotrophy and heterotrophy. **Autotrophy** can be defined as "self feeding"

What Do I Need to Know?

To get the most out of this chapter, please review the following terms from your previous courses in biology, anatomy, physiology, or chemistry or in previous chapters of this book as indicated in parentheses: active site of enzyme molecule (**2**), alcohol, aldehyde, denaturing a protein (**2**), energy of activation (**2**), gradient, hydrogen bond, inorganic compound (**2**), organic compound (**2**).

because autotrophic organisms obtain carbon atoms (the building blocks of organic molecules) from inorganic sources such as carbon dioxide (CO_2) in the environment. Green plants are one familiar example of autotrophic organisms. Autotrophs can be further divided into photoautotrophs and chemoautotrophs. **Photoautotrophs** are organisms that use sunlight as a source of energy. **Chemoautotrophs** are autotrophic organisms that obtain energy from chemical reactions involving inorganic substances such as nitrates and sulfates.

Heterotrophy can be looked at as "other feeding" because heterotrophic organisms get carbon atoms from the organic molecules present in other organisms. Humans are of course heterotrophs because we get our carbon atoms from the food we eat. **Photoheterotrophs** obtain energy from sunlight and convert it to chemical energy, and **chemoheterotrophs** obtain energy by breaking down organic compounds.

Nearly all infectious organisms are chemoheterotrophs, and so chemoheterotrophs will be our main focus in this chapter. Now let's look at some of the basic concepts of metabolism more closely.

Oxidation and Reduction Reactions

Metabolism can be broken down into two parts, catabolism and anabolism (**Figure 3.1**). **Catabolism** is the collective term for all the metabolic processes in which molecules are broken down to release the energy stored in their chemical bonds. **Anabolism** is the collective term for all the metabolic processes in which the energy derived from catabolism is used to build large organic molecules from smaller ones. All catabolic pathways involve *electron transfer*.

Electron transfer is part of the *oxidation* and *reduction* reactions. An oxidation reaction is a chemical reaction in which an atom, ion, or molecule *loses* one or more electrons. A reduction reaction is a chemical reaction in which an atom, ion, or molecule *gains* one or more electrons. Oxidation and reduction reactions always occur together — always — and for this reason the combination of an oxidation reaction and its reduction partner is referred to as a **redox reaction**. In many cases, oxygen is the acceptor for electrons in a redox reaction, but other substances can also accept them.

With our definitions of oxidation and reduction, we can say that when a substance is oxidized, it loses electrons, and when a substance is reduced, it gains electrons. The classic example of this is seen in the chemical reaction

$$Na^+ + Cl^- \longrightarrow \underset{\text{Sodium chloride}}{NaCl}$$

In this reaction, the sodium atom loses an electron and so becomes oxidized and the chlorine atom gains the electron lost by the sodium atom and so becomes reduced.

Respiration

Respiration means two different things in the biological sciences. At the macroscopic level, respiration is the exchange of carbon dioxide and oxygen in the lungs. At the cellular level, which is the level we are interested in, the term *cellular respiration* is used to describe catabolic processes, and these can be found in two forms, **aerobic respiration** (meaning metabolism that uses oxygen) and **anaerobic respiration** (metabolism that occurs without the use of oxygen). Whether a microorganism uses aerobic cellular respiration or anaerobic cellular respiration for catabolism determines the amount of ATP that the microorganism produces. We

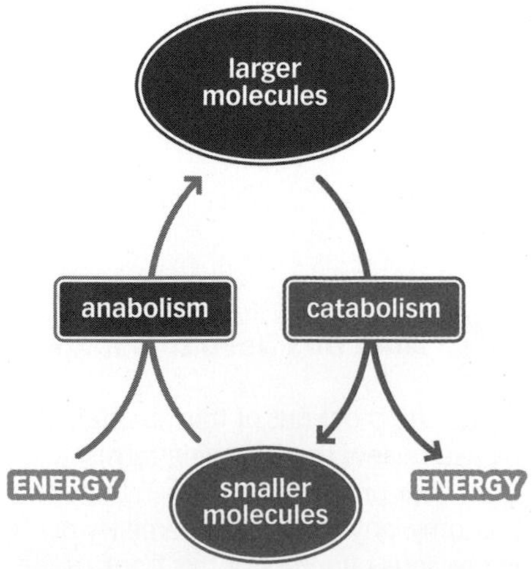

Figure 3.1 Metabolism is composed of catabolism and anabolism. In catabolism, large molecules are broken down and energy is released. In anabolism, large molecules are created from smaller ones, with energy consumed in the process.

will see below that anaerobic cellular respiration through glycolysis yields only 2 ATP molecules for every glucose molecule consumed, whereas coupling glycolysis with aerobic cellular respiration using the Krebs cycle and electron transport yields a total of 38 ATP molecules for every glucose molecule consumed. Because ATP is vitally important, organisms that use aerobic cellular respiration are able to grow (that is, increase in number) more quickly than those that use the anaerobic form.

It is important to note that some organisms can carry out metabolic processes either way, aerobically or anaerobically. These organisms are called **facultative anaerobes**. They grow well in the presence of oxygen, but when oxygen is absent, they can still grow, although not as well as with oxygen. Many pathogens are facultative anaerobes.

The chemical reaction for the complete oxidation of glucose by aerobic respiration is

$$\underset{\text{glucose}}{C_6H_{12}O_6} + \underset{\text{oxygen}}{6\,O_2} \longrightarrow \underset{\text{carbon dioxide}}{6\,CO_2} + \underset{\text{water}}{6\,H_2O} + \text{energy}$$

All the substances to the left of the reaction arrow are called either the reactants or the substrates of the reaction, and all the substances to the right of the reaction arrow are called the products of the reaction.

Metabolic Pathways

Nearly all chemical processes in living organisms consist of a series of chemical reactions called pathways. In these pathways the product of one reaction serves as the substrate for the next reaction. An example of a pathway is

$$A \xrightarrow{\text{Enzyme 1}} B \xrightarrow{\text{Enzyme 2}} C \xrightarrow{\text{Enzyme 3}} D \xrightarrow{\text{Enzyme 4}} E$$

In this example, A is the initial substrate and E is the final product of the pathway, with B, C, and D being intermediates. Notice that each step in the pathway is mediated by an enzyme. These enzymes facilitate, or promote the running of, certain reactions under physiological conditions and are required for the reactions to proceed at an appropriate speed.

Before we move on to our discussion of metabolic pathways, let's take a closer look at enzymes.

Keep in Mind

- Metabolism is the chemical process that provides or stores energy for organisms.
- Metabolism can be broken down into two parts: catabolism, in which molecules are broken down and energy is released, and anabolism, in which molecules are constructed and energy is consumed.
- Nearly all infectious organisms are chemoheterotrophs, obtaining energy by breaking down organic molecules.
- Oxidation–reduction (redox) reactions involve the transfer of electrons. When a molecule gives up an electron it is oxidized, and when a molecule obtains an electron it is reduced.
- Cellular respiration can be aerobic, in which oxygen is the final acceptor of electrons, or anaerobic, in which oxygen is not involved.
- Nearly all chemical processes consist of a series of chemical reactions known as pathways.

ENZYMES

Recall from Chapter 2 that enzymes are protein molecules and therefore have a distinctive three-dimensional shape. In Chapter 2, we learned that the shape of a protein is directly related to its function. If the shape is changed, the function can be inhibited or even stopped. Each enzyme is specific to its function. Enzymes are found in all living organisms, and most cells contain hundreds of enzymes, which are constantly being manufactured and replaced.

In chemistry, a catalyst is a substance that speeds up a chemical reaction but is not itself changed in any way by the reactions. Enzymes act as catalysts for the reactions that take place in organisms. This means that enzymes help get reactions started and help them to proceed.

Properties of Enzymes

Enzymes work in a chemical reaction by lowering the energy of activation, which is the energy required to start the reaction (**Figure 3.2**). In any metabolic reaction, a certain amount of energy (the energy of activation) is required for the substrate to be converted to product. The amount of energy needed is greater when no enzyme is present and less when enzyme is present. It is as if there is a boulder that wants to roll down a hill, but a bump at the edge of the hill keeps the boulder from moving. We can consider the bump as representing the activation energy needed for the rolling "reaction." This bump is high when no enzyme is present, but when we add an enzyme it lowers the bump. Then the boulder easily rolls over the smaller bump and down the hill.

Because enzymes are proteins, each enzyme molecule has a three-dimensional shape that results from the folding of the molecule. The shape of the molecule provides a site where the enzymatic activity takes place, called the *active site*, and it is here that the substrate fits into the enzyme (**Figure 3.3**). This is a precise fit based on the shape of the active site, which means that only one shape of substrate can fit at each active

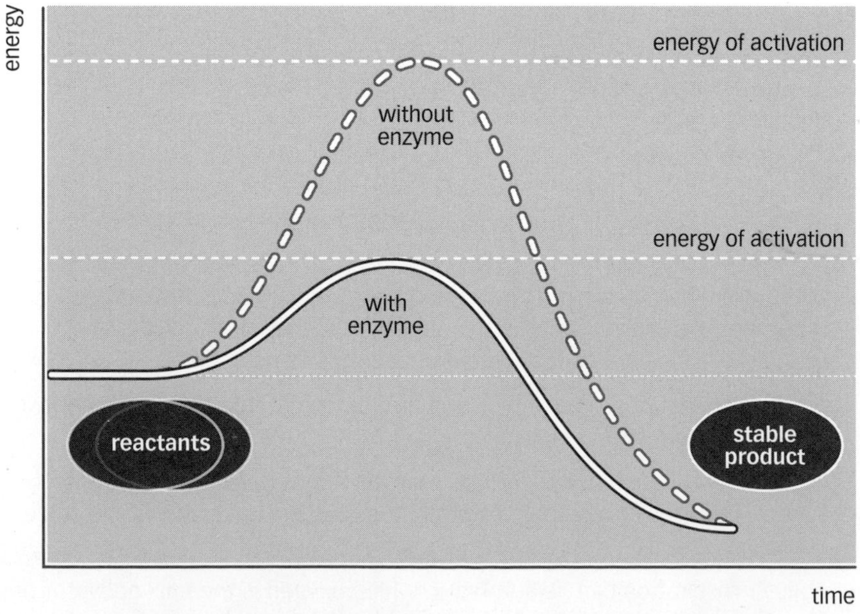

Figure 3.2 Energy of activation. A chemical reaction cannot take place unless there is enough activation energy available to start the reaction. Enzymes lower the amount of activation energy necessary to start the reaction.

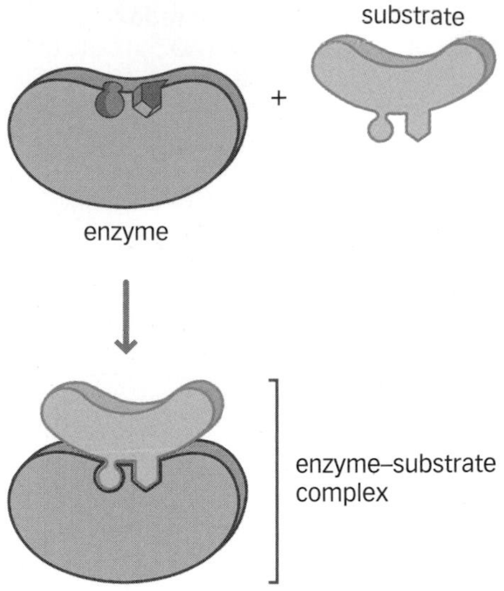

Figure 3.3 Active site on an enzyme. Only a substrate that fits exactly in the active site can combine with the enzyme. When the substrate binds to the active site, an enzyme–substrate complex is formed.

site. It is at the active site that the enzyme and substrate interact to form what is referred to as an **enzyme–substrate complex**. As a result of this binding to the active site, some of the chemical bonds in the substrate molecule undergo changes. These changes lead to the formation of the product of the reaction. In the hypothetical pathway shown on page 35, substrate A binds to enzyme 1, and this binding "changes" A to product B. After the change has taken place, A is released from the active site of enzyme 1 as product B. Product B now becomes the substrate of enzyme 2, and the reaction between B and enzyme 2 forms product C. This continues until the final product E is formed.

Enzymes are generally highly specific. This means that a given enzyme catalyzes only one type of reaction. In addition, most enzymes react with only one particular substrate. Therefore, in our hypothetical pathway, the only reaction that enzyme 1 catalyzes is the one in which the substrate A is converted to product B. This specificity is important because it allows a cell to function in an orderly way and use its energy stores efficiently. The shape of an enzyme molecule and the electrical charges found at the active site are responsible for the enzyme's specificity.

It should be noted that some enzymes can work on more than one substrate, but in these cases the enzymes always work in a particular type of reaction. For example, a proteolytic enzyme always degrades proteins because it reacts only with peptide bonds. Thus, a given proteolytic enzyme may react with, say, five or six different substrates, but every one of the substrates will be a protein.

Coenzymes and Cofactors

Many enzymes can catalyze a reaction only if other substances are present. These enzymes are referred to as apoenzymes (**Figure 3.4**) and require the help of some other substance before they can react with substrate. When the helper substance is an inorganic ion, such as magnesium, zinc, or manganese, it is called a **cofactor**. When the helper substance is a nonprotein organic molecule, it is called a **coenzyme**. Apoenzymes have an active site that does not fit the enzyme's substrate. It is therefore necessary for a cofactor or coenzyme to bind first to the active site. When the coenzyme or cofactor for the apoenzyme has bound to the active site, the binding changes the site's shape so that the substrate now fits (see **Figure 3.4**).

Metabolic reactions may require the presence of **carrier molecules** to carry hydrogen atoms or electrons in redox reactions. Coenzymes and cofactors can be used as carrier molecules. When a carrier molecule receives either hydrogen atoms or electrons, it becomes reduced. When the carrier molecule releases them, it becomes oxidized. Two coenzyme carrier molecules frequently encountered in biological reactions are **FAD** (**flavin adenine dinucleotide**) and **NAD^+** (**nicotinamide adenine dinucleotide**). Both are

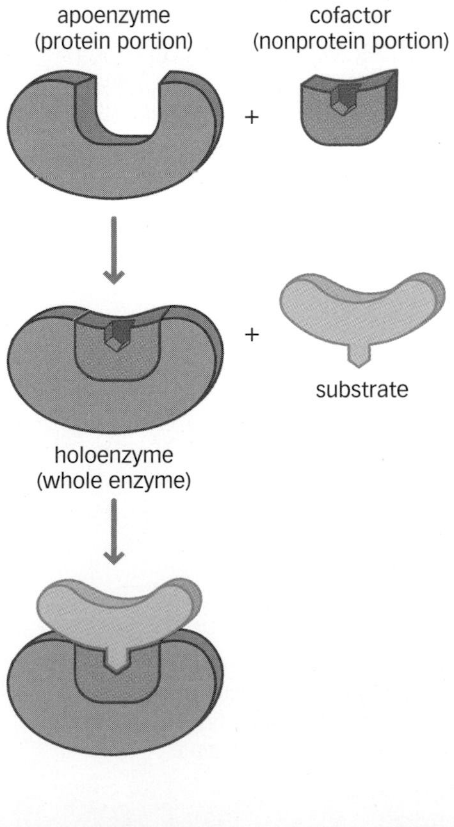

Figure 3.4 Apoenzymes and holoenzymes. With the type of enzymes known as apoenzymes a cofactor is needed in the active site for the substrate to fit properly. After the cofactor has bound to the apoenzyme, the combination is referred to as a holoenzyme. Note that the presence of the cofactor at the active site modifies the site so that the substrate can fit perfectly.

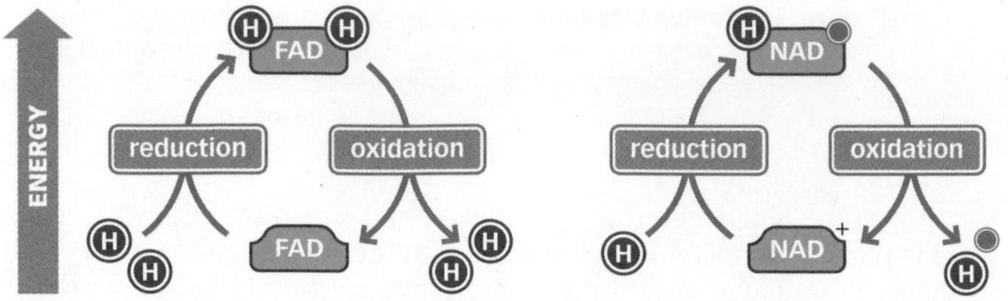

Figure 3.5 Coenzymes in redox reactions. The coenzyme FAD can carry two hydrogen atoms. The coenzyme NAD^+ can carry one hydrogen atom plus one electron. When coenzymes are reduced, they gain electrons, which increases their energy level. When the reduced forms of the coenzymes are oxidized, they lose electrons, which decreases their energy level.

Fast Fact

Many coenzymes are derived from vitamins, which is why vitamins are such an important part of our daily nutritional requirements.

vitamins and are represented in **Figure 3.5**. FAD can carry two hydrogen atoms and two electrons, and when it does it is called $FADH_2$, which is the reduced form of FAD. (Remember, reduction is the gain of electrons.) The coenzyme NAD is positively charged in its oxidized state (NAD^+) because it is lacking one electron. When it is reduced, it has gained a hydrogen atom and an electron and so becomes NADH. When these two carrier molecules are in their reduced state — $FADH_2$ and NADH — it is the electron that carries the energy that is being transferred from one molecule to another.

Enzyme Inhibition

Organisms cannot allow continuous enzyme activity because this requires a great deal of energy. It is also wasteful of components that may be hard to come by for the organism. In addition, continuous enzyme activity may result in the buildup of potentially harmful products. It is therefore necessary to be able to regulate enzyme activity. This is accomplished in two major ways: competitive inhibition and allosteric inhibition.

In **competitive inhibition**, an inhibitor molecule, which is a molecule similar in structure to the substrate for a given enzyme, competes with the substrate to bind to the enzyme's active site (**Figure 3.6**). Once the inhibitor molecule has bound to the active site, the substrate cannot bind. Because the inhibitor cannot react with the enzyme, the metabolic pathway is stopped. This competition is reversible and depends on the relative numbers of inhibitor molecules and substrate molecules present. If the number of substrate molecules is high and the number of inhibitor molecules is low, the active sites of only a few enzyme molecules are blocked, which leads to only a slight decrease in enzyme activity. If the number of inhibitor molecules is high and the number of substrate molecules is low, the active sites of most of the enzyme molecules are blocked, leading to a steep decrease in enzyme activity.

Fast Fact

Competitive inhibition is an example of how sulfa drugs work in inhibiting bacterial growth and is also how drugs such as AZT (azidothymidine) are used to treat HIV infections.

Allosteric inhibition also involves inhibitor molecules, but in this case the molecules do not block the active site. Instead they bind to another part of the enzyme, at a site called the *allosteric site* (**Figure 3.7**). Before an allosteric inhibitor binds, the enzyme's active site is the right shape to accept

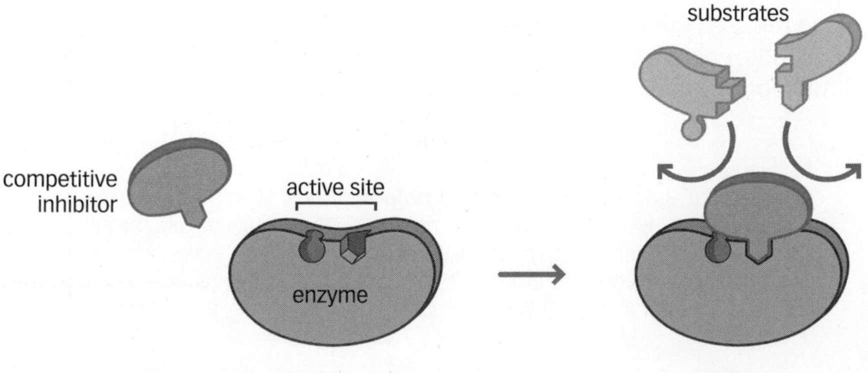

Figure 3.6 Competitive inhibition. The competitive inhibitor fits into the active site of the enzyme, preventing the substrate from binding to the enzyme.

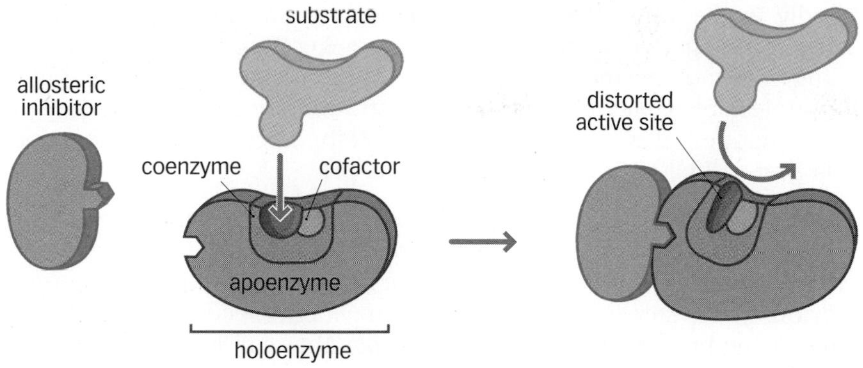

Figure 3.7 Allosteric inhibition. The allosteric inhibitor binds to a location on the enzyme that is different from the enzyme's active site. This binding changes the shape of the active site, and as a result enzymatic activity is stopped.

substrate. The binding of the inhibitor at the allosteric site changes the three-dimensional structure of the enzyme in such a way that the active site changes and can no longer fit properly with the substrate. As a result, there is no enzymatic activity. Some allosteric inhibitors bind reversibly and so can unbind and allow activity to resume. Others do not, however. When the binding is irreversible, the enzyme is permanently disabled.

Fast Fact

Some examples of irreversible allosteric inhibitors are lead, mercury, and other heavy metals. This is one of the reasons we don't use lead-based paint any more.

Feedback inhibition is also a regulatory mechanism for enzymatic activity and is used for many of the metabolic pathways found in the cell. With this type of inhibition, the final product in a pathway accumulates and begins to bind to and inactivate the enzyme that catalyzes the first reaction of the pathway (**Figure 3.8**). This inactivation is once again associated with a change in the three-dimensional structure of the enzyme molecule such that it no longer functions properly. Feedback inhibition is reversible, and when the level of end product decreases, the inhibition stops and the pathway begins to function again.

Factors That Affect Enzyme Reactions

Three major factors affect enzyme activity: temperature, pH, and concentration of substrate, product, and enzyme. Extremely high temperatures can break the hydrogen bonds that hold the enzyme in its three-dimensional shape, and this change in structure can inhibit enzyme activity. However, minor increases in temperature can help reactions occur more

Figure 3.8 Feedback inhibition. The inhibitor molecule is one of the end products of the metabolic pathway. If the inhibitor levels build up sufficiently, inhibitor molecules bind to an allosteric site on the enzyme, changing the shape of the active site and preventing the reactions of the pathway from continuing.

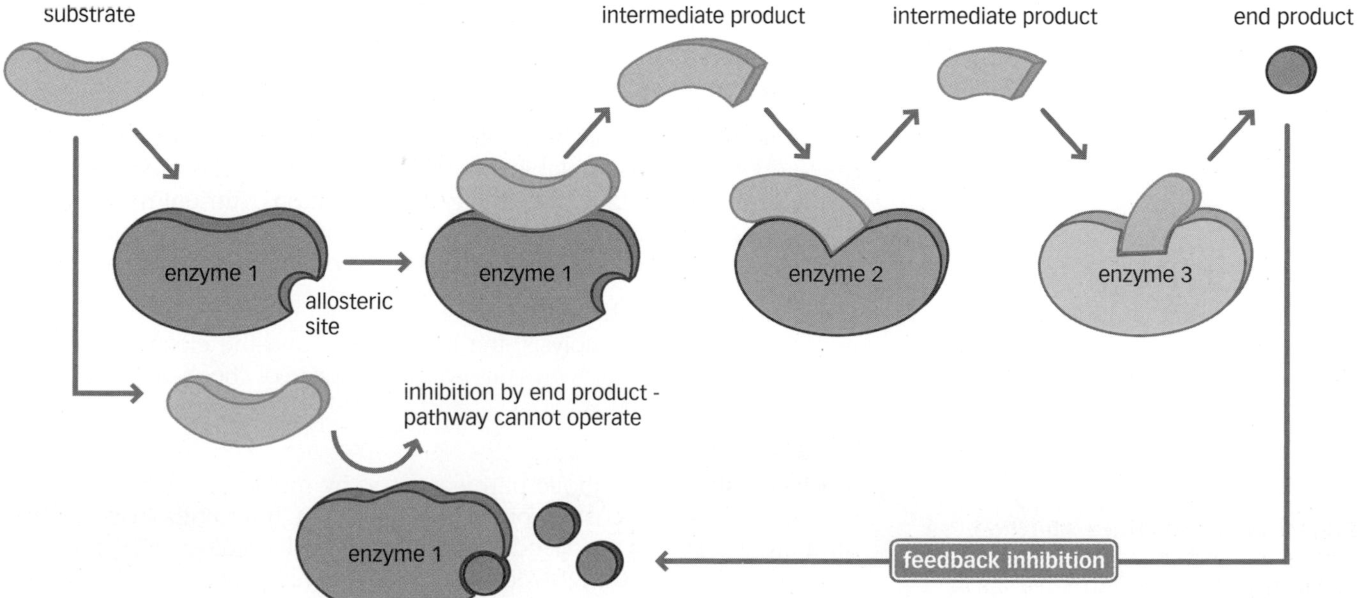

rapidly. By the same token, lower temperatures slow metabolic reactions and slow the growth of organisms. Changes in pH can alter the electrical charges in the enzyme molecule, and these changes in charges inhibit the molecule's ability to bind to its substrate. Very high or very low pH can denature the enzyme (which is a protein). Microbial enzymes function best at their optimal temperature and pH.

The concentrations of enzyme, substrate, and product all affect enzyme reactions. The job of an enzyme in a reaction is to lower the activation energy needed and thereby increase the rate at which the reaction runs. Obviously, the smaller the number of enzyme molecules available, the smaller the number of substrate molecules that can bind to enzyme at any instant and so the slower the reaction. If the substrate concentration is too low, that limits the number of enzyme–substrate complexes that can form at any instant and so also affects reaction rate. In some cases, once an enzyme reaction has run long enough to build up a quantity of product, the product can begin feedback inhibition, and so again the reaction rate is changed.

Fast Fact

Do you recall that egg salad sandwich you put in the refrigerator and forgot about? When you finally found it, it looked as though it had grown hair! The contamination that caused the sandwich to spoil is the result of metabolism. The lower temperature of the refrigerator causes the enzymes involved in metabolism to slow down, but it does not stop them completely, and so eventually the microbial growth spoils your lunch. You could have prevented the spoiling by freezing the sandwich, because at freezing temperatures the enzyme activity needed for metabolism is halted.

Keep in Mind

- Enzymes are proteins that work in metabolism by lowering the energy of activation.
- These proteins have a specific three-dimensional shape and complex with the substrate they act upon at a place on the molecule called the active site.
- Enzymes are highly specific and in some cases require cofactors or coenzymes to function.
- Enzyme function can be regulated by competitive inhibition or allosteric inhibition.
- Temperature, pH, and the concentration of substrate all affect the function of enzymes.

CATABOLISM

Now let's turn to the details of the various types of metabolic reaction seen in microorganisms. Here we will concentrate on catabolic processes, which are those that break down large molecules into smaller ones. After a detailed look at several types of catabolic reaction, we'll close the chapter with a brief look at anabolic reactions, which are those that assemble smaller molecules into larger ones.

Catabolic reactions can be looked on as *fueling reactions* because they make "fuel" (energy) available to an organism. This fuel is made available as the organism metabolizes the food it has eaten. Nutrient molecules in the food can be processed in several ways to release energy that is then stored in the high-energy bonds of ATP and other energy molecules.

There are several pathways by which most organisms release energy from nutrient molecules: glycolysis, the Krebs cycle, and the electron transport chain. Each of these pathways involves a sequence of chemical reactions.

Glycolysis

In **glycolysis**, the catabolic pathway used by most organisms, a carbohydrate such as glucose is broken down through a series of steps that ultimately result in the production of two molecules of ATP for each molecule of glucose involved (**Figure 3.9**). This process occurs in the cytoplasm and does not require oxygen.

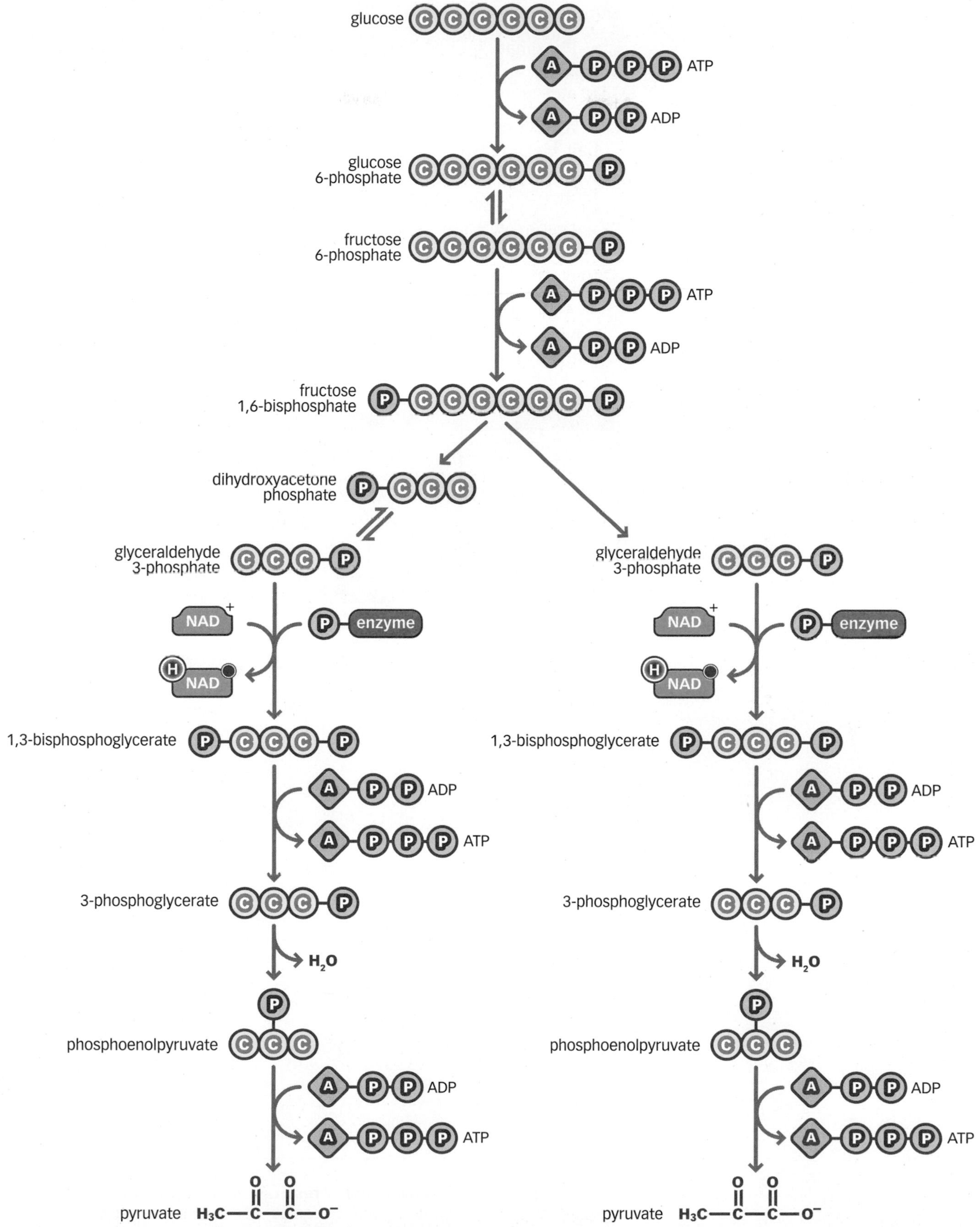

Figure 3.9 Glycolysis is a metabolic pathway involving a series of chemical reactions in which a glucose molecule is broken down into 2 three-carbon pyruvate molecules. This series of reactions yields 4 molecules of ATP, but the first steps of the pathway consume 2 ATP molecules, which makes the net gain of ATP through glycolysis 4 ATP – 2 ATP = 2 ATP. Glycolysis is used by both aerobic organisms and anaerobic organisms.

During glycolysis:

- Phosphate groups (PO_4^{3-}) are removed from ATP and transferred to substrates. This process is called **phosphorylation** and makes the substrates more energetic.
- After a series of steps, the 6-carbon glucose molecule is broken in half, yielding two 3-carbon molecules called pyruvate. These pyruvate molecules can be processed and moved into the Krebs cycle or into fermentation pathways (**Figure 3.10**).
- During glycolysis, two electrons are transferred to the carrier molecule NAD^+, converting it to NADH. The NADH then carries the electrons to electron-acceptor molecules.
- Although four ATP molecules are produced in glycolysis, there is a net gain of only two ATP molecules because the first steps of the pathway consume two ATPs.

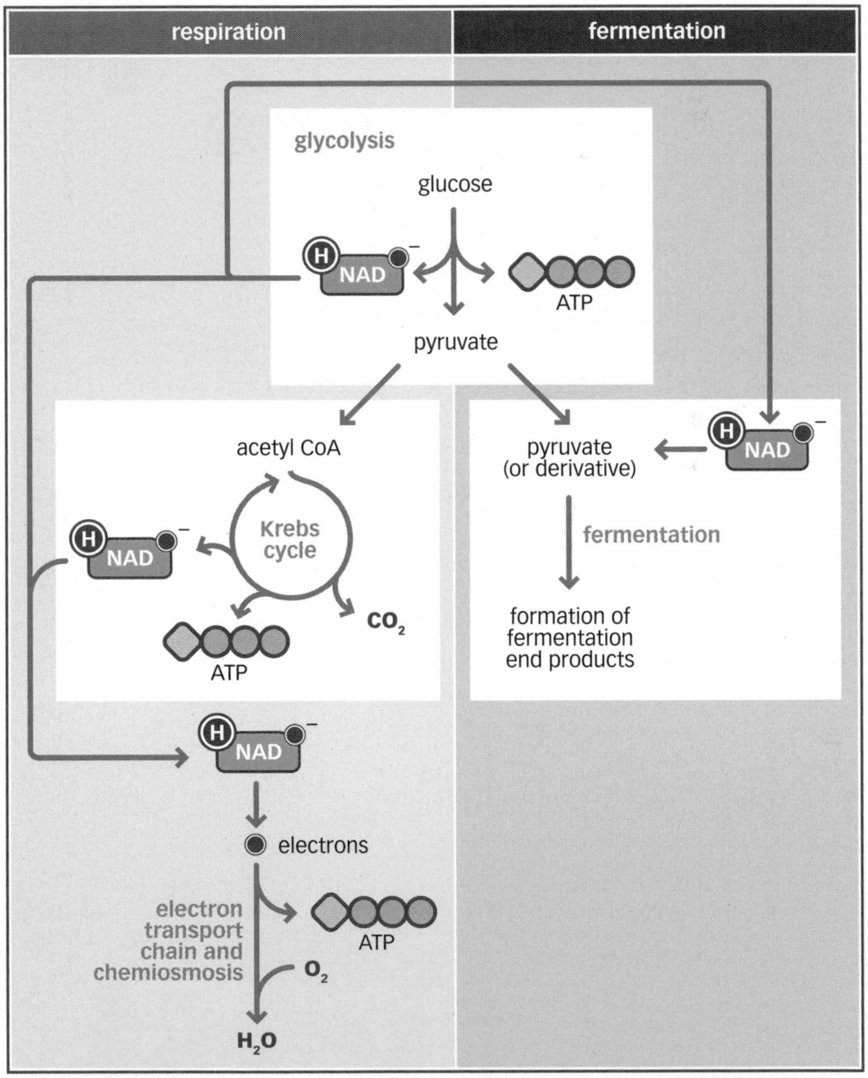

Figure 3.10 An overview of respiration and fermentation. The pyruvate produced in glycolysis can be used in either respiration or fermentation. Respiration involves both the Krebs cycle and the electron transport chain, with oxygen as the final electron acceptor. The Krebs cycle and electron transport yield 36 ATP molecules for each glucose molecule. Remember that glycolysis yields a total of 2 ATP molecules, so the total yield of ATP would be 38 molecules of ATP per molecule of glucose.

Let's take a moment to look at the mechanism of phosphorylation, which is the addition of a phosphate group to a molecule. When phosphate is added to a molecule, the energy of the molecule *increases*. In the first two steps of the glycolysis pathway, two molecules of phosphate are added to glucose, increasing its energy. This increase in energy is required to start the glycolytic pathway.

During the course of glycolysis, the energy released from the high-energy phosphorylated molecule as it loses its phosphate groups is used to form high-energy bonds between ADP and inorganic phosphate (P_i) via the reaction

$$ADP + P_i + \text{energy} \longrightarrow ATP$$

This reaction is reversible, which means that ATP can be broken down into ADP, inorganic phosphate, and energy:

$$ATP \longrightarrow ADP + P_i + \text{energy released}$$

The reversible nature of these reactions is a fundamental requirement for the survival of the cell because it allows the cell to recycle the ATP molecule.

As **Figure 3.10** indicates, the pyruvate produced in glycolysis can be metabolized further, either via the Krebs cycle in cellular respiration or via fermentation. Let's look at the Krebs cycle first.

The Krebs Cycle

The Krebs cycle is a catabolic pathway seen in aerobic cellular respiration in which the pyruvate produced in glycolysis is metabolized further (**Figure 3.11**). The cycle was named for the German biochemist Hans Krebs, who identified the steps in the 1930s. It is also referred to as the *tricarboxylic acid cycle* (TCA) or the *citric acid cycle*.

The cycle accepts only two-carbon molecules, and because pyruvate contains three carbons, it must be modified before it can enter the cycle. This modification involves the conversion of the three-carbon molecule to a two-carbon molecule through the removal of one carbon in the form of CO_2. This is a complex reaction involving the carrier molecule NAD^+ and the addition of a molecule called coenzyme A (CoA) to form the complex acetyl-CoA (**Figure 3.12**).

The Krebs cycle is a sequence of reactions in which hydrogen atoms are removed and their electrons are transferred to coenzyme carrier molecules. CO_2 is also given off during this cycle, and each step in the cycle is controlled by a specific enzyme. During the Krebs cycle, three important things occur:

1. Carbon is oxidized to CO_2.
2. Electrons are transferred to coenzyme carrier molecules that take the electrons to the electron transport chain (described below).
3. Energy is captured and stored when ADP is converted to ATP.

The Electron Transport Chain

Electron transport is the cellular process in which electrons are transferred to a final electron acceptor. In aerobic metabolism, that final electron acceptor is oxygen. In anaerobic metabolism, the final electron acceptor is some inorganic oxygen-containing molecule such as nitrate (NO_3^-), nitrite (NO_2^-), or sulfate (SO_4^{2-}). During electron transport, hydrogen atoms are transferred to NAD^+ carrier molecules, which transfer the

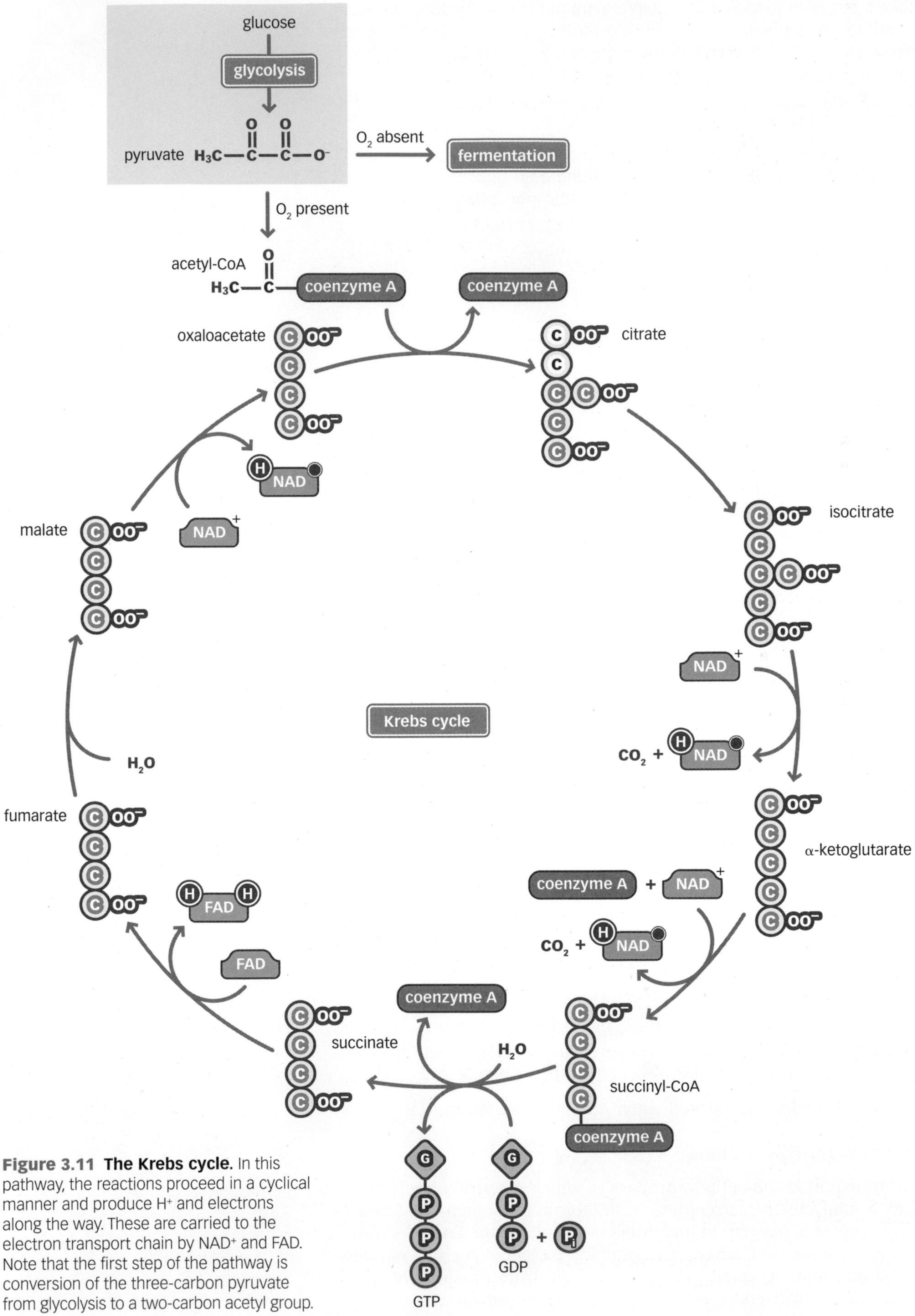

Figure 3.11 The Krebs cycle. In this pathway, the reactions proceed in a cyclical manner and produce H+ and electrons along the way. These are carried to the electron transport chain by NAD+ and FAD. Note that the first step of the pathway is conversion of the three-carbon pyruvate from glycolysis to a two-carbon acetyl group.

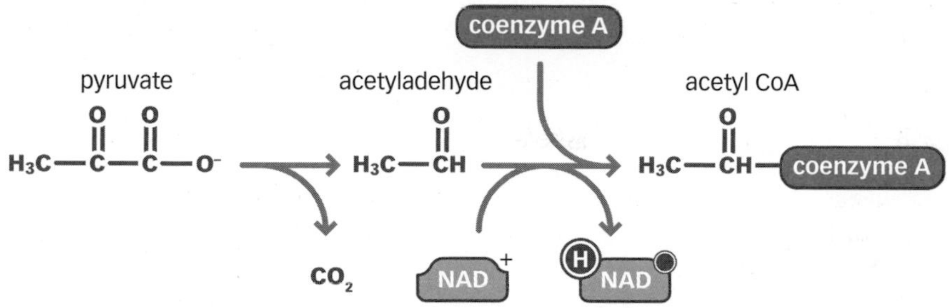

Figure 3.12 To enter the Krebs cycle, the three-carbon pyruvate from glycolysis must be modified to a two-carbon acetyl group by losing a carbon as CO_2. Once the acetyl group has combined with a molecule of coenzyme A, it is ready to enter the Krebs cycle.

hydrogen atoms to proteins; these proteins are found in a precise arrangement in the microbial cell membrane or in the inner membrane of the mitochondria in eukaryotic cells (**Figure 3.13**).

The transfer of hydrogen atoms from one molecule to another in the electron transport chain involves oxidation and reduction reactions. Each member of the chain becomes reduced as it picks up electrons. When it gives the electrons to the next molecule in the chain, it becomes oxidized and the accepting molecule becomes reduced (**Figure 3.14**). From the catabolism of a single molecule of glucose, there are 10 pairs of electrons transported to the electron transport chain by NAD^+ and an additional 2 pairs transported by FAD. When oxygen is the final electron acceptor in the chain it becomes reduced to the final form found in the H_2O molecule.

Electron transport differs from organism to organism, and some organisms can use more than one type. Keep in mind that during aerobic metabolism the electron transport chain also keeps the Krebs cycle turning by being a place for the reduced carrier molecule NADH to drop off

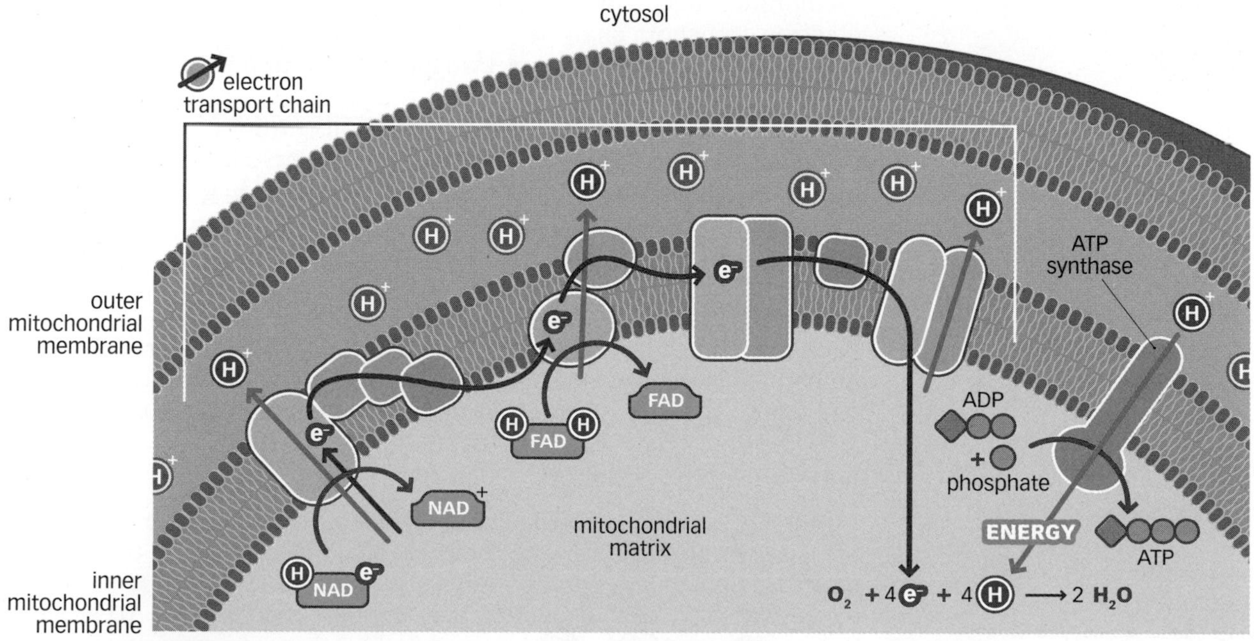

Figure 3.13 An illustration of the electron transport chain with oxygen as the final electron acceptor. This process is where the majority of ATP is produced.

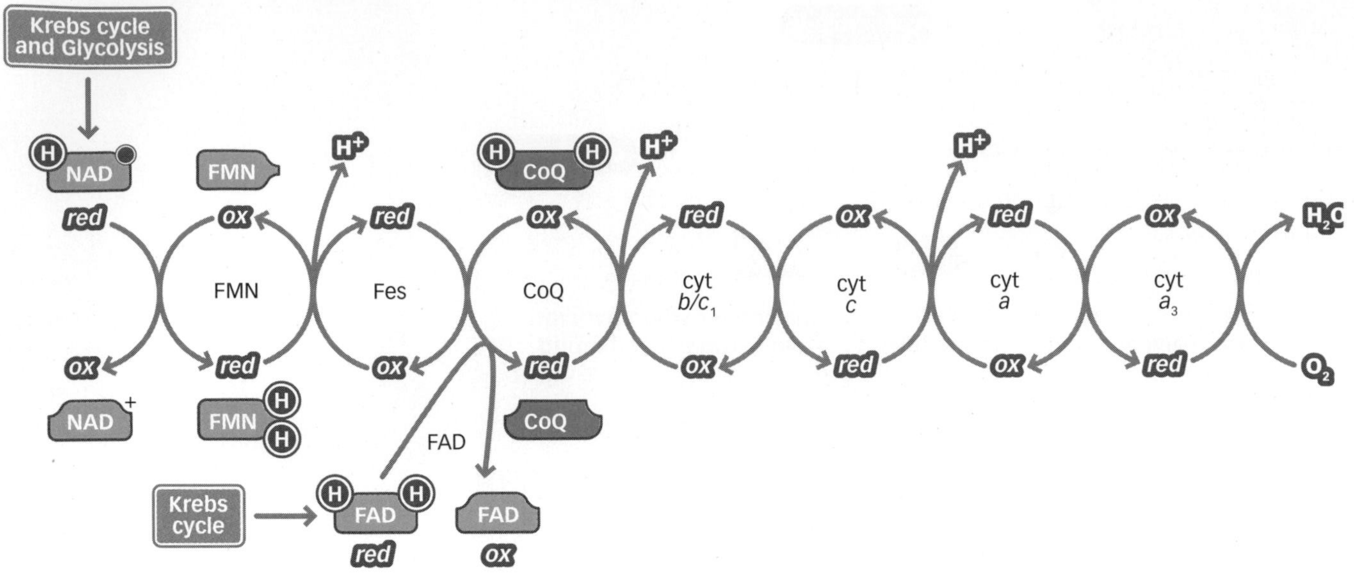

Figure 3.14 The electron transport chain. This is the final sequence of chemical reactions seen in aerobic respiration. In this series of steps, electrons and H^+ ions are passed between the intermediates in a sequence of redox reactions. In this case, the final electron acceptor is molecular oxygen, which then uses the acquired electrons to combine with hydrogen to form water, H_2O.

electrons. If the electron transport chain were blocked, the Krebs cycle would shut down.

We can think of the electron transport chain as a series of steps in a stairway. NADH enters at the top step and passes the electrons it is carrying to the next step. As these electrons descend the stairs, energy is released via a process called chemiosmosis.

Chemiosmosis

As electrons are transferred along the electron transport chain, protons (in this case hydrogens, which carry a positive electrical charge) are pumped out of the cell. In eukaryotic cells chemiosmosis occurs in the mitochondria, and these protons are pumped into the intermembrane space. Prokaryotes do not have mitochondria so their electron transport chain is located in the plasma membrane. Thus, in prokaryotes, the electron transport chain pumps protons across the plasma membrane. This causes the proton concentration outside the cell to be higher than the proton concentration inside the cell, causing a *concentration gradient* to form. Because it is a difference in proton concentrations, this particular concentration gradient is called the **proton motive force**. Whenever a concentration gradient exists, the particles tend to move from the region of high concentration to the region of low concentration. So, because the proton concentration is higher outside the cell, the proton motive force moves protons back into the cell. As the protons move through specialized proteins embedded in the membrane, energy is released and used to bind inorganic phosphate to ADP and form the high-energy molecule ATP.

Looking At The Numbers

During the aerobic catabolism of one glucose molecule, 10 pairs of electrons are transported by NAD^+, and each pair releases from the glucose molecule enough energy for 3 ATPs, yielding 30 ATPs. Each pair of electrons carried by FAD generates only enough energy for 2 ATPs. Since there are 2 pairs of electrons carried by FAD there is a gain of 4 ATPs by this carrier molecule. Include the net gain of 2 ATPs obtained from glycolysis and the 2 ATPs from the Krebs cycle and we see that aerobic respiration yields a total of 38 ATP molecules from one molecule of glucose.

Keep in Mind

- Catabolism is the metabolic process in which organic molecules are broken down to release energy.
- Catabolism can involve glycolysis, which occurs in the cytoplasm, as well as the Krebs cycle and electron transport, which occur at the bacterial plasma membrane.
- When glycolysis is linked to the Krebs cycle and electron transport, and oxygen is the final electron acceptor, 38 ATP molecules are produced for every molecule of glucose that is broken down.
- Aerobic cellular respiration uses oxygen as the final electron acceptor, whereas anaerobic cellular respiration uses an inorganic molecule other than elemental oxygen as the final electron acceptor.
- The Krebs cycle involves a series of chemical changes that generate the release of protons and electrons.
- Electrons and protons are carried from the Krebs cycle to the electron transport chain by reduced coenzymes such as NADH and $FADH_2$.
- In the electron transport chain, the protons and electrons are moved through a series of oxidation–reduction reactions that result in the formation of a concentration gradient called the proton motive force.
- Protons are moved by the proton motive force across the plasma membrane of the bacterial cell, and energy is released.
- The energy produced from protons moving across the plasma membrane is used to form ATP.

Fermentation

Fermentation is defined as the enzymatic breakdown of carbohydrates in which the final electron acceptor is an organic molecule. ATP is synthesized by substrate-level phosphorylation, which is not linked to the electron transport chain, and oxygen is not required. In fermentation, there is no Krebs cycle or electron transport chain (**Figure 3.15**). Fermentation does not increase the yield of ATP from what it is in glycolysis: two ATPs for every glucose molecule. This means that a huge amount of glucose must be consumed by organisms that use fermentation if they are to satisfy their energy requirements.

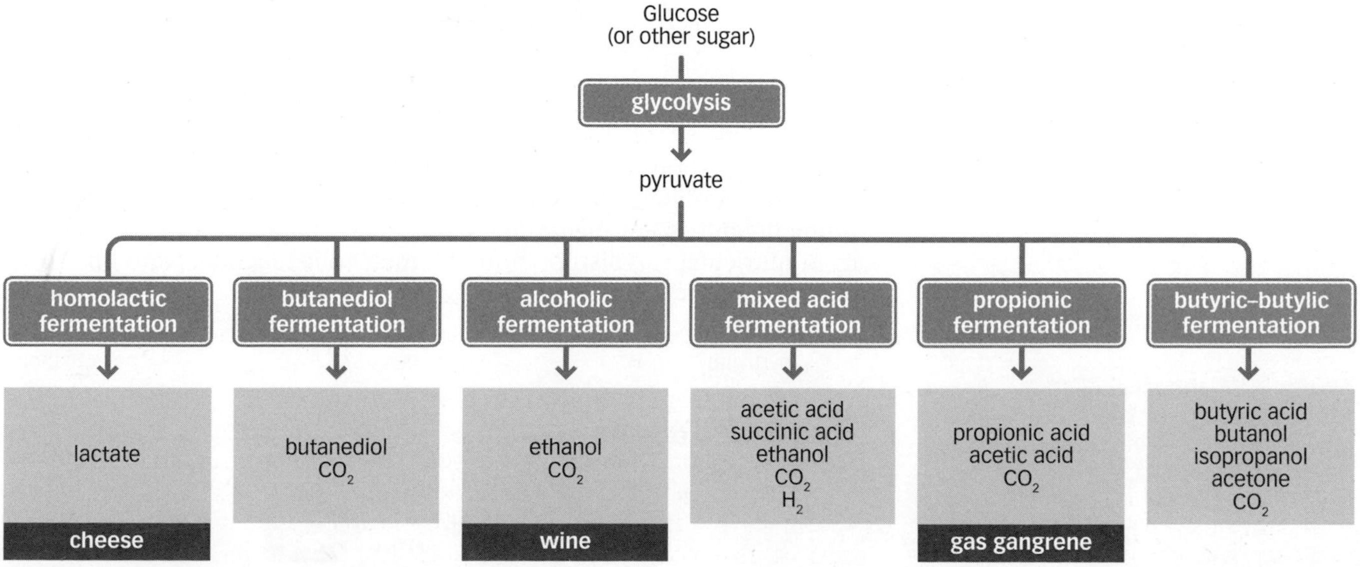

Figure 3.15 There are a variety of fermentation pathways possible for the pyruvate formed in glycolysis. Which pathway is used depends on which microorganism is doing the fermenting.

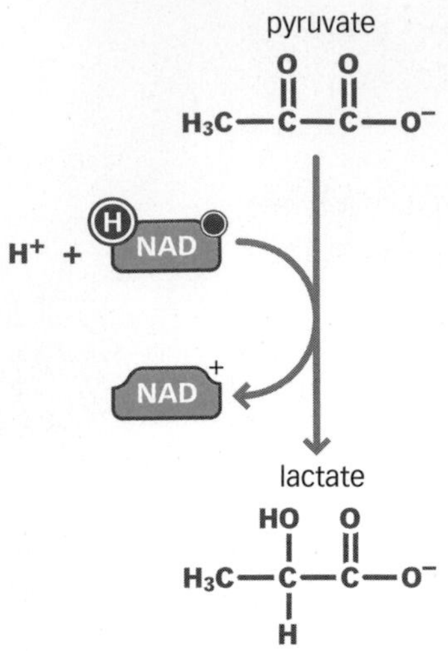

Figure 3.16 The homolactic fermentation pathway. Pyruvate obtained from glycolysis is reduced to lactate through a redox reaction with the carrier molecule NAD+.

Different microorganisms use different fermentation pathways, and in some cases these differences can be used diagnostically to aid in identifying microorganisms.

Two of the most important and common fermentation pathways are homolactic fermentation and alcoholic fermentation. Neither pathway stores energy in ATP as the pyruvate produced by glycolysis is catabolized. However, both fermentation pathways recycle the carrier molecule NAD^+ so that it can be used to continue glycolysis. This recycling is accomplished by removing the electrons from the reduced form, NADH, so that the resulting NAD^+ can go back to the glycolytic pathway and acquire more electrons.

Homolactic Fermentation

This is the simplest fermentation pathway for metabolizing pyruvate. The pyruvate molecule is converted directly to lactate with the reactions driven by the energy of the electrons from NADH (**Figure 3.16**). Note in **Figure 3.15** that the homolactic pathway is the only one not showing the gas CO_2 as a product. Unlike most other fermentation pathways, this pathway does not produce gas as a by-product. Lactic fermentation occurs in several forms of bacteria and is also seen in mammalian muscle cells. Also, as **Figure 3.15** indicates, it is a major method used in the making of cheese.

Alcoholic Fermentation

In alcoholic fermentation, CO_2 is released from pyruvate to form the intermediate acetaldehyde (**Figure 3.17**). This intermediate is quickly reduced to ethanol by electrons from NADH (and of course the NADH is oxidized to NAD^+). Alcoholic fermentation is rarely seen in bacteria but is very common in yeasts, as seen in the production of beer.

ANABOLISM

Anabolic reactions are classified as *biosynthetic reactions* because they are used to synthesize all the biological molecules needed by the cells of living organisms. Biosynthetic reactions form the network of pathways that produce the components required by the cell for growth and survival. These reactions are fueled by the energy stored in high-energy bonds in ATP (and those high-energy ATP bonds, remember, were created with energy released from nutrient molecules during the catabolic reactions described above). Using anabolic reactions, cells construct amino acids, nucleotides, carbohydrates, fatty acids, and all the other molecular biological building blocks they need (**Figure 3.18**). These anabolic pathways serve as a target for treatment strategies. As we will see in Chapter 19, the sulfonamide and trimethoprim antibiotics act by disrupting biosynthetic pathways.

Anabolic reactions cost energy and therefore must proceed in an orderly, efficient, coordinated fashion. Microorganisms cannot control their environment, and so any environmental change (in temperature or availability of nutrients) can disrupt or inhibit metabolic reactions, and this could be lethal for the microorganism.

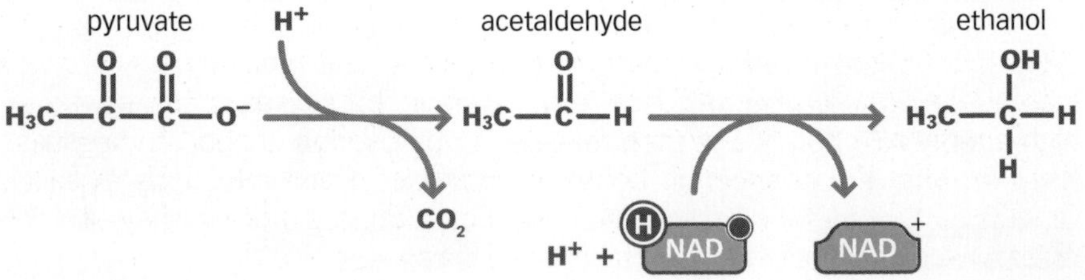

Figure 3.17 The alcoholic fermentation pathway. In this two-step process, a molecule of CO_2 is first removed from the pyruvate to form acetaldehyde. NADH then reduces the acetaldehyde to ethanol. At the same time, the NADH is oxidized to NAD^+.

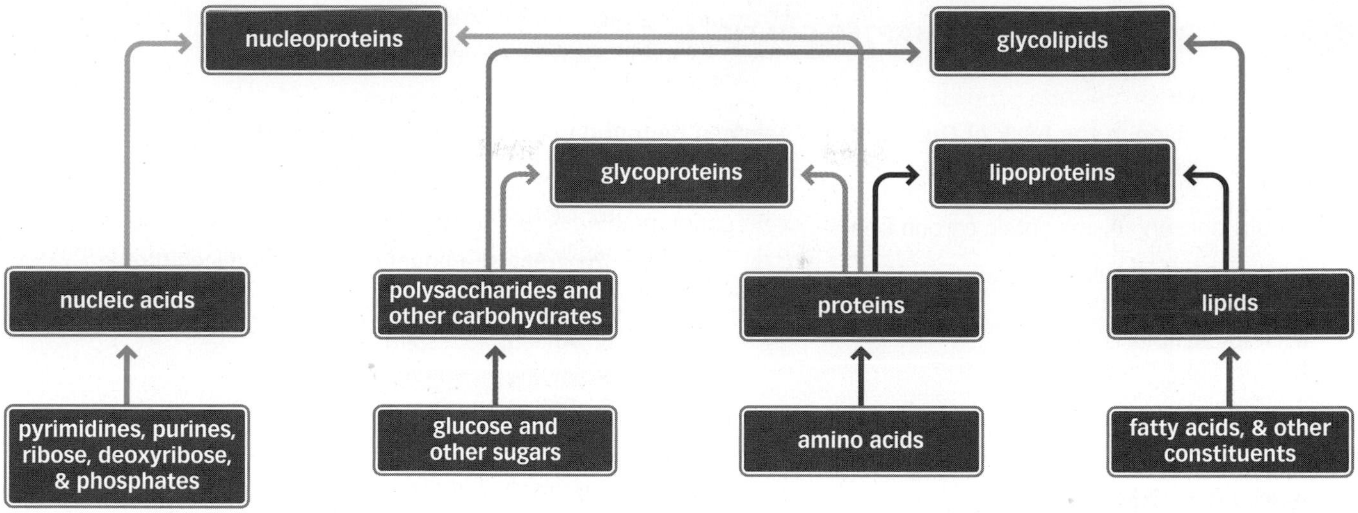

Figure 3.18 Biosynthesis involves the formation of complex biomolecules from simpler ones. This represents the anabolism arm of metabolism and requires the input of energy locked up in ATP.

Control of enzyme activity is the most common means by which organisms modulate the flow of materials through the catabolic and anabolic pathways. This ensures that the flow of carbons from the major substrates through the various pathways is adjusted to meet the biosynthetic demands of the cell. In biosynthetic pathways, reactions are also controlled through feedback inhibition.

SUMMARY

- Metabolism is the chemical process that provides or stores energy for the organism.
- Metabolism can be broken down into two parts, catabolism (breaking down molecules) and anabolism (building up molecules).
- Oxidation and reduction reactions involve the transfer of electrons.
- Nearly all chemical processes of the cell consist of a series of chemical reactions known as a pathway.
- Enzymes are proteins that speed up chemical reactions by lowering the energy of activation.
- Enzymes work on the basis of their three-dimensional shape. They are specific and in some cases require cofactors or coenzymes to function.
- Temperature, pH, and concentration of substrate all affect enzyme function.
- When oxygen is involved, catabolism occurs through glycolysis, the Krebs cycle, and electron transport.
- Aerobic metabolism requires oxygen and yields 38 ATPs from the breakdown of one molecule of glucose.
- Breakdown of one molecule of glucose without oxygen (anaerobic metabolism) yields only two molecules of ATP.

In this chapter we have briefly reviewed the essentials of metabolism. Metabolism is all about building molecules (anabolism) and breaking them down (catabolism). The energy released from catabolic processes is first stored in high-energy ATP bonds and then released to be used in anabolic processes. Keep in mind the connection between metabolism and infectious disease. Organisms that metabolize well increase their numbers more quickly, and increased numbers can enhance the infectious process.

SELF EVALUATION AND CHAPTER CONFIDENCE

Multiple Choice

Answers are given in the back of the book and help can be found in the student resources at:
www.garlandscience.com/micro

1. Autotrophic organisms obtain carbon from
   - A. Organic molecules
   - B. CO_2
   - C. Photosynthesis
   - D. The soil
   - E. Water
2. Heterotrophic organisms obtain carbon from
   - A. Organic molecules
   - B. CO_2
   - C. Photosynthesis
   - D. The soil
   - E. Water
3. Anabolism is the process in which
   - A. Molecules are broken down
   - B. Molecules are transformed into other more essential components.
   - C. Molecules are built up
   - D. Energy is decreased
4. Catabolism is the process in which
   - A. Molecules are broken down
   - B. Molecules are transformed into more essential components
   - C. Molecules are built up
   - D. Energy is decreased
5. Oxidation is defined as
   - A. Gaining an electron
   - B. Utilizing oxygen for metabolism
   - C. Destroying oxygen during metabolism
   - D. Losing an electron
   - E. None of the above
6. During a redox reaction
   - A. Electrons are lost
   - B. Electrons are multiplied
   - C. Electrons are gained
   - D. Electrons are gained and lost
   - E. Electrons are not used at all
7. During a reduction reaction a substance
   - A. Gains an electron and becomes more positively charged
   - B. Loses an electron and becomes more negatively charged
   - C. Gains an electron and becomes more negatively charged
   - D. Loses an electron and becomes positively charged
   - E. Neither gains nor loses an electron
8. The greatest amount of ATP is produced through
   - A. Fermentation
   - B. Anaerobic respiration
   - C. Aerobic respiration
   - D. Both A and B
   - E. None of the above
9. Organisms that can use either oxygen or no oxygen to metabolize are called
   - A. Aerobes
   - B. Obligate aerobes
   - C. Obligate anaerobes
   - D. Facultative aerobes
   - E. Facultative anaerobes
10. Complete oxidation of glucose yields
    - A. Carbon monoxide, water, and energy
    - B. Carbon dioxide and water
    - C. Carbon monoxide and energy
    - D. Carbon dioxide, water, and energy
    - E. None of the above
11. In enzymatic reactions the reaction material is the
    - A. Product
    - B. Enzyme
    - C. Substrate
    - D. Substrate and enzyme
    - E. Product and enzyme
12. Enzymes
    - A. Cause a reaction to happen
    - B. Cause a reaction to happen more slowly
    - C. Increase the activation energy
    - D. Decrease the energy of activation
    - E. Have no effect on the energy of activation
13. The active site of an enzyme is the result of the
    - A. Structure of the substrate
    - B. Structure of the product
    - C. Structure of the protein
    - D. Activation energy
    - E. Enzymatic pathway
14. In order to function, apoenzymes require
    - A. Cofactors
    - B. Coenzymes
    - C. Neither A nor B
    - D. B only
    - E. Either A or B

15. During competitive inhibition of enzyme function

A. The product competes with the substrate for the active site
B. ATP competes with the substrate for the active site
C. A molecule that is similar in structure to the substrate competes for the active site
D. Any molecule can compete with the substrate for the active site
E. A molecule similar to the substrate helps the substrate to bind to the active site

16. Allosteric inhibition occurs

A. At the active site
B. Because of the product
C. Because of the substrate
D. At a site away from the active site
E. Because of a molecule identical to the substrate

17. Both allosteric and competitive inhibition involve

A. Changes to the structure of the product
B. Changes to the structure of the substrate
C. Changes to the protein structure of the enzyme
D. Both A and B
E. None of the above

18. Fermentation respiration involves

A. Glycolysis only
B. Glycolysis and the Krebs cycle
C. Glycolysis, the Krebs cycle, and electron transport
D. Only the Krebs cycle
E. Only electron transport

19. During phosphorylation

A. ATP is broken down
B. ATP is converted to ADP
C. A phosphate is transferred to a substrate
D. Substrates become more energetic
E. All of the above

20. The net gain of ATP after glycolysis is

A. 4
B. 6
C. 2
D. 38
E. No ATP is produced in glycolysis

21. How many net ATP molecules are derived from fermentation?

A. 4
B. 2
C. 38
D. 16
E. None

22. Lactate fermentation

A. Is the most complex form of fermentation
B. Is a form of aerobic respiration
C. Is not involved in anaerobic respiration
D. Is used to make cheese
E. None of the above

23. The Krebs cycle only accepts

A. Single carbons
B. Three-carbon molecules
C. Pyruvate from the glycolytic pathway
D. Two-carbon molecules
E. Only products derived from fermentation

24. The final electron acceptor in aerobic respiration is

A. Water
B. Oxygen
C. Hydrogen
D. Both A and B

25. The majority of electrons are carried to the electron transport chain by

A. NAD
B. FAD
C. Glycolysis
D. The Krebs cycle
E. None of the above

26. In anaerobic metabolism, the final electron acceptor can be any of the following except

A. Nitrite
B. Sulfate
C. Nitrate
D. Oxygen

27. Biosynthetic reactions are part of

A. Glycolysis
B. Catabolism
C. Anabolism
D. Electron transport
E. The Krebs cycle

28. The number of ATP molecules that result from aerobic respiration is

A. 4
B. 38 for bacteria, <38 for eukaryote
C. 16
D. 36
E. 32

DEPTH OF UNDERSTANDING

Questions listed here require you to bring together the concepts you have learned in this chapter into a discussion format. This helps you to increase your depth of understanding of the material you have learned. Help can be found in the student resources at:
www.garlandscience.com/micro

1. Using your knowledge of metabolism and examples, describe the connection between anabolism and catabolism. Why is this connection so important?

2. Describe the relationship between glycolysis and the Krebs cycle.

3. "It is better to respire aerobically than anaerobically." Using what you have learned in this chapter, defend this statement.

An Introduction to Cell Structure and Host-Pathogen Relationships

Chapter 4

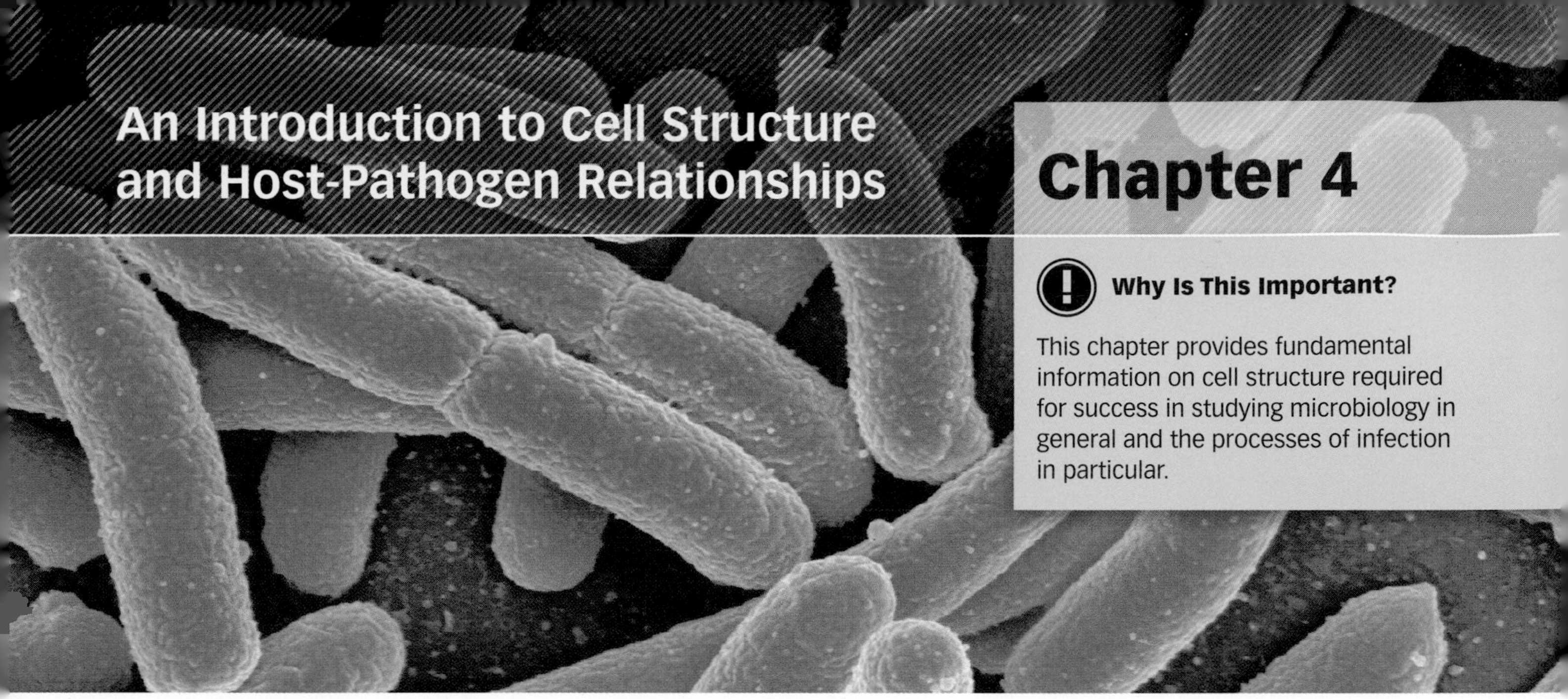

Why Is This Important?

This chapter provides fundamental information on cell structure required for success in studying microbiology in general and the processes of infection in particular.

OVERVIEW

In the preceding two chapters, we have reviewed first the fundamental chemical principles applicable to microbiology and then the basics of metabolism in microorganisms. These topics are important to our understanding of many of the subjects we discuss as we explore microbiology and infectious disease. In this chapter we look at the basic characteristics of bacterial cells, the concepts of pathogenicity and virulence, and the relationship between a pathogen and its host. The last section of the chapter is devoted to a review of the eukaryotic host cell, which is the target for infectious disease. Although this chapter focuses on bacteria, it is important to remember that viruses, fungi, and parasites can also cause infections. Each of these groups of microorganisms is discussed in detail in subsequent chapters (viruses in Chapters 12 and 13; parasites and fungi in Chapter 14).

We will divide our discussions into the following topics:

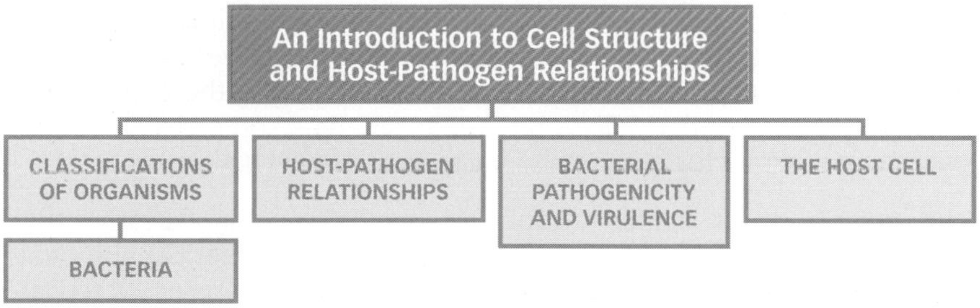

CLASSIFICATION OF ORGANISMS

All living organisms can be classified as either *prokaryotes* or *eukaryotes.* **Prokaryotes** are extremely simple cells that lack a nucleus and any other structures enclosed by a membrane. **Eukaryotes** are organisms made up of cells that do contain a membrane-enclosed nucleus as well as other membrane-enclosed structures outside the nucleus. All the membrane-enclosed structures are collectively referred to as **organelles**. We will look at the major differences between prokaryotic and eukaryotic cells later in this chapter.

Biologists classify organisms by their genus name and species name. For example, humans are referred to as *Homo sapiens* (genus *Homo*, species *sapiens*), the bacterium that causes **tetanus** is called *Clostridium tetani* (genus *Clostridium*, species *tetani*), and the bacterium that causes botulism is *Clostridium botulinum* (genus *Clostridium*, species *botulinum*).

What Do I Need to Know?

To get the most out of this chapter, please review the following terms from your previous courses in biology, anatomy, physiology, or chemistry or in previous chapters of this book as indicated in parentheses: ATP (**3**), chromosome, DNA (**2**), flagellum, gene, hydrophilic (**2**), hydrophobic (**2**), nosocomial infection, polar and nonpolar covalent bonds (**2**), RNA (**2**), transcription of genetic information, vesicle.

Fast Fact

Bacteria that cause infections in humans are referred to as pathogens.

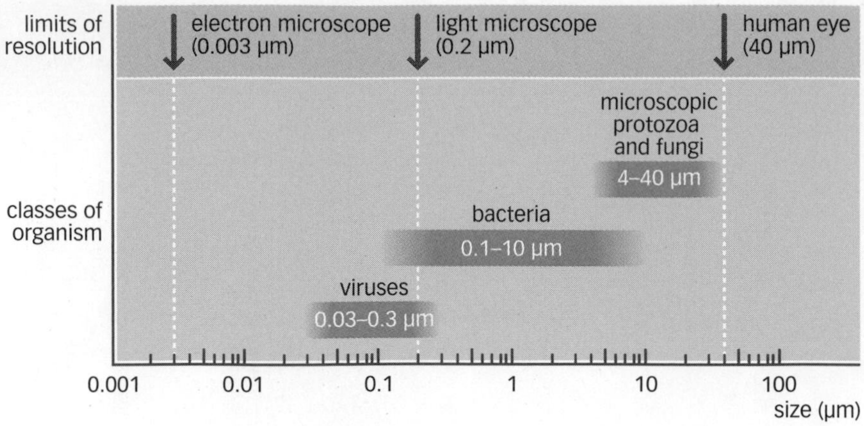

Figure 4.1 The relative sizes of microorganisms.

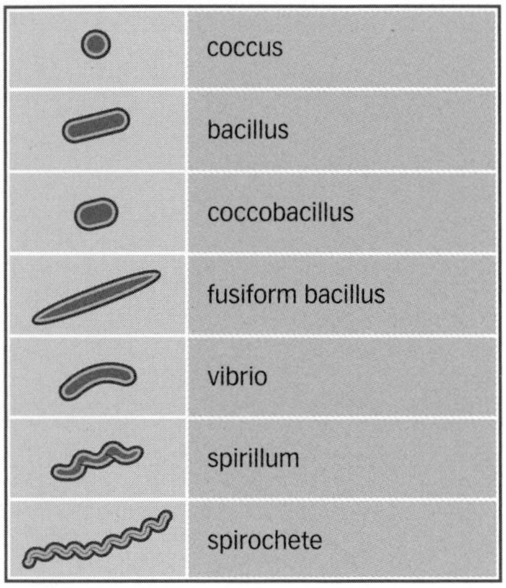

Figure 4.2 Bacterial shapes.

A genus (plural *genera*) can include one species or multiple species. For example, the genus *Homo* now has only one species, *sapiens*, whereas the genus *Clostridium* has several species, including *tetani* and *botulinum*. The convention for writing the genus and species names of organisms is to put the entire name in italics with the genus name capitalized and the species name in lower case.

BACTERIA

Bacteria are the product of about 3 billion years of natural selection and have emerged as immensely diverse and very successful organisms that colonize all parts of the world and its inhabitants. Our initial discussion of bacteria begins with a look at their size, their shape, and how clusters of them form identifiable multicell arrangements. We then examine the procedures used to stain bacteria so that they can be seen microscopically.

Size, Shape, and Multicell Arrangement

Bacteria are the smallest living organisms. Those that either colonize or infect humans range from 0.1 to 10 μm (1 μm = 10^{-6} meter) in the largest dimension. Some bacteria, such as those belonging to the genera *Rickettsia*, *Chlamydia*, and *Mycoplasma*, are as small as some viruses, as the overlap of the bars for viruses and bacteria in **Figure 4.1** shows.

Fast Fact

Some bacteria are referred to as pleomorphic, meaning that they can change their shape.

The major bacterial shapes are spheres, ovoids (egg-shaped rather than true sphere), straight rods, curved rods, and spirals. Spherical and ovoid bacteria are called **cocci** (singular **coccus**), and both types of rod are called **bacilli** (singular **bacillus**). Very short rods, which can be mistaken for cocci, are referred to as coccobacilli. Rod-shaped bacteria that have tapered ends are called fusiform bacilli. Spiral bacteria are called spirilla (singular **spirillum**) if the cell is rigid and **spirochetes** if it is flexible and undulating (**Figure 4.2**).

In addition to their distinctive shapes, bacteria can form distinctive multicell arrangements that are easily identified (**Figure 4.3**). One factor determining the shape of these multicell arrangements is the degree of stickiness of the organisms (which can vary depending on growth conditions). Cocci can form two-cell arrangements called **diplococci** (**Figure 4.4**). This arrangement is seen in bacteria such as *Streptococcus pneumoniae*, which causes respiratory infections, and *Neisseria gonorrhoeae*, which causes the sexually transmitted disease we call **gonorrhea**. In addition, cocci can form chains (**Figure 4.5**), as in *Streptococcus pyogenes*, and clusters (**Figure 4.6**), as occurs with staphylococcal organisms.

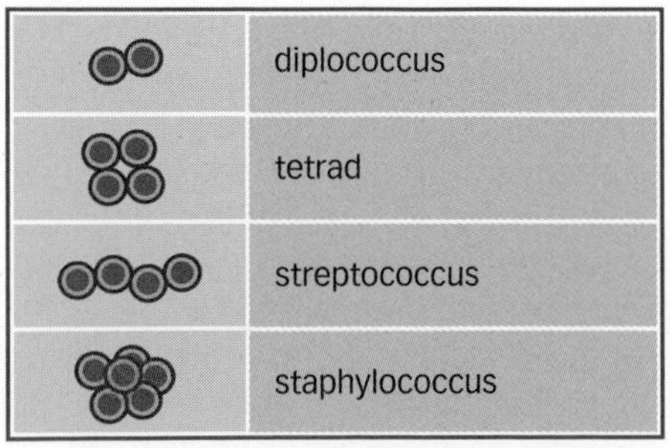

Figure 4.3 Multicell arrangements seen with spherical bacteria.

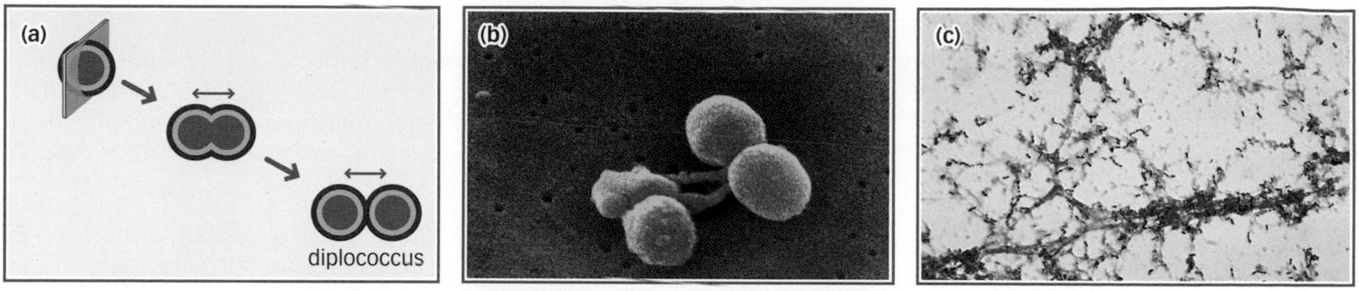

Figure 4.4 Arrangement of Diplococci. Panel **a**: Cocci can arrange in pairs based on the plane of division. Panel **b**: A scanning electron micrograph of *Streptococcus pneumoniae*. Panel **c**: A photomicrograph of *Streptococcus pneumoniae* grown from a blood culture.

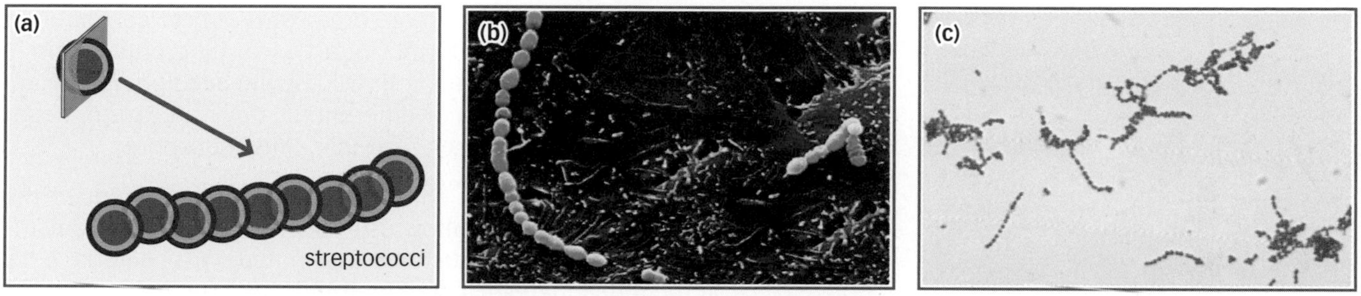

Figure 4.5 Arrangement of Streptococci. Panel **a**: Streptococci can grow in chains. Panel **b**: A scanning electron micrograph of chains of streptococci. Panel **c**: A sputum smear showing the typical chaining arrangement.

Although some bacteria are named for their distinctive shapes or multi-cell arrangements, these criteria cannot be used alone to identify particular bacteria. For an additional degree of classification, we use staining.

Keep in Mind

- All living organisms can be divided into prokaryotes or eukaryotes.
- Prokaryotes do not contain a nucleus or membrane-enclosed structures like those found in eukaryotes.
- Bacteria can be classified by genus and species.
- Further classification of bacteria can be based on size, shape, and arrangement.

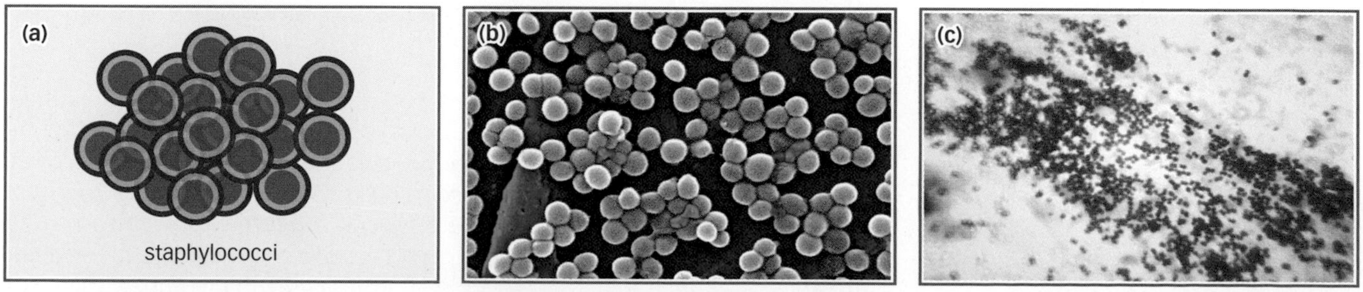

Figure 4.6 Arrangement of Staphylococci. Panel **a**: Staphylococci arrange in grape-like clusters. Panel **b**: A scanning electron micrograph of methicillin-resistant *Staphylococcus aureus*. Panel **c**: *Staphylococcus aureus* in a sputum sample from a patient with staphylococcal pneumonia.

Staining

In addition to being very small, microorganisms are essentially colorless. We therefore use stains to see them under a microscope. Staining is caused by dye molecules that bind to the microorganisms and give them color so that they are visible. Microorganisms can be categorized according to which stains color them and which do not.

The most commonly used stains, called basic dyes, are composed of positively charged molecules. These molecules are attracted to bacterial cells, which have an overall negative charge on their surface.

There are two main types of staining procedure that use basic dyes (**Table 4.1**). **Simple stains** consist of only one dye, and these stains are used to identify the shape and multicell arrangement of bacteria. Methylene blue, carbolfuchsin, safranin, and crystal violet are some of the most commonly used simple stains. **Differential stains** use two or more dyes to distinguish either between two or more organisms or between different parts of the same organism. The format of a typical differential stain is first the addition of the primary stain, followed by the decolorizing agent, and last the counterstain. The major differential stains are Gram, negative (capsule), flagella, Ziehl–Neelsen acid-fast, and endospore.

Simple stains let us see only the shape, size, and arrangement of the organisms, but differential stains allow us to begin classifying organisms and also show us many of the structures associated with bacteria. Let's take a look at the individual types of differential stain one by one.

Example	Result	Use
Simple Stain (single dye)		
Methylene blue Safranin Crystal violet	Uniform blue Uniform red Uniform purple	Shows size, shape, and multicell arrangement
Differential Stain (two or more dyes)		
Gram	Gram-positive: purple Gram-negative: pink Gram-variable: intermediate or mixed colors Gram-nonreactive: stain poorly or not at all	Distinguishes Gram-positive from Gram-negative organisms
Ziehl-Neelsen acid-fast	Acid-fast organisms retain red of carbolfuchsin; non-acid-fast organisms are colored blue with methylene blue	Distinguishes members of genera *Mycobacterium* and *Nocardia*
Negative	Stains background	Allows bacterial capsules to be seen
Flagella	Dyes or silver used to build up layers on flagella	Allows flagella to be seen
Endospore	Endospores retain malachite green color while cell counterstained with safranin is red	Used to identify endospores present in bacterial cells

Table 4.1 A Comparison of Staining Procedures for Bacteria

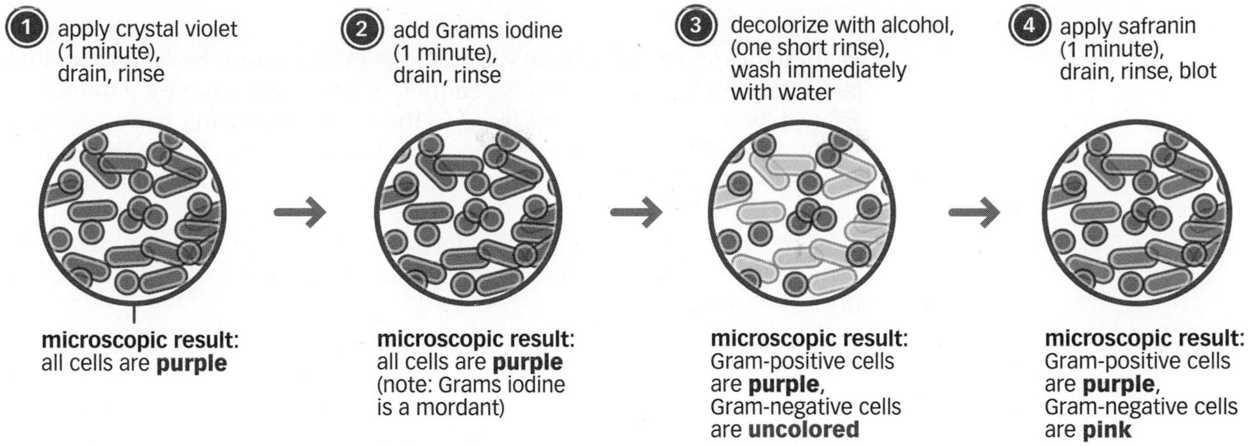

Figure 4.7 The steps involved in the Gram stain. Notice the color of the Gram-negative rods after the decolorizing and counterstaining steps.

The Gram Stain

The Gram stain, developed by the Danish physician Hans Christian Gram in 1884, can be used to separate most bacteria into four major groups: *Gram-positive*, *Gram-negative*, *Gram-variable*, and *Gram-nonreactive*. The Gram stain takes advantage of the differences in the cell walls of these groups of bacteria, which we will discuss in detail in Chapter 9. Most bacteria are either Gram-positive or Gram-negative.

The Gram staining procedure (**Figure 4.7**) initially uses the dye crystal violet, which is taken up by all bacteria. The cells are then treated with iodine as a *mordant* (a substance that sets the color and makes it permanent), which helps the Gram-positive cells retain the crystal violet. When alcohol is added in the next step, the Gram-negative cells lose their color but the Gram-positive cells retain the violet dye. Because the now-colorless Gram-negative cells would be invisible under a microscope at this point, they are counterstained with the red dye safranin. Any Gram-variable bacteria in the mix are recognizable by the way in which they stain unevenly. Gram-nonreactive bacteria do not stain and must be stained with a simple stain to be seen. Cell age is also a factor with old cells not staining properly.

The Negative (Capsule) Stain

The negative stain can be used to identify bacterial shapes, and in particular spirochetes. It can also be used to identify the presence of a **capsule**, which is a structure that surrounds certain bacterial cells. The capsule is important in bacteria that infect humans because it limits the access of antiseptics, disinfectants, and even antibiotics, thereby protecting the infecting bacteria. A capsule can also defeat a host's defense mechanisms. In fact, some bacteria can infect humans only if the bacteria are surrounded by a capsule.

The negative staining procedure uses dyes such as nigrosin and India ink to color the background surrounding encapsulated bacteria in a sample being tested, making the capsule very visible. The stain colors only the background and not the capsule itself. It is the coloring of the background that allows us to see the capsule. A second dye can be added to color the bacterium inside each capsule (**Figure 4.8**).

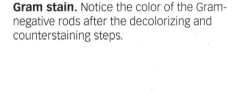

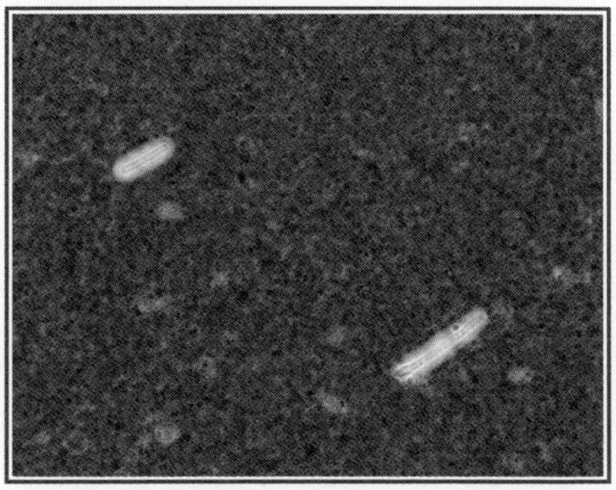

Figure 4.8 The negative stain. Staining the background with India ink makes the capsules visible. The bacteria shown here are encapsulated *Bacillus anthracis*.

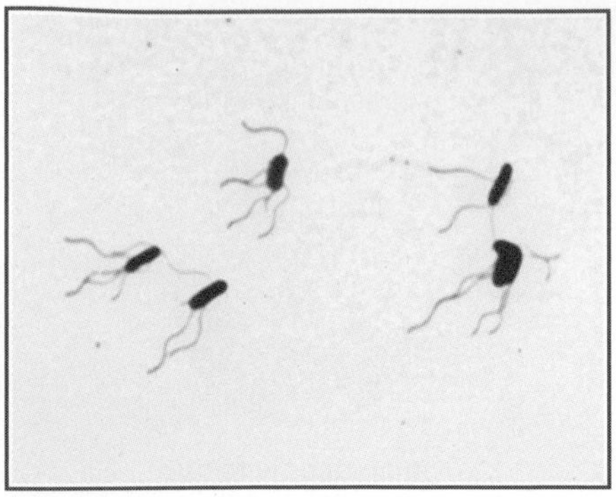

Figure 4.9 The flagella stain. A photomicrograph showing the flagella found on *Bordetella bronchiseptica*.

The Flagella Stain

Some types of bacteria have flagella, a property that allows the bacteria to be *motile* (capable of moving). Motility is an important part of the infectious process because it allows the invading organisms to move from the initial site of the infection. Because bacterial flagella are too thin to be seen under the microscope, a flagella stain is used to coat the surface of the flagella either with layers of dye or with metals such as silver. The coating makes the flagella visible (**Figure 4.9**). This staining procedure is very difficult to carry out and is very time-consuming, so it is not routinely done.

The Ziehl–Neelsen Acid-fast Stain

The Ziehl–Neelsen acid-fast staining procedure is used to detect *Mycobacterium tuberculosis*, the causative agent of tuberculosis, and *Mycobacterium leprae*, which causes leprosy. These bacteria have a cell wall that contains mycolic acid (a waxlike molecule) and lipids, and the presence of these substances in the cell wall makes the wall difficult to penetrate. Therefore, heat is used as part of the staining procedure to break down the mycolic acid and permit the entry of stain. Acid-fast bacteria stain poorly or not at all with the Gram stain.

The term *acid-fast* refers to whether or not a bacterium that is first stained and then washed with acid retains the staining dye. Dyed cells that do not lose their color when washed with acid are classified as acid-fast. An everyday use of *fast* in this sense is the "colorfast" label you sometimes see on clothing and linens. Products made from colorfast fabrics retain their color through numerous launderings (or so the manufacturers would have us believe).

The presence or absence of a cell wall containing mycolic acid is the underlying principle of Ziehl–Neelsen acid-fast staining because only this type of cell wall is acid-fast. A sample is treated first with heated carbolfuchsin, a red dye that can penetrate all cell walls, regardless of whether or not they contain mycolic acid. The next step is the addition of an acid–alcohol solution that removes the red color only from those cells in the sample in which the cell wall does not contain mycolic acid — in other words, only from cells that are not acid-fast. The acid washing does not disturb the red color of carbolfuchsin in any acid-fast cells. Because any cells in the sample that are not acid-fast are now colorless, the procedure is finished by counterstaining with methylene blue to give color to these cells (**Figure 4.10**).

The Endospore Stain

An **endospore** is a small, tough, dormant structure that forms in bacterial cells, and several types of bacteria can undergo sporulation (the process in which endospores are formed). Endospore formation involves gathering a copy of the genetic information of the bacterium together with other important chemicals and enclosing the collection inside a tough coating. As we will see in Chapter 9, bacteria that can undergo sporulation are particularly difficult to neutralize because the endospores are extremely resistant to antiseptics, disinfectants, and antibiotics.

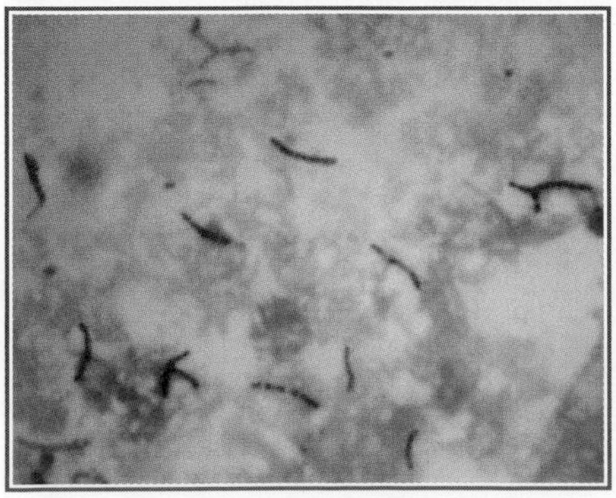

Figure 4.10 The Ziehl–Neelsen acid-fast stain. The vivid red shown here is typical of acid-fast bacteria, such as *Mycobacterium tuberculosis* and *Mycobacterium leprae*.

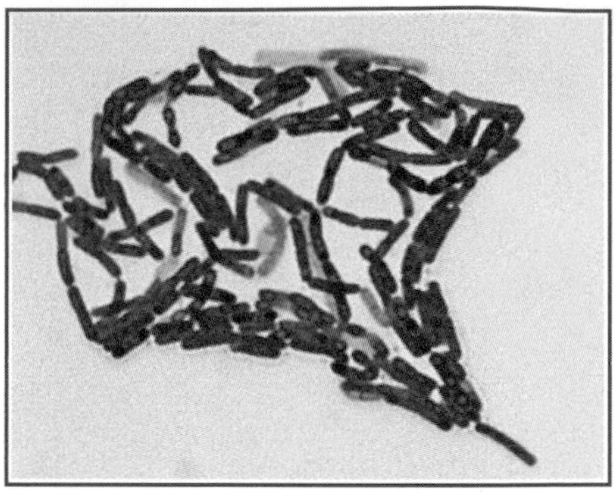

Figure 4.11 The endospore stain. This is a differential stain that shows green-dyed endospores inside red-dyed rod-shaped bacteria.

The endospore stain is a differential stain in which a sample is colored by heating with malachite green for 5 minutes. The endospore walls are so thick and resistant that extensive heating is required to make them permeable to the stain. The sample is then washed thoroughly with water for 30 seconds, and in this washing the dye is removed from all of each cell except the endospore, which stays green. A final step of counterstaining with safranin turns the non-spore part of each cell red (**Figure 4.11**). If the cells were stained with safranin only, the endospores would appear as colorless regions in the cell interior.

In the above discussions we have seen that bacterial cells have distinct sizes, shapes, and multicell arrangements and that the various types of bacteria can be differentiated by staining. In later chapters we will focus on the structures associated with bacteria in great detail. Now, however, let's examine the relationship between bacteria and their host cells.

Keep in Mind

- Staining is used to make organisms microscopically visible.
- There are two major types of stain: simple stains, which use one dye, and differential stains, which use more than one dye.
- The Gram stain is used to classify bacteria on the basis of their cell wall structure.
- Several stains, such as the capsule, flagella, and endospore stains, are used to identify structures associated with the bacterial cell.
- The acid-fast stain is used to confirm the identity of *Mycobacterium* species.

HOST–PATHOGEN RELATIONSHIPS

Pathogens are organisms that can infect humans and cause illness in them. Before we move into our examination of pathogens and the infectious diseases they cause, it's important to look at the relationship that we as hosts have with pathogens. Infectious diseases have been the major cause of human death and suffering throughout history, up to and including the present day. To a large extent, this is due to the human population's having become large enough to sustain pathogens and enhance their ability to multiply, which in turn has contributed to increased disease.

The relationship between humans and pathogens is relatively recent when looked at from an evolutionary perspective. Even though *Homo sapiens* have been around since about 300,000 years ago, it is only over the past several centuries that we can appreciate the effect of disease on our evolution. Consider, for example, the bubonic plague of the fourteenth century (also referred to as the Black Death), which led to the death of one-third to one-half of the population of Europe at that time. It is interesting to speculate on the effect of this epidemic in terms of what may have been lost from the human gene pool. We know of many important individuals who survived the plague and went on to affect the course of history, but it is likely that just as many potentially influential individuals died, and we will never know what they might have gone on to do. The same can be said today with regard to epidemics such as AIDS and also the newly emerging infectious diseases.

Fast Fact

Poverty, crowding, unsanitary conditions, and malnutrition are leading contributors to increased susceptibility to infection and disease, while war, famine, and civil unrest also increase the level of disease.

Infectious diseases are complex and involve a series of shifting interactions between pathogen and host. These interactions can vary, depending on the pathogen in question; relevant factors include the pathogen's ability to either overcome or evade host defenses, to increase in number, and to establish the infection. In addition, the pathogen must find a way to transmit the infection to new hosts. At the same time, the host uses its defenses to control or eliminate the pathogen. It is essential to keep in mind that the ability of a pathogen to infect or cause disease depends in large part on the host's susceptibility to infection. This susceptibility is usually associated with a diminished host defense.

Pathogenicity: A Matter of Perspective

We have defined a pathogen as an organism capable of causing disease. From the perspective of the organism, however, being pathogenic is simply a *strategy for survival*. In fact, the human body is host to a myriad of microorganisms, both externally and internally: from mouth to anus and from head to toe. Every cell exposed to the outside world interacts with a variety of microorganisms. Most of these microorganisms are harmless and in some cases perform useful functions in our lives by providing protection from pathogens, producing vitamins, and helping us to digest food. These microorganisms are classified as **mutualistic**, meaning that they depend on us for their survival and we live more comfortably with them than without them. If given the opportunity, however, some mutualistic organisms become opportunistic pathogens, as we discuss in the next section.

Opportunistic Pathogens and Primary Pathogens

Organisms that cause disease by taking advantage of a host's increased susceptibility to infection are called **opportunistic pathogens**. Under normal conditions, when the host is not especially susceptible, these organisms are not pathogenic. In contrast to opportunistic pathogens, primary pathogens are those that can cause disease in individuals who are healthy. Primary pathogens include the viruses that cause diseases such as colds and **mumps** and the bacteria that cause diseases such as typhoid fever, **gonorrhea**, tuberculosis, and **syphilis**. Primary pathogens have evolved mechanisms that allow them to overcome the defenses of the host and, once inside a host, to multiply greatly. Some primary pathogens are restricted to humans, whereas others infect both humans and other animal hosts.

Disease and Transmissibility

Infection by any pathogen, whether opportunistic or primary, requires that the pathogen (1) be able to multiply in sufficient numbers to secure establishment in the host and (2) be transmissible to new hosts. An interesting point to note is that the signs of a disease can help pathogens accomplish the objective of transmission. For example, coughing promotes the transmission of diseases such as influenza and tuberculosis from the respiratory system, and diarrhea spreads pathogenic bacteria, viruses, and protozoan parasites from the digestive tract of an infected host.

From the pathogen's standpoint, we can look at the death of an infected host as an inadvertent, unfavorable outcome. In fact, pathogens that are well adapted to humans usually spare the majority of those infected. The pathogens benefit from the illness produced because it permits easier transmission from one host to another. However, if a large number of hosts die, spreading the infection is no longer possible. For example, Ebola and HIV are both deadly viruses, but HIV is transmitted far more effectively, and is far more successful, because it progresses slowly. This allows infected hosts to survive long enough to transmit the disease. In contrast, Ebola kills its hosts so rapidly that outbreaks usually die out fairly quickly. So in most cases there is a subclinical resolution to the infection in which the pathogen causes damage but no disease. In this case there can be continued transmission, which is critical to the pathogen's survival.

Keep in Mind

- Pathogens are organisms that cause disease in humans.
- Infectious disease is a complex process that involves both the pathogen and the host.
- Pathogens can be classified as opportunistic (causing disease if the host's defenses are compromised in some way) or primary (causing disease even if the host's defenses are intact).

BACTERIAL PATHOGENICITY AND VIRULENCE

Most bacteria have the capacity to grow and survive under harsh conditions, and we know that only a small fraction of all the bacteria that exist are associated with humans either as normal flora or pathogens. Therefore, we can ask the question: What makes a bacterium pathogenic? As we touched on in Chapter 1, pathogens must be able to accomplish the following:

- A potential pathogen must be able to adhere to, penetrate, and persist in the host cell (the get-in-and-stay-in rule).
- It must be able to avoid, evade, or compromise the host defense mechanisms.
- It must damage the host and permit the spread of the infection.
- It must be able to exit from one host and infect another host.

Virulence refers to just how harmful a given pathogen is to a host. How virulent a given pathogen is depends on genetic factors of the pathogen. Most bacterial pathogens have to survive in two very different environments: the environment external to the host and the environment either on or in the host. Depending on which of these two environments a pathogen is in, some of its genes are active and others are not. For example, it would be a waste of resources for a bacterium to produce substances harmful to a host when it is not in or on the host. The genes for producing these substances are therefore turned off when the pathogen is in the external environment. The question is how the pathogen knows where it is, and which genes to turn on and which to turn off.

It is now well established that pathogens often carry clusters of genes whose activity results in the production of factors that increase its virulence. Bacteria typically have only one chromosome, and these virulence genes are carried either on that chromosome or on mobile genetic elements called **plasmids**. We will see in Chapter 11 that plasmids can move from one bacterium to another and cause previously harmless types to become virulent. Clusters of virulence genes are called **pathogenicity islands** because they occupy distinct regions on the chromosome or the plasmid, and all genes on the islands are involved with pathogenicity. Some virulence genes are regulated by quorum sensing, an environment-sensing mechanism that we discuss next.

Quorum Sensing

During **quorum sensing**, specialized proteins in a pathogen cell called *sensing proteins* relay information about the cell's environment to other proteins that regulate genes controlling the transcription of virulence genes. This environment-sensing mechanism is referred to as quorum sensing because it is based on cell population densities; that is, certain genes are expressed only when there is a sufficient population density. (A *quorum* is the minimum number of things that must be present for something to happen. For instance, a vote cannot take place in Congress unless a quorum of legislators is present in the room.) As an example, *Salmonella* bacteria, which cause food poisoning, do not secrete the toxins responsible for the signs of food poisoning until there are sufficient numbers — a quorum — of *Salmonella* present in the host. It is as if the bacteria do not want to tip their hand before they are present in large enough numbers to make the host sick. The scientific explanation is probably that the bacteria have evolved to delay the production of toxins as a means of hiding from the host defenses, which would have no trouble in dealing with small numbers of pathogens.

Quorum sensing occurs when bacteria in a host secrete into the host circulation small diffusible molecules that can be "sensed" by other bacteria. In a process called auto-induction, these other bacteria then wait till a sufficient increase in population has occurred.

There is a difference in the genetic involvement of quorum sensing between opportunistic pathogens and primary pathogens. For example, in *Pseudomonas aeruginosa*, which is an opportunistic pathogen, only 5 to 20% of its virulence genes are subject to quorum-sensing regulation. In a primary pathogen, such as *Staphylococcus aureus*, quorum sensing is used to control all of its virulence genes.

Biofilms

Whether in natural conditions such as soil or marine environments or on artificial surfaces such as the surfaces of prosthetic devices, catheters, or internal pace makers, bacteria adhere and grow as aggregated assemblies of cells referred to as a **biofilm**. A biofilm is important for health care because it can either impede or totally prevent the entry of antimicrobial agents and other molecules that are potentially toxic to bacteria. At the same time, a biofilm is able to capture and retain nutrients, thereby allowing the bacterial population to increase in number. The adhesion of a biofilm can also affect the host inflammatory response by attracting host defensive cells that attempt to engulf the biofilm. This results in what is referred to as "frustrated phagocytosis", causing the formation of gigantic cells that eventually form a tough collagen capsule that inhibits new blood vessel development and interferes with wound healing.

The body usually reacts to the implantation of a medical device by first coating it with a protein film (we will see in Chapter 22 that the same coating process occurs on our teeth). This protein film is composed of fibronectin, fibrinogen, albumin, immunoglobulins, and other proteins that serve as binding sites and receptors for bacteria. Through a series of steps, bacteria begin to accumulate on this protein film (which is then referred to as a *substrate layer*) and form a biofilm (**Figure 4.12**). Several things can happen at the level of the biofilm:

1. The bacteria can accumulate on the device and be absorbed into it.
2. The bacteria can detach from the device surface and move into the liquid surrounding the biofilm.

Fast Fact

The production of medical devices and tissue-engineered materials is an industry generating $170 billion per year, with 5 million prosthetic devices such as heart valves, orthopedic implants, and catheters implanted each year in the United States alone. Unfortunately, many nosocomial infections are associated with implants and indwelling catheters.

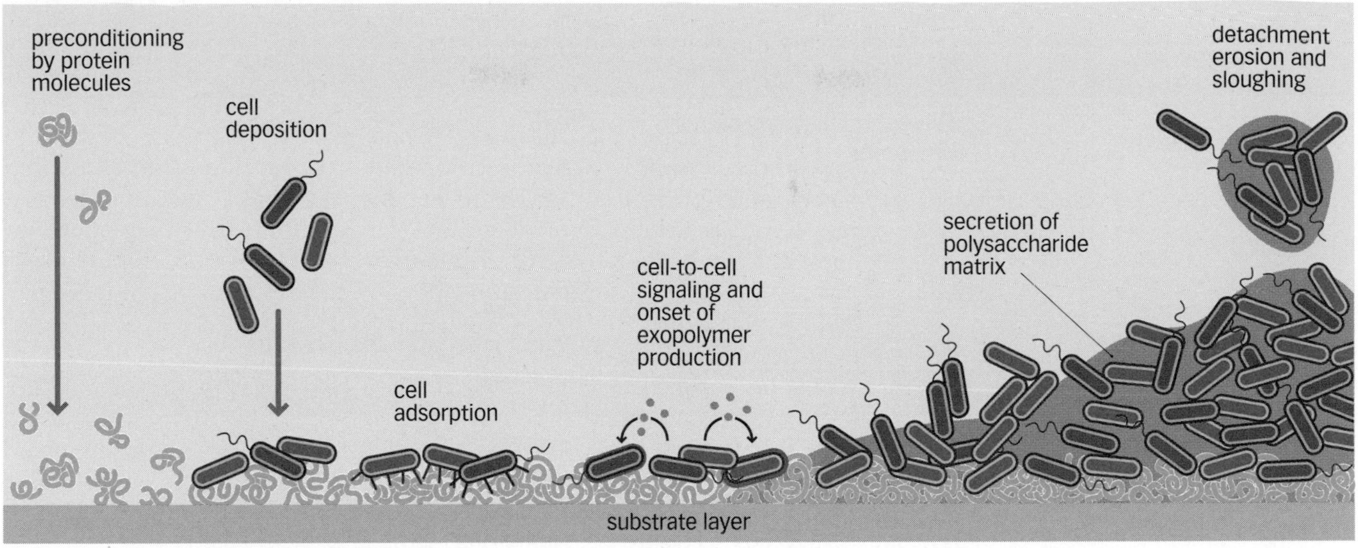

Figure 4.12 Development of a biofilm. This process starts when a film of proteins forms. Organisms adhere to these proteins and then aggregate with other organisms to produce the biofilm.

3. The biofilm can form a tenacious gelatinous mass that is resistant to host defenses.

4. Large pieces of the biofilm can detach from the device surface and spread to other locations. In some cases, these pieces can also be large enough to cause fatal thromboembolisms (clots).

Keep in Mind

- Infection by a pathogen requires that the pathogen get in, stay in, defeat the host defense, damage the host, and be transmissible.
- Virulence refers to how harmful a given pathogen is to a host.
- There are virulence genes that are carried on the bacterial chromosome or on extrachromosomal pieces of DNA called plasmids.
- Virulence genes can be regulated by quorum sensing.
- A biofilm can assist in the infectious process by inhibiting the exposure of bacteria to antibiotics and other molecules toxic to bacteria.

THE HOST CELL

In this section, we compare prokaryotic and eukaryotic cell structures. In addition, we examine the structures found in the eukaryotic cell in detail because the eukaryotic cell is the primary target for infectious diseases. Because many structures of the eukaryotic cell have important roles in the infectious process, we include a brief description of how some of these structures are involved in infection.

Prokaryotic Versus Eukaryotic Cell Structure

Bacteria are prokaryotic cells (having no membrane-enclosed nucleus), whereas the cells of humans, other animals, and fungi are eukaryotic (having a membrane-enclosed nucleus). **Table 4.2** lists the major differences between the two cell types. The eukaryotic cell is bigger and far more complex than the prokaryotic cell in almost every way. During the

Characteristic	Prokaryotic Cells	Eukaryotic Cells
Genetic Structures		
Genetic material	Usually found in single circular chromosome	Found in paired chromosomes
Location of genetic material	Nuclear region (nucleoid)	Membrane-enclosed nucleus
Nucleolus	Absent	Present
Histones	Absent	Present
Extrachromosomal DNA	In plasmids	In mitochondria and plasmids
Intracellular Structures		
Mitotic spindle	Absent	Present
Plasma membrane	Lacks sterols	Contains sterols
Internal membranes	Only in photosynthetic organisms	Numerous membrane-enclosed organelles
Endoplasmic reticulum	Absent	Present
Respiration (ATP)	At cell membrane	In mitochondria
Golgi	Absent	Present
Lysosomes	Absent	Present
Peroxisomes	Absent	Present
Ribosomes	70S	80S in cytoplasm and on endoplasmic reticulum, 70S in mitochondria
Cytoskeleton	Absent	Present
Extracellular Structures		
Cell wall	Peptidoglycan, LPS and teichoic acid	None in most eukaryotic cells; chitin in fungal cells
External layer	Capsule or slime layer	None in most eukaryotic cells; pellicle or shell in certain parasites
Cilia	Absent	Present in certain cell types
Pili	Present	Absent
Flagella	Present	Present in certain cell types
Reproduction		
Cell division	Binary fission	Mitosis or meiosis
Reproduction mode	Asexual	Sexual or asexual

Table 4.2 A Comparison of Eukaryotic and Prokaryotic Cells

following discussions, we will look at eukaryotic cell structures (**Figure 4.13**) by using the human cell as the example. However, keep in mind that there are many other types of eukaryotic cell.

Plasma Membrane

The plasma membrane is the outer layer of the eukaryotic cell. It is structurally similar to the plasma membrane of bacteria but with significant differences in the amounts of lipid and certain other components. For example, the eukaryotic plasma membrane contains cholesterol molecules that confer support and strength. The bacterial plasma membrane does not contain cholesterol because in prokaryotes this membrane is surrounded, supported, and protected by a strong outer cell wall.

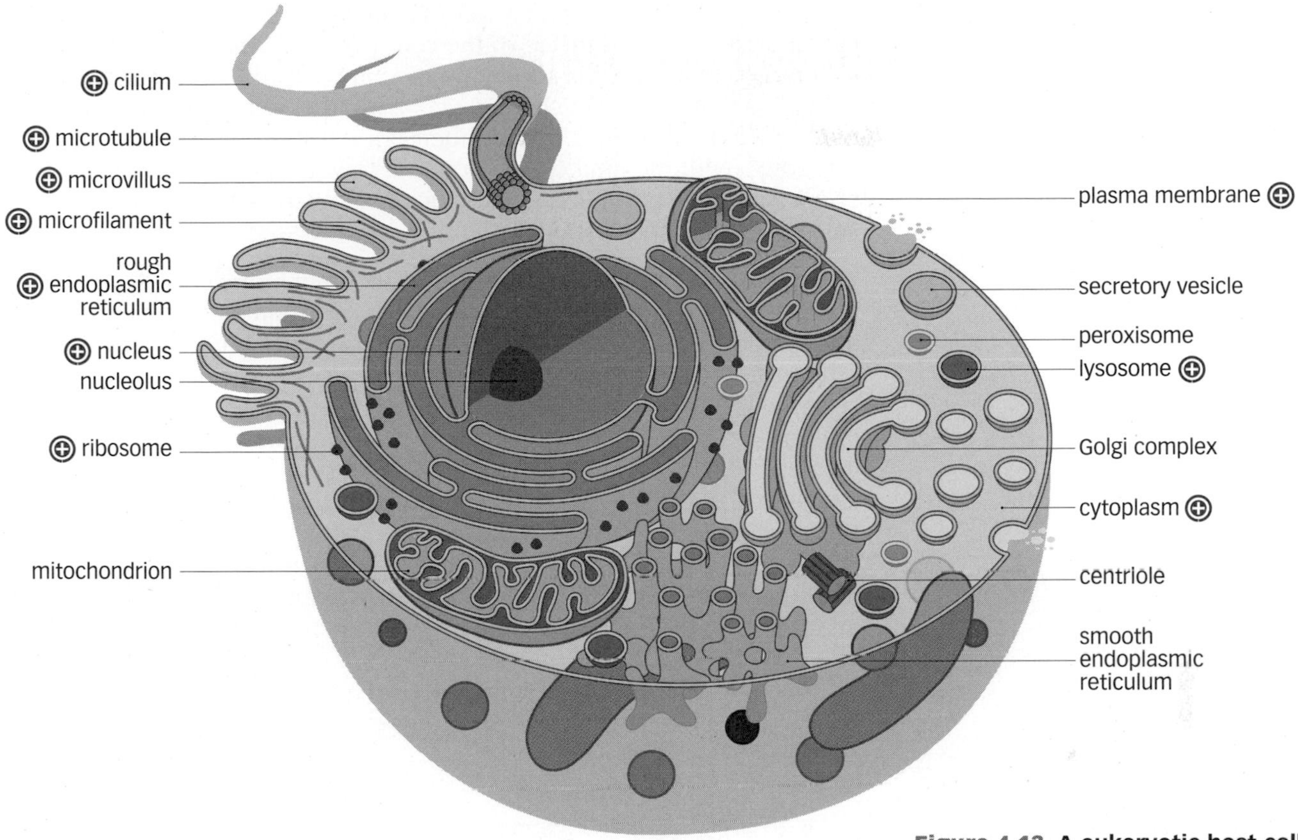

Figure 4.13 A eukaryotic host cell, which is much more complex than the prokaryotic cell. The internal structures of the host cell that are involved in the infectious process are indicated by ⊕.

The eukaryotic plasma membrane is made up of a phospholipid bilayer (**Figure 4.14**). Recall from Chapter 2 that lipids are not soluble in water and for this reason they are said to be *hydrophobic* (water-fearing). Lipids

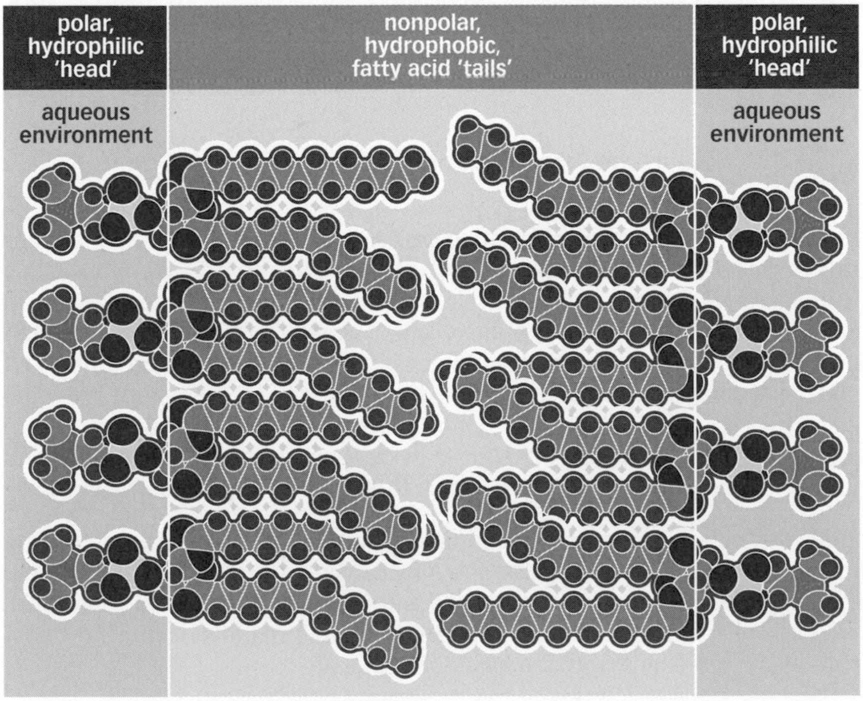

Figure 4.14 The plasma membrane of a eukaryotic cell. The structure is a phospholipid bilayer that has the polar heads of the bilayer facing the inside and outside of the cell. These polar heads are hydrophilic, a property that allows them to interact with the aqueous environments on both sides of the plasma membrane. The inner portion of the bilayer consists of the nonpolar, hydrophobic fatty acid tails that separate the aqueous environments inside and outside the cell.

therefore provide a perfect barrier between water on the outside of the cell and water on the inside of the cell. However, the cell must be able to interact with its environment. To accomplish this, the lipid is bound with phosphate ions, which are *hydrophilic* (water-loving). In the plasma membrane, lipid molecules with their bound phosphate groups align in two rows, with the lipid chains pointing toward the center of the membrane and the phosphate groups facing outward and forming the two surfaces of the membrane, as **Figure 4.14** shows.

The plasma membrane contains a variety of other molecules (**Figure 4.15**). In particular there are proteins involved in communication and transport and also in structural roles, connecting to the cytoskeleton of the eukaryotic cell. Some of these proteins act as receptors for signals sent by other cells; they also can serve as a site for virus attachment. These proteins are not stationary in the membrane but instead move freely through the lipid bilayer. (Because of this mobility the plasma membrane represented in Figure 4.15 is referred to as the *fluid mosaic model* of the membrane.) It is important for you to remember that the phospholipid bilayer structure is also seen on membrane-enclosed structures inside the cell. This structural characteristic allows the plasma membrane to interact with these internal structures during certain cellular functions (described below).

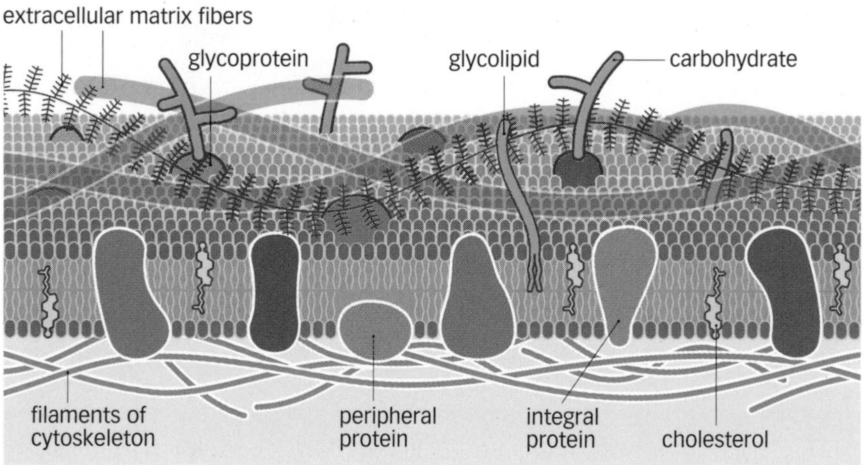

Figure 4.15 Components of the eukaryotic plasma membrane. The proteins and carbohydrate molecules are free-floating in the lipid bilayer. This model of the structure of the plasma membrane is called the fluid mosaic model.

The Role of the Plasma Membrane in Infection

Because the plasma membrane is the barrier between the inside and the outside of the cell, it must be breached if microorganisms are to gain entrance. This is particularly true for viruses. As we will see in Chapter 12, viruses are able to infect host cells that have a specific receptor for viral particles located on their plasma membrane. A viral particle attaches to these receptors on a host cell and thereby gains entry into the cell (**Figure 4.16**). A good example of this is the influenza virus, which binds to receptors called ICAM-1 receptors. These receptors are normally found on the surface of cells of the respiratory system, and therefore these cells are very susceptible to infection by the influenza virus.

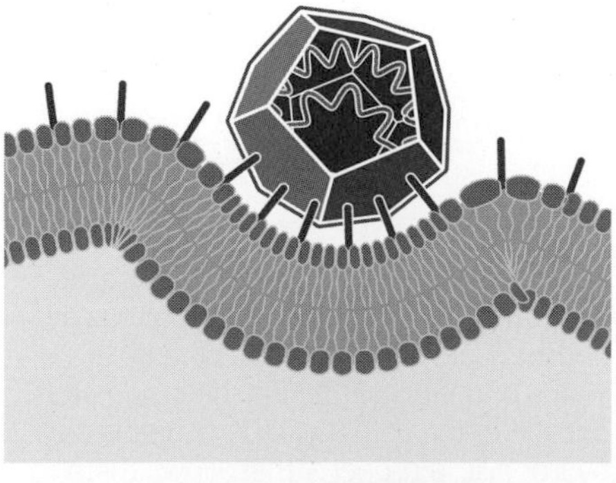

Figure 4.16 Rhinovirus (which causes the common cold) binding to the plasma membrane of a host cell. Viruses infect host cells by binding to the receptors on the plasma membrane.

In many viral infections, the viral particle acquires a part of the host's plasma membrane as the particle leaves the infected host cell. This piece of plasma membrane wraps around the virus and is referred to as the **viral envelope**. We will see that this envelope is important both in the infection of new host cells and in protection from the host defense.

Cytoplasm

The **cytoplasm** of the eukaryotic cell is all the volume of the cell that is inside the plasma membrane but outside the nucleus. It is made up of (1) a semifluid material consisting mainly of water plus a variety of dissolved substances, referred to as the cytosol, (2) membrane-enclosed structures called *organelles*, as noted at the beginning of this chapter, and (3) structures that are not bound by a membrane.

Cytoplasm is found in both prokaryotic cells and eukaryotic cells, but membrane-enclosed organelles are found only in the latter. Even though eukaryotic cells are much larger than prokaryotic cells, there is actually less cytosol in eukaryotic cells. This is because the cytosol of eukaryotic cells must share the extranuclear volume with many organelles and other structures.

The Role of the Cytoplasm in Infection

The cytoplasm is involved in a variety of infections, particularly viral infections. During a viral infection, the structures in the host cell are taken over and used by the virus. It is in the host cell's cytoplasm that the individual viral particles are manufactured and assembled.

To study the interior of the eukaryotic cell, we can separate the structures found in the cytoplasm into two groups: those not bound by a membrane and those bound by a membrane. Recall in what we discuss next that the structures named all float in the semifluid cytosol.

Cytoplasmic Structures Not Enclosed by a Membrane

In the eukaryotic cell, there are three major categories of cytoplasmic structure that are not enclosed by any membrane: *cytoskeleton structures*, *cilia*, and *ribosomes*.

The Cytoskeleton

The cytoskeleton of the cell is much like that of the skeleton of the body in that it gives the cell structural integrity. There are three cytoskeletal components: *microfilaments*, *intermediate filaments*, and *microtubules*. In addition to maintaining the shape of the cell, the filaments are also involved in determining how cells are joined together to form tissue.

Microfilaments are thin structures made up of molecules of the protein actin. They are solid structures and anchor the cytoskeleton to proteins in the plasma membrane (much like scaffolding). These small filaments give the cytoplasm a gel-like consistency.

Intermediate filaments are larger than microfilaments and are composed of a variety of different molecular subunits belonging to the family of proteins called keratins. These structures provide additional strength and stability to the cytoskeleton. They are also involved with positioning cells alongside one another in tissues.

Microtubules are hollow tubes made up of the protein tubulin. They are the largest of the cytoskeletal structures and are found in the cilia and flagella seen on some types of eukaryotic cell. Microtubules are involved in the movement of other structures that reside in the cytoplasm, in particular chromosomes during mitosis and meiosis.

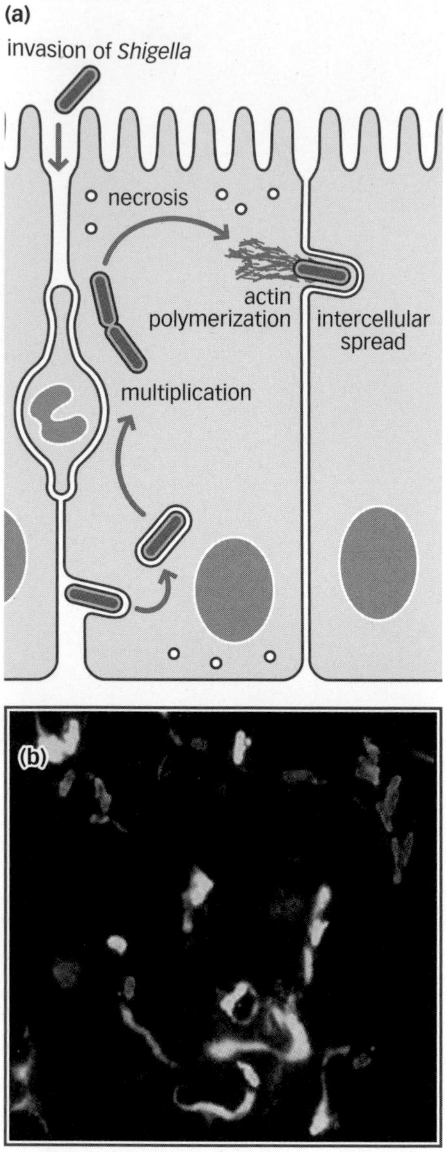

Figure 4.17 Intercellular infection by *Shigella* Panel **a**: Infection of the intestinal lining by the bacteria *Shigella*. After entering a cell that lines the intestinal lumen, *Shigella* uses the actin molecules of microfilaments in the host cell to move from one cell to the next. Panel **b**: A colorized photomicrograph of *Shigella*. Here *Shigella* organisms (stained red) are propelling themselves through the cytoplasm with their actin tails (stained green).

The Role of the Cytoskeleton in Infection

Many pathogens take advantage of the cytoskeleton as part of the infectious process. For example, *Shigella* bacteria, which cause serious infections in the digestive system, use the cytoskeleton as part of the infectious process. These organisms destroy the lining of the intestinal tract when they infect the cells forming that lining. As the infection spreads, the *Shigella* bacteria move laterally from one cell to the next by protruding a finger-like structure made up of host-cell microfilaments (**Figure 4.17**). The cytoskeleton also has an integral role in phagocytosis, which is one of the most important host defense mechanisms.

Cilia

Found only on eukaryotic cells, **cilia** (singular *cilium*) are made up of microtubules arrayed in an arrangement in which nine groups of microtubules form a ring encircling two central microtubules (a *9 + 2 arrangement*; **Figure 4.18**). The cilia on a cell project from the surface of the cell and are anchored to the plasma membrane. By moving in unison, cilia move liquids and secretions across the surface of the membrane.

There are many ciliated cells in the human body. A good example is the lower respiratory tract, which is lined with ciliated cells that work together with mucus-producing cells to move trapped particles upward and out of the respiratory tract.

The Role of Cilia in Infection

Cilia are involved both in the infection strategy of invading pathogens and in the host's defense against the invasion. Ciliated cells work with mucus-secreting cells to remove foreign materials, including microorganisms, from the respiratory tract, thereby preventing the "staying in" requirement for successful infection. In some respiratory diseases, such as **pertussis** (whooping cough) or **diphtheria**, pathogens attach to host ciliated cells as an initial part of the infection (**Figure 4.19**).

Flagella

4.1

As noted earlier, flagella are responsible for cell motility. They are composed of the globular protein tubulin. In humans, the only cells that contain flagella are sperm cells. The flagellum is anchored in the plasma membrane of the cell and uses a complex sequence of reactions, almost like motorized turning, at the anchor point. This action moves the flagellum and permits the whip-like motion seen in the flagella found on eukaryotic cells.

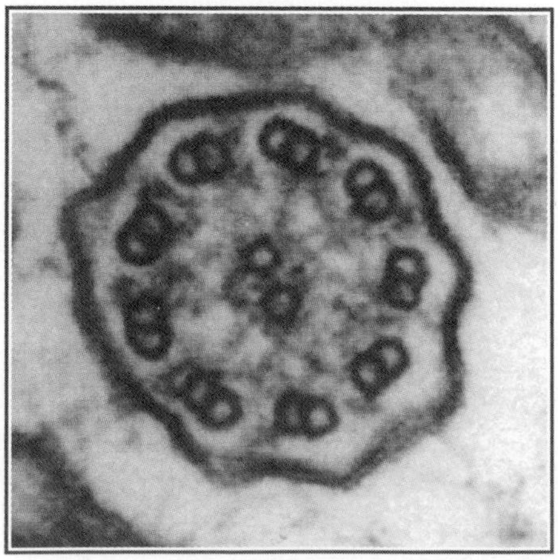

Figure 4.18 Photomicrograph of a cross section of a cilium associated with eukaryotic cells. Notice the 9 + 2 arrangement of the microtubules: a ring of nine groups surrounding two central ones. These cytoskeleton structures give cilia support and strength.

Figure 4.19 Scanning electron micrograph of ciliated cells of the respiratory tract being infected by *Bordetella pertussis*, the bacterium that causes whooping cough.

We will see in Chapter 9 that flagella are commonly found in the microbial world and have a significant role in the infectious disease process by allowing pathogens to move from one location to another in a host body.

Fast Fact

The bacterium *Neisseria gonorrhoeae*, which causes gonorrhea, can sometimes attach to sperm cells and "ride along" from one partner to the other to spread this sexually transmitted disease. However, this is not the primary method of infection with *Neisseria* pathogens.

Ribosomes

The third category of cytoplasmic structures not bound by a membrane is the **ribosome**, found in both prokaryotic and eukaryotic cells and responsible for the production of proteins. Recall from your introductory biology course that ribosomes are found either floating free in the cytosol or attached to the endoplasmic reticulum (discussed below).

The ribosomes in eukaryotic cells differ from those found in prokaryotic organisms, although both types of ribosome are made up of protein and a specific form of RNA (called ribosomal RNA). We will discuss the ribosomes of bacteria in Chapter 9.

The Role of the Eukaryotic Ribosome in Infection

Eukaryotic ribosomes are actively involved in viral infections, but not by choice. They are part of the cellular mechanisms that are taken over by the virus. All of the protein components of new viral particles are made by the host cell's ribosomes. Although ribosomes are not directly involved in the bacterial infectious process, they do have a role in the treatment of bacterial infections because the bacterial ribosome is one of the targets attacked by certain antibiotics. When this happens, the invading bacterium can no longer make protein and therefore dies. Because bacterial ribosomes are structurally different from those in human cells, this targeting is referred to as *selective toxicity.* This means that the antibiotic kills the bacteria but not the host cell. We will see in Chapter 19 that selective toxicity is an important consideration when developing and prescribing antibiotics.

Cytoplasmic Structures Enclosed by a Membrane

Now let's turn to those components of the cytoplasm that are enclosed by a membrane — the cell's **organelles**. It is important to keep in mind that the membrane enclosing any organelle is of the same type as the plasma membrane surrounding the whole cell — a phospholipid bilayer. This two-layer configuration makes it possible for organelles to fuse with one another and also with the plasma membrane.

Mitochondria

Mitochondria (singular *mitochondrion*) are the energy-producing elements of the eukaryotic cell. They are the organelles where most ATP is produced. There are large numbers of mitochondria in cells that are working and fewer in cells that are resting. A mitochondrion has two membranes enclosing it, and ATP is made on the folds (called *cristae*) of the inner membrane (**Figure 4.20**). Mitochondria also contain their own ribosomes and their own DNA, which replicates independently of the host cell.

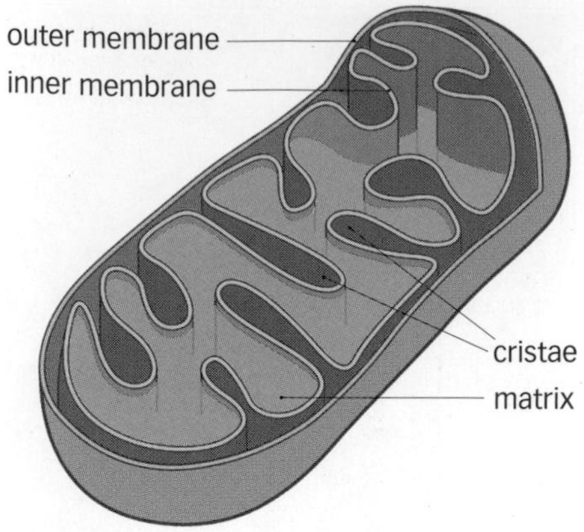

Figure 4.20 The mitochondrion is a double-membrane structure.

Mitochondria have many characteristics that are similar to those seen in bacteria: they replicate independently just as bacteria do, the mitochondrial chromosome is single and circular, and the ribosomes in mitochondria are different from those in the rest of the eukaryotic cell but the same as those seen in bacteria. The RNA and DNA polymerase molecules found in mitochondria are also structurally similar to those seen in bacteria. The mechanism of ATP production on the inner membrane of mitochondria is similar to that seen in bacteria. All these similarities between mitochondria and bacteria have fostered the **endosymbiotic theory**, which describes a process whereby early in evolution, bacteria and eukaryotic organisms had a symbiotic relationship — a relationship in which two organisms lived as one unit. Over time, the bacteria were integrated into the eukaryotic cells as mitochondria.

Endoplasmic Reticulum and Golgi Apparatus

Both the **endoplasmic reticulum** (ER) and the **Golgi apparatus** are systems of membranes that form numerous flattened sacs and platelike structures in the cytoplasm of the eukaryotic cell (**Figure 4.21**). These structures are not found in prokaryotes. The ER is the site where various cellular components are synthesized. It is sometimes associated with ribosomes. If ribosomes are attached to the ER, it is referred to as *rough endoplasmic reticulum*. ER to which no ribosomes are attached is called *smooth endoplasmic reticulum*. Rough ER is where proteins are produced; smooth ER is the site where lipids and other nonprotein components are produced. The ER moves synthesized materials either to the Golgi apparatus, where additional finishing steps are completed, or directly to the plasma membrane for transport out of the cell.

The Golgi apparatus has three functions: (1) modifying and packaging products that come from the ER, (2) renewing the plasma membrane, and (3) producing lysosomes (discussed below). Because both the ER and the Golgi apparatus are surrounded by the same type of phospholipid bilayer, the two organelles can interact with each other by fusing together. It is this mechanism that is used to move newly synthesized components from the ER to the Golgi apparatus.

The Role of the Endoplasmic Reticulum in Infection

In viral infections, the ER is the site of the biosynthesis and assembly of viruses. The ER is also associated with the adaptive immune response, a major host defense against infection that we study in Chapter 16.

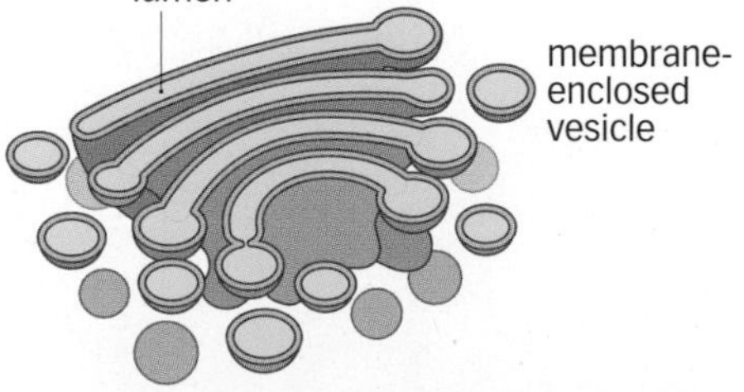

Figure 4.21 The Golgi apparatus is the cellular organelle in which molecules produced in the ER are completed. Notice the membrane-enclosed vesicles that have "pinched" off from the flat membrane sacs. Lysosomes are produced in the Golgi apparatus.

Lysosomes

Lysosomes are the organelles responsible for destroying invading microorganisms and any other foreign materials that get inside the cell. They are produced by the Golgi apparatus and bud off as vesicles containing powerful enzymes that destroy the invaders. Lysosomes are also responsible for recycling any host-cell components that are no longer needed or are no longer functioning properly. The enzymes in lysosomes break down these components and get rid of anything that is not recyclable. Any inhibition of lysosomal function can be a lethal event for the eukaryotic cell.

The Role of the Lysosome in Infection

Lysosomes have a pivotal role in *phagocytosis*, which is part of the innate immune response (Chapter 15). In this process, invading pathogens are enclosed by the host phagocytic cell's plasma membrane to form a vesicle. Once this membrane has fused with the lysosome membrane, the lysosomal enzymes destroy the pathogens.

Proteasomes

Proteasomes are organelles composed of ring structures stacked together (**Figure 4.22**). They function in the degradation of proteins. Proteins tagged with molecules of a small protein called **ubiquitin** are recognized and bound by the regulatory proteins of the proteasome and enter the core of the rings, where they are degraded into fragments. In many cases, these fragments are further degraded, and the useful components are reused by the cell.

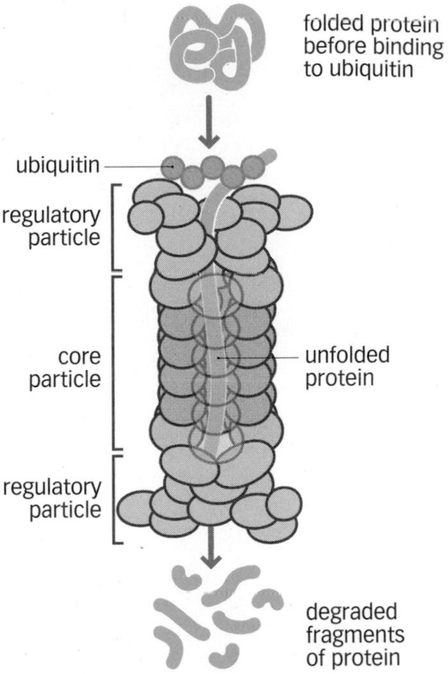

Figure 4.22 The proteasome is an organelle involved in protein degradation. When the proteins being degraded are from invading pathogens, the proteasome participates in the host's defense reactions.

The Role of the Proteasome in Infection

Proteasomes degrade proteins associated with pathogens and move them into the endoplasmic reticulum. Here the degraded proteins trigger the host's immune system to attack the pathogens.

Peroxisomes

Peroxisomes are the organelles responsible for the breakdown of fatty acids in the eukaryotic cell. Many of the by-products of this breakdown are poisonous to the cell, and one role of the peroxisomes is to get rid of these products. For example, hydrogen peroxide (H_2O_2), a substance extremely toxic to cells, is one fatty acid by-product. Peroxisomes contain enzymes that break H_2O_2 down to oxygen (O_2) and water (H_2O).

Nucleus

The nucleus is a structure unique to eukaryotic cells. It is not found in prokaryotic cells. It is an organelle and so is enclosed by a phospholipid membrane, in this case called the **nuclear membrane**, and contains a unique form of cytoplasm called **nucleoplasm**. There are one or more structures called **nucleoli** (singular *nucleolus*) in the nucleus, and it is in the nucleoli that ribosomal RNA is made.

The nucleus is where the DNA of the eukaryotic cell is stored. When the cell is not dividing to create new cells, the DNA is in the form of **chromatin**, which has a hairlike structure. When the cell is in the process of dividing, the chromatin condenses to form pairs of **chromosomes**.

The nuclear membrane is a *double* phospholipid bilayer (**Figure 4.23**). Pores in this membrane allow material to move into and out of the nucleus. The nucleus is the location of **transcription,** the process in which DNA is used as a template to produce RNA. Newly formed RNA uses the nuclear membrane pores to move from the nucleus to the cytoplasm.

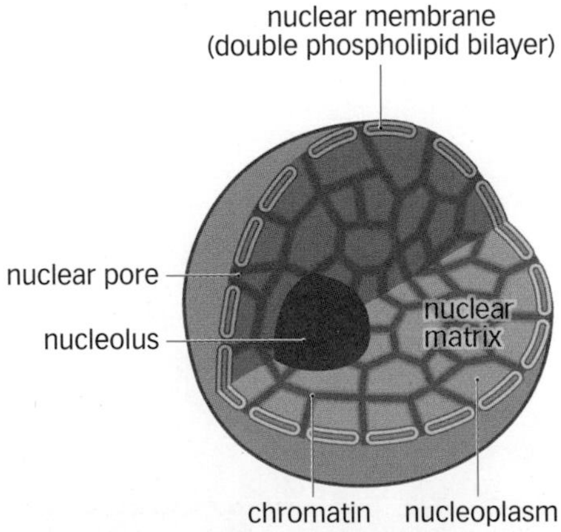

Figure 4.23 The nucleus of the eukaryotic cell, showing the double phospholipid bilayer and the nuclear pores. Notice the nucleolus, the region where ribosomal RNA is produced.

The Role of the Nucleus in Infection

The nucleus of the host cell is important in many infections, particularly those caused by viruses that contain DNA. For this type of virus to infect a host, the viral DNA must enter the host nucleus. As we will see in Chapter 12, this involves a fascinating series of steps that move the viral DNA from the host cytoplasm through the nuclear pores into the nucleus. In some cases, the viral DNA that enters the nucleus becomes incorporated into the host DNA, and the virus becomes **latent** in the host.

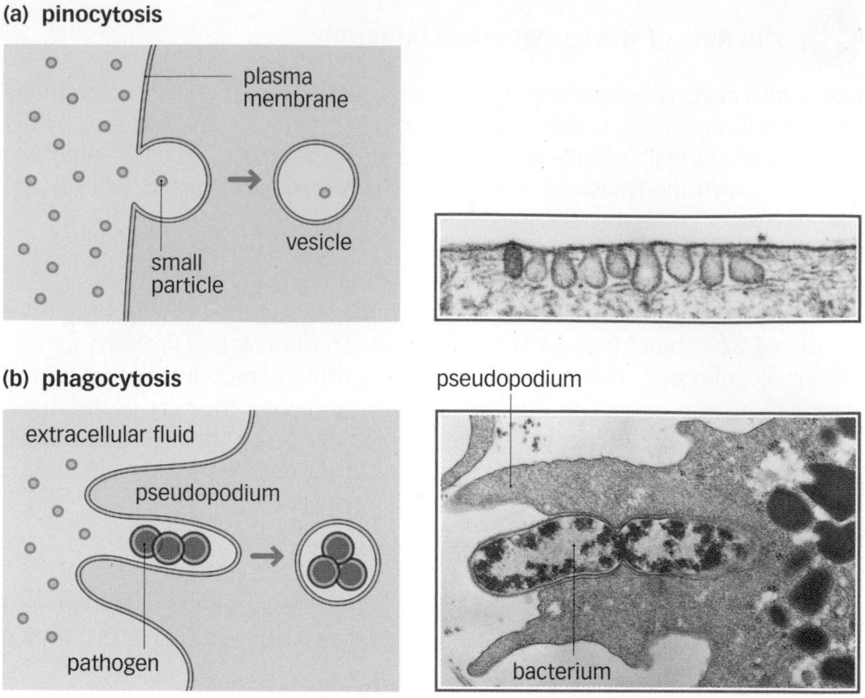

Figure 4.24 Pinocytosis and Phagocytosis. Panel **a**: Pinocytosis involves formation of a concave region in the membrane that flaps over the material to be taken in and creates a vesicle. Pinocytosis is seen when small molecules are brought into the cell. Panel **b**: Phagocytosis is the process by which large items are taken into the cell. This process involves the formation of pseudopodia that attach to and enclose the organism, as shown in the electron micrograph.

Endocytosis and Exocytosis

Before we finish looking at the host cell, it will be helpful to review the processes of **endocytosis**, the process by which a cell takes in materials, and **exocytosis**, the process by which a cell expels unneeded materials. Endocytosis can occur in three ways: *pinocytosis*, *phagocytosis*, and *receptor-mediated* endocytosis.

When a cell takes in material from the extracellular fluid via **pinocytosis** (**Figure 4.24a**), the plasma membrane of the cell will roll over anything that is in contact with it. This is seen with small molecules and results in the formation of a small vesicle that is like a bag surrounded by plasma membrane. This vesicle is then moved into the cytoplasm. **Phagocytosis** (**Figure 4.24b**), also involves the formation of a vesicle, but in this process the cell membrane is pushed outward to form pseudopodia (singular *pseudopodium*, "false foot") as shown in **Figures 4.24**b and **4.25**. The pseudopodia surround the material to be brought into the cell, forming a membrane-enclosed vesicle. As noted earlier, phagocytosis is part of the host defense mechanism, and so materials imported into the cell by this process are usually pathogens that the cell must destroy.

The third form of endocytosis is **receptor-mediated endocytosis**. In this process, receptors on the surface of the cell bind with the extracellular material that is to be brought into the cell. Then the plasma membrane begins to sink into the cell interior, as in pinocytosis, and eventually a vesicle forms (**Figure 4.26**).

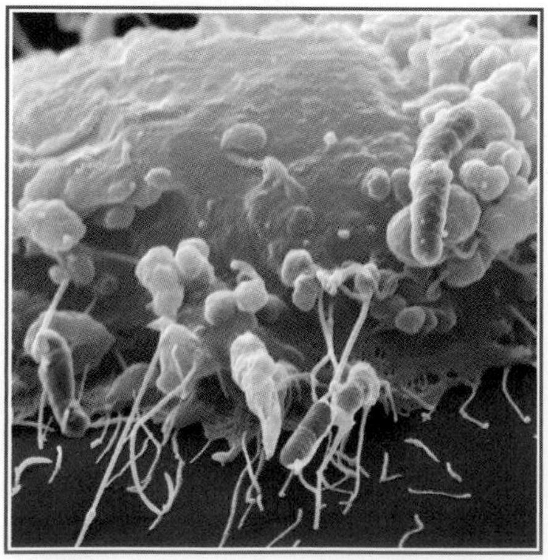

Figure 4.25 Phagocytosis.. A scanning electron micrograph of phagocytosis of bacteria (red).

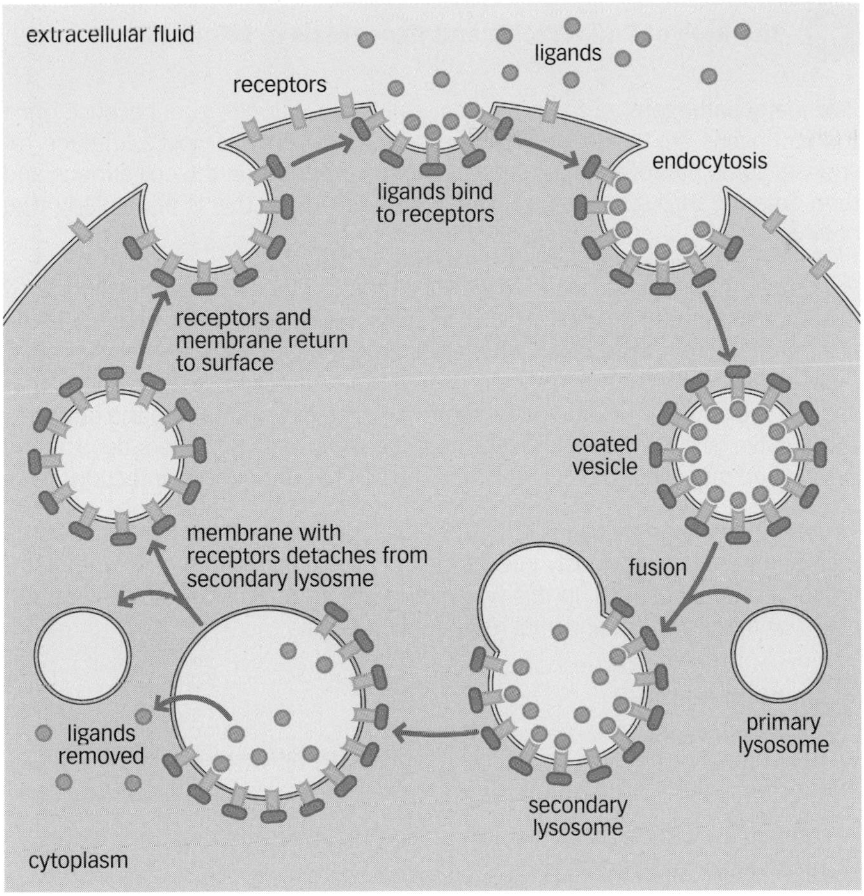

Figure 4.26 An illustration of receptor-mediated endocytosis. Notice the formation of the vesicle and the fusion of the vesicle with the lysosome. This fusion of vesicle and lysosome is possible because of the phospholipid bilayer structure of the membrane.

Exocytosis is essentially the reverse of endocytosis. Vesicles are formed by organelles such as the endoplasmic reticulum and the Golgi apparatus. These vesicles then move to the plasma membrane and, because of their phospholipid bilayer structure, fuse with the plasma membrane and release their components to the extracellular fluid (**Figure 4.27**).

Figure 4.27 Exocytosis. Panel **a**: An illustration of exocytosis in which vesicles pinch off from the Golgi apparatus and move to the surface of the cell. Here they become incorporated into the cell membrane. Panel **b**: A transmission electron micrograph of exocytosis.

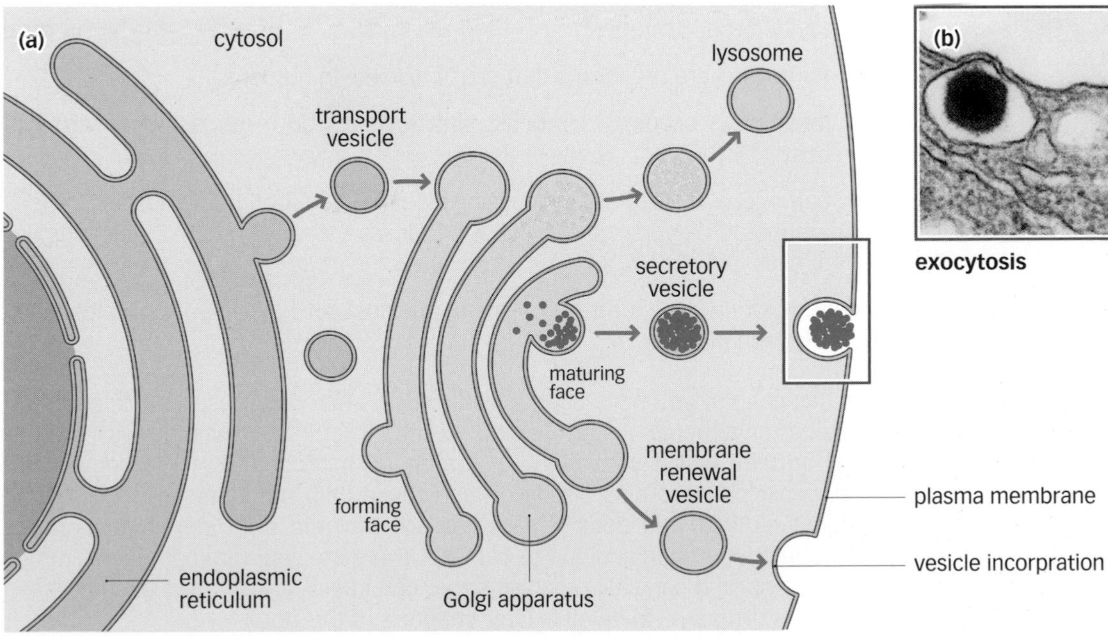

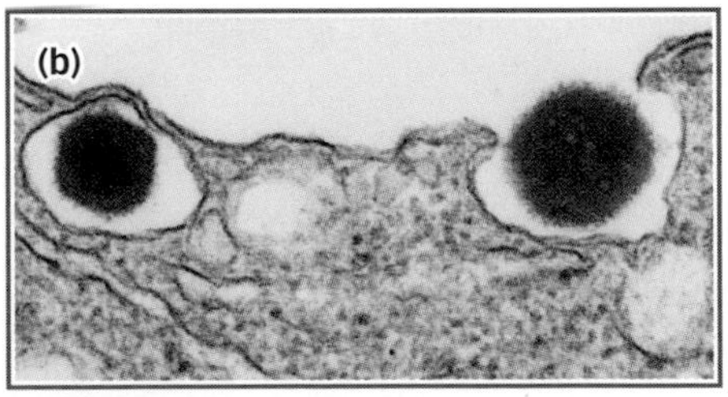

exocytosis

The Role of Endocytosis and Exocytosis in Infection

For many pathogens, entry into a host cell is the ultimate goal because once the pathogens are inside, they can be protected from the host's immune response. Some pathogens enter by binding to receptors on the cell surface and then entering through receptor-mediated endocytosis. This is particularly true of viruses.

For other pathogens, the host's defensive mechanism of phagocytosis is used for entry. As noted in our discussion of lysosomes earlier, in phagocytosis the plasma membrane of a host cell encloses a pathogen, forming a vesicle that fuses with a lysosome in the cell. The lysosome is filled with enzymes that destroy the pathogen. Some pathogens, however, are resistant to the enzymes and survive, and even grow, inside the phagocytic vesicle. Others destroy the vesicle and move into the cytoplasm of the cell to continue the infection.

Viruses that have envelopes use the envelope, which is essentially plasma membrane from a previously infected host cell, to fuse with the plasma membrane of a new host cell. In this way viral particles are released into the cytoplasm of the new host, and viral reproduction begins.

Keep in Mind

- The structure of the prokaryotic cell is distinctly different from that of the eukaryotic cell.
- Many of the structures of the eukaryotic cell have important roles in the infectious disease process.

SUMMARY

- All living organisms can be divided into prokaryotes or eukaryotes.
- Prokaryotes are very simple cells that do not contain a nucleus or cytoplasmic membrane-enclosed organelles like those seen in eukaryotic cells.
- Bacteria are classified by genus and species and have distinct sizes, shapes, and arrangements.
- There are several staining techniques that can be used to classify and identify different bacteria.
- Pathogens are organisms that cause disease in humans.
- Infection is a complex process that involves both the pathogen and the host.
- Pathogens can be primary (causing disease even though the host's defenses are intact) or opportunistic (causing disease when the host's defenses are diminished).
- Many of the structures of the eukaryotic host cell have important roles in the infection process.

In this chapter you have been introduced to bacterial sizes, shapes, and multicell arrangements. We also looked at the structures associated with the eukaryotic cell, which is the target of pathogenic microorganisms. Pathogen–host relationships were also discussed and these help us understand the events associated with infectious disease that are explored in the next two chapters. Keep in mind that in addition to bacteria, there are also viral pathogens, fungal pathogens, and parasitic organisms that can infect humans, and each of these topics is discussed in detail in later chapters of this book.

SELF EVALUATION AND CHAPTER CONFIDENCE

Multiple Choice

Answers are given in the back of the book and help can be found in the student resources at:
www.garlandscience.com/micro

1. Which of the following pairs is mismatched?
   - **A.** Gram-negative bacteria and negative stain
   - **B.** Alcohol and decolorizer
   - **C.** Acid–alcohol and decolorizer
   - **D.** Iodine and mordant
   - **E.** None of the above
2. Place the steps of the Gram stain in the correct order, using this code: 1 alcohol wash, 2 crystal violet stain, 3 safranin stain, 4 iodine stain.
   - **A.** 2-1-4-3
   - **B.** 2-4-1-3
   - **C.** 4-3-2-1
   - **D.** 1-2-3-4
   - **E.** 1-3-2-4
3. The color of Gram-positive bacteria after the addition of counterstain is
   - **A.** Colorless
   - **B.** Red
   - **C.** Brown
   - **D.** Purple
   - **E.** None of the above
4. Which of the following stains allows you to see the capsule of an encapsulated bacterium under the microscope?
   - **A.** Ziehl–Neelsen acid-fast
   - **B.** Spore
   - **C.** Negative
   - **D.** Gram
5. Virulence genes are carried on
   - **A.** The chromosome
   - **B.** The ribosomes
   - **C.** The endoplasmic reticulum
   - **D.** Plasmids
   - **E.** None of the above
6. Virulence genes are arranged into
   - **A.** Reservoirs
   - **B.** Pathogenicity islands
   - **C.** Clusters
   - **D.** Plasmids
   - **E.** Individual chromosomes
7. Auto-induction is a mechanism seen in
   - **A.** Biofilms
   - **B.** Plasmids
   - **C.** Quorum sensing
   - **D.** Pathogenicity islands
   - **E.** None of the above
8. Biofilms
   - **A.** Are aggregations of many bacterial cells
   - **B.** Form on a base layer of fibrin, albumin, and immunoglobulins
   - **C.** Protect bacteria
   - **D.** All of the above
9. Organisms that can cause disease even in healthy individuals are
   - **A.** Opportunistic pathogens
   - **B.** Variable pathogens
   - **C.** Directed pathogens
   - **D.** Primary pathogens
   - **E.** None of the above
10. Organisms that are commensal can become
    - **A.** Primary pathogens
    - **B.** Intermediate pathogens
    - **C.** Temporary pathogens
    - **D.** Opportunistic pathogens
    - **E.** None of the above
11. One definition of a pathogen is
    - **A.** A mutualistic organism
    - **B.** An opportunist
    - **C.** An organism capable of causing disease
    - **D.** An organism that uses biofilms
    - **E.** None of the above

DEPTH OF UNDERSTANDING

Questions listed here require you to bring together the concepts you have learned in this chapter into a discussion format. This helps you to increase your depth of understanding of the material you have learned. Help can be found in the student resources at:
www.garlandscience.com/micro

1. Describe the host cell structures that are involved in the infection process.
2. Describe the relationship between quorum sensing and infectious disease.
3. Explain how a biofilm might be involved in the development of a nosocomial infection.

PART II

Disease Mechanisms

PART I gave us a basic understanding of the infectious process and the chemistry we will need. In **PART II**, we look at the disease process, which is a fundamental topic for health care professionals who study microbiology.

Chapter 5 gives you a basic understanding of the disease process including detailed information on four of the five requirements for a successful infection that we were introduced to in **PART I** (getting in, staying in, defeating the host defense, and damaging the host).

In **Chapter 6** we discuss the fifth requirement, transmissibility. Equally importantly, this chapter also describes the compromised host, which is a major target for infection. Throughout these discussions the things you learned in **Chapter 5** will help in your understanding. In the last part of **Chapter 6** we take a brief look at epidemiology and its importance in how we view infectious disease.

Chapter 7 provides you with discussions of the etiology or cause of disease as well as how diseases develop. This includes topics such as communicable and contagious diseases and persistent infections. The topic of herd immunity, which has become increasingly important as new diseases continue to emerge and old diseases re-emerge, is included here. The chapter concludes by discussing the scope of infections, including topics such as toxic shock.

Chapter 8 gives you a look at emerging infectious diseases that are now becoming major problems for public health. In addition, this chapter talks about re-emerging diseases, which are those that were once controlled but are now becoming a threat to public health again. In this section of the chapter we take a close look at avian influenza.

When you have finished with **PART II** you will:

- Understand the tactics used by microbial pathogens to get into the host, remain in the host, defeat the defenses of the host and damage the host.
- See how infection can be spread from one person to another.
- Be aware of how individuals can be immunocompromised and at greater risk for infection.
- Have an understanding of epidemiology and how it can help us understand infection better.
- Understand the principles of disease including the etiology and classifications used to describe diseases.
- Have a better understanding of the importance of emerging infectious diseases and also how some diseases that were controlled are now re-emerging as public health problems.

It is important to remember that what we learn in **PART II** will provide essential information for understanding what we will find in **PART III**.

IMPORTANT NOTE: Beginning with the chapters in **PART II** we have added a new feature to help you understand the information you are learning about. It is called the **CLINICAL CORNER**. It will be located at the end of each chapter and consists of clinical type questions that help you put what you have learned into a clinical context. There are additional questions like these on the web site.

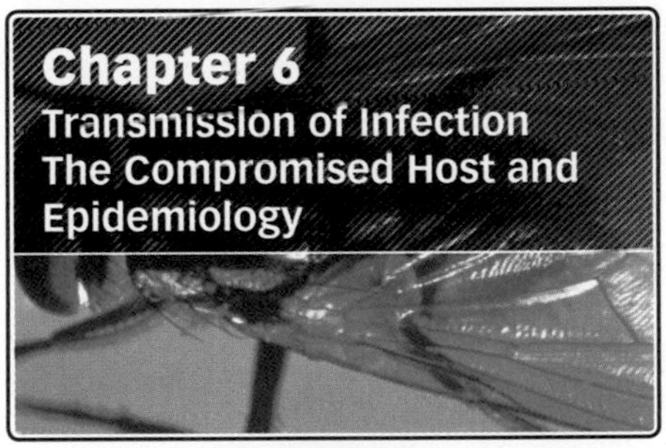

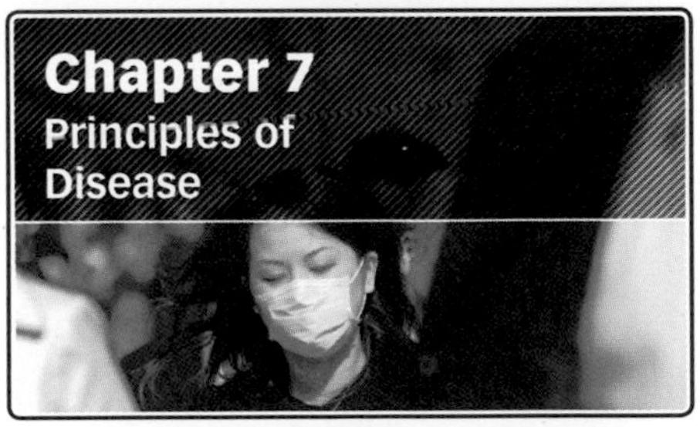

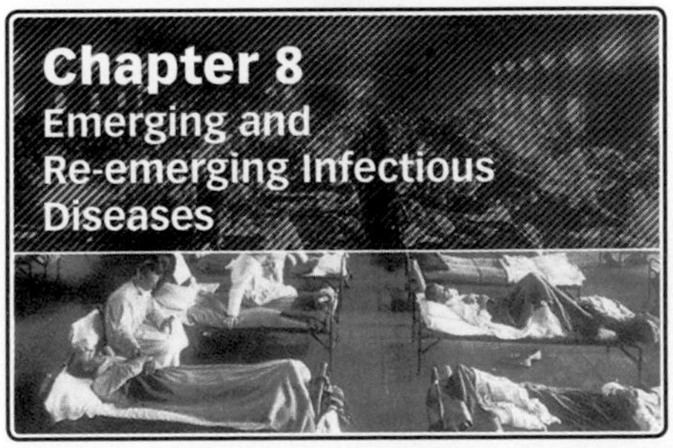

Requirements for Infection

Chapter 5

Why Is This Important?

This chapter introduces you to the mechanisms involved in the infectious disease process. What you learn here will be the foundation for the rest of your studies in microbiology.

OVERVIEW

This is a very important chapter because it is here that we begin to look at the fundamentals of the infectious disease process. Armed with the concepts from previous chapters, we are well prepared for these discussions about the specifics of what is involved in the infectious disease process. We will then build on this information as we move through future chapters. In this chapter, we examine four of the five requirements necessary for a successful infection, namely portals of entry (getting in), establishment (staying in), defeating the host defenses, and damaging the host.

We will divide our discussions into the following topics:

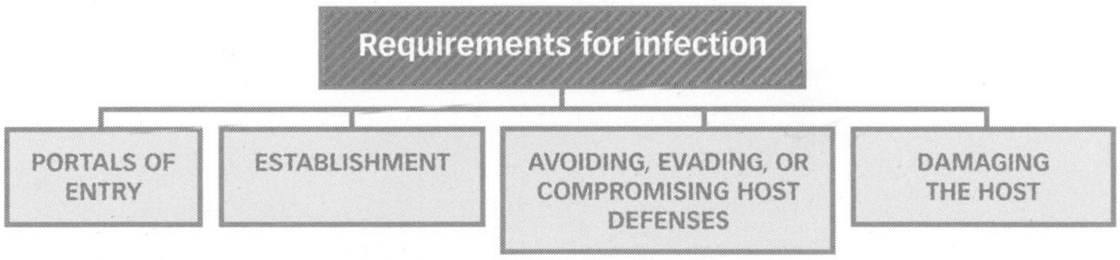

As we saw in PART I of this book, there are five fundamental requirements for a successful infection. Pathogens must be able to:

1. Enter the host (get in).
2. Have the ability to remain in a location while the infection gets established. We can call this process *establishment* (staying in), and this can include increasing the number of pathogenic organisms.
3. Avoid, evade, or compromise the host defenses (defeat the host defenses).
4. Damage the host.
5. Exit from the host and survive long enough to be transmitted to another host (transmissibility, covered in Chapter 6).

For pathogens to accomplish this, they use their **virulence factors**. These are the characteristics that the pathogens possess that allow them to thrive and survive in the host's environment. They are also the "weapons" that have a role in how virulent a microbe is and are responsible for many of the symptoms in the host. As we discuss the requirements for a successful infection in this chapter, you will see the important role of these virulence factors in the process.

What Do I Need to Know?

To get the most out of this chapter, please review the following terms from your previous courses in biology, anatomy, physiology, or chemistry or in previous chapters of this book as indicated in parentheses: cytokine, endocytosis (**4**), exocytosis (**4**), electrolyte, lipoprotein (**2**), lipopolysaccharide (**2**), lysis, lamina propria, phospholipid (**4**), Peyer's patches.

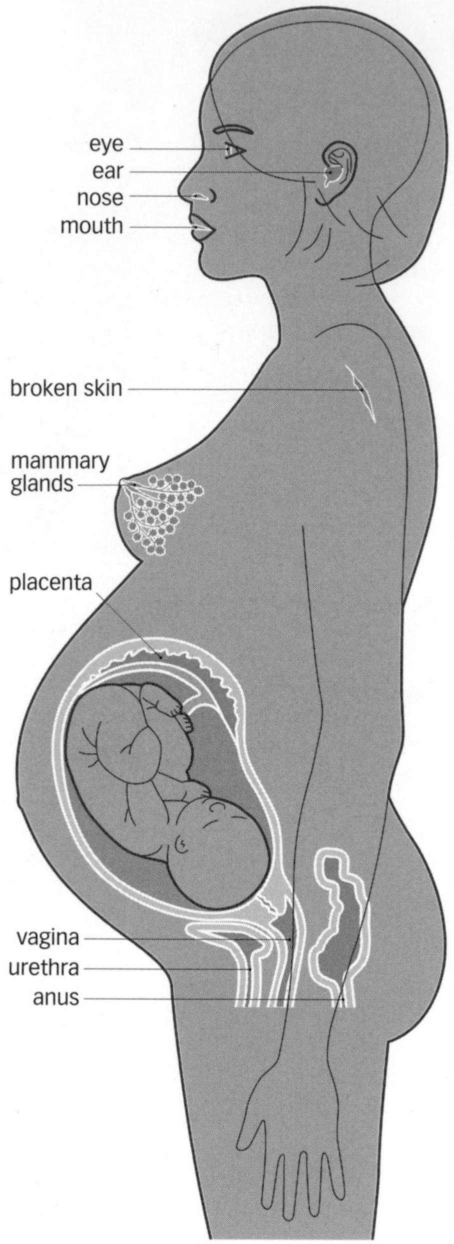

Figure 5.1 Portals of entry

PORTALS OF ENTRY (GETTING IN)

The entry of pathogens is a primary requirement for infection and it is relatively easy in humans because so much of the body is open to the outside world (**Figure 5.1**). Any point at which organisms can enter the body is called a **portal of entry**, and we can divide these portals into three categories: mucous membranes, skin, and parenteral routes (**Table 5.1**). The skin and mucous membranes are in direct contact with the exterior environment and are therefore in close proximity to potential pathogens. In contrast, pathogens that enter the body via the parenteral route take advantage of breaks in the body's barriers to gain access.

Mucous Membranes

Recall from your studies of anatomy that mucous membranes are located in areas of the body that are adjacent to the outside world. These membranes are found in the respiratory tract, the gastrointestinal tract, and the genitourinary tract.

Fast Fact

In total, there are about 400 square meters of mucous membrane surface area in the human body. That represents a lot of potential entry points.

Table 5.1 Portals of Entry for Some Common Pathogens

Portal of Entry	Pathogen	Disease
Respiratory-tract mucous membranes	*Streptococcus* species	Pneumonia
	Mycobacterium tuberculosis	Tuberculosis
	Bordetella pertussis	Whooping cough
	Influenza virus	Influenza
	Measles virus	Measles (rubeola)
	Rubella virus	German measles (rubella)
	Varicella-zoster virus	Chickenpox
Gastrointestinal-tract mucous membranes	*Shigella* species	Shigellosis (bacillary dysentery)
	Escherichia coli	Enterohemorrhagic disease
	Vibrio cholerae	Cholera
	Salmonella enterica	Salmonellosis
	Salmonella typhi	Typhoid fever
	Hepatitis A virus	Hepatitis A
	Mumps virus	Mumps
Genitourinary-tract mucous membranes	*Neisseria gonorrhoeae*	Gonorrhea
	Treponema pallidum	Syphilis
	Chlamydia trachomatis	Nongonococcal urethritis
	Herpes simplex virus	Herpes
	Human immunodeficiency virus	Acquired immunodeficiency syndrome
Skin or parenteral route	*Clostridium perfringens*	Gas gangrene
	Clostridium tetani	Tetanus
	Rickettsia rickettsii	Rocky Mountain spotted fever
	Hepatitis B and C	Hepatitis
	Rabies virus	Rabies
	Plasmodium species	Malaria

You can think of the enormous surface area of the mucous membranes as a border, analogous to the border between two nations. Fortunately, as we will see in Chapter 15, the body has a variety of powerful border defenses that prevent entry. Consequently, even though the surfaces of the respiratory, gastrointestinal, and genitourinary tracts are in contact with potential pathogens, the surfaces have means of protecting the body against the entry of microorganisms.

The Respiratory Tract

Of all of the portals of entry, this is probably the most favorable to pathogens (**Figure 5.2**). We live in a cloud of potentially dangerous microbial pathogens, and the respiratory tract facilitates entry through breathing. Organisms can be found on droplets of moisture in the air and even on dust particles, and many diseases, such as colds, pneumonia, tuberculosis, influenza, **measles**, and even **smallpox**, use this portal of entry. As we will see later in this chapter, the respiratory tract is also a very productive portal of exit that can be used to transmit pathogens through coughing or sneezing. (A **portal of exit** is any point at which microorganisms can leave the body.)

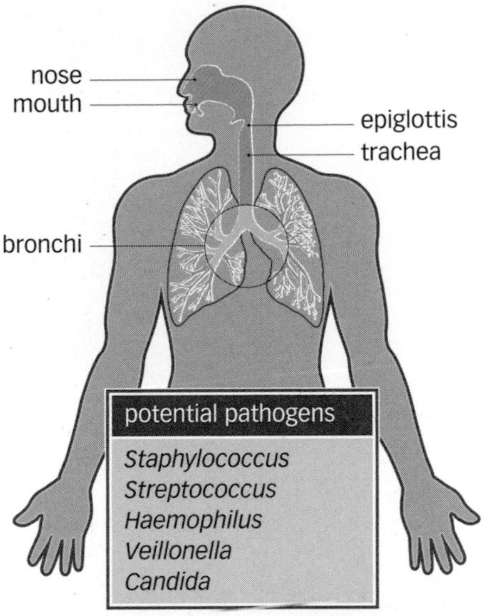

Figure 5.2 Mucous membranes of the upper respiratory tract are portals of entry for potential pathogens. The upper respiratory tract is the body's most accessible portal of entry because organisms are brought in through the breathing process.

The Gastrointestinal Tract

This system is also open to the outside world, and organisms can enter the body via the foods and liquids we eat and drink. The gastrointestinal tract has many protective barriers against pathogens, the most obvious of which is the production of stomach acid and bile. These substances are required for normal digestion but produce hostile environments that limit the survival of most pathogens. Still, there are many organisms that not only use this portal of entry but actually prefer or require it. For example, the polio virus uses the gastrointestinal tract as a required part of its infectious cycle. In addition, the tract is a preferred entry point for hepatitis A virus, the parasite *Giardia*, the bacterium *Vibrio cholerae*, and organisms that cause **dysentery** and typhoid fever (**Figure 5.3**).

One of the more interesting pathogens that use this portal of entry is *Helicobacter pylori*. This organism is carried in the gastric mucosa of one out of every two people in the world, and infection with this organism is a known risk factor for the development of gastroduodenal ulcers. For many years, it was thought that the acidity of the stomach (about pH 1.0) would preclude bacterial survival, but *H. pylori* has consistently been shown to

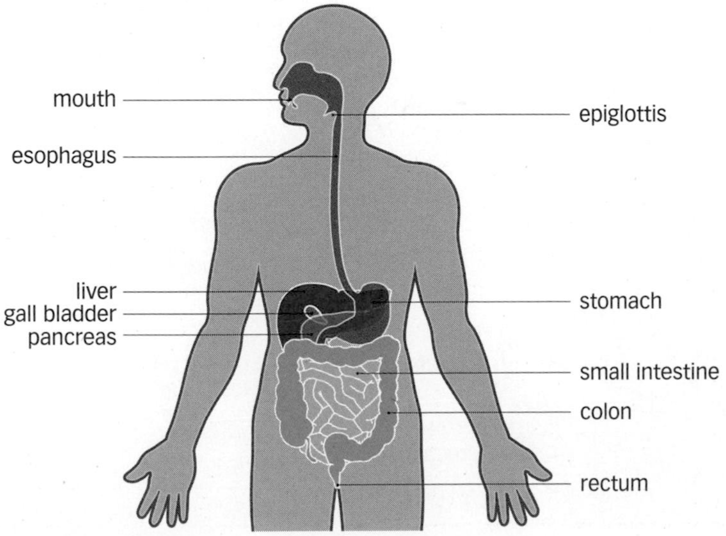

Figure 5.3 Mucous membranes of the gastrointestinal tract are portals of entry to the body. Microorganisms enter with food or water, but much of this system is an inhospitable environment for them. This is a preferred portal of entry for some pathogens such as *Salmonella*.

upper digestive tract	lower digestive tract
potential pathogens	potential pathogens
Haemophilus	*Escherichia*
Actinomyces	*Lactobacillus*
Treponema	*Clostridium*
Neisseria	*Enterococcus*
Corynebacterium	*Proteus*
Entamoeba	*Shigella*
Trichomonas	*Candida*
	Entamoeba
	Trichomonas

be associated with gastroduodenal lesions. It turns out that *Helicobacter* can produce a relatively alkaline atmosphere around itself that protects it during its journey in the stomach by neutralizing the acidic pH found there. The organism eventually makes its way to the mucus that lines the wall of the stomach and duodenum of the small intestine. Nestled in this mucus, it is protected and can begin the process of infection that results in the destruction of tissue and the formation of an ulcer.

The gastrointestinal tract is also a leading portal of exit for pathogens in feces. In fact, we will see throughout our discussions of infectious disease that the **fecal–oral route of contamination** has a major role in many infections, especially with Gram-negative bacteria, viruses, protozoa, and other parasites.

The Genitourinary Tract

The urinary and reproductive tracts are also open to the outside world, but unlike the respiratory and gastrointestinal tracts, they are more complicated with respect to entry.

Fast Fact

Urinary tract infections are easily treated, and no serious damage is done unless the pathogens find a way up into the bladder or the kidneys.

Urinary tract infections are more common in women than in men because of the anatomical relationship between the anus and urethra, which is much closer in women than in men (**Figure 5.4**). Because fecal material contains bacteria, it is easy for these organisms to find their way to the urinary tract.

Diseases of the reproductive tract are usually sexually transmitted and occur as a result of either abrasions or tiny tears in the tissues that routinely occur during sexual activity. Once the mucous membrane barrier is broken, pathogens gain entry. Conditions such as syphilis, gonorrhea, chlamydia, herpes, **genital warts**, and HIV infections are caused by pathogens that use this portal of entry. The genitourinary tract can also be used as a portal of exit through which these infections can be transmitted. Fortunately, this route is well protected by host defenses.

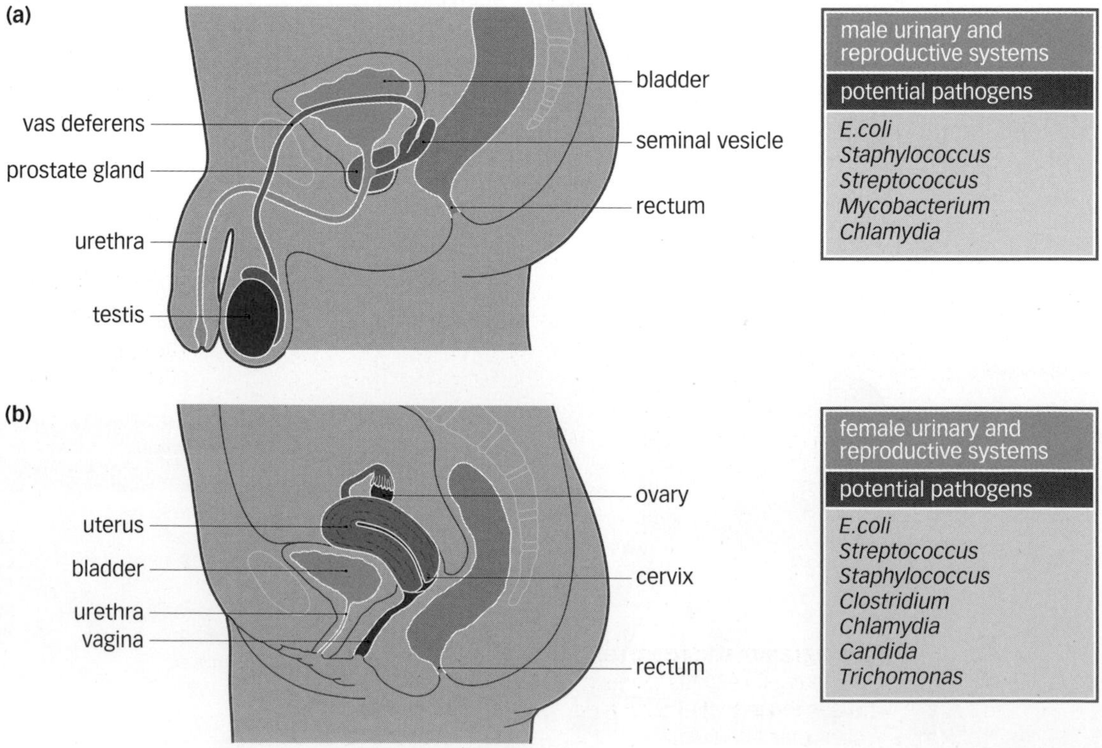

Figure 5.4 Genitourinary portals of infection. Panel **a**: The male urinary and reproductive tract. Panel **b**: The female reproductive and urinary tract. In both sexes, both tracts are portals of entry for pathogens. Urinary tract infections occur more frequently in females than in males because of the anatomical relationship between the anus and the urethra.

Skin

The skin is the largest organ in the body, and like the mucous membranes it has a vast surface area through which microorganisms may enter the body. However, unlike the case with the mucous membranes, the association between the skin and microorganisms does not depend on their being taken in through breathing, eating, or sexual activity. The microorganisms are already there. In fact, the skin is literally covered with many types of microorganism and easily accessible to many other types, including pathogens. Fortunately, the skin is impenetrable to most microorganisms. In fact, many bacteria, fungi, and some parasites, that live on the skin are completely harmless to the host (**Figure 5.5**). To initiate an infection, these organisms must find an opening, such as a hair follicle, a perspiration duct, or a break in the skin, so as to gain entry. We will see that these potential entry points are very well guarded.

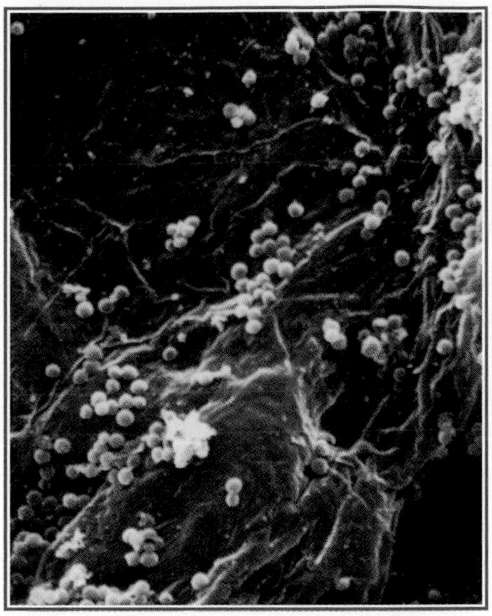

Figure 5.5 The skin and conjunctiva of the eye are in constant contact with microorganisms but present an impenetrable barrier to entry. These barriers must be compromised in some way in order for organism to enter. This scanning electron micrograph shows some of the microorganisms that inhabit the skin.

The Parenteral Route

Movement of organisms past the barrier of the skin requires a break in the barrier, and the portal of entry referred to as the **parenteral route** depends on such breaks. Things such as injections, which are routinely used in clinical applications, can easily become parenteral portals of entry any time that microorganisms are present in close proximity to the site of injection. Entry can also occur via insect bites (referred to as **vector transmission**), and many organisms, such as *Plasmodium* (the causative agent of **malaria**), use this as a way to enter the host.

Obviously, any cut or wound is going to allow the entry of skin-dwelling organisms, but the extent of the trauma can also have a role in the severity of the subsequent infection. Recall that the skin is made up of two basic layers, the epidermis and the dermis (**Figure 5.6**). Because the epidermis is made up of dead and dying cells, there is no access of blood to this

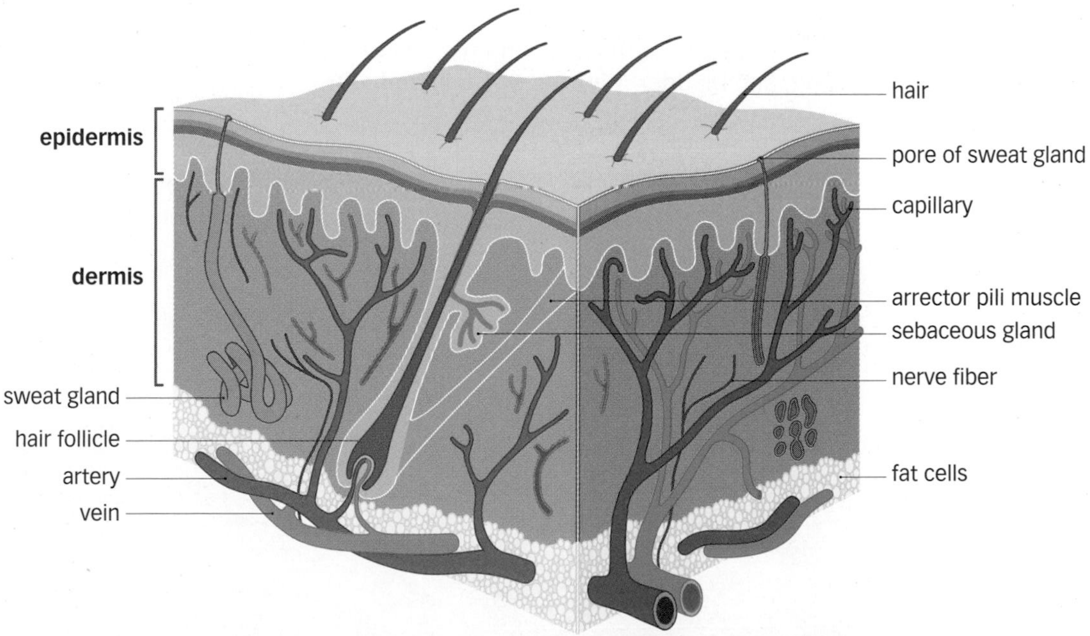

Figure 5.6 A diagrammatic representation of a cross section of the skin. Pathogens that gain access to the epidermis usually result in localized infections, whereas those that enter the dermis can cause systemic problems because of the availability of blood in this layer of the skin.

layer. Therefore, cuts or wounds limited to this layer are less likely to spread beyond the site of entry. In contrast, the dermis is associated with blood vessels, and cuts or wounds that involve this layer or go deeper are far more likely to cause more serious systemic infections. This is even more apparent when we look at surgical procedures, in which contaminating organisms can gain access deep into internal tissues.

Some organisms have *preferred* portals of entry, and only entry through the preferred portals will result in infection. For example, *Salmonella enterica* serovar Typhi (also known as *Salmonella typhi*) must be swallowed if it is to cause intestinal infections, whereas streptococci must be inhaled to cause pneumonia. However, many organisms can cause infection no matter what entry point is used. A case in point is *Yersinia pestis*, the organism that causes bubonic plague. This organism uses multiple entry points, and in the Middle Ages it wiped out one-third to one-half of the population of Europe.

ESTABLISHMENT (STAYING IN)

Entry into the host is just the beginning of the problems most pathogens face. After entry, a pathogen must find a way to stay in the host if it is to establish the focus of the infection. This task is very difficult for a variety of reasons. For instance, there can be physical obstacles to overcome. Let's use *Neisseria gonorrhoeae* as an example. If, after gaining access to the genitourinary tract, this microorganism does not have a way of adhering to the tissue, it might be flushed back out during urination. On top of this is the vast array of defenses the body has in place to destroy the organism.

In Chapter 9 we will look at the anatomical structures of bacteria and see how these structures can have a role in clinical pathogenesis. During these discussions, we will see that organisms such as *Streptococcus pneumoniae* are not infectious without a surrounding capsule that allows them to adhere to the body's tissue and inhibit the host defense. In point of fact, almost all pathogens have some means of attachment. Some Gram-negative organisms — for instance, *Escherichia coli* — use structures called **fimbriae** to attach to certain receptors on cells of the small intestine, colon, and bladder (**Figures 5.7**, **5.8**, and **5.9**).

Figure 5.7 A false color scanning electron micrograph of the surface of the colon mucosa with clusters of bacteria (possibly E. coli) attached.

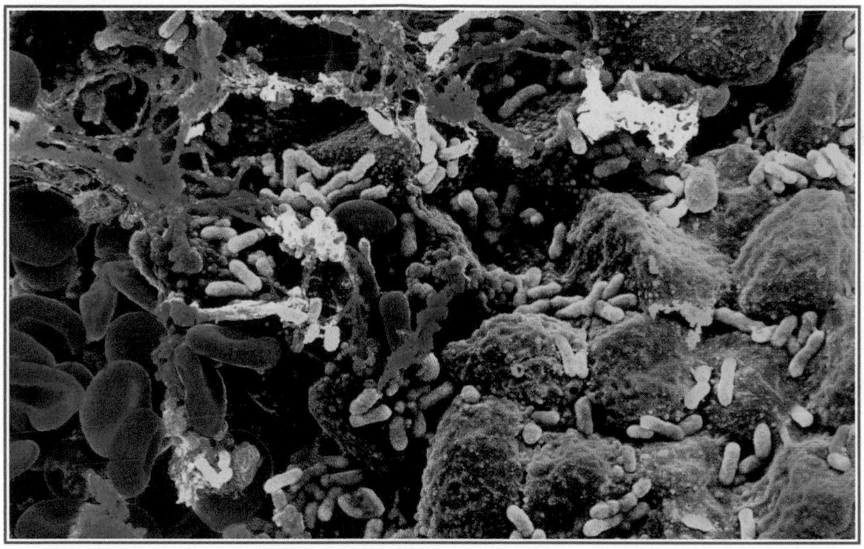

Figure 5.8 A scanning electron micrograph of bacteria (colored green) adhering to the surface of cells in the colon.

In many cases, pathogens use molecules called **adhesins**, which are glycolipids or lipoproteins located on the pathogen surface, as a means of adhering to tissue (**Table 5.2**). For example, *N. gonorrhoeae* can use fimbriae coated with adhesins to adhere to tissue in the genitourinary tract. Pathogens can also take advantage of "sticky" glycoproteins found on the surface of host cells.

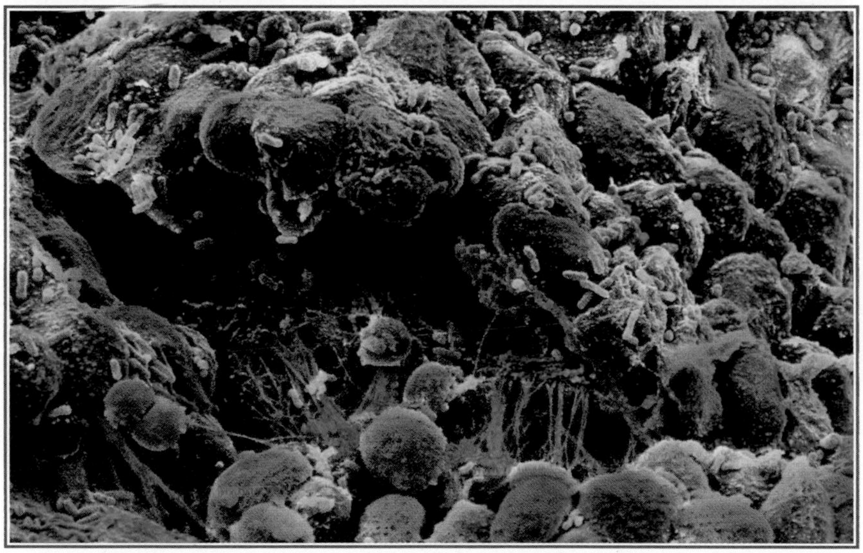

Figure 5.9 A colorized scanning electron micrograph of cells of the human bladder infected with bacteria. Rod shaped *E. coli* (colored yellow) are seen attached to the epithelial cells of the bladder (colored blue). Note the purple colored cells which have swelled and developed a rough surface due to this chronic bladder infection.

Let's look at a good example of adherence. The organism *Streptococcus mutans* has for a long time been accused of causing tooth decay. In fact, this is not strictly true because tooth decay actually starts with fluids produced by your oral tissues. These fluids form a **pellicle**, which is a protein film that coats your teeth. *S. mutans* adheres to this pellicle (**Figure 5.10**) and begins to produce the enzyme glycosyltransferase. The problem is

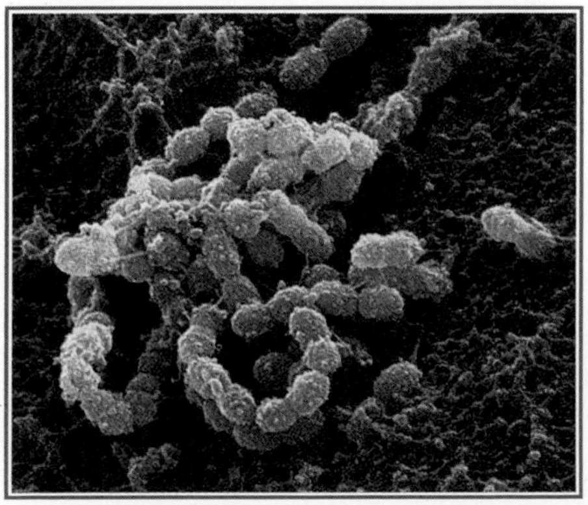

Figure 5.10 A colorized scanning electron micrograph of dental plaque. The yellow-green structures are *Streptococcus mutans* adhering to a biofilm that covers the teeth. Other organisms adhere to *S. mutans* and cause the development of a biofilm. The organisms in this biofilm produce the enzymes that can cause destruction of the tooth, resulting in the formation of a cavity.

Location	Bacterium	Disease	Mechanism of Adherence
Upper respiratory tract	*Mycoplasma pneumoniae*	Atypical pneumonia	Cell surface adhesion molecules bind to receptors on cells lining respiratory tract
	Streptococcus pneumoniae	Pneumonia	Adhesion molecules attach to carbohydrates on respiratory cells
	Neisseria meningitidis	Meningitis	Adhesion molecules on bacterial cell attach to respiratory cells
Genitourinary tract	*Treponema pallidum*	Syphilis	Bacterial proteins attach to cells in reproductive tract
	Neisseria gonorrhoeae	Gonorrhea	Adhesion molecules on bacterium attach to cells in reproductive tract
Gastrointestinal tract	*Shigella* species	Dysentery	Mechanism not known
	Escherichia coli	Diarrhea	Adhesin molecules on bacterial pili attach to gastrointestinal cells
	Vibrio cholerae	Cholera	Adhesin molecules on bacterial flagella attach to gastrointestinal cells

Table 5.2 Adherence Factors Associated with Infection

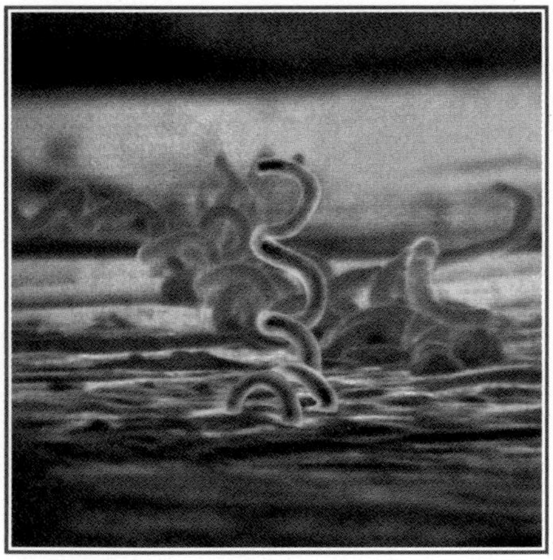

Figure 5.11 The spirochete *Treponema pallidum* "corkscrewing" into tissue.

Fast Fact

Drastic increases in pathogen numbers could be a cause for worry except for the fact that humans have developed tremendous defense mechanisms. Consequently, even though a pathogen could potentially proliferate from a single cell to 10^{21} cells in 24 hours, each antibody-secreting plasma cell in the human immune system can produce antibodies against that pathogen at the rate of 2,000 antibody molecules *per second*. More importantly there can be millions of plasma cells!

exacerbated when other organisms adhere to *S. mutans*, forming a biofilm, which is essentially a living coating on the teeth. This combination of organisms and the enzymes they produce causes the destruction of the tooth enamel, resulting in the formation of a cavity. The plaque the dentist removes from your teeth is made up of this complex of organisms. When you consider how hard the dentist has to work to pry this plaque from your teeth, you get a good idea of "establishment."

Some organisms, such as *Treponema pallidum* (which causes syphilis), avoid the need for protracted periods of adherence by boring into the tissues (**Figure 5.11**).

Increasing the Numbers

For pathogens, there is safety and success (that is, infectivity) in numbers. In fact, the doubling time for some bacteria can be as little as 20 minutes. This extremely high reproductive rate requires that the growth environment be satisfactory, and for most pathogens the tissues and fluids of the human body are an ideal environment. So the number of organisms required for successful infection can easily be achieved.

There is considerable variability among organisms with regard to the number required for success, and we can run specific experiments to establish the criteria for virulence for any given pathogen. The **lethal dose 50%** (**LD_{50}**) is the number of organisms required to kill 50% of the hosts, and the **infectious dose 50%** (**ID_{50}**) is the number of organisms required for 50% of the population to show signs of infection. Pathogens having the lowest LD_{50} and ID_{50} values are the most virulent. Using these types of information, we can categorize organisms according to virulence, as shown in **Figure 5.12**.

It is important to remember that bacteria divide by **binary fission** (one bacterium splits into two) and that a pathogen that has a low LD_{50} and a short doubling time could be extraordinarily dangerous, with the rapid increase in organisms quickly overwhelming a patient. Fortunately, attacking only rapidly growing bacteria is the way in which many

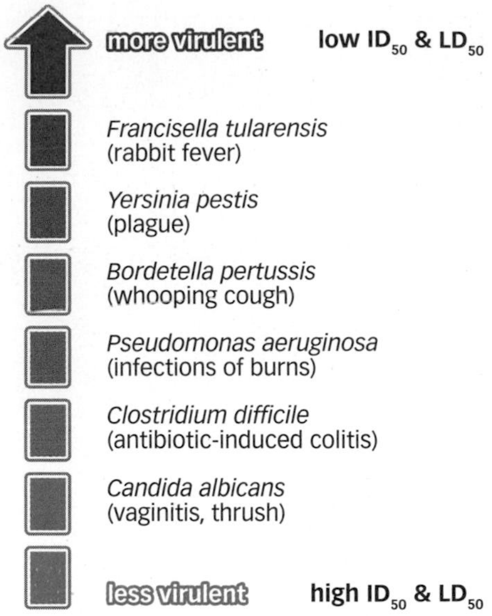

Figure 5.12 Degree of virulence attributed to different pathogens. This type of appraisal can be made after determining the LD_{50} and ID_{50} of the pathogens. Remember, the lower the LD_{50}, the more virulent the pathogen.

antibiotics work, thereby preventing negative outcomes from infection. Unfortunately, this story is changing, as we will see when we discuss the rapidly expanding resistance to antibiotics in Chapter 20.

In viral infections, the number requirement is easier to understand. A **lytic virus** is defined as one that functions by filling a host cell with viral particles called **virions** until the host cell bursts open and pours virions into the intercellular fluid. These new virions locate new host cells, and the process is repeated until there are no more host cells available. Infection with HIV is a good example. As the infection continues, the host white blood cells known as T4 lymphocytes, which are the viral targets, simply disappear (along with immune capability).

AVOIDING, EVADING, OR COMPROMISING HOST DEFENSES (DEFEAT THE HOST DEFENSES)

If a pathogen has managed to get into a host, stay in, and rapidly increase its numbers, it is on the way to causing a successful infection. Unfortunately for the pathogen, these steps are usually not enough because there are many ways in which the host can defend itself. Thus, the pathogen has to deal with the host's defenses. There are two basic ways in which this is accomplished; one involves the structure of the pathogen cells, which is a built-in (passive) defense, and the other involves attacking the host's defenses (an active defense).

The main structural defenses of pathogenic bacteria are capsules and cell wall components. In fact, as we noted earlier, encapsulation is required for some organisms to become virulent. For example, unless they are encapsulated, *S. pneumoniae* will not cause pneumonia and *Klebsiella pneumoniae* will not cause Gram-negative bacterial pneumonia. Other bacteria that are virulent only when encapsulated are *Haemophilus influenzae*, *Bacillus anthracis*, and *Yersinia pestis*.

Capsule and Cell Wall (Passive) Defenses

The bacterial capsule protects against phagocytosis. In humans, a first line of defense for the innate immune response is **phagocytosis**. In this process, cells known as **phagocytes** ingest pathogens and then destroy them. Pathogens can encapsulate themselves, covering their entire surface in a slimy capsule, which protects against phagocytosis (**Figure 5.13**). The capsular material seems to prevent a phagocyte from adhering to the surface of the bacterium. This adherence is required for the phagocyte to develop *pseudopodia*, "feet" that then surround the organism.

You might think that capsule protection would be all that the pathogen required to overcome the host defense, but that is not quite true. As discussed in Chapter 16, the host can use the adaptive immune response

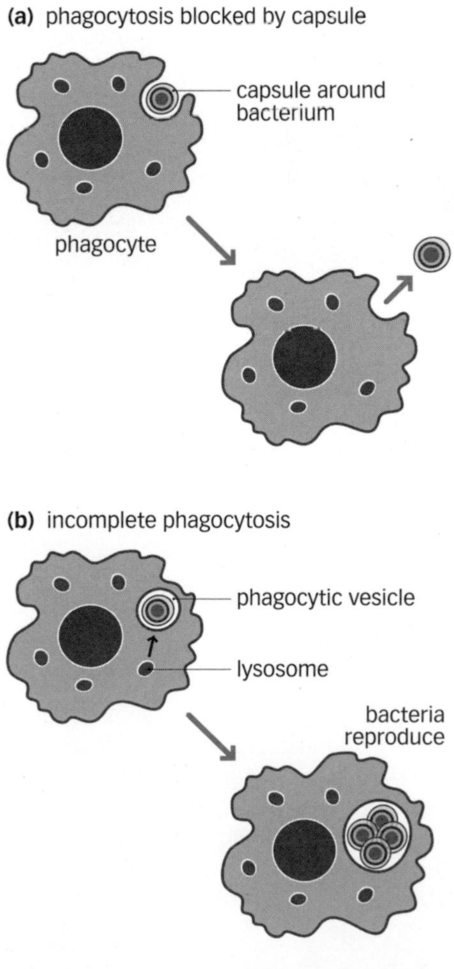

Figure 5.13 A diagrammatic representation of the inhibition of phagocytosis by an encapsulated bacterium. Encapsulation is a defense mechanism that pathogens have against the defenses of the host. Panel **a**: The capsule keeps the surface of the engulfing phagocyte from sticking to the bacterium, and the bacterium goes free. Panel **b**: Some bacteria not only resist being phagocytosed, but can even multiply once inside the phagocyte.

to produce antibodies against the capsule. When the antibody molecules bind to the capsule (a process known as **opsonization**, which we will discuss in detail in Chapter 16), they attract phagocytic cells that use the antibody molecules as bridges to adhere to the organism and phagocytose it.

A second structural defense available to bacteria is the bacterial cell wall. In Chapter 9, we will see that this wall is a complex structure that protects the bacteria from environmental pressure. The components of the wall can help to increase virulence by defending against host defenses. For instance, *Streptococcus pyogenes* incorporates **M proteins** into its cell wall. These proteins increase the virulence of this pathogen by increasing adherence to host target cells and by making the pathogen resistant to heat and to acidic environments. In addition, M proteins also inhibit phagocytosis and are intimately involved in the condition known as *toxic shock*. Fortunately, the host's immune response can provide antibodies and also fragments of complement proteins that opsonize these bacteria.

Another structural protection is seen in *Mycobacterium tuberculosis*, in which the cell wall is infused with **mycolic acid**, a waxy substance that inhibits phagocytosis and protects the bacterium against antiseptics, disinfectants, and antibiotics.

Enzyme (Active) Defenses

Capsules and cell wall components are passive measures used to defeat the host defenses. Bacteria also use active defenses against a host by producing extracellular enzymes that enable the infection to spread, and by killing off host defensive cells. The following enzymes are useful against the host defenses.

Leukocidins are enzymes that destroy white blood cells in the host. White blood cells are an important part of both the innate and the adaptive host defense systems. Two types of white blood cell are neutrophils and macrophages, which are powerful phagocytic cells. In addition, the white blood cells known as lymphocytes, which are responsible for the adaptive immune response to infection, are destroyed by leukocidins. Leukocidins are produced by staphylococcal and streptococcal pathogens, and it is easy to see how the ability to kill off host defenders can make these organisms dangerous.

Hemolysins are membrane-damaging toxins that disrupt the plasma membrane of host cells and cause the cells to lyse. These toxins can damage the plasma membrane of both red blood cells and white blood cells and are produced by a variety of bacteria, including staphylococcal species, *Clostridium perfringens* (which causes **gas gangrene**), and streptococcal species. Hemolysin produced by streptococcal bacteria is referred to as streptolysin and can be divided into different groups, such as group A and group O. The various groups of streptolysin differ from one another in the type of cell destruction that they cause. For example, streptolysin O is associated with β-hemolysis (complete destruction) of red blood cells.

Fast Fact

Knowledge about streptococcal bacteria and their kinase enzymes has saved the lives of many cardiac patients! Scientists have been able to make use of streptococcal bacteria in the lab to produce these enzymes, which are then used as a medical treatment to destroy blood clots.

Coagulase (**Figure 5.14a**) is a pathogen-produced enzyme that causes fibrin clots to form in the blood of a host. Clot formation can be used by both host and pathogen during an infection. The host can use clotting as part of the defense against infection. This clotting occurs in blood vessels around the site of the infection, thereby closing in the pathogens and preventing the spread of the infection. An example of bacterial "walling off" is a **boil** resulting from a localized staphylococcal infection. Here the organism will wall itself off to avoid the host defenses.

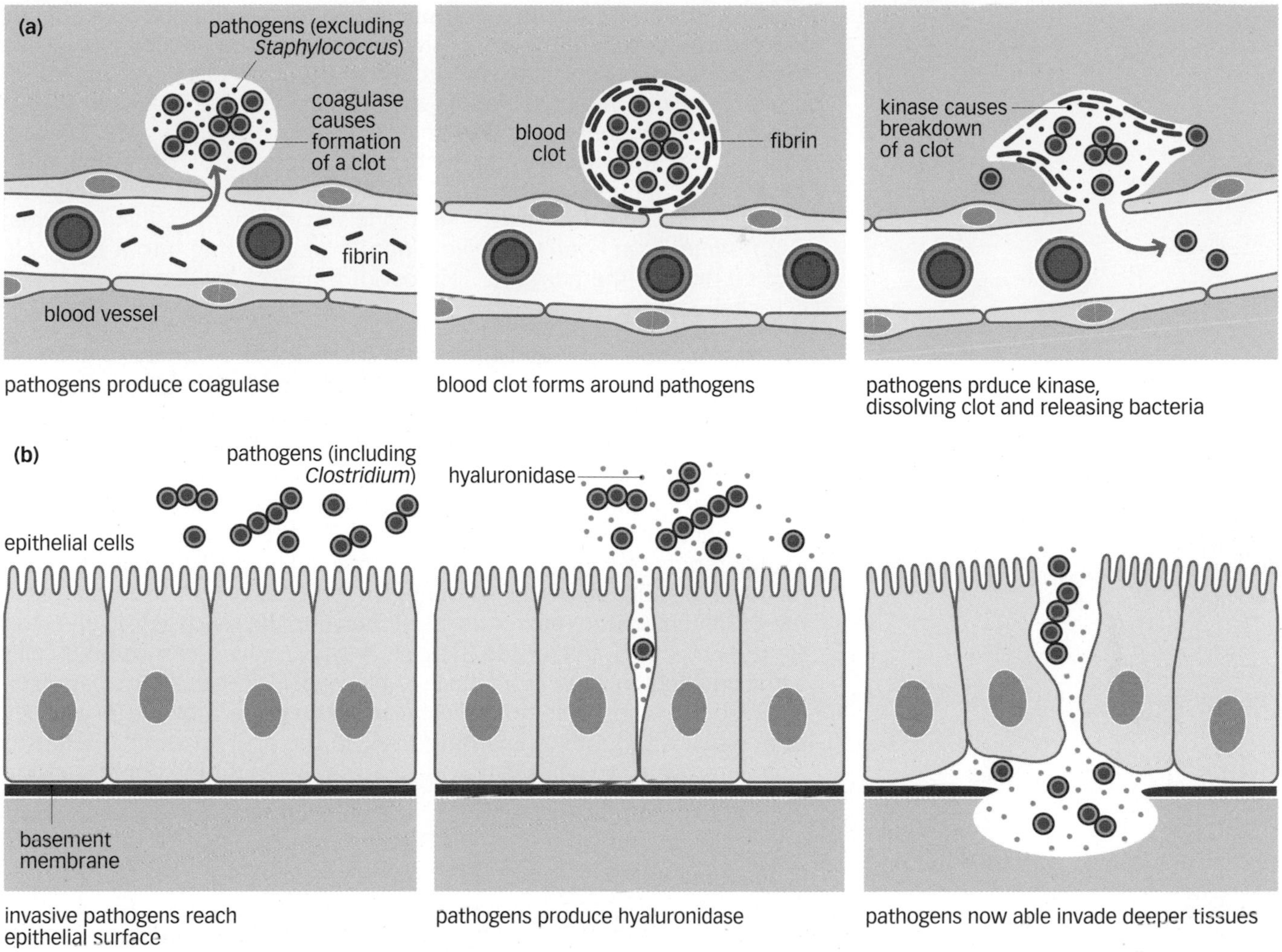

Figure 5.14 Some of the enzymes produced by pathogenic bacteria. Panel **a**: Pathogens can wall themselves off from host defenses using coagulase but can also use kinase to dissolve clots. Panel **b**: The enzyme hyaluronidase allows pathogens to invade deep tissues.

Kinases (**Figure 5.14a**) are enzymes that break down fibrin and dissolve clots. These enzymes can be used by a pathogen to overcome attempts by the host to wall off the infection, thereby ensuring its spread. Staphylokinase is produced by staphylococcal species, and streptokinase is produced by streptococcal species.

Hyaluronidase and collagenase are enzymes that break down connective tissue and collagen in a host, thereby allowing infections to spread (**Figure 5.14b**). Both are active, for instance, in the infection gas gangrene (caused by *C. perfringens*), with the result that the infection is usually widespread and usually involves the destruction of connective tissue.

The process of spreading out is fundamental for the increased virulence of many pathogens. Group A streptococci are a good example. These organisms do not infect any mammals other than humans. They use enzymes as described above to inhibit the clotting machinery or to break down clots formed by the host, allowing the pathogens to spread. There may be a genetic predisposition in humans that has a role in whether streptococcal infections are minor, such as strep throat, or major, such as **necrotizing fasciitis** (flesh-eating; **Figure 5.15**).

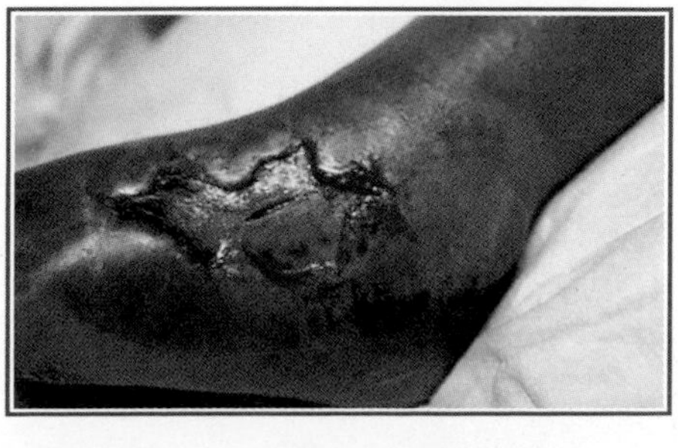

Figure 5.15 A patient with necrotizing fasciitis. This infection is caused by *Streptococcus pyogenes.*

There is one final tactic that pathogens use to defend themselves. They hide! Pathogens can use any of the tactics described above, but in the long run the host's innate and adaptive immune response will catch up with them. Therefore, being able to get *inside* a host cell, where the immune response cannot detect them, is the best possible defense. This is easy for viral pathogens because they are by definition **obligate** intracellular parasites. (*Obligate* means able to survive in only one environment, and for viruses that one environment is inside a host cell.) Bacteria, in contrast, have a harder time getting into a host cell, and so they let the host cell do most of the work by usurping the cell cytoskeleton.

Recall from Chapter 4 that the eukaryotic cell has a variety of fibers — microfilaments, intermediate filaments, and microtubules — that are part of the cellular cytoskeleton and are responsible for cellular support and intracellular movement. Bacteria use these filaments and tubules to penetrate and move around inside the infected cell. As an example, let's look at *Salmonella*. When this pathogen makes contact with a host cell, the contact causes the plasma membrane of the host cell to change its configuration. *Salmonella* produces a molecule called **invasin** that can change the structure of actin filaments in the cytoskeleton. The change in these filaments in turn moves the bacterium into the cell. It gets even better for the pathogen once inside the cell, because now it can use the cell's actin filaments to move from place to place inside the cell. The pathogen can also use a host cell molecule called **cadherin** to move from one cell to another without ever exposing itself to the host's immune defenses searching for it. We will see this mechanism again during our discussion of infections of the digestive tract.

Keep in Mind

- There are five requirements for a successful infection: get in, stay in, defeat the host defenses, damage the host, and be transmissible.
- Places at which pathogens enter the body are called portals of entry.
- The major portals of entry are the mucous membranes, the skin, and parenteral routes.
- Mucous membrane portals of entry are associated with the respiratory, digestive, and genitourinary tracts of the body.
- Establishment (the requirement of staying in) can be accomplished using adhesin molecules, which are glycolipids or lipoproteins. In addition, some pathogens take advantage of structures such as fimbriae to adhere to tissues.
- Criteria for virulence for a given pathogen can be gauged by the ID_{50} and LD_{50} of that organism.
- Pathogens can defeat a host's defenses in two ways: passively (by using structures such as the capsule) and actively (by attacking the host defense directly through the production of enzymes).

DAMAGING THE HOST

In the above discussions, we talked about some of the requirements for a successful infection. Now let's look at the damage that occurs to host cells during the disease process. Most of the damage associated with infection can be divided into two parts: damage that occurs because the bacteria are present, and damage that is a by-product of the host response. The damage directly attributable to the pathogen can be either direct or indirect.

Direct damage is the obvious destruction of host cells and tissues and is usually localized to the site of the infection. In direct damage, the host defense response is timely and potent, usually limiting the damage done. *Indirect damage* is seen in most serious infections and is much more dangerous to the host because it involves systemic disease. This type of distal pathology results from the production of bacterial toxins. These toxins (which can be very poisonous) are soluble in aqueous solutions and easily diffuse into, and move through, the blood and lymph systems. Thus, they can travel throughout the body quickly. This causes pathogenic changes far away from the initial site of infection.

Toxins can produce fatal outcomes in patients. They also have some common characteristics associated with them, such as fever, shock, diarrhea, cardiac and neurological trauma, and destruction of blood vessels. There are two types of toxin — exotoxins and endotoxins — and they are very different from each other.

Exotoxins

Exotoxins are toxins that are produced by pathogens and then leave the pathogen cells and enter host cells (**Figure 5.16**). They are among the most lethal substances known, with some exotoxins a million times more poisonous than the poison strychnine. An exotoxin is usually an enzymatic protein that is soluble in blood and lymph. Being soluble in these liquids, exotoxins rapidly diffuse into tissues and stop metabolic function in the host cell. They are so dangerous that they are usually produced by pathogens in the *proenzyme* form, which is inactive, and then activated only after having left the pathogen's cells. Many of the genes that encode these toxins are carried on *plasmids* in the pathogens, which make them even more dangerous because the plasmids make it possible to transfer genetic information from one bacterium to another (discussed in Chapter 11).

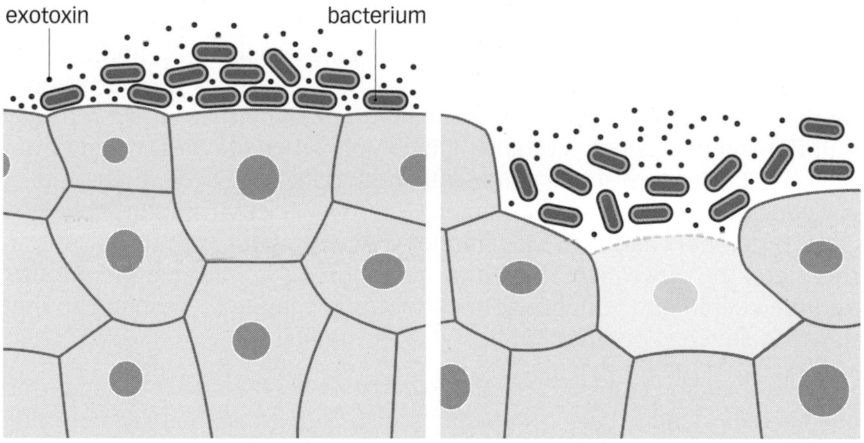

Figure 5.16 Exotoxins are produced by living pathogenic bacteria. These toxins then enter cells of a host and prevent those cells from functioning properly.

There are three types of exotoxin: **cytotoxins**, which kill cells that they come in contact with; **neurotoxins**, which interfere with neurological signal transmission; and **enterotoxins**, which affect the lining of the digestive system (**Table 5.3**). Let's look at some of the more dangerous exotoxins.

Exotoxin	Organism Producing Toxin	Action of Toxin	Symptoms
Anthrax (cytotoxin)	*Bacillus anthracis*	Increases vascular permeability	Hemorrhage and pulmonary edema
Enterotoxin (enterotoxin)	*Bacillus cereus*	Causes host body to lose electrolytes	Diarrhea
Botulinum (neurotoxin)	*Clostridium botulinum*	Blocks release of acetylcholine	Respiratory paralysis
Alpha (gas gangrene); (cytotoxin)	*Clostridium perfringens*	Breaks down cell membranes	Extensive cell and tissue destruction
Tetanus (neurotoxin)	*Clostridium tetani*	Inhibits motor-neuron antagonists	Violent skeletal muscle spasms (lockjaw), respiratory failure
Diphtheria (cytotoxin)	*Corynebacterium diphtheriae*	Inhibits protein synthesis	Heart damage, possible death weeks after apparent recovery
E. coli (enterotoxin)	*Escherichia coli* O157:H7	Hemolytic uremic syndrome	Destruction of intestinal lining, kidney failure
Bacillary dysentery (enterotoxin)	*Shigella dysenteriae*	Poisons host cells	Severe diarrhea
Vibrio (enterotoxin)	*Vibrio cholerae*	Causes excessive loss of water and electrolytes	Severe diarrhea; death can occur within hours
Erythrogenic (cytotoxin)	*Streptococcus pyogenes*	Causes vasodilation	Maculopapular lesions, as seen in scarlet fever

Table 5.3 Effects of Exotoxins

Anthrax Toxin

Anthrax toxin is a cytotoxin produced by the bacterium *B. anthracis*, a Gram-positive rod commonly found in the soil of pastures. The toxin is made up of three parts: an edema factor, a protective antigen, (which is a transmembrane factor also used for vaccine production); and a lethal factor. It is so dangerous that each bacterium produces the parts separately and assembles them on the bacterial cell surface in such a way that the complex is not yet toxic. The complex then leaves the bacterium and attaches to a host cell. Once attached to its target cell, the complex is endocytosed into a vesicle, where the low-pH environment causes a conformational change in the complex, converting it to the toxic form. In this state, the protective antigen forms a pore in the vesicle membrane, and the edema and lethal factors temporarily change their shape so that they can squeeze out into the host cell's cytoplasm.

Anthrax toxin interrupts the signaling capability of host macrophages and causes their death. It also interrupts the signaling capability of dendritic cells but does not kill them. However, even though they are still alive, the infected dendritic cells are no longer able to participate in the host defense against the infection. The importance of the loss of macrophage and dendritic cell function will become obvious in Chapters 15 and 16.

Fast Fact

As with many other toxins, anthrax has become important in discussions of potential biological weaponry (Chapter 28).

Diphtheria Toxin

Diphtheria toxin is a cytotoxin that was discovered in the 1880s and is produced by the bacterium *Corynebacterium diphtheriae*. It works by inhibiting protein synthesis and is produced in an inactive form, just as anthrax toxin is. After secretion from the bacterium, the diphtheria toxin is changed enzymatically into an active form. This toxin is very potent, and a single molecule of it is sufficient to kill a cell.

The structure of the diphtheria toxin seems to be the same as that seen in many other exotoxins. It is composed of two polypeptide chains, α and β. The β chain is responsible for binding to the target cell. This occurs through receptor molecules on the target cell and is followed by transport of the α chain across the cell membrane. Once inside the host cell, the α chain inhibits protein synthesis. Without the β chain, the toxin is harmless because it cannot gain entry into the host cell.

Botulinum Toxin

Botulinum toxin is a neurotoxin produced by the bacterium *Clostridium botulinum*, a Gram-positive anaerobic rod. There are seven types of botulinum toxin, and all of them inhibit the release of the neurotransmitter acetylcholine. This disrupts the neurological signaling of skeletal muscles and results in paralysis. The mechanism of action is fascinating and uses specialized proteins known as *snare proteins* to block neurotransmitter release. These proteins are found on vesicle and cytoplasmic membranes in the host cell and normally facilitate the fusion of vesicles to the cell membrane so that routine exocytosis can occur. The botulinum toxin clips these snare proteins off the vesicle membranes, thereby disrupting fusion and inhibiting the exocytosis of the neurotransmitter. Because it affects muscles required for respiration, the resulting paralysis can lead to the death of the patient. Botulinum toxin has become a favorite candidate for potential use as a biological weapon (Chapter 28).

Fast Fact

What would the cosmetic industry do without Botox® (a commercial brand of botulinum toxin) injections? No more wrinkles — because the muscles that would contract to form the wrinkles are actually paralyzed! But remember that the material they are injecting to get rid of those wrinkles is derived from one of the deadliest poisons ever known.

Tetanus Toxin

Tetanus toxin is a neurotoxin produced by the bacterium *Clostridium tetani*, a Gram-positive, obligate, anaerobic rod commonly found in soil. This toxin is closely related to botulinum toxin and causes a loss of skeletal muscle control by blocking the relaxation impulse. This loss of control results in uncontrollable convulsive muscle contractions. The condition known as **lockjaw**, in which the facial muscles contract uncontrollably, is a symptom of infection with *C. tetani*.

Vibrio Toxin

5.1

Vibrio toxin, also known as cholera toxin, is an enterotoxin produced by the bacterium *V. cholerae*. This toxin also consists of two polypeptide chains, with binding of the β chain to receptors on the target cell allowing the α chain to enter the cell. Once inside, the α chain induces the epithelial cells of the digestive system to release large quantities of electrolytes. The result is severe diarrhea and vomiting that can be lethal. One symptom of **cholera** seen in patients with advanced disease is what is called *rice-water stool*, in which the fecal material is mainly liquid with bits of mucus in it. Cholera is transmitted through the fecal–oral route of contamination and is an endemic problem in many parts of the world that are socioeconomically depressed. Treatment involves proper sanitation, killing of the organisms, and large-scale infusion of electrolyte solution.

Other Exotoxins

Exotoxin effects are also seen in scarlet fever. Here, *S. pyogenes* produces a cytotoxin that destroys blood capillaries, and the result of this capillary destruction is the characteristic rash seen with this disease. If not treated, this exotoxin can cause occult heart damage that may lead to death.

The bacterium *Staphylococcus aureus* produces an enterotoxin that affects the digestive system and can cause **toxic shock**. This condition causes significant loss of liquids from the body, a loss that can lower the blood volume and blood pressure to the point at which the patient first goes into shock and then dies.

Fast Fact

Staphylococcal organisms such as *S. aureus* can also cause a form of toxic shock that has been found to be associated with the use of tampons. The exact connection remains unclear, but researchers suspect that certain types of high-absorbency tampon provide a moist, warm environment where these organisms can thrive.

Fortunately, exotoxins are very *antigenic*, which means they are substances that stimulate a host to produce antibodies. It is therefore relatively easy to generate antibodies against these toxins. In fact, the DTaP vaccination that children receive is made up of the exotoxins from *C. diphtheriae* and *C. tetani* coupled with antigenic components of *Bordetella pertussis* (the organism that causes **whooping cough**). So that they can serve as vaccines, these toxins have been chemically treated to inactivate them. The treatment causes them to lose their toxicity but not their antigenicity (the ability to induce an antibody response). After this inactivation treatment, toxins are referred to as **toxoids**.

Endotoxins

Endotoxins are bacterial toxins that are part of the cell wall of Gram-negative bacteria. They are active only after a bacterium containing them has been killed. Once the bacterial cell is dead, the endotoxins leave the cell wall and enter the bloodstream of the infected host.

Endotoxins are very different from exotoxins (**Table 5.4**). For a start, endotoxins are a component of Gram-negative organisms, whereas most exotoxins are produced by Gram-positive organisms. Endotoxins are not nearly as toxic and are part of the bacterial cell wall. We will see in Chapter 9 that Gram-negative bacteria have an outer layer made up of lipoproteins, phospholipids, and lipopolysaccharides. While the organism is alive, this outer layer stays in place around the cell, but when the organism dies, the layer falls apart (**Figure 5.17**). The lipopolysaccharides of this layer contain a particular lipid called lipid A, which has endotoxin properties.

All endotoxins cause essentially the same symptoms: chills, fever, muscle weakness, and aches. However, large amounts of endotoxin can cause more serious problems, such as shock and **disseminated intravascular clotting** (**DIC**), a condition in which minor clotting occurs throughout the body. This minor clotting uses up the clotting elements so that they are not available when needed for serious blood loss situations.

Table 5.4 A Comparison of Exotoxins and Endotoxins

Property	Exotoxins	Endotoxins
Organism	Almost all Gram-positive	Almost all Gram-negative
Location	Extracellular, excreted by living organisms	Part of pathogen cell wall, released when cell dies
Chemistry	Polypeptide	Lipopolysaccharide complex
Stability	Unstable; denatured above 60°C	Stable; can withstand 60°C for hours
Toxicity	Among the most powerful toxins known (100 to 1 million times more lethal than strychnine)	Weak, but fatal at high doses
Effects	Highly specific, several types	Nonspecific; local reactions, such as fever, aches, and possible shock
Fever production	No	Yes, rapid rise to very high fever
Usefulness as antigen	Very good, long-lasting immunity conferred	Weak, no immunity conferred
Conversion to toxoid form	Yes, by chemical treatment	No
Lethal dose	Small	Large
Typical infections caused	Botulism, gas gangrene, tetanus, diphtheria, cholera, plague, scarlet fever, staphylococcal food poisoning	Salmonellosis, typhoid fever, tularemia, meningococcal meningitis, endotoxic shock

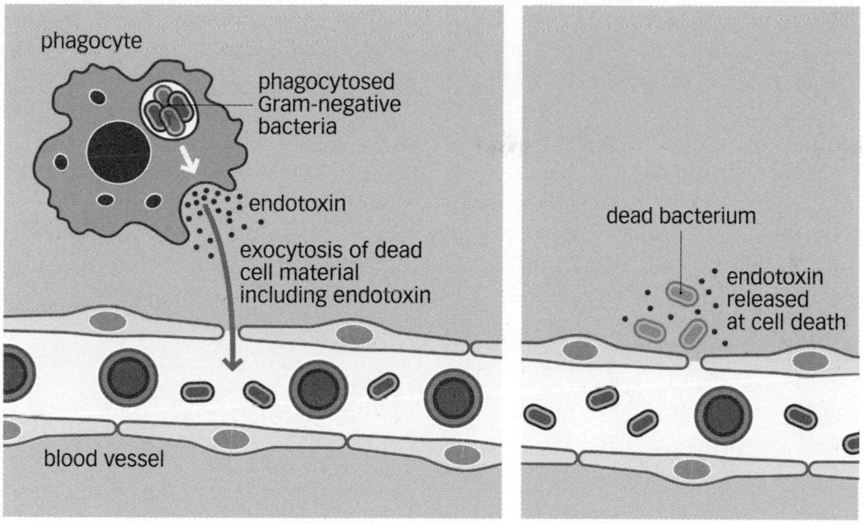

Figure 5.17 A diagrammatic representation of endotoxin release. Endotoxins are components of the Gram-negative cell wall, and on the death of the cell they are released to travel through the blood.

Endotoxins can sometimes elicit antibody production in the host, but in general this immune response is extremely weak because endotoxins are not very antigenic. As a result, the antibody response to them is usually poor. One of the problems with endotoxins is that because they are the result of cell death, they can contaminate materials and equipment used in clinics and hospitals, and remain toxic for long periods. This potential for long-term contamination makes the chance of endotoxins being transferred to patients a problem. Therefore, a test to determine whether there is endotoxin contamination of supplies or equipment was developed. This test is referred to as the ***Limulus* amebocyte lysate assay (LAL)** and takes advantage of the white blood cells of the horseshoe crab (*Limulus*), which are very different from human white blood cells. In the presence of endotoxin, *Limulus* white blood cells clot into a gel-like matrix that becomes turbid (cloudy). The degree of turbidity can be used as a measure of the amount of endotoxin contamination present.

Keep in Mind

- Damage to the host can be either direct or indirect. Direct damage is usually localized, and indirect damage is usually through the production of toxins.
- Exotoxins are extremely lethal substances produced by living cells (usually Gram-positive) and in most cases are proteins.
- Exotoxins can be cytotoxins (which kill cells), neurotoxins (which interfere with neurological signaling), or enterotoxins (which affect the lining of the digestive system).
- Exotoxins can cause the production of antibody.
- Endotoxins are contained inside the bacterial cell wall and are released on the death of the organism.
- Endotoxins are products of Gram-negative organisms, do not effectively cause the generation of antibody, and are less toxic than exotoxins.
- Lipid A, which is part of the Gram-negative phospholipid outer layer of the bacterial cell wall, has endotoxin properties.

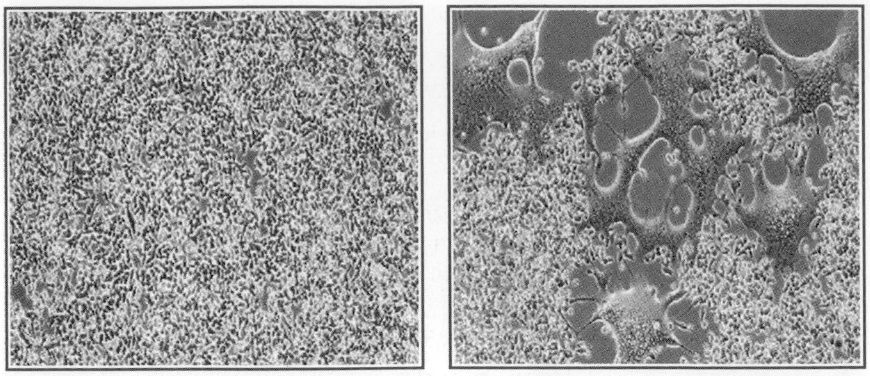

Figure 5.18 Cytopathic effects of viral infection caused by viral overload. The left panel shows uninfected cells, which grow to confluence (meaning that they completely cover the surface). The right panel shows the same cells after viral infection. The host cells are destroyed as the viral particles burst out and enter the surrounding cells, causing their destruction.

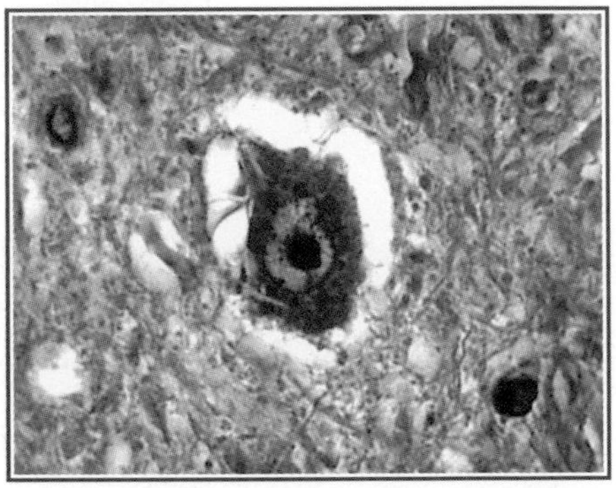

Figure 5.19 The cytopathology of the rabies virus, which produces large Negri bodies in the cell. These inclusion bodies contain newly formed viral particles.

Viral Pathogenic Effects

The pathogenic effects caused by viruses are discussed in detail in Chapters 12 and 13. For now, we can say that we classify virally caused cell damage or death as a **cytopathic effect** (**CPE**) and that the cytopathology associated with viral infections can occur in three ways.

The most obvious way is *viral overload*, a condition that causes the virus to explode out of the host cell and infect and lyse surrounding host cells (**Figure 5.18**). A second type of CPE occurs when host defense mechanisms identify and kill virally infected cells. This is categorized as a **cytocidal effect**. The third type of viral cytopathology occurs when a virus shuts down the host DNA, RNA, and protein synthesis and thereby forces the host cell to devote all its efforts to virus production. This is classified as a **non-cytocidal effect**.

Viral cytopathology can be identified microscopically. In some cases, **inclusion bodies** filled with virus become visible inside the cell. For example, **Negri bodies** are inclusion bodies seen in rabies viral infections (**Figure 5.19**) and can be diagnostic for the disease. Some viral infections cause breaks in the host chromosomes, and these breaks are easily identifiable. The formation and appearance of *syncytia* (gigantic cells formed as a number of infected host cells merge; this is discussed in Chapter 13) can also be a visual indication of viral infection (**Figure 5.20**). More subtle pathogenic effects, such as the production of hormones or interferon, are also signs of virally infected cells.

The pathogenic effects seen in diseases caused by fungi and parasites will be discussed in Chapter 14.

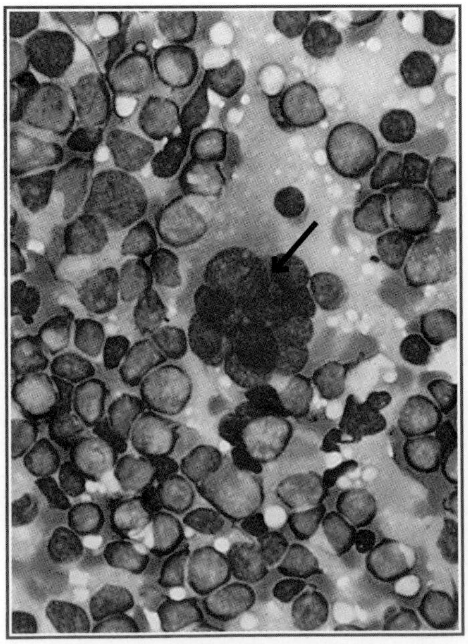

SUMMARY

- The requirements for infection include entry, establishment, avoiding host defenses, damaging the host, and exiting from the host.
- Portals of entry for pathogens include skin, parenteral routes, and the mucous membranes of the respiratory, gastrointestinal, and genitourinary tracts.
- Pathogens use virulence factors such as adhesins to establish themselves in the host.
- To avoid being killed by the host's defenses, pathogens use an array of virulence factors, including capsules, M proteins, mycolic acids, leukocidins,

Figure 5.20 Formation of syncytia (arrow) in cells infected with virus. Formation of these structures is one type of cytopathic effect seen in viral infections.

hemolysin, coagulase, kinases, hyaluronidase, and collagenase.

- Some bacterial pathogens cause damage to host cells by releasing exotoxins; these include specific types such as cytotoxins, neurotoxins, and enterotoxins.
- Endotoxins on the cell wall of Gram-negative bacteria cause damage to the host and are released and disseminated when the bacterial cell dies.

Overall, the basis for our discussion here was the ability of pathogens to invade a host, remain in that host, and defend against host defenses so as to cause disease. As you organize and reflect on the concepts in this chapter, keep in mind what you have learned in the previous chapters. We explore the structures of the bacterial cell in Chapter 9, and there you will be able to reaffirm the connection with the disease processes you have learned here and those structural components.

SELF EVALUATION AND CHAPTER CONFIDENCE

Multiple Choice

Answers are given in the back of the book and help can be found in the student resources at:
www.garlandscience.com/micro

1. You nicked yourself while shaving and it has become infected. Which of the following portals of entry did the pathogen most probably use?
   - **A.** Gastrointestinal tract
   - **B.** Skin
   - **C.** Respiratory tract
   - **D.** Genitourinary tract
   - **E.** Exotoxin tract

2. You overhear that the microbe causing an infection in your patient got into the body via the parenteral route. What does this actually mean?
   - **A.** It entered through the nervous system.
   - **B.** It entered and exited through the digestive system.
   - **C.** It entered through a break in the skin.
   - **D.** It entered by using cytotoxins.
   - **E.** It was inhaled through the respiratory system.

3. A pathogen has entered the body. All of the following will have a role in its establishment except
   - **A.** Using fimbriae to attach to cell receptors
   - **B.** Releasing several exotoxins to destroy host cells
   - **C.** Using adhesins to attach to tissues
   - **D.** Creating a biofilm on a body surface
   - **E.** Releasing endotoxin that will cause clotting

4. The LD_{50} of a pathogen is the number of organisms required to
   - **A.** Benefit 50% of the hosts
   - **B.** Infect 50% of the hosts
   - **C.** Kill 50% of the hosts
   - **D.** Produce lytic deaminase toxin
   - **E.** None of the above

5. Imagine you are working in a lab and read reports about two different bacteria. Organism A has an ID_{50} of 20 cells, whereas organism B has an ID_{50} of 100 cells. Which of the following conclusions would you make?
   - **A.** Organism A must have endotoxin.
   - **B.** Organism A could be considered more virulent than organism B.
   - **C.** Organism B could be considered more virulent than organism A.
   - **D.** Organism B has a portal of entry but organism A does not.

6. A bacterial toxin that causes damage to the plasma membrane of host red blood cells which results in lysis is
   - **A.** Coagulase
   - **B.** Leukocidin
   - **C.** Hyaluronidase
   - **D.** Collagenase
   - **E.** Hemolysin

7. A bacterial enzyme that breaks down connective tissue is
   - **A.** Coagulase
   - **B.** Hemolysin
   - **C.** Hyaluronidase
   - **D.** All of the above
   - **E.** None of the above

8. An invasin would be used by a microbe to
   - **A.** Change the nuclear structure of host cells
   - **B.** Inhibit the functions of host ribosomes
   - **C.** Increase the rate of bacterial division
   - **D.** Change the structure of actin filaments in host cells
   - **E.** Change the shape of microtubules in host cells

9. You are working in a clinical lab and have a culture of cells that are producing exotoxins. On the basis of this information, the organisms are most probably
   - **A.** Gram-negative bacteria
   - **B.** Viruses
   - **C.** Gram-positive bacteria
   - **D.** All dead
   - **E.** None of the above

10. Three types of exotoxin are
    - **A.** Febrile toxins, enterotoxins, and neurotoxins
    - **B.** Neurotoxins, cytotoxins, and febrile toxins
    - **C.** Cytotoxins, diarrhea toxins, and neurotoxins
    - **D.** Neurotoxins, cytotoxins, and enterotoxins
    - **E.** None of the above combinations

11. Botulism toxin is
    - **A.** A neurotoxin
    - **B.** A muscular toxin
    - **C.** A febrile toxin
    - **D.** An enterotoxin
    - **E.** None of the above

12. Many people refer to tetanus infection with the pathogen *Clostridium tetani* as lockjaw. Which of the following best explains why?
    - **A.** These bacteria produce a capsule that hardens the surface of the jaw.
    - **B.** These bacteria produce a toxin that causes jaw muscles to remain contracted.
    - **C.** These bacteria accumulate in the jaw and prevent proper jaw movement.
    - **D.** These bacteria produce an enzyme that breaks apart jaw muscle cells.

13. Which of the following would best describe endotoxins?
    - **A.** They are toxins found on Gram-negative cell walls.
    - **B.** They are toxins that are released by Gram-positive cells.
    - **C.** They are toxins that are produced by dead Gram-positive and Gram-negative cells.
    - **D.** They are toxins found on Gram-positive cell walls.
    - **E.** They are identical in structure to exotoxins.

14. You are caring for a new patient in the intensive care unit and see in his chart that he has disseminated intravascular clotting (DIC). On the basis of the information you learned in this chapter, the patient's disease is caused by
    - **A.** Exotoxins
    - **B.** Enterotoxins
    - **C.** Endotoxins
    - **D.** Vascular toxins
    - **E.** Neurotoxins

Q DEPTH OF UNDERSTANDING

Questions listed here require you to bring together the concepts you have learned in this chapter into a discussion format. This helps you to increase your depth of understanding of the material you have learned. Help can be found in the student resources at: www.garlandscience.com/micro

1. Compare the portals of entry used by bacteria and give examples of organisms that use each portal.

2. Discuss the mechanisms that bacteria use to avoid or defeat the host defenses.

3. Describe the properties of exotoxins and compare them with endotoxins.

Q CLINICAL CORNER

Help can be found in the student resources at: www.garlandscience.com/micro

1. *Neisseria meningitidis*, commonly known as meningococcus, is a Gram-negative bacterium that not only can cause meningitis but can also enter the bloodstream and cause a deadly infection of the blood. If the bacteria do invade the bloodstream, the death rate can be very high. Luckily, if it is detected early, doctors can administer antibiotics that kill the bacteria. However, once antibiotics have been administered and the bacteria are killed, the patient can often get much sicker before recovering.
   - **A.** Can you identify the virulence factor that might be responsible for this "turn for the worse" after antibiotic killing of these Gram-negative bacteria?
   - **B.** Imagine you must explain to the patient's concerned family what is happening inside the patient's body after antibiotic treatment. What would you say, in simple terms?

2. Imagine a new drug has been designed that targets and destroys the M protein of a deadly strain of *Streptococcus*. It is experimental and your very ill patient might be able to benefit from this new treatment. The doctor has presented the idea to the patient's family but needs a signed consent. The family is confused, however, about how this treatment will actually work. They've never heard of something called M protein. They turn to you to see whether you know how this might help. How would you explain this to them in simple terms they can understand? (Be sure to explain to them what you know about M protein, what it does, and what the benefits of destroying it would be.)

Transmission of Infection, the Compromised Host, and Epidemiology

Chapter 6

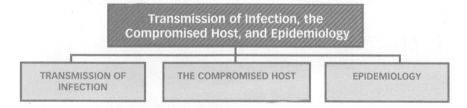

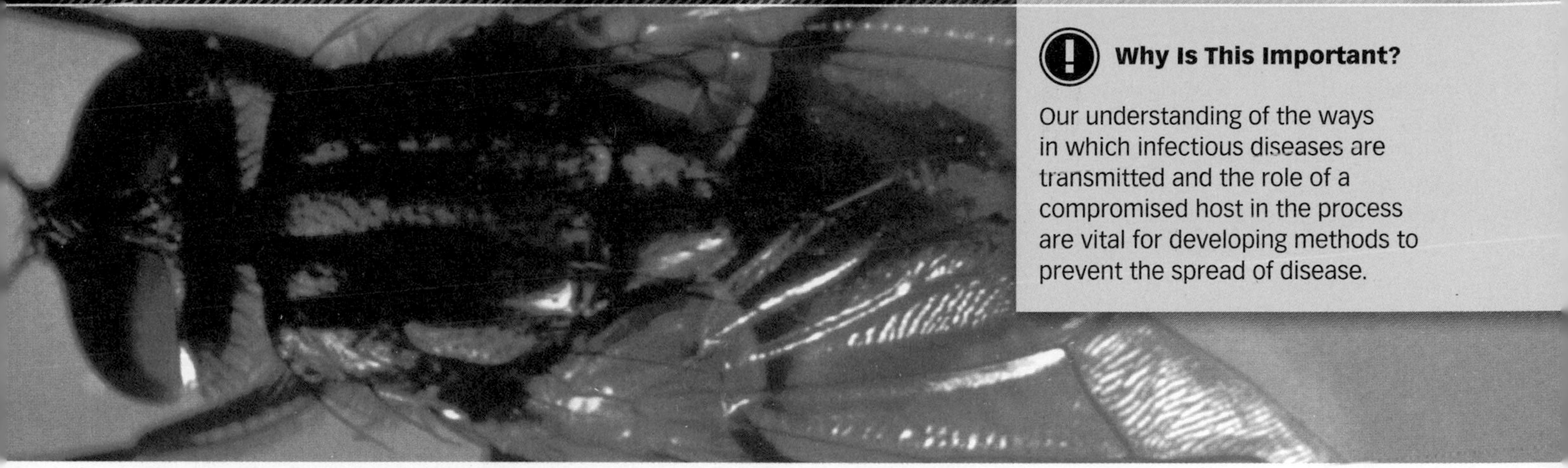

Why Is This Important?

Our understanding of the ways in which infectious diseases are transmitted and the role of a compromised host in the process are vital for developing methods to prevent the spread of disease.

OVERVIEW

In the previous chapter we discussed the principles involved in the infectious disease process. In this chapter we focus on the transmission of diseases by looking at where the pathogens responsible for infection are found and at the mechanisms by which those pathogens are transmitted from the environment to a host, and from one host to another. We also consider the compromised host, which is a very important and integral part of the infection process. The more compromised the host, the greater is the risk of successful infection. Last, we review some of the basic principles of epidemiology, the discipline that helps us to understand how infectious diseases are spread and thus to develop methods for prohibiting their spread.

We will divide our discussions into the following topics:

Transmission of Infection, the Compromised Host, and Epidemiology

- TRANSMISSION OF INFECTION
- THE COMPROMISED HOST
- EPIDEMIOLOGY

TRANSMISSION OF INFECTION

The transmission, or spread, of an infection is the final requirement necessary for a successful infection. We can best study this process by looking at two factors: **reservoirs** of infectious organisms (which are places where pathogens can grow and accumulate) and mechanisms of transmission (which are the various ways in which organisms move from place to place). We begin with reservoirs.

Reservoirs of Pathogens

There are three potential reservoirs where pathogens can accumulate: humans, other animals, and nonliving reservoirs. With human reservoirs, it is obvious that a sick person is a reservoir for the pathogens causing the infection and will continue to be a reservoir for as long as the infection continues. However, determining that a given individual is acting as a reservoir can be difficult because some infectious diseases are transmitted before the symptoms are manifested. Take measles as an example. In

What Do I Need to Know?

To get the most out of this chapter, please review the following terms from your previous courses in biology, anatomy, physiology, or chemistry or in previous chapters of this book as indicated in parentheses: fomite, neutropenia, zoonotic, morbidity, endemic, pandemic (**1**).

Fast Fact

Next time you are cleaning the litter box for Fluffy the cat, remember that the waste material you are dealing with can indirectly transmit pathogens such as *Toxoplasma gondii* from Fluffy to you! By the way, for those of you who carry the "pooper scooper," Fido's fecal material can also transmit pathogens indirectly.

a classroom full of second-grade students, by the time one student shows the symptoms — aches, fever, red spots — there is a good possibility that the rest of the class and the teacher have already been exposed.

Human reservoirs are also found in infected people who are carriers. These people can carry such diseases as AIDS, diphtheria, and typhoid fever but never show any signs of disease. However, they can readily infect others.

Our second type of reservoir of pathogens involves animals other than humans. Diseases that are transferred from animals to humans are called **zoonotic diseases**. Two well-known zoonotic diseases are **rabies**, in which the virus is transferred from a rabid animal to a human, and **Lyme disease**. Others are shown in **Table 6.1**. Zoonotic diseases usually occur after direct contact with the animal, but there are also ways in which indirect contact can cause infection. For instance, the waste material of a litter box, fur, feathers, infected meats, or vectors can indirectly transmit pathogens from an animal to a person.

Our third reservoir, nonliving, includes water, food, and soil. Of these three, water may be the most dangerous. In many underdeveloped countries there is poor sanitation, and in some cases little personal hygiene, which result in the fecal contamination of water. Because water is something that we cannot live without, it becomes the key component in the fecal–oral route of contamination. Many diseases are spread from this reservoir — typhoid fever and cholera, for instance — and these diseases are endemic in many parts of the world. In the case of food acting as a pathogen reservoir, contamination is part of the spoilage that occurs naturally. (Note that refrigeration only slows pathogen growth; it does not stop it.)

As far as soil is concerned, it is a normal habitat for many organisms that are potentially pathogenic, but these organisms must find a way to pass the physical barriers that protect the body if they are to cause disease. A case in point would be *Clostridium tetani*, a Gram-positive rod-shaped bacterium found in soil that can be transferred to humans through breaks in the skin, as, for example, in the "stepping on a rusty nail" scenario. In

Disease	Causative Agent	Mode of Transmission
Anthrax	*Bacillus anthracis*	Direct contact with infected animals; inhalation
Bubonic plague	*Yersinia pestis*	Flea bite
Lyme disease	*Borrelia burgdorferi*	Tick bite
Typhus	*Rickettsia prowazekii*	Louse bite
Rocky Mountain spotted fever	*Rickettsia rickettsii*	Tick bite
Relapsing fever	*Borrelia* species	Tick bite
Malaria	*Plasmodium* species	*Anopheles* mosquito bite
African sleeping sickness	*Trypanosoma brucei gambiense*	Tsetse fly bite
American sleeping sickness (Chagas' disease)	*Trypanosoma cruzi*	*Triatoma* species (kissing bug)
Rabies	Rabies virus	Bite of infected animal
Hantavirus pulmonary syndrome	Hantavirus	Inhalation of virus in dried feces and urine
Yellow fever	Flavivirus	*Aedes* mosquito bite

Table 6.1 Examples of Zoonotic Diseases

fact, the rust has nothing to do with it. It is the contamination of the nail with *C. tetani* that causes the infection. Recall from our discussions in Chapter 5 that this organism produces a neurotoxin that can be lethal.

Keep in Mind

- Transmission is the final requirement for a successful infection.
- Reservoirs are places where pathogens grow and accumulate.
- There are three potential reservoirs of infection; humans, animals, and nonliving reservoirs.
- Human reservoirs can be infected people who are carriers.
- Animal-to-human infections are referred to as zoonotic disease.
- Nonliving reservoirs are water, soil, and food.

Mechanisms of Transmission

There are three mechanisms that can be used to transfer infectious organisms: contact transmission, vehicle transmission, and vector transmission.

Contact transmission

In the mechanism known as **contact transmission**, a healthy person is exposed to pathogens by either touching or being close to an infected person or object. We can subdivide the contact transmission mechanism into three types: direct, indirect, and droplet. In *direct contact transmission*, there is no intermediary between the infected person and the uninfected person (**Figure 6.1**). This mechanism encompasses such things as touching, kissing, and sexual interactions. The diseases transmitted through direct contact include hepatitis A, smallpox, staphylococcal infections, **mononucleosis**, and of course sexually transmitted diseases.

Figure 6.1 shows that the direct-contact mechanism can also involve zoonotic disease. In that case, the infected organism is not a person but an animal.

Indirect contact transmission occurs through intermediates that are usually nonliving articles, such as tissues, handkerchiefs, towels, bedding, and contaminated needles, the latter easily transferring HIV and hepatitis B.

We refer to the nonliving intermediates that act as the agents of transmission by indirect contact as **fomites**.

Fast Fact

Contaminated needles have become a major factor in the spread of HIV among intravenous drug abusers, and many cities have instituted needle-exchange programs, where authorities swap clean needles for used ones to try to slow this type of transmission.

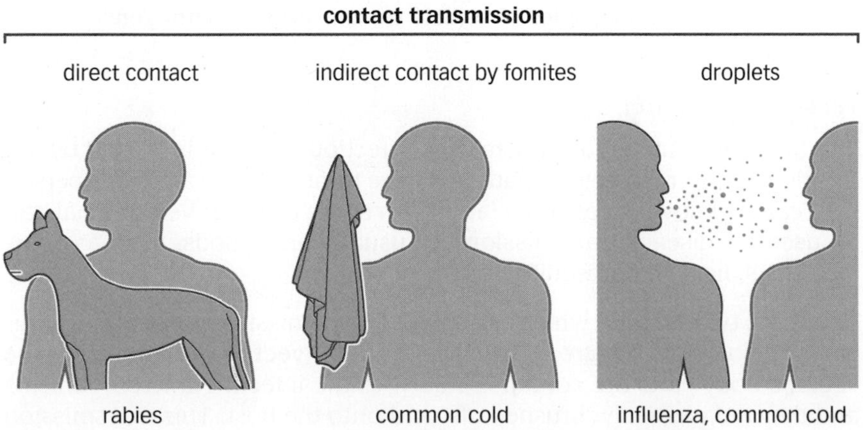

Figure 6.1 Examples of contact transmission of pathogenic organisms.

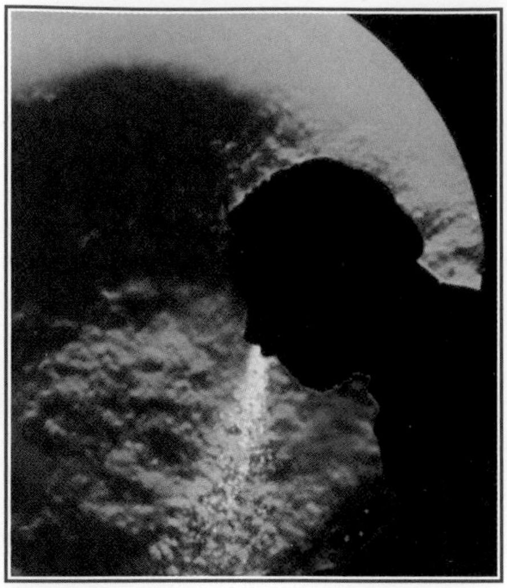

Figure 6.2 The sneeze, an example of droplet transmission.

Droplet contact transmission is seen in the transfer of respiratory diseases such as influenza and whooping cough. Droplet transmission can occur through sneezing (**Figure 6.2**), coughing, and even laughing. Although it is confined to short distances, the size of the droplet is important. Large droplets will fall to the ground quickly, but smaller droplets can stay airborne for long periods. The smaller the droplet, the more dangerous it is as an agent of disease.

Vehicle transmission

As the name implies, **vehicle transmission** involves pathogens "riding" along on supposedly "clean" components (**Figure 6.3**). For example, food can carry pathogenic organisms and is therefore considered a vehicle of disease transmission. Other possible vehicles are water, air, blood, body fluids, drugs, and intravenous fluids given to a hospitalized patient. The transmissibility is obvious for each of these vehicles, but let's look more closely at water, food, and air.

Food can transmit many infections that usually manifest themselves as food poisoning. Food contamination is usually the result of poor preparation or poor refrigeration, both of which are easily corrected. We have already mentioned water as a pathogen reservoir, and now we see it can also act as a vehicle. Obviously, proper sanitation can limit water's role as a vehicle of infection.

Perhaps the most difficult transmission vehicle to deal with is the air, because dust (which can contain huge numbers of microbes), microbial spores, and fungal spores can all use the air to travel from host to host. In fact, biological warfare strategies require ways to transform deadly pathogens into an aerosol form so as to achieve maximal transmission effectiveness.

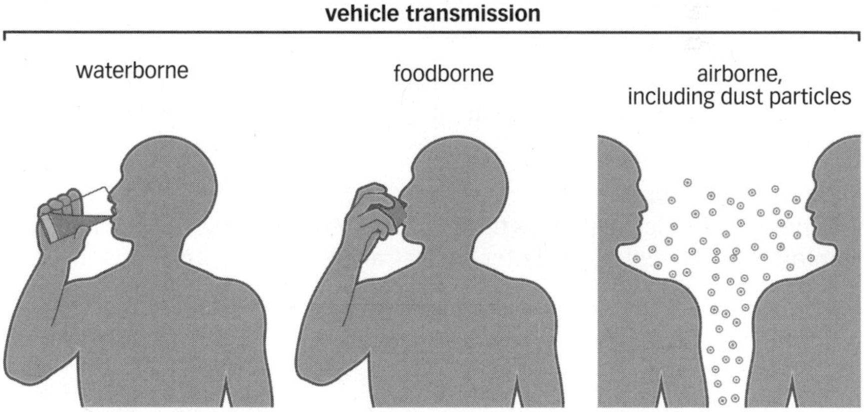

Figure 6.3 Examples of vehicle transmission of pathogens.

Vector transmission

The third mechanism for transmitting infectious disease is **vector transmission**. In this mechanism, pathogens are transmitted to a healthy person by a carrier known to be associated with some disease. Vectors that can be used for disease transmission are usually arthropods, such as fleas, ticks, flies, lice, and mosquitoes.

There are two ways in which this type of transmission occurs: mechanical and biological (**Figure 6.4**). In **mechanical vector transmission**, the vector's body parts are contaminated with the infecting microorganisms and they are passively "brushed off" and onto the host. This transmission

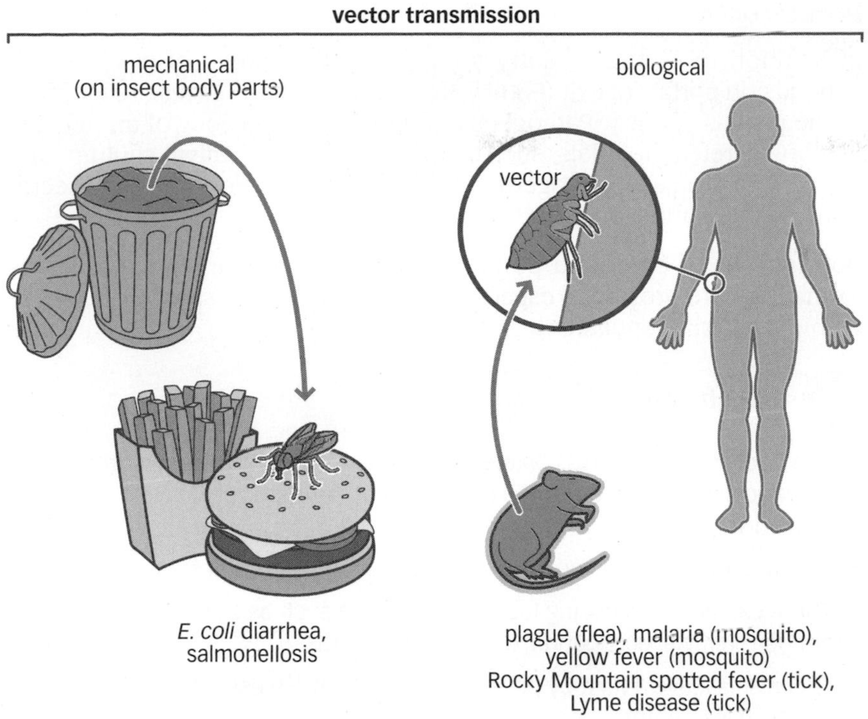

Figure 6.4 An illustration of vector transmission, which can be either mechanical (involving pathogens found on the body of insects such as houseflies) or biological (through the bite of an insect such as the flea or mosquito).

mode is seen with houseflies, which are known to frequent fecal material and can get pathogens from this source onto their feet, wings, or other body parts (**Figure 6.5**). *Biological vector transmission* occurs through a bite, as seen with, for instance, the fleas that had a primary role in the transmission of *Yersinia pestis* (the plague of the Middle Ages); ticks, which can transmit *Borrelia burgdorferi* (the causative agent of Lyme disease); and mosquitoes, which carry such pathogens as *Plasmodium* (the malaria organism) and West Nile virus.

Fast Fact

Some pathogens multiply in their vector and can arrive at the new host in larger numbers, which helps in establishing the infection.

Factors Affecting Disease Transmission

In addition to the overall health of the host, there are several other factors that can influence the disease process, including age, gender, lifestyle, occupation, emotional state, and climate.

Age affects the overall health of the host, in particular the decline in defensive capability, which translates to increases in disease as we age. Gender "bias" can be seen in certain infections; for instance, urinary tract infections are seen more in females than in males, and respiratory infections are seen more in males than in females. Obviously, lifestyle and occupation can lead to a predisposition to infection. However, things such as emotional state, although less recognizable, may also have a role in susceptibility to infection. Climate can have a role in infection in three ways: (1) there seems to be a greater incidence in respiratory infections associated with colder climates, (2) warmer climates allow longer periods when vectors are present, and (3) climate can effect changes in the variety of infectious organisms that may be present in a given location.

Figure 6.5 A common housefly can transmit pathogens through mechanical vector transmission.

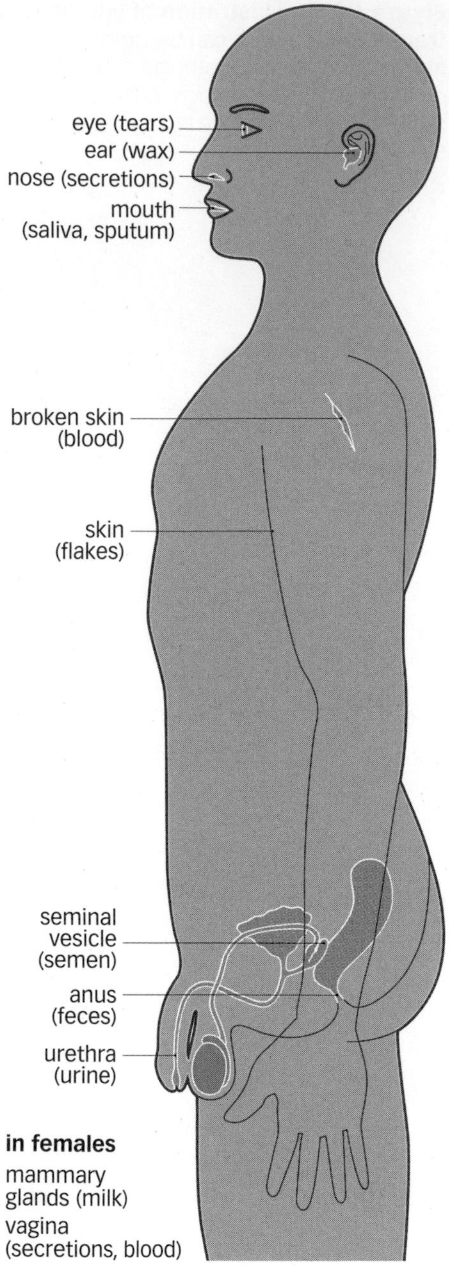

Figure 6.6 Portals of exit.

Portals of Exit

In addition to portals of entry (Chapter 5), the spread of disease also depends on portals of exit (**Figure 6.6**), which in many cases are identical to the portals of entry. Pathogens often exit from the body of an infected host in secreted materials, such as nasal secretions, saliva, sputum, and respiratory droplets. Pathogens can also exit in the blood, vaginal secretions, semen, urine, and feces.

Both portals of entry and portals of exit are important considerations for health care providers, especially in controlling the spread of disease within a patient population.

Keep in Mind

- Mechanisms of transmission are ways in which organisms move from place to place. There are three mechanisms: contact, vehicle, and vector transmission.
- Contact transmission can be direct, indirect, or by droplet transmission.
- There can be predisposing factors for disease such as the host's age, gender, lifestyle, occupation, or emotional status.
- Organisms leave a host by portals of exit, which are essentially the same as the portals of entry.

THE COMPROMISED HOST

In this section, we discuss one of the most important parts of infectious disease: the host. As we have repeatedly said, the spread of infectious diseases depends to no small extent on the hosts involved. The fact that we humans live in a cloud of potentially dangerous infectious organisms yet rarely get seriously ill illustrates that our defensive systems (Chapters 15 and 16) are very good at controlling the number and severity of infections we experience.

For most of this section, we consider the ability of a host to mount a defensive response against invading pathogens as a measure of that host's competence. As noted in Chapter 1, host defenses can be divided into innate and adaptive immune responses. These two powerful defensive systems can deal with most of the pathogens we come into contact with. As long as these mechanisms are in place and working adequately, infectious disease is in most cases only an annoyance. If these defenses are in some way compromised, however, the potential for damaging infectious disease increases. The relationship between host defense and infectious disease is dramatically illustrated by acquired immunodeficiency syndrome (AIDS), in which the loss of host defense leads to infections that were previously of no consequence but now cause the death of the host. (We discuss HIV in detail in Chapter 17.)

The loss of immunity in AIDS patients is obvious, but there are other situations in which the host's defense mechanisms can be compromised. There are specific kinds of compromise associated with particular risks for infectious disease. In addition to physiological compromise, meaning an actual impairment of the immune system, there are other risk factors for infection, including lifestyle, occupation, trauma, and travel. Aging is also a major factor in susceptibility to infection. For example, if you were looking at a 75-year-old previously healthy man who suffers a stroke, you normally would not have to think about infection in the central nervous system. However, if the patient is a 30-year-old with a depressed immune system and manifests the symptoms of a stroke, it might be important to consider an infectious disease as the potential cause.

We can list several groups of people who can be considered as having a compromised immune system and therefore being at increased risk of infection:

- People with AIDS
- People with genetic immunodeficiency diseases
- People undergoing chemotherapy
- Patients taking broad-spectrum antibiotics
- Transplant patients
- Burn patients
- Premature infants
- Newborn infants
- Health care workers
- The elderly
- Surgical patients
- Patients on artificial ventilators

We discuss immunodeficiency in detail in Chapter 17, covering the first two groups on this list. Here, we look at some of the other situations listed, keeping in mind that the defensive competence of the host has a major role in the extent and severity of many infectious diseases.

Neutropenia

The condition known as **neutropenia** is defined as a lower-than-normal number of neutrophils in the blood. As we will see, this type of white blood cell is a very important component of a host's innate immune response. The most common cause of profound neutropenia is the administration of **cytotoxic chemotherapy** for the treatment of the malignant tumors seen in cancer. The drugs used in this type of chemotherapy destroy rapidly growing cells such as those associated with the tumors, but as a side effect they render the patient temporarily immunoincompetent. Consequently, the chance of infection during the administration of these drugs is very high.

The infections associated with neutropenia are primarily bacterial and fungal. Bacterial infections can begin as soon as the neutrophil level decreases, but fungal infections occur only in people who have been neutropenic for long periods. Although neutropenia is a transitory compromise of host defense and disappears when chemotherapy ends, patients must be monitored carefully for the onset of infection. Neutropenic patients also risk infection with any invasive surgical procedure and whenever an indwelling catheter is necessary.

Gram-negative bacteria from either the patient's colon or the environment can cause life-threatening septicemias in neutropenic patients, and some of these infections can be especially difficult to deal with because of resistance to antibiotics. Gram-positive pathogens, such as staphylococci and streptococcal species (which are normally found on the skin and in the nasal cavity), may cause infections originating in either the bloodstream or central-venous catheters.

Infections caused by the fungus *Candida albicans*, which is commonly present on the mucosa of the human gastrointestinal tract and in the vagina, can cause **mucocutaneous candidiasis** in patients with neutropenia. The spores of *Aspergillus fumigatus*, a fungus found in many places and carried in the air, can cause **fungal pneumonia** in neutropenic patients.

Organ Transplantation

Unless a patient needing an organ transplant has an identical twin, there will always be immunological differences between the patient and a transplanted organ. These differences are immediately recognized by the recipient's immune system, and, depending on how closely the donor and recipient were matched, cause a reaction intended to destroy the transplanted organ. Consequently, transplant patients are placed on drug regimens designed to lessen the immune response against the organ to prevent rejection. These drugs reduce the chances of rejection by diminishing the patients' *overall* immune capability. However, this causes the patient to be more susceptible to infection.

Infections in organ-transplant patients require the administration of broad-spectrum antibiotics, and as we will see in Part IV the overuse of these antibiotics leads to superinfections and, more importantly, to increased bacterial resistance to antibiotics. This cycle of overuse of antibiotics and increased resistance is seen in many clinical situations and remains one of the most serious concerns of modern health care.

Burn Patients

These patients are at risk because of the loss of large areas of the primary physical barrier to infection, the skin. When skin is lost, there is a greater chance of **septicemia**. Infections with *Pseudomonas aeruginosa* are a particular problem in burn patients because this organism is very resistant to methods used to control bacterial growth, such as antiseptics and disinfectants, and is becoming increasingly resistant to antibiotic treatment. Burn patients are also very susceptible to fungal infections.

The categories just mentioned each refer to one group of patients, namely chemotherapy patients, organ-transplant patients, and burn victims. Now let us look at two types of infection, opportunistic and nosocomial, that can attack multiple groups of susceptible people.

Opportunistic Infections

Many infections are caused by microorganisms that are opportunists, and any infection acquired from such microorganisms is referred to as an **opportunistic infection**. To understand this phenomenon, we must first consider the normal bacterial flora that is part of us. As previously mentioned, there are many bacteria that are residents of the body and provide essential services to the host. For example, the large intestine is filled with organisms that are supposed to be there. These normal bacterial floras found in different areas of the body also prevent other organisms from inhabiting those areas. In some cases, the presence of bacteria in various regions of the body is associated with the production of *bacteriocins*, which are chemicals similar to antibiotics that inhibit the growth of any organisms other than the bacteria that produced them. Loss of these resident bacteria allows the entry and growth of pathogens, which then cause opportunistic infections.

It is interesting to note that many of the resident "harmless" organisms in the body also have the potential to be pathogenic given the right circumstances. This switch in character occurs when these bacteria move to a place in the body where they are normally not found. The best example of this is urinary tract infections. One of the primary causes of this type of infection is *E. coli* traveling from the large intestine, where it is part of the normal flora, to the urethra, where it is pathogenic.

The ability of pathogens to become established, or of resident bacteria to move to unfamiliar places and infect opportunistically, can be compounded by the improper use of antibiotics because such use destroys normal populations of resident bacteria, opening the way for both superinfections and opportunistic infections. In addition, improper antibiotic use fosters the development of resistant strains of bacteria. Similarly, improper use of drug therapy can result in the emergence of resistant bacteria in the hospital environment, paving the way for nosocomial infections.

Nosocomial Infections

As noted in Chapter 1, a **nosocomial infection** is any infection acquired in a hospital or other medical facility. Because they occur during medical treatment, nosocomial infections can affect not only patients but also medical workers. In the United States, the Center for Disease Control has estimated that as many as 2 million infections a year are acquired in hospitals, resulting in 90,000 deaths. The costs associated with these infections are estimated to be over $5 billion per year.

Any patient with a break in the skin (even a **bed sore**) is susceptible to a nosocomial infection, although these infections are usually associated with intravenous applications, urinary and other catheters, invasive tests, and surgical procedures. In fact, anything that allows the entry of organisms into the body is a potential source of nosocomial infections.

The same factors used when considering any other type of infection apply to hospital-borne infections. Therefore, hospitals must consider:

- The source of the infection
- The mode of transmission of the pathogen
- The susceptibility of the patient to infection
- Prevention and control

The most common sites of nosocomial infection are shown in **Figure 6.7**. Sources of the infection are usually the environment, other patients, hospital staff, or visitors. In addition, insects such as ants, roaches, and flies can cause the problem, and fomites such as trash and unsanitary toilet facilities can also be involved. Last, water supplies, respiratory equipment, plastic supplies, and catheters can all act as sources of nosocomial infections.

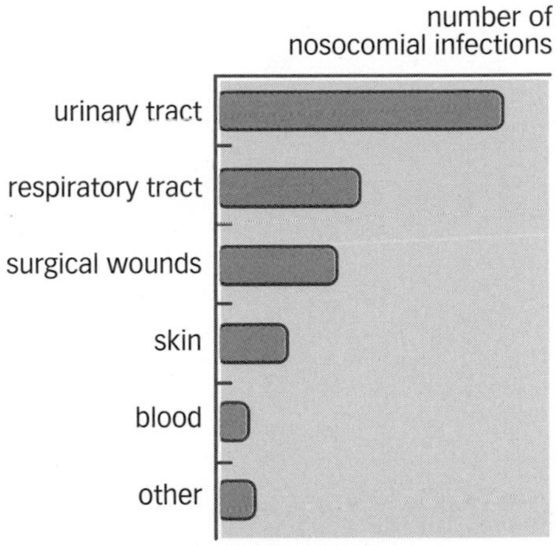

Figure 6.7 Relative frequencies of sites of nosocomial infections.

Fast Fact

In some cases, even soaps and cleaning solutions can become a problem if pathogens such as *Pseudomonas* have become resistant to the antibacterial compounds contained in the soap or other cleaning products.

Opportunistic organisms can have a role in nosocomial infections, and once again we must consider the patient's degree of immune competence. Any disease or medical procedure requiring convalescence in a hospital will have an effect on the strength and defensive competence of a patient. This combination of a compromised host and an opportunistic pathogen increases the possibility of a nosocomial infection, and the potential for serious infection increases.

Most nosocomial infections are caused by *Escherichia coli*, *Enterococcus* species, *Staphylococcus aureus*, or *Pseudomonas aeruginosa* (**Figure 6.8**). These organisms are ubiquitous in hospitals and can easily be moved from place to place and patient to patient by staff, visitors, and other patients. More importantly, many of these organisms are now resistant to antibiotics. For example, **MRSA** (methicillin-resistant *S. aureus*) and **VRSA** (vancomycin-resistant *S. aureus*) have become major medical problems because there are few alternative antibiotic treatments for these infections, which can be deadly in compromised hosts.

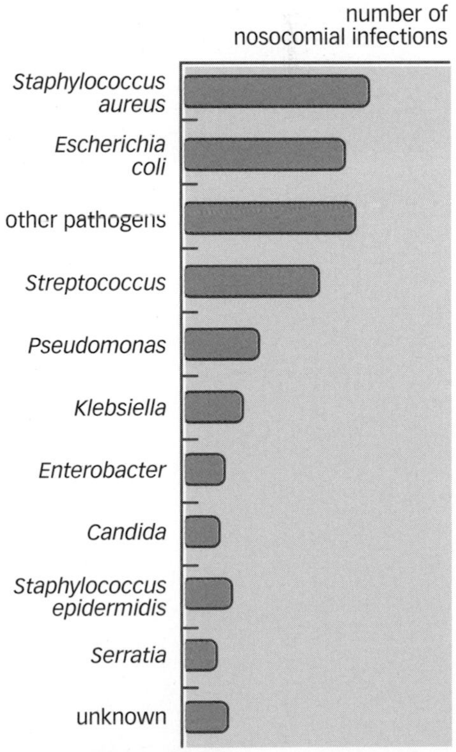

Figure 6.8 Common causative agents of nosocomial infections.

Of the equipment and procedures that contribute to nosocomial infections, catheters and respiratory equipment are the most likely to be involved. For example, nebulizers that administer oxygen gas and expand the passageways in the lungs can also distribute pathogens deep into the lung, where infection is much more serious. In addition, pathogens can grow in humidifiers and be transmitted to patients in aerosol form. All respiratory equipment must therefore be either disinfected or sterilized daily when in use and also between uses.

Infections can also easily result from contaminated catheters and tubing used for medical procedures, from loose connections in catheters or tubing, and from inadequate preparation of injection or surgical sites. A major concern is patients undergoing dialysis as there are many potential places for the initiation of a nosocomial infection. This procedure is essential for many people with renal disease, and dialysis clinics routinely reuse equipment for multiple patients daily.

Universal Precautions

In 1988, the Centers for Disease Control (CDC), concerned about the spread of HIV in hospitals, published a set of universal procedures requiring all medical facilities in the United States to conform to specific guidelines for patient care (**Table 6.2**). These procedures apply to all patients, and they include protocols that deal with blood, semen, vaginal secretions, and tissue specimens, and cerebrospinal, synovial, plural, peritoneal, pericardial, and amniotic fluids. They do *not* apply to feces, nasal secretions, sputum, sweat, tears, urine or vomit, as long as these substances do not contain blood.

1.	Wear gloves and gown if hands, skin, or clothing is likely to be soiled with patient's blood or other body fluids
2.	Wear nose/mouth mask and protective eyewear or chin-length plastic face shield whenever splashing or splattering of blood or body fluids is likely. A nose/mouth mask alone is not sufficient
3.	Wash hands before and after patient contact and after removal of gloves. Change gloves between patients
4.	Use disposable mouthpiece/airway for cardiopulmonary resuscitation
5.	Discard contaminated needles and other sharp items immediately in puncture-proof container. Needles must not be bent, clipped, or re-capped
6.	Clean spills of blood or contaminated fluids by (1) putting on gloves and any other barriers needed, (2) wiping up with disposable towels, (3) washing with soap and water, and (4) disinfecting with 1:10 solution of household bleach and water. Bleach solution should not have been prepared more than 24 hours beforehand

Table 6.2 CDC Recommendations for Preventing Nosocomial Infections

Preventing and Controlling Nosocomial Infections

Hospitals are very aware of the problems associated with nosocomial infections. Every hospital must have control programs in place if it is to be accredited by the American Hospital Association. These programs involve:

- Surveillance of nosocomial infections in patients and staff
- On-site microbiology laboratory plus standardized isolation procedures
- Standardized procedures for use of catheters and hospital equipment
- Proper decontamination and sanitary procedures
- Mandatory nosocomial-disease education programs
- In some cases, infection-control specialist on staff

Keep in Mind

- One of the most important parts of the infectious process is the host defense.
- Compromise of the host defense can be due to a variety of problems such as infection with HIV resulting in AIDS, congenital immunodeficiency, transplantation, chemotherapy, or other conditions that debilitate the patient.
- Some infections are caused by opportunistic pathogens, which are normally harmless but can be pathogenic if the right conditions exist.
- Infections that occur in hospitals are referred to as nosocomial infections.
- Universal precautions are specific guidelines for patient care, which provide proper procedures for dealing with blood, semen, vaginal secretions, and other samples of tissue and body fluids.

EPIDEMIOLOGY

Epidemiology is the study of the factors and mechanisms involved in the frequency and spread of diseases or other health-related problems, and is an important part of our understanding of disease. These health-related problems can vary from infection to cigarette smoking or lead poisoning, but we will limit our discussion to infectious diseases, which means those caused by infectious pathogens invading a formerly healthy host. Epidemiology can be used not only as a tool to study disease but also as a way to design methods for the control and prevention of diseases.

As we have seen in all our discussions so far, there is tremendous variability associated with disease. When you combine things such as contagiousness, latency, and virulence with the large population on our planet, it is easy to see the need for up-to-date information on and strategies for combating infectious disease.

Incidence and Prevalence

The **incidence** of a disease is defined as the number of new cases contracted within a set population in a specific period. This kind of information can provide us with a reliable indication of the spread of a particular disease. **Prevalence** is defined as the total number of people infected within a population at any given time. Prevalence data can be used to measure how seriously and for how long a particular disease affects the population (**Figure 6.9**).

Morbidity and Mortality Rates

The **morbidity rate** of a disease is the number of individuals affected by the disease during a set period divided by the total population. The **mortality rate** is the number of deaths due to a specific disease during a specific period divided by the total population.

Figure 6.9 A graphic representation of *incidence* and *prevalence*. Incidence is the number of new cases contracted within a set population in a specific time. Prevalence is defined as the total number of people infected within a population at any given time.

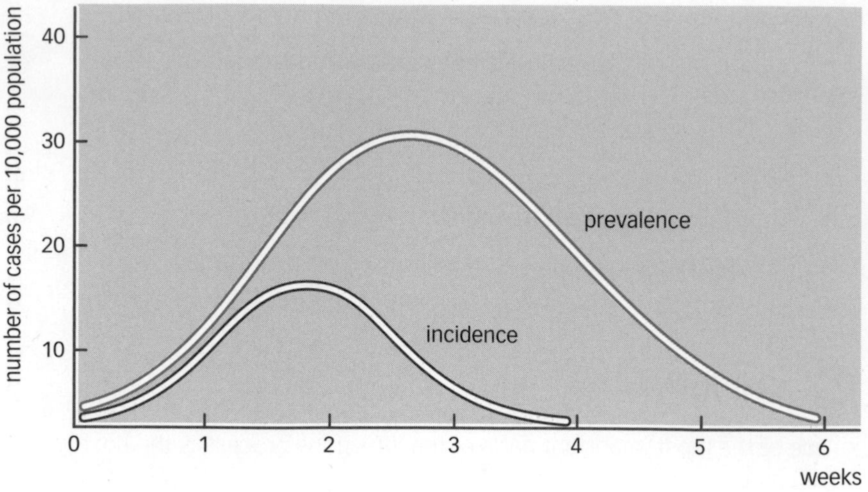

When epidemiological studies are used to examine parameters such as particular geographic areas and the degree of harm caused by a disease, they classify diseases as sporadic, endemic, epidemic, and pandemic. **Sporadic diseases** occur in a random and unpredictable manner and pose no threat to public heath. **Endemic** refers to diseases that are constantly in the population but in numbers too low to be a public health problem. As an example, the common cold is an endemic disease because there is always a percentage of any given population that has this infection. However, this percentage of the total population is low.

An **epidemic** occurs when the incidence of a disease suddenly becomes higher than the normally expected number of individuals affected by the disease. Over the past hundred years, for example, there have been several epidemics of influenza, such as the one that occurred in 1918 and killed millions. An epidemic causes morbidity and mortality rates to increase, and these increases signal a public health problem. It is important to note that there are several factors that contribute to the potential for the development of an epidemic. For example, the large populations that inhabit major cities allow the rapid spread of disease within a population. Increased aging of a population leads to a higher percentage of people who are immunologically or otherwise compromised and are therefore more predisposed to infection. Lack of herd immunity as a result of incomplete or inadequate vaccination is another factor in epidemics. Perhaps the most important factor of all in our twenty-first-century lives is the ability to travel easily and rapidly and thereby transmit diseases around the world.

When a disease occurs in epidemic proportions throughout the world, it is called a **pandemic**.

Epidemics are affected by the type of pathogen causing the disease and also by the mode of transmission. There are two basic types of epidemic, graphically represented in **Figure 6.10**: common-source outbreaks and propagated epidemics.

- A **common-source outbreak** is an epidemic that arises from contact with contaminated substances and most commonly occurs either when a water supply is contaminated with fecal material or when food is improperly prepared. Although common-source outbreaks affect large numbers of people, they quickly subside when the source of contamination has been identified and resolved.

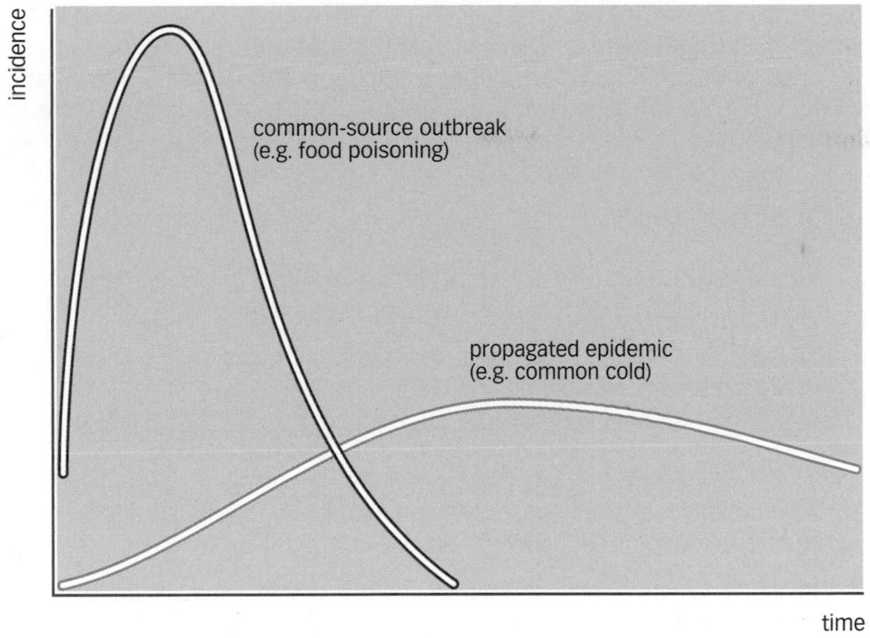

Figure 6.10 A graphic representation of the two types of epidemic: the common-source outbreak arising from contact with contaminated substances such as water supplies, and the propagated epidemic resulting from amplification of the number of infected individuals as person-to-person contact occurs.

- A **propagated epidemic** is one that results from amplification of the number of infected individuals as person-to-person contacts occur. Propagated epidemics stay in the population for long periods and are more difficult to deal with than common-source outbreaks.

Types of Epidemiological Study

There are two types of epidemiological study: descriptive and analytical. **Descriptive epidemiological studies** are concerned with the physical aspects of patients and the spread of the disease. These studies include data on the number of cases and on which segment of a population was affected. They also include the location and time of the infections, as well as the age, gender, race, marital status, and occupation of those affected. Careful examination of this type of information can allow epidemiologists to trace the outbreak of the disease and possibly identify the index case, which is the first person to have been infected with the disease. Identifying an index case is not mandatory for a successful descriptive epidemiological study. In some cases, it may be impossible to identify the index case because that person has recovered, moved away, or died.

In **analytical epidemiological studies**, the focus is on establishing a cause-and-effect relationship. These studies are always done in conjunction with a control group, which is a population known to be free of the disease being studied. A **retrospective analytical study** is one in which the records of patients who have already contracted the disease are studied. Using this type of already-available patient information, workers running retrospective studies can take into account a wide variety of factors that preceded an epidemic so that they can narrow down the potential causes. In contrast, **prospective analytical studies** do not allow the benefit of hindsight; in this type of study, data are analyzed as they are collected. In other words, prospective studies consider factors that occur as the epidemic proceeds.

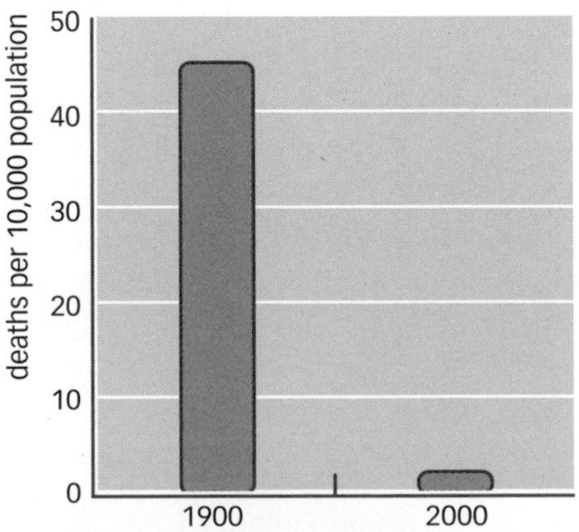

Figure 6.11 An illustration of the changes in deaths caused by infectious diseases over a century in the United States.

Health departments at the local and state levels require that doctors and hospitals report certain diseases. This type of information has been able to show how the effects of infectious diseases have changed over the years (**Figure 6.11**). In addition, some diseases are classified as **nationally notifiable** (**Table 6.3**), which means they must also be reported to the Centers for Disease Control in Atlanta, the clearing house for epidemiological research. Without question, the United States is one of the world's healthiest countries, which is due in no small measure to epidemiology.

Keep in Mind

- Epidemiology is the study of factors and mechanisms involved in the frequency or spread of diseases or health-related problems.
- Incidence describes the number of new cases contracted within a set population in a specific period.
- Prevalence is the total number of people infected within a population at any given time.
- The morbidity rate of a disease is the percentage of individuals affected by the disease during a set period.
- The mortality rate is the percentage of deaths due to a specific disease during a specific period.
- Diseases can be sporadic (occurring only occasionally), endemic (constantly in the population), epidemic (a higher than normal incidence of a disease), or pandemic (a worldwide epidemic).
- There are two types of epidemiological study: descriptive and analytical.
- Analytical epidemiological studies always contain a control group and can be retrospective or prospective.

Acquired immunodeficiency syndrome	Hemolytic uremic syndrome	Rubella syndrome, congenital
Anthrax	Hepatitis A	Salmonellosis
Botulism	Hepatitis B	Shigellosis
Brucellosis	Hepatitis C	Streptococcal disease group A, invasive
Chancroid	HIV infection	Streptococcal toxic shock syndrome
Chlamydia trachomatis	Legionellosis	*Streptococcus pneumoniae*, drug-resistant
Cholera	Malaria	*Streptococcus pneumoniae*, invasive
Coccidiomycosis	Measles	Syphilis, congenital
Cyclosporiasis	Meningococcal disease	Syphilis, contracted
Diphtheria	Mumps	Tetanus
Ehrlichiosis	Pertussis	Toxic shock syndrome
Arboreal encephalitis/meningitis	Plague	Trichinosis
Enterohemorrhagic *E. coli*	Psittacosis	Tuberculosis
Giardiasis	Q fever	Tularemia
Gonorrhea	Rabies, animal	Typhoid fever
Haemophilus influenzae, invasive	Rabies, human	Varicella (fatal)
Hansen's disease (leprosy)	Rocky Mountain spotted fever	West Nile encephalitis/meningitis
Hantavirus pulmonary syndrome	Rubella	Yellow fever

Table 6.3 Infectious Diseases that Are Nationally Notifiable in the United States

SUMMARY

- Transmission is the fifth requirement for a successful infection.
- There are three major reservoirs for infections: humans, animals, and nonliving.
- Transmission of infection can occur through contact, vehicular, and vector transmission.
- Contact transmission can be direct, indirect, or by droplets.
- Several factors, including age, life style, overall health, and emotional state, can influence the disease process.
- The portals of exit involved in the transmission of disease are the same as the portals of entry.
- One of the most important factors in the infection process is the immunocompetence of the host.
- Opportunistic infections are more likely in people with lowered immunity.
- Infections occurring in the hospital are referred to as nosocomial infections, and many of these are associated with antibiotic-resistant pathogens.
- There are universal precautions that are required for preventing nosocomial infections.
- Epidemiology is used to study the factors and mechanisms involved in the spread of disease. It can provide an important tool for developing strategies to combat infectious diseases.

On the basis of our studies up to now, we can say that we live in a fog of potentially harmful if not lethal microorganisms that, given the opportunity, can cause us great harm. You might be asking yourself how any of us are still left alive. The answer lies in later chapters, where we will learn about the elegant, lethal, and in some cases mind-boggling mechanisms that protect us from diseases every minute of every day. It is important to note that new infectious diseases are constantly emerging and that some diseases we thought had been eradicated are re-emerging. Therefore, it is important to examine some of these new threats, our main focus in Chapter 7.

SELF EVALUATION AND CHAPTER CONFIDENCE

Multiple Choice

Answers are given in the back of the book and help can be found in the student resources at:
www.garlandscience.com/micro

1. Reservoirs for pathogens include
   - **A.** Humans
   - **B.** Animals
   - **C.** Nonliving environments
   - **D.** Carriers
   - **E.** All of the above

2. Zoonotic diseases occur after direct contact with all of the following except
   - **A.** Animal feed
   - **B.** Soiled litter box
   - **C.** Feathers
   - **D.** Fur
   - **E.** Bite of rabid animal

3. Nonliving reservoirs include all of the following except
   - **A.** Food
   - **B.** Water
   - **C.** Blood
   - **D.** Soil

4. Which of the following is not a type of contact transmission mechanism?
   - **A.** Droplet
   - **B.** Indirect
   - **C.** Direct
   - **D.** Nosocomial

5. HIV transmitted by a contaminated needle is an example of
   - A. Direct contact transmission
   - B. Droplet contact transmission
   - C. Indirect contact transmission
   - D. Vector transmission
6. We refer to nonliving intermediates used in disease transmission as
   - A. Vectors
   - B. Fomites
   - C. Nosocomial agents
   - D. Zoonotic agents
7. Examples of vehicles for disease transmission include
   - A. Water
   - B. Air
   - C. Blood
   - D. Intravenous fluids
   - E. All of the above
8. Patients can become compromised in their ability to ward off infection because of
   - A. Age
   - B. Race
   - C. Gender
   - D. None of the above
9. Which of these factors influence whether or not a person becomes infected with a pathogen?
   - A. Lifestyle
   - B. Occupation
   - C. Age
   - D. Travel
   - E. All of the above
10. Nosocomial infections are
    - A. Seen only in the very young
    - B. Seen only in the elderly
    - C. Seen only in debilitated patients
    - D. Hospital-acquired infections
11. The most common site of a nosocomial infection is
    - A. The brain
    - B. The digestive system
    - C. The eyes
    - D. The urinary tract
    - E. None of the above
12. Likely sources of nosocomial infections include all of the following except
    - A. Catheters
    - B. Nebulizers
    - C. Bed sheets
    - D. Humidifiers
    - E. Contaminated needles
13. Universal procedures are required
    - A. Only when an infection occurs
    - B. Only in major hospitals
    - C. In all medical facilities in the United States
    - D. Only in nursing homes
14. The number of new cases of a disease contracted within a set population in a specific period is the definition of
    - A. Prevalence
    - B. Mortality rate
    - C. Morbidity rate
    - D. Incidence
    - E. None of the above
15. Diseases that occur in a random and unpredictable manner are referred to as being
    - A. Pandemic
    - B. Sporadic
    - C. Epidemic
    - D. Endemic
16. The total number of people infected within a population at any given time is the
    - A. Mortality rate
    - B. Prevalence
    - C. Incidence
    - D. Epidemic number
17. Diseases that are constantly in a population are called
    - A. Epidemic
    - B. Pandemic
    - C. Sporadic
    - D. Endemic
    - E. None of the above
18. Epidemics that stay in a population for long periods are called
    - A. Common-source outbreaks
    - B. Nosocomial infections
    - C. Propagated epidemics
    - D. Superinfections

DEPTH OF UNDERSTANDING

Questions listed here require you to bring together the concepts you have learned in this chapter into a discussion format. This helps you to increase your depth of understanding of the material you have learned. Help can be found in the student resources at:
www.garlandscience.com/micro

1. Compare and contrast the three major types of reservoir associated with infectious disease.

2. Discuss the predisposing factors that can contribute to the infectious disease process.

3. Evaluate the potential for infectious disease in burn patients, patients undergoing chemotherapy, and transplant patients.

CLINICAL CORNER

Help can be found in the student resources at:
www.garlandscience.com/micro

1. Eight children, all under the age of nine, are brought into the emergency room complaining of various degrees of diarrhea and vomiting. After initial examination it is learned that all of the children are from the same elementary school class and that the class was taken on a field trip to a nearby water park on the previous day.
   - **A.** What is the likely reason for their condition?
   - **B.** How would you confirm your suspicions?
   - **C.** What are the epidemiological possibilities?

2. Shady Grove retirement village has experienced an outbreak of bacterial pneumonia. This is a common problem seen in nursing home facilities. Eighteen of the 50 residents of the facility have severe symptoms and must be sent to the hospital.
   - **A.** What are the possible reasons why only 18 of the 50 residents became ill?
   - **B.** How could the disease have spread?
   - **C.** Explain whether this is a common-source outbreak or a propagated epidemic.
   - **D.** What are the main concerns you would have for the residents sent to the hospital?

Principles of Disease

Chapter 7

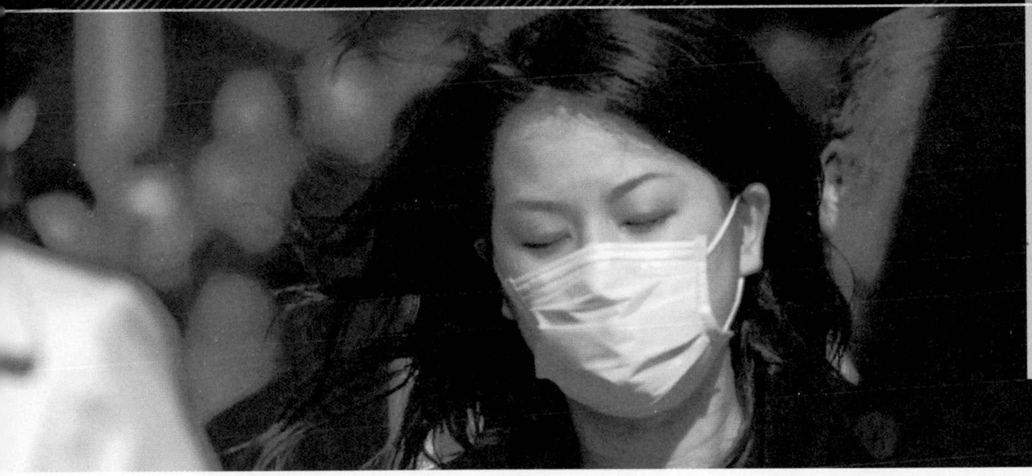

Why Is This Important?

This chapter introduces you to the principles of disease, and in particular how diseases are caused (the etiology), how they can be characterized, and the concepts of sepsis and shock. These concepts are important for developing an in-depth understanding of infections.

OVERVIEW

This is an important chapter because here is where we begin to look at the principles of infectious disease. Armed with the things we learned in the previous chapters, we are well prepared for these discussions, and the things we learn here will become part of the foundation for all of the other topics we examine. Our discussions will be broken down into three major parts. The first will examine how infectious disease can get started, which is referred to as etiology. Part two will examine how disease can develop, including communicability and contagiousness. The final discussion will be about infections that move into the systemic circulation.

We will divide our discussions into the following topics:

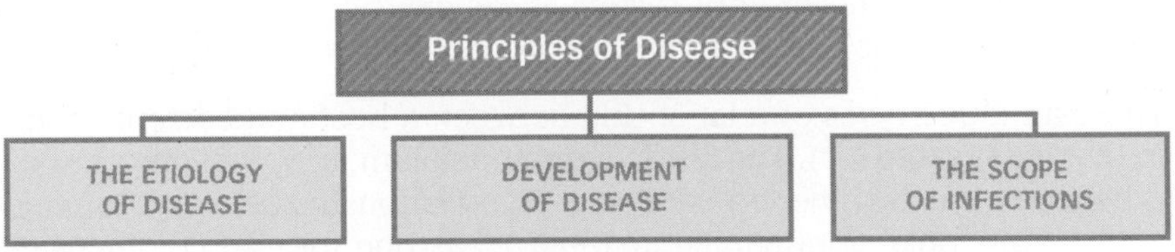

Now that we have looked at the primary requirements for a successful infection (Chapters 5 and 6), we can examine some of the principles associated with disease. **Table 7.1** lists several of the terms used when discussing disease. We can start off by defining **disease** as any negative change in a person's health. This change has a cause, and we refer to the cause of the disease as its **etiology**. Thus, for infectious disease, we can say:

- That the decline in health — in other words, the disease — is caused by microorganisms invading a host's body
- That these microorganisms are the etiological agents of the disease

One of the things we need to re-examine is the presence of normal microbial flora, which we mentioned in Chapter 6. There is no denying that we exist in an environment of microorganisms and that our bodies contain a variety of them that have a useful relationship with us. There are many microorganisms located in many places in the body that make up the normal **microbial flora** (**Table 7.2**). For example, the large intestine has the largest population of bacteria (in fact, fecal material is mostly bacteria). These organisms are supposed to be there because they provide

What Do I Need to Know?

To get the most out of this chapter, please review the following terms from your previous courses in biology, anatomy, physiology, or chemistry or in previous chapters of this book as indicated in parentheses: endocytosis (**4**), exocytosis (**4**), electrolyte, lipoprotein (**2**), phospholipid (**2**), lipopolysaccharide (**2**), lysis, lamina propria, Peyer's patches, cytokine.

Term	Defining Characteristic
Acute disease	Symptoms develop quickly but infection lasts only a short time
Chronic disease	Symptoms develop slowly and disease can last a long time
Subacute disease	Symptoms are between those of acute disease and those of chronic disease
Latent disease	Symptoms can continue to reappear long after initial infection
Local infection	Confined to a small area of the body
Focal infection	Initial site of a spreading infection
Systemic infection	Pathogens use blood or lymph to move around body
Bacteremia	Presence of bacteria in the blood
Septicemia	Organisms multiplying in the blood
Viremia	Presence but not multiplication of viruses in the blood
Toxemia	Presence of toxins in the blood
Primary infection	Initial infection
Secondary infection	Immediately follows primary infection; can be more dangerous than initial infection

Table 7.1 Some Common Terms Used When Describing Infections

necessary functions that are helpful to us. To use another example, the skin is covered with bacteria of many kinds, including normal resident bacteria, transient organisms, and even pathogens. As we saw in Chapter 6, however, unless these organisms can find a way through the skin (and some do), they are essentially harmless because the intact skin is such an impenetrable barrier. The mouth, nose, and throat are also places where large numbers of microorganisms are found, and once again many of these are harmless. Some are pathogenic, though, so it is important to remember that this area of the body is the most common portal of entry for bacteria. The respiratory system is readily accessible to potential pathogens from the nasopharynx and oropharynx.

There are three types of relationship between a host and bacteria living in that host (**Table 7.3**). The first is **commensalism**, in which one partner is benefited but the other is unaffected. The relationship between humans and the microbial flora named in Table 7.2 is one form of commensalism. The second type of host–bacteria relationship, **mutualism**, is one in which the host offers benefits to the bacteria and the bacteria offer benefits to the host. Some bacteria provide us with vitamins such as K and B, for instance, and we provide them with nutrients and a place to stay.

Table 7.2 Representative Normal Microflora

Region of the Body	Representative Microorganisms
Skin	*Staphylococcus epidermidis*, *Staphylococcus aureus*, *Propionibacterium acnes*, *Candida* species, and *Corynebacterium* species
Conjunctiva	*Staphylococcus aureus*, *Staphylococcus epidermidis*, and *Corynebacterium* species
Nose and throat	*Staphylococcus aureus* and *Staphylococcus epidermidis* in the nose; *Staphylococcus aureus*, *Streptococcus pneumoniae*, *Haemophilus*, *Corynebacterium*, and *Neisseria* species in the throat
Mouth	*Streptococcus* species, *Lactobacillus*, and *Corynebacterium*
Large Intestine	*Lactobacillus*, *Enterococcus*, *Escherichia coli*, *Enterobacter*, *Proteus*, *Klebsiella*, and *Corynebacterium* species
Urogenital tract	*Staphylococcus epidermidis*, *Enterococcus*, *Lactobacillus*, *Pseudomonas*, *Klebsiella*, and *Proteus* in the urethra; *Lactobacillus*, *Streptococcus*, and *Staphylococcus* in the vagina

Table 7.3 Host–Microorganism Relationships

Type of Relationship	Microorganism	Host	Example
Commensalism	Benefits	Neither benefits nor is harmed	Saprophytic bacteria that live off sloughed-off cells in the ear and external genitalia
Mutualism	Benefits	Benefits	Bacteria in colon
Parasitism	Benefits	Is harmed	Tuberculosis in lungs

The third type of relationship is called **parasitism**, and here one of the partners benefits at the expense of the other. This is the relationship we see with pathogens, which benefit at the expense of our health.

There are two points that are important to remember about host–bacteria relationships. First, microbial flora can protect us from disease. This is accomplished by **microbial antagonism**. Like a dog that marks its territory, bacteria that reside in certain areas of the body will fight to prevent "outsiders" (such as pathogens) from taking up residence. This prevention is easy for the resident bacteria to accomplish because they are well established in the host. For instance, our microbial flora competitively inhibits the growth of pathogenic interlopers by more efficiently using all the available nutrients and all the available oxygen, leaving none for the pathogens. In addition, the flora can acidify the region and make it less hospitable to any infringing pathogen. In fact, many bacteria, such as *Streptococcus* species and *Escherichia coli*, can produce bacteriocins, which are essentially localized bacterial antibiotics. These bacteriocins can kill invading organisms but do not affect the bacteria that produce them. Bacteriocins are specific for the bacteria that produce them, and for that reason they are useful as diagnostic tools for identifying bacteria.

The second important point about host–bacteria relationships is that many normally harmless resident bacteria are in fact opportunistic pathogens. Recall from the previous chapter that this type of pathogen is defined as one that is harmless in its normal location but can cause disease if it moves to a°on. One of the most prevalent of these is caused by *Escherichia coli*, which moves from the large intestine, where it is part of the normal microbial flora, to the urethra, where it is a pathogen. Urinary tract infections can occur repeatedly and become increasingly dangerous if the infection moves into the bladder or the kidneys. (We discuss this in more detail in Chapter 23.)

THE ETIOLOGY OF DISEASE

To understand the etiology of disease, it is important to mention one of the fundamental paradigms of microbiology, **Koch's postulates**. Robert Koch was a physician in the nineteenth century who studied anthrax in cattle. His studies were the first to close the loop between cause and effect for infectious disease and became a prerequisite for establishing the etiology of bacterial diseases. His postulates can be listed as follows:

- The same pathogen must be present in every case of the disease.
- The pathogen must be isolated from the sick host and purified.
- The pure pathogen must cause the same disease when given to uninfected hosts.
- The pathogen must be re-isolated from these newly infected hosts.

The procedure for testing the four postulates is illustrated in **Figure 7.1**.

These postulates allow us to place the responsibility for infection clearly and squarely on the organism that caused it, and they are still required today for determining the etiology of infections. Unfortunately, there are several exceptions, in particular the requirement for isolation and purification of the potential pathogen. There are many organisms that will not grow on artificial media and therefore cannot be purified. For example, the spirochete *Treponema pallidum* and the Gram-positive acid-fast rod *Mycobacterium leprae* (which causes **leprosy**) cannot be used in conjunction with Koch's postulates, and neither can *Rickettsia* organisms and viruses. In addition, some organisms can cause a variety of diseases, making the use of Koch's postulates impossible. Consequently, there are some diseases in which we can identify the etiological agent by using Koch's postulates, such as tuberculosis (etiological agent *Mycobacterium tuberculosis*) and Lyme disease (etiological agent *Borrelia burgdorferi*), and some in which we cannot use the postulates.

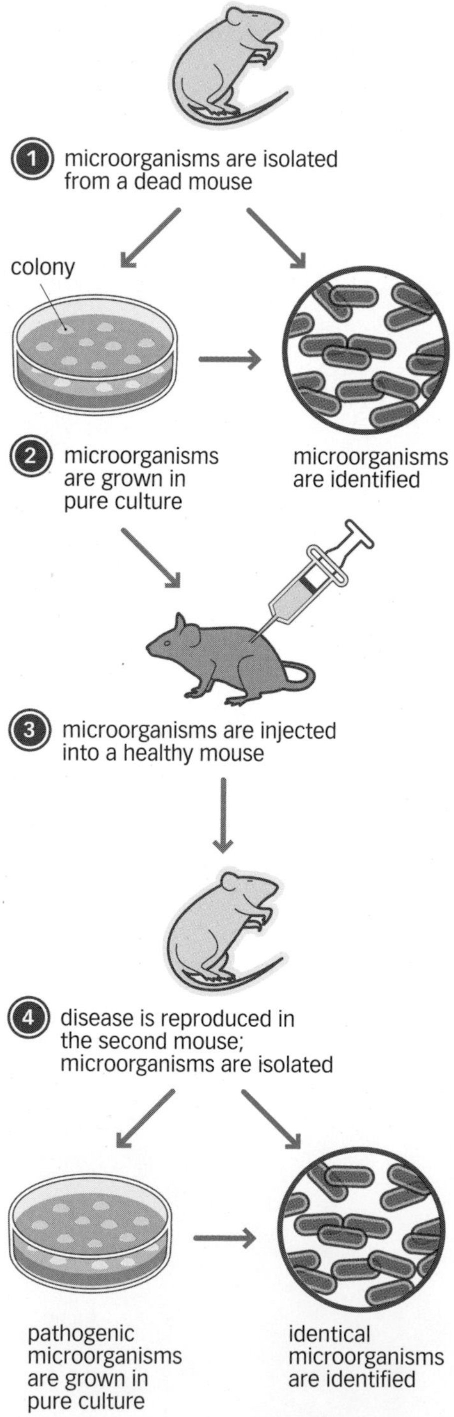

Figure 7.1 Koch's postulate test system. This test is used to determine the etiology of infectious diseases.

Keep in Mind

- The cause of a disease is referred to as the etiology.
- The body contains normal microbial flora made up of bacteria that are beneficial to the host.
- There are three types of relationship between bacteria and their hosts: commensalism, mutualism, and parasitism.
- Koch's postulates are an important way of evaluating the etiology of a disease.

DEVELOPMENT OF DISEASE

The course of a disease can be broken down into five specific periods (**Figure 7.2**). The first is called the **incubation period** and covers the time between the initial infection and the first symptoms of the disease. The length of the incubation period depends on the virulence of the pathogen: the lower the virulence, the longer this period will last (**Figure 7.3**). The second period is the **prodromal period**, during which the first mild symptoms appear. Once *major* symptoms are noted the disease has moved into the **period of illness**. It is here that the immune response is highest, and depending on the severity of the illness the patient may die during this period. The fourth period is called the **period of decline**, during which the symptoms subside. Although this period is an indicator of the end of the illness, it is also a time when the patient can acquire a secondary infection. This is particularly true of opportunistic infections and of nosocomial infections. In many cases, because of the debilitated condition of the patient, these secondary infections can be more dangerous than the initial problem. The final period is called the **period of convalescence**, during which the patient regains strength and proceeds to a full recovery.

Communicable and Contagious Disease

Some diseases are **communicable**, which means that they can spread from one person to another. Tuberculosis is one example. Others are non-communicable, which means they cannot spread from one person to another, with tetanus being a common example. In addition, communicable diseases that spread very easily, such as chickenpox or measles, are classified as **contagious**, meaning that they are communicable on contact with an infected individual.

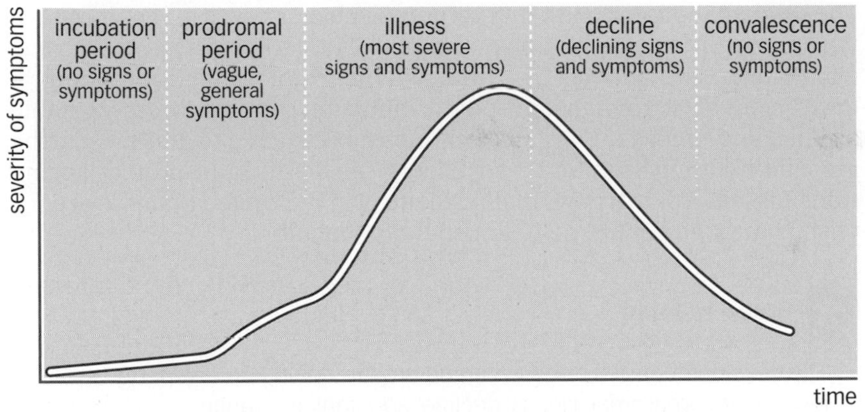

Figure 7.2 A graphical representation of the five periods of infection. The duration of the incubation period varies with the virulence of the pathogen, and secondary infections can occur during the period of decline. Because the patient is already debilitated, secondary infections can be more dangerous than the primary infection.

There are three methods available for full or partial control of communicable and contagious diseases: isolation, quarantine, and vector control.

Isolation means preventing an infected person from having contact with the general population. There are seven categories of isolation, which is usually done in a hospital. However, patients with diseases such as tuberculosis can be difficult to isolate before they have been diagnosed. As we saw in Chapter 1, in a busy emergency room, preference is given to individuals who are bleeding or who have chest pain. Someone with tuberculosis might be asked to be seated in a crowded waiting room for hours before being examined, diagnosed, and isolated. In a situation like this, the term *contagious* becomes easily understandable.

Quarantine involves separating from the general population healthy individuals (either humans or animals) who may have been exposed to a communicable disease. Quarantine usually lasts as long as the expected incubation times for the suspected disease and is lifted if symptoms do not present during that period. Although quarantine is the oldest method of dealing with communicable diseases, it is now generally used only for very severe diseases, such as cholera and yellow fever. In fact, quarantine is rarely used today because of the difficulty in enforcing it.

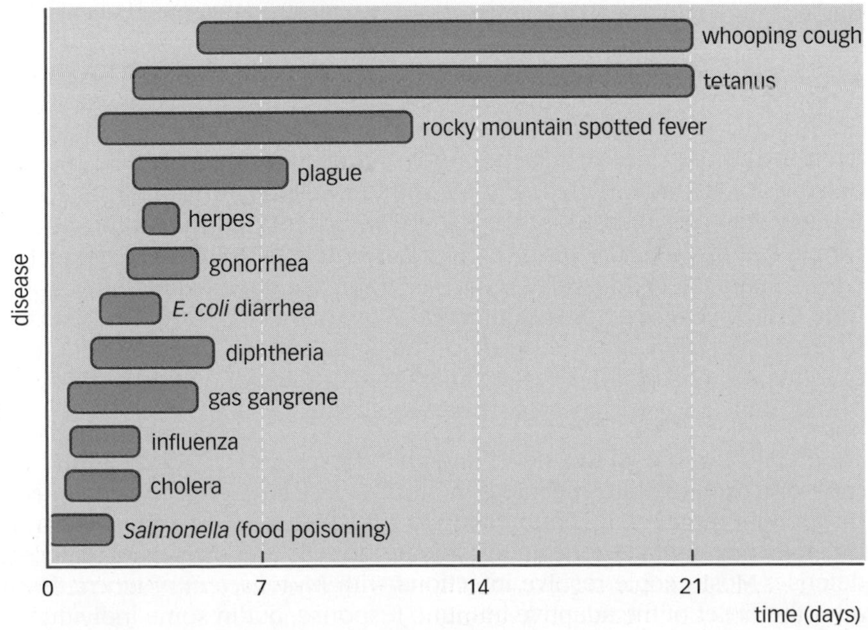

Figure 7.3 Durations of the period of incubation for selected organisms and diseases.

The third method for controlling communicable diseases is vector control, where **vector** is defined as any organism that carries disease pathogens from one host to another. The principal means of vector control are the destruction of vector habitats and the inhibition of vector breeding habits and feeding behaviors. In addition, measures such as window screens, mosquito netting, and insect repellents can be used to protect against transmission. For instance, malaria control in the United States is accomplished through control of the mosquito population.

Keep in Mind

- The development of a disease can be broken down into five periods: incubation, prodromal, illness, decline, and convalescence.
- Communicable diseases can be spread from one person to another.
- Contagious diseases are communicable on contact with an infected individual.
- Methods for the control of communicable and contagious diseases include isolation, quarantine, and vector control.

Duration of Disease

The duration of a disease can vary depending on several factors, in particular the overall health of the host. When we look at duration, there are four basic categories we can use. **Acute diseases** such as chickenpox and measles develop quickly but last for only a short time, whereas **chronic diseases**, such as **hepatitis**, mononucleosis, and tuberculosis, develop slowly but remain for long periods. Diseases such as **sclerosing panencephalitis** fall in the category of **subacute**. This disease has an insidious onset that can take 6–12 months to develop and can be fatal. **Latent diseases** are those that remain in the host after the initial signs and symptoms disappear and can be become reactivated after long periods. Latency is seen in several viral diseases, the most common example being **chickenpox**. The patient initially shows the classical signs of this pox infection (reviewed in Chapter 13), and after a while these disappear. However, the virus has taken up residence in the patient's neurons. At some point it can be reactivated and the patient will present with the rash identified as **shingles**. The reasons for reactivation of the virus are not yet understood, but they may involve stress or emotional anxiety.

Persistent Bacterial Infections

Some pathogenic bacteria are capable of maintaining infections in hosts, even in the presence of inflammatory and specific antimicrobial mechanisms as well as a perfectly good immune response (**Table 7.4**). Even though they are infected, some people with persistent infections show no clinical signs of the disease. *Mycobacterium tuberculosis*, *Salmonella enterica* serovar Typhi, and *Helicobacter pylori* are good examples of bacteria that can cause persistent infections, and there are many questions as to how a persistent infection can occur in spite of the innate and adaptive immune responses of the host. In an effort to understand this, let's look at two examples of persistent infections: tuberculosis and typhoid fever.

Tuberculosis is one of the oldest known diseases and affects one-third of the world's population. The infection starts at a site in a lung and can move throughout the lung. It is interesting to note that there is some evidence that this movement is performed by dendritic cells that are part of the host defense. Most people resolve infections with *Mycobacterium tuberculosis* after the onset of the adaptive immune response, but in some individuals

Pathogen	Disease	Site of Persistence
Mycobacterium tuberculosis	Tuberculosis	Granulomas and in other sites in macrophages
Salmonella enterica serovar Typhi	Typhoid fever	Macrophages in the bone marrow and possibly the gall bladder
Chlamydia species	Respiratory and cardiovascular disease; trachoma; genital tract infections and *Lymphogranuloma venereum*	Epithelial and endothelial cells
Helicobacter pylori	Gastritis, ulcers, and gastric cancer	Extracellular and possibly intracellular in the stomach
Neisseria gonorrhoeae	Genital tract infection, which can lead to pelvic inflammatory disease and infertility	Extracellular; intracellular at mucosal sites
Streptococcus pneumoniae	Pneumonia; acute otitis media; meningitis	Nasopharynx
Streptococcus pyogenes	Acute tonsillitis; pneumonia; endocarditis; necrotizing fasciitis	Nasopharynx
Haemophilus influenzae type B	Pneumonia; meningitis	Nasopharynx

Table 7.4 Selected Persistent Infections in Humans

the organisms are never completely cleared. Persistently infected hosts can harbor the pathogens for life, and in some cases the tuberculosis is reactivated later in life. This reactivation usually occurs in patients who have become immunocompromised through the aging process. It is for this reason that tuberculosis is seen in so many nursing homes.

In persistent tuberculosis infections, pathogens survive inside **granulomas**, which are bodies made up of host defensive cells, such as macrophages, T cells, B cells, dendritic cells, neutrophils, fibroblasts, and matrix components (**Figure 7.4**). Granulomas form as activated macrophages aggregate into gigantic cells similar to the syncytia seen in viral infections. The question is: How do the pathogens survive initial contact with cells that are programmed to phagocytose and kill them? This process is not completely understood, but it seems that the *Mycobacterium tuberculosis* organism gets taken into a phagosome in the normal fashion but then "remodels" this structure and prevents the development of an acidic, hydrolytic environment. In addition, the pathogens inhibit the formation of phagolysosomes by preventing the fusion of the phagosome with cellular lysosomes. Several genes have been identified in *M. tuberculosis* and seem to be involved in this process, but the events remain to be completely worked out.

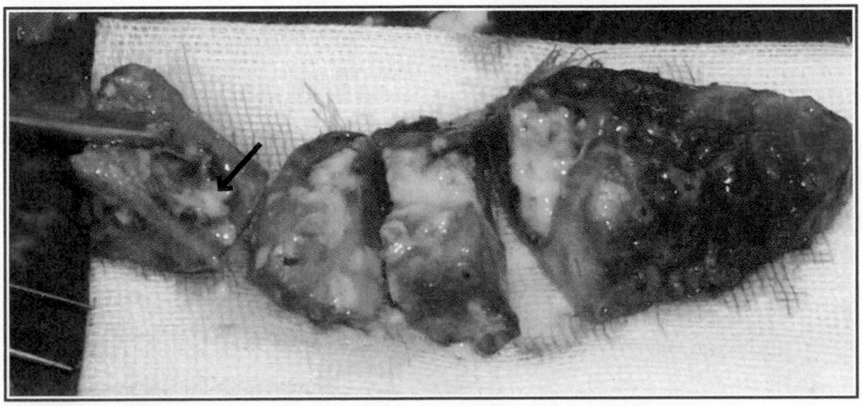

Figure 7.4 Gross pathology of human lung tissue granuloma removed during resection surgery. The arrow indicates a granuloma.

Typhoid fever is caused by *Salmonella enterica* serovar Typhi and can cause a variety of problems in the intestinal tract, such as localized gastritis. However, if this pathogen becomes systemic, it can cause typhoid fever. This disease starts with a localized infection and inflammatory response, usually in connection with Peyer's patches in the intestine. The *Salmonella* infect the phagocytic cells in the lamina propria of the intestine and gain access to the blood and lymph. Once in those liquids, they can spread to the liver and spleen and can become persistent in the gall bladder and bone marrow (**Figure 7.5**).

Salmonella infection is a major public health problem and is endemic in regions of the world in which there are unclean water supplies and a general lack of sanitation. It is also difficult to treat because the level of antibiotic resistance is rising. Between one and six per cent of people infected with *Salmonella enterica* serovar Typhi will become carriers who will shed large numbers of pathogens in their stool and urine for the rest of their lives but will never show symptoms of the infection (remember Mary from Chapter 1). The process by which *Salmonella* survives the host defensive response seems to be similar to the process by which *Mycobacterium* does: the invaders are phagocytosed, but the phagosomes never join with lysosomes to form phagolysosomes.

It is important to remember that several other mechanisms are used by pathogenic bacteria to survive a host's defensive responses and cause persistent infection. Some pathogens evade the innate response by enzymatically destroying the antimicrobial toxin nitric oxide as it is produced. Others, such as *Helicobacter pylori*, form megasomes inside a host macrophage by fusing together many phagosomes. Megasomes seem to protect the pathogens even though they are inside a phagocytic cell, but the reasons for this protection are still unclear. Some persistent pathogens subvert the adaptive immune response by blocking the activation

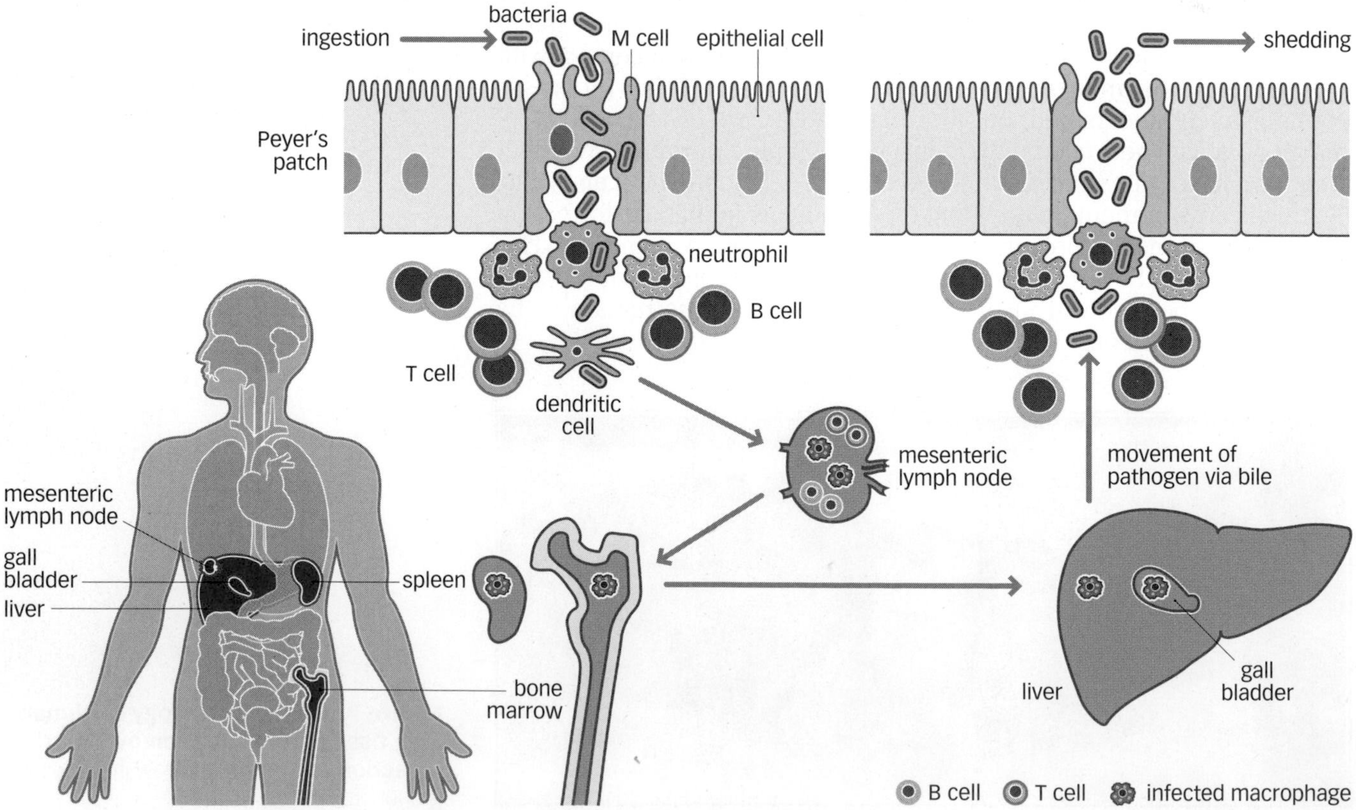

Figure 7.5 A schematic representation of persistent *Salmonella* infection in humans caused by *Salmonella enterica* serovar Typhi. Bacteria enter the intestinal tract by invading M cells (which are cells that are specifically localized in the digestive tract). This invasion is followed by inflammation, phagocytosis of bacteria by neutrophils and macrophages, and recruitment of immune T and B cells. In typhoid fever, *Salmonella* may use dendritic cells to disseminate to lymph nodes and deep tissues, which can lead to transport of the pathogens to the spleen, bone marrow, liver, and gall bladder. Re-infection and transmission occur by way of pathogens that are released from the gall bladder and re-enter the digestive system with bile.

of T cells, whereas others take advantage of genetic diversity (similar to that seen in influenza virus) to confuse the adaptive immune response. In fact, some pathogens change the level and type of lipopolysaccharide on their cells so that it will not bind as well to receptors on defensive cells, thereby lessening the host response against them.

Herd Immunity

One of the most important aspects of infection and the spread of infection is **herd immunity**, illustrated in **Figure 7.6**. This immunity can be conferred through vaccination or developed after infection. Let's look at smallpox as a way of understanding this concept. In years past, smallpox was a dangerous killer and had a hand in wiping out thousands. When a vaccination against the disease was developed and given widely, the population became immune to this infection. (Many older people carry the vaccination mark on an arm.) When the population had become immune, the disease essentially disappeared because there were no more potential hosts in the "herd". When smallpox was no longer a problem, vaccinations stopped. As a result, there are now lots of potential (non-vaccinated) targets in the herd and the infection can spread once again.

Herd immunity is a major and fundamental parameter for infectious diseases. However, it also presents a socioeconomic problem. Should we vaccinate against diseases that are no longer present in the population? What about **poliomyelitis**, a disease that killed and crippled thousands in the past but is today essentially nonexistent in the United States? What is the herd immunity like for this disease? What would happen if we stopped vaccinating against it? Keep in mind that polio is still prevalent in other countries in the world and therefore could resurface here at any time.

THE SCOPE OF INFECTIONS

Infections can be local or systemic. In many cases the host defenses wall off an infection and keep it a **local infection**, as occurs with boils or abscesses. From a host defense perspective, these infections are the easiest to deal with. However, when pathogens find a way to move from their original location in a host, a point called the **focus of infection**, they can become far more dangerous. An infection that has moved away from the focus of infection is called a **systemic infection** and is usually accomplished when pathogens gain access to the blood or lymph. Remember, the blood and lymph go everywhere in the body and interact with all the

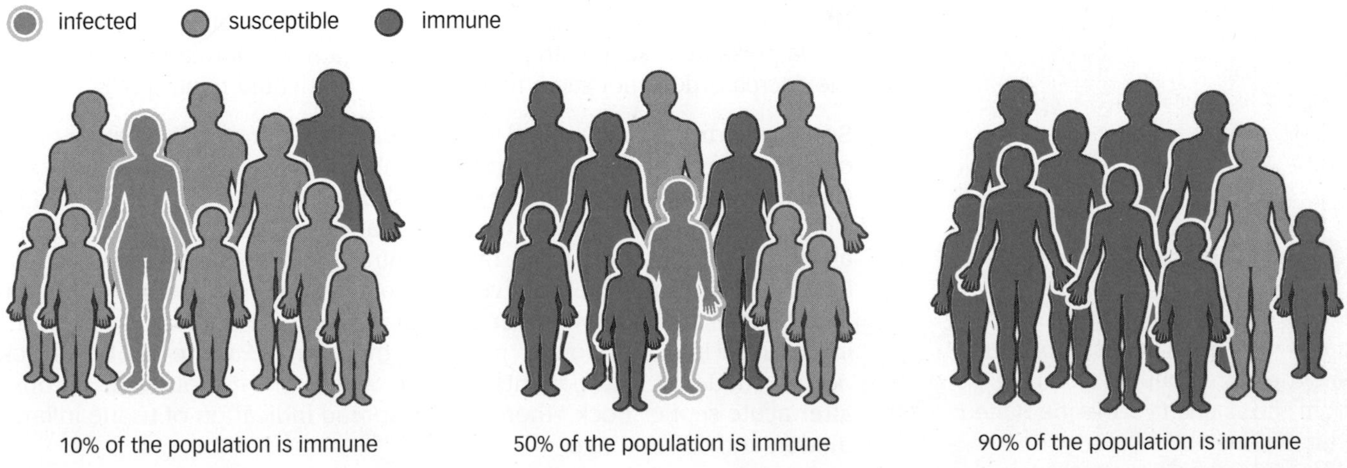

Figure 7.6 Herd immunity. Notice that the lower the percentage of immune individuals in the group, the greater is the number of potential targets and therefore the greater is the number of cases of disease.

organs. Bacteria in the blood are called a **bacteremia**, and if organisms are growing in the blood, the condition is referred to as a **septicemia**. When bacterial toxins move through the blood, the condition is referred to as **toxemia**, whereas viruses moving through the blood are called **viremia**.

Infections can also be primary, subclinical, or secondary. A **primary infection** is the one causing the initial acute onset of symptoms, whereas a **subclinical infection** is one in which symptoms do not appear even though the infection is ongoing (for example, polio and hepatitis A). Even though individuals with subclinical infections do not show signs of the disease or feel any symptoms, they are carriers who can infect others. A **secondary infection** is one that establishes itself in a host weakened by some primary infection. Secondary infections, such as *Pneumocystis* pneumonia, which is often seen in connection with AIDS, can be more dangerous than the primary infection because they take advantage of the weakened state of the host.

Toxic Shock and Sepsis

It is important that we take a minute here to discuss sepsis and toxic shock. These are two distinctly different clinical situations that arise as a result of infection. In **toxic shock**, there is massive leakage of plasma from the circulation that causes the blood pressure to plummet. This condition is fatal for between 30 and 70% of patients in which it occurs. It has been shown that toxic shock is caused by the activation of neutrophils, which occurs when the neutrophils come in contact with such bacterial surface proteins as the M proteins found on streptococcal species. In fact, when *Streptococcus pyogenes* begins to reproduce in the blood, it can cause streptococcal toxic shock syndrome (STSS), a condition that can rapidly cause death.

The mechanism by which STSS occurs is interesting and may provide researchers with possible methods to prevent a negative outcome. The M proteins of *S. pyogenes* are shed while the pathogens are growing in the blood. These proteins then bind to the plasma protein fibrinogen, which normally promotes clotting. The M protein–fibrinogen complexes then bind to and activate neutrophils that begin to release potent inflammatory molecules called heparin-binding proteins. It is these proteins that induce dynamic changes in vascular permeability in the host, changes that cause a rapid loss of fluids from the body, a catastrophic decrease in blood pressure, and difficulty in ventilating the lungs. In fact, strains of *S. pyogenes* that are rich in M proteins have been shown to be far more virulent than those containing only small amounts of these proteins.

The dynamic changes in the host just described result in a state of shock for the host. This type of shock is not the same as that seen in infections with Gram-negative endotoxins, however. Although a similar rapid decrease in blood pressure is seen with some Gram-negative endotoxins, the cause of the decrease does not seem to be related to activated neutrophils.

Sepsis syndrome is a general term indicating the presence of either pathogenic organisms or their toxins in the blood. There are two categories of sepsis: severe sepsis and acute septic shock. **Severe sepsis** is defined by signs of systemic inflammation and organ dysfunction accompanied by abnormal temperature, heart rate, respiratory rate, and leukocyte count as well as elevated levels of liver enzymes and altered cerebral function. Severe sepsis kills slowly over several weeks, and autopsy shows minimal tissue inflammation or necrosis. In contrast, **acute septic shock** occurs suddenly, and patients can die within 24–48 hours. At autopsy after acute septic shock, there is widespread indication of tissue inflammation and cell damage.

Fast Fact

Sepsis is the most common cause of death in hospital intensive care units, and there are about 225,000 deaths from sepsis each year in the United States alone.

Keep in Mind

- Disease can be acute, chronic, or subacute.
- Latent disease remains in the host after signs and symptoms have disappeared but can be reactivated after long periods.
- Pathogens such as *Mycobacterium tuberculosis* can cause persistent disease in which infections continue even though the host has a working immune defense.
- Herd immunity can limit the spread of infection.
- Infection can be localized or systemic and can be classified as primary (with acute initial symptoms), subclinical (without symptoms), or secondary (occurring after a primary infection).
- Infection can result in toxic shock or sepsis.

SUMMARY

- Normal microbial flora helps to protect against opportunistic pathogens.
- Etiology is defined as the cause of a disease.
- Koch's postulates can be used to evaluate and identify the etiology of a disease.
- Disease can be acute, chronic, sub-acute, or latent.
- Infection can be local or systemic.
- Relationships between bacteria and hosts can be described as commensal, mutual, or parasitic.
- There are five periods in the development of an infection: incubation period, prodromal period, period of disease, period of illness, and period of convalescence.
- Communicable diseases can be spread from one person to another.
- Contagious diseases are spread on contact with an infected person.
- Herd immunity can limit the spread of infection.
- Infection can cause septic shock or sepsis.

This chapter has taught us about how different infections get started, how diseases can develop and be characterized, and how infections can spread throughout the body. In addition, we learned about the importance of herd immunity and also about communicable and contagious diseases. Coupled with Chapter 5 on the requirements for infection and Chapter 6 on the transmission of infection, the material presented here completes our discussion of the infection process.

SELF EVALUATION AND CHAPTER CONFIDENCE

Multiple Choice

Answers are given in the back of the book and help can be found in the student resources at: www.garlandscience.com/micro

1. Etiology refers to
   - **A.** Viral infection
   - **B.** The results seen after a disease occurs
   - **C.** The cause of the disease
   - **D.** The portal of exit
   - **E.** None of the above

2. Microbial antagonism refers to
   - **A.** An increase in bacterial metabolism
   - **B.** A decrease in bacterial metabolism
   - **C.** An increase in infection symptoms
   - **D.** Protection by normal bacterial flora
   - **E.** Enhanced disease due to normal bacterial flora

3. The relationship in which one partner benefits at the expense of the other is called
   - **A.** Mutualism
   - **B.** Parasitism
   - **C.** Commensalism
   - **D.** Discordance
   - **E.** None of the above

4. Koch's postulates were the first to show the
   - **A.** Portals of entry for pathogens into the body
   - **B.** Portals of exit for pathogens leaving the body
   - **C.** Existence of exotoxins
   - **D.** Etiology of viral infections
   - **E.** Etiology of bacterial infections

5. The time between the initial infection and the onset of symptoms is called the
   - **A.** Disease period
   - **B.** Illness period
   - **C.** Incubation period
   - **D.** Prodromal period
   - **E.** Decline period

6. The first mild symptoms of diseases are seen in the
   - **A.** Illness period
   - **B.** Decline period
   - **C.** Prodromal period
   - **D.** Incubation period
   - **E.** None of the above

7. The period when major disease occurs is called the
   - **A.** Period of illness
   - **B.** Period of decline
   - **C.** Prodromal period
   - **D.** Period of convalescence
   - **E.** Major period

8. Chronic diseases
   - **A.** Develop quickly and subside quickly
   - **B.** Develop quickly but subside slowly
   - **C.** Develop slowly and subside quickly
   - **D.** Develop slowly and remain for a long time
   - **E.** Develop quickly and remain for a long time

9. Latent diseases
   - **A.** Develop quickly and subside quickly
   - **B.** Develop quickly and subside slowly
   - **C.** Remain in the host only when symptoms are present
   - **D.** Remain in the host after symptoms have gone but are able to be reactivated
   - **E.** Remain in the host after symptoms have gone but are never reactivated

10. Septicemia refers to pathogens growing in the
    - **A.** Focus of infection
    - **B.** Tissues
    - **C.** Urinary tract
    - **D.** Blood
    - **E.** None of the above

DEPTH OF UNDERSTANDING

Questions listed here require you to bring together the concepts you have learned in this chapter into a discussion format. This helps you to increase your depth of understanding of the material you have learned. Help can be found in the student resources at:
www.garlandscience.com/micro

1. Overuse of antibiotics can lead to antibiotic resistance and opportunistic infections. Why would these problems occur when antibiotics are our first choice for the treatment of bacterial infections?

2. There are three methods used to control communicable diseases. In today's world, which of them is most effective and why?

3. Tuberculosis has been referred to as both a persistent infection and a re-emerging infection. Justify this duality of classification.

CLINICAL CORNER

Help can be found in the student resources at:
www.garlandscience.com/micro

1. Your patient is a 78-year-old male who is in the hospital because of a serious bladder infection. He has been on antibiotic therapy for 4 days and seems to be getting better. Although he says he feels better, he is tired and not eating well. On the fifth day, you are surprised to find that he has been transferred to the intensive care unit. The doctor informs you that he has developed a severe upper respiratory infection and is in a grave condition. How will you explain to his wife and family what has happened?

2. Your neighbor has had frequent bouts of severe gastritis. She has been to the doctor and was diagnosed with a salmonella infection each time. After she was given antibiotics, the infection subsided but always seemed to reoccur. Finally the doctor explained that she needed to have her gall bladder removed. She is confused by all this and frightened by the possibility of surgery.

   A. How would you explain her condition to her?
   B. Why does the doctor want to remove her gall bladder?
   C. Why did the infection keep recurring even after antibiotic therapy?

3. Montrose elementary school is located in southern Florida, a region that sees a seasonal influx of migrant workers who help during the fruit picking season. Two weeks into the orange-picking season, there was an outbreak of measles in the second grade. Of the 26 students, only 4 came down with the infection and one of those was the daughter of a migrant worker.

   A. Explain why only four students got the measles.
   B. Why did the daughter of the migrant workers get infected?
   C. Why did the three other children who were not from migrant worker families get sick?

Emerging and Re-emerging Infectious Diseases

Chapter 8

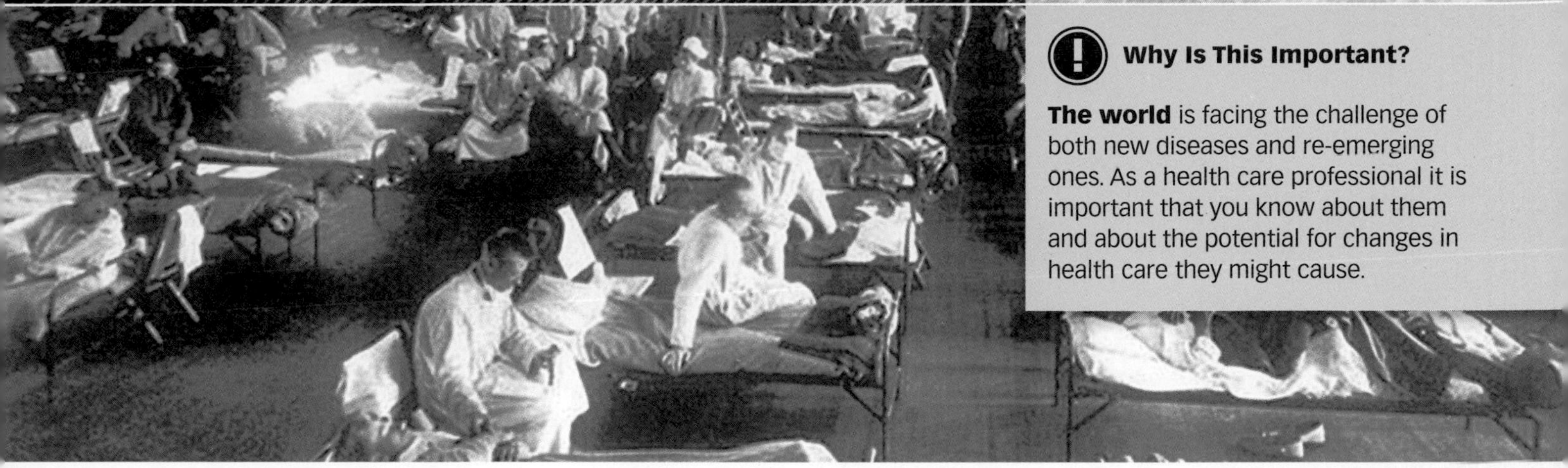

Why Is This Important?

The world is facing the challenge of both new diseases and re-emerging ones. As a health care professional it is important that you know about them and about the potential for changes in health care they might cause.

OVERVIEW

In this chapter we look at emerging infectious diseases, by which we mean those that have appeared over the past 30 years or so or are new to the United States. These diseases are important because they will have a great impact on health care in the future. Emerging infections such as SARS (severe acute respiratory syndrome) and West Nile virus not only make headlines but are also a wake-up call for the health-care community. They remind us that there are always new and potentially dangerous diseases cropping up in various populations. We will discuss re-emerging infectious diseases via a close look at influenza and tuberculosis. In the last part of the chapter, we look at two diseases that are emerging: avian influenza ("bird flu") and transmissible spongiform encephalopathies, which are diseases of the brain caused by abnormal proteins known as prions.

We will divide our discussions into the following topics:

Emerging and Re-emerging Infectious Diseases

- EMERGING INFECTIOUS DISEASES
- RE-EMERGING INFECTIOUS DISEASES
- AVIAN INFLUENZA
- PRIONS AND PRION DISEASES (TRANSMISSIBLE SPONGIFORM ENCEPHALOPATHY)

EMERGING INFECTIOUS DISEASES

"There will come yet other new and unusual ailments in the course of time. And this disease will pass away, but it later will be born again and be seen by our descendents."

Girolamo Frascatoro (speaking about syphilis)

Even though the quoted passage was written more than 450 years ago, it is just as valid today as we take our first look at emerging and re-emerging infectious diseases. When we look at infectious disease from a historical perspective, we can see that it has had a prominent role in shaping world events. *Yersinia pestis*, which caused the "Black Death," wiped out one-third to one-half of Europe in the Dark Ages. The measles virus contributed to the decline of the entire Aztec civilization, and many of the indigenous peoples of North and South America were decimated by smallpox brought by explorers from Europe. Smallpox was also used as the first biological weapon when blankets from smallpox victims were knowingly given to Native Americans by the British in pre-Revolutionary War days.

What Do I Need to Know?

To get the most out of this chapter, please review the following terms from your previous courses in biology, anatomy, physiology or chemistry or in previous chapters of this book as indicated in parentheses: apoptosis, coronavirus, fibrinolysis, lymphopenia, myalgia, T cells, viremia (**7**).

More recently, the ability of disease to affect history has continued. More than 30 new diseases have been identified in the past 30 years, including legionnaires' disease, HIV, hepatitis C, **Creutzfeldt–Jacob disease**, Nipah virus infection, several types of hemorrhagic fever, **SARS**, and avian influenza. In addition, we have had a re-emergence of "old favorites" such as tuberculosis and cholera that can be attributed to changes in ecology, migration, and mobility.

Emerging infectious diseases are those whose incidence in humans have either increased in the past 30 years or threaten to increase in the near future. These may be diseases that had previously been unseen in humans or had been seen only rarely (HIV/AIDS, SARS). The category also encompasses diseases that had previously been recognized but whose cause was unknown until recently (such as hepatitis C and the association between *Helicobacter pylori* and ulcers).

In addition to emerging infectious diseases, we are today faced with the problems presented by **re-emerging infectious diseases**. This category comprises diseases that were previously controlled but now have returned. Listings of emerging and re-emerging infectious diseases are given in **Tables 8.1** and **8.2**.

Along with the rise in new types of infectious disease, there has naturally been a concurrent rise in interest in them. In 1972 two medical researchers summed up the future of infectious diseases in this remarkably hopeful way: "The most likely forecast about the future of infectious disease is that it will be very dull." This quote must be taken in the context of what medical research had accomplished in the years before 1972. During that period, antibiotics had become commonplace, and we had conquered so many diseases that deaths from infectious disease had fallen dramatically (**Figure 8.1**), which actually led to an increase in life expectancy (**Figure 8.2**). In fact, this trend continues today except where HIV infection is widespread.

In the past 15 years, however, falling living standards and the decline of infrastructure in places such as the former Soviet Union have aided the re-emergence of some infectious diseases. As an example, there is currently speculation that there will be an influenza pandemic similar to the 1918 outbreak, which is believed to have killed 30 to 50 million people worldwide.

Twenty-five to thirty per cent of the approximately 60 million deaths that occur worldwide each year are caused by infectious disease, although many of the infections are rarely seen in the United States. Infectious

Table 8.1 Examples of Emerging Infectious Diseases

Disease	Infectious Agent	Year Recognized	Contributing Factors
Lassa fever	*Arenaviridae*	1969	Urbanization and consequent increased rodent population; increased nosocomial transmission
Ebola hemorrhagic fever	*Filoviridae*	1977	Unknown reservoir; nosocomial transmission
Legionnaires' disease	*Legionella pneumophila*	1977	Cooling and plumbing systems
Lyme disease	*Borrelia burgdorferi*	1982	Environments that favor tick and deer populations
AIDS	HIV	1983	Global travel, intravenous drug abuse, multiple sexual partners
Cholera	*Vibrio cholera* 0139	1992	New strain of bacteria with increased virulence
Hantavirus pulmonary syndrome	*Bunyaviridae*	1993	Encroachment into rodent territories

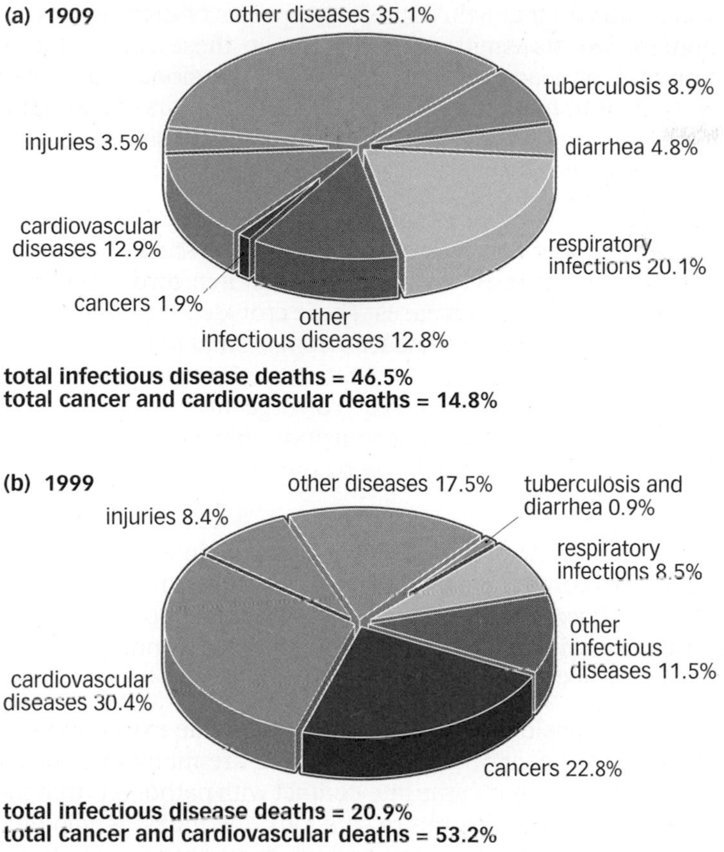

Figure 8.1 Proportions of total deaths from major causes in Chile. Panel **a**: 1909. Panel **b**: 1999. This country is a good illustration of the full transition from a developing nation to a developed status in the twentieth century. Notice the dramatic decrease in infectious diseases, which can be attributed to the development of antibiotics.

diseases, both established ones and emerging and re-emerging ones, will remain major global health threats that will affect society for the foreseeable future.

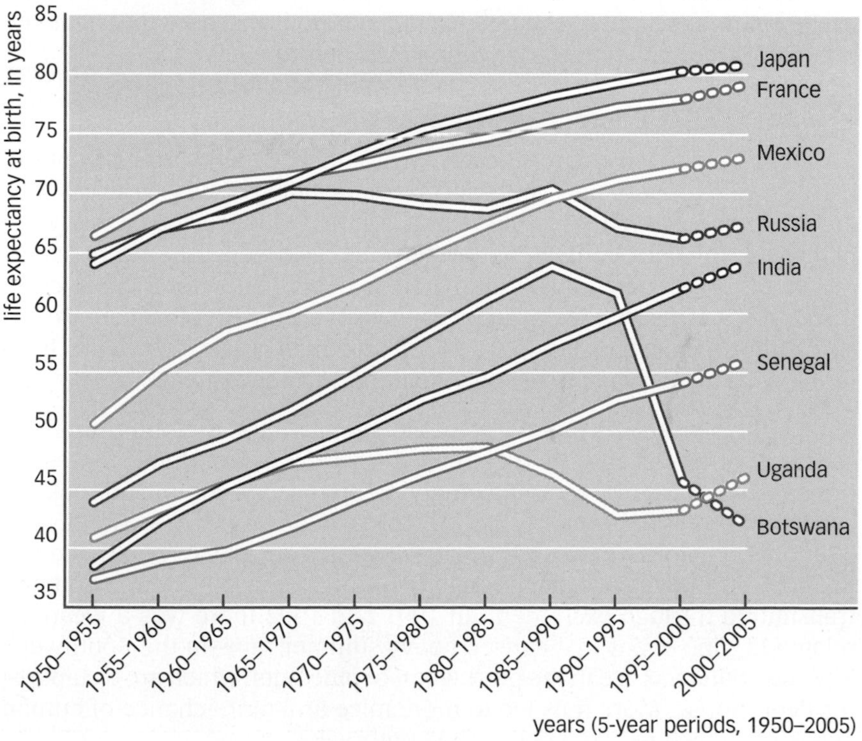

Figure 8.2 Changes in life expectancy at birth for both sexes in eight representative countries during the period 1950–2005. The trend toward longer life expectancies is a direct result of our success in controlling infectious disease.

It is estimated that 61% of the 1415 species of organisms known to infect humans are transmitted by animals. In these cases, the human is the terminal, or dead-end, host. However, occasionally a zoonotic infection will adapt to human-to-human transmission and in so doing will diversify away from its animal origin. Examples of this sequence are measles, smallpox, and HIV.

Throughout history, emerging diseases have followed a pattern of four transitions. The first was called a "crowd" transition and occurred when people began living close to one another in crowded places, allowing the easy transmission of diseases. These crowded conditions also allowed the movement of zoonotic diseases from animal to human. A good example is measles, which was confined to cattle 7000 years ago but then through a series of mutations was able to diverge into a pathogen that infects only humans. Although this first transition involves ancient history, it is still relevant today, as we will see below when we look at the development of SARS in China.

The second transition occurred in classical times as European civilization came into contact with other societies through either war or trade. This contact allowed various groups to exchange pools of infections and vectors. For example, in 430 BC during the Peloponnesian War, Thucydides wrote the first reports describing typhus.

The third transition coincided with worldwide exploration and colonization from the year 1500 onward. There are many examples of emerging diseases as humans came into contact with pathogens that they had never seen before. Cortez writes of smallpox and measles helping to defeat the Aztec empire. Captain Cook and his crew destroyed many Pacific island populations by bringing syphilis, measles, and tuberculosis along with them. Pacific islanders who had never encountered these pathogens had no defense against them.

The fourth transition is currently under way and can be classified as global urbanization. The increase in densely populated settlements coupled with poverty, social upheaval, air travel, long-distance trade, technological development, land clearance, and climate changes are contributing to the emergence and re-emergence of dangerous diseases that have the potential to spread more quickly than ever before.

Environment and Infectious Disease

As humans encroach on uncultivated environments, there are new contacts between humans, animals, and wild fauna (including disease vectors). This increased contact increases the risk of contact with previously unknown pathogens for both humans and other animals. Even if these infectious organisms develop in animals first, there is the potential for them to cross the species barrier. A good example is the establishment of piggeries close to the tropical forest in Malaysia. In 1988 the Nipah virus crossed over from fruit bats to the pigs and then began to infect the pig farmers.

Similarly, a Hantavirus that was a pathogen prevalent in South American rodents worked its way into rodents of the southwestern United States. As humans encroached into areas inhabited by these rodents, outbreaks of acute and sometimes fatal respiratory disease (now known as Hantavirus pulmonary syndrome) occurred, first observed in 1993. This previously unknown Hantavirus was maintained in the deer mouse population and transmitted through excrement. In 1991 and 1992 there was a weather-related El Niño event that caused heavy summer rains in the Southwest. This led to an increase in the production of pine nuts, which are a staple of the deer mouse. More nuts led to more mice and more chance of human exposure, which resulted in the 1993 outbreak.

This connected series of events is also seen with the dengue virus, which is spreading because its mosquito vector is now breeding rapidly in congested urban environments in developing countries. Lack of proper mosquito control and poor living standards set the stage for outbreaks of this disease. For Lyme disease, seen today in many regions of the United States, the vector is the deer tick (**Figure 8.3**). As the number of deer increases, the number of ticks increases and so, too, does the potential for increased outbreaks of Lyme disease.

Food-borne Infection Vectors

As the population grows worldwide, there is increased pressure to produce more meat. This has led to the emergence of infections that are transmitted from farm animals to humans. One well-known example is the 1986 outbreak of "mad cow" disease in Britain. This disease is caused by a prion, which is not even an organism (discussed below). Similarly, infectious pathogens such as *Salmonella* and *Escherichia coli* O157:H7 can cause fatal diseases in humans. This problem is compounded by the increasing demand for exotic and wild animals. SARS, for instance, is associated with such exotic animals as the palm civet and the raccoon dog.

Globalization and Transmission

Changing patterns in human behavior and a changing ecology on our planet contribute to the emergence of infectious disease in two ways:

- Increased opportunity for animal-to-human infection because of greater exposure
- Increased opportunity for the transmission from one human to another once a person is infected

Genetic changes in pathogens can occur through a process known as **re-assortment**. This is a type of gene "shuffling" in which an organism's genes rearrange themselves and cause changes in the characteristics of the organism. Re-assortment can also take place between different pathogens. Genetic re-assortment between pathogens could give rise to new, rapidly spreading strains of pathogens. For instance, as we will see later

Figure 8.3 The female Lone Star deer tick *Amblyomma americanum*, a vector for Lyme disease.

Fast Fact

Novel infectious disease can emerge in any part of the world at any time. For example, HIV and Ebola came out of Africa, avian influenza and SARS from China, Nipah from Malaysia, and Hantavirus from South America.

in this chapter, there is the potential for deadly avian strains of influenza to recombine genetically with strains that commonly infect humans, resulting in an influenza outbreak that could be comparable to or even worse than the pandemic of 1918.

Some of the most sinister infections are latent ones that have long incubation periods, such as HIV. By the time these diseases are recognized, the infection is already established in humans and out of control. The spread of latent pathogens is exacerbated by the marvel of modern air travel, which can disperse infections worldwide in a matter of days. The West Nile virus, for instance, which was unknown in the United States before 1999, arrived in the northeastern part of the country. It is difficult to pinpoint how it arrived, because it could have been in a bird that was blown a long way off course, in a mosquito that hitched a ride on an airplane, or in an infected traveler. Whichever is the case, it took less than four years for this virus to move from isolated counties in New York state to the Pacific coast of the United States.

If we combine all the factors mentioned above, especially travel, the potential for deadly infections to appear quickly worldwide is not small. Keep in mind that the immunodeficiency associated with HIV infection as well as increasing numbers of compromised hosts can amplify the entire process by presenting increasing numbers of potential targets.

Hurdles to Interspecies Transfer

It is important to remember that crossing the species barrier is not a simple task. A pathogen must overcome two major hurdles to replicate successfully in a human host:

- It must adapt in such a way that it can replicate in human cells. This can be a complex problem for the pathogen.
- It must be able to configure itself so that it can be easily transmitted from one human to another.

Many organisms, such as the Hantavirus, Nipah and avian influenza have overcome the first hurdle and have jumped from animals to humans, but they have not yet been able to overcome the second. One exception is HIV, which has successfully cleared both hurdles.

The changes required by a pathogen that allow it to clear these two hurdles require extensive genetic mutation, rearrangement, or re-assortment. Such changes are easier for viruses, which are prone to mutation because of the lack of fidelity in replication (especially RNA viruses).

Emerging Viral Diseases

The great preponderance of today's emerging infectious diseases are caused by viruses rather than by bacteria or other pathogens. Because viral replication requires the presence of particular specific factors in the host cell, it may be either the absence or the "foreignness" of these host factors that prevents virus replication in foreign species. If adjusting to this absence or foreignness requires only minor genetic change in the virus, there is an increased risk that the jump to a new host species will be made. All in all, given the extent of the obstacles facing the virus, it is amazing that a jump ever occurs. In fact, the number of viral emerging infectious diseases is quite small when you consider the number of viruses that exist in animal reservoirs.

The jump does occur, however, and RNA viruses are the best at it because they have developed the three basic strategies mentioned above to deal with the problem: mutation, recombination, and re-assortment (**Figure 8.4**).

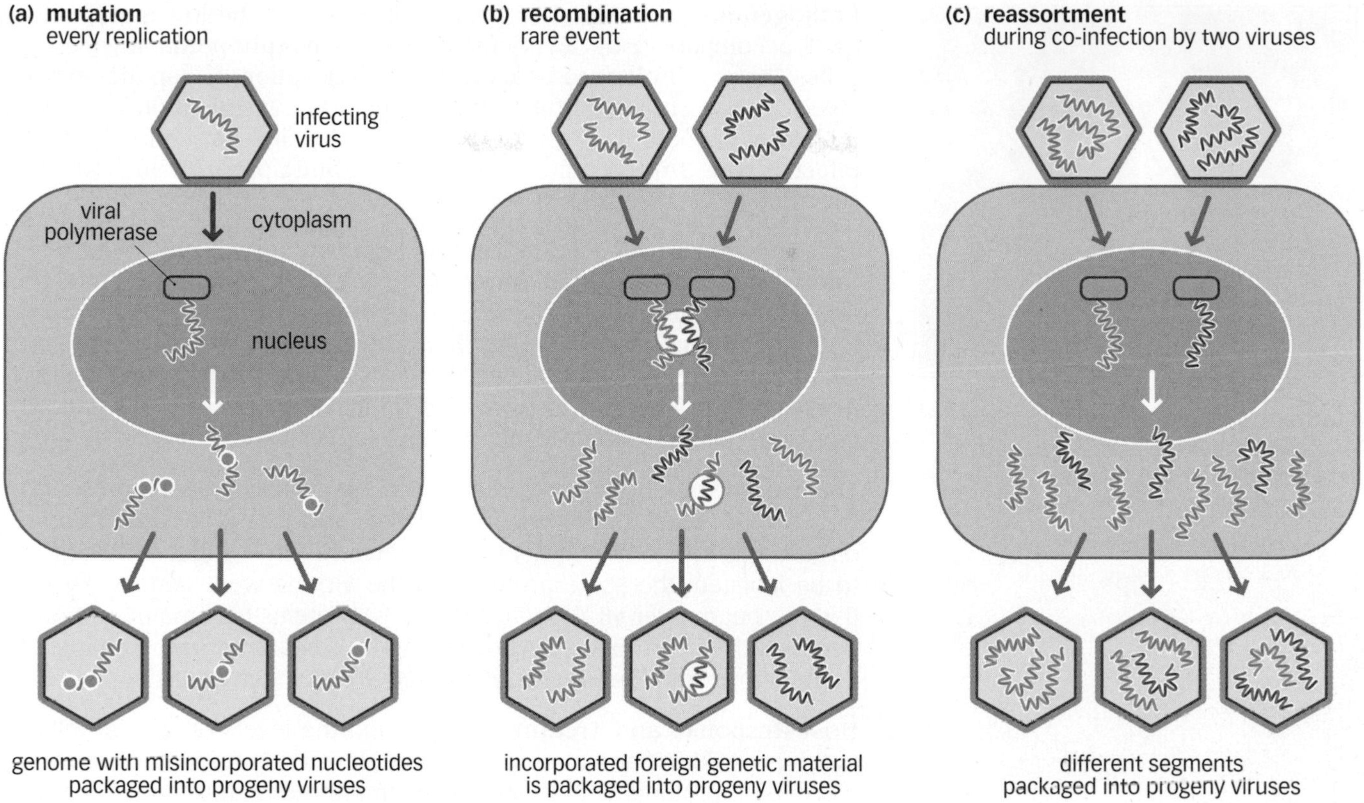

Figure 8.4 Three molecular mechanisms for generating viral diversity. Panel **a**: Single-point mutations incorporated into one or more positions as a result of the lack of proofreading by the viral polymerase. Panel **b**: During recombination, foreign genetic material is incorporated into the viral genome. Panel **c**: Re-assortment, in which whole gene segments can be swapped. All three mechanisms, which are not exclusive, may result in viruses with new biological properties such as host ranges and pathogenic potentials.

We know that both the Asian influenza strain of 1957 and the 1968 Hong Kong influenza pandemic arose through re-assortment of human and avian strains of the virus. Now there is increasing global concern that re-assortment between genes from the avian strain known as H5N1 and the human influenza virus will result in a repeat of the 1918 influenza pandemic.

Let's look at a few of the emerging viral diseases.

SARS (Severe Acute Respiratory Syndrome)

SARS was a previously unrecognized animal coronavirus that exploited what are called the "wet markets" of southern China, places where live animals such as palm civets and raccoon dogs are kept in close proximity to humans until sold. In these hospitable locations, the virus adapted to and became readily transmissible to humans during the 1990s. The course and spread of the original infection are a great example of how infectious disease spreads in the modern world. The first human case of SARS was seen in Guangdong Province, China, and was transmitted unknowingly from the patient to a local physician, who then traveled to Hong Kong for a wedding. He spent one night in a hotel in Hong Kong, during which time he transmitted the virus to 16 other guests, most of whom were foreigners. They in turn spread the virus, causing outbreaks in Hong Kong, Singapore, Vietnam, and as far away as Toronto.

It is thought that the first humans to be infected were workers involved with wild game, such as palm civets and raccoon dogs. It is likely that the virus tried many times to overcome the hurdles of crossing to a new species before it was finally successful in adapting to replication in humans. The second and perhaps more difficult hurdle — being transmissible from human to human — took longer but was also finally cleared. SARS can now be transmitted by droplet, aerosol, and fomite and is deposited on the respiratory mucosal epithelium, where it initiates infection.

Fast Fact

Within weeks, SARS had spread to more than 8000 people in 25 countries on 5 continents. It killed 744 people.

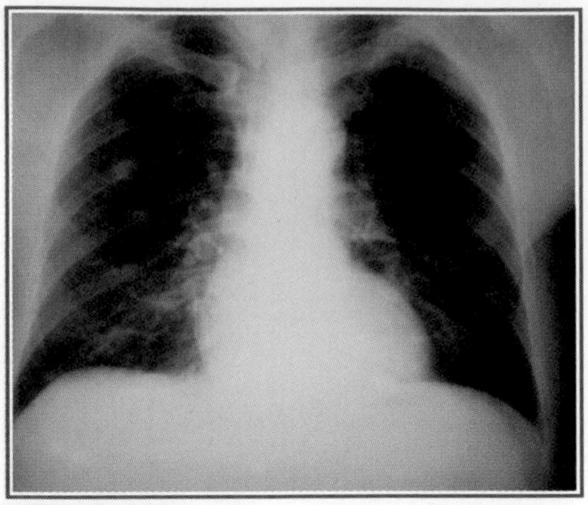

Figure 8.5 Chest X-ray of a patient with SARS (severe acute respiratory syndrome). The white shadow seen in the lungs is caused by the infection.

Pathogenesis of SARS. SARS causes infection of the lower respiratory tract accompanied by fever, **malaise**, and **lymphopenia** involving T cells. There is prolonged coagulation and elevation of hepatic enzyme levels. X-rays show infiltrates and sub-pleural consolidation consistent with viral pneumonia (**Figure 8.5**). Twenty to thirty per cent of patients infected with SARS require intensive care, and approximately 10% will succumb to the disease.

The pathogenesis of SARS is due to a high viral load in the lower respiratory tract. This viral load is low in the first 4 or 5 days of the infection and peaks around day 10. There is also a direct correlation between viral load and prognosis, with higher loads associated with poor prognosis. On about the 10th day of illness, the virus can be found in nasopharyngeal aspirates, feces, and serum. Poor clinical outcome is associated with continued uncontrolled viral replication.

The SARS virus enters host cells by attaching spike proteins present on the virion to receptors on the host plasma membrane (**Figure 8.6**). (We discuss this interaction in detail in Chapter 12.) In the first SARS viruses to be isolated, the spike proteins on the virions were genetically very diverse, but in later viral isolates the spike proteins became more homogeneous. This is probably because the virus is becoming more adapted to human hosts and therefore does not require as much mutation.

Host Response and Treatment. Although the levels of several inflammatory **cytokines** and **chemokines** are elevated in patients with SARS, there is a prolonged immunological impairment during the disease. Treatments vary; antiviral chemotherapeutics are effective if administered during the first few days of the infection. Attempts to develop a vaccine for SARS are under way, and it has also been shown that spikes isolated from the virion can be used to keep the virus from attaching to host cells, thereby preventing infection. So far, however, there is no vaccine for this disease.

West Nile Virus

The coronaviruses exemplified by SARS are not the only instance of emerging viral disease today. We also have arboviruses, which are those carried by arthropods. The best examples of this group of emerging viruses are the West Nile and dengue viruses. The arthropod vectors that carry these pathogens must get blood if they are to complete their life cycle, and it is during this feeding event that the virus is transmitted. There are more than 70 viruses in this group; 40 are mosquito-borne, 16 are tick-borne, and the rest have no known vector. Arboviruses are RNA viruses that are thought to have emerged in the human population as a result of continued deforestation in several African nations. These viruses most probably dispersed from the original sites via wind-blown mosquitoes, migrating birds, or infected travelers.

West Nile virus is a member of the Japanese encephalitis group and was first isolated in 1937 from the blood of a woman in the West Nile district of Uganda. Today, this virus is widespread throughout Africa, the Middle East, and parts of Europe, Russia, Asia, and Australia. It was not detected

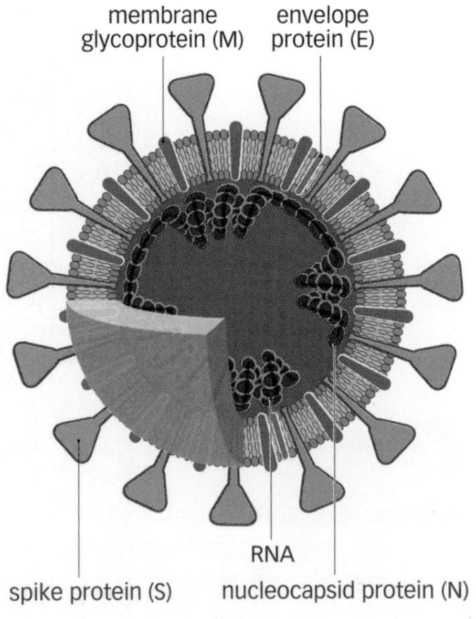

Figure 8.6 A schematic representation of the SARS coronavirus structure. Viral surface proteins (labelled M, E, and S) are embedded in a lipid bilayer envelope derived from the host cell. The single-stranded viral RNA is associated with the nucleocapsid protein (labelled N). The spike proteins are used to bind to receptors on host cells.

in the United States till 1999, when there was an outbreak in New York City. Epidemiologists believe that the virus arrived via an infected traveler from the Middle East. The rapid spread of this virus throughout the United States, Mexico, and the Caribbean in just five years is truly remarkable (**Figure 8.7**).

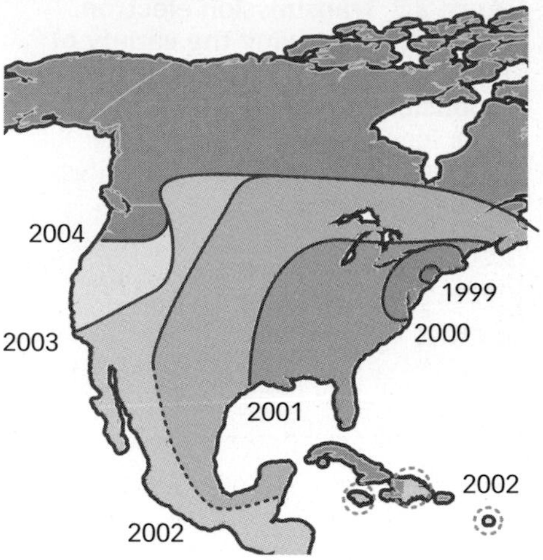

Figure 8.7 The approximate geographical distribution of West Nile virus in the Americas from 1999 to 2004. This was determined by virologic surveillance of dead birds reported to the Centers for Disease Control in the United States and the Department of Prevention and Health in Canada. The dashed lines indicate estimated range limits.

Birds are the primary host of the West Nile virus, and it is spread from bird to bird by mosquitoes. Humans and animals such as horses are incidental hosts; they can be infected by mosquitoes carrying the virus. The virus has been identified in more than 50 species of mosquito, but it is still unclear whether there is a specific mosquito species that transmits the virus to humans. Nearly all infections have resulted from mosquito bites, but transmission through blood transfusion and transplanted organs has occurred.

Pathogenesis Of West Nile Virus. The pathogenesis of West Nile infection is not well understood, but it is clear that the virus is transmitted to humans in the saliva of the mosquito. Most infected people are asymptomatic unless the infection causes an invasive neurological disease called **West Nile fever**, which occurs in 20–30% of those infected. There is a 2–14-day period associated with fever, headache, back pain, **myalgia**, and anorexia (weight loss). There may also be profound fatigue and skin rash. Severe infection can cause **myocarditis**, **pancreatitis**, and hepatitis, and 1 in 150 patients develop either **encephalitis** or **meningitis**. Approximately 10% of these latter patients will succumb to the disease.

Advanced age is a predictor of poor prognosis, and survivors of severe symptoms can have long-term neurological impairment. West Nile virus can also cause acute flaccid paralysis syndrome, from which complete recovery is uncommon. In the neurological conditions caused by West Nile virus, viral antigens are most commonly associated with neurons and neuronal processes. There is both a cellular and a humoral host defense response against this viral infection.

Fast Fact

A genetic mutation has been identified that increases a person's susceptibility to infection with West Nile virus. Ironically, it is the same mutation that seems to protect people from infection with HIV.

Viral Hemorrhagic Fever (VHF)

The emerging infectious diseases classified as viral hemorrhagic fevers include the conditions caused by the Ebola, Marburg, and yellow fever viruses. Some of these diseases, in particular Ebola and Marburg, are frequently fatal. VHF infections are some of the most exotic emerging diseases and are characterized by fever, bleeding, and **circulatory shock**. All of the viruses are single-stranded RNA types and carry a lipid envelope, but they vary in morphology (**Figure 8.8**).

Pathogenesis of Viral Hemorrhagic Fever. These viruses are transmitted in diverse ways, via both arthropod and rodent vectors. All of the hemorrhagic viruses have developed the ability to be transmitted from human to human, usually through direct contact with infected blood or fluids. Working with these viruses or with patients infected with them requires the highest levels of protection possible.

The hemorrhagic viruses have incubation periods of between 2 and 21 days, and the virulence of the infection depends on the virulence of the virus, the route of exposure, the viral load, and the competence of the host defense. The symptoms are fever, myalgia with prostration, **flushing**, and *petechial hemorrhaging* (the appearance of red spots on the skin). Hepatic dysfunction is common, and fatality rates average between 5 and 20% for all VHFs. However, for Ebola infection, the death rate is 50 to 90%.

VHF viruses target monocytes, dendritic cells, endothelial cells, hepatocytes, and adrenal cortical cells. They use endocytosis as the mechanism of penetration and cause varying degrees of cell destruction. All VHF

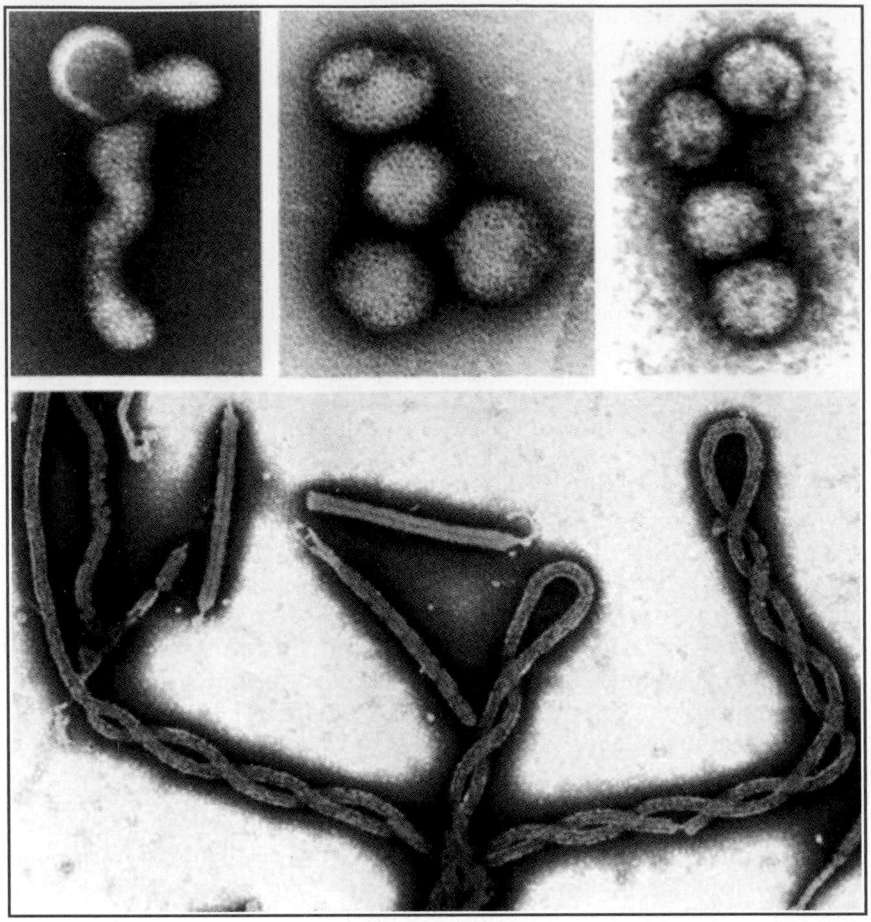

Figure 8.8 Transmission electron micrographs showing the variety of morphologies seen in the viruses responsible for viral hemorrhagic fever (VHF). Top left, Junin virus; top center, Rift Valley virus; top right, yellow fever virus; bottom, Ebola virus.

viruses target and impair host antiviral responses, a strategy that permits high levels of viremia and also immunosuppression (**Figure 8.9**). Sudden, severe shock-like symptoms are seen in all VHF fatalities.

The immunosuppression seen in VHF infections is caused by varying degrees of lymphocyte depletion and necrosis of the spleen and lymph nodes. Although lymphopenia occurs in all VHF infections, the cause remains a mystery. However, there is evidence in Ebola and Marburg of increased apoptosis of lymphoid cells. Rapid loss of lymphocytes results in failure to control viral replication and consequent enormous systemic viral burdens.

VHF infections in humans trigger the expression of several inflammatory mediators, including interferon and interleukins, but these agents can contribute to the disease process rather than inhibit it. However, it seems that the earlier and more robust a host defensive response, the greater the chance of survival. Although the mechanisms involved are yet to be worked out, it seems that there is a balance between the protective and deleterious cytokines produced during these infections. The result of tipping this balance can be either recovery or death.

The production of nitric oxide (NO), a powerful antimicrobial agent, also seems to have an important role in the pathogenesis of these infections. Increased blood levels of NO are associated with mortality because they cause increased apoptosis, tissue damage, and loss of vascular integrity, all of which may contribute to shock. NO is also an important mediator of hypotension, which is a prominent problem seen in VHF infections.

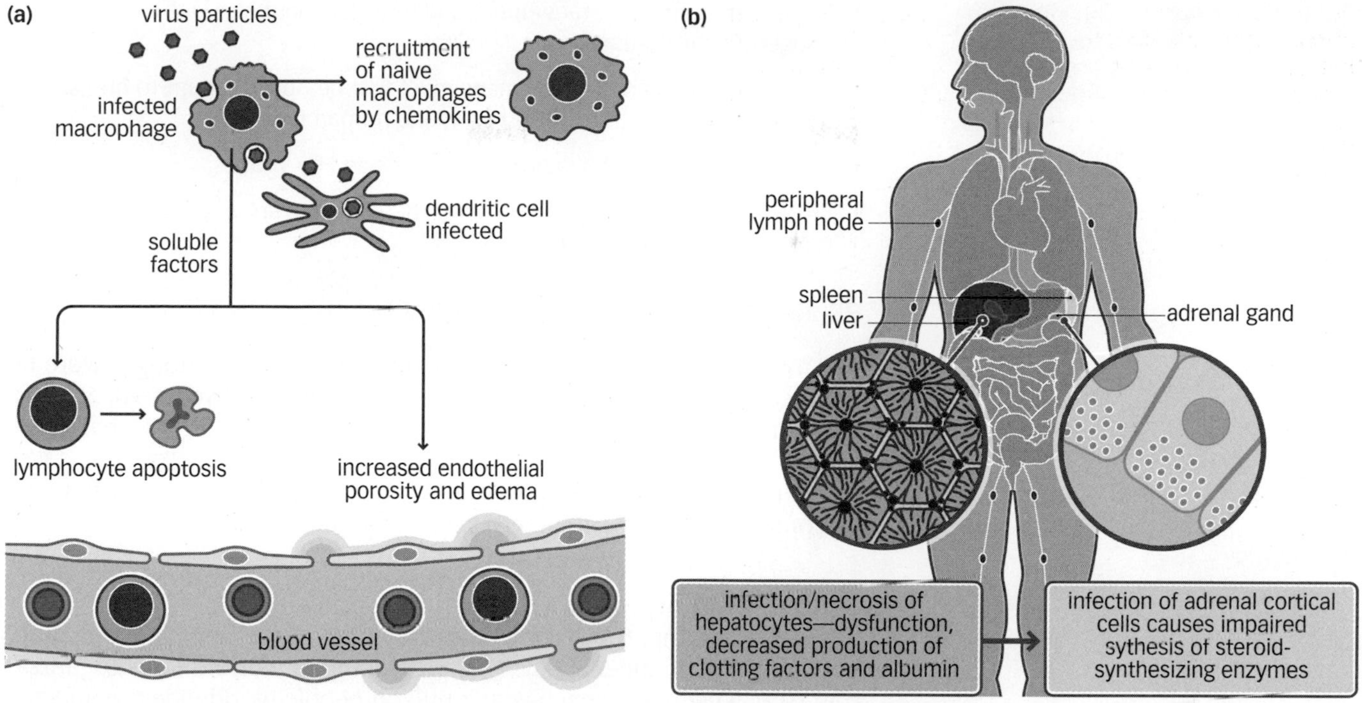

Figure 8.9 A model for VHF pathogenesis. Panel **a**: Virus spreads to regional lymph nodes, liver, and spleen. At these sites, virus infects macrophages (including Kupffer cells) and dendritic cells. Soluble factors are released and act both locally and systemically. This recruits more macrophages to the site, and these also become infected. This process continues and the host defenses begin to malfunction with a loss of lymphocytes. Panel **b**: Hemodynamic and coagulation disorders common to all VHF infections are exacerbated by infection of hepatocytes and adrenal cortical cells, which results in reduced synthesis of albumin by hepatocytes and leads to reduced plasma osmotic pressure. The result is edema and eventual hypotension.

Abnormalities in blood coagulation and *fibrinolysis* (the destruction of fibrin) during VHF infection result in **petechiae**, **ecchymoses** (bluish or purple patches on the skin), and mucosal hemorrhage. However, massive loss of blood is atypical, restricted largely to the intestinal tract, and not enough to cause death. Cutaneous flushing, rash, **thrombocytopenia**, and **DIC** (**disseminated intravascular coagulation**) characterize most VHF infections. This syndrome is characterized by an increased rate of blood coagulation, an increase so great that the large number of clots in the blood vessels can lead to organ failure. In humans, a large proportion of hemorrhagic manifestations are associated with fatal outcome.

Fortunately, outbreaks of VHF infections are very infrequent, small in size, and usually confined to remote areas. Heightened interest in bioterrorism has changed the perspective on these diseases, however, and consequently there is a renewed interest in vaccine development. There are as yet no vaccines available, and so far all attempts to develop one have failed.

Therapy for VHF Infection. Currently, there are no useful therapeutic strategies to deal with VHF infections. Early detection seems to give the greatest chance of success if the viral load can be kept low. However, it has been shown that even decreasing the viral load by 50% may not be sufficient to stop the disease. Remember, Ebola infections have a fatality rate of 50 to 90%. If left unchecked, VHF infection will move to hepatocytes and to adrenal cortical cells. This leads to the coagulation abnormalities seen with these infections and results in multiple organ failures similar to those seen in septic shock.

Fast Fact

Two new vaccines have recently been developed against Marburg and Ebola virus. These vaccines may someday protect health care workers and scientists who come in contact with these very dangerous viruses.

Keep in Mind

- Emerging infectious diseases are those whose incidence in humans has increased in the past 30 years or threatens to increase in the near future.
- The environment can contribute to emergence of disease when humans come into contact with previously unknown pathogens.

- Genetic mutation, rearrangement and re-assortment can give rise to more dangerous pathogens.
- Pathogens that can affect humans must first be able to adapt to humans and then be easily transmitted from one human to another.
- Most emerging infectious diseases are caused by viruses.
- SARS, West Nile virus and viral hemorrhagic fevers are examples of emerging infectious diseases.

RE-EMERGING INFECTIOUS DISEASES

As **Table 8.2** points out, a number of diseases we once thought were no longer a threat to humans have bounced back in recent years. All of them, from **cryptosporidiosis** to **yellow fever**, present important challenges to health care workers today. Coverage of all the re-emerging diseases listed in Table 8.2 would be beyond the scope of this chapter. Instead, let us consider two of them: tuberculosis and influenza.

Tuberculosis

It is estimated that 2 billion people (one-third of the world's population) are infected with tuberculosis (TB). According to World Health Organization (WHO) estimates, each year 8 million people worldwide are infected with TB, and 2 million die. In 2006, the Center for Disease Control and Prevention (CDC) reported 13,779 cases of active TB in the United States. In fact, *Mycobacterium tuberculosis* is still the leading killer of young adults worldwide. Minorities are affected disproportionately by TB, which occurs nine times more frequently among foreign-born individuals living in the United States than in people born here.

The development of effective antibiotics in the 1950s slowed the spread of TB for some years, but by the year 2000, the incidence of TB had begun to rise. Several factors, often interrelated, are behind the resurgence of TB:

- The HIV/AIDS epidemic: people with HIV are particularly vulnerable to infection with *M. tuberculosis* and to developing the active form of TB rather than the more common latent form.
- Increased poverty, intravenous drug abuse and homelessness: TB is rapidly spread in crowded shelters and prisons, where people are weakened by poor nutrition, drug addiction, and alcoholism while being constantly exposed to *M. tuberculosis.*

Table 8.2 Re-emerging Infectious Diseases

Disease	Infectious agent	Contributing factors
Cryptosporidiosis	*Cryptosporidium parvum*	International travel; contaminated water supplies
Diphtheria	*Corynebacterium diphtheriae*	Interruption of immunization program due to political changes
Influenza	Influenza virus	Genetic re-assortment
Malaria	*Plasmodium* species	Drug resistance; inadequate mosquito control
Pertussis	*Bordetella pertussis*	Refusal to vaccinate; decreased vaccine efficiency; waning immunity in adults
Rabies	Rhabdovirus	Breakdown in public health measures; travel; changes in land use
Tuberculosis	*Mycobacterium tuberculosis*	Antibiotic resistance; increased immunocompromised populations
Yellow fever	Flavivirus	Urbanization; insecticide resistance

- People from many other nations moving to the United States: increased numbers of people born in places where many cases of TB occur, such as Africa, Asia, and Latin America, are immigrating to the United States. In fact, TB cases among foreign-born nationals now living in the United States account for more than half of the US total.

- Increased numbers of residents in long-term care facilities: many elderly people whose general health is declining develop active TB from latent TB infections they have had for many years. Others are exposed to TB within the facility and because of lowered immune system protection they rapidly develop active TB infections.

- Failure of patients to take all prescribed antibiotics against TB: patients with TB who do not complete the required drug treatment can stay infectious for longer periods and spread the infection to more people. In addition, failure to follow the treatment protocols can result in the evolution of strains of *M. tuberculosis* that are resistant to the standard antibiotic treatments.

Fast Fact

In sub-Saharan Africa it is estimated that 50% of HIV-infected patients also have TB

Early detection is critical to prevent the spread of TB and to increase the chances of successful treatment. Because the initial signs are similar to those seen in other respiratory infections, it is important to look for signs such as fever, fatigue, weight loss, chest pain, and shortness of breath in conjunction with coughing as indications of potential TB.

We will cover the pathogenesis and treatment of TB in Chapter 21, which deals with diseases of the respiratory system.

Influenza

Unlike measles or smallpox, which are genetically stable, influenza is caused by an RNA virus that is continuously undergoing mutations that change the characteristics of the virus an infected host must fight off. When we couple this high mutation rate with the fact that this virus has a stable animal reservoir (aquatic birds), it becomes apparent that new epidemics and pandemics are likely to occur, and the eradication of the disease will be very difficult to achieve.

The influenza virus contains eight segments of RNA and has two major glycoproteins on its surface, hemagglutinin and neuraminidase, which are required for successful infection. There are at least 15 hemagglutinin subtypes and 9 neuraminidase subtypes. Human infection has been linked to hemagglutinins H1, H2, and H3 and to neuraminidases N1 and N2. All the other subtypes of the virus are found primarily in birds. As a result of changes in the surface glycoproteins, there have been devastating epidemics and pandemics in humans, and major *epizootics* (animal epidemics) in poultry, pigs, horses, seals, and even camels.

As we mentioned previously, there were three human influenza pandemics in the past century. The 1918 pandemic of **Spanish influenza** was the most highly contagious and deadly of the three and had a major global impact (**Figure 8.10**). It was caused by a virus containing H1 and N1 surface glycoproteins and was considered the most deadly in recorded history, killing an estimated 30–50 million people in *less than one year*! The outbreak in 1957, known as **Asian influenza**, was characterized by H2 and N2 glycoproteins. Because these proteins were different from those found in the Spanish virus, there was no herd immunity to the Asian variety. Recall from Chapter 7 that herd immunity results from having individuals in the population who are immune to the infection, thereby limiting the number of potential targets for infection. It is estimated that the Asian virus of 1957 caused 70,000 deaths in the United States, and many more

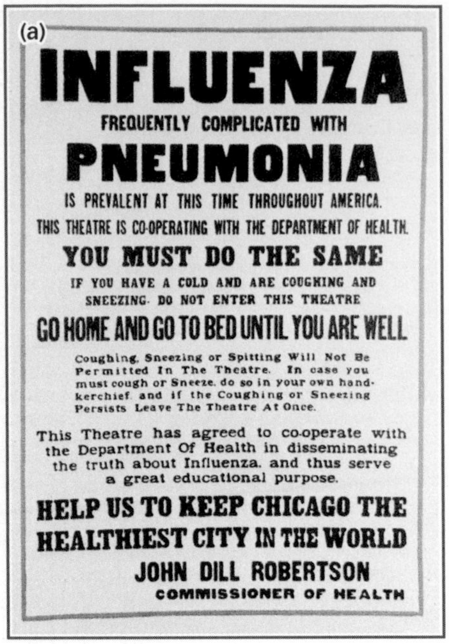

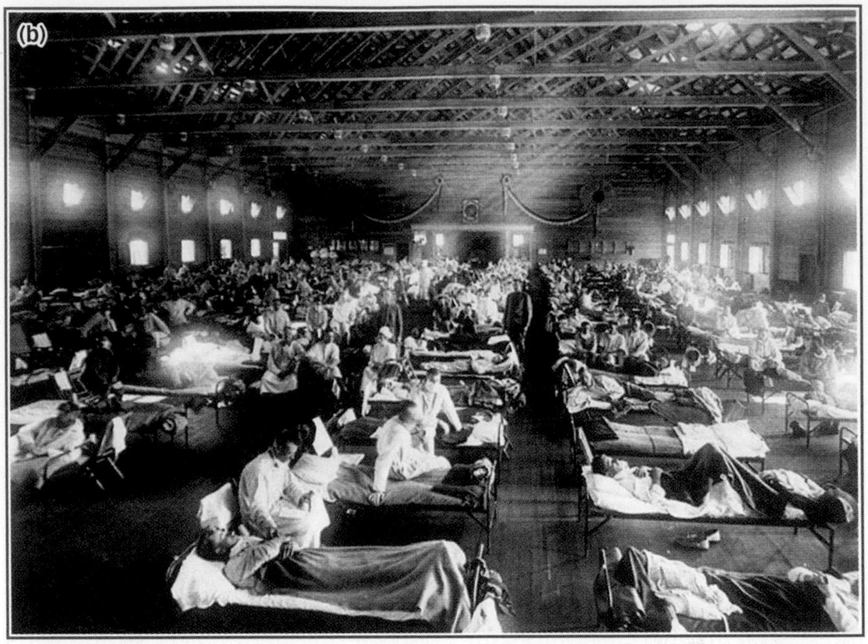

Figure 8.10 The 1918 influenza pandemic Panel **a**: A public health poster from the Spanish influenza pandemic of 1918. Panel **b**: Hospital wards filled with patients during the 1918 flu pandemic. Note the absence of any isolation facilities.

worldwide. In 1968, **Hong Kong influenza** killed an estimated 30,000 people in the United States. In this case, the number of deaths may have been kept low because this virus contained the same H2 and N2 glycoproteins as those found in the Asian virus, so that there may have been residual immunity in the population. An outbreak of the influenza known as swine flu emerged in Mexico and the United States in April of 2009 and rapidly spread to 74 countries around the world. In June 2009, there had been over 27,000 reported cases worldwide, including over 140 deaths, and the World Health Organization declared the outbreak a pandemic, making it the first influenza pandemic since the Hong Kong flu of 1968. Swine flu is subtype H1N1, the same as the devastating Spanish flu of 1918, and as with that virus it preferentially infects younger people. As yet it is only a moderate illness and most experience mild symptoms. A vaccine to combat the swine flu is in production at several pharmaceutical companies around the world.

Virulence Factors of Influenza

Though influenza does not usually result in high death rates, it remains a major public health problem that causes losses in work and school time. The outcome of the infection depends on both the virus and the health status of the host. A host's previous exposure can lead to partial protection, but in a naive host (one that has never been exposed to the virus) it is the virulence of the influenza subtype that determines the outcome of the infection. The virus has what are called **gene constellations**, clusters of genes that determine its virulence, and single mutations in these constellations can markedly affect the level of virulence of the strain. Given that influenza viruses are continuously undergoing mutations, there is a continuing potential for increased virulence in future strains.

We will cover the pathogenesis and treatment of influenza in Chapter 21.

Keep in Mind

- Some infectious diseases are re-emerging. They were previously thought to be controlled but now have returned.

- Tuberculosis is one of these re-emerging infectious diseases because of several contributing factors such as HIV infection, poverty and failure to follow recommended treatment protocols.
- Tuberculosis is becoming more resistant to antibiotic therapy.
- Influenza is another good example of a re-emerging infectious disease because of the variety of strains that continue to cause infection.
- Influenza virus contains gene constellations that help to determine its virulence.

AVIAN INFLUENZA

One of the most potentially devastating emerging infections is **avian influenza** — "bird flu." We have therefore chosen to separate this from the discussion of other emerging infections. Recall that viruses must be able to overcome two hurdles: they must be able to jump from some nonhuman species into humans, and they must be transmissible from human to human. The avian influenza virus has already overcome the first hurdle and at the time this book is being written there have been several hundred isolated cases in Asia. But the virus has not yet cleared the second hurdle and is as yet not easily transmitted from person to person.

The avian virus has the subtype designation H5N1 and is one of 15 subtypes that use birds as a reservoir. The H5N1 virus mutates at an extremely high rate and can acquire genes from other viruses. This is what makes it so potentially dangerous. Human influenza viruses have already overcome the hurdle of transmission from person to person. Thus if the H5N1 avian virus were to acquire these genes, it would be immediately transmissible. In point of fact, this may have already occurred.

In 1996 and 1997, the first cases of avian influenza were observed in humans. The virus had successfully jumped from poultry into humans. This led to the killing of more than 1 million birds, a measure that initially squelched the outbreak. Six years later, however, the H5N1 virus reappeared and was much more pathogenic, causing deaths in both birds and people. Although the human fatalities were small in number, they were scattered all across Asia, indicating that the infection was spreading. Today H5N1 is spreading at an alarming rate because migratory birds are carrying the virus in ever-expanding patterns. The only good news concerning H5N1 is that nearly all confirmed cases so far have resulted from direct contact with infected birds.

The H5N1 virus is more deadly than any of the viruses seen in previous influenza pandemics, including the devastating 1918 pandemic. It is estimated that H5N1 could have a 50% mortality rate, 10 times that seen with the 1918 pandemic. A 20% fatality rate would translate to a half million dead and 2 million hospitalized in the United States alone. Thus the effects of a pandemic having a 50% mortality rate are truly frightening.

Many of the scenarios now being described tell the tale. For example, if the United States is not prepared when this infection begins, the hospitals in New York City alone could see 300 to 400 cases of influenza a day, many of which will be patients in serious condition. Numbers of this magnitude will quickly overwhelm the health care system. Furthermore, the cost of a pandemic to the US economy could be several hundred billion dollars, a figure that does not include the health care costs, which could exceed 100 billion dollars.

Adding to the problem is the fact that the H5N1 strain is resistant to amantidine and rimantadine, two of the most effective antiviral drugs. This resistance is attributed to efforts in Asia to control the epidemic by treating birds with massive doses of these drugs. This approach did little to solve the problem, however, and in fact worsened it by fostering resistance. The drug Tamiflu® has some effect on the H5N1 stain but is in extremely short supply. Because every country is now clamoring for Tamiflu®, there is little optimism that enough can be produced in time to make a difference.

The last piece of the picture concerns the development of a vaccine. The "flu shot" currently given each year is composed of only a few of the more than 100 subtypes of influenza virus that have been identified. Since we know that H5N1 is the most dangerous subtype, concerted efforts to develop a vaccine for this subtype have been under way for several years. Unfortunately, so far these efforts have been unsuccessful.

Both the H5N1 avian virus and the human influenza virus can be transmitted to pigs. This presents the potential for an increased rate of gene re-assortment using the pig as an "incubator" (**Figure 8.11**). If the genes for transmissibility from human to human (which come from the human influenza virus) combine with avian influenza genes, the resulting virus would be extremely pathogenic and highly transmissible in humans. Perhaps more importantly, this combination will be one that has not been seen in humans. Therefore there is no herd immunity, yielding the

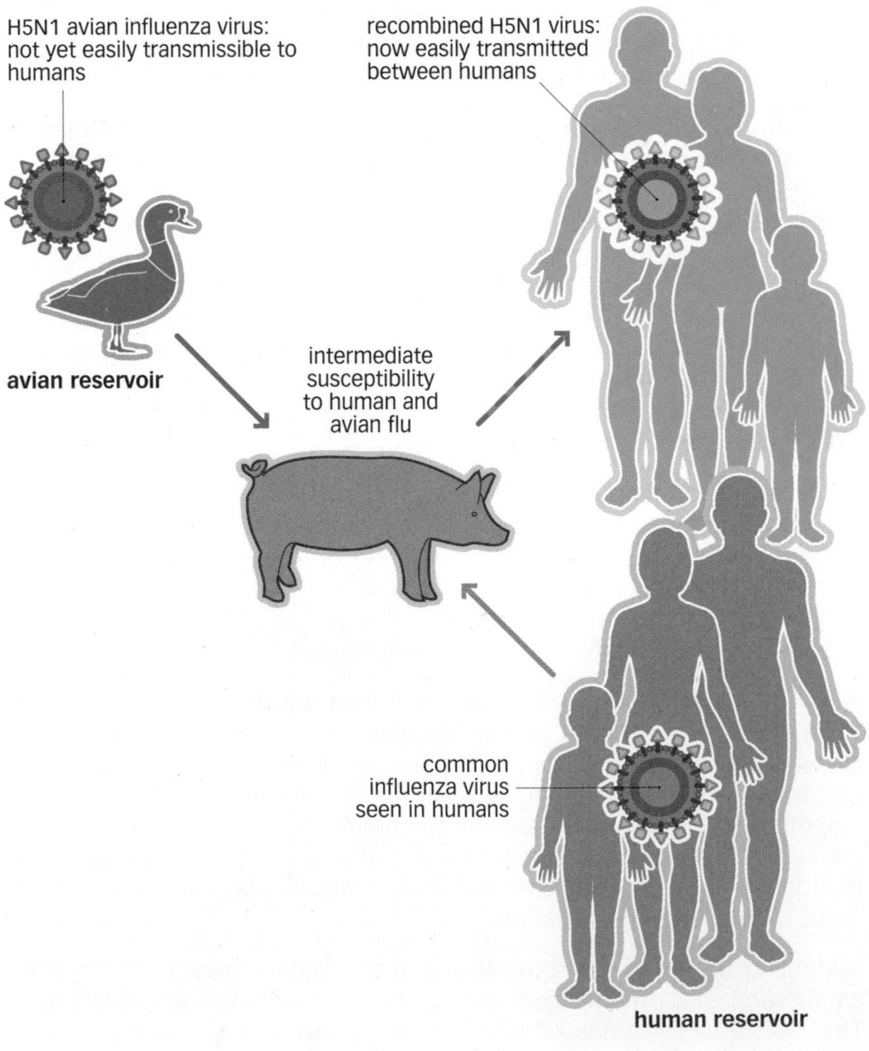

Figure 8.11 An illustration of how the avian influenza virus (H5N1) could become easily transmissible between human hosts. Both the avian virus and the influenza virus that is easily transmitted by humans can infect the pig. Re-assortment of genes within the pig could lead to an H5N1 virus that could easily move from human to human.

prospect of a widespread and potentially devastating epidemic. It must be stressed that the governments of the world are becoming more and more aware of the danger, and are working together to develop vaccines and treatments to deal with this potentially devastating problem.

Keep in Mind

- Avian influenza ("bird flu") has the potential to become a dangerous problem.
- The virus can jump from birds to humans but has as yet not become transmissible from human to human.
- If this infection becomes epidemic in the United States it could cripple the health care system.

PRIONS AND PRION DISEASES (TRANSMISSIBLE SPONGIFORM ENCEPHALOPATHY)

We have separated **prion** diseases from the other emerging infectious diseases because these infectious diseases are not caused by a microorganism at all. In 1982, Stanley Prusiner, a medical researcher, proposed the existence of diseases caused by infectious proteins called *prions*, and this work won him the Nobel Prize in Physiology or Medicine in 1997. It was hard to attribute disease to anything that is not an organism, but Prusiner and others showed that prions are resistant to inactivation by heat or radiation, one characteristic of living organisms. More important, enzymes such as RNase and DNase, which can destroy nucleic acids, have no effect on prions, indicating that prions contain no nucleic acids, another characteristic that disqualifies them from being called organisms. Study of prions led to the prion hypothesis.

The Prion Hypothesis

The *prion hypothesis* is centered on the premise that all normal nerve cells contain prions, proteins that are known as PrP^c (prion protein cellular) and found predominantly on the outer membrane of neurons. The function of PrP^c prions is not yet understood, but it has been proposed that they may help in some way to maintain the body's population of stem cells.

Infectious prions, designated **PrP^{sc}**, seem to be prions that become folded improperly and then cause normal PrP^c prions to fold improperly. The abbreviation PrP^{sc} stands for **prion protein scrapie**. (*Scrapie* is a neurological disease known for many years to affect sheep.) PrP^{sc} prions seem to aggregate into fibrous structures that disrupt the cell membrane and cause cell death.

Abnormal PrP^{sc} prions have a predominance of flat, β-pleated sheets in their structure, whereas normal PrP^c prions contain mostly helical folding (**Figure 8.12**). The abnormal prions are found inside the nerve cells rather than on the membrane and, as noted above, can convert PrP^c prions into the abnormal form. For reasons that are not yet understood, when PrP^c prions come into contact with PrP^{sc} prions, the former unfold and then refold abnormally

Abnormal PrP^{sc} prions are practically indestructible. They are not destroyed by normal cooking, and even the standard autoclave is not enough to deactivate them. Abnormal prions survive disinfectants and antiseptics and are resistant to strong alkali solutions for over an hour. They can also survive in soil for over three years and still be infective. To effectively inactivate abnormal prions requires treatment for one hour with bleach containing 2% chlorine or one hour of autoclaving in an alkali solution.

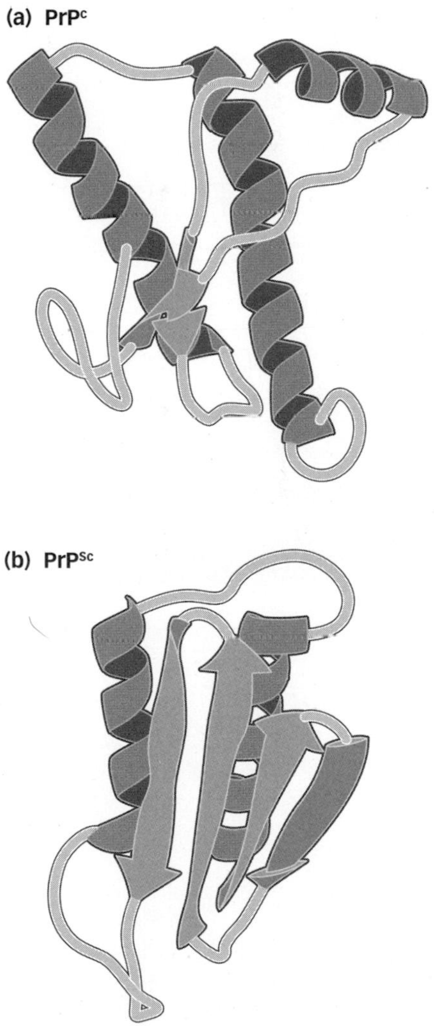

Figure 8.12 An illustration of the differences in structure between the normal and abnormal forms of prion protein. Panel **a**: the normal form, PrP^c. Panel **b**: the abnormal form, PrP^{sc}.

Prion Diseases

Infective (PrP^{sc}) prions can be ingested with prion-containing material. These prions can then pass through the intestinal wall rapidly and enter lymph nodes, where they incubate. Researchers believe the next step in infection to be that the abnormal prions are picked up by peripheral nerves and transported to the spinal cord and brain. Infectious prions can be transmitted between species, but when they do cross species the incubation time becomes significantly longer.

Prion diseases are referred to as **transmissible spongiform encephalopathy** (**TSE**). They are neurodegenerative diseases for which there is no treatment or cure. These infections seem to strike at random and there is no reliable test for them. TSE can affect humans as well as cattle and sheep. It is known as **mad cow disease** in bovines and **scrapie** in sheep, as noted above.

Prion disease was first observed in humans in the Fore people of New Guinea in the 1920s. It was the custom of these people to honor their dead by cannibalizing them. Women, infants, and children ate the brains and internal organs of the deceased, while men consumed the muscles. Women also spread the tissues of the dead over their skin. This material dried and remained on the skin for weeks.

In the 1920s a mysterious and horrifying disease spread among the Fore people, a disease that incapacitated and killed many of them. The men were rarely affected, and the disease was primarily found in women and children. The symptoms of the disease included lack of coordination, staggering, slurred speech, dramatic mood swings, paralysis, and death within one year. The Fore people called the disease **kuru** (to tremble with fear).

Fast Fact

In 1976, the Nobel Prize in Physiology or Medicine was awarded to D. Carlton Gajdusek for his work on kuru.

Autopsy showed that the brains of the afflicted were full of abnormally large deposits of protein called plaque. Study of the Fore people showed that kuru had a long incubation period (40 or more years) and that the infectious agent could survive cooking. When the practice of cannibalism was stopped, kuru disappeared among the Fore people.

Mad cow disease, which is similar to kuru, was first seen in England in 1984 and by the year 2000 there were 180,000 confirmed cases in cattle in that country. It was estimated at that time that as many as 2 million cattle in England could have been infected but were asymptomatic. Of these, 1.6 million could have already entered the food chain. The incubation time for cattle seems to be shorter than that observed with kuru and is estimated to be five years. The infection in cattle has been attributed to feed supplemented with sheep brains, a product used in an effort to increase the weight of the cattle. The belief is that the feed supplement was made from sheep infected with scrapie.

In 1996, the British government announced that mad cow disease had infected a human. Ten young patients showed the symptoms, and all of them degenerated clinically and died quickly. So far, more than 120 people have died of TSE in the United Kingdom and in continental Europe, with the incidence increasing by 20% a year. It is estimated that a few hundred to 150,000 people will become infected with TSE in the next few decades.

The Biology of TSE

The characteristics of transmissible spongiform encephalopathy are long incubation periods, spongiform effects in brain tissue, and protein deposits (called *plaque*) in the brain. There are no antibodies produced in response to TSE, and there is no inflammatory response observed during the infection.

Five forms of TSE are seen in humans: kuru, Creutzfeldt–Jacob disease (CJD), variant *CJD* (vCJD), **Gerstmann–Straüssler–Scheinker syndrome** (GSS), and **fatal familial insomnia** (FFI). These different forms affect different areas of the brain, as illustrated in **Figure 8.13**. We will focus our discussions on CJD and vCJD.

Creutzfeldt–Jacob Disease

This form of TSE was first identified in the 1920s and is the most common form in humans. The rate of infection with CJD is one person per million per year, with the infection predominantly seen in people aged 55 to 75 years. Signs are rapid progressive dementia, visual problems, speech abnormalities, muscle tremors, agitation, and depression. Death usually occurs within 12 months after signs first appear. The brain of those infected with CJD shows widespread spongiform changes and moderate protein deposits. The incubation time for CJD is estimated to be from 3 to over 20 years.

There are three forms of CJD: sporadic, familial, and iatrogenic. The sporadic form encompasses 85% of the cases seen, seems to occur spontaneously and has no identifiable risk factors. Familial CJD accounts for 10% of the cases and is genetically linked to mutations in the genes coding for normal prions (*Prnp*). The iatrogenic form of CJD is transmitted from person to person through medical treatments such as blood transfusions, cornea transplants, and dura mater grafts. It can also occur as a result of contaminated instruments or growth hormones obtained from the pituitary glands of infected individuals.

Variant CJD (vCJD)

This form of TSE was first reported in 1996 and has unique properties that distinguish it from CJD. With vCJD, patients are usually young (16 to 39 years old), whereas patients with CJD are usually over 50 years old. The first symptoms of vCJD are behavioral and include aggressiveness, anxiety, apathy, depression, and delusions. The first neurological symptoms, such as shaking, incontinence, and immobility, are not seen for the first six months. On autopsy, brain tissue from infected patients resembles that seen in kuru. vCJD is also different from CJD in that the plaque deposits found in the brain tissue of patients with vCJD are larger than those found in the brain tissue of patients with CJD.

In the United States, new regulations have been enacted to make sure that the blood supply is safe from infectious prions. These laws include banning donors who resided in the United Kingdom for three or more months between 1980 and 1996. Blood donors who have resided in continental Europe for five or more years or received blood transfusions in the United Kingdom between 1980 and the present are also banned from donating blood.

Keep in Mind

- Prions are proteins that cause transmissible spongiform encephalopathy in humans and mad cow disease in cattle.
- Kuru was one of the first human prion diseases described in detail.
- Transmissible spongiform encephalopathy comes in five forms in humans, but all have the same general characteristics: long incubation periods, spongiform defects in brain tissue, and plaque deposits in the brain.

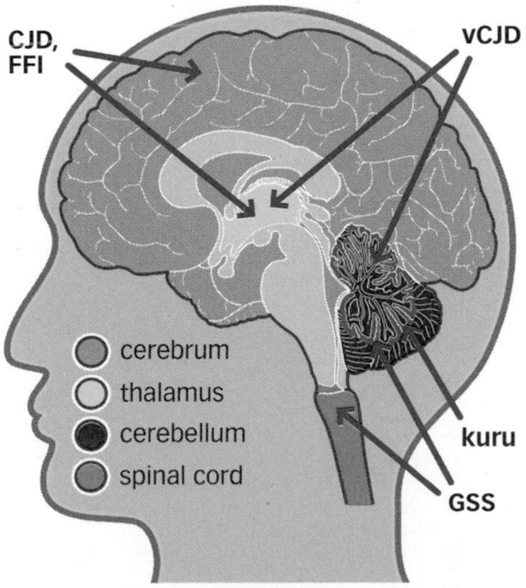

Figure 8.13 Different forms of prion disease affect different sites in the brain.

SUMMARY

- Emerging and re-emerging diseases will provide new and challenging problems for health care professionals.
- Genetic mutation, environmental changes, and travel can contribute to these types of infection.
- Most emerging diseases are caused by viruses.
- Re-emerging diseases are those that were thought to be under control but have returned. In many cases this is due to an increase in resistance to antibiotics.
- Avian influenza has the potential to become a very dangerous problem that could cripple the health care system of the United States.
- Prions are proteins that are not folded correctly and cause transmissible spongiform encephalopathy in humans and mad cow disease in cattle.

Emerging and re-emerging infectious diseases will have an increasing impact on health care, and some of these diseases could have devastating effects on the health care systems in the United States and in the rest of the world. The resulting problems will directly affect you as a health care professional. Therefore it is important for us to include these diseases in our study of microbiology.

SELF EVALUATION AND CHAPTER CONFIDENCE

Multiple Choice

Answers are given in the back of the book and help can be found in the student resources at:
www.garlandscience.com/micro

1. All of the following are emerging infectious diseases except
   - **A.** Creutzfeldt–Jacob disease
   - **B.** Legionnaires' disease
   - **C.** Mumps
   - **D.** SARS
   - **E.** Hepatitis C
2. Emerging infectious diseases are classified as those for which
   - **A.** The incidence has decreased in the past 50 years
   - **B.** The incidence has remained unchanged over the past 10 years
   - **C.** The incidence has decreased over the past 30 years
   - **D.** The incidence has increased over the past 30 years
   - **E.** None of the above
3. The Nipah virus originated in
   - **A.** Pigs
   - **B.** Chickens
   - **C.** Fruit bats
   - **D.** Monkeys
   - **E.** Rats
4. The Hantavirus originated in
   - **A.** North American rodents
   - **B.** Fruit bats
   - **C.** South American rodents
   - **D.** Monkeys
   - **E.** None of the above
5. New, rapidly spreading pathogens arise because of
   - **A.** Cultural changes
   - **B.** Agricultural improvements
   - **C.** Urban sprawl
   - **D.** Genetic re-assortment
   - **E.** All of the above
6. SARS emerged from
   - **A.** North America
   - **B.** Malaysia
   - **C.** Hong Kong
   - **D.** China
   - **E.** South America
7. HIV emerged from
   - **A.** North America
   - **B.** China
   - **C.** Japan
   - **D.** Africa
   - **E.** None of the above

8. Which of the following has achieved the ability to be transmissible from one human to another?
   A. Nipah virus
   B. HIV
   C. SARS
   D. Hantavirus
   E. B and C

9. SARS causes infection of the
   A. Digestive system
   B. Nervous system
   C. Respiratory system
   D. Skin
   E. Reproductive system

10. West Nile virus is transmitted by
   A. Rat fleas
   B. Mosquito saliva
   C. Fecal–oral contamination
   D. Ticks
   E. Sexual intercourse

11. West Nile virus causes infection in the
   A. Respiratory system
   B. Skin
   C. Digestive system
   D. Nervous system
   E. Urinary system

12. West Nile virus is initially spread from mosquitoes to
   A. Humans
   B. Birds
   C. Pigs
   D. Rodents
   E. None of the above

13. Viral hemorrhagic fever infections include all of the following except
   A. Ebola
   B. Marburg
   C. Yellow fever
   D. Measles

14. Which of the following is a characteristic seen in VHF infections?
   A. Rash
   B. Cutaneous flushing
   C. Thrombocytopenia
   D. Intravascular coagulation
   E. All of the above

15. The influenza genome is unique in that it is made of
   A. DNA only
   B. RNA only
   C. Eight segments of DNA
   D. Eight segments of RNA
   E. Both DNA and RNA

16. The influenza outbreak of 1918 was known as the
   A. Hong Kong flu
   B. Swine flu
   C. Spanish flu
   D. Asian flu
   E. Japanese flu

17. The virulence of influenza is determined by
   A. The health of the host
   B. Segments of the DNA chromosome
   C. Gene constellations
   D. DNA constellations
   E. None of the above

18. The subtype of the avian influenza virus referred to as bird flu is
   A. H1N1
   B. H5N2
   C. H3N1
   D. H5N1
   E. None of the above

19. Prion diseases are caused by
   A. Bacteria
   B. Viruses
   C. Fungi
   D. Abnormal proteins

20. In prion diseases, normal proteins
   A. Couple with infecting viruses
   B. Integrate into bacterial cells of the normal flora, making them infectious
   C. Refold into abnormal shapes
   D. Cause viruses to produce plaque proteins that infect the brain
   E. Remain normal throughout the infection

21. Ingested prions initially incubate in the
   A. Brain
   B. Spinal cord
   C. Digestive tract
   D. Lymph nodes
   E. None of the above

22. Transmissible spongiform encephalopathy was first seen in
   A. Great Britain
   B. France
   C. China
   D. New Guinea
   E. The United States

DEPTH OF UNDERSTANDING

Questions listed here require you to bring together the concepts you have learned in this chapter into a discussion format. This helps you to increase your depth of understanding of the material you have learned. Help can be found in the student resources at: www.garlandscience.com/micro

1. People wishing to donate blood must answer several questions before donating. One of these questions asks if they have lived in the United Kingdom for longer than three months between 1980 and 1996 or more than five years in Europe between 1980 and the present. Why would these questions be important, and why would a positive answer be reason for rejecting their donation?

2. What is the relationship, if any, between socioeconomic conditions and emerging infectious diseases?

3. Three influenza pandemics have occurred in the United States since 1918. Were they connected to one another in any way, and if so, how?

CLINICAL CORNER

Help can be found in the student resources at: www.garlandscience.com/micro

1. Mr Johnson, a 75-year-old male, was admitted to the hospital complaining of flu-like symptoms, shortness of breath and chest pain. He was producing large amounts of mucus that contained blood. On examination, he was diagnosed with tuberculosis. Mr Johnson's daughter tells you that her father had TB over 20 years ago. He was treated for it and has been healthy ever since.
   - **A.** What would you tell Mr Johnson's daughter about his current condition?
   - **B.** Should Mr Johnson be put on the antibiotic therapy routinely prescribed for TB?
   - **C.** Can you think of any tests that could help in the treatment of Mr Johnson?

2. Your patient is a 50-year-old white male who has been admitted to the psychiatric ward because of dramatic mood swings and bouts of uncoordinated movements. His speech is also slurred. He has been prescribed medication to stabilize his mood swings. During his admission interview he mentions that he has lived in England for most of his life and has never had any medical problems except for complications associated with a corneal transplant when he was 35 years old.
   - **A.** Is there another possible explanation for his condition?
   - **B.** How would you test for that other possibility?

PART III

Characteristics of Disease-Causing Microorganisms

PART I gave us a basic understanding of the infectious process and the chemistry we need. In **PART II**, we looked at the disease process. Having studied the disease process, our journey now takes us to a detailed discussion of the organisms that can infect us. We have divided our discussions into bacterial pathogens, viruses, and parasitic organisms and include chapters on growth requirements and bacterial genetics.

In **Chapter 9**, we thoroughly investigate the anatomy of the bacterial cell. In this chapter you will learn about the structures that make up bacteria and how they can be used in the infection process. You will see that many of these structures satisfy several of the five requirements for satisfactory infection, for example, getting in, staying in, and defeating the host defense that we learned about in **PART I** and **PART II**.

Chapter 10 teaches you about the requirements for growth of bacteria, and here you will see that many pathogenic bacteria have very strict requirements for the essentials that they require to divide and grow.

Chapter 11 gives you a brief description of bacterial genetics. Many students do not realize how important this information is. The chapter has four major themes: replication, gene expression, mutation, and transfer of genetic information. All of these subjects are intimately involved in the infection process of pathogens.

Chapters 12 and **13** are devoted to viruses. In **Chapter 12**, we look at the structures of viruses that infect us and find that there is a distinct cycle involved in infection. In **Chapter 13**, we look at the pathogenesis of viral infection and the development of vaccines.

Chapter 14 gives us an overview of protozoan parasites, worms, and fungi, each of which can infect humans.

When you have finished with **Part III** you will:

- Understand how the structures of the bacterial cell can provide assistance in the infection process and also help to defeat the defenses of the infected individual.
- See how bacterial structures are also targets for antibiotic treatment.
- Have an understanding of how important genetic mechanisms are to the infection process and how mutations and the transfer of genetic information can increase the severity and duration of an infection.
- See how variable the structure of viruses can be and how these structures have an important role in the process of viral infection.
- Realize how widespread and diverse protozoan and parasitic infections can be, and also how fungal infections are on the rise as a result of compromise of host defenses.

Once again, it is important to remember that the things we learn in **PART III** will provide essential information for understanding what we will talk about in **PART IV**.

Chapter 9
The Clinical Significance of Bacterial Anatomy

Chapter 10
Bacterial Growth

Chapter 11
Microbial genetics and Infectious Disease

Chapter 12
The Structure and Infection Cycle of Viruses

Chapter 13
Viral Pathogenesis

Chapter 14
Parasitic and Fungal Infections

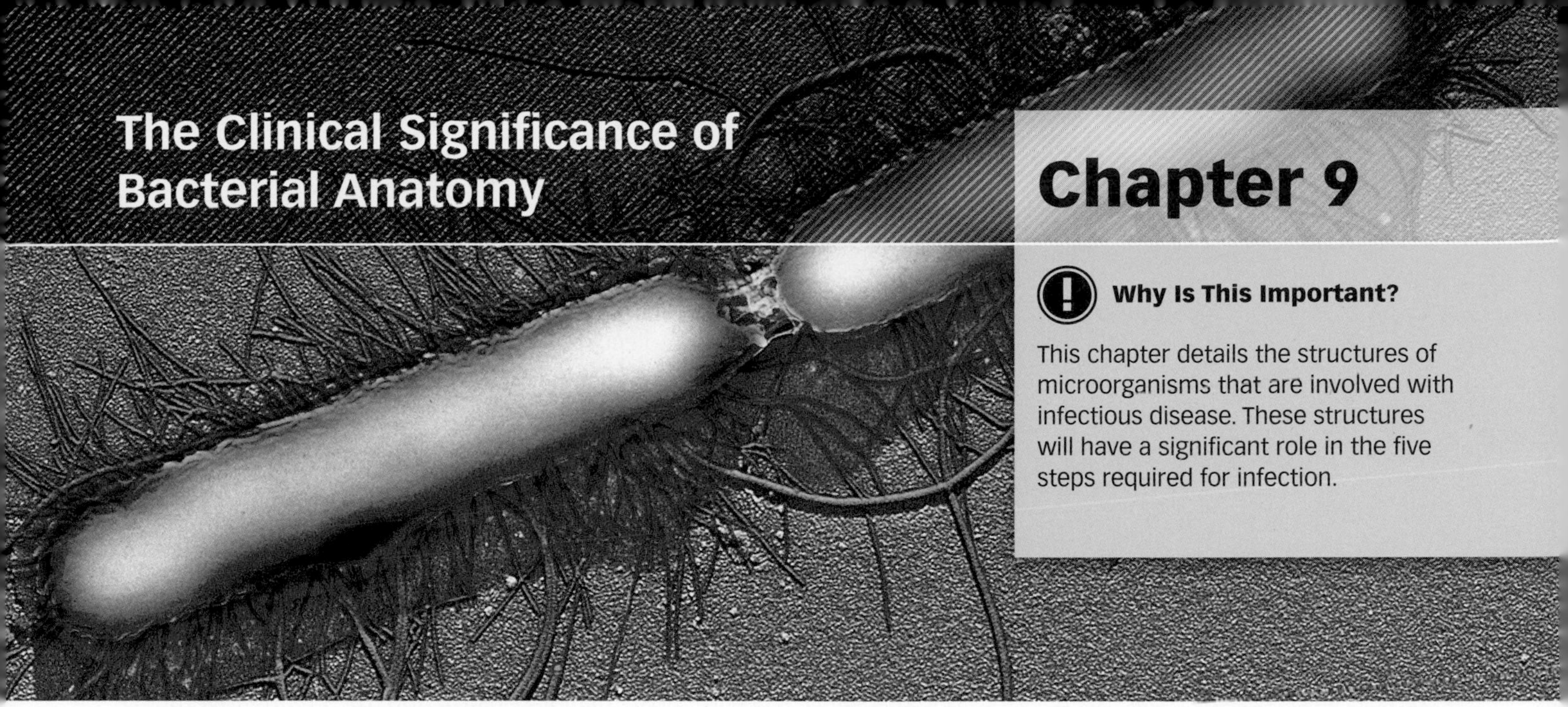

The Clinical Significance of Bacterial Anatomy

Chapter 9

Why Is This Important?

This chapter details the structures of microorganisms that are involved with infectious disease. These structures will have a significant role in the five steps required for infection.

OVERVIEW

In this chapter, we examine the anatomy of the bacterial cell. It is important that we understand the structures associated with bacterial cells because so many of them have an important role in the generation of disease. As you read this chapter, keep in mind what you learned in previous chapters, especially structure–function relationships. As we saw in PART I, at the molecular level the three-dimensional structures of proteins are directly responsible for their functions. The same can be said for the cell wall, capsule, fimbriae, pili, and flagella, which are all structures associated with the bacterial cell. The structure–function relationships of the bacterial cell show us how bacterial structures influence the development of disease. These structures will become core objectives for our studies and will be mentioned repeatedly throughout subsequent chapters.

We will divide our discussions into the following topics:

The Clinical Significance of Bacterial Anatomy

- THE BACTERIAL CELL WALL
- STRUCTURES OUTSIDE THE BACTERIAL CELL WALL
- STRUCTURES INSIDE THE BACTERIAL CELL WALL

As we have learned, the infectious process can be considered a "two-way street." Pathogenic bacteria use a variety of structures and substances to overwhelm host defenses, while the host uses nonspecific and specific defense mechanisms as well as antibacterial drugs to stop the infection. Our discussions in this chapter include not only a description of how the structures that pathogens use to overwhelm a host are relevant to the infectious process but also a description of how some of these structures can be targets for antibiotics. It is important to remember that the bacterial structures identified in this chapter have evolved over time. This evolution continues through genetic mutations such that the host-against-pathogen struggle is continuously unfolding.

What Do I Need to Know?

To get the most out of this chapter, please review the following terms from your previous courses in biology, anatomy, physiology, or chemistry or in previous chapters of this book as indicated in parentheses: amino acid (**2**), bacteremia (**7**), endotoxin (**5**), hydrophilic (**4**), hydrophobic (**4**), inclusion body (**5**), lyse, osmotic pressure (**4**), pathogenesis (**1**), plasmid (**4**), proton motive force (**3**), ribosome (**4**), septicemia (**7**), solute, sugar molecule general structure (**2**).

THE BACTERIAL CELL WALL

Because they are prokaryotic cells, bacteria face the problem of environmental stress because they may live independently in the outside world (that is, outside some hospitable host). Consequently, they must develop mechanisms to handle osmotic pressure, which can be looked at as the mechanism that keeps the outside out of the bacterial cell and the inside in. Failure to do so is a lethal event for the organism. To accomplish this, bacteria construct a complex cell wall. This wall is composed of several parts and is different in Gram-positive and Gram-negative bacteria. The primary structure, or "backbone," of the wall is **peptidoglycan**, a structure composed of repeating molecules of the sugars ***N*-acetyl glucosamine (NAG)** and ***N*-acetyl muramic acid (NAM)**, which together with small peptide chains form the meshwork of the cell wall. **Figure 9.1** shows the structure of NAG and NAM and how they bind together in repeating units. It is important to note that this binding is covalent and therefore very strong. This strong bonding is a major part of the mechanical strength of the cell wall and provides a barrier to osmotic pressure.

Building the Cell Wall

Let's take a look at how peptidoglycan is constructed, because the mechanisms of construction have implications for antibiotic therapy and antibiotic resistance. As with most cellular construction, the linking together of the NAG and NAM subunits is facilitated by enzymes. These proteins are responsible for placing the subunits in the proper orientation for promoting growth of the NAG/NAM chain. As we learned in Chapter 2, proteins are three-dimensional molecules whose folding is based on their amino acid sequence and ultimately the DNA sequence that codes for that particular protein. In the case of proteins that are enzymes, the distinctive molecular shape resulting from the folding is what allows the molecule to present its substrate-binding site, which is the "business end" of the molecule.

There are several enzymes, including *transglycosylase*, *transpeptidase*, *polymerase*, and *hydrolase*, that work together to construct a peptidoglycan backbone. There are three phases of peptidoglycan assembly: a cytoplasmic phase, a membrane-associated phase, and an extracytoplasmic phase. The events associated with each phase are dynamic and complex,

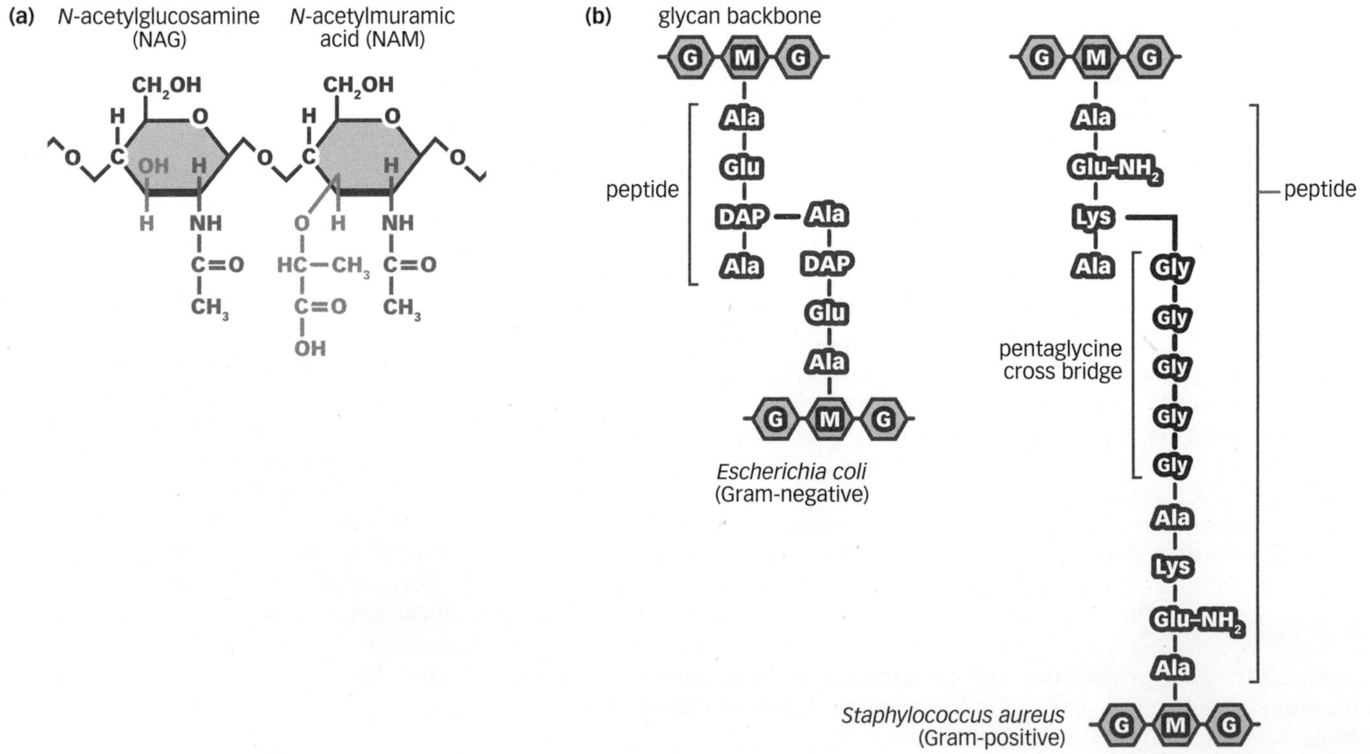

Figure 9.1 The Structural Components of the Bacterial Cell Wall. Panel **a**: *N*-acetylglucosamine (NAG) bound to *N*-acetyl muramic acid (NAM) is the basic building block of peptidoglycan. The blue lettering indicates where these two sugar molecules are the same; the gold lettering indicates where they differ. Panel **b**: The binding of peptidoglycan subunits (G representing NAG, and M representing NAM) with attached polypeptide chains. Note the difference between the Gram-positive and Gram-negative cell walls. Cross-linking in Gram-positive cells occurs between the amino acid lysine and a string of five glycine (Gly) molecules referred to as the pentaglycine crossbridge. The linkage in Gram-negative cells occurs between the amino acid alanine and diaminopimelic acid (DAP).

but understanding the basic events associated with each phase will give us an opportunity to understand the strength and durability of the cell wall and also how it can be a prime target for antibiotic therapy.

The Cytoplasmic Phase

During this phase, the NAG/NAM building blocks for peptidoglycan are fashioned in the cytoplasm of the cell. The cell wall must be constantly refurbished, especially when the cell is ready to divide. This requires that peptidoglycan layers be laid down both transversely (along the long axis of the cell) and vertically as the intracellular partition (septum) that divides the cell into the two daughter cells develops (**Figure 9.2**)

The NAG and NAM components of peptidoglycan are formed by enzymes that are encoded by the *mur*A–*F* genes. Although the chemistry associated with these reactions is too complex for this discussion, suffice it to say that the enzymes coded for by these genes attach five amino acids to each NAG and NAM molecule, hence the name *peptido* (peptide piece) *glycan* (sugar). Because peptidoglycan subunits must be assembled if a bacterial cell is to build new cell wall, this cytoplasmic phase is an important potential target for antibiotic therapy. In fact, the antibiotic **fosfomycin** targets *mur*A gene activity and thereby prevents peptidoglycan subunits from being produced.

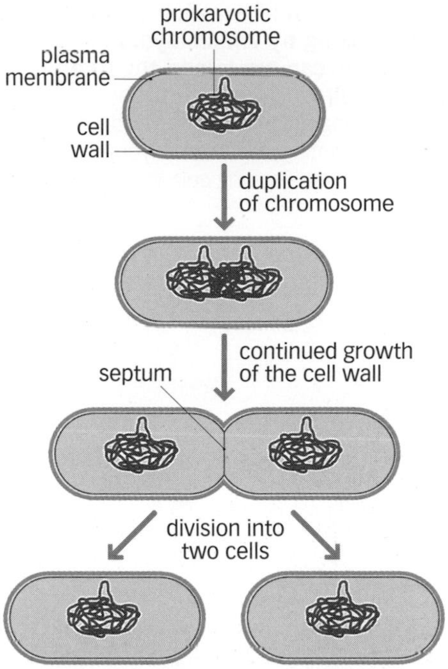

Figure 9.2 Prokaryotic cell division shows the development of new cell wall as one cell becomes two through the process of binary fission.

The Membrane-Associated Phase

In the membrane-associated phase of cell-wall building, specific enzymes link the NAG and NAM subunits with a lipid portion of the bacterial cell plasma membrane. Recall from Chapter 4 that the plasma membrane is the outer layer of the eukaryotic cell (in animal cells). In bacterial cells, however, it is the cell wall that is the outer layer, with the plasma membrane lying inside. This insertion of NAG and NAM into the plasma membrane is performed via the lipid carrier cycle. The first step in the cycle is the formation of a bond between the peptidoglycan and the side of the plasma membrane that faces the cytoplasm. This is efficient because the subunits were put together in the cytoplasm adjacent to this layer.

Fast Fact

The lipid carrier cycle is also a target for antibiotic therapy with drugs such as bacitracin and the antibiotic peptide mersacidin.

The Extracytoplasmic Phase

Because the cell wall grows outward from the cell, the new subunits must be moved from the inner side of the plasma membrane to the outer. This "flip-flop" is accomplished in the extracytoplasmic phase by a mechanism that is not yet understood. We do know that on arriving at the outer lipid layer of the plasma membrane, the peptidoglycan subunits react with a series of membrane-bound enzymes and it is these reactions that allow the orderly incorporation of the new subunits into the growing cell wall. As you might have worked out by now, these enzymes are also very efficient targets for antibiotic therapy, as we will see in Chapter 19.

The peptidoglycan molecules form a "meshwork" foundation for the cell wall, and the last step in our process is the binding together of the peptidoglycan layers to form the meshwork. This is accomplished with polypeptide chains, as shown in **Figure 9.3**. These connections can give the cell wall many layers and consequently increased strength. The amount of peptidoglycan in cell walls differs in Gram-positive bacteria (many layers) and Gram-negative bacteria (few layers).

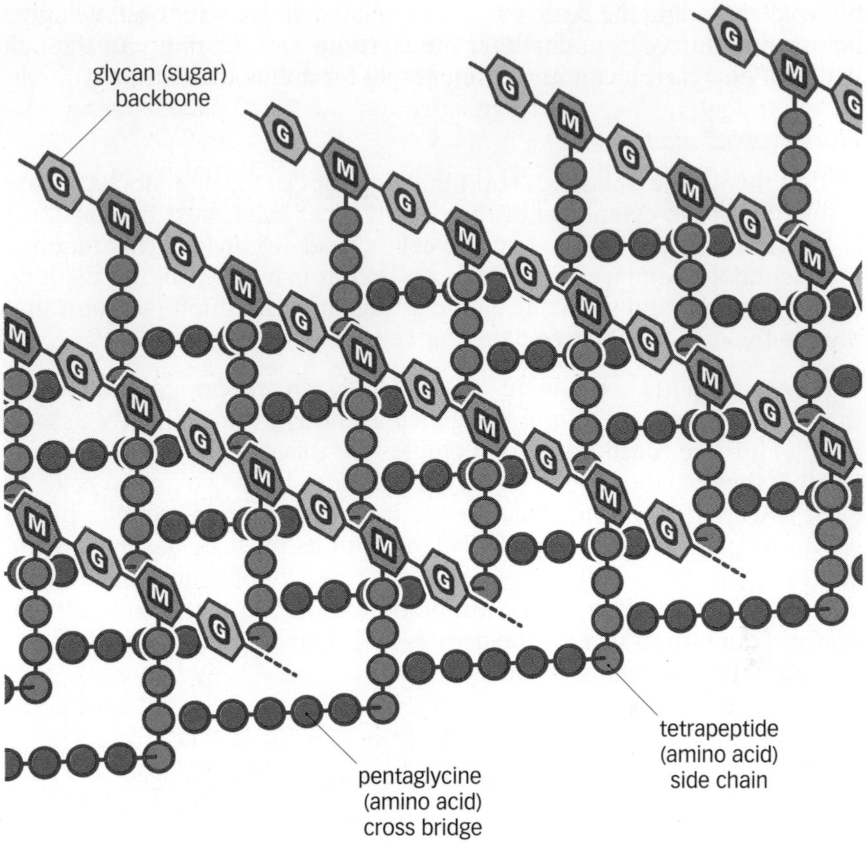

Figure 9.3 An illustration of the cross-linking by which layers of peptidoglycan are joined together during cell wall formation. The cross-linking by pentaglycine bridges and tetrapeptides between layers make peptidoglycan a strong cell wall component.

Keep in Mind

- Many of the structures associated with the bacterial cell wall are involved in the infection process.
- The cell wall is the protective outer structure of the bacterial cell.
- The cell wall is made of peptidoglycan, which is a series of repeating disaccharides and polypeptide pieces.
- The wall is thicker in Gram-positive bacteria than in Gram-negative ones.
- Peptidoglycan is constructed inside the cell and transported to the outside of the cell.
- The peptidoglycan forms a meshwork foundation for the cell wall and is a target for antibiotics.

Additional Cell Wall Components

Now that we have seen how peptidoglycan, the basic building block of the cell wall, is constructed, let's turn our attention to the other elements of the cell wall. In Chapter 4, we learned that, in most cases, bacteria can be divided into two major groups based on the differential Gram stain. This staining effect is predicated on the components of the cell walls. The Gram-positive cell wall is rich in peptidoglycan, with multiple layers of meshwork, whereas the Gram-negative cell wall contains very little peptidoglycan.

Wall Components in Gram-Positive Bacteria

In addition to many layers of peptidoglycan, the cell wall of Gram-positive bacterial cells also contains **teichoic acid** molecules. These are repeating subunits of sugar-phosphate molecules (either ribitol or glycerol) to which various other sugars and amino acids are attached. Teichoic acids come in two types (**Figure 9.4**a): those that go partway through the peptidoglycan layers (**wall teichoic acids**) and those that go completely though the peptidoglycan layer and link to the plasma membrane (**lipoteichoic acids**). It is important to note (1) that both types of teichoic acid protrude above the peptidoglycan layer and (2) that because these molecules are negatively charged, the Gram-positive bacterial cell has an overall negative charge.

Fast Fact

Wall teichoic acid can be involved with Gram-positive upper respiratory infections.

Wall Components in Gram-Negative Bacteria

The cell walls of Gram-negative bacteria are much more complex than those of Gram-positive bacteria and contain only a thin layer of peptidoglycan (**Figure 9.4**b). Consequently, the kind of environmental protection seen with Gram-positive bacteria is not available to Gram-negative organisms. However, the latter acquire additional protection through the presence of an outer membrane, which is also known as the **lipopolysaccharide layer** (LPS layer) because it is composed of lipids, proteins, and polysaccharides. Lipoprotein molecules are used to fasten the outer membrane of the cell to the peptidoglycan layer of the cell wall.

The structure of the outer membrane is unusual in that the lipid portion, although similar to the phospholipid bilayer seen in many cell types, has a unique outer layer. This layer is composed of lipopolysaccharides instead of the standard phospholipid molecules. It serves as a major barrier to the outside world and contains specialized proteins called **porins**. These proteins contain channels that vary in size and specificity, and they are responsible for the passage of molecules and ions into and out of the Gram-negative cell. The size of the channel in the porin has a role in determining which substances can be moved into or out of the cell.

The outer membrane also contains a variety of **translocation protein systems** that move substances out of the cell. Many of these translocation systems are found in the **periplasmic space**, which is the space between the plasma membrane and the outer membrane. This space is filled with a gel-like material and contains a variety of proteins secreted by the cell. Many of these proteins have important roles in the breakdown of nutrients and transport, whereas others are found only transiently in this space on their way to the outer membrane.

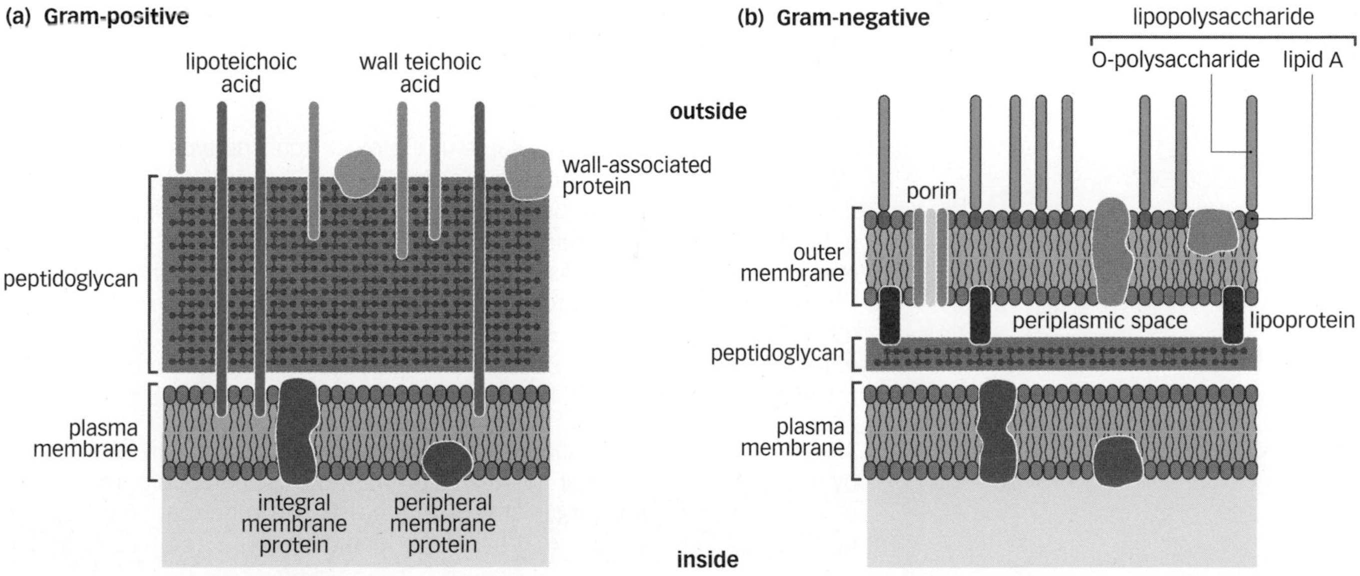

Figure 9.4 A comparison of bacterial cell walls. Panel **a**: The cell wall of a Gram-positive bacterium. Gram-positive cell walls are easy to identify because of the numerous layers of peptidoglycan. In addition to wall-associated proteins, the Gram-positive wall contains wall teichoic acid and lipoteichoic acid. The latter type spans the entire layer of peptidoglycan and fastens the cell wall to the underlying plasma membrane. The structure of this wall has a direct bearing on how certain antibiotics function and also on how the Gram stain is performed. Panel **b**: The Gram-negative cell wall contains very little peptidoglycan and no teichoic acids. One of the most prominent features of the Gram-negative wall is the outer membrane, which is composed of lipopolysaccharides, lipoproteins, and phospholipids. The Gram-negative cell wall contains endotoxin components that can cause toxic shock, and the O polysaccharides in the wall can be used diagnostically.

The difference between the Gram-positive cell wall and the Gram-negative cell wall is significant. **Table 9.1** compares the two types of cell wall. It is important to understand the components that make up these walls and how the two types of wall differ from each other, because these walls can have a role in the disease process and can also be targets for antibiotics.

It should also be noted that some bacteria, such as *Mycoplasma* (which can cause a mild pneumonia), have no cell wall. As a result, these bacteria have no distinct shape, are difficult to culture, and do not respond to antibiotics such as penicillin and cephalosporin, which use the cell wall as a target.

Table 9.1 Characteristics of Gram-Positive and Gram-Negative Bacteria

Characteristic	Gram-Positive Bacteria	Gram-Negative Bacteria
Peptidoglycan	Thick layer	Thin layer
Teichoic acid	Present	Absent
Lipids	Very little	Lipopolysaccharide layer
Outer membrane	No	Yes
Toxins	Exotoxins	Endotoxins
Sensitivity to antibiotics	Very sensitive	Moderately sensitive

Keep in Mind

- Gram-positive cells have many layers of peptidoglycan and also contain teichoic acid.
- Teichoic acid is a glycoprotein that can be found in two forms: wall teichoic acid and lipoteichoic acid.
- Gram-negative organisms have no teichoic acid and very little peptidoglycan but have an outer membrane layer made up of phospholipids and lipopolysaccharides.

Clinical Significance of the Bacterial Cell Wall

Gram-Positive Bacteria

As we discussed in Chapter 6, some of the most troubling types of respiratory infections are nosocomial ones caused by *Staphylococcus aureus*. These infections begin with colonization of the nasal passages by *S. aureus*. It has been determined that the wall teichoic acid of these Gram-positive organisms mediates the colonization of nasal epithelial cells and is an essential element for the infection to be successful. Sadly, these infections often lead to bacteremia (Chapter 7), where bacteria are found in the blood, and septicemia (Chapter 7), where the organisms are actively growing in the blood. These conditions can result in high rates of mortality.

Although colonization of the nasal passages with *S. aureus* can be inhibited by the application of the antibiotic **mupirocin**, the appearance of mupirocin-resistant Gram-positive organisms is on the rise. The gene that codes for the synthesis of wall teichoic acid has been identified and is referred to as the *tag* O gene. Strategies are in hand to develop new antimicrobials that target this gene and its products. In addition, it may be possible to use teichoic acid as an immunizing agent for the development of a vaccine that can be used in hospitals to protect against these Gram-positive infections.

The lipoteichoic acid component of the Gram-positive cell wall can also elicit an inflammatory response in the host, a response that is a by-product of Gram-positive cell death and fragmentation of the cell walls.

M protein. The Gram-positive organism *Streptococcus pyogenes* incorporates virulence factors into its cell wall. One of these factors is the M protein, which protrudes from the cell wall. This protein is required for infection, as illustrated by the fact that antibodies against it can inhibit infection. Unfortunately, this protein is highly susceptible to mutations, and there are currently more than 80 variants. Because antibodies against one variant do not bind sufficiently to others, the organism can use M-protein variants to defeat the adaptive immune responses of infected individuals.

Mycolic acid. The Gram-positive organism *Mycobacterium*, species of which cause tuberculosis (*M. tuberculosis*) and leprosy (*M. leprae*), synthesizes a waxy lipid called *mycolic acid* that is incorporated into the cell wall. In fact, as much as 60% of the cell wall of these organisms is mycolic acid, which makes the organisms extremely resistant to environmental stress. More importantly, the incorporation of mycolic acid into cell walls provides a barrier that prevents the actions of many antibiotics and many host defense mechanisms.

Fast Fact

The multiple drug resistance of *M. tuberculosis* has become a major health care concern.

Gram-Negative Bacteria

The complexity of the Gram-negative cell wall has a clinical role in a variety of ways because the outer membrane protects bacteria from antiseptics and disinfectants. More importantly, this membrane also provides a barrier against the uptake of antibiotics. Therefore, infections with Gram-negative bacteria are on the whole less sensitive to antibiotic treatment. The porin channels that run through the outer membrane exclude large antibiotic molecules but not smaller ones. As antibiotics are administered, bacterial genes encoding porin molecules can mutate and specify smaller openings, thereby inhibiting the exposure of the cell to even smaller antibiotic molecules.

Perhaps the most important clinical aspect of Gram-negative bacteria is the function of the outer membrane as an endotoxin. Many of the clinical effects associated with Gram-negative infections are due to the release of endotoxin molecules. There are two parts of the outer membrane that have clinical relevance.

- **Lipid A** anchors the LPS portion of the outer membrane to the phospholipid bilayer portion (**Figure 9.4b**). When lipid A is released as a by-product of bacterial cell death, it is referred to as an endotoxin. The body's defenses overreact to this endotoxin, and if enough of it is present, these defensive reactions can eventually have an adverse effect on host cells.

- **O polysaccharides** are carbohydrate chains that are part of the outer membrane located on the side of the membrane that faces the extracellular fluid. These molecules vary from one bacterial species to another and can be recognized by the immune system. In addition, they can be used as a diagnostic tool for the identification of certain bacteria. For example, in the name *E. coli* O157 H7, which causes a potentially lethal intestinal infection, the O indicates the O polysaccharide (designated 157) found on the outer membrane of this strain of *E. coli*.

Our examination of the bacterial cell wall configuration gives us a good idea of how this structure provides the environmental protection required for bacteria. We have also seen that the wall has important implications for infections and therapy. Because the major focus of our discussions is on clinical considerations, we now return our attention to the five requirements discussed throughout previous chapters: get in, stay in, defeat host defenses, damage the host, and be transmissible. There are structures associated with the bacterial cell that can help fulfill some of these requirements.

Keep in Mind

- The bacterial cell wall has clinical significance.
- Gram-positive teichoic acid is involved with the production of inflammation.
- Some Gram-positive bacteria incorporate virulence factors known as M proteins in their cell walls.
- *Mycobacterium* species are Gram-positive bacteria that incorporate mycolic acid (a waxy substance) in their cell walls, which makes them resistant to antibiotics.
- In Gram-negative organisms the outer membrane layer protects against disinfectants and antibiotics.
- The outer layer of Gram-negative organisms functions as an endotoxin.

STRUCTURES OUTSIDE THE BACTERIAL CELL WALL

There are five structures that can be found outside the cell wall of a bacterial cell. Not all bacteria will have all of these structures. Three of them — the *glycocalyx*, *fimbriae*, and *pili* — are primarily involved with adherence (the requirement of staying in), and the *flagella*, *axial filament* and also pili are involved with motility (which can be used to defeat host defenses and damage the host).

The Glycocalyx

Colonies of some pathogens grown on agar plates look wet and shiny. This wet look is there because each cell in the colony is surrounded by the **glycocalyx**, a sticky substance composed of polysaccharides and/or polypeptides. The polysaccharides and polypeptides are produced in the cytoplasm and secreted to the outside of the cell, where they become fastened to the cell wall in two distinct ways. If the molecules are loosely attached to the cell wall, the glycocalyx they collectively form is referred to as a **slime layer**. If the molecules produce a highly organized structure that adheres tightly to the cell wall, the glycocalyx is referred to as a **capsule**.

Both the capsule and the slime layer give the bacterial cell the ability to adhere to surfaces. For example, certain bacteria can grow on rocks in the middle of river rapids. The ability to adhere to surfaces in the presence of fast-moving water is due to the presence of a glycocalyx. In addition, the glycocalyx protects the organism from desiccation and in desperate times can be used as a source of nutrition. This structure also has an important role in pathogenesis.

Clinical Significance of the Glycocalyx

The glycocalyx is a prime factor in any successful invasion of a host by respiratory and urinary tract pathogens. After gaining entry to these anatomical systems, the invading bacteria face the possibility of being swept away by the natural defensive mechanisms of the host. This is where the adherence properties of the glycocalyx have a significant role in the infectious process.

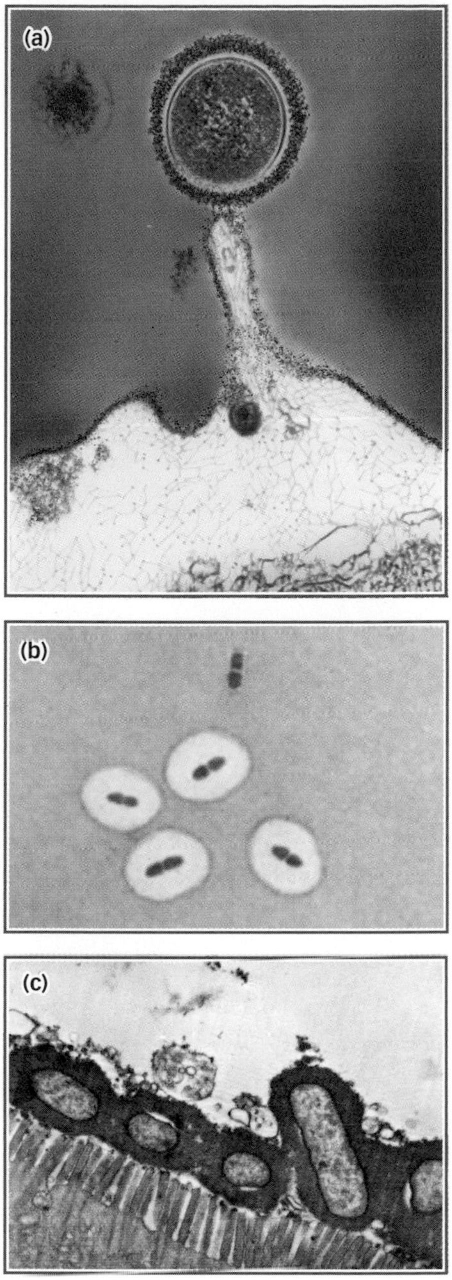

Figure 9.5 Bacterial capsules. Panel **a**. A colored transmission electron micrograph of a section through a *Streptococcus* bacterium attached to a human tonsil cell. The capsule form of the glycocalyx (thick red outer layer) can be seen tightly attached to the outside of the cell wall (thin yellow layer). Panel **b**: A photomicrograph of *Streptococcus pneumoniae* capsular swelling using the Neufeld-Quellung test. Panel **c**: An electron micrograph showing how capsules facilitate the attachment of bacteria to cells of the intestine.

The Slime Layer

The slime layer associated with bacteria is seen in some forms of dental decay. The development of this decay initially depends on the adherence of organisms to the biofilm located on the tooth surface. The slime layer permits the initial organisms to adhere to the tooth surface, and this population of initial invaders can be rapidly followed by a progression of organisms that adhere to one another. The final result is the breakdown of the dental surface.

The Capsule

The capsule form of the glycocalyx is clearly required for infection. For example, *Streptococcus pneumoniae* is not infectious unless it is encapsulated. The same holds true for *Bacillus anthracis* and *Klebsiella pneumoniae*. Examples of encapsulated bacterial cells are shown in **Figure 9.5**. As we will see in Chapter 11, transformation, a type of genetic transfer, was demonstrated with encapsulated and nonencapsulated *S. pneumoniae* and clearly showed that the development of a capsule conferred infectivity on this organism.

One of the primary reasons for the infectivity associated with the presence of a capsule is the inhibition of phagocytosis. As we will see in Chapter 15, this defense is nonspecific and requires a sequence of events to destroy invading organisms. The first step in this sequence is the adherence of the phagocytic cell to the bacterium. The presence of a capsule can prevent this initial step, allowing the bacterium to continue with the infection process.

Fimbriae and Pili

Fimbriae (singular *fimbria*) and **pili** (singular *pilus*) are two more cell-wall components involved in adherence. These sticky projections, which are shorter than flagella, are found on Gram-negative organisms, with the number of projections varying from species to species. Fimbriae (**Figure 9.6**a) and pili (**Figure 9.6**b) are composed of the same **pilin protein**

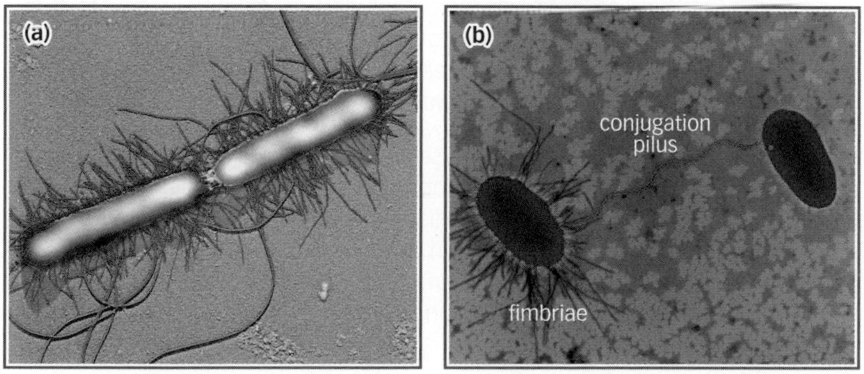

Figure 9.6 Structures exterior to the bacterial cell wall. Panel **a**: An electron micrograph showing a dividing *Proteus vulgaris* organism, which has both flagella and fimbriae. Notice the difference in length between these two structures. Fimbriae are made of pilin protein and are used for attachment; flagella are made of flagellin protein and are used for movement. Panel **b**: A transmission electron micrograph of a bacterial pilus during conjugation.

Fast Fact

Pili have been shown to have a specificity for host cell receptors that seems to be associated with specialized protein molecules found only at the tip of the pili.

subunits. Both structures are involved in adherence to host cells and to other bacterial cells. Although adherence seems to be the only function of the fimbriae, pili are also involved in the transfer of genetic material from a donor bacterium to a recipient bacterium through the process called conjugation (Chapter 11).

Clinical Significance of Fimbriae and Pili

Because of the adherence characteristic of the fimbriae and pili, they are important in fulfilling the "stay in" requirement. For example, *Neisseria gonorrhoeae* is infectious only if fimbriae are present. Adherence factors are obviously important for organisms that infect the urinary tract, because without this ability they could easily be flushed out of this area. Similarly, successful *E. coli* infections of the small intestine require the presence of fimbriae because the rapid movement of the intestinal contents would prevent the attachment of these organisms to the intestinal tissue.

In bacteria that use pili for adherence, the binding ability of the pili provides a possible target for vaccine development, because antibodies that bind to pili could block the binding to host cell surface receptors. Early studies on the binding of *Vibrio cholerae* have confirmed this possibility.

Pili also give bacteria a motility that is important for moving the organisms to different areas of the host. In Gram-negative organisms, this movement is a specialized twitching or gliding that occurs by retraction of the pili. A bacterium binds to nonspecific host cell surface determinants (such as proteins) and then retracts its pili, a motion that allows the bacterium to pull itself across the surface. The mechanism responsible for the motion is the detachment of pilin subunits, an action that pulls the cell forward, followed by pilin subunit reformation and elongation (the result is similar to the movement of a caterpillar). This ability to move also allows organisms to form biofilms, which help colonies become resistant to immune responses, antibiotics, and other therapies, leading to persistent infections.

Pili can also be used for **immune escape**, which is the ability to evade a host immune response. For example, *Neisseria* species use **phase variation**, a tactic in which the number of pili decreases after initial infection, taking away the target for antibodies. They can also use **antigenic variation** or **post-translational modification**, in which they change or mask the structure of the pili so that antibodies no longer recognize the *Neisseria* invaders. Lastly, in what is perhaps their most amazing tactic, they secrete fragments of pili, referred to as **S pili**,that bind to antibody molecules and essentially inactivate them. These strategies for use of the pili as both a pathogenic factor and a protective mechanism show the ingenuity that bacteria have evolved to function more efficiently and also to protect themselves.

Pili are also clinically important because they facilitate the transfer of genetic material from one bacterial cell to another (the process called conjugation, described in Chapter 11). This mechanism provides Gram-negative organisms with a method for genetic modification and self-defense. These genetic modifications can involve the production of dangerous toxins and the transfer of genes responsible for antibiotic resistance.

Axial Filaments

Axial filaments are flagellum-like structures that wrap around the bacterial cell and give it mobility (**Figure 9.7**). These filaments, often referred to as endoflagella, are found on spirochetes and are one of two structures exterior to the cell wall that are responsible for bacterial motility. They are confined to the space between the plasma membrane and the

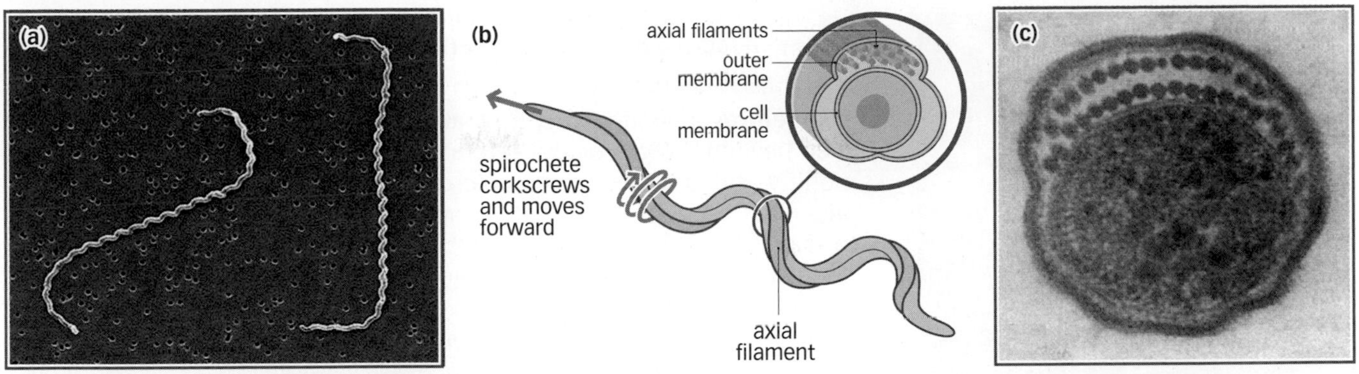

Figure 9.7 The structure of the axial filament. Panel **a**: A scanning electron micrograph of two *Leptospira interrogans* spirochetes. The filament runs inside the cell wall of this spirochete. Panel **b**: A diagrammatic representation of the axial filament wrapping around a spirochete. Panel **c**: A transmission electron micrograph of a cross section of a spirochete. Note the numerous axial filaments (dark circles).

outer membrane. Like flagella, axial filaments can produce a rotational movement. This rotation of all the filaments causes the entire organism to rotate in a corkscrew manner.

Clinical Significance of Axial Filaments

The corkscrew rotational motion created by axial filaments gives the bacterium the ability to bore through tissue. Evidence for this boring capability can be seen in the infectious process associated with *Treponema pallidum* (the causative agent of syphilis; **Figure 9.8**a) and *Borrelia burgdorferi* (the causative agent of Lyme disease; **Figure 9.8**b). In both cases, the infection involves movement of the organism from an external location through tissue, into the blood, and back into tissue. The design and structure of the axial filament facilitate this function.

Flagella

It is important to remember that the human body is home to a wide variety of bacterial organisms, which we refer to as *normal flora*. For example, the large intestine is home to a large number of microorganisms, and 99% of fecal material that leaves the body is composed of bacteria. In most cases, these organisms are not only completely harmless to us but may also provide us with important components, such as certain vitamins, as well as protection from pathogens.

Unfortunately, as we saw in Chapter 5, these same harmless bacteria can cause opportunistic infections if they move from the large intestine to other places in the body. For example, if *E. coli* moves from the large intestine, where it is harmless, to the urinary tract, it can cause a serious infection. Bacteria have developed complex structures called flagella that allow them to move rapidly from one location to another.

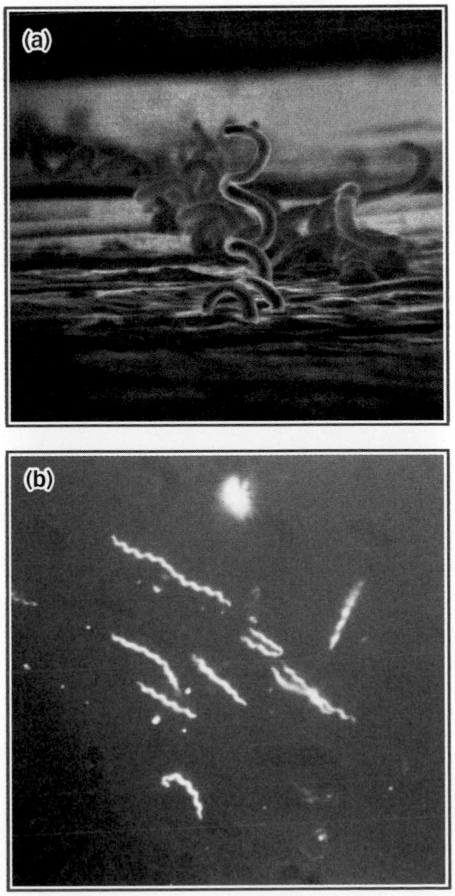

Figure 9.8 Bacterial axial filaments. Panel **a**: Electron micrograph of the organism *Treponema pallidum* (the causative agent of syphilis) invading rabbit epithelial tissue. The organism uses its axial filament to produce a corkscrew motion, enabling it to enter the tissue. Panel **b**: A dark-field micrograph showing the "corkscrew-shaped" bacterium *Borrelia burgdorferi*, which is the causative agent of Lyme disease.

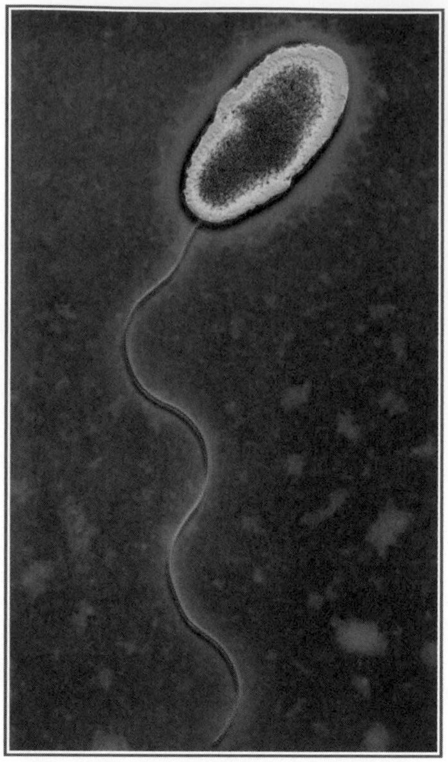

Figure 9.9 Colorized scanning electron micrograph of a rod-shaped bacterium with a prominent flagellum. 9.1

A **flagellum** is a long structure that extends far beyond the cell wall and even beyond the glycocalyx (**Figure 9.9**). This structure is used only for motility and is a classic example of the relationship between structure and function. Remember, in nature, structures are usually configured to maximize function. This reduces energy wastage, and as we saw in Chapter 3, energy is a very valuable and much-in-demand commodity.

Structure of the Bacterial Flagellum

The bacterial flagellum contains three parts: the *filament*, the *hook*, and the *basal body*. The *filament* is made up of molecules of the protein **flagellin** (**Figure 9.10**). These molecules attach to one another to form a chain that is twisted into a helical structure, giving the flagellum a hollow core.

The **hook** is the structure that links the flexible filament to the basal body (described in a moment), which anchors the structure to the bacterial cell interior.

Before we discuss the basal body, it is important to consider how the flagellum moves. This occurs through rotational movement that can be as fast as 100,000 rotations per minute and can move the organism up to 20 cell lengths in seconds. Although this movement rate may not seem like much, it equates to a 6-foot-tall human running at 80 or more miles per hour! With this in mind, it is easy to see that there is considerable stress and torque on the connection between the filament and the cell. It is at this junction that we find the basal body, which fastens the flagellum to the cell. The structure of the basal body and how it connects in Gram-positive and Gram-negative cells constitute another excellent example of structure matching function.

The *basal body* is a rod that has rings strategically fastened to it. Depending on which type of cell (Gram-positive or Gram-negative) the flagellum is attached to, there is either one or two pairs of rings. Remember that the cell wall of the Gram-positive cell has multiple layers of peptidoglycan, which confers strength and stability. Consequently there is only one pair of rings (an inner ring and an outer ring) on the basal body of a flagellum attached to this type of cell. This pair of rings is fastened to the plasma membrane, with the flagellum passing through the entire layer of peptidoglycan (**Figure 9.11**a).

Fast Fact

If we think about the energy required to propel a cell through a medium such as water or blood, we can understand the need for the propelling structure to be as energy-efficient as possible. To get a better picture of this process, remember that a bacterium moving through water is like a human moving through molasses. Obviously, moving through a viscous substance such as blood is even more difficult for a bacterial cell.

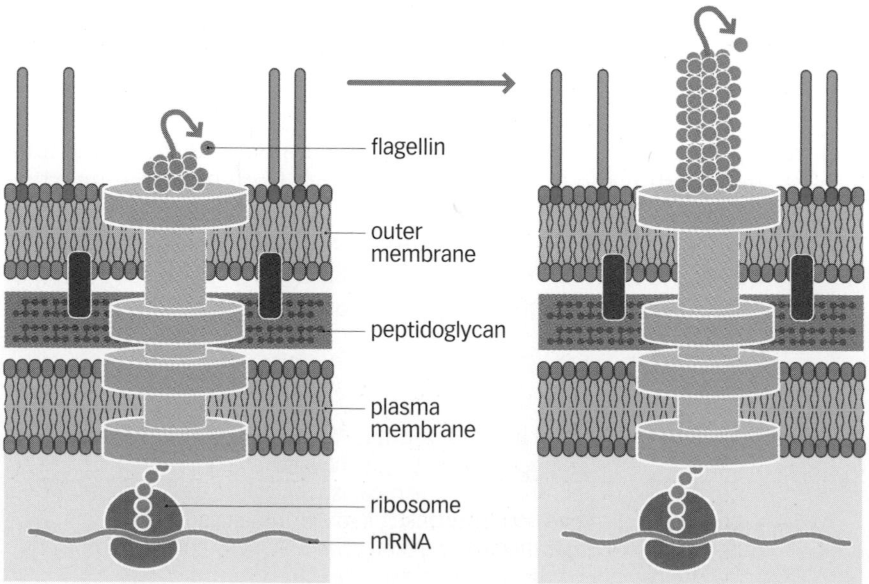

Figure 9.10 Deposition of flagellin during the construction of a bacterial flagellum. The protein is produced in the cytoplasm and moved through the developing flagellum, to be deposited at the top of the elongating filament.

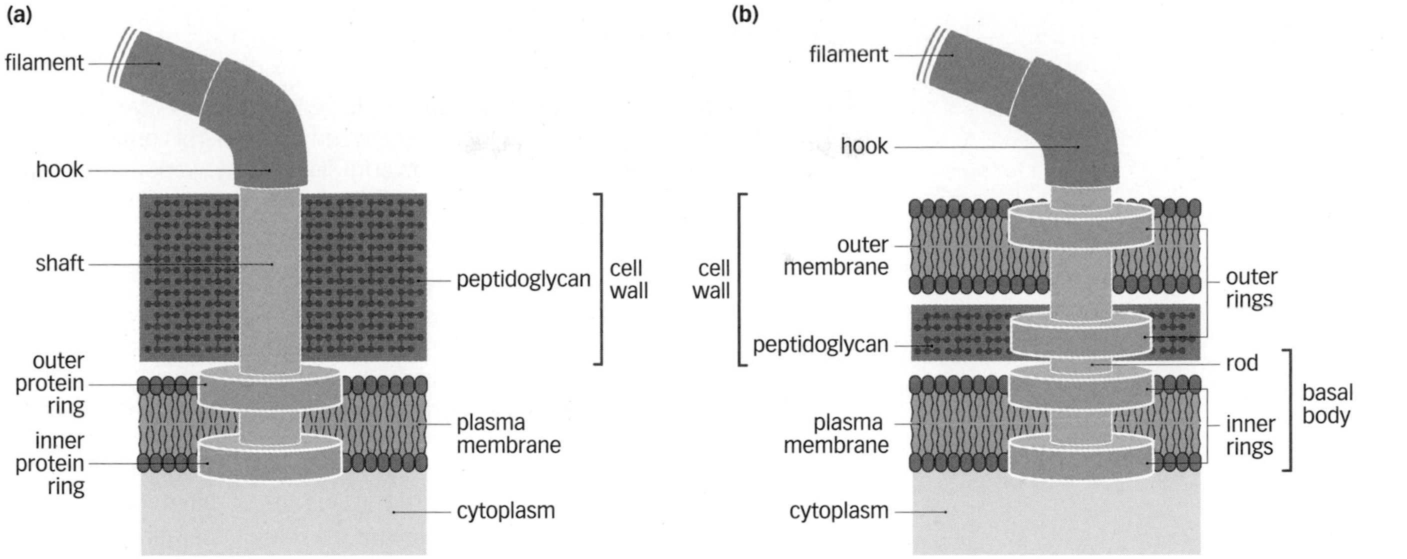

Figure 9.11 Bacterial flagella.
Panel **a**: Anchoring of the flagellum to a Gram-positive cell. The basal body of the flagellum on a Gram-positive organism has only one pair of rings, and these rings are both fastened to the plasma membrane. The many layers of peptidoglycan found in the Gram-positive cell allow this one pair to be sufficient for stability. Panel **b**: The anchoring of the flagellum to a Gram-negative cell requires two pairs of rings, with one pair binding to the plasma membrane and the other pair to the cell wall and outer membrane. The Gram-negative cell wall contains very little peptidoglycan and therefore requires additional stabilization.

Because the Gram-negative cell wall contains very little peptidoglycan, the point at which the flagellum connects to the cell needs reinforcement. This is accomplished by two pairs of rings on the basal body. **Figure 9.11**b shows that one pair is attached to the plasma membrane and the other is attached to the peptidoglycan/LPS layer of the Gram-negative cell.

Flagellar Configurations

Bacteria have four distinct patterns of flagella (**Figure 9.12**):

- **Monotrichous** — the most common form, in which the bacterium has one flagellum located at one end of the cell.
- **Amphitrichous** — in which the bacterium has two flagella, one at each end of the cell.
- **Lophotrichous** — in which the bacterium has two or more flagella located at the same end of the cell.
- **Peritrichous** — another common form, in which the entire cell is surrounded by flagella.

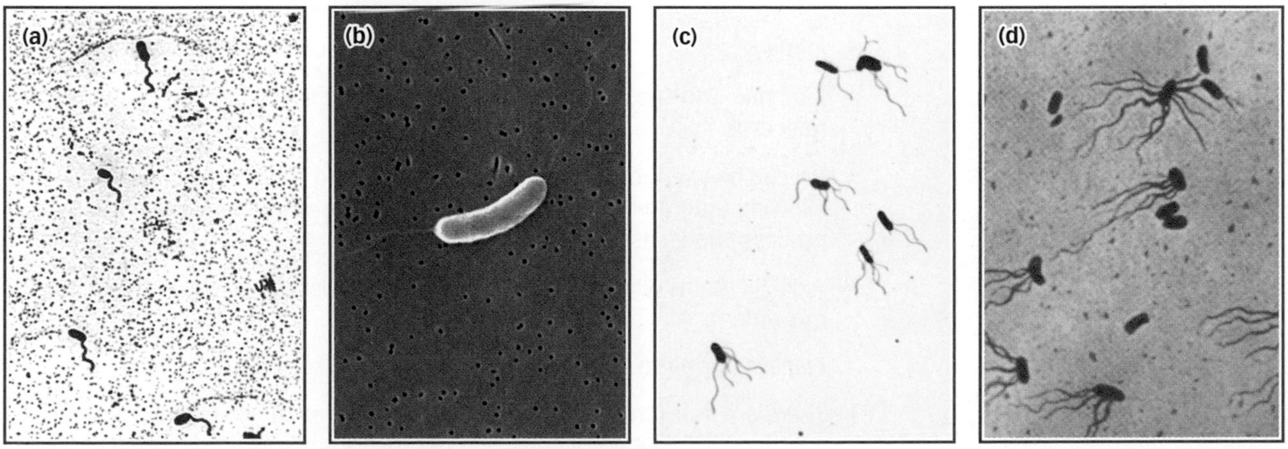

Figure 9.12 Photomicrographs of flagellar arrangements.
Panel **a**: Monotrichous. Panel **b**: Amphitrichous. Panel **c**: Lophotrichous. Panel **d**: Peritrichous.

Clinical Significance of Flagella

A good example of the clinical importance of flagella is *Helicobacter pylori*, the causative agent for some forms of gastric ulcers in humans. This organism lives in the folds of the stomach and uses powerful flagellar movements to propel itself into the heavy layer of mucus that protects the epithelial cells of the stomach from stomach acids.

In general, as we discussed previously, the movement of bacteria to places where they do not ordinarily reside results in opportunistic infection. In healthy individuals, when invading organisms enter regions already occupied by normal flora, the resident bacteria inhibit the growth of the newcomers. However, if resident populations are destroyed or diminished in number, as in the overuse of antibiotics, invading organisms that are motile because of the presence of one or more flagella can gain a foothold and initiate an opportunistic infection. Similarly, any organism that can move from tissue to the lymph or blood systems has the opportunity to travel throughout the body, generating a systemic infection.

Intracellular Movement

Any infection in which bacteria gain access to the inside of a host cell is good for the bacteria because then they are shielded from defenses such as phagocytosis and host immune reactions. Some pathogens — *Salmonella* is one example — use **cytoskeletal structures** of the host cell, such as actin fibers, not only to gain access to the host cell but also to move around the host cell and move from one host cell to another. This type of movement also occurs with *Shigella*, a bacterium associated with a type of dysentery (bloody diarrhea). Although *Shigella* is classified as nonmotile, it can use cytoskeletal structures of the host to move from cell to cell.

Keep in Mind

- Structures found outside the cell wall are all involved in the infection process.
- The glycocalyx can be found in either the capsule or slime-layer form and can satisfy the requirement of "staying in" as a result of their adherence to host cells.
- In many cases, an organism is not pathogenic unless it has a capsule.
- Fimbriae also satisfy the "stay in" requirement for infection by adhering to host cells.
- Pili can be used to adhere to host cells but also serve the function of allowing the transfer of genetic information between bacteria in the process known as conjugation.
- Axial filaments allow organisms to penetrate tissues by means of rotational movement.
- Flagella permit movement, which helps organisms stay in.
- There are four forms of flagella seen in bacteria: monotrichous, amphitrichous, lophotrichous, and peritrichous.
- The host cell's cytoskeleton can be used for intracellular movement during the infection.

STRUCTURES INSIDE THE BACTERIAL CELL WALL

The bacterial anatomy just described involves structures that deal with external environmental stress and, in many cases, pathogenic processes. In this section we discuss the six major structures inside the cell wall, all involved with intrinsic cellular functions: the *plasma membrane*, the *nuclear region*, *plasmids*, *ribosomes*, *inclusion bodies*, and *endospores*.

The Plasma Membrane

Just like the plasma membrane of eukaryotic cells (Chapter 4), the plasma membrane of a bacterium is a delicate flexible structure that holds in the internal cellular matrix of cytosol and organelles. The most important feature of any plasma membrane — bacterial or eukaryotic — is **selective permeability**. The membrane is constructed in such a way that only certain small molecules, like gases and hydrophilic molecules, can pass through. To understand the structure of this membrane fully, we must continue to think about structure–function relationships. The bacterial plasma membrane, like the cell wall that lies outside it, provides a barrier between the inside and the outside of the cell. However, one important difference between the two is that, unlike the cell wall, the plasma membrane is involved in the physiological functions of the bacterial cell, such as transport, secretion, DNA replication, and the generation of energy. It is the unique design of the membrane that allows it to participate in these functions.

Structure of the Plasma Membrane

The membrane is a phospholipid bilayer, with each layer composed of lipid molecules attached to phosphate groups (**Figure 9.13**). This two-layer arrangement gives the molecules a head-and-tail arrangement. The head region is hydrophilic, and the tail region is hydrophobic. With the two layers arranged such that the two tail regions are adjacent and the heads are pointing outward in opposite directions, we have the perfect structure for membrane function. Remember, the environment both inside and outside the cell is water-based, and there are components on both sides that need to be moved across the membrane. Because the two outward-facing sides of the bilayer are hydrophilic, they interface with water-soluble molecules on both sides of the membrane. The unique part of the membrane structure is that because the hydrophobic tails of both layers face each other, a barrier to water is produced.

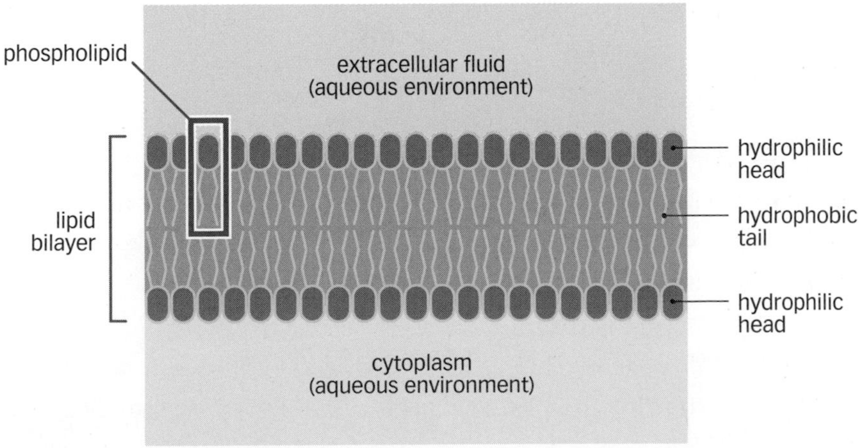

Figure 9.13 A diagrammatic representation of the phospholipid bilayer of the plasma membrane. The arrangement of the hydrophobic lipid tails provides a barrier between the outside and the inside of the cell.

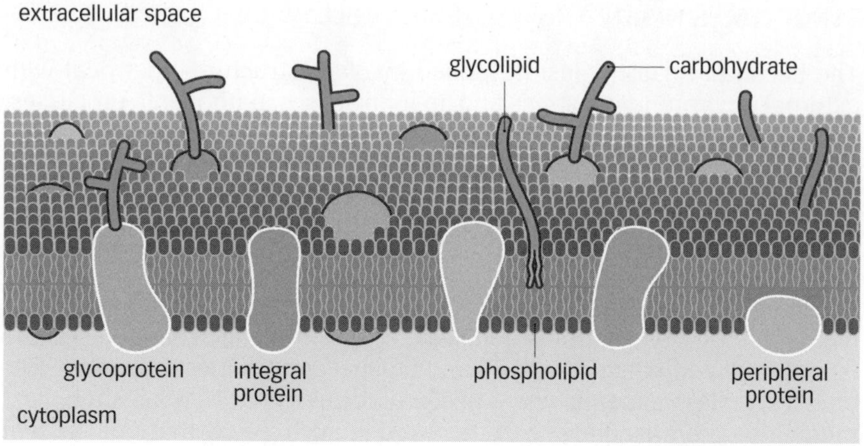

Figure 9.14 The bacterial plasma membrane, showing membrane proteins and other components. Some of the proteins extend through the entire membrane (integral proteins), whereas others are restricted to either the inner or outer layer (peripheral proteins).

The many functions required of the bacterial plasma membrane are performed by a variety of proteins that float in the phospholipid bilayer. There are two basic types of membrane protein (**Figure 9.14**). Peripheral proteins are found on either the inside or outside layer of the bilayer, whereas **integral proteins** penetrate the membrane completely and in some cases contain a pore that connects the interior of the cell to the external environment. Although all of the functions of these membrane proteins are important for cell function, we will concentrate our discussions on energy and transport.

Fast Fact

There are more than 200 different proteins in the plasma membrane of *E. coli*.

Energy Production

Recall from Chapter 3 that cellular energy, in the form of ATP, is required for all metabolic functions. To understand the magnitude of this requirement, think back to our discussion of the flagellum and how many rotations it produces. Each rotation requires energy. Unlike eukaryotic cells, in which mitochondria produce the energy required for cellular functions, bacteria use the plasma membrane. Here, the electron transport chain involves a series of membrane proteins that generate ATP through the sequential transfer of electrons and hydrogen ions from protein to protein. As we saw in Chapter 3, this forms a concentration gradient across the membrane and results in the proton motive force, a form of energy that can be either used immediately or stored in ATP by reactions involving the enzyme ATP synthase (**Figure 9.15**). (Remember from your chemistry

Figure 9.15 Membrane electron transport. This important membrane function is performed in bacteria by a series of compounds located in the plasma membrane. These compounds eject hydrogen ions (H^+) from the cell, thereby creating a concentration gradient that results in a proton motive force. As the concentration of hydrogen ions outside the cell increases, the ions begin to flow back into the cell and energy is released.

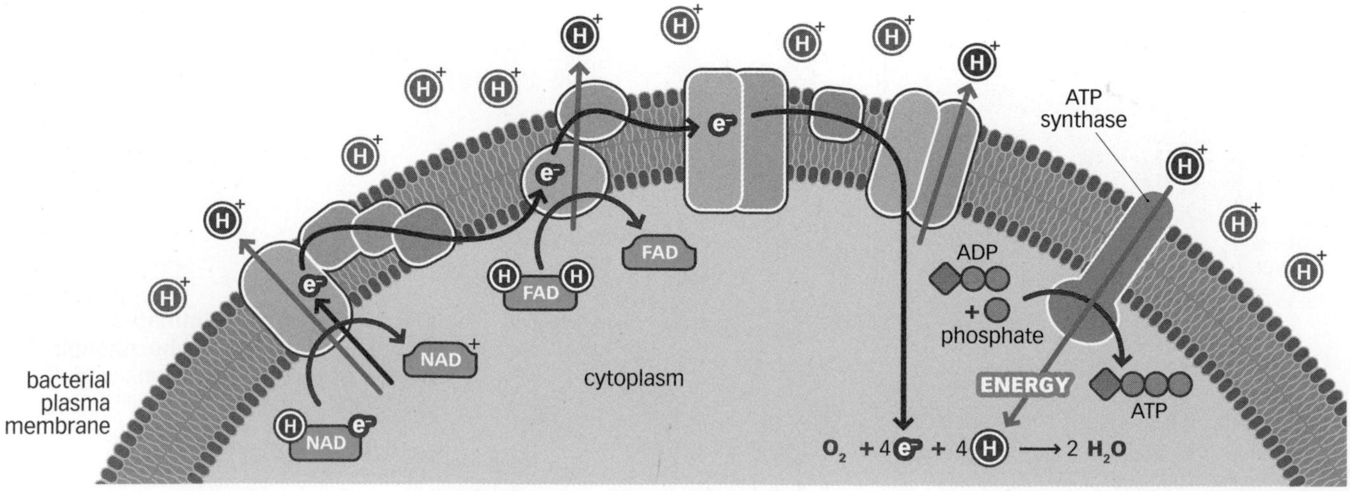

classes that a hydrogen ion, H^+, is nothing more than a single proton. Thus *proton* and *hydrogen ion* mean the same thing, so that we can call it a *proton* motive force but talk about hydrogen ions being moved.)

Membrane Transport

Transport across the plasma membrane should be viewed in the context of overall bacterial cell structure. As we have discussed, bacteria are under considerable environmental stress, and to alleviate this problem they surround themselves with a complex cell wall. The plasma membrane serves a similar function. It is a barrier between the cytoplasm and the exterior environment. Keep in mind that the cell wall is a meshwork structure that lets things percolate through, but it is the plasma membrane that regulates what gets into the cell and what does not.

Cells generate waste through everyday functions that must be removed from the cell, and they also require access to nutrients and other materials located in the extracellular fluid. The mechanisms of transport and the membrane proteins involved are designed to accomplish these functions. There are three types of membrane transport: *osmosis*, *passive transport*, and *active transport*.

Osmosis. Although the plasma membrane is selectively permeable, small molecules such as water can move across it. For the purposes of our discussion, we can define **osmosis** as follows: "water chases the concentration." To understand what it means to "chase a concentration," you must remember that the cell cytosol is composed of water and solutes (things that go into solution). For this discussion, the quantity of solutes dissolved in the cytosol determines the *concentration* referred to in our definition of osmosis. If the concentration of solutes inside the cell is less than that outside the cell, the water inside the cell "chases" the higher solute concentration and leaves the cell. When this happens, the cell loses water and shrivels up. This is referred to as **plasmolysis**.

Fast Fact

Although the cell wall keeps the cell safe from outside pressures, the structure of the peptidoglycan meshwork is such that water can move through it, thereby exposing the plasma membrane to osmotic pressure.

If the concentration of solutes is greater inside the cell than outside, the water outside chases the higher solute concentration and enters the cell. This influx of water causes the cell to expand and eventually lyse in a process called **osmotic lysis**.

When a cell is placed in a solution in which the solute concentration outside the cell is higher than the solute concentration inside the cell, as in **Figure 9.16**, the solution is referred to as **hypertonic**. When a cell is placed in a solution in which the solute concentration is lower than the solute concentration inside of the cell, as in **Figure 9.17**, the solution is called **hypotonic**. To keep these two similar terms straight, remember the phrase "**O**h n**o**, it's g**o**ing t**o** bl**o**w" (as in blow up, in other words, undergo osmotic lysis). All the o's in this expression will remind you of hyp**o**tonic. Once you remember that a hyp**o**tonic solution results in osmotic lysis, you also know that a hyp**er**tonic solution must elicit the opposite effect, plasmolysis.

It is important to note that cells are normally found in an **isotonic** environment, in which the solute concentrations inside and outside the cell are essentially equal. In this situation, water does not move in significant amounts in either direction.

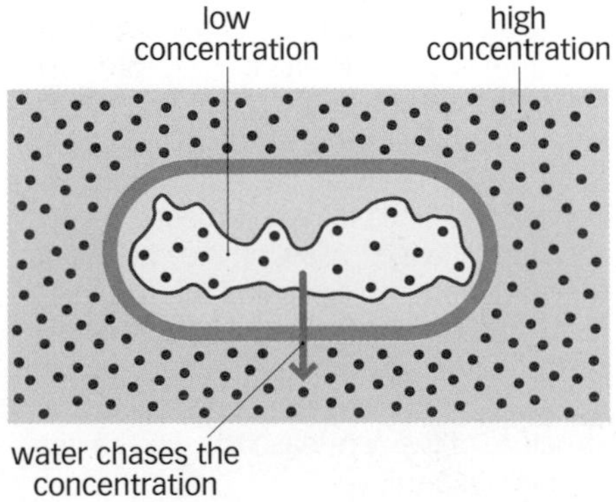

Figure 9.16 Plasmolysis. Plasmolysis occurs when cells are placed in a hypertonic environment, which means an environment in which the solute concentration is higher outside the cell than inside. Because water chases the solute concentration, it moves out of the cell and causes the plasma membrane to pull away from the cell wall and the cell volume to shrink.

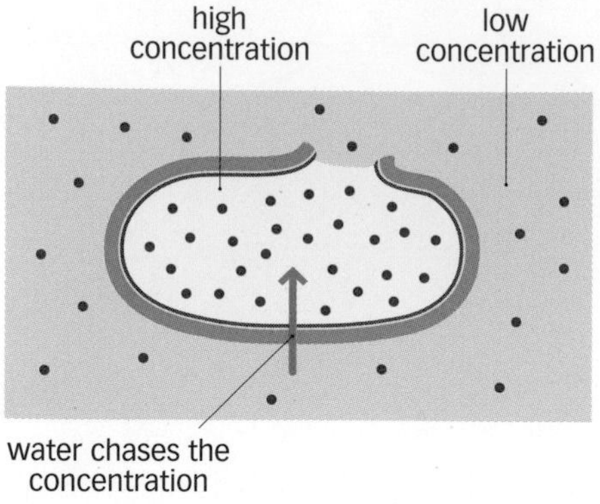

Figure 9.17 Osmolytic lysis. Osmotic lysis results when cells are placed in a hypotonic environment, which means an environment in which the solute concentration is higher inside the cell than outside. Because water chases the higher solute concentration, it moves into the cell and causes it to expand. This expansion can cause the cell to lyse (burst).

Passive transport. There are two types of **passive transport**: *simple diffusion* and *facilitated diffusion*. The most important thing to remember about both types of passive transport is that they do not require energy.

Simple diffusion is predicated on the development of concentration gradients. You can think of a concentration gradient as a truck going downhill with no brakes. The steeper the hill, the faster the truck goes. The higher a concentration gradient between two regions, the faster the solute moves from the region of higher concentration to the region of lower concentration. The movement of a solute by simple diffusion slows down as the concentration gradient becomes smaller but continues until the concentration of solute is the same on both sides of the membrane and there is no longer a gradient.

Simple diffusion occurs only with solute molecules that are either soluble in lipids or small enough to pass through the cell wall and the plasma membrane.

The second type of passive transport, **facilitated diffusion**, is similar to simple diffusion in that it requires no ATP. However, in facilitated diffusion, molecules are brought across the plasma membrane by carrier molecules called **permeases**. Permeases are specialized proteins found in the plasma membrane whose function is based on their three-dimensional shape.

The binding of a solute to a permease changes the three-dimensional shape of the protein (**Figure 9.18**). To see how this works, let's look at a situation in which the concentration of a particular solute is higher outside the cell than inside. A permease in the cell's plasma membrane has a specific shape when not bound to any solute. When a solute attaches to a binding site located on the side of the permease facing the extracellular fluid, the shape of the permease changes such that the opening facing the extracellular fluid closes and the side of the permease facing the cell interior opens to allow the solute to move into the cytoplasm. On release of the solute, the three-dimensional structure of the permease regains its original shape and the binding site on the extracellular side is again available.

If the solute molecule is too large to fit into the binding site of a permease, the cell secretes enzymes into the extracellular fluid, and these enzymes break the solute into smaller pieces that can then be accommodated by the carrier permease.

Active transport. The carrying of solute either into or out of a cell against the concentration gradient is known as **active transport**. It is the most common form of membrane transport but also the most "expensive"

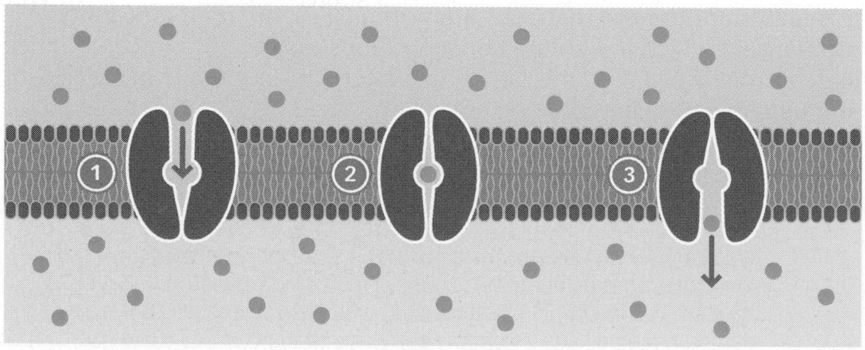

Figure 9.18 Facilitated diffusion. A permease assists in the transport of molecules across the plasma membrane. This is based on the three-dimensional structure of the permease (carrier) proteins. (1) In this simplistic illustration with no solute attached, the three-dimensional structure causes the end of the protein facing the outside of the cell to be open and the end facing the inside of the cell to be closed. (2) When the solute to be transported binds to the permease, its structure changes, allowing the end facing the cell exterior to close and the end facing the interior to open. (3) The solute is released from the permease and enters the cytoplasm, causing the three-dimensional structure of the permease to revert to its original shape.

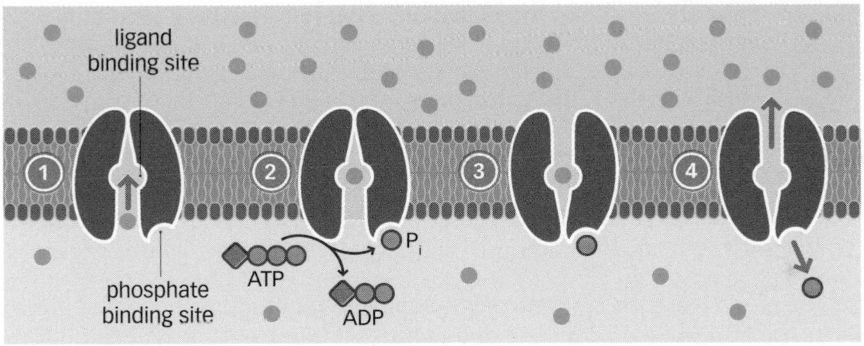

Figure 9.19 Active transport. This transport mechanism uses the energy in ATP to move a solute against its concentration gradient. In this drawing, that movement is from inside the cell to outside, even though the solute concentration is higher outside. Note that the carrier protein has one binding site for the solute to be transported and a separate binding site for an organic phosphate group (P_i) from ATP. As in facilitated diffusion, the three-dimensional changes that occur allow the transport of solute across the plasma membrane while ATP is broken down into ADP + P_i.

because it requires the expenditure of energy, either from ATP or from the proton motive force. Active transport uses specific carrier proteins found in the plasma membrane (**Figure 9.19**).

There are three types of active transport: *efflux pumping*, *ABC transport systems*, and *group translocation*.

Efflux pumping involves proteins that are part of what is called the *superfamily of transporters* and is seen in both prokaryotic and eukaryotic membrane transport. Efflux pumping employs a "revolving door" mechanism in which membrane pumps bring in certain molecules and expel others at the same time (**Figure 9.20**). The energy source for these membrane pumps is not ATP but rather the proton motive force that comes directly from electron transport. Although this process consumes energy, it is efficient because it accomplishes two functions (intake and output) at the same time.

The second type of active transport, **ABC transport systems**, is a complex system that involves several proteins. (*ABC* stands for **A**TP-**b**inding **c**assette.) During ABC transport, the molecule to be transported forms a complex with a binding protein on the outside of the plasma membrane and is then handed off to a complex of proteins located in the plasma membrane. Together these proteins then transport the molecule into the cytoplasm.

The third type of active transport, **group translocation**, is unique to bacteria and involves a clever way of making sure that molecules transported into the cell remain inside. Let's use glucose as an example. Glucose is the preferred sugar for many microorganisms because it is the easiest to breakdown (Chapter 3) and most efficient source of energy for them. With this in mind, it is easy to see that once a glucose molecule is inside a bacterial cell it must be kept there. As the concentration of glucose inside the cell increases, a concentration gradient forms. This gradient allows glucose molecules inside the cell to move by simple diffusion back across the cell's plasma membrane and out of the cell. To prevent this migration out of the

Fast Fact

As we shall see in Chapter 20, efflux pumping is routinely used by bacteria to expel antibiotic molecules.

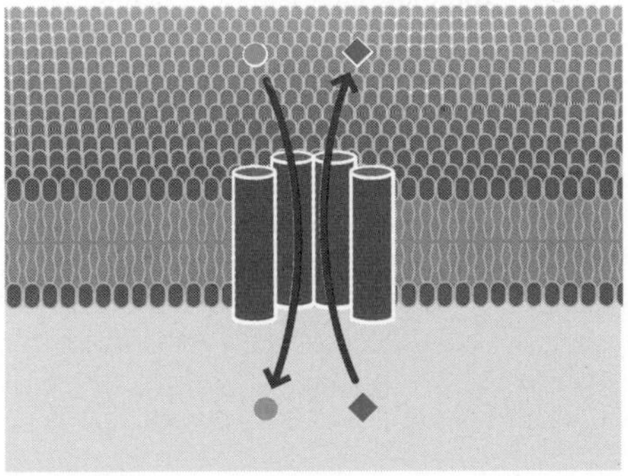

Figure 9.20 Efflux pumping, a type of membrane transport. The membrane protein pump brings one type of molecule into the cell as another type of molecule is simultaneously transported out of the cell. Bacteria regularly use this mechanism as a protection against antibiotics.

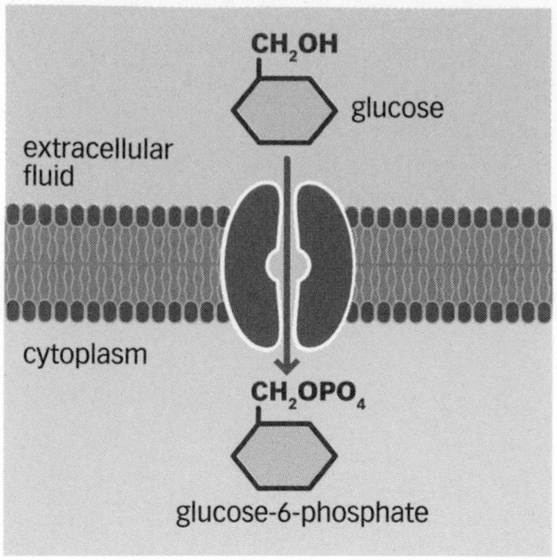

Figure 9.21 Group translocation, a type of transport seen only in bacteria. The molecule being transported into the bacterial cell is chemically altered as it crosses the plasma membrane. This process requires energy in the form of phosphoenolpyruvate. Alteration of the molecule avoids the buildup of a concentration gradient of the original molecule, with the result that the molecule cannot leave the cell by simple diffusion. This mechanism is often used for molecules such as glucose, which are in high demand by the cell.

cell, bacteria use a *phosphotransferase* enzyme to attach a phosphate group to the glucose molecule once it enters the cell (**Figure 9.21**). This modification of the glucose molecule removes it from the diffusion equilibrium reaction so that it stays inside the cell and is available as an energy source.

As with all other forms of active transport, energy is required to bind the phosphate to the glucose. In group translocation, the energy source is the high-energy molecule **phosphoenolpyruvate** (PEP) instead of ATP.

Secretion

Up to now, we have talked about the methods by which bacteria bring in molecules from the outside. The plasma membrane is also involved with **secretion**, which involves moving substances from the inside of the cell to the extracellular fluid. Secretion involves several plasma membrane proteins that act in a specific sequence. The exact interactions are not yet understood, although it has been shown that protein molecules destined to be moved out are recognized through a signal sequence of about 20 amino acids. It is thought that this sequence helps the protein molecule to interact with the plasma membrane and is clipped off as the molecule crosses the plasma membrane and exits from the cell.

Clinical Significance of the Plasma Membrane

Although the plasma membrane does not contribute any pathogenic qualities to a bacterium, it is a primary target for therapeutic strategies. Obviously, any damage to this membrane inhibits the transport of essential components into the cell, not to mention the ability to produce energy and replicate DNA. In a broader sense, loss of integrity of the bacterial plasma membrane causes the outside to come in or the inside to come out, and either is lethal for the organism.

In Chapter 19, we will see that the bacterial plasma membrane is one of the five primary targets for antibiotics. However, because of the similarity to plasma membranes in eukaryotic cells, this target does not demonstrate "selective toxicity," and antibiotics directed against it can have major side effects.

Keep in Mind

- The plasma membrane surrounds the cytoplasm in bacterial cells.
- This membrane is composed of a phospholipid bilayer, which is selectively permeable.
- The membrane contains many different types of protein that are either localized on the inner or outer side of the membrane (peripheral proteins) or span the entire membrane (integral proteins).
- The membrane in bacteria is involved with energy production (ATP is formed here) and also in the replication of the bacterial chromosome.
- Membrane transport can involve osmosis and passive transport (which require no energy) or active transport (which requires ATP).

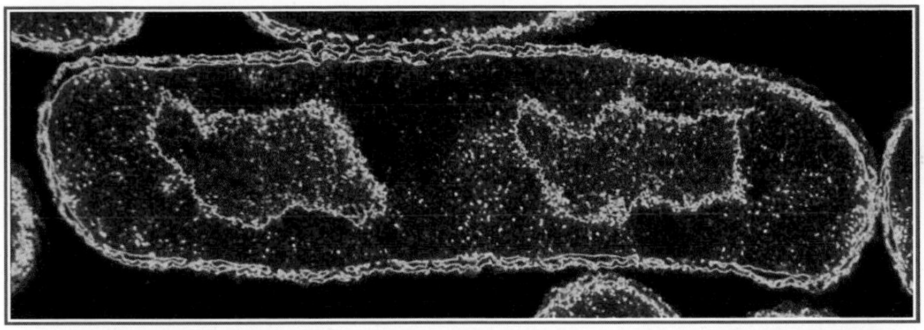

Figure 9.22 This colorized transmission electron micrograph shows an *E. coli* organism in the process of cell division. The bright red area inside the cell is the nucleoid region where DNA is localized.

The Nuclear Region

Because bacteria are prokaryotes, they have no nucleus. Therefore we refer to the region of the cell containing DNA as the **nuclear region** (also known as the nucleoid region). It is this region that is the most discernible area inside the cell wall (**Figure 9.22**). Bacteria usually have only one circular chromosome, which contains all of the genetic information required by the organism. It is supercoiled and associated with specific positively charged proteins that stabilize it. We discuss DNA in detail in Chapter 11.

Plasmids

Plasmids are extrachromosomal pieces of DNA that can be seen separated from the main DNA structure (**Figure 9.23**). The size of a plasmid ranges from 0.1% to 10% of the size of the chromosome, and some bacteria carry more than one plasmid. Plasmids often carry genes for toxins and resistance to antibiotics, and they can be transferred from one cell to another through pili during the process of conjugation.

Clinical Significance of DNA and Plasmids

Because DNA is the genetic blueprint for the organism, any disruption or damage to it can be a lethal event. It is therefore one of the five targets for antibiotic therapy. Like the plasma membrane, DNA is not a selectively toxic target because the DNA of pathogenic microbes is essentially the same as the DNA of the hosts they infect. However, several therapeutic strategies are routinely used for viral infections. These employ chemicals that are reactive with nucleic acids (DNA or RNA).

The clinical significance of plasmids is not only as a potential therapeutic target but, more importantly, as a direct pathogenic factor in antibiotic resistance. Plasmids carry genes for toxins that are in many cases responsible for the clinical symptoms of an infection. Equally important is the fact that many of the genes for resistance to antibiotics are carried on plasmids, and the ease of movement of plasmids from one bacterial cell to another is a major cause of the spread of antibiotic resistance among bacteria.

Ribosomes

Recall from Chapter 4 that ribosomes are nonmembrane-enclosed organelles involved in protein synthesis. More active bacteria contain more ribosomes. For example, *E. coli* can have anywhere between 7000 and 25,000 ribosomes.

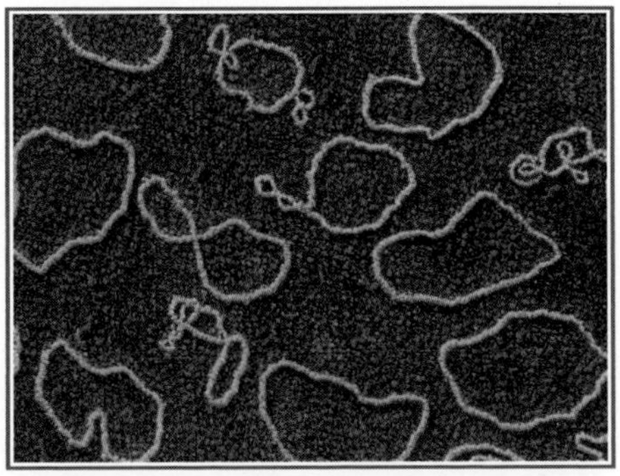

Figure 9.23 A color-enhanced electron micrograph showing bacterial plasmids.

In bacteria each ribosome is composed of two subunits that remain apart until a messenger RNA molecule approaches. At that time, the two subunits enfold the mRNA in a very specific way, and the process of protein synthesis, referred to as **translation**, takes place (Chapter 11).

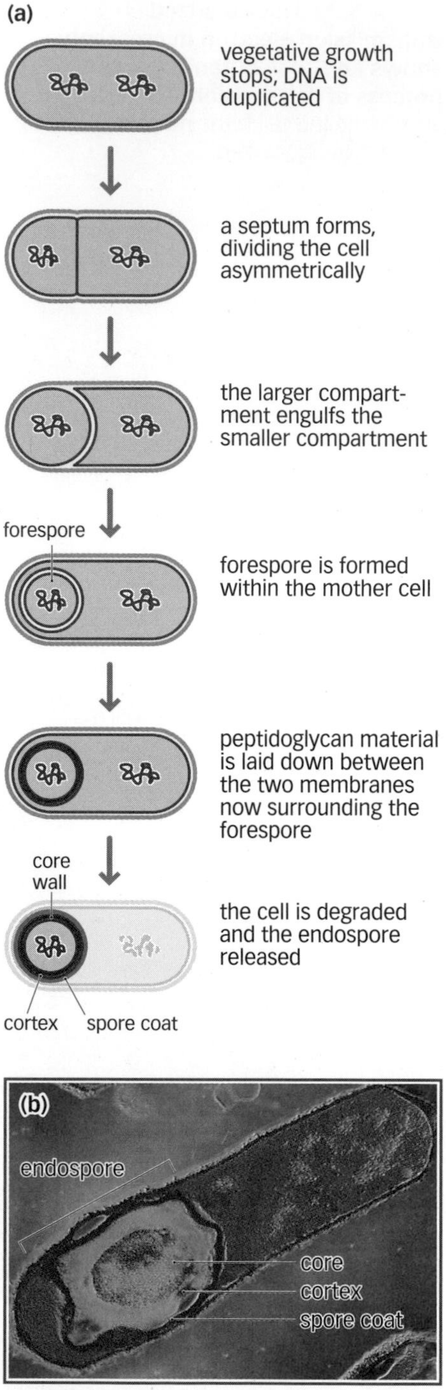

Figure 9.24 Endospore formation. Panel **a**: The sequence of events in sporulation, the process by which a bacterial endospore is formed. Panel **b**: Colorized electron micrograph of a bacterium undergoing sporulation.

Fast Fact

Endospores have been found that are 25 million years old.

Clinical Significance of Ribosomes

Because protein synthesis is required for any cell to live, any disturbance or inhibition of this process can be lethal. Because of this necessity, ribosomes are another of the five major targets for antibiotic therapy. As we shall see in Chapter 19, several antibiotics target different parts of the ribosome so as to disrupt or damage protein synthesis. Pathogens have developed several efficient ways of resisting these antibiotics and prevent the loss of protein synthesis (Chapter 20).

Inclusion Bodies

Other structures found in the interior of a bacterial cell are **inclusion bodies**, which are membrane-enclosed organelles used to store materials that the cell may require. There are several types of inclusion body, including those for glycogen, which is a stored form of the energy-rich glucose molecule (Chapter 2). Some organisms — the species *Corynebacterium* is one example — store phosphates in inclusion bodies called **metachromatic granules**.

Endospores

An amazing characteristic of bacteria is the ability to form **endospores** through a process called **sporulation**. An endospore forms whenever a bacterium is exposed to too great an environmental stress. The process is restricted mainly to Gram-positive rods. Keep in mind, however, that one Gram-negative organism, *Coxiella burnetii*, can also form endospores.

Once an endospore has formed, it is easily identifiable in the cytoplasm of the cell (**Figure 9.24**) and confers on the cell a kind of dormancy. Organisms can return from the endospore state to the vegetative (growing) state through the process of **germination**.

The endospore is extremely resistant to heat, desiccation, toxic chemicals, antibiotics, and ultraviolet irradiation. It can also survive long periods of boiling. In the endospore state, bacteria can survive for extraordinary lengths of time, but a pathogen that undergoes sporogenesis will still be a pathogen once it germinates.

Sporogenesis involves several steps (**Figure 9.24**). The first step is replication of the bacterial chromosome, followed by sequestering of the copy along with some cytosol. This volume becomes cut off from the rest of the cell as a wall referred to as a **septum** forms. The larger part of the cell then wraps itself around this newly formed smaller volume, forming what is called a **forespore**. The forespore at this point is composed of two membranes, one on top of the other. As sporogenesis continues, large amounts of peptidoglycan begin to be deposited between the two membranes of the forespore. Finally the rest of the original cell deteriorates and degrades, leaving the bacterial genetic information protected inside the endospore.

Germination of the endospore into a vegetative cell occurs when the environmental stress has subsided. At this time, the endospore accepts water molecules, swells, and cracks. The water also activates metabolic reactions involving dipicolinic acid and calcium, two of the few chemicals packaged in the endospore.

Clinical Significance of Sporogenesis

Endospores represent a significant clinical challenge. Because they are so resistant, organisms that can produce them can continue to cause serious problems no matter what we do. For example, *Clostridium botulinum* is a serious problem for the food industry, and the botulinum toxin produced by this organism can cause significant infections. Because of the resistance of endospores to heat, simple cooking cannot deal with the problem. For this reason, when items are sterilized, heat is combined with high pressure. This combination effectively deals with both vegetative *C. botulinum* cells and *C. botulinum* endospores.

Sadly, endospores have also become a tool for terrorists, as demonstrated in the anthrax attacks that occurred through the mail in 2001. *Bacillus anthracis* is commonly found in fields where cattle graze. This organism forms endospores that can be manipulated in specific ways (referred to as machining) to produce a highly dispersible endospore powder. It was this type of material that was found on the letters involved in the 2001 anthrax incidents that killed five people and injured 17. The machined anthrax endospores were extremely small (3.5 μm or less in diameter) and therefore remained airborne for a long time. This small size also allowed them to penetrate deeper into the lungs, making the infection much more serious. It must be remembered that the endospore form of these potential "weapons" is resistant to antibiotics, making it difficult to prevent environmental contamination that can lead to infection.

Table 9.2 presents an overview of the clinical significance of all the bacterial-cell structures described in this chapter.

Keep in Mind

- There is no nucleus in bacteria: the DNA is found freely floating in the cytoplasm in an area referred to as the nuclear region.
- Bacteria can contain extrachromosomal structures known as plasmids.
- Plasmids are very important in clinical disease because they can carry genes for antibiotic resistance as well as genes for the production of toxins.
- Ribosomes of bacteria are different from those found in eukaryotic cells and are consequently a selective target for antibiotic therapy.
- Endospore formation can protect bacteria from environmental pressures and are also very resistant to antibiotics and disinfectants.
- After environmental pressures subside, endospores can undergo germination into the original bacteria.
- Endospore formation makes pathogens extremely difficult to deal with.

Table 9.2 Clinical Relevance of Bacterial Cell Structures

Structure	Direct Clinical Relevance	Target for Antibiotics
Cell wall	Yes	Yes
Glycocalyx	Yes	No
Fimbriae	Yes	No
Pili	Yes	No
Axial filaments	Yes	No
Flagella	Yes	No
Plasma membrane	No	Yes
DNA	No	Yes
Ribosomes	No	Yes
Inclusion bodies	No	No
Endospores	Yes	No

SUMMARY

- Many of the structures associated with the bacterial cell wall are involved in the infection process.
- The cell wall is a meshwork made up of layers of peptidoglycan. There are many layers in Gram-positive organisms and few in Gram-negative organisms.
- The cell wall is a primary target for attack by antibiotics.
- Some Gram-positive bacteria incorporate M protein or mycolic acids, which are virulence factors.
- Gram-negative cells contain an outer layer made up of lipoproteins, lipopolysaccharides, and phospholipids that protect against antibiotics and function as an endotoxin.
- The five structures found outside the cell wall — capsules, fimbriae, flagella, axial filaments, and pili — are involved in the infection process.
- The plasma membrane surrounds the cytoplasm and is the place where the replication of DNA and the production of ATP take place. It is also a target of antibiotics.
- Bacteria have no nucleus, and the DNA floats freely in the cytoplasm.
- Plasmids are extrachromosomal structures made of DNA that contain the genes for toxins and antibiotic resistance. These plasmids can be transferred from one bacterial cell to another.
- Endospore formation protects bacteria from environmental pressure and also from antibiotics and disinfectants. This structure can have a major role in clinical settings.

In this chapter, we have looked at the anatomy of bacterial cells. It is important to keep in mind what we have learned as we continue our discussions, in particular how these structures are involved in the infectious process. When we look at bacterial genetics, antibiotics, and resistance, these anatomical structures will be revisited.

Q SELF EVALUATION AND CHAPTER CONFIDENCE

Multiple Choice

Answers are given in the back of the book and help can be found in the student resources at:
www.garlandscience.com/micro

1. The bacterial cell wall is important because it
   - **A.** Is required for infection
   - **B.** Protects the cell from the environment
   - **C.** Is used for DNA replication
   - **D.** Is used for mitochondrial attachment
   - **E.** Can be porous if necessary
2. The primary substance making up the bacterial cell wall is
   - **A.** Phospholipid
   - **B.** Carbohydrate
   - **C.** *N*-acetyl glucosamine
   - **D.** Protein
   - **E.** Peptidoglycan
3. The three phases of peptidoglycan assembly are
   - **A.** Membrane-associated phase, cytoplasmic phase, flagellar phase
   - **B.** Membrane-associated phase, extracytoplasmic phase, lipid phase
   - **C.** Membrane-associated phase, peptidoglycan phase, lipid phase
   - **D.** Membrane-associated phase, cytoplasmic phase, extracytoplasmic phase
   - **E.** None of the above
4. Peptidoglycan is made up of
   - **A.** *N*-acetyl glucosamine only
   - **B.** Repeating *N*-acetyl glucosamine molecules
   - **C.** *N*-acetyl muramic acid only
   - **D.** Repeating *N*-acetyl muramic acid molecules
   - **E.** Repeating *N*-acetyl glucosamine and *N*-acetyl muramic acid molecules
5. The growing cell wall is linked to the plasma membrane of the cell during
   - **A.** The cytoplasmic phase
   - **B.** The extracytoplasmic phase
   - **C.** The membrane-associated phase

D. The replication phase
E. The lipid phase

6. Gram-positive cell walls differ from Gram-negative cells in that the Gram-positive ones contain

A. Proteins
B. *N*-acetyl muramic acid
C. *N*-acetyl glucosamine
D. Phospholipids
E. Teichoic acid

7. Wall teichoic acids

A. Are found only in Gram-negative cells
B. Go completely through the cell wall
C. Go only halfway through the cell wall
D. Are part of the plasma membrane
E. Are attached to the nucleus of the cell

8. Lipoteichoic acids

A. Are found only in Gram-negative cells
B. Go completely through the cell wall
C. Go only halfway through the cell wall
D. Are made up of proteins

9. Teichoic acids

A. Have no clinical significance
B. Are involved in sexually transmitted diseases
C. Lead to skin lesions
D. Are involved in respiratory infections
E. Are the primary cause of bacterially induced headaches

10. Gram-negative organisms differ from Gram-positive organisms in that the Gram-negative ones

A. Contain teichoic acids
B. Contain specialized forms of peptidoglycan
C. Have three cell wall layers
D. Have an outer membrane
E. Contain specialized carbohydrate molecules in the cell wall

11. Gram-negative organisms contain

A. Large amounts of peptidoglycan
B. Only wall teichoic acid
C. Only lipoteichoic acid
D. Translocation proteins
E. None of the above

12. Mycoplasma is an organism that

A. Has a unique cell wall
B. Has a cell wall similar to that found in Gram-negative bacteria
C. Has no cell wall
D. Has three layers of cell wall
E. Losses its cell wall as it grows

13. M proteins

A. Contribute to infections
B. Are considered virulence factors
C. Are found on *Streptococcus pyogenes*
D. All of the above
E. None of the above

14. Mycolic acid

A. Can be found in the cell wall of all Gram-positive bacteria
B. Is found only in Gram-negative bacteria
C. Is found in all bacteria
D. Is found only in certain Gram-positive bacteria
E. Is part of the plasma membrane

15. Lipid A is

A. Part of the Gram-positive cell wall
B. Part of the exotoxins of Gram-positive bacteria
C. Part of the endotoxins of Gram-negative bacteria
D. Part of the plasma membrane of all bacteria
E. None of the above

16. O polysaccharides are

A. Found on all bacteria
B. Not part of the lipopolysaccharide found in Gram-positive cells
C. Used to identify certain bacteria
D. Found only on Gram-positive cells
E. Found in the plasma membrane of all bacteria

17. All of the following are associated with the exterior of the bacterial cell except

A. Fimbriae
B. Phospholipid bilayers
C. Pili
D. Flagella
E. Axial filaments

18. Slime layers are

A. Part of the fimbriae
B. Part of the cell wall
C. Part of the glycocalyx
D. Part of the flagella
E. Not associated with bacteria

19. Capsules are

A. Part of the slime layer
B. Part of the glycocalyx
C. Part of the cell wall
D. Part of the plasma membrane
E. None of the above

20. The presence of a capsule

A. Makes bacteria noninfective
B. Causes bacteria to become more infective

C. Causes bacteria to divide more slowly
D. Causes bacteria to become more susceptible to treatment
E. Has no effect on pathogenicity

21. Fimbriae are involved with

A. Getting into the host
B. Staying in the host
C. Defeating the host's defenses
D. Organisms being transmitted from host to host
E. Organizing the structures of the bacterial cell wall

22. Pili are responsible for which of the following?

A. Immune escape
B. Movement of bacteria
C. Transfer of genetic information
D. All of the above

23. Axial filaments are

A. The same as flagella
B. Confined to areas adjacent to the surface of the cell
C. The same as fimbriae
D. Similar to pili
E. Always peritrichous

24. Flagella seen in Gram-positive bacteria are

A. Attached to the cell wall
B. Attached to both the plasma membrane and the cell wall
C. Anchored in the cytoplasm
D. Attached to the nuclear membrane

25. Flagella located around an entire bacterial cell are called

A. Circumferential flagella
B. Amphitrichous flagella
C. Peritrichous flagella
D. Monotrichous flagella
E. Lophotrichous flagella

26. The plasma membrane is composed of

A. A single phospholipid layer
B. A phospholipid bilayer
C. Only lipids
D. Lipid A
E. None of the above

27. Integral proteins

A. Are found on either side of the plasma membrane
B. Go all the way through the membrane
C. Connect the plasma membrane to the cell wall
D. Are found only in the nuclear membrane of bacteria
E. None of the above

28. Membrane transport can occur by all of the following processes except

A. Passive transport
B. Facilitated diffusion
C. Active transport
D. Plasmolysis
E. Osmosis

29. Simple diffusion involves

A. Hypotonicity
B. Hypertonicity
C. A concentration gradient
D. Active transport
E. The cell wall of the organism

30. Group translocation involves

A. ATP
B. PEP
C. ADP
D. None of the above

31. Plasmids are

A. Specialized parts of the chromosome
B. Parts of bacterial mitochondria
C. Found in the cell wall
D. Extrachromosomal pieces of RNA
E. Extrachromosomal pieces of DNA

32. Bacterial ribosomes

A. Are identical to those found in eukaryotes
B. Are involved in protein synthesis
C. Are enclosed by their own individual membrane
D. Are made up of three subunits
E. Are made up of four subunits

33. Bacterial ribosomes are clinically important because they

A. Cause infection
B. Produce plasmids
C. Are targets for antibiotic therapy
D. Are resistant to antibiotics
E. Are part of endotoxins

34. Bacterial endospores are

A. Resistant to heat
B. Sensitive to heat
C. Part of the division process seen during bacterial growth
D. Targets of antibiotics
E. None of the above

DEPTH OF UNDERSTANDING

Questions listed here require you to bring together the concepts you have learned in this chapter into a discussion format. This helps you to increase your depth of understanding of the material you have learned. Help can be found in the student resources at:
www.garlandscience.com/micro

1. Using your knowledge of the structures outside the bacterial cell wall, discuss the clinical relevance of the flagella and the glycocalyx.

2. Compare and contrast the Gram-positive and Gram-negative cell wall.

3. Describe the clinical advantages and disadvantages associated with endospore formation.

CLINICAL CORNER

Help can be found in the student resources at:
www.garlandscience.com/micro

1. A nosocomial infection has broken out in the hospital. The organism responsible is a Gram-positive endospore-forming bacillus. You are asked to devise a strategy to deal with this problem.
   A. What is the most important consideration?
   B. Should you recommend changes in hospital disinfection procedures?
   C. Should you consider any recommendations about patient therapy protocols?
2. Your patient is suffering from an intestinal infection. After several days on broad-spectrum antibiotics without improvement, the patient was admitted to the hospital. A routine lab workup reveals that the pathogen responsible for the infection is a Gram-negative bacillus.
   A. What other characteristics of the organism would be useful for you to know about?
   B. What other information would be helpful for developing a plan for the treatment of the patient?

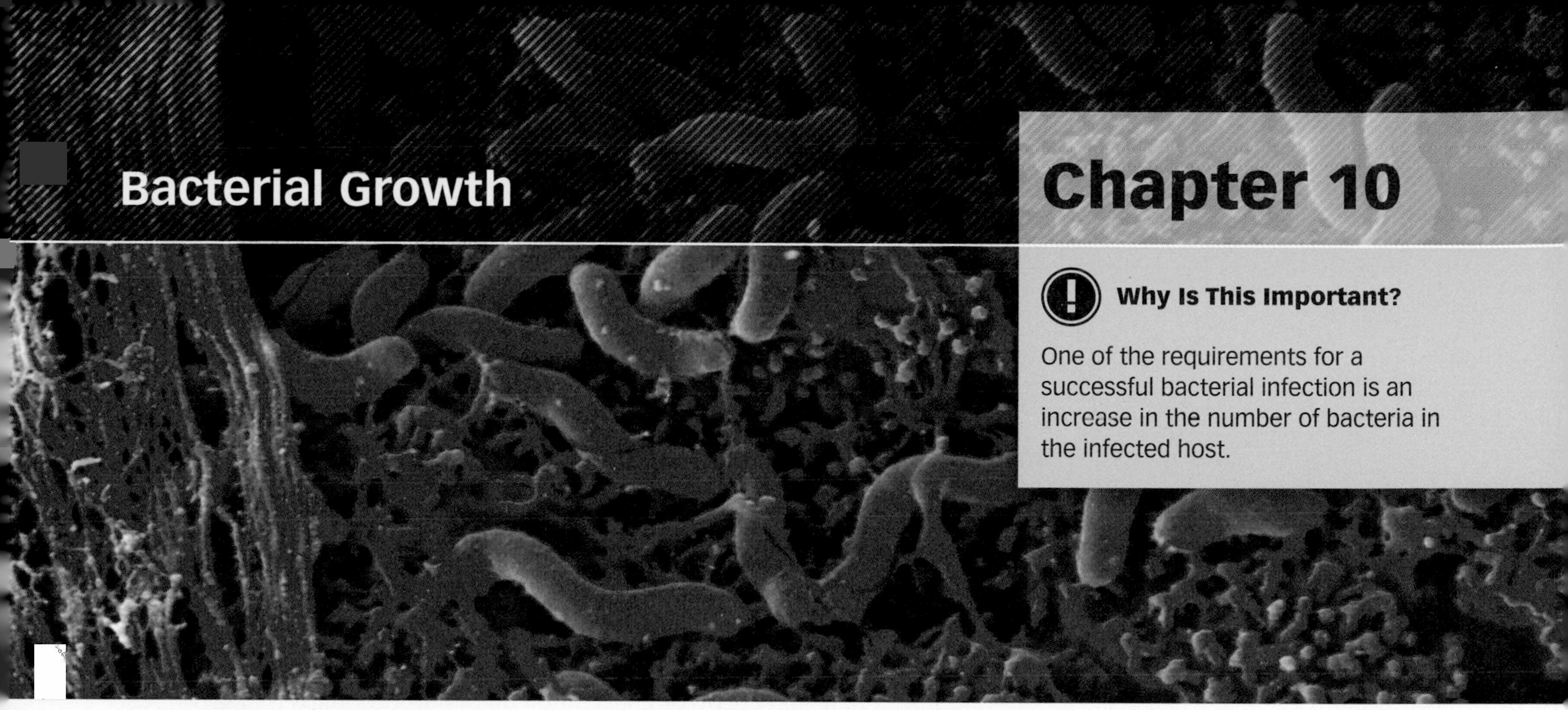

Bacterial Growth

Chapter 10

Why Is This Important?

One of the requirements for a successful bacterial infection is an increase in the number of bacteria in the infected host.

OVERVIEW

In this chapter, we examine the characteristics of bacterial growth. This information is important because infectious organisms have specific growth requirements, and understanding the requirements for growth will help us to understand the infectious process better. Recall that one of the requirements for a successful infection is being able to defeat the host defense. One way to accomplish this is to overwhelm it numerically. As we examine bacterial growth, you will also learn about specific techniques used by clinical microbiologists that have a key role in identifying and diagnosing bacterial diseases.

We will divide our discussions into the following topics:

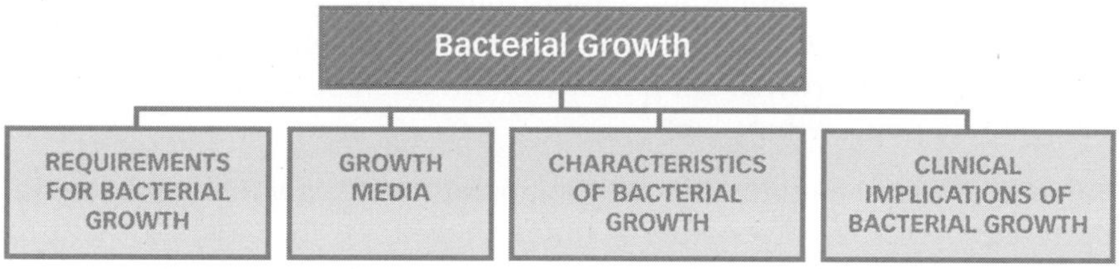

In Chapter 4 you learned that bacteria have specific sizes, shapes, and staining reactions, and in Chapter 5 you learned that most bacteria use the process of binary fission ("splitting in half") for growth. In Chapter 9 we examined the structures of bacterial cells and the functions associated with these structures. During those discussions, several things became apparent about the relationship between bacteria and their environment. For example, bacteria develop cell walls, and in some cases capsules or slime layers, that protect them from the environment. In addition, the structure–function relationships seen in membrane transport proteins showed you how bacteria can import the nutrients and other components needed for growth. We will build on these concepts as we explore bacterial growth in this chapter.

What Do I Need to Know?

To get the most out of this chapter, please review the following terms from your previous courses in biology, anatomy, physiology, or chemistry or in previous chapters of this book as indicated in parentheses: beta hemolysis, binary fission, chemical buffer, chemoautotroph (**3**), chemoheterotroph (**3**), genetic engineering, facultative anaerobe, hemolysins, hypertonic (**9**), hypotonic (**9**), incubation, *in vivo*, logarithms, MRSA, nitrogen fixation, osmotic lysis (**9**), osmotic pressure, pH (**2**), slope of a line, VRSA.

REQUIREMENTS FOR BACTERIAL GROWTH

How well bacteria grow depends on the environment in which the organisms live, and we can divide growth requirements into two major categories, physical and chemical

Physical Requirements for Growth

The physical requirements for growth fall into three classifications: temperature, pH, and osmotic pressure. Each of these factors can markedly, and in some cases lethally, affect bacteria.

Temperature

Remember that bacteria are ubiquitous throughout the Earth. In fact, they can be found in every conceivable environment as well as at every temperature. On the basis of the latter characteristic, one way of classifying bacteria is to separate them according to the temperature ranges at which they grow best (**Figure 10.1**).

Fast Fact

Psychrophiles are not pathogenic to humans. However, many of these organisms can cause spoilage of refrigerated food, which can result in illness.

Bacteria known as **psychrophiles** grow at cold temperatures. (The prefix *psychro-* comes from the Greek word meaning "cold.") There are two subsets of psychrophiles. One subset, which has no special name, is made up of psychrophiles that grow at temperatures from 0 to 15°C and are found in the deep oceans and in arctic environments. The other subset, referred to as **psychrotrophs**, is made up of psychrophiles that grow best between 20 and 30°C.

Mesophiles are bacteria that grow best at moderate temperatures, in the range between 25 and 40°C. Most common types of bacteria, including human pathogens, are mesophiles.

Thermophiles are bacteria that grow only at temperatures above 45°C. As with psychrophilic bacteria, there are two subsets of thermophiles: those that grow between 45 and 60°C and those that grow above 80°C. Members of the latter group are called either **extreme thermophiles** or hyperthermophiles. No thermophilic bacteria are pathogenic to humans.

Fast Fact

For most human pathogens, the optimal growth temperature is approximately 37°C (normal body temperature).

The **minimum growth temperature** is the lowest temperature at which an organism grows. For example, the minimum growth temperature for a thermophile is 45°C. If you place a thermophilic bacterium in an environment below this temperature, there will be no growth. The maximum growth temperature is the highest temperature at which growth occurs, and the **optimal growth temperature** is the temperature at which the highest rate of growth occurs.

Table 10.1 lists the optimal growth temperatures for a variety of bacteria, correlated with their generation times (how long it takes for them to divide). Note that some pathogens such as *Mycobacterium tuberculosis*

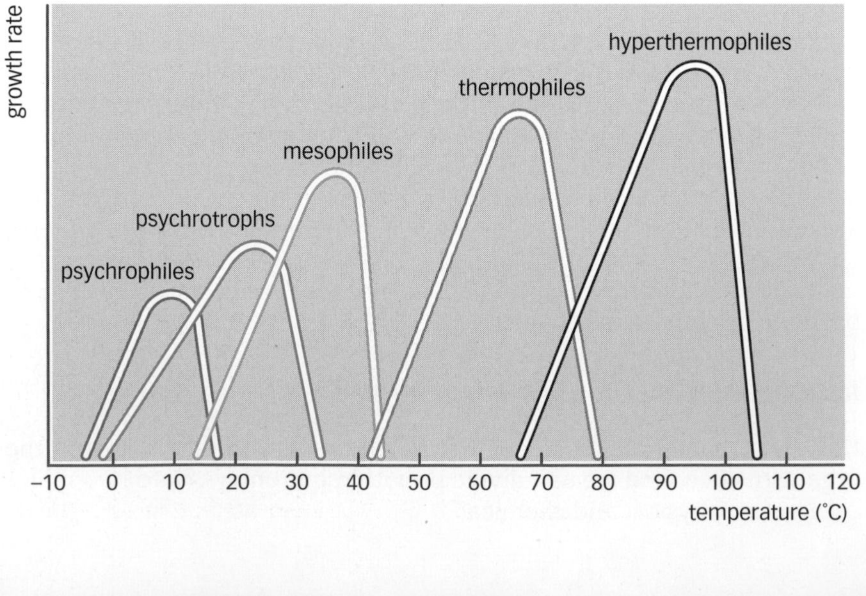

Figure 10.1 Bacteria can be classified according to temperature ranges at which they grow best. Psychrotrophs are considered a subset of the major group psychrophiles, and hyperthermophiles are considered a subset of the major group thermophiles. Note that there is a range of temperatures for each group and that in some cases neighboring ranges overlap. Most of the bacteria that are harmful to humans are in the mesophilic group.

Microorganism	Disease	Optimal Growth Temperature (°C)	Generation Time (hours)
Escherichia coli	Diarrheal diseases, opportunistic infections, and urinary tract infections	40	0.35
Staphylococcus aureus	Skin, respiratory, and other infections	37	0.47
Pseudomonas aeruginosa	Nosocomial infections	37	0.58
Clostridium botulinum	Botulism	37	0.58
Mycobacterium tuberculosis	Tuberculosis	37	12.0

Table 10.1 Comparison of Optimal Growth Temperatures and Generation Times for Several Pathogenic Bacteria

and *Treponema pallidum* have exceptionally long generation times. These slow growth rates make organisms of this type difficult to culture. In contrast, pathogens such as *Staphylococcus aureus* as well as opportunistic pathogens such as *Escherichia coli* have short generation times. Keep in mind that growth rates are important because the number of pathogens can have a direct effect on the disease process.

Let's consider how temperature can affect growth. To do this, we have to think about protein structure (discussed in Chapter 2). Remember that proteins are integral parts of the bacterial cell and can function as structural components, membrane transport mechanisms, and, perhaps most importantly, as enzymes. Because the three-dimensional structure of a protein is the key to its function, any change in this structure can have drastic effects on function. Remember also that the three-dimensionality of the protein is due to hydrogen bonds that form during the development of secondary and tertiary structure. Because they are weak, these hydrogen bonds are easily broken as the temperature rises. Consequently, exposure to high temperatures can destroy the three-dimensional structure of a protein, thereby destroying its ability to function in the bacterial cell.

An interesting observation is that variable temperature requirements are seen in certain diseases. For example, *Treponema pallidum*, the organism that causes syphilis, prefers temperatures that are on the low side of normal body temperature. Therefore lesions of this disease are initially seen on the lips, tongue, and genitalia, where the temperature of the body is slightly cooler than 37°C (**Figure 10.2**). This is also true for leprosy, which is caused by the organism *Mycobacterium leprae* and is initially manifested on the extremities of the body, in particular the face, ears, hands, feet, and fingers (**Figure 10.3**).

Fast Fact

Although bacteria can deal with acidic conditions fairly well, alkaline environments inhibit the growth of most of them. For this reason, disinfectants usually contain either ammonia or ammonia-like compounds because these substances are alkaline.

pH

Bacteria grow in a wide range of pH values. In Chapter 2 we saw that pH represents a measurement of the acidity or alkalinity of an environment, and ranges from 1.0 (very acidic) to 14.0 (very alkaline). Although most

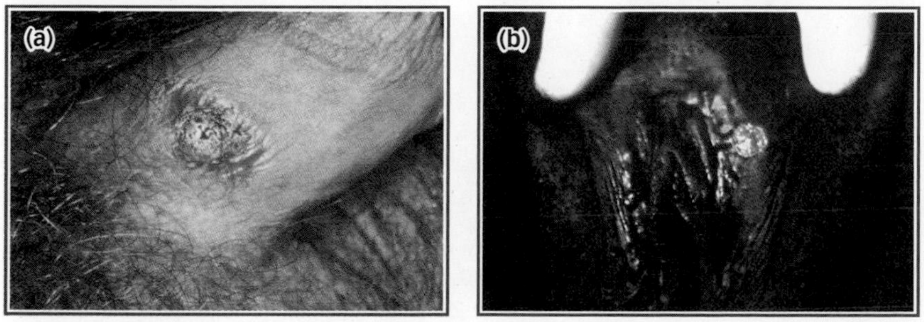

Figure 10.2 Indicators of syphilis. The appearance of chancroids on the penis (panel **a**) or on the external female genitalia (panel **b**) are indicators of syphilis, the infection caused by the bacterium *Treponema pallidum*. Although this microorganism is mesophilic, it initially grows best on the external genitalia, where the temperature is slightly cooler than within the body.

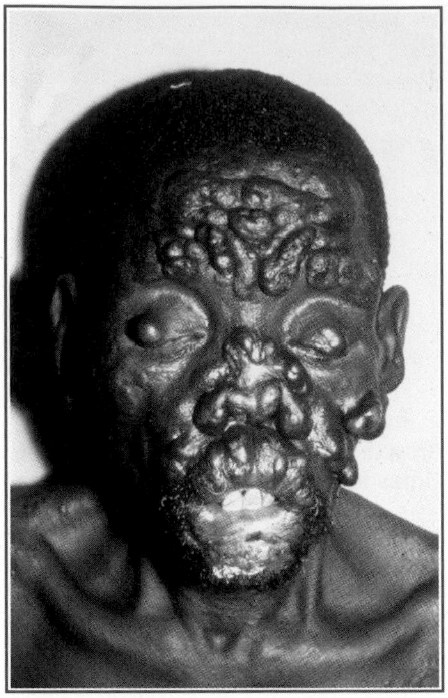

Figure 10.3 Leprosy *Mycobacterium leprae*, the etiological agent of leprosy, is a mesophilic organism that prefers to grow at the lower end of this temperature range. Consequently, it is seen on the ears, lips, and nose.

bacteria prefer the neutral pH of 7.0, some have adapted so that they are able to grow in environments in which the pH is far from 7.0. For instance, **acidophiles** are bacteria that grow at extremely low pH values.

Helicobacter pylori is not an acidophile but has been implicated as a cause of stomach ulcers and is found in the stomach, where the pH can be as low as 1.0. However, it should be noted that *H. pylori* survives for long periods in this low-pH environment by moving into the heavy mucous layer that coats and protects the digestive epithelium of the stomach (**Figure 10.4**). In fact, many bacteria, including *H. pylori*, produce chemical buffers to protect against pH changes in their environment.

The mechanisms by which changes in pH affect bacteria are similar to those we discussed for temperature. Just as hydrogen bonds are disrupted by elevated temperatures, increased quantities of free hydrogen ions (H^+) in low-pH environments or of free hydroxyl ions (OH^-) in high-pH environments also disrupt the hydrogen bonding that gives proteins their three-dimensional structure. As we have emphasized, changes in protein structure lead to changes in protein function, and these changes can be lethal for bacteria.

Osmotic Pressure

Osmotic pressure is the pressure exerted on bacteria by their surroundings, and this can affect bacterial growth. One of the major agents exerting such pressure is water. Not only do bacteria get nutrients in aqueous solution, but the bacterial cytoplasm is also 80% water.

Osmotic pressure can be used to inhibit bacterial growth. For example, food can be preserved with high salt concentrations, which create a hypertonic environment, causing organisms that might spoil the food to undergo plasmolysis. However, even this method of food preservation can be overcome by bacteria classified as **halophiles**. (Recall from your chemistry courses that *solutes* are frequently ions of salts. *Halo-* is a prefix meaning "salt," and so halophiles are bacteria that "love" being in

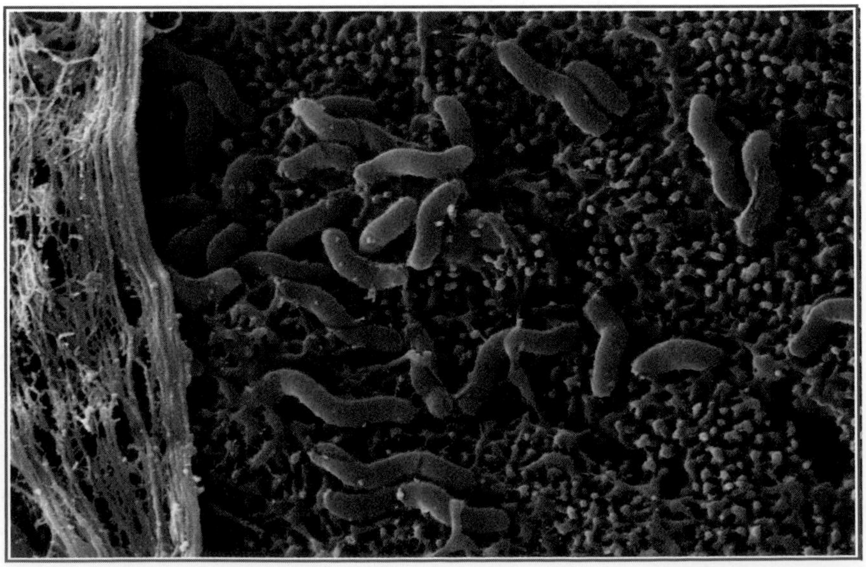

Figure 10.4 A color-enhanced scanning electron micrograph of *Helicobacter pylori*, the cause of gastric ulcers, growing on the surface of the human intestine. This organism is not an acidophile but can withstand very low pH. In the stomach, *H. pylori* survives by burrowing into the mucous layer that protects the cells of the stomach.

a high-salt, or high-solute, environment.) Halophilic organisms can be divided into three categories: obligate, facultative, and extreme. **Obligate halophiles** *require* high salt concentrations for growth, **facultative halophiles** can live with or without high salt concentrations, and **extreme halophiles** can grow in the presence of *very high* levels of salt.

In summary, remember that the main physical factors that affect bacterial growth — temperature, pH, and osmotic pressure — are relatively clear-cut. We can also consider these physical parameters in the context of pathogenic bacteria and infection. The normal human body temperature falls within the range required by pathogenic bacteria that are mesophilic. Although most bacteria grow best in a pH range from 6.5 to 7.5, acidophiles can survive even in the highly acidic environment of the stomach (pH 1.0). As far as osmotic pressure is concerned, remember that the same requirements for osmotic pressure exist for the cells of our bodies. Consequently, pathogens rarely have to deal with hypertonic or hypotonic environments during the infectious process.

Chemical Requirements for Growth

When we look at the chemical requirements for bacterial growth, we see that they are almost as variable as bacterial species themselves. **Table 10.2** summarizes several of the core chemicals required for bacterial growth. Let's look at some of them in detail.

Carbon

The biosphere of the planet Earth is *carbon-based*, which means that biological molecules all contain carbon. Therefore all living things require carbon atoms for the construction of biological molecules (carbohydrates, lipids, proteins, and nucleic acids). In fact, except for water, carbon is the most important chemical requirement for bacterial growth.

There are two basic ways in which living organisms obtain carbon. The first is through the breakdown of preexisting molecules that contain carbon atoms, which are then used for construction of new molecules. This "recycling" process is very common in biological systems, and organisms that use it are referred to as *chemoheterotrophs*.

The second way in which organisms obtain carbon is from CO_2 molecules, and these organisms are referred to as *chemoautotrophs*. Pathogenic bacteria are chemoheterotrophs.

Nitrogen

Many of the cellular tasks carried out by bacteria, as well as some of the molecules that make up bacterial cells, require nitrogen. This element is involved in protein synthesis and is an integral part of the structure of amino acids. Nitrogen is also part of the structure of DNA and RNA molecules.

Table 10.2 Chemical Requirements for Bacterial Growth

Chemical	Function
Carbon, oxygen, hydrogen	Required for cell structures
Nitrogen	Required for making bacterial amino acids and nucleic acids
Sulfur	Required for making some bacterial amino acids
Phosphorus	Required for making bacterial nucleic acids, membrane phospholipid bilayer, and ATP
Potassium, magnesium, calcium	Required for functioning of certain bacterial enzymes
Iron	Required for bacterial metabolism

Bacteria obtain nitrogen in a variety of ways, including the decomposition of existing proteins and from the ammonium ions (NH_4^+) found in organic materials. Nitrogen can also be obtained through nitrogen fixation.

Sulfur

Bacteria must have sulfur to make amino acids and some vitamins. This element is acquired primarily by decomposing existing proteins in which one or more of the component amino acids contain sulfur. Sulfur can also be procured in the sulfate ion form (SO_4^{2-}) and from H_2S, which is a by-product of many biochemical reactions carried out by bacteria.

Phosphorus

The element phosphorus is essential for the synthesis of nucleic acids, AMP, ADP, and ATP molecules (it's the P in these compounds). It is also a major chemical requirement for the development of the plasma membrane (remember the phospholipid bilayer).

Bacteria obtain the phosphorus they need by cleaving ATP to ADP plus inorganic phosphate. They can also use phosphorus in the form of phosphate ions (PO_4^{3-}).

Organic Growth Factors and Trace Elements

In addition to the major chemical requirements we have discussed, bacteria also use organic growth factors, such as vitamins B_1, B_2, and B_6. Bacteria cannot synthesize these growth factors, so they must be obtained from the environment. In addition, bacteria also require access to the elements potassium, magnesium, and calcium, which function as enzyme cofactors, and such trace elements as iron, copper, molybdenum, and zinc.

The functions of several growth factors are listed in **Table 10.3**.

Table 10.3 Growth Factors and their Functions

Growth Factor	Function
Amino acids	Components of bacterial proteins
Heme	Functions in electron transport system in bacteria
NADH	Electron carrier
Niacin (nicotinic acid)	Precursor of bacterial NAD^+ and $NADP^+$
Pantothenic acid	Component of bacterial coenzyme A
para-Aminobenzoic acid	Precursor of folic acid, which is involved in the bacterial metabolism of carbon compounds and in nucleic acid synthesis
Purines, pyrimidines	Components of bacterial nucleic acids
Pyridoxine (vitamin B_6)	Used in the synthesis of bacterial amino acids
Riboflavin (vitamin B_2)	Precursor of bacterial FAD
Thiamine (vitamin B_1)	Used in some bacterial decarboxylation reactions

Oxygen

Oxygen would seem to be one of the most important chemical requirements for bacterial growth. However, many bacteria not only do not require oxygen for growth, they actually die in the presence of oxygen. Although we know from Chapter 3 that using oxygen as a final electron acceptor in aerobic metabolism maximizes the yield of ATP, oxygen also has a "dark side" because during normal bacterial respiration, a small number of oxygen molecules can assume a superoxide free radical form (O_2^-). These free radicals are very unstable and steal electrons from other molecules, a process that causes the electron-deficient molecules to steal electrons from other molecules. This cascading free radicalization of molecules eventually leads to the death of the bacterial cell.

Two types of bacteria can grow in the presence of oxygen: **aerobes**, which are bacteria that *require* oxygen for growth, and *facultative anaerobes*, which can grow with or without oxygen. Both aerobic bacteria and facultative anaerobic bacteria produce an enzyme called **superoxide dismutase** (**SOD**) that can convert free-radical oxygen to molecular oxygen and peroxide:

$$O_2^- + O_2^- + 2H^+ \xrightarrow{\text{SOD}} H_2O_2 + O_2$$

Unfortunately, the product H_2O_2 (hydrogen peroxide) contains the peroxide anion O_2^{2-}, which is just as toxic to the bacteria as free radical oxygen. (This toxic peroxide anion is the active component in antimicrobial agents such as hydrogen peroxide and benzoyl peroxide.) To avoid this toxic anion, bacteria have developed two enzymes to neutralize it. The enzyme **catalase** converts hydrogen peroxide to water and oxygen:

$$2H_2O_2 \xrightarrow{\text{catalase}} 2H_2O + O_2$$

The enzyme **peroxidase** converts hydrogen peroxide not to water plus molecular oxygen, but rather to just water:

$$H_2O_2 + 2H^+ \xrightarrow{\text{peroxidase}} 2H_2O$$

Fast Fact

Exposing obligately anaerobic organisms to oxygen kills them.

We divide bacteria into three major groups based on oxygen use: **obligate aerobes,** which require oxygen for growth, **obligate anaerobes,** which cannot survive in the presence of oxygen, and the facultative anaerobes mentioned earlier, which grow either with or without oxygen. Remember that because ATP production is greater under aerobic conditions, facultative anaerobes grow *better* in the presence of oxygen. Many of the bacteria that, like *E. coli*, make up our normal flora as well as many pathogenic bacteria are facultative anaerobes.

There are two smaller groups of bacteria that are classified according to oxygen use. **Aerotolerant bacteria** can grow in the presence of oxygen but do not use it in metabolism, and **microaerophiles** are aerobic bacteria but require only low levels of oxygen for growth.

Although many bacteria are facultative anaerobes, several are obligate anaerobes. For example, *Clostridium perfringens* is an obligate anaerobe that causes **gas gangrene**, a condition in which there is significant tissue destruction (**Figure 10.5**). Because this organism is an obligate anaerobe, one of the most effective methods of treatment is simple debridement (removal of the affected tissue), because exposing the *C. perfringens* to oxygen in the air destroys the bacteria.

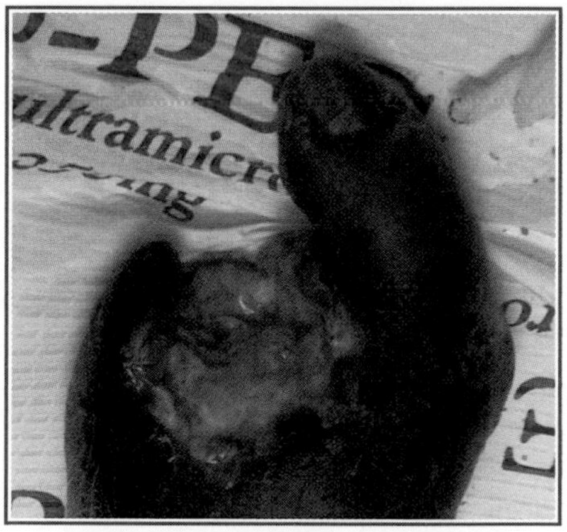

Figure 10.5 Gas gangrene in the left foot. *Clostridium perfringens*, an obligate anaerobe, causes this highly destructive tissue infection. Exposure of this bacterium to oxygen is a lethal event for the bacterium.

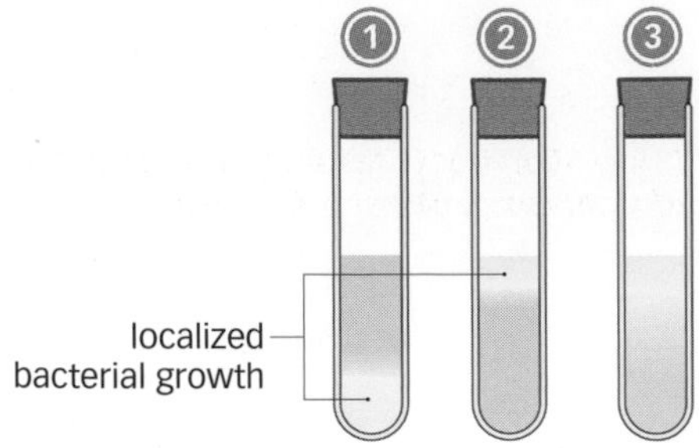

Figure 10.6 Sodium thioglycolate tubes are used for the incubation of bacteria with specific oxygen requirements. Tube 1 shows the growth of an obligate anaerobe - note the absence of growth in the top portion of the broth where oxygen is present. Tube 2 shows the growth of an obligate aerobe - note the growth is only in the top portion of the tube where oxygen is present. Tube 3 shows the growth of a facultative anaerobe – note the uneven distribution of growth from top to bottom (more growth at the top).

Growth of Anaerobic Organisms

Special growth media and incubation are required for anaerobic bacteria to grow. Considering these requirements is especially important when working with anaerobic bacteria that could easily be misidentified if not cultured properly.

Two methods are used for culturing anaerobic bacteria. The first uses a medium called **sodium thioglycolate**. This medium forms an oxygen gradient such that the farther into the medium we go, the less oxygen there is. When bacteria with different oxygen requirements are compared using this medium (**Figure 10.6**), obligate anaerobes will always grow in the area of the tube where oxygen is absent, whereas obligate aerobes grow only where the oxygen concentration is highest. Facultative anaerobes, which can grow either with or without oxygen, grow throughout the sodium thioglycolate medium even though the oxygen concentration decreases steadily from top to bottom of the tube.

The second method for culturing anaerobic organisms uses what is called a **GasPak™ jar**. This incubation container provides an oxygen-free environment that is developed through a series of chemical reactions, as shown in **Figure 10.7**. Because the jar is totally devoid of oxygen, only obligate and facultative anaerobic organisms can grow in it.

Keep in mind that organisms that do not use oxygen make less ATP, which causes them to grow more slowly. Therefore incubation times for anaerobes must be longer than those for aerobes. Abbreviated incubation times could cause anaerobic bacteria to be missed, leading to mistakes in diagnosis and treatment.

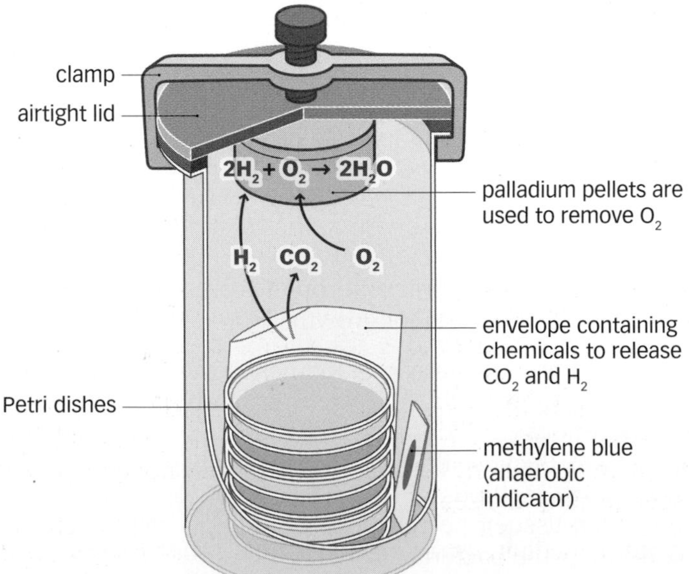

Figure 10.7 The GasPak™ jar is used for incubating anaerobic organisms. The oxygen-free environment results from the use of palladium pellets that catalyze the removal of oxygen through the production of water (as shown in the equation written just under the jar lid). An indicator of the presence of oxygen is included to ensure that the environment is oxygen-free during incubation.

Table 10.4 summarizes the temperature, pH, and oxygen requirements of bacteria.

Table 10.4 Bacterial Requirements for Temperature, pH, and Oxygen

Requirement and bacterial type	Description
Temperature	
Psychrophile	Grows well at 0°C and optimally at 15°C or lower
Psychrotroph	Can grow at 0–7°C but optimally between 20 and 30°C; maximum 35°C
Mesophile	Optimal growth between 25 and 45°C
Thermophile	High temperatures between 55 and 65°C
Hyperthermophile	Optimal growth between 80 and 130°C
pH	
Acidophile	Can grow between pH 0 and 5.5
Neutrophile	Growth between pH 5.5 and 8.0
Alkalophile	Optimal growth at pH 8.5 and 11.5
Oxygen	
Obligate aerobe	Dependent on the presence of oxygen
Facultative anaerobe	Does not require oxygen but grows better in its presence
Obligate anaerobe	Dies in the presence of oxygen
Microaerophile	Requires oxygen levels below 2–10%
Aerotolerant anaerobe	Grows the same with or without oxygen

Keep in Mind

- Bacterial growth has an effect on disease.
- Successful bacterial growth depends on the physical and chemical environment.
- Physical requirements include temperature, pH, and osmotic pressure.
- Chemical requirements include carbon, nitrogen, sulfur, phosphorus, organic growth factors, and trace elements.
- Bacteria have different requirements for oxygen: obligate aerobes require oxygen to grow, whereas obligate anaerobes are killed in the presence of oxygen and facultative anaerobes can grow with or without oxygen.

GROWTH MEDIA

As noted above, bacteria need a variety of growth factors. Any bacterium that requires not one or two but rather a large number of these growth factors is said to be a **fastidious bacterium**. Fastidious bacteria are usually slow growing, and to grow them in the laboratory, the growth medium must provide all of the essential growth factors. Also, laboratory conditions never exactly reproduce the conditions found in nature, and so you must keep in mind that any laboratory results may not completely mimic what is seen in clinical disease.

There are two types of medium used to grow bacteria: *chemically defined* and *complex*.

Ingredient	Quantity
Carbon and energy sources	
Glucose	9.1 g
Starch	9.1 g
Sodium acetate	1.8 g
Sodium citrate	1.4 g
Oxaloacetate	0.3 g
Salts	
Potassium phosphate dibasic	12.7 g
Sodium chloride	6.4 g
Potassium phosphate monobasic	5.5 g
Sodium bicarbonate	1.2 g
Potassium sulfate	1.1 g
Sodium sulfate	0.9 g
Magnesium chloride	0.5 g
Ammonium chloride	0.4 g
Potassium chloride	0.4 g
Calcium chloride	0.4 g
Ferric nitrate	0.006 g
Amino acids	
Cysteine	1.5 g
Arginine and proline (each)	0.3 g
Glutamic acid and methionine (each)	0.2 g
Asparagine, isoleucine, and serine (each)	0.2 g
Cystine	0.06 g
Organic growth factors	
Calcium pantothenate	0.02 g
Thiamine	0.02 g
Nicotinamide adenine dinucleotide	0.01 g
Uracil	0.006 g
Biotin	0.005 g
Hypoxanthine	0.003 g
Reducing agent	
Sodium thioglycolate	0.00003 g
Water	1.0 liter

Table 10.5 An Example of a Chemically Defined Medium

Chemically Defined Media

A **chemically defined growth medium** is one for which the chemical composition is precisely known. An example is shown in **Table 10.5**. Because all the ingredients in chemically defined media are precisely known, these media can be used for the laboratory analysis of compounds produced by specific bacteria.

Complex Media

In contrast to chemically defined media, complex media contain not only numerous ingredients of known chemical composition but also digested proteins and extracts derived from plants or meat. Such media are referred to as complex because the exact chemical composition of these digests and extracts is not known. **Table 10.6** lists a recipe for a complex medium. Don't be puzzled by the fact that the table shows an exact amount for each ingredient. Even though the manufacturer knows the *amount* of each extract and digest added to the medium, the *chemical composition* of each is not known. (Contrast this with the chemically defined medium of Table 10.5, where each ingredient is pure and thus its chemical composition is known.)

In complex media, the energy, carbon, sulfur, and nitrogen required for bacterial growth are all provided by proteins. These proteins are digested into fragments referred to as peptones. These fragments are more easily used than the intact protein molecules, which are relatively insoluble in water. In addition, the meat or plant extracts in complex media provide many of the vitamins required for optimal growth. A complex medium is usually referred to as a nutrient and is available both as a liquid (**nutrient broth**) and as a solid with agar added (**nutrient agar**). (**Agar** is a polysaccharide found in marine algae. It is used as a thickener or solidifier in the manufacture of jams, jellies, and — of most interest to us — laboratory growth media.)

Media as a Tool for Identifying Pathogens

In an ill patient, pathogens must be identified before a medical diagnosis can be made and effective treatment options can be explored. There are several ways that media can be used to help establish the presence of pathogenic bacteria and then identify them. These methods all use selective and differential media. A **selective medium** is one that contains ingredients that prohibit the growth of some organisms while fostering the growth of others. A **differential medium** is one that contains ingredients that can differentiate between organisms (**Table 10.7**).

It is informative to look at several media that have a role in identifying pathogenic bacteria. The selective medium bismuth sulfate agar selects for the growth of the bacterium *Salmonella enterica* serovar Typhi, which causes typhoid fever. This disease is a serious epidemic infection associated with poor sanitation and the fecal–oral route of infection. Although normally rare in the United States, this pathogen has been found in contaminated shellfish harvested off the coast of the United States. The bismuth sulfate in this nutrient agar selects for the growth of *S. enterica* serovar Typhi by inhibiting the growth of all other bacteria. Consequently, fecal samples from patients who present with symptoms consistent with typhoid fever can be inoculated onto this agar; if growth occurs, a preliminary diagnosis can be formulated.

The selective medium Brilliant green uses a dye to inhibit the growth of most microorganisms and selects for the genus *Salmonella*. Another selective medium, Sabouraud's agar, takes advantage of a low pH (5.6) to select for the growth of fungi while inhibiting the growth of bacteria. This type of medium has become increasingly important because the number of fungal infections has risen as a result of AIDS.

Many selective media are also differential. These media use specific ingredients to differentiate between organisms. As an example, *Escherichia coli*, which can be an opportunistic pathogen, is an enteric Gram-negative rod that is difficult to distinguish from other enteric Gram-negative rods. However, when grown on **eosin methylene blue (EMB)** agar, the colonies of *E. coli* exhibit a bright and easily identifiable green sheen created as the *E. coli* ferment lactose (**Figure 10.8**). Because other enteric bacteria do not ferment this sugar in the same way, the growth of *E. coli* is easily distinguishable from that of other enteric organisms that could grow on this medium. From a clinical perspective, because many urinary tract infections are caused by *E. coli*, EMB can be used to preliminarily identify *E. coli* as a source of the infection. (EMB is also a selective medium because the eosin and methylene blue dyes inhibit the growth of Gram-positive organisms.)

Ingredient	Quantity
Water	1 liter
Peptone	5 g
Beef extract	3 g
Sodium chloride	8 g
Agar	15 g

Table 10.6 Recipe for a Complex Medium

Medium	Use	Interpretation of Results
MacConkey medium	Culture and differentiation of bacteria based on their ability to ferment lactose	Lactose fermenters form red to pink colonies; non fermenters form colorless or transparent colonies
Eosin methylene blue (EMB) agar	Isolation, culture, and differentiation of Gram-negative bacteria	Lactose fermenting bacteria form green metallic sheen; non fermenting bacteria form colorless or light purple colonies
Triple sugar iron agar	Differentiation of Gram-negative bacteria on the basis of their fermentation of glucose, sucrose, and lactose and on their production of H_2S gas	Red slant/red butt, no fermentation; yellow slant/red butt, glucose fermentation; yellow slant/yellow butt, glucose and lactose fermentation; butt turns black, H_2S produced
Blood agar	Culture of fastidious bacteria and differentiation of hemolytic bacteria	Partial digestion of blood, alpha hemolysis; complete digestion of blood, beta hemolysis; no digestion of blood, gamma hemolysis

Table 10.7 Examples of Differential Media

Mannitol salt agar (MSA) is another selective/differential medium that is useful for identifying Gram-positive organisms. The medium uses a high salt concentration (7.5%) to select for *Staphylococcus* species while inhibiting the growth of other bacteria. In addition to this selection, the mannitol sugar in MSA permits differentiation between species of *Staphylococcus*. As an example, *S. aureus* can be differentiated from other staphylococcal species because it readily ferments mannitol. When grown on MSA,

Figure 10.8 Differential media. An eosin methylene blue (EMB) plate inoculated with *E. coli*, showing the characteristic green sheen that results from the vigorous fermentation of either lactose or sucrose.

S. aureus presents a bright yellow zone around the growth (**Figure 10.9**). Preliminary identification of this pathogen in clinical isolates is easily accomplished with MSA.

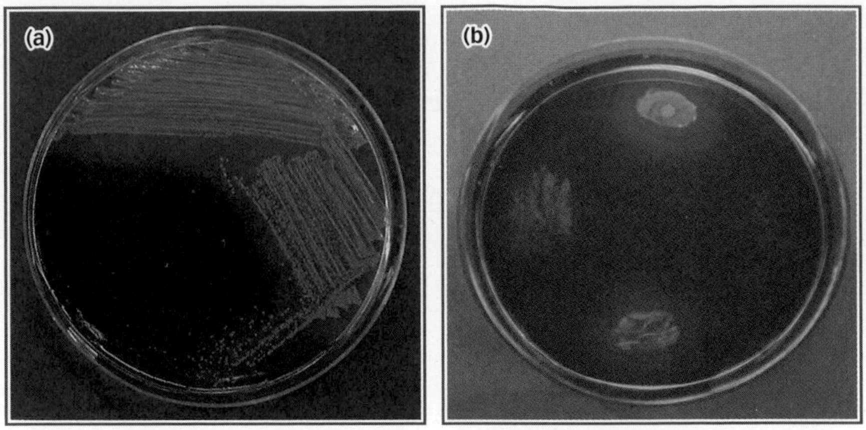

Figure 10.9 An example of selective and differential media. Panel **a**: Mannitol salt agar was streaked for isolation with *Staphylococcus aureus*. Results show an easily identifiable yellow zone around the growth. This medium's high salt concentration (7.5%) makes it selective for Gram-positive *Staphylococcus* organisms. Panel **b**: Mannitol salt agar inoculated with (clockwise from top) *Staphylococcus citreus* showing fermentation of mannitol (yellow zone), *Staphylococcus epidermidis* showing growth but no fermentation of mannitol (no yellow zone), *Staphylococcus aureus* showing fermentation of mannitol (yellow zone), and *Staphylococcus saprophyticus* showing growth but no fermentation of mannitol (no yellow zone).

Another example of a clinically useful differential medium is blood agar. This medium is not selective because blood is a nutrient for most microorganisms. The medium is used to differentiate between organisms on the basis of the hemolysis of red blood cells. Many pathogenic organisms produce hemolysins as part of the infectious process. These hemolysins can be separated into three groups (**Figure 10.10**) on the basis of how completely they destroy red blood cells and they can be used as preliminary diagnostic tools:

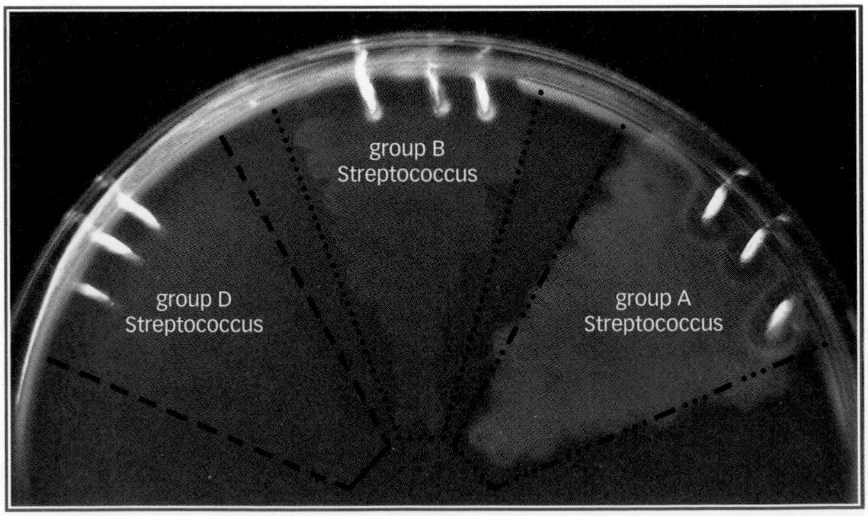

Figure 10.10 Hemolysis. This blood agar plate has been inoculated with three types of Streptococcal organisms: Group D which shows gamma hemolysis, Group B which shows alpha hemolysis and Group A which shows beta hemolysis.

- **Gamma hemolysis** shows no destruction of red blood cells around the bacterial growth.
- **Alpha hemolysis** shows incomplete hemolysis of red blood cells, resulting in a greenish halo around the bacterial growth.
- **Beta hemolysis** causes complete destruction of red blood cells, resulting in a clear area around the bacterial growth. This type of hemolysis is produced either by streptolysin O, an enzyme that is antigenic (it can elicit an immune response) or by the enzyme streptolysin S, which is nonantigenic.

Keep in Mind

- Identification of organisms is important for the study of infectious disease and requires that organisms be grown in a laboratory setting.
- Medium is used to grow bacteria.
- There are two types of medium: chemically defined and complex.
- Selective medium contains ingredients that prohibit the growth of some organisms while fostering the growth of others.
- Differential medium contains ingredients that can differentiate between organisms.

CHARACTERISTICS OF BACTERIAL GROWTH

Bacteria divide primarily by binary fission, in which a parent cell divides into two daughter cells (**Figure 10.11**). Because each division is one generation, the time interval between divisions is called the **generation time** (see **Table 10.1**).

Generation times vary between bacterial species and are heavily influenced by environmental pH, oxygen level, the availability of nutrients, and temperature. With proper levels of nutrients present, the number of bacteria in a colony increases exponentially from generation to generation.

Fast Fact

For an organism with a generation time of 20 minutes, incubation for 24 hours yields 10^{21} organisms.

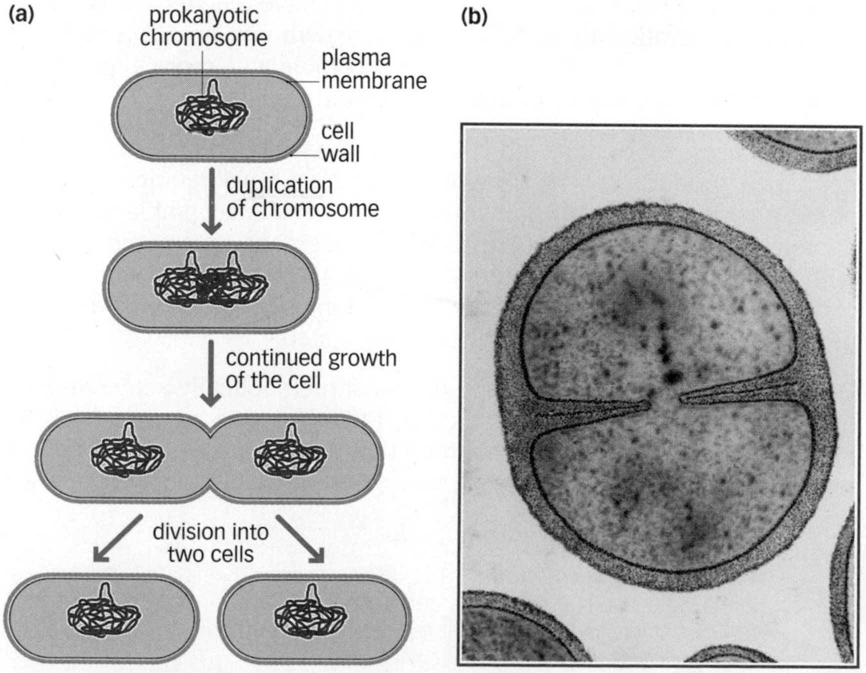

Figure 10.11 Bacterial cell division. Panel **a**: A graphic illustration of binary fission, the most common type of bacterial growth. Panel **b**: A colorized electron micrograph of a dividing cell.

The human body is a very suitable incubator for bacteria because of its constant temperature, excellent environmental conditions, and large supply of available nutrients in tissues and blood. Knowing this, it is easy to understand how quickly an invading bacterial population can multiply. If we magnify the problem by looking at toxin production or antibiotic resistance, the difficulty in dealing with infectious organisms becomes even more apparent.

The Bacterial Growth Curve

As shown in **Figure 10.12**, there are four phases of bacterial growth. We can examine each of these by using bacterial numbers and functions. During the **lag phase** of bacterial growth, the bacteria are adjusting to their environment and may have to synthesize enzymes to utilize the nutrients available in the environment. In this phase, little if any binary fission occurs, indicated by the fact that the growth curve is horizontal. How long this phase lasts varies between bacterial species and also depends on culture conditions.

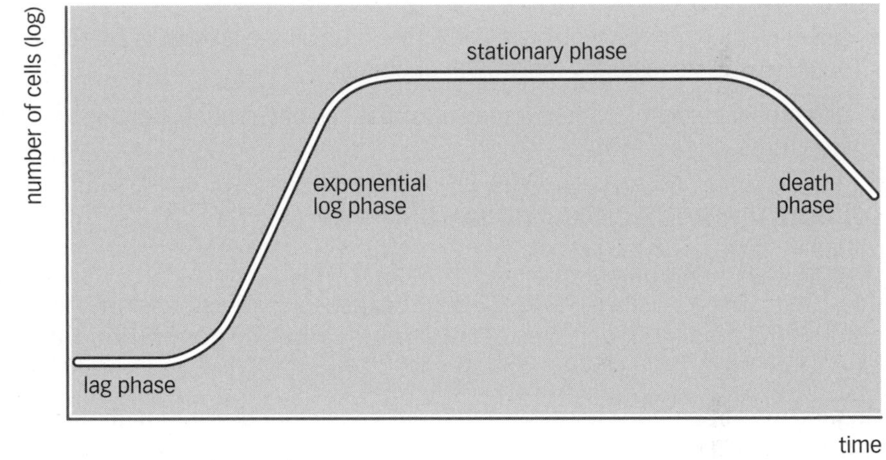

Figure 10.12 The bacterial growth curve. Specific functions and numbers of cells differ at each phase. During the lag phase, there is essentially no cell division. In the log (exponential) phase, the growth rate is exponential, and in the stationary phase the number of dying organisms equals the number produced during binary fission. The death phase, which inversely correlates with the log phase, is also referred to as the logarithmic decline phase.

During the **log phase** (log as in logarithmic; also known as the exponential phase) the number of bacteria doubles and increases exponentially (1 becomes 2, 2 become 4 and so on) and will have reached the constant minimum generation time. This level of growth can be sustained only while environmental conditions remain favorable and, more importantly, only if an adequate supply of nutrients remains available.

In some cases, especially with bacteria that have been genetically engineered to produce specific products, the conditions required for log-phase growth can be artificially maintained. This is accomplished using a **chemostat,** a device that permits the replacement of depleted medium with fresh, nutrient-rich medium (**Figure 10.13**). Under these conditions, organisms can be kept in the log phase for long periods during which the product in question can be continually harvested.

During the log phase, bacteria are the most metabolically active and are therefore most sensitive to antibiotics, radiation, and adverse environmental changes. Recall from Chapter 9 that when a bacterial cell is dividing, cell wall production is at its peak. Therefore antibiotics that target

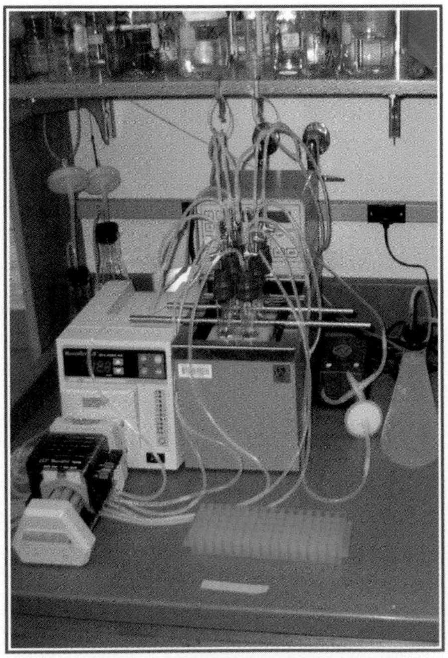

Figure 10.13 A chemostat. This instrument can maintain organisms in the log phase of growth for extended periods. This is accomplished by removal of wastes and constant replacement of nutrients. This technology is widely used by the pharmaceutical industry.

the synthesis of peptidoglycan are extremely effective during this phase of growth. So are antibiotics and/or treatments that target DNA replication or the transcription of RNA. In addition, because the translation of proteins is maximal during this phase, treatments targeting the bacterial ribosome are also highly effective. We will examine these antibiotic target interactions in more detail in Chapters 19 and 20.

At the point where nutrient levels begin to fall and waste levels (resulting from high rates of cell division) begin to rise, a **stationary phase of growth** begins. This phase can be characterized as the phase in which the number of cells dying is essentially equal to the number being produced through cell division. This phase is relatively short because it is predicated on the availability of nutrients, which continue to disappear as the growth curve shifts to the last phase.

The final phase of the bacterial growth curve — the **death phase** — represents a continuous decline in the number of dividing cells. This decline is caused by exhaustion of the nutrient supply as well as collapse of the environment due to the build-up of toxic waste materials. Notice in **Figure 10.12** how the death phase inversely parallels the log phase. Because of this relationship, the death phase is sometimes called the **logarithmic decline phase**.

Measurement of Bacterial Growth

Because the severity of bacterial infections can be correlated with the number of bacteria, it is important to be able to determine the number of organisms. This can be accomplished by either direct or indirect methods.

One direct method for determining the bacterial population of an infecting medium is filtration, which can be used to examine water samples for bacterial contamination. As we have previously mentioned, a favorite route of infection for many pathogens is the "fecal–oral" route. These types of infection are routinely seen in developing countries where sanitation is poor and as a result human waste may be commingled with available water supplies. Filtration is routinely used to examine water supplies for contamination with pathogenic bacteria. This testing generates the **fecal coliform count**, which is used as an indicator of the level of contamination. The setup is shown in **Figure 10.14**.

A water sample is passed through a filter in which the holes are small enough to exclude bacteria. The filter is then placed on a selective medium. After incubation, potentially pathogenic bacteria filtered out of the water form visible colonies on the filter. With this method, contaminated water sources can be easily and inexpensively identified and bacterial concentrations measured.

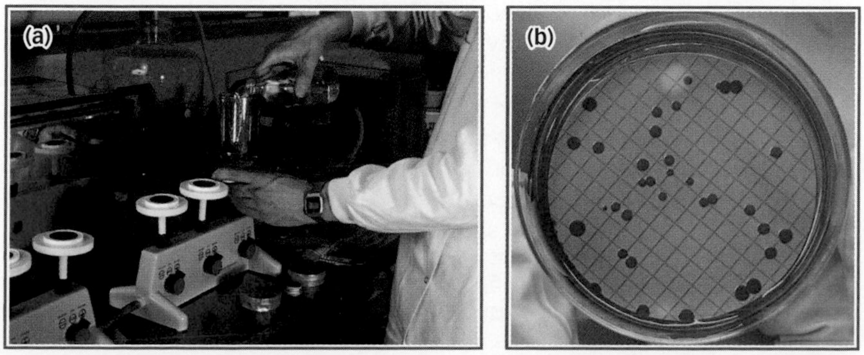

Figure 10.14 Filtration testing. Panel **a**: A water filtration system used to detect and quantify the level of fecal contamination in a water supply. Panel **b**: After water has been passed through it, the filter is placed on a nutrient agar plate and incubated. The number of colonies that grow indicates the level of contamination of that water sample.

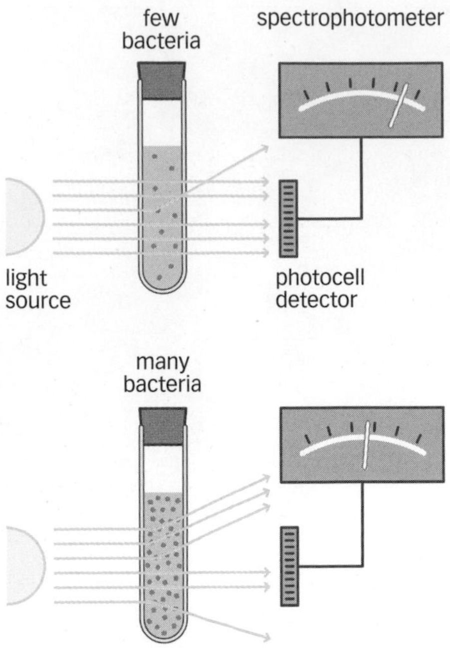

Figure 10.15 An illustration of spectrophotometry, an indirect method of measuring the number of bacteria in a sample. As the number of bacteria increases, more and more of the light is deflected from the pathway to the light detector. How much light is lost from the pathway as the beam passes through the turbid sample is registered on the meter and is compared with standard curves to arrive at the number of bacteria per milliliter of sample.

The most common indirect method for measuring bacterial populations is *spectrophotometry*. As illustrated in **Figure 10.15**, a *spectrophotometer* is an instrument that measures light at specific wavelengths. A liquid containing bacteria becomes turbid as the number of bacteria increases, and when a turbid liquid is placed in the path of light entering a spectrophotometer, some of the light is deflected from the path and is therefore not measured by the light detector. How much of the light is deflected is a measure of how turbid the liquid is. The turbidity is then used in conjunction with a standard curve as a method for determining the number of bacteria.

Direct and indirect methods for measuring bacteria are described in **Table 10.8**.

CLINICAL IMPLICATIONS OF BACTERIAL GROWTH

Many factors affect the rate at which bacterial populations grow. For example, generation time for a given bacterial species is not one unchanging number. Instead, this number can fluctuate depending on the nutrients available to the bacteria, and this variation influences the slope of the log phase of the bacterial growth curve. Because this slope is directly related to growth rate, a change in slope means a change in growth rate.

In addition, many pathogenic bacteria are fastidious in laboratory settings. For example, *Mycobacterium tuberculosis*, which grows well *in vivo*, requires a much longer time to grow in a laboratory, and *Mycobacterium leprae* (the causative agent of leprosy) cannot be grown in a microbial culture at all and must instead be grown *in vivo*.

Some pathogenic bacteria have stringent medium requirements. A good example is *Haemophilus influenzae*, which inhabits the mucous membranes of the upper respiratory tract, mouth, vagina, and intestinal tract. This bacterium is ordinarily fastidious but also requires medium supplements, such as the X factor of blood hemoglobin and the cofactor nicotinamide adenine dinucleotide. These substances are liberated from blood by heating, and the resulting brown agar is referred to as **chocolate agar** (**Figure 10.16**).

Because many bacteria have specific requirements for growth, it is easy to understand how these requirements can affect the diagnosis and treatment of diseases. If clinical specimens are not handled and cultured properly, the presence of some pathogenic bacteria can be missed, leading to errant diagnoses and potential exacerbation of the disease. The problem can be compounded by improper isolation of clinical specimens. It should therefore be noted that samples taken from different parts of the body require different methods of isolation. For example, sputum samples are intrinsically important for diagnosis of respiratory diseases and can be collected by coughing or catheterization. However, needle aspirations are used for blood and cerebrospinal fluids. **Table 10.9** lists several methods for collecting clinical samples.

In the end, it is interesting to speculate on how many diseases may go undiagnosed even today because of the unknown growth requirements of their causative agents.

Type	Method and Process	Characteristics and Limitations
Direct		
	Direct cell counts	Used to determine total number of cells; can be used for those bacteria that cannot be cultured
	Direct microscopic count	Rapid, but at least 10^7 cells/ml must be present to be counted effectively. Counts include living and dead cells
	Cell-counting instruments	Coulter counters and flow cytometers count total cells in dilute solutions. Flow cytometers can also be used to count organisms to which fluorescent dyes or tags have been attached
	Viable cell counts	Used to determine the number of viable bacteria in a sample, but includes only those that can grow in given conditions. This requires an incubation period of about 24 hours or longer. Selective and differential media can be used to enumerate specific species of bacteria
	Plate count	A time-consuming but technically simple method that does not require sophisticated equipment. Generally used only if the sample has at least 100 cells/ml
	Membrane filtration	Concentrates bacteria by filtration before they are plated; thus can be used to count cells in dilute environments
	Most probable number	Statistical estimation of likely cell number; it is not a precise measurement. This method can be used to estimate numbers of bacteria in relatively dilute solutions
Indirect		
	Measuring biomass	Biomass can be correlated with cell number
	Turbidity	Very rapid method; used routinely. A one-time correlation with plate counts is required in order to use turbidity for determining cell number
	Total weight	Tedious and time-consuming; however, it is one of the best methods for measuring the growth of filamentous microorganisms
	Chemical constituents	Uses chemical means to determine the amount of a given element, usually nitrogen. Not routinely used
	Measuring cell products	Methods are rapid but must be correlated with cell number. Frequently used to detect growth, but not routinely used for quantification
	Acid	Titration can be used to quantify acid production. A pH indicator is often used to detect growth
	Gases	Carbon dioxide can be detected by using a molecule that fluoresces when the medium becomes slightly more acidic. Gases can be trapped in an inverted Durham tube in a tube of broth

Table 10.8 Methods Used to Measure Bacterial Growth

Approximately 10% of the infections seen in humans are caused by bacteria, but more than 80% result from virus infections. Because viruses are obligate intracellular parasites, they have growth requirements that are very different from those for bacteria. We will examine the growth of viruses in detail in Chapter 12.

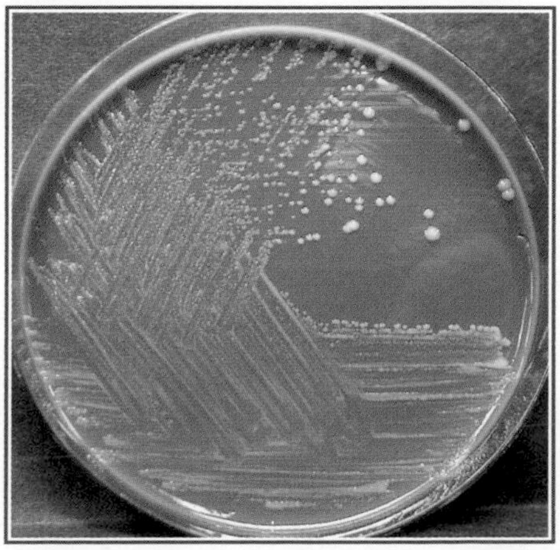

Figure 10.16 Chocolate agar is used for growing fastidious bacteria such as *Haemophilus influenzae* (which does not cause influenza). To undergo cell division and increase population numbers, this organism requires certain substances found in blood. These substances are released by cooking the blood, which gives the medium a brown color.

Type or Location of Specimen	Collection Method
Skin, accessible mucous membrane (including eye, outer ear, nose, throat, vagina, cervix, urethra) or open wound	Sterile swab brushed across the surface; care should be taken not to contact neighboring tissue
Blood	Needle aspiration from vein; anticoagulants are included in the specimen tube
Cerebrospinal fluid	Needle aspiration from subarachnoid space of spinal column
Stomach	Intubation, which involves inserting a tube into the stomach, often via a nostril
Urine	In aseptic collection, a catheter is inserted into the bladder through the urethra; in the "clean catch" method, initial urination washes the urethra, and the specimen is midstream urine
Lungs	Collection of sputum either dislodged by coughing or acquired via a catheter
Diseased tissue	Surgical removal (biopsy)

Table 10.9 Methods Used To Collect Clinical Specimens

Keep in Mind

- Bacteria grow by binary fission in which one cell divides into two.
- Different bacteria have different generation times.
- The bacterial growth curve has four phases: lag, log, stationary, and death.
- The number of organisms can be determined directly through methods such as direct microscopic counting, filtration, and automated cell counting.
- Indirect measurement of bacterial numbers involves methods such as spectrophotometry, total weight, and measurement of cell products.
- Samples taken from different parts of the body are collected with different methods.

SUMMARY

- Bacterial growth is an important part of the infection process.
- Growth of bacteria requires proper physical and chemical environments.
- Different bacteria have different requirements for oxygen. Obligate anaerobic organisms cannot grow in the presence of oxygen, aerobic organisms cannot grow without it, and facultative anaerobes can grow with or without oxygen.
- Identification of organisms is important for the study of bacteria and the treatment of disease.
- Media, either complex or chemically defined, can be used to grow bacteria in a laboratory setting.
- Selective medium prohibits the growth of some organisms while fostering the growth of others.
- Differential medium contains ingredients that can differentiate between organisms.
- Most bacteria grow by the process of binary fission.
- Different bacteria have different generation times.
- The bacterial growth curve has four phases: lag, log, stationary, and death.
- Different methods are used to take samples from different parts of the body.

In this chapter we have examined the characteristics of and requirements for bacterial growth. Remember that there are both physical and chemical requirements and that bacterial growth can be subdivided into four basic phases: lag, log, stationary, and death. Bacteria carry out different functions in each phase, and the log phase (where the bacteria are undergoing cell division at the highest rate), can be the most important for targeting by inhibitors of bacterial growth, such as antibiotics. We saw that bacteria have generation times, which can vary on the basis of factors such as nutrient and incubation requirements. We also learned that selective and differential media, as well as specific measuring methods, can be important in the identification and isolation of pathogenic bacteria.

SELF EVALUATION AND CHAPTER CONFIDENCE

Multiple Choice

Answers are given in the back of the book and help can be found in the student resources at:
www.garlandscience.com/micro

1. Psychrophiles are organisms that grow
   - **A.** Between 0 and 15°C
   - **B.** Between 45 and 60°C
   - **C.** At room temperature
   - **D.** Between 25 and 40°C
   - **E.** Above 80°C

2. Most human pathogens are
   - **A.** Psychrophiles
   - **B.** Psychrotrophs
   - **C.** Thermophiles
   - **D.** Extreme thermophiles
   - **E.** None of the above

3. Minimum growth temperature is
   - **A.** The highest temperature at which an organism will grow
   - **B.** The lowest temperature at which an organism will grow
   - **C.** The temperature at which there will be minimum growth
   - **D.** The temperature at which there will be the greatest amount of growth
   - **E.** The temperature at which cells will grow the best

4. Optimal growth temperature is
   - **A.** The highest temperature at which an organism grows
   - **B.** The lowest temperature at which an organism grows
   - **C.** The temperature at which the highest rate of growth occurs
   - **D.** The temperature at which bacteria grow poorly
   - **E.** None of the above

5. Acidophilic bacteria grow best at
   - **A.** pH 7
   - **B.** pH 2
   - **C.** pH 9
   - **D.** pH 14
   - **E.** None of the above

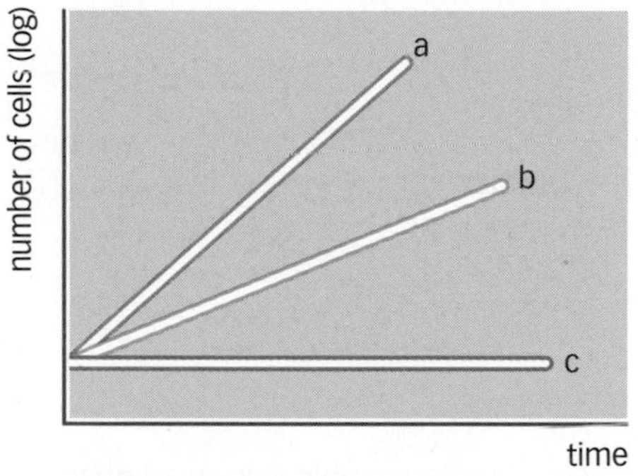

6. In the figure above, which line best depicts a facultative anaerobe in the absence of O_2?
   - **A.** Line a
   - **B.** Line b
   - **C.** Line c

7. In the figure above, which line shows the growth of an obligate aerobe incubated anaerobically?
   - **A.** Line a
   - **B.** Line b
   - **C.** Line c

8. In the figure above, which line best illustrates the growth of a facultative anaerobe incubated aerobically?
   - **A.** Line a
   - **B.** Line b
   - **C.** Line c

9. In the figure above, which line best depicts *Neisseria gonorrhoeae* when growing inside the human body?
   - **A.** Line a
   - **B.** Line b
   - **C.** Line c

10. The term facultative anaerobe refers to an organism that

A. Is killed by oxygen
B. Doesn't use oxygen but tolerates it
C. Uses oxygen or grows without oxygen
D. Requires less oxygen than is present in air
E. Prefers to grow without oxygen

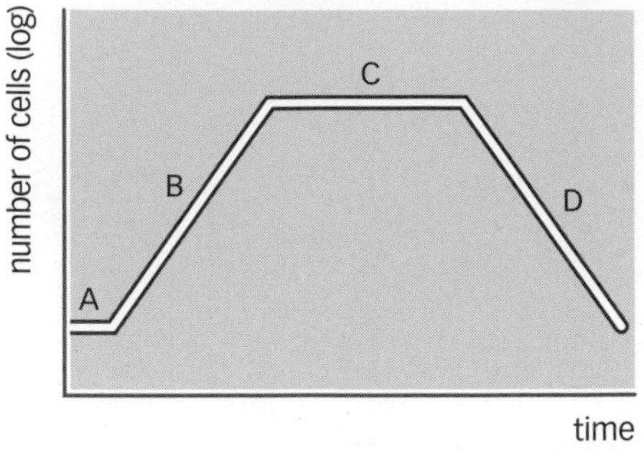

11. In the figure above, which section shows a growth phase where the number of cells dying equals the number of cells dividing?

A. B
B. A
C. D
D. C
E. A and C

12. During which growth phase are Gram-positive bacteria most susceptible to penicillin?

A. Lag phase
B. Death phase
C. Log phase
D. Stationary phase
E. The culture is equally susceptible during all phases.

13. Which of these organisms produce catalase and superoxide dismutase?

A. Obligate anaerobes
B. Aerobes
C. Aerotolerant anaerobes
D. Halophiles
E. Extreme halophiles

Medium A	Medium B	Medium C
Na_2HPO_4	Tide detergent	Glucose
KH_2PO_4	Na_2HPO_4	Peptone
$MgSO_4$	KH_2PO_4	$(NH_4)_2SO_4$
$CaCl_2$	$MgSO_4$	KH_2PO_4
$NaHCO_4$	$(NH_4)_2SO_4$	Na_2HPO_4

14. Which medium or media in the table above is/are chemically defined?

A. A
B. B
C. A and B
D. A and C
E. None of the above

15. Which enzyme catalyzes the reaction: $2H_2O_2 \rightarrow 2H_2O + O_2$?

A. Peroxidase
B. Superoxide dismutase
C. Catalase
D. Oxidase
E. None of the above

16. An organism that can grow both with and without oxygen is called

A. An obligate anaerobe
B. An obligate aerobe
C. A dual purpose organism
D. A facultative aerobe
E. A facultative anaerobe

17. Mannitol salt agar (MSA) is a medium that is

A. Selective only
B. Differential only
C. Both selective and differential
D. Neither selective nor differential
E. Used for culturing organisms from the ocean

18. Eosin methylene blue (EMB) medium

A. Differentiates organisms by glucose fermentation
B. Is selective because of the lactose it contains
C. Differentiates organisms by lactose fermentation
D. Is not selective
E. Is not differential

19. Gamma hemolysis on a blood agar medium shows

A. Complete destruction of red blood cells
B. Partial destruction of red blood cells
C. Complete destruction of platelets
D. No destruction of red blood cells
E. Partial destruction of platelets

20. Logarithmic decline is another way of describing which phase of bacterial growth?

A. Lag
B. Log
C. Stationary
D. Death

DEPTH OF UNDERSTANDING

Questions listed here require you to bring together the concepts you have learned in this chapter into a discussion format. This helps you to increase your depth of understanding of the material you have learned. Help can be found in the student resources at:
www.garlandscience.com/micro

1. When you arrive at the pool, there is a sign indicating it is unsafe to swim because of coliform contamination. Using what we have learned in this chapter, discuss how this decision was arrived at.

2. Discuss the relationship between bacterial growth and pathogenicity and the relationship between bacterial growth and treatment for infections.

3. Compare and contrast organisms that are obligate anaerobes and those that are facultative anaerobes.

CLINICAL CORNER

Help can be found in the student resources at:
www.garlandscience.com/micro

1. Mr Rodriguez is in the hospital because of an intestinal infection. He has been having bouts of diarrhea and vomiting for two days and has become seriously dehydrated.
   - **A.** What samples would you take from this patient and how would you do so?
   - **B.** What tests would you perform on these samples?
2. Amy Smith has been coughing for a week and is producing a greenish mucus when she coughs. Examination of her breathing indicates she may have a lower respiratory infection, which could be pneumonia.
   - **A.** What samples would you collect from Amy, and how would you do so?
   - **B.** What physical and chemical requirements would be necessary to grow this organism, and what tests would you perform on the samples you collect from her?

Microbial Genetics and Infectious Disease

Chapter 11

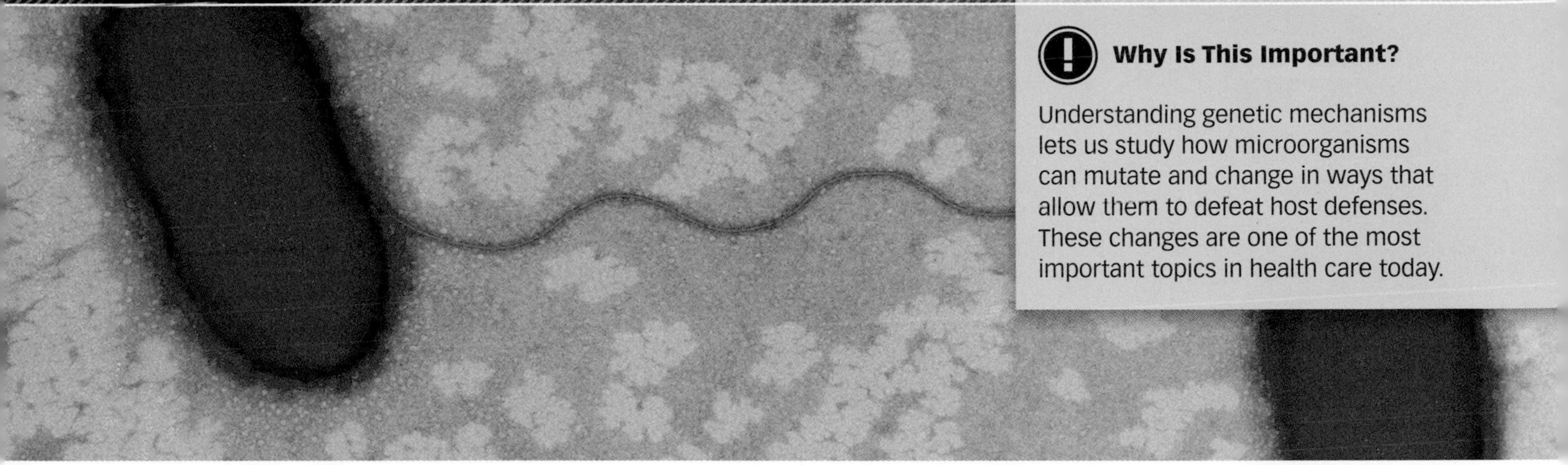

Why Is This Important?

Understanding genetic mechanisms lets us study how microorganisms can mutate and change in ways that allow them to defeat host defenses. These changes are one of the most important topics in health care today.

OVERVIEW

One of the most difficult problems in medicine today is antibiotic resistance. Pathogenic organisms become resistant to antibiotics through mutation, a process that causes changes in the genetic information of the organisms. These changes can be transferred from one organism to another, which means that microbes have the capacity to exchange genetic information. The genetic changes resulting from these exchanges can make harmless organisms dangerous and pathogenic organisms lethal. To understand pathogenesis and virulence, we must be familiar with microbial genetics. In this chapter, we begin to acquire this familiarity by looking at the structure of the nucleic acids DNA and RNA. We then examine the processes of replication, gene expression, mutation, and gene transfer and end the chapter by taking a look at the relationship of genetics to pathogenesis and virulence.

We will divide our discussions into the following topics:

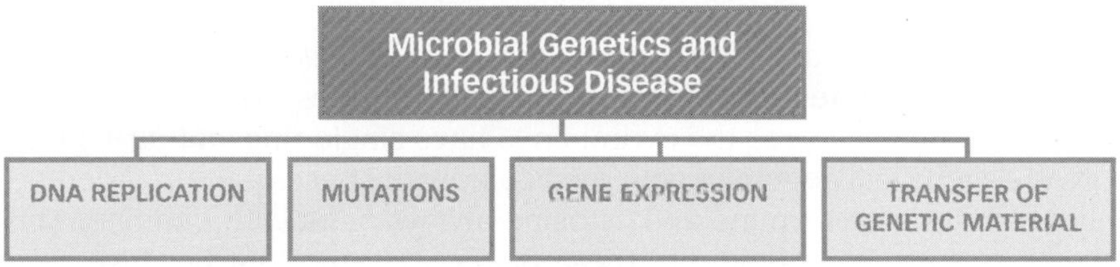

What Do I Need to Know?

To get the most out of this chapter, please review the following terms from your previous courses in biology, anatomy, physiology, or chemistry or in previous chapters of this book as indicated in parentheses: active site of protein (**2**), chromosome (**4**), complementary base pairing (**2**), deamination (**2**), dehydration synthesis (**2**), gene (**4**), gene-chromosome relationship, membrane transport (**4**), nucleic acid (**2**), pathogenesis (**1**), phenotype, pilus (**9**), plasmid (**9**), primary, secondary, tertiary protein structure (**2**), ribosome (**4**), stereochemistry, virulence (**1**).

THE STRUCTURE OF DNA AND RNA

If we are to understand the mechanisms discussed in this chapter, we must first review some of the fundamental information about nucleic acids. In the following sections we examine the structure of DNA and RNA.

Deoxyribonucleic Acid (DNA)

As we begin our discussions of DNA structure, remember that, in nature, structure and function are tightly connected to each other. The function of DNA is as a blueprint for all of the components of the cell which can be faithfully passed on to subsequent generations of cells. It is the structure of DNA that allows this to be accomplished. This structure–function relationship maximizes the ease of replication and makes gene expression, defined as the process of transcription (DNA to RNA) and translation (RNA to protein), efficient and error-free. When we consider the structure of DNA, we must first examine its components and how they orient themselves together.

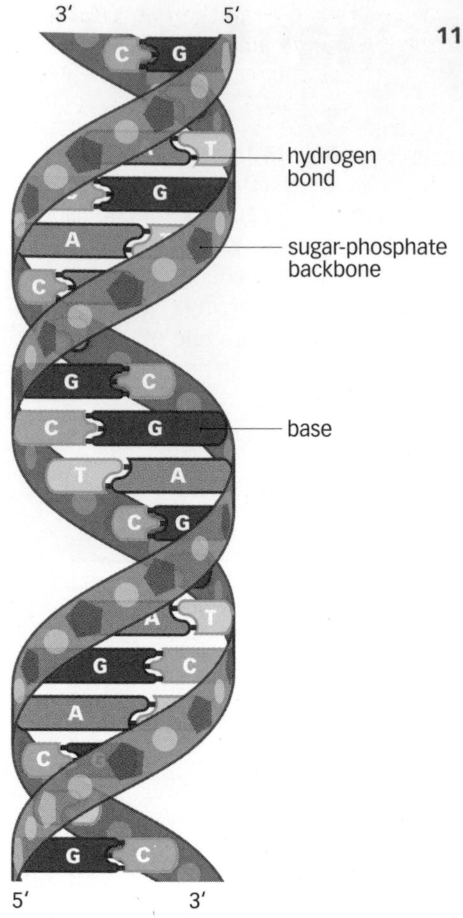

11.1 **Figure 11.1 The helical structure of DNA contains two strands that are anti-parallel.** There are three hydrogen bonds between the guanine and cytosine bases but only two between adenine and thymine.

DNA stands for deoxyribonucleic acid and is a double-stranded helical molecule made up of nucleotides (**Figure 11.1**). Nucleotides are combinations of a phosphate, a sugar (which in DNA is **deoxyribose**) and a nucleotide base (**Figure 11.2**). The two polynucleotide strands of DNA are complementary and wind around each other to form a double helix. In this structure, the bases project inward but are still accessible because the helical nature of the molecules naturally produces grooves (a major and a minor groove) between the twists of the helix.

It is important to note that these components bind together in a very specific way chemically, and that binding permits the correct and precise orientation of the nucleotide. It is this orientation that permits the building of the strands of DNA (**Figure 11.3**). Nucleotides join to each other when the 3' hydroxyl group of the sugar of one nucleotide joins to the 5' hydroxyl group of another nucleotide (**Figure 11.4**). These linkages impart inherent *polarity* and *structural orientation* to the growing chain, which is required for the proper addition of bases.

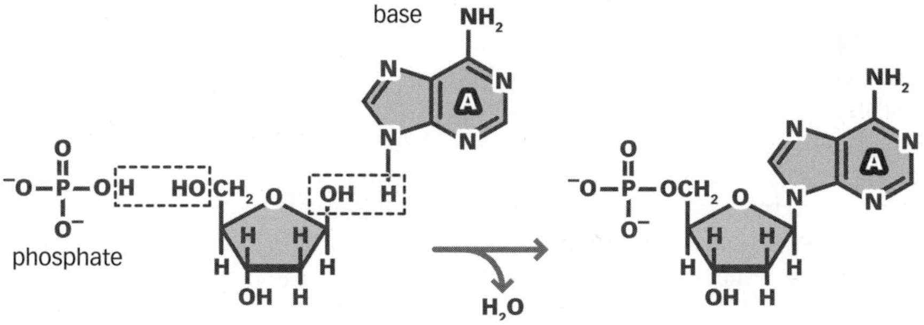

Figure 11.2 Formation of a nucleotide. The phosphate, deoxyribose sugar, and base are assembled through the process of dehydration synthesis (removal of water).

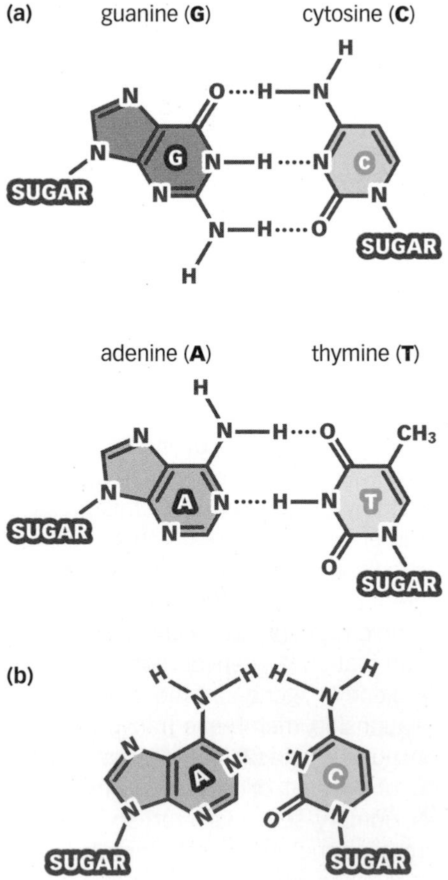

The bases in DNA are of two types: the **purines** (*adenine* and *guanine*) and the **pyrimidines** (*thymine* and *cytosine*). Purines are large double-ring structures, whereas the pyrimidines have single rings (**Figure 11.5**). DNA has a helical geometry governed by how the bases pair up. Adenine always pairs with thymine and cytosine always pairs with guanine. The size, structure, and binding of these bases are important because they affect this helical geometry. The chemistry of the way in which these bases bond together requires that one of the strands be oriented upside down relative to the other. This is referred to as *anti-parallel* and confers the proper orientation necessary for bonding of the bases. Proper base pairing also keeps the sugars and phosphates in alignment so that there is no distortion of the helical DNA structure. This causes the bases to "stack" on top of each other. All of these elements help to ensure the thermodynamic stability of DNA. Consequently, any mismatched base pairing becomes chemically unstable. Remarkably, this chemical instability is one of the ways in which errors are identified and corrected.

Figure 11.3 Hydrogen bonding between the bases of DNA. Panel **a**: The structures of the nucleotide bases give them the ability to form the two hydrogen bonds between the A and T bases and three bonds between the C and G bases. This is important for the fidelity of DNA replication. In fact, mismatching of bases is structurally inhibited (panel **b**).

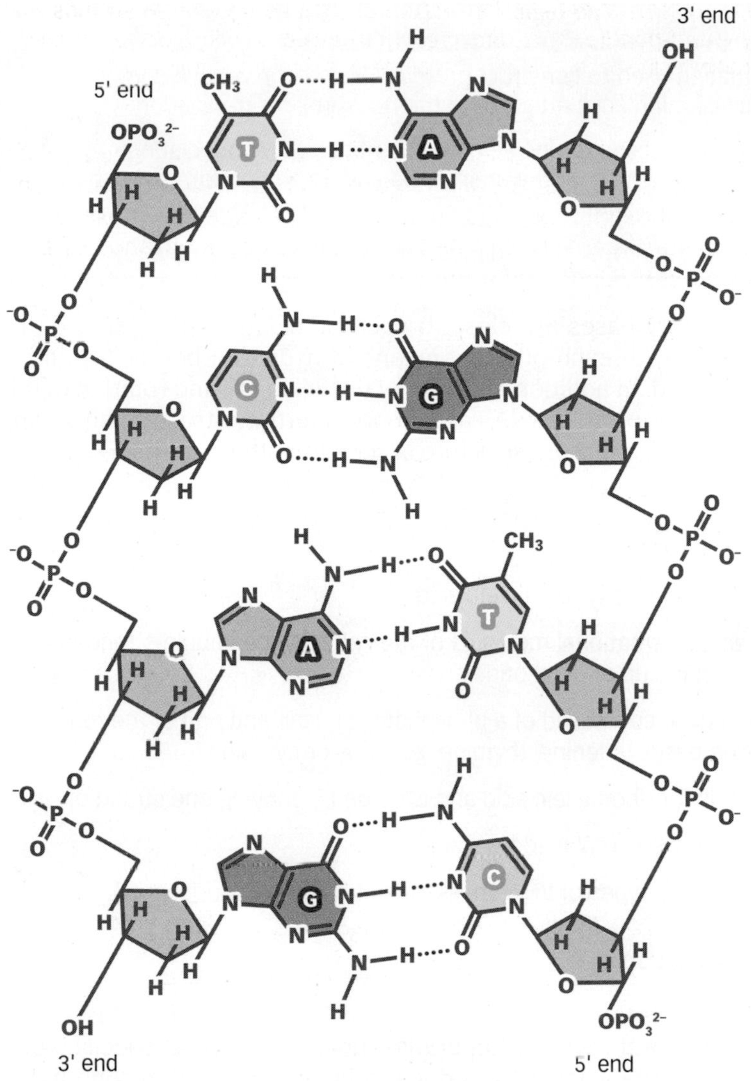

Figure 11.4 Detailed structure of a polynucleotide polymer. Notice the pairing between purine and pyrimidine as well as the phosphodiester linkages of the phosphate backbone. Many of the functions of DNA replication and repair are based on the stereochemistry of this structure.

Ribonucleic Acid (RNA)

The second nucleic acid we are interested in when studying genetic mechanisms is ribonucleic acid, which differs from DNA in the following ways:

- RNA contains the sugar ribose rather than deoxyribose.
- The bases in RNA are adenine, cytosine, guanine, and uracil, and the base pairings are adenine with uracil, cytosine with guanine.
- RNA is usually found in a single-stranded form. However, RNA can fold on itself and form areas that are in a double-stranded form (see the discussion below).

RNA functions in three ways (**Table 11.1**). It is found as **messenger RNA** (mRNA), a form containing information derived from DNA that is used for construction of proteins. In the form of **transfer RNA** (tRNA), it carries amino acids to the ribosome where protein is being constructed. Finally, as **ribosomal RNA** (rRNA), it helps in maintaining the proper shape of the ribosome and the orientation of the protein under construction.

As we examine each of these roles of RNA, keep in mind the complementary nature of base pairing described in Chapter 2 — adenine with uracil, cytosine with guanine — because this pairing is very important in constructing exactly the right proteins needed by a cell.

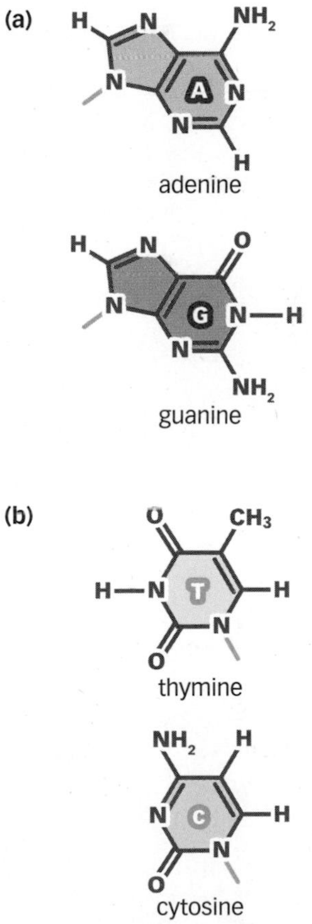

Figure 11.5 The structures of purines and pyrimidines. Panel **a**: The structure of the purines adenine and guanine. Panel **b**: The structure of the pyrimidines thymine and cytosine. The orange line shows the bond attaching it to the deoxyribose sugar.

Type of RNA	Properties and Function
Messenger	Transcribed from DNA; carries information used to construct proteins; uses a three-base combination called a codon that codes for a specific amino acid; attaches to the ribosome for translation
Transfer	Found in cytoplasm and used to pick and transfer amino acids to the ribosome; has a distinct three-dimensional shape with an attachment arm for the amino acid and an anti-codon region that guarantees the proper orientation at the ribosome
Ribosomal	Combines with specific proteins to form the proper three-dimensional configuration of the ribosome

Table 11.1 Types of RNA

If complementary bases are close to each other on an RNA strand, the bases can pair with each other by means of hydrogen bonds, forming a loop in the strand. In addition, because of the flexibility and rotation of this single-stranded molecule, RNA can also form tertiary structure by folding on top of itself. This tertiary structure is seen in both tRNA and rRNA.

Keep in Mind

- DNA stands for deoxyribonucleic acid.
- DNA is the informational molecule of the cell and is a double-stranded molecule made up of nucleotides.
- A nucleotide is composed of a phosphate, a sugar, and one of the four nucleotide bases (adenine, thymine, guanine, or cytosine).
- RNA stands for ribonucleic acid and is made by copying one strand of DNA.
- In RNA, thymine is replaced by uracil.
- There are three types of RNA: messenger, transfer, and ribosomal.

DNA REPLICATION

DNA replication is the process by which the DNA is copied, and is a carefully controlled and regulated operation involving several specific components and mechanisms. It is a critical cellular procedure accomplished with remarkable accuracy, at astounding speed.

DNA Separation and Supercoiling

One of the characteristics of a double-helical molecule is the potential for **supercoiling**, which occurs when the helix twists around itself. An analogy is a coiled phone cord that can become coiled over again and again. Although the chemical and physical reasons for supercoiling are beyond the scope of this discussion, it is important to note that before strands can be unwound and separated for replication or transcription (a process we discuss below), the DNA supercoiling must be relaxed. This is accomplished by **topoisomerase**, a remarkable enzymatic protein that unwinds the supercoils by first breaking the DNA chain so that the supercoil relaxes, and then precisely resealing the break. Once the supercoiling has been relaxed, the enzyme **helicase** will unwind and separate the chains.

There are two basic requirements for replication:

- An ample supply of each of the nucleotides adenine, thymine, cytosine, and guanine
- A **primer:template junction**. Once the double-stranded DNA has been unwound, each unwound single DNA strand is called a template. A portion of this template is then paired with a short segment of RNA called a primer. The point at which the primer and template join is called a primer:template junction (**Figure 11.6**).

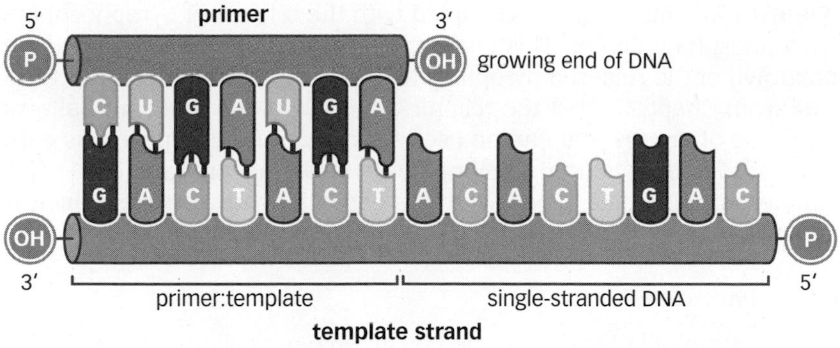

Figure 11.6 The general structure of a primer:template junction. The primer region, which is complementarily base paired to the DNA strand, provides a free 3' hydroxyl group for addition of the appropriate DNA base.

Because DNA replication proceeds in only one direction, the primer:template junction gives the DNA polymerase a place to which the next base can be attached. Remember, this binding is between the 3' end of one base and the 5' end of the next nucleotide. This one-way addition of bases is a universal feature of both DNA and RNA synthesis.

Elongation of bases from the 3' end is required because of the stereochemistry of the bases. The template portion of the primer:template junction dictates which base will be added to the elongating strand. As can be seen

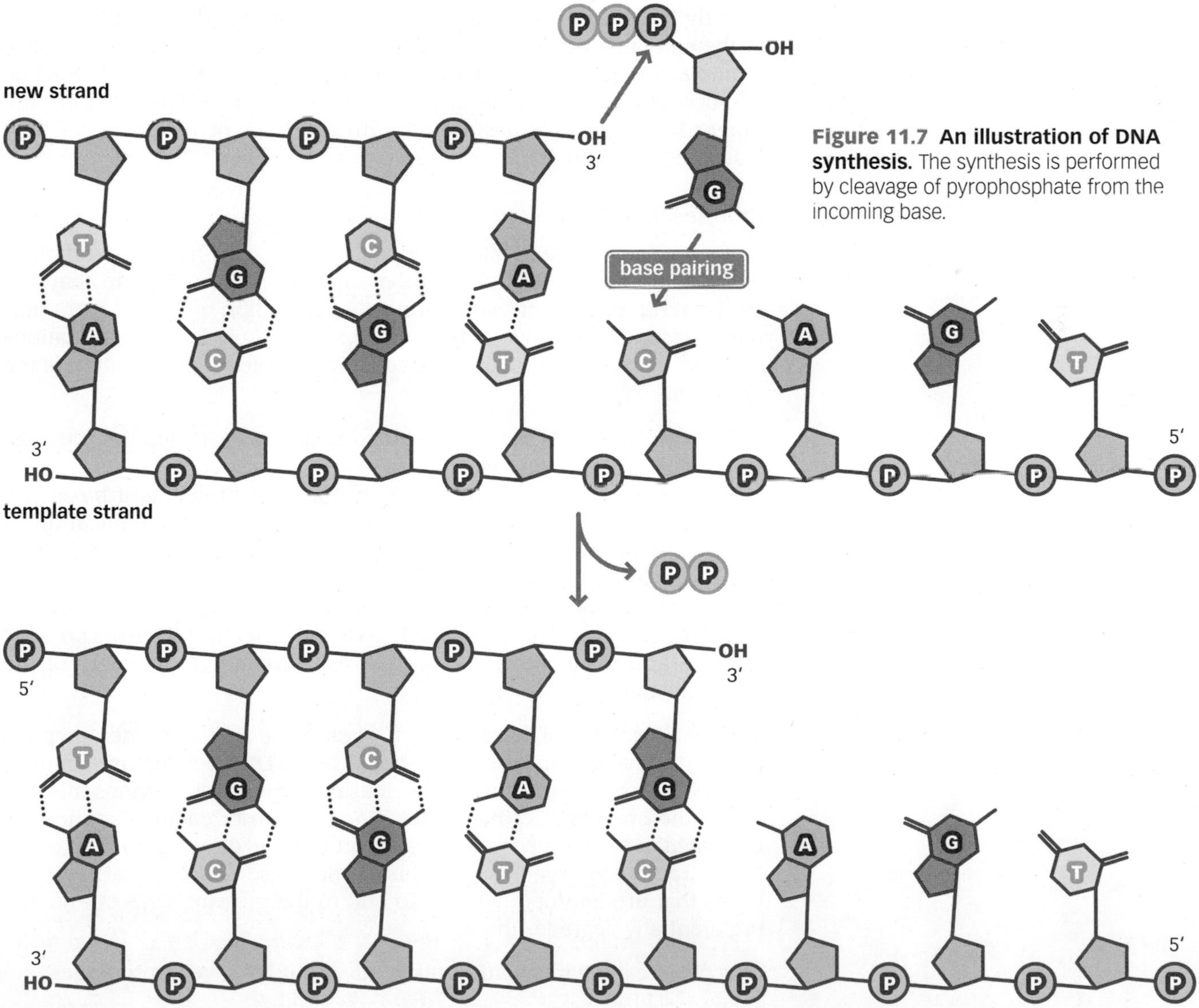

Figure 11.7 An illustration of DNA synthesis. The synthesis is performed by cleavage of pyrophosphate from the incoming base.

in **Figure 11.7**, this reaction is coupled with the release of pyrophosphate (a two-phosphate chain). This energetic reaction couples the immediate breakdown of the released pyrophosphate into two individual molecules. Recall from Chapter 2 that the "clipping" of phosphate molecules allows the release of energy that can be used for cellular functions. In this case the released energy is used to elongate the growing strand of DNA. This coupling process makes DNA synthesis essentially irreversible, which is important because DNA must be as stable a molecule as possible.

DNA Polymerase

Fast Fact

An important and interesting characteristic of DNA polymerase is that it will never pick up RNA bases, even though there may be more than 1000 times as many available.

The replication of DNA is performed by an enzyme called **DNA polymerase**. Remember that enzymes have *active sites* that are used in enzymatic reactions (Chapter 3), and this type of activity is used by DNA polymerase to form new strands of DNA using the primer:template junction as a guide. The enzyme takes any of the four bases and attempts to pair them. This allows the enzyme to work incredibly fast because it does not have to wait for the right base to appear at the active site.

DNA polymerase can bind more than one base at a time at the primer:template junction. It should be noted that there are several types of DNA polymerase that perform specific functions and have different levels of processivity (that is, how many bases they can add at a time), ranging from a few to more than 50,000 bases per binding event. Consequently, the rate of DNA synthesis is directly correlated to the processivity of the particular polymerase. It takes about 1 second for the polymerase to find and bind to the primer:template junction. Once bound, the actual addition of a nucleotide is in the millisecond range. Furthermore, the longer the enzyme is associated with the DNA being elongated, the faster the addition of nucleotides occurs.

Proofreading By DNA Polymerase

Since the same genetic information is passed down from generation to generation of cells, it is vital to the survival of the organism that this information remain correct. Fortunately, replication of DNA is extraordinarily accurate. However, there are always going to be some mutations because these genetic changes have a pivotal role in the evolution of the organism.

As we consider what we have learned so far, we can see that the system of DNA replication, which is based on the structure and the proper pairing of bases, can be fraught with potential problems if bases are added incorrectly. It has been shown that during DNA replication, it is *possible* for incorrect base pairing to occur at a frequency of about once in every 10^5 pairings. This frequency is significantly higher than the actual *observed* error for replication, which is approximately once in every 10^{10} pairings. The difference between the number of errors possible and the actual number is due to the "proofreading" capability of the DNA polymerase.

This proofreading of bases added takes place at the primer:template junction active site and is possible because the DNA polymerase contains an *exonuclease* component. Exonuclease enzymes are enzymes that can attack the open ends of molecules. The DNA proofreading exonuclease is located in the DNA polymerase and attacks the growing end (3') of the DNA strand being synthesized. This exonuclease is strongly attracted to bases that are improperly trying to add to the growing DNA strand and preferentially degrades them.

This proofreading mechanism takes advantage of several properties:

- Chemical structure. DNA polymerase activity slows down if an improper nitrogenous base tries to attach to the growing DNA strand. This is due to the chemical structure of the base (simply put, the incoming base does not fit correctly).
- Any improper binding changes the configuration of the primer:template junction in the active site of the polymerase, causing the junction to slide to the exonuclease active site (**Figure 11.8**).
- At the same time, this improper chemical structure causes an increase in the exonuclease activity, which speeds up the removal of the improper base when it arrives at this site.
- Once the "wrong" base has been removed, the primer:template configuration changes once again and causes the junction to leave the exonuclease active site and move back into the polymerase active site.

One of the most important aspects of this proofreading mechanism is that the corrections are made without having to disengage the DNA polymerase. If removal were required, the process could be very time-consuming because the polymerase would first have to disengage and then re-engage once the correction had been made.

The Replication Fork

The **replication fork** is the location along a DNA double helix where replication is going on. It is easily recognizable because, in this fork, the double helix is unwound and the strands are separated from each other. It is important to note that the fork is moving along the DNA as the strands are unwound. When we examine the elements involved at this location, we can get a good idea of the overall process of DNA replication. However, before we examine these events, we need to review three facts:

- Both strands of DNA are replicated at the same time.
- The two strands are antiparallel (**Figure 11.9**).
- Each strand is replicated by the addition of bases to the 3' end of the strand. This presents a dilemma for the polymerase because the antiparallel alignment of the strands makes it impossible to replicate both of them simultaneously.

In DNA replication, one strand is in perfect position for the addition of bases to its 3' end. We call this the **leading strand**, and that is how it is labeled in Figure 11.9. Looking at the leading strand, it is easy to understand that the polymerase can add bases and move toward the advancing replication fork (in effect, the polymerase on this strand is "chasing" the fork). In contrast, because of the antiparallel alignment of the two strands of DNA, the DNA polymerase replicating the other strand — called the **lagging strand** — is moving away from the fork. This movement of the enzyme away from the replication fork necessitates replicating the lagging strand in pieces, which are called **Okazaki fragments** (after the scientist who discovered them).

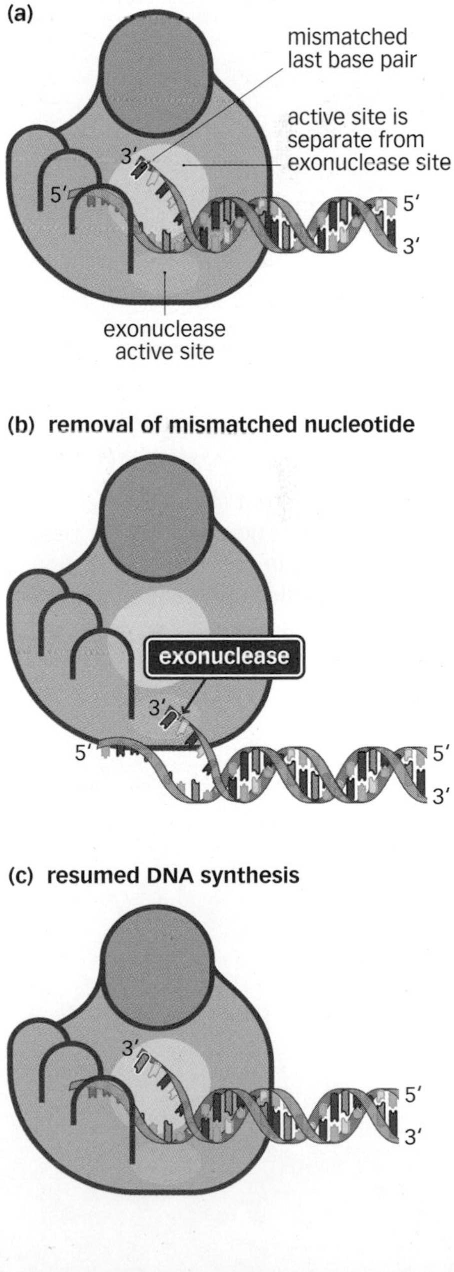

Figure 11.8 A representation of the proofreading capability of the DNA polymerase, which roughly resembles a curled hand. Mismatching of base pairs at the active site of the enzyme causes replication to slow down (panel **a**), and movement of the growing strand from the active site to the exonuclease part of the enzyme (panel **b**). Here, the mismatched base is removed and the strand moves back into the active site (panel **c**).

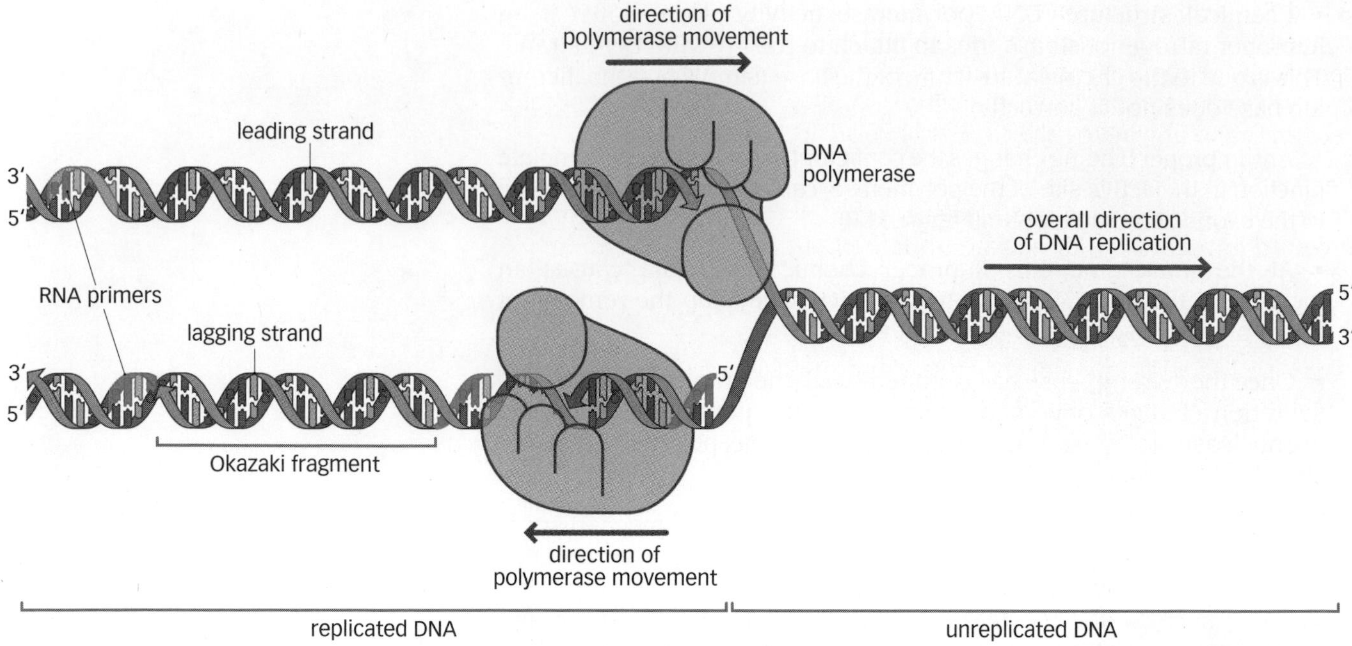

11.2

Figure 11.9 The replication fork. This is the site where DNA replication is occurring. The direction of replication in the leading (continuous) strand is opposite to the direction of replication in the lagging (discontinuous) strand. Note the RNA primer (blue) that is present at the start of each Okazaki fragment.

Remember that for bases to be added to a growing DNA strand, there has to be a primer:template junction that the polymerase can use. This is no problem for the leading strand because the one primer segment at the 3' end is all that is required for replication for this strand. In contrast, for the lagging strand, the DNA polymerase has to wait for the replicating fork to move far enough to expose a substantial segment of the template DNA. Therefore, each new Okazaki fragment needs to have its own primer so as to form a primer:template junction.

Fast Fact

Primase is different from the RNA polymerase that normally transcribes DNA into RNA, in that its *only* function is to synthesize short primer segments on the lagging strand.

Formation of the junctions is accomplished by a form of RNA polymerase called **primase**, which can synthesize RNA without a 3' end being present (this synthesis can occur at any place along the DNA sequence). As the replication fork moves, the lagging strand of DNA elongates, and a primase molecule attaches to the strand and synthesizes a small piece of RNA. This RNA becomes the primer part of the primer:template junction.

Once the short RNA primer segments have been synthesized, the primer is ready for the DNA polymerase to begin adding bases, which it does until it reaches the end of the previously synthesized Okazaki fragment. (Okazaki fragments are usually between 1000 and 2000 nucleotides long.) When the synthesis of an Okazaki fragment has finished, an enzyme called **RNAase H** removes the primer RNA. The gap that results from the missing primer is then easily filled in by DNA polymerase by using the 3' end of the previous Okazaki fragment as a primer to be added onto. Once the gap is filled, the ends of the DNA pieces are linked together by the enzyme **DNA ligase**.

Initiation and Termination of Replication

Initiation of DNA replication in bacteria begins at a specific spot on the chromosome that is called the **origin of replication**. It is thought that there are two components that control this initiation. The first is the **replicator sequence**, a specific set of DNA sequences that includes a long string of A-T base pairs. Remember that the hydrogen bonding between

A and T is weaker than that between G and C, so this stretch is more easily opened. The second component in initiation is the **initiator protein**, which specifically recognizes the replicator sequence of DNA. The series of events involved in initiation includes the initiator protein binding to the replicator sequence and unwinding the DNA strands adjacent to that site.

Replication will terminate when the entire DNA chromosome has been copied; it also requires a specific set of events in which the components involved in replication dissociate from the DNA. At this point, the replicated circular chromosomes will still be linked together (**Figure 11.10**). If the daughter cells that result from cell division are to acquire one of these chromosomes, they must be separated from each other. This separation is accomplished by topoisomerase, the same enzyme involved in relaxing the supercoils in DNA before replication.

Keep in Mind

- DNA is faithfully replicated so that the same genetic information is passed on from generation to generation.
- The enzymes topoisomerase and helicase unwind and separate the strands of DNA to be replicated.
- DNA polymerase copies each of the parental strands so that each daughter cell will contain a chromosome made up of a parental and daughter strand.
- DNA polymerase has a proofreading capability to prevent mistakes during replication, and it replicates in only one direction (from the 3' end of the strand).
- The replication fork is the site at which replication is occurring.
- At the replication fork there is a leading strand, which is replicated continuously, and a lagging strand, which is replicated in pieces known as Okazaki fragments.
- Replication is initiated at a site on the DNA called the origin of replication and proceeds until the entire chromosome has been copied.

THE GENETIC CODE

Before we can examine gene expression, we have to have an understanding of the genetic code. The information contained in DNA that is used for gene expression (transcription and translation) is based on a four-letter alphabet (A, T, G, and C) which uses combinations of three letters called **codons**. These codons code for specific amino acids. If you were to list all the possible three-letter combinations using the four available letters A,

Figure 11.10 Replication and separation of two completed chromosomes. The Initiator binds to the replicator sequence of the chromosome. After replication is complete, the two identical chromosomes are separated by the topoisomerase enzyme in a reaction similar to that seen in the relaxation of supercoiled DNA.

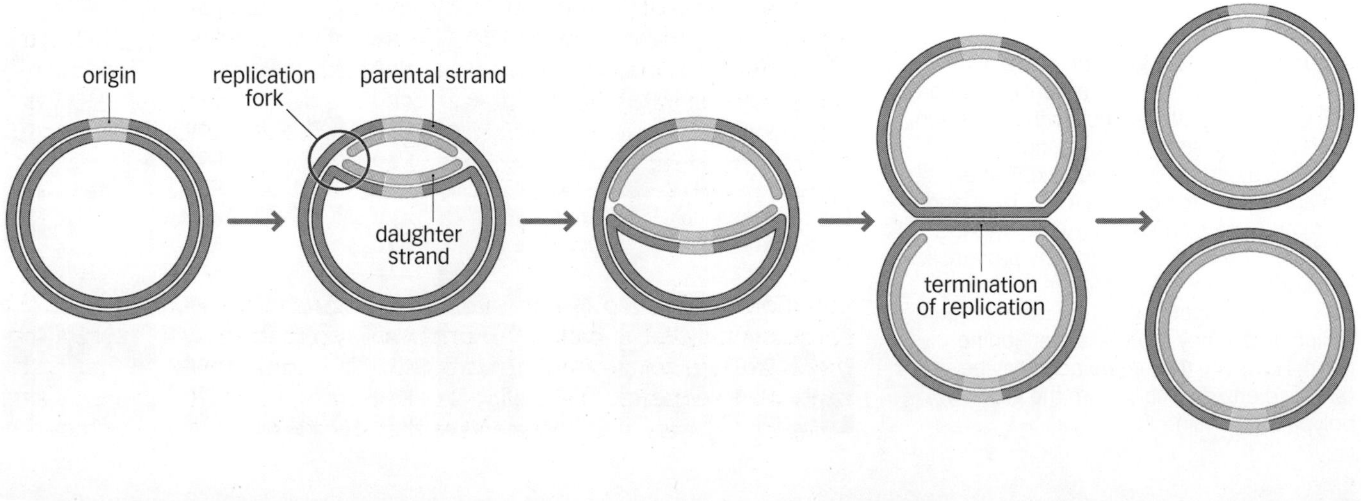

T, G, and C, you would come up with 64 possible codons (**Figure 11.11**). However, only 20 amino acids are used to make all the proteins found in all organisms. We use the term **degenerate** to describe this redundant feature of the genetic code.

It is thought that the genetic code evolved in this degenerate fashion as a way to minimize the potentially devastating effects of mutations. Consequently, if you look at the combinations shown in **Figure 11.11**, you see that changes in the third letter of the codon do not change the selected amino acid. The "insignificance" of a last letter change in the codon is part of the **wobble hypothesis** that was originally put forth by Francis Crick. In addition, notice the three **stop codons**, UAA, UAG, and UGA. These stop codons are important in the proper and precise construction of proteins (a role we shall look at below).

There are three rules that govern the arrangement and use of codons in mRNA. First, codons are always read in only one direction. As in DNA structure, we can use the chemical structure designations of the nucleotide base to orient our understanding. When we do this, we see that the orientation goes from 5' to 3'. This is important because the movement of the message through the ribosome during protein synthesis must be in an orderly fashion. Second, the message is translated in a fixed **reading frame**, which is set by an **initiation codon**. Third, there is never any overlap or gap in the code. If this were to occur, it would throw off the reading frame and cause aberrant changes in protein structure.

Keep in mind that any deviations in the DNA sequence, such as a mutation, would ultimately affect the structure of the protein that is coded for.

GENE EXPRESSION

Now that we have examined DNA replication and the genetic code, we can turn our attention to the expression of genes. A gene is a segment of DNA that codes for a functional product, and the production of that product based on the information contained in the DNA is called gene expression.

As we have seen, replication faithfully ensures that the DNA of one cell is available for the next generation. Any changes (mutations) that occur as a parent DNA strand is replicated are therefore carried in subsequent cell generations. This is important because changes in the genetic information can help organisms become and remain more pathogenic. As we

first position (5' end)	second position: U		second position: C		second position: A		second position: G		third position (3' end)
U	UUU	phe	UCU	ser	UAU	tyr	UGU	cys	U
U	UUC	phe	UCC	ser	UAC	tyr	UGC	cys	C
U	UUA	leu	UCA	ser	UAA	stop	UGA	stop	A
U	UUG	leu	UCG	ser	UAG	stop	UGG	trp	G
C	CUU	leu	CCU	pro	CAU	his	CGU	arg	U
C	CUC	leu	CCC	pro	CAC	his	CGC	arg	C
C	CUA	leu	CCA	pro	CAA	gln	CGA	arg	A
C	CUG	leu	CCG	pro	CAG	gln	CGG	arg	G
A	AUU	ile	ACU	thr	AAU	asn	AGU	ser	U
A	AUC	ile	ACC	thr	AAC	asn	AGC	ser	C
A	AUA	ile	ACA	thr	AAA	lys	AGA	arg	A
A	AUG	met	ACG	thr	AAG	lys	AGG	arg	G
G	GUU	val	GCU	ala	GAU	asp	GGU	gly	U
G	GUC	val	GCC	ala	GAC	asp	GGC	gly	C
G	GUA	val	GCA	ala	GAA	glu	GGA	gly	A
G	GUG	val	GCG	ala	GAG	glu	GGG	gly	G

Figure 11.11 The genetic code. The three nucleotides in the messenger RNA are designated as first, second, or third position. Each set of these three nucleotides specifies a particular amino acid. There are also three stop codons (UAA, UAG, and UGA) and one start codon (AUG). The AUG is always the start codon for any protein. If this codon is not present in the message, translation cannot begin. Therefore the first amino acid in any protein is methionine (Met). However, this amino acid may be removed enzymatically from the growing polypeptide chain.

examine the expression of genes into functional proteins, we will see that the mechanisms of expression (1) involve specific interactions between proteins, RNA, and DNA and (2) are highly regulated.

There are two parts to gene expression, *transcription* (construction of RNA from a DNA template) and *translation* (construction of protein by using RNA instructions).

Transcription

As we mentioned briefly in Chapter 4, **transcription** is the process whereby RNA is made from DNA. This process is similar to DNA replication in that one strand of DNA serves as a template for the production of RNA. However, there are several important differences:

- RNA synthesis does not require a primer:template junction.
- RNA does not remain base-paired to the DNA template once transcription is complete.
- Unlike DNA polymerase, RNA polymerase is a poor proofreader. This makes RNA synthesis less accurate than DNA replication.
- Transcription copies only certain portions of a DNA strand, whereas replication copies all of it.

The process of transcription proceeds through three steps, namely *initiation*, *elongation*, and *termination*, in which different events occur (**Figure 11.12**).

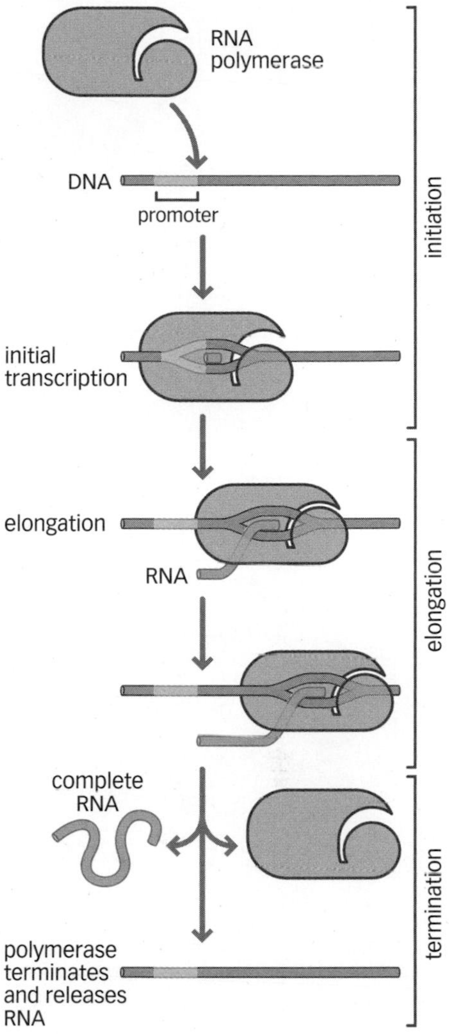

Figure 11.12 The three phases of transcription. Initiation, elongation, and termination result in the synthesis of a complete RNA strand. Notice the "claw" shape of the RNA polymerase, which binds to the promoter region of the DNA strand to begin transcription.

11.3

11.4

During *initiation*, a DNA sequence called the *promoter* initially binds the RNA polymerase. The formation of this complex causes a specific conformation to form. This configuration allows the initiation phase to continue; the DNA strands are separated in front of the complex, producing what is referred to as a "bubble" in the DNA (**Figure 11.13**) in which RNA is being made. It is important to remember that the bubble is only as large as needed for the production of RNA, and the separated strands of the DNA are immediately rejoined right behind the polymerase. Just as we saw in the replication of DNA, RNA bases are added only at the 3' end of the growing strand of RNA.

After about 10 RNA bases have been added, the polymerase transitions from the initiation phase to the *elongation* phase. During elongation, the RNA polymerase "multitasks" in that it unwinds the DNA ahead of it, adds bases to the growing end of the RNA strand and re-anneals (closes) the DNA strands behind. During this process, the polymerase also does limited proofreading as well.

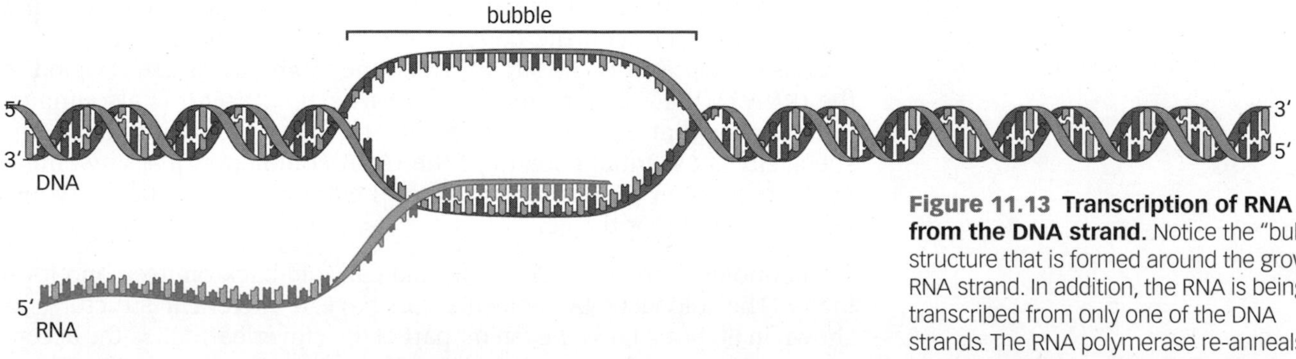

Figure 11.13 Transcription of RNA from the DNA strand. Notice the "bubble" structure that is formed around the growing RNA strand. In addition, the RNA is being transcribed from only one of the DNA strands. The RNA polymerase re-anneals (closes) the strands of DNA directly after transcription.

Of all the steps in the synthesis of RNA, the *termination* event is the least well understood. For our purposes, it is enough to say that when the polymerase reaches a segment of the DNA that signals the end of the required RNA segment it detaches from the DNA.

Translation

Now that we have seen how the informational molecule called RNA is assembled, we need to examine **translation**, the process by which a cell makes proteins by "reading" the information contained in its RNA. This process is directly affected by any errors in either the DNA used as the template for the newly created RNA or in the RNA itself.

Translation uses the sequence of nucleotides in messenger RNA to generate a specific sequence of amino acids and thereby manufacture a protein needed by the cell. Like transcription, translation is a very highly conserved function seen in all cells. Translation is also "expensive" in that it requires a high energy investment. In a rapidly dividing cell, up to 80% of all the energy is invested in the making of new proteins by translation.

Translation requires all three types of RNA — messenger, transfer, and ribosomal. Let's look at these different RNA molecules, remembering that each is transcribed from one strand of DNA during transcription.

Messenger RNA in Translation

The protein-coding region of mRNA includes an **open reading frame** (ORF). Each ORF indicates the start of a specific protein's amino acid sequence. Translation starts at the ORF and goes from the 5' to the 3' end of the mRNA strand. The mRNA contains a "start" codon where protein assembly begins. At the end of the synthesis, stop codons on the mRNA stop the process.

Recall from Chapter 4 that **ribosomes** are non-membrane-enclosed cytoplasmic structures made up of protein and ribosomal RNA. They are the site of protein production in the cell. Prokaryotic mRNA molecules contain a sequence of nucleotides serving as a binding site that "recruits" a ribosome to the mRNA. The ribosome and mRNA join at this binding site through complementary base pairing. This allows the mRNA to be properly oriented along the ribosome, with the start position of the mRNA aligned.

Transfer RNA in Translation

Transfer RNA molecules function as adapters between mRNA attached to a ribosome and the amino acids being added to the growing protein chain. This bridging function is elegantly accomplished because each tRNA attaches to a specific amino acid and to a specific codon in the mRNA. Each tRNA molecule is about 75–95 nucleotides long, and every tRNA ends with the codon 3'-CCA-5'. This codon is important because it represents the site at which the amino acids are coupled to the tRNA by the enzyme aminoacyl-tRNA synthetase. It is also important to note that each tRNA is specific for only one amino acid. This specificity is essential because if the tRNA could carry just any amino acid, the number of potential errors in the amino acid sequence would be catastrophic for the cell.

As mentioned earlier, a tRNA molecule can fold back on itself and form loops. The folded tRNA molecule can have a cloverleaf structure, as shown in **Figure 11.14**. The "stem" part of the cloverleaf, called the acceptor arm, is where the amino acid specific for that particular tRNA binds. The loop farthest from the acceptor arm is called the **anticodon loop** and

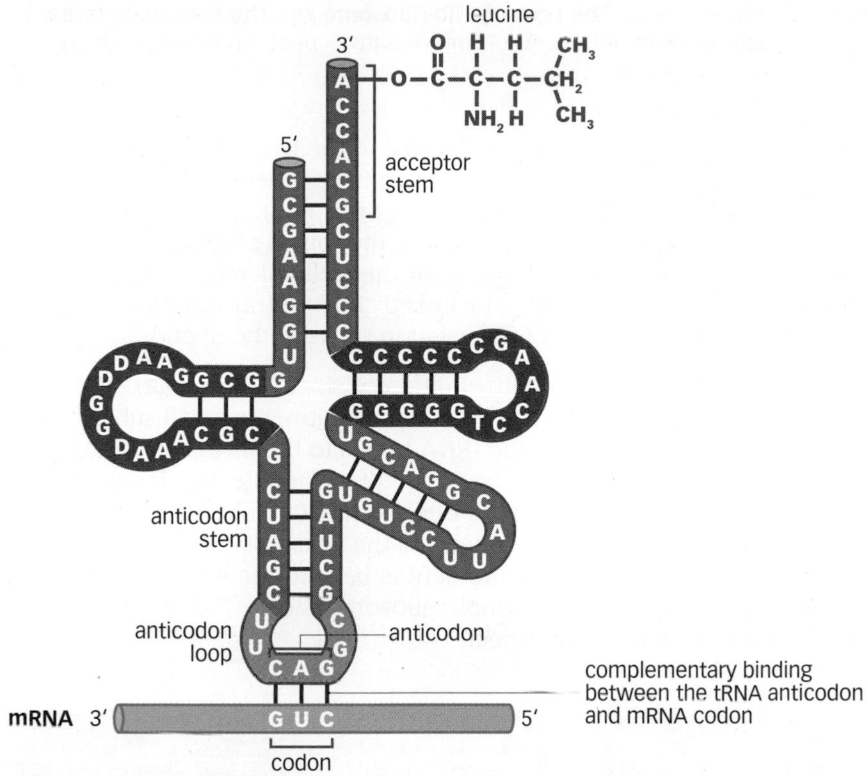

Figure 11.14 A diagram of the transfer RNA (tRNA) molecule. The acceptor arm sequence of 3'ACC that is present on all tRNAs signifies the amino acid-binding section of this carrier molecule. Also note the anticodon region, which uses complementary base pairing with the messenger RNA to ensure that the correct amino acid is placed in the growing polypeptide chain.

is the most important loop because it is where codon recognition occurs. This whole region of the cloverleaf is referred to as the **anticodon region** and uses complementary base pairing to ensure that the correct amino acid is placed in the protein sequence encoded by the mRNA.

Let's examine how tRNA picks up the amino acid. This may be the most important reaction in protein synthesis because the bond joining the amino acid to the tRNA is a high-energy bond. The energy released in breaking this bond will then be used to form the peptide bonds between amino acids as a protein is created.

Attaching an amino acid to tRNA is a two-step process. Step one is coupling of the amino acid to AMP. The second step is transfer of the amino acid to the 3' end of the tRNA. These reactions require the presence of enzymes called **aminoacyl-tRNA synthetases**, with each amino acid having its own synthetase. Because there are 20 amino acids, most organisms have 20 different synthetase enzymes.

This process of specifically coupling amino acids has a high rate of fidelity and is associated with an interaction between the acceptor arm and the anticodon loop of the tRNA molecule. In fact, the incorrect amino acid is attached on fewer than 1 in 1000 occasions. This specificity is very important because the ribosome is unable to discriminate right from wrong amino acids. It blindly accepts any tRNA if there is proper codon–anticodon recognition. The specificity of translation is therefore due to aminoacyl-tRNA synthetases.

The Ribosome in Translation

As already noted, the ribosome contains the machinery that directs protein synthesis. It is composed of three molecules of rRNA and more than 50 proteins and is a highly efficient machine that can add 2–20 amino acids per second. In prokaryotes, the ribosome attaches to mRNA as the latter is being made. Recall that RNA polymerase moves along the DNA strand

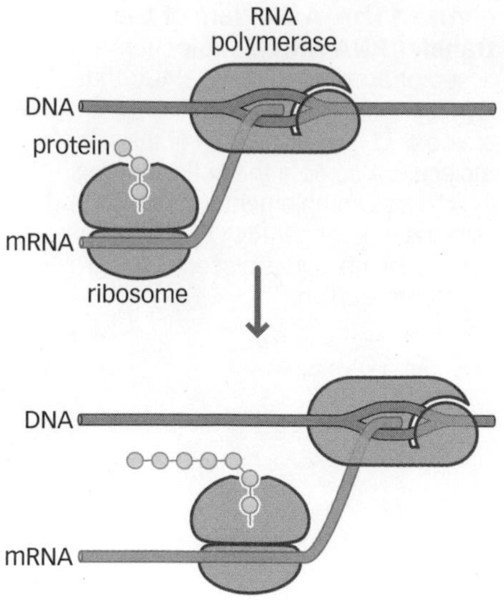

Figure 11.15 The prokaryotic ribosome and the RNA polymerase can work in unison. As the mRNA is transcribed, ribosomes begin to translate protein.

and transcribes the information encoded in the DNA base sequence into a strand of mRNA. As this mRNA leaves the polymerase molecule, it can be immediately bound to a ribosome (**Figure 11.15**). In fact, as one ribosome moves along the mRNA strand, another binds onto the same strand right behind the first. The elegance of this cellular engineering permits transcription and translation to be linked because transcription occurs in the 5' to 3' direction, whereas translation starts at the 5' end.

The ribosome has two subunits of different sizes (**Table 11.2**). The large subunit is responsible for peptide bond formation; the small subunit contains the *decoding center*, where tRNA bonds to the mRNA (it is here that codon–anticodon recognition occurs). We designate the large subunit as the 50S subunit and the small one as the 30S subunit. The S stands for Svedberg units, a measurement of the mass of the subunit (a discussion of the Svedberg measurement is beyond the scope of this text). Interestingly, the mass of the whole ribosome is only 70S rather than the 50S + 30S = 80S you might expect.

Table 11.2 The Structure of Ribosomes

Property	Value
Overall size	70S
Small subunit	30S
Proteins in small subunit	Approximately 21
Size of RNA in small subunit	16S (1500 bases)
Large subunit	50S
Proteins in large subunit	Approximately 34
Size of RNA in large subunit	23S (2900 bases) and 5S (120 bases)

The 50S subunit is made up of a 5S rRNA and a 23S rRNA, which are coupled with many different proteins. The 30S subunit also contains a variety of proteins but contains only one 16S rRNA molecule. It is important to once again recall that the combination of rRNA molecules and protein molecules confers on these subunits a three-dimensional shape (**Figure 11.16**).

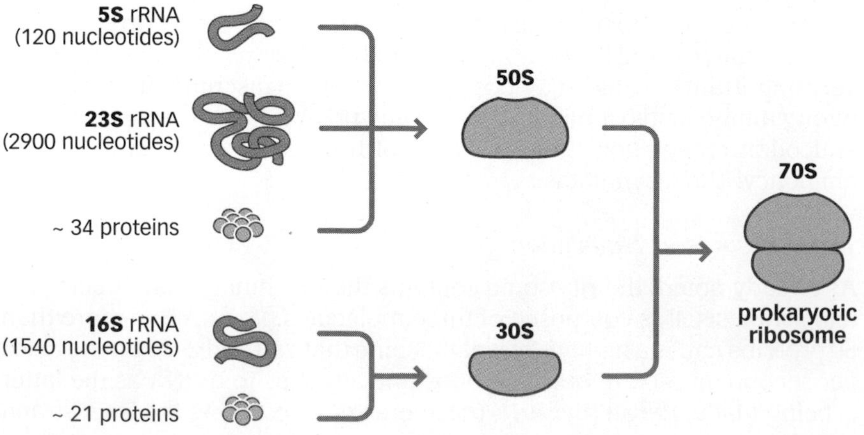

Figure 11.16 Composition of the prokaryotic ribosome. Notice the difference in weights (S units) between the intact ribosome and each of the subunits

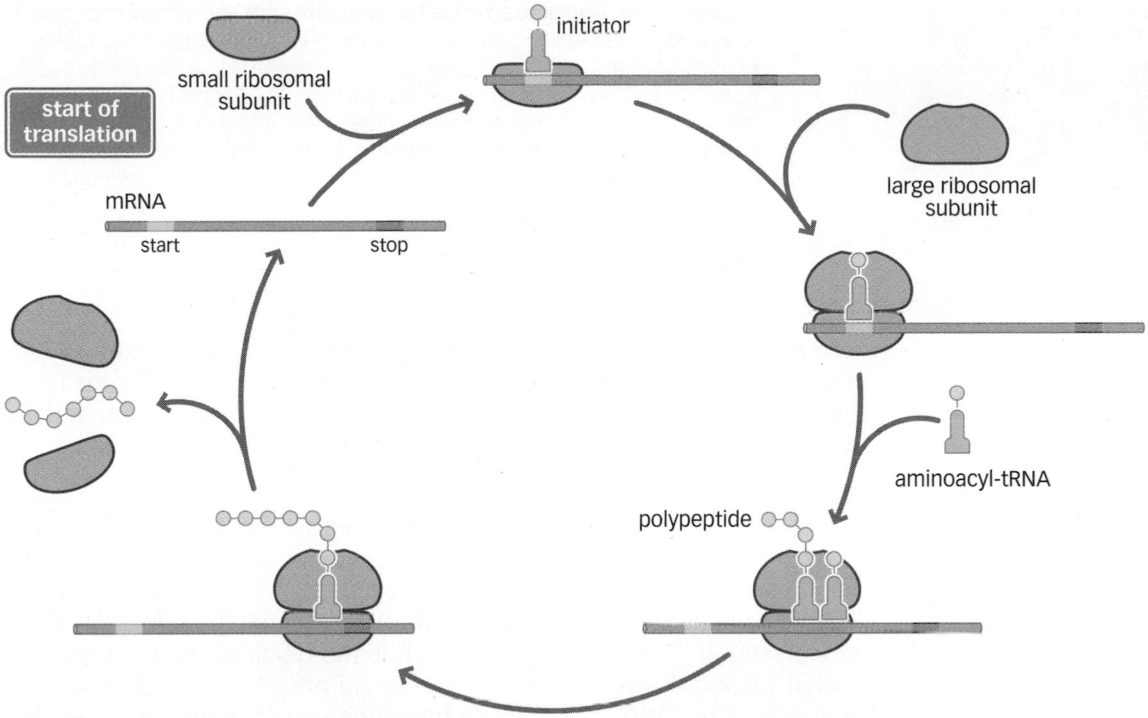

11.5

Figure 11.17 An overview of the events associated with translation in prokaryotes. The binding of the first amino acid (the initiator), which is always methionine (Met), occurs without the large subunit. In addition, notice that the entire translation complex falls apart after the translation of the protein has been completed. It is the stop codon that causes this to occur.

Translation occurs in cycles. During each cycle, the large and small ribosome subunits associate with the mRNA, translate the information contained in the mRNA base sequence, and then dissociate from it (**Figure 11.17**). As mentioned above and illustrated in **Figure 11.18**, more than one ribosome can attach to the same mRNA molecule. When this occurs, the complex is referred to as a **polyribosome** (or polysome).

Formation of Peptide Bonds in Translation

Recall that proteins are composed of sequences of amino acids linked to one another by peptide bonds. Although the ribosome catalyzes the formation of these bonds, they are actually formed on the tRNA molecule attached to the ribosome. The reaction that bonds each amino acid being added to the growing peptide chain is called the **peptidyl transferase reaction**. Keep in mind that the tRNA carrying an amino acid contains energy that can be used to create the peptide bonds in a protein being synthesized. The energy released when the amino acid is released from tRNA is available for the formation of that peptide bond.

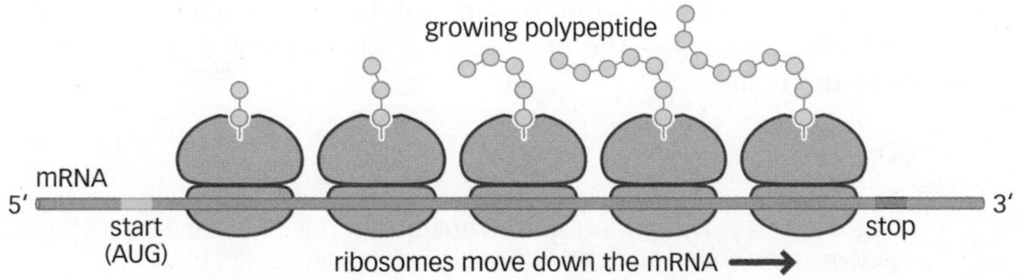

11.6

Figure 11.18 A polyribosome. Ribosomes can complex with messenger RNA as it is being transcribed. Therefore, many ribosomes moving down the same mRNA can be translating the same protein. This is important if large amounts of the specific protein are required by the organism.

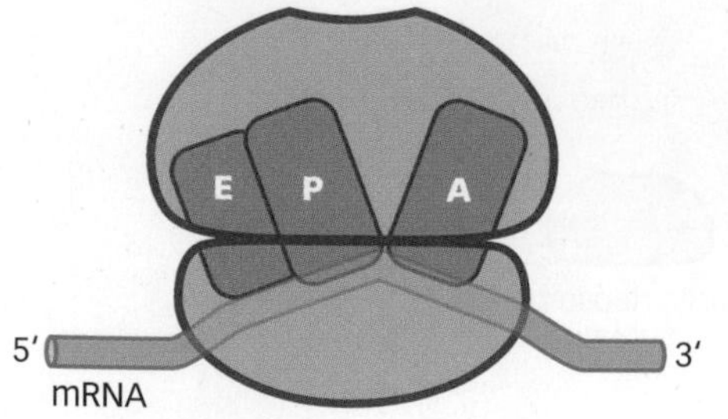

Figure 11.19 There are three binding sites on the ribosome: A, P, and E. As the ribosome moves along the message, tRNA molecules will enter at the A site, then move from there to the P and eventually the E site, where they exit from the ribosomal complex. Each site spans both the large and small subunits. This complex is dependent upon three-dimensional shape, and it is this shape that allows translation to proceed normally. We will see in later chapters that the shape of the ribosome is a primary targeting strategy for antibiotics.

Fast Fact

The shape of the ribosome is a primary targeting strategy for antibiotics.

The ribosome has three binding sites for tRNA molecules (**Figure 11.19**). The A site is for tRNA carrying a single amino acid. The **P site** is for peptidyl-tRNA, which is holding on to the growing peptide chain, and the **E site** is for tRNA that is in the process of exiting from the ribosome.

The ribosome is honeycombed with tunnels that facilitate the movement of the components involved in protein synthesis. Both the decoding center on the small subunit and the site of peptide bond formation on the large subunit are buried deep in the interior of the ribosome. The mRNA strand enters through one tunnel and exits through another, both located in the small subunit. The entry tunnel, which is the result of the three-dimensional folding, is wide enough for a single mRNA strand only, so that there can never be problems with bunching up of the mRNA molecules. Between the mRNA entry and exit tunnels is an area of the ribosome accessible to incoming tRNA molecules. This area is the location of the A and P sites. Finally, there is a separate tunnel through the large subunit that is the exit path for the newly synthesized polypeptide chain, and the width of this tunnel restricts premature folding of the polypeptide.

Just as with transcription, there are three stages of translation: initiation, elongation, and termination.

Translation Initiation

Initiation of translation requires three things: (1) recruitment of the ribosome to the mRNA to form what is referred to as the **translational apparatus**, (2) placement of a methionine tRNA and its amino acid on the P site (note that this first tRNA does not go to the A site first), (3) precise positioning of the ribosome over the mRNA start codon (**Figure 11.20**). This last step is the most crucial because it establishes the open reading frame. If this positioning step is off by even one base position, the synthesized protein will be completely inaccurate. The result of this initiation stage is the formation of an intact 70S ribosome situated at the start codon of the mRNA strand.

Translation Elongation

After initiation, three things must occur in order for the amino acids to be added to the methionine joined to the mRNA during initiation (**Figure 11.21**). First, a tRNA carrying the next amino acid coded for in the mRNA base sequence is loaded into the A site (remember that this placement requires matching at the anticodon region). Second, a peptide bond is formed between this amino acid and the amino acid in the P site. Third, each tRNA moves — the one at the A site to the P site, the one at the P site to the E site.

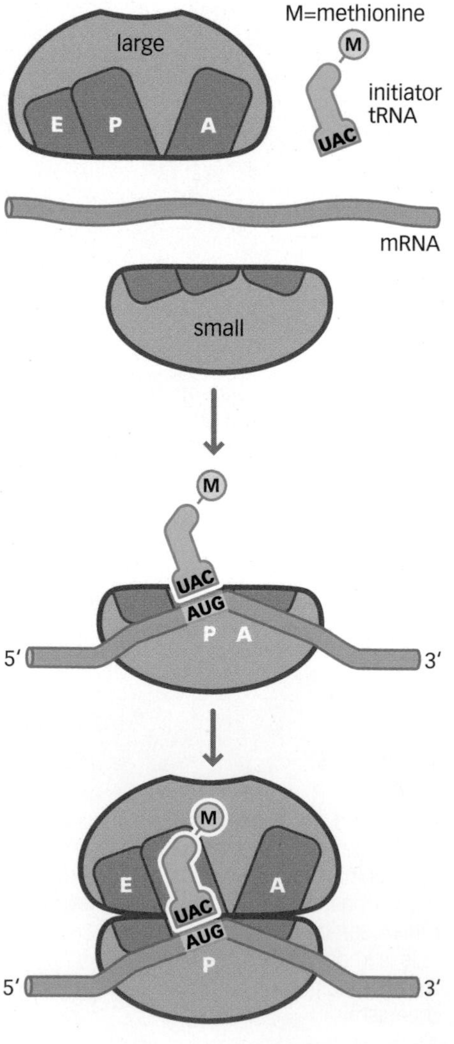

Figure 11.20 The initiation of translation. Notice that the polypeptide always begins with the tRNA for Methionine binding at the P site.

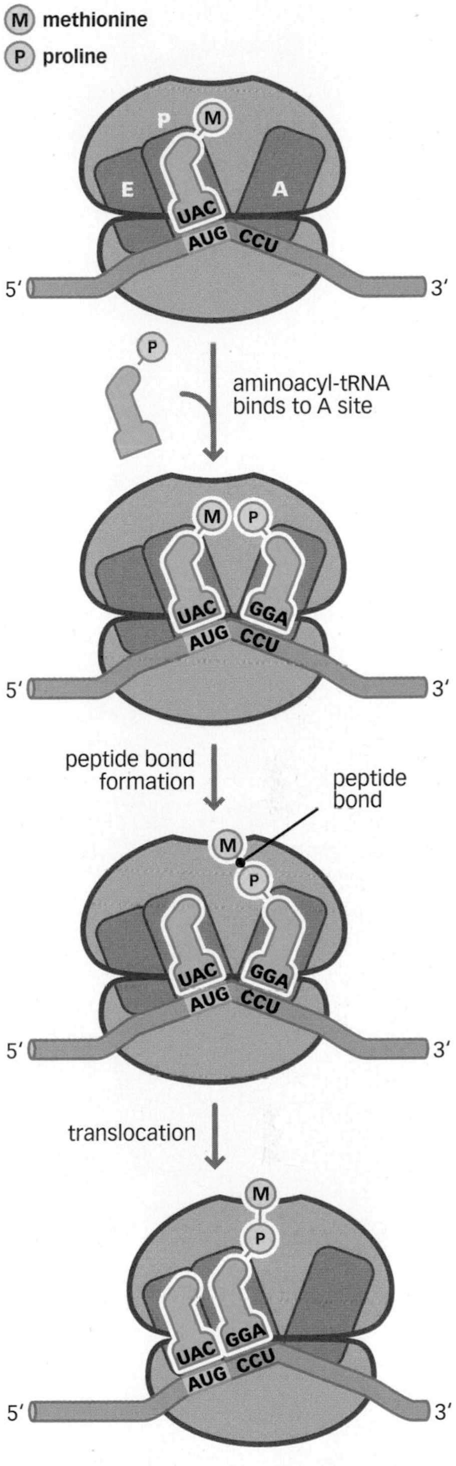

Figure 11.21 A summary of the steps involved in translation. The ribosome moves along the message, causing the tRNA molecules to move from site to site. Also take note that the peptide bond is formed between the newly arrived amino acid in the A site and the amino acid in the P site. Recall that the energy needed to form this bond is carried by the charged tRNA entering the A site.

It is interesting to note that the ribosome has several methods for preventing binding of the wrong amino acids in a sequence. The first of these proofreading methods is seen at the site where the mRNA codon and the tRNA anticodon interact. Here, additional hydrogen bonds form to help orient and hold the codon and anticodon in place. If there is disagreement in base pairing between the mRNA codon and the tRNA anticodon, these hydrogen bonds cannot form. Furthermore, proteins that regulate elongation help prevent the incorrect binding of amino acids. Last, there is the process of accommodation, which ensures that the correct tRNA is used. Accommodation has to do with the strain placed on the tRNA anticodon when it is aligning with the mRNA strand. Only the correct tRNA–amino acid complex can withstand the strain of interacting with the mRNA codon. The wrong tRNA is too strained to remain in place and so is automatically dissociated from the ribosome (**Figure 11.22**)

Translation Termination

Translation continues until a stop codon enters the A site. These codons are recognized by specialized proteins that cause the translation complex (ribosome plus newly synthesized peptide chain) to break down. At this time, the peptide chain is released from the ribosome and begins to form secondary and tertiary structure. In addition, the large and small subunits of the ribosome dissociate from the mRNA and begin to recirculate in the cytoplasm of the cell in search of a new mRNA strand.

Keep in Mind

- The genetic code is based on combinations of three letters called codons.
- Each codon codes for a specific amino acid.
- There are three stop codons and one start codon in the code.

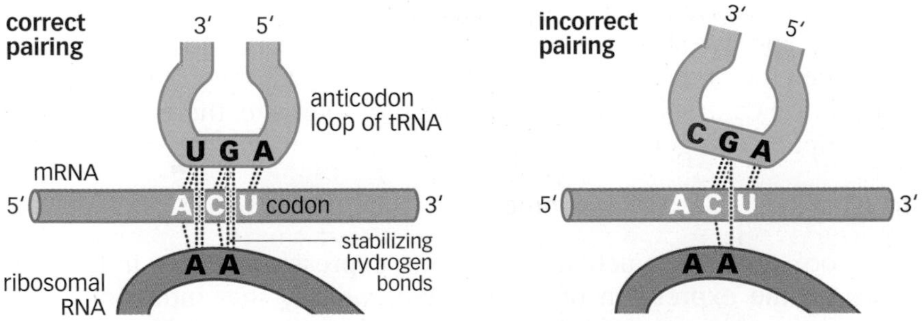

Figure 11.22 One of the mechanisms that ensures the correct pairing of the messenger RNA codon and the anticodon region of the transfer RNA. Note that the stabilizing hydrogen bonds that form between the ribosomal RNA and the anticodon region of the transfer RNA can only form if there is complementary binding between the mRNA codon and the anticodon region of the tRNA. If there is a mismatch, these bonds do not form and the strain involved causes the mismatched tRNA to break away from the ribosome.

- Gene expression is the process of making a functional product based on the genetic information contained in the DNA and consists of transcription and translation.
- Transcription proceeds through three steps: initiation, elongation, and termination.
- Translation uses messenger RNA, transfer RNA, and ribosomal RNA and occurs at the ribosome of the cell.
- The ribosome is made up of two subunits that contain RNA and protein.
- Amino acids are brought to the ribosome by transfer RNA molecules, which are specific for certain amino acids.
- While at the ribosome, peptide bonds form between adjacent amino acids.
- Translation stops when a stop codon enters the ribosome.

REGULATION OF GENE EXPRESSION

Now that we have examined DNA structure, the genetic code, transcription, and translation, it is important to put into perspective the expression of genes. This is an important mechanism used by bacteria to develop resistance to antibiotics. As we have seen, genes code for functional proteins, and the synthesis of these proteins is highly regulated and energetically expensive. Because bacteria are required to be energetically frugal, not all genes in a bacterium are expressed all the time. For example, in *Escherichia coli*, about 80% of genes are classified as **constitutive**, meaning they are always "on" (being expressed). The other 20% are genes that are expressed only when required.

Gene expression is regulated by mechanisms that either increase or decrease the rate at which a gene is expressed. There are various stages where this regulation can occur, but the most common is during transcription. There are two types of gene expression at this level, both controlled by regulatory proteins. Positive regulation (*induction*, discussed in a moment) involves protein *activators*, which increase the rate at which the DNA in a gene is transcribed to mRNA. Negative regulation (*repression*, also discussed below) uses protein *repressors* to either decrease the transcription rate for a given gene or prevent any transcription of that gene.

These activators and repressors are DNA-binding proteins that recognize two sites on the DNA double helix at or near the genes they control. The promoter site is the place where RNA polymerase binds to the DNA to begin transcribing specific genes. This site is adjacent to a region of the DNA strand referred to as the **operator site**, which is where the regulatory proteins bind. We shall see later that these sites overlap to a degree such that when the operator site is occupied by a regulatory protein, the RNA polymerase cannot bind properly to the promoter site, thereby inhibiting transcription and gene expression.

Induction

Let's look first at the activation of gene expression, which induces or turns on the expression of a gene. We can say that induction turns "on" genes that are off. The best example of positive regulation of gene expression (that is, induction) is the ***lac* operon** (**Figure 11.23**).

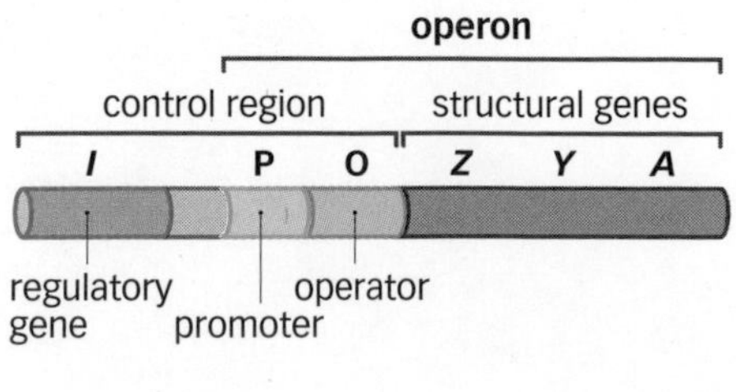

Figure 11.23 The lactose operon. The lactose operon contains the regulatory elements O and P and the structural genes *Z*, *Y*, and *A*. The regulatory gene, *I*, is located nearby. Expression of the structural genes is controlled by *I*, P, and O.

An operon is a set of genes that is regulated. The *lac* genes code for enzymes that enable bacteria to utilize the sugar lactose. It is a system that employs both repressor and inducer mechanisms.

Before we look at the details of this system, it is important to think about one of the most pressing problems faced by any biological organism: energy. Bacteria have to have ways to get it and they cannot waste it. As we mentioned above, making any protein requires energy; if that protein is not used, its synthesis could be a waste of energy. In bacteria, as in other cells, one of the best energy-yielding materials is glucose. This is because glucose can easily be converted to energy. As we saw in Chapter 3, if broken down using oxygen, one molecule of glucose can provide 38 ATPs. So glucose is preferred because it yields high levels of energy quickly and inexpensively. Therefore, the genes coding for enzymes that break down glucose are constitutive (always on), even if glucose is not present. You might think this is wasteful on the part of the bacteria, but glucose breakdown is important enough for this to be overlooked.

So what does the cell do if no glucose is present? It uses whatever it can, for example lactose. However, because lactose is used only if glucose is not present, the genes that code for enzymes that break down lactose are normally off. Therefore, the genes for lactose are said to be inducible (off but can be turned on). There are three structural genes used for the breakdown of lactose, namely *lac* Z, *lac* Y, and *lac* A. These lie adjacent to one another on the *E. coli* chromosome. The promoter for these genes is located at the 5' end of the *lac* Z gene. The mRNA from these genes is polycistronic, meaning that it contains the information for all three proteins on one message. The *lac* Z gene codes for β-galactosidase, an enzyme that cleaves lactose into galactose and glucose. The *Y* gene codes for permease, which is a membrane transport protein for lactose. Last, the *lac* A gene codes for transacetylase, which rids the cell of toxic by-products of galactose.

It is important to remember that *lac* genes are only expressed when there is no glucose present. In all other cases, they are "off" or repressed. There are two regulatory proteins involved in this repression: the *lac* repressor protein and the *lac* activator called CAP (for catabolic activator protein).

The repressor protein for the *lac* operon is encoded by the *lac I* gene and is expressed constitutively, meaning that it is always being produced. This makes sense because the cell does not want to use lactose as long as glucose is present, and having the genes to utilize lactose turned on would be a waste of energy. This repressor is therefore always being manufactured. The repressor protein and the CAP proteins have the capacity to bind to the DNA, and this binding occurs at the **operator site**. This site is a segment of the DNA adjacent to the *lac* **promoter region** (see **Figure 11.23**).

The *lac* repressor protein binds to the DNA in such a way as to prevent the transcription of these genes unless lactose is present and glucose is absent. This repressor protein actually binds to the operator region of the DNA and in doing so overlaps part of the promoter region. In this way it physically interferes with the binding of the RNA polymerase, resulting in no message being made (**Figure 11.24**).

The CAP protein activates the *lac* operon, but only if the environment does not contain glucose. In essence, the CAP protein "recruits" the RNA polymerase to come to the DNA-binding site and then interacts with it in such a way that the configuration of the polymerase changes so that it binds properly and mRNA can be made.

For the genes of the *lac* operon to be turned on, the repressor must first be inhibited. This occurs through a mechanism called allosteric control.

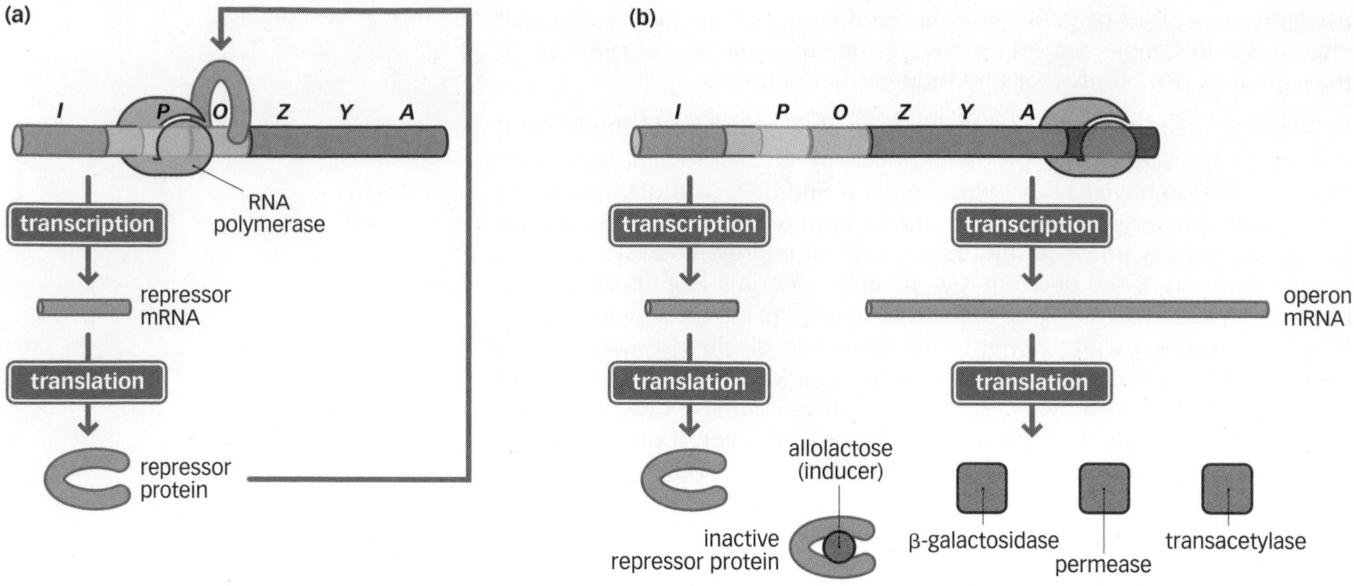

Figure 11.24 The operation of the lactose operon. Panel **a**: The lactose operon in the "off" (repressed) position. Notice that the I gene repressor protein is able to bind to the operator segment of the operon and prohibit transcription of the structural genes. Panel **b**: Induction (turning "on") of the lactose operon occurs when allolactose (a fragment of lactose) binds to the repressor protein. This changes the three-dimensional configuration of the repressor protein such that it can no longer bind to the operator gene. Therefore the RNA polymerase in combination with CAP protein can bind to the promoter region, and transcription of the structural genes can proceed.

Recall that the three-dimensional structure of proteins can be affected in a variety of ways. For example, binding of a molecule at one site can change the capacity to bind another molecule at a separate site on the same protein. This is the mechanism used for induction of the lactose operon.

Although there is continuous repression of the lactose genes when glucose is present, the expression of *lac* genes is "leaky" and a few transcripts are usually made such that there is always a low level of β-galactosidase present. This enzyme permits the entry of small amounts of lactose into the cell, which is then converted to **allolactose**. Allolactose binds to the lactose repressor protein and changes its shape such that it can no longer bind to the operator site on the DNA (see **Figure 11.24**). It is important to note that the binding site for allolactose on the repressor protein is away from the active site that binds to the DNA molecule. Consequently, when allolactose is no longer present, the repressor protein can return to its original shape and bind again to the operator region of the DNA, effectively reestablishing repression of the operon.

The CAP protein acts in a similar fashion. When glucose is present, the levels of cyclic AMP (cAMP) are low. Remember, that cAMP is a product that accumulates as available energy decreases. It serves as a signal to the cell that alternative methods for energy production are required. As glucose disappears, the level of cAMP rises and it binds to the CAP protein. This causes a change in the three-dimensional shape such that the CAP–cAMP complex now binds to the DNA. This in turn recruits the RNA polymerase to the promoter site for transcription of the *lac* operon to begin. The binding of cAMP to the CAP protein is also allosteric. The protein configuration will therefore return to normal when cAMP levels decline and the CAP will "fall off" the DNA. Without the CAP protein, RNA polymerase will not bind efficiently to the promoter region of the DNA. This CAP protein has been shown to be involved with more than 100 genes in *E. coli* and combines with a wide variety of protein partners.

Repression

In addition to the induction (turning on) of genes that we have discussed above, there are also cellular mechanisms that repress (turn off) genes. This is also a very important factor for the energy-conscious cell, because protein synthesis costs energy, and the synthesis of excess amounts of protein is wasteful. Consequently the cell can regulate the expression of genes in a "negative" fashion. This is similar to feedback inhibition and occurs post-transcriptionally. For an example of this repression we can look at the synthesis of the amino acid tryptophan. In this situation, tryptophan is made until it begins to accumulate in excessive amounts. At this point the tryptophan actually becomes a co-repressor of its own synthesis.

This occurs because the cell produces a repressor protein specific for the production of tryptophan, and like the repressor for lactose it is always produced. However, this protein is produced in a configuration that cannot bind to DNA and repress the production of messenger RNA. Excess tryptophan will bind to this repressor protein and change the shape such that the repressor will now bind to the DNA and repress the production of tryptophan (**Figure 11.25**). This is an elegant mechanism because when

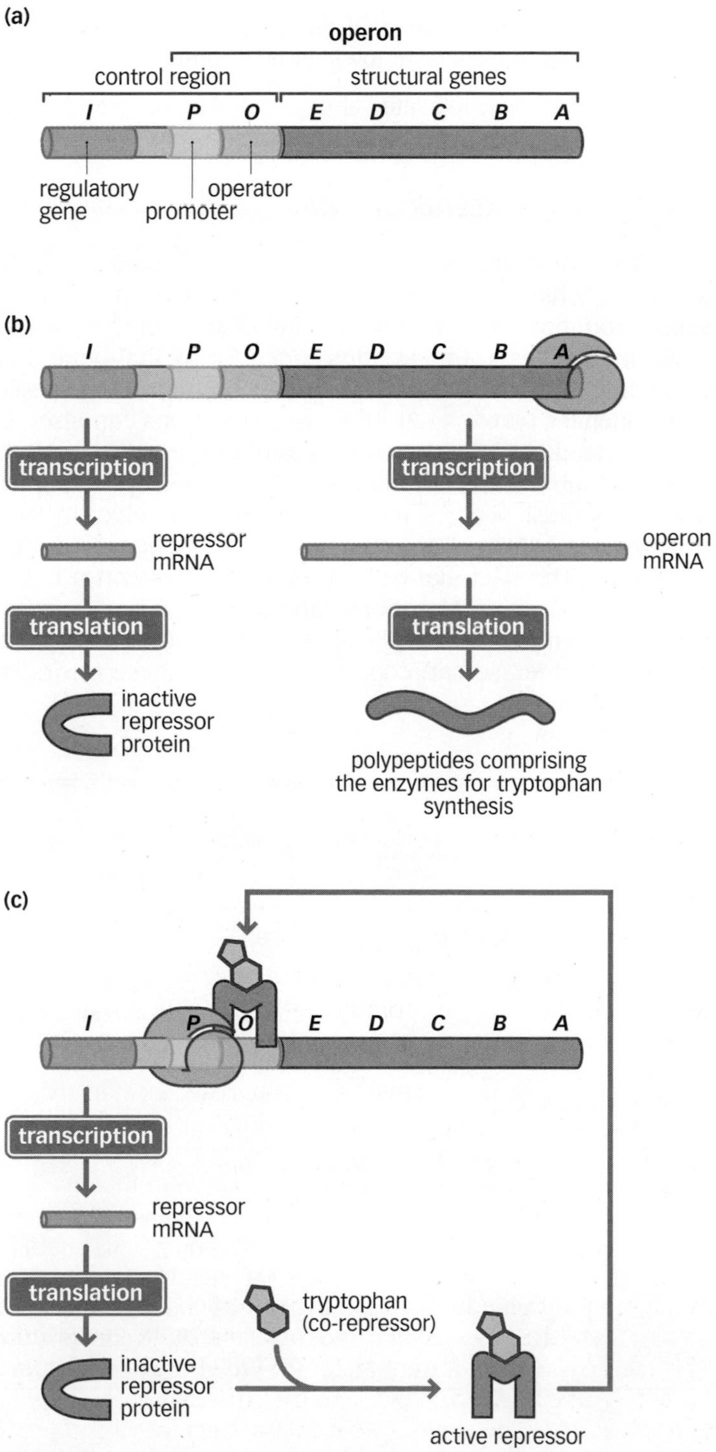

Figure 11.25 The tryptophan operon. Panel **a**: The tryptophan operon is a repressible operon (that is, it is "on" and can be turned "off"). This repression mechanism is a form of "feedback" inhibition. Panel **b**: The repressor protein for tryptophan is always produced, but it has a three-dimensional shape that does not fit onto the operator region of the DNA. The message for tryptophan is therefore always made and tryptophan continues to be produced. Panel **c**: When tryptophan is produced in excess, some of it will begin to bind to the repressor protein. This will change the shape of the repressor such that it now binds to the operator gene, thereby blocking the transcription of the structural genes for tryptophan. When the amount of tryptophan decreases, there is none available to bind to the repressor protein; it thus re-forms its original shape, which cannot bind to the operator, and tryptophan production resumes.

the amount of tryptophan decreases, there is none left to bind to the repressor. This causes the configuration of this protein to change back to its original shape such that it can no longer bind to the DNA, and tryptophan is again produced.

Keep in Mind

- The expression of a gene is carefully regulated.
- Genes can be constitutive (always "on"), inducible ("off" and can be turned "on"), or repressible ("on" but can be turned "off").
- Regulatory proteins control induction and repression through binding on the DNA at the site known as the operator site.
- An operon is a set of structural genes that share a common promoter and operator and are regulated together by a control gene.
- Gene expression is regulated at the level of mRNA production. When induced, message is made; when repressed, message is not made.

MUTATION AND REPAIR OF DNA

One of the most important topics in bacterial genetics is the process of mutation, which can cause changes in the DNA sequence. As we have seen throughout our discussions, any change in the DNA sequence will result in changes in the proteins coded for by that gene. In many cases, these changes can have adverse affects on the cell and could have lethal consequences (**Table 11.3**). However, mutations can also be responsible for increased virulence as well as antibiotic resistance. Remember also that although the level of mutations has to be kept to a minimum, some mutations must occur if the organism is to evolve. In fact, adaptation of a species, which is a cornerstone of evolution, is directly related to mutations. The bacterial cell therefore depends on a balance between mutation and repair. Mutations can occur because of errors in the accuracy and fidelity of DNA replication; although the proofreading capability of DNA polymerase can cope with most of these errors, some escape correction.

Table 11.3 Types of Mutation

Type of Mutation	Effect
Point mutation	
Single base change in DNA with no change in amino acid	No effect on the protein. Called a "silent" mutation
Change in the DNA sequence that results in a change in the amino acid sequence	Causes a change in the protein that can cause significant alteration of protein function
Change in DNA sequence that creates a premature stop codon	Produces truncated and non-functional protein
Frameshift mutation	
Deletion or insertion of one or more bases into the DNA sequence	Changes the entire sequence of codons and greatly alters the amino acid sequence. Transposition is a form of insertion that can cause a frameshift and change the genetic makeup of a bacterium

Mutations can result from chemical damage to DNA, and more importantly insertions that are generated by **transposons**. For these reasons, there are multiple overlapping safety systems that enable the bacterial cell to cope with changes in DNA. In the following discussion, we examine some of these systems.

Replication Errors

The simplest mutations are switches of one base for another. These are called **missense** or **point mutations** because they affect only one base. There are two kinds of point mutation: transitions, in which one pyrimidine is switched for the other, or one purine is switched with the other, and transversion, in which a pyrimidine is switched for a purine. More drastic mutations are those that involve insertions or deletions of bases in the DNA sequence. These may be the result of transposition of genes from one place to another on the chromosome, or they may result from erroneous recombination. When more than one base is involved, as in insertion or deletion, the mutation is referred to as a **frameshift**.

The rate of **spontaneous mutations** ranges from 1 in 10^{-6} to 1 in 10^{-11} per round of replication. However, there are certain sections of the chromosome called "hot spots" that seem to be more susceptible to mutation and therefore have a higher rate of spontaneous mutations. Unless corrected, any mismatch that occurs during replication can become a permanent change and will be propagated through subsequent generations.

It is important to note that there are also **suppressor mutations** that can reverse the primary mutation and reestablish the construction of a functional protein (a case of "two wrongs making a right"). These suppressor mutations can be *intragenic* (occurring in the same gene) or *intergenic* (occurring in separate genes).

How DNA Damage Occurs

DNA can be damaged by hydrolysis, by deamination (the loss of an amino group, $-NH_2$), and by chemicals called **mutagens**. For example, water can cause the deamination of cytosine, which causes it to become uracil. Although uracil is not normally found in DNA, it can bind to adenine and cause a mutation that is then carried from generation to generation.

DNA can also be damaged by alkylation, oxidation, and radiation. In alkylation, a G-C base pair can be changed to an A-T pair after replication. An example of damage by oxidation is the oxidation of guanine to a derivative that can bind to adenine as well as to cytosine. After replication, the G-C pair is replaced by A-T.

Gamma radiation and ionizing radiation cause double-strand breaks in the DNA, which are very difficult to repair. In most cases, these types of radiation are lethal to bacteria. Ultraviolet radiation can also cause DNA damage through the formation of thymine dimers, which are two thymine bases on the same strand of DNA bound to each other. When these bases link together on the same strand, they cause DNA polymerase to stop at that site, and replication stops.

In addition, mutations can be caused by compounds that can slip between two bases on the DNA strand and cause errors in replication. Examples of these compounds are acridine and ethidium, which can cause shifts in the reading frame and negatively affect transcription and translation. There are also *base analogs* that are so similar to the DNA bases that they are mistakenly placed in the growing DNA strand during replication. **Figure 11.26** provides two good examples of this.

Figure 11.26 Analogs of nucleotide bases can become erroneously inserted into growing DNA strands. Notice the structural similarity between the chemical 2-aminopurine and the base adenine (panel **a**), as well as that between 5-bromouracil and thymine (panel **b**).

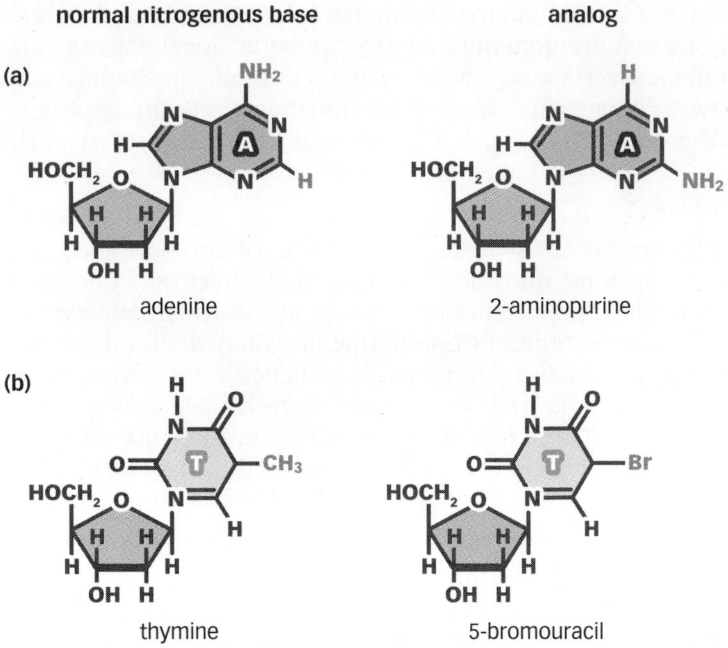

Repair of DNA Damage

The most frequently used method for repairing damaged bases in DNA is the removal and repair of altered bases. There are two principal mechanisms. In the first, base excision, a damaged base is removed (in other words, *excised*) from a DNA double helix. After the base is removed, DNA polymerase fills in the gap, and DNA ligase repairs the break in the strand.

The enzymes involved in base excision have different specificities that recognize different types of damage. These enzymes diffuse along the DNA double helix, each scanning for a specific type of damage. X-ray crystallographic studies have shown that when a damaged base is detected by an enzyme, the base is flipped away from the helix and into a "pocket" on the enzyme molecule. What is amazing about all this is that this flipping out of the base does not cause major distortion of the DNA helix, reaffirming the flexibility of DNA.

The second repair mechanism is called **nucleotide excision** (**Figure 11.27**). Unlike the enzymes involved in base excision, those involved in nucleotide excision do not recognize any particular type of damage. Instead, the repair enzymes look for distortions in the double helix. These distortions trigger a chain of events leading to the removal of short single-strand sections of DNA surrounding the distortion. As with base excision, nucleotide excision is followed by the actions of DNA polymerase and DNA ligase to fill in the gap.

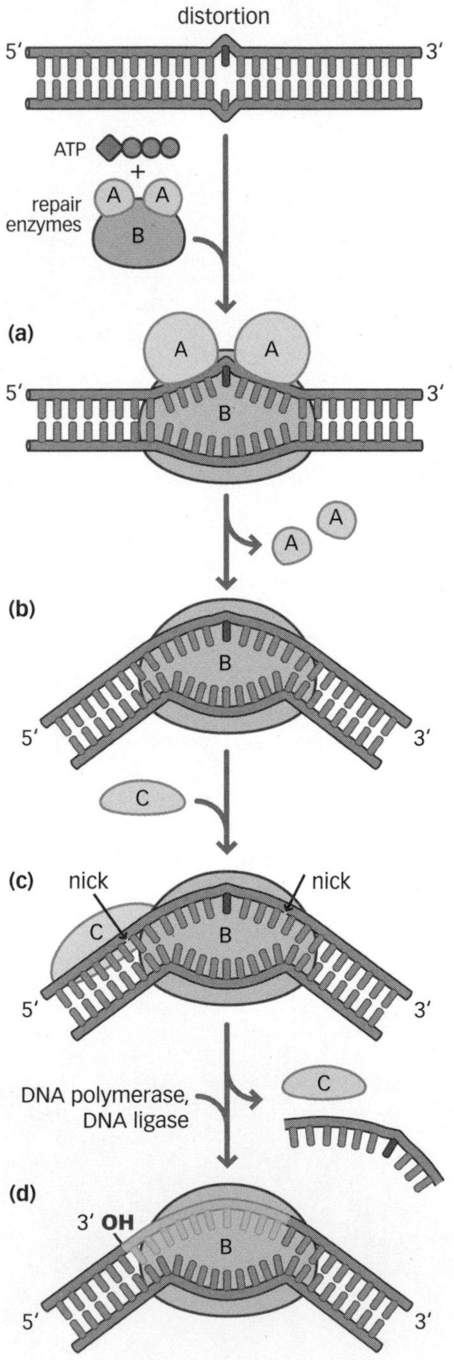

Figure 11.27 Nucleotide Excision Repair. Panel **a**: This repair process uses proteins (A) to "scan" DNA looking for chemical distortions (remember the stereochemistry of the DNA molecule). Panel **b**: An enzymatic protein (B) "melts open" the area around the distortion. This is followed by "nicking" of the strands on each side of the distortion by protein (C) (panel **c**). Lastly, DNA polymerase fills the gap and DNA ligase reseals the area (panel **d**).

Another DNA repair mechanism is called **photoreactivation**. This mechanism unlinks the damaging thymine dimers formed when DNA is exposed to ultraviolet radiation. Photoreactivation is accomplished by an enzyme called photolyase. In the dark, this enzyme binds to the dimer. When the DNA is then exposed to light, the photolyase becomes activated and breaks the thymine–thymine bond (**Figure 11.28**).

Keep in Mind

- Mutations have an important role in the infection process because pathogens can become resistant to antibiotics through mutation.
- Bacteria depend on a balance between mutation and repair.
- Mutations can result from transposition of genes in the chromosome, point mutations, or frameshift mutations.
- Suppressor mutations can reverse the primary mutation.
- DNA can undergo spontaneous damage from hydrolysis or deamination and can also be damaged by chemicals called mutagens.
- Repair of damaged DNA can be accomplished by excision repair or nucleotide excision systems in the cell.

TRANSFER OF GENETIC INFORMATION

In the last part of this chapter, it is very important that we examine how bacteria can transfer genetic information. This discussion is based on the fact that bacteria can "shuffle" their genes. We refer to this as genetic **recombination**. In previous chapters, as well as in those that follow, we see that there are many factors that can influence the virulence and pathogenicity of bacteria. Recall from Chapter 9 that the presence of fimbriae or a capsule can make organisms more virulent, and the capacity to form endospores can protect organisms. In addition, many disease symptoms result from the bacterial production of toxins. Each of these traits is genetically controlled, and the transfer of these genetic determinants can make harmless bacteria dangerous.

There are four ways in which genetic recombination can occur. Three of the four mechanisms — *transformation*, *transduction*, and *conjugation* — are transfers of genetic material from one bacterium to another. The fourth, *transposition*, is a way in which the genes in the DNA of a bacterium are "shuffled" into a different order (in the way that shuffling a deck of cards rearranges their order).

Figure 11.28 Photoreactivation. This is a mechanism that can repair thymine dimers. There are two reactions that take place in this repair mechanism, the action of the enzyme DNA photolyase, which occurs in the dark, and the second reaction, which requires visible light.

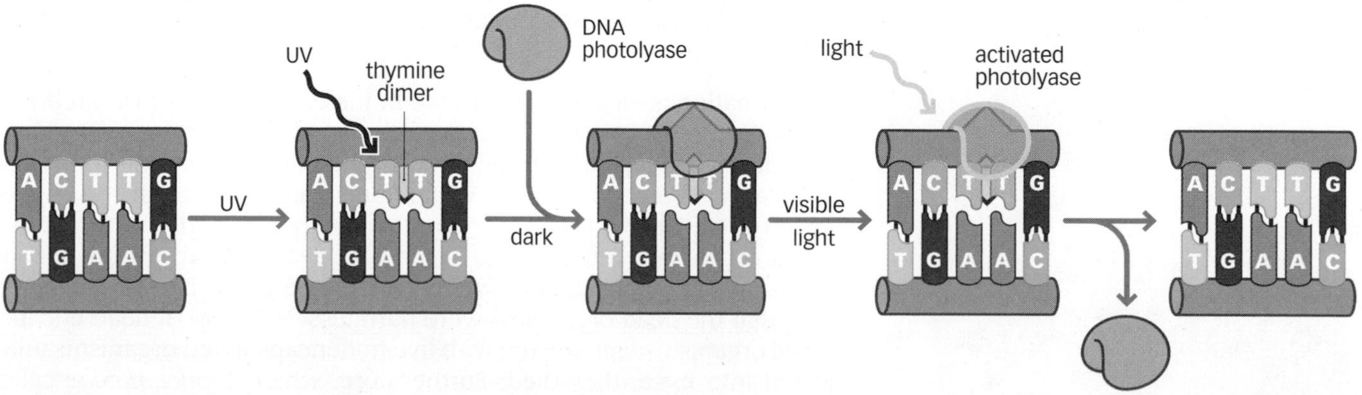

Transposition

In the recombination process called **transposition**, genetic elements called **transposons** move from one place on a bacterial chromosome to another. (An analogy would be taking a card from one place in the deck and putting it back in another place in the deck.) This rearrangement is accomplished in a random fashion, so that the result can be either detrimental to the bacterium or beneficial. If changes to DNA result in negative consequences for the cell, the end result is usually the death of that cell. Therefore negative consequences are rarely, if ever, seen. In contrast, any genetic change that enhances a bacterium's ability to survive and grow is selected for and maintained throughout subsequent generations.

Transpositions are the most common cause of new "mutations" in many organisms. They may be important causes of mutations that lead to genetic diseases in humans. In bacteria, transposons carry genes encoding proteins that promote resistance to one or more antibiotics. Even viruses seem to use methods similar to transposition to integrate into host chromosomes (discussed in Chapter 12). Transposons move in and out of DNA strands by using cleavage and rejoining mechanisms akin to cutting and pasting. These mechanisms involve excision of the transposon from its initial location in the DNA followed by integration into a new DNA site.

The rest of our gene-transfer discussion involves the transfer of genetic information from one cell to another. There are three ways in which this transfer can happen: *transformation*, *transduction*, and *conjugation*.

Transformation

The major point to remember about **transformation** is that it involves naked DNA, which can be released from cells that have died. This DNA is taken up by a bacterial cell and recombines with that cell's DNA. Therefore traits that are encoded on these DNA pieces can become part of the recipient cell's genetic repertoire.

Fast Fact

Many biotechnology companies routinely use transformation to genetically engineer bacteria.

For this to happen, the recipient cell must be "competent," which means that the wall of this cell can accommodate the uptake of large molecules such as pieces of DNA. Some bacteria are naturally competent, whereas others can become competent after chemical treatment.

The mechanism that conveys competence to a bacterial cell is not well understood but may involve Ca^{2+} ions, which seem to shield the electrical charge on the naked DNA pieces, allowing them to move through the membrane.

Transformation is an inefficient process because only a small percentage of competent cells take up naked DNA from their surroundings, and those that do, take up only a small percentage of the total amount available to the cell.

Transformation is elegantly illustrated in the experiments of Dr. Fredrick Griffith (**Figure 11.29**). He used two strains of *Streptococcus pneumoniae*, one that was encapsulated and caused pneumonia, and one that was nonencapsulated and harmless. When nonencapsulated organisms were injected into mice, no disease occurred. In contrast, injection with encapsulated organisms caused pneumonia and death. When encapsulated organisms that had been killed were injected there was no disease, indicating that the dead organisms were harmless. However, if dead encapsulated organisms were mixed with live nonencapsulated organisms and injected into mice, they died. Furthermore, when *S. pneumoniae* cells

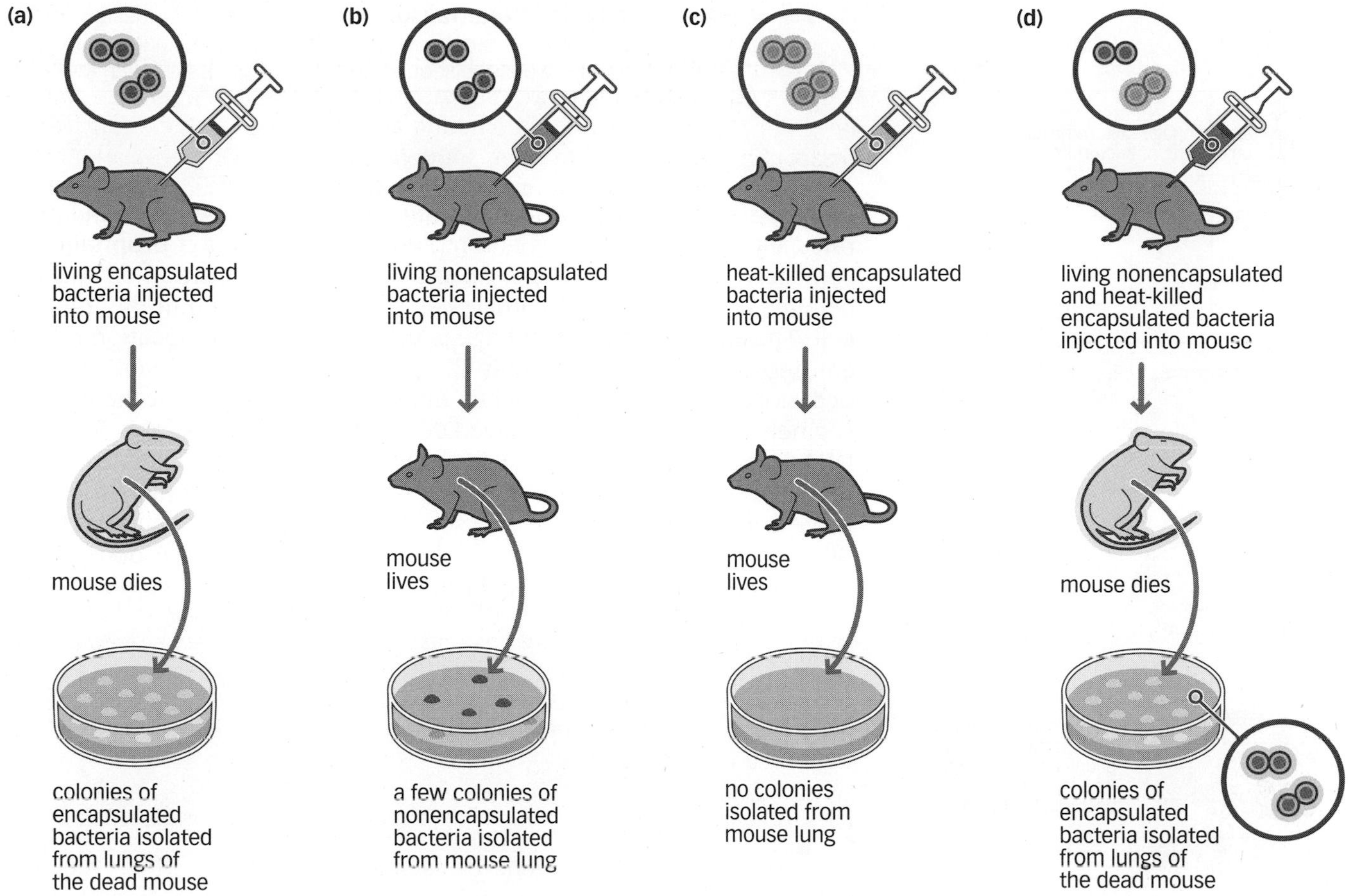

Figure 11.29 An illustration of the Griffith experiment that demonstrated transformation.

were isolated from these dead mice, they were encapsulated. Clearly, something from the dead organisms (naked DNA) had been taken up by the live organisms. Keep in mind that this observation was made decades before the structure and function of DNA were completely elaborated.

Transduction

Transduction is a common event in both Gram-positive and Gram-negative organisms, and involves the transfer of genetic information using a bacterial virus (known as a bacteriophage). Transduction permits the movement of genetic information from one infected host to another followed by recombination of the newly arrived DNA into the host chromosome. This integration into the new host DNA actually uses the same "cut and paste" mechanisms as transposition.

Transduction can be random, in which case it is referred to as **generalized transduction**, or specific, in which case it is referred to as **specialized transduction**. In generalized transduction (**Figure 11.30**), after a phage infects a bacterium, a phage-encoded enzyme called DNase is activated. This enzyme cleaves DNA contained in the phage into the proper size so that it can be enclosed in new viral particles as the infection proceeds. The DNase also chops the bacterium's chromosome into small pieces, and some of these pieces also get enclosed in viral particles. Although these latter viral particles contain bacterial DNA instead of viral DNA, they are still capable of "infecting" other bacterial cells. When the infected cell lyses (ruptures) and releases its viral particles, some contain viral DNA and others contain bacterial DNA. When one of the virus particles containing bacterial DNA infects a new host cell, the bacterial DNA it carries recombines with the newly infected host cell DNA in the recipient's chromosome.

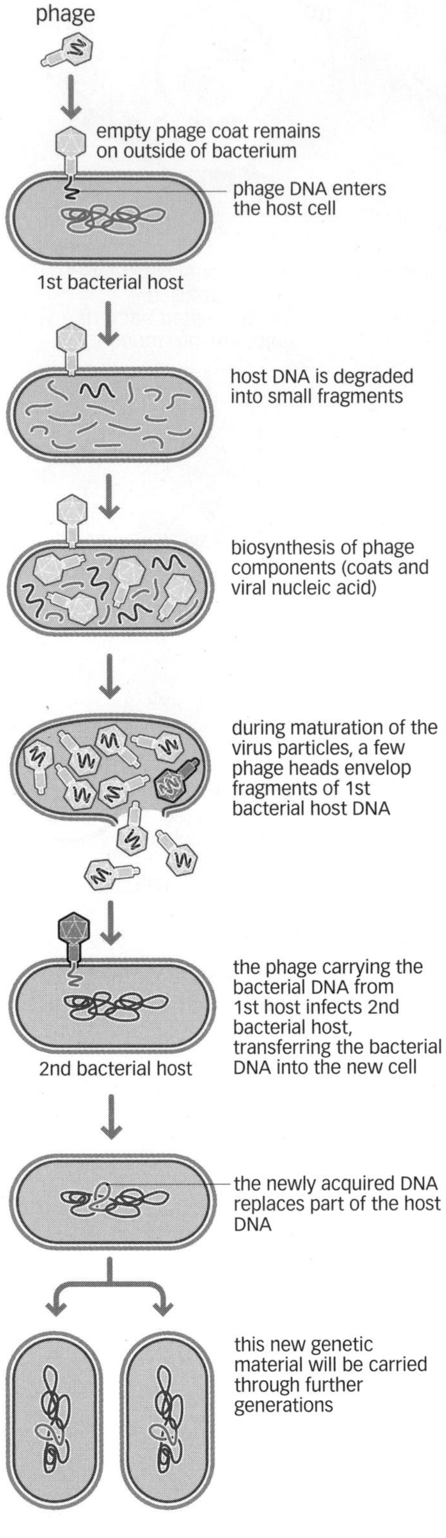

Figure 11.30 Generalized transduction. Follow the color designations of the DNA as the process proceeds. The outcome is recombination ("gene shuffling") between the sequences of DNA from host 1 with the chromosome of host 2.

In specialized transduction, the phage DNA initially becomes incorporated into an infected cell's chromosome and is called a **prophage**. At some later time, the prophage can excise itself from the chromosome and become enclosed in viral particles as they are produced in the infected cell. The excision of the prophage is usually in a precise location on the chromosome and involves only the prophage DNA. However, excision occasionally happens in a slightly different position, so that the excised segment contains both prophage DNA and bacterial DNA. This bacterial DNA is incorporated into viral particles, just as in generalized transduction. The infected bacterium then lyses and releases the viral particles — some containing viral DNA, and some containing donor bacterial DNA. When one of the latter infects a recipient bacterium, the donor DNA recombines with the recipient chromosome.

Fast Fact

The genes that code for production of the diphtheria toxin are transferred by transduction.

Conjugation

Our final mechanism for gene transfer is **conjugation**, which requires direct contact between donor and recipient cells. This important mechanism occurs in both Gram-positive and Gram-negative bacteria. In the former, direct contact involves the cell walls of two bacteria sticking together. In the latter, the process is more elaborate, so that is where we focus our discussion.

The major element involved in any type of conjugation is the plasmid. Recall that a plasmid is a genetic element that is separate from the bacterial chromosome and can replicate independently.

In Gram-negative bacteria, conjugation is facilitated through the formation of a bacterial "sex" pilus on the donor cell (**Figure 11.31**). This structure is a conduit through which the plasmid DNA travels from donor to recipient. Although the pilus is at times relatively long, as shown in **Figure 11.31**, it

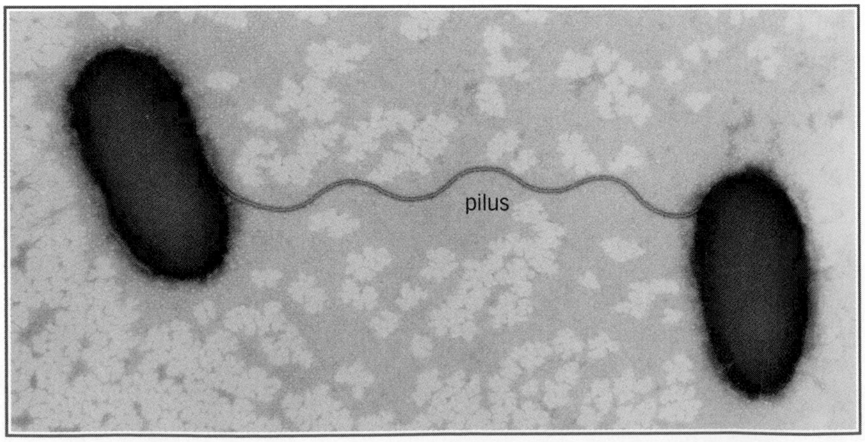

Figure 11.31 Bacterial pilus during conjugation. Although the connection shown here is relatively long, it is thought that in addition to permitting the passage of DNA between cells, the pilus may also act as a "grappling hook," pulling the recipient cell closer to the donor cell.

does not remain that way during conjugation. It is thought that although the pilus acts as a bridge, it also functions like a grappling hook that pulls donor and recipient close together.

Let's examine the sequence of events associated with conjugation in *Escherichia coli*. There are four steps in the process, as described in **Figure 11.32**.

- The sex pilus of the donor cell recognizes specific receptor sites on the cell wall of the recipient.
- An enzyme in the donor cell causes the plasmid DNA double helix to unwind, beginning at a specific site called the origin of transfer.
- One of the two single strands of plasmid DNA stays in the donor cell and the other moves from donor to recipient across the sex pilus.
- The single strand of plasmid DNA remaining in the donor and the single strand now in the recipient are both replicated. Because of the complementary nature of DNA, after this replication is complete, the donor and the recipient contain identical plasmids.

At the completion of this transfer, the recipient cell becomes a donor and can conjugate with another recipient cell. Some of the traits that can be transferred on plasmids are listed in **Table 11.4**.

It should be noted that cells can also spontaneously lose their plasmid, and when this happens the cell is referred to as "cured."

Conjugation can have several outcomes for the recipient cell. The plasmid produced after a transferred single strand is replicated in the recipient can remain in the cytoplasm as a plasmid, capable of being transferred to another recipient.

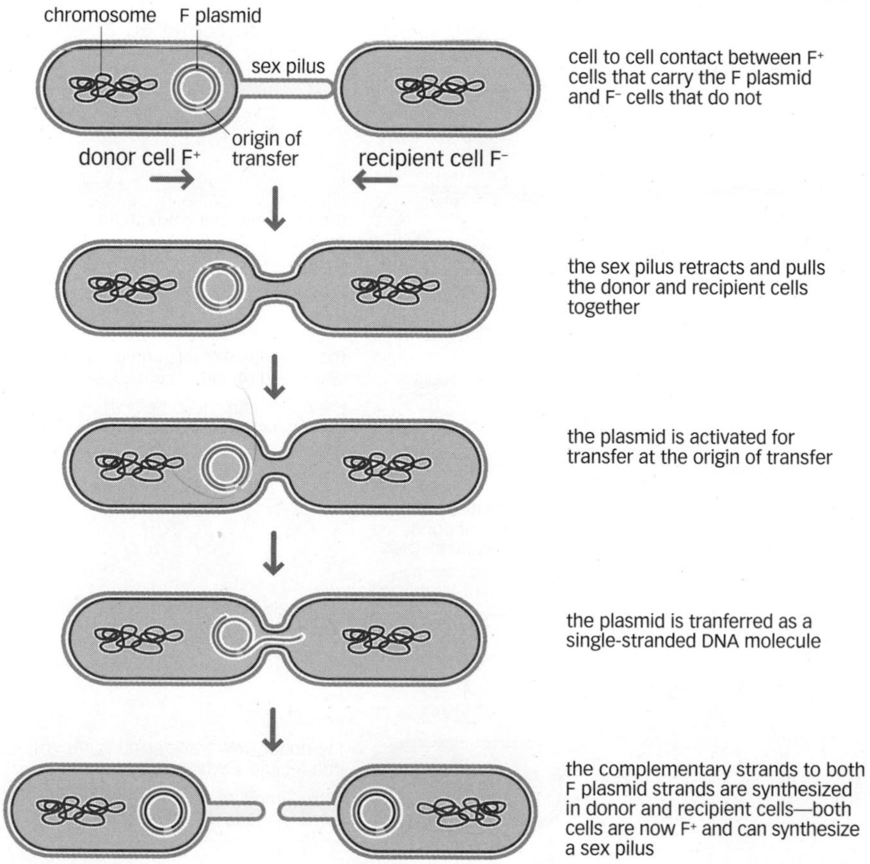

11.7

Figure 11.32 The steps involved in conjugation. Remember that only one strand of the plasmid is passed from the F^+ donor cell to the F^- recipient cell. DNA replication results in complete plasmids in both donor and recipient cells, which causes the F^- cell to become F^+.

Table 11.4 Traits that are Coded for by Plasmids

Trait	Examples of Organisms with the Trait
Antibiotic resistance	*E. coli*, *Salmonella* species, *Neisseria* species, *Staphylococcus* species, *Shigella* species and many more
Synthesis of pilus	*E. coli*, *Pseudomonas* species
Utilization of unusual nutrients	Dissimilation plasmids found in many species such as *Pseudomonas*
Increased virulence	*Yersinia enterocolitica*
Toxin production	*Corynebacterium diphtheriae* and others
Antibiotic synthesis	*Streptomyces* species

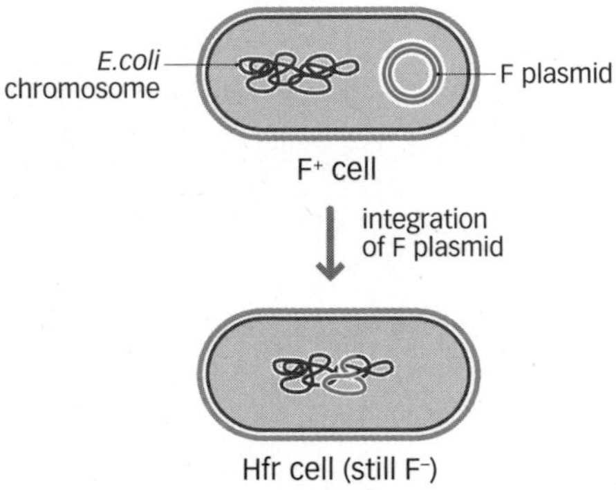

Figure 11.33 Formation of an Hfr recipient cell by conjugation. Conjugation can result in the formation of an Hfr recipient cell if the plasmid DNA that is received from the F⁺ donor cell becomes incorporated into the chromosome of the F⁻ cell. Note that in this case the recipient cell remains F⁻.

Alternatively, the replicated plasmid in the recipient can become incorporated into the cell's chromosome (**Figure 11.33**). This occurs because of the presence of *recombination sites* on the chromosome, which are similar to the insertion sites seen in transposition (discussed above). As a result of the incorporated plasmid, the recipient chromosome becomes larger than it was before conjugation. A bacterial cell that harbors a plasmid integrated into its chromosome is referred to as an **Hfr cell** (the letters stand for **high frequency of recombination**). The way in which genetic material is transferred from this type of cell is illustrated in **Figure 11.34**. The first difference is that most recipient cells that receive DNA transfer from Hfr cells cannot then become donor cells and therefore cannot pass the information they receive to other cells. A second difference is that the DNA from an Hfr cell integrates into the recipient cell's chromosome not by adding to the chromosome but rather by *replacing* the existing DNA segment of the recipient, which is then enzymatically destroyed. In other words, some genes in the recipient chromosome are replaced by genes from the donor.

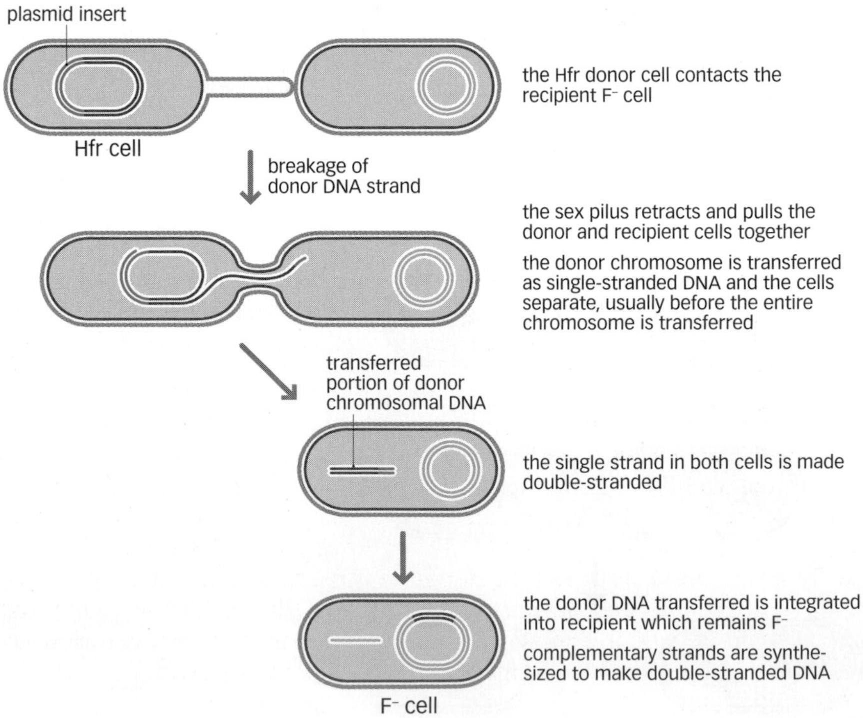

Figure 11.34 Conjugation between an Hfr cell and a F⁻ recipient cell results in the complete replacement of recipient segments of DNA with donor DNA. Although the replaced segments are part of the host chromosome, they are degraded. Interestingly, the recipient cell remains F⁻.

Table 11.5 summarizes the types of horizontal genetic transfer we have discussed.

Transfer Mechanism	Effect
Transformation	Transfer of naked DNA. First demonstrated by experiments with *S. pneumoniae*. Requires the recipient cell to be "competent." Occurrence is very rare. Traits that are transferred depend on the uptake of the DNA by the recipient cell
Transduction	Transfer of genetic material by a bacterial virus. Two types: specialized and generalized. In specialized, only genes that are near the integrated prophage are transferred. In generalized, the host chromosome is fragmented and random genes can be incorporated into new phage particles, resulting in random transfer of genes
Conjugation	Transfer of genetic material contained in plasmid. When recipient genetic material remains in plasmid form, recipient can become donor. When recipient material is integrated into host chromosome, cell becomes Hfr. Recipient conjugating with Hfr cannot become a donor
F^+	Cell can transfer plasmid DNA to a recipient F^- cell
Hfr	In an Hfr cell, the host chromosome has a prophage integrated into it. When this cell is involved in conjugation it can transfer host cell DNA sequences that are adjacent to the integrated prophage

Table 11.5 Horizontal Genetic Transfer Mechanisms

Keep in Mind

- Genetic recombination occurs in bacteria through transposition, transformation, transduction, or conjugation.
- Transposition is a specific form of recombination in which genetic elements called transposons move from one place in the chromosome to another.
- Transformation involves the uptake of naked DNA from one cell to another.
- Transduction is caused by a virus transferring pieces of DNA from one cell to another.
- Conjugation occurs when DNA is moved from a donor cell (designated F^+) to a recipient cell (designated F^-).
- Each of the transfer mechanisms causes genetic recombination in the recipient cell and thus can be important in making a pathogen more dangerous.

GENETICS AND PATHOGENICITY

Although the discussions in this chapter are complex, they are important for understanding disease processes because some of the mutations that occur during DNA replication and the various processes of gene expression can lead to increased virulence and antibiotic resistance in bacteria.

Perhaps the most obvious genetic connection to virulence and pathogenesis is the transfer of genetic material. Through recombination, organisms that may be harmless to humans can change characteristics and become clinically dangerous. For example, the genes for toxins, which in many cases are the main causes of disease progression, are coded for by genes found on plasmids. As we have seen, this information is easily transported from one bacterial cell to another through conjugation. Furthermore, certain plasmids known as **dissimilation plasmids** can contain genetic information that allows organisms to become more resistant to disinfectants and more able to adapt to otherwise destructive environments.

SUMMARY

- DNA is the informational molecule of the cell.
- DNA is a double-stranded molecule made up of nucleotides (which consist of a phosphate, a deoxyribose sugar and one of the four bases adenine, thymine, guanine, or cytosine).
- Nucleotides are bound together through complementary base pairing.
- RNA is a single-stranded molecule and contains uracil instead of thymine. It can be found in the form of messenger RNA, transfer RNA, or ribosomal RNA.
- DNA is faithfully replicated by the enzyme DNA polymerase.
- DNA polymerase has a proofreading capability that prevents errors in replication.
- Replication occurs at the replication fork (a separation of the DNA strands) and is continuous on one strand and discontinuous (made in pieces) on the other.
- DNA is transcribed into RNA by the enzyme RNA polymerase.
- The genetic code is based on codons, which are combinations of three nucleotides.
- Gene expression is the process of making a functional product by transcription and translation; it is highly regulated.
- Mutations are changes in the DNA and are important in infectious disease because they can lead to antibiotic resistance.
- Genetic recombination can occur in bacteria through transposition, transformation, transduction, or conjugation.

In this complex chapter, we have looked at four major topics — DNA replication, gene expression, mutations, and the transfer of genetic material. We have spent so much time on these topics because pathogenic bacteria use these methods to infect their hosts successfully. One of the most obvious examples is the development of antibiotic resistance. Mutations can bring about this resistance, and the genes for antibiotic resistance are located on these easily transferable genetic elements. We shall see in later chapters that antibiotic resistance may be the most important problem facing health care today. As we move through subsequent topics, keep in mind the things we have discussed in this chapter and how they relate to the issues that follow.

SELF EVALUATION AND CHAPTER CONFIDENCE

Multiple Choice

Answers are given in the back of the book and help can be found in the student resources at:
www.garlandscience.com/micro

1. A nucleotide is composed of
   - **A.** A phosphate and a nucleotide base
   - **B.** A nucleotide base
   - **C.** A sugar and a nucleotide
   - **D.** A phosphate and a sugar
   - **E.** A phosphate, a sugar, and a nucleotide base

2. In complementary base pairing
   - **A.** A always binds to C
   - **B.** T always binds to G
   - **C.** A always binds to G
   - **D.** T always binds to C
   - **E.** A always binds to T

3. The purines are
   - **A.** Adenine and thymine
   - **B.** Guanine and cytosine
   - **C.** Adenine and cytosine
   - **D.** Adenine and guanine
   - **E.** None of the above pairs is correct

4. Supercoiled DNA is "relaxed" by
   - **A.** DNA polymerase
   - **B.** DNA ligase
   - **C.** RNA polymerase
   - **D.** Topoisomerase
   - **E.** Ligase

5. Transcription is the process in which DNA is copied into all of the following except
   - **A.** Messenger RNA
   - **B.** Ribosomal RNA
   - **C.** A new DNA molecule
   - **D.** Transfer RNA

6. In RNA, uracil replaces
   - **A.** Adenine
   - **B.** Guanine
   - **C.** Thymine
   - **D.** Cytosine
   - **E.** Cyclic AMP

7. The genetic code uses codons, which are combinations of
   - **A.** Two bases
   - **B.** Three bases
   - **C.** Four bases
   - **D.** Five bases
   - **E.** None of the above

8. The genetic code is referred to as degenerate because
   - **A.** The codons are unstable
   - **B.** It is prone to errors
   - **C.** Codons code for only one amino acid
   - **D.** More than one amino acid is coded for by the same codon
   - **E.** Codons can vary in the number of bases they contain

9. The genetic code includes
   - **A.** Several start codons
   - **B.** Several start codons and one stop codon
   - **C.** Several start codons and several stop codons
   - **D.** One start codon and several stop codons
   - **E.** None of the above

10. A gene is best defined as
    - **A.** A segment of DNA
    - **B.** Three nucleotides that code for an amino acid
    - **C.** A transcribed unit of DNA
    - **D.** A sequence of nucleotides in RNA that codes for a functional product
    - **E.** A sequence of nucleotides in DNA that codes for a functional product

11. Which of the following statements is false?
    - **A.** The lagging strand of DNA is started by an RNA primer
    - **B.** DNA polymerase joins nucleotides in one direction around the bacterial chromosome
    - **C.** DNA replication proceeds in one direction around the bacterial chromosome
    - **D.** Multiple replication forks are possible on a bacterial chromosome
    - **E.** The leading strand of DNA is made continuously

12. The base that is added to the growing DNA strand is determined by
    - **A.** The base itself
    - **B.** The primer portion of the primer template junction
    - **C.** The template portion of the primer template junction
    - **D.** The DNA polymerase
    - **E.** The RNA polymerase

13. The "processivity" of DNA polymerase has to do with
    - **A.** Reading the genetic code
    - **B.** The fidelity of RNA transcription
    - **C.** The speed of DNA polymerase
    - **D.** The speed of translation
    - **E.** The speed of transcription

14. "Proofreading"
    - A. Permits fewer mistakes in the DNA
    - B. Permits fewer mistakes in the RNA
    - C. Is done by DNA polymerase
    - D. Is done by RNA polymerase
    - E. All of the above

15. The replication fork is the place where
    - A. RNA polymerase attaches to the DNA
    - B. DNA replication is interrupted
    - C. DNA replication is ongoing
    - D. DNA replication ends
    - E. Ribosomes attach to the DNA

16. The leading strand of DNA
    - A. Is made first
    - B. Is made last
    - C. Is made in pieces
    - D. Is made continuously
    - E. Is made in the 5'→3' direction

17. Okazaki fragments are found on
    - A. The leading strand
    - B. The lagging strand
    - C. In some places on both DNA strands
    - D. Only where the DNA has been damaged
    - E. Where transcription ends

18. Okazaki fragments are joined together by the enzyme
    - A. DNA polymerase
    - B. Helicase
    - C. Topoisomerase
    - D. RNA polymerase
    - E. Ligase

19. Which of the following is NOT a product of transcription?
    - A. Messenger RNA
    - B. A new strand of DNA
    - C. Ribosomal RNA
    - D. Transfer RNA
    - E. None of the above

20. The steps involved in transcription are
    - A. Initiation
    - B. Elongation
    - C. Termination
    - D. A and C only
    - E. All of the above

21. Transcription results in the formation of
    - A. Messenger RNA
    - B. Ribosomal RNA
    - C. Transfer RNA
    - D. All of the above
    - E. Only A and B

22. Specific transfer RNA molecules
    - A. Carry new bases for DNA replication
    - B. Carry any amino acids
    - C. Are involved with transfer of plasmids
    - D. Carry only one specific amino acid
    - E. Carry only two amino acids at a time

23. The ribosome is the site where
    - A. Transcription occurs
    - B. Replication occurs
    - C. Translation occurs
    - D. All of the above
    - E. None of the above

24. Ribosomes are composed of
    - A. DNA and protein
    - B. DNA, RNA, and protein
    - C. Ribosomal RNA and protein
    - D. Ribosomal RNA, transfer RNA, and protein
    - E. Ribosomal RNA, messenger RNA, and protein

25. The ribosome
    - A. Has one site used for translation
    - B. Has two sites for translation
    - C. Has three sites for translation
    - D. Has four sites for translation
    - E. Is not involved in translation

Codon on mRNA and corresponding amino acid			
UUA	Leucine	UAA	Nonsense
GCA	Alanine	AAU	Asparagine
AAG	Lysine	UGC	Cysteine
GUU	Valine	UCG, UCU	Serine

26. (Use the table above.) If the sequence of amino acids coded for by a strand of DNA is serine-alanine-lysine-leucine, what is the order of bases in the sense strand of DNA?
    - A. 3'-TCTCGTTTGTTA-5'
    - B. 3'-UGUGCAAAGUUA-5'
    - C. 3'-AGACGTTTCAAT-5'
    - D. 5'-TGTGCTTTCTTA-3'
    - E. 5'-AGAGCTTTGAAT-3'

27. (Use the table above.) What is the sequence of amino acids coded for by the sequence of bases 3'-ATTACGCTTGCA-5' in a strand of DNA
    - A. Leucine-arginine-lysine-alanine
    - B. Asparagine-cysteine-valine-serine
    - C. Transcription would stop at the first codon
    - D. Asparagine-arginine-lysine-alanine
    - E. Can't tell

28. Genes that are always "on" are referred to as

A. Active
B. Inducible
C. Repressible
D. Constitutive
E. None of the above

29. Genes that are "off" but can be turned on are referred to as

A. Constitutive
B. Repressible
C. Inducible
D. Capable
E. Armed

30. According to the operon model, for the synthesis of an inducible enzyme to occur, the

A. Repressor must bind to the operator
B. End product must not be in excess
C. Repressor must not be synthesized
D. Substrate must bind to the enzyme
E. Substrate must bind to the repressor

31. Point mutations are errors in

A. The reading frame of the RNA
B. The reading frame of the DNA
C. A single base in the RNA
D. A single base in the DNA
E. None of the above

32. Errors in the DNA

A. Can be fixed by suppressor mutations
B. Can never be fixed
C. Are of no consequence to the protein
D. Can be fixed by mismatch repair
E. Are fixed by RNA polymerase

33. Mutagens are chemicals that

A. Cause mutations
B. Repair mutations
C. Destroy the DNA polymerase
D. Improve proofreading
E. Increase the number of mutations

34. Thymine dimers are caused by

A. X-rays
B. Ionic radiation
C. Ultraviolet radiation
D. Abnormal thymine molecules
E. Replication errors

35. Conjugation differs from reproduction because conjugation

A. Transcribes DNA to RNA
B. Transfers DNA vertically, to a new generation of cells
C. Replicates DNA
D. Transfers DNA horizontally, to cells in the same generation
E. None of the above

36. Transposition is

A. A form of conjugation
B. A type of bacterial "sex"
C. Seen only in Gram-positive bacteria
D. Seen only in Gram-negative bacteria
E. A specific form of recombination

37. Transformation is

A. A type of genetic transfer
B. Performed by "competent" cells
C. Done with uptake of "naked" DNA
D. Different than conjugation
E. All of the above

38. Transduction

A. Is carried out by a virus
B. Is facilitated by a pilus
C. Involves naked DNA
D. Is not seen in Gram-negative cells
E. Is not involved in genetic recombination

39. For conjugation to occur, one of the cells must have

A. DNA fragments
B. A latent virus
C. A plasmid
D. An F factor plasmid
E. A large enough chromosome

40. Information that makes organisms more resistant to disinfectants and antiseptics is found on

A. F factor plasmids
B. Transposons
C. Transducing phage
D. Dissimilation plasmids
E. R factor plasmids

DEPTH OF UNDERSTANDING

Questions listed here require you to bring together the concepts you have learned in this chapter into a discussion format. This helps you to increase your depth of understanding of the material you have learned. Help can be found in the student resources at: www.garlandscience.com/micro

1. Discuss the arrangement and orientation of DNA and how this affects replication.
2. Explain how an inducible operon works including the genes and molecules that are involved.
3. Explain how point and frameshift mutations affect the protein's structure and function.

CLINICAL CORNER

Help can be found in the student resources at: www.garlandscience.com/micro

1. A recent outbreak of diphtheria has occurred at the local junior college. When students were examined it was found that both those infected and those not infected were harboring the *Corynebacterium diphtheriae* organisms.

   A. Why were some students sick and others not?
   B. What would you recommend be done to help with this outbreak and to prevent future outbreaks?

2. You are working in Dr Richard's bacteriology laboratory. The work he is doing involves potentially pathogenic microorganisms that could cause respiratory infections. Consequently, all work with these organisms is done under a sterile hood. You have been instructed to wipe down the surfaces of the work area under the hood thoroughly and turn on the ultraviolet light when you are finished. Although these instructions are part of the standard operating procedures for all laboratory personnel, you know that the last technician to work in at this hood became sick.

   A. Why do you have to turn on an ultraviolet light when you finish working?
   B. Will the ultraviolet light alone be effective at preventing contamination?
   C. Why did the other technician become sick?

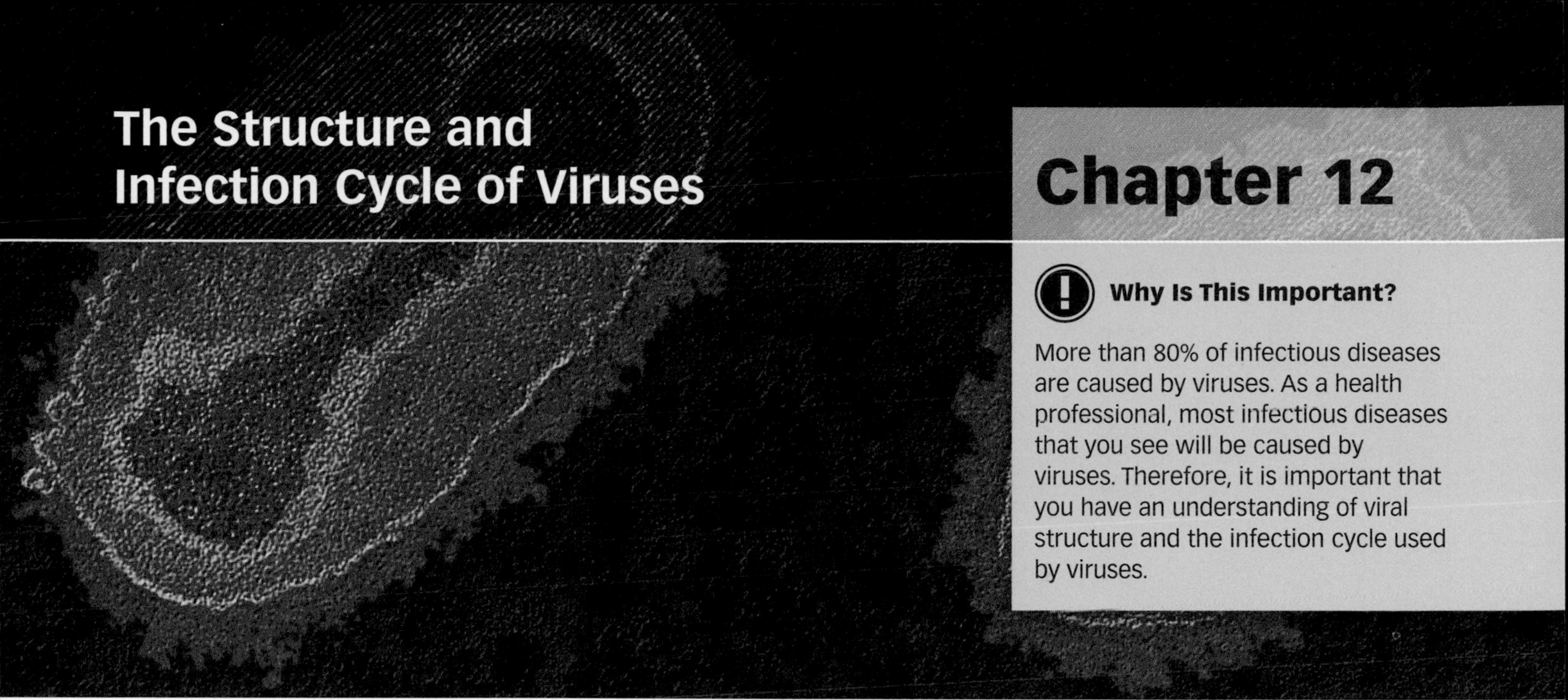

The Structure and Infection Cycle of Viruses

Chapter 12

Why Is This Important?

More than 80% of infectious diseases are caused by viruses. As a health professional, most infectious diseases that you see will be caused by viruses. Therefore, it is important that you have an understanding of viral structure and the infection cycle used by viruses.

OVERVIEW

Recall that in PART III (Chapters 9–14) we are looking at the characteristics of disease-causing microorganisms. In Chapter 9, we surveyed bacterial structures and related them to potential for infection, and in Chapter 10 we looked into what bacteria require for growth and infectivity. In Chapter 11, we examined the genetics of bacteria in an effort to understand how they can undergo changes that can make them more infectiously formidable. However, only about 10–15% of all infectious diseases are caused by bacteria, with the vast majority of the remaining 85% being caused by viruses. As you begin Chapter 12, therefore, it is important to remember that viruses have only one "goal": a productive infection. Productive in this case means making as many copies of themselves as possible so that they can continue infecting host cells. In this chapter, we examine the structure of viruses and the events associated with the viral infection cycle. Then, in Chapter 13, we concentrate on the pathogenesis of viral diseases.

We will divide our discussions into the following topics:

The Structure and Infection Cycle of Viruses

- VIRUS STRUCTURE
- THE INFECTION CYCLE
- SPREAD OF VIRUSES

Viruses are defined as **obligate intracellular parasites**. This means that they cannot "live" outside a cellular host. Keep in mind that the term *live* in this context refers to being able to multiply and produce new viral particles.

VIRUS STRUCTURE

To understand the infection process associated with viruses, we have to examine the structure of viruses. Viruses are "simple" particles, known as virions, that are composed of proteins and genetic material. There are two types of virus, DNA viruses and RNA viruses, a division based on the type of nucleic acid that they contain. This genetic material is enclosed in a protein coat called a **capsid**, and some viruses will also wrap themselves in an envelope structure made up of lipids obtained from internal membranes or the plasma membrane of the infected host cell.

What Do I Need to Know?

To get the most out of this chapter, please review the following terms from your previous courses in biology, anatomy, physiology, or chemistry or in previous chapters of this book as indicated in parentheses: apical surface of cell, basolateral surface of cell, electrostatic interaction, endocytosis (**4**), genome (**4**), glycoprotein (**2**), hydrophilic (**4**), hydrophobic (**4**), integral protein (**9**), pathogenesis (**1**), phospholipid bilayer (**4**), polypeptide, progeny, proteolytic enzyme (**3**), replication (**11**), transcription (**4**, **11**), translation (**11**), vesicle (**4**).

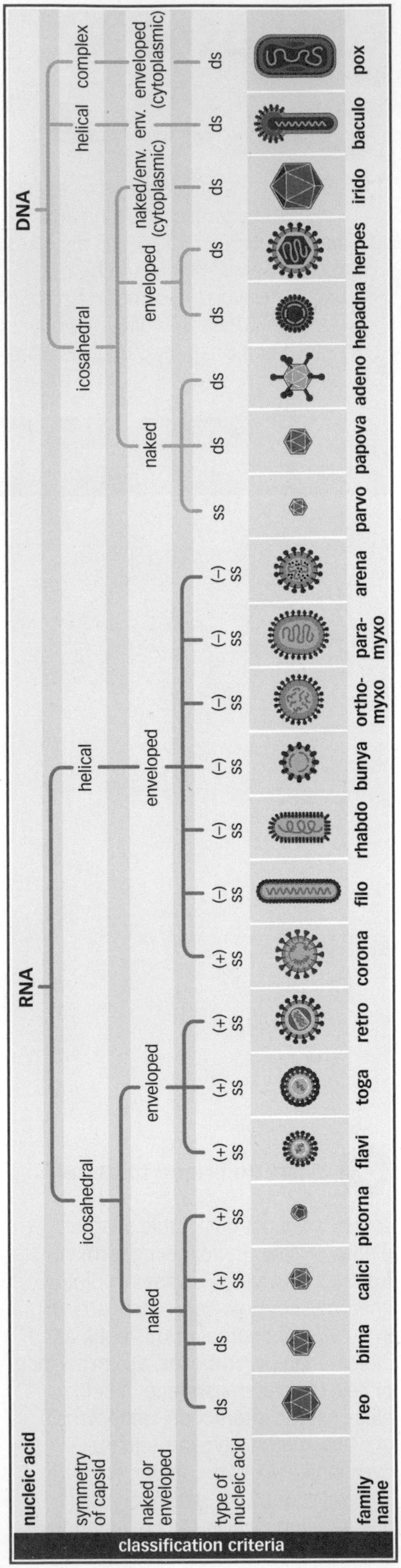

Figure 12.1 Classification of animal viruses. This illustration shows a summary of the major characteristics of the 22 families of animal viruses that infect humans. Note the shapes of these viruses. ds stands for double-stranded nucleic acid; ss stands for single-stranded nucleic acid.

Viruses come in many sizes and shapes (**Figure 12.1**), with both size and shape determined by the molecules from which they are constructed. Please look this figure over carefully so that you can get a good idea of the structures and basic characteristics of viruses.

The Virion

A single viral particle is referred to as a **virion**. When considering the structure of a virion, we have to consider two basic problems. First, the virion must be strong enough to withstand the environment in which it must survive until it successfully infects a host cell. This will depend on the strength of the capsid. Second, the virion must be able to shed this same protective protein coat easily on entry into a host cell. The overall design of the virion is based on the ability to meet these two challenges.

Close examination of the virion shows that the protein coat is what protects the viral genome from chemical agents and physical conditions such as pH or temperature. The presence of a lipid envelope surrounding the virion can also protect viruses from agents that cannot penetrate lipids. The assembly of virions includes very stable intermolecular interactions that help protect the genome, but these interactions are also reversible so that the virus can disassemble the structure after entering the host cell.

Viral Nomenclature

Viral architecture is usually described from the polypeptide chain (the smallest biochemical unit) to the overall virion, and some of the terms used to describe virus structure are listed in **Table 12.1**. A widely used method for studying viruses is electron microscopy, but even with this very powerful tool, detailed study of the viral structures is very difficult. Therefore, viruses have been studied by computerized image analysis and three-dimensional reconstruction methods. A second method, X-ray crystallography, has also proved to be excellent for observing the three-dimensional structure of virions.

Table 12.1 Viral Structures

Term	Synonym	Definition
Capsomere		Protein molecule forming capsid
Capsid	Protein coat	Protein shell surrounding nucleic acid
Nucleocapsid		Nucleic acid plus capsid
Envelope	Viral membrane	Phospholipid bilayer with embedded glycoproteins surrounding capsid in enveloped virus
Virion	Viral particle	Complete infectious viral structure: nucleic acid plus capsid for non-enveloped virus; nucleic acid plus capsid plus envelope for enveloped virus

The Capsid

Capsids are built from identical protein subunits called **capsomeres**, which are arranged to provide maximal contact and bonding among the subunits. The bonding of these subunits occurs in such a way that there is structural symmetry. All except the most complex viruses have either **helical symmetry** or icosahedral symmetry.

Helical Viruses

Helical viruses have either a rod shape (straight, relatively rigid) or a filamentous shape (relatively flexible, curved or coiled). A classic example of a helical RNA virus is the influenza virus (**Figure 12.2**). The RNA in this virus is segmented, with each segment enclosed in its own helical capsid held in place by multiple proteins called **nucleoproteins**. This virus also has a second layer of protein called **matrix protein** just inside the envelope.

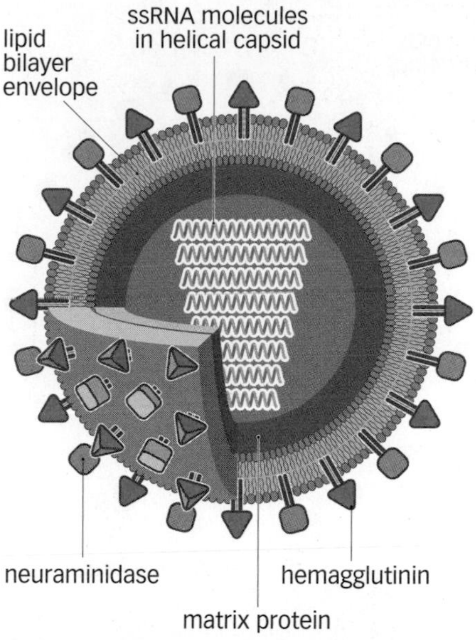

Figure 12.2 A diagrammatic representation of the influenza virus in cross section. The single-stranded RNA is in eight segments, each enclosed in a helical capsid and surrounded by a lipid bilayer envelope. Note the neuraminidase and hemagglutinin molecules embedded in the envelope.

Icosahedral Viruses

The shape of icosahedral viruses is derived from 20 triangular faces that make up the capsid. This shape gives the capsids of these non-enveloped viruses 12 points (**Figure 12.3**). There are two types of icosahedral viruses, *simple* and *complex*. In the *simple icosahedral virus*, the capsid is made up of a large number of identical three-polypeptide capsomeres. A good example is the poliovirus, which is a picornavirus. This virus is one of the smallest and simplest of the viruses that infect humans and is composed of only 60 copies of four proteins.

Poliovirus uses the digestive tract in some of its infection strategies, and therefore it must be able to withstand the harsh environment that it encounters there. Survival is possible because of the very stable protein–protein interactions that take place in the capsid. Keep in mind the functional dichotomy of the capsid: even though poliovirus must be strong enough to withstand harsh environmental conditions, the protein coat is still designed to disassemble when contact is made with a receptor on a host cell.

Fast Fact

The name picornavirus comes from *pico* meaning small, and *rna*, meaning an RNA genome.

In *complex icosahedral viruses*, the complexity comes from additional proteins and lipids surrounding the capsid. The best example of this type of virus is the enveloped herpesvirus (**Figure 12.4**), in which the virion contains three groups of proteins: the proteins of the capsid; those forming a structure called a **tegument**, which is a protein layer located between the capsid and the envelope; and glycoproteins that form spikes on the surface of the envelope.

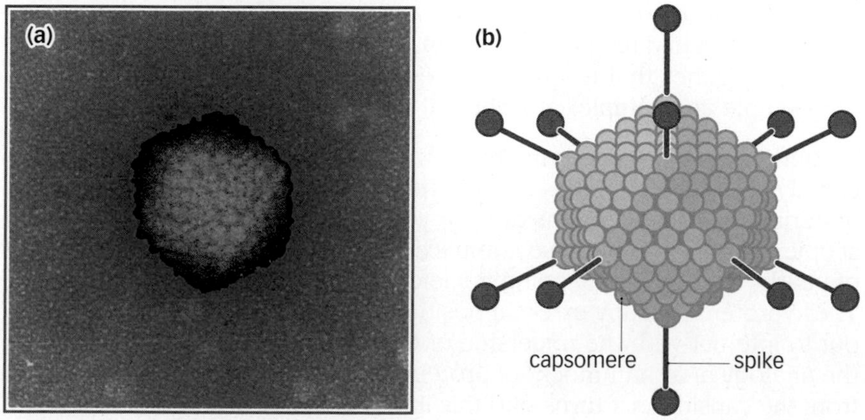

Figure 12.3 Icosahedral viruses. Panel **a**: Colorized transmission electron micrograph of adenovirus, a non-enveloped simple icosahedral virus. Panel **b**: A diagrammatic representation of the non-enveloped complex icosahedral virus. The spikes on the vertices of the capsid are glycoproteins used to attach the virion to receptors on a host cell.

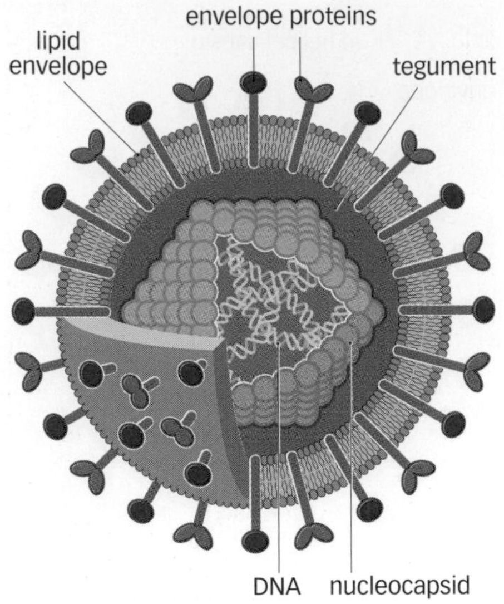

Figure 12.4 An illustration of the herpes virus, which has an enveloped complex icosahedral structure. Notice that this virus has three distinct groups of proteins: those of the capsid, the tegument, and the envelope.

Complex Viruses

Viruses that do not have either helical or icosahedral symmetry are categorized as **complex viruses**. The best example is the variola (pox) virus, which has a smooth rectangular shape encased in an envelope (**Figure 12.5**).

Viral Envelopes

Many viruses that infect humans and other animals are enveloped. Envelopes are formed when viral glycoproteins and oligosaccharides associate with membranes of the host cell. As we will see later, movement of a virion out of an infected host can occur in such a way that the virion gets wrapped in host membranes. Envelopes vary in size, morphology, complexity, and composition, but the foundation of all envelopes is the phospholipid bilayer. The precise amount of lipid in a viral envelope varies depending on the virus and on the mechanism by which it captures lipid from the host membrane.

Envelope Glycoproteins

The glycoproteins in viral envelopes are integral membrane proteins that are firmly embedded in the lipid bilayer of the envelope. This embedding is facilitated by domains (sections) of the membrane proteins called *spanners*, which associate with the envelope bilayer in such a way that part of the protein faces inward toward the virus capsid and part faces outward from the envelope. On the exterior of the membrane, the glycoproteins can form spikes or other surface structures that the virion can use to attach to a host cell.

Genomic Packaging

The goal of any virus is a productive infection. To achieve this, the virus must move its nucleic acid into a host cell, and every structure of the virus has been formed to accomplish this task. The way in which the nucleic acid is packaged has an important role in the infection. There are three ways in which viruses package their nucleic acid: (1) directly in the capsid, (2) enclosed in specialized proteins, and (3) enclosed in proteins from the host cell. Let's look at the first two of these mechanisms in more detail.

Enclosing the nucleic acid in a capsid is the simplest way to package the viral genome, and the attachment occurs at the inner side of the protein coat. This method of using the same proteins to package the genetic material and build the protein coat makes very effective use of the small number of genes that the virus carries. We see this type of packaging in both simple and complex icosahedral viruses.

In some viruses the genetic material is associated with specialized proteins called *nucleic acid–binding proteins*. The interactions between the genetic material and these proteins are not yet understood, but electron microscopic examination of these complexes gives the appearance of a string of pearls, with the string being nucleic acid and the pearls being protein. These proteins occupy external positions on the nucleic acids and reach out to interact with the underside of the capsid proteins. One could use the analogy of a "hammock" of protein holding the genome and hanging from the capsid. As it turns out, this is a very stable arrangement.

Keep in Mind

- Viruses are obligate intracellular parasites because they cannot "live" outside a host cell.
- Viruses come in a variety of sizes and shapes.
- Viruses contain either DNA or RNA but never both.
- The nucleic acid in a virion is surrounded by a protein coat called a capsid.
- The capsid is made up of repeating protein subunits known as capsomeres.
- Some viruses are surrounded by an envelope composed of viral glycoproteins and oligosaccharides complexed with host cell membranes.
- Viral nucleic acids are packaged (1) directly in the capsid, (2) enclosed in viral proteins, or (3) enclosed in host proteins.

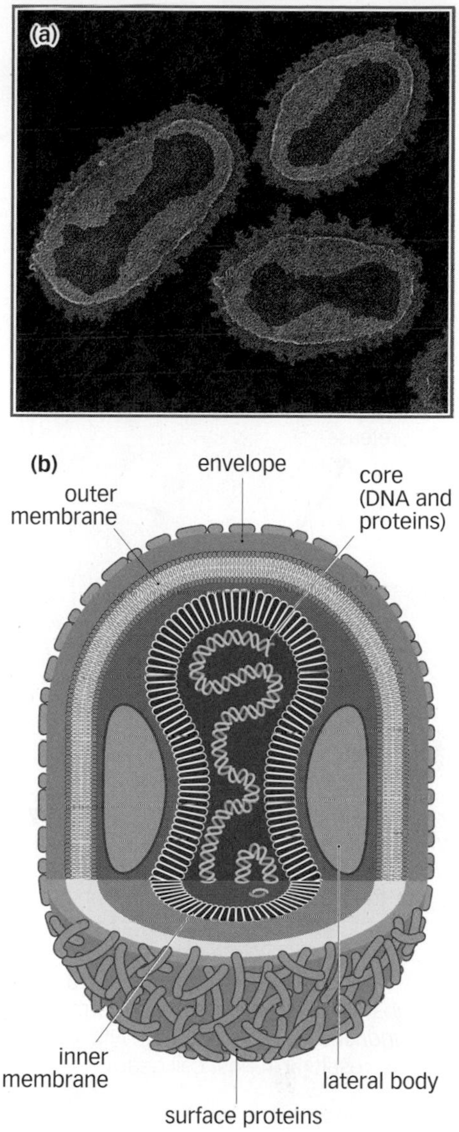

Figure 12.5 The structure of the variola virus. Panel **a**: Colorized electron micrograph of the complex virus variola, the agent of smallpox. Panel **b**: Diagrammatic cross section of the variola virus, showing significant structural features.

THE INFECTION CYCLE

The stages of the infection cycle of viruses have been worked out, and a great deal of information has been accumulated for each step. In this chapter our primary focus is on animal viruses, by which we mean those that infect eukaryotic (animal) cells. Besides animal viruses, there are also viruses that infect bacteria, called either **bacteriophages** or simply phages. A great amount of information about the infection cycle of animal viruses has been derived from studies on bacteriophages.

Lytic versus Lysogenic Infection

During the infection cycle of many viruses, the host cell is used to produce more virions, and when the host cell is completely filled with new virions the host cell simply bursts. This is called lysis and is the hallmark of the **lytic infection cycle** seen in animal virus infections.

However, there is an alternative infection strategy called the **lysogenic cycle**. In this cycle the viral genome becomes incorporated into the host cell's DNA and can remain this way for an extended period. This causes a **latent infection** in which there is no new virus made and no increase in infection. During lysogeny, the viral DNA insert becomes part of the host cell's genetic material and is carried there through the subsequent generations of host cell division. At some point, the integrated viral DNA (referred to as a provirus) can be induced to exit the host genome. When this occurs, the virus begins the lytic cycle resulting in the death of the host cell. The lytic and lysogenic cycles are illustrated in **Figure 12.6**.

So how is the choice made between lytic and lysogenic infections? It seems that the decision is based on the overall health of the host cell. Remember that the most important thing for the virus is to produce as many new virions as possible. The healthier the host cell, the greater will be the probability that there are a large number of additional healthy host cells available when new virus is released by lysis of the originally infected host cell. Large numbers of prospective host cells means more chance of new infections and therefore more new virus. Consequently, when a virus encounters a population of healthy host cells, lytic infection is what usually happens.

In an unhealthy host, the number of healthy host cells is too low to allow successful continuation of the infection. In this case, the lysogenic cycle is more favorable to the virus because the virus can await better conditions when it is integrated into the host chromosome.

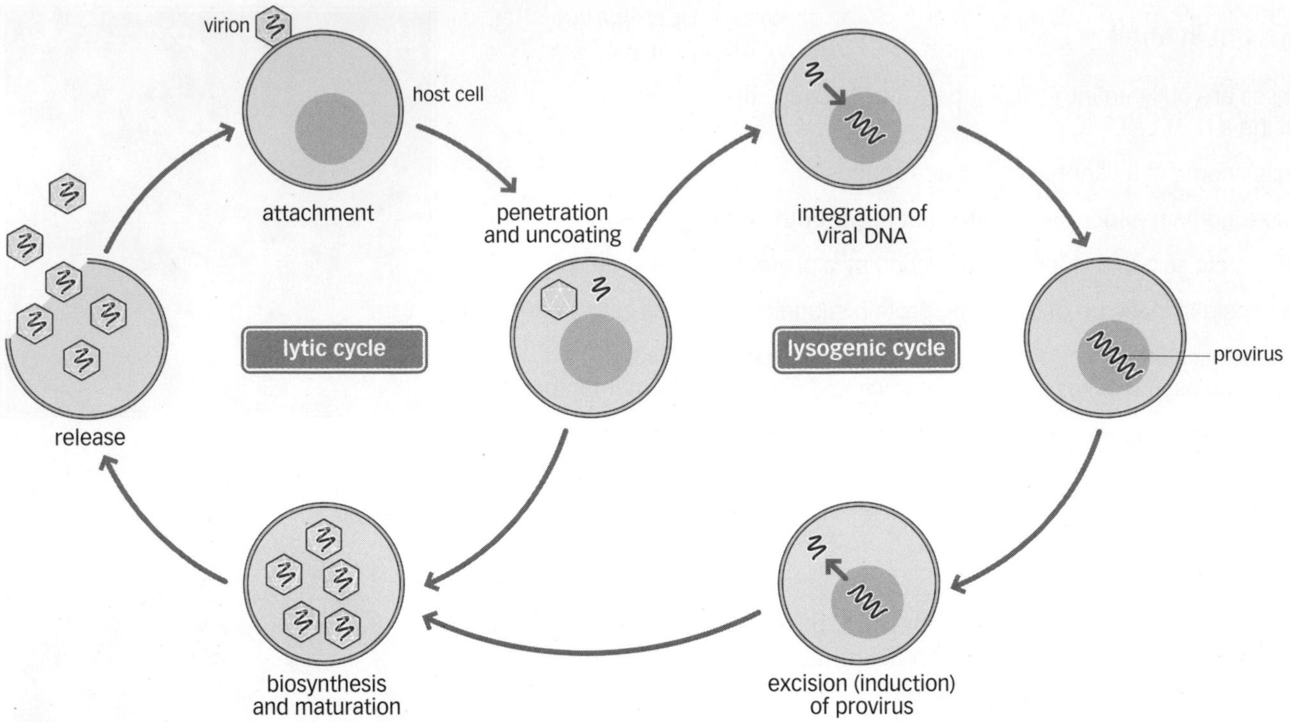

12.1

Figure 12.6 Infection of host cells by viruses can result in lytic infection or in lysogenic infection. In a lytic infection cycle, the host is killed as it bursts and releases large numbers of newly formed virus. In a lysogenic infection cycle, the DNA of the virus is first integrated into the host chromosome and remains there until *induced*, at which time the lytic cycle begins, resulting in host cell death.

For animal viruses the lytic infection cycle contains the following steps: *attachment*, *penetration*, *uncoating*, *biosynthesis*, *maturation*, and *release*. Let's look at each of these in detail.

Attachment

Attachment occurs as a virion binds to specific receptors on a host cell. For some viruses, only one receptor interaction is required, whereas for others there needs to be a co-receptor involved. Receptor/virus interactions are important for understanding the pathogenesis of viruses and also for examining potential blocking mechanisms that might be useful for therapy.

Virus entry into the host cell is not a passive process but instead relies on the virus "usurping" normal cell functions such as vesicular transport. Remember that the capsid is made up of capsomeres, which are three-dimensional proteins. In some cases the interaction between the virion and the host cell's receptor initiates a conformational change in the capsomeres that prepares the virion for the uncoating step. In other viruses, the host cell's receptor is more like a tether that holds the virion in place next to the host cell until the virion has gained entry.

When Virus Meets Host

Viral interactions with host cells occur through random collisions and are essentially governed by chance. Consequently, the concentration of viral particles is important in determining whether an infection occurs. The production of large numbers of viral progeny in each infected host cell increases the chance that the number of random encounters with new host cells will be high enough to continue the infection. It is interesting to note that viruses can be either "promiscuous" or highly specific when it comes to attachment to a host cell. The presence of a receptor makes identification of the host cell easy, but binding to the receptor may not be enough to cause infection. In addition to having the proper receptor, the host cell must also be *permissive*, which entails having the cellular components necessary to produce new virions.

Viral infections most often occur at the apical surface of epithelial cells, and such infections are usually localized. In contrast, viruses that are transported to the basolateral surface of the host cell and released into the underlying tissues usually spread to other sites.

Many viruses attach only to specific areas of the host plasma membranes, called **lipid rafts**. These areas of the membrane are rich in cholesterol, fatty acids, and other lipids, and consequently they are more densely packed and less watery, making them more reliable for stable attachment. These areas of the membrane are also the release site for many viruses, such as Ebola and HIV type 1.

Many different plasma membrane molecules on the host cell can serve as receptors for virus attachment. Some viruses can attach to more than one type of receptor molecule, and some receptors can be shared by different types of virus. In fact, for some viruses, the receptor determines the host range.

Initial virus–receptor interactions are probably electrostatic. However, these initial reactions are followed by high-affinity binding, which occurs because of the conformational interactions between virion and receptor.

In some cases the host's response to the attachment of virus can amplify the attachment step. For example, rhinovirus, which causes upper respiratory infections (including 50% of common colds), uses a glycoprotein host cell receptor called **ICAM-1**. It turns out that this is a common adhesion molecule normally involved in a variety of host cell inflammatory responses. Ironically, inflammation seen with upper respiratory infections caused by rhinovirus increases the number of ICAM-1 molecules, which facilitates more binding to host cells by the virus. Consequently, the initial defensive response of the host to rhinovirus infection leads to more infection by the virus!

Fast Fact

It has been determined that, in many cases, a single rhinovirus–receptor binding event is all that is required to mediate the entry of the virus into the host cell.

Some viral infections require the involvement of co-receptors. These molecules are host components that interact with the virus and the primary receptor to allow continuation of the infection cycle. Co-receptors have been shown to be involved in HIV type 1, picornavirus, adenovirus, and herpesvirus infections. Viruses that require co-receptors cannot successfully complete the infection without these molecules.

Receptor Binding

Examination of non-enveloped viruses gives us a good idea of receptor binding mechanisms. For these viruses, attachment is between the host cell's receptor and structures on the viral capsid. For example, picornaviruses have icosahedral capsids built from 60 protein subunits and arranged in 12 pentamers. It is the points of these pentamer structures that bind to host cell receptors. Adenoviruses, which are much larger than picornaviruses, also have icosahedral capsids but these are much more complex structures. These viruses have fibers that protrude from each of the pentamers, and these fibers work in conjunction with the base of the pentamer to attach the virus to the host cell's receptor (**Figure 12.7**).

For enveloped viruses, remember that the lipid envelope surrounding the virus originates from host cells previously infected by the virus. During the assembly of these viruses, specific viral proteins are inserted into these envelopes. It is the attachment site on one or more of these viral envelope proteins that binds to specific host cell receptors. Two of the most studied enveloped viruses are influenza and HIV type 1. In the influenza virus, hemagglutinin (HA) is a viral glycoprotein found in the viral envelope that binds to the host cell's receptor (sialic acid). Each of the HA molecules consists of a long helical stalklike structure anchored into

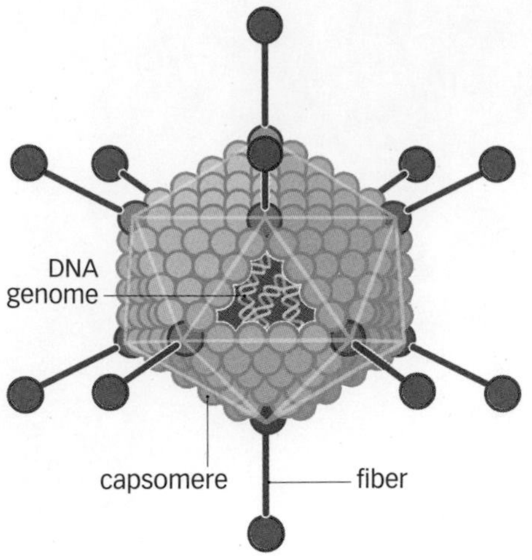

Figure 12.7 An illustration of adenovirus showing the complex icosahedral capsid. Notice the fibers that extend from each of the pentamers. These fibers, which are similar to spikes seen on many viruses, are used for attachment of the adenovirus to receptors on host cells.

the membrane and topped by a large globular protein that binds to sialic acid. Together the stalk and globular protein are referred to as a *spike*. For HIV type 1, attachment is accomplished by similar spikelike structures that are located on the outer surface of the envelope. These spikes bind to the CD4 receptor found on the helper T cell population of lymphocytes. With HIV, there is also a requirement for co-receptor binding.

Penetration and Uncoating

Once attached to the host cell, the virion must gain access to the interior of the cell in the penetration step. However, the viral genetic material is still enclosed in either a capsid or a capsid and envelope, and so the next step in the infection cycle must be uncoating. In some cases, uncoating is a simple process that occurs at the plasma membrane (**Figure 12.8**). In others, it is a very complex mechanism that includes endocytosis through the plasma membrane and a "docking" event at the nucleus. In still other cases, the nucleus of the host cell is involved in uncoating the virus.

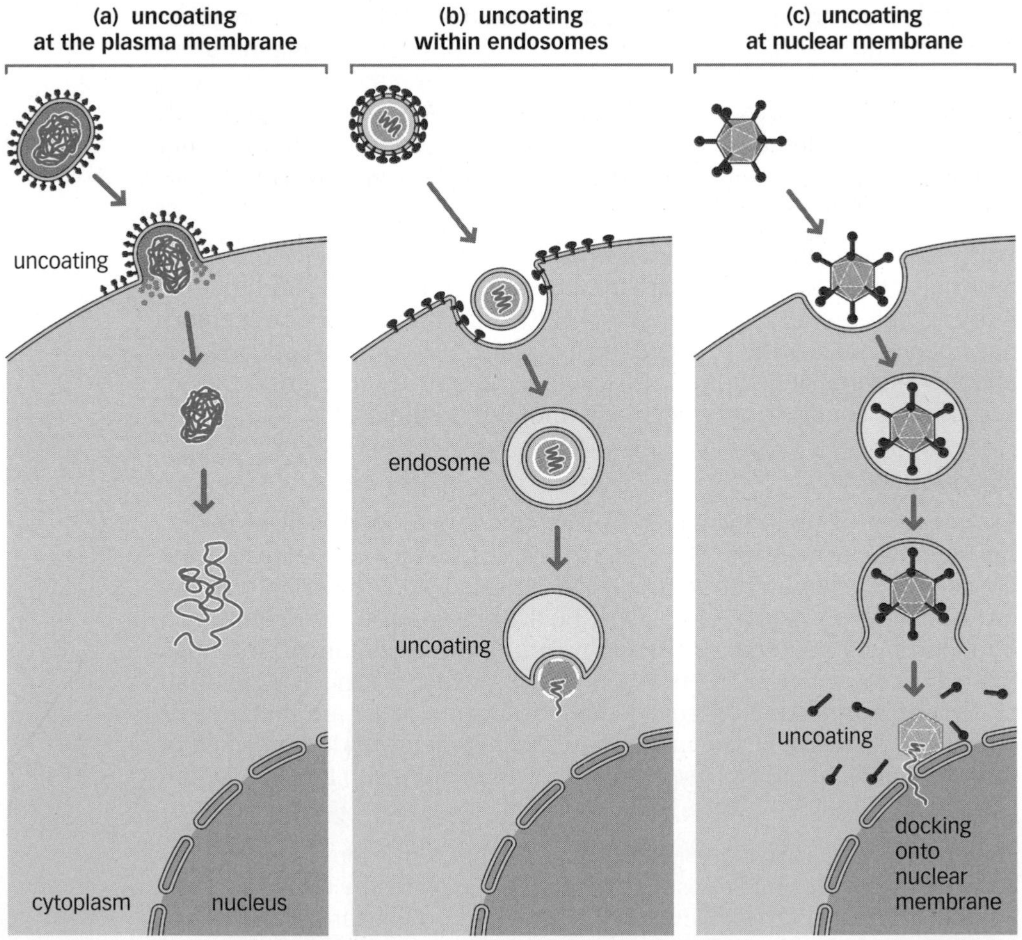

Figure 12.8 Three virus uncoating strategies. Panel **a**: Uncoating at the plasma membrane. Panel **b**: Uncoating in endosomes. Panel **c**: Uncoating at the nuclear membrane. In the first two, the viral genome is released into the cytoplasm. In the last example, uncoating occurs at the nucleus, which causes the release of the viral genome directly into the nucleus.

Penetration and Uncoating by Non-enveloped Viruses

Even the smallest viral particle is too big to move across the host plasma membrane without help. Non-enveloped viruses take advantage of normal cellular mechanisms, such as receptor-mediated endocytosis, to gain entry into the host cell. Using receptor-mediated endocytosis, a virion binds to a receptor on the host cell's plasma membrane, causing a pit to form in the membrane. The membrane then encloses the virion to form a vesicle called an **endosome**. When first produced, these vesicles, referred to as *early endosomes*, travel through the host cytoplasm and eventually fuse with or become *late endosomes*, which are endosomes that have been in the cell for a while. The inside of a late endosome is more acidic than the inside of an early endosome, and this acidic environment begins uncoating the virion. The late endosome then fuses with lysosomes in the cytoplasm, and uncoating is completed. Keep in mind that the phospholipid bilayer structure of the plasma membrane facilitates the fusion of the endosomes with one another and with lysosomes.

Some non-enveloped viruses form a pore in the host plasma membrane to gain entrance to the cytoplasm. Binding of the virus to the host cell's receptor causes conformational changes to the capsomeres such that they become hydrophobic and are attracted to the plasma membrane. The hydrophobic part of the virus inserts itself into the membrane, forming a pore that allows the viral genetic material to enter the cytoplasm of the host cell.

Penetration and Uncoating by Enveloped Viruses

With enveloped viruses, penetration of the virus into the host cell is relatively straightforward because the envelope is essentially a plasma membrane and can fuse with the host plasma membrane. This fusion is mediated by specialized fusion proteins, and the fusion is believed to result in the formation of a large opening called a fusion pore in the host cell. This opening permits the virion to move across the plasma membrane and into the host cell.

In some cases, virion–receptor binding causes conformational changes in the host cell that expose the fusion proteins. In other instances, fusion requires the presence of co-receptor molecules. Perhaps the best example of the need for co-receptor molecules is seen in HIV type 1 infections. In these infections, not only does HIV require specific binding at the CD4 receptor of the helper T cell, but other host cell proteins are also required to complete the fusion event. In fact, all of the co-receptors for HIV are normal cell surface receptors for small molecules called chemotactic cytokines. Several of these co-receptors have been identified and it seems that all of them are required for the fusion to occur.

Cytoplasmic Transport of Viral Components

We have chosen to mention cytoplasmic transport as part of the penetration and uncoating process because viral infections involve *compartmentalization*. In this process, DNA or RNA synthesis of viral genomes as well as synthesis of the new capsids and other viral proteins occurs in specific locations of the host cell. After synthesis, these components are finally moved to other specialized sites for final assembly of the intact virions. Therefore, there need to be mechanisms that allow the movement of viral components through the host cell's cytoplasm. After the virus uncoats, there are two ways in which this is done. The first uses membrane-enclosed vesicles in conjunction with host cell cytoskeletal structures such as microtubules. The second method employs direct association of viral components with cellular transport mechanisms.

In some cases there are actually specialized lipids and proteins called *chaperones* that will facilitate the movement of virus through the host cell's cytoplasm. The actual methods employed by these chaperones are not yet understood, but they seem to be associated with maintaining stabilization of the three-dimensional structure of viral components during the journey through the cytoplasm.

Transport of the Viral Genome into the Nucleus

Some viruses must ultimately enter the host cell's nucleus for replication of the viral genome. Recall from Chapter 4 that the structure of the nucleus in eukaryotic cells involves a double phospholipid bilayer membrane. Genomes of DNA viruses, as well as the DNA synthesized by *reverse transcriptase* (discussed below), must penetrate this double membrane for a successful infection to occur. This transport is accomplished using *import pathways* that are routinely used by host cell proteins, and it involves pores that are part of the nuclear membrane.

Schemes for entry into the nucleus depend on particular viruses. For example, DNA viruses are too large to get into the nucleus intact, so the capsid of these viruses moves through the cytoplasm and docks on the outside of the nucleus and then delivers the viral genome into the nucleus.

There are RNA viruses such as the retroviruses that require entry into the nucleus. These viruses carry the enzyme **reverse transcriptase**, which can convert RNA into DNA. The synthesis of retrovirus DNA into a *pre-integration complex* (a DNA copy of the viral RNA that can integrate into the host chromosome) actually begins in the cytoplasm and takes 4–8 hours. This complex then moves into the nucleus and integrates with the host chromosome (**Figure 12.9**). It turns out that the movement of this pre-integration complex actually correlates with the breakdown of the nuclear membrane during mitosis. The timing of this host cell mitotic event is neatly connected to two important viral requirements: first, the entry into the nucleus, which is made easy because the nuclear membrane is breaking down during mitosis, and second, integration into the host chromosome, which is in the process of replicating.

Keep in Mind

- Viruses can go through a lytic cycle in which the host cell fills with virions and bursts.
- In some cases, viruses cause a lysogenic infection in which the viral genome becomes integrated with the host cell's chromosome.
- Host cell receptors used by viruses represent a small fraction of the cell membrane proteins.
- Many different viruses can share the same receptor.
- Virus–receptor interactions facilitate the infection process by enhancing virus entry into host cells.
- For enveloped viruses, penetration occurs through a fusion event between the viral envelope and the host cell's plasma membrane.
- This fusion is catalyzed by a specialized viral glycoprotein called a fusion protein, whose function is highly regulated.
- Once fusion has occurred, the virus is released into the cytoplasm of the host cell, where various mechanisms allow the uncoating of the virus.
- Penetration mechanisms of non-enveloped viruses are thought to be through host cell endocytotic pathways that are routinely used by host cells for the normal importation of molecules.

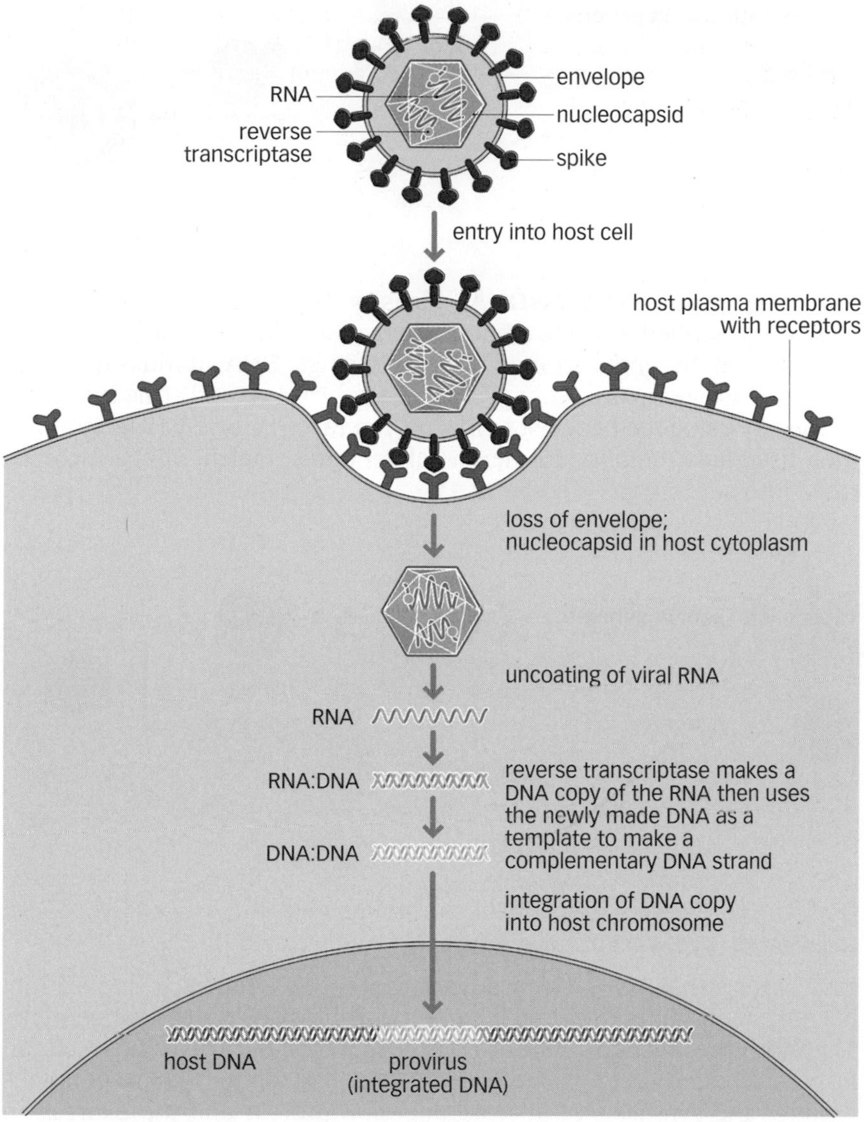

Figure 12.9 Illustration of the integration of proviral DNA into the host cell chromosome after infection with a retrovirus. Notice the action of reverse transcriptase as it copies viral RNA into DNA.

Biosynthesis

Now that we have seen how viruses attach, penetrate, and uncoat, we need to examine the way in which new virions are made. This process involves the synthesis of new viral protein components as well as new viral genomes. As we have seen, viruses contain either DNA or RNA, and the genomes of both types of virus can be either single-stranded or double-stranded. Let's look at how each of these viral types makes the protein and genome components required for new virions.

DNA Viruses

Double-stranded DNA (dsDNA) viruses. For these viruses, the mRNA required to make viral proteins can be produced by transcribing one strand of the viral DNA. This can be accomplished by using either host or viral RNA polymerase molecules in a manner consistent with host cell RNA synthesis. The replication of new DNA viral genomes also follows the replication mechanisms seen in the host cell (**Figure 12.10**).

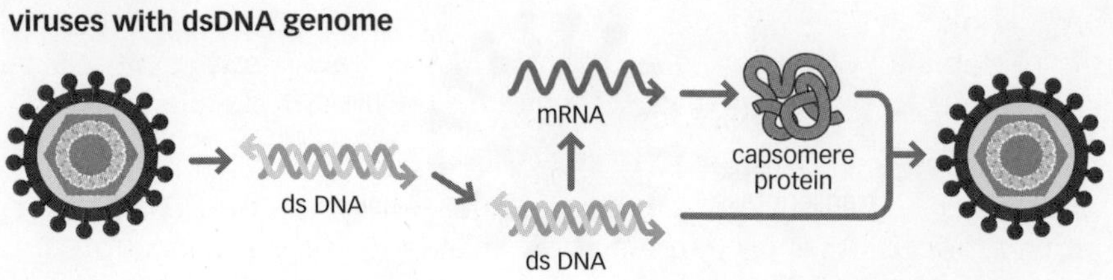

Figure 12.10 Transcription and replication strategy of double-stranded DNA (dsDNA) viruses. In this case one of the strands of the viral DNA is used to transcribe a message (mRNA) which can be used to make capsomere proteins. These capsomere proteins assemble into the capsid. The double stranded viral DNA is replicated using the host DNA polymerase to make viral genomes. The capsid and viral genome are used to construct new virus.

12.2

Single-stranded DNA (ssDNA) viruses. Here the DNA strand must still be transcribed to mRNA, but in these viruses mRNA cannot be transcribed from the single-stranded DNA. The single DNA strand is therefore first used as a template for copying a second DNA strand. This is accomplished by using the host cell's DNA polymerase. This new DNA strand is then used as a template for new viral genomes, which will be incorporated into new virions (**Figure 12.11**).

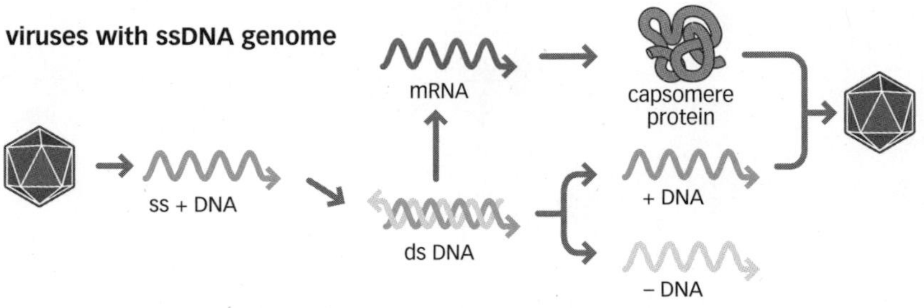

Figure 12.11 Transcription and replication of single-stranded DNA (ssDNA) viruses. Single stranded DNA viruses must make a template strand of their genome (light green). This template strand is transcribed into mRNA which is used to make viral capsomere proteins. The template strand is also used to make new viral genomes (dark green) which can be incorporated into the newly constructed capsids. Multiple genomic strands can be made from these templates but only the viral ssDNA (dark green) is incorporated into new virus.

12.3

Replication of DNA Virus Genomes

DNA viruses have a decided advantage when it comes to replication, because their genome is in the same configuration as the host genome. Therefore, the mechanisms that are used for replication of the host cell can also be used for viral replication. However, replication of these genomes requires the synthesis of at least one viral protein and the expression of several viral genes. Consequently, viral DNA synthesis cannot begin immediately after infection but must wait until these viral components have been made in sufficient quantity.

Once viral DNA synthesis begins, there are many cycles of replication during which large numbers of viral DNA molecules begin to accumulate. Replication of viral DNA is performed by host cell machinery, which avoids the need for the virus to devote any of its limited genetic capacity to coding for unnecessary enzymes. In the case of latent DNA viruses (see discussion below), the amount of DNA replicated is much smaller and there is no accumulation of viral DNA.

Remarkably, DNA synthesis in the host cell is inhibited by the virus, so that all polymerases and proteins that are involved in replication can concentrate on viral DNA synthesis. In fact, adenovirus and herpesvirus actually induce the arrest of the entire cell cycle of the host cell so that viral DNA replication can be maximized.

In DNA virus infections, specialized sites form in the host cell. These distinct replication compartments are essentially "viral factories." They contain both the DNA templates and also the viral replication machinery. This localization of the components needed for viral DNA synthesis at a limited number of cellular locations facilitates exponential viral DNA replication. This arrangement also increases the local concentration of the proteins involved in replication and allows each newly synthesized DNA to serve as an immediate template for the next molecule.

The same type of concentration phenomenon is useful for viral gene expression because large numbers of DNA templates and the proteins required for transcription are all concentrated in the same location. It is important to note that viral replication compartments do not form randomly but are actually formed as a result of viral "colonization" of specialized areas of the host cell.

DNA replication in latent viruses. A latent virus (one that incorporates its genome into the host chromosome) does not require maximal replication of viral genomes or the large-scale production of mature virus. In infections with a latent virus such as Epstein–Barr virus, there is expression of only a small number of viral genes and replication of a limited number of viral genomes during the period when the virus is latent. The replication of latent viral genomes also begins at a different location on the genome from that associated with replication in lytic infections. As a latent infection is established, the Epstein–Barr viral DNA also becomes increasingly methylated. In some way this seems to repress transcription, which in turn inhibits the expression of viral genes and allows the virus to remain in a latent state.

DNA Virus Transcription

The next part of the biosynthesis step for animal viruses involves transcription, in which a newly made viral DNA molecule acts as a template for the creation of viral mRNA. Transcription of viral DNA is performed by the host cell's RNA polymerase. (An exception to this statement involves the poxviruses, which bring their own RNA polymerase into the host cell.) Expression of the genes in DNA viruses occurs in a sequential and reproducible order. Viral enzymes, regulatory proteins, and structural proteins that will be needed by the virus are made only after viral DNA synthesis has begun. This is different from gene expression in RNA viruses, in which all the genes are expressed continuously so that enzymes, regulatory proteins, structural proteins, and new viral RNA are all being made at the same time.

With double-stranded viral DNA, transcription is easy because the DNA is in the same conformation as the host cell's DNA (each one is a double helix of A, C, G, and T bases). Therefore, the transcription of viral DNA begins as soon as it reaches the host nucleus. With some dsDNA viruses, such as adenovirus, herpesvirus, and papillomavirus, transcription continues after replication of new viral DNA because this gives more templates for transcription. In contrast, transcription with single-stranded DNA viruses is more complicated because the single DNA strands must first be converted to double strands to serve as templates for transcription.

Viral genes are transcribed at very high rates. This is necessary because large quantities of proteins must be produced so as to make as many new virions as possible within an infection cycle. Although the rate of transcription is high, there is considerable regulation of the process. In most cases this regulation is performed by regulatory proteins of the host cell. However, some viruses do bring along their own regulatory proteins. The carefully regulated process of DNA virus transcription ensures the coordinated production of the structural proteins required to make new virions. This regulation also serves another important function in that it prevents the overproduction of components that are not needed, making the infection process very efficient.

Fast Fact

Antiviral drugs that inhibit viral DNA synthesis also inhibit the production of proteins and therefore the construction of new virions.

One of the hallmarks of DNA virus transcription is the coordination of transcription with the synthesis of viral DNA. It is as if the virus does not make the proteins needed to build the new virions until there is something to put inside them (that is, the newly synthesized viral DNA).

One last point about the transcription of viral DNA is important. On infection with a DNA virus, host cell transcription is shut down, and part of the transcriptional machinery is "stolen" for use by the virus. This offers the virus several advantages. For instance, inhibition of host cell function can allow host resources that were required for these functions to be devoted exclusively to the needs of the virus. In addition, there is no competition between the transcripts of the host and those of the virus when it comes to translation. These advantages amplify the primary goal of the virus, which is a productive infection.

Keep in Mind

- Replication of DNA viral genomes requires several proteins that are also seen in replication of the host cell.
- Replication of viral DNA genomes ranges from the simple synthesis of both strands of a linear double-stranded DNA virus to mechanisms that replicate the DNA genome in pieces (as seen in herpesvirus).
- Viral DNA replication occurs at specialized sites in the host cell, which contributes to the efficiency and productivity of the viral infection.
- Transcription of viral DNA is performed by the host cell's RNA polymerase (except for poxviruses, which bring their own RNA polymerase).
- Viral genomes are transcribed at a very high rate, to make as many new virions as possible.
- For latent DNA viruses, special mRNA sequences inhibit the lytic cycle and allow the viral DNA to integrate with the host chromosome.

RNA Viruses

For RNA viruses, making the components needed for new virions is more complicated. Host cells do not possess RNA-dependent polymerase enzymes, which are required to make viral mRNA and replicate new viral RNA genomes. Therefore, single-stranded RNA viruses must carry with them genes that encode these novel enzymes.

RNA viruses are classified using a (+) or (–) strand designation for their genomes. Before we look at illustrations of RNA biosynthesis, it will be helpful to revisit the concept of complementary base pairing. Here we can use one strand as a *template* for another. When there is a C in one strand, the template strand will have the complementary base, G. In RNA viruses an A is paired with a U. Let's look at an example.

Original strand	⟶	AUGACCAGUACC
Template strand	⟶	UACUGGUCAUGG
New strand	⟶	AUGACCAGUACC

Notice that the template strand is used to make more copies of the original strand. This template mechanism is used for RNA viruses as described below.

Double-stranded RNA (dsRNA) viruses. Double-stranded RNA virus genomes contain both a (+) and a (–) strand. During infection, the (–) strand is first copied into mRNA by a viral RNA polymerase to produce viral proteins. This newly synthesized strand is then used as a template to make a double-stranded genome, which will be placed into new virions (**Figure 12.12**).

viruses with dsRNA genome

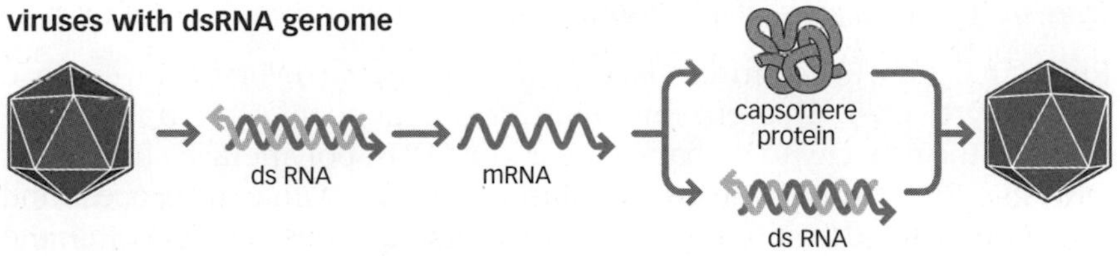

12.4

Figure 12.12 Transcription and replication of double-stranded RNA (dsRNA) viruses. In this case, one of the strands of the viral genome (dark blue) is a messenger RNA strand. It is immediately used to make viral capsomere protein which can form a capsid. This same strand is then used to make the complementary second strand of the virus (light blue) and the double strands are incorporated into capsids to form new virus.

(+) Single-stranded RNA (+ssRNA) viruses. In this case, the (+) strand is essentially already mRNA and as such it can be directly translated into viral proteins by the host cell's ribosomes. Genome replication takes two steps. First the (+) strand is copied into a (–) strand. Then this (–) strand is used as a template to produce more (+) strand genomes to place into new virions (**Figure 12.13**).

viruses with +ssRNA genome

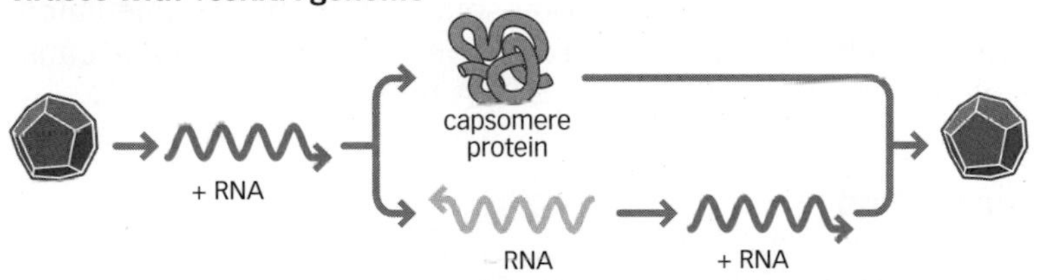

12.5

Figure 12.13 Replication and transcription strategies of (+) strand single-stranded RNA (+ ssRNA) viruses. Recall that single stranded RNA viruses are designated (+) or (–). This is an illustration of a + ssRNA virus. In the case of (+) stranded RNA viruses the genome strand (dark blue) is a messenger RNA which can immediately be used to make capsomere proteins. This strand is also used to make a template strand (light blue) which can be used to produce more (+) strands that will be incorporated into new virus.

(–) Single-stranded RNA viruses (– ssRNA). In viruses that contain (–) strand RNA, things are more complicated. This (–) strand cannot be directly used as mRNA. Therefore, it must first be copied into a (+) strand by viral RNA polymerase, which is brought in with the virus. This (+) strand copy is mRNA and can be used for the synthesis of viral proteins. The production of genomes for new virions goes through a two-step sequence in which the (–) strand is copied to a (+) strand and this (+) strand is then used as the template for new (–) strand genomes that are packaged into new virions (**Figure 12.14**).

viruses with –ssRNA genome

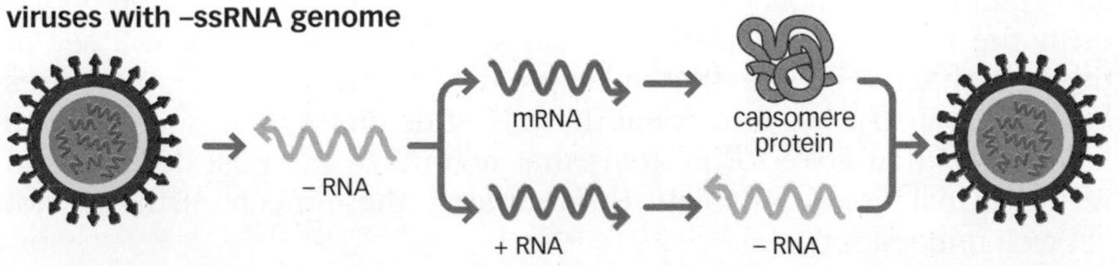

12.6

Figure 12.14 Replication and transcription of (–) strand single-stranded (– ssRNA) RNA viruses. In this case, the (–) single strand (light blue) is not a messenger RNA. Therefore a template (dark blue) must be made in order for capsomere protein to be made. This template is also used to make new genomic (–) strand RNA to be placed in the capsid of new virus.

Fast Fact

Reverse transcriptase was first discovered independently by David Baltimore and Howard Temin in 1970, and both men were subsequently awarded the Nobel Prize in Physiology or Medicine for this work.

Retroviral Transcription and Integration

Retroviruses are RNA viruses that contain the enzyme reverse transcriptase (see **Figure 12.9**). This enzyme is an RNA-dependent DNA polymerase that has the capacity to go backwards! Most DNA polymerases transcribe DNA into RNA, but reverse transcriptase is able to reverse the process and transcribe viral RNA into DNA. Because this enzyme is carried within the retrovirus, it is ready to use immediately on infection of a host cell.

The two RNA strands in the retrovirus are held together by multiple regions of complementary base pairing, most of which occurs at the ends of each of the RNA strands. The viral RNA is complexed with two molecules of a specific host cell tRNA that serves as a primer for the initiation of reverse transcription. It turns out that only certain host cell tRNAs can serve this initiation function and only these tRNAs are assembled into new virions as the infection proceeds.

There are between 50 and 100 copies of the reverse transcriptase enzyme in one retrovirus virion. Because it is carried in with the virus during infection, it can begin to function immediately after the virus has uncoated and nucleotides have become available. Reverse transcription occurs in the cytoplasm of the host cell, and the newly copied DNA molecules are transported from the cytoplasm into the nucleus.

About 4–8 hours after the initial infection of the host cell, newly synthesized viral DNA transcripts have begun entering the nucleus, and integration of this DNA into the host cell's chromosome commences immediately. After the viral RNA has been copied, it is automatically degraded because there is no longer any need for the viral RNA template.

Keep in Mind

- RNA virus replication is more complicated than DNA virus replication.
- RNA viruses have either single-stranded or double-stranded RNA.
- Single strands can be (–) or (+).
- In all cases, RNA viruses use a template strand of RNA to make new viral genomes.
- Retroviruses are RNA viruses that contain the enzyme reverse transcriptase.
- Reverse transcriptase can convert RNA into DNA, which can than be integrated into the host cell's chromosomes.

Viral Control of Translation

Viruses are completely reliant on the host cell's translational machinery for translating mRNA templates to proteins. In fact, viral infection often results in modification of this machinery so that viral RNA is translated preferentially.

During viral infections, the host cellular translation apparatus is altered. Alterations in the translation apparatus of an infected host cell can signal host defenses that a viral infection is under way. Two host genes code for enzymes that prevent any RNA, whether viral or host, from associating with ribosomes. This effectively stops all translation. Obviously the cell will die without protein synthesis, but, as we will see in the next chapter, cell death may be preferable to viral infection. Viruses have developed ways to combat these host defenses, and many viruses have genes that code for proteins that neutralize the host defense and restore translational capability. Unfortunately, these mechanisms are not yet well understood.

Maturation

The next step in the infection cycle is maturation. Up to now we have seen how the viruses attach and penetrate host cells and how new components (capsids and genomes) are synthesized. In this section, we will see that the maturation step of the infection cycle involves the movement of newly made viral components to specific sites in the host cell (referred to as *intracellular trafficking*) where the *assembly* of the new virions takes place.

Intracellular Trafficking

Some viral components are synthesized in the host cytoplasm, and some are synthesized in the host nucleus. They then have all to be brought to one site for assembly into new virions. Before looking at the assembly step, let's look at some general features of intracellular trafficking and follow some viral components as they move around in a eukaryotic host cell.

If we consider the distances that newly synthesized viral components must travel in human terms, that travel could involve many "cell miles" to get to assembly sites. Simple diffusion mechanisms are not able to cover these distances. Instead, newly synthesized viral components are transported through the cell by means of microtubules, a process that requires a considerable expenditure of energy. However, this is of no concern to the virus because the energy is all supplied by the infected host cell!

Assembly sites in the cell are determined by several factors, such as whether the virions will have an envelope, the type of genome (DNA or RNA) and the mechanism of genome replication. Because viral envelopes are derived from the host plasma membrane (modified by the insertion of viral proteins), many enveloped viruses assemble at sites adjacent to the plasma membrane. Other viral envelopes are derived not from the host's plasma membrane but rather from organelle membranes, and in these cases the virion's assembly site is near the organelle.

In contrast to enveloped viruses, non-enveloped viruses assemble in the host nucleus. Therefore, all of the structural proteins for these virions must be transported from the cytoplasm, where they are constructed, to the nucleus. Viral structural proteins seem to enter the host nucleus through cellular pathways that are normally used to import host nuclear proteins.

Viral proteins travel from their site of assembly to the cell surface through a series of membrane-enclosed compartments and vesicles. The first step is the transport of correctly folded viral proteins from the ER to the Golgi apparatus (**Figure 12.15**). In the Golgi, the viral proteins are sorted according to "delivery addresses" in their protein sequences, using transport vesicles that bud from one compartment and move to the next. Because of the fusion that occurs between phospholipid bilayer membranes, it is easy for viral components to move in vesicles and use the fusion with other membranes to release these components into the lumen. When the transport vesicle encounters its target structure, docking occurs.

Fast Fact

Vesicular movement from compartment to compartment means that viral proteins are never again exposed to the cytoplasm of the host cell, thereby avoiding cytoplasmic enzymes that might degrade them.

Just as protein components are synthesized and transported to assembly sites, the newly synthesized viral genomes must also be moved to the site of final assembly. For enveloped viruses, this site is near the host plasma membrane; for non-enveloped viruses, it is at the host nucleus. During this movement, the newly constructed genome becomes dormant because it must not be allowed to start transcription or replication mechanisms. Doing so would either slow down or inhibit the assembly of new virions.

Packaging of most DNA viral genomes takes place in the nucleus of the infected cell. However, for some DNA viruses and many RNA viruses, synthesis of the viral genome and assembly of the virions take place in the cytoplasm.

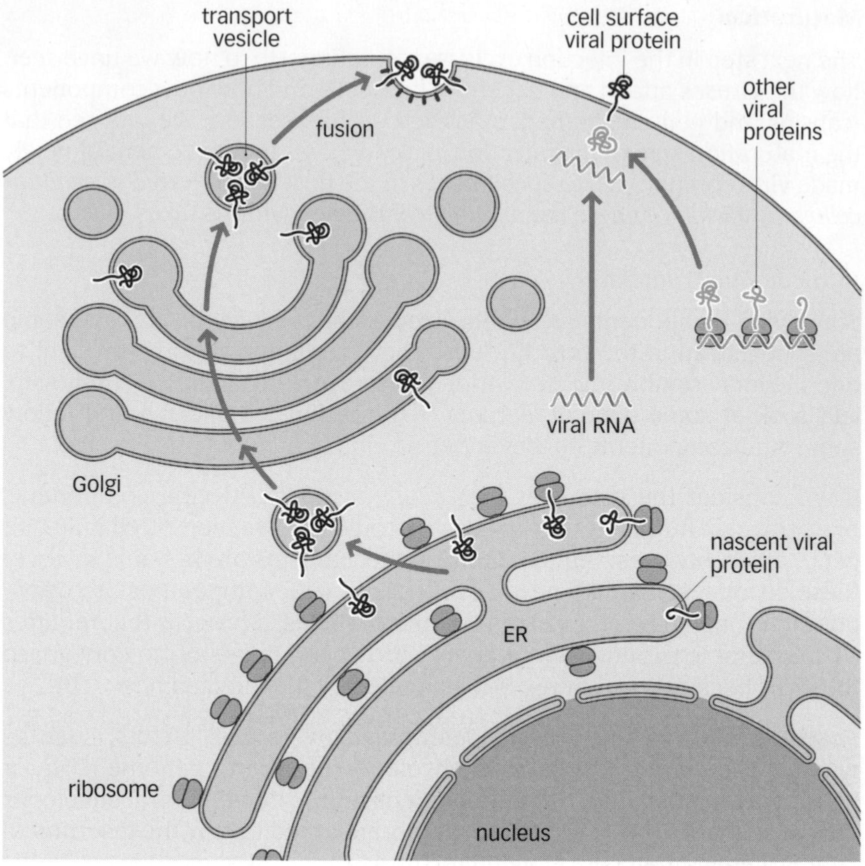

Figure 12.15 A diagrammatic illustration of localization of viral proteins to the plasma membrane. The viral glycoproteins (in red) are translocated from ribosomes into the lumen of the endoplasmic reticulum (ER). They then travel to the plasma membrane by vesicular transport and are united with other viral proteins and new viral genomes.

Keep in Mind

- Viruses are completely reliant on host cell machinery for translation.
- During infection, translation of host cell proteins is suppressed and viral RNA is preferentially translated.
- Intracellular trafficking is crucial for viral reproduction.
- Intracellular trafficking requirements can be quite complex, with transport of viral macromolecules over long distances in the cell occurring at different times during the infection cycle.
- Assembly of different viral components occurs at different sites and requires that all viral components be sorted.
- Some viruses interfere with normal transport of proteins in the host cell and completely disrupt the host cell's secretory pathways.
- Perhaps the most important consequence of viral transport is the formation of microenvironments inside the host cell that contain high concentrations of structural proteins and viral genomes. This leads to maximal assembly of new virions and continuation of a progressive infection.

Assembly

All virions must complete a common set of assembly reactions to ensure reproductive success. For non-enveloped viruses these reactions are:

- Formation of structural subunits for the capsid
- Assembly of the capsid
- Association of viral genome within the capsid

For enveloped viruses (**Figure 12.16**), the sequence is:

- Formation of structural subunits for the capsid
- Assembly of the capsid
- Association of viral genome within the capsid
- Assembly of viral envelope glycoproteins

Assembly is a remarkable process that requires exquisite specificity as well as the coordination of multiple reactions. It must be efficient and irreversible. The structure of the virion determines how it is assembled and also affects the mechanism of entry into the next host cell.

Assembly of capsid subunits. Recall that the viral capsid is made up of protein molecules called capsomeres. These capsomeres are assembled first, and there can be differences in how this process works between RNA viruses and DNA viruses. There are several mechanisms for forming these structural subunits, and in some cases other proteins may be involved in the process. Some viruses use common cellular techniques to produce capsid protein subunits that then fold into the proper three-dimensional structure.

Interestingly, the number of subunits produced is always far in excess of the number required. This is because the subunits must "find" one another, and this "search" has to occur in the host cell's cytoplasm, which is filled with irrelevant host proteins. Remember, if the capsids cannot be assembled, the infection will fail. The chances of random interactions between viral subunits are increased by locating the production of capsid subunits at distinct assembly sites and producing more of the subunits than are necessary. Together these increase the chances for successful virion assembly.

In some viruses, the assembly of capsids can be assisted by host chaperone proteins. These proteins facilitate viral protein folding by preventing the improper association of subunits. In some cases, these chaperones participate in the formation of the capsid subunits.

Assembly of the viral genome. Perhaps the most important part of assembling a new virion is the placement of the viral genome inside the capsid. If this step is either incomplete or inaccurate, the potential for continued infection is compromised. There are two assembly mechanisms, *concerted assembly* and *sequential assembly*. In **concerted assembly**, the virion is assembled while the viral genome is being synthesized (the two events take place *in concert*). In **sequential assembly**, the viral genome is inserted into already-formed capsids. This latter type of assembly requires a mechanism in which the genome can be pushed or pulled into the capsid. It also requires a portal of entry into the capsid. When you factor in errors in capsid development or difficulties in pushing or pulling the genome into the capsid, it is easy to see that some capsids wind up empty. It turns out that these empty capsids have an important role as a viral defense against the host immune response.

One important question about filling capsids is this: how does the capsid recognize viral nucleic acid in a sea of host nucleic acids? The answer is that the viral nucleic acid can be identified by a specific nitrogenous base sequence referred to as the packaging signal.

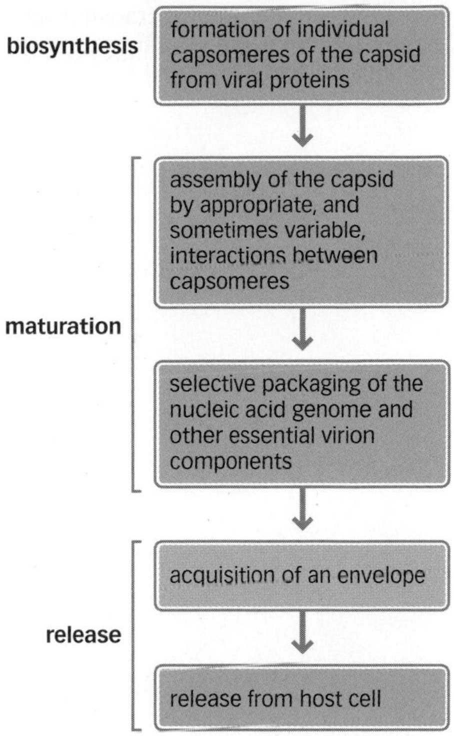

Figure 12.16 A hypothetical pathway of virion assembly and release of an enveloped virus.

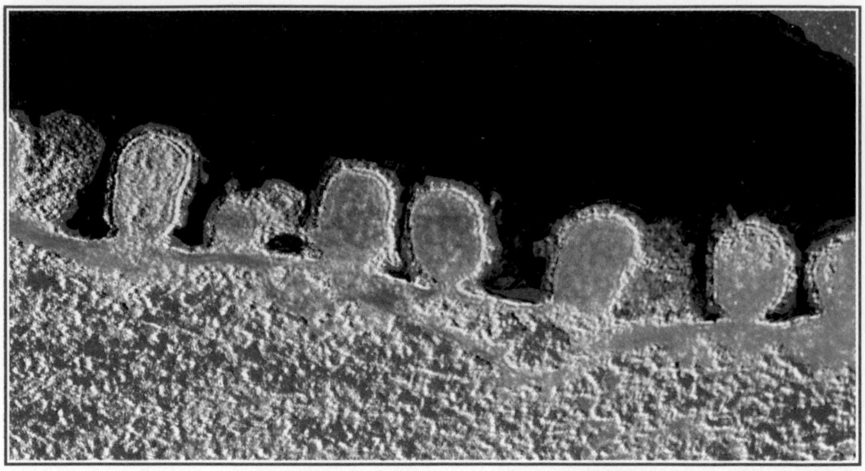

Figure 12.17 A colorized transmission electron micrograph of influenza virus budding from the host cell.

Release

New virions are released from the host cell in one of two ways: by lysis or by budding off from the host cell (**Figure 12.17**). When release is by lysis, the host cell dies immediately. When release is by budding, the host cell remains alive for a while and intermittently releases newly formed virions.

For non-enveloped viruses, release usually results in the death and destruction of the host cell, and some of these viruses will use viral proteins to kill their host. One example is poliovirus, which uses a viral protein that increases the permeability of the host cell membrane and allows an enhanced release of virions. Other viruses will actually cause the collapse of the cytoskeletal proteins, disrupting the integrity of the plasma membrane. Because the viruses shut down cellular biosynthesis at the start of the infection, the host cell cannot repair the damaged cytoskeletal structures.

Budding from the Plasma Membrane

The mechanisms by which enveloped virions bud off from a host plasma membrane are not completely understood. However, it is known that this process has four steps: bud formation, bud growth, fusion of the bud membrane, and separation from the host. In the formation and growth steps, the viral bud forms and begins to push the host plasma membrane out. Many viral proteins are required to make this happen, but the exact mechanisms are not understood.

Membrane fusion reactions pinch off the viral bud from the host plasma membrane. This is also a complex mechanism that is not well understood, but it has been shown that host proteins are involved.

Last, in the least understood of these processes, the viral bud uses what seem to be exocytosis mechanisms to free itself from the infected host cell. Suffice it to say that this simple process of exiting from the infected host cell without killing it involves a complex series of reactions that include both viral and host cell components.

It is important to point out that in some viral infections, virions are completely assembled but remain non-infectious. These virions are referred to as *immature*, and proteolytic enzyme processing must be employed to convert them to mature infectious virions. These conversion reactions are performed by virally encoded enzymes and take place either very late in the assembly process or after the release of the virions from the host cell.

Fast Fact

Proteolytic proteins responsible for the conversion of immature virions into mature infectious virions are very good potential targets for antibiotic therapy because without their function, viral particles will remain immature and non-infectious.

Keep in Mind

- All virions complete a set of assembly reactions.
- Capsomere proteins are assembled first.
- The number of capsomeres produced is always far in excess of the number required for the number of virions to be assembled.
- Some viruses use host cell proteins called chaperones to assemble viral capsids.
- Viral genomes are added to virions that are being assembled in one of two ways: concerted assembly or sequential assembly.
- New virions are released through lysis, which kills the host cell, or by budding from the host cell, which allows the host cell to survive for a period.
- Some viruses are released in an immature non-infectious state and must be activated enzymatically before they can infect host cells.

SPREAD OF VIRUSES

Before we leave the infection cycle, it is important to discuss how viruses spread from host cell to host cell. Remember, viruses have only one goal — to continue the infection. For a newly made virion, the next host cell target can be "right next door" (a neighboring cell) or "body miles" away. In either case, as we have seen, virions are built to withstand the trip. However, they are susceptible to host defense mechanisms along the way. Exposure to host defenses does not occur if the next host cell is next door (**Figure 12.18**). A good example of this neighborliness is the herpesvirus, which seems to spread from cell to cell in epithelial tissue and also between neurons, using specialized cellular junctions called tight junctions.

Viruses can also spread through the formation of **syncytia** (**Figure 12.19**), which are multinucleate masses formed by the fusion of many infected cells into one gigantic cell. A syncytium allows virions to move to places far from where they were synthesized without leaving the confines of the infected syncytium and being exposed to host defenses. The presence of a syncytium is easily visible by light microscopy and can be used as a diagnostic tool for the identification of virus-infected cells.

In another very clever trick, some viruses produce decoy virions, which are either empty capsids or non-infectious virions, and release large numbers of them from infected cells. The decoys confuse and distract the host defenses so that the real virus particles can achieve infection of new host cells. In the same way, some viruses keep certain host cell proteins, such as major histocompatibility proteins, in the virion as a type of camouflage.

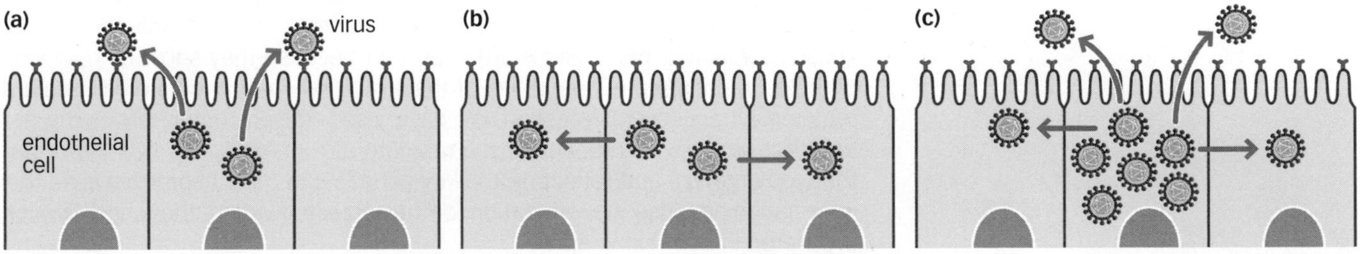

Figure 12.18 The spread of viral particles. Panel **a**: The movement of virions through the apical surface of the host causes infection of neighboring cells. Panel **b**: The spread of viral particles from the basolateral surface of the host cell can cause infection of neighboring cells and can also keep the virus from being identified by the host's immune system. Panel **c**: Some viruses use both extracellular and cell-to-cell contact.

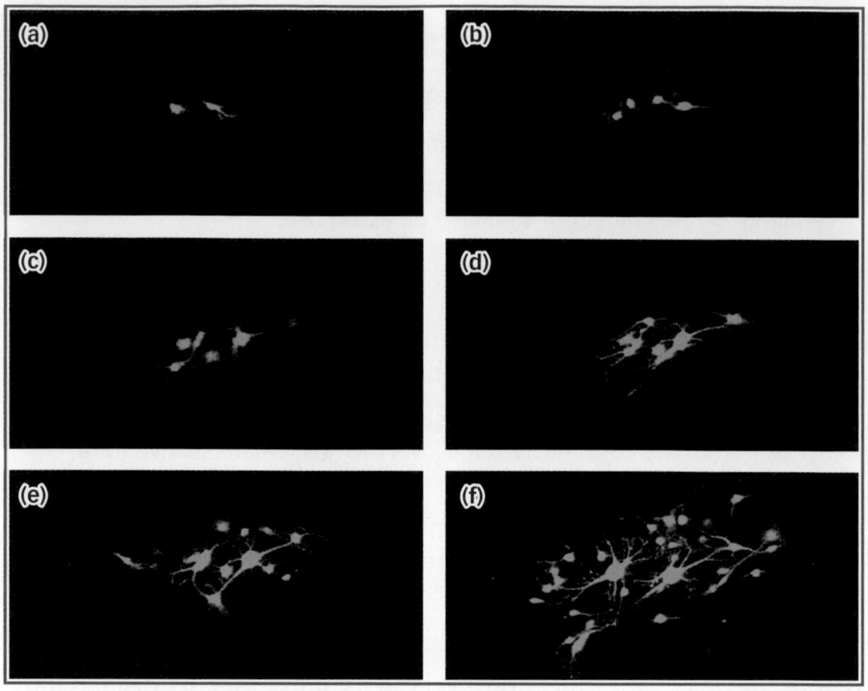

Figure 12.19 The cell-to-cell spread of virus through the formation of syncytia. Notice that, over time, the amount of virus (shown in green) becomes more and more noticeable. This is because the generation of syncytia (giant cells filled with virus) allows adjacent cells to become infected and incorporated into the growing syncytia. Notice that by the 87th hour (last frame), the infection has become widespread. Remember, the formation of syncytia allows the virus to be protected from discovery by the host's immune system.

SUMMARY

- Viral structures include nucleic acids (either DNA or RNA), and a capsid protein coat made of capsomere subunits.
- Some viruses are surrounded by envelopes composed of viral glycoproteins, oligosaccharides, and host cell membrane lipids.
- Viruses may lyse the host cells by using a lytic infection cycle or they may employ a lysogenic infection cycle in which the viral genome is incorporated into the host cell's DNA, resulting in a latent infection.
- A lytic infection cycle involves the steps of attachment, penetration, uncoating, biosynthesis, maturation, and release.
- Host cell receptors are used to facilitate virus attachment and entry into host cells.
- Synthesis of new viral components involves complex replication mechanisms of the viral DNA or RNA.
- Intracellular trafficking and assembly of viral components in the host cell occur during the maturation stage of the viral infection cycle.
- Newly assembled viruses may be released through lysis of the host cell or by budding from the surface of the host cell.

In this chapter, we have focused on viral structure, the mechanisms by which viruses are constructed, and the events associated with the infection cycle. These events and mechanisms are important because they show us how viral particles are made and how many of the structural components of viruses are required for successful infection. We have also referred repeatedly to the notion that productive infection is the ultimate aim of any virus. It is important that you organize and reflect on the key concepts in this chapter because this information provides the foundation for the subject of viral pathogenesis, which is covered in the next chapter.

SELF EVALUATION AND CHAPTER CONFIDENCE

Multiple Choice

Answers are given in the back of the book and help can be found in the student resources at:
www.garlandscience.com/micro

1. Viruses contain
   - **A.** Either RNA or DNA
   - **B.** Both RNA and DNA
   - **C.** No nucleic acids
   - **D.** The RNA of host cells they infect
   - **E.** Only proteins

2. Viral structures can contain any of the following except
   - **A.** Capsomeres
   - **B.** A capsule
   - **C.** Viral nucleic acid
   - **D.** An envelope
   - **E.** A capsid

3. Viral envelopes are formed
   - **A.** When the virus synthesizes them
   - **B.** When the virus is first replicated
   - **C.** When the virus leaves the host cell
   - **D.** When the virus enters the host cell

4. Bacteriophage are viruses that
   - **A.** Infect other viruses
   - **B.** Infect human cells
   - **C.** Infect plant cells
   - **D.** Infect bacteria
   - **E.** Are no longer able to infect

5. A lytic virus has infected a patient. Which of the following would best describe what is happening inside the patient?
   - **A.** The virus is causing the death of the infected cells in the patient
   - **B.** The virus is not killing any cells in the host
   - **C.** The virus is incorporating its nucleic acid with that of the patient's cells
   - **D.** The virus is slowly killing the patient's cells
   - **E.** The virus is infecting cells and then releasing only small amounts of virus

6. A lysogenic virus
   - **A.** Kills its host immediately
   - **B.** Forms plasmids in the host cell
   - **C.** Releases massive numbers of virus all at once
   - **D.** Incorporates its nucleic acid with that of the host chromosome
   - **E.** Causes host cell to become lytic

7. Imagine you are a virologist studying how a type of virus attaches to its host cells. You have found a virus attachment area on the host cell membrane that seems to be dense and full of fatty acid molecules. This area on the host cell is most likely to be
   - **A.** A lipid shaft
   - **B.** An envelope
   - **C.** A lipid raft
   - **D.** A spike
   - **E.** A capsomere

8. The common cold virus (rhinovirus) uses which of the following for attachment?
   - **A.** ICAM molecules
   - **B.** Hemagglutinin
   - **C.** Capsids
   - **D.** Capsomeres
   - **E.** Host cell nucleic acids

9. The influenza virus uses which of the following for attachment?
   - **A.** Capsids
   - **B.** Protein coats
   - **C.** ICAM molecules
   - **D.** Hemagglutinin
   - **E.** Host cell nucleic acids

10. Non-enveloped animal viruses penetrate host cells by
    - **A.** Injection of nucleic acid into the host cell
    - **B.** Wrapping themselves in host cell lipids
    - **C.** Endocytosis
    - **D.** Exocytosis
    - **E.** None of the above

11. Enveloped viruses penetrate their host cells by
    - **A.** Endocytosis
    - **B.** Exocytosis
    - **C.** Injection of nucleic acid into the host cell
    - **D.** Fusion of their envelope to the host cell membrane
    - **E.** Uncoating their nucleic acid at the exterior of the host cell membrane

12. During the events associated with attachment and penetration, all of the following are true except
    - **A.** Viruses never share host cell receptors
    - **B.** Virus–receptor interactions facilitate the infection process
    - **C.** Host cell receptors used by viruses represent a large portion of cell membrane proteins
    - **D.** For enveloped viruses, fusion occurs between the viral envelope and the host cell membrane
    - **E.** Fusion is facilitated by specialized fusion proteins

13. Animal viruses can have any of the following types of nucleic acids except
    - A. Double-stranded RNA
    - B. Single-stranded RNA
    - C. Single strands of RNA and DNA
    - D. Single-stranded DNA
    - E. Double-stranded DNA

14. RNA viruses synthesize their nucleic acid in the host cell
    - A. Nucleus
    - B. Cytoplasm
    - C. Ribosomes
    - D. Both A and B
    - E. Both A and C

15. The enzyme reverse transcriptase
    - A. Reverses the sequence of DNA
    - B. Converts DNA to RNA
    - C. Converts RNA to DNA
    - D. Converts messenger RNA to transfer RNA
    - E. None of the above

16. Which of the following enzymes is/are involved with the integration of viral DNA into the host chromosome?
    - A. Reverse transcriptase
    - B. Integrase
    - C. Polymerase
    - D. A only
    - E. A and B

17. Construction of viral proteins is performed by
    - A. The host cell's ribosomes
    - B. Proteins brought into the host cell by the virus
    - C. A combination of host cell and viral proteins
    - D. The Golgi apparatus of the host cell
    - E. All of the above depending on the virus

18. The assembly of virus requires all of the following except
    - A. Formation of capsomeres
    - B. Formation of capsids
    - C. Formation of viral ribosomes
    - D. Association of viral genomes with the capsid
    - E. Coordination of multiple assembly reactions

19. Newly formed virus can be released from the cell by which of the following?
    - A. Budding
    - B. Cell division
    - C. Cell lysis
    - D. A and C
    - E. A only

20. Viruses can spread through which of the following methods?
    - A. Fusion of the host cell with a neighbor cell
    - B. The formation of syncytia
    - C. The rupture of the host cell
    - D. All of the above
    - E. None of the above

DEPTH OF UNDERSTANDING

Questions listed here require you to bring together the concepts you have learned in this chapter into a discussion format. This helps you to increase your depth of understanding of the material you have learned. Help can be found in the student resources at:
www.garlandscience.com/micro

1. Discuss the structures associated with an enveloped virus and how each has a role in the infection process.

2. One of the mechanisms used to design vaccines against viral infections is to introduce free host cell receptor molecules. How would this influence the viral infection cycle?

3. Compare and contrast the lytic and lysogenic infection cycles.

CLINICAL CORNER

Help can be found in the student resources at:
www.garlandscience.com/micro

1. Imagine you are administering an admitted HIV patient's drug cocktail treatment for the day. The patient has become curious about how the drugs he's taking actually help him. He points to one of the drugs and says he read on the Internet that it is a reverse transcriptase inhibitor but doesn't know what that means.

   A. Explain to him what reverse transcriptase is.
   B. Then explain what the reverse transcriptase inhibitor does.
   C. Explain what overall effect this will have on the virus.

2. Your friend is feeling the onset of a cold and runs to the store to buy some zinc nasal spray that she saw in an advertisement. She hopes that it will work. She calls you to ask about what she is reading on the box, because you are the only person she knows in the medical field. It says that the zinc in the spray has been clinically tested, and "binds to ICAM-1 receptors on cells" to block the cold virus and "shorten the duration" of the infection.

   A. On the basis of this brief description, which stage in the virus infection cycle is this treatment targeting?
   B. Explain why this treatment may actually shorten the time for which your friend is sick.

Chapter 13

Viral Pathogenesis

Why Is This Important?

Because viral infections are so prevalent in humans, health care professionals must understand the pathogenic mechanisms used by these pathogens.

OVERVIEW

Recall from Chapter 12 that most infections are caused by viruses. In that chapter, we learned about the mechanisms responsible for the viral infection cycle. With this information in mind, we can now turn our attention to viral pathogenesis. It is important to keep in mind the events associated with the infection cycle from the last chapter as we look at viral pathogenesis.

We will divide our discussions into the following topics:

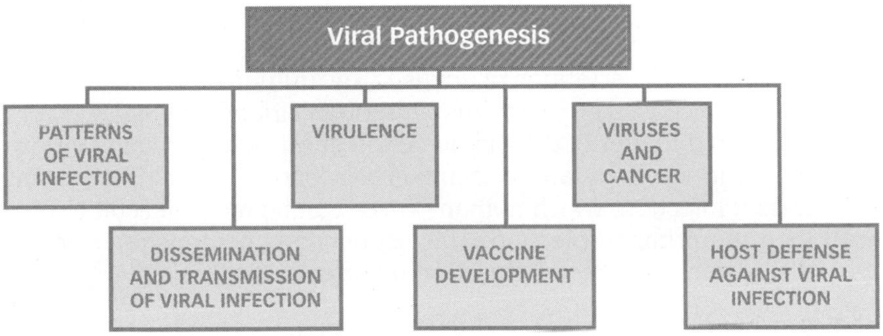

What Do I Need to Know?

To get the most out of this chapter, please review the following terms from your previous courses in biology, anatomy, physiology, or chemistry or in previous chapters of this book as indicated in parentheses: acute infection (**7**), antibody titer, antigen-presenting cell, apical membrane (**12**), apoptosis, basement membrane (**12**), basolateral surface (**12**), choroid plexus, chromatin (**4**), chronic infection (**7**), cytopathic effect (**5**), dermis, diapedesis, endosome (**4**), epidermis, fenestrated, genome (**4**), hepatocyte, herd immunity (**7**), IgA/IgG/IgM immunoglobulins, immunogenic, incubation period (**7**), latent infection (**12**), MHC, persistent infection (**7**), sclera, serotype, vesicle (**4**), viremia (**5**), virion (**12**), virulence (**4**).

PATTERNS OF VIRAL INFECTION

Viral infections can be *acute* (rapid and self-limiting) or *persistent* (long-term). There are various types of persistent infection. For example, *latent* viral infections are essentially extreme versions of persistent infections. In contrast, *slow* and *transforming* infections are more complicated types of persistent infection.

Some viruses kill their host cells rapidly to yield the maximum number of virions for continuation of the infection. These viruses are referred to as **cytopathic viruses** because they cause cytopathology in the host. Other viruses produce virions but do not cause cytopathology; these are called **non-cytopathic viruses**. Still others produce no virions and no cytopathology even though there is an infection.

There is considerable variation in incubation (the appearence of symptoms) periods among viruses, with some periods as short as days and others, such as HIV, as long as years (**Table 13.1**). During the incubation period, the virus is replicating and the host is beginning to respond.

Table 13.1 Incubation Periods of Some Common Viral Diseases

Viral Disease	Incubation Period
Influenza	1–2 days
Common cold	1–3 days
Acute respiratory disease (adenoviruses)	5–7 days
Herpes	5–8 days
Enterovirus disease	6–12 days
Poliomyelitis	5–20 days
Measles	9–12 days
Smallpox	12–14 days
Chickenpox	13–17 days
Mumps	17–20 days
Mononucleosis	1–2 months
Hepatitis A	15–40 days
Hepatitis B and C	2–5 months
Rabies	1–3 months
Papilloma (warts)	2–5 months
AIDS	1–10 years

Acute Infections

These infections are the best-understood infections and involve the rapid production of virions followed by rapid resolution and elimination of the infection by host defenses (**Figure 13.1**). Two excellent examples of this type of infection are influenza (caused by influenza virus) and the common cold (caused by rhinovirus). Although virions and infected cells are cleared in a matter of days in acute infection, some progeny virions can remain in the body and disseminate virions to other tissues. This can cause reinfection, which in many cases can be more difficult to deal with than the primary infection. This sequence is demonstrated in the varicella-zoster (chickenpox) infection (**Figure 13.2**).

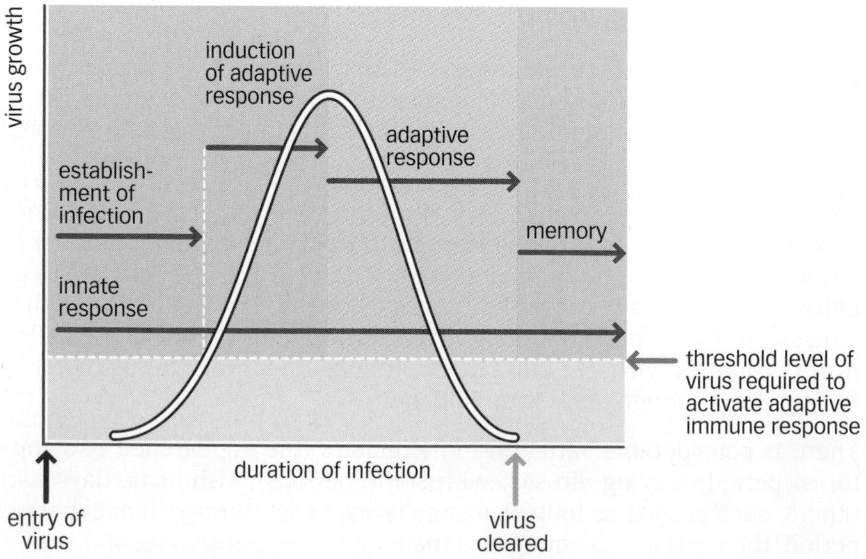

Figure 13.1 A graphic representation of the course of a typical acute viral infection. In this graph, viral growth is shown as a function of how long the virus is in the body. Notice that during the early stages of the infection, only the innate defenses are at work (left area). When the level of virus reaches a certain point referred to as the threshold, the adaptive immune response is mobilized. In acute infections the virus is normally cleared in four to five days. After clearance, there is immunological memory (Chapter 16).

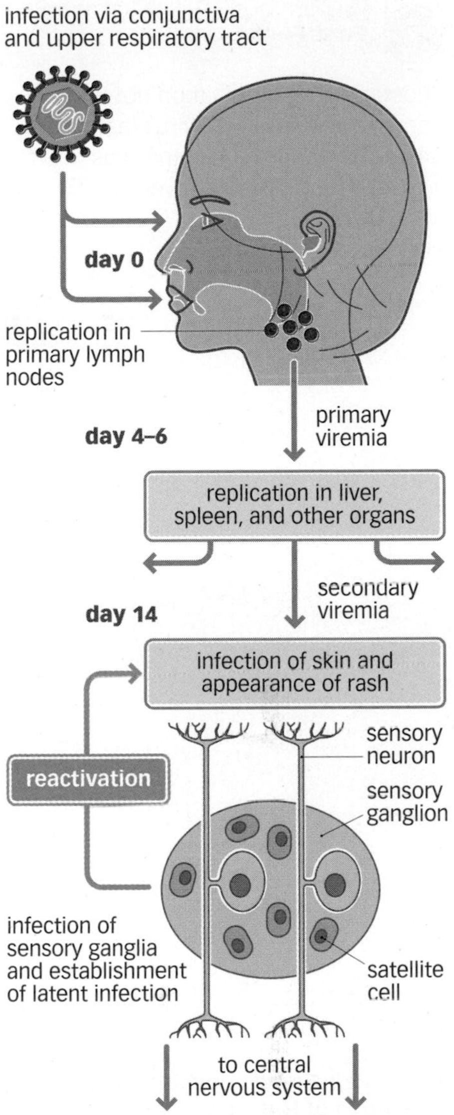

Figure 13.2 A model of viral infection with varicella-zoster (chickenpox) virus. The infection is initiated either in the conjunctiva of the eye or in the mucosa of the upper respiratory tract. It then moves quickly to regional lymph nodes, where it can infect T cells. About four to six days later, infected T cells move into the blood, causing a primary viremia. These cells move to the liver, spleen, and other organs, causing a second round of infection and then back into the blood, causing a secondary viremia. It is during this period that the skin lesions appear. The virus is not finished once the lesions disappear, however. It now moves into sensory ganglia in the peripheral nervous system, where it becomes latent. If reactivated at a later date, the patient will exhibit skin lesions called shingles.

Some acute infections are asymptomatic. For example, 90% of people infected with the polio virus are asymptomatic. In another example, 95% of the population of the United States has antibody titers to varicella-zoster but fewer than 50% of these people have ever had symptomatic disease. This is because the innate immune response limits and contains most acute viral infections.

Fast Fact

Asymptomatic hosts can still transmit the virus to others.

Acute viral infections are severe public health problems and are usually associated with epidemics such as those associated with measles, influenza, and polio. The main problem with these viral infections is the incubation period. In many cases, by the time people feel sick and begin to show symptoms, they have already transmitted the infection. In fact, by the time that symptoms appear in a patient, the infection is essentially over for that patient. This delay in the appearance of identifiable symptoms makes it very difficult to control the transmission of these diseases in crowded populations such as those found in schools, military bases, and nursing homes.

Antigenic Variation

In most cases, hosts that survive acute infections are immune to re-infection for life. That is the reason we vaccinate against many viral diseases. We will see in later chapters that the adaptive immune response comes with memory, and it is this memory that confers long-term protection from viral re-infection. However, some diseases escape this restriction. For example, rhinovirus (which causes about 50% of all common colds) and influenza virus continue to plague us after we have survived an infection. With viruses such as these, re-infection occurs because of structural changes in the virions. Much of the host defense against viral infection revolves around the recognition and destruction of virally infected cells and the elimination of free virions. The adaptive response has, for any given virus, a *specific* memory based on the structure of the virus. Once that structure changes, adaptive memory is no longer effective.

For viruses such as rhinovirus and influenza virus, the virions essentially "change their spots." That is, they use amino acid substitutions to change the protein configuration to a form that the adaptive response has not "seen" and therefore cannot respond to. Antigenic variation can vary within related viruses. For example, both rhinovirus and poliovirus are picornaviruses, but polio does not undergo structural changes. In fact, there are over 100 serotypes of rhinovirus but only 3 serotypes of poliovirus. This is why the polio vaccine made in the 1950s, which comprises all three serotypes of the virus, is still effective today but there is no vaccine for the common cold.

Fast Fact

Substantial recombination between genes of the Influenza virus leads to antigenic shift, which is the reason why we need new flu shots every year.

Changes in virion structure are referred to as **antigenic variation**. We will see in Chapter 16 that any protein that can induce an immune response is classified as an antigen. When the immune system is triggered by a particular antigen, the response against that antigen is very specific, but when antigenic variation occurs, the response to that antigen is changed and becomes either less effective or nonexistent.

Antigenic variation can occur in two forms. **Antigenic drift** involves a slight change in the virion structure, resulting from mutations, and occurs after the infection has begun. **Antigenic shift** involves major changes in the structure of the virion as a result of the acquisition of new genes either from co-infection or through recombination events.

13.1

13.2

Persistent Infections

Persistent infections (**Table 13.2**) are caused when the defenses of the host are either modulated or completely bypassed. In persistent infections, virions or other viral products continue to be produced for months or even years. This type of infection can have two variations. A persistent infection that is eventually cleared is referred to as a chronic infection, and those that last a lifetime are called latent.

Changes in the immune response to a viral infection can induce a persistent infection. For this reason, many individuals who are receiving immunosuppression therapy after they have undergone an organ transplant come down with persistent infections.

It has been shown that a host's immune system responds only to certain viral peptides present on infected cells. Because this response is so narrow, it is easily bypassed, and any change in these immunodominant peptides causes the infected host cells to become essentially invisible to the immune response. Viruses that can mutate their immunodominant peptides are called *cytotoxic T lymphocyte escape mutants*. These mutants are very important in HIV pathogenesis and arise because of error-prone viral replication and the constant selective pressure that comes from exposure to the immune response. If cytotoxic T lymphocyte escape mutations occur early, a persistent infection occurs; if the mutations do not occur early, there is no persistent infection.

Table 13.2 Some Persistent Viral Infections of Humans

Virus	Site of Persistence	Consequence
Adenovirus	Adenoids, tonsils, lymphocytes	None known
Epstein–Barr virus	B cells, nasopharyngeal epithelia	Lymphoma, carcinoma
Hepatitis B virus	Liver, lymphocytes	Cirrhosis, hepatocellular carcinoma
Hepatitis C virus	Liver	Cirrhosis, hepatocellular carcinoma
Human immunodeficiency virus	$CD4^+$ T cells, macrophages, microglia	AIDS
Herpes simplex virus types 1 and 2	Sensory and autonomic ganglia	Cold sore, genital herpes
Papillomavirus	Skin, epithelial cells	Papillomas, carcinomas
Polyoma virus BK	Kidney	Hemorrhagic cystitis
Polyoma virus JC	Kidney, central nervous system	Progressive multifocal leukoencephalopathy
Measles virus	Central nervous system	Subacute sclerosing panencephalitis, measles inclusion body encephalitis
Rubella virus	Central nervous system	Progressive rubella panencephalitis
Varicella-zoster virus	Sensory ganglia	Zoster (shingles), postherpetic neuralgia

Cytotoxic T lymphocytes (CTLs) and CTL escape mutations may also have a significant role in hepatitis C infections. More than 70 million people in the world are infected with this virus, but CTL responses by a host are effective in less than 30% of cases. An insidious persistent infection remains in most of the rest, and after several years this can lead to serious liver damage and often to fatal hepatocellular carcinoma.

Killing of Cytotoxic T Lymphocytes

One of the host defenses against viral infection is the CTL. It is programmed to kill cells that are infected with virus (discussed in Chapter 16). However, a host's CTLs do not function until they have been activated by a host stimulatory signal that is given once an antigen has been sensed. The function of activated CTLs is to kill the invaders, but sometimes the tables get turned, so that when a host's CTLs attack a virally infected target, it is the CTLs that are killed rather than the target. This is a fascinating viral defense mechanism that occurs through a membrane receptor called **Fas** located on the surface of the activated CTLs. Fas is a member of the tumor necrosis factor (TNF) family of cytokines. When Fas interacts with viral components on the infected cell it causes these surface receptor molecules to aggregate abnormally (a process known as trimerization). The trimerization in turn triggers a cascade of events that wind up killing the activated CTLs. Some viruses increase the amount of Fas on the surface of the infected cell as a way of killing host CTLs. The Fas system is a normal component of a host cell and is routinely used to eliminate activated CTLs in the host when the infection is over.

To avoid a CTL response by a host, some viruses infect only tissues that have either limited or reduced immunosurveillance ability. The best example of this is papillomavirus, which causes skin warts (**Figure 13.3**). These infections occur in terminally differentiated outer skin layers, where there is no local immune response. This same strategy is seen in the central nervous system and the vitreous humor of the eye, both of which are areas that are normally sequestered from the immune system.

Latent Infections

Recall that latent infections are one type of persistent infection. Latent infections have three general characteristics:

- Absence of a productive infection, in other words, no large-scale production of virions
- Reduced or absent host immune response
- Persistence of an intact viral genome so that productive infections can occur later

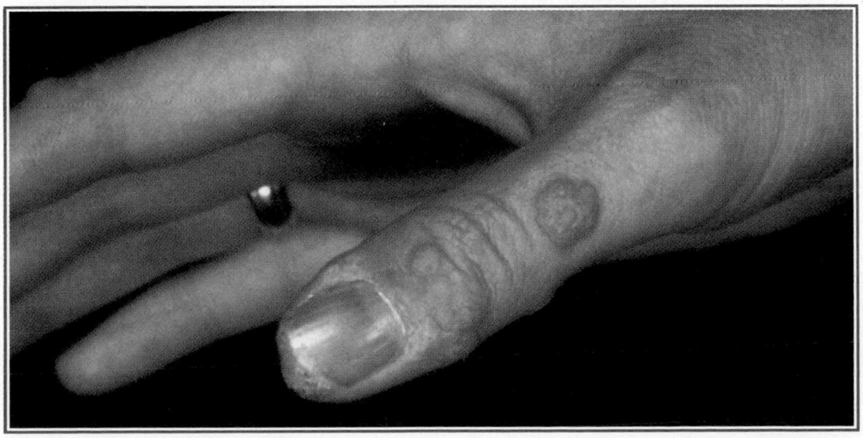

Figure 13.3 Seed warts caused by infection with papillomavirus.

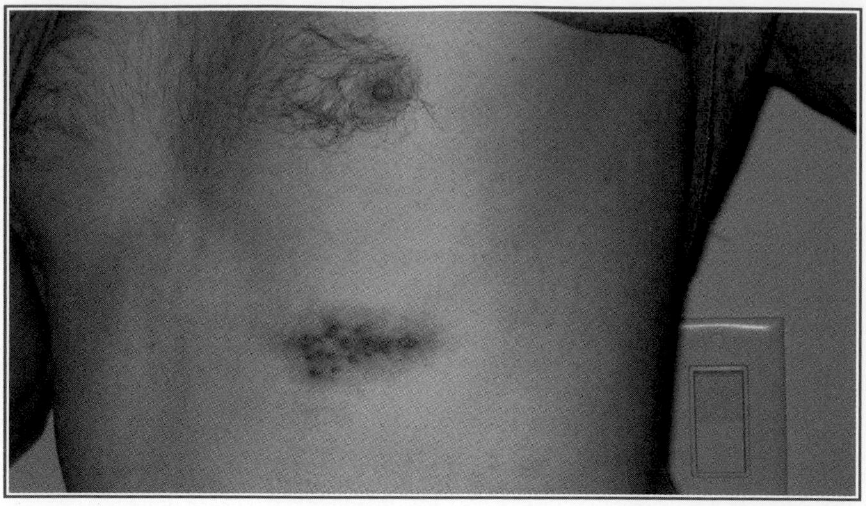

Figure 13.4 Shingles, a reaction to reactivation of latent varicella-zoster virus.

Several viruses are latent, but perhaps the best examples are herpes simplex and varicella-zoster. Both use host neurons as a vehicle for their latency, and the genomes of both of these viruses remain as extrachromosomal elements in the nucleus of infected host cells. It is important to point out that, as we saw in the last chapter, the genomes of some latent viruses can integrate into the host's chromosome.

Latent viruses have the ability to be reactivated years after they enter a host. As an example, varicella-zoster can cause chickenpox in a child, and shingles (**Figure 13.4**) in that same individual many years later. The reasons for reactivation are not completely understood but may have to do with trauma, stress, or any condition that indicates to the virus that the host cell is no longer a suitable place to stay. Reactivation can also be a way of establishing new or improved latency by moving the virus to non-infected neurons.

Slow Infections

These types of infection are lethal. They are usually associated with fatal brain infections and show signs such as ataxia (loss of motor control) and/or dementia. Slow infections are variants of persistent infections and consequently signs may not be seen until years after the primary infection. Once signs appear, death usually follows very quickly. It should be noted that viral diseases such as measles, certain polyoma infections, and HIV can establish slow infections with severe nervous system pathology in the end stages of the disease.

Keep in Mind

- Viral infections can be acute or persistent.
- Cytopathic viruses kill host cells quickly, whereas non-cytopathic viruses do not cause cell death even though virions are being produced.
- Viruses differ in their incubation periods.
- Most acute viral infections result in lifelong immunity.
- The process by which virions change structure is called antigenic variation.
- Persistent infections last for long periods and can be chronic or latent.
- In latent viral infections there is no large-scale production of virus, but these infections can be reactivated later, with large numbers of virions being produced.

DISSEMINATION AND TRANSMISSION OF VIRAL INFECTION

In this section we look at how viral infections are disseminated within the infected host and transmitted from one host to another. There are three basic requirements for successful infection. First, there must be sufficient virus present in the host's body. Second, the virus must have access to susceptible and permissive host cells. When we speak of *susceptibility*, we mean that the host cell must have the appropriate receptor for viral attachment. *Permissiveness* requires that the host cell contain the gene products used by the virus for a successful infection. This is an important consideration because accessibility to nonsusceptible or nonpermissive host cells is not enough to cause infection. Third, the host immune response must be ineffective, at least initially.

Fast Fact

Empty capsids are perfect decoys for the immune system, but they also tie up host-cell receptor sites that the intact virus could use.

In Chapter 12, we commented about hardships that the virus faces during the trip to a potential host. Many viruses avoid these problems by using insect vectors (the mosquito is one example of a viral vector) to avoid exposure to the environment.

So the question is: how many virions in a host are enough? In theory, it takes only one virion to get the infection initiated, but in practice the required number is considerably higher and depends on the virus, the site of infection, and on the age and health of the host. As you learned in Chapter 12, many virions are empty capsids without genomes.

Viral Dissemination

We can use entry points in the host body as starting points for exploring dissemination. Some of the common sites of viral entry are shown in **Figure 13.5**.

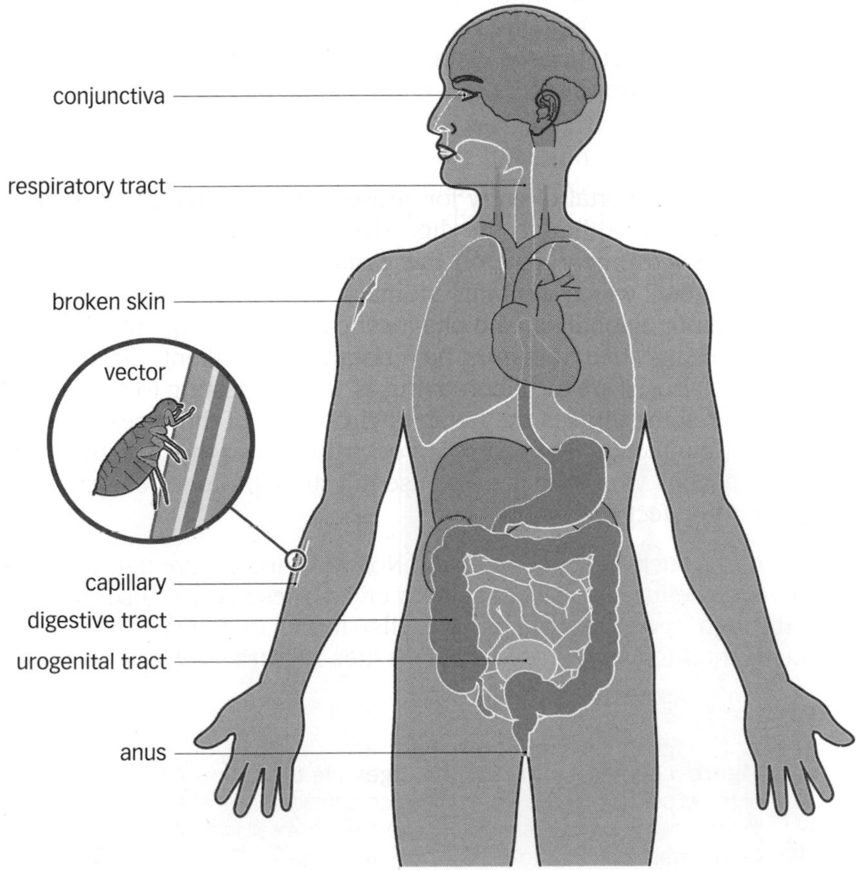

Figure 13.5 Portals of viral entry into a host. Areas that have access to the outside world, such as the respiratory system, make the best portals of entry. Note that the skin is impermeable to virions unless it is broken by events such as a scratch, needle stick, vector bite, or other injury.

Table 13.3 Viral Dissemination Using the Central Nervous System

Pathway	Viruses
From peripheral neurons to neurons of central nervous system	Poliovirus, yellow fever virus, Venezuelan encephalitis virus, rabies virus, reovirus (type 3 only; type 1 disseminated by viremia), herpes simplex virus types 1 and 2, pseudorabies virus
From olfactory neurons to neurons of central nervous system	Herpes simplex virus, coronavirus
From blood to neurons of central nervous system (*hematogenous* route)	Poliovirus, coxsackievirus, arenavirus, mumps virus, measles virus, herpes simplex virus, cytomegalovirus

Nervous System

Because neurons are found throughout the body, many viruses use the neuronal network to disseminate through a host's body (**Table 13.3**). For certain viruses, the neuron is the target. In rabies, for instance, both mature virions and nucleocapsids have been identified in the axons and dendrites of neurons. For other viruses, such as polio, neurons are not the primary target (although polio virions are sometimes found on a patient's neurons). Virions move through neurons via microtubule structures. Obviously, virions disseminated from neurons connected to the spinal cord (viral meningitis) and brain (viral encephalitis) can have devastating effects on the host.

Viruses can also go from the central nervous system into the periphery of the body. When they arrive in the periphery, they cause local infections, such as the **cold sore** seen with herpes (**Figure 13.11**). Fortunately for humans, the direction of viral dissemination is most frequently from the central nervous system to the peripheral system and not the other way around.

Dissemination via Organs

The liver, spleen, and bone marrow are all good candidates for systemic viral infection. In the liver, virions enter the organ via the blood that is being filtered there. The presence of the infected blood leads to viral infection of the liver's Kupffer cells followed by translocation through these cells to the hepatocytes. This process engenders a potent inflammatory response that destroys liver tissue, resulting in hepatitis.

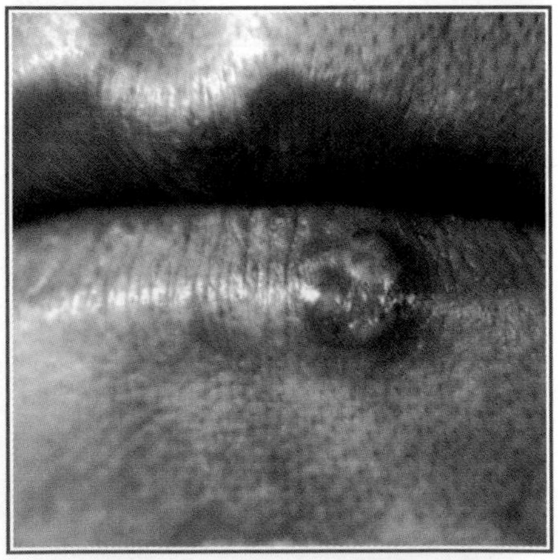

Figure 13.11 The "cold sores" that are a common sign of oral herpes. About 90% of cold sores are caused by herpes simplex type 1 infections.

Recall from your study of anatomy that in some parts of the brain, the capillary epithelium is fenestrated with a sparse basement membrane. Viruses such as mumps virus can pass through this area and move into the choroid plexus, where cerebrospinal fluid is being made. From here, the viruses can be disseminated throughout the central nervous system.

Another dissemination tactic used by viruses is to adhere to blood vessel walls. This way the virions can invade such organs as the pancreas, renal glomerulus, and colon with relative ease. In fact, some viruses — the herpes simplex, yellow fever, and measles viruses are examples — move out of the blood and into the tissues during diapedesis, which occurs as part of the inflammatory response.

Remember that virions (viral particles) can remain localized or can disseminate systemically to other tissues. For example, the rhinovirus begins as a localized respiratory infection, whereas the measles virus moves from the respiratory system to other tissues. When a virus infects other organs it is said to be a systemic virus. Directional movement is an important factor in viral dissemination. So if the virion is released from the apical membrane of the infected host cell, it will establish a localized limited infection. However, if the virion is released from the basement membrane of the cell into underlying tissue, it can spread systemically.

The bloodstream is the best route by which systemic viruses can disseminate; this is referred to **hematogenous dissemination** (**Figure 13.12**). Recall from Chapter 7 that *viremia* is defined as a condition in which virions are present in the blood. If the virions replicate in the blood, the condition is called **active viremia**. Although most systemic viruses spend very little time in the blood (about 1–60 minutes), some, such as hepatitis B and C, can spend years traveling through the blood. This makes sense because the target for these viruses is the hepatocytes of the liver, which is the major blood-filtering organ in the body.

Figure 13.12 Dissemination of virus using the blood. Notice the target organs for certain viruses.

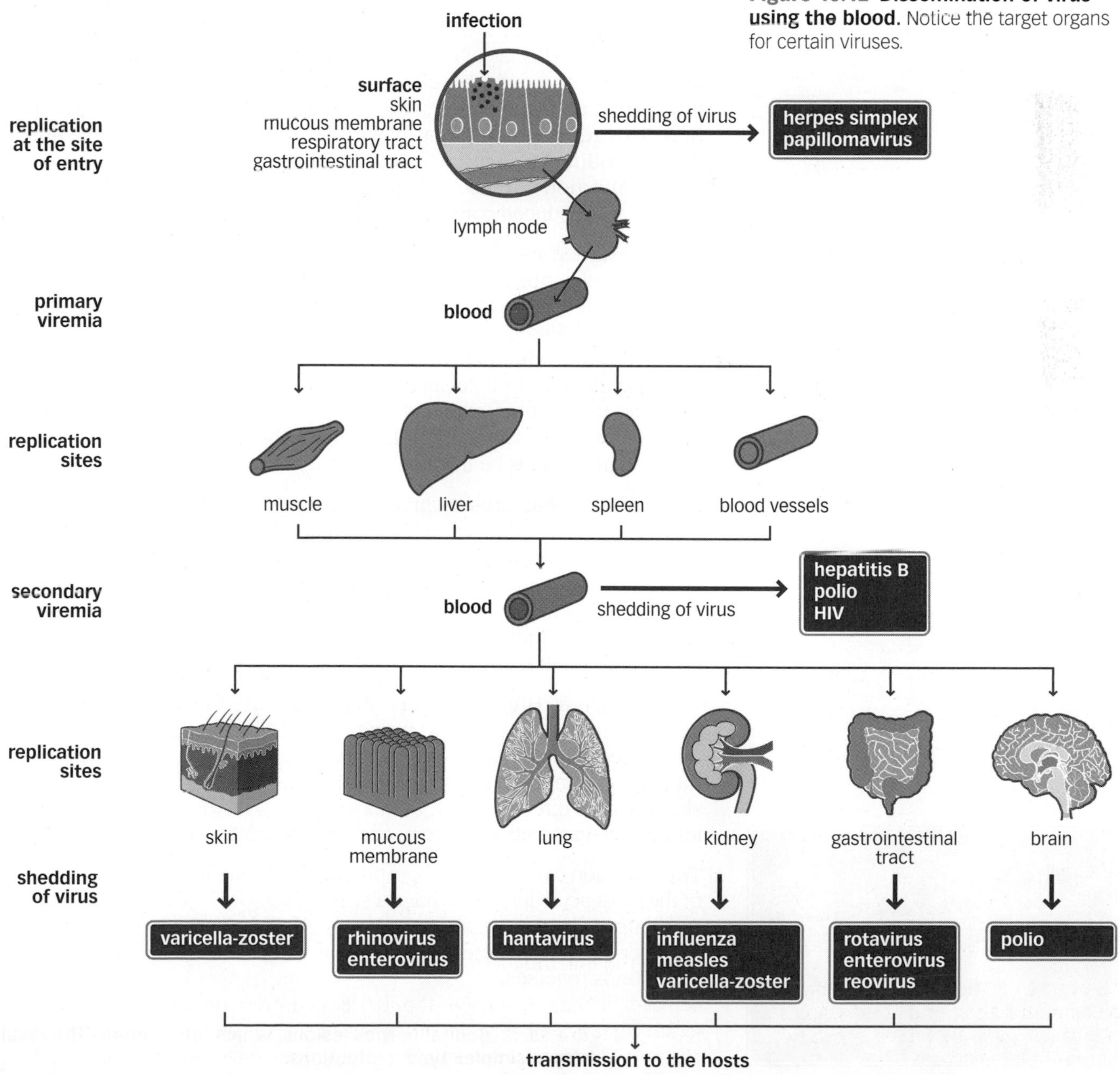

Viral Transmission

Now that we have seen the various ways in which viruses disseminate, we turn next to transmission. Remember, the difference between these two terms is that dissemination refers to spread of the virus through the tissues of the body and transmission refers to spread of the virus from one person to another.

There are two general patterns of virus transmission. The first is perpetuation of the infection in only one species, as is seen with measles and hepatitis A, both of which occur only in humans. The second is perpetuation of the infection by transmission from other animals to humans. Rabies and influenza are two examples of this interspecies pattern. Recall that diseases caused in this manner are referred to as *zoonotic diseases*.

Some viruses that can withstand harsh environments are transmitted by the fecal–oral route of infection, and some are acquired from *fomites*, which are inanimate objects contaminated with disease-causing organisms. (The singular of *fomites* is *fomes*, as in "A computer keyboard used by several persons can easily become a fomes for the rhinovirus.") Transmission can also be facilitated by poor techniques employed by health care workers. This is called **iatrogenic transmission**. Viruses that cause acute infections must be efficiently transmitted. Efficient transmission means that enough virus must be produced to permit a productive infection. This requirement is easily satisfied in acute infections because of the huge numbers of virions released. In contrast, viruses that cause persistent infections do not require efficiency in transmission because the virions are produced continuously over many years.

As you learned in Chapter 12, release is usually required for the transmission and continuation of a productive infection. For respiratory-tract infection, the sneeze is probably the best transmission method; however, transmission also occurs to a smaller degree in coughing, in laughing, and even in normal exhalation. It has been estimated that the volume of air expelled in a single sneeze can contain more than 20,000 droplets, whereas the volume expelled in a typical cough contains only a few hundred droplets. Obviously infectivity in this situation requires that the exposed individuals be in close proximity to the aerosol.

Keep in mind that viruses can also be transmitted on hands, on fomites such as tissues, and in saliva. The fecal–oral route of contamination seen in so many bacterial infections is also a common route for the spread of viral infections, and viral transmission via urine (known as *viruria*) can contaminate food and water supplies. In fact, several human viruses that replicate in the kidneys are shed into the urine, although this is not the most desirable transmission vehicle.

Viruses such as HIV, herpes, and hepatitis B are found in semen and are sexually transmitted. Genital herpes virus (**Figure 13.13**) and papillomavirus infect the genital mucosa, causing lesions that can be transmitted through genital secretions. Viral transmission can also occur from the skin through direct contact, as occurs with poxvirus, herpesvirus, papillomavirus, varicella-zoster virus and even Ebola virus.

Transmission can also be geographically and seasonally influenced. Many of the most acute infections have a striking seasonal variation. For example, respiratory viral infections are seen more often in the winter than in the summer months, whereas for viral infections of the digestive tract the situation is reversed.

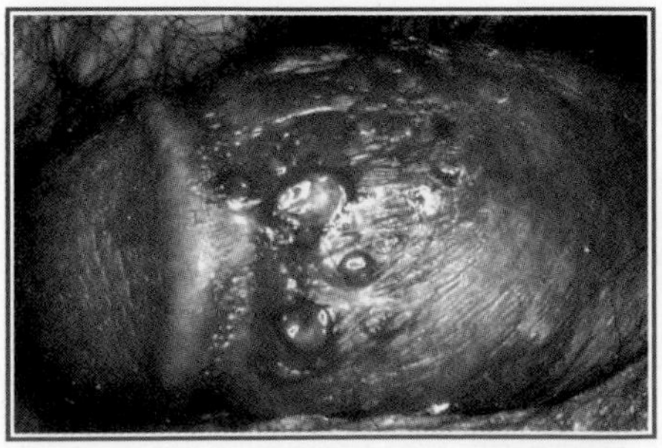

Figure 13.13 Genital herpes lesions, which are primarily the result of herpes simplex type 2 infections.

Transmission via the Respiratory Tract

Respiratory infections can be spread from individual to individual by coughing or sneezing (**Figure 13.14**), as well as by contact with saliva. Large droplets of fluid are found in the nose and smaller droplets in the lungs. When these droplets are expelled from an infected individual, the larger ones fall to the ground more quickly than the smaller ones. However, some small droplets stay airborne for very long periods. Furthermore, inhaling the smaller droplets can increase the risk of a more severe infection because they can more easily find their way into the alveoli of the lungs.

Figure 13.14 The sneeze is the main mechanism by which respiratory tract virus is spread from one person to another. Notice how large the volume occupied by droplets can be. Small droplets can circulate in the air for prolonged periods, increasing the potential for viral transmission.

Transmission via the Epidermis

Many systemic viruses leave a telltale reminder — in the form of a skin rash — of when and where they leave blood vessels (**Table 13.4**). All of the skin lesions seen in systemic viral infections are the result of viral destruction of host cells. Two types of viral marker appear on the skin. The types called macular lesions (red spots on the skin) and papules (small solid elevated lesions) result from inflammation of the dermis. The types called lesions and pustules result from the transmission of virus from the capillaries to the skin surface.

Table 13.4 Viruses That Cause Skin Rashes in Humans

Virus	Disease	Rash Type
Coxsackievirus A16	Hand-foot-and-mouth disease	Maculopapular
Measles virus	Measles	Maculopapular
Parvovirus	Erythema infectiosum	Maculopapular
Rubella virus	German measles	Maculopapular
Varicella-zoster virus	Chickenpox, zoster	Vesicular

Some lesions form in the mucosal tissue of the mouth and throat. For example, measles forms vesicles in the mouth known as Koplik's spots (**Figure 13.15**), which begin to ulcerate before the familiar red-spot skin lesions appear. In fact, by the time spots appear on the skin, the infection is on the wane. This means that by the time a patient develops the spots, they have already transmitted the virus to anyone who has been in close contact and has no immunity to it.

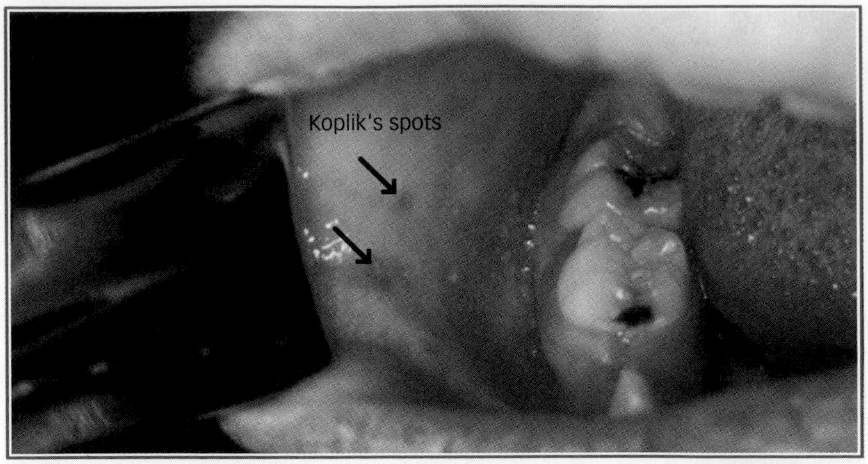

Figure 13.15 Koplik's spots on the mucosa lining the mouth interior are the earliest signs of infection by the measles virus.

Fast Fact

Recent studies have shown that aggressive treatment of infected mothers can dramatically reduce the incidence of babies born with the HIV virus.

Fetal Infection

Viremia in pregnant women can expose the fetus to dangerous virions. With rubella, for instance, the fetal infection rate during the first trimester is more than 80%. About 1 in 5 babies can be infected by HIV *in utero* from infected mothers, and some viral transmission can even occur during breast feeding.

Keep in Mind

- Viruses can be disseminated through the respiratory, digestive, and urogenital tracts as well as the eyes, skin, and nervous system.
- Transmission of the viral infection is required for the infection to be successful.
- For adequate transmission there must be adequate numbers of virions and permissive host cells and an ineffective host defense response.
- Viruses are transmitted through the fecal–oral route of contamination, by fomites, and also by health care workers (this is known as iatrogenic transmission).
- Viral transmission can be influenced seasonally and geographically.

VIRULENCE

Recall from Chapter 4 that *virulence* refers to the capacity of an infectious organism to cause disease. In general, virulent viruses cause significant disease, and non-virulent, or *attenuated*, viruses cause little or no disease. As we will see later in this chapter, attenuated virus can be used for vaccination. There are three ways to measure the virulence of viruses; two of these are LD_{50} (lethal dose 50%), a measure of how much virus is required to cause the death of 50% of infected individuals, and ID_{50} (infectious dose 50%), a measure of how much virus it takes to infect 50% of a population. A third way, called the $\mathbf{PD_{50}}$ (paralytic dose 50%), indicates how much virus is needed to paralyze 50% of infected individuals. Although all these measurements may seem to be somewhat arbitrary, they are important indicators of the potential virulence of viruses.

Virulence varies from one virus to another and can be directly affected by the route of infection and by the age, the health, and in some cases the gender of the host. In addition, alterations in the ability of the virus to replicate can affect virulence. For example, a genetic mutation that decreases the number of virions produced in an infected host cell can lower the virulence of the virus. In addition, viral cytopathology genes can be mutated and turned off. This limits the destructiveness of the virus and thereby lowers its virulence. Last, any change in a host cell's function required by the virus can affect virulence.

Virulence and Host Susceptibility

As we have mentioned previously, virulence is affected by the health of the intended host. There are two types of host, susceptible and immune. A susceptible host is one who can be infected and can also transmit the disease. This will depend on how efficient the virus infection is and how prevalent the disease is in the prospective population. In groups where most individuals have been immunized against a particular viral disease, that disease cannot take hold. As we saw previously, this is referred to as "herd" immunity. In addition, if the potential populations all have healthy immune response capabilities, any infection will be resolved quickly and the transmission of the infection will be limited. This is amply demonstrated in nursing homes, where the population has limited immunity. In this case viral infections that would be easily cleared by younger people spread with devastating quickness.

Interestingly, some viral infections are milder in the elderly. This may be due to physical as well as physiological changes that take place with age. For example, the alveoli of elderly individuals are smaller than those of young people, allowing less area for infection. In addition, the muscles of elderly people are no longer strong enough to propel viral particles long distances during coughing and sneezing.

Fast Fact

In developing countries where many individuals are malnourished and have poor immune function, measles can be 300 times as lethal as it is in the United States. In these cases, the Koplik spots become massive, and mortality can be as high as 50%.

Gender may also have a role because males seem to be slightly more susceptible to viral infections, and pregnant women seem to be more susceptible to hepatitis A, B, and E, for reasons that are not clear. Last, the physical condition of the host seems to have a significant role in viral disease.

VACCINE DEVELOPMENT

As we will see in Chapter 19, there are very few drugs effective against viral infections. Consequently, the most effective strategy in dealing with viruses is prevention through vaccination. With many viral diseases, the memory property of the adaptive immune response translates into lifelong immunity. In addition, immunization increases herd immunity. Together these help to control serious diseases that could easily reach epidemic proportions.

Let us look at smallpox as an example of how vaccination works. This is a disease that was perhaps the most devastating the world has ever seen. It caused the death of more than 300 million people in the twentieth century alone. Yet it was the first disease to be eradicated through vaccination. Edward Jenner, a rural physician determined that, for some reason, milkmaids did not seem to contract smallpox. He noticed that these women had poxlike lesions on their hands, lesions that came from cowpox, a harmless virus related to the smallpox virus. In 1796 — that's right, more than 200 years ago — Jenner took pus from the hand lesions of milkmaid Sarah Nelmes and placed the pus under the skin of a young boy named James Phipps. Then in what would today be considered an unethical, immoral, and definitely illegal decision, he infected young James with smallpox. James did not get sick, however! It was the beginning of the end for smallpox and the beginning of the era of vaccinations for humans.

Fast Fact

Mary Wortley Montagu, after seeing smallpox inoculations in Turkey where her husband was ambassador, brought the idea of vaccination for smallpox back to England and had her children inoculated before Jenner's work took place.

Since Jenner's time, many vaccines have been developed and are routinely administered (**Table 13.5**). Vaccines can be remarkably effective in limiting viral diseases, as demonstrated most convincingly by the polio and measles vaccines, which have essentially eradicated these diseases in the United States (**Figure 13.16**). Unfortunately, vaccines can have side effects that can be unpredictable.

Vaccines can be broadly classified into three groups:

- A **live attenuated vaccine** is made up of intact viral particles that have been rendered non-infectious through some form of treatment (usually chemical). Examples are the MMR (measles, mumps, and rubella) vaccine routinely given to children and the oral polio vaccine. Because these vaccines are composed of infectious virions, there is the potential danger of accidental infection of individuals receiving the vaccination.
- An **inactivated vaccine**, also called a **killed vaccine**, is composed of virus that is either dead or non-infectious. This type is potentially safer than a live attenuated vaccine because there are no infectious virions present, or there are virions present but they have been rendered non-infectious.
- A **subunit vaccine** is composed of immunogenic parts of the virus and is usually derived through the use of genetic engineering and recombinant DNA techniques. This is the safest type of vaccine because there are no intact virions present, but this type can also be less effective than live attenuated vaccines or inactivated vaccines, depending on the viral proteins that have been engineered.

Table 13.5 Viral Vaccines Licensed in the United States

Disease or Virus	Type of Vaccine	Population Vaccinated	Schedule
Adenovirus	Live, attenuated, oral	Military recruits	One dose
Hepatitis A	Inactivated whole virus	Travelers, other high-risk groups	0, 1, and 6 months
Hepatitis B	Yeast-produced recombinant surface protein	Universal in children, exposure to blood, sexual promiscuity	0, 1, 6, and 12 months
Influenza	Inactivated viral subunits	Elderly and other high-risk groups	Two-dose primary series, then one seasonal dose
Measles	Live attenuated	Universal vaccination of infants	12 months; 2nd dose, 6-12 years
Mumps	Live attenuated	Universal vaccination of infants	Same as measles, given as MMR[a]
Polio (inactivated)	Inactivated whole viruses of types 1, 2, and 3	Changing: commonly used for immunosuppressed for whom live vaccine cannot be used	2, 4, and 12–18 months; then 4–6 years
Polio (live)	Live, attenuated, oral mixture of types 1, 2, and 3	Universal vaccination; no longer used in United States	2, 4, and 6–18 months
Rabies	Inactivated whole virus	Exposure to rabies, actual or prospective	0, 3, 7, 14, and 28 days after exposure
Rubella	Live attenuated	Universal vaccination of infants	Same as measles, given as MMR
Smallpox	Live vaccinia virus	Certain laboratory workers; military personel	One dose
Varicella	Live attenuated	Universal vaccination of infants	12–18 months
Yellow fever	Live attenuated	Travel in areas where infection is common	One dose every 10 years

[a]MMR: measles, mumps, and rubella.

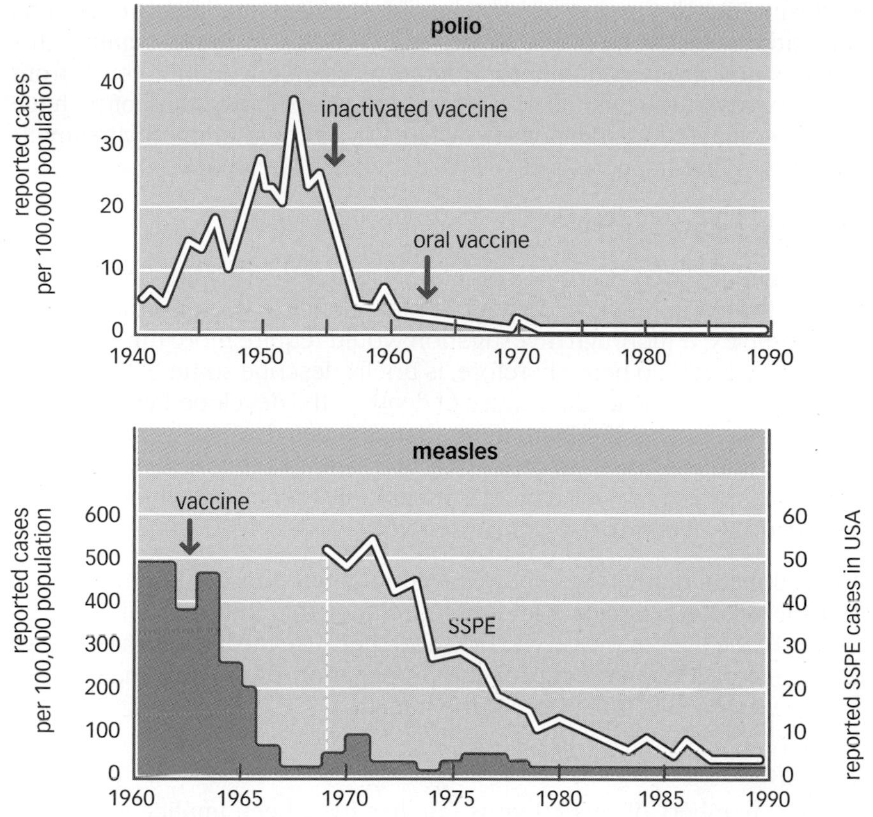

Figure 13.16 Profiles of vaccination success for polio, measles, and SSPE (subacute sclerosing panencephalitis, a specific form of measles) viruses in the United States.

Vaccination can be active or passive. In **active immunization**, the antigen representing the infectious agent is administered and causes the onset of an immune response. In **passive immunization**, an already-formed antiviral product, such as antibody, is administered.

Table 13.6 shows the basic and fundamental requirements for an effective vaccine. It is estimated that the cost of developing an approved vaccine is in the hundreds of millions of dollars. Consequently, there are not many companies willing to risk the cost of development. This problem is compounded by the fact that accidental infection during vaccination can bring about considerable litigation and expense. Therefore, in general, there is little effort to make vaccines for many viral diseases. Another factor is the economics of producing vaccines for diseases that do not affect industrialized nations.

Fast Fact

Economics dictate that if a disease infects only people too poor to pay for vaccination, the vaccine will not be made.

Requirement	Comments
Safety	The vaccine must not cause disease. Side effects must be minimal
Induction of protective immune response	Vaccinated individual must be protected from illness due to pathogen. Proper innate, cellular, and humoral responses must be evoked by vaccine
Practical issues	Cost per dose must not be prohibitive. The vaccine should be biologically stable (no genetic reversion to virulence; able to survive use and storage in different surroundings). Vaccine should be easy to administer (oral delivery preferred to needles). The public must see more benefit than risk

Table 13.6 Requirements of an Effective Vaccine

In closing our discussion of vaccines, let's revisit smallpox. In previous generations, everyone in the United States was vaccinated against this disease, and the resulting herd immunity essentially eliminated it from this country. In the past 30 or so years, however, how many of us have been vaccinated for smallpox? How large is our herd immunity to smallpox now? What happens if the disease makes a comeback?

VIRUSES AND CANCER

Our discussion would not be complete without mentioning how viruses are implicated in the development of malignancies. Because this topic is very complex, a thorough investigation would require more time than we have. All we can do here, therefore, is briefly describe some of the salient points. Cancer is the leading cause of death in the developed world. In the United States alone, there are more than 500,000 deaths a year from this disorder. Some viruses can cause cancer in animals (one example is Rous sarcoma in chickens), but cancer is essentially associated with mutations that result in uncontrolled cellular growth.

It is estimated that viruses are involved in about 20% of human cancers, and there is a clear cause-and-effect relationship between viruses and cancer of the liver or the cervix. However, because the induction of malignancy is not a requirement for the propagation of viruses, malignancy can be viewed as a side effect of the infection.

Oncogenic Viruses

Specific members of several virus families have been implicated both in cancers in humans and in experimentally induced cancers in laboratory animals. Retroviruses, which were initially called RNA tumor viruses, have been shown to be able to inactivate genes that suppress tumor formation.

Fast Fact

There is now a vaccine for HPV infection that is recommended in the United States for young girls; it may prevent the development of cervical cancer caused by infection with this virus.

Several human cancers are associated with infection by one of five viruses: Epstein–Barr, hepatitis B, hepatitis C, human lymphotropic virus and human papillomavirus (HPV). In many of these infections, it seems that viral proteins or transcriptional controls override the mechanisms that normally ensure that cells divide only when necessary. Such overriding leads to the uncontrolled cell growth that we know by the generic term *cancer*.

Keep in Mind

- Virulence refers to the capacity of a virus to cause disease.
- Virulence varies from one virus to another and can be affected by the route of infection, by the age and health of the host, and in some cases by the sex of the host.
- Susceptible hosts can be infected and transmit the disease, but immune hosts cannot be infected.
- Vaccines have been very effective in limiting viral diseases.
- Vaccines can be composed of live attenuated virus, inactivated virus, or subunits of the virion (parts of the virion that can elicit an immune response).
- There are strict requirements for vaccines, including minimal side effects coupled with maximum protection from infection.
- Oncogenic viruses can induce cancer in animals.
- There is clear evidence for human papillomavirus as a cause of cervical cancer.

HOST DEFENSE AGAINST VIRAL INFECTION

13.3

One of the cornerstones of human health is the presence of a fully functional immune system. It keeps us safe from myriad infectious organisms and diseases. This concept is well illustrated in the battle against viral infections. In point of fact, we live most of our lives in a "cloud" of infectious organisms, and it is the combination of innate and adaptive immunity that keeps us safe. The battle between virus and host is a fascinating one in which we humans have developed weapons to defeat viruses, and viruses have developed ways to defeat our weapons. In fact, the genomes of successful pathogenic viruses code for many products that modify or block almost every step we take to defend ourselves. Put another way, for every host defense, there is a viral offense. We examine our defense mechanisms against viral infection in Chapters 15 and 16.

SUMMARY

- Viral infections can be acute or persistent.
- Persistent infections last longer and can be chronic or latent.
- Cytopathic viruses kill host cells when they release new virions, whereas non-cytopathic viruses do not cause cell death even though they are releasing new virions.
- Latent viral infections do not produce large numbers of virions, but these infections can be reactivated later and release virions.
- Viruses can be disseminated (move to other parts of the infected host's body) through the respiratory, digestive, and urogenital tracts as well as the nervous system.
- Transmission of virus from one host to another can be through the fecal–oral route of infection, fomite transmission, and iatrogenic mechanisms (by health care workers).
- Virulence varies from one virus to another and can be affected by the age of the host, the health status of the host, and the route of infection.
- Vaccination has been very effective in limiting viral diseases.
- Vaccines can contain live attenuated virus, inactivated virus, or viral subunits.
- Some oncogenic viruses have been implicated in the development of malignancies.

In this chapter, we have looked at the pathogenesis of viral disease and explored the diverse and complex ways in which viruses can disseminate through the body or be transmitted from one host to another. Keep in mind that pathogenic viruses have only one objective: a productive infection. Even latent and persistent viral infections begin this way, but they produce virions for the life of the host cell rather than all at once at the expense of the host cell. Protection against viral infection is through vaccination, and there are many safe and effective vaccines that are routinely administered as a way of preventing viral infection.

SELF EVALUATION AND CHAPTER CONFIDENCE

Multiple Choice

Answers are given in the back of the book and help can be found in the student resources at:
www.garlandscience.com/micro

1. Viral infections can be any of the following except
   - A. Slow
   - B. Latent
   - C. Acute
   - D. Temporary
   - E. Persistent

2. Acute infections are represented by
   - A. Slow production of virus but rapid resolution of the infection by host defense
   - B. Rapid production of virus and slow resolution of infection by host defense
   - C. Rapid production of virus and rapid resolution of infection by host defense
   - D. Slow production of virus and slow resolution of infection by host defense
   - E. None of the above

3. A person who is asymptomatic for viral infection
   - A. Is not infected
   - B. Is showing symptoms of the infection
   - C. Is infected but shows no symptoms of the infection
   - D. Has only a mild infection
   - E. Has an acute infection

4. Major changes in the structure of the virus are referred to as
   - A. Viral shifts
   - B. Antigenic drift
   - C. Antigenic shift
   - D. Antigenic shuffling
   - E. Viral camouflage

5. Cytotoxic T cells can be killed by
   - A. The virus capsid
   - B. Fas molecules on the virus
   - C. Fas molecules on the infected host cell
   - D. Fas molecules released by the virus
   - E. None of the above

6. Latent infections have all of the following characteristics except
   - A. The absence of a productive infection
   - B. A reduced or absent immune response
   - C. The viral genome remains intact
   - D. The presence of a productive infection
   - E. No large-scale production of virus

7. Latent viruses
   - A. Can never be reactivated
   - B. Occur only in adults
   - C. Can integrate into the host chromosome
   - D. Destroy the host chromosome
   - E. None of the above

8. All of the following help to determine viral pathogenesis except
   - A. Ability of the infection to kill cells
   - B. Interactions of the virus with the target cells
   - C. The host response to the infection
   - D. B and C
   - E. None of the above

9. Which of the following is most important for a successful viral infection?
   - A. Icosahedral capsids
   - B. Permissive host cells
   - C. Viral envelopes
   - D. Genetic variability
   - E. None of the above

10. When virus is released from the basement membrane of a host cell it will
    - A. Cause a localized infection
    - B. Not infect other cells
    - C. Cause a systemic infection
    - D. Be deactivated
    - E. None of the above

11. Iatrogenic transmission of virus is caused by
    - A. Ingestion of virally contaminated food
    - B. Mosquitoes
    - C. Health care workers
    - D. Family members
    - E. Coworkers

12. The target site for rabies virus is
    - A. The digestive epithelium
    - B. The conjunctiva of the eye
    - C. The muscle cells of the heart
    - D. The neuron
    - E. Both A and C

13. Hepatitis is caused by
    - A. Viral infection of keratinocytes

B. Inflammation of the liver in response to viral infection
C. Viral infection of the liver
D. B and C only
E. All of the above

14. Viral shedding is

A. Part of the recovery process
B. Part of the transmission of viral infection
C. Destruction of the viral particle
D. Used to describe the uncoating of a virus
E. None of the above

15. Viral vaccines are usually composed of

A. Virulent virus
B. Non-virulent virus
C. Attenuated virus
D. B and C
E. All of the above

16. PD_{50} defines the

A. Lethal dose of a particular virus
B. Paralytic dose of a virus
C. The pandemic dose of a virus
D. The prior dose of a virus
E. None of the above

17. Virulence of a virus can be affected by which of the following?

A. The route of infection
B. The age of the host
C. The sex of the host
D. The ability of the virus to replicate
E. All of the above

18. Koplik spots are seen in

A. Herpes infections
B. Hepatitis infections
C. Polio
D. Measles

19. The first successfully tested viral vaccine was for

A. Polio
B. Smallpox
C. Rubella
D. Measles
E. Mumps

20. The vaccine that is most potentially dangerous

A. Consists of live attenuated virus
B. Contains killed virus
C. Contains inactivated virus
D. Is a subunit vaccine
E. None of the above

21. All of the following are requirements for an effective vaccine except

A. The vaccine must be safe
B. The vaccine must be made from killed pathogens
C. The vaccine must induce a protective response
D. The vaccine must be stable
E. The vaccine should be inexpensive if possible

22. Oncogenic viruses

A. Cause acute infections
B. Are genetically unstable
C. Cause tumors to develop
D. Are lytic viruses that kill the host cells
E. Do not cause tumors to develop

DEPTH OF UNDERSTANDING

Questions listed here require you to bring together the concepts you have learned in this chapter into a discussion format. This helps you to increase your depth of understanding of the material you have learned. Help can be found in the student resources at:
www.garlandscience.com/micro

1. From the perspective of the virus, which type of infection is more valuable, the acute infection or the latent infection? Defend your choice.

2. Discuss the affects of herd immunity on viral infection and how this affects transmission of viral disease.

3. Using what you have learned, design the best vaccine against viral infection.

CLINICAL CORNER

Help can be found in the student resources at:
www.garlandscience.com/micro

1. Viruses such as smallpox have become of interest to terrorists as a bioweapon. As a result, military personnel are routinely vaccinated against smallpox.
   A. Why would smallpox be a good choice for a bioweapon?
   B. What precautions could be taken to prevent a terrorist attack with these weapons, and would these precautions be effective?

2. Millicent's grandmother has lived at the Shady Grove retirement home for more than six years and Millicent visits her as often as possible. At her last visit she found out that two of her grandmother's friends had come down with viral pneumonia and one had died. Fortunately, her grandmother seemed to be fine.
   A. Should she be worried about her grandmother? If so why?
   B. Should she be worried about her own health?

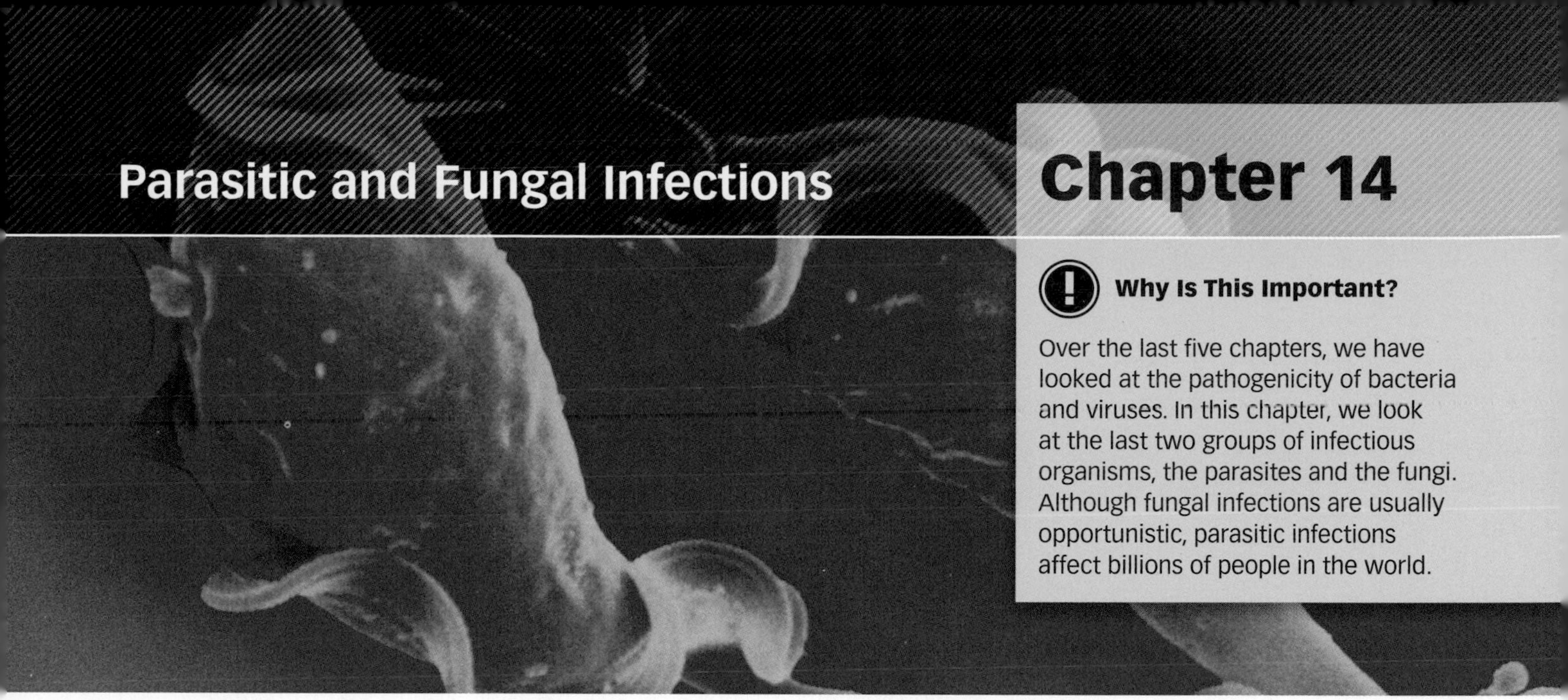

Chapter 14

Parasitic and Fungal Infections

Why Is This Important?

Over the last five chapters, we have looked at the pathogenicity of bacteria and viruses. In this chapter, we look at the last two groups of infectious organisms, the parasites and the fungi. Although fungal infections are usually opportunistic, parasitic infections affect billions of people in the world.

OVERVIEW

In this chapter, we complete our discussion of the organisms that can cause disease in humans. We begin by looking at protozoan parasitic pathogens, which account for millions of infections in the world. The parasite-caused disease malaria, for instance, infects more than 500 million people in the world, and more than 2 million people die from this infection each year. Next, we look at helminths, the parasites more commonly known as worms, and the diseases they cause. Then, in the last part of the chapter, we look at diseases caused by fungi. We have chosen to look at fungal infection last because they are for the most part opportunistic and rarely bother us if we have intact and functioning defenses. However, fungal infections can be very dangerous for individuals who are immunocompromised.

We will divide our discussions into the following topics:

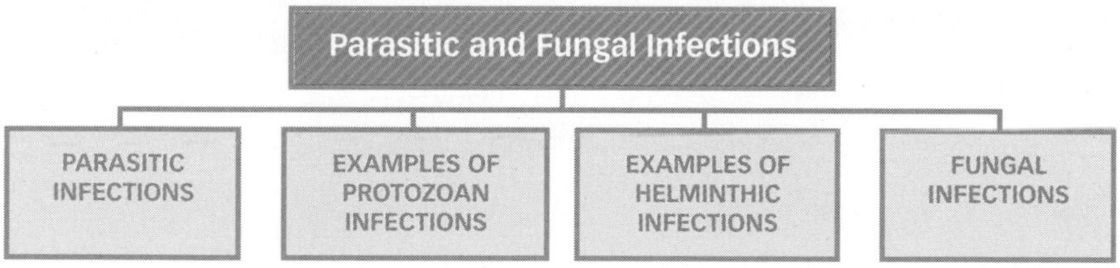

PARASITIC INFECTIONS

Parasites can be divided into two groups, protozoans and helminths. Protozoans, both the parasitic members of this group and the free-living members, are microscopic, single-celled eukaryotes. Helminths, both the parasitic and free-living members, are macroscopic, multicellular worms that possess differentiated tissues and complex organ systems. The helminths called *roundworms* can vary in length from less than 1 mm to a meter or more, and those known as *tapeworms* vary in length from 1 cm to 10 meters or more. Most protozoans and helminths that are free-living (in other words, not parasitic) have an important role in ecology and are incapable of infecting humans. The disease-causing parasites live in synergy with and depend on their infected host for survival.

What Do I Need to Know?

To get the most out of this chapter, please review the following terms from your previous courses in biology, anatomy, physiology, or chemistry or in previous chapters of this book as indicated in parentheses:anaerobic respiration (**3**), arthropod, cecum, cell-mediated immune response, chancre, commensal (**5**), diploid, endocytosis (**4**), endogenous, eosinophil, eosinophilia, etiology (**5**), facultative anaerobe (**3**), fusiform (**4**), haploid, heterotroph (**3**), humoral antibody response, ileum, *in vivo*, jejunum, lumen, lymphadenopathy, lymphocyte, meiosis, mesenteric vessel, microaerophilic (**10**), mitotic division (**4**), neutrophil, obligate intracellular parasite (**12**), oocyst, opsonization, pathogenesis (**4**), peristalsis, phagocytosis (**4**), pinocytosis (**4**), sequestration, T-cell response, transmission vector (**6**), zygote.

Fast Fact

It has been estimated that as many as 25–50% of Americans harbor parasitic worms.

Significance of Parasitic Infections

In the industrialized world, we hardly ever think about parasitic infections. However, they remain among the major causes of human misery and death in the world (**Table 14.1**). For example, more than 500 million people are infected with malaria, a disease caused by a parasite, and more than 2 million people, mostly children, die of this disease each year. Perhaps more incredibly, 2.5 billion people live in areas of the world where malaria is endemic. *Plasmodium falciparum* is the deadliest of the four species of the genus *Plasmodium* that cause malaria, but *Plasmodium vivax* is the most widespread species. All of the *Plasmodium* species are becoming resistant to the drugs used to treat malaria.

Sadly, changes in mosquito control have increased the incidence of malaria a hundredfold in recent years, and in Africa the disease is out of control. Only about 100 cases of malaria are reported in the United States each year, and these occur in immigrants and in people who have traveled to areas of the world where the disease is prevalent.

Two other important carriers of parasitic diseases are *Entamoeba* and *Trypanosoma*. Members of the genus *Entamoeba* are intestinal parasites that infect 10% of the world's population; and two to three per cent of Americans are infected with this organism. In Latin America, the parasite

Table 14.1 The Prevalence of Parasitic Infections

Disease	Estimated Population Affected
Amebiasis	10% of the worlds population
Annual deaths	40,000–110,000
Giardiasis	200 million
Malaria	500 million
Population at risk	2.0 billion
Annual deaths	2-3 million
Leishmaniasis	12 million
African trypanosomiasis	100,000
Population at risk	50 million
Annual deaths	5000
American trypanosomiasis	24 million
Population at risk	65 million
New cases per year	60,000
Schistosomiasis (blood fluke)	200 million
Population at risk	600 million
Annual deaths	0.5–1.0 million
Clonorchis (liver fluke) infection	13.5 million
Paragonimiasis (lung fluke)	2.1 million
Ascaris infection	1.3 billion
Hookworm infection	1.3 billion
Enterobius vermicularis	400 million
Cestodiasis	65 million

Trypanosoma cruzii infects an estimated 16 million people every year. This infection can cause heart and gastrointestinal lesions that can be very serious. A related species, *T. brucei*, causes sleeping sickness, one of the most lethal diseases in humans.

In the United States, parasitic infections such as toxoplasmosis, giardiasis, trichomoniasis, and pinworms are well known. In fact, toxoplasmosis is thought to infect one-third of the population of the world! Although this infection is usually asymptomatic, it can be dangerous for pregnant women and infants.

Parasitic Protozoans

Parasitic protozoans cause a variety of infections affecting large numbers of people throughout the world. In this section we examine some of the more important types of parasitic protozoans.

Morphology, Classification, and Physiology

The parasitic protozoans vary in size, ranging from 2 to 100 μm in diameter, and contain membrane-bound nuclei and cytoplasm. The cytoplasm of these organisms is divided into an inner form, called *endoplasm*, and an outer form, called *ectoplasm*. The endoplasm is involved with nutrition and metabolism, whereas the ectoplasm is concerned with movement of the organism. Protozoan organisms can be classified on the basis of their method of locomotion, with some using flagella and others using the classical pseudopodial movement. The mode of reproduction and type of locomotion are used to classify protozoan organisms into several groups (**Table 14.2**).

Table 14.2 Classes of Protozoans

Class	Organelles of Locomotion	Method of Reproduction
Rhizopods	Pseudopods	Binary fission
Ciliates	Cilia	Binary fission
Flagellates	Flagella	Binary fission
Sporozoans	None	Schizogony/gametogony

Rhizopods are free-living protozoans, but many rhizopod species form commensal relationships in the human intestines. **Ciliates** move through the use of cilia and are for the most part rarely parasitic. Several genera of **flagellates** are parasitic organisms. *Leishmania*, for instance, invades the blood and causes skin lesions as well as chronic illness. The flagellate *Trichomonas vaginalis* (**Figure 14.1**) inhabits the human genitourinary tract, and *Giardia intestinalis* (**Figure 14.2**) inhabits the gastrointestinal tract. Both of these infections cause mild disease, but *Giardia* can be serious in certain people who have allergies. Two of the most important parasitic diseases, malaria and trypanosomiasis, are caused by members of the class **Sporozoa**, which uses both asexual and sexual reproduction during the infection cycle.

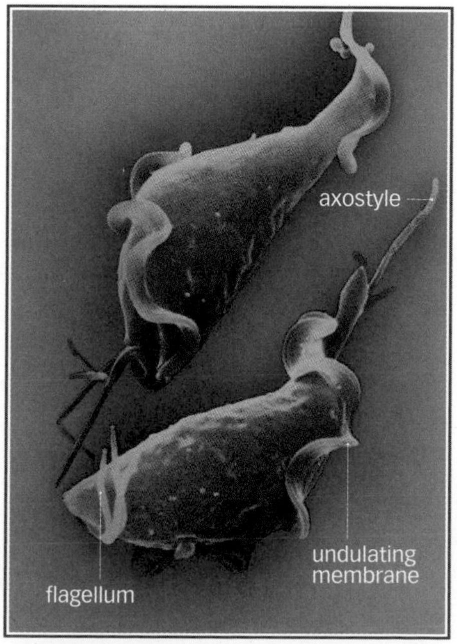

Figure 14.1 Scanning electron micrograph of *Trichomonas vaginalis*, a protozoan parasite associated with genitourinary tract infections. Notice the axostyle, which may be used for tissue attachment and destruction.

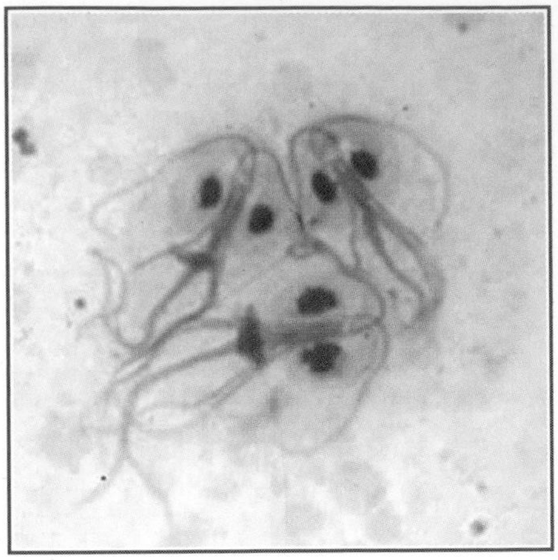

Figure 14.2 Light micrograph of the trophozoite form of *Giardia intestinalis* (formerly known as *lamblia*), which causes giardiasis. This protozoan has eight flagella and two very prominent nuclei. *Giardia* can also be found in a cystic form for protection from the environment.

Most infectious protozoans are facultative anaerobes and heterotrophs that engulf food into digestive vacuoles either through pinocytosis or through phagocytosis. Protozoans have highly developed reproductive and protective systems, and some can form cysts as a way of protecting themselves from their environment and also as a mechanism to move from host to host. Reproduction of many protozoan organisms is by simple binary fission, but some go through a cycle of simple fission (called **schizogony**) followed by a sexual reproductive phase called **gametogony**.

Parasitic Helminths

In addition to protozoan parasites, humans can also be infected with parasitic helminths.

Morphology, Classification, and Physiology

Helminths are worms, and all of them, both parasitic and free-living, are elongated, bilaterally symmetrical animals of various lengths. The body of the worm is covered by a tough acellular cuticle, which may be smooth or may possess ridges, spines, or tubercles. At the anterior end of some helminths there can be suckers, hooks, or plates, which are used for attachment.

All helminths have differentiated organs, including primitive nervous and excretory systems and highly developed reproductive tracts. None have circulatory systems. There are three classes of parasitic helminth that infect humans (**Table 14.3**), and these classes can be distinguished from one another by (1) the configurations of the body and alimentary tract, (2) the nature of the reproductive system, and (3) the need for one or more hosts to complete the life cycle.

The three classes of helminth that infect humans are:

- **Nematodes** (roundworms), which have a cylindrical body and an alimentary canal that goes from the anterior mouth to the posterior anus. There are two types of parasitic nematode: those that dwell in the gastrointestinal tract and use only one host to complete their life cycle, and those that infect blood and tissues and use multiple hosts to complete their life cycle.

Table 14.3 Characteristics of Helminths That Infect Humans

Characteristic	Nematode (Roundworm)	Cestode (Tapeworm)	Trematode (Fluke)
Morphology	Spindle-shaped	Head with segmented body (proglottids)	Leaf-shaped with oral and ventral suckers
Sex	Separate sexes	Hermaphroditic	Hermaphroditic (except *Schistosoma*)
Alimentary tract	Tubular	None	Blind pocket
Intermediate host	Variable[a]	One[b]	Two[c]

[a]Tissue nematodes have intermediate hosts; intestinal nematodes do not.
[b]*Diphyllobothrium* has two intermediate hosts.
[c]*Schistosoma* has one intermediate host.

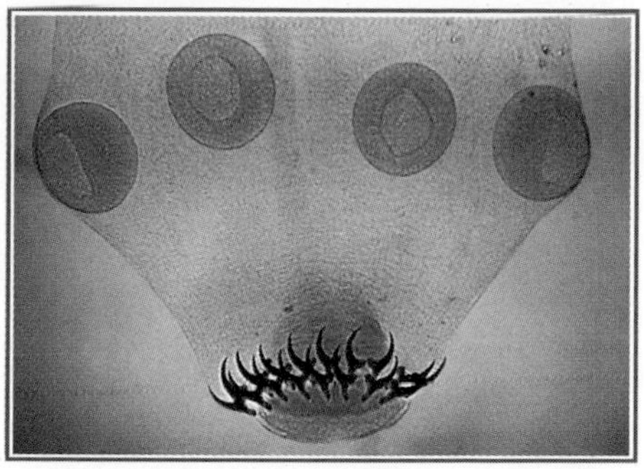

Figure 14.3 The rostellum of a helminth. These hooks, which can come in one or two rows, are found on the scolex of the worm and are used to attach and stay fixed to the tissue. Notice the four suckers above the rostellum.

- **Cestodes** (tapeworms), which have a flat, ribbon-shaped body. At the anterior end of the body is a head, which contains suckers and frequently has hooks for attachment (**Figure 14.3**). In these worms the "neck" region generates reproductive segments called **proglottids**, each of which contains both male and female gonads. These worms have no digestive tract, and nutrients are absorbed across their cuticle. Some of these helminths use one host to complete their life cycle, and others use two.

- **Trematodes** (flukes), which are leaf-shaped and have a blind branched alimentary tract (**Figure 14.4**). (A blind tract is one that has an opening at one end only.) They have two suckers, an oral sucker through which nutrients are taken in and waste material is regurgitated, and a distal sucker responsible for attachment.

Helminthic parasites are nourished by the ingestion or absorption of bodily fluids, by lysed tissue, or by the intestinal contents of the host. Respiration in these worms is primarily anaerobic, but some larval stages require access to oxygen.

Their tough cuticle and the enzymes they secrete protect helminths from host defense responses. In fact, the trematode *Schistosoma* will incorporate some of the host antigens into its cuticle as a way of protecting itself from host immune responses. Though the life span of most of these worms is usually short (weeks or months), some species, such as hookworms and flukes, can survive in the host for decades.

Fast Fact

Most of the energy required by parasitic helminths is used for reproduction, because the number of eggs that some of these worms produce is 250,000 per day under ideal conditions.

Life Cycles and Transmission Pathways of Protozoans and Helminths

The life cycles and transmission mechanisms of parasitic protozoans and helminths are variable, depending on the organism. There is also a difference with regard to the number of hosts that must be infected to complete the parasite's life cycle: in some cases, parasitic organisms use a single host; in other cases multiple hosts are required.

Parasites That Use a Single Host

Many parasites require only a single host to complete their life cycle, and transmission of these parasites from one host to another depends on the ability of the parasites to survive in the external environment. Some examples of this are shown in **Table 14.4**. For example, the flagellate

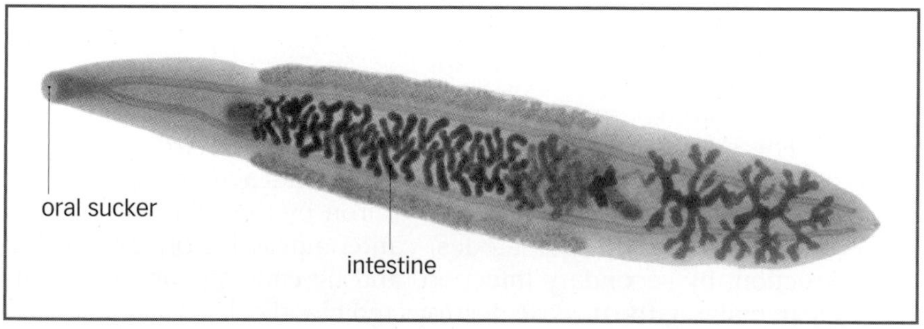

Figure 14.4 Light micrograph of the Asian liver fluke *Clonorchis sinensis*, which like all flukes has a leaf-like shape and an oral sucker. This fluke has an incomplete digestive system.

Organism	Infective Form	Mechanism of Transmission	Distribution
Trichomonas vaginalis (flagellate protozoan)	Trophozoite	Venereal (direct)	Worldwide
Entamoeba histolytica (rhizopod protozoan)	Cyst/trophozoite	Fecal–oral (direct)	Worldwide
Ascaris lumbricoides (helminth)	Eggs	Fecal–oral (indirect)	Areas of poor sanitation
Plasmodium falciparum (sporozoan protozoan)	Sporozoite	*Anopheles* mosquito	Tropical and subtropical regions

Table 14.4 Transmission and Distribution of Representative Parasites

protozoan *Trichomonas vaginalis* can survive outside a host for only a few hours and requires direct genital contact through sexual intercourse to be transmitted. In contrast, the rhizopod protozoan *Entamoeba histolytica* lives in the human gut and produces strong cysts that are passed in the stool and transmitted through the fecal–oral route of contamination. These cysts can survive in the environment for long periods and may eventually contaminate food or drinking water.

In the case of the parasitic helminth *Ascaris lumbricoides*, this worm produces highly resistant eggs that are passed in the stool of the infected host. These eggs are not immediately infectious and must mature in the soil for a period. Consequently, this parasite cannot be directly transmitted from host to host.

Parasites That Use Multiple Hosts

A few protozoans and many helminths need more than one host to complete their life cycle. Usually there are two hosts, the **definitive host**, in which asexual reproduction occurs, and the **intermediate host**, in which sexual development occurs. In some cases, such as the beef tapeworm *Taenia saginata*, both hosts are vertebrates (with humans being the definitive host and cattle the intermediate host). However, in parasites that live in the blood and tissue of humans, it is more common to find blood-feeding arthropods serving as intermediate hosts and transmitting vectors.

The most important example of a parasite that uses multiple hosts is the sporozoan protozoan *Plasmodium*, the causative agent of malaria (described below), which is transmitted by the *Anopheles* mosquito. In this case, the mosquito is the organism in which sexual reproduction occurs and *Anopheles* is therefore the definitive host, making the human an intermediate host. The areas where malaria is endemic are restricted by the availability of mosquito vectors, an availability that depends on a warm climate. Consequently, in tropical and subtropical climates, the transmission of malaria is constant and intense.

Pathogenesis of Parasitic Infection

Pathogenesis of both protozoan and helminthic parasitic diseases is quite variable. For example, the protozoan *Giardia* and the worm *Strongyloides* both interfere with the absorption of nutrients across the intestinal mucosa. Hookworms can cause a loss of iron by their feeding mechanisms, and schistosomes (trematodes) can compromise organ function by obstruction, by secondary infection, and by causing carcinomatous changes. In malaria, the disease destroys red blood cells.

Because most worms cannot increase their numbers while in a host, the severity of a helminthic infection is related to the number of worms acquired by repeated infection over time — the smaller the number, the greater the chance of asymptomatic infection. However, many worms are long-lived, and repeated infections with these worms will drive up the number in the host. The resulting large worm loads result in increased disability of the host.

It is important to note that, in addition to the damage caused by the invading organisms, the immunological defense of a host against these parasites can also cause extensive tissue damage and clinical symptoms. For example, allergic and anaphylactic cutaneous reactions are seen in response to hookworms, whipworms, and *Ascaris*, and fever and lymphadenopathy are associated with the response to *Schistosoma* larvae.

Keep in Mind

- Parasites can be protozoans or helminths (although not all protozoans and not all helminths are parasites).
- Parasitic infections affect hundreds of millions of people throughout the world and cause millions of deaths each year.
- Protozoan parasites can be divided into four groups: rhizopods, ciliates, and flagellates, and sporozoans.
- Parasitic helminths (worms) are bilaterally symmetrical animals that vary in length.
- There are three classes of helminth that infect humans: nematodes (roundworms), cestodes (tapeworms), and trematodes (flukes).
- Some parasites have a life cycle that involves a single host, whereas others use more than one host.
- Pathogenic mechanisms for both protozoan and helminthic infections vary and depend on the specific parasite.

EXAMPLES OF PROTOZOAN INFECTIONS

In this section, we look at diseases caused by several protozoan parasites. In each case we discuss the pathology and treatment of the disease.

Sporozoans

Sporozoans are a unique class of intracellular protozoans that alternate between sexual and asexual reproduction during the infectious cycle. Asexual reproduction occurs by schizogony, and sexual reproduction is by gametogony. Two of the most important sporozoan infections, malaria and toxoplasmosis, are common in humans and affect one-third of the world's population! These infections alone kill or debilitate a million newborn infants and children every year.

*Malaria (*Plasmodium *species)*

Malaria is a febrile illness caused by a parasitic infection of human red blood cells. As noted earlier, it results from infection by the protozoan *Plasmodium* and is transmitted through the bite of the *Anopheles* mosquito. This disease is found throughout the world and in particular in areas with warm climates. There are four species of *Plasmodium* that can cause malaria symptoms to different degrees, with *P. falciparum* being the most pathogenic and the dominant species found in the tropics. Spread of the disease depends on the density and feeding habits of the mosquito

Fast Fact

It is estimated that 2 billion people live in areas endemic for malaria, and 25–50% of this group are believed to carry the parasite, resulting in more than 2 million deaths a year from malaria.

vectors, and mortality is largely restricted to infants and immunocompromised adults. In some areas, the transmission of malaria can be seasonal, and in these areas the infection can be seen in people of all ages. In the United States, about 100 cases are reported each year, and these occur mostly in immigrants coming from endemic areas and people who travel to these areas. Clinical symptoms are usually seen within six months of returning from these areas.

Life cycle of *Plasmodium*. The sexual cycle of *Plasmodium* begins when a female mosquito ingests circulating male and female *Plasmodium* **gametocytes** from the blood of an infected person. In the gut of the mosquito, the male gametocyte fertilizes the female gametocyte, and the resulting zygote penetrates out of the gut wall of the mosquito and forms an oocyst. Inside this oocyst, thousands of **sporozoites** are formed, and the cyst eventually ruptures (**Figure 14.5**), releasing the sporozoites into the body cavity of the mosquito. Some of these sporozoites will penetrate the salivary glands of the mosquito, and this is the form responsible for infection of humans. One life cycle takes about one to three weeks to complete, depending on temperature and humidity.

The asexual reproductive cycle begins in the human host when sporozoites from the saliva of the biting mosquito are injected into the human bloodstream and begin to circulate. Within 30 minutes, they find their way to the liver and attach to hepatocytes. Once inside a hepatocyte, each sporozoite can produce 2000–40,000 daughter cells called **merozoites**. The infected hepatocytes begin to rupture within two weeks and release the merozoites, which then attach to specific receptors on red blood cells and are carried into the cells by endocytosis. This stage of the infection is referred to as the **ring stage** because of the appearance of the merozoites in the red blood cells (they form a ringlike structure). Within 72 hours, the infected red blood cells begin to rupture, releasing merozoites (**Figure 14.6**). Some of the released merozoites continue to

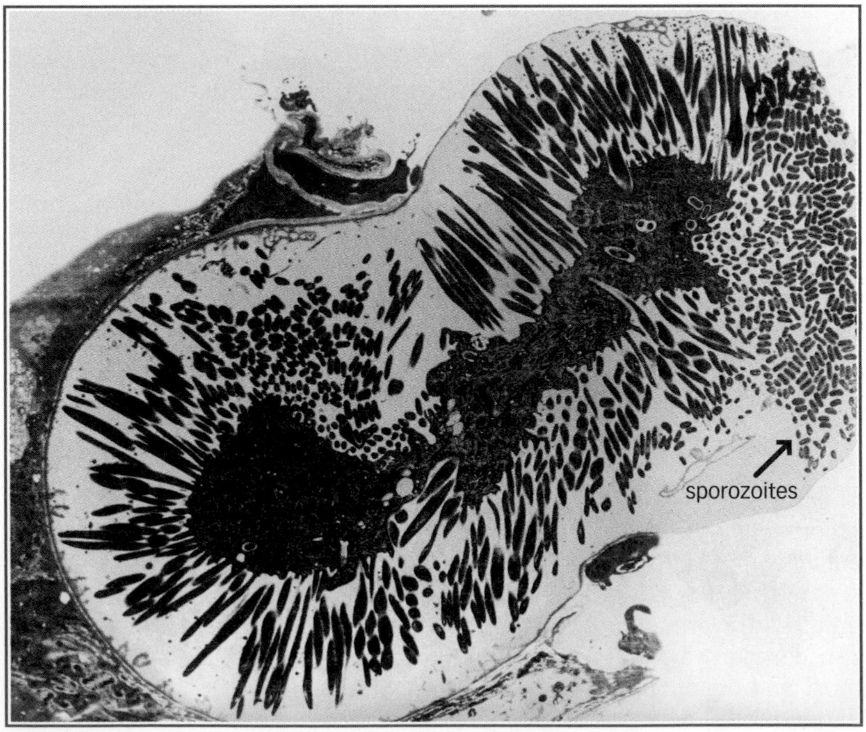

Figure 14.5 Malaria parasites. Colored transmission electron micrograph (TEM) of malaria sporozoites in a mosquito (*Anopheles* species) gut. The sporozoites (*Plasmodium* species, purple) are seen bursting from an oocyst (white/yellow) at upper right.

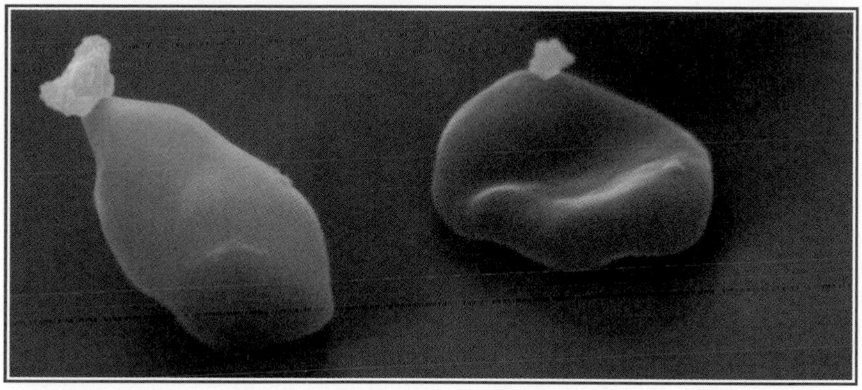

Figure 14.6 A photomicrograph of *Plasmodium falciparum* (yellow) emerging from red blood cells.

invade and destroy red blood cells, whereas others transform into the gametocyte form. This gametocyte form is not capable of destroying red blood cells but circulates in the peripheral blood until it is ingested by a mosquito, and the infectious cycle begins again (**Figure 14.7**).

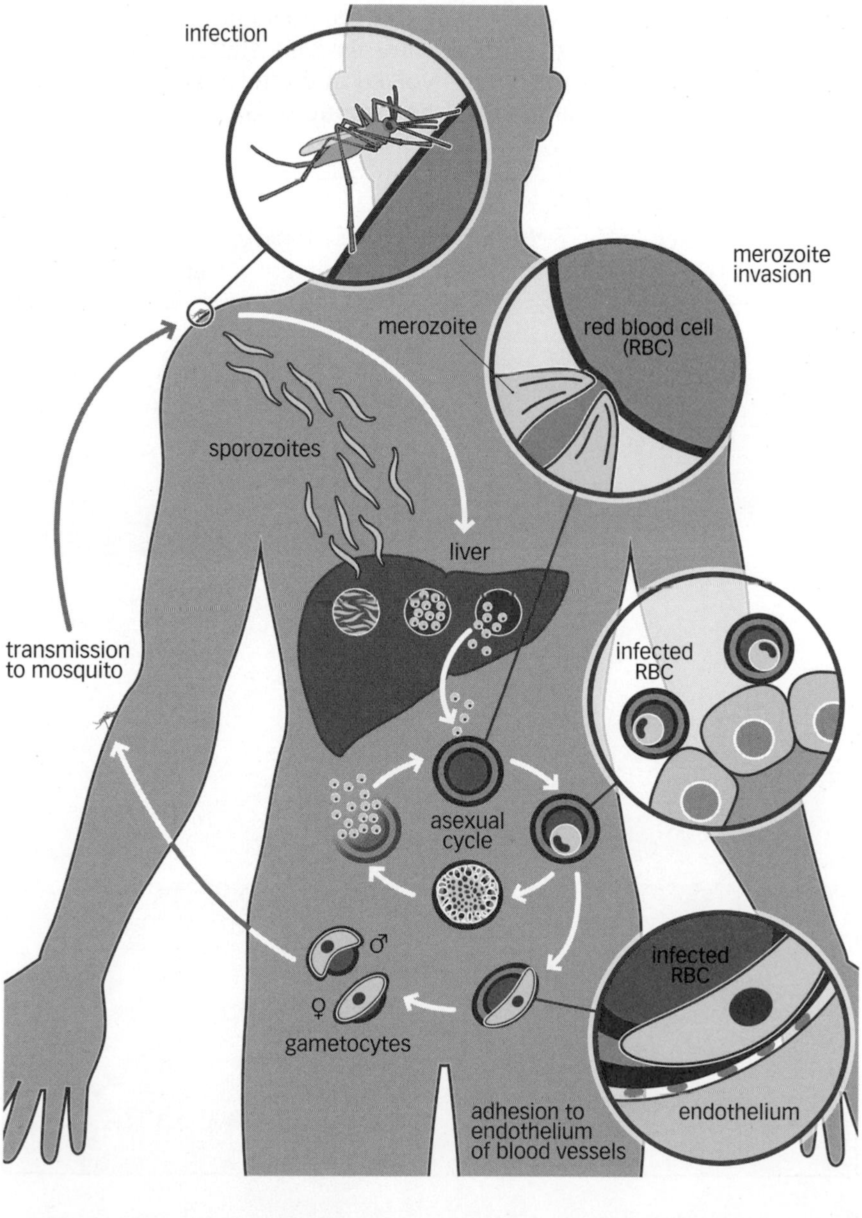

Figure 14.7 The life cycle of malaria. Keep in mind that the mosquito is the definitive host and that sexual reproduction occurs in this vector. In contrast, humans are the intermediate host and the parasite undergoes asexual reproduction here.

Pathogenesis of malaria. Several symptoms occur during this disease, including fever, anemia, and circulatory changes. Fever is the hallmark of malaria and seems to be initiated by the rupture of red blood cells, which in turn liberates new parasites to continue the infection. However, the actual mediators of the fever have yet to be identified. As it turns out, temperatures above 40°C destroy mature parasites, and at these temperatures the sporulation of merozoites eventually becomes synchronized, with the result that fever occurs about every 48 hours.

The anemia seen in malaria is caused by the destruction of red blood cells and is accompanied by a depression of marrow function and sequestration of red blood cells in the enlarged spleen. When the destruction of red blood cells is severe, the patient gets hemoglobinuria, which causes the urine to become dark. This is why malaria is sometimes referred to as "black water fever." One common circulatory change seen during the infection is hypotension, which occurs as the high fever causes the blood vessels to dilate. During infections with *P. falciparum*, red blood cells stick to the walls of capillaries and impair microcirculation in the organs, especially the brain (**Figure 14.8**). Thrombocytopenia (low levels of platelets) is common in malaria and seems to be due to the shortened life span of platelets.

Fast Fact

As a malaria infection continues, the host defense response to the infection can cause decreased renal function.

The incubation period between the bite of the mosquito and the onset of disease is about two weeks, and the clinical manifestations vary depending on the species of *Plasmodium* involved in the infection. The cycle of the disease (referred to as **malarial paroxysm**) begins with a *cold stage*,

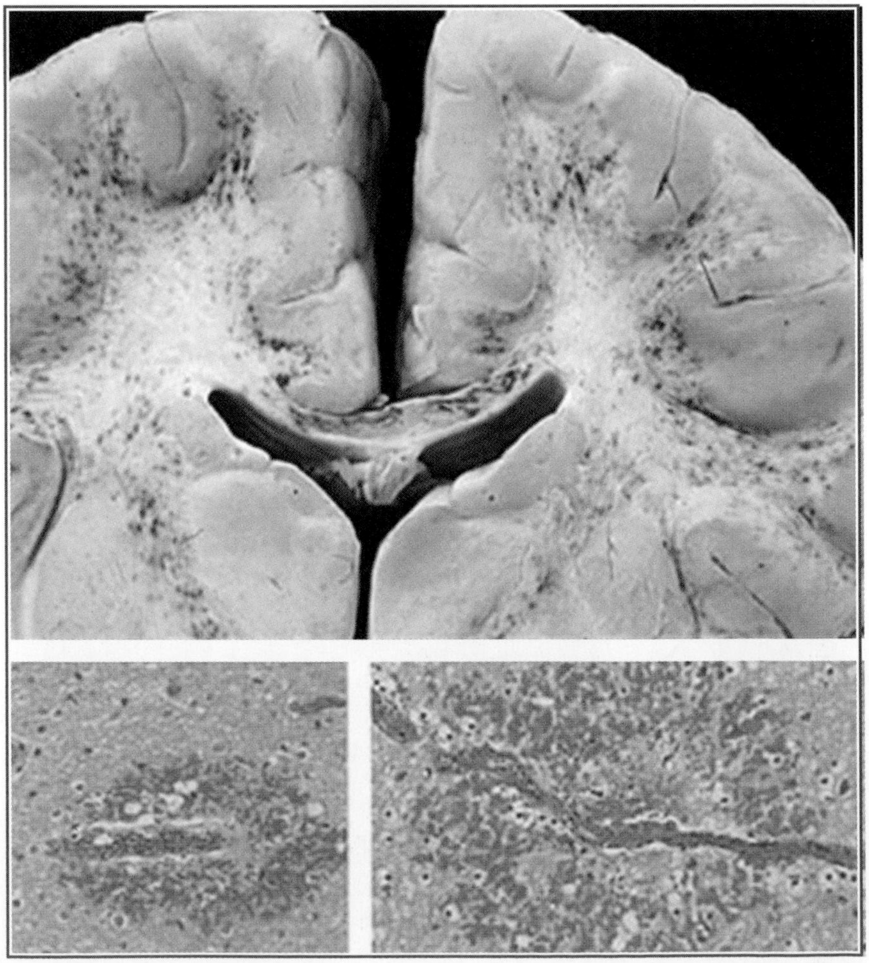

Figure 14.8 The pattern of sequestration and microvascular pathology seen in cerebral malaria. Notice the multiple petechial hemorrhages in the cortical white matter, and also the classical ring shape of hemorrhage surrounding a parasitized vessel (bottom left and right).

which lasts for 20–60 minutes. During this period the patient experiences continuous shaking and chills. The body temperature increases for 3–8 hours, and the *hot stage* begins as the shaking and chills subside. During this period, the temperature can reach 40–41.7°C and cause profuse sweating. This sequence of recurring hot and cold stages leaves the patient exhausted, but well, until the next cycle begins. By the second or third week of infection, this cyclical sequence becomes repetitively synchronized.

Eventually, the paroxysms diminish and finally disappear as the parasites in the blood disappear. In the most deadly form of malaria, caused by *Plasmodium falciparum*, the central nervous system can become involved, and this form is called *cerebral malaria*. It is marked by delusions, convulsions, paralysis, coma, and rapid death. Acute **jaundice**, renal failure, and coma are also common. The mortality for cerebral malaria is 80%, with most patients succumbing in as little as three days.

Treatment of malaria. The treatment of this infection rests on two factors: which species of *Plasmodium* is causing the infection and the immune status of the patient. If the infecting organism is *P. falciparum*, the most dangerous of the species, treatment must be started as soon as possible. Regardless of which species is involved, successful treatment requires the destruction of all forms of the parasite (which are called the *erythrolytic schizont form*, the *hepatic schizont form*, and the *erythrogenic gametocyte*). This breaks the transmission cycle. At present, there is no single drug that affects all three forms. Treatments used for malaria are listed in **Table 14.5**.

*Toxoplasmosis (*Toxoplasma gondii*)*

This infection is caused by the parasitic protozoan *Toxoplasma gondii*, which is an obligate intracellular sporozoan. The definitive host for this parasite is the domestic housecat, and transmission can occur through the ingestion of oocysts present in the fecal material of infected cats. The major morphological forms of this parasite are oocysts, trophozoites, and tissue cysts.

Life cycle of *Toxoplasma*. The life cycle of *Toxoplasma* begins in the intestine of the cat. Here the trophozoite form of the parasite enters the epithelial cells of the ileum, with entry into the cells being aided by specialized enzymes secreted by the parasite for this purpose. The trophozoites remain in cellular vacuoles and undergo schizogony into merozoites. This change causes the epithelial cell to rupture and release the parasites. The merozoites then differentiate into female and male gametocytes and begin gametogony. Millions of oocysts are released each day for two to three weeks, and these oocysts mature in the external environment and are stable in the soil for months.

Table 14.5 Treatments Used for Malaria

Form of the Parasite	Clinical Goal	Drug of Choice
Erythrocytic schizont	Treatment of clinical attack	Chloroquine; quinine, antifolates or sulfonamides for resistant *P. falciparum*
	Suppression of clinical attack	Chloroquine; antifolates and sulfonamides for resistant *P. falciparum*
Erythrocytic gametocyte	Cure of relapsing malaria; cure of falciparum malaria; control of mosquito population	Chloroquine; primaquine
Hepatic schizont	Cure of relapsing malaria	Primaquine

In the intermediate host (humans), sporozoites are released from disrupted oocysts and enter macrophages present in the host's blood, using these cells to travel to all the organs of the body. Eventually the macrophages rupture and release new parasites, which invade adjacent host cells and begin another turn of the asexual portion of the cycle. In the brain, heart and skeletal muscles, trophozoites produce a membrane that surrounds and protects the tissue cyst. These cysts can eventually hold more than 1000 organisms and persist for the life of the host.

Pathogenesis of toxoplasmosis. In the primary infection, the proliferation of trophozoites leads to the death of host cells, which initiates an immune response. Normally this response controls the infection, but in patients that are immunodeficient there can be continuous tissue necrosis. During this defensive response, extracellular parasites are killed and intracellular multiplication is inhibited. Serious disease can lead to the inhibition of cell-mediated immunity, and intense host defense can also exacerbate the pathology associated with this infection.

Treatment of toxoplasmosis. No treatment is usually required unless the symptoms are severe and persistent, or unless vital organs are involved. In the United States, the most common treatment is a regimen of combinations of pyrimethamine and sulfonamides.

Rhizopods

The parasitic protozoans known as rhizopods are amebas and they are therefore the most primitive of all protozoans. Rhizopods multiply by simple binary fission and move by using the structure known as a *pseudopod*. This motion involves projecting the relatively solid ectoplasm outward and following with the streaming of the inner endoplasm. These organisms can also produce a chitin wall around themselves for protection, which is referred to as a cyst and can survive in the external environment for long periods.

*Amebiasis (*Entamoeba histolytica*)*

Most rhizopods are free-living and do not infect humans. However, there are several species, including *Entamoeba*, that are obligate intracellular parasites of the human alimentary canal and cause the disease called *amebiasis*. These parasites are passed from host to host as cysts via the fecal–oral route of contamination. Only *E. histolytica* produces disease in humans, and it has been subdivided into two genetic forms, *E. histolytica*, which is always pathogenic, and *E. dispar*, which is a harmless commensal organism.

The infectious dose for this parasite is approximately 1000 or more organisms. However, the ingestion of a single cyst can also cause an infection. Infection rates are higher in warmer climates, and *Entamoeba histolytica* is thought to produce more deaths worldwide than any other parasitic disease except malaria and **schistosomiasis**. In the United States, it is estimated that one to five per cent of the population harbor *Entamoeba*, but the majority form is *E. dispar* rather than *E. histolytica*. Amebiasis has been on the rise in the United States, especially in institutionalized individuals, on Indian reservations, and among AIDS patients and migrant workers. It is also found among individuals who have traveled to parts of the world where this parasite is prevalent.

Life cycle of *Entamoeba histolytica*. *E. histolytica* can be found in either the trophozoite form or the cyst form. The trophozoites are microaerophilic and dwell either in the lumen or the wall of the colon, where they feed on the bacteria found there as well as the tissue cells. They can multiply rapidly in the environment of the gut, and when diarrhea occurs the trophozoites are passed in the watery stool. Electron micrographs show that the trophozoites contain microfilaments and an external glycocalyx, along with cytoplasmic projections that are thought to be important for attachment to host cells. Trophozoites usually encyst themselves before they leave the gut. These cysts can survive temperatures of 55°C as well as the chlorine concentrations usually found in municipal water supplies. Cysts are also able to survive the acidic conditions found in the stomach of infected individuals.

Fast Fact

Humans are the principal host and reservoir for *Entamoeba histolytica*, and infected humans can pass a staggering 45 million cysts a day in their feces!

Pathogenesis of amebiasis. The initial amebiasis infection is the result of direct person-to-person contact through the fecal–oral route, whereas systemic amebiasis usually occurs only after the parasite has colonized the colon. During the infection, the parasite produces several virulence factors and extracellular proteinases. The trophozoite form adheres to the target cell and then releases a protein that causes membrane lesions in the host cell that result in cellular cytolysis. In most cases, tissue damage is minimal, and the host remains essentially asymptomatic. However, *Entamoeba histolytica* can create a portal of entry in the intestinal mucosa through which bacteria and viruses can readily enter and either contribute to the ulceration process or spread to other parts of the body.

After passage through the stomach, the cyst reaches the small intestine. Here the wall of the cyst disintegrates, releasing a quadri-nucleated (four nuclei) parasite, that divides to form four trophozoites that move to the colon, where they attack the epithelial cells lining the colon and produce small mucosal ulcerations. Large numbers of trophozoites can also be seen at the margins between the ulcerations and the adjacent normal mucosa. These lesions spread laterally in the submucosal layer and eventually compromise blood flow, causing necrosis followed by fibrosis. There are rare instances in which the lesions occur in the brain, liver, lung, or spleen; in these areas, abscesses can form.

Symptoms are usually diarrhea, flatulence, cramping, and abdominal pain. The diarrhea is intermittent and can be accompanied by bouts of constipation that can last for months to years. During the episodes of diarrhea, it is common to find blood in the stool. Fulminating amebic dysentery is the most virulent amebiasis and is usually seen in debilitated individuals as well as during pregnancy. Here the onset is abrupt, with a high fever, severe cramping, and profuse diarrhea.

Treatment of amebiasis. Treatment involves treating the symptoms through blood and fluid replacement. The drug of choice for eradication of the parasite is metronidazole, which is effective against all forms of amebiasis. In addition, efforts toward the sanitary disposal of feces and good personal hygiene can help to prevent this infection.

Flagellates

Like the rhizopod parasites, flagellates are widespread in nature. They multiply by binary fission and use flagella for locomotion. Depending on the organism, the locomotion structure is either a single flagellum or multiple flagella. Only four flagellate protozoans cause disease in humans. *Trichomonas* and *Giardia* are noninvasive and inhabit the lumen of the genitourinary or gastrointestinal tracts. These organisms do not need an intermediate host and result in low morbidity in the host. In contrast, *Leishmania* and *Trypanosoma* are invasive blood and tissue parasites that produce high morbidity and frequently lethal disease in humans. These parasites require an intermediate host for transmission.

*Trichomoniasis (*Trichomonas vaginalis*)*

Trichomonas infection is a sexually transmitted disease that in females produces the condition known as vaginitis, the symptoms of which are pain, discharge, and **dysuria**. The infection can last for weeks to months and may cause prostatitis or urethritis in men. It is estimated that 3 million women in the United States and 180 million worldwide acquire this infection each year. In fact, 25% of sexually active women will be infected at some time in their lives, and this rate is 70% for prostitutes. The peak period of prevalence in women is between 16 and 35 years of age, but the infection can also be transmitted to newborn infants during passage down the birth canal. The *Trichomonas vaginalis* trophozoite has a rounded anterior end and a pointed posterior end, and measures about 7 μm across (see **Figure 14.1**). It has five flagella and contains an **axostyle**, which is a microtubule believed to be used for attachment and which may also cause tissue damage in a host.

Life cycle of *Trichomonas*. *Trichomonas* does not form cysts, but the trophozoite form can survive outside a host for one to two hours. In urine, semen, and water, the trophozoite is viable for up to 24 hours.

Pathogenesis of trichomoniasis. Direct contact of the parasite with the epithelial cells of the genitourinary tract results in the destruction of these cells and an inflammatory response accompanied by petechial hemorrhage. The exact mechanisms of pathogenicity are not understood, but *Trichomonas* is not invasive and does not produce a toxin. Interestingly, changes in the vaginal environment (such as pH, hormonal, and microbial) can affect the severity of the pathological changes that occur during *Trichomonas* infection.

The infection causes a persistent vaginitis with clinical symptoms that can last for months. These include discharge, itching or burning, dysuria, and in some cases disagreeable odor. In mild cases, there is reduced vaginal and mucosal tissue damage. However, in severe cases there can be hemorrhaging and extensive tissue erosion.

Fast Fact

Men infected with *Trichomonas* are usually asymptomatic.

Treatment of trichomoniasis. Oral metronidazole administered either in a single dose or over seven days cures 95% of these infections. However, this therapy should not be administered in the first trimester of pregnancy. Sexual partners should also be treated for the infection.

*Trypanosomiasis (*Trypanosoma *species)*

This infection is caused by the protozoan parasite *Trypanosoma*. This flagellate parasite uses an insect, the *tsetse* fly (*Glossina* species; **Figure 14.9**), as a vector with which to infect humans, and there are several morphological changes in the parasite during the cycling between insect and human (similar to what happens in the malaria life cycle). This protozoan parasite is motile and fusiform, with a blunt posterior end and a sharp anterior end. It moves in a spiral fashion.

There are two forms of trypanosomiasis: the African form, which is called **sleeping sickness**, and the American form, which is referred to as **Chagas' disease**. We confine this discussion to sleeping sickness and discuss Chagas' disease in Chapter 25, which covers infections of the blood. Sleeping

Figure 14.9 The tsetse fly is the vector for *Trypanosoma* (the etiological agent of African sleeping sickness). Notice that the fly has become engorged with blood.

sickness is confined to central Africa, where 50 million people live and 10,000 to 20,000 of them get this disease each year. Although the tsetse fly is the vector, it is believed that the reservoir for this pathogen is humans.

Life cycle of *Trypanosoma*. *Trypanosoma* reproduces by longitudinal binary fission. There are three subspecies — *T. brucei gambiense*, *T. brucei rhodesiense*, and *T. brucei brucei* — and all of them undergo morphological changes as they cycle from insect to human host. In the mammalian host, they multiply extracellularly and eventually invade the blood. They also have the ability to change their antigens, expressing dozens to hundreds of variations, which makes it difficult for the host's immune system to respond effectively.

Pathogenesis of trypanosomiasis. Trypomastigotes (one of the morphological forms of the parasite) are deposited by the bite of the tsetse fly and begin to multiply, causing localized inflammation. This develops into a chancre from which organisms spread into the blood and lymph of the host, causing lymphadenopathy and recurrent parasitemia. The host response in the form of IgM antibodies causes destruction of the parasites by opsonization, and the trypomastigotes disappear from the blood. Amazingly, they reappear three to eight days later with different antigenic markers. These reappearances become less frequent but can last for years.

During parasitemia, the *Trypanosoma* trypomastigotes localize in the small blood vessels of the heart and the central nervous system, causing endothelial proliferation and perivascular infiltration of plasma cells and lymphocytes. In the brain, the infection can cause hemorrhage and **demyelinating panencephalitis**. During recurrent bouts of parasitemia, the patient experiences fever, tenderness in the lymph nodes, skin rash, headache, and impaired mental status. Bouts of fever can last for years before gradual problems with the central nervous system appear. Eventually, alertness diminishes, attention wavers and the patient must be prodded to eat or talk. Speech becomes indistinct, tremors develop, and loss of sphincter control brings about the final stage, which includes coma and death.

Treatment of trypanosomiasis. Because of the involvement of the central nervous system, agents that cross the blood–brain barrier, such as melarsoprol (an arsenic compound), must be used. If there is no central nervous system involvement, pentamidine or eflornithine is effective, and the cure rate is high with recovery being complete.

Keep in Mind

- Protozoan parasitic diseases can be caused by several classes of protozoan, including sporozoans, rhizopods, and flagellates.
- Sporozoan diseases include malaria (caused by *Plasmodium* species) and toxoplasmosis (caused by *Toxoplasma gondii*).
- One disease caused by a rhizopod protozoan is amebiasis (caused by *Entamoeba histolytica*).
- Flagellate diseases include trichomoniasis (*Trichomonas vaginalis*) and trypanosomiasis (*Trypanosoma* species).

EXAMPLES OF HELMINTHIC INFECTIONS

Recall from earlier in the chapter that there are three types of parasitic helminth that cause disease in humans: nematodes, cestodes, and trematodes. Let's take a closer look at each of them now.

Parasite	Human Disease
Enterobius vermicularis (pinworm)	Enterobiasis
Ascaris lumbricoides (large roundworm)	Ascariasis
Necator americanus (hookworm)	Hookworm infections

Table 14.6 Some Intestinal Nematodes That Cause Human Disease

Intestinal Nematodes

The group of roundworms known as nematodes has two subgroups: intestinal nematodes and tissue nematodes. Here we look at the intestinal variety first. These parasitic roundworms have a fusiform body consisting of a tough acellular cuticle, a tubular alimentary canal, and a muscular layer. There are male and female forms, and the female can produce thousands of offspring. Eggs must incubate outside the host before they are infectious, and this maturation involves the development of a larval form.

There are several intestinal nematodes that routinely infect humans (**Table 14.6**), including pinworms (*Enterobius vermicularis*), whipworms (*Trichuris trichiura*), and large roundworms, such as *Ascaris lumbricoides* and *Strongyloides stercoralis*. These worm infections produce discomfort, malnutrition, anemia, and occasionally death. They also cause embarrassment because they can unexpectedly exit the body from the anus, nose, mouth, or ears (**Figure 14.10**).

It is interesting to note that, with intestinal nematodes, the severity of the disease depends on the level of adaptation to the host. The more adapted the worm becomes, the less severe the symptoms of the infection. Conversely, the less adapted the worm is to the host, the more serious the disease. This relationship can also be viewed from the perspective of worm *load* (that is, the number of worms) in that infections resulting from smaller worm loads are normally asymptomatic, whereas those resulting from larger worm loads cause symptomatic disease. Overall, the immune response is slow to develop in parasitic infections caused by intestinal nematodes.

Fast Fact

Nematode worms infect 25% of the entire human race.

We will only discuss infections caused by two of these nematodes, *Enterobius* and *Ascaris*.

Enterobiasis

The pinworm *Enterobius vermicularis* is a ubiquitous parasite of humans and the cause of the condition known as *enterobiasis*. It is estimated that more than 200 million people (a large percentage of which are children) are infected with this worm every year. It is most frequently found in the temperate regions of Europe and North America and is relatively rare in the tropics. *Enterobius* is entirely restricted to humans, and infection can be readily transmitted in places where large numbers of children gather together (such as nurseries, child care facilities, and orphanages).

Figure 14.10 *Ascaris lumbricoides* worms. These worms can exit from the anus, nose, mouth, or ears.

Pathogenesis of enterobiasis. These pinworms lie attached to the mucosa of the cecum portion of the large intestine. The female migrates down the colon and through the anal canal (**Figure 14.11**), and deposits about 20,000 sticky eggs on the perianal skin, as well as on bedclothes and linens. These eggs are near maturity at the time they are deposited, and scratching of the perianal area results in adherence of the eggs to fingers and eventual transfer to the oral cavity. In addition, the eggs can be shaken into the air (when making the bed) and inhaled or swallowed.

Eggs hatch in the upper intestine, and the larvae begin the migration down to the cecum. This cycle takes about two weeks and causes sleep deprivation and daytime irritability in children. In addition, the skin abrasion due to scratching can occasionally cause cutaneous bacterial infections. Female worms will occasionally migrate into the genitourinary tract of women.

Treatment of enterobiasis. Several agents, including mebendazole and pyrantel pamoate, can effectively treat this infection. However, recurrence is common.

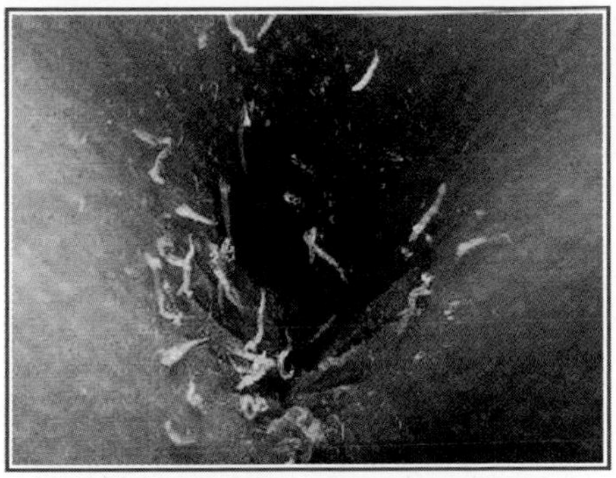

Figure 14.11 Pinworms (*Enterobius vermicularis*) leave the anus to lay eggs. These eggs are very sticky and can become affixed to the fingers during scratching of the irritation, as well as to bedclothes.

Ascariasis

Ascaris lumbricoides is the largest and most common of the intestinal nematodes, measuring 15–40 cm in length. The female can lay 250,000–500,000 eggs per day; these are very resistant to environmental stress and can be viable in temperate climates for up to six years. The medical condition is called *ascariasis*, and the infection is maintained by small children who defecate indiscriminately and pick up mature eggs from the soil while playing. These eggs can also contaminate food, and in dry, windy areas they can become airborne, then inhaled or swallowed.

Fast Fact

More than 1 billion people, including 4 million Americans, are infected with *Ascaris* annually, and it is estimated that a phenomenal 25,000 tons of *Ascaris* eggs are moved into the environment every year.

Pathogenesis of ascariasis. Adult *A. lumbricoides* parasites live in the human small intestine, and eggs are passed into the feces. The eggs require about three weeks to embryonate, and this process must occur in the soil before the eggs can become infectious. Once the infectious eggs have been ingested, they proceed to the larval stage. The larvae of the worms penetrate the intestinal mucosa of the human host and invade the liver. Here the larvae are still small enough to exit through the hepatic vein, be carried to the right side of the heart and subsequently be pumped into the lung. By the time they reach the pulmonary capillaries, they are too large to pass through the left side of the heart and remain in the lung. Eventually, the larvae will rupture into the alveolar spaces and be coughed up and swallowed, regaining access to the intestine.

Clinical ascariasis may result either while the *A. lumbricoides* larvae migrate from host liver to host lung as just described or when the larvae establish themselves in the intestinal lumen. The symptoms include fever, coughing, wheezing, and shortness of breath. However, if the worm load is small the patient will be asymptomatic. Worms can be vomited up or passed in the stool during episodes of fever and can be observed crawling out of the anus, nose, mouth, or ear. Heavy worm loads can cause malabsorption of fat, protein, carbohydrates, and vitamins. In addition, there can be abdominal pain and obstruction of the bile duct and pancreatic ducts. Worm loads of 50 are common in this infection and as many as 2000 worms have been recovered from a single child.

Treatment of ascariasis. Albendazole, mebendazole, and pyrantel pamoate are very effective in dealing with this infection. Sanitation is also important in preventing its spread.

Tissue Nematodes

These parasites induce disease through their presence in the tissue, blood, and lymph systems of the host's body. There are four major types of nematode that use humans as their definitive hosts, and the adults live for years in subcutaneous tissues and lymphatic vessels. Here they discharge live offspring called **microfilariae**, which circulate in the blood or in subcutaneous tissue until they are ingested by a specific blood-sucking insect.

Trichinosis

The nematode *Trichinella spiralis* lives in the duodenal and jejunal mucosae of flesh-eating animals, particularly swine and bears. In the intestinal mucosae, the tiny male worm couples with the larger female and from this one insemination the female can give birth for 4–16 weeks, generating up to 1500 larvae. The larvae enter the host's vascular system and are distributed throughout the body. Larvae that penetrate tissue other than skeletal muscle disintegrate and die, but those that find their way into skeletal muscle will continue to grow and become encapsulated over a period of several weeks. Once encapsulated, the larvae can remain viable for 5–10 years. The muscles most often invaded are highly vascularized muscles, such as the eye muscles, tongue, deltoid, pectoral, intercostal, diaphragm, and gastrocnemius.

The disease trichinosis is widespread in carnivores, with swine being the most often involved. Only two types of *Trichinella* show a high level of pathogenesis in humans. Human infection results from eating undercooked meat, and it is estimated that there are more than 1.5 million Americans carrying either live *Trichinella* or dead encysted larvae in their musculature. Between 150,000 and 300,000 cases are reported in the United States each year, but they are for the most part asymptomatic.

Pathogenesis of trichinosis. Pathogenic lesions related to the presence of larvae can be found in striated muscle, muscle of the heart and also in the central nervous system. The invaded muscle cells undergo host defensive measures including basophilic degranulation, which causes the surrounding area to be infiltrated by neutrophils, lymphocytes, and eosinophils. These eosinophils destroy the circulating larvae and slow the development of new larvae.

One or two days after a host has ingested tainted meat, the worms mature, and adult worms can perforate the intestinal mucosa, causing nausea, abdominal pain, and diarrhea. In serious infections these symptoms may persist for days and render the patient prostrate. Larval invasion starts one week later and initiates the long period of disease, which can last for six weeks. Worm load is important here because ten or fewer worms will cause no symptoms. However, a worm load of 100 or more can cause significant disease, and 1000 to 5000 can be lethal.

The most prominent symptoms are fever, muscle pain, tenderness, and weakness. In severe disease there can also be pulmonary dysfunction. If the heart is involved, there can be tachycardia or congestive heart failure, and central nervous system manifestations involve encephalitis and meningitis.

Treatment of trichinosis. In patients with severe edema, myocardial involvement, or disease of the central nervous system, the use of corticosteroids is required. Mebendazole and albendazole halt the production of new larvae, but in severe infection the destruction of larvae by the host defenses may cause the onset of dangerous hypersensitivity.

Cestodes

These long, ribbonlike helminths, commonly known as **tapeworms**, represent the largest and most repulsive of the intestinal parasites. As noted at the beginning of this chapter, they lack vascular and respiratory systems and are devoid of a gut or body cavity. Nutrients are absorbed directly across the cuticle surrounding the organism. The adult has three parts: a head, called the **scolex**; a regenerative neck region; and a long, segmented body.

The scolex of certain cestodes is equipped with four muscular sucking disks, whereas other species have only two. These disks serve to attach the worm to the intestinal mucosa of the host. In some species of cestode, the scolex can have a retractable **rostellum** armed with crowns of chitinous hooks (**Figure 14.12**).

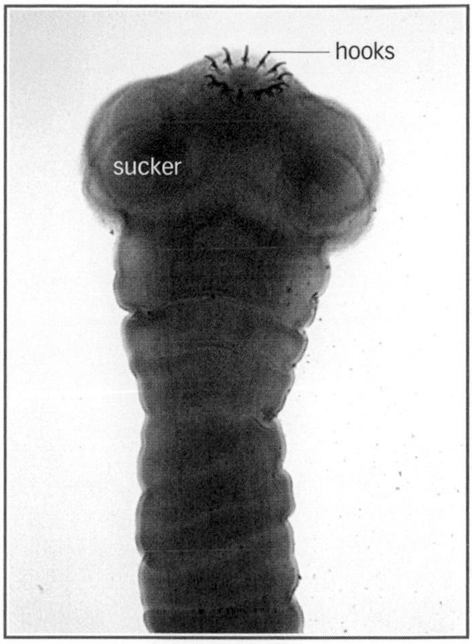

Figure 14.12 The scolex of a tapeworm showing the suckers used to obtain nutrients and also the rostellum bearing the hooks used for attachment.

The neck region of the worm is where the proglottid segments are generated, and each segment is a self-contained hermaphroditic reproductive unit containing male and female gonads. Sexual reproduction, fertilization, and maturation occur as a proglottid moves farther from the neck, and the proglottid eventually ruptures and releases eggs. The development of all except one type of cestode requires passage through more than one intermediate host.

Pathogenesis of cestode infection. The severity of these infections depends on whether the patient is the primary or intermediate host. If a patient is the primary host, the tapeworm stays in the lumen of the gut, causing only minor symptoms. If the patient is the intermediate host, the larval stages of the worm will cause tissue invasion, with frequently serious disease. We will use the cestode *Taenia saginata* as an example.

Taenia saginata inhabits the human jejunum and can live there for 25 years, growing to a length of 10 meters (more than 30 feet; **Figure 14.13**). Mature *Taenia* can have six to nine terminal proglottids, each containing about 100,000 eggs, which break free and either crawl through the anal canal of the host or are passed in the stool. When these proglottids reach the soil, they disintegrate and release their eggs, which can survive in the soil for months. If the eggs are ingested by cattle, the resulting larvae will penetrate the intestinal wall and be carried by the blood to striated muscle of the tongue, diaphragm, or hindquarters, where they transform into a **cysticercus**, giving a "mealy" appearance to the meat.

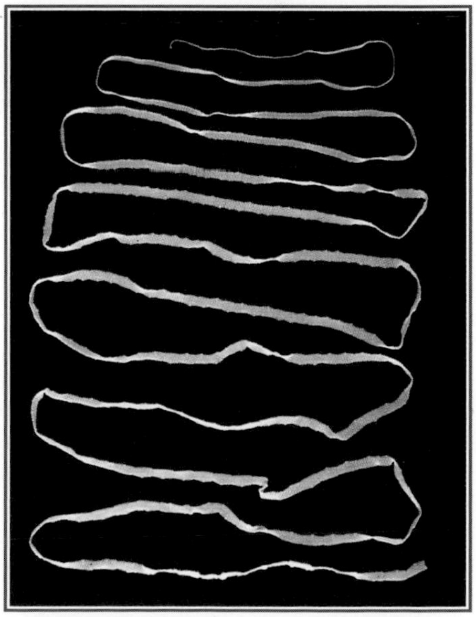

Figure 14.13 An illustration of the size of a tapeworm (recall that some can be 10 meters in length). Notice the segmented body of the worm which broadens at the distal end. These segments are the proglottids.

Humans are infected when they eat improperly prepared meat, but most patients are asymptomatic and become aware of their infection only by observing proglottids being passed in their stool or visible on their bedclothes. In some cases, there can be some gastric discomfort, nausea, diarrhea and weight loss.

Treatment of cestode infection. The drugs of choice for these infections are praziquantel and niclosamide, which either paralyze or kill the worm. Peristalsis then pushes the worm out of the host.

Trematodes

Adult trematodes, also known as **flukes**, live for decades in human tissue and blood vessels, producing progressive damage to vital organs. These worms have bilateral symmetry, vary in length from a few millimeters to several centimeters, and, as noted at the beginning of the chapter, possess two deep suckers. One of these suckers surrounds the oral cavity, and the other is located on the ventral surface of the worm.

There are two major categories of fluke, based on the reproductive systems: **hermaphrodites** and **schistosomes**. Eggs are excreted from the human host, and if they reach water, they hatch and release ciliated larvae called **miracidia**. These larvae find and penetrate snails. The snail is the intermediate host, and it is here that the larvae undergo reproduction into tail-bearing larvae called **cercariae**, which are continuously released from the snail for several weeks. What happens next depends on whether the species is a hermaphrodite or a schistosome. If it is a schistosome, the cercariae shed their tails and invade the skin of humans. If the species is a hermaphrodite, the cercariae encyst in or on an animal or plant, which is the second intermediate host. Here the larvae develop into the **metacercariae** form. Humans become infected when they eat the sea animal or plant contaminated with the metacercariae.

Although many types of fluke infect humans, there are three major groups that are most involved: the lung flukes (*Paragonimus*, various species), the liver fluke (*Clonorchis sinensis*), which is hermaphroditic, and the blood flukes (*Schistosoma*, various species). We will look at the pathogenesis of these three.

Paragonimiasis

Pathogenesis of paragonimiasis. There are several species of *Paragonimus* that infect humans, and this lung fluke causes more than 5 million infections, mostly in the Far East. This infection is caused by eating infected crabs and is not a problem if the shellfish are cooked properly.

Presence of the adult worm in a human host causes **eosinophilia**, inflammation, and eventually the formation of a fibrous capsule that surrounds one or more parasites. An infected individual may have as many as 25 of these capsules, which eventually swell and erode into the bronchioles of the lung, causing expectoration of brownish eggs, blood, and inflammatory exudate. If the capsules rupture in the pleural cavity, chest pain can result. Eventually the capsules form cystic rings and calcify and can resemble tuberculosis lesions on X-ray. Adult flukes in the intestine can cause pain, bloody diarrhea, and occasional cutaneous masses, and in 1% of cases (mostly children) there can be brain invasion, causing epilepsy and paralysis.

Treatment of paragonimiasis. The disease responds well to praziquantel or bithionol.

Clonorchiasis

Pathogenesis of clonorchiasis. The trematode *Clonorchis*, the liver fluke, forms metacercariae that encyst on freshwater fish. When the infected fish is eaten by a human host, larvae are released into the duodenum and ascend to the common bile duct, where they mature over the course of 30 days. Migration of the larvae from the duodenum may produce fever and chills, as well as mild jaundice, eosinophilia, and enlarged liver. Adult worms cause **epithelial hyperplasia**, inflammation, and fibrosis around the bile ducts, but with a low worm load the patient will be mostly asymptomatic. However, repeated infection can produce worm loads of up to 1000, that can lead to bile stones and bile-duct carcinoma. These flukes can occasionally move to the pancreas and cause obstruction of the pancreatic duct and **acute pancreatitis**.

The adult form of *Clonorchis*, the liver fluke, can survive for 50 years in a human host, feeding on mucosal secretions. In addition to humans, rats, cats, dogs, and pigs can also be definitive hosts.

Treatment of clonorchiasis. Praziquantel and albendazole are effective for this infection. Prevention requires thorough cooking of fish as well as sanitary disposal of human feces.

Schistosomiasis

There are five species of *Schistosoma* that infect humans, and 200–300 million people are infected with this blood fluke, from Africa to the Middle East, Southeast Asia and the Caribbean. About 1 million people die from this infection each year.

Pathogenesis of schistosomiasis. The fluke has a cylindrical body (**Figure 14.14**); there are separate males and females, which copulate and stay conjoined for life. *Schistosoma* couples first mate in the portal vein of a host and then use their suckers to ascend the mesenteric vessels, traveling against the flow of blood, until they reach the ascending colon. Here they lay eggs in the submucosal venules (between 300 and 3000 eggs, depending on the species of schistosome) every day for the remainder of their life, which can be as long as 35 years.

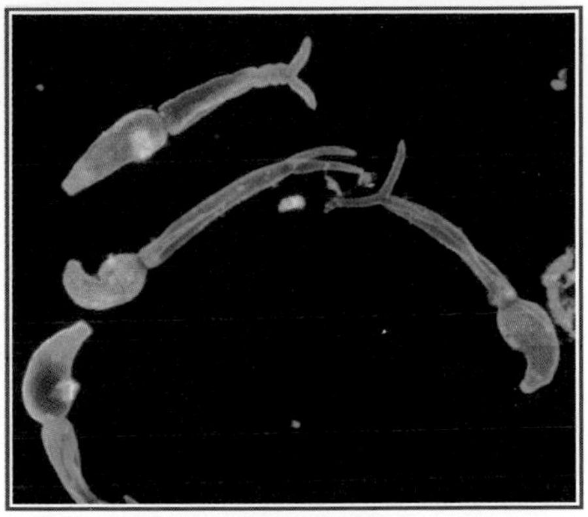

Figure 14.14 A micrograph of four Schistosoma blood flukes.

The eggs deposited closest to the mucosal layer rupture into the lumen of the bowel or bladder and are excreted to the outside. If they reach fresh water, they hatch quickly into the miracidia form, which invades snails. Once in a snail, the miracidia transform into thousands of fork-tailed cercariae, which can penetrate human skin. The cercariae spend one to three days in the skin and then enter the small blood vessels. From there they move into the systemic circulation, on to the gut and through the intestinal capillaries to the portal vein, where they mature to the adult form.

Fast Fact

Schistosomiasis has become the single most important helminthic infection in the world, with 200 million people infected in 74 countries.

This infection is so widespread worldwide that there is extensive morbidity. It continues to be a problem because of the ongoing practice of disposing of human excrement into fresh water. Although most of those infected will have low worm loads of less than 10 and will be asymptomatic, heavier worm loads result in serious clinical disease and death.

Some species of *Schistosoma* can cause bladder infections in their hosts, with progressive obstruction leading to renal failure and uremia. If the infection moves to the bowel, patients experience abdominal pain, diarrhea, and blood in the stool. If eggs reach the central nervous system, epilepsy or paraplegia can result.

Treatment of schistosomiasis. There is no specific treatment for this infection, but treatment with corticosteroids may limit more severe infection. In late stages, therapy is directed at interrupting the deposition of eggs by killing or sterilizing adult worms.

Keep in Mind

- Infections with helminths can be caused by intestinal nematodes, such as *Enterobius* and *Ascaris*, and by tissue nematodes, such as *Trichinella spiralis*.
- Nematodes cause tissue, blood, and lymph infections.
- Cestodes (tapeworms) are the largest intestinal parasites and have a scolex, which incorporates both muscular sucking disks and in some cases attachment hooks called a rostellum.
- Trematodes (flukes) can infect the blood, liver, and lungs.

FUNGAL INFECTIONS

In this section, we look at the pathogenicity of fungi. It is important to remember that the fungi associated with the body are in most cases commensal organisms that live harmoniously with humans. Also keep in mind that these organisms have important roles in the environment and in the production of many nutritional components used by humans.

Consequently, when we look at fungi from a clinical perspective, they are for the most part viewed as opportunistic pathogens. Our discussion here deals with the structure of fungi and the pathogenic mechanisms used by these organisms, but we reserve discussion of the clinical diseases in which fungi are involved for the chapters on infections of specific body systems (Chapters 21–26).

Mycology is the study of fungi, which are eukaryotes and can be either unicellular (the **yeasts**) or multicellular (the **molds**). Infections with fungi are usually either subacute or chronic with relapsing features. Acute fungal infections, such as those seen in bacterial or viral diseases, are uncommon.

Fungal Structure and Growth

The fungal cell possesses typical eukaryotic structures, such as a nucleus containing nucleoli and linear chromosomes. Fungal cytoplasm contains an actin cytoskeleton and organelles such as mitochondria, endoplasmic reticulum, and Golgi bodies. The plasma membrane of fungi differs from that of bacteria in that it incorporates the sterol ergosterol, which helps make it stronger. Each fungal cell is surrounded by a cell wall that is different from the cell wall in bacteria. In the fungal cell, this wall is composed of the polysaccharides mannan, glucan, and chitin. The mannan is found on the surface and in the structural matrix of the wall, where it is linked to proteins. It is these mannan–protein associations that make up the antigenic determinants of the fungal cell. Glucans are polymers of glucose, and some glucans form the fibrils that, in association with chitin, increase the strength of the cell wall. Chitin, which is composed of long, unbranched polymers of *N*-acetylglucosamine, is an inert, water-insoluble, rigid substance. It is the major component in cell walls of certain fungi and allows these fungi to form stable aerial hyphae.

Fungi utilize heterotrophic metabolism, requiring carbon for growth and obtaining nutrients from decaying organic matter. There is considerable metabolic diversity in these organisms. Most fungi are obligately aerobic, and although there are some facultative anaerobic forms, there are no obligately anaerobic fungi.

Fungi reproduce either asexually or sexually. The asexual reproductive elements are called **conidia**. Asexual reproduction involves mitotic division and is associated with the production of budding structures. For sexual reproduction, fungi use spores, which are classified into ascospores, zygospores, and basidiospores. In sexual reproduction, haploid nuclei of the donor and recipient cells fuse to form a diploid nucleus, which may then divide by meiosis. It is during this fusion that genetic recombination occurs in fungi.

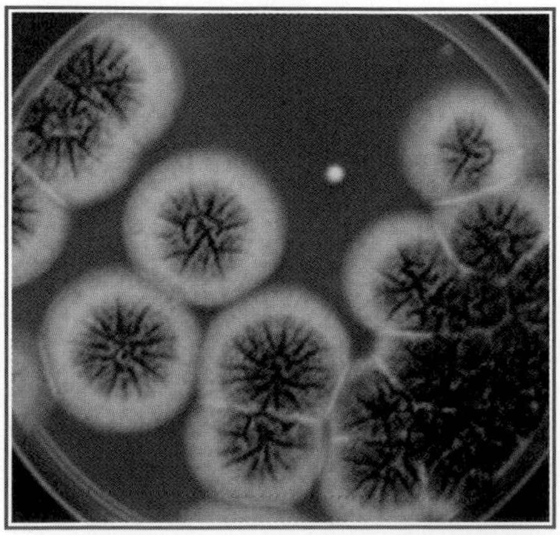

Figure 14.15 The mold form of fungi.

The sizes of fungi vary widely, and a single cell may vary from microscopic to a macroscopically visible structure. Growth can occur in colonies or in some of the most complex multicellular, colorful, and beautiful structures seen in nature. In this chapter we limit our discussions to fungi of clinical importance.

Yeasts and Molds

Fungi can occur in two forms: as a *mold*, which is multicellular (**Figure 14.15**), or as *yeast*, which is unicellular (**Figure 14.16**). The simplest form of growth seen in fungi is by budding, in which a new cell projects from an existing cell. These buds are called **blastoconidia** and occur in yeasts. Fungi can also form **hyphae** (seen in molds), which are tubelike extensions of the cytoplasm that have thick parallel cell walls. As the hyphae

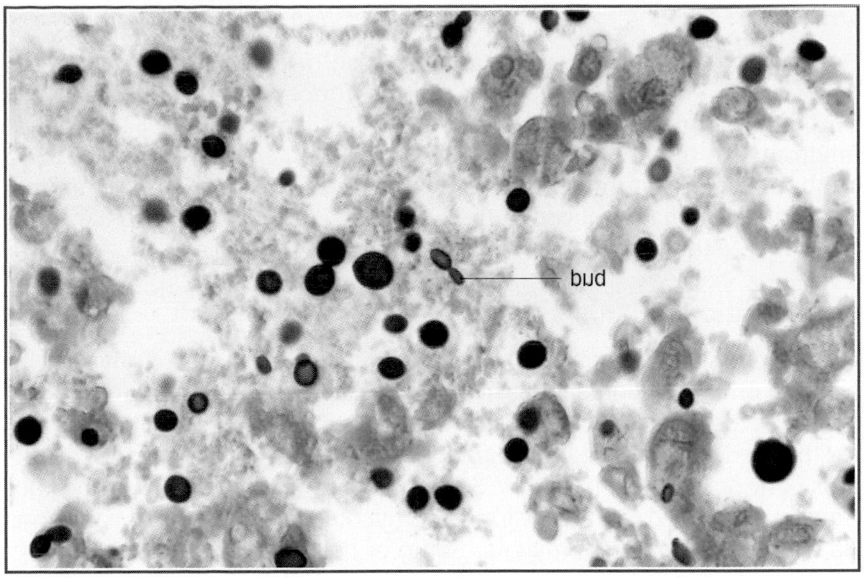

Figure 14.16 An example of the yeast form of fungi. Yeasts are single cells and they can divide by budding off new cells. Notice the buds that are forming on the individual yeast cells.

of the mold extend, they can intertwine to form a **mycelium**. Most molds form **septa**, or crosswalls, within the hyphae (**Figure 14.17**); however, some molds are nonseptate (without septa). Both septate and nonseptate hyphae contain multiple nuclei.

A portion of the mycelium will "root" itself in the nutrient medium (such as soil) and become an anchor, while the rest of the hyphae become aerial as they push upward. In molds, it is the aerial hyphae that contain the reproductive structures. The reproductive conidia and spores of molds and the structures that bear them come in a variety of shapes and sizes and have a variety of relationships to the hyphae. Looking at the morphology and development of these structures is the primary way of identifying the medically important fungi.

Conidia may arise either directly from the hyphae or on a specialized stalk known as the **arthroconidium**. Conidia that form on arthroconidia are delicately attached and break off and disseminate when disturbed.

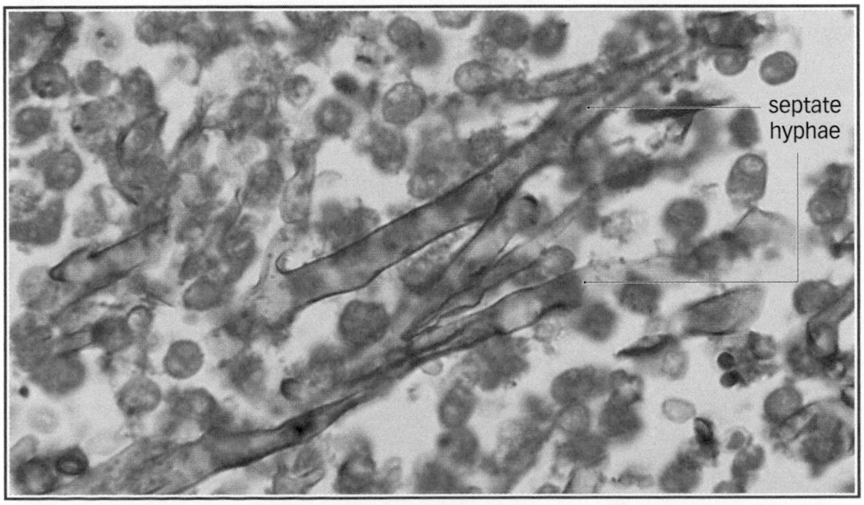

Figure 14.17 A photomicrograph of septate hyphae. Remember that not all forms of fungi will form these separations between cells.

Dimorphism

Some fungi can grow in either a yeast form or a mold form depending on the environment. This is referred to as **dimorphism**. The yeast phase of dimorphic growth requires conditions similar to those of the *in vivo* environment (that is, 35–37°C) and an enriched source of nutrients. In contrast, the mold form requires only ambient temperatures and minimal nutrients. Conidia produced in the mold form may be infectious and can disseminate the fungi.

Classification of Pathogenic Fungi

The kingdom Fungi has four subgroups called divisions (equivalent to phyla in other kingdoms). Classification of fungi depends primarily on the nature of the sexual spores and on the septation of the hyphae. However, for the medically important fungi, ribosomal RNA typing is used to distinguish one species from another. For now, most of the medically important fungi are assigned to the division Ascomycota, with a few in the divisions Basidiomycota and Zygomycota. Some of the medically important fungi are described in **Table 14.7**.

Keep in Mind

- Fungi are mostly harmless commensal organisms that cause no problems for humans.
- Fungi are eukaryotes that have the sterol ergosterol incorporated in their plasma membrane and the polysaccharides mannan, glucan, and chitin in their cell walls.
- Fungi are heterotrophic, metabolically diverse, and either aerobic or facultatively anaerobic.
- Fungi reproduce either sexually or asexually.
- Fungal growth can be in a mold or yeast form, but some fungi are dimorphic and can grow in either form depending on the environmental conditions.
- Medically important fungi are classified and distinguished by ribosomal RNA typing.

Table 14.7 Classification of Medically Important Fungi

Genus	Typical Growth	Formation of Hyphae	Sexual Form	Division	Medical Classification
Aspergillus	Mold	+	?	Ascomycota	Opportunistic pathogen
Blastomyces	Dimorphic	+	?	Ascomycota	Systemic
Candida	Dimorphic	+	?	Ascomycota	Opportunistic pathogen
Coccidioides	Dimorphic	+	?	Ascomycota	Systemic
Cryptococcus	Yeast		+	Basidiomycota	Systemic
Histoplasma	Dimorphic	+	+	Ascomycota	Systemic
Mucor	Mold	+	+	Zygomycota	Opportunistic pathogen
Pneumocystis	Cysts[a]		?	Ascomycota	Opportunistic pathogen
Rhizopus	Mold	–	+	Zygomycota	Opportunistic pathogen
Sporothrix	Dimorphic	+	?	Ascomycota	Subcutaneous
Trichophyton	Mold	+	+	Ascomycota	Superficial

[a]Tissue form but cannot be grown in culture.

In addition to being typed by their ribosomal RNA, medically important fungi can be categorized according to the types of tissues they parasitize and the diseases they produce. We will look at the fungi in this latter context during our discussions. Using this method, we can come up with four general categories of fungal infection: superficial mycoses, mucocutaneous mycoses, subcutaneous mycoses, and deep mycoses.

Superficial Mycoses

Any disease caused by a fungus is called a **mycosis**, and **superficial mycoses** are fungal infections (many of which are caused by species of the genus *Trichophyton*) that do not involve tissue response. Examples include:

Piedra — a colonization of the hair shaft characterized by nodules affixed to the hair. The nodules can be either white or black, and the infection is caused by various species of fungi.

Tinea nigra — brown or black superficial skin lesions found mostly on the palms. This condition is usually seen in women less than 20 years of age and is caused by the fungus *Hortaea werneckii*.

Pityriasis — caused by yeasts of the genus *Malassezia*, which are commonly found on the skin. Overgrowth leads to dermatitis characterized by redness of the skin, itching, and sloughing off of skin cells. The symptoms are caused by hypersensitivity to the fungus. The infection can lead to localized lesions on the chest and back, and more severe forms of the infection can lead to **folliculitis** (mostly seen in immunocompromised patients).

Tinea capitis, which occurs on the scalp and eyebrows. Folliculitis, which is an infection of the hair follicle, is common in this condition.

Favus. This causes hair loss due to permanent destruction of the hair follicle. The result is bald spots associated with crusty scarred skin.

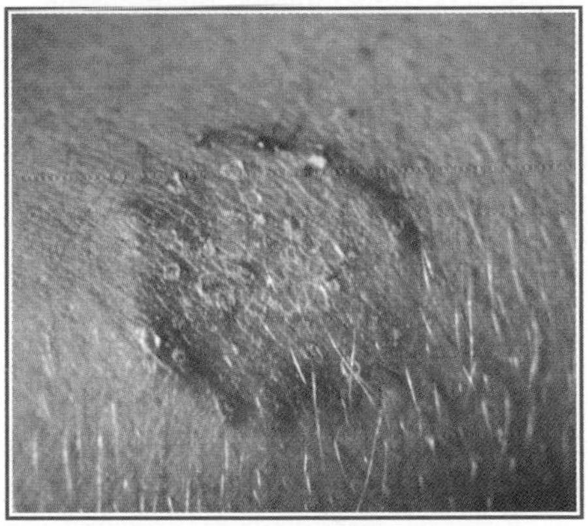

Figure 14.18 An example of ringworm, tinea corporis.

Cutaneous and Mucocutaneous Mycoses

These fungal infections are associated with the skin, eyes, sinuses, oropharynx, external ears, or vagina. Examples of these infections include:

Ringworm. This presents as skin lesions characterized by red margins, numerous scales and reddish itching skin (**Figure 14.18**). The infection is restricted to the stratum corneum of the skin and causes the development of an inflammatory response. Classification of ringworm is based on the location of the lesions:

- **Tinea pedis**, which is found on the feet (**Figure 14.19**) and spaces between the toes

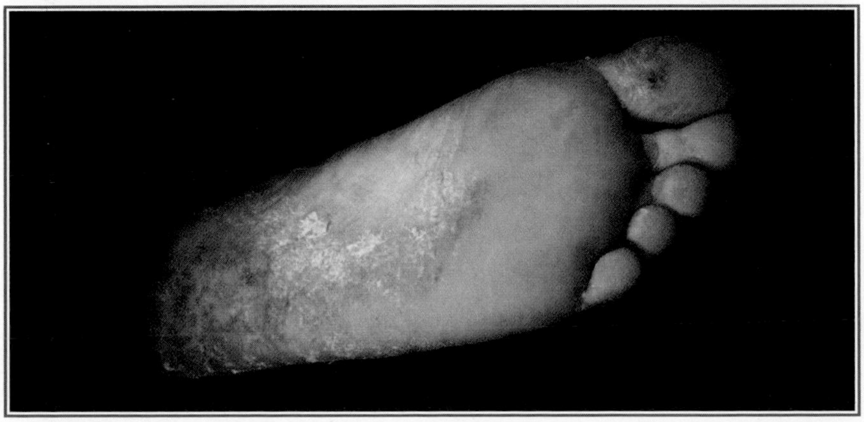

Figure 14.19 An example of athlete's foot, tinea pedis.

- **Tinea corporis**, which is found between the fingers, in wrinkles in the palm, and on scaly skin
- **Tinea cruris**, which causes lesions on hairy skin around the genitalia

Onychomycosis. This is a chronic infection of the nailbed and nail (**Figure 14.20**), and is most commonly found in toenails. The infection expands from the periphery to the center of the nail and can cause the distal part of the nail to rise up and crumble.

Mucocutaneous candidiasis. This is caused by the yeast *Candida albicans*, which commonly colonizes the mucous membranes. The infection is associated with the loss of immunocompetence. There are two clinical types of candidiasis. The first is *thrush*, in which there is fungal growth in the oral cavity. This is one of the first indications of acquired immune deficiency syndrome (AIDS). The second form of candidiasis is *vulvovaginitis*, which results in the accumulation of a dry, white, crumbling material in the vaginal canal and often causes irritation and erythema. This infection can in many cases be associated with changes in hormone balance.

Fast Fact

Some eyedrops contain preservatives that fungi can use for growth, thereby increasing the potential for fungal infection of the eyes.

Hyperkeratosis. This causes extended scaly areas on the hands and feet.

Keratitis. Involves the colonization or infiltration of corneal epithelium; it can occur after surgery, the use of corticosteroids, or careless application of contact lenses. The affected eyes can become ulcerated and scarred.

Subcutaneous Mycoses

These mycoses present as localized traumatic primary infections of the subcutaneous tissues. Infection provokes the innate immune response and increases in the eosinophil population. The infection can cause the development of cysts and granulomas. Although not life-threatening, these growths can sometimes cause disfigurement. There are several types of subcutaneous mycosis, including:

- *Sporotrichosis*, which occurs after traumatic implantation of fungal organisms.

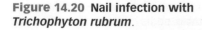

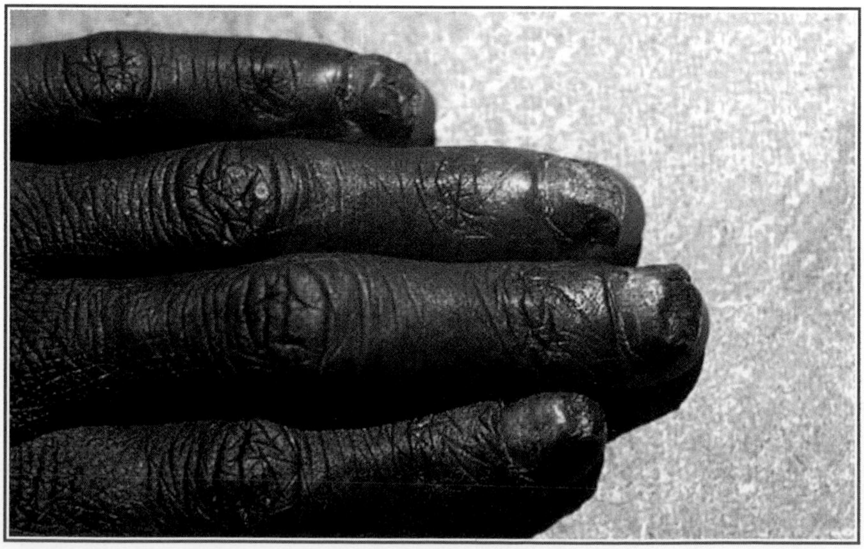

Figure 14.20 Nail infection with *Trichophyton rubrum*.

- *Paranasal conidiobolomycoses*, an infection of the submucosa of the paranasal sinuses with the formation of granulated fibrotic tissue filled with eosinophils. The infection progresses slowly but can be severe.

- *Zygomatic rhinitis*, in which areas of the mucosa become grayish black (similar to a blood clot). The fungus invades the tissue through the arteries and causes thrombosis (blood clotting). Severe cases of this infection can involve the nervous system, leading to meningitis and unilateral blindness.

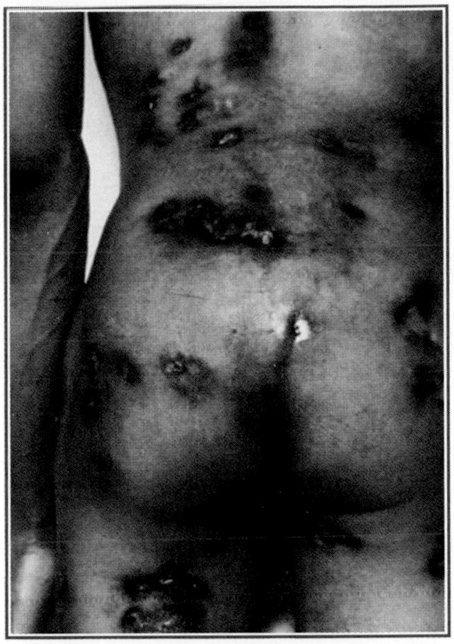

Figure 14.21 Surface lesions are characteristic of a disseminated fungal infection.

Deep Mycoses

Deep mycoses can be either *localized* or *systemic*. These infections are usually seen in immunosuppressed patients. There have been increases in opportunistic systemic infections in immunocompromised patients who have AIDS, cancer, or diabetes and also in intravenous drug abusers. Deep mycoses can be acquired by inhalation of fungi or fungal spores, as well as through enteric invasion. Infection can also be caused by contaminated surgical instruments, catheters, and hypodermic needles. If the infection remains localized in the deep tissues and organs, it is referred to as *systemic mycosis*. However, if it moves from the initial site through the blood or lymph system, it is referred to as *disseminated mycosis*. *Secondary cutaneous mycoses* are disseminated infections that move to the skin (**Figure 14.21**).

Some examples of deep mycoses are:

- **Coccidiomycosis** is caused by members of the genus *Coccidioides*. The primary infection enters the body through the respiratory system. The infection leads to fever, bronchial pneumonia, and erythema. Conjunctivitis may also occur. In most patients, coccidiomycosis resolves spontaneously as a result of the host's defensive systems. However, in a small percentage of cases there can be systemic spread to skin, bones, and visceral organs, which often results in a fatal outcome.

- **Histoplasmosis** caused by the fungus *Histoplasma capsulatum*, is associated with the monocyte/macrophage components of the innate immune response and is often associated with immunodeficiency. Macrophages containing the fungal cells multiply in great numbers and give rise to granuloma formation. These granulomas necrotize and become calcified. If this infection becomes disseminated (**Figure 14.22**), the result is frequently fatal.

- **Aspergillosis** is also associated with reduced immune competence. There are several species of *Aspergillus* that can cause infection, with the worst form being invasive. In this case, the fungal mycelium grows between epithelial cells and disseminates through the blood to the lungs. This results in acute pneumonia with high fever. The mortality rate for this type of aspergillosis infection is very high, and death occurs in one to three weeks.

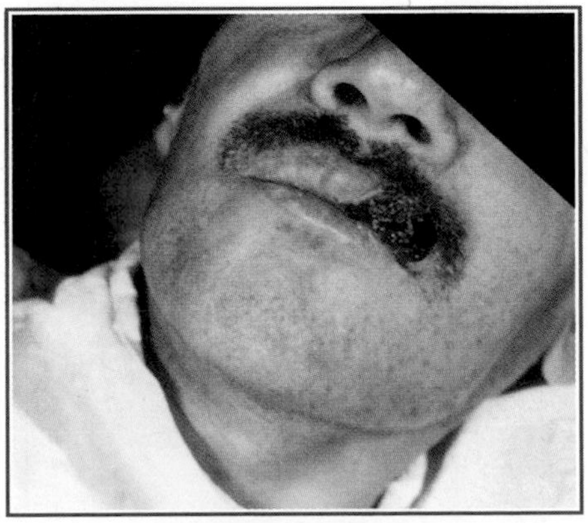

Figure 14.22 A skin lesion associated with a *Histoplasma capsulatum* disseminated infection.

Pathogenesis of Fungal Infections

We all have regular contact with fungi on a daily basis. In fact, thousands of fungal spores are inhaled or ingested every day. Most fungi are so well adapted to us that they are part of our normal microbial flora. Clinical fungal infections are very uncommon, but when they occur, progressive systemic fungal infections are some of the most difficult of all infectious diseases to diagnose and treat. This is especially true in immunocompromised individuals.

Fungal infections can be acquired from the environment through the inhalation or ingestion of infectious conidia from molds. Some of these molds are ubiquitous, but others are limited to specific geographical areas and can infect only individuals who travel through these areas. Keep in mind that pathogenic fungi constitute only a tiny fraction of the total number of fungi found in nature. Endogenous fungal infections are restricted to a few yeasts (primarily *Candida*) that have the ability to colonize by adherence to host cells; if given the opportunity to do so, these yeasts will invade more deeply.

The pathogenesis of fungal infections can be divided into three categories: *adherence*, *invasion*, and *tissue injury*.

Adherence

Several fungal species, particularly yeasts, are able to adhere to the mucosal surfaces of the gastrointestinal tract and the female genital tract. This adherence usually requires a surface adhesion molecule on the fungi and a receptor on the target cell. For example, in *Candida albicans* the mannan seems to be the adhesion molecule, which attaches to fibronectin receptors on target cells.

Invasion

This is a very important part of the infectious process, and some fungi are introduced to tissues through mechanical breaks in the skin. However, fungi that infect the lung must produce conidia that are small enough to be inhaled past the defenses of the upper respiratory system. For example, *Coccidioides immitis* conidia are between 2 and 6 μm in diameter. This small size allows the conidia to remain suspended in the air for long periods, a condition that allows them to reach the bronchioles of the lung directly and initiate pulmonary coccidiomycosis.

Some dimorphic fungi can be triggered by temperature change to undergo a metabolic shift and change their morphology into a more invasive form. For example, the yeast *Candida albicans* can convert and grow hyphae that invade tissues and spread the infection.

Tissue Injury

Fungi do not produce extracellular products that damage tissues in the way that bacteria produce exotoxins, for instance. Some fungi do produce exotoxins in the environment but do not do so *in vivo*. In fact, the primary tissue injury seen in fungal infections is associated with the host inflammatory and immune response stimulated by the prolonged presence of fungi during the infection.

Host Defense Against Fungal Infection

The defenses seen with fungal infection primarily involve phagocytosis and the T cell-mediated activation of macrophages.

Phagocytosis

Most people have a high level of resistance to fungal infection, and this is due to the ability of neutrophils to kill any fungal hyphae that reach the tissues. However, fungi that are dimorphic and cause mild to severe disease are resistant to phagocytosis. For example, *Coccidioides immitis* has a cell wall component that is anti-phagocytic, and *Candida albicans* can bind to complement in such a way as to inhibit opsonization. *Histoplasma capsulatum* is like some bacteria in that it grows inside phagocytic cells.

The Adaptive Immune Response

Remember that (1) most fungi will not cause even a mild infection in immunocompetent individuals and (2) even if a fungal infection begins, it will be short-lived. This is due to the adaptive immune response and is the province of the T cell component of that response. This statement is supported by the fact that progressive fungal disease is seen only in immunocompromised patients, in which the T cell response is missing.

The humoral antibody response is always seen in a fungal infection, but it does not seem to have as much of a role in its eradication as the T cell response. One exception to this is *Cryptococcus neoformans*, which is the only encapsulated fungus. Recall from Chapter 5 that encapsulated organisms can resist innate nonspecific defensive responses and that antibody can coat the capsule and allow phagocytosis to take place. Although the antibody acts in this fashion with *Cryptococcus*, it is still the T cell that does most of the work to remove this fungal problem.

Let's look briefly at the sequence of events associated with defense against a fungal infection:

- When hyphae or yeast cells reach deep tissues in a host, either they are killed by phagocytosis or they can resist phagocytosis, using some of the mechanisms described above.
- The surviving fungal cells continue to grow. If they are dimorphic, they convert to yeast or to hyphal structures, which may not be killed as effectively by macrophages.
- Continued presence of the organism will initiate an immune response in which T cells interact with the fungi causing activation of macrophages and increased production of interferon-γ. These activated macrophages destroy the fungi and control the infection.

The pathogenic potential of fungi is limited and usually affects only those persons who are in some way immunocompromised. We will explore some of the pathogenic problems seen with fungi as we look at infections in the different systems of the body (Chapters 21–26).

Keep in Mind

- Medically important fungi can be divided into four categories of mycoses (diseases caused by fungi).
- Superficial mycoses do not involve tissue responses and include infection of the hair shafts and superficial skin.
- Mucocutaneous mycoses are associated with the skin, eyes, sinuses, oropharynx, external ears, or vagina.
- Subcutaneous mycoses are localized infections of the subcutaneous tissues.
- Deep mycoses can be localized or systemic, and are usually restricted to patients who are immunocompromised.
- The pathogenesis of fungal infections involves adherence, invasion, and tissue injury.
- Primary host defense against fungal infection involves phagocytosis and the T cells of the adaptive immune response.

SUMMARY

- Parasitic infections in humans can be caused by protozoans and helminths.
- These infections affect hundreds of millions of people around the world, and millions die from these infections each year.
- Protozoan parasites include rhizopods, ciliates, flagellates, and sporozoans.
- Parasitic helminths are found in three classes: nematodes, cestodes, and trematodes.
- Medically important fungi can be divided on the basis of the types of infection they cause: superficial mycoses, subcutaneous mycoses, mucocutaneous mycoses, or deep mycoses.
- Deep mycoses are the most serious fungal infections and can be either localized or systemic.

In this chapter we have looked at the pathogenesis of parasites and fungi. This completes our examination of the different types of infectious organism. Taken together with what we have learned in the previous chapters, we now have a good grounding in the disease process and in the different pathogens that are involved in the infectious disease process.

SELF EVALUATION AND CHAPTER CONFIDENCE

Multiple Choice

Answers are given in the back of the book and help can be found in the student resources at:
www.garlandscience.com/micro

1. Protozoan parasites are
   A. Multicellular
   B. Prokaryotic
   C. Macroscopic
   D. Unicellular
   E. Worms

2. *Plasmodium falciparum*
   A. Causes sleeping sickness
   B. Is a helminth
   C. Causes the mildest form of malaria
   D. Is the most dangerous of the malaria-causing parasites
   E. Causes dysentery

3. Which of the following invade the blood and cause chronic illness?
   A. Rhizopods
   B. Flagellates
   C. Ciliates
   D. Cestodes

4. Protozoan parasites can be all of the following except
   A. Aerobic
   B. Facultatively anaerobic
   C. Anaerobic
   D. Heterotrophic

5. Sexual reproduction in protozoans is called
   A. Schizogony
   B. Binary fission
   C. Budding
   D. Gametogony

6. The body of a worm is covered by a
   A. Cell wall
   B. Membrane
   C. Cuticle
   D. Scolex
   E. None of the above

7. Nematodes are
   A. Tape worms
   B. Flukes
   C. Roundworms
   D. A type of fungus
   E. Single-celled eukaryotes

8. Trematodes are
   A. Tapeworms
   B. Flukes
   C. Roundworms
   D. A type of fungus
   E. Cestodes

9. The severity of a helminth infection is directly related to
   - A. The type of helminth
   - B. A preexisting bacterial infection
   - C. The size of the worm
   - D. The number of worms that are present
10. The organism that causes malaria is transmitted by
   - A. The bite of a fly
   - B. The bite of a tick
   - C. The bite of a mosquito
   - D. The respiratory system
   - E. The fecal–oral route of contamination
11. The intermediate host for *Plasmodium* species is
   - A. A fly
   - B. A mosquito
   - C. A human
   - D. A tick
   - E. There is no intermediate host
12. Merozoites of *Plasmodium* are found in
   - A. The saliva of a mosquito
   - B. The human intestinal tract
   - C. The intestinal tract of a mosquito
   - D. Human hepatocytes
   - E. None of the above
13. All of the following are true about flagellates except
   - A. They are widespread in nature
   - B. They use flagella for locomotion
   - C. They require an intermediate host for transmission
   - D. All flagellates cause disease in humans
14. *Enterobius vermicularis* is a
   - A. Tapeworm
   - B. Cestode
   - C. Trematode
   - D. Pinworm
   - E. None of the above
15. Which of the following systems in a parasitic helminth is not greatly reduced compared with free-living helminths?
   - A. Nervous system
   - B. Digestive system
   - C. Locomotion
   - D. Reproductive system
   - E. All are reduced
16. Transmission of helminthic diseases to humans is usually by
   - A. The genitourinary route
   - B. The respiratory route
   - C. Vectors
   - D. The gastrointestinal route
   - E. All of the above
17. Cercariae, metacercaria, miracidia, are stages in the life cycle of
   - A. Sporozoans
   - B. Nematodes
   - C. Cestodes
   - D. Trematodes
18. Which one of the following does not belong with the others?
   - A. *Entamoeba*
   - B. *Trypanosoma*
   - C. *Cryptosporidium*
   - D. *Giardia*
   - E. *Plasmodium*
19. What do tapeworms eat?
   - A. Intestinal bacteria
   - B. Intestinal contents
   - C. Red blood cells
   - D. Host tissues
   - E. All of the above
20. Ringworm is caused by a
   - A. Protozoan
   - B. Nematode
   - C. Trematode
   - D. Cestode
   - E. Fungus
21. Mycology is the study of
   - A. Parasites
   - B. Worms
   - C. Fungi
   - D. Insects
   - E. The structure of bacterial colonies
22. Fungal plasma membranes contain
   - A. Peptidoglycan
   - B. Chains of *N*-acetylglucosamine
   - C. Cholesterol
   - D. Ergosterol
   - E. Mannan
23. Fungal cell walls contain
   - A. Mannon
   - B. Ergosterol
   - C. Glucan
   - D. A and C
   - E. B and C
24. The branching structures seen in fungi are called
   - A. Spores
   - B. Conidia
   - C. Hyphae
   - D. Blastoconidia
   - E. None of the above

25. Superficial mycoses

A. Involve tissue destruction
B. Are systemic infections
C. Are associated with hair shafts
D. Are infections of the liver
E. Are infections of the intestines

26. All of the following are true about mucocutaneous candidiasis except

A. It is caused by *Candida albicans*
B. It is an infection of the mucous membranes
C. It can be seen as thrush
D. It is a deep mycosis
E. It can be seen as vulvovaginitis

27. Tinea capitis is

A. A subcutaneous mycosis
B. A systemic infection
C. A form of ringworm
D. Found on the feet and between the toes

28. Successful host defense against fungal infection

A. Is restricted to phagocytosis
B. Is restricted to the humoral antibody response
C. Is restricted to the T cell response of adaptive immunity
D. Is a combination of phagocytosis and the production of antibody
E. Is a combination of phagocytosis, the production of antibody, and the T cell response

29. The most commonly seen yeast infections are caused by

A. *Histoplasma*
B. *Aspergillus*
C. *Penicillium*
D. *Saccharomyces cerevisiae*
E. *Candida albicans*

DEPTH OF UNDERSTANDING

Questions listed here require you to bring together the concepts you have learned in this chapter into a discussion format. This helps you to increase your depth of understanding of the material you have learned. Help can be found in the student resources at:
www.garlandscience.com/micro

1. Describe the life cycle of *Plasmodium falciparum* and why we in the United States do not see many cases of this infection.

2. Fungal infections are usually opportunistic. Explain how this relates to patients with (a) Acquired Immunodeficiency Syndrome (AIDS) and (b) patients who are undergoing prolonged treatment with antibiotics.

CLINICAL CORNER

Help can be found in the student resources at:
www.garlandscience.com/micro

1. Ronald Johnson has just received a kidney transplant. With the exception of the kidney problem, he has been in relatively good health until now. After the transplant, he has been put on drugs that suppress his immune response. During his stay in the hospital, he has come down with *Pneumocystis* pneumonia.

A. Explain how this could have happened.

B. Could the administration of the immunosuppressive drugs have had a role in his infection?

2. You are working in a large county hospital emergency room. Two men have been brought in from a homeless shelter. Both are complaining about shortness of breath, but initial exam has ruled out heart attack. When you examine them, you notice that one of the men has oral thrush. Both men have a long history of alcoholism and have been homeless for several years.

A. What does the oral thrush indicate?

B. Why would one of the men have this condition but not both?

C. What would you be concerned about in each of these patients?

3. You are working as a physician's assistant in a large family practice. Your patient is a 44-year-old female in apparent good health. She complains of a chronic cramping in the abdomen, excessive flatulence, and intermittent diarrhea. She has also lost weight for no apparent reason. She explains that she had traveled extensively in Africa for several months more than two years ago.

A. Based on her symptoms and what she has told you, what kind of problem do you think she may have?

B. What would be the easiest way to find out what may be happening to her?

PART IV

Host Defense

Up to now we have looked at the infection process in a variety of ways that focused on the organisms that infect us. Here in **PART IV** we will look at how we defend ourselves from infection.

Chapters 15 and **16** provide you with information on host defense. In **Chapter 15** we take a detailed look at the innate immune response, which is the first response to any infection. We will see that in many cases, this powerful protective response is enough to overcome an infection. Infection will also initiate the adaptive immune response, which is discussed in **Chapter 16**. The adaptive response is very powerful and very specific. It involves the production of antibody and a potent cellular response to infection. The adaptive response also involves the gift of memory, which in many cases provides us with lifelong immunity to infection.

After **Chapters 15** and **16** describe how we protect ourselves from infection, **Chapter 17** shows you how things can go wrong with these protective systems and how that loss of protection can have catastrophic consequences for the host. Failure of our immune response results in a greater possibility of infection and also in the development of autoimmune diseases. In this chapter, we will also take a detailed look at HIV infection and AIDS, which remain a significant problem for the health care community.

When you have finished **PART IV** you will:

- See how the host defends itself against infection by using the innate immune response, which is nonspecific and the first to fight when infection occurs.
- Know how the adaptive immune response not only gives us powerful protection against infection but also has the gift of memory. Therefore if this infection has been seen before, the response to it will be very fast and very powerful.
- Understand **Chapter 16**, which will show you why vaccinations are so important in providing long-lasting immunity to many of the most dangerous infections.
- Have an understanding of how the immune response can be ineffective, fail completely, and even turn on the host.

As usual, it is important to continue to use what you find in each part as a foundation for those chapters to come. Host defense will be an important consideration in everything we have described previously and in what we are about to discuss in chapters to come.

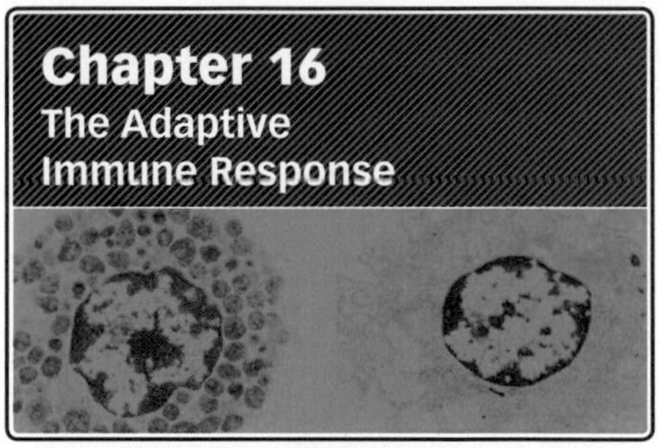

Chapter 15

The Innate Immune Response

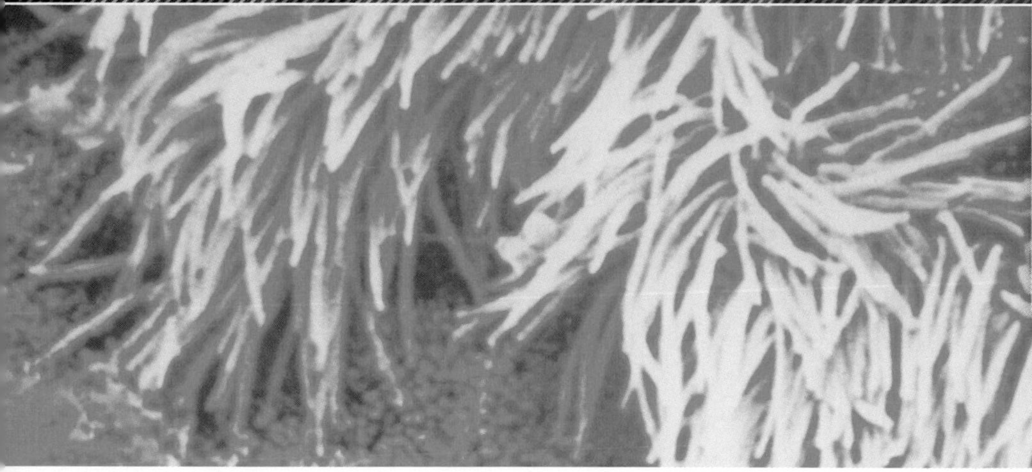

Why Is This Important?

This chapter introduces you to the innate immune response, which includes all the nonspecific host defenses that are of paramount importance in fighting off infectious disease.

OVERVIEW

In this chapter, we begin to examine the ways in which a host defends itself against microbial infection. During this discussion, it is important that you keep in mind what you learned in Chapters 5 and 6 about compromised hosts and about the mechanisms that pathogens use to undermine our defenses. The interaction between pathogen and host is one of the cornerstones of infectious disease. We will see in this chapter that there are several natural barriers to infection. There are also highly effective and a lethal nonspecific defense mechanisms that hosts can employ as guards against infection. Along with the specific immune reactions that we will learn about in the next chapter, these nonspecific defense mechanisms help to protect us against a variety of potentially dangerous infections.

We will divide our discussions into the following topics:

The Innate Immune Response

- THE FIRST LINE OF DEFENSE IN THE INNATE IMMUNE RESPONSE
- THE SECOND LINE OF DEFENSE IN THE INNATE IMMUNE RESPONSE
- GENETIC SUSCEPTIBILITY TO INFECTION

The body can defend itself against infection by using two types of immune response, the innate and the adaptive. The **innate immune response** is available to us when we are born and is nonspecific, where *nonspecific* means that this response can react against any infection or pathogen. In contrast, the **adaptive immune response** (Chapter 16) is specific, meaning that it responds against a specific pathogen. The adaptive immune response also has the gift of "memory," which allows it to remember any pathogen it reacted against in the past and to respond quickly and powerfully if that pathogen returns.

In this chapter, we look at the nonspecific innate immune response. This response can be divided into two parts: (1) mechanisms that use barriers to prevent infectious organisms from entering the body's environment, and (2) mechanisms that destroy any infectious organisms that manage to break through those barriers. Barriers can be looked on as the first line of nonspecific host defense, and the destruction of pathogens can be looked on as the second line of nonspecific defense.

What Do I Need to Know?

To get the most out of this chapter, please review the following terms from your previous courses in biology, anatomy, physiology, or chemistry or in previous chapters of this book as indicated in parentheses: apoptosis, basement membrane (**13**), ciliated cell, dendrite, effector cell, endocytosis (**4**), endotoxin (**5**), free radical (**2**), hematopoietic, immunoglobulin, lamina propria (**5**), ligand, lymph (**5**), macrophage, monoblast, mucous membrane (**5**), mucus (**5**), myeloid stem cell, parasite (**14**), phagocyte, promonocytes, Peyer's patches (**5**), reticular fiber, sebaceous gland.

The best way to look at host defense mechanisms is to consider what is happening in an infected patient. As we saw in previous chapters, bacterial and viral infections can result in cell and tissue damage. Host innate immune responses are triggered when this damage occurs, and a series of chemical messages are sent out from the damaged site signaling the injury. These messages can be chemotactic, meaning they give off a chemical "scent." If we use the analogy of crumbs in the forest, defensive cells follow these chemical signals — the "crumbs" — back to the site of the injury. As the initial-responder cells arrive at the damaged site, they release chemicals that amplify the defense against the cause of the injury. This complex "ballet" of interactions works to maximize the defensive efforts such that the infection is brought under control as quickly and efficiently as possible.

THE FIRST LINE OF DEFENSE IN THE INNATE IMMUNE RESPONSE

There are several natural barriers to infection. The principal functions of these barriers are not directly related to defense. Thus, their defense functions can be looked at as a secondary effect. Just their presence and normal function in the body inhibit entrance of and colonization by infectious organisms. We can divide these barriers into two types: mechanical and chemical.

Mechanical Barriers

Skin

The most obvious mechanical barrier presented to microorganisms trying to get inside the body is the skin. Recall from your anatomy courses that the skin is a semi-watertight, impermeable barrier made up of epidermis and dermis. The outer layer, the epidermis, is composed of tightly packed dead and dying cells that contain keratin, a tough protein that gives the cells strength and "semi-waterproofs" them. Aside from the places where glandular structures and hair follicles are located, the epidermis is essentially impenetrable to most microorganisms (**Figure 15.1**). Because there is no access to blood or lymph in the epidermal layer, any intrusion of bacteria into this layer will be localized and not systemic.

Entry through the epidermal layer requires breaking the barrier — for example, by a cut, burn, or insect bite. Of these intrusion routes, burns are the most deadly from the perspective of infection because large patches of the epidermis can be lost. The skin is constantly occupied by organisms such as *Staphylococcus* and *Streptococcus* (**Figure 15.2**), and a

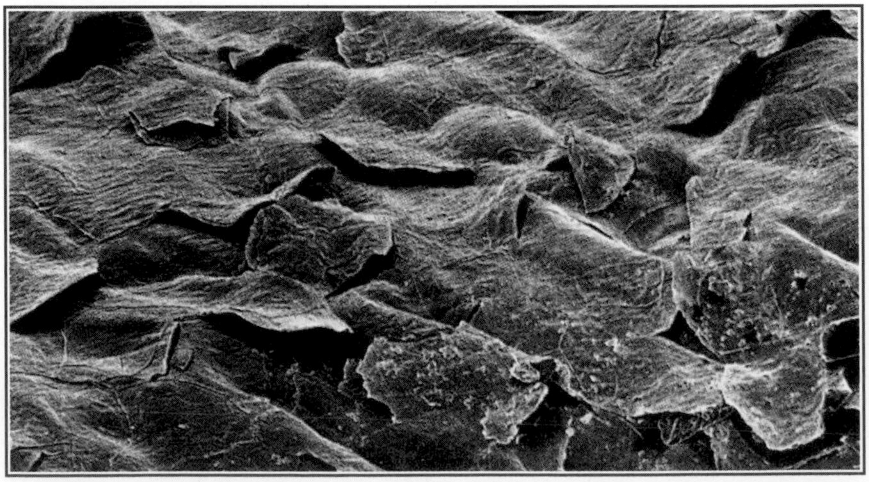

Figure 15.1 A scanning electron micrograph of the surface of human skin. Notice how tightly the epidermal cells are packed together. These cells are also keratinized, providing a very strong barrier to entry into the body.

Figure 15.2 Scanning electron microscopy shows microorganisms on the surface of the skin. Notice the depression in the center, which is a pore. Any break in the skin will allow these organisms immediate access to the interior of the body.

loss of large sections of this barrier allows these organisms to enter the dermis and underlying tissue. Because the dermis and tissue underlying it are vascularized, the chances of systemic infection are greatly increased when skin has been burned away. In addition, immunological incapacitation and debilitation can accompany serious burn injury, thereby compounding the potential for severe infectious disease.

Mucous Membranes

Recall from Chapter 5 that the major portals of entry for infectious organisms — the respiratory, gastrointestinal, and genitourinary tracts — all contain mucous membranes. Although the primary function of the mucus produced by these membranes is to keep tissue moist, it also traps microorganisms. The respiratory tract is a good example of how this works. Although the lower respiratory tract is essentially sterile, the upper respiratory tract is constantly exposed to pathogenic bacteria and viruses that enter through the nasal and oral cavities.

Mucus helps defend against these intrusions through an elegant mechanism known as the **mucociliary escalator**. The lower respiratory tract is lined with ciliated cells (**Figure 15.3**) and goblet cells. The goblet cells produce mucus, which traps any microorganisms that have entered the tract. Then the ciliated cells rhythmically move this mucus up to the oral cavity (hence the "escalator" metaphor), where it is either swallowed or expectorated.

Fast Fact

The mucociliary escalator normally moves a small amount of mucus at a rate of about 1–3 cm per hour, but coughing or sneezing accelerates the movement.

Mucus also has a role in the gastrointestinal tract. Produced in copious amounts in the stomach, it coats the stomach wall to protect it against the acidic fluids needed for digestion. Some microorganisms use this mucous coating for their own purposes. For example, *Helicobacter pylori*, a bacterium associated with stomach ulcers, settles into the mucous coating as part of the process of ulcer formation.

Other Mechanical Barriers

Several other barriers help keep microorganisms from establishing a foothold. Like those discussed above, these barriers also have a role in normal bodily functions in addition to their (secondary) role in inhibiting microbial growth.

The lacrimal apparatus. Recall from Chapter 5 that the eye is a portal of entry for microbes. This structure is an *immunologically protected site*, meaning that the immune system is not permitted to work here. This is a protective mechanism because the immune system can in some cases turn against the body in the form of autoimmune disease (Chapter 17). You would think that because host defense is reduced in this location, the eye would be a good region for microbial entry, but in fact few infections occur here. This is because of tears, which are produced in the lacrimal glands and constantly flush any foreign particles across the eye

Figure 15.3 Ciliated cells found in the lower respiratory tract are part of the mucociliary escalator. These cells, shown here by scanning electron microscopy, work with mucus-producing goblet cells to trap and remove organisms and also any other materials that enter the respiratory tract.

Fast Fact

Iron is required as a cofactor for many metabolic functions that take place in bacteria and is therefore a requirement for successful infection.

and into the lacrimal canal (**Figure 15.4**). The lacrimal apparatus increases the flow of tears whenever any irritant enters the eye. In addition to the flushing action, tears also contain three important components: IgA, an immunoglobulin; lysozyme, an enzyme that breaks down microbial cell walls; and **lipocalin**, which has recently been shown to have antimicrobial activity by inhibiting pathogens as they scavenge for iron.

Saliva. This secretion is produced by the salivary glands of the mouth and is similar to tears. The main functions of saliva are to cleanse the teeth and mucous membranes of the mouth and prepare food for digestion. However, this liquid can also inhibit microbial growth because, like tears, it contains both lysozyme and the immunoglobulin IgA. It is important to remember that although saliva can reduce the extent of microbial colonization, the mouth is always a very busy place when it comes to microbial traffic.

Epiglottis. This flap of tissue at the back of the throat is a barrier that keeps food from entering the respiratory tract. In this way, organisms that enter through the fecal–oral route of contamination are kept away from the respiratory tract. Infection of the epiglottis can be a life-threatening problem.

Chemical Barriers

The body also secretes chemical substances that act as barriers to microbial entry and growth. Like the mechanical barriers described above, each of these secreted substances is a passive barrier that provides protection against infection ancillary to its primary physiological function.

Sebum. Recall from your earlier courses that the sebaceous glands produce an oily secretion called sebum that prevents hair from drying out and forms a protective film over the skin. Sebum contains unsaturated fatty acids, which inhibit bacterial and fungal growth, and organic acids that make the environment of the skin naturally acidic (pH about 3–5) and inhibitory to bacterial growth. The combination of these chemical factors and dead bacterial cells on the skin can cause body odor and influence certain skin conditions, such as acne. In some cases, antibiotics are prescribed for acne but in many cases these drugs are not warranted. This unnecessary use of antibiotics can increase both the chances of opportunistic pathogenicity and the creation of drug-resistant pathogens.

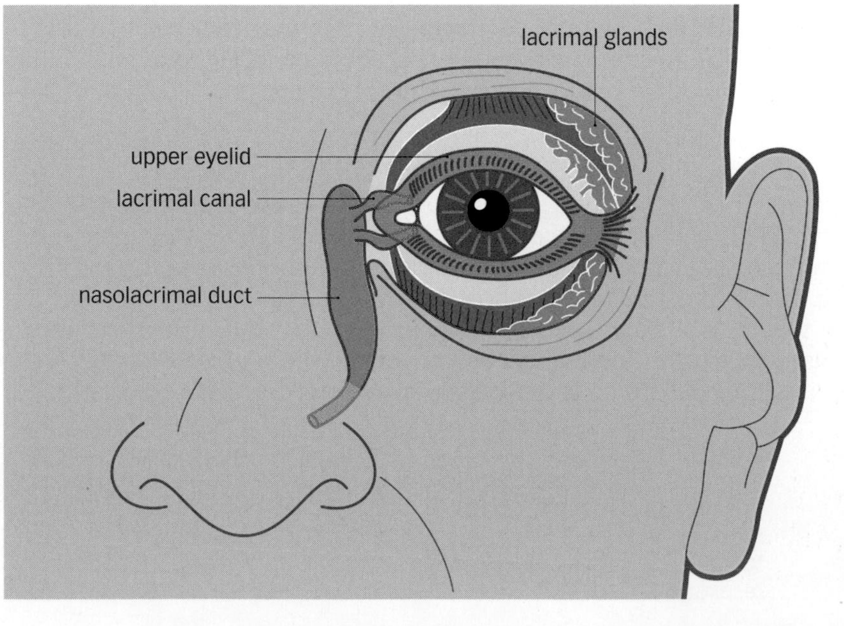

Figure 15.4 The lacrimal apparatus is the primary mechanical barrier protecting the eye. Tears, which contain the antimicrobial enzyme lysozyme, are constantly produced in the lacrimal gland and flush across the eye, taking organisms and debris with them. The tears are collected in the nasolacrimal duct. Any irritant that enters the eye increases the rate at which tears are produced.

Perspiration. Perspiring is a continuous natural mechanism that regulates body temperature and eliminates wastes. The effect of perspiring is like that seen in the lacrimal apparatus of the eye. Because we are continuously perspiring, there is a flushing action by which the perspiration we produce clears many organisms from the surface of the skin. When we perspire profusely, this action is magnified.

Just as tears do, perspiration contains the enzyme lysozyme, which breaks down microbial cell walls.

Gastric Juice. Stomach acids and enzymes, collectively called *gastric juice*, break down the stomach contents. This combination of substances produces a harsh chemical environment that is not conducive to microbial growth. However, bacteria hidden in food particles are protected from this environment and can therefore thrive in the digestive tract. The bacterium *Helicobacter pylori* and certain viruses use the stomach and its very acidic environment as part of the infection process.

Urine. In Chapter 5, you learned that the genitourinary tract is a portal of entry for many infectious organisms. Secretions coupled with lactic acid produced by resident bacteria make this environment acidic, a condition that prevents colonization by many potentially pathogenic bacteria. The acidity of urine inhibits most microbial growth in the tract, and the flushing action of this body fluid keeps microbes from attaching to tissues. It is interesting to note, however, that low pH conditions are more prevalent in the female genitourinary tract than in the male tract. Because an acidic environment does allow fungal organisms to grow, yeast infections are more common in women than in men.

Transferrins. In addition to lipocalin, the iron-binding substance contained in tears, the body produces another class of substances, called **transferrins**, that also bind iron as a defense mechanism. Iron is required as a cofactor for many metabolic functions that take place in bacteria. Human blood contains transferrins, and bacterial growth is competitively inhibited when the body's transferrin molecules capture iron and prevent bacteria from using it. **Table 15.1** summarizes the various fluids acting as the barriers, both mechanical and chemical, involved in the innate immune response's first line of defense.

In this section we have seen that many of the body's anatomical and physiological mechanisms can prevent the entry and growth of infectious microorganisms. Keep in mind, however, that defense is not the primary function of these mechanisms. Instead, the defensive results are just passive side effects of normal functions.

Table 15.1 Secretions That Function as Barriers in the Innate Immune Response

Secretion	Barrier Type	Defense Function
Saliva	Mechanical	Washes microbes from teeth and gums; contains the antibacterial enzyme lysozyme and immunoglobulin IgA
Tears	Mechanical	Contains lysozyme and IgA; constant flushing across eyeball keeps microbes from attaching
Perspiration	Mechanical	Flushes organisms from skin surface; contains lysozyme
Stomach acid	Chemical	Digests microbes
Bile	Chemical	Inhibits growth of most microbes
Urine	Chemical	Contains lysozyme; acidity inhibits most microbial growth; flushing action keeps microbes from attaching
Sebum	Chemical	Low pH inhibits growth of some bacteria

Keep in Mind

- There are two types of immune response that protect the host: the innate immune response and the adaptive immune response.
- The innate immune response is a nonspecific response that responds immediately to any type of infection.
- The first line of defense in the innate immune response consists of mechanical and chemical barriers.
- Mechanical barriers include the skin, mucous membranes, the lacrimal apparatus, saliva, and the epiglottis.
- Chemical barriers include sebum, perspiration, gastric juice, urine, and transferrins.

THE SECOND LINE OF DEFENSE IN THE INNATE IMMUNE RESPONSE

Let's look now at the second line of defense in the innate immune response. This second line defends against infection by means of one cellular mechanism — *phagocytosis* — and four chemical mechanisms — *inflammation*, *fever*, *the complement system*, and *the production of interferons*. Some of these second-line defenses are among the most potent and lethal seen in nature. As we review these processes, keep in mind that they are all interrelated and work together to magnify the overall innate immune response.

Toll-Like Receptors

Before we get into a discussion of the cells involved in nonspecific host defense, it is important to talk about how cells can differentiate between *self* and *nonself*. An **antigen** is any substance that triggers an immune response in a host body, and the host defense provides protection against nonself antigens found on foreign invading organisms. How do our cellular defenses know what is nonself? The answer to this question is simple: **Toll-like receptors** (TLRs). These molecules are located on the surface of defender cells and are a required part of the innate immune response. They bind to antigens found on pathogens.

15.1 The first TLR to be discovered was TLR-4, which recognizes and binds with lipopolysaccharide (LPS), a substance found on Gram-negative bacteria. Subsequently, other TLRs were found, and there are now 13 of them that seem to be involved in the innate immune response. These receptors are the way in which our defenses "see" the microbial world, and they are used to activate the innate response. **Table 15.2** shows that TLRs react to a variety of microbial structures.

A TLR is activated as soon as it binds to a target antigen. This activation triggers the host defense cell associated with the TLR to release inflammatory substances, chiefly **tumor necrosis factor** (**TNF**). For example, TLR-4 is found on the surface of dendritic cells. When it recognizes LPS on the surface of Gram-negative bacteria, it triggers the release of substances that cause inflammation. TLR-4 signaling stimulates the dendritic cells to remodel their actin cytoskeleton so as to mediate transient increases in antigen-dependent endocytosis. This increased importing of substances into the dendritic cells enhances antigen presentation, which as we will see in the next chapter is required for an adaptive immune response to develop.

Recall from the beginning of this discussion that TLRs are the answer to the question of how the body distinguishes self from nonself. One of the most interesting questions that have yet to be answered in this area, however, is just *how* TLRs do this.

Fast Fact

Toll-like receptors were initially discovered by accident during studies on the growth of insects.

Toll-like Receptor	Ligand Bound
TLR-1	Lipoproteins
TLR-2	Bacterial lipoproteins
TLR-3	Double-stranded RNA
TLR-4	Lipopolysaccharide, some viral proteins
TLR-5	Flagellar protein
TLR-6	Lipoteichoic acid
TLR-7	Single-stranded viral RNA
TLR-8	Single-stranded viral RNA
TLR-9	Bacterial DNA
TLR-10	Unknown
TLR-11	Toxoplasma profiling
TLR-12	Unknown
TLR-13	Unknown

Table 15.2 Some Toll-like Receptors and the Nonself Components They Bind

As we mentioned above, the second-line host defense comprises one cellular and four chemical components. The cellular component is **phagocytosis**, the cellular process in which pathogens, damaged host cells, cell debris, and other foreign materials are removed from the body. The chemical components are **inflammation**, **fever**, **complement**, and **interferon**. Phagocytosis is performed by white blood cells, which are also known as leukocytes. There are several types of white blood cell, and they are all derived from stem cells located in the bone marrow (**Figure 15.5**).

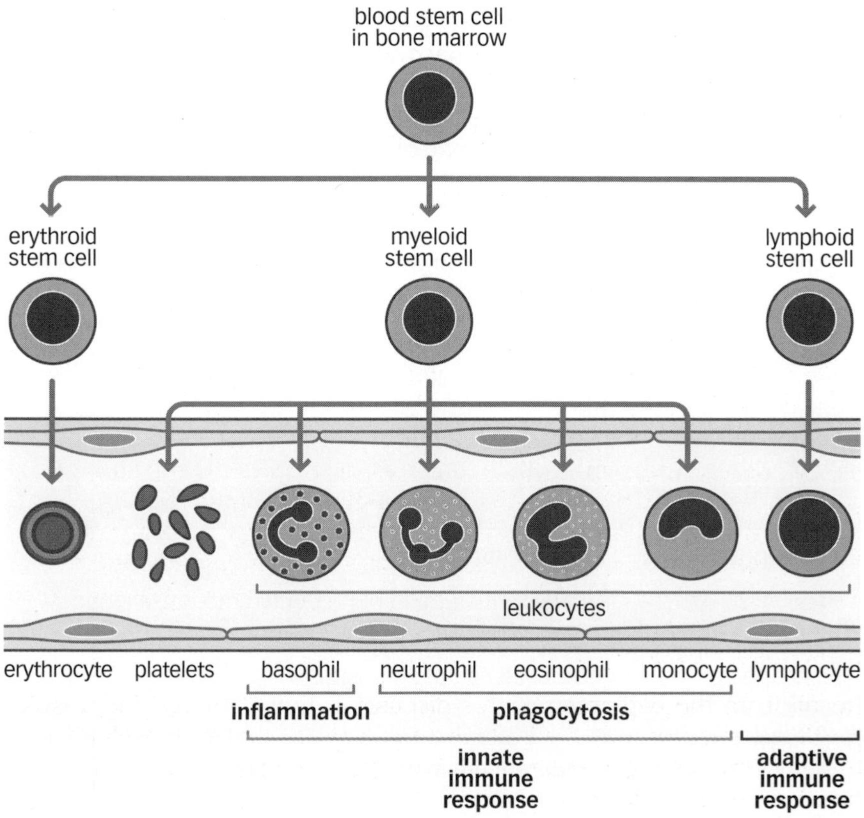

Figure 15.5 The differentiation of cell types in the blood. Stem cells in the bone marrow give rise to a variety of cell types known collectively as the blood's formed elements. Erythrocytes are red blood cells. Red blood cells and platelets are shown on the left. The remaining cells are the five types of white blood cells. Basophils, neutrophils, eosinophils, and monocytes are involved in the innate immune response. Lymphocytes function in the adaptive immune response.

The innate immune response relies heavily on white blood cells, and the number of these cells can be directly correlated to the stage of disease. Blood analysis usually includes a **complete blood count** (CBC), in which the total number of white blood cells is identified. In addition, a **differential analysis** is carried out, and this test gives the percentage of each type of white blood cell. Taken together, these two pieces of information can help in understanding the infectious process. For example, whether the CBC increases or decreases depends on the nature of the infection. It increases in cases of pneumonia, meningitis, appendicitis, and gonorrhea but decreases in *Salmonella* infections, pertussis, and some viral infections. No matter whether the CBC increases or decreases, though, it always returns to normal as the infection subsides. Consequently, in many cases, a CBC either higher or lower than normal tells the health care provider that an infection is under way, and a return to the normal level indicates that the infection is subsiding.

White blood cells can be broken down into two major groups, **granulocytes** and **agranulocytes**. The types are easily differentiated from each other microscopically by cellular appearance and by the presence or absence of visible granules (see **Figure 15.5**).

Granulocytes

These cells have a granular cytoplasm and multilobed nuclei, and they can be divided into three types: neutrophils, basophils, and eosinophils. Each of these cells has a different role in the innate immune response (**Table 15.3**).

15.2 **Neutrophils** are phagocytic cells that guard skin and mucous membranes; they represent about 70% of the white blood cell population. They protect against bacterial and fungal infection by "sensing" the site of the infection, migrating to it, and destroying the infectious organisms.

Neutrophils are derived from bone marrow and mature there. This process takes about two weeks, and then the mature neutrophils move from the bone marrow into the blood. At any given time, there are about 20 times as many neutrophils in the bone marrow as in the blood, kept in reserve in case of infection. In some infections, this reserve is exhausted, and in these cases immature neutrophils that cannot function at maximum level are moved out of the bone marrow. Neutrophils use TLRs to detect lipoproteins, peptidoglycan, and lipoteichoic acid from Gram-positive pathogens, and endotoxins and LPS from Gram-negative organisms.

15.3 Neutrophils circulate in the blood for about 6–10 hours but can remain in tissues in an inactive state for 2–6 days. The passage of these cells from blood to tissue is known as **diapedesis**. In this process, the neutrophils squeeze between the cells lining capillary walls (**Figure 15.6**). However,

Table 15.3 White Blood Cells Involved in the Innate Immune Response

Cell Type	Function
Granulocytes	
Neutrophils	Phagocytosis
Basophils	Release histamine during inflammation; allergic responses
Eosinophils	Defense against parasites
Agranulocytes	
Monocytes	Differentiate into macrophages at infection site

before they can leave the blood, the neutrophils have to settle down at the exit site. This is accomplished by the process of **margination**, which is quite amazing. The vascular environment becomes activated locally in the area of damaged tissue, and epithelial cells lining the vessels begin to secrete a substance called **selectin**. This substance causes a series of binding interactions with receptors on the surface of the neutrophils, causing them to slow down and begin to roll along the vessel wall. As the cells roll along, the binding events continue, eventually causing the neutrophils to stop. Once they have come to a stop, they begin to respond to chemotactic gradients that guide them through the vessel wall and into the tissue.

15.4

Although neutrophils have a wide range of toxic mechanisms for fighting invading organisms, neutrophil functions are tightly controlled. Overproduction of these toxic substances can lead to catastrophic consequences for the host. Examples are: septic shock; disseminated intravascular coagulation leading to respiratory distress syndrome; hypoxia; and fatal circulatory collapse.

Neutrophils are programmed for apoptosis (programmed cell death) and have a relatively short life span. They undergo apoptosis within hours after they move into extravascular tissues. The mechanism for this programmed cell death is not yet understood, but it seems that loss of phagocytic activity seems to be a prelude to the apoptotic event. Dead neutrophils are eliminated from the tissue by macrophages, which are agranular white blood cells that we will discuss in more detail in a moment.

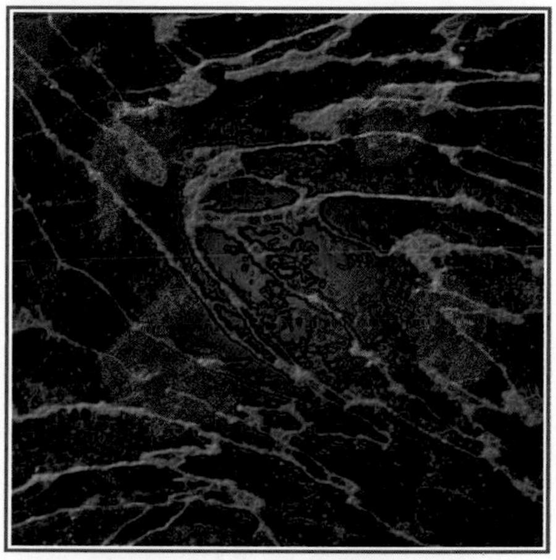

Figure 15.6 White blood cell movement. Computer-enhanced confocal light micrograph of white blood cells (red) moving through the intact walls of a blood vessel, a process known as diapedesis. The membranes of the cells in the blood vessel wall are green.

15.5

Basophils, our second type of white blood cell, are derived from hematopoietic progenitor cells in the bone marrow and have a life span of only several days. They mature in the bone marrow and circulate in small numbers in the blood. They are recruited into tissues as needed and carry the receptors for the immunoglobulin IgE. After linking to IgE, the basophils release their cytoplasmic granules. These granules contain histamine, a chemical that is involved in allergic reactions but can also increase and magnify innate immune responses.

Basophils can be activated by bacteria, viruses, and parasites, all of which they detect using TLRs. When basophils are stained with a basic dye, such as methylene blue, they stain a distinct purple in which the cytoplasmic granulation is very visible.

The white blood cells known as **eosinophils** are normally found in very small numbers in the blood and are easily identified by the orange color they display when stained with acidic dyes. Eosinophil numbers increase greatly in cases of parasitic infection and in allergic reactions. Although their involvement in allergy is not clear, we know they are the primary defense in parasitic infections, during which they produce powerful toxic enzymes. **Figure 15.7** illustrates this event dramatically. These cells also modulate the inflammatory response and may be involved in the detoxification of foreign substances.

Agranulocytes

These white blood cells have granulation in the cytoplasm, but it is so small that the cytoplasm looks agranular under a microscope. There are three types of cell in this group, namely **monocytes**, **macrophages**, and **lymphocytes**. Because lymphocytes are involved in the adaptive immune

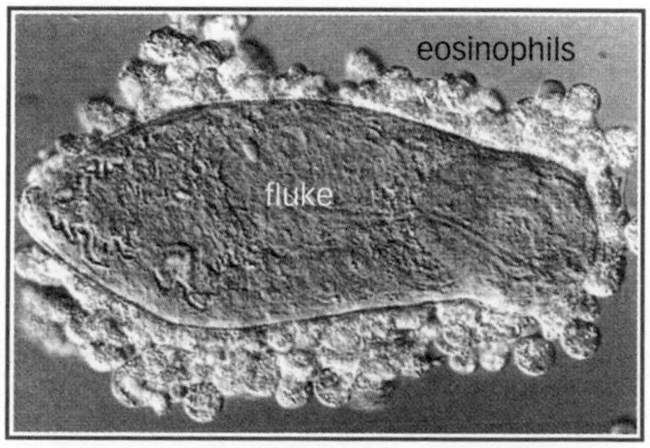

Figure 15.7 Eosinophils attacking the larval stage of a liver fluke. These white blood cells will discharge toxic enzymes that tear holes in the organism.

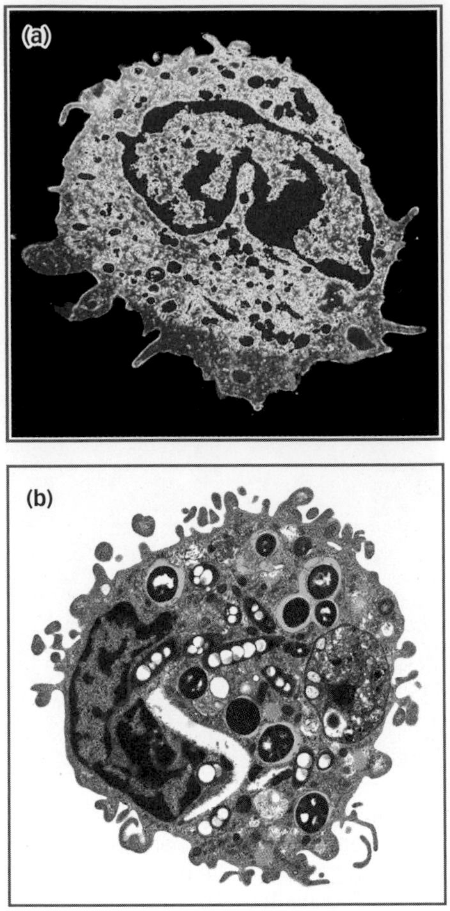

Figure 15.8 Colorized transmission electron micrographs showing the maturation (differentiation) of a monocyte into a macrophage. Panel **a**: Monocytes have no phagocytic capability, but after differentiation into macrophages (panel **b**) they are the most phagocytic of all white blood cells. Notice that the macrophage shown here has phagocytized several *Legionella* bacteria (white areas in purple).

response, we discuss these cells in Chapter 16. Three types of macrophage exist: those that differentiate from monocytes, those that move freely through tissues (referred to as *wandering macrophages*), and those that remain stationed in certain tissues (*resident macrophages*).

Monocytes are mononuclear white blood cells derived from stem cells. They normally circulate in the blood in a nonphagocytic form. However, when there is an infection, they are called to the site by chemotactic factors released from damaged tissue and from neutrophils already joined in battle at the site. Once monocytes reach the site, they differentiate and become macrophages (**Figure 15.8**). In this form, they are the most phagocytic of all white blood cells and can engorge themselves on bacteria as well as tissue debris.

Monocytes are released from the bone marrow into the blood within 24 hours of maturation. When they are released, they already possess migratory and chemotactic ability. Normally the number of monocytes circulating in the blood is quite small. During an infection, however, this number increases markedly, and the monocytes begin to adhere to vessel walls and migrate out into the tissues, where they differentiate into macrophages.

Macrophages are responsible for recognizing, engulfing, and destroying bacteria, fungi, and parasitic worms. They are also involved in removing tumor cells, virus-infected cells, and normal cells that have undergone apoptosis. Equally importantly, macrophages function in wound healing, tissue repair, and bone remodeling. As we will see in Chapter 16, they also function as antigen-presenting cells in adaptive immune responses. Some of the common features of macrophages are listed in **Table 15.4**.

The bone marrow stem cells destined to become macrophages go through a series of differentiation steps from monoblasts to promonocytes to monocytes and finally to macrophages. The body is filled with macrophages that wander (**wandering macrophages**) and those that remain fixed in a given tissue (**resident macrophages**). Resident macrophages are called by different names but are found in every organ system (**Figure 15.9**). They maintain their position in the tissues through interactions with the tissue reticular fibers and form a network known as the **mononuclear phagocytic system** (formerly called the **reticuloendothelial system**).

Table 15.4 Common Features of Macrophages

Feature	Description
1	Located adjacent to basement membranes of epithelial and endothelial cells
2	Contain high levels of enzymes
3	Adhere to plastic
4	High levels of phagocytic and endocytic activity
5	Contain surface receptors for complement, Toll-like receptors and regulatory receptors

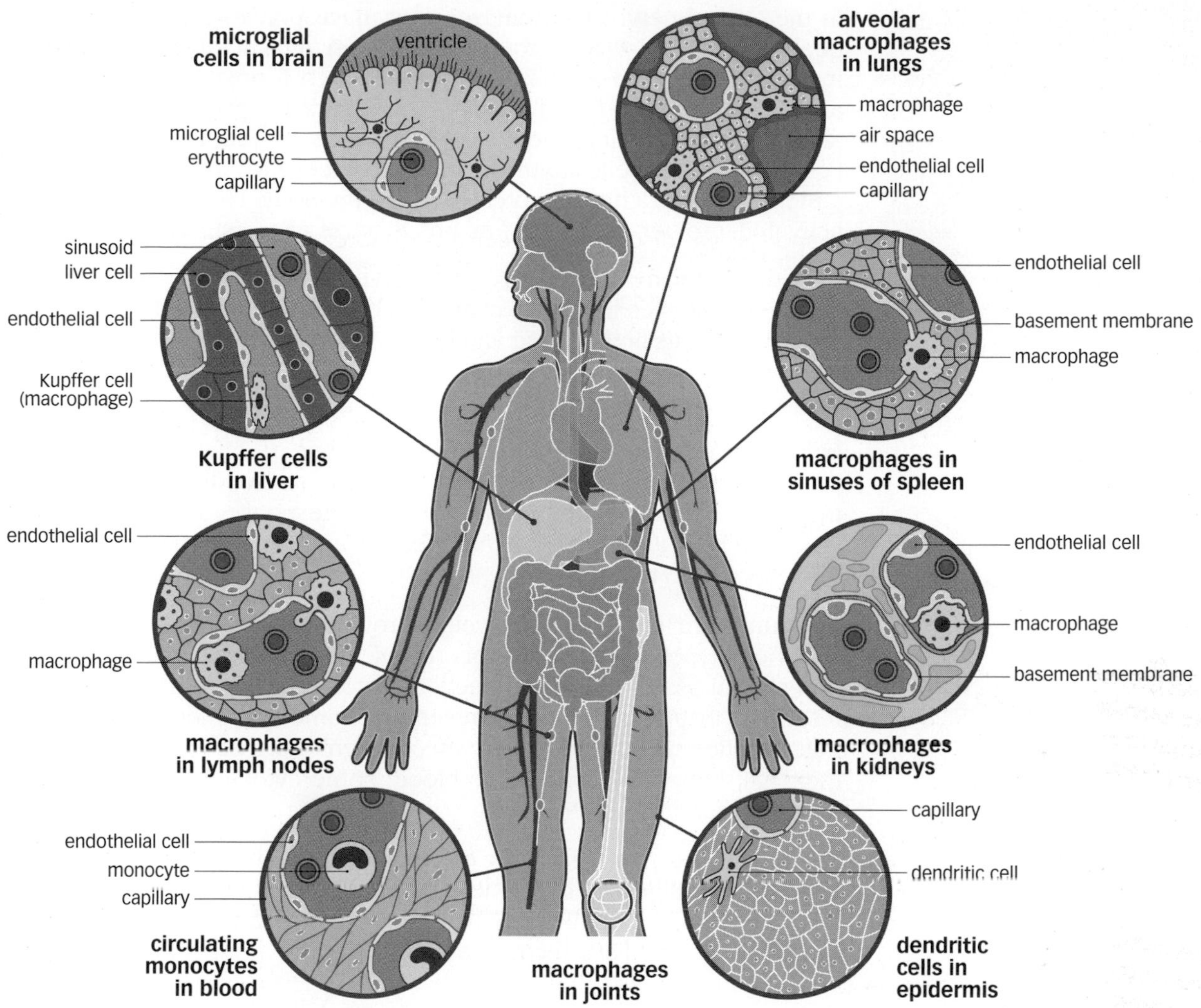

Figure 15.9 Resident macrophages are found in every one of the body's organ systems. They sometimes are given specific names, such as Kupffer cells (the liver's resident macrophages) and microglial cells (the brain's resident macrophages).

Let's look now at some resident macrophages and their functions in the liver, lungs, and central nervous system. In the liver, resident macrophages are called **Kupffer cells** (**Figure 15.10**). They reside in the sinusoids of the liver, where they are anchored to the endothelium by long cytoplasmic projections. The liver contains the largest population of macrophages in the body. This makes sense because the liver is responsible for the filtration of the blood supply, which could be filled with microbes in the event of a systemic infection. The liver is also the site of routine clearance of defunct red blood cells, which are cleared and recycled by Kupffer cells.

In humans, there are two populations of resident macrophages in the lungs: alveolar and interstitial. **Alveolar macrophages** are the first-line defenders against inhaled microbes that enter the lung. These macrophages are extremely active and highly phagocytic cells that kill with great efficiency. **Interstitial macrophages** are found in the stroma of the lung and are smaller and less phagocytic than alveolar macrophages. It is important to keep in mind the route of entry here and the need to protect the alveolar spaces, where gas exchange occurs. We will see in Chapter 21 that there is a variety of infections that can affect the lung, many of them potentially fatal.

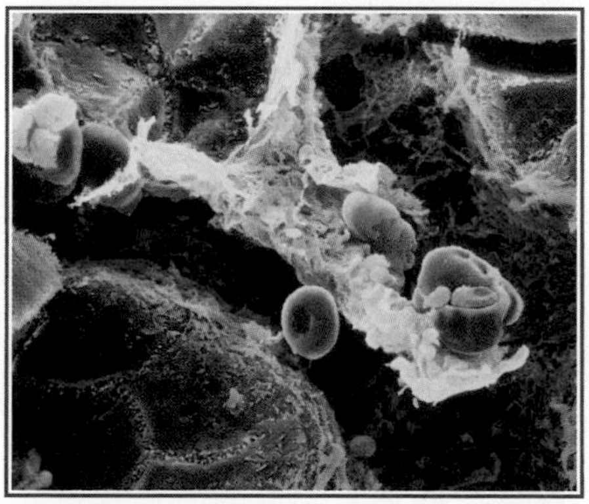

Figure 15.10 A scanning electron micrograph of the resident macrophages known as Kupffer cells (yellow), which are found in the liver.

In the central nervous system, resident macrophages are called **microglial cells**. These cells are found in the brain and spinal cord, representing about 10% of brain tissue. There are two forms of microglia. The amoeboid forms are found traveling through developing brain tissue and also in diseased brains, and the ramified forms are found in normal brain tissue. Microglial cells are the first line of defense for the central nervous system, which is a highly protected site in the body. These cells are easily activated, proliferate rapidly, and phagocytize aggressively.

In addition to the liver, lungs, and nervous system, we also find resident macrophages lining the chambers of the heart and attached to the endothelial cells lining blood and lymph vessels.

Cytokines and Chemokines

Before we look more closely at the second line of defense in the innate immune response, it will be helpful to discuss two types of chemical mediator that are an important part of the response. These factors are produced both at the onset of an infection and throughout the course of the response to that infection.

Cytokines are low-molecular-weight proteins that are released by a variety of cell types in the body. This release can be in response to activation stimuli associated with infection. Cytokines induce innate immune responses by binding to specific receptors. Some cytokines affect the cells that produce them, and others affect neighboring cells. Depending on their half-life and access to the blood, some cytokines can also affect cells in other areas of the body.

There are two major families of cytokines, the hematopoietin family, which includes growth hormones and interleukins, and the tumor necrosis factor family. Both are involved in the innate and adaptive immune responses. When TLRs recognize a pathogen, a variety of cytokines are released in response.

We can summarize the characteristics of cytokines as follows:

- They are secreted from white blood cells when the body has been invaded by pathogens.
- They regulate inflammation and the immune response.
- They react with specific receptors on their target cells and alter the activity of those cells.
- Each cytokine has overlapping functions that act as a network to either induce or inhibit the effects of other cytokines.
- The action of a cytokine depends on the concentration of that cytokine.

Some of the major cytokines produced in the inflammatory response are shown in **Figure 15.11**.

Chemokines are cytokines that appear in the earliest part of an infection and "attract" defensive cells to the site of infection. Chemokines are released by many types of immune-system cell and guide defensive cells to the site of trauma. Some chemokines are also involved in angiogenesis (building new blood vessels) when the body is repairing damaged tissues.

There are two broad groups of chemokines, the *CC* group, which bind to CC receptors, and the *CXC* group, which bind to CXC receptors. CC chemokines promote the migration of monocytes, lymphocytes, and other cells and can also induce monocytes to differentiate into resident macrophages. Newly formed resident macrophages then move to

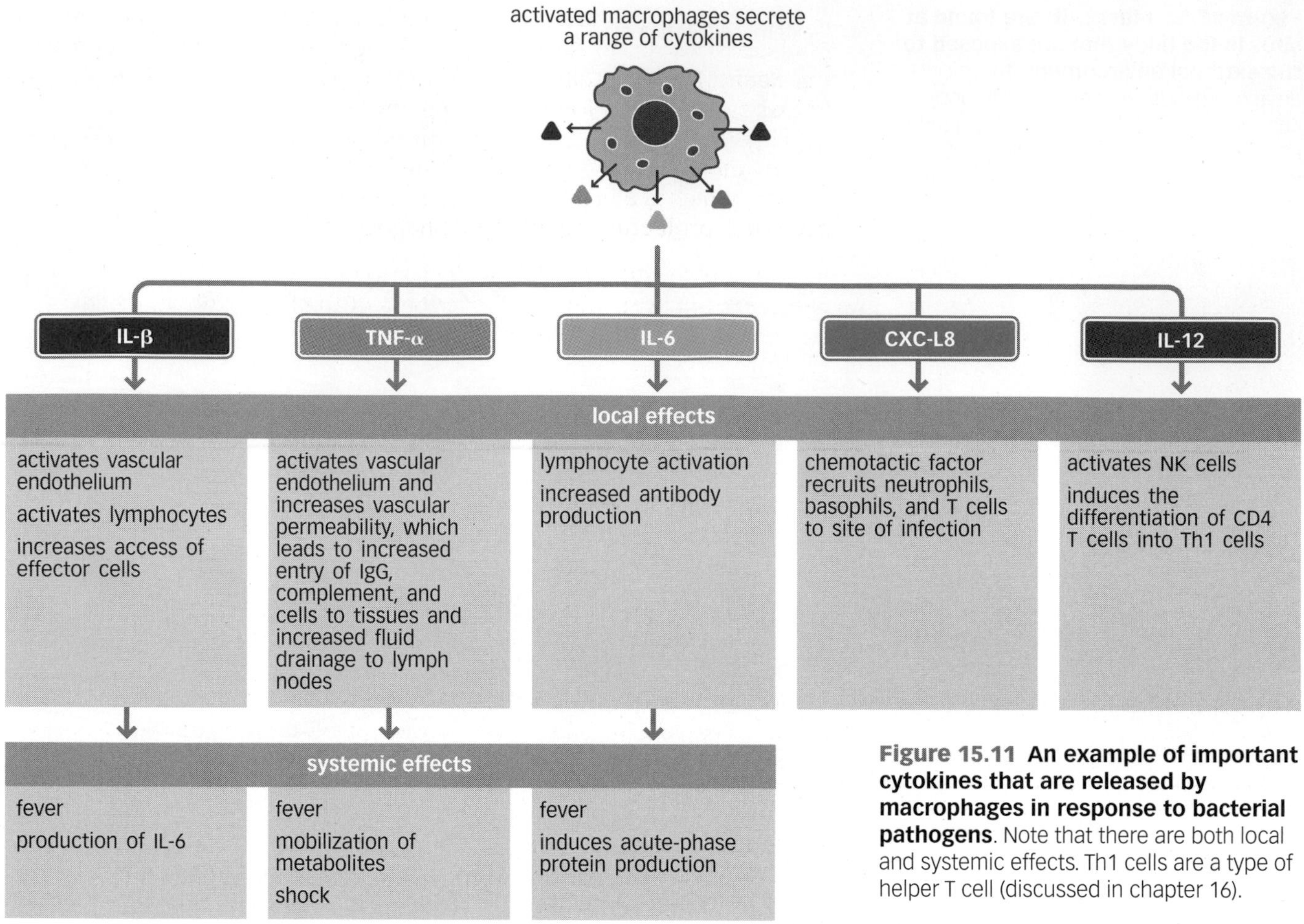

Figure 15.11 An example of important cytokines that are released by macrophages in response to bacterial pathogens. Note that there are both local and systemic effects. Th1 cells are a type of helper T cell (discussed in chapter 16).

specific locations in the body, where they become a part of the mononuclear phagocytic system. CXC chemokines are responsible for the migration of neutrophils and lymphocytes to the site of infection, and also promote the diapedesis of these phagocytic cells from the blood vessels into the infected tissue.

Chemokines are released into the environment around the infection in response to the presence of bacterial invaders, viral invaders, or tissue damage due to physical injury. They also have a role in the destruction of pathogens. For example, neutrophils exposed to CXC-L8 and the cytokine TNF release oxygen atoms in the free-radical form and nitric oxide, two powerful antimicrobial toxins that cause cellular lysosomes to release their contents.

Fast Fact

Antimicrobial toxins destroy pathogenic organisms and also break down damaged tissue, a process that results in the formation of the thick yellow substance we know as pus.

Mast Cells, Dendritic Cells, and Natural Killer Cells

There are three other types of cell that are very important in the innate immune response: mast cells, dendritic cells, and natural killer cells.

Mast Cells

Mast cells are derived from stem cells in the bone marrow. Often referred to by the alternative name *sentinel cells*, mast cells are responsible for allergic reactions and responses to parasitic infections. They are found throughout the body but most commonly in tissues that are exposed to the external environment, such as the skin, respiratory tract, and digestive tract (**Figure 15.12**).

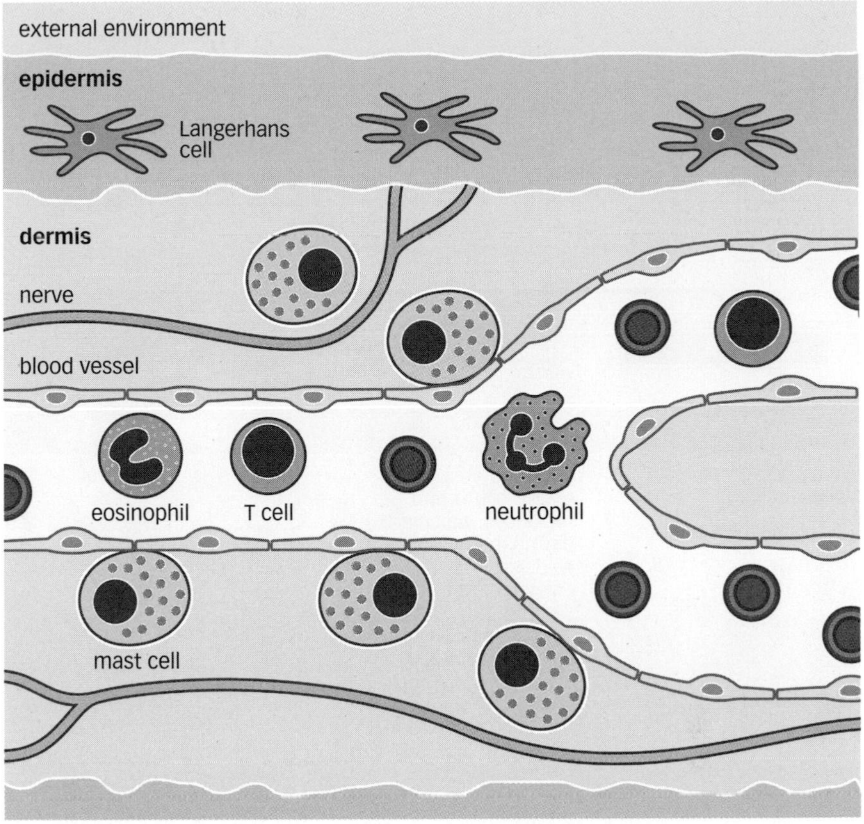

Figure 15.12 Mast cells are found at sites in the body that are exposed to the external environment. These cells are located close to blood vessels and can regulate vascular permeability during an infection. They also recruit key cells required for the innate immune response.

Mast cells have three distinct properties. First, they can rapidly and selectively produce mediators that work in host defense. Second, because they are positioned for long periods adjacent to blood vessels, they can enhance the recruitment of effector cells that respond to infection. Third, they can influence the adaptive immune response, as we will see in Chapter 16. Mast cells leave the bone marrow in an immature form and differentiate into mature cells when they arrive at tissue sites. Interestingly, mature mast cells that are resident in the tissues can proliferate there. This replacement capability in the tissues makes possible a sustained level of effector cells during infection.

When they are directly activated by the presence of invading pathogens, mast cells produce a variety of mediators (**Table 15.5**) that cause alterations in vascular function and cellular recruitment (**Figure 15.13**). In addition,

Table 15.5 Mediators Released by Mast Cells

Mediator	Function
Histamine, serotonin	Alter vascular permeability
Heparin, chondroitin sulfate	Enhance cytokine and chemokine function; enhance angiogenesis
Tryptase, carboxypeptidase, other proteases	Remodel tissue; recruit effector cells
Cytokines	Induce and regulate inflammation and angiogenesis; recruit effector cells; enhance angiogenesis
Chemokines	Recruit effector cells, including dendritic cells; regulate innate immune response

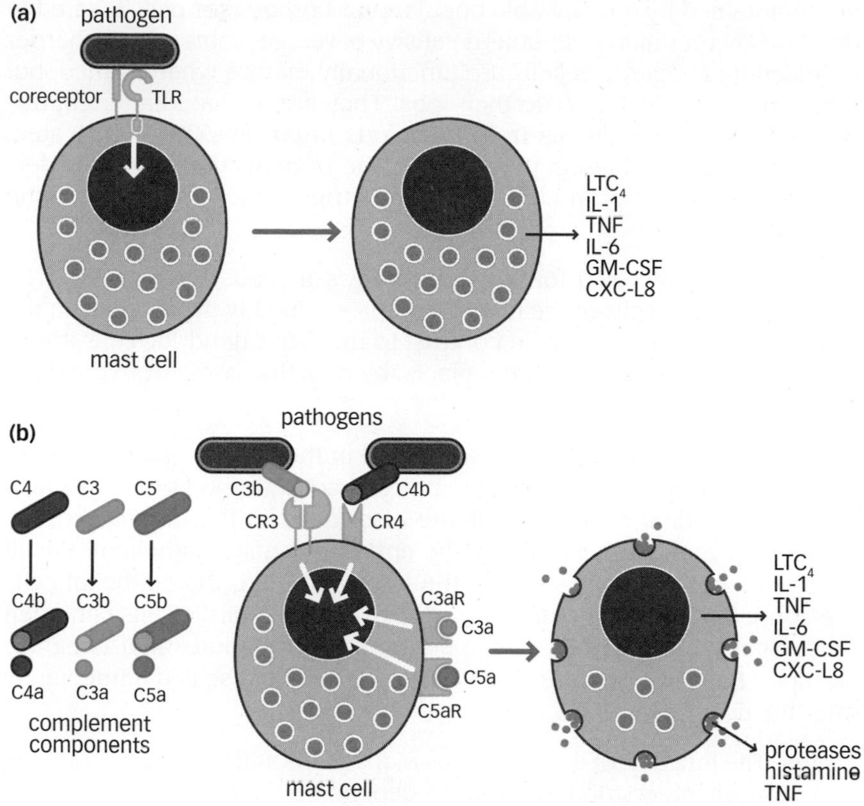

Figure 15.13 Two ways in which mast cells are activated by pathogens. Panel **a**: Direct activation by the pathogen leads to the production of several mediators but in most cases does not cause degranulation and release of histamine. Panel **b**: Indirect activation by components of the complement system. In this situation, degranulation occurs, and histamine and other non-specific defense mediators are released.

mast cells can reposition themselves during tissue repair and can initiate and maintain an effective adaptive immune response. Using TLRs, mast cells can distinguish between different types of pathogen and generate a highly selective response to bacterial, viral, fungal, or parasitic infections.

Mast cells can also be activated indirectly through interactions with the complement system (discussed below). The combination of complement system and mast cells provides crucial initial signals for the recruitment of more cells to fight an infection. Mast cells can also recruit and activate dendritic cells (see below).

As with a variety of other defense mechanisms, the response by mast cells can be potentially damaging to the host. As a result of their close proximity to blood vessels and their ability to produce large and sustained amounts of inflammatory cytokines, mast cells have been implicated in allergic diseases, rheumatoid arthritis, vasculitis, and atherosclerosis.

Dendritic Cells

Dendritic cells regulate both the innate and the adaptive immune system, depending on the local environment in which they are found. They are called dendritic cells because they have long membranous extensions that resemble the *dendrites* of nerve fibers. The cells are continuously produced in the bone marrow and move from the marrow to all tissues. Dendritic cells are strategically located in mucosal tissues associated with routes of pathogen entry such as the oral, respiratory, and genital mucosae. They have different effects depending on the tissue in which they are located, the microbial environment, and the presence or absence of inflammation. Let's look next at some of the areas in which we find dendritic cells.

Skin and mucous membranes. In the skin, dendritic cells are called **Langerhans cells**. They are located in the basal layers of the epidermis, where they connect to one another to form a network. Langerhans cells are maintained by a renewable population of progenitor cells located in the skin. This permits continuous defensive coverage of this primary barrier to infection. Langerhans cells are functionally mature when formed, but they must be activated to do their jobs. They are activated after capturing and processing signals from infectious organisms. Once activated, the Langerhans cells migrate from the site of activation in the epidermis to regional lymph nodes, where they trigger the adaptive immune response.

Dendritic cells are also found in the body's mucous membranes. The function of these cells in the membranes is essentially the same as in the skin. One difference is that, in contrast to the skin's dendritic cells, those in the mucous membranes are replaced by cells that come from the bone marrow.

Intestines. Here, dendritic cells are found in the Peyer's patches, where they are close to **M cells**, which are antigen-collecting cells in the Peyer's patches. Intestinal dendritic cells are also found in the lamina propria, where they are distributed along the entire intestinal epithelium. Recall from your earlier courses that in the intestinal tract, the epithelial cells are connected with tight junctions. Remarkably, dendritic cells can open these tight junctions and extrude dendrites into the intestinal lumen to "sample" the bacteria there. The tight junctions are resealed immediately after the dendritic process is withdrawn.

Lungs. The function of dendritic cells in the pulmonary spaces is complex and not well understood. It seems to depend on the context and form in which the infectious agent is acquired. More importantly, this area of the body is constantly bombarded with inhaled antigens. Most, of course, are harmless, but some are infectious. The processes on pulmonary dendritic cells must discriminate between the two. During an inflammatory response, pulmonary dendritic cells are recruited into the airway epithelium to deal with pathogens entering the respiratory tract.

Lymphoid tissues. Dendritic cells in lymphoid tissue are functionally mature but less phagocytic. Lymphoid dendritic cells can produce inflammatory cytokines and chemokines quickly once an infection has taken place, usually within two or three hours. They use TLRs to identify non-self antigens, and their placement in lymphoid tissue is strategic for their function. They are located in areas of the lymph nodes where they can load up on antigens that are flowing in with lymph fluid and then move into the areas of the nodes that are filled with T cells. Here they prime antigen-specific T cells as part of the adaptive immune response.

Natural Killer Cells

The cells known as **natural killer cells** (NK cells) were initially thought to be restricted to the immune surveillance of tumor cells. More recent research, however, has shown that they are also involved in the host response to microbial pathogens. Remember that the response to infection comes in two parts, the innate response and the adaptive response. Innate host defense occurs first and must hold the fort until the more powerful adaptive response gets going (a period of days). NK cells fulfill the early need for the destruction of infected cells.

NK cells are a unique population found in peripheral tissues and the blood. They are normally seen as large granular cells that have no antigen receptors. There seem to be different kinds of NK cell found in different tissues. NK cells are descended from bone marrow stem cells and use the same margination and diapedesis mechanisms used by other cells that migrate into tissues during the innate response.

NK cells can kill tumor cells, virus-infected cells, bacteria, fungi, and parasites. The first evidence that they have functions in addition to tumor-cell surveillance came from observation of patients who had selective deficiencies of NK cells. These patients had all of the other necessary innate defenses but still presented with frequent and recurrent viral infections, especially varicella-zoster and herpes simplex. It has also been shown that the NK response is diminished in HIV infections.

NK cells are involved in the innate immune response in two ways: through target-cell killing and through the production of cytokines. Target-cell killing is mediated by the instigation of apoptosis in the target cell. This is accomplished by the triggered, directional release of enzymes called **perforin** and **granzyme**. The perforins polymerize in the target cell membrane and form a pore, which the granzyme then uses to enter the target cell and stimulate apoptosis.

When activated, NK cells produce cytokines, such as **TNF** and **granulocyte–macrophage colony stimulating factor** (which recruits large numbers of additional phagocytic cells to fight the infection). In addition, NK cells respond to cytokines released during the innate response to infection, such as interleukin-2 and interleukin-12. This complex amplification relationship is illustrated in **Figure 15.14**.

It is becoming apparent that the NK response is a protective mechanism that takes place in the interval before the more powerful adaptive immune response is ready to begin. This is nicely demonstrated when we look at host defense over the course of a viral infection (**Figure 15.15**).

Unlike all other cells involved in the immune response, NK cells do not use TLRs. The question is, then, how do NK cells know friend (normal cell) from foe (infected or tumor cell)? The recognition process for NK cells, which is very interesting and somewhat complex, takes advantage of two types of receptor on the surface of the NK cell. One of these is an *inhibitory receptor*, and the other is an *activation receptor*. The activation receptor activates the killing effect, and the inhibitory receptor inhibits killing by blocking activation. The fate of the target pathogen cell depends on its interaction with both of these receptors.

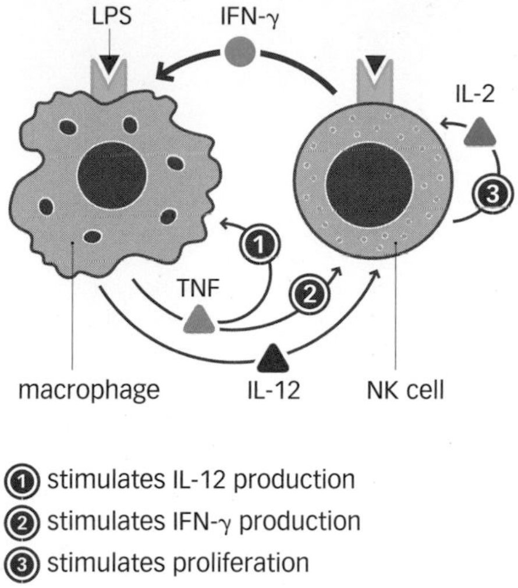

Figure 15.14 The interaction between NK cells and macrophages is a good example of the amplification seen in the innate immune response. The gamma interferon (IFN-γ) produced by NK cells amplifies the production of the cytokines tumor necrosis factor (TNF) and interleukin 12 (IL-12) by macrophages. The elevated levels of these cytokines then causes NK cells to produce more interleukin-2 (IL-2). Production of IL-2 increases the number of NK cells.

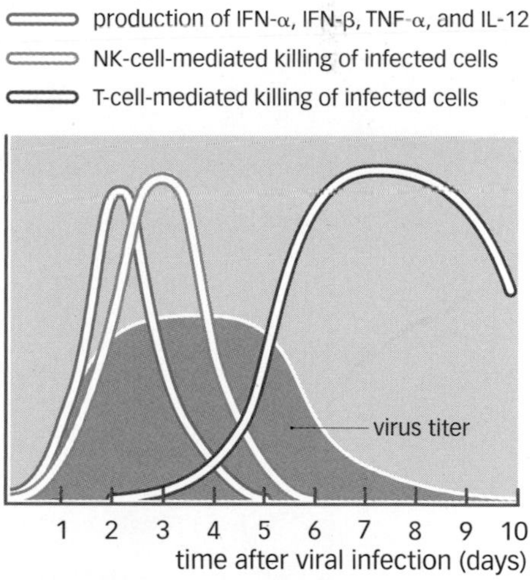

Figure 15.15 Host response to viral infection. The earliest response is from the innate immune system: first the production of interferons (IFN), interleukins (IL), and tumor necrosis factor (TNF) and then the killing of infected cells by NK cells. Notice that after the initial rise, the level of infection (shown as the titer of virus) stays constant even though NK cell activity attacks infected cells. The infection does not decrease until the adaptive response (killing by T cells) is initiated on about day 6.

Keep in Mind

- The second line of defense in the innate immune response incorporates cellular and chemical mechanisms that actively defend the body.
- Cells that are involved in defending the body carry Toll-like receptors that identify molecules associated with antigens.
- The cellular components of the second line of defense are white blood cells.
- These white blood cells can be granular (neutrophils, basophils, and eosinophils) or agranular (monocytes).
- The process of diapedesis describes the movement of these white blood cells out of the blood and into the infected tissues.
- Cytokines and chemokines are chemical mediators that are produced at the onset of and throughout the course of the response to an infection.
- There are three types of cell that have important roles in the innate immune response: mast cells, dendritic cells, and natural killer cells.

As we mentioned at the outset of this chapter, the second line of host defense can be broadly broken down into five distinct mechanisms: **phagocytosis**, **inflammation**, **fever**, the **complement system**, and **interferon**. Each of these nonspecific mechanisms has characteristics that give it a unique role in host defense. Although we will examine them one by one, it is important to keep in mind that the five mechanisms are highly integrated and that interactions between them permit amplification of the host defense to infection.

Phagocytosis

15.6

Phagocytosis, the cellular mechanism in the second-line of defense, is carried out primarily by neutrophils and macrophages. (These two types of white blood cell are collectively called **phagocytes**, which mean any immune-system cell that conducts phagocytosis.) When an infection begins, the traumatized tissue produces chemokines, and the presence of chemokines in the body signals neutrophils to move immediately to the damaged site, as noted earlier. Monocytes are the second cells to arrive at the scene; as they arrive, they differentiate into macrophages, which are more phagocytic than the neutrophils.

Phagocytosis consists of five phases (**Figure 15.16**):

- **Chemotaxis** is the release of chemokines so as to attract phagocytes to the infected area or damaged tissue. Once some phagocytes reach the damaged site, they also release chemokines, thereby bringing even more phagocytes to the site and amplifying the effectiveness of the host defensive response. As phagocytes get closer to the site, the chemokine concentration becomes higher and higher, causing the phagocytes to move more quickly to the site. Chemotaxis is a very interesting process. Once chemokine molecules released by damaged tissue cells reach a phagocytic cell, receptors on one side of the phagocytic cell are occupied by the chemokine molecules, and this causes the cytoskeleton of the cell to rearrange itself and lean toward that side. This distortion in turn causes the cell to move forward. During this process, the matrix proteins of the cytoskeleton form at the front of the cell and disassemble at the rear. This pattern continues as the cell moves toward the chemotactic source (the damaged cells that triggered the initial chemokine release). As this occurs, the phagocyte surface receptors bound to the chemokine molecules must be freed of the molecules so that they can repeat the process. This happens through internalization of the receptor–chemokine complex. Once inside the phagocytic cell, the complex is broken apart, and free receptors then return to the cell surface, ready to bind more chemokine molecules.

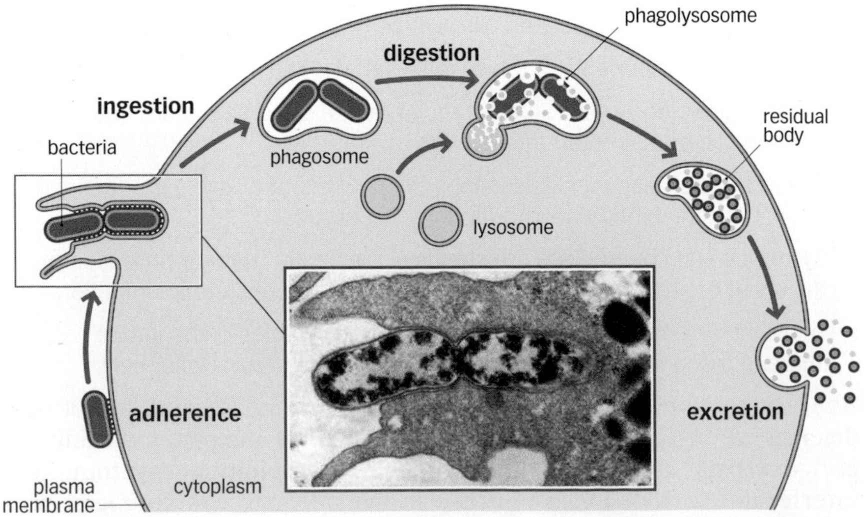

Figure 15.16 Four of the five phases of phagocytosis: adherence, ingestion, digestion, and excretion. The fusion of phagosomes with lysosomes enables the destruction of the pathogen cells. (However, several microorganisms have evolved ways to defeat this process.)

- Adherence, the second phase of phagocytosis, occurs as the plasma membrane of the phagocytic cell comes into contact with a pathogen cell. This involves a binding of structures such as glycoproteins on the pathogen to surface receptors on the phagocyte. Recall from Chapter 5 that bacterial capsules can inhibit this attachment, making the pathogen cell resistant to phagocytosis.

- Ingestion describes the process in which the pathogen is taken inside the phagocytic cell. This process involves the formation of pseudopodia, which are projections that extend from the phagocyte and surround the pathogen (see **Figure 15.16**). A pseudopod can engulf a pathogen only if receptors on the pseudopod surface bind to the pathogen surface. This binding also allows the pseudopod to fit tightly around the pathogen (**Figure 15.17**) and eventually forms a vesicle called a **phagosome** (see **Figure 15.16**). This phagosome, like any other vesicle, can then move farther into the phagocytic cell and eventually fuses with lysosomes in the cell.

- Digestion of the pathogen begins when the lysosome releases its contents. Recall from Chapter 4 that the lysosome is a vesicle that is filled with such enzymes as lysozyme, lipase, protease, and ribonuclease in a low-pH (4.0) environment. Phagosomes fuse with lysosomes by virtue of their similar membrane structure to form **phagolysosomes**. This fusion allows the enzymes of the lysosome to come into contact with the bacterium and destroy it. This process can occur in as little as 30 minutes. Macrophages use oxygen derivatives such as hydrogen peroxide and hypochlorite ions for the digestion phase. (Hypochlorite ions are the ions found in laundry bleach.) All oxygen derivatives work by destroying the pathogen's plasma membrane. If a pathogen cell is too large to be ingested, toxic compounds from the lysosome will be released onto the pathogen while it is outside the macrophage, and this can damage the surrounding host tissues.

- Excretion, the final phase of phagocytosis, takes place after digestion of the pathogen. The phagolysosome, now filled with pathogen fragments and referred to as a **residual body**, moves to the surface of the phagocyte and discharges the debris. Before this discharge takes place, the phagocytic cell will scavenge any useful molecules to be recycled.

Phagocytosis is a powerful tool in the host defense response. The infectious process is a tug-of-war between pathogen and host, however, and in the case of phagocytosis, bacteria have evolved ways to defeat this host defense. As we saw in Chapter 5, some bacteria produce enzymes called leukocidins that destroy white blood cells, thereby removing the phagocytic threat. Other bacteria have capsules that interfere with the adherence phase of phagocytosis.

Many bacteria can also resist the digestion phase. For example, *Yersinia pestis*, the organism that causes plague, has a tough capsule that resists enzymatic digestion, and as a result *Yersinia* actually multiplies in the phagolysosome. Other pathogens that resist digestion are *Mycobacterium tuberculosis*, *Mycobacterium leprae* and the parasite *Leishmania*. Some bacteria produce toxins that destroy the membrane of the phagolysosome, thereby releasing the toxins from the lysosome into the cytoplasm of the phagocytic cell, killing it.

There are several other ways in which phagocytosis can be compromised. For example, patients who are receiving chemotherapy and/or radiation for cancer have deficient phagocytic responses. Acquired immunodeficiency syndrome (AIDS) also causes a loss of this response, and even

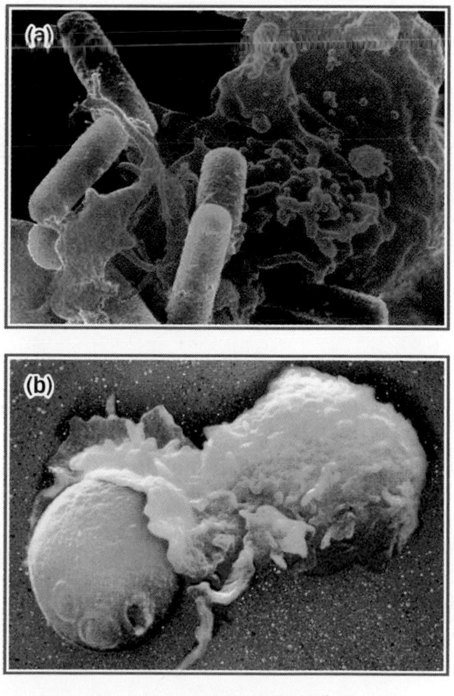

Figure 15.17 Phagocytosis. Panel **a**: Colored scanning electron micrograph of a polymorphonuclear white blood cell or leukocyte, attacking *Bacillus cereus* bacteria. The leukocyte (orange), part of the body's immune system, is attaching to and engulfing the *Bacillus* cells (blue, rod- shaped). Panel **b**: A scanning electron micrograph showing phagocytosis of a yeast cell.

drug treatments used after organ transplantation will decrease the ability of the innate immune system to respond.

The function of phagocytosis is to destroy invading microorganisms. However, any microorganisms that can resist destruction while inside a phagocyte can remain hidden from the powerful and specific adaptive immune response.

As noted earlier, phagocytosis is the cellular mechanism of the second line of defense in the innate immune response to infection. It is closely correlated, however, with the four chemical mechanisms of the second line of defense, which we now examine.

Inflammation

15.7

Inflammation is a normal physiological response to body trauma. It is beneficial because it helps to destroy infectious agents and also participates in the repair and replacement of damaged tissue. Irrespective of the cause, from bacterial infection to a burned hand or broken bone, an inflammatory response involves redness, pain, heat, and swelling. Vasodilation, an increase in the diameter and permeability of blood vessels, is associated with each of these symptoms.

Vasodilation

The cornerstone of inflammation, vasodilation is a localized rather than systemic reaction. Because vasodilation causes an increased amount of blood to flow into the injured area, the area becomes redder and warmer. The swelling and pain characteristic of inflammation are also related to vasodilation because the increased vascular permeability causes fluid to escape from the vessels into the tissues. This causes swelling, and the swelling puts pressure on nerve endings, signaling pain.

Vasodilation occurs in response to a variety of chemical signals sent from damaged tissues. Four representative signals are histamine, kinins, prostaglandins, and leukotrienes.

- Histamine is found in many cell types, including mast cells and basophils, and also in connective tissue.
- Kinins are chemotactic factors that are released from damaged tissue and draw phagocytes to the site of injury.
- **Prostaglandins** intensify the effects of both histamine and kinins, resulting in a magnification of the overall response. Prostaglandins also help in the migration of phagocytes through the walls of blood vessels.
- **Leukotrienes** are produced by mast cells. They increase vascular permeability and promote the adherence phase of phagocytosis.

Vasodilation also helps to deliver clotting elements to the site of injury. These elements prevent the spread of infection by walling off the site. This walling-off mechanism is part of the process that can lead to the formation of abscesses, such as those seen in boils.

Phagocyte Migration

One of the most fascinating aspects of the inflammatory response is the migration of phagocytes. Without the ability to leave the blood, there would be little that these very important cells could do to prevent further infection or repair tissue damage. The increased blood flow that results from vasodilation leads to increased numbers of phagocytes moving to the site of injury. However, being able to get there is only part of the story. As discussed earlier, phagocytes must stop at the site and leave the blood vessel. This part of the inflammatory process causes margination, in which cells stick to the vessel walls at the site of the infection, and diapedesis, in which the phagocytes leave the blood and move into the tissues (see **Figure 15.6**). This egress from the bloodstream is facilitated by the localized increase in permeability of the vessel walls seen during vasodilation. Diapedesis occurs rapidly, and phagocytes can be seen leaving the bloodstream as early as two minutes after arrival at the site of injury. At this point, destruction of the invading organisms begins and does not stop until they are gone.

The Acute-Phase Response

A response to pathogen invasion that is related to inflammation but is seen only in acutely ill patients is the *acute-phase response*. It involves the production of proteins called **acute-phase proteins**, including cytokines, fibrinogen (used in clotting), and kinins (which enhance vasodilation). The response begins when phagocytes somewhere in the body ingest pathogens, an event that stimulates the synthesis and secretion of several cytokines. One of these cytokines is **interleukin 6** (**IL-6**), which causes the liver to produce acute-phase proteins.

The presence in the body of these acute-phase proteins initiates a nonspecific host defense that is distinct from the classical inflammatory response and in some ways similar to the specific defenses seen with the adaptive immune response. This acute-phase response seems to recognize foreign substances just as the adaptive immune response does, but acts like an early inflammatory response that takes place before the appearance of antibody (antibody molecules are specific proteins produced during the adaptive immune response).

The best-understood acute-phase proteins are **C reactive protein** (**CRP**), which binds to phospholipids, and **mannose-binding protein** (**MBP**), also known as *mannose-binding lectin*, which binds to the mannose sugars found in many bacterial and fungal membranes. These two acute-phase proteins coat pathogens and by doing so make the pathogens more readily taken up by phagocytic cells. In addition, the proteins activate the complement system (see below) and also stimulate the production of chemotactic factors that attract more phagocytes to an injury.

Fever

Fever is a systemic rise in body temperature that often accompanies and augments the effects of inflammation. The normal body temperature can vary to some degree but is usually 37°C (98.6°F). The clinical definition of fever is an oral temperature above 37.8°C (100.5°F) or a rectal temperature above 38.4°C (101.5°F). A fever in which the body temperature rose to 43°C (109.4°F) would cause death.

Body temperature is controlled in the hypothalamus of the brain, and many pathogenic infections can reset body temperature, causing a fever. In some cases, fever accompanies certain immune responses, such as that seen after vaccination, but most types of tissue injury will elicit a fever reaction.

Fever is caused by chemicals called **pyrogens**. There are two forms, *endogenous pyrogens*, which are produced by the host, and *exogenous pyrogens*, which are produced by the invading pathogens but cause fever in the host. Bacterial toxins and even just the cytoplasm from ruptured pathogen cells act as exogenous pyrogens. **Interleukin 1** (**IL-1**) is the best-known endogenous pyrogen. This pyrogen resets the hypothalamus temperature control within 20 minutes of release. It moves through the systemic circulation into the hypothalamus and triggers the release of prostaglandin, which resets the body temperature (**Figure 15.18**). The increased body temperature continues for as long as IL-1 is present. As the infection subsides, there are fewer phagocytes secreting IL-1 and the levels of prostaglandin drop, causing the thermostat in the hypothalamus to be reset to normal. This is referred to as the *crisis phase* of the fever.

Fast Fact

In years past, fever was looked at as a negative reaction, and the standard treatment protocol was to administer aspirin to bring the body temperature down to normal. Fever is now looked at positively and, unless it goes too high, is left alone as part of the defense response.

Keep in mind that fever is a beneficial response because it raises the body temperature to levels that inhibit the growth of many bacteria. Fever also inactivates many bacterial toxins by changing their three-dimensional shape. The fever response can cause the release of **leukocyte endogenous mediator** (LEM), a factor that lowers plasma iron concentration; without iron, pathogens do not grow well. In addition, the elevated body temperature increases the speed at which host defenses work. Last, fever makes you feel ill, and this feeling takes away your energy and forces you to rest, allowing all available energy to be used for fighting off the infection.

Unfortunately, a fever that causes body temperature to go too high can be dangerous for the host. Furthermore, as long as body temperature is elevated, there is vasoconstriction. This inhibits the movement of phagocytes to the site of infection, thereby countering one of the most important parts of the innate immune response. Unchecked fever can also increase the rate of metabolism by 20%, making the heart work harder. Prolonged high temperature can cause denaturation of proteins, inhibition of nerve impulses, and electrolyte imbalance due to the loss of water. These changes can lead to hallucinations, convulsions, coma, and eventually death. Therefore, if fever goes beyond 40°C antipyretic medications are given to the patient to counter the effects of pyrogens in the body.

The Complement System

Although the responses we have discussed so far are very effective in dealing with infection and tissue damage, they are not alone in the innate immune response arsenal. Another system — the complement system — also participates in our defense. This system not only has a lethal capacity but also enhances and magnifies other parts of the innate immune response. About 30 serum proteins are involved, some of them functioning in a cascade sequence that amplifies as it progresses because more

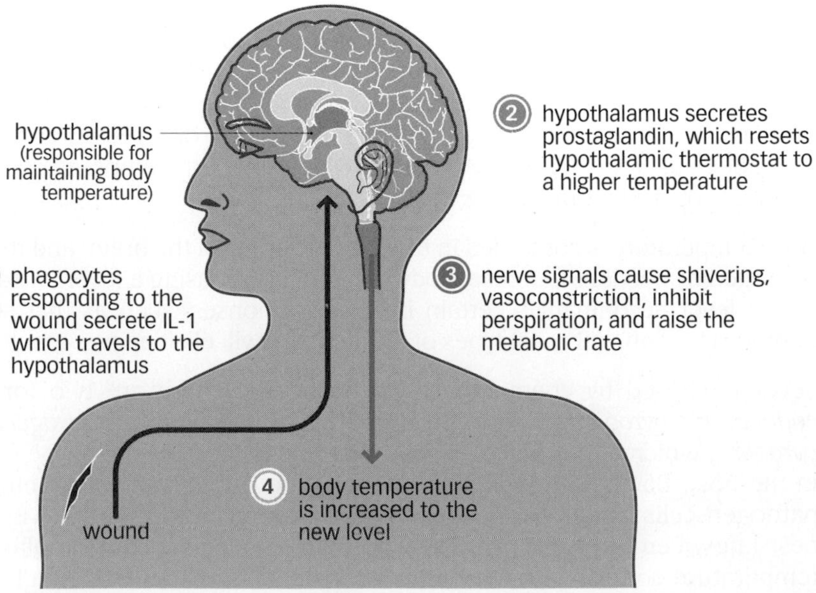

Figure 15.18 The development of a fever. Interleukin-1 (IL-1) is secreted by phagocytic cells at the injury site. IL-1 travels to the brain and influences the hypothalamus to release prostaglandins, which then reset the body temperature. The hypothalamus signals nerves to cause vasoconstriction and shivering and inhibit perspiring, all of which increase body temperature. The fever continues until IL-1 is no longer produced by phagocytic cells, indicating that the infection is under control.

product is formed with each reaction. The proteins are produced in the liver and then circulate in the blood in inactive forms. As the cascade proceeds, the proteins are activated. The proteins not directly involved in the cascade either activate or inhibit the response.

The complement system functions in a variety of ways. Its proteins can trigger and modulate the inflammatory response, and they also act as opsonins and chemotactic factors. However, the major function of this system is the lysis of bacterial cells and viral envelopes. This breaking down is accomplished through the production of a **membrane attack complex** (**MAC**) that essentially "punches" a hole in the plasma membrane of the bacterium or the envelope of a virus (**Figure 15.19**). The complement system is activated as soon as an invading organism is detected.

The complement response is a complex series of enzymatic interactions between complement proteins that can follow any one of three pathways: classical, alternative, or lectin-binding.

Activation of the Classical Pathway

The classical pathway is activated by antigen–antibody complexes. The presence of antibodies means that the host has previously seen the infecting organism and has already generated an adaptive immune response to it. The antibody molecules bind to antigens associated with the pathogens.

The proteins of the complement system are numbered in the order in which they were discovered, beginning with C1. The classical pathway is illustrated in **Figure 15.20**. The C3 protein shown in the drawing is referred to as the *nexus* of the complement system because once step 3 has finished, the three complement-activating pathways all follow the same sequence. In other words, it is only in the first two steps that the pathways differ from one another.

Activation of the Alternative Pathway

The alternative pathway works with pathogens that have never before infected the host. Because the infectious organism has not been seen before, it takes several days for the adaptive immune system to respond maximally. During this time, if the only available pathway were the classical one, we would be without the benefit of the very powerful complement system.

The alternative pathway (also referred to as the **properdin pathway**) is activated by contact between LPS and endotoxins on the pathogen surface and three factors found in the blood — **factor B**, **factor D**, and **factor P** (**properdin**). Once all these substances have combined as shown in **Figure 15.21**, complement protein C3 is attracted to the complex, and we are again at the nexus of the pathways.

Although less efficient than the classical pathway, the alternative pathway is still very useful in the early stages of infection.

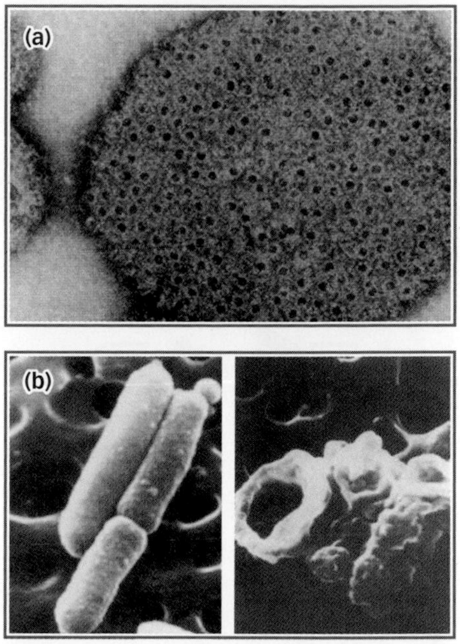

Figure 15.19 Complement mediated bacterial lysis. Panel **a**: A transmission electron micrograph of a pathogen cell after attack by the complement membrane attack complex. Note the numerous holes produced in the membrane by the complex. Panel **b**: 'Before and after' scanning electron micrographs of a rod-shaped pathogen, clearly showing the cytolysis caused by the membrane attack complex.

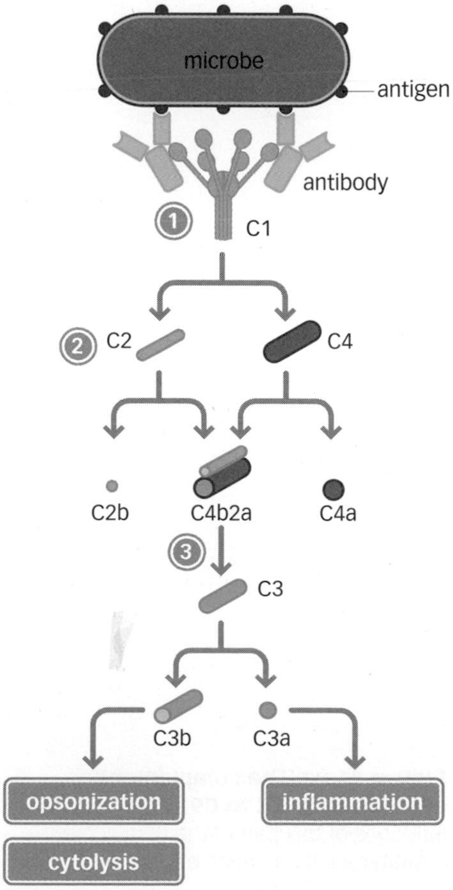

15.8 **Figure 15.20 Activation of the complement system via the classical pathway**. Step 1: binding of antibody with antigen causes the complement protein C1 to bind to the antibody–antigen complex. Step 2: this complex then enzymatically cleaves proteins C2 and C4 into two fragments each. Step 3: fragments C2a and C4b join and react with protein C3 to launch the rest of the pathway, culminating in the formation of the membrane attack complex. After step 3, the classical, alternative, and lectin-binding pathways are identical.

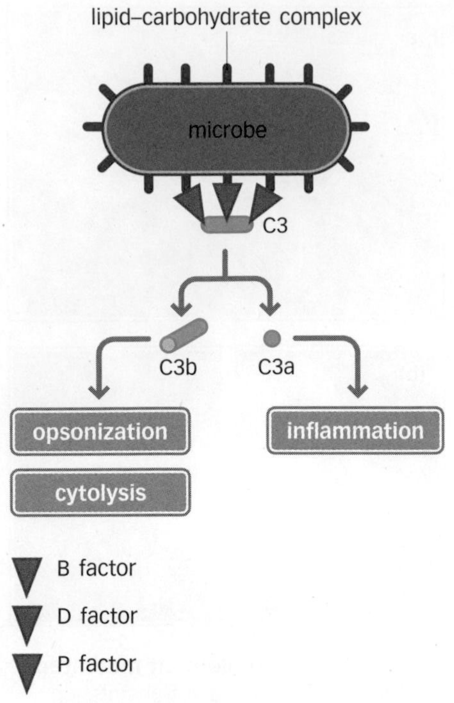

Figure 15.21 Activation of the complement system via the alternative pathway, which requires no antibodies. Carbohydrates on the surface of the pathogen, and also endotoxins, interact with factors B, D, and P in the blood. These interactions attract complement protein C3 and launch the cascade.

Activation of the Lectin-Binding Pathway

The third and perhaps most important complement-activating pathway — the **lectin-binding pathway** — is stimulated by the carbohydrate mannose. The pathway involves **mannose-binding protein** (**MBP**), a protein produced in the liver. This protein binds to the carbohydrate mannose, which is found on pathogens such as *Candida*, HIV and influenza virus, many bacterial pathogens, and also parasites such as *Leishmania*. More important than where mannose is found, however, is the fact that it is not found on normal healthy cells. MBP is activated when it binds to mannose, and it enzymatically cleaves complement protein C3, just as in the classical and alternative pathways, initiating the rest of the complement sequence as described next.

C3 and Beyond

The interactions of the remaining proteins of the complement system involve conformational changes and enzymatic cleavages that result in the formation of the membrane attack complex. In addition, the cascade can also influence other parts of the nonspecific host defense. As seen in **Figure 15.22**, at the nexus of the system, protein C3 is cleaved into C3a and C3b. The C3a fragment interacts with mast cells, causing them to release histamine, which amplifies the inflammatory response. The C3b fragment can act as an opsonin which coats pathogens, making them more susceptible to phagocytosis. It can also attract protein C5, which on binding to the complex is cleaved into two fragments. Here again we see the amplification of the host defense: the C5a fragment reinforces the production of histamine by mast cells, and the C5b fragment interacts with C6, causing the cascade to continue through to C9.

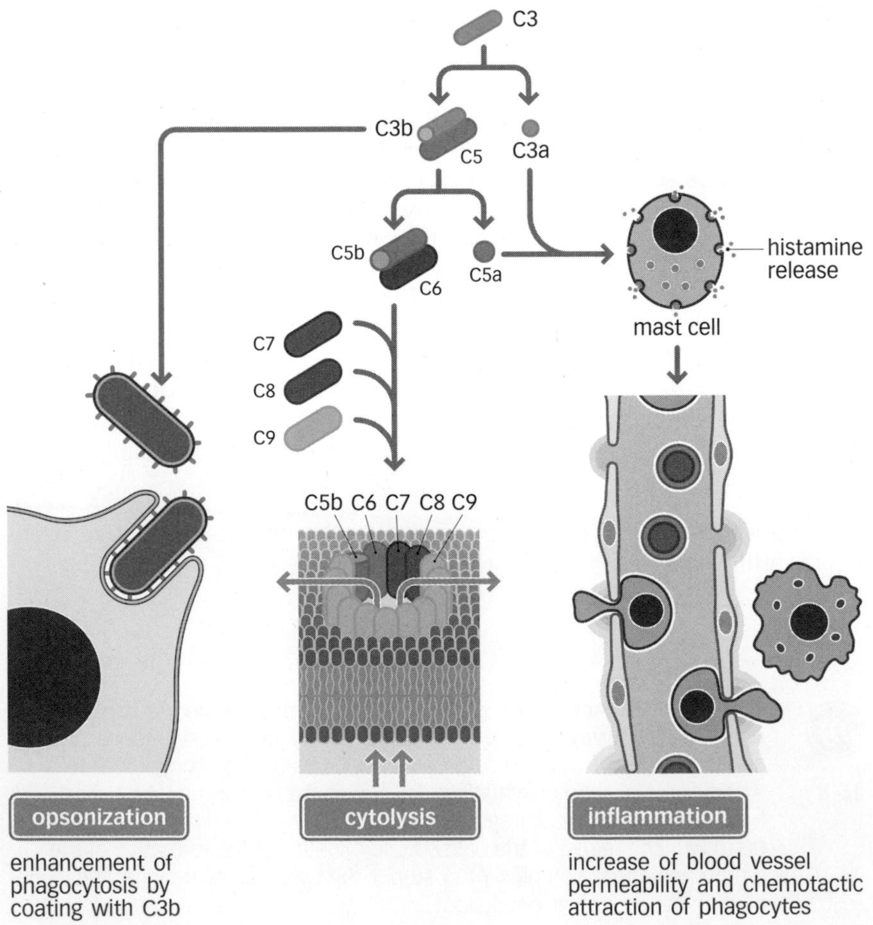

Figure 15.22 The complement cascade from C3 to C9. Note the influence of this part of the complement system on inflammation and phagocytosis.

The combination of proteins C5 to C9 is the membrane attack complex, which creates holes in a pathogen's cell membrane, thereby destroying the cell (see **Figure 15.19**). The same effect can be seen in enveloped viruses: the envelope is destroyed through this membrane attack mechanism. Without its envelope, the virus can no longer infect a host cell.

As we have come to expect by now, some organisms have evolved defenses against the complement response. One defense strategy involves encapsulation: the capsules contain large amounts of sialic acid, which discourages the formation of membrane attack complexes. In another approach, certain Gram-negative organisms lengthen surface glycolipid complexes so as to prevent membrane attack by complement. Some Gram-positive bacteria release enzymes that destroy C5a fragments, to inhibit the amplification of phagocytosis.

It is important to note that the complement system is lethal and can adversely affect normal host cells. This potential for damage to the host is controlled by having complement proteins produced in an inactive form and activating them only when an infection occurs. Furthermore, there is in the antibody–antigen reaction a specificity that prevents spurious binding and initiation of the classical pathway. In fact, most host cells contain membrane-bound proteins that bind to and inactivate complement proteins, thereby interrupting any aberrant cascade. In addition, human cells routinely replace surface membranes at a high rate. Therefore, any membrane attack complex formed inadvertently is shed off the membrane or endocytosed into the cell and degraded.

Some people are genetically deficient and cannot produce complement proteins, a condition that makes them prone to infections (**Table 15.6**). C3 deficiencies are the worst because this protein is the nexus of all the activation pathways. Without C3, the entire complement system and all its beneficial responses are lost. Deficiencies in any of the proteins (C5 to C9) making up the membrane attack complex also result in recurrent infections, especially with *Neisseria* species.

Interferon

It is easy to see how bacterial infections can be handled through the inflammatory response and phagocytosis. Infection by viruses is a different problem, however, because viruses are obligate intracellular parasites, meaning that they are obliged to enter the host cell as part of the infectious process. When outside a host cell, they are subjected to all the host defenses we have been discussing. Recall from Chapter 12 that the goal of any virus is to infect a host cell, and this goal carries with it the requirement to become essentially invisible to host defenses. The production of interferons is one of the host responses to viral infection.

Deficiency	Condition
C3	Severe recurrent infections
C1, C2, C5	Less severe recurrent infections
C6, C8	Gonococcal infections
C6	Meningococcal infections
C1, C2, C4, C5, C8	Systemic lupus erythematosus

Table 15.6 Conditions Associated with Complement Deficiency

Interferon is a protein produced by virus-infected host cells and released from those cells. The interferon then moves to as yet uninfected host cells and prompts them to make antiviral proteins that prevent virus from replicating. The main effect of **interferons** (IFNs) is that they make a host cell incapable of being infected by a virus. During studies aimed at isolating and purifying IFNs, it was found that (1) many species produce them and (2) they are species-specific. They are not *virus*-specific, however, and are produced in response to any viral infection.

Different interferon molecules are produced in different tissues, and there are three major forms found in humans: *alpha*, *beta*, and *gamma* (**Table 15.7**). The alpha and beta forms are very similar and bind to the same receptor on target cells. Consequently, both are referred to as type I. In contrast, the gamma form is distinctly different and is classified as type II. Although there are these various forms, we often speak in terms of the singular *interferon*, in which case we mean all the forms in general.

The alpha forms (of which there are more than a dozen) are produced in monocytes and macrophages, whereas the beta form is made in fibroblasts. Both are synthesized immediately after infection and protect uninfected neighboring cells. This protection occurs as IFN molecules bind to surface receptors on neighboring cells, a process that stimulates the cells to transcribe genes that code for **antiviral proteins** (**AVPs**) (**Figure 15.23**). These AVPs are inactive until they come into contact with double-stranded RNA. Recall that this form of RNA is not normally seen in host cells; it appears only in cells infected by certain RNA viruses.

Fast Fact

Interferons were originally thought to be the "magic bullet" both for viral diseases and for some malignancies. Sadly, they have not fulfilled this promise.

Gamma IFN is produced by T lymphocytes and NK cells. It also appears later in the course of an infection, when the adaptive immune reactions begin. Gamma IFN protects against viral infection, as do the other IFNs, but it can also "heal" macrophages and neutrophils that are infected and re-stimulate their phagocytic activity. For this reason, gamma IFN was originally referred to as *macrophage-activating factor*.

Initially, the amounts of IFN produced by cells were too small for therapeutic use. This hurdle was overcome with the advent of biotechnology (Chapter 27), when the genes for IFN were cloned and large amounts of pure IFN were produced in the laboratory. Interferon was approved in 1986 as a treatment for a rare leukemia called hairy cell leukemia, and it has proved to be effective in keeping this malignancy under control. In fact, if taken off IFN therapy, 90% of these patients relapse. Interferon is also useful in chronic granulomatous disease, a congenital condition in which patients do not produce gamma IFN and have no phagocytic activity in their neutrophils. This condition leads to repetitive serious infections and a life span of about 10 years. Treatment with gamma IFN increases the life expectancy of these patients. IFN has also been approved for the treatment of several viral diseases, such as genital warts and hepatitis C.

Table 15.7 Summary of Interferons

Class	Cell Source	Stimulated By	Effects
Alpha interferon (IFN-α)	Leukocytes	Virus infection	Stimulates production of antiviral proteins in uninfected cells
Beta interferon (IFN-β)	Fibroblasts	Virus infection	Same as those seen with IFN-α
Gamma interferon (IFN-γ)	T lymphocytes and natural killer cells	Virus infection and antigenic stimulation	Kills infected cells and activates destruction of tumors

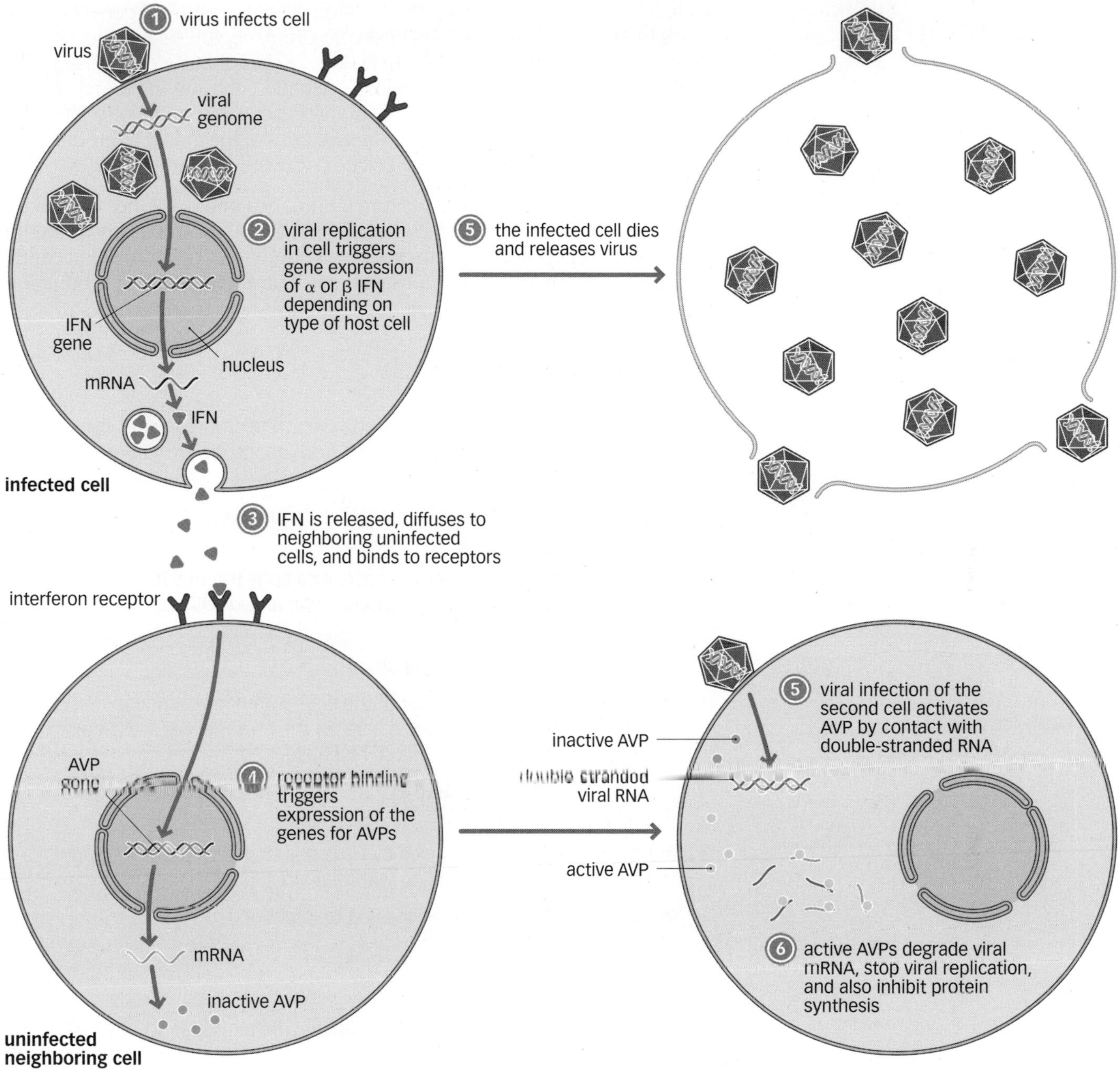

Figure 15.23 The mechanism of action of α and β interferons. Only uninfected cells neighboring an infected cell are protected. Interferons cannot save cells that are already infected.

It is important to point out that IFN is not a cure but only a treatment. Furthermore, the stability of recombinant IFN is poor, and some adverse side effects are seen with IFN treatment, including nausea, vomiting, fatigue, weight loss, and, in some cases, damage to the central nervous system. IFN can also increase the fever response, and in high concentrations it damages the heart, bone marrow, liver, and kidneys.

GENETIC SUSCEPTIBILITY TO INFECTION

Before we leave our discussion of the innate immune response, it is important to put into perspective the fact that there can be a genetic predisposition to infectious disease. In fact, in some cases, genes that are protective for one bacterial disease may increase susceptibility to another. Genetic abnormalities can involve receptors or other components of the defense network we have been discussing. For example, genetic variability in TLRs can result in serious problems with host defense. Variations in

TLR-2 have been shown to be associated with leprosy, and genetic variations in TLR-4 are seen in cases of severe Gram-negative systemic disease. In addition, genetic variability in TLRs can contribute to increased susceptibility to anthrax and tuberculosis. Genetic variation in receptors on T cells and NK cells are associated with increased susceptibility to infection by *Plasmodium*, HIV, and hepatitis C virus. Loss of the ability to produce the cytokine TNF-α can increase the risk of staphylococcal infections.

Variation in the genes that code for chemokines and cytokines are linked to at least 20 infectious diseases and syndromes. At least six variants in chemokine receptors and cytokines are known to contribute significantly to AIDS. Furthermore, variations in the genes coding for IFN production can lead to overproduction, with concomitant host pathogenesis.

Keep in Mind

- Phagocytosis is the cellular response in which microorganisms, damaged host cells, and cellular debris resulting from apoptosis are removed from the body.
- Chemotaxis is the release of chemicals by cells involved in infection that attract phagocytic cells to an infected area.
- Inflammation is a physiological response to body trauma. It involves vasodilation, resulting in redness, pain, heat, and swelling.
- Fever is a systemic rise in body temperature that often accompanies and augments the effects of inflammation.
- The complement system is part of the innate immune response that destroys bacteria by "punching a hole" in the bacterial cell wall. It also enhances other parts of the innate response.
- Complement can be activated in three ways: the classical, alternative, or lectin-binding pathways.
- Interferon is a protein produced by virus-infected cells that can protect neighboring cells from infection with that virus.
- There can be a genetic predisposition to infection.

SUMMARY

- The innate immune response is a nonspecific response to any type of infection.
- This response is composed of both mechanical and chemical barriers. There are cellular and chemical responses that are part of the innate response, including phagocytosis, inflammation, complement fixation, fever, and the production of interferon.
- Cellular responses are carried out by white blood cells.
- Host defense cells use Toll-like receptors to identify nonself antigens.
- Diapedesis is the process whereby white blood cells leave the blood vessels and migrate into tissues to reach the site of the infection.
- A variety of cytokines and chemokines are produced throughout the innate response.
- Phagocytosis is one of the most important parts of the innate response and involves attachment, ingestion, and digestion.
- Inflammation involves vasodilation and results in redness, pain, heat, and swelling.

- Complement fixation involves a cascade of proteins that result in the destruction of a bacterial cell.
- The complement cascade can be initiated in three ways: the classical pathway, the alternative pathway, and the lectin-binding pathway.
- There can be a genetic predisposition to infection.

In this chapter we have learned that in the innate immune response there are mechanical and chemical barriers constituting a first line of defense, and also powerful cellular and chemical mechanisms in a second line of defense. Keep in mind that all these barriers and mechanisms work together to maximize the nonspecific response to any infection. In addition to these powerful innate immune defenses, we also have a pathogen-specific response called the adaptive immune response. Like the innate mechanisms we have discussed here, adaptive mechanisms are also lethal. However, as we will see in the next chapter, the adaptive immune response is an exquisitely sensitive and specific response that also has the gift of memory.

Q SELF EVALUATION AND CHAPTER CONFIDENCE

Multiple Choice

Answers are given in the back of the book and help can be found in the student resources at:
www.garlandscience.com/micro

1. Nonspecific defense is
   - **A.** The body's defenses against any kind of pathogen
   - **B.** The body's lack of resistance to infection
   - **C.** The body's defense against a particular pathogen
   - **D.** None of the above
2. The innate response includes all of the following except
   - **A.** Phagocytosis
   - **B.** Inflammation
   - **C.** Production of antibody
   - **D.** Production of interferon
   - **E.** Activation of complement
3. Which of the following is not a mechanical barrier to protect the skin and mucous membranes from infection?
   - **A.** Lysozyme
   - **B.** Tears
   - **C.** Layers of cells
   - **D.** Saliva
   - **E.** Gastric juice
4. The function of the mucociliary escalator is to
   - **A.** Kill microorganisms
   - **B.** Remove microorganisms from the upper respiratory tract
   - **C.** Remove microorganisms from body cavities
   - **D.** Remove microorganisms from the lower respiratory tract
   - **E.** All of the above
5. Tears contain
   - **A.** Lipocalin
   - **B.** Lysozyme
   - **C.** IgA
   - **D.** All of the above
   - **E.** None of the above
6. Perspiration inhibits bacteria because
   - **A.** It contains mucus
   - **B.** It contains IgA
   - **C.** It contains lysozyme
   - **D.** It flushes them away
   - **E.** Both C and D are correct
7. Toll-like receptors are responsible for all of the following except
   - **A.** Recognition of structures on bacterial pathogens
   - **B.** Recognition of viral double-stranded RNA
   - **C.** Recognition of flagellar proteins
   - **D.** Recognition of host cell DNA
   - **E.** Recognition of peptidoglycan
8. All of the following cells are involved in the innate response except
   - **A.** Neutrophils
   - **B.** Eosinophils
   - **C.** Basophils
   - **D.** Monocytes
   - **E.** None of the above

9. Neutrophils attach to the vascular linings and move out of the blood and into the tissues in a process known as
   - A. Intravascular clotting
   - B. Selection
   - C. Diapedesis
   - D. Margination
   - E. None of the above

10. Margination is the process in which white blood cells
    - A. Separate from red blood cells
    - B. Leave the blood vessels
    - C. Produce selectin
    - D. Slow down, stop, and attach to vessel walls
    - E. Speed up and attach to vessel walls

11. The primary phagocytes in the blood are
    - A. Basophils
    - B. Eosinophils
    - C. Lymphocytes
    - D. Monocytes
    - E. Neutrophils

12. The most phagocytic white blood cells are
    - A. Neutrophils
    - B. Macrophages
    - C. Basophils
    - D. Monocytes
    - E. Lymphocytes

13. Macrophages located in the liver are called
    - A. Alveolar macrophages
    - B. Dendrites
    - C. Microglial cells
    - D. Kupffer cells
    - E. None of the above

14. Macrophages located in the central nervous system are called
    - A. Alveolar macrophages
    - B. Dendritic macrophages
    - C. Microglial cells
    - D. Kupffer cells
    - E. None of the above

15. The characteristics of cytokines include all of the following except
    - A. Regulation of inflammation response
    - B. Secretion from white blood cells
    - C. Reaction with specific receptors on target cells
    - D. Phagocytic activity
    - E. Having overlapping functions with other cytokines

16. Mediators released by mast cells include
    - A. Histamine
    - B. Serotonin
    - C. Cytokines
    - D. Proteases
    - E. All of the above

17. Dendritic cells found in the skin are called
    - A. Langerhans cells
    - B. Kupffer cells
    - C. Microglial cells
    - D. None of the above

18. Natural killer cells are
    - A. Restricted to immune surveillance of tumors
    - B. Part of the adaptive immune response
    - C. Involved in both surveillance and response to pathogens
    - D. Restricted to destruction of pathogens

19. The sequence of phases in phagocytosis is
    - A. Ingestion, chemotaxis, adherence, digestion, excretion
    - B. Digestion, adherence, chemotaxis, ingestion, excretion
    - C. Chemotaxis, ingestion, adherence, digestion, excretion
    - D. Chemotaxis, adherence, digestion, ingestion, excretion
    - E. Chemotaxis, adherence, ingestion, digestion, excretion

20. Phagolysosomes are formed during which phase of phagocytosis?
    - A. Ingestion
    - B. Chemotaxis
    - C. Digestion
    - D. Adherence
    - E. Excretion

21. Redness, pain, heat, and swelling are hallmarks of
    - A. Phagocytosis
    - B. Vasoconstriction
    - C. An anti-inflammatory response
    - D. An inflammatory response
    - E. Intravascular clotting

22. The acute-phase response is seen only in
    - A. Recovering patients
    - B. Patients with no visible signs of infection
    - C. Acutely ill patients
    - D. Patients who are immune
    - E. Patients who are immunodeficient

23. Fever is caused by chemicals known as
    - A. Pyretics
    - B. Intravascular clotting factors
    - C. Pyrogens
    - D. Pyrotechnics
    - E. None of the above

24. The complement system is activated by
    - A. The alternative pathway
    - B. The restriction pathway
    - C. The lectin-binding pathway
    - D. A and C
    - E. None of the above

25. The classical pathway for activation of the complement system requires
    - A. A phagocytic response
    - B. Mannose-binding ligands
    - C. Factor D
    - D. Antigen–antibody complexes
    - E. Properdin

26. The alternative pathway for activation of the complement system is initiated at protein
    - A. C1
    - B. C2
    - C. C6
    - D. C3
    - E. C1–C2–C4 complex

27. The combination of complement proteins C5 to C9 is known as
    - A. The terminal complex
    - B. The defense complex
    - C. The membrane defense complex
    - D. The membrane attack complex
    - E. None of the above

28. Gamma interferon is produced by
    - A. T lymphocytes
    - B. T lymphocytes and NK cells
    - C. T lymphocytes, NK cells, and neutrophils
    - D. NK cells
    - E. NK cells and neutrophils

DEPTH OF UNDERSTANDING

Questions listed here require you to bring together the concepts you have learned in this chapter into a discussion format. This helps you to increase your depth of understanding of the material you have learned. Help can be found in the student resources at:
www.garlandscience.com/micro

1. Describe the chemical barriers associated with the innate immune response.

2. The complement system is extremely important in the nonspecific defense response. It can be said that this system is an amplification system. Defend this statement.

3. While chopping an onion, you inadvertently cut your finger. Recalling what you have learned in this chapter, explain the chemical and cellular defense responses that occur at the site of this trauma.

CLINICAL CORNER

Help can be found in the student resources at:
www.garlandscience.com/micro

1. Your patient is a 35-year-old male who presents with a history of infections. These infections are primarily due to staphylococcal and streptococcal pathogens. His blood work shows a high titer of antibody against staphylococci and streptococci but he is currently infected with a strain of *Neisseria*. Further evaluation of his blood work indicates he has no titer to these organisms and also that he has little to no C3 protein in his blood. As part of his history he has told you that his father was also prone to repeated infections.

A. How will you explain his situation to him?
B. Should he be put on broad-spectrum antibiotics?
C. What do you tell him if he asks whether he will continue to have repeated infections?

2. Mr Edison is recovering from a kidney infection. He has been receiving antibiotic therapy and his symptoms have diminished markedly. He has returned to the clinic because he has noticed blood in his urine. The doctor tells Mr Edison that in some cases the response to an infection can have side effects that are as bad as the infection. After the doctor leaves, Mr Edison asks you to explain what the doctor meant.

A. What would you tell him?
B. Which of the innate responses would be the most involved in this bleeding event?

Chapter 16

The Adaptive Immune Response

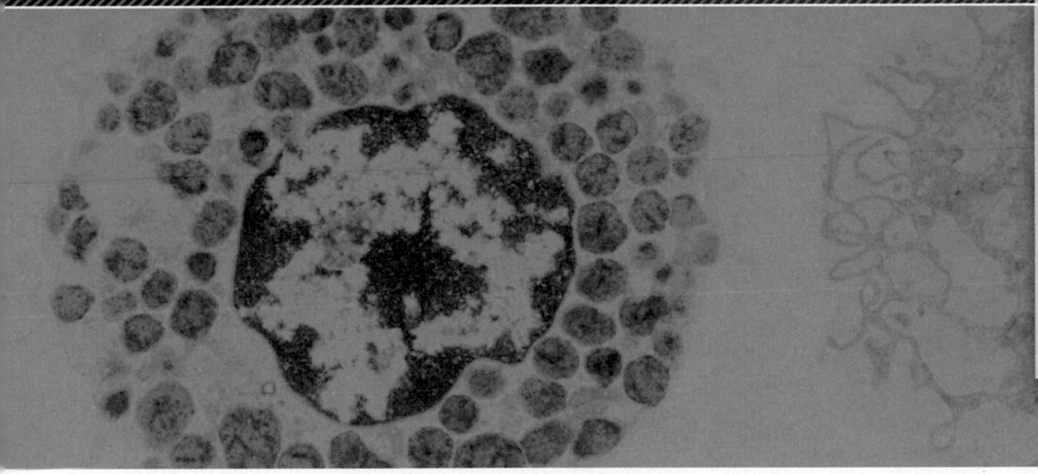

Why Is This Important?

The adaptive immune response is a very powerful system that protects us from a multitude of infectious organisms. It has the gift of memory, which provides a more powerful reaction if the same pathogen is seen again. Without the adaptive immune response we would not survive.

OVERVIEW

In this chapter, we look at the adaptive immune response, the second level of host defense (recall that the innate response was the first level). In Chapter 15, we examined the nonspecific defenses associated with the innate immune response. We learned that a host has a formidable array of innate defense mechanisms that in many cases are more than enough to handle potential infections. However, the adaptive immune response also has a powerful role in the defense of the host. This response involves lymphocytes and also has the "gift" of memory. That means that when a host is infected by a pathogen, the adaptive immune response not only defeats the infection but also remembers the pathogen. When that pathogen invades the host again, even decades after the first infection, the pathogen will be defeated even more quickly. Memory is the reason that vaccinations work and one of the major reasons that we are able to survive even though surrounded by microorganisms that could repeatedly infect us. As we look at the adaptive response, you will see that it is like a carefully choreographed ballet in which the dancers all depend on one another in giving a powerful performance.

We will divide our discussions into the following topics:

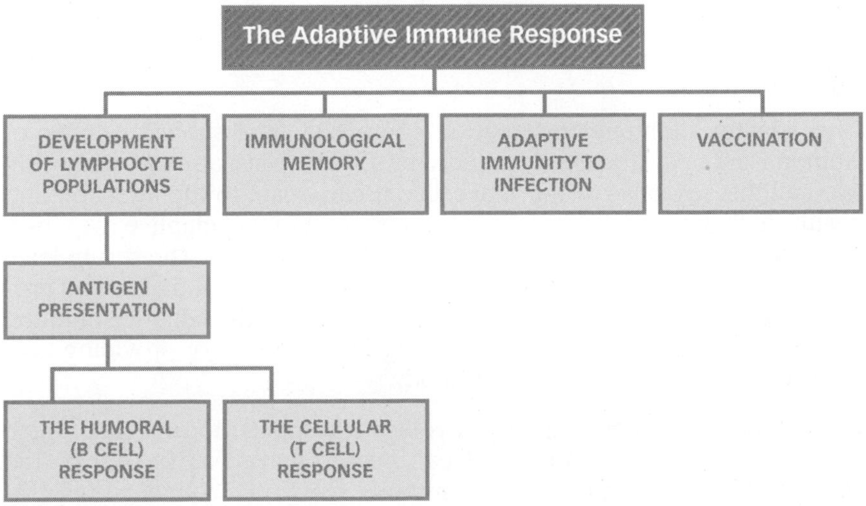

What Do I Need to Know?

To get the most out of this chapter, please review the following terms from your previous courses in biology, anatomy, physiology, or chemistry or in previous chapters of this book as indicated in parentheses: apoptosis (**8**), fenestrated, regional lymphoid structures, lymph fluid (**5**), peripheral tissue, peripheral lymphoid structures, strategic lymphoid structures, efferent lymphatic vessels, inflammatory cells (**15**), lymph nodes (**8**).

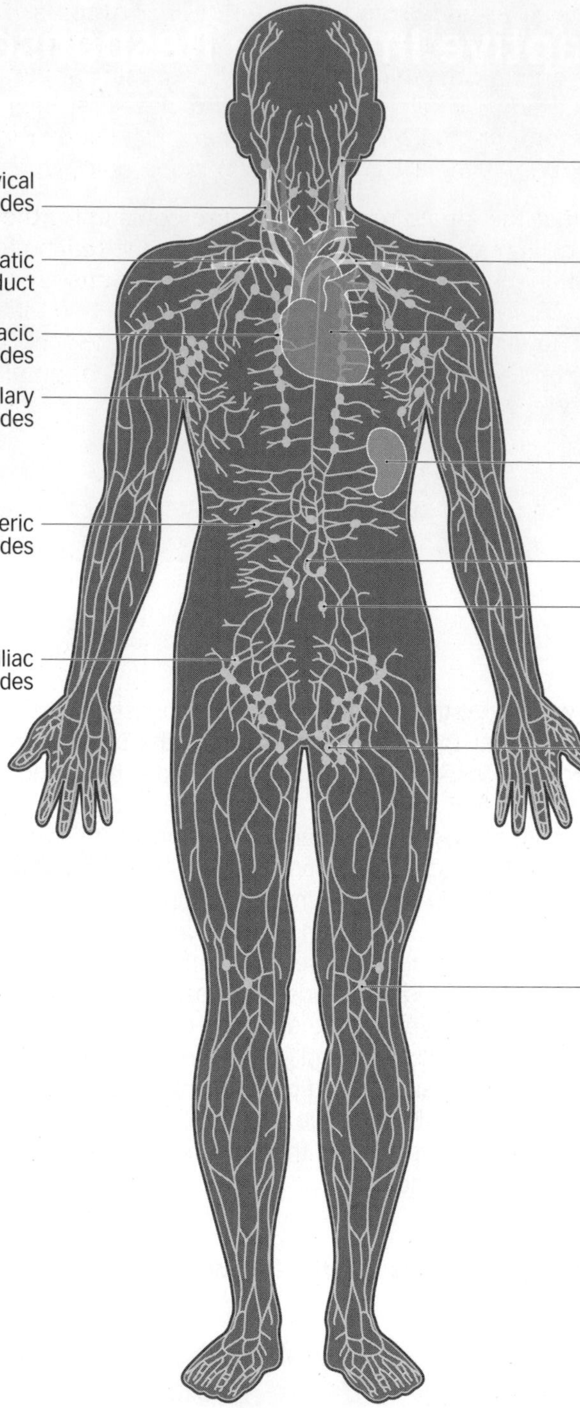

Figure 16.1 An illustration of the human lymphatic system. Notice how this system covers the entire body. It is the strategic placement of these lymphoid structures that makes it possible for the adaptive immune response to deal with potential pathogens from any place that is involved with infection.

INTRODUCTION

The adaptive immune response is composed of a *specific* cellular or humoral (antibody) response produced by a host against a *specific* pathogen. We call this response *adaptive* because it can adapt to the infection and results in lifelong immunity. In the humoral part of the adaptive response, an antibody is produced against the invading pathogen. The serum level of this antibody is called a **titer** and can be an indicator of the level of protection against that pathogen. In the cellular part of the adaptive immune response, T lymphocytes react by killing infected cells or providing help to B cells for the production of antibody.

The adaptive immune response involves a response to **antigen**, which is defined as any substance that can be recognized as foreign by the adaptive response and cause an immune reaction. To understand this response, we need to understand the difference between self antigens and nonself antigens. Perhaps the simplest way to understand this is to realize that beginning early in life, the adaptive immune system takes an inventory of all the molecules in the body and classifies them as "self". From that time on, any molecules that appear in the body that are not part of the inventory stimulate an adaptive immune reaction.

Keep in mind that the innate response is a prerequisite for the adaptive response, and cells that participate in innate immune reactions are also involved in the adaptive response. The adaptive response is associated with the lymphatic system of the body, which is laid out so that it covers the entire body (**Figure 16.1**). The strategic location of the lymphoid structures, in particular the lymph nodes, permit the identification of potential invaders from any location in the body.

Components of the Adaptive Response

The cells involved in adaptive immunity arise in the bone marrow from stem cells (**Figure 16.2**). Of the cells that participate in the early steps

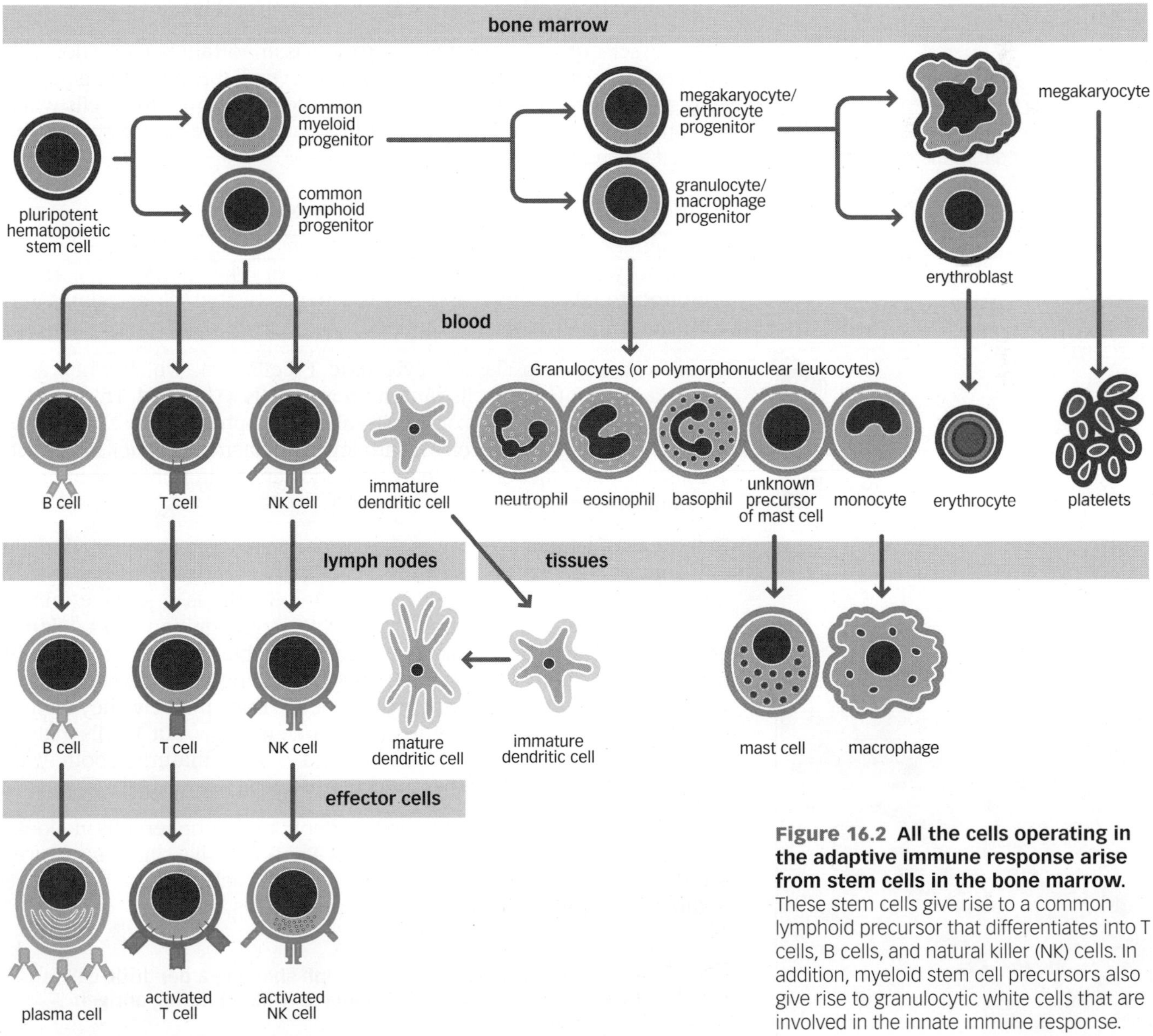

Figure 16.2 All the cells operating in the adaptive immune response arise from stem cells in the bone marrow. These stem cells give rise to a common lymphoid precursor that differentiates into T cells, B cells, and natural killer (NK) cells. In addition, myeloid stem cell precursors also give rise to granulocytic white cells that are involved in the innate immune response.

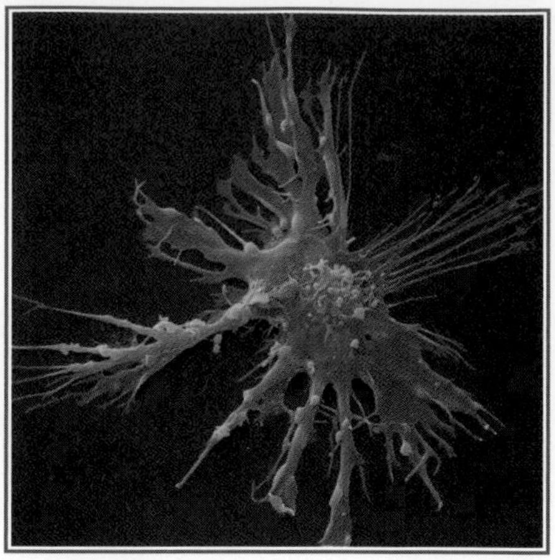

Figure 16.3 A scanning electron micrograph of a dendritic cell, which is the major initiator of the adaptive immune response. Note the extensive elongated projections, which greatly increase the amount of surface area available for antigen presentation.

of adaptive immunity, dendritic cells (**Figure 16.3**) are some of the most important. These cells are specialized to take up antigens, process them, and then present them to **T lymphocytes** (which are known as **T cells**). Note the large number of long projecting arms in the dendritic cell in **Figure 16.3**. These projections give the cell an enormous surface area to which antigens can attach and thus be presented.

Immature dendritic cells migrate from the bone marrow into the blood and then from blood into tissues. Here they use phagocytosis and macropinocytosis (which involves engulfing large amounts of fluids) to capture antigens and deliver them to T cells. Dendritic cells are constantly gathering up antigens, both self and, during infection, nonself. The dendritic cells migrate from the tissues to regional lymphoid structures (such as lymph nodes), carrying the antigens attached to their surface. It is in lymphoid structures such as the lymph nodes that nonself antigens are presented to lymphocytes and trigger the adaptive response (**Figure 16.4**).

Fast Fact

Tolerance prevents the immune system from reacting against self. (We will see in Chapter 17 that this safeguard can fail and autoimmune diseases can occur.)

Carrying self antigens and presenting them is important because doing so induces a **tolerance** to those antigens. However, if the immature dendritic cells are carrying nonself antigens, they release a signal to T cells when presenting the antigens. This results in an adaptive response against the nonself antigens that were presented.

Differentiation and Maturation of Lymphocytes

The common lymphoid precursor cell shown in **Figure 16.2** gives rise to T lymphocytes, B lymphocytes and also natural killer (NK) cells. The **B lymphocyte** (also known as a *B cell*), when activated, differentiates into a *plasma cell*, which secretes antibody.

16.1

16.2

T cells come in two classes: **cytotoxic T cells** (*Tc*), which when activated can kill infected cells, and **helper T cells (Th1 and Th2)**, which can activate B cells, cytotoxic T cells, and macrophages. Most T cells are small and relatively featureless, with few cytoplasmic organelles. In fact, T cells have no functional activity until they encounter an antigen and a co-stimulatory signal.

Both types of lymphocyte can mount a specific response against virtually any antigen because as a B cell or a T cell matures, it acquires a unique antigen receptor. The antigen receptor for B cells is a surface immunoglobulin molecule that has two identical recognition sites (**Figure 16.5**a). The antigen receptor for a T cell resembles an immunoglobulin molecule but there is only one antigen-binding site (**Figure 16.5**b). Although B cells and T cells are both formed in bone marrow, they mature in different places. B cells mature in the bone marrow, but T cells mature in the thymus (hence the designation B and T). After maturing, both types of lymphocyte move into other parts of the body.

Lymphocytes circulate through blood, lymph, and peripheral tissue looking for antigens that fit their receptors. Peripheral lymphoid structures are specialized to trap antigen-bearing dendritic cells so that they can be "examined" by lymphocytes for potential antigens.

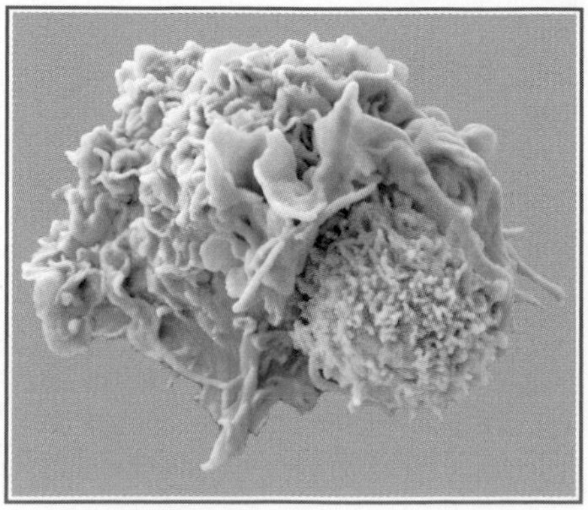

Figure 16.4 An electron micrograph showing a dendritic cell (blue) interacting with a lymphocyte (yellow) during antigen presentation.

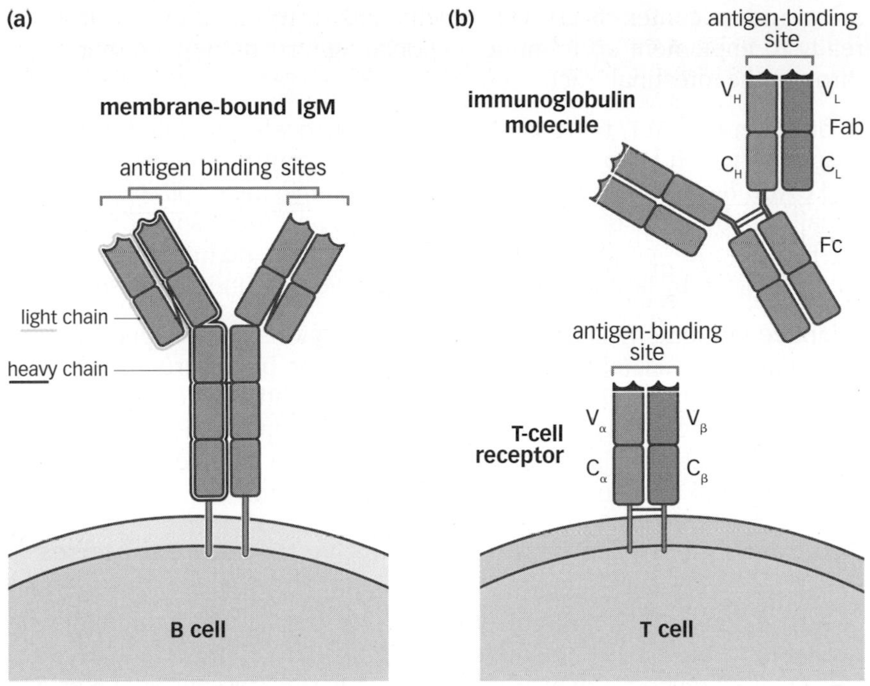

Figure 16.5 Lymphocyte antigen receptors. Panel **a**: An illustration of IgM bound to the membrane of a B cell; this is a B-cell receptor. Note that the molecule is bivalent, having two antigen-binding sites. Panel **b**: The T-cell receptor is similar to an immunoglobulin molecule but has only one binding site for antigen.

Keep in Mind

- The adaptive immune response is specific and involves both a cellular and a humoral (antibody) component.
- Antigens are substances that can elicit an adaptive immune response.
- Cells involved in adaptive immunity arise from stem cells in the bone marrow.
- The two lymphocyte types involved in the adaptive response are T cells and B cells.
- Both T cells and B cells have receptors for antigen.
- For B cells the antigen receptor is an immunoglobulin molecule, and for T cells it is an immunoglobulin-like molecule.

Strategic Lymphoid Structures

As we saw in earlier chapters, there are several areas of the body that are portals of entry for pathogens, where immune protection is essential. In all of these areas we find lymphoid tissue. In the gastrointestinal tract, for example, the adaptive immune response is focused in the **GALT** (*g*ut-*a*ssociated *l*ymphoid *t*issue), which includes the tonsils, adenoids, appendix, and Peyer's patches of the small intestine. The Peyer's patches are the most important part of the GALT because of the antigen-collecting **M cells** found in them. M cells are unlike any other part of the small intestine because no nutrient absorption takes place here. These cells have no villi; instead they are fenestrated (**Figure 16.6**). Under each M cell, there

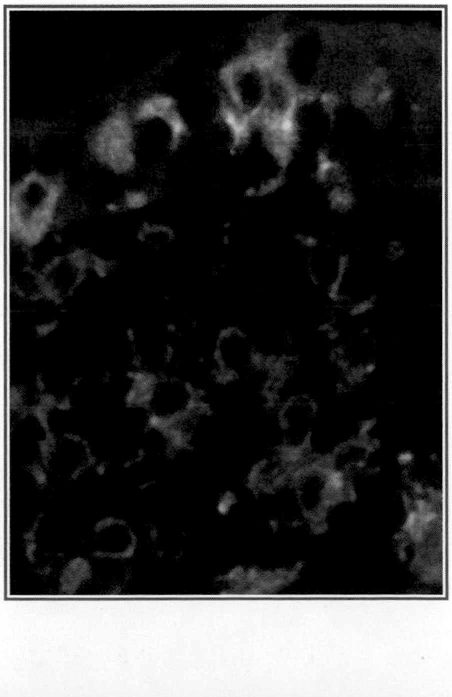

Figure 16.6 A micrograph of a Peyer's patch. Some of the epithelial cells (dark blue) are M cells that form pockets where T cells and B cells (green) accumulate.

are germinal centers filled with B cells and surrounded by T cell areas ready to implement an immune response against pathogens that enter through the intestinal tract.

In addition to GALT, there is also **MALT** (*m*ucosal-*a*ssociated *l*ymphoid *t*issue), in which lymphocytes are located adjacent to mucosal surfaces, and **BALT** (*b*ronchial-*a*ssociated *l*ymphoid *t*issue) in the respiratory system. Recall that both the body's mucosal surfaces and the respiratory system are major portals of entry for infectious organisms and must therefore be well protected by both the innate and adaptive immune responses.

Adaptive immune mechanisms in all these regions are designed to trap potential antigens (either by dendritic cells or by macrophages) and induce an adaptive response by presenting the antigens to lymphocytes. Of equal significance is the fact that peripheral lymphoid tissues also produce important signals that allow lymphocytes to survive and continue migrating until they encounter their specific antigen.

Relationship between Innate and Adaptive Immune Responses

Recall from Chapter 15 that the innate immune response is the first line of host defense. It is very powerful but nonspecific and is immediately available to the host. This immediate availability is important because the adaptive response, which is equally powerful but specific, takes five to seven days to appear once a host has been invaded by a pathogen. The innate response must deal with the infection until the adaptive response is ready. In fact, the innate response helps to regulate the adaptive response by providing cytokines and chemokines needed for the activation and maintenance of the adaptive immune response.

Dendritic cells are good examples of the interrelationship between the innate and adaptive immune responses. These phagocytic cells are resident in most tissues, particularly at epithelial surfaces, and carry Toll-like receptors. Therefore they are great "watchdogs" for potential pathogens. When they detect antigens, they phagocytose and degrade them (the innate portion of their work) and then move to the nearest lymphoid tissue. The antigen recognition and processing causes the dendritic cell to mature into an **antigen-presenting cell** (**APC**). When activated, these APCs begin to produce the cytokines that influence the adaptive immune response.

Macrophages also have antigen presentation ability. During an infection, monocytes move to the site and differentiate into macrophages. These cells phagocytose the pathogens and move to regional lymphoid structures, where they present the antigens to T cells.

Keep in Mind

- There are lymphoid structures such as GALT, MALT, and BALT that are strategically located in major portals of entry used by pathogens.
- The adaptive immune response is interrelated with the innate immune response through antigen presentation by cells called antigen-presenting cells (APCs).

DEVELOPMENT OF LYMPHOCYTE POPULATIONS

Both B cells and T cells are formed in the bone marrow, and some B cells mature there. Most T lymphocytes, however, mature in the thymus. This organ is prominent in the young but atrophies by the time we reach puberty. It was difficult for scientists to accept the fact that the thymus atrophies even though in many cases the T cell response lasts essentially for life. As it turns out, the T cell response is maintained through long-lived T cells that can occasionally divide to maintain the response capability.

The situation for B cells is slightly different because they mature in the bone marrow, which remains functional for life. There is therefore always a supply of mature B cells readily available.

The stages of lymphocyte development are marked by successive rearrangements of antigen receptor genes. In addition, maturation requires signals from the microenvironment in which the lymphocyte develops (bone marrow for B cells and thymus for T cells).

Clonal Selection of Lymphocytes

Clonal selection, illustrated in **Figure 16.7**, is the process by which some lymphocytes are destroyed and others are allowed to mature. This selection process takes place in the bone marrow for B cells and in the thymus for T cells. As shown in the top panel of **Figure 16.7**, a precursor cell produces lymphocytes, each with specificity for a different antigen. This wide range of specificities is the result of a genetic mechanism called rearrangement, which occurs during lymphocyte development. This rearrangement generates millions of variants of the genes that code for the antigen receptors on the surface of the lymphocytes. Because each lymphocyte has a unique specificity and because there are billions of lymphocytes in the body, the "repertoire" of antigen receptors is vast.

Clonal deletion (second panel in **Figure 16.7**) is a part of clonal selection. In this step, those members of the newly formed lymphocyte group that react with self antigens are eliminated (in other words, *deleted*). The exact mechanism for clonal deletion is not yet understood, but because these cells would react against the host they must be removed.

All the lymphocytes remaining after deletion will mature, and each will be specific for a different antigen. When one member of this group encounters its particular antigen, it binds that antigen and becomes *activated*. Lymphocyte activation gives rise to a *clone* of identical lymphocytes specific for that antigen (bottom panel of **Figure 16.7**). Only those lymphocytes that encounter their specific antigen are activated. This limitation prevents indiscriminate activation of the immune response because indiscriminate activation can have dangerous autoimmune consequences for the host.

Clonal selection is one of the central principles of adaptive immunity and has three important consequences:

- It enables a limited number of gene segments to rearrange and thereby generates a vast number of different antigen receptors.
- Each receptor is specific for a different antigen.
- Because genetic rearrangement is irreversible, all progeny of a lymphocyte that is a member of a clone will have that same receptor.

Survival of Lymphocyte Populations

16.3

Each day the bone marrow produces millions of new B lymphocytes. The survival of these lymphocytes is determined by signals sent out from peripheral lymphoid tissue and received by the antigen receptors on the lymphocytes. These signals cause the B lymphocytes either to become activated and proliferate, or to die. Eventually, those B lymphocytes that never get stimulated undergo apoptosis. This helps to keep the size of the overall B lymphocyte population relatively constant.

For T cells, survival signals come from specialized epithelial cells in the thymus during development. Signals can also come from dendritic cells in the peripheral lymphoid tissues. Once they leave the thymus, T cells migrate to the lymph nodes. After being presented with their antigen, T cells in the lymph nodes stop migrating and become activated. They

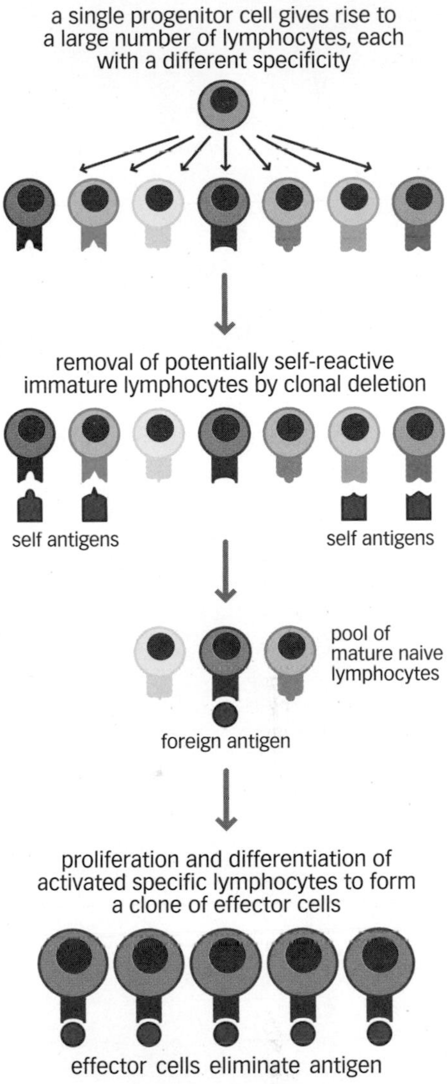
a single progenitor cell gives rise to a large number of lymphocytes, each with a different specificity
removal of potentially self-reactive immature lymphocytes by clonal deletion
self antigens
self antigens
pool of mature naive lymphocytes
foreign antigen
proliferation and differentiation of activated specific lymphocytes to form a clone of effector cells
effector cells eliminate antigen

Figure 16.7 Clonal selection, a process in which each lymphoid progenitor gives rise to a large number of lymphocytes, each bearing a distinct antigen receptor. Lymphocytes bearing receptors for self are lost through the process of clonal deletion before they become fully mature. When a naive lymphocyte binds to its particular antigen, the lymphocyte proliferates and differentiates to form a clone of cells all specific for that particular antigen.

become larger and increase their numbers up to fourfold every 24 hours for three to five days. Therefore, one naive T lymphocyte can give rise to thousands of daughter cells of the same specificity, and each of these can differentiate into an *armed effector T cell*. These changes also affect the surface adhesion molecules on these cells so that they can either migrate to sites of infection or remain in the lymph nodes. Activated T lymphocytes have a limited life span and although most will eventually undergo apoptosis, a significant number will become **memory cells**, which are long-lived lymphocytes able to respond faster and more powerfully when the antigen is encountered again.

Responses to antigen by both B lymphocytes and T lymphocytes require not only the signals that result from antigen binding to lymphocyte receptors but also a second **co-stimulatory signal**. This signal is generated from, and regulated by, cells of the innate immune response. Contact with antigen without a co-stimulatory signal leads to automatic inactivation of the lymphocyte, a state called **anergy**. In this state, the lymphocyte is in a *refractory mode* and cannot be activated. This is part of the tolerance required to prevent autoreactivity toward self.

Lymphoid Tissues

Once lymphocytes leave the thymus or bone marrow, they are carried by the blood into peripheral lymphoid tissues. These tissues, which include the lymph nodes, are organized into distinct areas where T cells or B cells are found. We can use the lymph node to illustrate this organization (**Figure 16.8**). B cells in the lymph node are found in areas called **follicles**, which are located in the outer cortex. T cells are found in zones that surround these follicles in what are called **paracortical areas**.

Normally, a lymphocyte circulates continuously through the lymphoid tissue by way of the blood and lymph fluid until the lymphocyte either encounters its specific antigen or dies. When a lymphocyte dies, it is replaced by a new one.

In peripheral lymphoid tissue, the fate of the lymphocytes is controlled by their antigen receptors. In the absence of an encounter with their specific antigen, B cells must be stimulated by a weak engagement of their antigen receptors; they die if this does not happen. Although large numbers

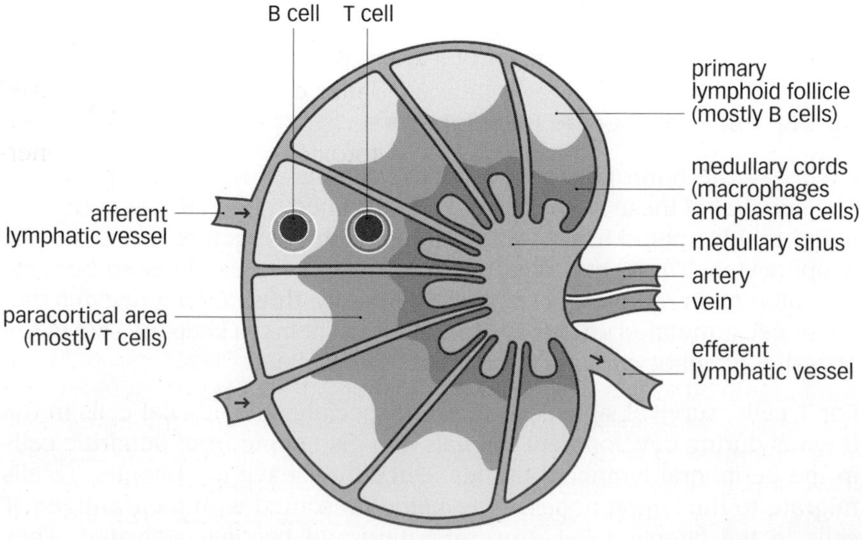

Figure 16.8 A diagrammatic representation of the structure of the lymph node. Notice how B and T cells are located in different regions of the node.

of B cells leave the bone marrow every day, most die soon after they arrive in the peripheral lymphoid tissue. However, the total number in the body remains constant through replacement.

T cells leave the thymus fully mature but in smaller numbers than B cells leaving the bone marrow. T cells generally have a longer life span and are thought to be self-renewing in peripheral lymphoid tissue.

Keep in Mind

- T cells mature in the thymus; B cells mature in the bone marrow.
- Clonal selection is the process by which some lymphocytes are allowed to mature and others are deleted from the body.
- Lymphocytes continue to circulate through lymphoid tissue by way of the blood or lymph until they either encounter their specific antigens or die.
- T cells are naive until they come in contact with their specific antigens, at which time they become armed effector T cells.

ANTIGEN PRESENTATION

For antigen presentation to occur, the antigen recognized by a T lymphocyte receptor must be bound to the **major histocompatibility complex** (**MHC**) molecules found on APCs. To initiate a response, the T-cell receptor has to recognize both the antigen and the MHC molecule, a combination referred to as the *antigen–MHC complex*. This requirement for the proper recognition of the complex is referred to as **MHC restriction** and is necessary for the regulation of the response.

As noted earlier in this chapter, there are two basic types of T cell — helper T cells and cytotoxic T cells. The two types are differentiated from each other by which surface proteins are expressed. The surface protein called CD4 is found on helper T cells, and the surface protein called CD8 is found on cytotoxic T cells.

There are also two types of MHC: class I and class II. These two types differ in structure and in the way in which they are expressed on the body's cells. Cytotoxic T cells (CD8) bind to class I MHC molecules, and helper T cells (CD4) bind to class II MHC molecules. Because T cells do not bind directly to an antigen but rather to an antigen–MHC complex, we can view class I and class II MHC molecules as CD8 and CD4 T cell receptor ligands.

Antigens are delivered to the cell surface of APCs by MHC molecules. As we mentioned above, there are two classes of MHC — I and II — and each delivers antigens from different cellular compartments to the APC surface. Antigens from the cytoplasm of the APC cell bind to class I MHC molecules and are presented to CD8 (cytotoxic) T cells. Antigens generated in vesicles of the APC cells, as in phagocytosis, are bound to the class II MHC molecules and are presented to CD4 (helper) T cells. 16.4

For either class of MHC molecule to function properly, the antigen–MHC complex must be stable at the APC cell surface. This stability permits long-term display of the complex, which is required for effective presentation of the antigen to a T cell. If the antigen is removed from the MHC molecule while on the surface of the APC cell, or if an MHC molecule shows up on the surface of the APC without an antigen, the entire MHC molecule undergoes a conformational change and is quickly degraded. This is a failsafe mechanism for the adaptive immune response that protects against aberrant reactions that could damage the host.

T Cell Response to Superantigens

Many T cells can respond to what are called **superantigens**. These are distinct classes of antigens produced by many different pathogens that provoke responses helpful to the pathogen rather than the host. Superantigens do not need to be bound to MHC molecules to be recognized by T cells. These antigens can bind to the *outside* of MHC molecules that have already bound to antigens. This type of antigen presentation causes massive overproduction of cytokines by CD4 helper T cells. These cytokines cause systemic toxicity and suppression of the adaptive immune response.

Keep in Mind

- For antigen presentation to occur, antigen must be bound to MHC molecules found on antigen-presenting cells (APCs).
- There are two types of MHC molecule, class I and class II.
- CD8 cytotoxic T cells bind to class I MHC molecules, and CD4 helper T cells bind to class II MHC molecules.
- Antigens from the cytoplasm of the APC bind to class I MHC molecules, and antigens generated in vesicles of the APC are bound to class II MHC molecules.
- T cells that bind to superantigens can backfire on the host and cause suppression of the response and also systemic toxicity.

Fast Fact

One well-known family of superantigens are the staphylococcal enterotoxins, which cause food poisoning and also toxic shock.

THE HUMORAL (B CELL) RESPONSE

We can divide the adaptive immune response to infection into two parts: humoral and cellular. The **humoral response** is carried out by B lymphocytes, whereas the **cellular response** is carried out by T lymphocytes. In this section, we look at the humoral response. The cellular response is the subject of a later section of this chapter.

Many pathogens that cause infectious disease multiply in the extracellular spaces of the host's body. Attacking this type of problem is the task of the humoral immune response. This response involves the production of antibody and is the province of B cells. However, in most cases, activation of B cells requires help from T cells, and it is the CD4 helper T cells that are involved (hence the designation helper). Activation of B cells causes them to proliferate and differentiate into plasma cells that produce large amounts of specific antibody. It is important to note that, rather than differentiating into plasma cells, some of the activated B cells will become memory cells.

Antibodies produced by plasma cells are found in the blood and extracellular spaces and contribute to immunity in three major ways: neutralization, opsonization, and complement activation (**Figure 16.9**). In the neutralization pathway, the antibodies neutralize toxins and viruses and also prevent bacterial attachment (the "staying in" requirement for infection). In the opsonization pathway, the antibodies facilitate the uptake of pathogens by phagocytic cells. In the complement activation pathway, the antibodies activate the classical pathway of the complement system (as we saw in Chapter 15).

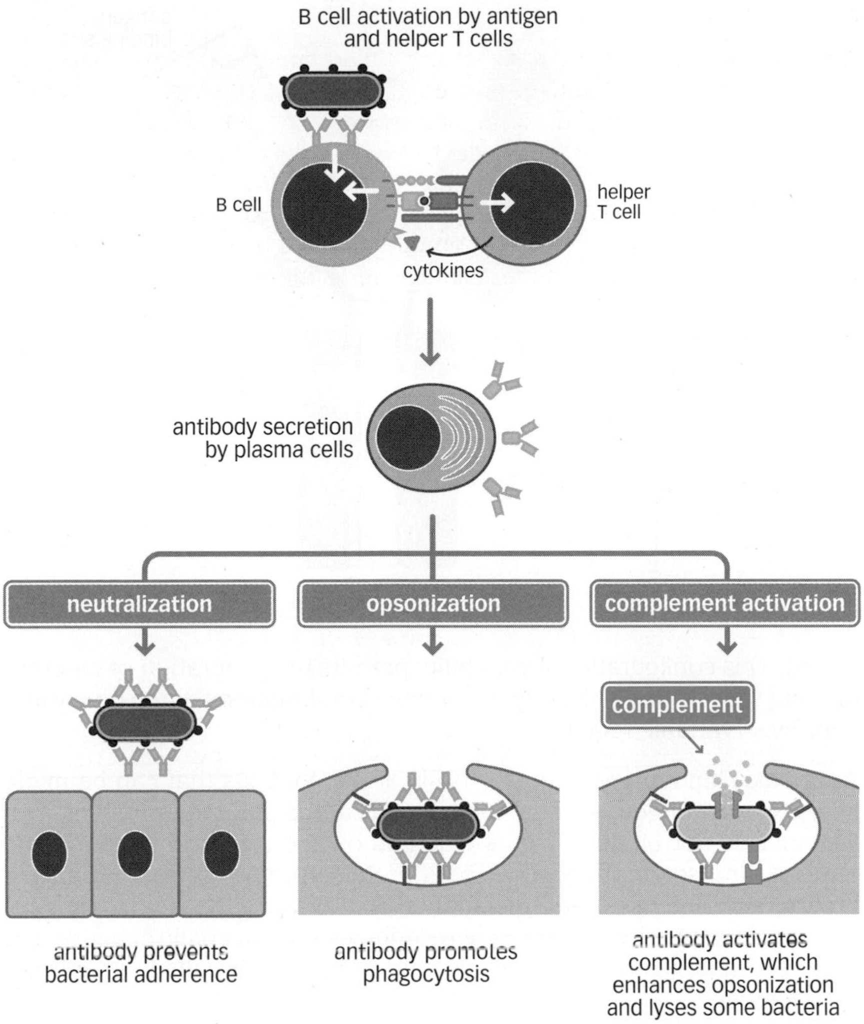

Figure 16.9 The three mechanisms by which antibody protects a host. Notice in the top panel the complex interactions between surface markers of both cells, an interaction in which T and B cells must recognize both parts of the antigen–MHC complex before the B cell can proliferate and differentiate into a plasma cell.

The Immunoglobulin Molecule

As noted earlier, the antigen receptors on B cells are immunoglobulin molecules. Every immunoglobulin molecule has a Y shape and is composed of four polypeptide chains, two *light* chains and two *heavy* chains (**Figure 16.10**). The light and heavy refer to the molecular weights of the chains. When we look at **Figure 16.10** we can see that the four amino-terminal ends (one on each of the four chains) make up the **variable region** of the molecule, which contains the **antigen-binding site**. The remainders of the two heavy chains make up what is referred to as the constant region. It is the constant region that denotes the antibody isotype. The light chains also have constant and variable regions.

The effector functions of immunoglobulin molecules such as complement fixation are associated with their constant region. Immunoglobulin molecules are flexible, and the fact that there are two identical antigen-binding sites is important because it allows cross-bridging of antigens. The binding resulting from cross-bridging is very stable and also acts as a trigger for B cell activation.

Parts of the variable region of each immunoglobulin molecule are so variable that they are called **hypervariable regions**. When the variable regions of the heavy and light chains are paired up, the hypervariable regions of each are brought together and create a hypervariable binding site at the tip of each "arm" of the molecule. The strength (or weakness) of this binding site is determined by how the polypeptide chains are

Figure 16.10 The immunoglobulin molecule is a four-peptide chain structure made up of two light chains bonded by disulfide linkages to two heavy chains. Note the constant and variable regions in the molecule. These regions arise from the three-dimensional structure of the polypeptides. The variable regions of the heavy and light chains combine to form the antigen-binding sites of the molecule. Because there are two binding sites, the molecule is referred to as *bivalent*.

16.5

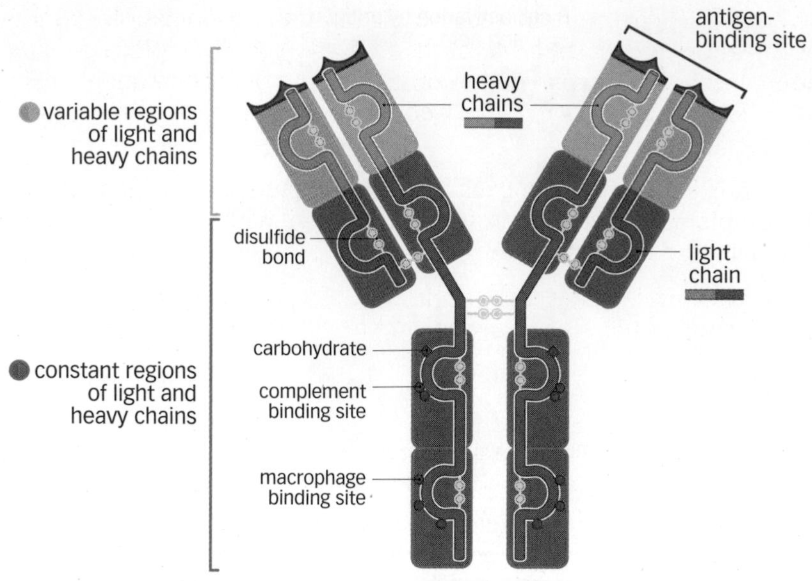

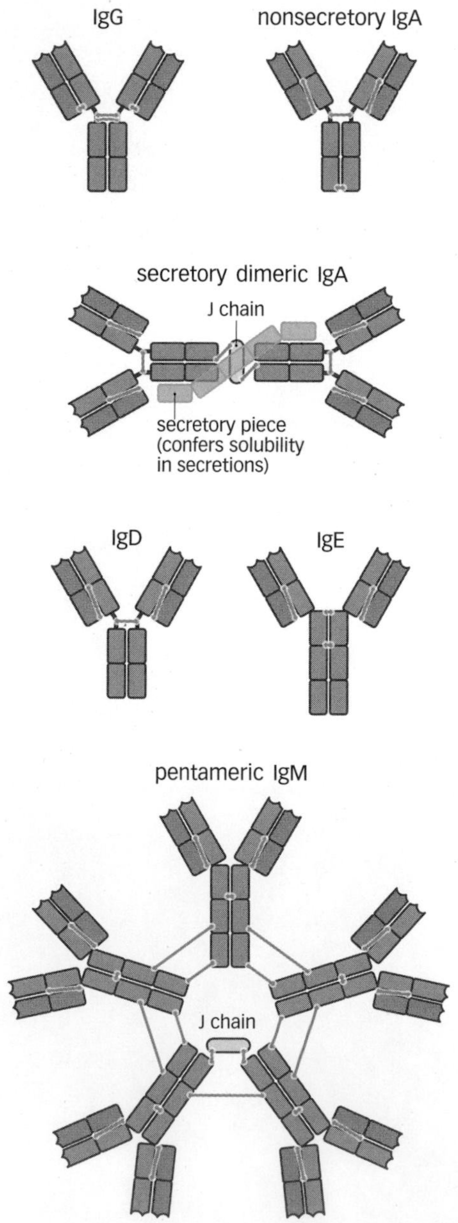

folded. This configurational variability permits the generation of different binding specificities by simply varying the combinations of the heavy and light hypervariable regions.

Antibodies bind antigens on the basis of the contacts that can be made between the antigens and the amino acids at the antibody's binding site. Whether or not binding occurs depends on the size and shape of the antigen, and some antigens such as proteins are too large to fit in the binding site. In these cases, the antibody binds along the side of the antigen structure. Antibody generally recognizes only a small region on the surface of a large antigen molecule, and this region is called the **epitope** of the antigen.

The binding of an antibody molecule to an antigen involves hydrophobic forces and electrostatic forces (in which electrical charges on various parts of the antibody and antigen molecules either attract or repel one another). In addition, hydrogen bonds help to hold the two molecules together.

Immunoglobulin Isotypes

There are five immunoglobulin isotypes: IgG, IgA, IgM, IgD, and IgE (**Figure 16.11**). In humans, there are four subclasses of IgG (IgG1, IgG2, IgG3, and IgG4) and two subclasses of IgA (IgA1 and IgA2). The five isotypes are sometimes designated with Greek letters: γ (gamma) for IgG, α (alpha) for IgA, μ (mu) for IgM, δ (delta) for IgD, and ϵ (epsilon) for IgE.

The constant region of any immunoglobulin molecule (see **Figure 16.10**) has three main functions:

- The constant region is recognized by specialized receptors expressed on cells such as macrophages and neutrophils. Recall that some pathogens produce capsules that inhibit phagocytosis (Chapter 5), but

Figure 16.11 Immunoglobulin isotypes. Notice that all the molecules have the same four-chain structure but differ in the length of the Fc region and in the disulfide linkages (shown in gold in all molecules). IgA can dimerize to form secretory IgA and IgM is able to form a pentameric (five-sided) structure.

such capsules can be rendered useless through opsonization. During opsonization, the binding site of the immunoglobulin molecule binds to the capsule, leaving part of the constant region of the immunoglobulin molecule exposed. It is this exposed portion that is recognized by the phagocytic cell.

- Antigen–antibody complexes can initiate the classical pathway of the complement system through interaction with the complement protein C1 and the constant region of the antibody molecule.
- The constant region is involved in delivering antibody to places such as breast milk, mucous secretions, and tears.

Distribution and Function of Immunoglobulins

Because pathogens find their way to most sites in the body, antibodies must be able to do the same. Most antibody molecules are distributed by diffusion, but specialized transport mechanisms are required to bring antibody to the epithelial surface linings of the intestine and lungs.

Fast Fact

Antibody at the surface of mucosal linings is important because it can neutralize a potential pathogen or promote its elimination before a significant infection can get started.

The first antibody to be produced in a humoral response is always IgM, which usually does not bind tightly to its antigen. The capacity of IgM to form a pentameric structure in which there are 10 binding sites (see **Figure 16.11**) helps in binding to antigen. Even though each individual binding site of IgM binds less tightly, because there are many binding sites they add up to a greater binding strength. IgM is usually found in the blood and in smaller amounts in the lymph. It is one of the best activators of the complement system because its pentameric shape makes five constant regions available for complement binding. IgM has a prime role in reactions against bloodborne pathogens. It is also found in the pleural cavity and the pleural spaces, where it protects against environmental pathogens.

IgG, IgA, and IgE antibodies are smaller than IgM molecules and can easily diffuse out of the blood and into the tissues. IgG is the principal isotype in the blood and extracellular fluid and is very effective for opsonization and activation of complement. IgA is the principal antibody found in secretions, especially those of the epithelial lining of the intestine and the respiratory tract. However, IgA is not very efficient at opsonization and only weakly activates the complement cascade.

The IgE isotype is found in very low levels in the blood and extracellular fluids, but it is bound tightly by mast cells just below the skin and mucosa and also along blood vessels in the connective tissue. When antigen is bound by the IgE on mast cells, the IgE triggers these cells to release powerful chemical mediators that induce reactions such as coughing, sneezing, and vomiting.

Fast Fact

Violent physical reactions such as coughing and sneezing are actually beneficial to the host because they can expel infectious agents from the body.

IgD is found in very small amounts (about 0.2% of the total serum immunoglobulin) and is found in the lymph and blood as well as on the surface of B cells. IgD has no known function in serum, but it is involved as an early antigen receptor on the surface of B cells.

Transport proteins can bind to the Fc fragments of immunoglobulins and carry particular isotypes across epithelial barriers during a process called **transcytosis** (**Figure 16.12**). This form of transport is most frequently seen with the IgA molecule, which can be transported across the epithelial layer of the gut or bronchi. For this to happen, IgA is secreted as a dimer held together with a protein called a **J chain** and a specialized **secretory piece** (shown in **Figure 16.11**). Dimeric IgA uses the J chain to bind to a receptor located on the basal side of the epithelial cell, and is internalized. It then moves through the cytoplasm in a transport vesicle,

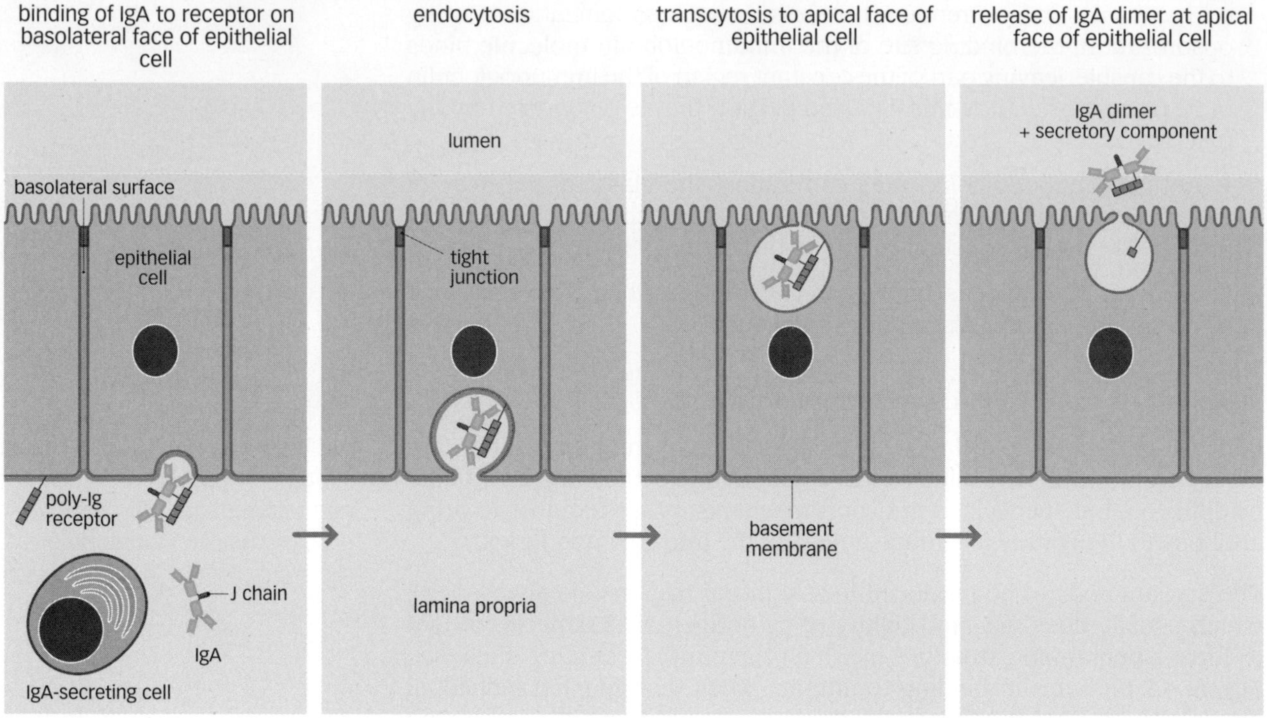

Figure 16.12 Transcytosis of dimeric IgA across an epithelial layer. This process is mediated by a specialized transport protein known as the poly-Ig receptor.

to the surface of the cell facing the lumen. As the vesicle fuses with the membrane, the IgA is deposited in the mucus. The IgA molecule is held in place in the mucus by mucins that bind to the carbohydrate portions of the secretory piece.

Newborn infants are especially vulnerable to infection, and maternal IgA in colostrum is transferred to the gut of the infant, where it provides protection from newly encountered bacteria until the infant can synthesize its own. Newborn protection also results from maternal IgG crossing the placenta.

IgG and IgA neutralize many dangerous bacterial toxins. Neutralization occurs because antibody binds to the toxin molecule and thereby blocks the molecule's ability to bind to its cellular receptor site. This prevents the toxin from gaining entry into the cell. To neutralize toxins, antibody must be able to diffuse into tissues and rapidly bind the toxins. As a result of its ability to diffuse rapidly and because of its strong binding characteristics, IgG is the principal neutralizing antibody. However, IgA can also neutralize toxins at the mucosal surfaces of the body.

IgG and IgA antibodies also inhibit the infectivity of viruses. For example, the hemagglutinin structure found on the envelope of the influenza virus is a target for antibody. Antibody against hemagglutinin can prevent influenza infection by preventing the virus from binding to receptors on the host target cell. In a similar fashion, antibody can block the adherence of bacteria to host cells. In fact, IgA, which is secreted onto the mucosal surface of the intestinal, respiratory, and reproductive tracts, is particularly important in preventing infection by inhibiting the adherence of organisms such as *Salmonella* and *Neisseria*. In tissues, it is IgG that provides this function.

Another important function of antibody is activating a variety of cells that have receptors for the constant region of the antibody molecule. Non-phagocytic cells, such as NK cells, basophils, and mast cells, are triggered to secrete stored mediators when these receptors for the constant region are engaged. Examples of this receptor binding are dramatically illustrated in parasitic infections in which the parasites are too large to be engulfed. Here, the phagocytic cells attach to the surface of the antibody-coated parasite; the lysosomes of the phagocytic cells move to the cell surface, fuse, and release their contents directly onto the parasite.

Activation of Basophils and Mast Cells by IgE

A substantial amount of IgE is bound to mast cells and basophils. Mast cells can be activated to release their granules and also to secrete lipid inflammatory mediators and cytokines. This occurs via specific IgE and IgG receptors, but only when the surface IgE is bound to antigen (**Figure 16.13**). Incredibly, this release occurs in just seconds. Degranulation of mast cells releases stored histamine, which causes a local increase in blood flow and vascular permeability. This quickly leads to an accumulation of fluid and blood proteins, including antibody, in the surrounding tissue. Shortly thereafter, there is an influx of neutrophils, macrophages, and lymphocytes into the area.

B Cell Activation by T Cells

B cells have an antigen receptor that is a surface immunoglobulin molecule containing two identical binding sites. This receptor transmits signals to the interior of the B cell when it binds to antigen. The receptor also brings the bound antigen into the cell by endocytosis. Once inside the B cell, the antigen begins to be degraded; it is combined with class II MHC molecules and returned to the surface of the B cell. This antigen–MHC complex is recognized by T cells, which then activate the B cell to proliferate and differentiate into a plasma cell.

A good example of cooperation between T and B cells is the complex set of reactions that occurs when antibody is produced against viruses. In this case, a B cell can directly bind a viral particle, which is then ingested by endocytosis and degraded. The viral peptides are then coupled to class II MHC molecules that move to the surface of the B cell so that the viral antigen can be presented to the helper T cell. This presenting to the T cell results in activation of the B cell and production of antibody against that virus. This way of attacking a viral particle is effective only while the particle is in the open (that is, not yet in a host cell). Once the particle has infected a host cell, it can no longer be taken up by the B cell receptor. However, if memory B cells are produced, the antibody response to subsequent infection by the same virus will be fast enough to prevent infection of host cells.

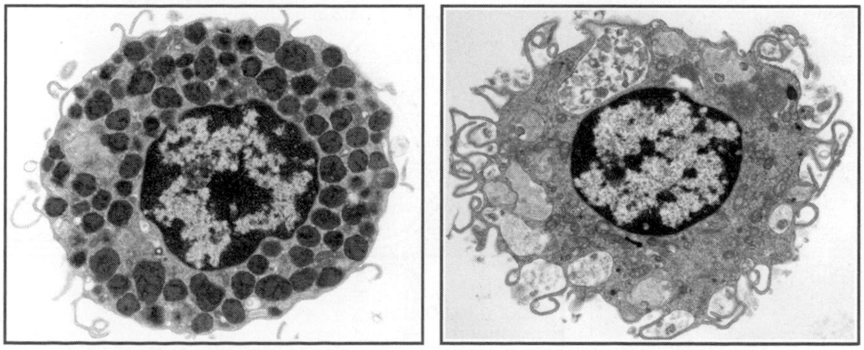

Figure 16.13 Release of inflammatory mediators by mast cells. Notice the dark granules containing these mediators in the resting mast cell at the left. Cross-linking of the IgE molecules on the surface of the cell leads to a rapid release of the mediator molecules, as shown in the activated cell at the right.

All mature naive B cells express both IgM and IgD on their surface. After a B lymphocyte is activated, its IgD disappears, and only IgM remains. IgM is always the first immunoglobulin isotype to be produced in a humoral immune response, and this part of the response is known as the **primary immune response**. In the **secondary immune response**, isotype switching occurs, and in most cases IgG is produced in quantity while IgM levels trail off (**Figure 16.14**). This secondary response occurs when an antigen is seen again; it is faster and more powerful than the primary response. Helper T cells (carrying the CD4 protein) regulate both the production of antibody and also which isotype of immunoglobulin will be produced.

Summary of Cooperation between T and B Cells

Remember, for the humoral response to work, each B cell must find a helper T cell to cooperate with. This is accomplished by a trapping mechanism that operates in the peripheral lymphoid tissue. The process, which is summarized below, is truly elegant.

- When antigens make their way into the body, they are captured and processed by antigen-presenting cells (APCs), especially dendritic cells, that then migrate to local lymph nodes lodging in the paracortical areas of the nodes.
- Naive T cells circulating in the body are continuously passing through these paracortical areas and "sampling" the APCs.
- Once a naive T cell passing through the paracortical area of a node is presented with an antigen specific for the T cell's receptor, the T cell becomes "armed" and remains trapped in the lymph node.
- When B cells enter the lymph nodes, they first pass through the paracortical areas on their way to the follicles.
- If an armed effector T cell is trapped in the paracortical areas, a B cell also becomes trapped and it is here that T and B cells cooperate. The agents responsible for trapping the cells in the paracortical area are adhesion molecules produced by the cells.
- After receiving help from T cells in the paracortical area of the lymph node, B cells migrate to the follicle region of the node. It is here that B cells proliferate and differentiate into plasma cells, which produce antibody.

The plasma cells created in this process have a variety of life spans. Some survive only days, but others receive signals from stromal cells in the bone marrow that allow them to survive for long periods. The presence of these long-lived plasma cells in the body provides a source of long-lasting antibody protection.

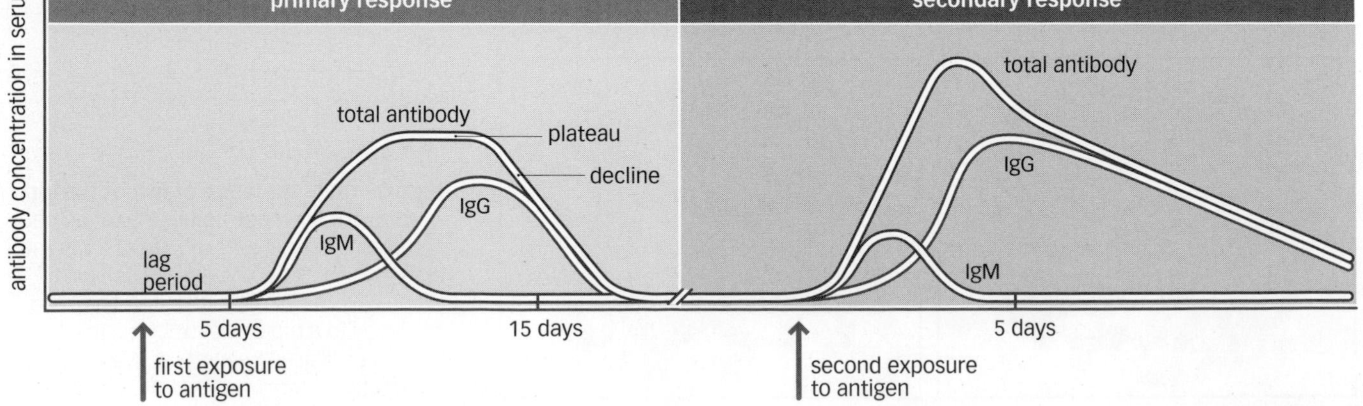

Figure 16.14 The primary and secondary response to an antigen shown as the level of antibody produced. Notice the production of different amounts and isotypes of antibody produced.

Keep in Mind

- B cells are responsible for antibody production.
- Activation of B cells causes them to proliferate and differentiate into plasma cells, which are end-stage cells that continuously produce antibody.
- Antibody protects the host through neutralization, opsonization, and complement activation.
- The immunoglobulin molecule is composed of two heavy and two light chains.
- There are five types of immunoglobulin molecule (IgG, IgA, IgE, IgD, and IgM), and each of these has different functions in protecting the host.
- B cells must cooperate with T cells to differentiate into plasma cells through what is referred to as T cell and B cell cooperation.
- T cell and B cell cooperation is carried out in conjunction with class II MHC molecules.

THE CELLULAR (T CELL) RESPONSE

Now that we have examined the humoral (B cell) response to antigens, let's turn our attention to the **cellular adaptive immune response**, which is generated by T cells. Recall that T cells that have not encountered antigen are referred to as *naive* T cells, and it is these cells that continually circulate between the blood and the peripheral lymphoid tissue. As they circulate, they receive periodic survival signals from encounters with self-antigen–MHC complexes as well as cytokines.

Fast Fact

The activation of naive T cells is referred to as the primary cell-mediated immune response. It provides for both the "arming" of T cells and the development of memory T cells.

T cells detect antigens associated with various types of pathogen and respond in a variety of ways. After encountering an antigen that is specific for its antigen receptor, a naive T cell becomes an *armed effector* T cell. This transformation occurs in cytotoxic T cells and helper T cells as follows:

- **Cytotoxic T cells.** Some pathogens that invade a host multiply in the cytoplasm of infected host cells. These pathogens are degraded, and the resulting antigens are carried to the cell surface by class I MHC molecules and presented to cytotoxic (CD8) T cells. This results in proliferation of these T cells, which then look for and kill infected host cells expressing that antigen.

16.6

- **Helper T cells.** Some pathogens that invade a host multiply in intracellular vesicles. Antigens from these pathogens as well as any antigens and miscellaneous toxins floating in the extracellular fluid are carried to the surface of the APC by class II MHC molecules. There they are presented to helper (CD4) T cells. These helper T cells then differentiate into either **Th1 cells** or **Th2 cells** (**Figure 16.15**). The Th1 helper T cells are involved with pathogens that accumulate inside vesicles in macrophages and dendritic cells, while the Th2 helper T cells deal with extracellular pathogens.

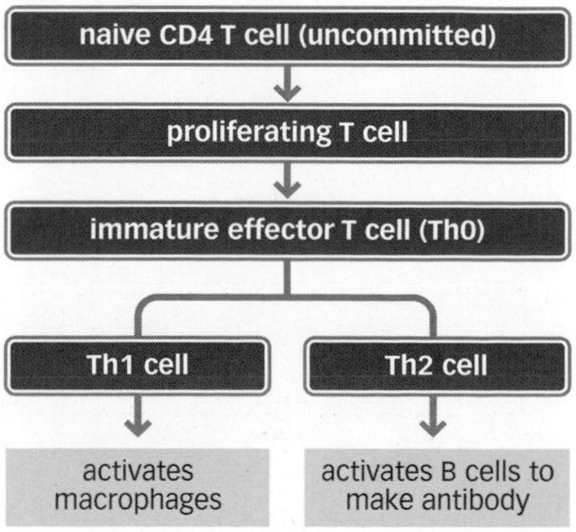

Figure 16.15 Differentiation of T cell populations. When stimulated, naive T cells proliferate and become immature effector T cells, which can be thought of as Th0 cells. Upon further stimulation, these cells will mature into either T helper cells that activate macrophages (called Th1 cells), or T helper cells that activate B cells to make antibody and also enhance macrophage function (called Th2 cells).

Production of Armed Effector T Cells

Naive T cells circulating in the lymphoid tissues sample the antigen–MHC complexes on the APCs. This lymphocyte migration depends on cell–cell interactions mediated by adhesion molecules on the surface of the T cells. As they migrate through a lymph node, naive T cells transiently bind to each APC. This binding allows the T cell to "crawl" along the APC and sample every antigen present. If the T cell finds an antigen that fits into the T cell antigen receptor, a conformational protein change occurs immediately. This change stabilizes the binding of the T cell to that site on the APC. The association of these two cells can last for several days, during which time the T cell proliferates. The progeny of the T cell also adhere to the APC and differentiate into armed T cells. Arming of T cells can be accomplished by dendritic cells, macrophages, and B cells.

Arming of T Cells by Dendritic cells

16.7 A dendritic cell secretes a chemokine that attracts naive T cells to come and "sample" antigens that the dendritic cell has phagocytosed. Dendritic cells that have never encountered an infection eventually move to the local lymphoid tissue and display self antigens they have acquired from self tissue, so as to reinforce the tolerance that keeps autoimmunity under control.

Dendritic cells present viral, fungal, and bacterial antigens and can also differentiate between the different classes of pathogens. In addition, they can present antigens, either as part of a transplantation rejection process or as part of the allergic response.

Arming of T Cells by Macrophages

Recall that macrophages also present antigens to T cells. As described in Chapter 15, macrophages are highly phagocytic and can destroy invading organisms. However, many pathogens have developed ways to avoid this, so the adaptive immune response must help in combating this problem. Resting macrophages have few or no class II MHC molecules on their surface and do not express co-stimulatory signals, but these molecules are induced to be expressed when the macrophage ingests a microorganism.

Because macrophages normally scavenge dead and dying cells of the host, it is important they do not activate T cells against self components that could initiate an autoimmune response. Fortunately, macrophages have a variety of receptors that recognize foreign molecule patterns (for instance Toll-like receptors, mannose receptors, and complement receptors). This is important because having these receptors permits distinguishing between infectious antigens and innocuous host proteins.

T cell arming occurs when a macrophage engulfs a microorganism, degrades it using cellular lysosomes, and generates antigens that become complexed with class II MHC molecules. This causes the production of co-stimulatory molecules that allow the arming process to continue.

Arming of T Cells by B Cells

B cells can also present antigens to T cells. Recall that the antigen receptor for B cells is a surface immunoglobulin that can readily bind to antigens. When this binding occurs, the B cell's receptor and its bound antigen are internalized by endocytosis. The antigen then is bound by class II MHC molecules and moved back to the surface of the B cell. B cells have high levels of class II on their surface, but as a safety measure they do not express co-stimulatory molecules. Instead, production of co-stimulatory signals must be induced.

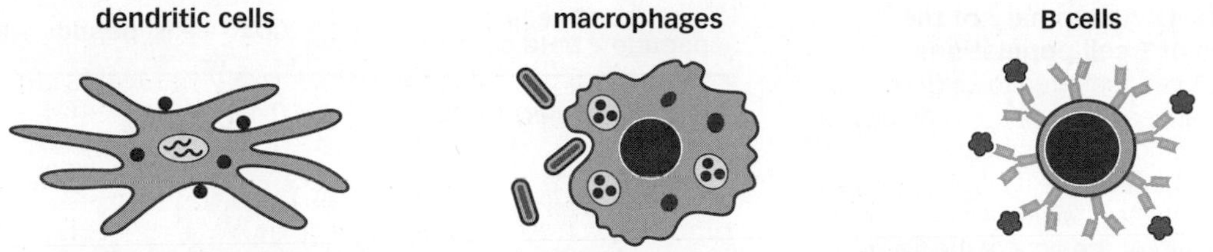

	dendritic cells	macrophages	B cells
antigen uptake	macropinocytosis and phagocytosis by tissue dendritic cells	phagocytosis	antigen-specific receptor (immunoglobulin)
MHC expression	low on tissue dendritic cells high on dendritic cells in lymphoid tissues	inducible by bacteria and cytokines	constitutive increases on activation
co-stimulator delivery	constitutive by mature, nonphagocytic lymphoid dendritic cells	inducible	inducible
antigen presented	peptides viral antigens allergens	particulate antigens intracellular and extracellular pathogens	soluble antigens toxins viruses
location	ubiquitous throughout the body	lymphoid tissue connective tissue body cavities	lymphoid tissue peripheral blood

Figure 16.16 Properties of the three groups of antigen-presenting cells

The overall importance of B cells as APCs has yet to be worked out.

In summary, the cellular adaptive immune response can be primed by three types of APC (**Figure 16.16**):

- Dendritic cells are the most efficient because they are optimally equipped to present a wide variety of antigens to naive T cells.
- Macrophages stimulate T cell responses to pathogens that the macrophages have taken up but are unable to eliminate.
- B cells can present fragments of antigens that were bound to immunoglobulin receptors on the surface of the B cells.

Properties of Armed Effector T Cells

The three classes of armed effector T cells specialized to deal with different pathogens — cytotoxic cells, Th1 helper cells, and Th2 helper cells — are summarized in **Figure 16.17**. Most of these cells leave the lymphoid tissue when they are activated, and they enter the blood via the thoracic duct. The initial binding of an armed T cell to its target is mediated by nonspecific adhesion molecules in a manner similar to that seen with naive T cells and APCs. However, the number of adhesion molecules is two to four times higher on armed T cells. This allows an armed T cell to bind more tightly to its target and remain bound long enough to release effector molecules. For example, helper (CD4) T cells must remain in contact with macrophages for long periods. In contrast, cytotoxic (CD8) T cells attach, kill, detach, reattach, kill, and so on, relatively quickly. The killing event somehow changes the cytotoxic T cell so that it detaches and moves directly to the next target and reattaches in a manner that is not yet understood.

The functions of T cells are determined by the effector molecules they produce. These molecules fall into two basic categories: *cytotoxins* released by cytotoxic (CD8) T cells, and *cytokines* released primarily by helper (CD4) T cells. Water-soluble cytokines and membrane-associated molecules often act in combination to mediate the effects of T cells

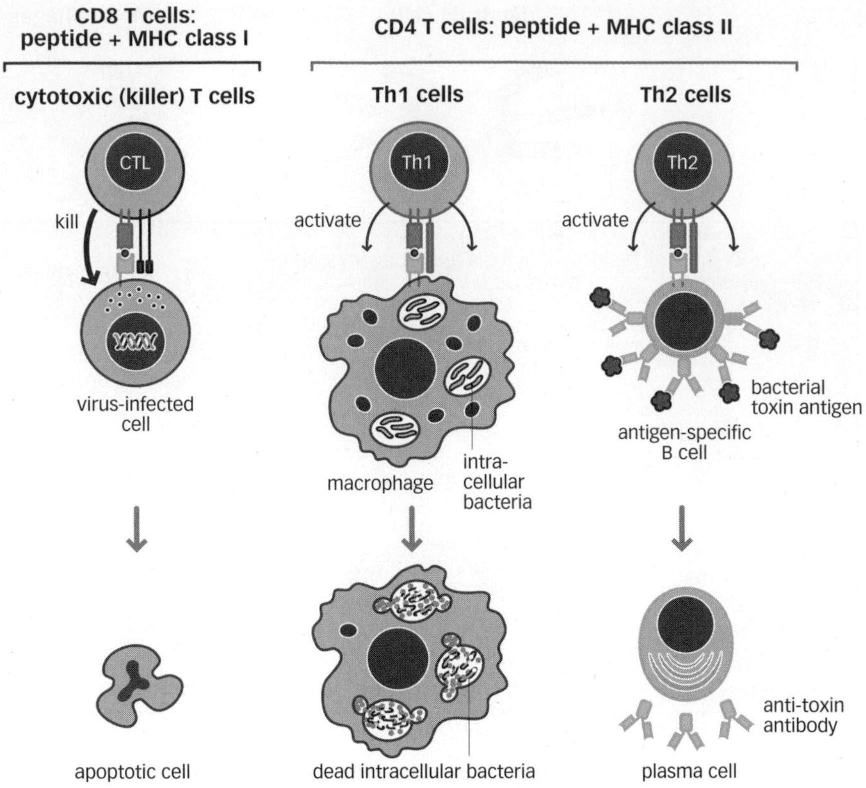

Figure 16.17 A summary of the functions of T cell populations. Cytotoxic T cells, referred to as CTLs, recognize antigen–MHC class I complexes and kill infected cells. T helper cells designated Th1 activate macrophages to kill organisms they have ingested through phagocytosis. At the right of the figure Th2 helper cells are shown activating B cells to produce antibody.

on their target cells. For example, CD40 is very important for helper T cell function. It is induced on Th1 and Th2 cells and delivers activation signals to B cells and macrophages. If the cytokine is produced by a lymphocyte, it is referred to as a **lymphokine** or, more frequently, as an **interleukin**. A summary of T cell cytokines and their functions is shown in **Figure 16.18**.

Fast Fact

Cytotoxic T cells also have a role in the destruction of transformed or malignant cells. Because transformation of cells can happen through genetic mutations, there is the possibility that cancerous cells may routinely appear in the body; the cytotoxic T cells keep them from developing into a tumor.

T Cell Cytotoxicity

All viruses and some bacteria multiply in the cytoplasm of infected cells. Once inside the host cell, the pathogen is no longer susceptible to antibody, and the elimination of the organism becomes the responsibility of the cytotoxic T cell. Because it must occur without the destruction of any healthy tissue, this elimination must be powerful but *accurately* targeted. Cytotoxic T cells kill their targets by programming them to undergo apoptosis. This programming takes only about five minutes, but the infected cell may take up to several hours to die.

The principal mechanism of action of cytotoxic T cells is the calcium-dependent release of preformed specialized lytic granules. These granules are modified lysosomes containing several proteins that are expressed in cytotoxic T cells. Three of these proteins, namely **perforin**, **granzyme** (protease enzymes), and the peptide **granulysin** (which induces the onset of apoptosis), are responsible for the death of the target cell. During the killing event, perforin creates a pore in the target cell membrane through which the other proteins gain access to the cytoplasm of the target cell.

16.8

16.9

Cytotoxic T cells are selective and repetitive killers of target cells. Because their effects are narrowly focused, they can kill an infected cell that is surrounded by healthy tissue without causing any tissue damage. These killer cells also release cytokines, such as interferon and tumor necrosis factor, which are part of the innate immune response. This is another example of cooperation between the innate and adaptive immune responses to maximize host defense.

cytokine	T cell source	effects on				
		B cells	T cells	macrophages	hematopoietic cells	other somatic cells
interleukin-2 (IL-2)	naive, Th1, some CD8	stimulates growth and J-chain synthesis	growth	–	stimulates NK cell growth	–
Interferon-γ (IFN-γ)	Th1, CTL	differentiation of B cells	Inhibits Th2 cell growth	activation, ↑MHC class I and class II	activates NK cells	antiviral ↑ MHC class I and class II
lymphotoxin (LT, TNF-β)	Th1, some CTL	inhibits	kills	activates, induces nitric oxide production	activates neutrophils	kills fibroblasts and tumor cells
interleukin-4 (IL-4)	Th2	activation, growth ↑MHC class II induction	growth, survival	inhibits macrophage activation	↑growth of mast cells	–
interleukin-5 (IL-5)	Th2	differentiation of B cells	–	–	↑eosinophil growth and differentiation	–
interleukin-10 (IL-10)	Th2 (human: some Th1), T_{reg}	↑MHC class II	inhibits Th1	inhibits cytokine release	co-stimulates mast cell growth	–
interleukin-3 (IL-3)	Th1, Th2 some CTL	–	–	–	growth factor for progenitor hematopoietic cells (multi-CSF)	–
tumor necrosis factor-α (TNF-α)	Th1, some Th2 some CTL	–	–	activates, induces nitric oxide production	–	activates microvascular endothelium
granulocyte-macrophage colony-stimulating factor (GM-CSF)	Th1, some Th2 some CTL	differentiation	inhibits growth ?	activation, differentiation to dendritic cells	↑production of granulocytes, macrophages and dendritic cells	–
transforming growth factor-β (TGF-β)	CD4 T cells (T_{reg})	inhibits growth IgA switch factor	inhibits growth, promotes survival	inhibits activation	activates neutrophils	inhibits/ stimulates cell growth
interleukin-17 (IL-17)	CD4 T cells macrophages	–	–	–	stimulates neutrophil recruitment	stimulates fibroblasts and epithelial cells to secrete chemokines

Figure 16.18 A summary of the effects of T cell cytokines on different cell populations. This figure describes some of the effects that different T cell cytokines have on different cell populations. The major activity of the cytokines are depicted in red. CTL, cytotoxic lymphocyte; NK, natural killer; CSF, colony stimulating factor.

Macrophage Activation by T cells

As we saw in Chapter 5, some bacteria can proliferate in macrophages by inhibiting the binding of phagosomes with lysosomes to form phagolysosomes. These pathogens can be eliminated when the macrophage has been activated by an armed Th1 helper cell. This is referred to as **macrophage activation**. An example of this activation can be seen in infections with *Pneumocystis carinii*, which is a common cause of death in individuals with AIDS. In this case, activation of the macrophage by armed Th1 cells promotes the killing of *Pneumocystis* pathogens.

Part of the macrophage activation process is the production of oxygen free radicals and nitric oxide, both potent antimicrobial substances. Activation also enhances the fusion of phagosomes to lysosomes. In addition, macrophage activation amplifies the overall immune response by increasing class II MHC molecules and cytokine receptors on the surface of the macrophage. The presence of these receptors makes macrophages better APCs.

Keep in Mind

- The cellular portion of the adaptive immune response is performed by T cells.
- Cytotoxic T cells (CD8) are activated by antigen in association with class I MHC molecules.
- Cytotoxic T cells kill infected host cells.
- CD4 helper T cells can be divided into Th1 and Th2 subpopulations.
- Th1 CD4 helper T cells are involved with pathogens that accumulate inside vesicles in macrophages and dendritic cells.
- Th2 CD4 helper T cells deal with extracellular pathogens.
- Armed effector T cells can be produced through interactions with dendrites, macrophages, or B cells.

IMMUNOLOGICAL MEMORY

One of the most important parts of the adaptive immune response is the development of **immunological memory**, which is the ability of the adaptive immune system to respond rapidly and effectively to pathogens that have been encountered previously. This occurs because of the persistence of antigen-specific lymphocytes and can be seen in both B cell and T cell responses. Immunological memory is long-lived and is seen after infection or vaccination.

Memory results from a small but persistent population of specialized memory cells. Although most of these cells are at rest, a small percentage is dividing at all times. The stimulus that causes these resting memory cells to divide remains unclear, but it is known that interleukins help to maintain the memory T cell population and also maintains the memory cytotoxic T cell population.

Differences between Memory and Effector T Cells

After either infection or immunization of a host, the number of host T cells reactive to a given antigen markedly increases as effector T cells are produced. As time goes by and the infection subsides, the number of T cells decreases to a persistent level. This level is 100–1000 times higher than the initial number before activation, with the increase being due to memory T cells. These memory cells are more sensitive to restimulation by the same antigen than are naive T cells, and the memory cells produce cytokines more quickly and vigorously. Keep in mind that memory B cells are also produced as part of the adaptive immune response.

Now let's add immunological memory to what we have learned about the adaptive response:

- A pathogen elicits an adaptive immune response.
- This stimulates the production of antibody and effector T cell responses that eliminate the pathogen.
- When the infection is over, most armed effector T cells and antibody levels slowly decline.

- Memory T and B cells remain. These cells are able to initiate a response to recurrence of infection with the same pathogen. This memory response occurs more quickly than the initial response and is more powerful.

Keep in Mind

- One of the most important properties of the adaptive immune response is the development of memory.
- The adaptive response causes memory to occur for both T and B cells.
- Memory causes a quicker and more powerful response to antigens that have been seen before.
- The adaptive response can be divided into a primary phase and a secondary phase.
- The primary response is predominantly with IgM and not very powerful. However, the secondary response is much more powerful than the primary, with IgG being the predominant class of antibody formed.

ADAPTIVE IMMUNITY TO INFECTION

When we consider how the cells and molecules of the innate and adaptive responses work together, it is easy to view the host defense as an integrated system that eliminates and controls an infectious agent and also provides long-lasting protection. Remember that the innate response has a role in the early response to infection. In fact, the innate response can be considered a prerequisite without which the adaptive response cannot occur. This is because antigen-specific lymphocytes of the adaptive immune response must be activated by co-stimulatory molecules found on or produced by cells which are involved in the innate response.

The course of an infection can be broken down into stages as shown in **Figure 16.19**. The establishment of the infection takes into account many of the things we have discussed, such as routes of entry, modes of transmission, and the numbers of pathogens. Only when pathogens have established a site of infection in the host does disease occur. Little damage is done unless the invading pathogen is able either to spread from the original site of infection or to secrete toxins that can spread to other parts of the body.

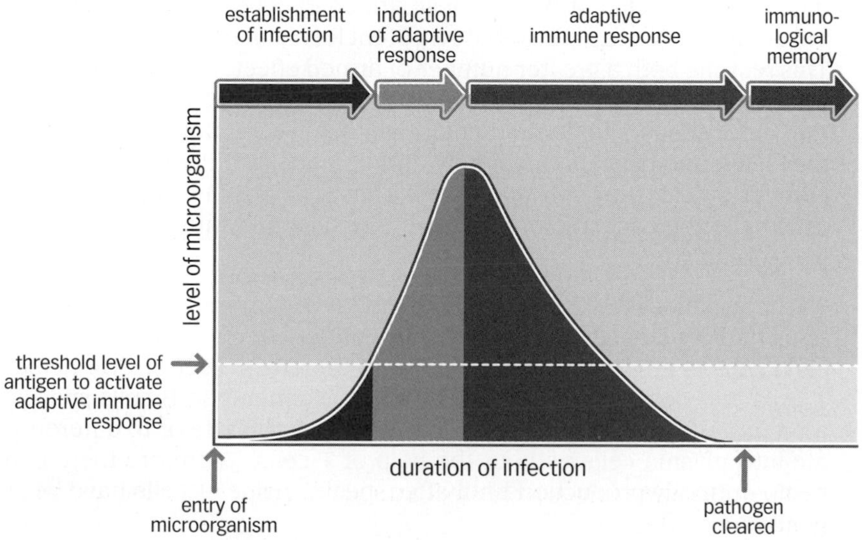

Figure 16.19 The course of a typical acute infection that is cleared by the immune response.

Although most pathogens are repelled by innate immune mechanisms, once they have established a stable site of infection, only the adaptive response can get rid of them. This process begins when APCs present antigens to T cells. Antigen-specific armed effector T cells and antibody-secreting B cells are generated by clonal selection and differentiation over several days. Armed effector T cells followed by antibody are released into the blood and recruited to the site of infection. Normally, there is little damage to uninfected tissue during the response to the pathogen, but in some cases collateral damage to uninfected tissue can be very serious. Furthermore, some pathogens such as *Mycobacterium tuberculosis* may never be completely cleared and persist in the host in a latent form.

Course of the Adaptive Response 16.10

The complex set of interactions that make up the adaptive immune response can be summarized as follows and are illustrated in **Figure 16.20**:

- A crucial effect of infection is the activation of potential APCs that reside in most tissues. These cells take up antigens through interactions with their Toll-like receptors. They become activated and mature into APCs. As part of this process, the APCs increase the synthesis rate of class II MHC molecules and begin to express co-stimulatory molecules on their surface. APCs carrying antigen move from the site of infection and enter peripheral lymphoid tissue. Here they initiate the adaptive response.

- Once APCs arrive in the lymphoid tissue, their only job is to activate antigen-specific naive T cells. These naive T cells are able to recognize the antigen presented on the surface of the APCs and become activated. This activation causes them to divide and mature into armed effector T cells, which then re-enter the circulation.

- The full activation of naive T cells takes four to five days and is accompanied by marked changes in the homing behavior of these cells. Armed cytotoxic T cells must travel from lymph nodes or other peripheral lymphoid tissues, where they became activated so as to attack and destroy infected cells. The same is also true for armed effector helper Th1 cells, which also leave the lymphoid tissue to activate macrophages at the site of infection.

- By the peak of the adaptive response, most of the T cells are specific for the infecting pathogen because of several days of proliferation and differentiation. One or two of these armed effector T cells encountering antigen in tissues can initiate a potent local inflammatory response. This recruits both a greater number of armed effector T cells and many more nonspecific inflammatory cells to the site. Armed effector T cells that enter tissue and do not find antigens there are rapidly lost. Either they leave the tissue immediately and go back into the blood, or they commit suicide through apoptosis. This is important because these cells have the capacity to damage the tissues and must be prevented from doing so.

- The production of antibody is essential in controlling many infections and develops in lymphoid tissues under the direction of helper T cells. This is predominantly the work of the Th2 subset of CD4 helper T cells and occurs in the lymphoid tissue. Remember, B cells specific for a protein antigen cannot be activated to proliferate or to differentiate into plasma cells without the help of T cells. Therefore there can be no antibody production until after specific helper T cells have been generated.

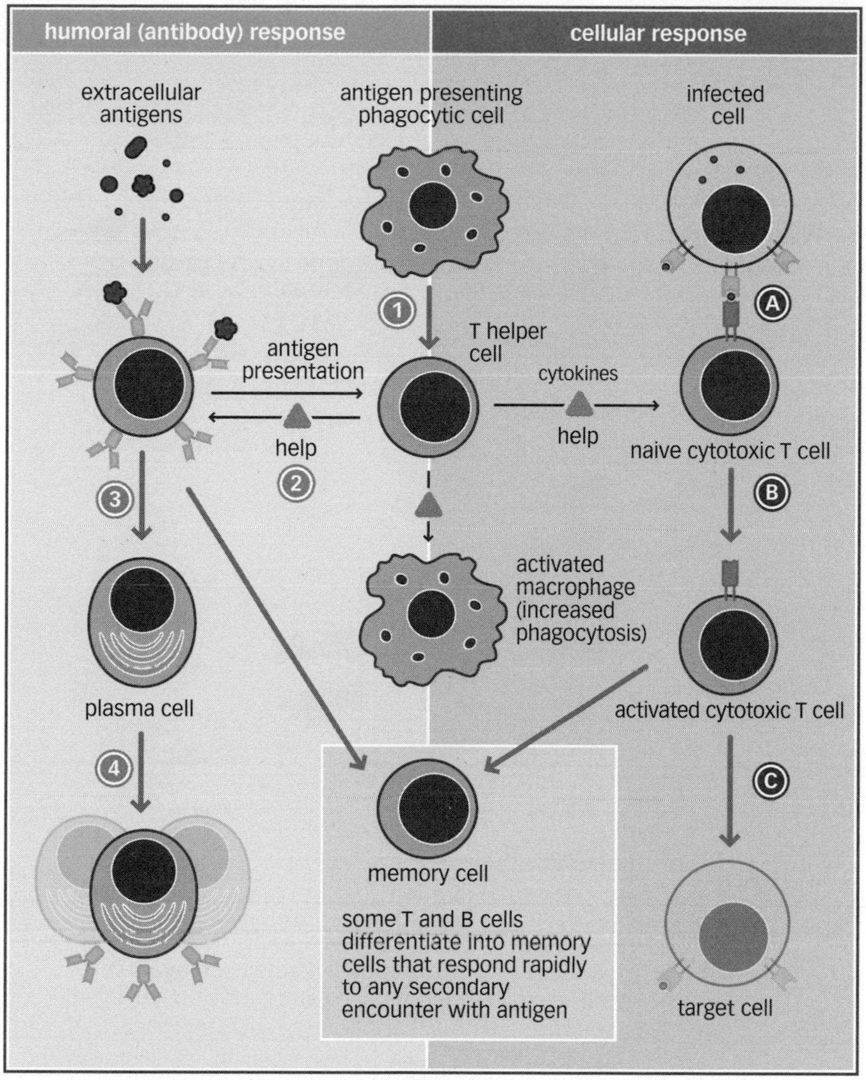

Figure 16.20 Cellular cooperation in the adaptive immune response. This illustration shows the cellular cooperation required for both the humoral (antibody) response and the cellular immune response. The left side of the figure shows the development of an antibody response. In step **1**, antigen is presented to the helper T cell by either a B cell or an antigen presenting cell such as a macrophage. Remember that this interaction is dependent upon MHC class II recognition. In step **2**, the helper T cell gives the B cell permission to differentiate into a plasma cell (step **3**) which produces antibody against the antigen that was presented (step **4**). On the cellular reaction side, T cells identify infected cells through MHC antigen complexes on their surface (step **A**) and become activated. The helper T cell produces cytokines that cause the activated T cell to differentiate into a cytotoxic T cell (**B**) which then kills the target cell (step **C**). Note that both B cells and cytotoxic T cells will become memory cells.

- Antibody responses are sustained in the lymph nodes and in the bone marrow. In these sites, plasma cells secrete antibody directly into the blood via the efferent lymph flow for distribution to the rest of the body. Plasma cells in the lymph nodes live only two to four days and then undergo apoptosis. However, in the bone marrow, plasma cells can live for long periods (months to years).

Figure 16.21 summarizes the different types of infection in humans and the ways in which they can be either eliminated or held in check by the adaptive immune response. Once the infection has been repelled by

	infectious agent	disease	humoral immunity				cell-mediated immunity	
			IgM	IgG	IgE	IgA	CD4 T cells (macro-phages)	CD8 killer T cells
viruses	varicella zoster	chickenpox						
	Epstein–Barr virus	mononucleosis						
	influenza virus	influenza						
	polio virus	poliomyelitis						
intra-cellular bacteria	*Rickettsia prowazekii*	typhus						
	mycobacteria	tuberculosis, leprosy						
extra-cellular bacteria	*Staphylococcus aureus*	boils						
	Streptococcus pneumoniae	pneumonia						
	Neisseria meningitidis	meningitis						
	Corynebacterium diphtheriae	diphtheria						
	Vibrio cholerae	cholera						
fungi	*Candida albicans*	candidiasis						
protozoa	*Plasmodium* spp.	malaria						
	Trypanosoma spp.	trypanosomiasis						
worms	schistosome	schistosomiasis						

Figure 16.21 A summary of the mechanisms used against infections by different pathogens. The pathogens are listed in order of increasing complexity. Red shading indicates known mechanisms of defense, and yellow shading indicates protective immunity. Clear or pale shading indicates that the mechanisms are either less well understood or not understood at all.

the adaptive immune response, two things happen. First, effector cells remove the specific stimulus that recruited them. Second, those effector cells undergo "*death by neglect*" (because there is no more war, the army simply fades away). Fortunately for us, not all the effector cells disappear. Some of them are retained as memory T and B cells that keep us safe should we encounter the same pathogen again.

VACCINATION

Now that we have an understanding of how the adaptive response functions, we can look at vaccination, which is essentially an artificially derived infection. Vaccination is the process of administering weakened or dead pathogens to a healthy person with the intent of conferring immunity. The term was coined by Edward Jenner and adopted by Louis Pasteur.

The first vaccine was derived from a virus, called cowpox virus, that affected cattle. This vaccine could also provide a degree of immunity to a related virus, smallpox virus. The smallpox virus causes a contagious and deadly infection in humans.

Fast Fact

The first vaccine-preventable disease was smallpox, with the last reported case in 1977. Polio is also a vaccine-preventable disease, but cases are still seen in children who have not been immunized.

Fast Fact

A vaccine against the major forms of human papilloma virus (HPV) that are connected with cervical carcinoma, has been approved for young women in the United States.

In an effort to protect against infectious disease, vaccination primes the immune system with an **immunogen**. This process can be called immunization and the immunogen can be called a vaccine. Most vaccines such as MMR (which protects against measles, mumps, and rubella) or DTaP (which protects against diphtheria, tetanus, and pertussis) are administered during childhood. However, some vaccines can be given after the patient has already contracted the disease, as in experimental vaccines against AIDS, cancer, and Alzheimer's disease. In fact, the first rabies vaccine was given by Pasteur to a young girl who had been bitten by a rabid dog.

Vaccination can have risks associated with it. For example, vaccines can contain adjuvants, which are chemicals designed to boost the immune response. In some people, these adjuvants can cause adverse reactions. The same situation can exist for vaccines that contain preservatives. Moreover, vaccines composed of weakened pathogens may cause a small percentage of vaccinated individuals to become infected.

SUMMARY

- The adaptive immune response is specific and involves both cellular and humoral responses.
- T cells and B cells are involved in the adaptive immune response.
- Both T cells and B cells have receptors for antigen.
- There are lymphoid structures strategically placed in major portals of entry.
- The adaptive immune response is connected to the innate immune response.
- T cells mature in the thymus, and B cells mature in the bone marrow.
- Clonal selection and deletion are processes that allow some lymphocytes to mature while others are deleted from the body.
- T cells are initially naive and become armed effector cells after encountering their specific antigen.
- Antigen presentation involves combining antigen with class II MHC molecules.
- CD4 helper T cells recognize class II MHC molecules, whereas CD8 cytotoxic T cells recognize class I MHC molecules.
- B cells are responsible for antibody production.
- B cells differentiate into plasma cells that produce antibody.
- There are five types of antibody molecule: IgG, IgM, IgA, IgD, and IgE.
- T cells direct the production of antibody.
- CD8 cytotoxic T cells kill specifically identified targets and remember them through the development of memory cells.
- CD4 helper T cells can be divided into two groups, Th1 and Th2, each with a different helper function.
- The adaptive immune response can be divided into a primary phase and a secondary phase.

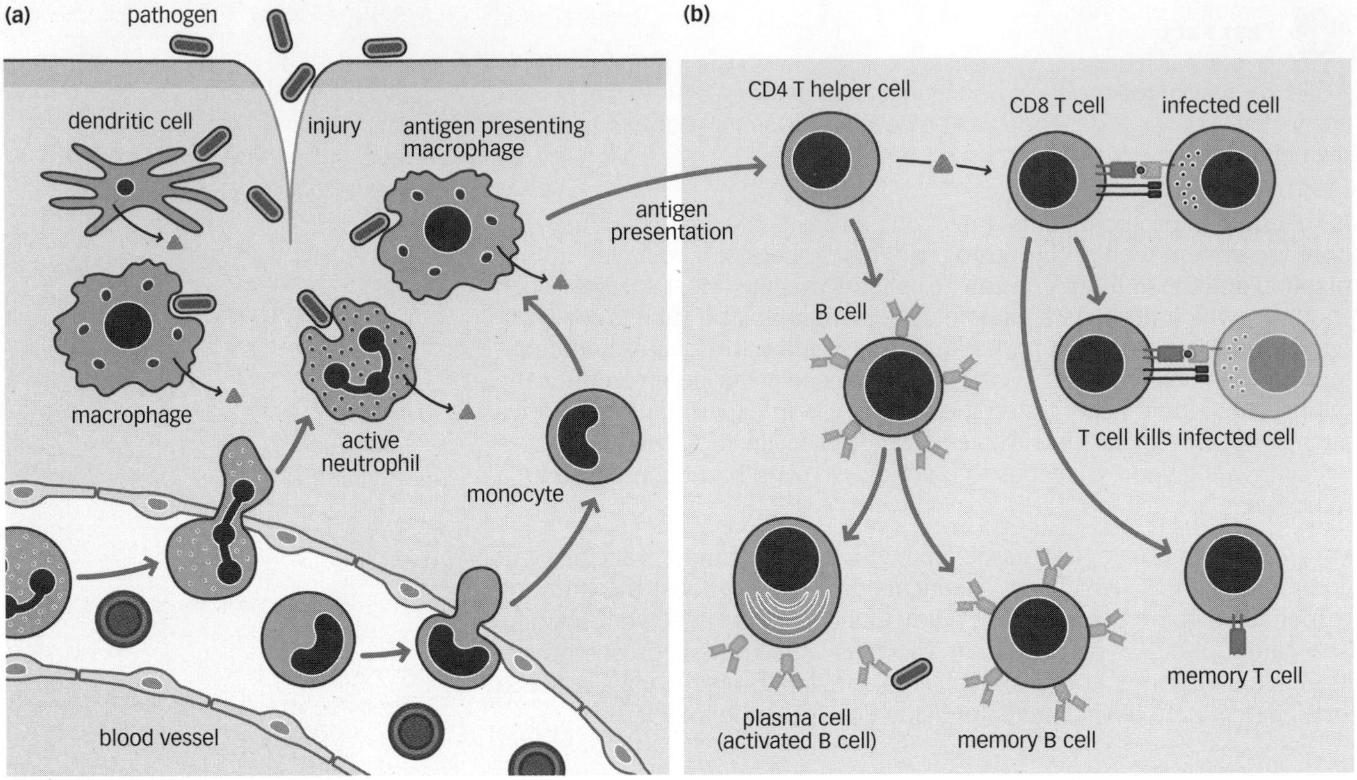

Figure 16.22 Relationship between the innate immune response (panel a) and the adaptive immune response (panel b) during infection. The connection is due to antigen presenting cells such as macrophages which are part of the innate response that interact with helper T cells to initiate an antibody response. Do not forget that these responses are dependant upon recognition of MHC class II molecules.

In this chapter we have explored the adaptive immune response, noting that it is intimately connected to the innate immune response we examined in the previous chapter (**Figure 16.22**). The adaptive response is called upon when infections overwhelm or evade the capacity of the innate response to deal with them, and get out of hand. This response is specific and results in the development of memory. It is among the most potent and lethal responses in nature and therefore it can cause severe problems if it goes wrong, over-reacts or, worse, is not available at all. In the next chapter, we look at what happens when the immune response is missing or functions incorrectly.

SELF EVALUATION AND CHAPTER CONFIDENCE

Multiple Choice

Answers are given in the back of the book and help can be found in the student resources at:
www.garlandscience.com/micro

1. The antibody response is part of the
   - **A.** Humoral response
   - **B.** The cellular response
   - **C.** The phagocytic response
   - **D.** The propagation response
   - **E.** None of the above

2. The amount of antibody to a specific pathogen found in serum is referred to as the
   - **A.** Level of antibody
   - **B.** Antibody concentration
   - **C.** Antibody titer
   - **D.** Antibody minimum
   - **E.** Antibody maximum

3. Specificity is seen in each of the following except
   - **A.** The humoral response
   - **B.** The cellular response
   - **C.** The adaptive immune response
   - **D.** The innate response
   - **E.** Antigen–antibody complexes

4. The adaptive response relies upon distinguishing
   - **A.** Complete from incomplete antigens
   - **B.** Proteins from lipid antigens
   - **C.** Carbohydrates from protein antigens
   - **D.** Self from nonself antigens
   - **E.** Carbohydrates from lipids

5. Presentation of antigen is done by
   A. Macrophages
   B. Dendritic cells
   C. Monocytes
   D. Macrophages and dendritic cells
   E. All of the above

6. Antibody is produced by
   A. T cells
   B. B cells
   C. Plasma cells
   D. Macrophages
   E. Dendritic cells

7. T cells mature in the
   A. Bone marrow
   B. Liver
   C. Lymph nodes
   D. Thymus
   E. Thyroid

8. B cells mature in the
   A. Bone marrow
   B. Liver
   C. Lymph nodes
   D. Thymus
   E. Thyroid

9. There are two classes of T cells called
   A. Antigen presenting and suppressor
   B. Suppressor and killer
   C. Cytotoxic and helper
   D. Suppressor and cytotoxic
   E. None of these pairs is correct

10. M cells are found in all of the following except
    A. The intestine
    B. The Peyer's patches
    C. The GALT
    D. The MALT
    E. None of the above

11. A mature antigen-presenting cell
    A. Is older than other cells
    B. Has recognized antigens
    C. Is able to phagocytose proteins
    D. Has recognized and processed antigens
    E. Is able to recognize antigens better than an imma ture cell

12. Clonal selection involves all of the following except
    A. Rearrangement of gene segments
    B. Specific antigen receptors
    C. Reversible genetic rearrangement
    D. Irreversible genetic rearrangement
    E. Passage of genetic rearrangement to progeny

13. After antigen presentation, T cells in the lymph nodes
    A. Immediately leave the node
    B. Become activated and leave the node
    C. Become activated and remain in the node
    D. Become inactivated and remain in the node
    E. Become inactivated and leave the node

14. Anergy is
    A. Lymphocyte activation by B cells
    B. Lymphocyte inactivation by B cells
    C. Lymphocyte inactivation due to increased co-stimulatory signals
    D. Lymphocyte inactivation due to lack of co-stimulatory signals
    E. Another name for apoptosis

15. The thymus
    A. Grows larger when puberty is reached
    B. Becomes populated with dendritic cells when puberty is reached
    C. Atrophies when puberty is reached
    D. Becomes filled with activated T cells when puberty is reached
    E. None of the above

16. T cells are found in which part of the lymph node?
    A. Stroma
    B. Follicles
    C. Paracortical areas
    D. Capsular areas
    E. Both A and B

17. Interaction of T cells with self-antigen–MHC causes
    A. Maintenance of the T-cell population
    B. Apoptosis of T cells
    C. Activation of T cells
    D. Inactivation of T cells
    E. None of the above

18. The B-cell receptor is
    A. Produced by plasma cells
    B. Found in the cytoplasm of B cells
    C. Found on membranes of plasma cells
    D. Found on membranes of B cells
    E. Only seen on immature B cells

19. The T-cell receptor
    A. Recognizes antigens
    B. Recognizes antigens linked to MHC molecules
    C. Recognizes MHC molecules only
    D. Recognizes small pieces of antigen associated with MHC molecules
    E. None of the above

20. Class I MHC molecules present antigen to
    A. Phagocytic cells
    B. B cells

C. Helper T cells
D. Cytotoxic T cells
E. All of the above

21. Class II MHC molecules present antigen to

A. Phagocytic cells
B. B cells
C. Helper T cells
D. Cytotoxic T cells
E. All of the above

22. Superantigens are

A. Extra-large proteins
B. Recognized by T cells after being bound to MHC molecules
C. Recognized by T cells without being bound to MHC molecules
D. Presented by special antigen-presenting cells
E. Presented only by dendrites

23. Antibody molecules are bivalent because they have

A. One binding site
B. One attachment site for macrophages
C. Two identical binding sites
D. Two binding sites that recognize different antigens
E. Four identical binding sites

24. Antibody generally recognizes a specific

A. Receptor
B. T cell
C. Antigen
D. Epitope
E. B cell

25. The antibody molecule that is found in colostrum (mother's milk) is

A. IgG
B. IgA
C. IgD
D. IgE
E. All of the above

26. T cells that have not been presented with antigen are referred to as

A. Armed
B. Effector
C. Cytotoxic
D. Primed
E. Naive

27. Immunological memory

A. Allows protection on re-exposure to a previous pathogen
B. Is associated with both T and B cells
C. Is the responsibility of long-lived T cells
D. Is the responsibility of bone marrow B cells
E. All of the above

28. Vaccines were first administered by

A. Pasteur
B. Salk
C. Jenner
D. Sabin
E. None of the above

DEPTH OF UNDERSTANDING

Questions listed here require you to bring together the concepts you have learned in this chapter into a discussion format. This helps you to increase your depth of understanding of the material you have learned. Help can be found in the student resources at: www.garlandscience.com/micro

1. Describe the events which lead up to arming of T cells after a pathogen has broken the barrier of the skin.
2. Compare the maturation of T lymphocytes with that of B lymphocytes.
3. Describe the cellular events associated with the adaptive immune response.

CLINICAL CORNER

Help can be found in the student resources at: www.garlandscience.com/micro

1. Your patient is a 33-year-old male who is complaining about the flu. He is a nurse in a large hospital and is immunized aganst flu every year, just to be sure he is protected. However, during the past two years he has come down with flu-like symptoms immediately after getting the shot.

A. Why would he come down with flu-like symptoms right after receiving the immunization?
B. Why has this only begun to occur within the past two years?

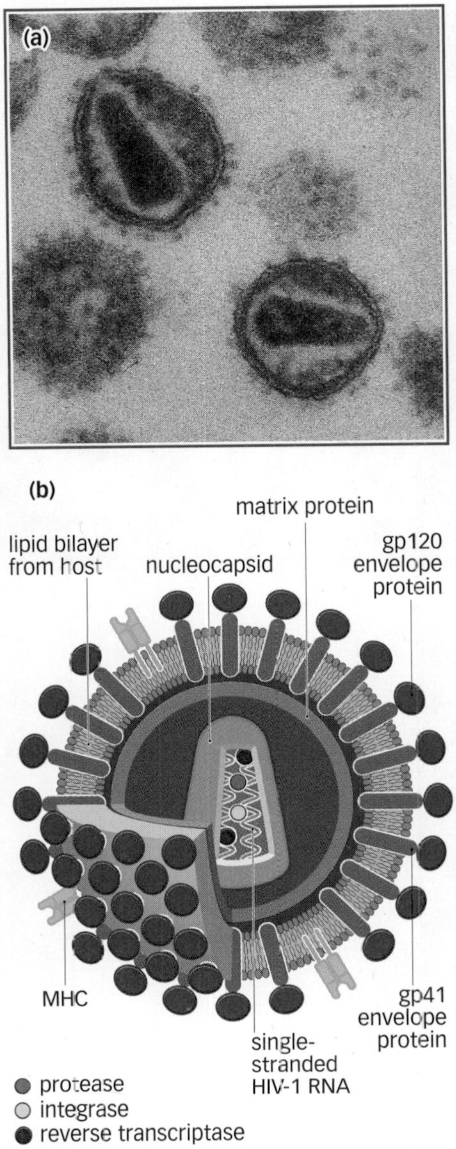

Figure 17.1 The virion of human immunodeficiency virus (HIV-1). Panel **a**: A transmission electron micrograph of HIV. Panel **b**: An illustration of the structures of the HIV virion. Notice that the reverse transcriptase, integrase, and viral protease enzymes are packaged in the virion. Although for clarity only one molecule of viral integrase and protease are shown, the virion actually contains many molecules of each enzyme. Note the nucleocapsid structure and envelope proteins gp120 and gp41 which are also found as part of this virus. Also shown (green) are MHC molecules acquired from the host during budding.

ACQUIRED IMMUNODEFICIENCY SYNDROME (AIDS)

Pneumocystis Pneumonia — Los Angeles. In the period October 1980 – May 1981, 5 young men, all active homosexuals, were treated for biopsy — confirmed Pneumocystis carinii *pneumonia at three different hospitals in Los Angeles, California. Two of the patients died....*

M.S. Gottlieb, H. M. Schancker, P. F. Tan et al. (1981) (Centers for Disease Control) *Morb. Mortal. Wkly Rep*. **30**, 250–252.

This prosaic posting in a weekly mortality report was the first warning of what was to come. There was an editorial note included with this entry that also proved to be prophetic: "*Pneumocystis* pneumonia is almost exclusively limited to immunosuppression patients. ... The occurrence of this disease in these five previously healthy individuals is unusual. ... The fact that these patients are homosexuals suggests an association between some aspects of a homosexual lifestyle or disease acquired through sexual contact."

This editorial was written in 1981, and by 1992 AIDS had become the major cause of death in individuals 25–44 years of age in the United States. Recent estimates from the World Health Organization are that 20 million people have died from the epidemic and more than 40 million are currently infected worldwide. Furthermore, infection by the human immunodeficiency virus (HIV), the virus that causes AIDS, is rising more rapidly in eastern Europe and central Asia than in the rest of the world. One-third of persons currently infected are between 15 and 24 years of age, and the majority of them are *unaware* they are infected.

HIV is an enveloped retrovirus (**Figure 17.1**). Retroviruses contain the enzyme reverse transcriptase, which as we saw in Chapter 12 allows it to copy its RNA genetic material into a DNA copy that can then integrate into a host chromosome. There are two major forms of this virus: HIV-1 is the predominant form seen throughout the world, and HIV-2 is the form mostly seen in Africa.

Cellular Targets of HIV

The first phase of the viral infectious cycle is attachment, and HIV uses a complex of two glycoproteins — **gp120** and **gp41** — in the viral envelope to bind receptors on helper (CD4) T cells, dendritic cells, and macrophages. For the infection to be successful, co-receptors are also required, and two of these co-receptors have been identified. One is the chemokine receptor known as CXCR4, which binds to the chemokine CXCL12 and causes the virus infection to form syncytia (giant cells). In contrast, some variants of HIV use the co-receptor CCR5, which binds to the chemokines CCL3, CCL4, and CCL5 and does not result in the formation of syncytia. Several other co-receptors have been found in places such as the brain and the thymus but are not yet well characterized. People who have mutations in the genes that code for these receptors are not susceptible to infection with HIV.

Fast Fact

Viral entry into target cells can also occur through endocytosis, when the virus is complexed with anti-HIV antibody. Therefore, even when the virus is being attacked by the adaptive immune response it can still infect.

Failures of the Immune Response

Chapter 17

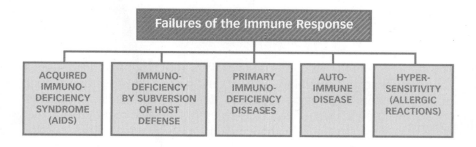

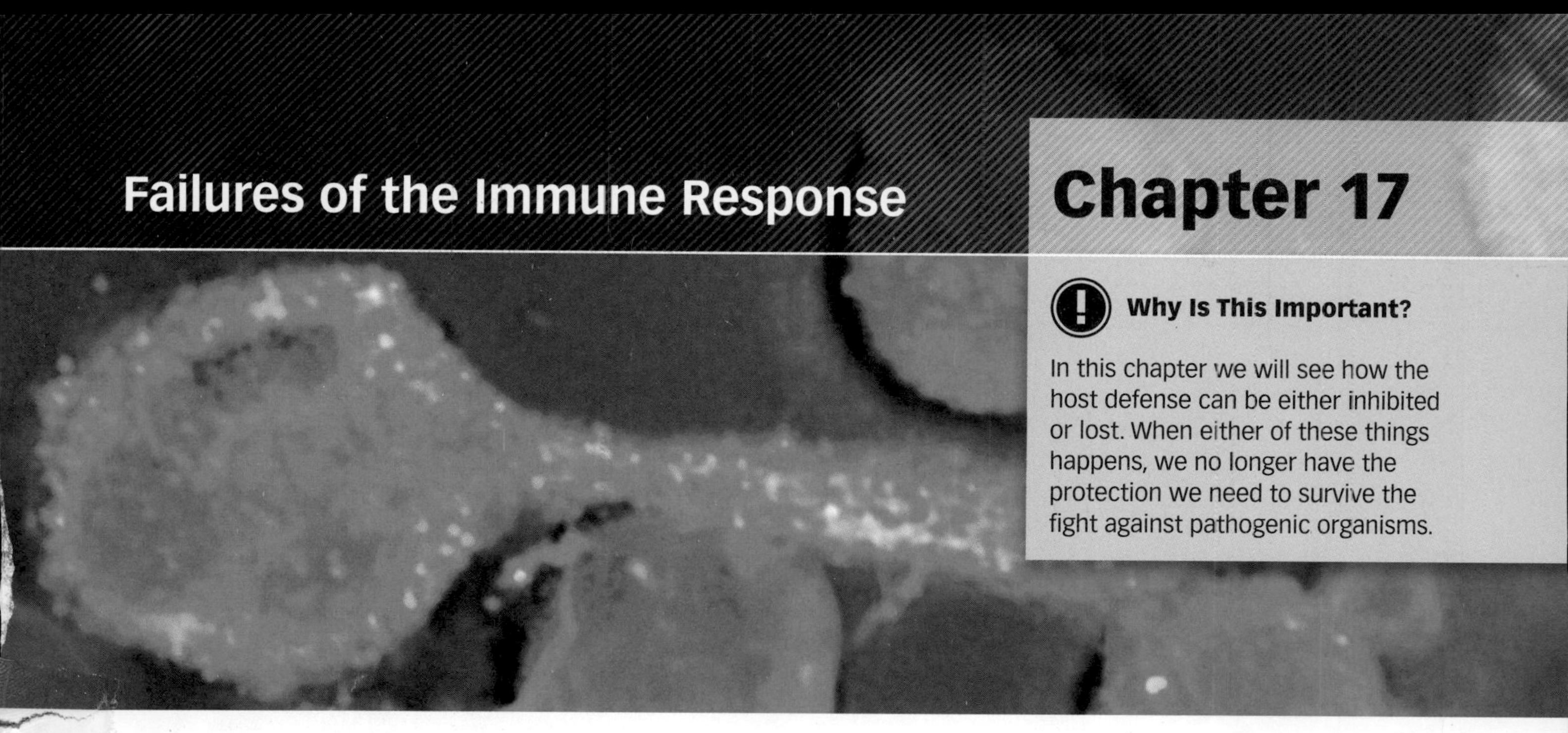

Why Is This Important?

In this chapter we will see how the host defense can be either inhibited or lost. When either of these things happens, we no longer have the protection we need to survive the fight against pathogenic organisms.

OVERVIEW

In the preceding two chapters, we saw that the host defense relies on two mechanisms to survive against relentless pathogenic assault. The innate, non-specific, immune response protects against most of these infections, and the powerful and specific adaptive immune response can also be mobilized. Recall that both of these systems are elegantly designed to deal with keeping the body safe. In this chapter, we examine what happens when the host defense either does not function properly or is completely absent. The discussion is broken down into three parts. In the first, we discuss the loss of host defense, which can occur in a variety of ways, including infection with HIV virus, pathogenic mechanisms designed to defeat the host immune response, and genetic abnormalities that cause deficiencies in the response. In the second part, we focus on what happens when the immune system turns on the host in the form of autoimmunity. The chapter concludes with a look at hypersensitivity reactions, which are immunological responses to allergens that can result in damage to the host.

We will divide our discussions into the following topics:

- Failures of the Immune Response
  - ACQUIRED IMMUNODEFICIENCY SYNDROME (AIDS)
  - IMMUNODEFICIENCY BY SUBVERSION OF HOST DEFENSE
  - PRIMARY IMMUNODEFICIENCY DISEASES
  - AUTOIMMUNE DISEASE
  - HYPERSENSITIVITY (ALLERGIC REACTIONS)

The adaptive immune response in any host can be subverted either indirectly or directly, and we will look at both ways in this chapter. The indirect route to subversion is via acquired immunodeficiency syndrome, more commonly known as AIDS. The direct routes occur when a pathogen either avoids or subverts the host immune response and when the host has a genetic immunodeficiency disease.

What Do I Need to Know?

To get the most out of this chapter, please review the following terms from your previous courses in biology, anatomy, physiology, or chemistry or in previous chapters of this book as indicated in parentheses: apoptosis (**8**), chromatin (**2**), commensal organism (**4**), endocytosis (**4**), neutropenia (**6**), pathogenesis (**4**), phagocytosis (**15**), syncytium/syncytia (**5**), venule, viremia (**7**), virion (**12**).

Table 17.1 The Estimated Quantity of HIV Particles in Body Fluids

Fluid	Estimated Quantity of Virus (Infectious Particles per Milliliter)
Plasma	1–5000
Intestinal mucosal secretions	1–5000
Tears	<1
Ear secretions	5–10
Saliva	<1
Sweat	None detected
Feces	None detected
Urine	<1
Vaginal–cervical fluid	<1
Semen	10–50
Breast milk	<1
Cerebrospinal fluid	10–10,000

Modes of HIV Transmission

HIV is transmitted through sexual contact and also via the blood. The efficiency of transfer is based on the concentration of viral particles in the fluid being exchanged (**Table 17.1**), with the highest viral loads being found in peripheral blood monocytes, blood plasma, and cerebrospinal fluid. Semen and genital secretions are also sources of the virus.

In the United States, HIV transmission is highest among homosexuals, whereas in the rest of the world, transmission is highest among heterosexuals. Other sexually transmitted diseases can contribute to HIV transmission, probably because infected and ulcerated genital tissue permits direct access of the virus to the blood. Although sexual activity is still the leading mode of transmission, intravenous drug abuse is the next most common form (**Figure 17.2**). Children are also at risk because the virus can be transmitted across the placental barrier as well as in breast milk.

Fast Fact

Until recently, the transmission from infected mother to fetus was extremely high. However, the administration of drugs to infected mothers has drastically limited this mode of transmission.

Except for direct access through a needlestick injury or blood transfusion, HIV makes entry into the lamina propria when the mucosal epithelial cell layer is broken. Infection relies on sufficient viral numbers, and infected cells are the richest source of the virus. Extrusion of virus from infected

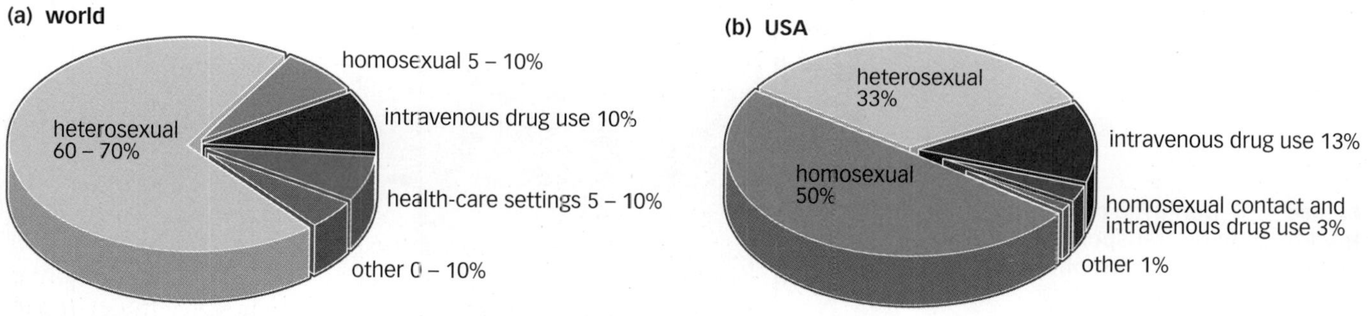

Figure 17.2 Modes of transmission of HIV. Panel **a**: Worldwide figures. Panel **b**: Figures for the United States.

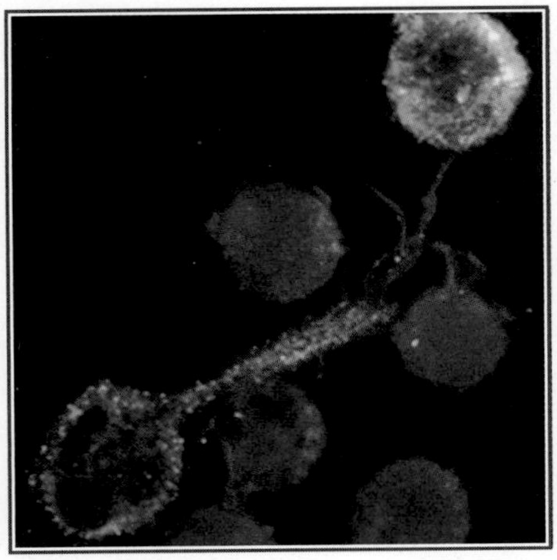

Figure 17.3 Lymphocytes infected with HIV. In this photomicrograph two lymphocytes are already infected with HIV and producing virus (stained green). Two of the other lymphocytes (colored blue) are beginning to show signs of infection (the green dots).

cells can be observed in tissue culture (**Figure 17.3**). Cells of the mucosal immune system, including M cells in the bowel and dendritic antigen-presenting cells in the vagina and cervical epithelium, are prime targets for the initial infection. In fact, it has been shown that dendritic cells can bind to the HIV-1 envelope proteins with high affinity and hold them stably for several days until susceptible T cells come along and are infected (**Figure 17.4**). In sexually active males, cells lining the penis can become infected by contact with virus from infected macrophages or Langerhans cells in the cervix or intestinal mucosa of an infected partner.

As with all other retroviruses, reverse transcription of HIV results in the copying of viral RNA to DNA. Retroviruses can use the viral enzyme integrase to integrate viral DNA into the host cell's chromosome. The HIV genetic material comprises only nine genes flanked by long terminal repeat sequences (called LTRs), which are required for the virus to integrate into the host chromosome.

Virus replication is initiated by transcription of the inserted viral DNA (**Figure 17.5**) and depends on transcription factors present in the host cell. It is important to note that the reverse transcriptase that converts viral RNA to DNA has no proofreading ability, and therefore many mutations occur. It is these mutations that account for the virus's ability to escape drug therapy and develop rapid resistance.

Course of Infection

The pathology of HIV infection consists of three phases (**Figure 17.6**): the acute phase, the asymptomatic phase, and the symptomatic phase with development of AIDS.

The **acute phase** begins in the first few days after the initial infection. During this time, virus is produced in large quantities by infected lymphocytes in the lymph nodes, resulting in lymphadenopathy and flulike symptoms. This initial viremia is greatly reduced by ten weeks after initial contact. The decrease is most probably a result of the actions of cytotoxic (CD8) T cells, which kill infected targets. Support for this hypothesis is demonstrated by the increase in the number of cytotoxic T cells seen during this period. The helper T-cell target population, which is initially

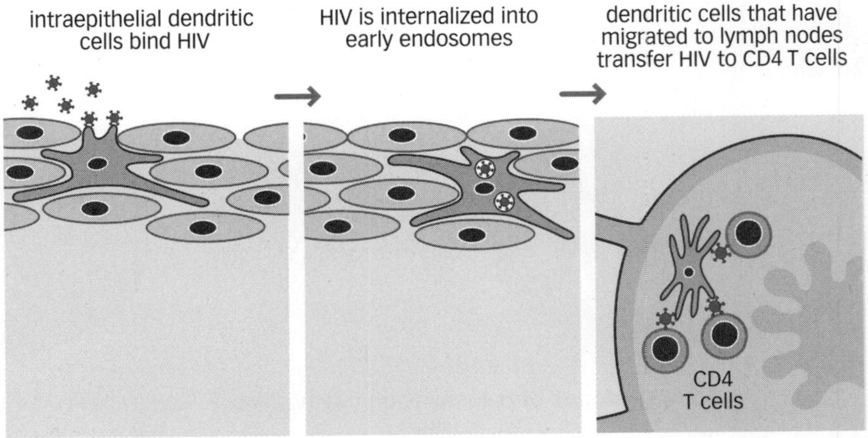

intraepithelial dendritic cells bind HIV
HIV is internalized into early endosomes
dendritic cells that have migrated to lymph nodes transfer HIV to CD4 T cells
CD4 T cells

Figure 17.4 Dendritic cells initiate the HIV infection by transporting HIV from mucosal surfaces to lymphoid tissue. The viral particles gain access to dendritic cells either at the site of mucosal injury or possibly on portions of the dendritic cell that protrude between the epithelial cells.

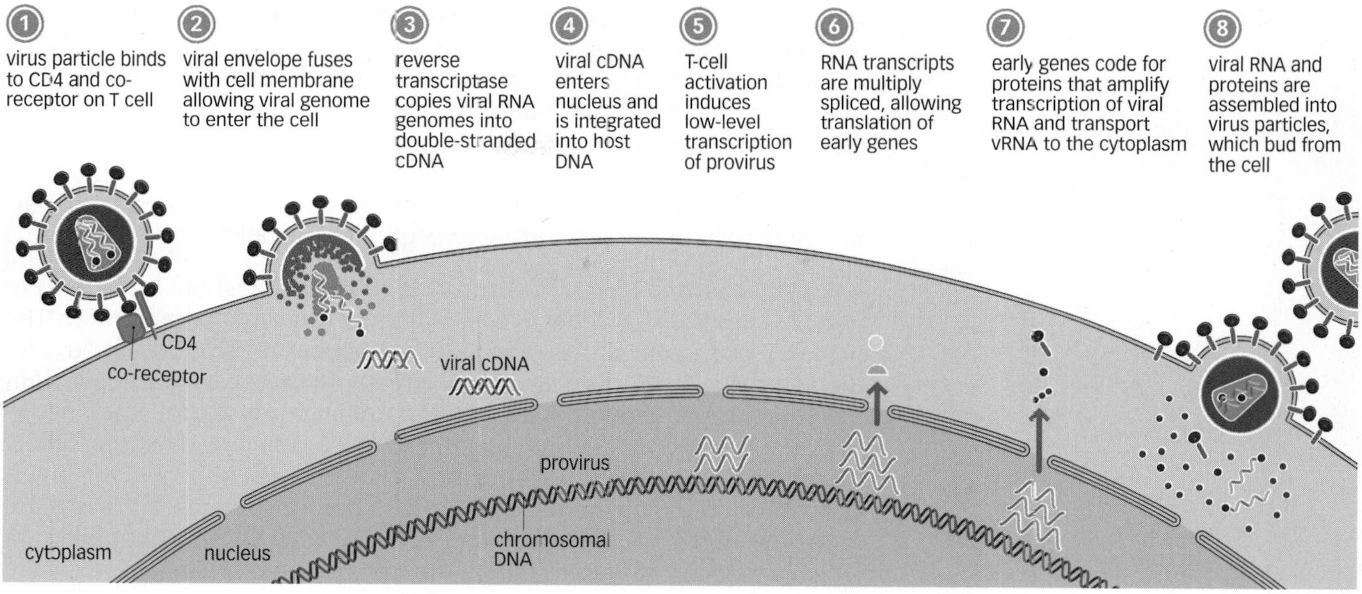
1 virus particle binds to CD4 and co-receptor on T cell
2 viral envelope fuses with cell membrane allowing viral genome to enter the cell
3 reverse transcriptase copies viral RNA genomes into double-stranded cDNA
4 viral cDNA enters nucleus and is integrated into host DNA
5 T-cell activation induces low-level transcription of provirus
6 RNA transcripts are multiply spliced, allowing translation of early genes
7 early genes code for proteins that amplify transcription of viral RNA and transport vRNA to the cytoplasm
8 viral RNA and proteins are assembled into virus particles, which bud from the cell
CD4
co-receptor
viral cDNA
provirus
cytoplasm
nucleus
chromosomal DNA

Figure 17.5 The sequence of events associated with HIV infection.

17.1

17.2

decreased, rebounds by the end of this phase to near-normal numbers. During this phase, the virus population is homogeneous, indicating that little mutation is occurring.

The asymptomatic phase of HIV infection usually begins three to four months after the initial infection. By this point, the viremia is very low, with only occasional small bursts of virus being released. The degree of viremia seen during this phase is a predictor of how fast the disease will progress in an individual — the greater the viremia seen here, the faster the progression of the disease.

The population of helper T cells declines at a steady rate during the asymptomatic phase. There are three ways in which this may happen: direct killing of infected cells by the virus, increased induction of apoptosis in

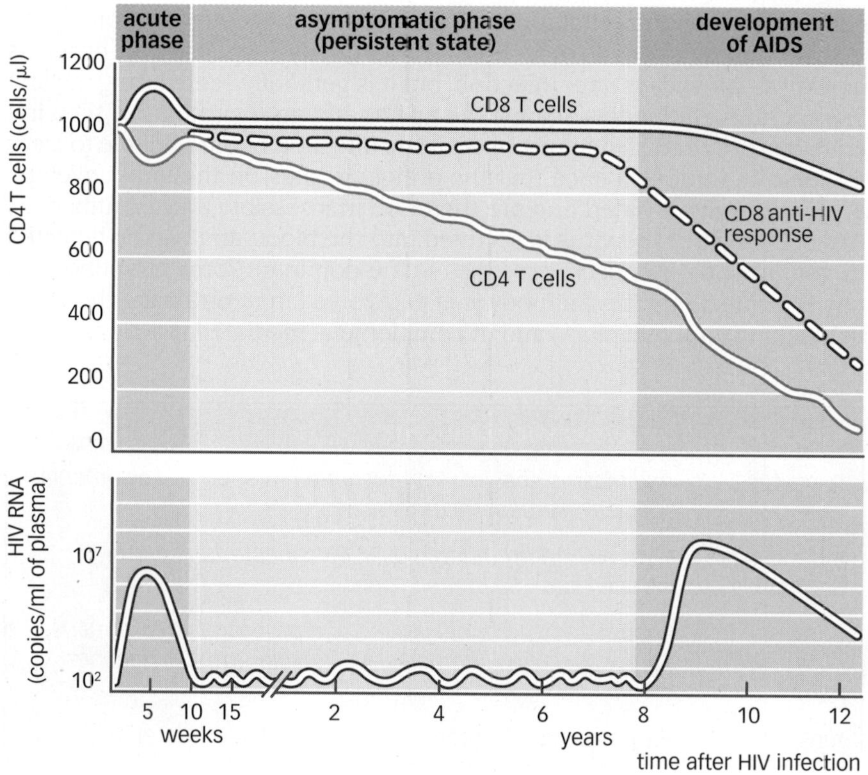
acute phase
asymptomatic phase (persistent state)
development of AIDS
CD4 T cells (cells/μl)
1200
1000
800
600
400
200
0
CD8 T cells
CD8 anti-HIV response
CD4 T cells
HIV RNA (copies/ml of plasma)
10⁷
10²
5 10 15
weeks
2 4 6 8 10 12
years
time after HIV infection

Figure 17.6 A schematic diagram of events occurring during the three phases of HIV infection. Notice there is an initial drop in the CD4 T cell population in the acute phase of the infection. This corresponds to the increase in amount of HIV present in the blood. In the asymptomatic phase there is little increase in viral numbers and a strong cytotoxic CD8 T cell response (solid purple line). In fact the majority of this response is directed against HIV infected T cells (dashed purple line). Eventually the level of cytotoxic T cells targeting HIV infected T cells declines and there is a concomitant increase in infectious HIV in the blood.

infected cells, and killing of infected helper T cells by cytotoxic T cells. This is a protracted phase that can last for years and is referred to as **clinical latency**, with virus replication continuing at a low rate, mainly in the lymph nodes. It is thought that cytotoxic T cells are the reason that the infection does not progress more rapidly during this time, but these cytotoxic T cells eventually begin to disappear. In addition, the virus population becomes more heterogeneous as this phase progresses, indicating that mutations are occurring with greater frequency.

The **symptomatic phase** is the end stage of the infection, and it is here that the infected individual develops the clinical symptoms of AIDS. The helper T cell count drops below 200 per cubic milliliter of blood, and the cytotoxic T cell number is also greatly decreased. Viral replication increases in the lymph nodes, and the architecture of the lymph nodes begins to deteriorate. During this phase there is increased susceptibility to opportunistic infection (**Table 17.2**) as well as generalized lymphadenopathy and the appearance of lesions. The most common lesions are *thrush*, a fungal infection of the oral cavity, and **hairy leukoplakia**, in which white patches are found on the tongue and oral mucosa.

Response to HIV Infection

Approximately 10% of HIV-infected individuals will progress to AIDS in the first three years after the initial infection. More than 80% of infected individuals show signs of clinical disease within 10 years. The remaining 20% are free of disease for long periods (more than 20 years in some cases), and a small percentage of infected individuals will never move past the asymptomatic phase.

The virus replicates best in activated T cells. Because infection activates T cells, the greater the number of opportunistic infections that occur, the greater is the number of T cells activated; the more T cells that are activated, the larger is the quantity of virus particles produced, leading to more infected cells, leading to more opportunistic infections, and so on.

The response to any viral infection is a combination of the humoral (B cell) portion and the cellular (T cell) portion of the adaptive response, and it is no different for HIV infection. The antibody response can occur as quickly as a few days after infection, but it is generally seen during the first few months. Antibodies against the gp120 and gp41 proteins of the viral envelope are produced in response to the infection but are unable to clear it. There is some evidence that the antigenic sites on the viral envelope are heavily glycosylated and are therefore inaccessible to the antibodies. Antibody against the virus is secreted into the blood and can be detected in genital and other secretions. IgG is the dominant form of antibody in any HIV infection. This antibody is also involved in neutralizing the virus, in blocking viral receptors, and in complement-mediated reactions.

Fast Fact

The primary protective mechanism of antibody neutralization of virus is difficult for HIV because the epitopes of the virus are hidden.

Table 17.2 The Types of Opportunistic Pathogen That Infect Patients with AIDS

Infections	Opportunistic Pathogens
Parasites	*Toxoplasma* species, *Cryptosporidium* species, *Leishmania* species, *Microsporidium* species
Intracellular bacteria	*Mycobacterium tuberculosis*, *Mycobacterium avium intracellulare*, *Salmonella* species
Fungi	*Pneumocystis (carinii) jiroveci*, *Cryptococcus neoformans*, *Candida* species, *Histoplasma capsulatum*, *Coccidioides immitis*
Viruses	Herpes simplex, cytomegalovirus, varicella-zoster

It has been shown that infected individuals all have a cellular adaptive response to AIDS, a response in which cytotoxic T cells are programmed to recognize essentially all HIV-1 proteins. This makes sense because there is a good correlation between the cellular response, low viral load, and slower disease progression. The obvious question therefore is why the cellular adaptive response does not control and defeat the infection. Sadly, there is as yet no clear-cut answer to this question.

Dynamics of HIV Replication in Patients with AIDS

The dynamics of HIV production during infection are truly astonishing. The minimum rate of release into the blood based on one cycle of infection per infected cell per day is an astronomical 10^{10} virions. When this high rate is coupled with the high mutation rate seen in this virus, it can be predicted that every possible mutation at every position in the viral genome can occur numerous times *each day*! It is estimated that the genetic diversity of HIV produced in a single infected person is greater than all the diversity that would be seen in a worldwide epidemic of influenza (influenza virus is one of the most genetically diverse viruses in the world).

On the basis of the way in which HIV infection progresses, even if we had an effective drug to treat the infection, it could take three to five years of continuous treatment to eliminate the virus. This is only an estimate because the virus can move to different parts of the body, and virus that finds its way to the brain can remain undetected. Sadly, the side effects of the current treatments are so severe that a patient could not withstand such a protracted period of treatment.

Major Tissue Effects of HIV Infection

Lymphoid Organs

Most HIV virus is trapped in germinal centers, which are areas of the lymph node that consist of networks of follicular dendritic cells. Early in the asymptomatic phase of the infection, the patient develops palpable lymphadenopathy (swollen lymph nodes) as a result of follicular hyperplasia and cell proliferation. Later, in the intermediate stage of the symptomatic phase of the disease, the lymph nodes begin to deteriorate as a result of cell death. In the advanced stage of the symptomatic phase of disease, lymphoid tissue is almost completely destroyed.

Nervous System

About one-third of HIV patients are diagnosed with neurological disorders at some point during the infection. HIV enters the nervous system early in the infection but does not seem to infect neurons. Nearly two-thirds of all patients infected with HIV-1 get sub-acute encephalitis, referred to as **AIDS dementia**. Several opportunistic infections that result from the immunodeficiency associated with AIDS can be neuropathological.

Gastrointestinal System

Advanced disease shows damage to the gastrointestinal system manifested in the form of diarrhea and chronic malabsorption of nutrients from food, which results in significant weight loss.

Other Systems

HIV infects the lungs, heart, and kidneys as well as joint fluids. In addition, HIV infection is associated with cancer. This latter finding makes sense because the immune system is responsible for "tumor surveillance," and when the immune system is disrupted, there can be an increased

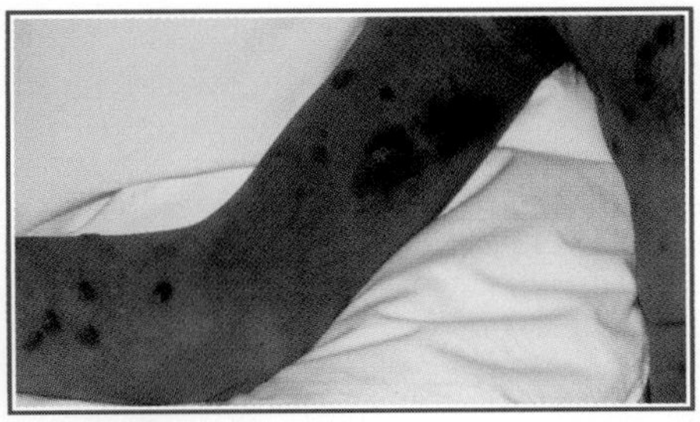

Figure 17.7 Kaposi's sarcoma in a young man infected with HIV-1.

incidence of malignancies. In addition, HIV infection leads to high levels of cytokine production, which might activate oncogenic viruses and promote *angiogenesis* (the vascularization required for tumor growth). Two malignancies are prevalent in patients with AIDS: Kaposi's sarcoma and B cell lymphoma.

Kaposi's sarcoma (**Figure 17.7**) is a tumor that contains many cell types. In immunocompetent individuals, this condition presents as a nonaggressive and rarely lethal malignancy, but in HIV-infected males the sarcoma is very aggressive, affecting both mucocutaneous and visceral areas of the body. Kaposi's sarcoma occurs in about 20% of HIV-infected men but in only 2% of infected women.

The second form of cancer prevalent in patients with AIDS, **B cell lymphoma**, is 60–100 times more common in patients with AIDS than in the healthy population. It is found in many areas of the body, including the lymph nodes, intestinal tract, central nervous system, and liver.

We have chosen to look at HIV infection and AIDS in detail, but remember that immunodeficiency can occur in other ways as well.

Keep in Mind

- Infection with HIV can lead to acquired immunodeficiency syndrome (AIDS).
- The target of the HIV virus is the helper T cell (CD4).
- HIV requires the presence of co-receptors to infect the T cell successfully.
- HIV is transmitted sexually or through the blood.
- In the United States the highest transmission rates are found among homosexuals, whereas in the rest of the world the highest transmission is among heterosexuals.
- HIV uses the enzyme reverse transcriptase to convert its RNA genome into DNA that can be integrated into the host chromosome.
- There are three phases of HIV infection, acute, asymptomatic, and symptomatic, with the development of AIDS.
- High rates of mutation by HIV make the infection very hard to treat.

IMMUNODEFICIENCY BY SUBVERSION OF HOST DEFENSE

Throughout Part II of this book, we mentioned ways in which a pathogen can subvert host defense responses. We can consider this subversion a type of host defense failure and therefore a topic for this chapter on failures of the immune response. Keep in mind that for the pathogen, the aim is to spread the disease to new hosts. To do this, many pathogens take an approach that can be described as "keep quiet and don't provoke too vigorous an immune response." In addition, they don't want to kill the host too quickly, because this would cause an end to the infection. Consequently, pathogens have developed a variety of strategies for avoiding destruction by the host defense. Here we look at several of these mechanisms, such as antigenic variation, latency, resistance to host defenses, and suppression of the immune response.

Antigenic Variation

This mechanism is particularly important for extracellular pathogens in defeating the humoral (B cell) arm of the adaptive response. There are three ways in which antigenic variation occurs:

- Many pathogens exhibit a wide variety of antigens. The various antigenic varieties of a given pathogen are said to be serotypes of the pathogen, a name that reflects the fact that serotypes are differentiated from one another by serological testing. For example, there are 84 known serotypes of *Streptococcus pneumoniae*, with each type containing different polysaccharides in the capsule surrounding the organism. Because of the large number of serotypes for this pathogen, it is possible to be repeatedly infected by *S. pneumoniae*.

- The second form of antigenic variation is more dynamic and involves antigenic drift and antigenic shift. In **antigenic drift**, the pathogen takes advantage of mutations to change the epitopes of the antigen such that the epitopes are not recognized by a host's cytotoxic T cells. For instance, a person infected with the influenza virus develops protection by producing antibody against the viral surface protein called hemagglutinin. Influenza virus infecting that person a second time undergoes antigenic drift, creating modified hemagglutinin molecules that are structurally different from the original molecule. These modified surface proteins on the virus allow it to evade the antibody specific for the original form of hemagglutinin. In **antigenic shift**, there is a reassortment (a shuffling) of the viral RNA genome, and this reassortment leads to major changes in viral proteins. The effect on the host is the same as with antigenic drift: the host's immune system does not recognize the changed proteins. Major pandemics of influenza resulting in widespread and often fatal disease result from antigenic shift.

- The third form of antigenic variation involves programming genetic rearrangements. For example, African trypanosomes, which are protozoan parasites, can change major surface antigens repeatedly within the infected host. In this case, the chronic cycle of new antibody development and clearance of the pathogen results in repeated bouts of inflammation, which eventually cause neurological damage. This leads to the classical coma symptoms seen in this infection. The same antigenic variation is seen during the course of a malarial infection with *Plasmodium*, or food poisoning caused by *Salmonella*.

Latency

In some cases, a pathogen adopts a hide-and-wait strategy in which it stops replicating until the host's immune response has diminished. This strategy is called **latency**. While a pathogen is latent in a host, the pathogen does not replicate and therefore does not trigger an immune response. A good example is herpes simplex type I virus, which initially infects the epithelial cells, spreads to sensory neurons, and then becomes latent. Factors such as sunlight, hormones, or even bacterial infections can reactivate the latent virus, which then travels down the axon of the neuron and re-infects the epithelium. It is the destruction of the infected epithelial cells that produces the "cold sore" representative of this infection.

There are two reasons the latent herpes simplex virus can safely stay in the neuron. First, it is quiescent (inactive) while there, and second, the neuron expresses very few class I MHC molecules, which makes it harder for cytotoxic T cells to recognize and attack the infected cell. Low expression of class I MHC is a protective mechanism that limits the risk that neurons, which do not regenerate well, will be attacked in an autoimmune fashion by a defective cytotoxic response. This makes the neuron a good spot for persistent viral infections such as chicken pox (varicella-zoster).

Fast Fact

Although latent herpes simplex can be reactivated repeatedly, shingles, a painful condition resulting from latent varicella-zoster virus infection, usually occurs only once in a lifetime.

Latency is also seen in B cells infected with Epstein–Barr virus. In most cases, the infection passes without even being noticed, but in some cases it can be severe enough to result in the medical condition known as *infectious mononucleosis*. In rare cases, the virus can lead to development of *Burkitt's lymphoma*. The Epstein–Barr virus expresses a viral protein called EBNA-1, which should be antigenic enough to stimulate an immune response. However, this protein interacts with the host cell's proteasome in such a way that the protein is not degraded into peptides that could be presented to T cells.

Table 17.3 gives examples of viral subversion mechanisms used to defeat the host immune response.

Resistance to Host Defense

Although we have looked at mechanisms that pathogens use to defeat the host defense several times in our previous discussions, it is important to include them here as well because they are examples of failures of host defense. We can list these as follows:

- Some pathogens are resistant to phagocytosis because of capsules, M proteins and other mechanisms.
- Some pathogens — *Mycobacterium tuberculosis* is one example — grow in macrophages by blocking the fusion of the phagosome with the lysosome.
- Some pathogens can escape from the phagosome after phagocytosis and use the host cell's actin to move into adjacent cells (an example is *Listeria monocytogenes*).
- The protozoan parasite *Toxoplasma gondii* isolates itself by generating its own vesicle that does not fuse with lysosomes in the host cell. This way the pathogenic proteins are kept away from the host cell's MHC molecules and cannot be presented to T cells.
- The spirochete *Treponema pallidum* (which causes syphilis) is believed to avoid antibody recognition by coating itself with molecules from host cells until after it invades tissues, at which time it is less accessible to antibody molecules. How this occurs is not yet completely understood.

Table 17.3 Examples of Mechanisms Used by Viruses to Subvert Host Defense

Viral Strategy	Specific Mechanism	Result	Virus
Inhibition of humoral immunity	Virally encoded Fc receptors	Blocks effector function of antibodies bound to infected cells	Herpes simplex
	Virally encoded complement receptor	Blocks complement system	Herpes simplex
Inhibition of inflammatory response	Virally encoded chemokine receptor	Advantage to virus not yet understood	Cytomegalovirus
	Viral inhibition of adhesion molecules	Inhibits interaction of inflammatory cytokines	Vaccinia virus
Blocking of antigen presentation	Inhibition of expression of MHC class I	Blocks recognition of infected cells by cytotoxic T cells	Herpes virus
Immunosuppression of the host	Virally encoded IL-10	Inhibits Th1 T cells and reduces interferon production	Epstein–Barr virus

Suppression of the Immune Response

Many pathogens can suppress a host's immune response. One group capable of this suppression is the staphylococcal organisms that produce enterotoxins and toxic shock syndrome (Chapter 5). These toxins also cause the production of cytokines that suppress the immune response.

Some pathogens cause mild or transient immunodeficiency during acute infection. Although the exact mechanisms are not understood, this form of host suppression seems to affect dendritic cells, causing them to become unresponsive and thereby disabling the T cell response through a lack of antigen presentation. This is important because the patient then becomes more susceptible to secondary infections, which are in many cases more dangerous than the primary infection.

Keep in Mind

- Immunodeficiency can occur through subversion of the host defenses.
- Three major strategies are used to subvert or defeat the host defenses: antigenic variation, resistance to host defense, and suppression of the immune response.
- Antigenic variation can occur through antigenic drift or antigenic shift.
- While a pathogen is latent, it does not multiply and does not trigger an immune response.
- Latency is usually associated with viral infections.
- Resistance to host defense can involve several mechanisms that inhibit phagocytosis.
- Some pathogens produce toxins that suppress the host's immune response.

PRIMARY IMMUNODEFICIENCY DISEASES

Up to now, we have seen that immunodeficiency can result in a variety of ways. Now let's look at primary immunodeficiencies that result from genetic abnormalities. The diseases classified as primary immunodeficiency diseases are characterized by an immune system that does not function properly. They can occur when parts of the immune response are defective because of mutations in one or more of the genes responsible for the response. Primary immunodeficiency diseases present as overwhelming infections in young children. The type of infection that occurs can indicate the type of primary immunodeficiency causing the problem. For example, recurrent infections by **pyogenic (pus-forming) bacteria** points to a defect in antibody production, in the complement system, or in the phagocytosis mechanism. In contrast, recurrent fungal or viral infections suggest that the defect is in reactions mediated by T cells.

Inherited immunodeficiency diseases are caused by recessive gene defects and were first reported in 1952. These diseases can affect both the development and function of T cells and B cells (**Table 17.4**). There can also be defects in phagocytic cells, complement, or cytokine receptors, any of which will result in a lack of proper host defense.

B Cell Defects

Defects in antibody production can result in severe and repeated infection with encapsulated bacteria. Recall from Chapter 15 that the bacterial capsule is a protective component that defeats phagocytosis. A host

Name of Deficiency	Specific Abnormality	Immune Defect	Susceptibility
Severe combined Immunodeficiency syndrome (SCID)	X-linked SCID	No T cells	General infections
	Autosomal SCID A DNA repair defect	No T cells or B cells	General infections
DiGeorge syndrome	Thymic aplasia (the thymus does not develop)	Variable numbers of T and B cells	General infections
MHC class II deficiency	Lack of expression of MHC class II	No CD4 T cells	General infections
Wiskott–Aldrich syndrome	X-linked defective gene	Impaired T cell activation, defective anti-polysaccharide response	Encapsulated extracellular bacterial infections
X-linked agammaglobulinemia	Loss of specific enzyme activity	No B cells	Bacterial and viral infections
Phagocyte deficiencies	Many different causes	Loss of phagocyte function	Bacterial and fungal infections
Complement deficiencies	Many different causes	Loss of specific complement components	Bacterial infections especially *Neisseria* species
Ataxia telangiectasia	Defective enzyme gene	Reduced numbers of T cells	Respiratory infections

Table 17.4 Common Primary Immunodeficiency Syndromes

deals with the presence of encapsulated bacteria by producing antibody that reacts with components of the capsule. The result is opsonization of the organism, which facilitates the elimination of the pathogen. Without the ability to produce capsule-binding antibody, the host is unable to defeat the encapsulated pathogens.

The same is true of neutralizing antibody, which protects against attachment of viruses and also neutralizes toxins. Antibodies can bind to viruses and prevent the virus from binding to the host cell, which prevents the infection. Defects in anti-viral antibody production can result in more viral infections. If a host has a B cell defect that prevents the formation of neutralizing antibody, the toxins produced by pathogens can cause severe problems.

Many of the immunodeficiency diseases classified as B cell defects have been identified, and most are due either to defects in the development of B cells or to defects in the activation of the humoral response. There is also a transient humoral deficiency that occurs during the first 6–12 months of life. Newborn infants have the same level of antibody as their mother because of the transplacental transfer of maternal IgG to the fetus. However, these maternal molecules are eventually catabolized, and the level of antibody in the newborn infant decreases after birth until the infant can begin to produce its own antibody at about six months of age. Consequently, the infant has low levels of antibody from three months to one year of age and is more susceptible to infection during that time.

Fast Fact

In people with B cell defects, infections are usually treated with antibiotics and also by infusion of immunoglobulin collected from large donor pools.

T Cell Defects

Defects in T cell function can result in **severe combined immunodeficiency syndrome** (**SCID**). In these cases there is no T cell function and therefore no cellular adaptive immune response. In addition, T cell defects eventually result in a lack of a humoral (B cell) adaptive response because, as we saw in the previous chapter, the helper T cell is directly involved in regulating the production of antibody. Some cases also exhibit a class II MHC deficiency in which there is no T cell development in the thymus.

In rare cases there can be deficiency in class I MHC, resulting in chronic respiratory infections and skin ulcerations. For example, in *DiGeorge syndrome* the thymic epithelium fails to develop properly, and therefore the thymic environment cannot support T cell maturation. Defects in cytokine production or activation can also result in immunodeficiency disease.

The SCID known as *Wiskott–Aldrich syndrome* causes impaired T cell function, reduced numbers of T cells, and a failure of B cell response to encapsulated bacteria. In this disease, the defect is in the actin cytoskeleton of the cell, which is required for cooperation between T cells and B cells. In these cases, there is no T cell cytotoxic response or T cell help. Bone marrow transplants or gene therapies can be used to treat these types of immunodeficiency disease, but these therapies are very difficult and rarely succeed.

Defects in Accessory Cells

Primary immunodeficiency diseases can also be the result of defects in the host defenses other than B cell and T cell defects. For instance, defects in the complement cascade can lead to an increase in the occurrence of infectious disease (**Figure 17.8**). Defects in phagocytic cells can cause widespread bacterial infections. Any deficit in the number of phagocytic cells is associated with severe immunodeficiency. In fact, a total absence of neutrophils (neutropenia) is fatal. For inherited neutropenia, the only effective treatment is a successful bone marrow transplant.

Some immunodeficiencies occur because of genetic defects in the production of the adhesion molecules required for the margination (rolling and stopping) of white blood cells in the blood vessels. However, most of the defects in white blood cells affect their ability to kill intracellular bacteria or ingest extracellular bacteria (**Table 17.5**). Among the best-known of these diseases is **chronic granulomatous disease**, in which the white blood cells cannot produce the superoxide radical, so that their antibacterial activity is seriously impaired. Another immunodeficiency disease involving white blood cells is **Chediak–Higashi syndrome**, which involves a defect in the gene encoding the protein involved in the formation of intracellular vesicles. This problem results in a failure to fuse phagosomes with lysosomes.

Figure 17.8 A summary of primary immunodeficiency diseases resulting from defects in components of the complement system. Deficiencies in the classical pathway lead to immune complex disease. Defects in the mannose binding lectin pathway are associated with increased bacterial infections in early childhood. Deficiencies in the early components of the alternative pathway lead to increased infection with pyogenic pathogens and *Neisseria* species. Perhaps the most problematic deficiencies are associated with defects in C3 which cause inhibition of the formation of the membrane attack complexes and essentially shut down the complement pathway.

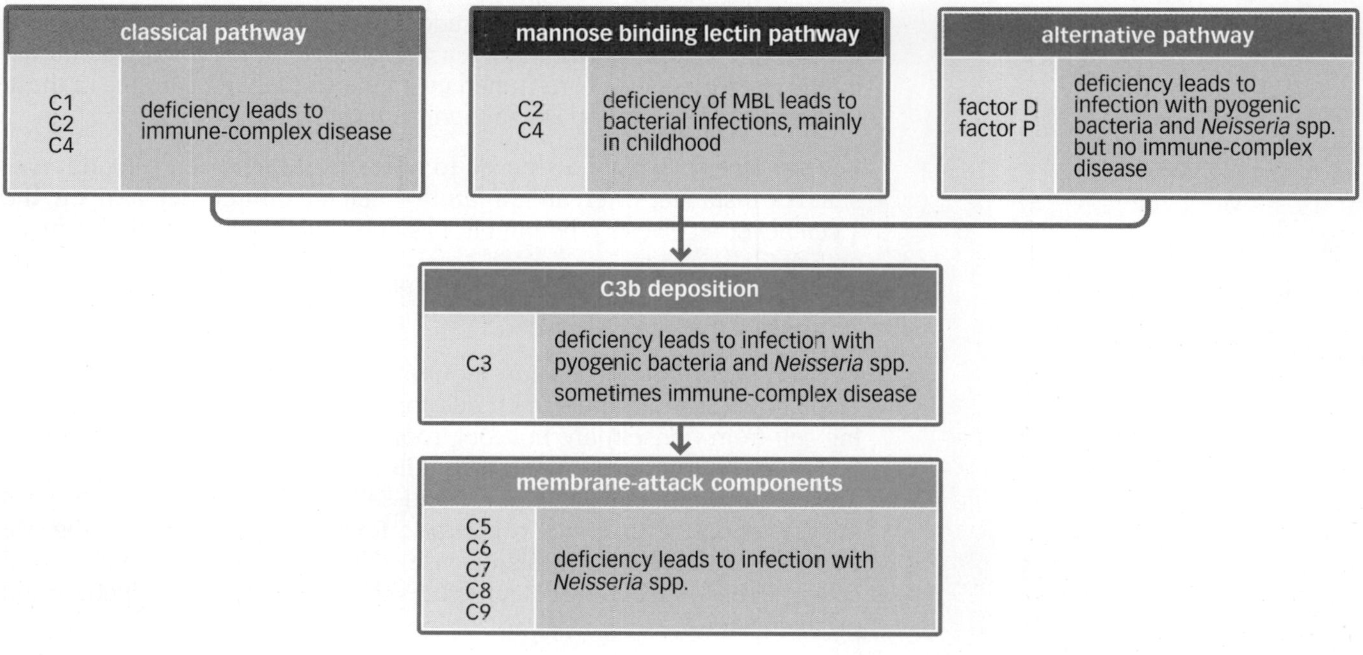

Type of Defect/ Syndrome	Associated Infections
Leukocyte adhesion deficiency	Widespread pyogenic bacterial infections
Chronic granulomatous disease	Intracellular and extracellular infections
Myeloperoxidase deficiency	Defective intracellular killing of pathogens
Chediak–Higashi syndrome	Intracellular and extracellular infection

Table 17.5 Defects in Phagocytic Cells Associated with Persistent Bacterial Infections

Keep in Mind

- Primary immunodeficiency can be caused by mutations in immune response genes.
- Primary immunodeficiency is usually identified by overwhelming infections in young children.
- Inherited immunodeficiency is caused by recessive gene defects.
- Defects in B cells can result in severe and repeated infections with encapsulated bacteria.
- Defects in T cell function result in severe combined immunodeficiency syndrome (SCID).
- Primary immunodeficiency can also be the result of defects in host defense other than T cells or B cells.

AUTOIMMUNE DISEASE

As we have seen so far in this chapter, a host defense can be subverted either indirectly via acquired immunodeficiency syndrome (AIDS), or directly via subversion by an invading pathogen or via genetic immunodeficiency diseases. No matter what the subversion mechanism is, the end result is that the host suffers from a lack of immunity that in many cases is fatal. In addition to these types of immune-system failure, an immune system can also fail by reacting against self components. This can be a devastating problem for the host because of the lethality of immune defense mechanisms. Let's now take a look at the causes of the phenomenon generally known as *autoimmune disease*.

Development of Tolerance

To understand how a body develops immune responses against itself, it is important to remember tolerance. Tolerance to self is part of the reason that autoimmune reactions should not happen. Despite these tolerance safeguards, some self reactions do occur, and they can manifest themselves as autoimmune disease. One of the mechanisms that permits a tolerance of self is constant antigen concentration. Self antigens are always present in essentially constant concentrations. In contrast, foreign antigens are introduced suddenly into the system, and the concentration of these antigens increases rapidly and exponentially during infection. Lymphocytes are "tuned in" to the sudden appearance of these foreign antigens and recognize them as nonself.

Several other clues are also used to differentiate between self and nonself. For instance, when an immature T cell encounters self antigen, the T cell never receives a co-stimulatory signal and therefore never becomes activated.

Triggers for Autoimmunity

In point of fact, concentration-monitoring mechanisms, absence of co-stimulation, and all other methods for inducing tolerance (differentiating self from nonself) are not foolproof, which means that autoimmune responses occasionally occur. We still do not know the exact events involved in initiating this type of anti-self response, but some triggers are apparent. For example, **rheumatic fever** is an autoimmune disease triggered by infection. In this disease, antibodies against streptococcal species cross-react with host structures that are similar to antigens found on the bacteria.

Disease	T Cells	B Cells	Antibody
Systemic lupus erythematosus	T cell help for autoantibody production	Presents antigen to T cells	Pathogenic autoantibody produced
Diabetes	CD8 cytotoxic cells	Presents antigen to T cells	None
Myasthenia gravis	Helper T cell response for antibody production	Antibody secretion	Pathogenic autoantibody produced
Multiple sclerosis	Pathogenic	Presents antigen to T cells	Autoantibody present but role is unclear

Table 17.6 Parts of the Immune Response Involved in Autoimmune Disease

Multiple parts of the immune response are usually involved in autoimmune reactions (**Table 17.6**). For example, in autoimmune diseases such as **systemic lupus erythematosus** (SLE; **Figure 17.9**) and **myasthenia gravis**, autoantibodies (antibodies against self) are the damaging factor. SLE involves autoantibodies against DNA when it is in the form of chromatin in a cell that is not undergoing mitosis, and autoantibodies in myasthenia gravis react against the receptors for the neurotransmitter acetylcholine, a reaction that results in muscle weakness.

In many cases, autoimmunity results in chronic disease because it is impossible to rid the host of the self antigen that triggers the response. This results in a vicious cycle in which the disease destroys host tissue containing the self antigen, which frees more self antigens into the circulation, which are reacted against, producing more autoantibody, which destroys more tissue, and so on (**Figure 17.10**).

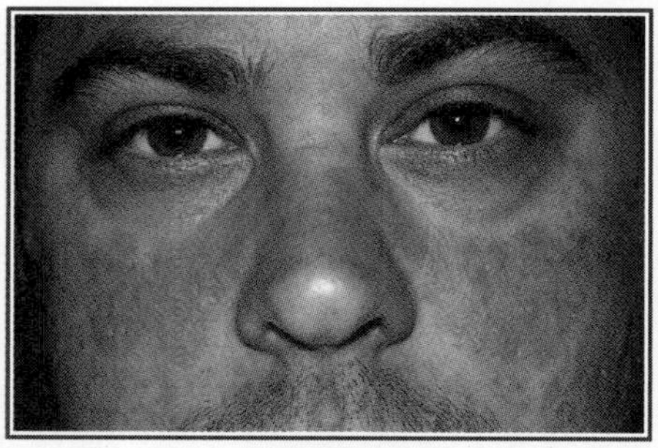

Figure 17.9 The characteristic skin rash seen in systemic lupus erythematosus.

Even though T cells and B cells that react strongly against self are eliminated in the thymus and bone marrow, respectively, not all self antigens are expressed in these two locations when the T cells and B cells are developing. There are antigens associated with the body's immunologically protected sites — brain, eyes, testes, and uterus — that do not induce an immune response but can be the target of autoimmune attack. For example, myelin basic protein from the brain is involved in the autoimmune reaction that causes **multiple sclerosis**.

circulating B cell binds self antigens released from injured cells

B cell is activated by a T cell specific for self peptide

B cells differentiate into plasma cells, secreting large amounts of self-antigen-specific antibody

at sites of injury, self-antigen-specific antibody initiates an inflammatory response, causing more cell injury

more B cells bind self antigens, amplifying the cycle of tissue damage

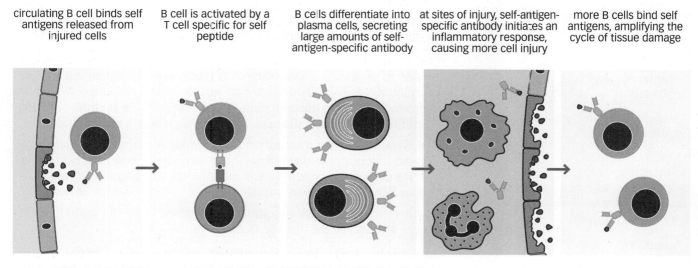

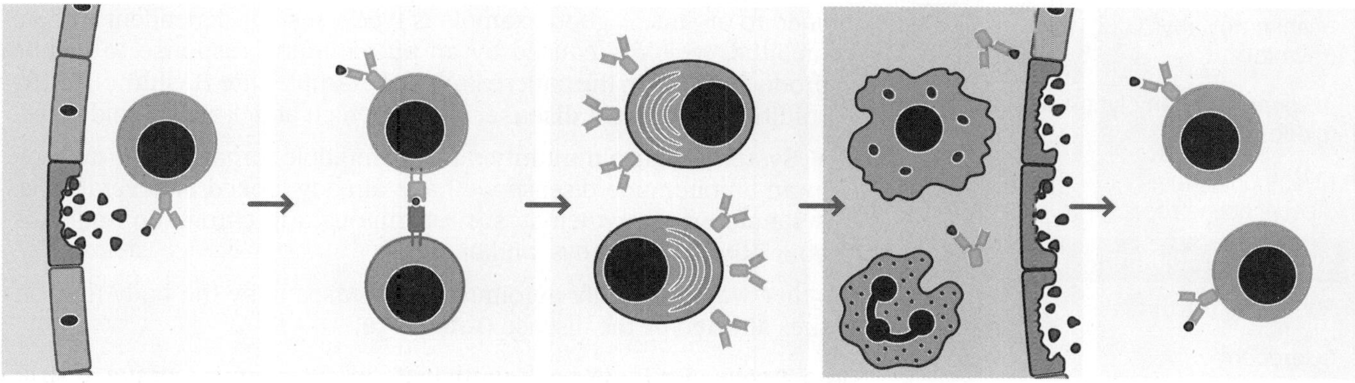

Figure 17.10 How autoimmune-mediated inflammation leads to the release of self antigens from damaged tissues, a release that further activates the autoimmune response to self.

Regulation of Autoimmunity

Specialized T cells called *regulatory T cells (Treg cells)* are involved in controlling the autoimmune response. These specialized T cells, which can regulate other T cells, B cells, and antigen-presenting cells, carry the cell surface marker CD25. Treg cells arise in the thymus in response to relatively strong recognition of self antigens and may also be involved in regulating normal immune reactions against pathogens. These cells regulate the effects of other T cells.

Yet another autoimmune control mechanism involves the cessation of the immune response. Lymphocytes have built-in limits when it comes to proliferation and survival. These limitations naturally restrict normal responses as well as autoimmune responses. In the end, all immune responses result in the death of most of the responding cells (except memory cells), through activation-induced cell death. This occurs through apoptosis and is part of the natural tendency of responding cells to die.

Causes of Autoimmunity

It is clear that some people are genetically predisposed to autoimmune disease. There is also some preliminary evidence that autoimmunity can be influenced by environmental factors. In addition, drugs and toxins can also cause autoimmunity. In these cases, it is thought that drugs or toxins react with self antigens in such a way that they form derivatives that are perceived as foreign by the immune system.

Fast Fact

Procainamide, a medication used to treat cardiac arrhythmias, can induce the production of autoantibodies resembling those present in systemic lupus erythematosus.

Infection can also result in an autoimmune reaction. Inflammation from an infection leads to tissue destruction, and self-reactive lymphocytes can become activated as the level of tissue destruction rises. In addition, some pathogens can upset the body's regulation of the immune response either by preventing apoptosis or by secreting their own cytokines. Molecular mimicry can occur when pathogens express proteins or carbohydrate antigens that resemble host molecules, and this can cause autoimmunity. In most of these cases, the autoimmune response is only transitory: when the infection is over, the autoimmunity ends. However, in individuals who are genetically predisposed to autoimmunity, the disease can become chronic.

Mechanisms of Autoimmunity

Autoimmune diseases vary in severity, in effector mechanisms, and in the tissue affected. These diseases can be divided into two types (**Table 17.7**):

- In **organ-specific autoimmunity**, the reaction against self is confined to organs. A good example is type 1 insulin-dependent diabetes mellitus, which is caused by an autoimmune response to insulin-producing cells in the pancreas. Other examples are **Hashimoto's thyroiditis** and **Graves' disease**, both of which attack the thyroid gland.
- **Systemic autoimmunity** affects multiple organs. For example, in an autoimmune disease we have already looked at several times, systemic lupus erythematosus, ubiquitous anti-chromatin antibodies attack the skin, kidneys, and brain.

Table 17.7 Classification of Some Common Autoimmune Diseases

Organ-specific Autoimmune Diseases
Type 1 diabetes mellitus (β cells of the pancreas)
Goodpasture's syndrome (kidneys)
Multiple sclerosis (central nervous system)
Graves' disease (thyroid)
Hashimoto's thyroiditis (thyroid)
Autoimmune hemolytic anemia (blood cells)
Myasthenia gravis (acetylcholine receptors)
Systemic autoimmune diseases
Rheumatoid arthritis
Scleroderma
Systemic lupus erythematosus
Sjögren's syndrome

Another way to classify autoimmune diseases is by the body tissue or tissues affected by the disease (**Table 17.8**).

Both T cells and B cells can participate in autoimmune tissue destruction. The production of autoantibodies requires T cell help, and some of the diseases caused by autoantibodies are shown in **Table 17.9**. Cytotoxic T cells can also be involved in autoimmune disease by directly

Disease	Autoantigen	Consequence
Autoimmune hemolytic anemia	Rh blood-group antigens	Destruction of red blood cells
Autoimmune thrombocytopenic purpura	Platelets	Abnormal bleeding
Goodpasture's syndrome	Basement membrane collagen	Glomerulonephritis, pulmonary hemorrhage
Acute rheumatic fever	Streptococcal cell wall antigens	Arthritis, myocarditis, scarring of the heart valves
Systemic lupus erythematosus	DNA, histone proteins, ribosomes	Glomerulonephritis, vasculitis, rash
Rheumatoid arthritis	Unknown synovial joint antigen	Joint inflammation and destruction
Insulin-dependent diabetes mellitus	Pancreatic β-cell antigen	β-cell destruction
Multiple sclerosis	Myelin basic protein	Brain invasion by CD4 T cells

Table 17.8 Autoimmune Diseases Classified by Immunopathogenic Mechanism

Syndrome	Antigen	Consequence
Graves' disease	Thyroid-stimulating hormone receptor	Hyperthyroidism
Myasthenia gravis	Acetylcholine receptor	Progressive weakness
Insulin-resistant diabetes mellitus	Insulin receptor	Hyperglycemia, ketoacidosis

Table 17.9 Autoimmune Diseases Caused by Autoantibodies

damaging tissues: this is what happens in insulin-dependent diabetes mellitus (**Figure 17.11**), rheumatoid arthritis, and multiple sclerosis.

Keep in Mind

- In autoimmune reactions, the immune system reacts against the host.
- Tolerance to self antigens prevents the development of autoimmune disease.
- Many autoimmune diseases are the result of production of autoantibodies, but multiple parts of the immune response are usually involved.
- Specialized regulatory T cells (Treg cells) control the development of an autoimmune response.
- Some people are genetically predisposed to autoimmune disease.
- Autoimmunity can be organ-specific such as type 1 diabetes mellitus, which affects cells in the pancreas, or systemic, affecting many organs.

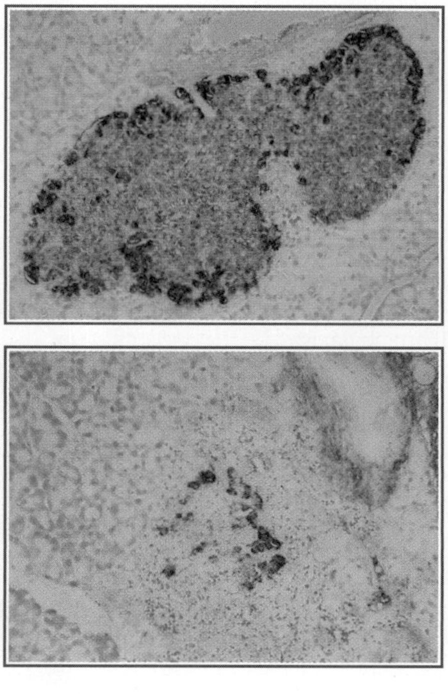

Figure 17.11 Tissue destruction in insulin-dependent diabetes mellitus, an autoimmune disease. Pancreatic cells have been stained to make insulin molecules visible (dark brown). Top: Cells from a healthy person, with the many dark brown spots indicating normal levels of insulin-producing cells. Bottom: Cells from a diabetic patient. Notice the absence of dark brown, indicating a loss of insulin-producing cells.

HYPERSENSITIVITY (ALLERGIC REACTIONS)

An allergic reaction to some substance such as pollen, peanuts, dust-mite feces, or animal dander is a form of immune response that is usually not life-threatening but can sometimes produce serious tissue injury and, in asthma and anaphylaxis, even death. Allergies can cause discomfort and distress for the patient as well as lost time from work or school. Allergic responses are the result of a person's becoming hypersensitized to a particular substance. In fact, the terms hypersensitivity and allergy mean the same thing.

There are four main types of hypersensitivity reaction. Type II hypersensitivity reactions are associated with sensitivity to drugs. In this case, drugs bind to the surface of cells and serve as targets for anti-drug IgG antibodies that cause destruction of the cells. This occurs in only a small percentage of the population, and the reason it occurs is still unclear.

Type III hypersensitivity occurs with soluble antigens that combine with antibodies to form immune complexes. These complexes bind to mast cells and other leukocytes and create a localized inflammatory response with increased vascular permeability at the site. This called an **Arthus reaction**. Type III reactions are also seen systemically in cases of serum sickness. Serum sickness occurs if large amounts of foreign antigen are injected into the host. Examples of this problem are seen when horse serum containing antibody against a pathogen is used as a treatment for infection. This type of hypersensitivity can also be seen when antigen is not effectively cleared by the host's immune response. For example, in subacute bacterial endocarditis or chronic viral hepatitis, pathogens are constantly generating new antigens and the adaptive immune response fails to clear them completely from the system. This causes immune complex disease with injury to small blood vessels in many tissues and organs.

Type IV hypersensitivity reactions are not mediated by antibody but result from the reactions of antigen-specific T cells.

Here we focus our discussion on type I allergic reactions. This type of allergic reaction occurs when the immunoglobulin IgE responds to antigens called **allergens**. These allergens trigger the activation of IgE-binding cells (mast cells and basophils) located in tissues exposed to these allergens. Some common type I allergic reactions are shown in **Table 17.10**.

Type I allergic reactions are mediated by IgE antibody molecules. These antibodies are produced by plasma cells located in lymph nodes that drain the site of allergen entry, and also by plasma cells in inflamed tissue. Certain antigens and routes of antigen presentation favor the production of

Table 17.10 Examples of IgE-Mediated Allergic Reactions

Syndrome	Common Allergens	Route of Entry	Response
Systemic anaphylaxis	Drugs, serum, venoms, peanuts	Intravenous directly or after oral absorption	Edema, increased vascular permeability, tracheal occlusion, circulatory collapse, death
Acute urticaria (wheal-and-flare)	Animal hair, insect bites, allergy testing	Through the skin	Local increases in blood flow and vascular permeability
Allergic rhinitis (hay fever)	Pollens, dust-mite feces	Inhalation	Edema and irritation of the nasal mucosa
Asthma	Dander (cat), pollens, dust-mite feces	Inhalation	Bronchial constriction, airway inflammation, increased production of mucus
Food allergy	Tree nuts, peanuts, shellfish, milk, eggs, fish	Oral	Vomiting, diarrhea, itching, hives, anaphylaxis (rare)

IgE, and allergens selectively stimulate the production of Th2 cells, which regulate the IgE response. Most human allergies are caused by inhaled small, water-soluble proteins carried on dry particles such as pollen grains or dust-mite feces (**Figure 17.12**). Once it comes in contact with the mucosa of the airways, the allergen is eluted from the carrier particle and diffuses into the mucosa, which allows presentation at a low dose.

In the respiratory mucosa, allergens encounter dendritic cells that take up and process allergens and become activated. This activation causes the dendritic cell to move to regional lymph tissue, where they present the allergens and produce the required co-stimulatory molecules that favor the production of Th2 cells.

It has been estimated that as many as 40% of people in the Western United States show a tendency to mount IgE responses to environmental allergens. This hypersensitivity to environmental allergens is referred to as **atopy**, and these individuals are called *atopic*. They have higher levels of eosinophils than normal and are more susceptible to such allergic diseases such as hay fever, rhinitis, and asthma. There is some evidence that this susceptibility may be genetic.

The prevalence of atopic allergies in general, and of asthma in particular, is explained by four environmental factors that predispose a person to atopy:

- Lack of exposure to infectious diseases in early childhood
- Environmental pollution
- Allergen levels
- Dietary changes

These predisposing factors led to the formulation of what is called the **hygiene hypothesis**, which suggests that a less hygienic environment can help protect against atopic allergies. For instance, early population of the gut by commensal bacteria such as lactobacilli or infection by gut pathogens is associated with a reduced prevalence of atopic allergies. In contrast, there is some evidence that children who have had attacks of bronchiolitis associated with viral infection are more prone to later development of asthma.

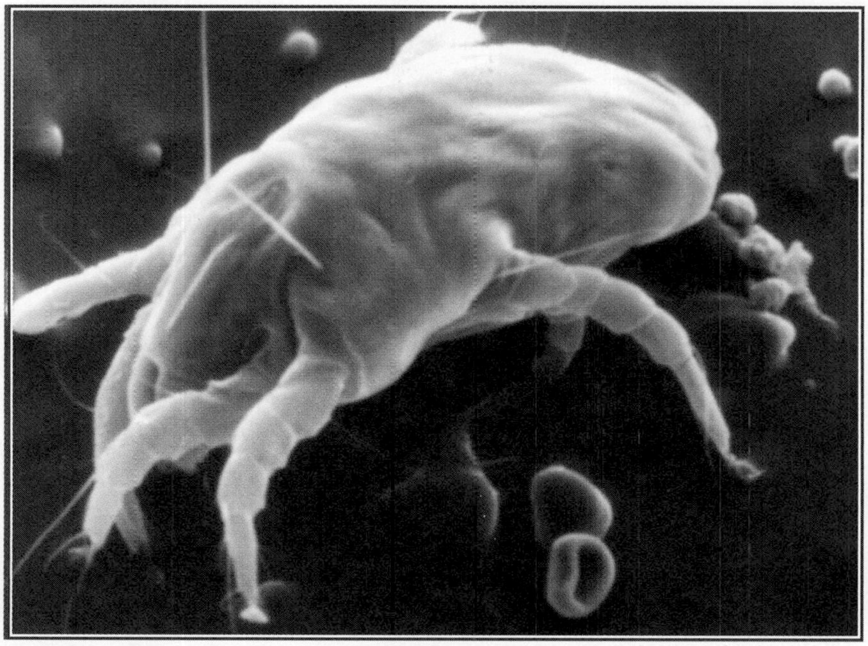

Figure 17.12 Scanning electron micrograph of *Dermatophagoides pteronyssimus* with some of its fecal pellets.

Effector Mechanisms in Allergic Reactions

Once a person has been exposed to an allergen, mast cells induce inflammatory reactions by secreting mediators and by synthesizing prostaglandins, leukotrienes, and cytokines. Symptoms range from the sniffles seen in hay fever to the life-threatening circulatory collapse seen in systemic anaphylaxis (**Figure 17.13**). The immediate allergic reaction is followed by a more sustained response that leads to the recruitment of more effector cells that contribute to the immunopathology.

Most IgE is found on the surface of mast cells, basophils, and activated eosinophils and is bound there by high-affinity receptors. Mast cells are highly specialized, prominent residents of mucosal and epithelial tissues. They are located in the vicinity of small blood vessels and venules, where they guard against invading pathogens and continuously express the IgE receptor on their surface. When a person previously sensitized to an allergen is exposed to it again, the allergen cross-links with an IgE molecule bound to a receptor on a mast cell. This causes degranulation of the mast cell, and within seconds a variety of inflammatory mediators are released.

Some of the mediators cause immediate increases in local blood flow, vessel permeability, and enzyme production leading to tissue destruction. Others contribute to smooth muscle contraction, increased secretion of mucus, and the influx and activation of leukocytes.

Eosinophils are also part of the allergic reaction; because of their destructive potential, they are usually under tight control. They are found in tissues, especially the connective tissue lying immediately under the respiratory, digestive, and genitourinary epithelia. Eosinophils have two functions in allergic reactions. Activation induces them (1) to release highly toxic granules and free radicals that kill microorganisms and cause significant tissue damage in allergic reactions and (2) to produce chemical mediators, prostaglandins, leukotrienes, and cytokines that enhance and amplify the allergic reaction. Eosinophils are strictly regulated at the level of origination in the bone marrow, and few are even produced unless there is an infection. They are also regulated at the site of infection, and the receptor for IgE is never present until the eosinophil is activated.

Fast Fact

Both eosinophils and basophils cause inflammation and tissue damage in allergic reactions and induce mast-cell degranulation.

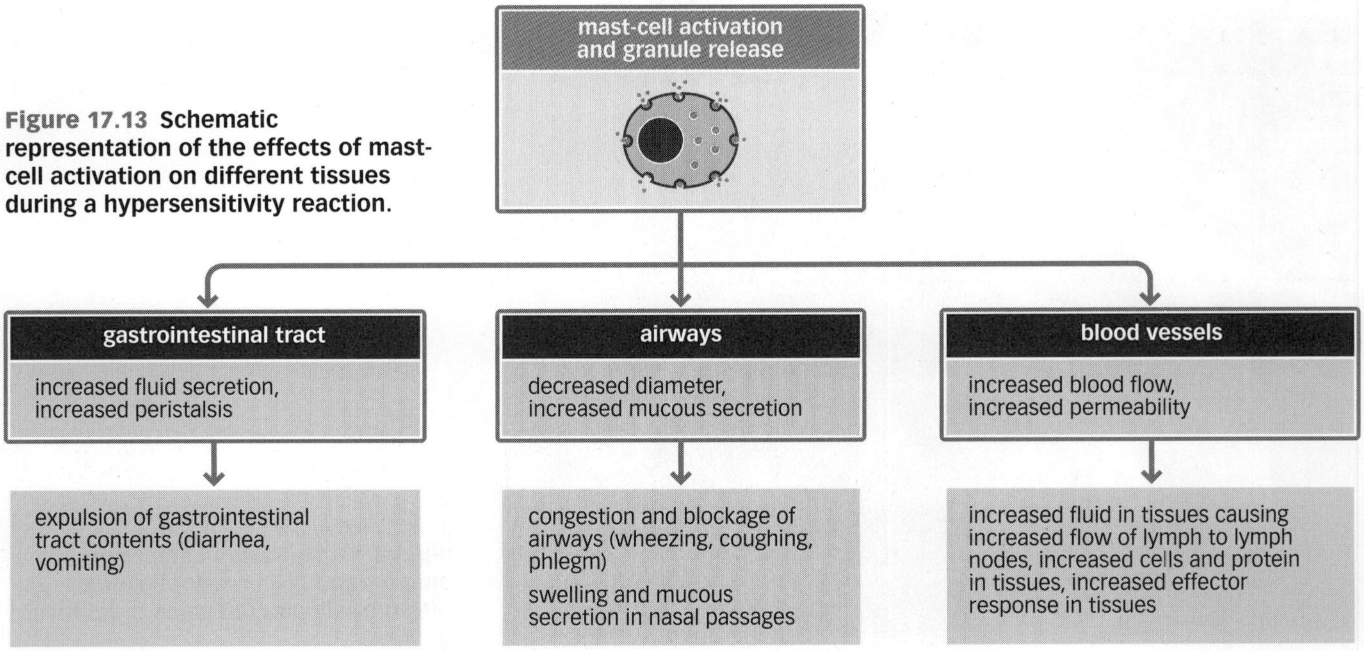

Figure 17.13 Schematic representation of the effects of mast-cell activation on different tissues during a hypersensitivity reaction.

Phases of Allergic Reactions

Allergic reactions can be divided into immediate and late-phase responses. IgE-mediated mast cell activation is an example of the immediate reaction that occurs within seconds after a person has been exposed to an allergen. In contrast, the late-phase reactions can take 8–12 hours to develop. The immediate reaction is due to the release of histamine and preformed mediators. Late-phase reactions are due to elaboration of the mediators (such as leukotrienes, chemokines, and cytokines) associated with smooth muscle contraction, sustained edema, and tissue remodeling. As an example, the wheal-and-flare reaction seen in positive skin tests for allergies has an immediate phase that is seen within a minute and lasts for about 30 minutes, whereas the late-phase reaction occurs about eight hours later in the form of a widespread edematous reaction that can persist for hours (**Figure 17.14**).

Clinical Effects of Allergic Reactions

The clinical effects of allergic reactions vary according to the site of mast cell activation. This depends on three variables: (1) the amount of allergen-specific IgE present in the mast cells, (2) the route by which the allergen is introduced into the body, and (3) how much allergen enters the body. If allergen is either deposited directly into the blood or rapidly absorbed from the gut, connective-tissue mast cells associated with the blood vessels become activated and the result is systemic anaphylaxis. This is a potentially fatal reaction caused by a widespread increase in vascular permeability, leading to a catastrophic loss of blood pressure, airway constriction causing breathing difficulties, and swelling of the epiglottis, which can cause suffocation. This is referred to as **anaphylactic shock** and can be seen in people who are allergic to substances such as penicillin and peanuts. An immediate injection of epinephrine (adrenaline) is required to control this reaction.

Allergic rhinitis is characterized by intense itching, sneezing, and a local edema that leads to blocked nasal passages and nasal discharge. A similar reaction is seen in **allergic conjunctivitis**, the condition that results when allergens enter the eye. In contrast, *allergic asthma*, which is more serious, is triggered by the activation of submucosal mast cells in lower airways. This leads to immediate bronchial constriction (in seconds) and an increased production of fluids and mucus, which makes breathing difficult and can be life-threatening. Asthma is due to a chronic inflammation of the airways that results in the continued presence of increased numbers of Th2 T cells, neutrophils, and other white blood cells.

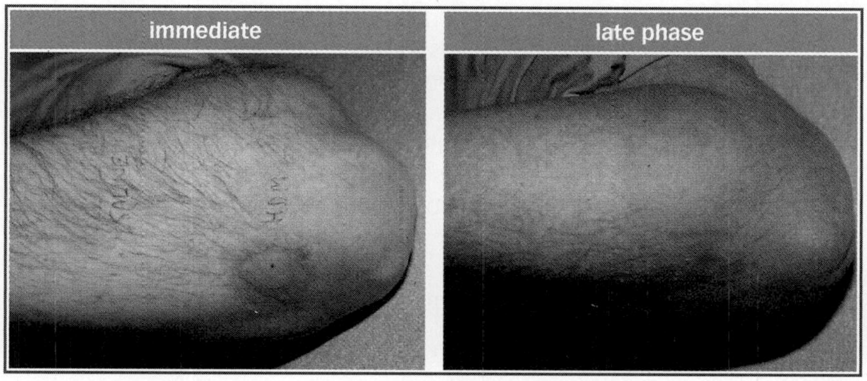

Figure 17.14 Early and late phases of an allergic reaction. The immediate wheal-and-flare reaction (left) is seen within a minute or two after antigen is injected under the skin and lasts for up to 30 minutes. The late-phase edematous reaction (right) develops approximately eight hours later.

Keep in Mind

- An allergic response is a type of immune response.
- This response is generated against allergens (antigens) such as peanuts, pollen, dust-mite feces, or animal dander.
- Allergic responses are usually not life-threatening but can sometimes cause anaphylaxis, a condition that can be fatal.
- There are four types of hypersensitivity response.
- Approximately 40% of people in the western United States show a tendency to mount IgE responses to environmental allergens.

SUMMARY

- The immune system can break down and in some cases fail. It can also turn against the host.
- There are three ways the adaptive immune response can fail: viral infection (HIV) leading to AIDS, subversion of the immune response by pathogens, and genetic immunodeficiency.
- HIV is a latent virus that carries the enzyme reverse transcriptase. This enzyme allows the viral RNA to be converted to DNA, which can integrate into a host cell chromosome.
- There are three strategies that are used to subvert or defeat the host defense: antigenic variation, resistance to host defense, and suppression of the immune response.
- Resistance to host defense can involve several mechanisms including the inhibition of phagocytosis.
- Primary immunodeficiency involves genetic abnormalities.
- Autoimmune disease occurs when the adaptive immune response attacks the host.
- Tolerance to self antigens prevents the development of autoimmune disease.
- Specialized regulatory T cells (Treg cells) control the development of an autoimmune response.
- Autoimmunity can be organ-specific or systemic.
- An allergic response is a type of immune response against antigens that are referred to as allergens.
- There are four types of hypersensitivity response.

In this chapter, we have examined ways in which our immune system can fail. We have seen that failure of host defense can result from infections with HIV, subversion of the defense mechanisms involving strategies developed by certain pathogens, and genetic diseases in which parts of the immune response do not develop. We also have looked at what can happen when the immune response mistakenly reacts against self in the form of autoimmunity and the damage that can result from the defense mechanisms responsible for allergies. Even in view of the problems that can occur, we are universally better off with an immune system than we would be without one. Given the number of pathogens that attack us on a regular basis and require our host defenses to clear the infections, it is amazing how seldom we have to worry about the failures of our immune system.

SELF EVALUATION AND CHAPTER CONFIDENCE

Multiple Choice

Answers are given in the back of the book and help can be found in the student resources at:
www.garlandscience.com/micro

1. All of the following are examples of failures of the immune response except
   A. Autoimmunity
   B. Primary immunodeficiency
   C. Acquired immunodeficiency syndrome (AIDS)
   D. Inflammation
   E. All of the above
2. HIV makes its entry through all of the following except
   A. The mucosal system
   B. A needlestick injury
   C. The respiratory system
   D. M cells of the digestive system
3. The number of T cells drops to below 200 per cubic millimeter in which phase of the HIV infection?
   A. The acute phase
   B. The developmental phase
   C. The symptomatic phase
   D. The asymptomatic phase
   E. None of the above
4. The major mode of HIV transmission is
   A. Intravenous drug abuse
   B. Sexual transmission
   C. Transplacental transfer
   D. Breast feeding
   E. Blood transfusion
5. The initial symptom of HIV infection is
   A. Loss of T cells
   B. Lymphadenopathy
   C. Appearance of thrush
   D. Significant weight loss
   E. High viral load
6. Antigenic variation is caused by which of the following?
   A. Bacterial endotoxins
   B. B cell lymphoma
   C. Antigenic drift
   D. Viral exotoxins
   E. All of the above
7. Latency is most often seen in
   A. Bacterial infections
   B. Fungal infections
   C. Viral infections
   D. Patients with suppressed immune responses
   E. Latency is seen in all infections
8. Pathogens can resist host defenses by all of the following except
   A. Resistance to phagocytosis
   B. Enhancing the fusion of phagosomes to lysosomes
   C. Escaping from the phagosome
   D. Growing inside the phagosome
9. Primary immunodeficiencies include all of the following except
   A. Severe combined immunodeficiency disease (SCID)
   B. DiGeorge syndrome
   C. Myasthenia gravis
   D. Wiskott–Aldrich syndrome
   E. Chronic granulomatous disease
10. Defects in antibody production can result in
    A. Increased fungal infections
    B. Infrequent infections with non-encapsulated bacteria
    C. Frequent infections with non-encapsulated bacteria
    D. Infrequent infection with capsulated bacteria
    E. Frequent infection with capsulated bacteria
11. Autoimmune diseases include all of the following except
    A. Chronic granulomatous disease
    B. Myasthenia gravis
    C. Multiple sclerosis
    D. Systemic lupus erythematosus
12. Organ-specific autoimmune diseases include all of the following except
    A. Graves' disease
    B. Goodpasture syndrome
    C. Type 1 diabetes
    D. Systemic lupus erythematosus
    E. Myasthenia gravis
13. Autoimmunity is prevented by which of the following?
    A. The innate response
    B. Tolerance
    C. Amplification
    D. Antibody
    E. Phagocytic cells
14. Which of the following immunoglobulin molecules is associated with hypersensitivity and allergy?
    A. IgG
    B. IgA
    C. IgH
    D. IgE
    E. None of the above
15. Which cells are most prominently involved in hypersensitivity reactions?
    A. Mast cells
    B. Neutrophils
    C. Monocytes
    D. Eosinophils
    E. None of the above

DEPTH OF UNDERSTANDING

Questions listed here require you to bring together the concepts you have learned in this chapter into a discussion format. This helps you to increase your depth of understanding of the material you have learned. Help can be found in the student resources at: www.garlandscience.com/micro

1. Compare and contrast acquired immunodeficiency, primary immunodeficiency, and subversion of the host defenses, and discuss their benefit to pathogenic microbes.

2. Discuss the role of T cells during the acute, asymptomatic, and symptomatic phases of HIV infection.

CLINICAL CORNER

Help can be found in the student resources at: www.garlandscience.com/micro

1. You are an emergency medical technician and are sitting in a restaurant with your wife and children. The place is crowded and you can't help hearing the woman at the next table asking the waiter whether there are peanuts in the dish she is ordering. The waiter assures her that there are none. You are about to finish and ask for your check when you notice that the woman at the next table is choking and has a bluish tint to her skin. Immediately you sprint for your car and dig out an epinephrine syringe. The man sitting at the table with the woman is trying to lift her up saying that she needs the Heimlich maneuver. You push him out of the way and jab the needle into her thigh. Within minutes she is breathing more easily and her color is returning. The man is grateful but indignant that she would have been okay if he could have used the Heimlich maneuver.
   - **A.** Explain to him why he is wrong.
   - **B.** What do you think caused this to happen?
   - **C.** How could you prove your suspicions?
2. Your patient, Richard Parks, is a 26-year-old male who is in the clinic because of problems with lethargy and loss of vigor. He works in construction and recently has had no desire to even go to work. Although he eats several meals a day, he is losing weight and he is always thirsty. You have ordered blood work and it shows a high level of triacylglycerols and a glucose value of 135 / 100ml. You have to tell Mr Parks that he has type 1 diabetes.
   - **A.** Can you explain how this happened?
   - **B.** What will you tell him about the immune response and its role in this disease?

PART V

Control and Treatment

In the first three parts of this book, we concentrated on the disease process and on the characteristics of organisms that can infect us. In **PART IV**, we looked at our defense against infection. Here in **PART V**, we turn our attention to how humans use antiseptics, disinfectants, and antibiotics to control pathogens.

Chapter 18 describes how we can use disinfectants to limit the contamination of inanimate surfaces and antiseptics to limit contamination of the skin. A variety of chemical treatments can be used for these purposes, and we look at the major categories of these chemicals as well as drawbacks that may be associated with their use.

In **Chapter 19**, we look at antibiotics. This chapter is especially important because of our dependence on these chemical treatments for fighting infectious diseases. This chapter gives us a description of selective toxicity, a process that is required for all successful antibiotics. In addition, we examine the different categories of antibiotics and the specific targets that they attack. We divide our discussion into antibiotics that are available for bacterial, viral, fungal, and parasitic infections.

Chapter 20 is one of the most important chapters because it looks at antibiotic resistance, which may be the biggest problem in health care today. Here we examine the causes for resistance and the mechanisms used by pathogens to inhibit these treatments.

When you have finished with **PART V** you will:

- Have a working knowledge of how the growth of potential pathogens can be inhibited through the use of antiseptics and disinfectants.
- Have learned about the targets used for antibiotic therapy and the different types of antibiotics.
- Know a variety of antibiotics that are used for bacterial, viral, fungal, and parasitic infections, with an emphasis on how potentially toxic the antibiotics might be to patients.
- Understand the problems associated with the overuse of antibiotics and how antibiotic resistance occurs.
- Be familiar with the potential problems that result from pathogens becoming resistant to antibiotic therapy.

To understand these chapters well, you will need to depend on what you have learned in the first four parts of this book.

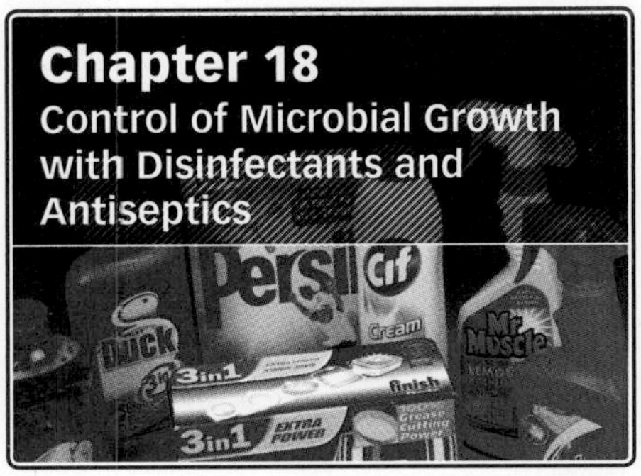

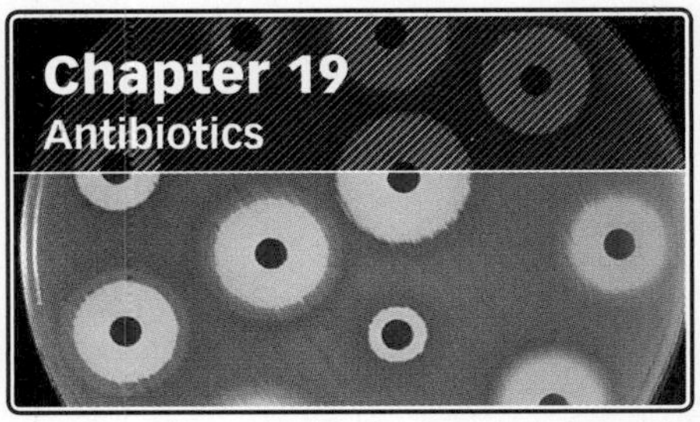

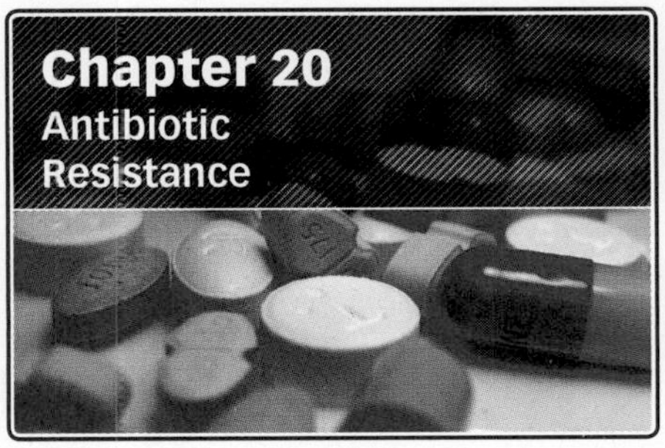

Control of Microbial Growth with Disinfectants and Antiseptics

Chapter 18

Why Is This Important?

Infection control through the use of disinfectants and antiseptics is essential to keep infections from spreading, particularly in the hospital setting.

OVERVIEW

As we saw in previous chapters, we humans live in constant contact with dangerous and potentially dangerous microorganisms. Because they can produce serious illnesses, it is important to know how we can defeat potential pathogens. In the chapters of PART IV we learned about our defense mechanisms. Here in PART V we will learn about methods we use to rid ourselves of infectious organisms. We begin here in Chapter 18 with a discussion of antiseptics, which are chemicals that we can use to prevent the growth of pathogens on our bodies, and disinfectants, which we commonly use to clean inanimate objects. We use disinfectants to keep down the level of microorganisms on those objects, thereby lowering the chances of our picking up some infection as we handle them. We will see that the targets for these chemicals are the same as those used for antibiotics, which will be discussed in Chapter 19.

We will divide our discussion into the following topics:

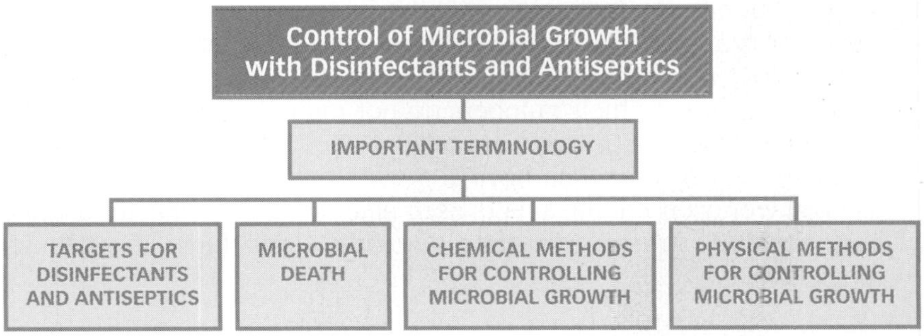

What Do I Need to Know?

To get the most out of this chapter, please review the following terms from your previous courses in biology, anatomy, physiology, or chemistry or in previous chapters of this book as indicated in parentheses: agar plate (**10**), denatured protein (**2**), halogen (**2**), hydrogen bond (**2**), hydrophilic/hydrophobic (**4**), hydroxyl radical (**2**), logarithmic growth curve (**10**), lysis (**12**), mutagen (**11**), nitrocellulose, nosocomial infection (**6**), oxidize (**3**), oxidizing agent, sepsis (**5**), thermophilic organism (**10**), virion (**12**).

IMPORTANT TERMINOLOGY

To enable us to understand how disinfectants and antiseptics can control the growth of microbes, it is important to become familiar with some of the terminology used to describe these substances and their use. Keep in mind that some treatments can be used for both *disinfection* and *antisepsis*, with the difference being that disinfection is associated with inanimate objects and antisepsis is associated with human tissue and skin. The following definitions give a basic understanding of the processes used for control of microbial growth.

- **Sterilization** involves the removal of all microbes, including bacteria, viruses, fungi, and spores. It should be noted that sterilization does not destroy prions, which are the infectious proteins we discussed in Chapter 8 that cause spongiform encephalopathies such as Creutzfeldt–Jacob disease.

- **Aseptic** is a term used to describe an environment or procedure that is free from contamination by pathogens. For example, surgeons and laboratory technicians use aseptic techniques to avoid contamination.

- Disinfection refers to the use of chemical or physical agents such as heat, alcohol, or ultraviolet radiation, which kill or inhibit the growth of microorganisms, including pathogens. It is important to note that some disinfecting agents do not affect spores. Therefore any organism that has the ability to form a spore can be very difficult to deal with. In some cases, the same chemical can be used on both an inanimate object, in which case it is referred to as a **disinfectant**, and on the skin and in tissue, in which case it is an **antiseptic**.

- **De-germing** is the removal of microbes from a surface by mechanical means. For example, the act of scrubbing when washing your hands is a de-germing mechanism. Another example is the use of an alcohol pad to prepare the skin for injection.

- Sanitization refers to the disinfection of places (restrooms) or items (dishes) used by the public. Sanitization is not sterilization, although the same techniques may be used: steaming; high-pressure, high-temperature washing; and scrubbing. Sanitization is used to reduce the number of pathogenic organisms present to a number that meets accepted public health standards.

- **Pasteurization** uses heat to kill pathogens and is most often seen in the food-processing industry. This method does not sterilize but is used to reduce the number of pathogens and also the number of organisms that can cause spoilage of the food. Milk, fruit juices, wine, and beer are routinely pasteurized.

- The two suffixes *-static* and *-cidal* are used on words that describe agents that either kill (-cidal) or inhibit the growth (-static) of organisms. For example, a *bacteriocidal* agent kills any bacteria exposed to it, but a *bacteriostatic* agent does not. When bacteria are in a bacteriostatic environment, their numbers do not multiply but the organisms do not die. Once the bacteria are taken out of the bacteriostatic environment, they resume their normal growth pattern. These suffixes are used on words describing agents used against bacteria (bacteriostatic, bacteriocidal), viruses (virustatic, virucidal), and fungi (fungistatic, fungicidal).

TARGETS FOR DISINFECTANTS AND ANTISEPTICS

With these terms in mind, let's examine the action of these agents. To do this, we have to think about the vulnerabilities of microorganisms. The discussion in Chapter 9 will be very helpful here. Recall that in that chapter we focused on the anatomy of bacterial cells. Because bacteria are single-celled organisms with a relatively simple anatomy, it is easy to see which targets in the cell could result in their death. For example, the cell wall of a bacterium is crucial for keeping the outside out and the inside in. Thus, damage to this structure could result in the death of the organism. The same is true for the plasma membrane of the bacterium, because this structure is responsible for enclosing the cytoplasm and is involved with DNA replication and ATP production. Therefore, loss or damage to the bacterial plasma membrane is a lethal event. Finally, any inhibition of protein synthesis or alteration in protein structure can result in disastrous consequences for the organism. As we will see, these targets are also some of the same ones that are used for antibiotic therapy (discussed in Chapter 19).

The Cell Wall

As mentioned above, the cell wall of bacteria maintains the integrity of the cell. Several chemical agents damage this barrier by inhibiting its synthesis, by digesting it, or by breaking down its structure. Any microorganism that loses the integrity of the cell wall becomes very fragile and susceptible to lysis.

The Plasma Membrane

This structure is composed of a phospholipid bilayer that also contains a variety of proteins. As we learned in Chapter 9, this membrane is selectively permeable, which means it has the ability to selectively allow some things to enter and leave the cell. When the plasma membrane of a bacterial cell is disrupted, the cell loses its selective permeability, and this change in permeability leads to the death of the cell.

Surfactants are chemicals that reduce the surface tension of solvents, such as water, by decreasing the attraction between molecules forming the surface layer of the solvent. Surfactants are very effective for disrupting the plasma membrane of cells because, like the phospholipid bilayer, surfactant molecules are also polar molecules that have hydrophobic and hydrophilic regions. Surfactant molecules bind to and penetrate the bilayer and cause openings to form in the plasma membrane, resulting in the lysis of the cell (**Figure 18.1**).

In addition, recall that some viruses have envelopes composed of the plasma membranes from host cells they have infected. Damage to this envelope can cause the virus to lose the capacity to infect a host cell.

Protein Structure and Function

Proteins are among the most important molecules in the microbial cell. They are responsible for structural elements of the cell and are also the cell's enzymes, which are important for the physiology and metabolism of the cell. Recall that proteins have a three-dimensional shape and that this shape is directly related to the function of the protein. If this shape is changed in any way, the protein is said to be *denatured*, and its function can be inhibited or eliminated, resulting in the death of the cell. Denaturation involves breaking of the hydrogen bonds and other bonds that hold the three-dimensional shape of the protein together. When these bonds are broken, the protein unfolds and is inactivated.

Heat and strong solvents, such as alcohols and acids, break the hydrogen bonds and can result in total denaturation and coagulation of the protein. In addition, metallic ions can inhibit enzymatic function by blockading the active site on the protein (**Figure 18.2**).

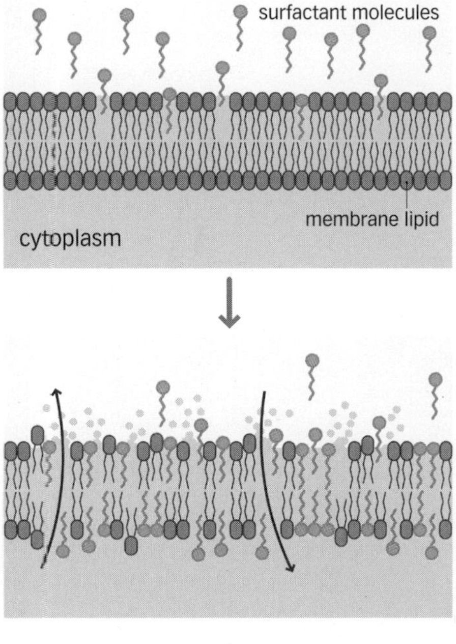

Figure 18.1 The mode of action of surfactants which affect the plasma membrane of cells by inserting into the lipid bilayer, disrupting it and creating abnormal channels that cause leakage from the cell and ultimate death.

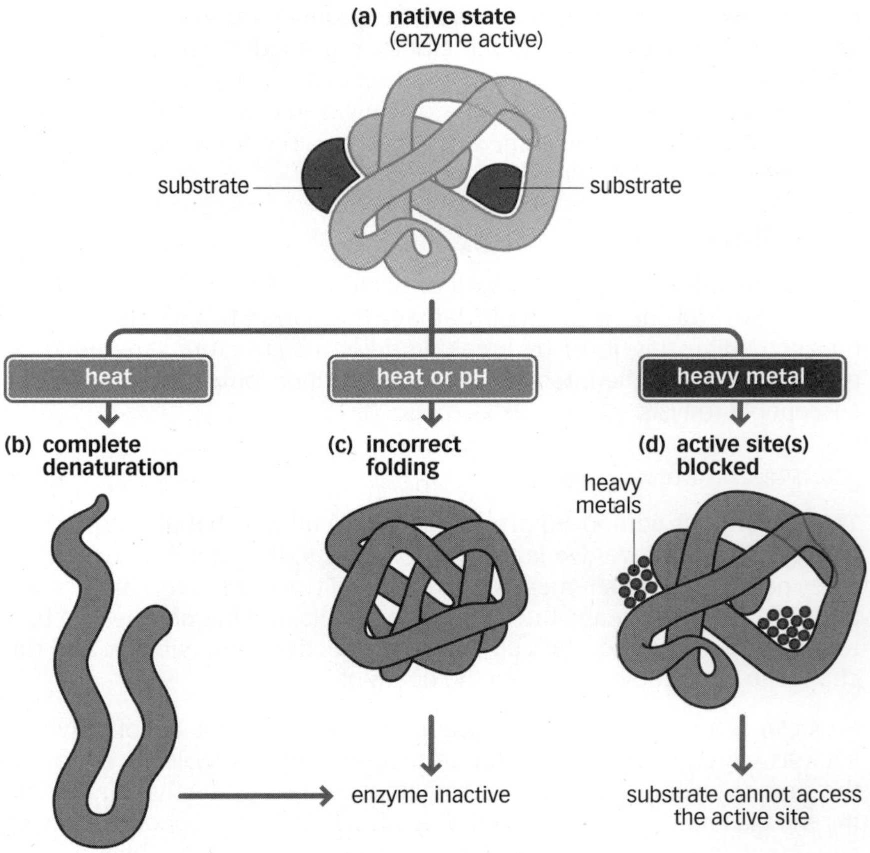

Figure 18.2 The effects of heat, pH, and heavy metals on protein structure. Panel **a**: The native functional configuration of the protein. Panel **b**: Denaturation of the protein, involving complete unfolding of the protein. Panel **c**: Incorrect folding of the protein. Panel **d**: Interference with the binding site of a protein enzyme.

Nucleic Acid Synthesis

Like proteins, nucleic acids are required for cell survival. The synthesis of DNA and RNA (reviewed in Chapter 11) is one of the most important functions of the cell, and any agent that can disturb this synthesis is a powerful antimicrobial compound. An **antimicrobial** is defined as a substance that inhibits the growth of a microbial organism. Some antimicrobial agents bind irreversibly to DNA, preventing gene expression (transcription and translation), whereas others are *mutagenic* and cause lethal mutations in gene sequences. Some chemicals, such as formaldehyde and ethylene oxide, interfere with DNA and RNA function, but the most powerful agent for disrupting nucleic acids is radiation. In point of fact, irradiation with gamma rays, ultraviolet radiation and X-rays causes mutations that can result in permanent inactivation of the nucleic acids and the death of the organism.

MICROBIAL DEATH

In reality, the death of a microbe such as a bacterium is hard to identify because there are no apparent signs of it. In complex organisms, the failure of a system necessary to the wellbeing of the organism is an indication of death, but in bacteria, which have simple structures, these systems do not exist. The difficulty in identifying bacterial death is compounded by the fact that many lethal compounds kill bacteria without changing what they look like under the microscope. Even a lack of movement by the bacteria cannot be a definitive sign of death. Therefore, it is necessary to develop special requirements to define microbial death.

The most efficient way to determine microbial death is to determine whether the organism can reproduce when moved from the antimicrobial environment to an environment that is suitable for growth.

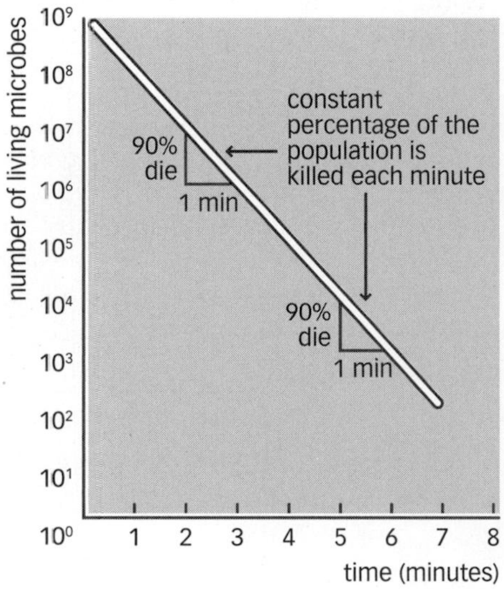

Figure 18.3 A plot of microbial death. The killing of organisms involves a constant percentage over time (in this case 90% per minute).

If the organism cannot reproduce even in the most suitable growth environment, it is dead. The permanent loss of reproductive capability even under optimum conditions has become the accepted definition of microbial death.

One of the techniques used to evaluate the efficacy of an antimicrobial agent is to calculate the microbial death rate, which is constant over time and under a particular set of conditions (**Figure 18.3**). You can see from the numbers on the vertical axis of the figure that the death rate is logarithmic (just like the logarithmic phase seen in the bacterial growth curve described in Chapter 10).

Factors That Affect the Rate of Microbial Death

There are several factors that can affect the death of microbes, particularly in a clinical setting.

- **Numbers.** The greater the number of organisms, the longer it will take to kill all of them. This is a matter of accessibility of the disinfectant or antiseptic to the organisms. If there are large numbers of organisms present, it will take time for the agent to reach each of the organisms.
- **Duration of exposure.** The duration of exposure can also vary depending on the accessibility of the agent to the organism as well the microbe in question. This time requirement also applies to radiation treatment, which can take a long period to be accomplished.
- **Temperature.** The lower the temperature at which the microbes are treated, the longer it will take to kill all of them.
- **The environment.** The environment in which the microbes to be treated exist is particularly important in health care because many pathogens will be associated with organic materials such as blood, saliva, bodily fluids, or even fecal material, which inhibit the accessibility of the antimicrobial agents to the organisms.
- **Endospore formation.** Of all the factors that affect the rate of microbial death, endospore formation may be the most important. Endospores are resistant to many of the agents routinely used to inhibit microbial growth, and spore-forming organisms can evade destruction by these agents. This can be especially important when considered in the context of nosocomial infections (discussed in Chapter 6).

Keep in Mind

- The cell wall, plasma membrane, proteins, and nucleic acids of microbes are targets for disinfectants and antiseptics.
- The permanent loss of reproductive capability has become the accepted definition of microbial death.
- Several factors can affect the rate of death, including the number of organisms, the duration of exposure, temperature, environment, and the ability to form a spore.

Figure 18.4 gives an overview of the methods used to control microbial growth. As you can see from this figure, there are three major methods of control: chemical, physical, and mechanical removal.

CHEMICAL METHODS FOR CONTROLLING MICROBIAL GROWTH

In this section we examine the use of chemicals as a means of controlling microbial growth. Many chemicals can kill microbes, but they can also be harmful to humans. Therefore, if they are to be useful as disinfectants or antiseptics, they must also be safe to use.

The Potency of Disinfectants and Antiseptics

There are several factors to consider when evaluating chemical methods for controlling microbial growth, because the effectiveness of antiseptics and disinfectants can be affected by such factors as time, temperature, and concentration. Let us agree to use the general term **chemical agent** when discussing a topic that is about both antiseptics and disinfectants. Using this general term, we can say that the death rate of organisms subjected to a chemical agent can be accelerated by, for example, increasing the temperature, as noted above. An increase of 10°C in the temperature at which a microbial specimen is treated will roughly double the rate of the chemical reaction, thereby increasing the potency of the chemical agent. The same holds true for changes in pH, which can either increase or decrease an agent's potency.

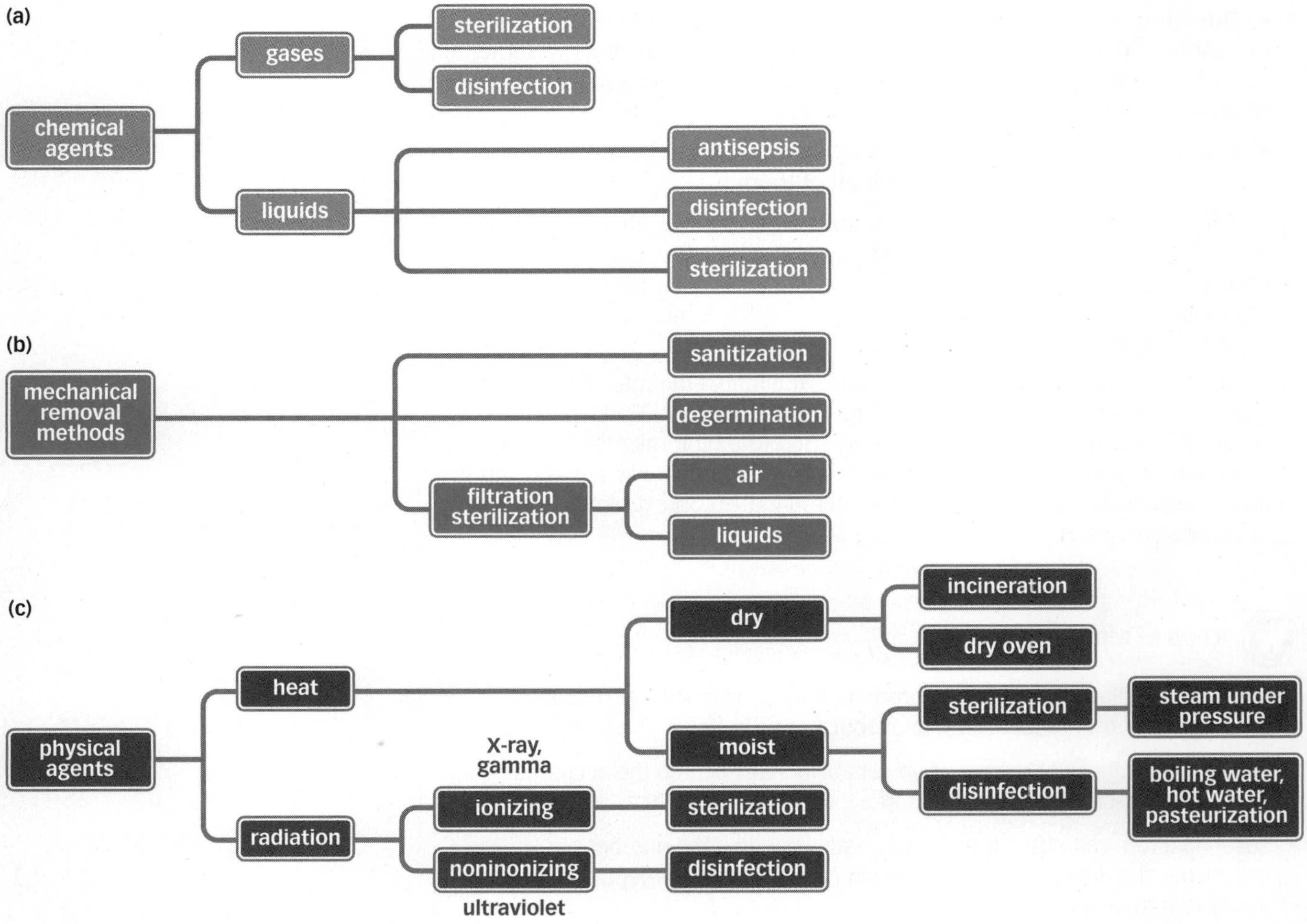

Figure 18.4 An overview of microbial control mechanisms. There are three methods that can be used: Chemical (Panel **a**), mechanical (Panel **b**), and physical (Panel **c**).

For most chemical agents, increases in the concentration of the agent will result in an increase in its potency: higher concentrations will be microbicidal, whereas lower concentrations will be microbistatic. This does not hold true for alcohol when used as an antimicrobial, however. In this case, an increase in the concentration does not increase the killing efficacy but actually hinders it. For alcohol to be effective, it must have some water associated with it for the effective penetration and denaturation of proteins. Consequently, 70% alcohol is better for microbial control than 99% alcohol.

Evaluation of Disinfectants and Antiseptics

There is no completely satisfactory method for evaluating antimicrobial chemical agents, but there are several tests that can be used, and we will now look at three of them. One is to compare the agent with phenol, a test known as the **phenol coefficient**. The second method is the *disk method*, which is a rapid test for efficacy that uses tiny disks of filter paper, and the last is called the **use dilution method**.

The Phenol Coefficient

Phenol was first used as a disinfectant by Joseph Lister in 1867 to reduce infection during surgery. It is still the benchmark disinfectant that others are compared with, and the comparison is reported as a phenol coefficient. Usually *Salmonella typhi* and *Staphylococcus aureus* are the two organisms on which the disinfectants are compared. A chemical agent with a phenol coefficient of 1.0 has the same effectiveness as phenol. Any number greater than 1.0 indicates an efficiency greater than that of phenol, and any number less than 1.0 indicates efficiency less than that of phenol. Phenol coefficients are reported separately for the two test organisms. As an example, the household disinfectant Lysol® has a coefficient of 5.0 against *Staphylococcus aureus* but only 3.2 against *Salmonella typhi*, whereas ethyl alcohol has a phenol coefficient of 6.3 against both.

The Disk Method

This simple method for determining the efficacy of a chemical agent against a particular microbe uses tiny disks of filter paper soaked in the agents to be tested. First an agar plate is inoculated with the microbe against which the various agents are being tested. Then the disks, each soaked in a different antimicrobial agent, are placed at various positions on the inoculated agar. Inhibition of growth around each disk (or lack thereof) is easily identifiable by a clear zone called a **zone of inhibition** (**Figure 18.5**). The sizes of the zones seen with different chemicals are not comparable because they may reflect differences in concentration of the chemicals and also differences in the diffusion rates of the molecules (larger chemical molecules on one disk move away from their disk more slowly than smaller chemical molecules on another disk move away from their disk, resulting in a smaller zone of inhibition around the disk containing the larger molecule). This method also cannot distinguish between a microbicidal agent and one that is microbistatic. In both cases there is inhibition of growth during the disk test, but if you were to subculture the organisms from the zone of inhibition surrounding the microbistatic agent, the organisms would resume growing.

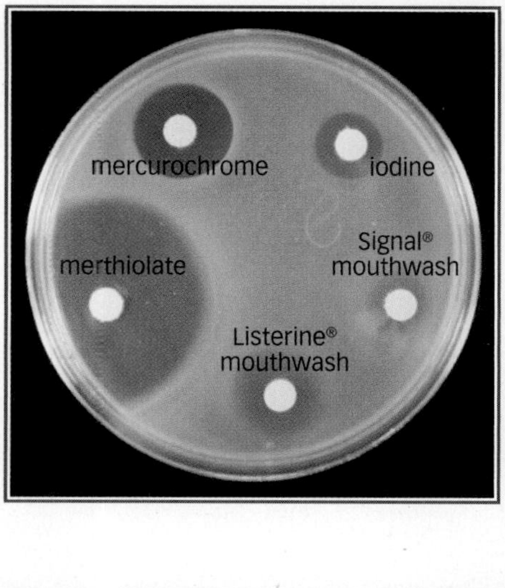

Figure 18.5 The disk method for evaluating potential antimicrobial agents. Notice the clear zones (called zones of inhibition) around some of the disks. The size of the zone of inhibition around a disk depends on the concentration of the agent and on the rate at which it diffuses away from the disk.

The Use Dilution Method

This evaluation method is more time-consuming than the two just described. However, it can tell you whether an antimicrobial agent is bacteriostatic or bacteriocidal, and it uses standardized preparations of certain bacteria. A series of solutions of different concentrations of the disinfectant are prepared, ranging from very dilute to very concentrated. Broth cultures of the test organism are dried down on stainless steel cylinders, and each cylinder is dipped for 10 minutes into one of the solutions. The cylinders are then removed from the solutions, rinsed with water to remove any remaining chemical agents, and placed into a tube of growth medium. The tubes are incubated and observed for the presence or absence of bacterial growth.

Fast Fact

Many microbiologists feel that the use dilution method is a more meaningful test than the phenol coefficient test.

Selecting an Antimicrobial Agent

Many chemicals are tested as possible antimicrobial agents, but it is important to remember that many of them have distinct and inherent potential side effects that would disqualify them from this use. In addition, some chemical agents are better for certain uses than others. In general, there are several qualities that are considered when deciding which antimicrobial agent to use:

- It must be effective against all types of infectious organisms without destroying tissues.
- It should be effective even in the presence of organic materials, such as bodily fluids, blood, and fecal material.
- It should be stable and, if possible, inexpensive.

As we discussed above, there are several ways to inhibit the growth of microbial organisms, such as destruction of the cell wall or plasma membrane, denaturation of proteins, and inhibition of nucleic acid synthesis. Antimicrobial chemical agents — both disinfectants and antiseptics — kill by participating in all these types of reaction. There are many different chemicals that are used for disinfection and antisepsis (**Table 18.1**), and these can be grouped into three types: those that affect proteins, those that affect membranes, and those that affect viruses.

As we discussed above, proteins are some of the major structural and enzymatic molecules of the cell. They can be denatured by destroying their three-dimensional shape, and this destruction of shape destroys their functional capability. It is important to note that mild treatments, such as low temperature or treatment with dilute acids or alkalis, denature proteins only temporarily, and when the agents are removed (or the temperature is allowed to return to room temperature), the protein will refold to its normal structure and be functional once again (**Figure 18.6**). Consequently, most antimicrobial chemical agents are used in strong enough concentration and for periods that guarantee permanent denaturation of proteins, which is a lethal event for the organism.

Chemical agents can also have a role in the control of viral pathogens by inactivating the ability of the virus to infect or to replicate. This can be accomplished by either destroying the proteins associated with the virion or destroying the replication, or gene expression, of the virus such that it can no longer replicate. Detergents, alcohols, and other agents that denature proteins can affect the protein found on the capsid of virions, and the envelopes of viruses are susceptible to agents that act on lipids. Furthermore, alkylating agents, such as ethylene oxide and nitrous acid, can act as mutagens for viral nucleic acid, thereby inhibiting the replication and proliferation of virions.

Fast Fact

Some viruses are difficult to inactivate even after being exposed to chemical agents and can remain infective after treatment.

Agent	Use	Mode of Action
Antiseptics		
Alcohol (60–85% ethanol or isopropanol in water)[a]	Skin	Lipid solvent and protein denaturation
Phenol-containing compounds (hexachlorophene, triclosan, chloroxylenol, chlorhexidine)	Soaps, lotions, cosmetics, body deodorants	Disrupts cell membrane
Cationic detergents, especially quaternary ammonium compounds (benzalkonium chloride)	Soaps, lotions	Interact with phospholipids of membrane
Hydrogen peroxide (3% solution)	Skin	Oxidizing agent
Iodine-containing iodophor compounds in solution (Betadine®)	Skin	Iodinates tyrosine residues of proteins
Silver nitrate	Eyes of newborn infant to prevent blindness due to infection by *Neisseria gonorrhoeae*	Protein precipitant
Disinfectants and sterilants		
Alcohol (60–85% ethanol or isopropanol in water)[a]	Disinfectant and sterilant for medical instruments and laboratory surfaces	Lipid solvent and protein denaturant
Cationic detergents (quaternary ammonium compounds)	Disinfectant for medical instruments, food, and dairy equipment	Interact with phospholipids
Ethylene oxide (gas)	Sterilant for temperature-sensitive laboratory materials such as plastics	Alkylating agent
Formaldehyde	3–8% solution used as surface disinfectant, 37% (formalin) or vapor used as sterilant	Alkylating agent
Glutaraldehyde	2% solution used as high-level disinfectant or sterilant	Alkylating agent
Hydrogen peroxide	Vapor used as sterilant	Oxidizing agent
Iodine-containing iodophor compounds in solution (Wescodyne®)	Disinfectant for medical instruments and laboratory surfaces	Iodinates tyrosine residues
Ozone	Disinfectant for drinking water	Strong oxidizing agent
Peracetic acid	0.2% solution used as high-level disinfectant or sterilant	Strong oxidizing agent
Phenolic compounds	Disinfectant for laboratory surfaces	Protein denaturant

[a]Alcohols, hydrogen peroxide, and iodine-containing iodophor compounds can act as antiseptics, disinfectants, or even sterilants depending on concentration, length of exposure, and form of delivery.

Table 18.1 Antiseptics, disinfectants, and sterilants

Types of Chemical Agent

Chemical agents are used more than physical means (described below) for disinfection, antisepsis, and preservation. Keep in mind that chemical agents — both disinfectants and antiseptics — affect microbial cell walls, plasma membranes, proteins, or nucleic acids. Remember also that the effect of chemical agents varies with temperature, the length of exposure, the amount of contaminating organic material present, pH, and the concentration and stability (freshness) of the agent.

Chemical agents tend to either destroy or inhibit the growth of enveloped viruses, bacteria, fungi, and protozoans but have trouble dealing with protozoan cysts and bacterial endospores. There are eight major categories of chemical agents used as antiseptics and disinfectants: phenol and

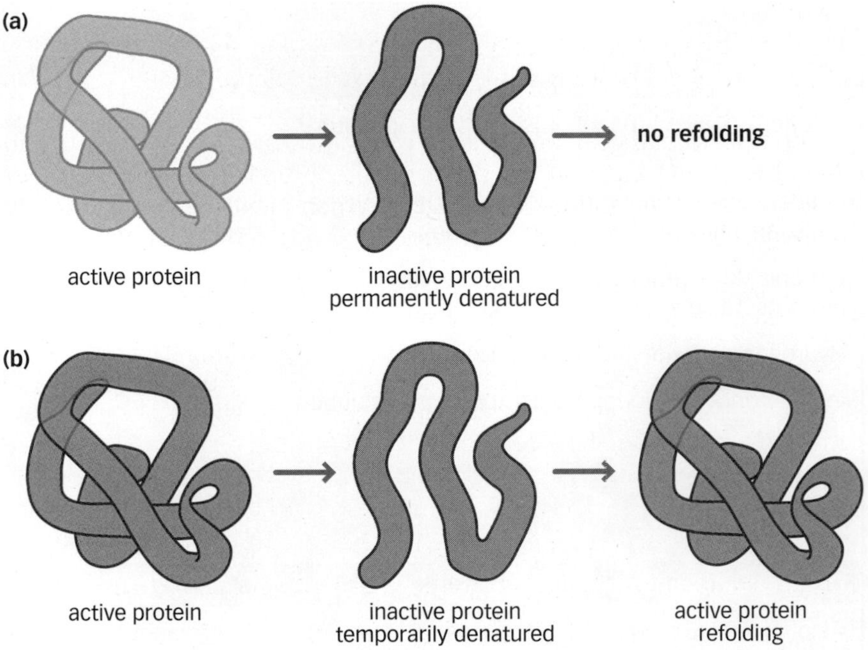

Figure 18.6 Denaturation of proteins can be either permanent or temporary. Panel **a**: Treatment with concentrated antimicrobial agents usually leads to a permanently denatured protein molecule. Panel **b**: Treatment with dilute antimicrobial agents can lead to denaturation that is only temporary. Once removed from the antimicrobial environment, the protein molecule reverts to its normal three-dimensional configuration.

phenolic compounds, alcohols, halogens, oxidizing agents, surfactants, heavy metals, aldehydes, and gaseous agents. (Some antimicrobial chemical agents combine one or more of these.) Let's look at these various categories one at a time.

Phenol and Phenolic Compounds

Phenolic compounds are derived from phenol by adding halogens or organic functional groups to the phenol ring (**Figure 18.7**). Many of these derivatives have greater efficacy and fewer side effects than phenol. In fact, natural oils such as pine oil and clove oil are phenolic compounds that can be used as antiseptics.

Bisphenolics are composed of two covalently linked phenolic compounds (the prefix *bis-* means "two"). The bisphenolic *ortho*-phenylphenol is the active ingredient in Lysol, and the bisphenolic triclosan is found in diapers, garbage bags and cutting boards. Phenols and phenolic compounds are low-level to intermediate-level disinfectants and antiseptics that denature proteins and disrupt the plasma membrane of microbial cells. More importantly, they are very effective in the presence of organic material, such as vomit, pus, saliva, and feces, and can remain active for prolonged periods. They are therefore commonly used as disinfectants in health care settings and laboratories.

Fast Fact

The bisphenolic hexachlorophene was once a popular disinfectant in houses and hospital nurseries but was found to cause brain damage in infants. It is now available only by prescription and is used in hospital nurseries only in cases of severe staphylococcal contamination.

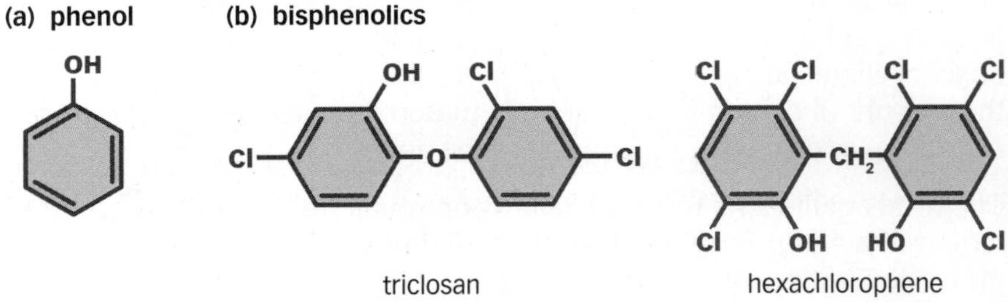

Figure 18.7 Phenolic compounds. Panel **a**: The phenol molecule. Panel **b**: Bisphenolics are molecules synthesized from the phenol molecule; they have greater antimicrobial efficacy than phenol and fewer side effects.

Alcohols

Alcohols are bacteriocidal, fungicidal, and virucidal for enveloped viruses but have no effect on fungal spores and bacterial endospores. Alcohols are intermediate-level disinfectants, with the most commonly used being isopropanol (rubbing alcohol, technically known as propan-2-ol) and ethanol (drinking alcohol). Of these two, isopropanol is slightly better than ethanol as an antimicrobial agent. When alcohol is used to carry other antimicrobial chemicals, the solution is referred to as a **tincture**.

Alcohol denatures proteins and also disrupts the plasma membrane of microbial cells. It has the benefit of evaporating, thereby leaving no residue, and it is routinely used as a de-germing agent to prepare sites for injection. The evaporative effect of alcohol can be a problem if the liquid evaporates before there has been sufficient time to kill all of the organisms in the area being treated.

Halogens

Four members of the chemical family known as *halogens* have antimicrobial activity: iodine, chlorine, bromine, and fluorine. These are intermediate-level antimicrobial chemical agents that are effective against bacteria and fungal cells, fungal spores, some bacterial endospores, protozoan cysts, and many viruses. They can be used alone or in combination with other elements, but the mechanism of action of halogens is still not completely understood. It is thought that they oxidize and/or denature enzymatic proteins in the microbes they attack.

Iodine is a well-known antiseptic and is used medically as a tincture or as an **iodophor**, which is an organic compound that incorporates iodine in such a way that the iodine is slowly released from the organic molecule during a chemical reaction. The advantage of iodophor antiseptics over tincture of iodine is that the former are longer lasting and non-irritating to the skin. **Betadine** is an example of an iodophor and is routinely used to prepare skin for surgery and injection as well as to treat burns.

Chlorine is found in drinking water, in swimming pools, and in wastewater from sewage treatment plants. It is also a major ingredient in disinfectants such as chlorine bleach and is used to disinfect kidney-dialysis equipment. **Chloramines**, which are combinations of chlorine and ammonia, are used in wound dressings, some skin antiseptics, and some municipal water supplies. Chloramines are less effective than chlorine as disinfectants/antiseptics, but they release their chlorine atoms more slowly and therefore last longer.

Oxidizing Agents

Oxidizing agents are high-level disinfectants and antiseptics that prohibit bacterial metabolism. They are very effective against infections of deep tissues because they release hydroxyl radicals ($\bullet OH^-$), which kill anaerobic organisms. Consequently, oxidizing agents are routinely used in deep puncture wounds. The three most commonly used are hydrogen peroxide, ozone, and peracetic acid.

Hydrogen peroxide (H_2O_2) is a common household antiseptic, but contrary to popular belief it is not useful for the treatment of open wounds because the enzyme catalase, which is released from damaged tissues, quickly neutralizes it. As we learned in Chapter 10, aerobic and facultatively anaerobic organisms can make catalase, but the amount of peroxide used as a disinfectant is more than sufficient to overwhelm the bacterial production of this enzyme.

Ozone (O_3) is a very reactive form of oxygen generated when O_2 is exposed to electrical discharge. Instead of chlorine, some Canadian and European cities use ozone generated in this way for water treatment. However, ozone is expensive to produce and difficult to maintain at the proper concentration in water.

Peracetic acid (CH_3-CO-O-OH, the peroxide form of acetic acid) is an extremely effective sporicide that can be used to sterilize surfaces. It is not affected by organic contaminants and leaves no toxic residue. Peracetic acid is used by food processors and medical personnel to sterilize equipment.

Surfactants

There are two common surfactants: soap and detergents. One end of a soap molecule is composed of fatty acid and is therefore hydrophobic; the other end is ionic and is therefore hydrophilic. When you wash your hands, the hydrophobic end of the soap molecule breaks down oily deposits into tiny drops that easily mix with water and are washed away. Soaps are good de-germing agents but poor antimicrobial agents. However, they can be made more potent by adding antimicrobial chemicals such as triclosan.

Detergents are positively charged organic surfactants that are more soluble in water than soap. The most popular detergents are **quaternary ammonium compounds** (**QUATS**), which contain the ammonium cation NH_4^+ with the hydrogen atoms replaced by functional groups or hydrocarbon chains (**Figure 18.8**). QUATS are considered low-level disinfectants/antiseptics, but their advantage is that they are odorless, tasteless, and harmless to humans (except at high concentrations). They are therefore used in many industrial and medicinal applications, such as mouthwashes.

QUATS function by disrupting the plasma membrane of microbial cells, which causes the cell to lose essential internal ions such as K^+. QUATs are bacteriocidal (especially for Gram-positive bacteria), fungicidal, and virucidal against enveloped viruses. They are not useful for non-enveloped viruses, mycobacteria, or bacterial endospores, and they are inhibited by the presence of organic contaminants.

Fast Fact

Pseudomonas species grow very well in the presence of QUATS.

Heavy Metals

The ions of heavy metals such as arsenic, zinc, mercury, silver, and copper are inherently antimicrobial, and mercury and silver have been used in clinical situations. However, the use of mercury was abandoned when toxicity was discovered, and only silver is still occasionally used. The mechanism of action is through protein denaturation because these ions interact with the sulfur atoms in the amino acid cysteine. This amino acid forms disulfide linkages in proteins and helps to maintain their three-dimensional shapes (for a reminder see Chapter 2).

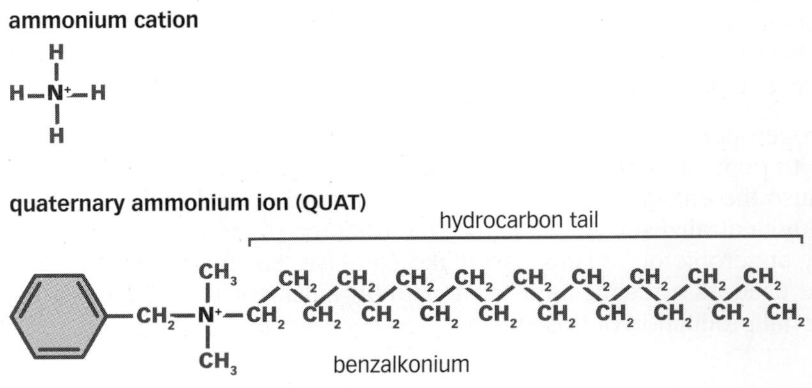

Figure 18.8 Quaternary ammonium compounds (QUATS) are surfactants in which the hydrogen atoms of ammonium ions are replaced by other functional groups.

Heavy metals are low-level bacteriostatic agents. At one time, it was routine to use silver nitrate cream to treat the eyes of newborn infants as a way to prevent blindness brought on by infection of the eyes with *Neisseria gonorrhoeae*, which could enter the eyes in the birth canal. Today, however, antibiotic creams, which are less irritating and can also kill *Chlamydia trachomatis*, are used instead for this purpose. Nevertheless, silver nitrate is still used in surgical dressings, burn creams, and catheters.

Aldehydes

Aldehydes are compounds containing a terminal –CHO group. There are two highly reactive aldehydes used as antimicrobials: glutaraldehyde, which is used in liquid form, and formaldehyde, which is used in both liquid form and gaseous form. Aldehydes function by cross-linking organic functional groups, and these cross-linking reactions denature proteins and also inactivate nucleic acids. Hospitals and research laboratories use 2% solutions of glutaraldehyde, which effectively kills bacteria, viruses, and fungi. In fact, treatments for 10 minutes will disinfect most objects, and treatment for 10 hours will sterilize. Glutaraldehyde is less toxic than formaldehyde but it is also more expensive.

Health care workers use a 37% formaldehyde solution called **formalin** to disinfect isolation rooms, exhausts, cabinets, surgical instruments, and dialysis machines. However, formaldehyde is an irritant for mucous membranes and has also been found to be carcinogenic.

Gaseous Agents

Many items cannot be sterilized with heat or chemicals (some examples are plastics, suture materials, heart–lung machine components, and dried or powdered foods). These items can be sterilized effectively by highly reactive antimicrobial and sporicidal gases such as ethylene oxide, propylene oxide, and β-propiolactone.

Gases rapidly penetrate and diffuse into any space, and over time (14–18 hours) they denature proteins and also DNA by cross-linking organic functional groups. In fact, they kill everything they come in contact with and do so without causing any damage to inanimate objects. Ethylene oxide is the most frequently used gaseous sterilizing agent and is found in hospitals and dentists' offices. Large chambers for use with ethylene oxide can be found in hospitals and are used for sterilizing instruments and equipment that might be sensitive to heat.

Fast Fact

Chlorine gas has been used to decontaminate large spaces such as the post offices that were contaminated during the anthrax event in 2001.

The primary disadvantages of using gaseous agents are that they are explosive, poisonous, and potentially carcinogenic. Disinfection with gas also takes a considerable time to complete because of the need for continuous cleanup.

Keep in Mind

- The potency of disinfectants and antiseptics can be affected by time, temperature, and concentration.
- The effects of disinfectants and antiseptics can be increased by increasing the temperature.
- Three tests — phenol coefficient test, disk method, and use dilution — are used to evaluate disinfectants and antiseptics.
- There are eight major categories of chemical agents used as antiseptics and/or disinfectants.

Type	Temp (°C)	TDT (min)
Non-spore-forming bacteria	58	28
Vegetative stage of spore-forming bacteria	58	19
Fungal spores	76	22
Yeasts	59	19
Viruses		
Non-enveloped	57	29
Enveloped	54	22
Protozoan trophozoites	46	16
Protozoan cysts	60	6

Table 18.2 Average Thermal Death Times (TDTs) for Selected Microorganisms

PHYSICAL METHODS FOR CONTROLLING MICROBIAL GROWTH

Physical methods for controlling microbial growth have been used for centuries. Ancient Egyptians dried food to preserve it, and Europeans used heat to can food 50 years before Pasteur worked out why heat was a necessary part of the process. In addition to drying and heating, there are many other physical methods used to control microbial growth, including cold, filtration, osmotic pressure, and radiation. Let's look at all of these in greater detail.

Heat

Elevated temperatures are usually lethal to most pathogenic microbes. Two types of heat are used to control microbial growth: moist and dry. Moist heat comes from hot water, boiling water, or steam, and the range of temperatures effective for killing microbes runs from 60°C to 135°C. Dry heat is hot air with low moisture content, such as the air in an oven. Temperatures for dry heat used as an antimicrobial range from 160°C to several thousand degrees. Hot air ovens are used for glassware, metallic instruments, powders, and oils. Exposure to a temperature between 150°C and 180°C for two to four hours ensures the destruction of spores as well as vegetative cells.

Moist heat can achieve the same level of effectiveness as dry heat in a much shorter time and at a lower temperature. Moist heat denatures proteins, which halts microbe metabolism and causes death. Dry heat dehydrates microbial cells, and the absence of water then inhibits metabolism. However, the loss of water can also stabilize some proteins present in microbes, and in these cases the object being treated must be exposed to the dry heat for a longer time so that the proteins are denatured.

When the temperature of dry heat is very high, cells are oxidized (burned to ashes). The flame of a Bunsen burner reaches 1870°C at its hottest point, for instance, and furnaces/incinerators operate at 800–6500°C. Direct exposure to such temperatures ignites and reduces microbes to ash and gases. An example of this is flaming the inoculating loop used to inoculate cultures so as to maximize aseptic technique.

Adequate sterilization with heat depends on both the temperature and the length of time for which the heat is used. Normally, high temperatures are associated with short treatment times. These two variables are used to calculate the **thermal death time** (TDT), which is defined as the shortest length of time needed to kill all organisms at a specific temperature (**Table 18.2** and **Table 18.3** show the TDTs for selected microorganisms

Organism	Temperature (°C)	TDT (min)
Moist heat		
Bacillus subtilis	121	1
Clostridium botulinum	121	10
Clostridium tetani	105	10
Dry heat		
Bacillus subtilis	121	120
Clostridium botulinum	120	120
Clostridium tetani	100	60

Table 18.3 Thermal Death Times (TDTs) for Various Endospores

Method	Characteristics and Typical Uses
Pressurized steam	Achieves temperatures above 100°C, which are necessary to destroy endospores. Used to sterilize surgical instruments and other items that can be sterilized by steam. Also used to process canned food
Boiling	Fast, reliable, and inexpensive way to destroy most bacteria and viruses. Does not affect endospores
Pasteurization	Significantly decreases the numbers of heat-sensitive microorganisms, including pathogens and those that spoil food. Does not affect endospores. Routinely used for milk, fruit juices, beer, and wine

Table 18.4 Moist Heat Methods for Controlling Microbial Growth

and bacterial endospores). Temperature and length of exposure time can also be used to calculate the **thermal death point** (TDP), which is defined as the lowest temperature required for killing all the organisms in a sample in 10 minutes.

There are three ways of using moist heat for controlling microbial growth: boiling, pressurized steam, and pasteurization (**Table 18.4**). When pressurized steam is used, it is not the pressure that kills, but the high temperatures that result from increased pressure. A good example of this is the **autoclave**, which is found in any health care facility. The autoclave is like a pressure cooker. It has a complex network of valves, ducts, and gauges that regulate and measure pressure and conduct steam into the pressurized chamber. Sterilization occurs when steam condenses to liquid water on the objects to be sterilized and the hot water gradually raises their temperature. Autoclaves routinely use pressures of 15 pounds per square inch, which yields a temperature of 121°C for the water, and are superior for sterilizing heat-resistant materials such as glassware, surgical dressings, some types of rubber glove, instruments, liquids, paper, some types of medium (but not all), and heat-resistant plastics. Autoclaves are not useful for substances that repel water, such as waxes, powders, oils, most plastics, and media containing ingredients that break down when heated.

Boiling water is easy to use in homes and clinical settings but it does not kill heat-resistant cells. It is therefore effective only for disinfection, not for sterilization. Immersing an object in boiling water for 30 minutes will kill most non-spore-forming pathogens, including *Mycobacterium tuberculosis* and *Staphylococcus* species. Boiling can also disinfect drinking water, but the disadvantage of boiling is that once the object is removed from the boiling water, re-contamination can occur.

The last heat-associated mechanism that we will discuss is **pasteurization**. As noted earlier in the chapter, pasteurization is used to reduce microbial load and destroy pathogens while preserving flavor and nutritive value, but this process does not sterilize.

Pasteurization can be accomplished in two ways: via the *flash method*, which involves exposure to a temperature of 71.6°C for 15 seconds, or via the *batch method*, which uses a temperature of 63–66°C for 30 minutes. The flash method is preferable because it is more effective against *Coxiella* and *Mycobacterium* organisms. In addition, longer exposure to a high temperature can destroy flavor and nutritive value. These treatments kill 97–99% of bacteria but do not affect endospores or non-pathogenic lactobacilli, micrococci, or yeasts.

Fast Fact

Pasteurized milk is not sterile: it still contains more than 20,000 organisms per milliliter.

The primary reason for pasteurization is to prevent the transmission of milk-borne diseases, such as those caused by *Salmonella* species (intestinal food poisoning), *Campylobacter jejunum* (acute intestinal disease), *Listeria monocytogenes* (listeriosis), *Coxiella burnetii* (Q fever) and *Mycobacterium tuberculosis* (tuberculosis).

Fast Fact

Even when stored in a refrigerator, *Clostridium botulinum* has been shown to grow and produce exotoxin while embedded deep in food where anaerobic conditions prevail.

Refrigeration, Freezing, and Freeze-Drying

Cold temperature retards the growth of microorganisms by slowing the rate of enzymatic reactions, but it does not kill the microorganisms. Refrigeration is used to delay the spoilage of food by keeping the temperature at about 5°C. Many bacteria and molds can continue to grow at this temperature, so refrigeration is useful only for a limited period.

Freezing at –20°C can preserve food but does not sterilize it. This low temperature slows the metabolic rate of microorganisms to such a degree that there is no growth or spoilage of food. Freezing can also be used to preserve microorganisms, but temperatures much lower than those used to freeze food must be used. The organisms to be preserved are usually frozen in glycerol or protein to prevent the formation of ice crystals that could puncture cell membranes, then placed in dry ice (the solid form of carbon dioxide, –78°C) or liquid nitrogen (–180°C).

Freeze-drying, also known as **lyophilization**, can also be used to preserve cells. This is the process used to produce instant coffee and requires the removal of water from the organisms. Lyophilization is used for long-term storage of organisms and also for ease of shipping and handling. Organisms are rapidly frozen in liquid nitrogen and then subjected to high vacuum, which pulls out the water molecules. The containers holding the frozen, dehydrated organisms are then sealed under vacuum, and the organisms are viable in this state for years. The addition of water is all that is required to restart the growth process.

Filtration

Filtration is useful for sterilizing liquids that would be destroyed by heat. The process involves passing the liquid through *membrane filters*, which contain pores that are too small to allow for the passage of microorganisms (**Figure 18.9**). Membrane filters are usually made of nitrocellulose. These filters can be made with specific pore sizes ranging in diameter from 25 μm to less than 0.025 μm. Membrane filters are relatively inexpensive, do not clog very often, and can be purchased either sterile or non-sterile (the user autoclaves the latter form before use). They can be used for growth media, drugs, and vitamins and in some cases they are used for commercial food preparation, such as the filtration of beer. Filtration is used to sample and test water samples for contamination, in particular fecal coliform contamination.

Filters can also be used to purify air. HEPA (high-efficiency particulate air) filters are routinely found in ventilation systems, such as operating rooms, burn units, and clean rooms of laboratories, where microbial control is required and sterility is important. These same filters are used in facilities where dangerous organisms are used, such as the Centers for Disease Control (CDC) in Atlanta. However, in this case the filters are used to keep organisms from escaping into the outside air. Air filters are usually soaked in formalin before disposal, to kill any trapped organisms.

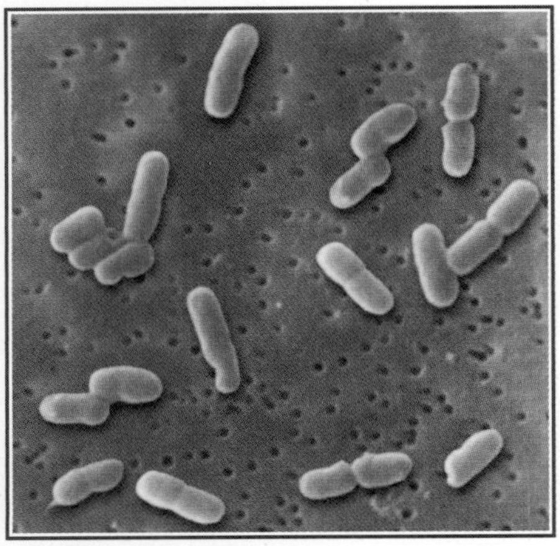

Figure 18.9 Colorized scanning electron micrograph of bacteria sitting on a membrane filter. The diameter of the pores in the filter are smaller than the diameter of the bacterial cells, so the bacteria cannot pass through the filter.

Osmotic Pressure

This technique is another that has been used in food preservation for many decades. As we learned in Chapter 9, high concentrations of salt (or sugar or other substances) will create a hypertonic medium that draws water from the organism through osmosis, leading to plasmolysis and death. High concentrations of sugar are used in making preserves, such as jams and jellies (notice that the word *preserve* fits here), and high concentrations of salt have been used for a long time to "cure" meats and produce pickles. Keep in mind that some halophilic (salt-loving) organisms thrive in these conditions and can spoil these products.

Radiation

Radiation can be defined as energy that is emitted from atomic activities and dispersed at high velocity through matter or space. There are three types of radiation involved with controlling microbial growth: gamma rays, X-rays, and ultraviolet rays. When a cell is bombarded with radiation, the cell's molecules absorb some of the energy, leading to changes in the structure of the cell's DNA (**Figure 18.10**). The radiation can be one of two types:

- **Ionizing radiation**, which changes atoms in a cell's molecules to ions by causing electrons to be ejected from the atoms. Gamma rays, X-rays, and high-speed electron beams are forms of ionizing radiation. DNA is very sensitive to this type of radiation, which means that in DNA exposed to the radiation there is large-scale mutation and breakdown of chromosomal elements.
- **Non-ionizing radiation**, which is best seen with ultraviolet radiation. This excites atoms but does not ionize them, and it leads to abnormal bonds within molecules, such as the formation of thymine dimers (Chapter 11).

Ionizing radiation has become safer to use in recent years and is very useful because it sterilizes without heat or chemicals. All ionizing radiation can penetrate liquids and most solid materials, but gamma rays are the most penetrating. Although irradiation has been used in food preservation for many years, there are still stigmas attached to its use. In fact, any use of radiation in food preservation must be clearly stated on the food label. Flour, meat, fruits, and vegetables are routinely irradiated to kill microorganisms, parasitic worms, and insects.

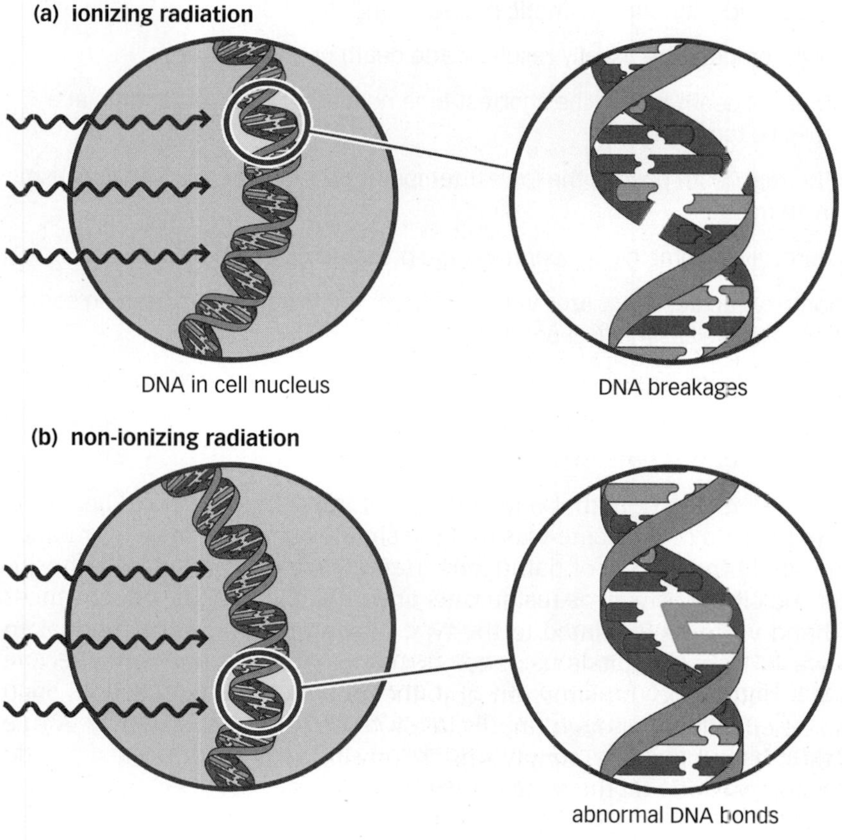

Figure 18.10 Disinfection with radiation. Panel **a**: Ionizing radiation can bombard and penetrate a cell, dislodging electrons from the DNA molecules in the cell and causing the DNA strand to break. Panel **b**: Non-ionizing radiation enters the cell, strikes the DNA molecules, and excites them, causing mutations and the formation of abnormal bonds in the double helix.

Fast Fact

Bacterial spores are 10 times more resistant to radiation than vegetative cells, but increasing the duration of exposure can eventually kill spores as well.

Although radiation can cause cancer, there are no side effects associated with the irradiation of food. In fact, it has been estimated that the irradiation of just 50% of the meat and poultry in the United States would result in 900,000 fewer cases of infection, 8500 fewer hospitalizations and 350 fewer deaths per year. Radiation is currently approved for the reduction of *E. coli* and *Salmonella* in beef and of *Trichinella* in pork.

Sterilization of medical products by radiation has become a rapidly expanding area. Drugs, vaccines, plastics, syringes, gloves, and even tissue that is to be used for grafting and heart valves are all sterilized using radiation. The main drawback to this technique is potential radiation poisoning of the operators who perform the sterilization.

Ultraviolet radiation is non-ionizing radiation that disrupts cells by generating free radicals, which then bind to the cells' DNA, RNA, and proteins. This radiation is a powerful killer of fungal cells and spores, bacterial cells, protozoans, and viruses.

Ultraviolet irradiation is used for disinfection, not sterilization. Germicidal lamps can cut down on microbes by as much as 99%, and the lamps are used in hospital rooms, operating rooms, schools, food preparation areas, and dental offices. In addition, disinfection of air with ultraviolet radiation has been effective in reducing postoperative infection, preventing the droplet transmission of infectious organisms and curtailing the growth of microorganisms in food preparation. It can also inhibit the growth of organisms in water, vaccines, drugs, plasma, and tissues used for transplantation. The major disadvantages of ultraviolet irradiation are poor penetration and damaging effects seen over long exposure to human tissues: retinal damage, cancer, and skin wrinkling.

Keep in Mind

- Physical methods for controlling microbial growth include heat, cold, desiccation, filtration, osmotic pressure, and radiation.
- High temperature usually results in the death of microbes.
- Thermal death time is the shortest time needed to kill all organisms at a specific temperature.
- Thermal death point is the lowest temperature required to kill all organisms in 10 minutes.
- Autoclaves combine temperature and pressure for sterilization.
- Ionizing radiation sterilizes without heat or chemicals and is used on some foods as well as medical products.
- Ultraviolet irradiation is used for disinfection but not for sterilization.

A Word About Hand Washing

It is of interest to note that one of the most important historical discoveries in the field of medicine was that the simple act of washing your hands can inhibit the spread of pathogens. Hand washing is required in hospitals and clinics as well as restaurants and bars. Much of the effectiveness of hand washing is related to the type of soap used and the time taken to wash. Many hospitals use bacteriocidal soaps that are very effective at preventing the transmission of pathogens, but even household soap can be effective if enough time is taken to do the job thoroughly. A rule of thumb for many is to simply sing "happy birthday" as you wash and do not stop washing till the song is over.

SUMMARY

- The cell wall, plasma membrane, proteins, and nucleic acids are targets for disinfectants and antiseptics.
- Several factors affect the rate of death in bacteria, including the number of organisms present, the duration of treatment, temperature, environment, and the ability to form an endospore.
- The potency of antiseptics and disinfectants can be affected by temperature and concentration.
- Three tests are used to evaluate disinfectants and antiseptics: the phenol coefficient test, the disk method, and the use dilution test.
- There are eight major categories of chemical agent used as antiseptics and/or disinfectants.
- Heat, cold, desiccation, filtration, osmotic pressure, and radiation are physical methods used to control microbial growth.

In this chapter, we have seen that there are many ways to control microbial growth, either through chemical or physical means, and that many of these methods are routinely used in commercial applications. It is important to keep in mind that control of microbial growth depends on many factors and that we can establish a hierarchy with regard to the susceptibility of microorganisms to these methods (**Figure 18.11**). We have also discussed ways to control microbial growth outside the body. In the next chapter we look at how antibiotics are used to control microbial growth inside the body.

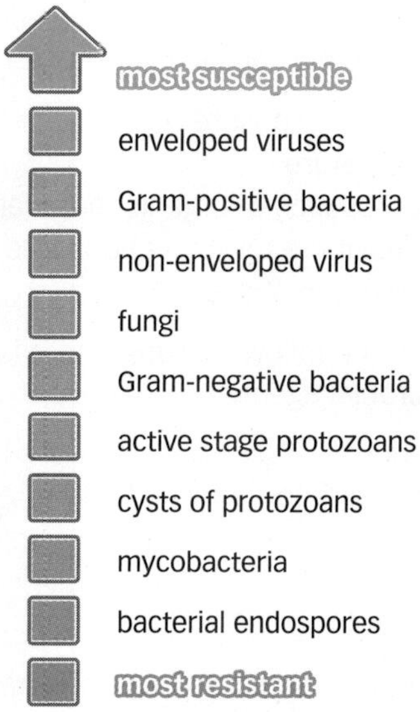

Figure 18.11 Relative susceptibility of selected microbes to antimicrobial agents.

SELF EVALUATION AND CHAPTER CONFIDENCE

Multiple Choice

Answers are given in the back of the book and help can be found in the student resources at:
www.garlandscience.com/micro

1. Which of the following best describes the pattern of microbial death?
   - **A.** All the cells in a culture die at once.
   - **B.** The cells in a population die at a constant rate.
   - **C.** All of the cells in a culture are never killed.
   - **D.** The pattern varies depending on the antimicrobial agent.
   - **E.** The pattern varies depending on the species.
2. Bacterial death will result from damage to which of the following structures?
   - **A.** Plasma membrane
   - **B.** Proteins
   - **C.** Nucleic acids
   - **D.** Cell wall
   - **E.** All of the above
3. Sterilization involves
   - **A.** Removal of all microbes and spores
   - **B.** Removal of all microbes but not endospores
   - **C.** Killing only pathogens
   - **D.** Bacteriostatic agents
   - **E.** All of the above
4. Which of the following is a direct result of heat?
   - **A.** Breaking hydrogen bonds
   - **B.** Cell lysis
   - **C.** Denaturing enzymes
   - **D.** Breaking sulfhydryl bonds
   - **E.** All of the above
5. Which of these factors affect the rate of bacterial cell death
   - **A.** The number of organisms present
   - **B.** The time of exposure
   - **C.** The environment
   - **D.** Endospore formation
   - **E.** All of the above
6. The bacteriocidal versus the bacteriostatic nature of a disinfectant can be determined with which of the following test methods?
   - **A.** The phenol coefficient
   - **B.** The disk method
   - **C.** The thermal death point
   - **D.** The use dilution method
   - **E.** The thermal death time

7. The effectiveness of chemical disinfectants varies with each of the following except
   - A. Length of exposure
   - B. Temperature
   - C. The presence of organic material
   - D. The ability to form an endospore
   - E. None of the above

8. Which of the following substances is the least effective antimicrobial agent?
   - A. Phenol
   - B. Cationic detergents
   - C. Soap
   - D. Alcohol
   - E. Iodine

9. Which of the following are unaffected by alcohol?
   - A. Bacteria
   - B. Viruses
   - C. Fungi
   - D. Endospores
   - E. None of the above

10. Oxidizing agents target
    - A. Cell membranes
    - B. Metabolic pathways
    - C. The cell wall
    - D. Protein synthesis
    - E. None of the above

11. In the table below, which compound was the most effective against *Staphylococcus*?

Disinfectant	Zone of Inhibition (mm)
A	0
B	2.5
C	10
D	5

Table A Disk-Diffusion Test with *Staphylococcus* Gave the Results Shown

   - A. A
   - B. B
   - C. C
   - D. D
   - E. Can't tell

12. QUATS target
    - A. Metabolic pathways
    - B. The cell wall
    - C. The plasma membrane
    - D. Protein synthesis
    - E. Endospores

13. Which one of the following is the best advertisement for a disinfectant?
    - A. Kills *Staphylococcus aureus*
    - B. Kills lipophilic viruses
    - C. Kills *E. coli*
    - D. Kills *Pseudomonas*
    - E. All are equal

14. All of the following are involved in the control of microbial growth except
    - A. Disinfection
    - B. Antisepsis
    - C. Neutralization
    - D. Pasteurization
    - E. Sanitization

15. All of the following are chemicals used in disinfection except
    - A. Aldehydes
    - B. Ultraviolet irradiation
    - C. Phenolic compounds
    - D. Halogens
    - E. Surfactants

16. Pasteurization kills
    - A. All microorganisms
    - B. Only endospores
    - C. Only pathogens
    - D. Gram-positive organisms preferentially
    - E. Gram-negative organisms preferentially

17. Refrigeration at –20°C
    - A. Sterilizes food
    - B. Preserves food for a period of time
    - C. Causes bacterial cell death through changes in osmotic pressure
    - D. Has no effect on microbial growth

18. Which of the following can sterilize?
    - A. Pasteurization
    - B. Autoclaving
    - C. Refrigeration
    - D. Freezing
    - E. None of the above

DEPTH OF UNDERSTANDING

Questions listed here require you to bring together the concepts you have learned in this chapter into a discussion format. This helps you to increase your depth of understanding of the material you have learned. Help can be found in the student resources at:
www.garlandscience.com/micro

1. Why does the endospore present such a problem when thinking about controlling the growth of bacteria?

2. Compare and contrast the effects of dry and moist heat.

3. Compare the benefits of ionizing and non-ionizing radiation as methods for controlling microbial growth.

CLINICAL CORNER

Help can be found in the student resources at:
www.garlandscience.com/micro

1. It was already a very busy day in the infectious disease ward of the hospital. Patients were being admitted and discharged and others were being sent for tests. It is your responsibility to see that the rooms that are being vacated are cleaned and ready to receive new patients. You have a new nursing assistant to help you with this.

   A. How will you explain the importance of the job at hand?
   B. What will you tell your assistant about the disinfection procedures that you will use?

2. Hospital standard operating procedures require that you wash your hands regularly, even though most of the time you are wearing rubber gloves when working with patients.

   A. Why is washing your hands so important?
   B. What effect does wearing rubber gloves have on the need for washing your hands?
   C. Is hand washing in the hospital the same as at home?

Antibiotics

Chapter 19

Why Is This Important?

Antibiotics have been used for decades and have changed the landscape of health care. Since they were discovered, the number of deaths due to infection has been drastically reduced.

OVERVIEW

Perhaps one of the most important scientific findings in history was the discovery that microbes protect themselves from other microbes. This simple principle led to the development of antibiotics, which are chemicals that we can take from one organism and use to protect ourselves from other organisms. Infectious diseases that once swept through the human population unabated were suddenly shut down with the simple administration of these chemotherapeutic substances, known as antibiotics. Because of antibiotics, people began to think that infectious disease was a thing of the past. We will focus in this chapter on the development of antibiotics and their functions, but keep in mind that overuse and improper use of these important medications has brought about a dangerous rise in the resistance of organisms to antibiotics, a topic we discuss in Chapter 20.

We will divide our discussion into the following topics:

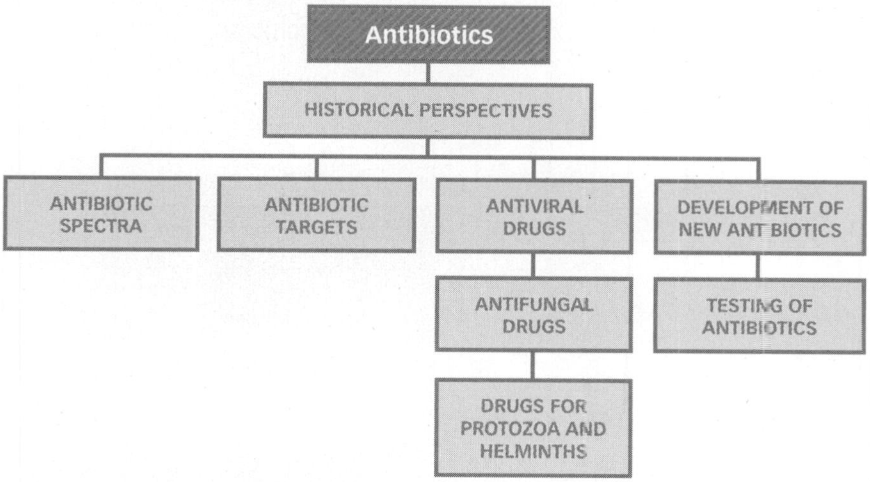

What Do I Need to Know?

To get the most out of this chapter, please review the following terms from your previous courses in biology, anatomy, physiology, or chemistry or in previous chapters of this book as indicated in parentheses: *competitive inhibition* (**2**), *constitutive gene* (**11**), *disk method* (**18**), *double bond* (**2**), *eukaryotic cell* (**4**), *exogenous*, *lipid A* (**5**), *in vivo*, lipopolysaccharide layer (**9**), log phase of bacterial growth (**10**), microbicidal versus microbistatic (**18**), obligate intracellular parasite (**12**), open reading frame (**11**), peptidoglycan (**9**), porins (**9**), prokaryotic cell (**4**), replication fork (**11**), stationary phase of bacterial growth (**10**), virulence factor (**4**), zone of inhibition (**18**).

HISTORICAL PERSPECTIVES

The discovery of the first antibiotic was an accident. Alexander Fleming was working with *Staphylococcus aureus* in 1928, and while plating this organism he accidentally allowed the fungus *Penicillium* to contaminate his plate. He subsequently observed that the plate had a uniform growth of *S. aureus* except where the fungi were growing. In this area, Fleming saw a clearly defined region where there was no bacterial growth; this

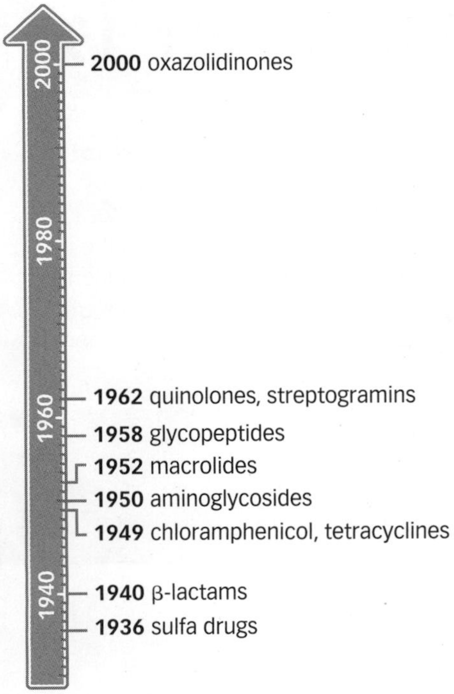

Figure 19.1 A timeline for the introduction of new classes of antibiotics into clinical practice. Note that no new naturally occurring antibiotics were discovered between 1962 and 2000.

was what eventually became referred to as a *zone of inhibition*. This accidental discovery was the beginning of an industry that has become central to successful patient care. It is estimated that more than 80 million prescriptions are filled in America each year, with about 12,500 tons of antibiotics being produced annually. Of this quantity, 25% or more is fed to livestock because the use of antibiotics in livestock can increase the rate of weight gain of the animal by 3–5%. Administration of antibiotics has permitted successful outcomes in many diseases for which there were previously no treatments available. In the United States from 1900 to 1980, mortality from infectious diseases dropped from 797 per 100,000 persons to 36 per 100,000 persons. However, the implementation of antibiotics has been a relatively slow process, as shown in **Figure 19.1**.

We also see from **Figure 19.1** that there have been no major discoveries of natural antibiotic substances for several years. As we will see below, efforts have now shifted to modifying existing antibiotics and searching for potential antibiotics in new places.

Antibiotics Are Part of Bacterial Self Protection

Before we get to the chemical and functional characteristics of the therapeutic chemicals we call antibiotics, it is important to understand that many of them are produced by microorganisms as part of their survival mechanism. These substances keep other organisms away and protect the supply of nutrients and oxygen. The microorganisms that produce these substances use elegant molecular mechanisms to control the production of these toxic molecules, thereby preventing self-destruction. **Table 19.1** shows microbial sources of some common antibiotics.

Table 19.1 Microbial Sources of Some Antibiotics

Microorganism	Antibiotic
Bacteria	
Streptomyces various species	Amphotericin B
	Chloramphenicol
	Erythromycin
	Kanamycin
	Neomycin
	Nystatin
	Rifampin
	Streptomycin
	Tetracyclines
	Vancomycin
Micromonospora various species	Gentamicin
Bacillus various species	Bacitracin
	Polymyxins
Fungi	
Penicillium various species	Griseofulvin
	Penicillin
Cephalosporium various species	Cephalosporins

Naturally produced antibiotics are products of secondary metabolic pathways. These are pathways that are not turned on all the time because continuous production of these destructive molecules could adversely affect the organism producing them. The organisms producing antibiotic substances must therefore develop ways that ensure they can protect themselves. There are several ways in which the organism can do this. For example, some bacteria restrict the production of these chemicals to the stationary phase of bacterial growth. As you remember from Chapter 10, in the stationary growth phase, bacteria are not actively dividing. Consequently, the bacteria are not susceptible to damage caused by antibiotics that impair cell wall construction or interfere with the metabolic functions associated with logarithmic growth. It is no coincidence that bacteria produce these toxic molecules during the stationary phase, because it is a time in the life of the bacteria when nutrients and other growth requirements are at a low level and competition for them is intense.

Other bacteria use mechanisms that keep the intracellular concentrations of the antibiotics they are producing at low levels. This is accomplished by tightly regulating the rates at which these molecules are produced and the rates at which they are exported. In these bacteria, the antibiotic molecules can be exported in an inactive form that will not affect the organism and then become activated by extracellular enzymes. Some antibiotic-producing microorganisms modify their own cell walls or polymerase molecules to ensure their safety. Interestingly, the genes responsible for these modifications are clustered with the antibiotic-producing genes so that the functions of these two sets of genes are integrated: when the genes that control antibiotic production are turned on, the modifications to potential self targets are made automatically.

Keep in Mind

- Penicillin was discovered accidentally by Alexander Fleming.
- There have been no major discoveries of natural antibiotics for several years.
- Many microorganisms produce toxic substances as part of their survival mechanism and have developed methods to protect themselves from the antibiotics they produce.

ANTIBIOTIC SPECTRA

There are a number of diverse molecules that inhibit the growth of microbes. The first molecules that inhibited bacterial growth were natural products isolated from specific microorganisms. Over time, and as a result of increasing bacterial resistance, these natural molecules have been modified, and several types of semi-synthetic antibiotics have been derived. The original natural molecules used by humans as antibiotics had a very narrow spectrum of activity. For example, penicillin activity was restricted to Gram-positive organisms and was essentially ineffective against Gram-negative organisms. As we perfected the ability to chemically modify the structure of some antibiotics, it became possible to broaden the spectrum of activity. Today, antibiotics are classified as either broad-spectrum drugs or narrow-spectrum drugs.

There are a variety of antibiotics that have inhibitory effects on bacteria. **Table 19.2** lists the major groups, their spectra and their side effects. It should be remembered that many of these groups contain generations of molecules that have been and are being continually modified.

Drug	Primary Effect	Spectrum	Side Effects
Ampicillin	Cidal	Broad (Gram +, some Gram –)	Allergic responses, diarrhea, anemia
Bacitracin	Cidal	Narrow (Gram +)	Renal injury if injected
Carbenicillin	Cidal	Broad (Gram +, many Gram –)	Allergic responses, nausea, anemia
Cephalosporins	Cidal	Broad (Gram +, some Gram –)	Allergic responses, thrombophlebitis, renal injury
Chloramphenicol	Static	Broad (Gram +, Gram –; *Rickettsia* and *Chlamydia*)	Depressed bone marrow function, allergic reactions
Ciprofloxacin	Cidal	Broad (Gram +, Gram –)	Gastrointestinal upset, allergic responses
Clindamycin	Static	Narrow (Gram +, anaerobes)	Diarrhea
Dapsone	Static	Narrow (mycobacteria)	Anemia, allergic responses
Erythromycin	Static	Narrow (Gram +, mycoplasma)	Gastrointestinal upset, hepatic injury
Gentamicin	Cidal	Narrow (Gram –)	Allergic responses, nausea, loss of hearing, renal damage
Isoniazid	Static	Narrow (mycobacteria)	Allergic reactions, gastrointestinal upset, hepatic injury
Methicillin	Cidal	Narrow (Gram +)	Allergic responses, renal toxicity, anemia
Penicillin	Cidal	Narrow (Gram +)	Allergic responses, nausea, anemia
Polymyxin B	Cidal	Narrow (Gram –)	Renal damage, neurotoxic reactions
Rifampin	Static	Broad (Gram –, mycobacteria)	Hepatic injury, nausea, allergic responses
Streptomycin	Cidal	Broad (Gram +, Gram –; mycobacteria)	Allergic responses, nausea, loss of hearing, renal damage
Sulfonamides	Static	Broad (Gram +, Gram –)	Allergic responses, renal and hepatic injury, anemia
Tetracyclines	Static	Broad (Gram +, Gram –; *Rickettsia* and *Chlamydia*)	Gastrointestinal upset, teeth discoloration, renal and hepatic injury
Trimethoprim	Cidal	Broad (Gram +, Gram –)	Allergic responses, rash, nausea, leukopenia
Vancomycin	Cidal	Narrow (Gram +)	Hypotension, neutropenia, kidney damage, allergic reactions

Table 19.2 Properties of Some Common Antibiotics

Before we examine the extensive array of antibiotics and how they work, let's focus on penicillin. The structure of this molecule has been the template for the development of an entire group of antibiotics (more than 50 so far). In its native form, penicillin is composed of a *core* four-sided ring structure called the *beta (β)-lactam ring* (**Figure 19.2**). This structure is also referred to as the *nucleus* of the penicillin molecule. All derivatives of penicillin contain this ring structure, and derivatives of penicillin such as ampicillin or amoxicillin contain a specific structure, referred to as a side chain and designated generically by the letter R, attached to the ring. As you chemically change the side chain, you can change the antimicrobial activity of the penicillin derivative as well as the resistance to stomach acid and overall half-life in the body.

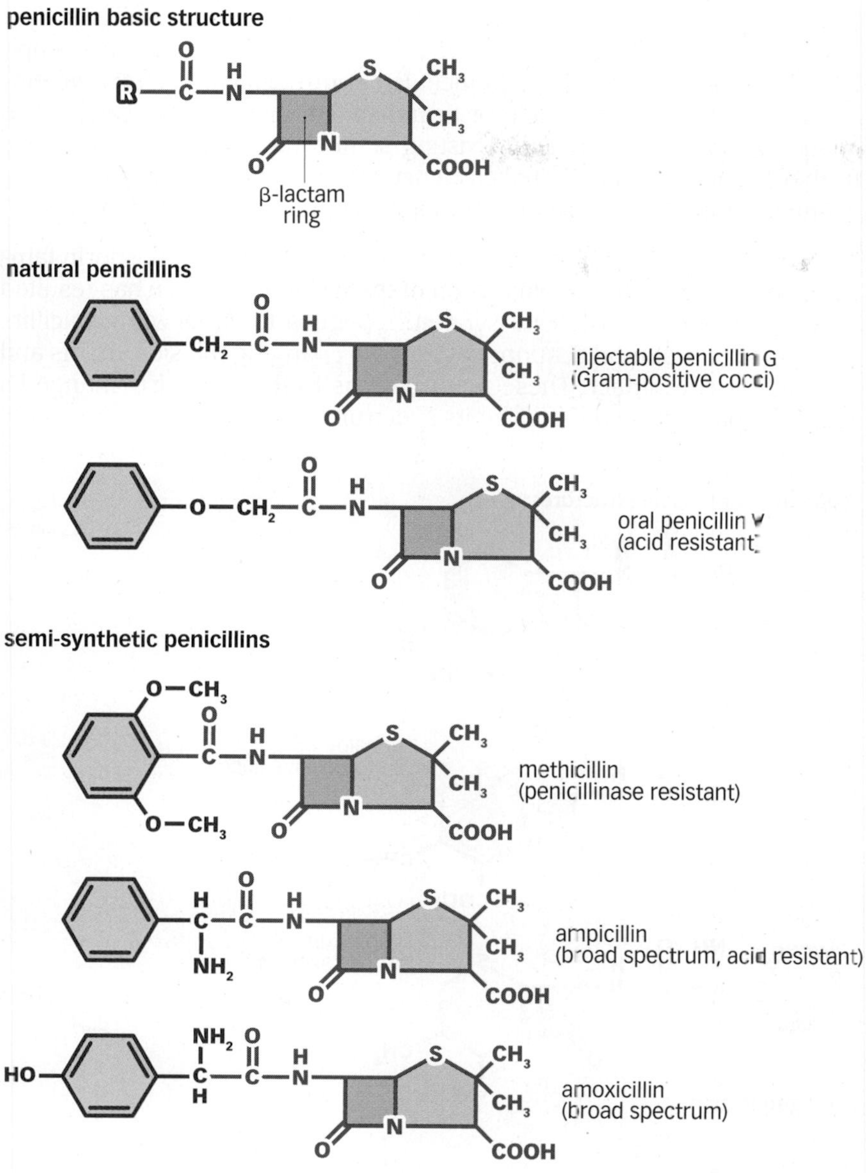

Figure 19.2 Natural and semi-synthetic forms of penicillin. Natural forms of penicillin (G and V) have narrow spectra of activity and variable resistance to acids and β-lactamase enzymes. Note that modifications of the side group (shown in red) on the β-lactam ring can change not only the spectrum of activity but also the susceptibility to acids and enzymes.

Figure 19.2 shows several examples of this kind of manipulation. Here we see that penicillin can be found in two natural forms called **G** and **V**. Both of these forms are naturally produced by the mold *Penicillium*. If we look at the four alternative forms of penicillin shown in this figure, we see that all of them have the same core structure of the β-lactam ring.

However, ampicillin, methicillin, and amoxicillin all have distinctly different side chains attached to the nucleus. These examples are referred to as semi-synthetic forms of penicillin because the modifications can be made in a laboratory. Modifying side chains has become an important part of the search for new antibiotics. Chemists can create and modify side chains that are attached to the core structure, thereby producing new semi-synthetic forms of penicillin. It is important to note that changing a side chain does not guarantee that a functional antibiotic will be formed. In some cases the newly derived molecule may have an increased spectrum of activity (that is, the number of different organisms it affects). In other cases, the synthetic side chain may result in a diminution or loss of activity.

Natural penicillin has a very narrow spectrum and reacts against only a small group of Gram-positive bacteria. Modifying penicillin to form ampicillin broadens the spectrum to include Gram-negative bacteria, whereas modifying it to carbenicillin or ticarcillin can broaden the spectrum to include *Pseudomonas* species. Existing semi-synthetic penicillins can be further modified to increase the efficiency of inhibiting bacterial growth. For example, ampicillin can be further modified to mezlocillin or azlocillin.

The same kind of manipulation can be seen with the cephalosporin family of antibiotics, where modification of the natural molecule has resulted in several generations of semi-synthetics (**Figure 19.3**). As with penicillin, you will note that modifications result from changing the side chains and leaving the core intact. These modifications to the core also change its reactivity patterns and therefore its spectrum.

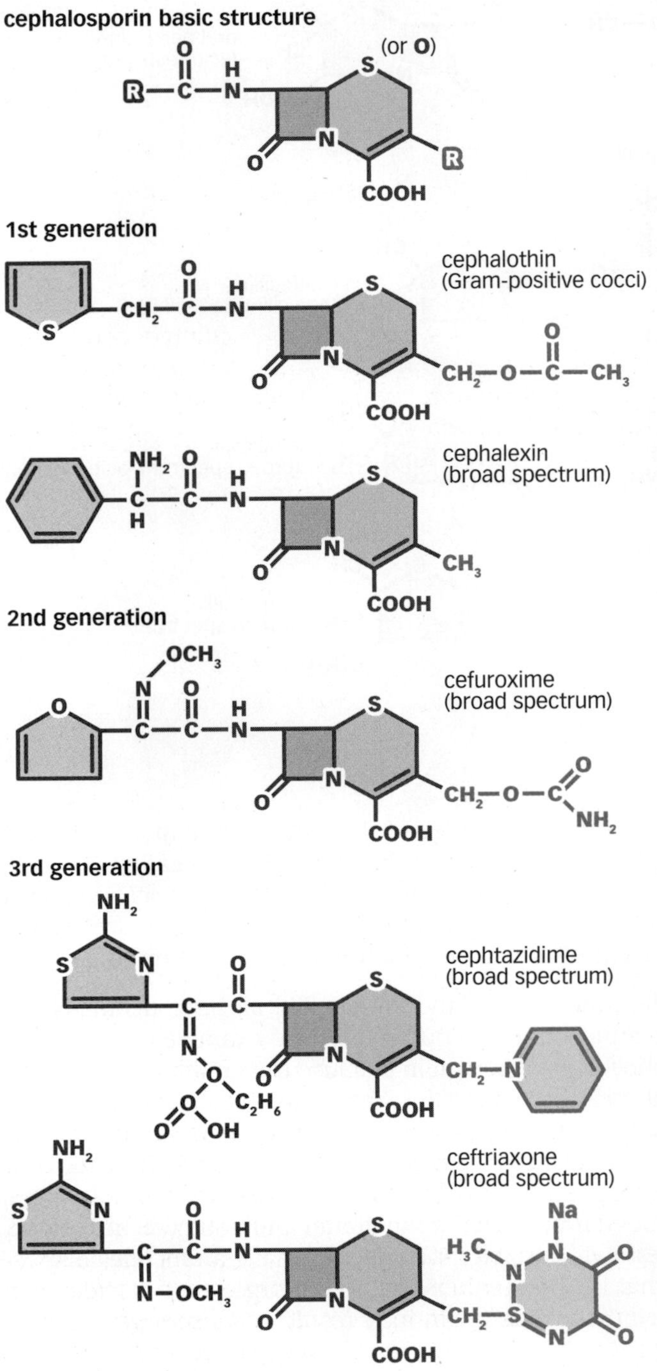

Figure 19.3 Chemical modifications of cephalosporin. The nucleus of the drug remains unchanged, while the side chains (shown in red and blue) change. We call these successive modifications *generations* of the drug.

Bacteria do not stand idly by as new forms of antibiotic are synthesized by medical researchers. In fact, bacteria are constantly finding ways to counteract these molecules. One of the most important defense mechanisms that bacteria use is the production of the enzyme β-lactamase. This enzyme cleaves open the β-lactam ring of the penicillin molecule, thereby inactivating the molecule. Because all forms of penicillin, either natural or semi-synthetic, contain this ring structure, bacteria that have evolved the ability to produce β-lactamase have become resistant to penicillin. As mentioned earlier, resistance to antibiotics has become one of the most important problems in patient care today and will be discussed in the next chapter.

One possible way to overcome this enzymatic resistance mechanism is to combine the penicillin drug with some molecule that protects the penicillin molecules or in some way diminishes or impedes β-lactamase activity. Potassium clavulanate (clavulanic acid) is a good example. Potassium clavulanate has no antibacterial activity but it can inhibit the activity of β-lactamase, thereby depriving the bacteria of this major resistance mechanism and allowing the antibiotic to function normally.

Several of these combinations have been developed, including:

Amoxicillin + potassium clavulanate = **Augmentin®** or **Timentin®**

Imipenem + cilastatin = **Primaxin®**

Ampicillin + sulbactam = **Unasyn®**

As you can see, the usefulness of antibiotics has become dependent on overcoming the mechanisms that microorganisms develop to escape antibiotic therapy. The development of resistance, which can be unbelievably fast, necessitates the continued search for new forms of old antibiotics and the identification and/or development of new ones. This is a continuous race, the loss of which could have dramatic and potentially catastrophic clinical consequences.

Keep in Mind

- Antibiotics have spectra of reactivity, either broad or narrow.
- Chemical modification of natural antibiotics can broaden their spectrum.
- Penicillin is composed of a core ring structure known as the β-lactam ring.
- Natural penicillin is found in two forms, G and V, which both have a narrow spectrum of activity.
- Some bacteria produce an enzyme called β-lactamase, which breaks the lactam ring of the penicillin molecule and so inhibits the activity of penicillin, causing the bacteria to become resistant to penicillin.

ANTIBIOTIC TARGETS

One of the fundamental criteria in the selection of an antibiotic for medical use is **selective toxicity**, meaning that the antibiotic should be destructive to the disease-causing organism but have no effect on the human host. The obviousness of this restriction is contradicted by several facts. First, the human body is based on homeostasis, the balance of chemicals and chemical reactions. The introduction of exogenous chemicals, especially in large quantities, can disrupt this homeostasis with harmful results. Unfortunately, the first antibiotic to be discovered, penicillin, was the most selectively toxic. This was unfortunate because it gave a false sense of security regarding antibiotic molecules, suggesting that all of them could be used without regard to toxicity. In fact, many chemicals that are useful in restricting bacterial growth are inherently toxic and cannot be used therapeutically. In addition, many antibiotic molecules can be toxic if administered at concentrations that are too high. The inherent toxicity of potential antibiotic molecules has necessitated an extensive testing program for these substances, a program that can take years and cost millions of dollars. We look at the testing and strategies for development of new antibiotics later in this chapter.

Our discussion of antibiotics here is based on the targets they affect. When we think of potential targets for antibiotic activity, we can use what we learned in Chapter 9 about the structures associated with bacteria. Antibiotic targets can be subdivided into five major groups:

- The bacterial cell wall
- The bacterial plasma membrane
- Synthesis of bacterial proteins
- Bacterial nucleic acids
- Bacterial metabolism

Figure 19.4 shows each of these targets and some of the major antibiotics that target them. We can now look at each of these targets and the antibiotic mechanisms that affect them in some detail.

The Bacterial Cell Wall

The most appealing target for antibiotics is the bacterial cell wall, because it is found in bacteria but not humans. It therefore meets the criterion of selective toxicity. Recall that the bacterial cell wall is a structure built to keep the outside out and the inside in, making it a necessary component for the survival of the bacteria. This wall is found in both Gram-positive and Gram-negative bacteria and is a complex network of linked molecules (Chapter 9).

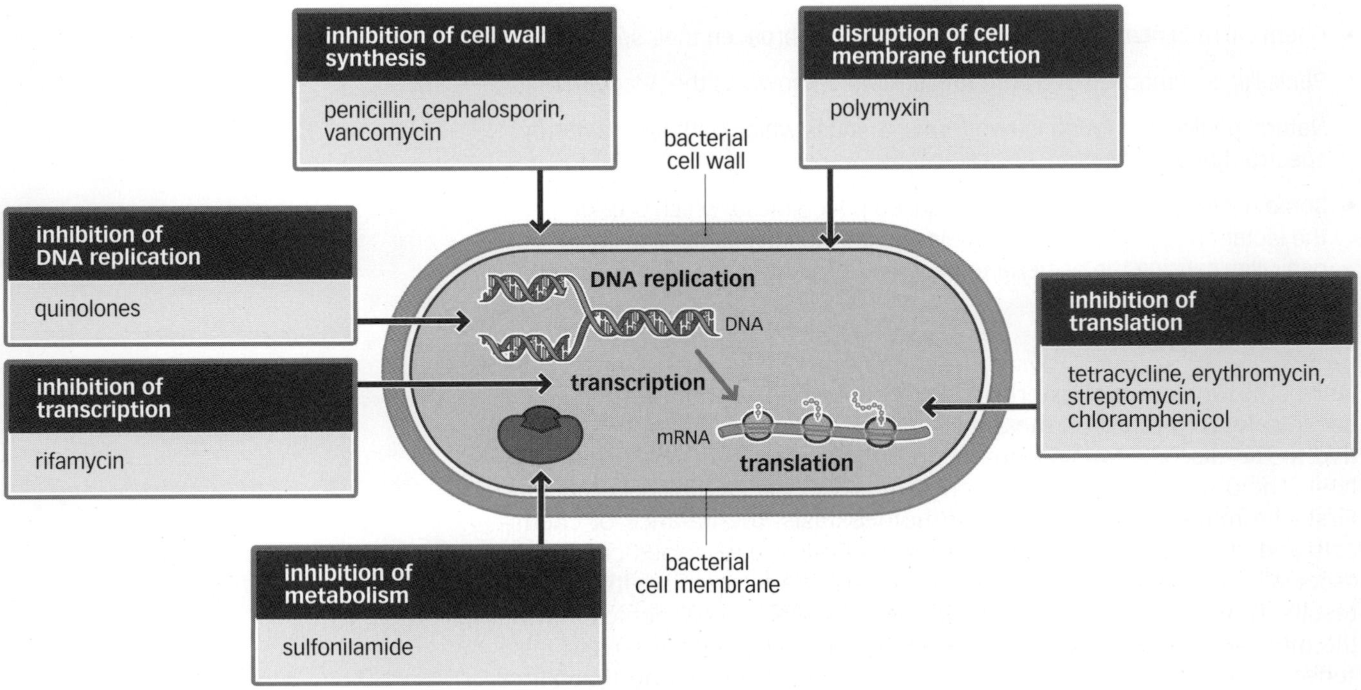

Figure 19.4 A summary of the targets used for the development of antibiotics.

Penicillin

Many enzymatic interactions take place during the construction of cell wall components. The enzymes involved in these processes can therefore be used as targets of antibiotic molecules. As we have learned, the cell wall is composed of layers of peptidoglycan, which is made up of repeating units of *N*-acetylglucosamine (NAG) and *N*-acetylmuramic acid (NAM). The NAG and NAM molecules are cross-linked through the activity of transglycosylase and transpeptidase enzymes. Many antibiotics that target the cell wall act by inhibiting the activity of these two enzymes. The result is that the cell wall is not properly cross-linked; it is therefore weak and unable to withstand the environmental pressures that are always present.

There are also **penicillin-binding proteins (PBPs)** in the cell walls of bacteria that function in the construction of the cell wall. They are called penicillin-binding proteins because the β-lactam ring of penicillin binds to these proteins. There can be large numbers of PBPs in a bacterial cell wall. For example, the cell wall in one *Escherichia coli* cell has more than 2500 of these molecules (although only about 300 can be used by penicillin to kill the cell). It should be noted that even though PBPs are useful as targets of antibiotic activity, any mutation in the bacterial genes that code for them will result in the organism's becoming resistant to penicillin. It may seem odd that the bacteria would have a protein that would bind to a potentially harmful agent, but this type of interaction is often seen in the microbial world, especially in interactions between viruses and their hosts, and may represent natural co-evolution.

During active growth of bacteria, new cell wall is continuously being built. It is at this time that the activity of penicillin is most effective, because penicillin prevents the cross-linking of the NAG and NAM units and thereby prevents the formation of an intact cell wall. Consequently, the more rapidly the bacteria are dividing, the more devastating is the effect of penicillin. Although both Gram-positive and Gram-negative cells contain peptidoglycan, the amounts are markedly less in Gram-negative bacteria. For this reason, these bacteria are normally less sensitive to penicillin than are Gram-positive organisms. As noted earlier, the reactivity of penicillin against Gram-negative bacteria has been enhanced by synthetically modifying penicillin G or V so that it will affect Gram-negative organisms.

Fast Fact

The penicillin molecule has undergone a number of synthetic modifications, and there are now five categories of the drug, based on narrow versus broad spectrum and the reactivity to the organism *Pseudomonas aeruginosa*.

Cephalosporins

Cephalosporin antibiotics act similarly to the penicillins, by preventing the construction of a stable cell wall. However, the cephalosporins differ from penicillin in that they have a much greater effect on Gram-negative bacteria (i.e. they are naturally broad-spectrum antibiotics). One major reason for their success is that they are not susceptible to the β-lactamase enzymes that inactivate penicillin. There are multiple generations of cephalosporins, as **Figure 19.3** shows; with more than 70 versions now in use, they are one of the most widely prescribed antibiotics.

The mechanism of action of this antibiotic is similar to that of penicillin, but cephalosporin has the capacity to penetrate through the porin molecules found in the outer layer of Gram-negative bacteria. As the side chains of the cephalosporin molecule continue to be modified, the spectrum of reactivity continues to increase. Cephalosporins have an excellent safety record with regard to adverse reactions and are therefore frequently used both preoperatively and postoperatively. However, the frequent use of these antibiotics in hospital settings has increased the number of bacteria that are resistant to cephalosporins.

Carbapenems

Like penicillin and cephalosporin, the carbapenems also contain a β-lactam ring as part of their structure and inhibit the synthesis of bacterial cell walls. However, these molecules differ from penicillin in that they have a double bond in the β-lactam ring. The presence of this double bond prevents β-lactamase from cleaving the ring. Carbapenems have a very broad spectrum of antibacterial activity, and two have been approved for clinical use in humans: imipenem and meropenem. Both of these antibiotics are useful against *Pseudomonas* species.

Primaxin is a combination of imipenem and a molecule called cilastatin. In the human body, the imipenem is antibacterial and the cilastatin prevents premature destruction of the antibiotic molecule by the kidneys, allowing the drug to remain active longer. Primaxin has become the antibiotic of choice for many nosocomial infections.

Monobactams

Unlike penicillin, monobactam molecules contain a different ring structure. Bacteria have developed resistance to antibiotics that have a β-lactam ring structure, as a result of their ability to produce enzymes such as β-lactamase that attack the ring. An antibiotic that has a different ring structure cannot be recognized by β-lactamase. The monobactams are therefore very effective in overcoming bacterial resistance. The monobactam aztreonam, for instance, is active against Gram-negative organisms only, and, more importantly, it has good reactivity against *E. coli* and *Pseudomonas* species, both of which are becoming very dangerous hospital infection agents.

Glycopeptide Antibiotics

Glycopeptide antibiotics, such as vancomycin and teicoplanin, have antibacterial activity. Teicoplanin differs structurally from vancomycin and has not been approved for use in the United States, although it is used in other countries. Both vancomycin and teicoplanin are derived from *Streptomyces* organisms. When discovered, they were found to have serious side effects, but that toxicity level has been reduced in recent years through improvements in purification. Both vancomycin and teicoplanin inhibit cell wall synthesis by forming a complex with the substrates that make up the wall's peptidoglycan molecules. However, neither can penetrate the porins of Gram-negative cells. They therefore have a very narrow spectrum, restricted to Gram-positive organisms. It should be noted that the glycopeptide antibiotics work on different parts of peptidoglycan synthesis from those affected by penicillin, making the use of these antibiotics in concert potentially attractive for enhanced protection.

Vancomycin has achieved a new status as the appearance and magnitude of bacterial antibiotic resistance have increased. The antibiotic methicillin was formerly the treatment of choice against pathogenic *S. aureus*. When this pathogen became resistant to methicillin, it was referred to as **MRSA** (**m**ethicillin-**r**esistant ***S****taphylococcus* ***a****ureus*). Vancomycin is now used as the last line of antibiotic defense against MRSA as well as against certain streptococci, enterococci, and other pathogens. As we will see in Chapter 20, there are now *S. aureus* strains that have also become resistant to vancomycin (referred to as **VRSA**). These organisms are considered to be very dangerous.

Isoniazid and Ethambutol

Isoniazid and ethambutol are used against bacteria that have modified their cell walls for further protection against environmental conditions and host cell defenses. The *Mycobacterium* species are a good example.

The pathogens that cause tuberculosis and leprosy are members of this genus. As we saw in Chapter 9, the cell walls of these organisms are modified by the incorporation of mycolic acids, which are waxy components that add extra protection [illegible] make these organisms uncommonly resistant to most antibiotics.

Isoniazid, however, is [illegible]ctive against these organisms because of a mechanism that seem[illegible]ibit the synthesis of mycolic acid. Ethambutol, which is given in c[illegible]h isoniazid, inhibits the incorporation of mycolic acid into the [illegible]acterial cell wall. It should be noted that ethambutol alone [illegible]ot very effective against mycobacterial species. The combinat[illegible]zid, ethambutol, and rifampin is now the treatment of cho[illegible]losis, with the administration of these combinations of dru[illegible]ng the potential for the development of resistance.

> **Fast Fact**
> Isoniazid and ethambutol are effective only against organisms that contain mycolic acid in their cell wall.

[illegible]ntibiotics

[illegible]e antibiotics, such as bacitracin, are used topically for super[illegible]ections of Gram-positive organisms, such as *Staphylococcus* and [illegible]*coccus* species. The mechanism of action of these antibiotics is [illegible]nhibition of binding between the *N*-acetylglucosamine and *N*-acetyl[illegible]uramic acid subunits, thereby preventing the synthesis of linear strands of peptidoglycan.

The Bacterial Plasma Membrane

The plasma membrane in bacteria is involved with membrane transport, DNA replication, the production of ATP, and other important physiological functions. It is therefore a prime target for antibiotics because any disruption of this membrane will destroy the bacteria's ability to survive. Unfortunately, the structure of a bacterial plasma membrane (the phospholipid bilayer) is remarkably similar to that found on eukaryotic host cells. This similarity of membranes does not allow antibiotics that attack this bilayer to be selectively toxic. Consequently, although there are antibiotics that target the plasma membrane — the antibiotic polymyxin B is one example — they do not strictly meet the requirements associated with selective toxicity.

Synthesis of Bacterial Proteins

On the basis of what we have discussed about the structure and function of proteins and their importance to all living cells, it is easy to understand how any disruption in the production of these molecules could be devastating to a bacterial cell. As we saw in Chapter 11, proteins are assembled at a ribosome in combination with messenger RNA, and assembly of a protein begins with the formation of an intact ribosome from two ribosomal subunits (**Figure 19.5**). Here amino acids are linked together through peptide bond formation. Because ribosomes are found in both prokaryotic and eukaryotic cells, the selection of protein synthesis as a target for antibiotic therapy against bacteria would, at first glance, seem to break the rule of selective toxicity. However, the ribosomes of prokaryotes are not the same as those of eukaryotes. Therefore, antibiotics that target the synthesis of proteins in bacteria do meet the criterion of selective toxicity. However, in the eukaryotic cell, the mitochondria contain ribosomes that are unique to mitochondria. These ribosomes are the same as the ribosomes in prokaryotic cells. Consequently, there can be some antibiotic interference in normal eukaryotic cell function if antibiotics that interrupt protein synthesis are given in excessive amounts.

To understand the effects of specific types of antibiotic on protein synthesis, let's briefly review the events that occur at the ribosome (for detailed information, see Chapter 11). Recall that the intact 70S ribosome

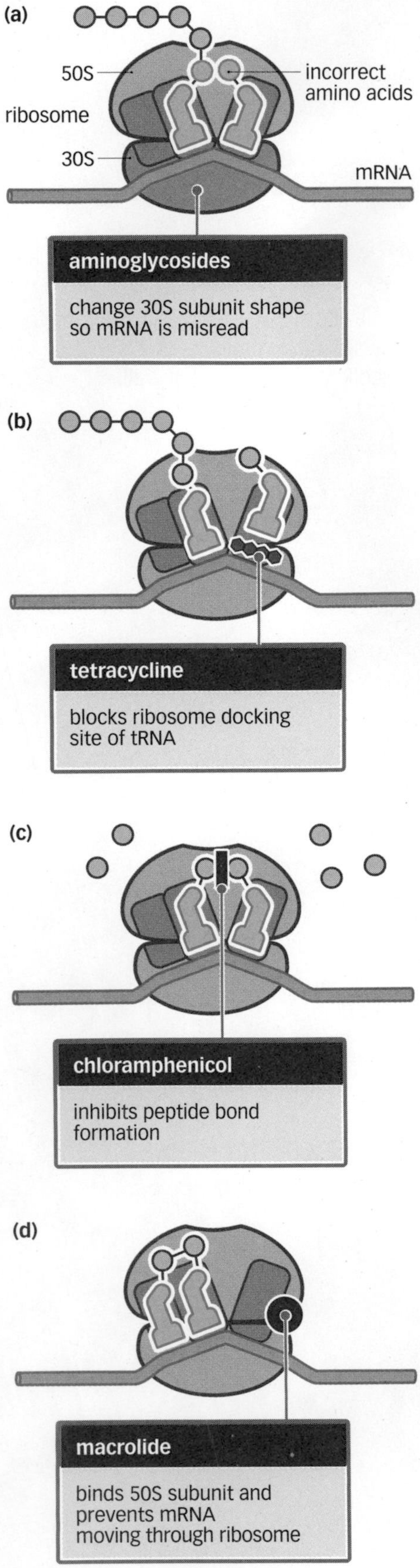

Figure 19.5 Antibiotic targets of the bacterial ribosome. The bacterial 70S ribosome is composed of two subunits, 30S and 50S. Several targets of antibiotics are located on these subunits. Some mechanisms of antibiotic inactivation involve improper orientation of the mRNA, inability to form peptide bonds, or inhibition of peptide elongation.

is composed of two parts, a small 30S subunit and which join together when they encounter mRNA. Eac ribosomal RNA and proteins. The 30S subunit contains a S subunit, is rRNA, and the 50S subunit has a 23S segment of rRNA. ntains has a three-dimensional configuration, and it is via this three ality that sites for protein synthesis form on the intact ribosome that

The A and P sites on the ribosome are where the transfer RNAs that specific amino acids "sit down." When both the A site and the P site occupied by tRNA molecules carrying amino acids, the amino acids are in close proximity to each other and will join together through the formation of a peptide bond. The three-dimensional configuration of the ribosome also permits a precise reading of the mRNA, without which a correct functional protein could not be synthesized.

Both the 30S and the 50S subunit have been examined by X-ray crystallography, and their three-dimensional structures have been determined. Knowing the exact three-dimensional structure has given researchers a new insight into the mechanism of action of antibiotics that target ribosomes. It has been shown that the mRNA threads through two tunnels in the 30S subunit. These tunnels are interrupted by a space that is about six nucleotides long and protrudes into the space between the 30S and 50S subunits. It is here that the A and P sites are located and therefore here that the tRNAs sit down with their attached amino acids. Transfer RNA molecules are accompanied to the sites by chaperone proteins that assist in the orientation of the RNA molecules at the site, and these chaperone proteins have come to be of interest as possible new targets for antibiotic activity. The peptide being constructed at the ribosome is growing as amino acids are added to each other. X-ray analysis has shown that the growing peptide extends from and leaves the ribosome through an "exit" tunnel.

Many antibiotics act at different sites on the bacterial ribosomes to inhibit protein synthesis. For example, spectinomycin, the aminoglycosides kanamycin and streptomycin, and tetracycline all target the 30S subunit of the ribosome, whereas clindamycin, chloramphenicol, linezolid, erythromycin, clarithromycin, and azithromycin all target the 50S subunit. Some antibiotics interfere in the process of peptide elongation by blocking the translocation of the growing peptide chain from the A site to the P site (spectinomycin); others interfere with the decoding process of the message (paromomycin); and still others upset the accuracy of the translation (streptomycin). Chloramphenicol totally blocks the binding of tRNA to the A site of the ribosome.

Both synthetic and natural forms of erythromycin act by blocking the approach to the peptide exit tunnel on a bacterial ribosome and also by blocking assembly of the 50S subunit. Erythromycin is safe and effective in both children and adults and has been routinely used in both in-patient and out-patient settings. The synthetic forms of erythromycin known as azithromycin and clarithromycin have expanded spectra and are used for respiratory infections because they have a longer half-life and cause less irritation of the gastrointestinal tract.

Several *Streptomyces* species make a pair of antibiotics called pristinamycin and streptogramin, which work synergistically to block translation at the 50S subunit. These antibiotics have been synthetically modified and make up the antibiotic known as Synercid®. Synercid has recently been approved for the treatment of vancomycin-resistant enterococcal (VRE) infections, which are becoming a great cause for concern, as we will see in the next chapter.

Fast Fact

There are now three generations of synthetic erythromycin-related antibiotics, which are selectively toxic because of the structural differences between prokaryotic and eukaryotic ribosomes.

Tetracyclines, which have been used since the late 1940s, are bacteriostatic and also target the ribosome. Their mechanism of action is to block the arrival of the tRNA at the A site. Unfortunately, these antibiotics have been in use for so long that many bacteria have developed resistance to them, and their use has steadily declined over the years.

Aminoglycoside antibiotics, which have also been around since the 1940s, are used in combination with other antibiotics and target the 16S RNA portion of the 30S ribosomal subunit. There have been several generations of these compounds, such as gentamycin and tobramycin, and although they are potent against Gram-negative organisms, they are not very effective against Gram-positive bacteria. They have been used in combination with β-lactam antibiotics to fight *Pseudomonas* infections, but they produce significant renal toxicity and ototoxicity (damage to hearing).

Linezolid is the first totally synthetic antibiotic to be introduced and used clinically. Linezolid blocks protein synthesis at the ribosome by occupying the P site and blocking the formation of peptide bonds. It can be taken orally and is very active against Gram-positive organisms. More importantly, it is also very reactive against vancomycin-resistant enterococci.

Bacterial Nucleic Acids

Perhaps the most obvious target of antibiotic therapy would be the nucleic acids DNA and RNA. Because these molecules have critical roles in the reproduction of bacteria, they would be excellent targets: any disruption in their function will result in the death of the bacteria. The main difficulty in using nucleic acids as targets is selective toxicity. DNA and RNA are universal components associated with all life on Earth, and the structure of DNA and RNA in bacteria is no different from the structure of these molecules in humans. In fact, over the years a variety of potential antibiotics have been developed for this target and they have indeed been found to be unusable. However, two families of synthetic compounds, the rifamycins and the quinolones, have been found to be somewhat effective in attacking bacterial nucleic acids while not harming the same molecules in humans.

Quinolones block either DNA replication or DNA repair. Replication involves a variety of protein molecules, including the enzymes topoisomerase and gyrase. These enzymes are involved with making small cuts in the supercoiled structure of DNA so that it can uncoil, unwind, and separate (Chapter 11). As we have discussed before, these enzymes are proteins and therefore have a three-dimensional structure that is very precise and critical for proper function. Any disruption of this structure will result in the loss of that function. This makes them very valuable targets for antibiotics that can disrupt the three-dimensional structure. Perhaps more importantly, the topoisomerase molecules found in bacteria are different from those found in eukaryotic organisms, making this enzyme an excellent target for any antibiotics that work by disrupting DNA replication.

Fluoroquinolones such as levofloxacin and ciprofloxacin are used against both Gram-positive and Gram-negative organisms to treat urinary tract infections, osteomyelitis, and community-acquired pneumonia and gastroenteritis. Ciprofloxacin has made the headlines as the primary antibiotic used against infections caused by *Bacillus anthracis* (anthrax). We discuss this disease and the subject of bioterrorism in Chapter 28. Fluoroquinolone antibiotics block the movement of the replication fork.

The rifamycins were originally isolated from *Streptomyces*. These antibiotics bind to the RNA polymerase molecule and disrupt its three-dimensional shape, rendering the polymerase molecule unable to function properly. It is interesting to note that this binding occurs away from the active site of the polymerase molecule. Once the activity of this molecule is blocked, there can be no transcription and therefore no protein synthesis, which is a lethal event. The rifamycin known as rifampin is the only antibiotic with this mechanism of action and is used only in combination therapy because resistance develops rapidly if the antibiotic is used alone. As mentioned earlier, rifampin is used in combination therapy of tuberculosis. All the rifamycins are considered selectively toxic because the sensitivity of eukaryotic RNA polymerase is 100 times less than that of bacterial RNA polymerase.

Fast Fact

Mycobacterium will develop resistance to rifampin through mutation of the RNA polymerase β subunit such that it continues to function but is no longer identifiable to the antibiotic as a target.

Bacterial Metabolism

Two other targets for inhibition of bacterial growth are (1) the production of nucleic acid precursors and (2) one or more metabolic pathways that occur at the plasma membrane. Fortunately, there are several pathways that are exclusive to bacteria, and interruption of these pathways selectively inhibits bacterial growth but has no effect on eukaryotic organisms. Perhaps the best example of this is the metabolism of folic acid, a molecule needed for nucleic acid synthesis (**Figure 19.6**). One of the intermediates in the pathway is *para*-aminobenzoic acid (PABA). Sulfa drugs competitively inhibit the function of the enzyme that incorporates the PABA molecule into the folic acid metabolic pathway. It is referred to as competitive inhibition because the sulfa molecule is remarkably similar in structure to the PABA molecule. X-ray diffraction studies have confirmed this similarity.

The enzyme simply gets fooled into incorporating the sulfa molecule into the folic acid structure instead of the PABA. Incorporation of sulfa instead of PABA stops the pathway and is a lethal event for bacteria because

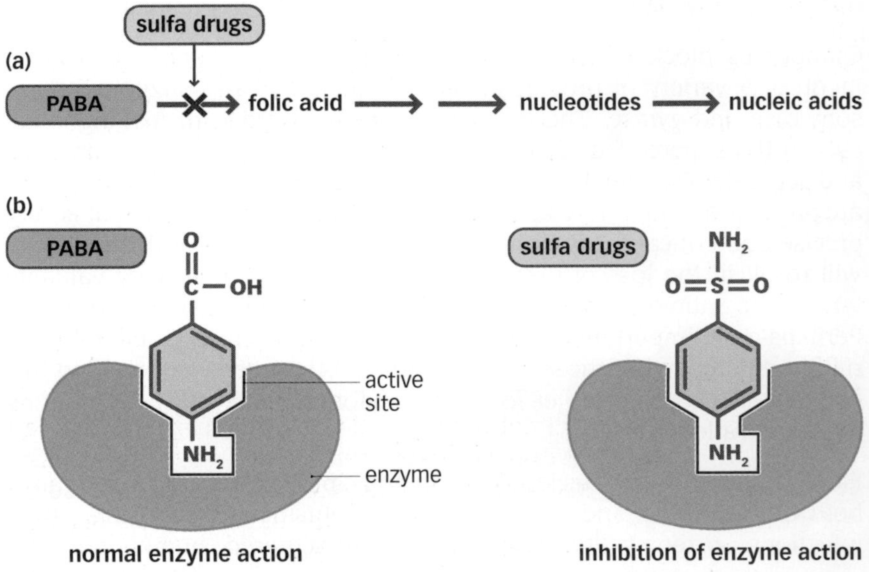

Figure 19.6 Competitive inhibition of metabolism. This figure shows the metabolic pathway used to generate nucleic acids. One of the intermediates in this pathway is *para*-aminobenzoic acid (PABA). Panel **a** shows that sulfa drugs can inhibit this pathway. Panel **b** illustrates the reason for this blockade. As you can see, the structure of sulfa drugs is very similar to the PABA molecule and will competitively inhibit the pathway.

they cannot survive without folic acid. This inhibition is selectively toxic because unlike bacteria, which must synthesize folic acid, we obtain folic acid through our diet.

Sulfa drugs have been in use longer than any other form of antibacterial inhibitor. The sulfa drug sulfamethoxazole is usually used in combination with trimethoprim for urinary tract infections and *Pneumocystis* infections that occur secondarily to AIDS. Both of these drugs block a step in folic acid metabolism. Sulfamethoxazole blocks the biosynthesis of folate, whereas trimethoprim blocks intermediates required for the synthesis of DNA. Unfortunately, because these drugs have been used for such a long time, bacterial resistance has continued to increase, decreasing their usefulness.

Keep in Mind

- Antibiotics must satisfy the criterion of selective toxicity, so they must react against the invading organism but not against the host.
- Most antibiotics have some side effects.
- There are five targets for antibiotics: the bacterial cell wall, the bacterial plasma membrane, protein synthesis, metabolic inhibition, and nucleic acids.
- Nucleic acids and the plasma membrane are the least selectively toxic targets because they are very similar in bacteria and in host cells.

ANTIVIRAL DRUGS

Viruses present a different set of problems for antibiotic therapy because they are obligate intracellular parasites. Many of the drugs that would eliminate the virus would be potentially dangerous to non-infected cells as well. Therefore, selective toxicity is difficult to maintain when searching for antiviral antibiotics. Because many viruses are difficult if not impossible to grow, it is also difficult to test potential antiviral drugs. In addition, many acute viral infections have such a short duration they are essentially over before antibiotics could be of any therapeutic use. This is compounded by the lack of rapid tests that can differentiate between various viral infections. Furthermore, successful antiviral drugs must eliminate all virions because the escape of even one virion from an infected cell into the host's blood could quickly restart the infectious cycle.

In the early development of potential antiviral antibiotics, the technique called *blind screening* was employed. This involved looking for any chemical that might inhibit viral infection. As we will discuss below, the advent of new molecular techniques using recombinant DNA as well as the understanding of many viral life cycles has made the use of blind screening obsolete. The first antibiotic to be used against viruses was the sulfa drug derivative known as thiosemicarbazone, in the 1950s. In the 1960s amantadine (Symmetrel®) was developed for use against influenza A. **Table 19.3** lists a variety of antiviral drugs. Let's look at some of the commercially available ones and their characteristics.

Acyclovir

Acyclovir is a specific, nontoxic drug that is highly effective against both genital and oral herpes simplex infections and has been used with some success in the treatment of varicella-zoster (chickenpox and shingles). It was discovered in 1974 but not widely used until the mid-1980s and can be taken intravenously or orally or used topically. Acyclovir is a nucleoside analog of guanine and is produced as a **prodrug**, which means a drug that must be activated enzymatically once in the patient's body. Because this activation is accomplished by enzymes found only in infected cells,

Target/Drug Examples	Comments/Characteristics
Viral uncoating	
Amantadine and rimantadine	Used to treat influenza A infections
Nucleic acid synthesis	
Nucleoside analogs [acyclovir, ganciclovir, ribavirin, zidovudine (azidothymidine, AZT), didanosine (2',3'-dideoxyinosine, ddI), lamivudine (2',3' -dideoxy-3' -thiacytidine, 3TC)]	Primarily used to treat infections caused by herpesviruses and HIV; they do not cure latent infections. The drugs are converted within eukaryotic cells to a nucleotide analog; virally encoded enzymes are prone to incorporate these, resulting in premature termination of synthesis or improper base pairing of the viral nucleic acid. Acyclovir is used to treat active herpes simplex virus (HSV) and varicella-zoster virus (VZV) infections. Ganciclovir is used to treat cytomegalovirus infections in immunocompromised patients. Ribavirin is used to treat respiratory syncytial virus (RSV) infections in newborn infants. Combinations of nucleoside analogs such as zidovudine (AZT), didanosine (ddI), and lamivudine (3TC) are used to treat HIV infections
Nonnucleoside polymerase inhibitors (foscarnet)	Primarily used to treat infections caused by herpesviruses. They inhibit the activity of viral polymerases by binding to a site other than the nucleotide-binding site. Foscarnet is used to treat ganciclovir-resistant cytomegalovirus (CMV) and acyclovir-resistant herpes simplex virus (HSV)
Nonnucleoside reverse transcriptase inhibitors (nevirapine, delavirdine, efavirenz)	Used to treat HIV infections. They inhibit the activity of reverse transcriptase by binding to a site other than the nucleotide-binding site and are often used in combination with nucleoside analogs
Assembly and release of viral particles	
Protease inhibitors (indinavir, ritonavir, saquinavir, nelfinavir)	Used to treat HIV infections. They inhibit protease, an essential enzyme of HIV, by binding to its active site
Neuraminidase inhibitors (zanamivir, oseltamivir)	Used to treat influenza virus infections

Table 19.3 Some Examples of Antiviral Drugs

acyclovir is selectively toxic. The mechanism of action is blockade and termination of viral DNA replication. New variations of acyclovir, such as Valtrex® and Famvir®, are very effective and widely used.

Ganciclovir

This drug is a derivative of acyclovir and was developed to treat cytomegalovirus infections. It was originally given intravenously, and at that time the side effects were so severe that it was used only in life-threatening illnesses such as AIDS. Recently, a less toxic oral derivative has been developed and is now used effectively for cytomegalovirus infections in immunocompromised hosts.

Fast Fact

Foscarnet has been found to be toxic to both kidney and bone marrow and is therefore a drug of last resort for life-threatening infections.

Foscarnet

Foscarnet is an inhibitor of DNA replication used to treat herpes infections. This drug inhibits the binding site of hepatitis B viral DNA polymerase and also HIV reverse transcriptase.

Ribavirin

This drug was developed in 1972 and, like acyclovir, is a nucleoside analog. It is highly toxic and has been very controversial. Even though it has been around for many years, its mechanism of action is still unclear. It has been used in the United States for Lassa fever and hantavirus infections.

Amantadine

As mentioned above, amantadine has been around for some time. It was the first highly specific potent antiviral agent used for influenza A viral infections. It targets viral protein and inhibits uncoating of the viral particle. Unfortunately, the influenza A virus frequently mutates the protein target of amantadine, thereby becoming resistant to this drug. Influenza B does not contain the target protein and is therefore unaffected by amantadine.

Fast Fact

Amantadine has side effects associated with the central nervous system, which have fortuitously been shown to be of some use in relieving symptoms of Parkinson's disease. In point of fact, more amantadine is sold for diseases of the central nervous system than for viral infections.

ANTIFUNGAL DRUGS

Fungi have always been relatively innocuous organisms from the perspective of infectious diseases. However, the emergence of diseases such as AIDS and other infections that render a host immunocompromised have led to increased secondary fungal infections. We will concentrate on specific fungal diseases in Part VI of this book; here we limit our discussion to antibiotics that are currently available for use against fungal infections.

Identification of drugs that can be used for fungal infections has been difficult because of selective toxicity issues. Selective toxicity in bacteria takes advantage of the differences between prokaryotes and eukaryotes. Because fungi are eukaryotes, however, many of the common targets that are attacked by antibiotics generate serious side effects in the host.

Polyenes

Polyenes are produced by the soil bacterium *Streptomyces*. Polyene antibiotics, such as amphotericin B, increase the permeability of the plasma membrane of fungi by interacting with sterols in the membrane. For many years, amphotericin B was used for systemic fungal infections such as histoplasmosis and coccidiomycosis. However, this antibiotic must be used with caution because of its renal toxicity.

Azoles

Azoles such as imidazole and triazole inhibit the production of sterols in fungi. Clotrimazole and miconazole, which are derivatives of imidazole, are sold without a prescription and are routinely used topically for the treatment of athlete's foot and vaginal yeast infection. The broad-spectrum derivative ketoconazole can be taken orally and is somewhat less toxic than the polyene amphotericin B for systemic fungal infections. The least toxic forms of azoles are fluconazole and itraconazole, which have become widely used for systemic fungal infections.

Griseofulvin

This drug is produced by a species of the fungus *Penicillium*. Although it is administered orally, griseofulvin is effective superficially for fungal infections of the skin (ringworm), hair, and nails. The drug seems to react with keratin and blocks the formation of microtubules, which in turn inhibits mitosis in the fungi.

Other Antifungal Antibiotics

The antifungal antibiotic flucytosine interferes with DNA and RNA synthesis. This drug is taken up preferentially by fungi but has a high level of toxicity toward kidney and bone marrow that limits its use. Pentamidine, which seems to bind to fungal DNA, is used in the treatment of *Pneumocystis* pneumonia, a fungal pneumonia that is commonly seen in immunocompromised individuals, particularly AIDS patients. A more complete list of antifungal drugs is given in **Table 19.4**.

Drug Target/Drug	Comments
Plasma membrane	
Polyenes (amphotericin B, nystatin)	Binds to ergosterol, disrupting the plasma membrane and causing leakage of the cytoplasm. Amphotericin B is very toxic but is the most effective drug for treating life-threatening infections; newer lipid-based emulsions are less toxic but very expensive. Nystatin is too toxic for systemic use, but it can be used topically
Azoles Imidazoles (ketoconazole, miconazole, clotrimazole) Triazoles (fluconazole, itraconazole)	Inhibit synthesis of sterol, which is essential to fungal cell membrane structure
Allylamines (naftifine, terbinafine)	Inhibits an enzyme in the pathway of ergosterol synthesis. Administered topically to treat dermatophyte infections. Terbinafine can be taken orally
Cell division	
Griseofulvin	Used to treat skin and nail infections. Taken orally for months; concentrates in the dead keratinized layers of the skin; taken up by fungi invading those cells and inhibits their division. Active only against fungi that invade keratinized cells
Nucleic acid synthesis	
Flucytosine	Used to treat systemic yeast infections; enzymes within yeast cells convert the drug to 5-fluorouracil, which inhibits an enzyme required for nucleic acid synthesis; not effective against most molds; resistant mutants are common

Table 19.4 Characteristics of Antifungal Drugs

Keep in Mind

- Viruses present problems for antibiotic therapy because they are intracellular parasites: drugs that kill infected host cells may also damage healthy host cells.
- Fungal infections have become more prevalent since the appearance of AIDS.
- Because fungi are eukaryotes (as are host cells), side effects of antifungal drugs can be serious.

DRUGS FOR PROTOZOA AND HELMINTHS

As we saw in Chapter 14, one of the most widespread and devastating diseases, malaria, which is caused by a parasite, was often overlooked in developing nations because the incidence of this disease was low in those places. And what was true for malaria has been true for many other parasitic infections: the development of drugs for them has lagged behind the development of antibacterial and antiviral drugs because parasitic infections do not occur at high frequency in developed nations. In other words, there is no money to be made in development of these types of drug.

There are two widely used anti-parasitic drugs: quinine and metronidazole. Quinine as a treatment for malaria has been used since the 1600s, and over the centuries it has been chemically modified into several synthetic forms, such as chloroquinine, for the treatment of malaria. As in bacteria, malarial resistance to chloroquinine can occur, and other quinine derivatives such as mefloquine must be substituted. Another derivative, diiodohydroxyquin, is used for the treatment of several intestinal amebic diseases but has been found to be toxic to the optic nerve.

Metronidazole is one of the most widely used anti-protozoan drugs. It is sold under the name Flagyl® and is the drug of choice for diseases such as vaginitis resulting from *Trichomonas vaginalis* infection, giardiasis and amebic dysentery. This drug interferes with anaerobic metabolism and has also been shown to be effective against some anaerobic bacteria.

Anti-helminthic drugs, like anti-protozoan drugs, have been largely ignored until recently because the affected populations were not found in developed countries. However, the popularity of sushi in the developed world has led to an increase in tapeworm infestations, and increased world travel has led to an increased occurrence of helminth infections in travelers. Niclosamide, which inhibits the production of ATP, is the first choice of treatment for these types of infection. The broad-spectrum anti-helminthic praziquantel is also effective against tapeworms, with its mechanism of action being the increase in permeability of plasma membranes. This is the drug of choice for fluke diseases such as schistosomiasis, in which the effect on the worm is the induction of muscle spasms. These spasms apparently expose antigenic sites for attack by the host immune system.

Mebendazole is used for treatment of several common intestinal helminthic infections, such as pinworm (*Enterobius vermicularis*) and ascariasis (*Ascaris lumbricoides*). The mode of action of mebendazole is disruption of microtubule formation, which affects the motility of the worm. There are also drugs that paralyze the worm, such as pyrantel pamoate and ivermectin. This paralysis induces the worm to exit from the body. **Table 19.5** lists anti-protozoan and anti-helminthic drugs.

Keep in Mind

- Drugs that are useful against parasitic protozoans and helminths have been slow to be developed because diseases caused by these organisms do not often occur in developed countries.
- Metronidazole is one of the most widely used anti-protozoan antibiotics.
- Mebendazole is used to treat several types of intestinal worm infection.

DEVELOPMENT OF NEW ANTIBIOTICS

As we have seen, antibiotics are natural products produced by microbes that can inhibit the growth of other microbes. Since the discovery of penicillin more than three-quarters of a century ago, the search for new antibiotics has continued in earnest. It is becoming safe to assume that most of the naturally occurring antimicrobial products produced by known organisms have already been found. This leaves us with the synthetic modification of the already existing compounds. Although this development of new generations of antibiotics has worked well so far, disease-causing microbes continue to develop resistance, and in some cases they do so very quickly. Therefore, the major question is how we find new compounds that are antimicrobial and also possess the necessary selective toxicity. There are several approaches that are being taken to deal with the problem.

One of the most promising is to identify novel microbial structures or functions that can be used as potential new targets when dealing with a microorganism that has developed resistance to all current drugs available. This type of analysis is often done in conjunction with DNA mapping, which can identify genes that code for essential products that can be used as targets.

Table 19.5 Characteristics of Anti-protozoan and Anti-helminthic Drugs

Causative Agent/Drug	Comments
Intestinal protozoa	
Iodoquinol	Mechanism unknown. Poorly absorbed but taken orally to eliminate amebic cysts in the intestine
Nitroimidazole, metronidazole	Activated by the metabolism of anaerobic organisms. Interferes with electron transfer and alters DNA. Does not reliably eliminate the cyst stage. Metronidazole is also used to treat infections caused by anaerobic bacteria
Quinacrine	Mechanism of action is unknown, but it may be due to interference with nucleic acid synthesis
Plasmodium* (malaria) and *Toxoplasma	
Folate antagonists (pyrimethamine, sulfonamide)	Interferes with folate metabolism. Used to treat toxoplasmosis and malaria
Quinolones (chloroquine, mefloquine, primaquine, quinine)	The mechanism of action is not completely clear. Chloroquine is concentrated in infected red blood cells and is the drug of choice for preventing or treating the red blood cell stage of the malarial parasite. Its effects may be due to inhibition of an enzyme that protects the parasite from the toxic by-products of hemoglobin degradation. Primaquine destroys the liver stage of the parasite and must be used to treat relapsing forms of malaria. Mefloquine or quinine is used to treat infections caused by chloroquine-resistant strains of the malarial parasite
Trypanosomes and *Leishmania*	
Eflomithine	Used to treat infections caused by some types of *Trypanosoma*. It inhibits the enzyme ornithine decarboxylase
Heavy metals (melarsoprol, sodium stibogluconate, meglumine antimonite)	These inactivate sulfydryl groups of parasitic enzymes, but they are very toxic to host cells as well. Melarsoprol is used to treat trypanosomiasis, but the treatment itself is often lethal. Sodium stibogluconate and meglumine antimonite are used to treat leishmaniasis
Nitrofurlimox	Widely used to treat acute Chagas' disease; it forms reactive oxygen radicals that are toxic to the parasite as well as the host
Intestinal and tissue helminths	
Avermectins (ivermectin)	Ivermectin causes neuromuscular paralysis in parasites. It is used to treat infections caused by *Strongyloides* and tissue nematodes
Benzimidazole (mebendazole, thiabendazole, albendazole)	Mebendazole binds to tubulin of helminths, blocking microtubule assembly and inhibiting glucose uptake. It is poorly absorbed in the intestine, making it effective for treating intestinal, but not tissue, helminths. Thiabendazole may have a similar mechanism, but it is well absorbed and has many toxic side effects. Albendazole is used to treat tissue infections caused by *Echinococcus* and *Taenia solium*
Phenols (niclosamide)	Absorbed by cestodes in the intestinal tract, but not by the human host
Piperazines (piperazine, diethylcarbamazine)	Piperazine causes a flaccid paralysis in worms and can be used to treat infections caused by Ascaris. Diethylcarbamazine immobilizes filarial worms and alters their surface, which enhances killing by the immune system. The resulting inflammatory response, however, causes tissue damage
Pyrazinoisoquinolines (praziquantel)	A single dose of praziquantel is effective in eliminating a wide variety of trematodes and cestodes. It is taken up and ultimately causes tetanic contractions in the worm
Tetrahydropyrimidines (pyrantel pamoate, oxantel)	Pyrantel pamoate interferes with neuromuscular activity of worms, causing a type of paralysis. It is not readily absorbed from the gastrointestinal tract and is active against intestinal worms including pinworm, hookworm, and *Ascaris*. Oxantel can be used to treat *Trichuris* infections

New Targets for Bacteria

We currently have the skills to map bacterial chromosomes efficiently. Analysis of these structures can identify open reading frames in the DNA sequence that are seen only in prokaryotes and are conserved over generations. These areas of the bacterial DNA could code for essential components that could then be used as targets.

Fast Fact

So far, there have been hundreds of gene products that have been identified as potential targets for antibiotic therapy.

In addition to mapping bacterial chromosomes, other routes hold promise for identifying new targets in drug-resistant bacteria. One method takes advantage of messenger RNA. We can easily identify this RNA in bacteria and use a technique called microarray analysis to look for the most abundant copies of specific messages. If these copies are abundantly produced over a variety of growth conditions, it is possible that the products they code for are required for the organism to survive and could be potential new targets for antibiotic therapy.

Molecular biological techniques can also be used to look for *auxiliary targets*, which mean targets that may be associated with old targets but are identified by new antibiotics. As an example, *Mycobacterium tuberculosis* is an organism that produces a waxy outer coat that can prevent the successful use of certain drugs. If it were possible to identify, target, and inactivate the proteins that assemble that waxy coat, the organism would be susceptible to antibiotics that could then easily make entry into the cell. In a similar fashion, RNA helicase proteins required for proper folding of the RNA molecule can be targeted such that the folding of the RNA molecule is absent or incorrect, thereby destroying normal RNA function.

The bacterial ribosome can also be re-examined for new targets. X-ray crystallography studies of the bacterial ribosome have shown that it is composed of a series of channels through which mRNA and newly formed proteins are constantly moving. Blockade of these channels would be devastating to the bacteria. A separate, but similar, approach would be to target bacterial efflux pumps, which are used as a means of resistance and function by "pumping" antibiotics out of the cell (we discuss this in detail in the next chapter).

Other new targets could be virulence factors or microbial survival mechanisms. One example could be targeting the production of the lipid A component of lipopolysaccharide layer in the outer layer of Gram-negative organisms (Chapter 9). Because lipid A is intimately associated with the toxicity of these organisms, elimination of this component could negate toxicity. A second example would be the targeting and inactivation of proteins that organisms such as *Listeria*, *Salmonella*, and *Yersinia* use to avoid destruction by phagocytic enzymes.

Peptide Fragments

The approach to looking for new molecules that have antibacterial activity has become more synthetic and molecular. This is possible because of new techniques in chemistry that permit the rapid and efficient production of synthetic molecules and fragments. We can couple our understanding of new genetically defined potential targets (described above) with these new molecules and look for a fit. Simply stated, computer analysis of bacterial genomes can be coupled with computer-generated construction of chemical fragments without initially working on the bacteria themselves. This permits the rapid screening of potential antibacterial compounds, following up on only the most promising in further studies.

There are several antibiotic compounds that are composed of peptide fragments that act on the bacterial plasma membrane, such as bacitracin, gramicidin S, and polymyxin B. Similarly, there are peptides, such as the

marginins produced by frogs and the **defensins** produced by humans, that have antibacterial activity. So far, there are over 500 known peptides produced by multicellular organisms that can serve as antibiotics and have been shown to act by insertion into the phospholipid plasma membrane. Even some Gram-positive bacteria produce these peptides, which in these organisms are called **lantibiotics**. The peptides drosocin and apidaecin have antibacterial activity and are produced by insects.

The search for new and potent antibiotics will continue as long as microorganisms become resistant to those currently in use. The problem is that the only thing organisms have to do is develop resistance, while we must develop new and clinically safe antibiotics. In this race we are the tortoises, as you will see below.

New Targets for Viruses

Molecular technologies are also being used for the development of drugs that can be used for viral infections. Mechanism-based (target-based) screening is useful because it permits a high throughput of possible compounds and can be aided with bioinformatics, combinatorial chemistry, and computer-based design. High-throughput screening aided by robotics can be used in the initial evaluation of as many as 50,000 compounds in a single day. In addition, pharmaceutical companies can place all the compounds they synthesize into "libraries" for future testing. Some companies have libraries with as many as 500,000 compounds in them.

Combinatorial chemistry uses computer programs to produce all possible combinations of a basic set of modular components and can generate thousands of compounds in a day. These can be combined into testing programs in the search for potential "hits", by which we mean reactivity to bacterial pathogens. If the molecular structure of a potential target is known (this is usually defined by X-ray crystallography), computer programs can select the library compounds that will fit with those targets, thereby identifying compounds with antibiotic potential.

Fast Fact

Although high-throughput techniques permit the rapid identification of potentially useful drugs, they do not give information on the potential toxicity or side effects of these compounds. Intensive and expensive testing is required before the drug can be considered for therapeutic use in humans.

Genetic sequencing, also known as genome sequencing, is another way in which potential compounds can be selected. Many viral genomes have been completely sequenced, and high-density arrays of DNA fragments can be used to assess which genes are expressed. In fact, more than 10,000 unique sequences can be put on a single microscope slide. The identification of gene sequences that are always switched on (constitutive) gives insight into which genes code for possibly unique targets.

The Cost of Research and Development

Even with the latest breakthroughs in molecular technology described above, the costs of drug development are not trivial. Although these techniques can help in rapidly identifying thousands of compounds, there still has to be a set sequence of rigorous and expensive testing that occurs before a new drug can be brought to market. In fact, there can be thousands of promising compounds that, after testing, yield only one candidate drug with real potential, and that compound may fail in tests for toxicity, allergic effects, mutagenicity, carcinogenicity, and effectiveness *in vivo*. It is estimated that the cost of a new drug can range from $100 million to $500 million and take as long as 5–10 years to reach the market place.

Keep in Mind

- Most of the naturally occurring antibiotics have already been found.
- There are several approaches being taken to provide new antibiotics,

including identifying new targets on bacteria and viruses, and the development of peptide fragment antibiotics.

- The development of new antibiotics is expensive ($100 million to $500 million) and time-consuming (5–10 years).

TESTING OF ANTIBIOTICS

Because of the risks to our health, the development of new antibiotics must follow a very stringent and highly regulated pathway. These rules and regulations have been put in place to protect us from inadequately researched and poorly manufactured drugs that could be harmful. Remember, any chemical can be antipathogenic, but selective toxicity is required to keep the public safe. However, the need for product safety is a two-edged sword in that it requires a great deal of time and money to produce an effective and safe antibiotic.

Any new anti-pathogen compound that shows promise has to go through the following steps to win approval from the federal Food and Drug Administration:

- The compound is first tested against a panel of pathogen strains, some of which are resistant to antibiotics. If positive against these strains, the new compound moves on to the next step.
- It is tested in infected animals to see whether they are cured by the compound. During this testing, side effects are also evaluated.
- It is compared with standard antibiotics that are currently in use, to determine whether they are more effective.
- Last, clinical testing must be performed in humans. This is the most expensive and time-consuming part of the procedure.

There are several testing systems that can be used to evaluate new compounds. Perhaps the most widely used test for antibiotics is the Kirby–Bauer test, which is similar to the disk method described in Chapter 18 and is shown in **Figure 19.7**.

As with the evaluation of disinfectants (Chapter 18), an agar plate is covered with a known pathogen. Tiny filter-paper disks that have been impregnated with known concentrations of the compound being tested are placed on the agar. After incubation, any disk that inhibits pathogen growth is easily identifiable by the zone of inhibition surrounding the disk.

In addition to testing various concentrations for a single new compound, the disk method is also used to compare the relative effectiveness of different compounds. In these tests, it is important to note that a larger zone of inhibition does not necessarily indicate a more powerful compound, because there can be differences in diffusion rates. Comparison

Fast Fact

The cost of generating new antibiotics coupled with the inherent "lottery" nature of succeeding has forced many pharmaceutical companies to abandon the development of new antibiotics.

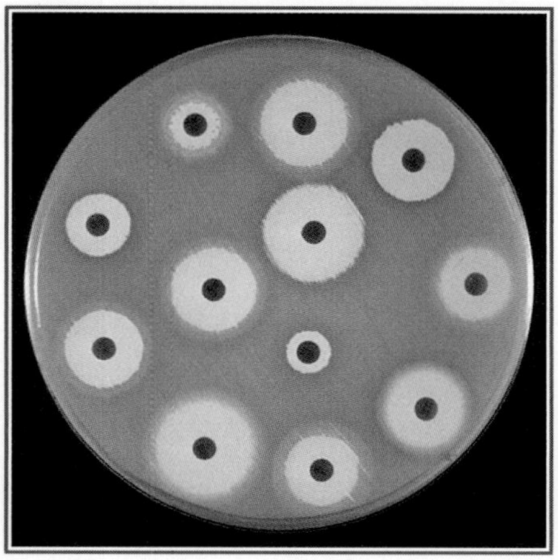

Figure 19.7 The Kirby–Bauer Test for testing potential antibiotic drugs. This plate shows an agar layer with continuous bacterial growth. When tiny discs containing either different compounds or different concentrations of the same compound are placed on the agar, the compounds diffuse out. If they are pharmacologically active, they form a zone of inhibition in which bacteria do not grow. Remember, unless the antibiotic concentrations and diffusion rates are comparable, the efficiency of one antibiotic cannot be compared with another using this test. Also remember that this test does not distinguish between microbistatic and microbicidal compounds.

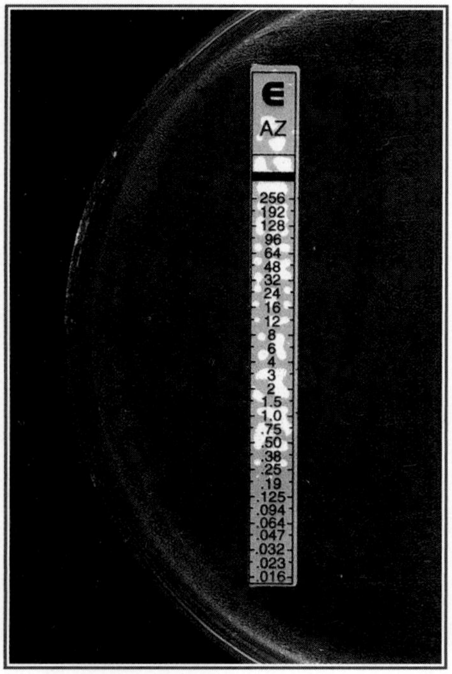

Figure 19.8 The E test uses a quantitative scale that can be used to determine the MIC (minimum inhibitory concentration) of the compound being tested. This picture shows the drug azithromycin; the strip contains a gradation of antibiotic with the strongest at the top and the weakest at the bottom. The minimum inhibitory concentration of antibiotic that will inhibit the bacteria is determined by where the growth of the bacteria starts, or the area of the growth where the ellipse meets the strip (in this case 0.25).

of similar compounds can be done by measuring the zones and referring to a standardized table for that type of drug and that concentration. When testing specific organisms by the disk method, the organisms can be classified as sensitive, intermediate, or resistant to the compound or compounds being tested. Although this is a simple and inexpensive test, it is inadequate for most clinical purposes.

A more advanced diffusion test known as the **E test** permits the determination of the **minimal inhibitory concentration** (**MIC**), which is defined as the lowest antibiotic concentration that prevents visible pathogen growth. This test employs plastic-coated strips that contain gradients of antibiotic concentrations. After incubation of pathogen-coated plates, the MIC can be read from the scale printed on each strip (**Figure 19.8**).

Although the disk method and the E test can show you which compounds inhibit pathogen growth, there is no way to determine whether a compound being tested is microbicidal or microbistatic. The **broth dilution test** is used for this purpose and determines the *minimal bactericidal concentration (MBC)* of a compound. This procedure involves incubating a specific organism in a sequence of wells containing decreasing amounts of the antibiotic compound being tested. Microbes from wells that do not show growth can be recultured in nutrient broth medium containing none of the test compound. Growth in this medium indicates that the compound being tested inhibited the growth of the organism but did not kill it. Therefore it was microbistatic. If no growth occurs in the reculture, the test compound killed the organism and is classified as microbicidal.

These types of dilution test are highly automated, and additional testing can use colorimetric methods to indicate reactivity and test for serum concentrations of the test compound. This is important because many drugs are toxic at high concentrations, making it essential to establish and monitor the levels for use.

Keep in Mind

- The development of antibiotics is highly regulated by government authorities such as the Food and Drug Administration in the United States.
- There are several testing systems used to evaluate new compounds.
- The most widely used test for antibiotics is the disk method described in Chapter 18.
- The E test is a diffusion test that determines the MIC (minimum inhibitory concentration) of antibiotics.
- The broth dilution test is used to determine the MBC (minimum bactericidal concentration) of an antibiotic.

SUMMARY

- There have been no new discoveries of natural antibiotics for several years.
- Microorganisms produce toxic chemicals as part of their natural defense.
- Antibiotics can be broad or narrow spectrum.
- Chemical modification of natural antibiotics can broaden their spectrum.
- Some bacteria produce an enzyme called β-lactamase that inhibits the reactivity of penicillin.
- Antibiotics must be selectively toxic.
- Most antibiotics will have side effects.
- The five targets for antibiotics are the cell wall, the plasma membrane, the ribosome, nucleic acids, and metabolic synthesis pathways.
- Viruses present problems for antibiotic treatment because they are obligate intracellular parasites.
- Antifungal drugs have serious side effects because fungi are eukaryotic cells.
- Drugs against parasitic protozoans and helminths have been slow to be developed because these infections occur mostly in underdeveloped countries.
- The development of new drugs is an expensive and time-consuming process.

With all the improved methods for drug discovery and testing, one fundamental problem remains. All pathogenic organisms can become resistant to new drugs very quickly. The problem of drug resistance has been referred to as one of today's greatest medical threats. We discuss antibiotic resistance in the next chapter.

SELF EVALUATION AND CHAPTER CONFIDENCE

Multiple Choice

Answers are given in the back of the book and help can be found in the student resources at:
www.garlandscience.com/micro

1. The first antibiotic discovered was
   - A. Streptomycin
   - B. Quinine
   - C. Sulfa drugs
   - D. Penicillin
2. Most of the available antimicrobial agents are effective against
   - A. Viruses
   - B. Fungi
   - C. Bacteria
   - D. Protozoa
   - E. All of the above
3. Which of the following does not belong with the others?
   - A. Monobactam
   - B. Cephalosporin
   - C. Bacitracin
   - D. Streptomycin
   - E. Penicillin
4. Which of these antimicrobial agents has the fewest side effects?
   - A. Penicillin
   - B. Chloramphenicol
   - C. Tetracycline
   - D. Erythromycin
   - E. Streptomycin
5. Which of the following methods of action would be bacteriostatic?
   - A. Inhibition of cell wall synthesis
   - B. Inhibition of RNA synthesis
   - C. Competitive inhibition with folic acid synthesis
   - D. Injury to plasma membrane
   - E. None of the above
6. Which of the following antimicrobial agents is recommended for use against fungal infections?
   - A. Amphotericin B
   - B. Penicillin
   - C. Bacitracin
   - D. Cephalosporin
   - E. Polymyxin

7. More than half of our antibiotics are
   A. Produced by Fleming
   B. Produced by bacteria
   C. Produced by fungi
   D. Synthesized in laboratories
   E. None of the above
8. Which of the following drugs is NOT used primarily to treat tuberculosis?
   A. Sulfonamide
   B. Rifampin
   C. Isoniazid
   D. Ethambutol
   E. None of the above
9. The antimicrobial drugs with the broadest spectrum of activity are
   A. Aminoglycosides
   B. Macrolides
   C. Chloramphenicol
   D. Lincomycin
   E. Tetracyclines
10. Which of the following organisms would most probably be sensitive to natural penicillin?
   A. *Streptococcus pyogenes*
   B. *Penicillium*
   C. Penicillinase-producing *Neisseria gonorrhoeae*
   D. *Mycoplasma*
11. *Streptomyces* bacteria produce all of the following antibiotics except
   A. Erythromycin
   B. Nystatin
   C. Kanamycin
   D. Rifampin
   E. Bacitracin
12. Broad-spectrum antibiotics
   A. React only with Gram-positive bacteria
   B. React only with Gram-negative bacteria
   C. React only with *Pseudomonas*
   D. React with Gram-positive bacteria, Gram-negative bacteria and *Pseudomonas*
   E. React only with large bacteria
13. A bacteriostatic antibiotic
   A. Inhibits bacterial growth by killing the organism
   B. Increases the electrical charge of the organism
   C. Damages the bacterial plasma membrane
   D. Inhibits bacterial growth but does not kill the organism
   E. None of the above
14. The difference between penicillin and ampicillin is
   A. The β-lactam ring
   B. The type of carbohydrates associated with the drug
   C. The side chains affixed to the core ring structure
   D. All of the above
15. β-Lactamase is
   A. A ring structure seen in semi-synthetic penicillin
   B. Found only on cephalosporin
   C. A chemical that enhances the effect of antibiotics
   D. An enzyme that cleaves the ring structure of penicillin
   E. None of the above
16. All of the following are targets for antibiotics except
   A. The cell wall
   B. Bacterial ribosomes
   C. The glycocalyx
   D. The plasma membrane of the bacteria
   E. Nucleic acids
17. Carbapenem antibiotics have
   A. The same structure as penicillin
   B. Different side chains from penicillin's
   C. The same ring structure as penicillin
   D. A different ring structure from penicillin's
   E. Three rings in the core
18. The antibiotic isoniazid is
   A. Similar to ethambutol
   B. Never used with ethambutol
   C. Used with rifampin
   D. Used with ethambutol and rifampin for the treatment of tuberculosis
   E. None of the above
19. Protein synthesis is
   A. Not a target for antibiotics
   B. Not a selectively toxic target
   C. A selective target
   D. A target of last resort for antibiotic therapy
   E. Used only for viral infections
20. All of the following target the ribosome except
   A. Streptomycin
   B. Tetracycline
   C. Penicillin
   D. Chloramphenicol
   E. Erythromycin
21. Sulfa drugs target
   A. Metabolism
   B. The cell wall
   C. The bacterial plasma membrane
   D. Ribosomes
   E. DNA
22. Antiviral drugs
   A. Are all selectively toxic
   B. Only affect infected cells

C. Only affect free viral particles
D. Must eliminate all viral particles to be effective
E. Only effect uncoating of viral particles

23. Acyclovir

A. Is effective against herpes infections
B. Is selectively toxic
C. Blockades viral DNA replication
D. All of the above
E. Only A and C above

24. All of the following are antifungal drugs except

A. Polyenes
B. Bacitracin
C. Azoles
D. Griseofulvin
E. Flucytosine

25. Which of the following is used for protozoan infections?

A. Griseofulvin
B. Bacitracin
C. Polyenes
D. All of the above
E. None of the above

26. Which of the following antimicrobial peptides is naturally produced by frogs?

A. Lantibiotics
B. Defensins
C. Marginins
D. Drosocin

27. Which of the following tests gives the MBC?

A. The Kirby–Bauer test
B. The E test
C. The broth dilution test
D. B and C above
E. A above only

DEPTH OF UNDERSTANDING

Questions listed here require you to bring together the concepts you have learned in this chapter into a discussion format. This helps you to increase your depth of understanding of the material you have learned. Help can be found in the student resources at: www.garlandscience.com/micro

1. Discuss the differences between natural and semi-synthetic penicillin, using structure and function as the foundation of your answer.

2. Evaluate the targets used for antibiotics with regard to selective toxicity.

3. Using what you know about the bacterial cell wall, describe how bacteria could develop resistance to antibiotics that attack this target.

CLINICAL CORNER

Help can be found in the student resources at: www.garlandscience.com/micro

1. You are working in the drug development section of a large university and have found a new compound with properties similar to penicillin's. When bacteria come in contact with the compound they stop growing.

   A. How would you test the effectiveness of this new compound?
   B. What are the most important questions that you need to answer about this compound?

2. Your patient has an infection with Gram-positive staphylococci. She has been on cephalosporin for seven days and has shown little improvement. The doctor has switched her to a combination of streptomycin and penicillin. She does not understand why this is necessary. How do you explain it to her?

   A. Why did the initial treatment not seem to work?
   B. What will the benefits of the new drug therapy be?
   C. Do you think that this switch will make a difference?

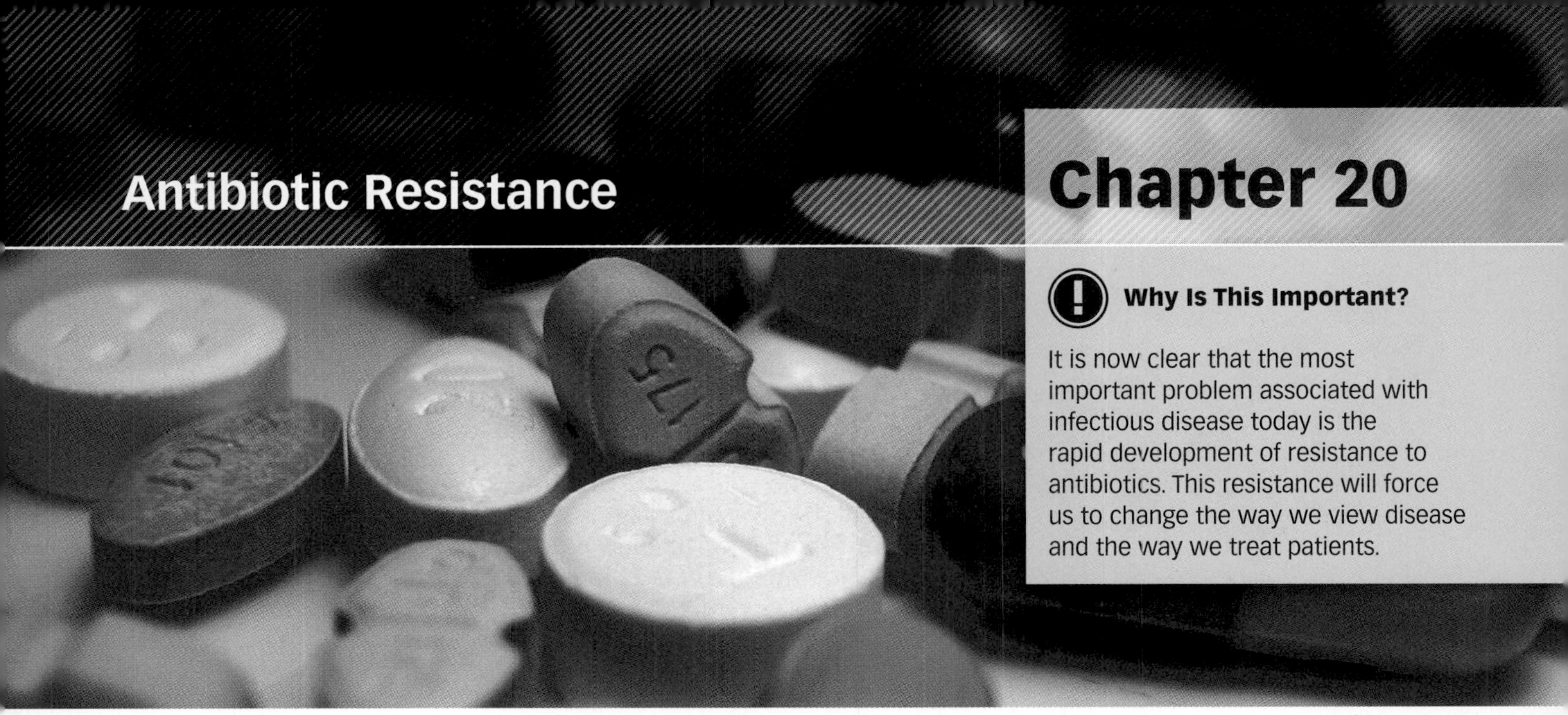

Antibiotic Resistance

Chapter 20

Why Is This Important?

It is now clear that the most important problem associated with infectious disease today is the rapid development of resistance to antibiotics. This resistance will force us to change the way we view disease and the way we treat patients.

OVERVIEW

This chapter is dedicated to bacterial resistance to antibiotics. Even though modifying existing antibiotics can increase the life spans of those antibiotics, resistance is inevitable. In other words, resistance to antibiotics is a matter not of "if" but of "when." Resistance has become such a widespread and alarming problem it is important that health care professionals understand why and how this resistance occurs.

We will divide our discussions into the following topics:

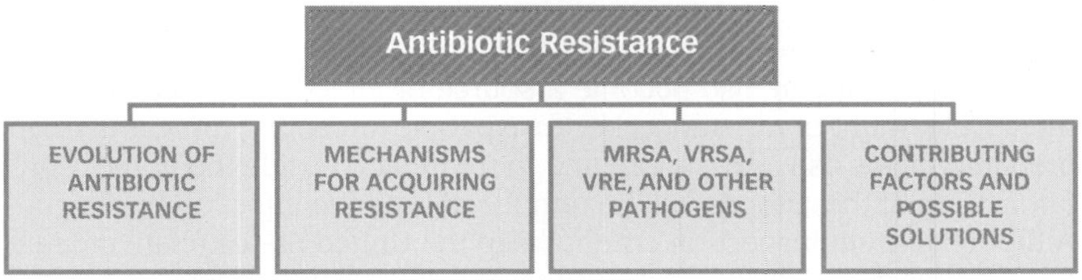

BACKGROUND

Although antibiotics have alleviated the impact of diseases, their use has not been without consequence. Several factors have a role in the development of antibiotic resistance, and you already have a feeling for some of these from our discussions in previous chapters. Because bacteria have very short generation times, they can quickly grow into large populations. Consequently, the potential for rapid spontaneous mutation is considerable. Among the mutations that occur are those that allow the survival of an organism exposed to lethal antibiotics, and these mutations will be selected for (**Figure 20.1**). As you can see from the figure, those bacterial cells that have developed resistance to a given antibiotic are not killed off when treated with the drug. These resistant cells continue to divide, and the resulting population will be completely resistant. The combination of mutation and evolutionary pressure has given rise to a rapid increase in populations of bacteria that have become resistant to antibiotics.

The development of resistant strains of bacteria can be further exacerbated by today's technologies and sociological structures. For instance, today a person can travel anywhere in the world within 24 hours and bring

What Do I Need to Know?

To get the most out of this chapter, please review the following terms from your previous courses in biology, anatomy, physiology, or chemistry or in previous chapters of this book as indicated in parentheses: active transport (**4**), β-lactam ring (**19**), β-lactamase (**19**), constitutive gene (**11**), efflux pumping (**19**), enteric, etiology (**5**), inducible gene (**11**), nucleus of antibiotic molecule (**19**), operon (**11**), periplasmic space, plasmid (**9**), repressible genes (**11**), transposon (**11**).

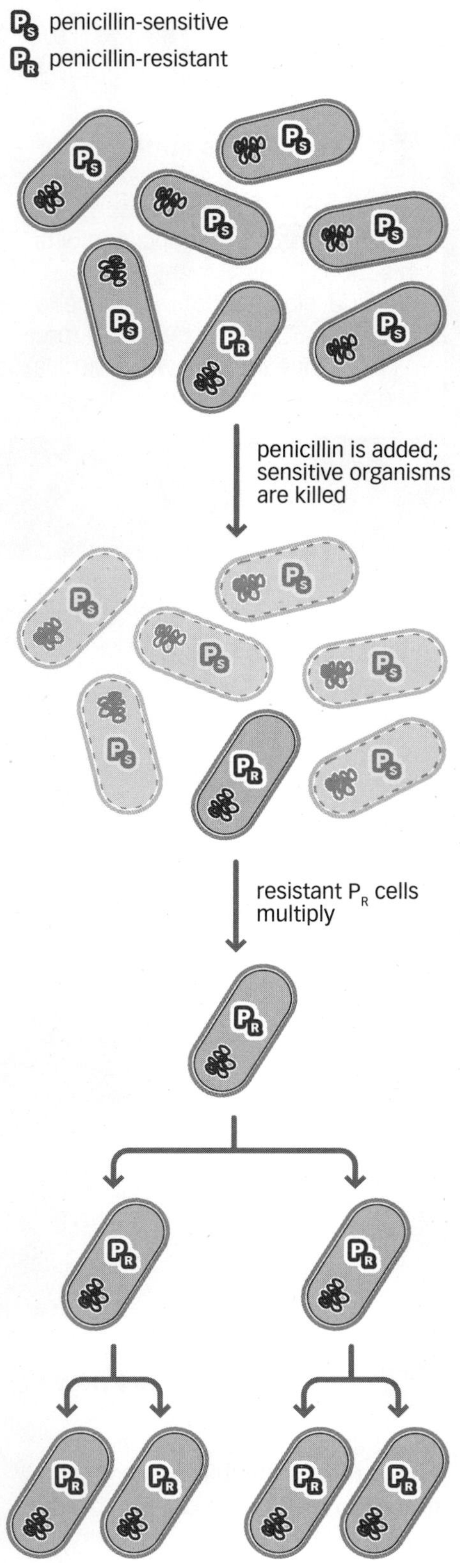

Figure 20.1 Development of antibiotic resistance. Resistance is an evolutionarily favorable mutation. In the initial population, most of the bacterial cells are sensitive (P_S) to the antibiotic being used for treatment, but one is resistant (P_R) to the drug. When the antibiotic is applied, the P_S cells die but the P_R cell survives and reproduces. Eventually the population is made up entirely of P_R cells. Any mutation that permits the survival of the organism will be selected for and passed on from generation to generation, as indicated here.

resistant bacteria with them. That travel can take place in an enclosed space (plane, train, or car) in association with several or many other people. In a plane, for instance, an individual infected with a resistant bacterium will be seated in an enclosed space, in which the air is continuously recirculated, for several hours. The problem increases because the individuals who may become secondarily infected by the resistant bacterial strain are usually making connections to other planes, where the process can be repeated. The SARS outbreak of 2003 was a good example of this type of dissemination by travel. In that case, the corona virus responsible for this severe respiratory disease was moved from a small rural village in mainland China to more than 20 countries in a matter of days.

Similarly, there are more large cities in the world today than in eras past. Large numbers of people in relatively small areas means that passing a contagious disease or antibiotic-resistant pathogen from one person to another is more easily accomplished than in areas where populations are spread out. This can be further compounded by the fact that many of these large urban populations are seen in developing countries, where sanitation is poor. In these situations, organisms such as the bacterium *Vibrio cholerae* can easily cause epidemics of cholera, and *Vibrio* organisms have become resistant to antibiotics, making these outbreaks even more dangerous.

The food we eat can also become a source of infection that could affect the development of resistance, because people are eating more and more prepared meals as well as more and more meals outside the home. Contamination of these foods will go unnoticed until the infection has begun. Although current inspection criteria keep the United States relatively safe from foodborne infections, many infections still occur as evidenced by the outbreaks of *Escherichia coli* O157 in spinach and lettuce seen in 2006, and *Salmonella* in peaunt butter in 2009. In addition, importations of food products from other countries, where safety standards may not be as stringent, present us with additional opportunities for the movement of hazardous bacteria. As people become sick from foodborne infections, the use of antibiotics increases and so does the development of resistance.

Perhaps the most important social change that affects bacterial resistance is the increase in the number of people who are immunocompromised. As you will remember from our discussions in Chapter 16, the adaptive immune system is responsible for our very survival. In Chapter 17, we saw that there are individuals who are immunocompromised for a variety of reasons. This increase in immunocompromised individuals has necessitated the increased use of antibiotics, which has in turn fostered the development of resistance.

Another source for development of increased bacterial resistance is emerging and re-emerging diseases, which we discussed in Chapter 8. The name *emerging diseases* encompasses infectious diseases that have not been seen before, while re-emerging diseases are those caused by organisms that have become resistant to treatment with antibiotics that were formerly successful. Bacterial diseases in the emerging category include toxic-shock-syndrome forms of *Staphylococcus aureus*, ulcer-causing *Helicobacter pylori*, and *Legionella pneumophila*. An example of a re-emerging disease is tuberculosis caused by *Mycobacterium tuberculosis*. **Table 20.1** lists some of these emerging diseases.

Table 20.1 Some Emerging Bacterial Diseases Since 1977

Year	Agent	Comment
1992	*Bartonella henselae*	Cat-scratch disease; bacillary angiomatosis
1992	*Vibrio cholerae* O139	New strain, epidemic cholera
1989	*Ehrlichia chaffeensis*	Human ehrlichiosis
1983	*Helicobacter pylori*	Peptic ulcer disease
1982	*Borrelia burgdorferi*	Lyme disease
1981	Toxin-producing strains of *Staphylococcus aureus*	Toxic shock syndrome
1977	*Campylobacter jejuni*	Enteric pathogen, global distribution

Keep in Mind

- Mutations that confer resistance to antibiotics are selected for by bacteria.
- Travel and modern technological advances have increased the spread of antibiotic resistance.
- Re-emerging diseases are occurring as a result of antibiotic resistance.
- The single most important factor in the rise of antibiotic resistance is the increase in the number of immunocompromised people.

EVOLUTION OF ANTIBIOTIC RESISTANCE

There was a considerable delay between the discovery of the first antibiotics and their first clinical use. As we discussed in Chapter 19, the clinical success of these substances led to ever-increasing efforts to discover new antibiotics, coupled with a great emphasis on the modification of existing drugs. The intent was to find or develop antibiotics with broader spectra of reactivity. However, in the recent past, that effort has been fueled by the need to overcome the development of bacterial strains that are resistant to treatment with antibiotics. It is important to note how quickly the development of resistance can occur (**Table 20.2**). Several groups of antibiotics — the sulfonamides, streptomycin, erythromycin, and vancomycin — were in use for many years before resistance to them was observed. In contrast, resistance to penicillin was observed only three years after it was widely used.

For some semi-synthetic forms of penicillin, such as ampicillin, the length of time before resistance was observed was relatively long (deployed in 1961, resistance observed in 1973). Other semi-synthetic forms, however, such as methicillin, lasted only a year before resistance was observed. In some cases, the short interval for some antibiotics is directly related to increased use. Methicillin is a good example of this. This modified penicillin had a broad spectrum of activity and was so effective that it became overprescribed. Remember, the more an antibiotic is used, the greater the occurrence of resistance. In fact, the therapeutic life span of a drug is based on how quickly resistance develops.

Table 20.2 Evolution of Resistance to Antibiotics

Antibiotic	Year Deployed	Resistance Observed
Sulfonamides	1930s	1940s
Penicillin	1943	1946
Streptomycin	1943	1959
Chloramphenicol	1947	1959
Tetracycline	1948	1953
Erythromycin	1952	1988
Vancomycin	1956	1988
Methicillin	1960	1961
Ampicillin	1961	1973
Cephalosporins	1960s	late 1960s

Table 20.2 also leads us to a discussion of perhaps the most important contributing factor for the development of antibiotic resistance: overuse. Take as an example upper respiratory infections, including the common cold, which represent one of the most prevalent clinical problems we face. It has been estimated that upwards of 60% of all upper respiratory infections are of viral etiology. That said, almost anyone seeing a doctor for flu-like symptoms will be given a prescription for an antibiotic. In fact, most of us would wonder about a visit to our doctor that did not result in the proffering of the "script." Because many antibiotics are not effective against viruses, taking the prescribed antibiotic does essentially nothing for relieving our symptoms, which thanks to our immune system would subside naturally within a week. In addition, taking the antibiotic can have a negative effect by destroying the normal flora of our bodies. As we have seen, the bacteria that make up these resident populations have a critical role in keeping under control organisms that can become opportunistic pathogens. Patients who, feeling better, decide not to finish the course of treatment, patients who use outdated drugs, and patients who use antibiotics that were prescribed for something else all contribute to the problem by allowing the invading bacteria to survive while the normal flora are destroyed.

The time it takes organisms to develop resistance to an antibiotic varies, but there is no doubt that the more an antibiotic is used, the more quickly resistance occurs. That is why hospital settings are outstanding reservoirs for the acquisition of resistance. Hospitals have all the necessary ingredients, including a population of people who for one reason or another have had their health compromised and a high concentration of organisms, many of which can be extremely pathogenic. The hospital is also the place where large amounts of different antibiotics are constantly in use. Depending on the patient, these may include drugs such as vancomycin, which is considered an antibiotic of last resort. Because increased use leads to resistance, the hospital is a place where resistance can develop rapidly, and resistance to antibiotics of last resort can be clinically devastating.

Transfer of resistance from one species of bacteria to another can also be easily accomplished in a hospital setting. As we know from Chapter 11, bacteria have several mechanisms for moving genes from one organism to another. Health care workers and employees in the hospital who do not follow infection control protocols can also aid bacteria in increasing resistance.

It has been shown that bacterial plasmids containing genes for resistance can integrate into the chromosome of a recipient bacterium at specific sites, and these sites can be referred to as *resistance islands* where resistance genes accumulate and are consistently maintained.

Microorganisms that produce antibiotic substances have genes that code for mechanisms to protect the microorganisms from accidentally committing suicide. These are referred to as *autoprotective mechanisms*. Usually these mechanisms involve membrane-bound protein export pumps, referred to as transmembrane pumps, that pump out the freshly produced antibiotic so that it does not accumulate to levels that would kill the organism producing it. The genes that code for these pumps are closely linked to genes that code for the antibiotic substances. Because of this linkage, there is a combined effect: as the genes for antibiotic production are turned on, so are the pump genes. In fact, there is some evidence to suggest that the antibiotic production machinery and the pumps are colocated on the membrane and are kinetically coupled.

Members of the bacterial genus *Streptomyces*, which produce antibiotics such as erythromycin, use three mechanisms for self-protection. Some streptomycetes modify the structure of their 50S ribosomal subunit (a target for erythromycin) in such a way that it is no longer "seen" by the antibiotic they are producing. Other streptomycete species use membrane protein-exporting pumps powered by ATP, and still others produce the antibiotic in an inactive form that is non-lethal while in the cell. After it has been exported out of the cell, it is converted to its active form by extracellular enzymes.

The production of vancomycin employs similar protective mechanisms. The streptomycetes that produce this substance reprogram their peptidoglycan layer to make it unsusceptible to attack by the self-produced vancomycin. In addition, these organisms produce vancomycin only during the stationary growth phase, thereby limiting the number of growing cells that could be targeted by the antibiotic.

Fast Fact

Bacteria that are resistant to more than one antibiotic are becoming more and more prevalent, and they present dangerous clinical challenges.

MECHANISMS FOR ACQUIRING RESISTANCE

As we discuss the mechanisms that bacteria use to become antibiotic-resistant, you will notice that these are the same mechanisms that antibiotic-producing bacteria use to protect themselves (**Figure 20.2**):

- Inactivation of the antibiotic
- Efflux pumping of the antibiotic
- Modification of the antibiotic target
- Alteration of the pathway

In some ways, these protective mechanisms represent countermeasures to the methods we use in our search for potential antibiotics. For instance, we choose targets that would be lethal for the organism, while the bacteria develop ways to change those targets so they are no longer identifiable. Let's look at each of these mechanisms in more detail.

Inactivation of Antibiotic

Obviously, if bacteria in an infected patient being treated with an antibiotic can inactivate the antibiotic in some way, the bacteria will become resistant to the antibiotic. Inactivation usually involves the enzymatic breakdown of the antibiotic molecules. We learned in Chapter 19 about β-lactamase, the enzyme produced by bacteria in response to antibiotics

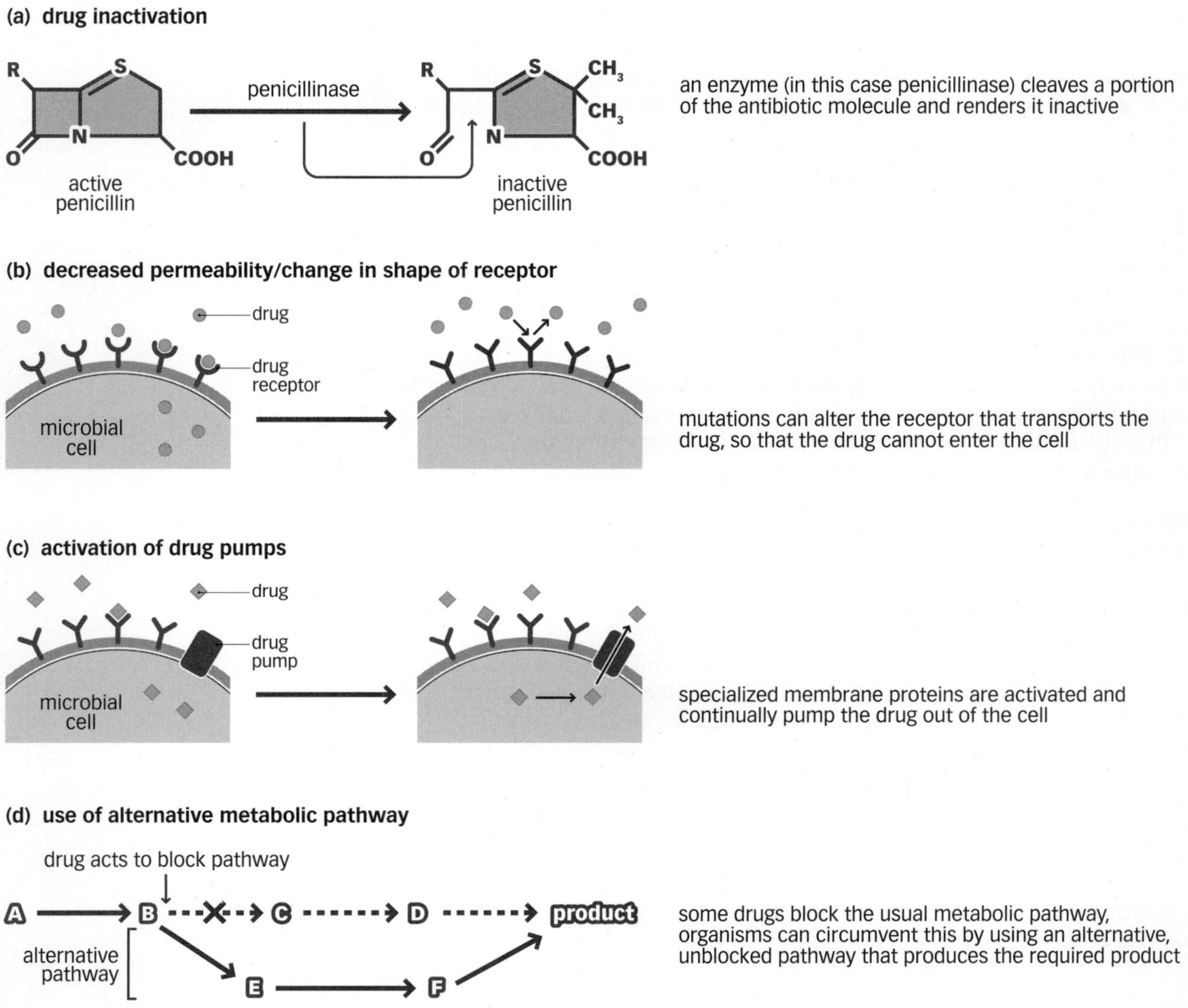
(a) drug inactivation
penicillinase
active penicillin
inactive penicillin
an enzyme (in this case penicillinase) cleaves a portion of the antibiotic molecule and renders it inactive
(b) decreased permeability/change in shape of receptor
drug
drug receptor
microbial cell
mutations can alter the receptor that transports the drug, so that the drug cannot enter the cell
(c) activation of drug pumps
drug
drug pump
microbial cell
specialized membrane proteins are activated and continually pump the drug out of the cell
(d) use of alternative metabolic pathway
drug acts to block pathway
A
B
C
D
product
E
F
alternative pathway
some drugs block the usual metabolic pathway, organisms can circumvent this by using an alternative, unblocked pathway that produces the required product

Figure 20.2 Mechanisms of resistance. Organisms that are resistant can use a variety of mechanisms to defeat the drug, including drug inactivation (panel **a**), alteration of the antibiotic transport receptor (panel **b**), increased elimination through efflux pumping (panel **c**), or the use of alternative metabolic pathways (panel **d**).

that have a β-lactam ring as their core. The lactamase hydrolyses that ring structure in penicillins, cephalosporins, and carbapenems. There are more than 190 forms of β-lactamase, and these enzymes are usually secreted into the bacterial periplasmic space and attack the antibiotic as it approaches its target. Efforts to obtain antibiotics that are not affected by these enzymes led to the development of cefotaxime and ceftazidime. Additionally, as we discussed in Chapter 19, another way to inhibit the destruction of the β-lactam ring in an antibiotic molecule is to couple the antibiotic with lactamase inhibitors such as potassium clavulanate (Augmentin® and Timentin®) and sulbactam (Unasyn®).

Regulation of β-Lactamase Activity

Two bacterial genes are involved with β-lactamase activity: the *Amp C* gene in enteric bacteria and the *bla Z* gene in *Staphylococcus aureus*. These genes can be embedded in the bacterial chromosome, or they can be carried on either plasmids or transposons. They are usually turned off and turned on (induced) only in the presence of molecules containing a β-lactam ring. As we look at a couple of examples of how this mechanism works, keep in mind what we learned about gene expression and inducible and repressible genes in Chapter 11.

Example 1: *Escherichia coli*. In *E. coli*, there are four genes involved in the production of β-lactamase. All of these genes are from the *Amp* gene group, which controls the expression of the *Amp C* gene, which codes for β-lactamase. The *Amp G* gene codes for a bacterial protein that binds to fragments of peptidoglycan created when the antibiotic destroys the bacterial cell wall. The Amp G protein transports these fragments through the cytoplasm and induces the transcription of the *Amp C* gene for β-lactamase. Therefore, when the bacterial cell wall begins to be destroyed, the gene for β-lactamase is turned on to stop the destruction.

Example 2: *Staphylococcus aureus*. In some cases, *S. aureus* will use the *bla Z* lactamase genes for antibiotic resistance. This is also an inducible system; it is normally off and not induced until penicillins or cephalosporins are present in the bacterial cell. In *S. aureus*, the production of β-lactamase involves three things — sensing, signal transduction and transcription — and three genes and two proteins are involved. All three genes are negatively regulated by the bla 1 repressor protein, which binds to the promoter region of the DNA and stops transcription. Other proteins embedded in the bacterial plasma membrane signal the detection of the β-lactam ring in the antibiotic molecules. These signals cause the production of a protein that cleaves apart the bla 1 repressor protein bound to the promoter, thereby allowing the transcription of β-lactamase, the enzyme that will destroy the antibiotic.

These two examples show how the production of resistance factors such as β-lactamase can be genetically regulated. This is in keeping with the fastidious nature of most bacteria in that they don't make things they don't need. In addition, these examples give an indication of the sophisticated nature of some bacterial resistance mechanisms.

Efflux Pumping of Antibiotic

The second major route of drug resistance is through the use of bacterial pumps known as **efflux pumps**. This is an active transport mechanism that requires ATP and is mediated by the efflux pumps, which transport proteins. These pumps are located in the bacterial plasma membrane but can also be found in the outer layer of Gram-negative organisms. They work by keeping the concentration of antibiotic in the bacterial cell below levels that would destroy the cell. Active transport is required because the concentration of antibiotic is high outside the cell as the infected patient ingests antibiotic. Therefore movement of antibiotic out of the bacterial cells is movement against a concentration gradient. Although efflux pumps are active against β-lactams and fluoroquinolones, their greatest activity is seen against tetracyclines.

Like the antibiotics they protect against, efflux pumps can be classified as narrow-spectrum or broad-spectrum, with the broad-spectrum pumps conferring resistance against more than one type of antibiotic. The concept of efflux pumps is not unique to bacteria, and many species of bacteria use them not only for protection against antibiotics but also for everyday physiological functions. The genes that code for efflux pumps are located on plasmids and transposons and as such can be readily acquired by non-resistant bacteria to transform them into resistant species.

Classification of Efflux Pumps

There are four mechanisms by which efflux pumps operate. Three of them use the principle of *counterflow*, in which the antibiotic is pumped out of the bacterial cell at the same time cations such as H^+ and Na^+ are pumped into the cell. Bacteria also use these pumps to remove antiseptic and disinfectant substances (in particular the quaternary ammonium compounds). The fourth pump mechanism is a one-way transport that depends solely on ATP expenditure and does not simultaneously import cations as antibiotic molecules are moved out of the cell. This form of efflux pump is relatively rare, and it is thought that although this ATP-consuming pump is required for normal physiology, it is co-opted into antibiotic efflux activity by the bacteria when necessary.

It has been shown that although transmembrane efflux pumps found on Gram-positive bacteria function independently of one another, there can be "partner" proteins associated with efflux pumps on Gram-negative bacteria. These partner proteins may become targets in the search for new antibiotics, because inhibition of a partner protein may keep the efflux pump from working.

Some Gram-negative bacteria contain efflux pumps in both the inner membrane and the outer layer, with a connecting protein between them. This three-protein sequence is seen in efflux pumps that work on tetracycline, ciprofloxacin, chloramphenicol, and β-lactams. *Pseudomonas* has four families of these three-protein sequence pumps, and they all overlap functionally, making it one of the most antibiotic-resistant organisms.

Fast Fact

E. coli alone has over 35 different genes that code for drug efflux pumps.

Regulation of Efflux Pumps

As with all other cellular mechanisms, the operation of efflux pumps is carefully regulated. This is especially true for any mechanism that joins the outside of the cell to the inside, because things that are brought into the cell or moved from the cell may have drastic implications for cell survival. The best-studied antibiotic pumping system is that used for tetracycline in Gram-positive and Gram-negative organisms.

These efflux pumps are regulated by the *Tet* gene family and are another example of the regulation of gene expression we discussed in Chapter 11. The *Tet R* gene codes for a repressor protein that binds to the *Tet* operon. This restricts the transcription of the *Tet A* gene, which codes for the efflux-pump protein that pumps out tetracycline. When no tetracycline is present in a bacterial cell, the protein is not made. However, when tetracycline is present, it binds to the Tet R protein, which is then inhibited from repressing the *Tet A* gene. The efflux-pump protein is then made in large quantities and sent to the bacterial plasma membrane, where it pumps tetracycline out of the cell, simultaneously importing needed cations. This same type of regulation occurs in *Pseudomonas*, *Neisseria gonorrhoeae*, and *E. coli*. The bacterium *Bacillus subtilis* uses the same system for the efflux of fluoroquinolones, chloramphenicol, and doxorubicin antibiotics.

It should be noted that in addition to membrane efflux pumping, some bacteria reduce the permeability of their membranes as a way of keeping antibiotics out. This is accomplished by turning off the production of porin and other membrane channel proteins. This mechanism is seen in *E. coli* O157:H7 as well as in *Pseudomonas*, which use this decreased membrane permeability for resistance to streptomycin, tetracycline, and sulfa drugs.

Modification of Antibiotic Target

Modification of some component of a bacterial cell that is the target of an antibiotic is the third way in which bacteria escape antibiotic activity. This is a very interesting mechanism because the bacteria must change the structure of the target but the modified target must still be able to function. Remember, the proper three-dimensional structure of proteins is required for them to function properly, and any change in the structure can render a protein nonfunctional. Therefore, the modifications made to any targets in a bacterial cell must be such that the target can still function but has become insensitive to drug activity.

This can be achieved in two ways: mutation of the gene that codes for the target protein or importing a gene that codes for a modified target that performs the required function but is not recognized by the attacking antibiotic. A good example of this is seen with MRSA (methicillin-resistant *Staphylococcus aureus*). Recall from Chapter 19 that bacteria have PBPs (penicillin-binding proteins) in their plasma membranes and that these proteins are targets for the penicillin family of antibiotics. MRSA has acquired the *mec A* gene, which codes for a PBP that is three-dimensionally different from the PBP usually found in *S. aureus* cells and is consequently less sensitive to penicillin antibiotics. MRSA is essentially resistant to all β-lactam antibiotics, cephalosporins, and carbapenems, making it a very dangerous pathogen, particularly in burn patients.

Fast Fact

The incidence of MRSA in hospitals in the United States has risen to levels of 20–40%.

The production of insensitive PBPs is an example of operon function at the genetic level. The gene coding for an insensitive PBP is always kept switched off by a repressor protein. Through a series of reactions, the antibiotic fragments the repressor protein, and the absence of the intact repressor protein then allows the insensitive PBP to be made. Because the PBP does not attach to any penicillin molecules, the bacterial cell wall can be constructed correctly even in the presence of the antibiotic. Through the natural selection of generations of bacterial growth, this type of antibiotic resistance can accumulate to very high levels. For example, when MRSA was treated with the fluoroquinolone ciprofloxacin, resistance to the drug increased from 5% to more than 85% in a year!

Another example of PBP target modification can be seen with *Streptococcus pneumoniae*, an organism that can make as many as five different types of PBP. Apparently this is not the product of simple mutation but rather the ability of the bacterium to rearrange or shuffle these genes. This is referred to as genetic *plasticity* and permits increased resistance.

Modification of Target Ribosomes

The ribosomes in bacterial cells are a primary target for antibiotics, and several antibiotics affect this target in different ways. For example, as noted in Chapter 19, erythromycin and its descendents azithromycin and clarithromycin attack the 23S ribosomal RNA of the 50S subunit. This family of antibiotics has been routinely used for respiratory infections. However, resistance to these drugs has increased markedly in the past 20 years. This resistance is the result of modification of the 23S RNA, a modification that makes the RNA no longer sensitive to the antibiotics. Some organisms use target modification in conjunction with production of efflux pumps, making the resistance even more effective.

Modifying ribosomes to make them insensitive to antibiotics is also a suicide-prevention mechanism used by organisms that produce antibiotics. For example, there is a gene in erythromycin-producing bacteria called *Erm E*. Being constitutive, it is always on and protects these bacteria from the erythromycin they produce. This same type of gene has been found in pathogenic bacteria that were originally not resistant to erythromycin but have developed resistance over the years. In these bacteria, however, the gene is inducible, not constitutive. The presence of low levels of erythromycin causes the gene to be expressed, which produces a protein that changes the three-dimensional shape of the ribosome. This change in shape closes off the site that erythromycin attacks, making the antibiotic ineffective.

A similar effect is seen in bacteria that are resistant to antibiotics such as gentamycin. Resistant bacteria produce enzymes that modify the 16S rRNA of the 30S subunit of the ribosome such that this target can no longer be attacked by the antibiotic. There are over 30 forms of these modifying enzymes, and the genes that code for them are found on plasmids. Because plasmids are easily transferred from one bacterial cell to another, the genetic information for synthesizing these gentamycin-resistant enzymes is also easily transferred. The result is the evolution of strains of gentamycin-resistant bacteria.

Alteration of a Pathway

Some of the drugs used to control the growth of bacteria focus on competitive inhibition of metabolic pathways as a mechanism of action (recall our discussion of the sulfa drugs in Chapter 19). Bacteria can overcome this method of control by changing to an alternative pathway. This is illustrated in **Figure 20.2**d. Because metabolic pathways can parallel one another it is easy to circumvent the blockade of an established pathway by a drug. This alternative pathway still achieves the required outcome.

MRSA, VRSA, VRE, AND OTHER PATHOGENS

There are currently several antibiotic-resistant bacteria that are considered clinically dangerous. It is important that we discuss them in some detail because they are rapidly becoming what can only be described as a health care nightmare.

MRSA (methicillin-resistant *S. aureus*) and VRSA (vancomycin-resistant *S. aureus*) are what can be called *professional pathogens*, meaning that they are very virulent in humans. These bacteria contain three or four *resistance islands*, located on the chromosome, plus 26–28 additional gene clusters located on mobile genetic elements, such as plasmids, which can be moved to other bacterial cells. In fact, genes for antibiotic resistance make up approximately 7% of the total *S. aureus* genome. In contrast, *Bacillus subtilis*, a non-pathogenic organism, has no resistance genes! In MRSA there are several specific resistance genes that have been identified. Some of these genes are associated with β-lactamase activity; some are associated with erythromycin resistance, some with the modification of aminoglycosides and some with the operation of efflux pumps (**Table 20.3**).

As mentioned in Chapter 19, VREs are vancomycin-resistant enterococci. One species, *Enterococcus faecalis*, accounts for more than 90% of all vancomycin-resistant bacteria. Resistant enterococcal strains are the leading cause of endocarditis and are common pathogens in patients with

Table 20.3 Resistance Genes of MRSA

Protein	Gene	Antibiotic Resistance
Bleomycin resistance protein	*ble O*	Bleomycin
PBP2'	*mec A*	β-Lactams
β-Lactamase	*bla Z*	β-Lactams
rRNA methylase	*erm A*	Erythromycin, pristinamycins
O-Nucleotidyl transferases	*ant 4'; ant 9'*	Aminoglycosides
Acetylase–phosphotransferase	*aacA-aphD*	Aminoglycosides
Tet M efflux protein	*tet M*	Tetracyclines
Qac A	*qac A*	Antiseptics

indwelling catheters. Very few antibiotics are effective against VRE, and current treatment for VRE infection involves the administration of Synercid® or a combination of oxazolidinone and linezolid.

Genetic resistance to vancomycin involves five tandem genes (meaning that they are located together) that work in sequence to change the structure of peptidoglycan such that it is no longer affected by the antibiotic. One of the major problems with these resistance genes is that they are easily transferred by plasmids or transposons; therefore, resistance to vancomycin can rapidly spread.

Table 20.4 lists some of the resistance mechanisms used by bacteria against commonly prescribed antibiotics. It should be noted that some antibiotics elicit multiple resistance mechanisms. Furthermore, this table lists the association of drug resistance to mutation and plasmid

Table 20.4 Bacterial Resistances to Various Classes of Clinically Used Antibiotics

Antibiotic	Structural Class	Target	Mutant/Plasmid	Efflux	Porin	Inact.	Target Alteration
Ampicillin	Penicillin	E	+/+	Yes	Yes	Yes	Yes
Ceftriaxone	Cephalosporin	E	+/+	Yes	Yes	Yes	Yes
Imipenem	Carbapenems	E	+/+	Yes	Yes	Yes	Yes
Fosfomycin	Phosphonic acid	E	+/+		Yes	Yes	
Gentamicin	Aminoglycoside	R	+/+	Yes		Yes	Yes
Chloramphenicol	Phenylpropanoid	R	+/+	Yes		Yes	Yes
Tetracycline	Polyketide (II)	R	+/+	Yes		?	Yes
Erythromycin	Macrolide	R	+/+	Yes		Yes	Yes
Clindamycin	Lincosamide	R	+/+			Yes	Yes
Synercid	Streptogramin	R	+/+	Yes		Yes	Yes
Telithromycin	Ketolide	R	+/+	Yes		Yes	Yes
Ciprofloxacin	Fluoroquinolone	D	+/+	Yes			Yes
Vancomycin	Glycopeptide	E	+/+				Yes
Sulfisoxazole	Sulfonamide	M	+/+				
Trimethoprim	–	M	+/+				
Rifampin	Ansamycin	P	+/+			Yes	Yes
Fusidic acid	Steroid	T	+/+	Yes			Yes
Linezolid	Oxazolidinone	R	+/–				Yes
Novobiocin	Coumarin	D	+/+				Yes
Isoniazid	–	M	+/–				
Pyrazinamide	–	M	+/–				
Nitrofurantoin	Nitrofuran	M	+/–			Yes	
Polymyxin	Peptide	E	+/–	Yes			Yes
Capreomycin	Peptide	R	+/–			Yes	Yes
Mupirocin	Pseudomonic acid	T	–/+				Yes

*D, replication; E, cell wall; M, metabolism; P, RNA polymerase; R. ribosome; T, translation.

localization. Remember, any resistance capability that is located on a transmissible element, such as a plasmid, can efficiently and in some cases rapidly transfer that resistance to other bacteria.

There is no doubt that MRSA, VRSA, and VRE are big problems for health care and getting bigger each day. In fact, VRSA strains are now found throughout the world. There also seem to be new mechanisms for resistance. For example, the cell walls of some resistant bacteria have increased in thickness. For some MRSA and VRSA strains, there are no longer any antibiotic treatments available. These are the ultimate hospital pathogens. They are genetically *flexible*, meaning they develop resistance to new antibiotics very quickly. It has been suggested that the only way to stop these organisms may be through increased education for health care providers and strict enforcement of hospital infection control protocols.

Fast Fact

It is estimated that over 80 tons of trimethoprim and sulfamethoxazole are placed in animal feed in the United States every year.

Furthermore, bacteria that are part of the normal flora of the body are becoming more dangerous. The best example is *E. coli*, which is part of the normal flora of the large intestine but has become more involved with urinary tract infections. Antibiotic-resistant *E. coli* infections are now being seen in countries throughout the world, and this bacterium has become more and more resistant to antibiotics such as trimethoprim and sulfamethoxazole.

CONTRIBUTING FACTORS AND POSSIBLE SOLUTIONS

A variety of factors have a role in the development of drug resistance. Perhaps the most underestimated is the success of so many of our antibiotics. Since the discovery of these drugs, the world health care landscape has changed drastically, not only for the patient but also for the health care worker. Antibiotics have given the physician powerful weapons that really work to alleviate problems, and patients have become dependent on the use of these drugs. This doctor–patient–drug relationship is most clearly seen in the case of common viral infections. As noted earlier in this chapter, even though many antibiotics will not work in these infections and even though the patient will recover normally in a matter of days without any medication, there is an expectation by the patient that antibiotics have to be prescribed, and in too many cases the physician acquiesces to those wishes. The result is overprescription of antibiotics that are not required. This problem is further compounded by patients who feel better (as they normally will anyway) and stop using the drug prematurely, which causes the normal microbial flora to be killed and opportunistic pathogens to take their place.

Another ingredient in the rise of resistance is the development and overprescription of broad-spectrum antibiotics. These drugs make it much easier for the health care community to deal quickly with sick patients. Rather than go through the trouble of identifying the cause of the infection and then selecting the appropriate narrow-spectrum antibiotic, it is much easier to use a broad-spectrum drug that will kill a large variety of organisms. A good example of this is seen with cephalosporins, broad-spectrum antibiotics that have no side effects.

Overuse of broad-spectrum antibiotics permits the *superinfection effect*, in which pathogens occupy areas where normal microbes have been killed. In these cases, the antibiotics have essentially compromised the patient. An example of a superinfection pathogen is *Clostridium difficile*, which can establish itself in the intestinal tract as part of a superinfection. This organism is very resistant to antibiotics, and patients with this infection are difficult to treat. **Figure 20.3** illustrates how superinfections like this occur.

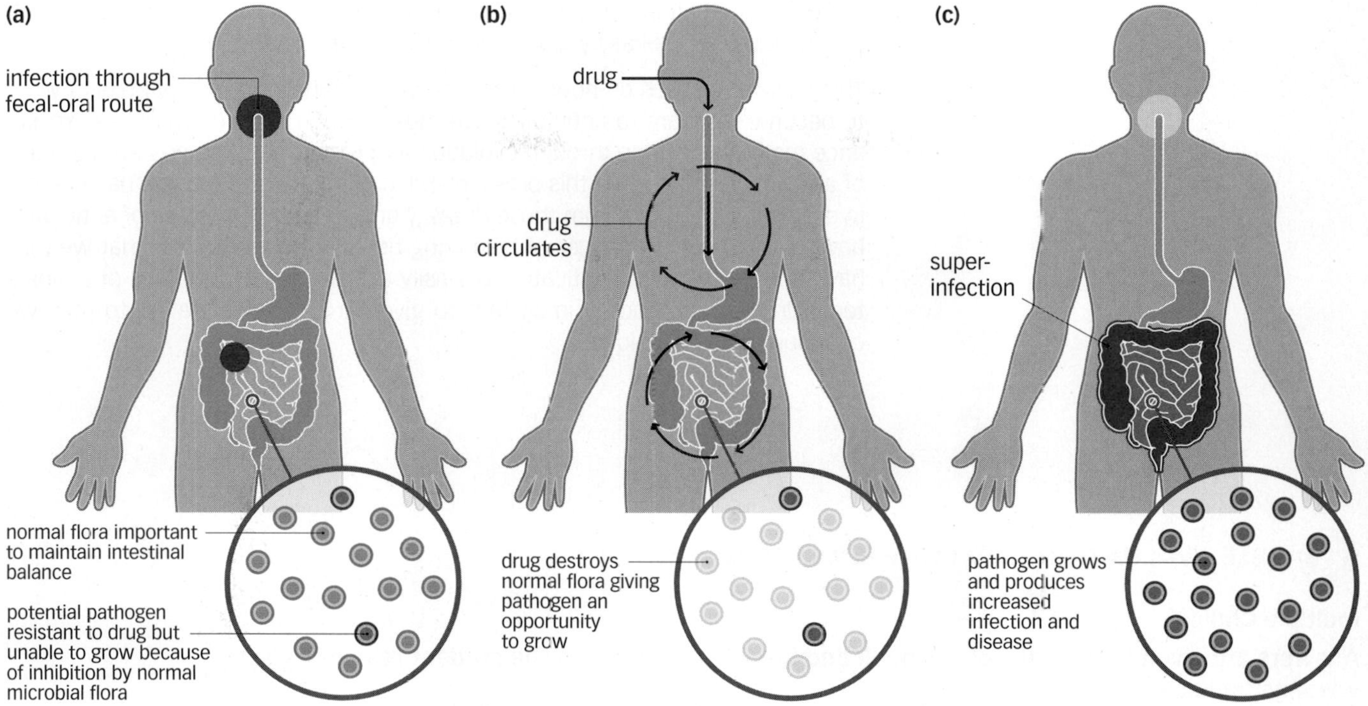

Figure 20.3 Development of superinfection. The destruction of normal bacterial flora allows potentially pathogenic bacteria that would normally be prohibited from growing to take over, resulting in disease.

When you couple the overuse of antibiotics with improper adherence to hospital infection control protocols, the difficulty of finding new antibiotics and the ease of worldwide travel, the specter of global antibiotic resistance can be alarming. However, there are things that we can do to control drug resistance. **Table 20.5** lists six guidelines that can be used to increase the useful life span of drugs.

Keep in Mind

- The time required to become resistant is relatively short for most antibiotics.
- There have been no new natural forms of antibiotics discovered for several years, so the "new" antibiotics are derivatives of antibiotics already in use.
- Bacterial resistance occurs as a result of inactivation of the antibiotic, pumping the antibiotic out of the cell, modifying the target of the antibiotic, or using an alternative pathway from that inhibited by the antibiotic.
- Resistance may be decreased by proper use of the antibiotic, by rotation or cyclical patterns of use, or by the use of combinations of antibiotics.

Point	Guideline
1	Optimal use of all antibacterial drugs
2	Selective removal, control, or restriction of classes of antibacterial agents
3	Use of antibacterial drugs in rotation or cyclic patterns
4	Use of combination antibacterial therapy to slow the emergence of resistance
5	Evaluation of routes of resistance
6	Implementation of global changes

Table 20.5 Guidelines for Extending the Useful Life of Antimicrobial Drugs

SUMMARY

- Mutations that confer resistance are selected for by bacteria.
- Travel and modern technology have increased the spread of antibiotic-resistant pathogens.
- One of the most important contributing factors to antibiotic resistance is the increase in the number of immunocompromised people.
- The time it takes to develop resistance to antibiotics is relatively short.
- Resistance to an antibiotic can occur through inactivation of the antibiotic, pumping the antibiotic out of the cell, modifying the target of the antibiotic, or using alternative metabolic pathways.

- Resistance can be reduced by rotational use or cyclical patterns of use for antibiotics as well as by using combinations of different antibiotics together.

This chapter focuses on how pathogenic bacteria have found a variety of ways to become resistant to antibiotics. We must keep in mind (1) that these resistance methods develop through evolutionary pressure that comes from the use of antibiotics and (2) that this pressure will continue as long as the bacteria are in danger of extinction from these drugs. Our overuse and misuse of antibiotics have placed them in danger of becoming useless. Given the fact that we may have exhausted the identification of easily accessible new sources of antibacterial drugs, it is critically important to give careful consideration to how we continue to use antibiotics.

SELF EVALUATION AND CHAPTER CONFIDENCE

Multiple Choice

Answers are given in the back of the book and help can be found in the student resources at:
www.garlandscience.com/micro

1. All of the following are involved in the spread of antibiotic resistance *except*
   - **A.** Travel
   - **B.** Overuse of antibiotics
   - **C.** Specific prescriptions for antibiotics
   - **D.** Improper use of antibiotics
   - **E.** None of the above

2. Resistance to antibiotics is facilitated by which of the following?
   - **A.** The antibody response
   - **B.** Host immunity
   - **C.** Frequency of use
   - **D.** The inflammatory response

3. All of the following are mechanisms of resistance *except*
   - **A.** Activation of the antibiotic
   - **B.** Efflux pumping
   - **C.** Modification of the target structure
   - **D.** Inactivation of the antibiotic
   - **E.** None of the above

4. The second most often used mechanism in development of antibiotic resistance is
   - **A.** Target modification
   - **B.** Efflux pumping
   - **C.** Destruction of the antibiotic
   - **D.** Enhancement of antibiotic activity

5. MRSA stands for which of the following?
   - **A.** Microbial-resistant *Streptococcus aureus*
   - **B.** Microbial-resistant *Staphylococcus aureus*
   - **C.** Methicillin-reactive *Staphylococcus aureus*
   - **D.** Methicillin-resistant *Staphylococcus aureus*

6. MRSA organisms are resistant to which of the following?
   - **A.** Penicillin
   - **B.** Cephalosporins
   - **C.** Carbapenems
   - **D.** None of the above
   - **E.** All of the above

7. Resistance to antibiotics seen at the level of the ribosome is caused by
   - **A.** Efflux pumping
   - **B.** Destruction of anti-ribosomal antibiotics
   - **C.** Changes in the shape of the ribosome
   - **D.** None of the above

8. Increased use of antibiotics can be attributed to the following
   - **A.** An increasing number of large cities
   - **B.** Emerging infectious diseases
   - **C.** Increased levels of immunodeficiency diseases
   - **D.** All of the above

9. The best way to deal with antibiotic resistance is to use
   - **A.** More antibiotics
   - **B.** Less antibiotics
   - **C.** Combinations of antibiotics
   - **D.** None of the above

10. The useful life of antibiotics can be extended by
    - **A.** Increasing the doses
    - **B.** Using more broad-spectrum antibiotics
    - **C.** Using combinations of antibiotics
    - **D.** None of the above

DEPTH OF UNDERSTANDING

Questions listed here require you to bring together the concepts you have learned in this chapter into a discussion format. This helps you to increase your depth of understanding of the material you have learned. Help can be found in the student resources at:
www.garlandscience.com/micro

1. It has been said that antibiotic resistance may be the greatest threat in medicine today. Defend this statement.

2. Using your knowledge of microbial genetics, describe the rise of antibiotic resistance.

3. Compare the following to determine which is the most important for development of bacterial resistance: destruction of the antibiotic, target modification, or efflux pumping.

CLINICAL CORNER

Help can be found in the student resources at:
www.garlandscience.com/micro

1. One of the attending physicians that see patients on your floor is well known for prescribing very broad-spectrum antibiotics for all of the patients he sees. He has told the patients that it is the quickest and easiest way to get them well.

   A. What is wrong with this approach?
   B. What would be the best way to prescribe antibiotics?

2. Your next-door neighbor just got a flu shot and then developed a runny nose coupled with sneezing and coughing. She went to see her doctor, who prescribed a regimen of penicillin to be taken for seven days. After three days, she felt better and stopped taking the drugs. She mentions to you that she stopped taking them so if she got sick again she could use the rest of the medicine and would not have to pay to see the doctor and get a new prescription.

   A. What should you tell her about her idea of saving what was left of the prescription for the next time she felt sick?
   B. Do you think the prescription was appropriate under the circumstances?

PART VI

Microbial Infections

As we have progressed through this book, we have built a strong foundation for understanding microbiology and infectious disease. **PART I** gave us a basic understanding of the infectious process and the chemistry we need. In **PART II**, we looked at the infection process. In **PART III** we had a detailed discussion of the organisms that can infect us. There we divided our discussions into bacterial pathogens, viruses, fungi, and parasitic organisms and included chapters on growth requirements and bacterial genetics. In **PART IV** we examined ways in which the body can defend itself against infection. **Part V** showed us how we can control microbial growth using disinfectants, antiseptics, and antibiotics as well as ways in which pathogens can defeat our attempts to control their growth.

In the next set of chapters, **PART VI**, we will look at infections of several of the body systems. In addition to the importance of these topics, these chapters can also be used as a form of **ready reference**. In each chapter of **PART VI** we will discuss the infection process for a body system and then divide our discussion into bacterial, viral, fungal, and parasitic infections that affect that system. The coverage of each infection will include a brief discussion of the pathogenesis of the infection and the treatments recommended for it. Although we will briefly discuss the anatomy of these systems, it will be helpful for you to review your anatomy and physiology textbooks so that you can more easily understand the mechanisms that we will be looking at.

In **Chapters 21**, **22,** and **23**, we look at the three most often infected systems: the respiratory, digestive, and genitourinary systems. Because these systems are open to the outside world, they are very disposed to infection, and many of the infections seen in these three systems are routinely seen in clinical settings. **Chapters 24** through **26** give us an understanding of the infections we see in the nervous system, the blood, and the skin and eyes.

When you have finished with PART VI you will:

- See that each system of the body can be involved in infection.
- Understand how the respiratory, digestive, and genitourinary systems are both major portals of entry for pathogens and major portals of exit that can transmit pathogens from one person to another.
- Understand the mechanisms and pathogenesis of some of the more common bacterial, viral, fungal and parasitic infections seen in clinical settings as well as some of the treatments used for these infections.

In order to get the most out of this part of the book, it is important to incorporate the basic concepts of *getting in, staying in, defeating the host defenses, damaging the host,* and *being transmissible* when you look at the various infections covered in **PART VI**. In addition, think about control mechanisms and the consequences of antibiotic resistance in the infections discussed in these chapters.

Chapter 21
Infections of the Respiratory System

Chapter 22
Infections of the Digestive System

Chapter 23
Infections of the Genitourinary System

Chapter 24
Infections of the Central Nervous System

Chapter 25
Infections of the Blood

Chapter 26
Infections of the Skin and Eyes

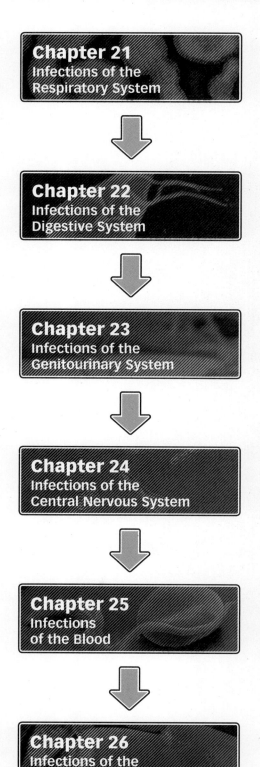

PART VI

[illegible]

As we have progressed through this book, we have [illegible] for understanding microbiology and infectious diseases. PART I gave us a basic understanding of the infectious process and the chemistry we need. In PART II we looked at [illegible]. In PART III we had a detailed discussion of the organisms [illegible]. There we divided our discussions into bacterial, [illegible], and parasitic organisms, and included chapters [illegible] and back[illegible]. In PART IV we examined ways in which the body can defend itself against infection. Part V showed us [illegible] antibiotics, and [illegible] ways in which [illegible] to combat these organisms.

In the next set of chapters, PART VI, we will look at [illegible] body systems. [illegible]

In Chapters 2[illegible] [illegible]

When you have finished PART VI you will [illegible]

[illegible]

In order to [illegible]

Infections of the Respiratory System

Chapter 21

Why Is This Important?

The respiratory system is the most commonly infected of all the systems of the body. As a health care provider you will see more respiratory infections than any other form of routine infection.

OVERVIEW

In this chapter, we discuss infections of the respiratory system. As we learned in Part II, this is one of the major portals of entry into the body for infectious organisms. We can divide the respiratory system into two tracts, upper and lower, on the basis of the structures and functions found in each part. However, the dichotomy between the upper and lower respiratory tracts can also be seen in the type of infections that occur in each region. We will see that infections of the upper respiratory tract, which includes the nasal cavity, sinuses, pharynx, and larynx, are fairly common and are usually nothing more than an irritation. In contrast, infections of the lower respiratory tract are more dangerous and can be very difficult to treat.

In this chapter, we organize our discussion of respiratory infections in two ways: (1) according to whether they occur in the upper or lower tract, and (2) according to whether they are bacterial, viral, or fungal.

We will base our discussions on the following topics:

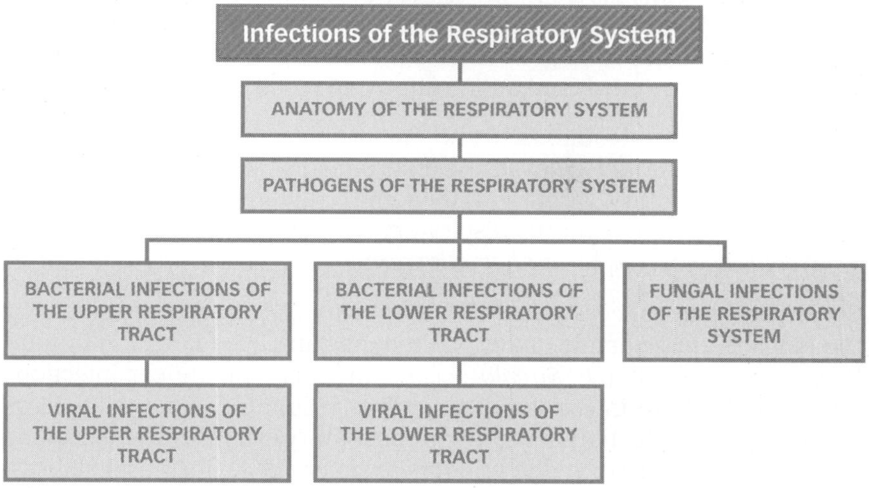

ANATOMY OF THE RESPIRATORY SYSTEM

The respiratory system is the most accessible system in the body. The simple and required act of breathing brings in clouds of potentially infectious bacteria, viruses, and even opportunistic fungi. As we examine

What Do I Need to Know?

To get the most out of this chapter, please review the following terms from your previous courses in biology, anatomy, physiology, or chemistry or in previous chapters of this book as indicated in parentheses: alveolus, antigenic drift and shift (**8**), arthroconidium (**14**), bronchiole, bronchus, chemotactic response (**15**), conidium (**14**), eustachian tube, fomites (**6**), mastoid cavity, mononuclear phagocytic system (**15**), M protein (**9**), nasopharynx, parenchyma, phagolysosome (**15**), pharynx, repressor protein (**11**), septate fungus (**14**), sinus, superinfection (**6**), trachea, uvula, viremia (**5**), vital capacity of lung, zoonotic disease (**6**).

some of the infections caused by these pathogens, remember that there are a variety of host defense mechanisms associated with the respiratory tract, including physical barriers, chemical barriers, and the innate and adaptive immune responses.

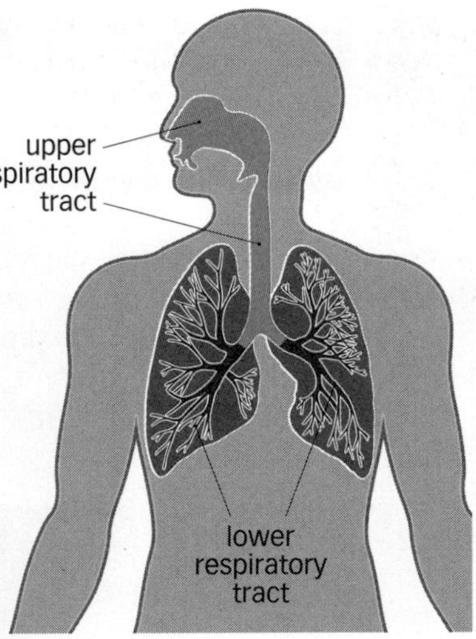

Figure 21.1 The upper and lower respiratory tracts.

We can divide the respiratory system into an upper tract and a lower tract (**Figure 21.1**). The upper tract, the primary entryway into the system, is continuously exposed to potential pathogens drifting around in the external environment. In contrast, the lower respiratory tract, which begins with the bronchi and continues into the lung, is essentially a sterile environment. When this area is infected, pneumonia can result.

PATHOGENS OF THE RESPIRATORY SYSTEM

The list of bacterial organisms that infect the respiratory system is extensive and includes *Streptococcus pneumoniae*, *Haemophilus*, group A streptococci, *Mycoplasma pneumoniae*, *Chlamydia pneumoniae*, *Bordetella pertussis*, and *Mycobacterium tuberculosis*. The upper respiratory tract is also a portal of entry for such viral pathogens as influenza, parainfluenza, respiratory syncytial virus, and rhinovirus (a cause of the common cold). Vaccination has eliminated many infections of the respiratory tract, but some, such as diphtheria and pertussis (whooping cough), are still seen in underdeveloped parts of the world. Some of the microbial sources of respiratory infection are shown in **Figure 21.2**.

Many respiratory infections are caused by pathogens that move from one human to another, with the infection spreading as the pathogens circulate within a community. This circulation can be exacerbated because some pathogens, such as *Streptococcus pneumoniae*, exist in some individuals as part of the normal flora of the upper respiratory tract. Other respiratory pathogens are acquired from animal sources, and so the infections they cause are classified as zoonotic infections. For example, Q fever stems from farm animals that spread the organism *Coxiella burnetii*, and the respiratory infection called psittacosis, caused by *Chlamydia psittaci*, is spread from parrots and other birds to humans.

Water can also be a source of respiratory infections. The prime example is legionellosis, a form of pneumonia caused by *Legionella pneumophila*. This organism can survive in water over the range of temperatures from 25°C to 45°C. If water contaminated with this organism is aerosolized as a mist (as is done in building-wide air-conditioning systems, for instance), droplets can be inhaled, and an infection can result.

Fungi, which are ubiquitous in the environment, are also a source of respiratory infection. However, these organisms do not normally cause infection unless the patient is in some way immunocompromised. Some of the more dangerous respiratory fungal infections are caused by *Aspergillus* species and *Pneumocystis (carinii) jiroveci*.

Some pathogens are restricted to certain sites. *Legionella* infecting only the lungs is one example. Other pathogens can cause infection in multiple sites. For example, *Streptococcus* can cause middle ear infections and sinusitis (both infections of the upper respiratory tract) as well as pneumonia (an infection of the lower respiratory tract). Four areas of the upper respiratory tract — the middle ear, mastoid cavity, nasal sinuses, and nasopharynx — are frequent sites of infection.

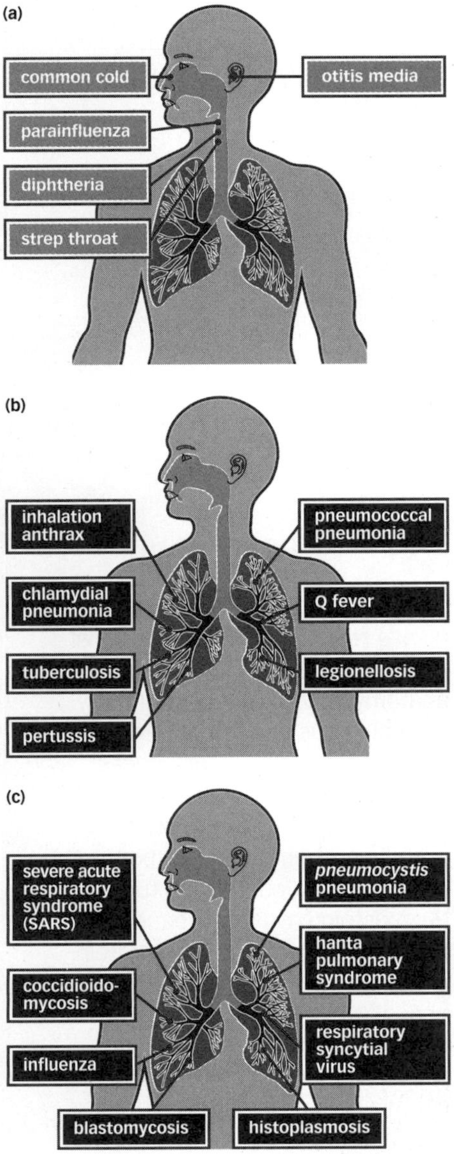

Figure 21.2 Common infections of the respiratory system. Panel **a**: Some of the more common infections and symptoms associated with the upper respiratory tract. Panels **b** and **c**: Infections seen in the lower respiratory tract.

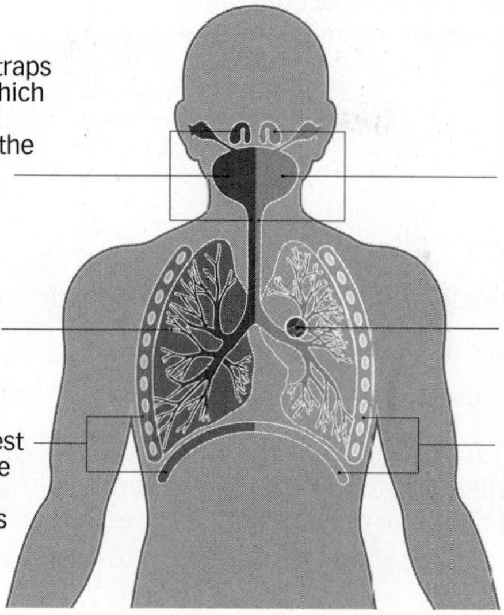
normal
inhibited
mucociliary escalator traps debris and bacteria, which are then moved to the esophagus and out of the respiratory system
viral infections can disrupt the function of the respiratory epithelium, which allows bacteria to enter the respiratory system
alveolar macrophages are an important part of the host defense
a bronchial carcinoma can obstruct a bronchus and infection can arise behind the obstruction
the muscles of the chest wall and diaphragm are important in coughing and clearing secretions from the respiratory system
trauma or abdominal surgery can affect the ability to cough

Figure 21.3 Some of the defenses of the respiratory tract and how they can be inhibited. Loss of these defenses is directly correlated to increased pathogen activity.

Keep in mind that there are significant defenses in the respiratory system (**Figure 21.3**). For example, the ciliated pseudostratified columnar epithelium of the upper respiratory tract, known as the *mucociliary escalator* (**Figure 21.4**), traps pathogens and rhythmically moves them up and out of the system. In addition, the act of coughing is a mechanical process that forcefully eliminates organisms from the respiratory system. These defenses are powerful barriers in the upper respiratory tract. In the lower respiratory system, alveolar macrophages in the alveoli of the lungs help protect against infection. It is important to remember that compromise of any of these defenses can lead to a predisposition to respiratory-system infection.

Bacteria that Infect the Respiratory System

Figure 21.5 provides an extensive list of bacteria that are involved with infections of the respiratory system. They can be divided into several groups as follows:

- Those that cause otitis media (middle ear infections), sinusitis and mastoiditis

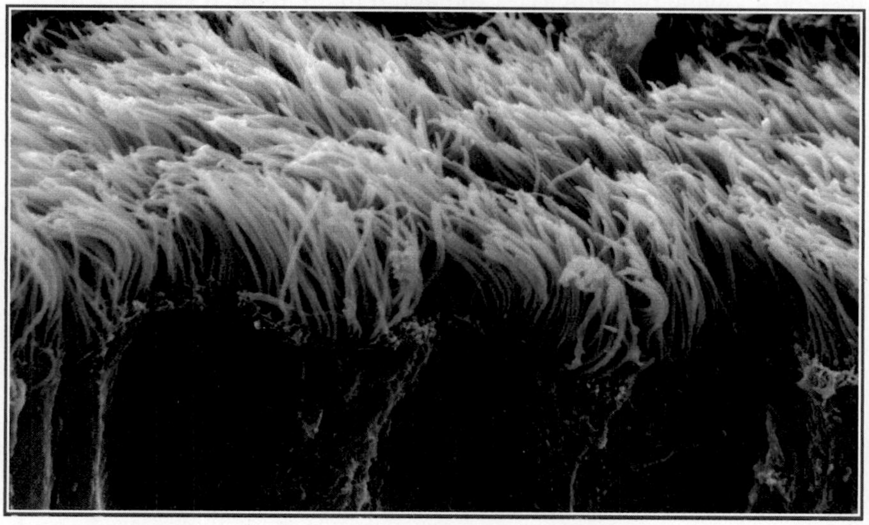

Figure 21.4 Ciliated cells are important barriers used to keep pathogens from entering the lower respiratory tract.

organism	treatment
otitis media, sinusitis, mastoiditis	
Streptococcus pneumoniae	amoxycillin, erythromycin
Haemophilus influenzae	amoxycillin, ciprofloxacin
Staphylococcus aureus	flucloxacillin, erythromycin, ciprofloxacin
group A streptococcus	amoxycillin, benzylpenicillin
pharyngitis	
group A streptococcus	benzylpenicillin, amoxycillin
group C streptococcus	benzylpenicillin, amoxycillin
Corynebacterium diphtheriae	benzylpenicillin, erythromycin (antitoxin)
Chlamydia pneumoniae	erythromycin, tetracycline
Mycoplasma pneumoniae	erythromycin, tetracycline
community-acquired pneumonia – typical	
Streptococcus pneumoniae	amoxycillin, erythromycin, cefuroxime
Haemophilus influenzae	amoxycillin, cefuroxime, ciprofloxacin
Staphylococcus aureus	flucloxacillin, erythromycin
community-acquired pneumonia – atypical	
Chlamydia pneumoniae	erythromycin, tetracycline
Chlamydia psittaci	erythromycin, tetracycline
Mycoplasma pneumoniae	erythromycin
Coxiella burnetii	erythromycin, tetracycline
Legionella pneumophila	erythromycin (and rifampicin)
hospital-acquired pneumonia	
Citrobacter spp.	gentamycin, ciprofloxacin, imipenem
Enterobacter spp.	gentamycin, ciprofloxacin, imipenem
Pseudomonas aeruginosa	gentamycin, ciprofloxacin, imipenem
Staphylococcus aureus	flucloxacillin (and gentamycin)
MRSA	vancomycin (and gentamycin)

Figure 21.5 Bacteria that can infect the respiratory system and antibiotics commonly used against them. Gram-positive bacteria are shown in purple and Gram-negative bacteria are shown in pink.

- Those that cause pharyngitis
- Those responsible for typical and atypical community-acquired pneumonia
- Those that cause hospital-acquired (nosocomial) pneumonia

We have included an example of the morphology and Gram reaction for each of the organisms listed, as well as commonly used antibiotics for infections caused by these organisms.

Keep in Mind

- The respiratory system is the most accessible system in the body and is therefore a major portal of entry for pathogens.
- The upper respiratory tract is continuously exposed to pathogens, whereas the lower respiratory tract is essentially sterile.
- Many infections of the lower respiratory tract result in pneumonia.
- The respiratory system is also a good portal of exit, and so it is easy for a pathogen to use this system to spread infections from person to person.
- Fungal infections of the respiratory system are usually restricted to individuals who are immunocompromised.

BACTERIAL INFECTIONS OF THE UPPER RESPIRATORY TRACT

Otitis Media, Mastoiditis, and Sinusitis

The middle ear, mastoid cavity, and sinuses are all connected either directly or indirectly to the nasopharynx. The ciliated epithelial cells that line both the sinuses and the eustachian tube push out bacteria that are trapped in mucus. In middle ear and sinus infections, it is likely that a virus, such as respiratory syncytial virus, initially invades the ciliated epithelium and destroys the cells involved with the mucociliary escalator, thereby allowing bacteria to invade and remain in the respiratory tract.

Mastoiditis is an uncommon problem but is very dangerous because of the proximity of this cavity to the nervous system and large blood vessels.

Pharyngitis

A variety of bacteria can cause infection in the pharynx. The classic form of pharyngitis is **strep throat** (**Figure 21.6**), caused by *Streptococcus pyogenes*. Recall from earlier discussions that this organism has associated with it an M protein that inhibits phagocytosis. In addition, proteins such as hemolysin O and S, DNase, streptokinase, and pyrogenic toxins produced by this pathogen account for the symptoms seen with pharyngitis. Group A streptococci (such as *S. pyogenes*) can also cause abscesses on the tonsils, a development that may require the removal of these lymphoid organs. *S. pyogenes* can also cause other infections that are potentially dangerous, in particular scarlet fever and toxic shock syndrome (**Figure 21.7**).

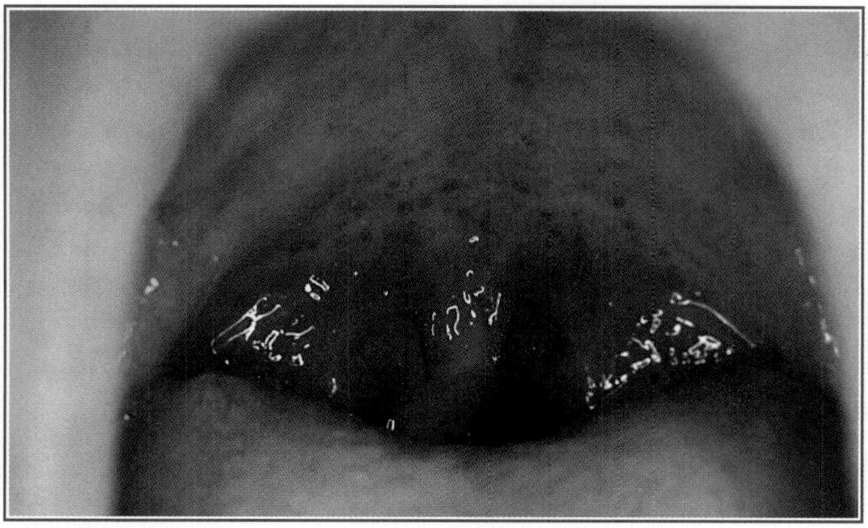

Figure 21.6 An example of strep throat, which is caused by *Streptococcus pyogenes*. Notice the redness and edema of the oropharynx, and the petechia (small red spots) on the soft palate.

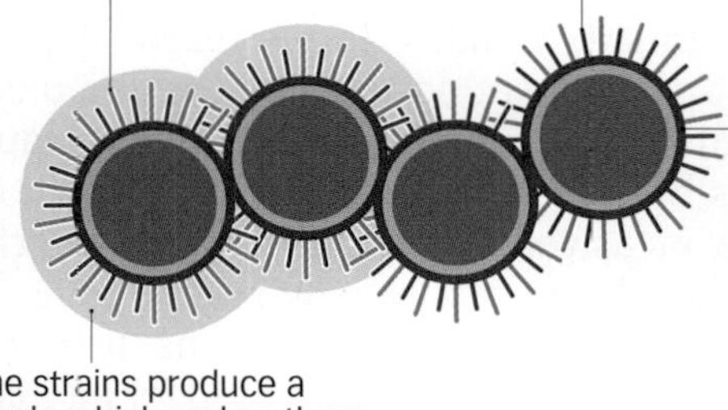
group-specific carbohydrate (e.g. A)
M protein is a major virulence factor, with antiphagocytic properties
extracellular products
streptococcal pyrogenic toxin which causes:
• scarlet fever rash
• toxic shock syndrome
Streptolysins (hemolysins)
• streptolysin O
• streptolysin S
Streptokinase
some strains produce a capsule which makes them resistant to phagocytosis

Figure 21.7 The properties of streptococcus organisms, including the carbohydrates that define the groups, the M protein associated with the cell wall of these bacteria, and some of the extracellular products they produce that increase their pathogenicity.

Scarlet Fever

Scarlet fever is a disease that is caused by Group A streptococci and can sometimes occur in people who have strep throat. The disease is usually seen in children under the age of 18 years.

Pathogenesis of scarlet fever — The symptoms of scarlet fever usually begin with the appearance of a rash that appears as tiny bumps on the chest and abdomen. This rash can spread over the entire body and usually appears redder in the armpits and groin areas. The rash normally lasts two to five days. In addition the patient may experience flushing and a very sore throat, which can be accompanied by yellow or white papules. There can be a fever of 101°F or higher and lymphadenopathy in the neck region. Headache, body aches, and nausea can also occur.

Treatment of scarlet fever — Treatment of scarlet fever involves the use of antibiotic therapy.

> **Fast Fact**
>
> The removal of tonsils and adenoids was once a common procedure until medical workers realized that these organs served an important protective function.

Diphtheria

Diphtheria is an infection caused by the toxin produced by *Corynebacterium diphtheriae*, a potent inhibitor of protein synthesis. The infection is localized and initially presents as severe pharyngitis accompanied by a plaque-like pseudomembrane in the throat (**Figure 21.8**). The life-threatening aspects of the infection are due to toxemia, which can involve multiple organ systems, including acute myocarditis. Diphtheria

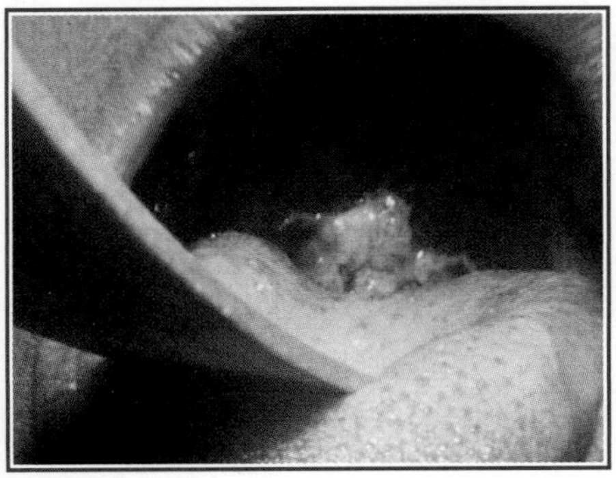

Figure 21.8 The pseudomembrane seen in diphtheria infections. In severe cases, this membrane can cover the trachea.

is transmitted by droplet aerosol, by direct contact with the skin of an infected person, or, to a lesser degree, by fomites (inanimate objects that can harbor disease-causing organisms).

Vaccination against diphtheria, part of the DTaP (diphtheria, tetanus, pertussis) vaccine protocol, is very effective, and infection is rare when vaccination is in place. In fact, fewer than 10 cases are reported in the United States each year, with the highest incidence seen in migrant workers or immigrants who have not been vaccinated. Diphtheria still occurs frequently in some parts of the world, particularly where socioeconomic conditions do not permit vaccination. For example, when a vaccination program existed in the Soviet Union, there were normally fewer than 200 reported cases of diphtheria per year. However, when the numbers of children vaccinated dropped, adults who had been exposed passed the infection to unvaccinated children and the number of cases increased enormously, with more than 47,000 cases and 1700 deaths between 1990 and 1995.

Pathogenesis of diphtheria — *Corynebacterium diphtheriae* is a small, Gram-positive bacillus that appears in V and L forms resembling Chinese letters (**Figure 21.9**). This arrangement occurs because of a unique process of cell division referred to as "snapping." The process causes the cells to arrange themselves both parallel and perpendicular to one another. This organism is poorly invasive, and the effects of the infection are mainly due to the exotoxin produced. As we saw in Chapter 5, this toxin has a classical configuration of two polypeptide chains, with the B chain used for entry into the target cell and the A chain used for inhibiting protein synthesis.

Local effects are seen as epithelial cell necrosis accompanied by inflammation, and the pseudomembrane that forms is composed of a mixture of fibrin, leukocytes, and cell debris. The size of this membrane varies from small and localized to extensive, in which case it can cover the trachea. Diphtheria can also be systemic, causing acute myocarditis. Iron seems to have a role in how much toxin is produced, and the *Tox* genes that code for this toxin are regulated by operons like those we learned about in Chapter 11.

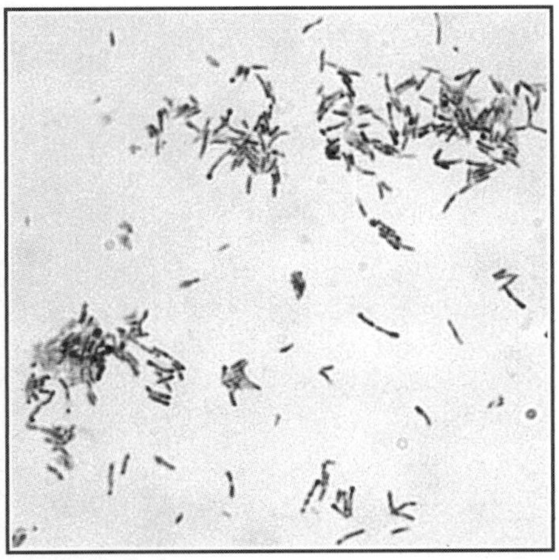

Figure 21.9 A photomicrograph showing the organization of *Corynebacterium diphtheriae* in the characteristic Chinese-letter pattern. This pattern results from snapping, which is the unique form of cell division seen in this organism.

Incubation takes two to four days and usually presents as pharyngitis or tonsillitis accompanied by fever, sore throat, and malaise. Patches of pseudomembrane can be seen on tonsils, uvula, soft palate, or pharyngeal walls and may extend downward into the larynx and trachea. In uncomplicated cases, the infection will resolve and the membrane will be coughed up after five to ten days. However, complications caused by respiratory obstruction can result in suffocation, and systemic infection can result in myocarditis during the second or third week of the infection. Diphtheria can also cause nonrespiratory infections, particularly of the skin, demonstrated by the formation of simple pustules or chronic nonhealing ulcerations.

Treatment of diphtheria — The most important treatment is neutralization of the toxin and elimination of the organism. Toxin neutralization is the more critical and should be done as quickly as possible. Antitoxin can be used to neutralize free toxin but has no effect on toxin that has become fixed to target cells. *Corynebacterium diphtheriae* is sensitive to many antibiotics, including penicillin, cephalosporin, erythromycin, and tetracycline.

VIRAL INFECTIONS OF THE UPPER RESPIRATORY TRACT

Rhinovirus Infection (the Common Cold)

There are several hundred serotypes of rhinovirus, and fewer than half of them have been well characterized. The 50% that have been characterized are all picornaviruses (**Figure 21.10**), which are extremely small, single-stranded RNA viruses that do not have envelopes. The optimum temperature for the growth of picornaviruses is 33°C, which happens to be the temperature in the nasopharynx. In fact, there is some thought that this lower temperature is why these viruses localize there.

Pathogenesis of rhinovirus infection — This virus uses the glycoprotein ICAM, which is an adhesion molecule, as a receptor to infect host cells. It is known as the common cold virus because it is the major cause of the mild upper respiratory infections that affect people of all ages but especially older children and adults. This infection is seen at any time of the year and is usually epidemic in the early fall and spring. Rhinovirus infections are rarely seen in the lower respiratory tract.

The incubation period for rhinovirus infection is approximately two to three days, and the acute symptoms can last for three to seven days. There is little damage to the mucosal layer of the upper respiratory tract during this infection, but some data do suggest that the infection causes an increase in the production of bradykinin, which may cause excessive secretion, vasodilation, and sore throat.

Treatment of rhinovirus infection — There is no specific therapy or treatment for these infections. Because of the large number of serotypes of this virus, it is unlikely that an effective vaccine will ever be developed. However, there is a possibility that soluble ICAM receptors (discussed in Chapter 13) may be a way of inhibiting viral attachment.

Parainfluenza

There are four types of parainfluenza virus: 1, 2, 3, and 4. This virus is enveloped and belongs to the paramyxovirus group. It is a single-stranded RNA virus, but the viral genome is not segmented as it is in the influenza virus. Like the influenza virus, the parainfluenza virus contains hemagglutinin and neuraminidase, and it is this similarity that makes the transmission and pathology of the parainfluenza viruses similar to those of the influenza virus.

There are also differences between these two viruses, one being that parainfluenza virus replicates in the cytoplasm of the host cell, whereas influenza virus replicates in the nucleus. In addition, parainfluenza is genetically more stable, and there is very little mutation seen with this virus. Consequently there is little antigenic drift and no antigenic shift.

The parainfluenza virus can cause serious problems in infants and small children, especially between the ages of one and three years. Overall, it is responsible for 15–20% of nonbacterial respiratory infection that requires hospitalization in infancy and childhood. There is only a transitory immunity to reinfection, but the infection becomes milder as the child ages.

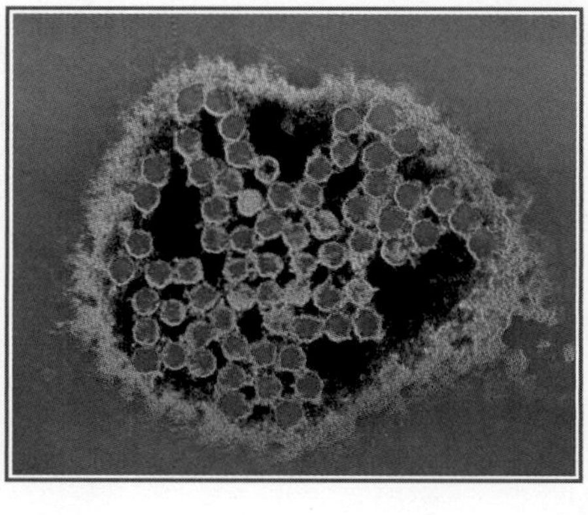

Figure 21.10 A colorized electron micrograph of rhinovirus, which is the etiologic agent of the upper respiratory infection known as the common cold.

Pathogenesis of parainfluenza infection — The onset of infection with parainfluenza may be abrupt and appear as an acute spasmodic **croup** (hoarseness and a barking cough). This usually begins as mild and progresses over one to three days to involve the lower respiratory tract. The duration of the illness can be between four and twenty-one days but is usually seven to ten days. Of the four types of parainfluenza virus, types 1 and 3 are the most clinically relevant. Type 1 is the major cause of laryngotracheitis (**acute croup**) in infants and young children. It can also cause severe upper respiratory illness, pharyngitis, and **tracheobronchitis** in all age groups. Outbreaks usually occur in the fall of the year.

Parainfluenza type 3 is the major cause of severe lower respiratory infection in infants and young children, often causing bronchitis and pneumonia in children less than one year of age. Infections with this virus can occur throughout the year, and it is estimated that 50% of all children are exposed to this virus during their first year of life.

Treatment of parainfluenza infection — There is currently no method of treatment or control for this viral infection except supportive care.

Keep in Mind

- Bacterial infections of the upper respiratory tract can involve otitis media, mastoiditis, sinusitis, and pharyngitis.
- Viral infections of the upper respiratory tract include the common cold (rhinovirus) and parainfluenza.

BACTERIAL INFECTIONS OF THE LOWER RESPIRATORY TRACT

Bacterial Pneumonia

Of all of the infections of the lower respiratory tract, pneumonia is one of the most serious. This infection of the lung parenchyma can be divided into two types, *community-acquired* and *nosocomial*, and each type can be caused by a variety of organisms. Nosocomial pneumonia occurs approximately 48 hours after a patient is admitted to the hospital and is usually associated with *Staphylococcus aureus* or with Gram-negative bacteria such as *Pseudomonas aeruginosa*. This infection can be particularly difficult to deal with if the staphylococcal organism is resistant to antibiotics.

In contrast, community-acquired pneumonia usually presents as a lobar pneumonia accompanied by fever, chest pain, and the production of purulent sputum. It is important to note that there can also be cases of atypical pneumonia characterized by coughing without the production of sputum. Atypical pneumonia can be caused by a variety of organisms, including *Mycoplasma pneumoniae*, *Chlamydia pneumoniae*, *Chlamydia psittaci*, *Coxiella burnetii*, and *Legionella pneumophila*. In some cases, bacterial pneumonia can progress to the production of lung abscesses.

Pathogenesis of nosocomial bacterial pneumonia — Ironically, a hospital setting is one of the most clinically dangerous, mainly because of the large number of infected patients in one location (many of whom are immunocompromised), each carrying different pathogens, many of which have developed antibiotic resistance (Chapter 20). Among these resistant bacteria are *Enterobacter* species, *Pseudomonas aeruginosa*, and methicillin-resistant *Staphylococcus aureus* (MRSA).

The problem is compounded by the fact that the hospital is filled with debilitated patients. Debilitation causes an increase in proteolytic enzyme activity in the saliva of these patients, which contributes to the rapid turnover of the fibronectin layer that covers the epithelium of the pharynx. This layer is associated with normal bacterial flora; when it is lost, the area can become colonized with opportunistic pathogens that can then be aspirated into the lungs and cause pneumonia (**Figure 21.11**).

Pathogenesis of community-acquired bacterial pneumonia — This infection usually occurs after the aspiration of pathogens such as *Pneumococcus* in great enough numbers to overwhelm the resident defenses present in the lungs. The establishment of an infection in the lungs depends not only on the number of pathogens entering the lungs but also on the competence of the mucociliary escalator to keep them out. The classical lobar pneumonia resulting from infection with *Pneumococcus* has four stages:

- ***Acute congestion*** occurs where local capillaries become engorged with neutrophils that are part of the host defense response.
- During the ***red hepatization*** stage, red blood cells from the capillaries flow into the alveolar spaces.
- In the ***grey hepatization*** stage, large numbers of dead and dying neutrophils are present, and degenerating red cells are seen.
- In the last stage, called ***resolution***, the adaptive immune response begins to produce antibodies that control the infection.

Treatment of bacterial pneumonia – The treatment of bacterial pneumonia depends on the severity of the infection and the type of organism causing the infection. The leading cause of bacterial pneumonia is *Streptococcus pneumoniae*, which is treated with penicillin, amoxicillin-clavulanate (Augmentin®) and erythromycin. Other antibiotics used to treat bacterial pneumonia include cefuroxime (Ceftin®), ofloxacin (Floxin®) and trimethoprim-sulfamethoxazole.

Chlamydial Pneumonia

This infection is caused by *Chlamydia pneumoniae*, which is found throughout the world and is responsible for 10% of pneumonias and 5% of bronchitis cases. The infection occurs throughout the year and is spread through person-to-person contact. There have been reports of this form of pneumonia in both community-acquired and nosocomial forms, and there are more infections with this organism in the elderly.

Figure 21.11 Mechanisms associated with nosocomial pneumonia. Panel **a**: Mucosal surfaces are coated with fibronectin and normal flora. Removal of this layer allows Gram-negative bacteria to colonize the oropharynx in significant numbers. Panel **b**: Some of the contributory factors for nosocomial pneumonia.

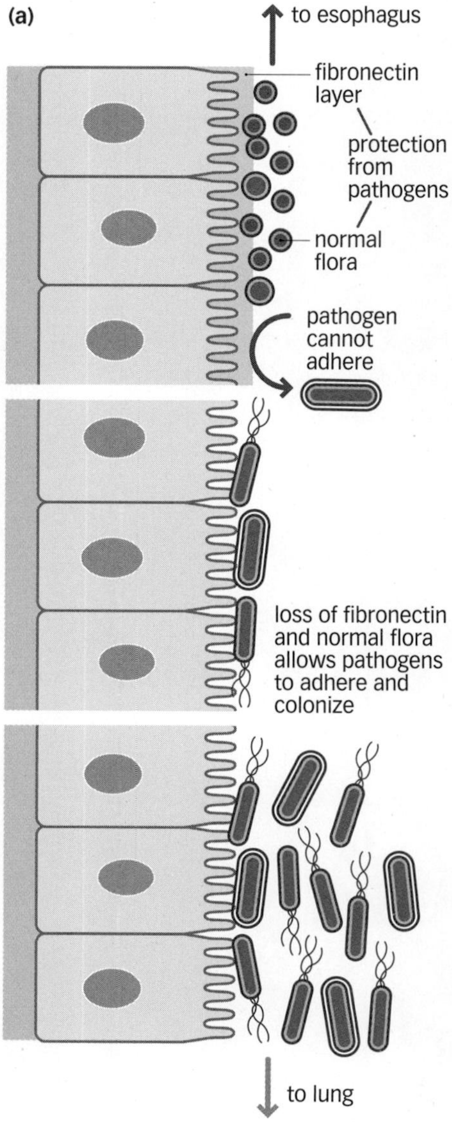

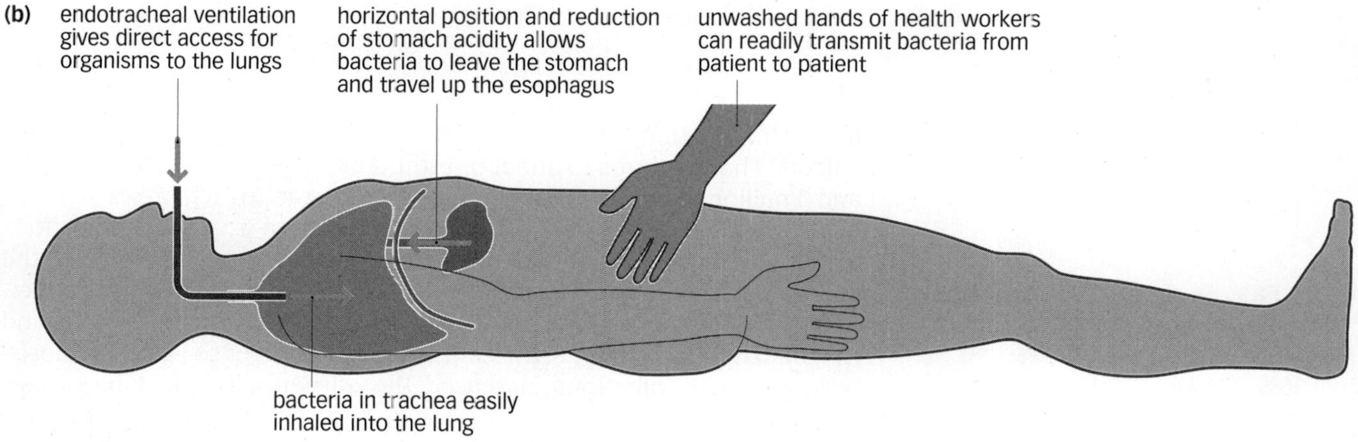

Pathogenesis of chlamydial pneumonia — The infection can present as either pharyngitis, lower-respiratory-tract infection, or both, and is clinically similar to *Mycoplasma* pneumonia (see below). An initial pharyngitis lasting one to three weeks is replaced by a persistent cough that can last for weeks. There may be an association between chlamydial pneumonia and vascular endothelial disease such as atherosclerosis.

Treatment of chlamydial pneumonia — Tetracycline or erythromycin will alleviate the symptoms of this infection.

Mycoplasma Pneumonia

This is a mild form of pneumonia that accounts for about 10% of all pneumonias. It is often referred to as *walking pneumonia* because there is no need for hospitalization during this infection. The most common age for patients with this infection is between five and fifteen years, and it causes approximately 30% of all teenage pneumonias. It is caused by *Mycoplasma pneumoniae*, a bacterium that lacks a cell wall. The infection is acquired by droplet transmission, and the infectious dose required to cause infection is fewer than 100 pathogens, making it very easy to contract. *Mycoplasma pneumoniae* is found throughout the world, especially in temperate climates.

Pathogenesis of *Mycoplasma* pneumonia — The incubation period is between two and fifteen days, and the infection has an insidious onset, with fever, headache, and malaise occurring for two to four days before the appearance of respiratory symptoms. Infection involves the trachea, bronchi, and bronchioles and may extend down to the alveoli.

The organism initially attaches to the cilia and microvilli of the cells lining the bronchial epithelium. This attaching interferes with the ciliary action, with the result being detachment of the mucosal layer and subsequent inflammation and appearance of **exudate**. The inflammatory response is initially composed of lymphocytes, plasma cells, and macrophages, which may infiltrate and thicken the walls of the bronchioles and alveoli. The organism can be shed in upper respiratory secretions for two to eight days before symptoms appear and for up to fourteen weeks afterward.

This infection causes a mild tracheobronchitis with fever, cough, headache, and malaise. It can occur with a sore throat, and some patients also experience symptoms of otitis media.

Treatment of *Mycoplasma* pneumonia — Erythromycin or tetracycline is the usual treatment. The use of either drug can shorten the clinical symptoms, but the organism may be in the nasopharynx for long periods after the symptoms have subsided. Azithromycin, clarithromycin, and most quinolones are also effective.

Tuberculosis

It is estimated that 1.7 billion people worldwide are infected with tuberculosis. That amounts to about one-third of the population of the world, and 3 million people die of tuberculosis each year. In fact, *Mycobacterium tuberculosis* is responsible for more deaths than any other infectious agent. The development of effective antibiotics in the 1950s slowed the spread of tuberculosis for some years, but by the year 2000 the incidence of tuberculosis had begun to rise. The appearance of AIDS and HIV infection has had a significant role in the increase of tuberculosis, because these infections increased the efficiency of the tuberculosis transmission cycle.

Fast Fact

In sub-Saharan Africa it is estimated that 50% of HIV-infected patients also have tuberculosis.

Poverty and poor socioeconomic conditions are breeding grounds for tuberculosis, and drug resistance has become an increasingly dangerous problem. One of the major reasons is non-compliance with the therapy regimen. Because the treatment of tuberculosis can require the daily administration of antibiotics for long periods (at least six months), many patients will stop taking the drugs. This can lead to the evolution of drug-resistant strains of *Mycobacterium tuberculosis*.

Early detection is critical to prevent the spread of tuberculosis and to increase the chances of successful treatment. Because the initial symptoms are similar to those seen in other respiratory infections, it is important to look for symptoms such as fever, fatigue, weight loss, chest pain, and shortness of breath in conjunction with coughing as an indication of potential tuberculosis.

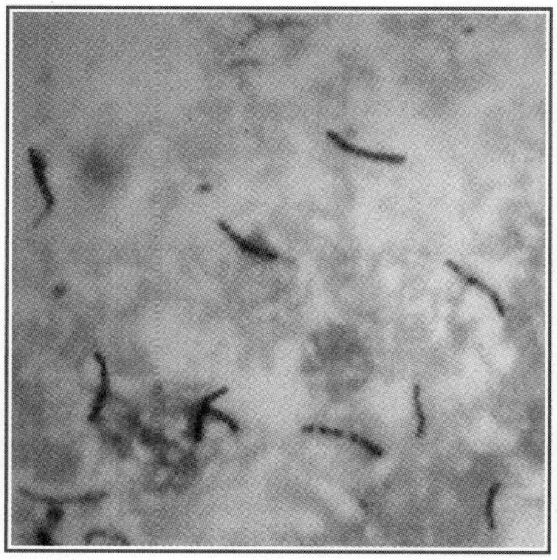

Figure 21.12 An acid-fast stain of *Mycobacterium* bacteria. Because of the mycolic acid that is part of the cell wall of this bacterium, heat is required to stain it.

M. tuberculosis is an obligate aerobic rod-shaped bacillus that is acid-fast and forms no spores (**Figure 21.12**). Recall from Chapter 9 that *Mycobacterium* produces mycolic acid as part of its cell wall, and it is this component that makes it difficult to Gram stain and also protects the pathogen from antibiotic therapy and host defenses.

Pathogenesis of tuberculosis — For most healthy people, tuberculosis is a self-limiting disease that will be dealt with by the host defense. However, if cell-mediated immunity is in some way compromised or inefficient, tuberculosis can be serious. *Mycobacterium* is a problem for the host defense because of its unique cell wall, which interferes with macrophage function and with T cell activation. In point of fact, when *Mycobacterium* is ingested by macrophages, it inhibits the formation of the phagolysosome and eventually escapes into the cytoplasm of the macrophage. Here the bacterium will increase in number and eventually spread to the lymph nodes, where it will enter the blood and distribute throughout the body. The cell wall components of *Mycobacterium* attract T cells and macrophages to the site of the infection, and there is an uncontrollable release of enzymes and metabolites that destroy tissues, leading to necrosis. Necrosis in the lung liquefies and spreads to adjacent areas of the lung, a migration that causes the cycle to continue (**Figure 21.13**).

There are two basic types of tuberculosis: *primary*, which follows initial exposure to the pathogen, and *secondary*, which can occur years later. Let's briefly examine the difference between primary and secondary tuberculosis. Primary tuberculosis occurs when the host encounters *M. tuberculosis* for the first time. At this time, organisms will find their way to the alveoli and a localized inflammatory response will develop. This will involve the phagocytosis of the bacilli by macrophages and neutrophils; however, the pathogens will not be killed during this process. Instead they will be transported by these white cells to the regional lymph nodes and will continue to divide intracellularly. At this time a cell-mediated immune response will begin, including a delayed-type hypersensitivity reaction to tuberculin protein. This can lead to a positive tuberculin (PPD) skin test reaction.

If the primary lesion is not contained by host defense, tubercles will form. A tubercle is composed of aggregates of enlarged macrophages filled with bacteria. This tubercle can be surrounded by fibroblasts and lymphocytes. Frequently, the center of the tubercle will undergo caseous necrosis, which may later calcify. When this occurs, these calcifications are referred to as **Ghon complexes**; they are readily seen on X-ray.

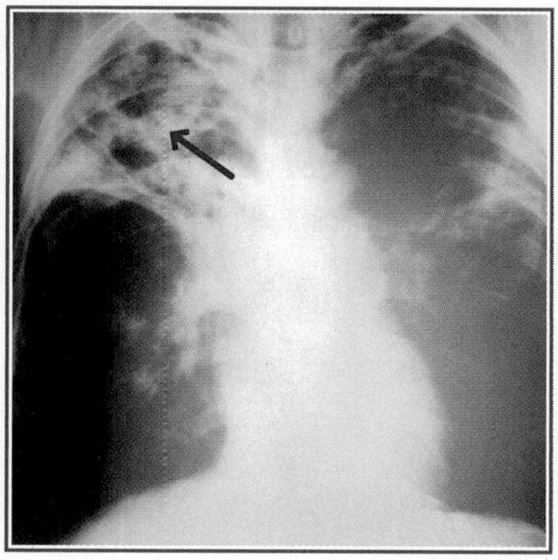

Figure 21.13 An X-ray of the lungs of a patient with advanced tuberculosis. The arrow shows infected areas containing tubercles filled with live *Mycobacterium*. These areas of the lungs are no longer functional.

At this point the primary infection becomes quiescent and asymptomatic in about 90% of patients. In the remaining 10%, the infection can evolve into clinical disease with bacilli spreading through the lymphatic channels and bloodstream and via the bronchi and gastrointestinal tract. These events result in tuberculous meningitis, miliary (disseminated) tuberculosis or both. If the localized tubercles discharge their contents into the bronchus, they can be aspirated and distributed to other parts of the lungs.

Secondary tuberculosis is usually due to the reactivation of old lesions or the gradual progression of primary tuberculosis into a chronic form. This recurrence of disease occurs in a small percentage of patients whose initial symptoms have subsided. Secondary tuberculosis usually manifests itself in the apices (top) of the lungs and occurs within two years of the primary infection. However, secondary tuberculosis may evolve many decades later when innate resistance begins to diminish. There is currently a debate as to whether the secondary infection is the result of a breakdown of preexisting tubercles, resulting in the release of viable pathogenic bacilli or the result of a new infection. The current thinking is that secondary infection with tuberculosis is more likely to be due to the release of organisms that have been there since the primary infection. However, the patient's ability to maintain an immune defense also has a role.

Treatment of tuberculosis — Treatment usually consists of a triple therapy involving isoniazid (INH), pyrazinamide (PZA), and rifampicin (RFP). These three drugs are taken once a day for two months, followed by a regimen of INH and RFP for nine more months. If the tuberculosis strain is drug-resistant, the initial regimen will also include ethambutol.

Patients must be monitored carefully because compliance with the drug therapy is very important. Compliance by patients can be difficult because of the toxicity of the drugs and the consequent side effects, the most serious of which is liver toxicity.

It has been shown that resistance to INH is most common, and in difficult cases, ciprofloxacin and clarithromycin have also been used.

Directly observed therapy (DOT) for tuberculosis — Directly observed therapy (DOT) for treatment of tuberculosis is used as a way of preventing multi-drug-resistant tuberculosis. It involves the delivery of scheduled doses of tuberculosis medication by a health care worker. The health care worker directly administers, observes, and documents the patient's ingestion or injection of the medication. The purpose of DOT for tuberculosis treatment is to ensure that patients receive the medication required to prevent the spread of tuberculosis and to prevent multi-drug-resistant tuberculosis. Patients who are required to receive DOT or are recommended for this treatment are listed in **Table 21.1**.

Pertussis (Whooping Cough)

This infection of the lower respiratory tract is spread by airborne droplets produced by patients in the early stages of illness. It is highly contagious and infects 80–100% of exposed susceptible individuals. Consequently, the spread of pertussis is rapid in schools, hospitals, offices, and homes — in other words, just about anywhere. The pathogen *Bordetella pertussis* is not found in animals and does not survive in the environment. Therefore, the reservoir for this pathogen is humans, and it has been shown that previously immunized individuals are becoming an important reservoir for this infection. Because the symptoms can be similar to those of a cold, many infected adults will spread the infection to places like schools and nurseries without knowing it. Mortality is highest in infants (70%) and in children under 1 year old.

DOT is required for the these patients	DOT is recommended for these patients
HIV patients co-infected with tuberculosis	Congregate living
Homeless	Children
A history of criminal incarceration	Adolescents
Psychiatric disorders	Elderly with cognitive impairment
Cognitive dysfunction	Recently immigrated indiv duals
Drug abuse or misuse	Patients who seem to have difficulty in understanding
History of non-adherence to medication therapies	
Physiological resistance to one or more anti-tuberculosis drugs	
Persistently positive test results	
History of refusal of medical care	

Table 21.1 Directly Observed Therapy

Immunization against pertussis was begun in the 1940s and continues today as part of DTaP vaccination. Once use of this vaccine became widespread, the incidence of pertussis dropped from approximately 250,000 cases a year to below 1 in 100,000 people. Unfortunately, pertussis appears to be making a comeback of sorts since 1980, with epidemics occurring every 3 to 4 years. In addition, the greatest numbers of infections are being seen in 10 to 20 year olds, which may be due to people who were not immunized because of fears associated with vaccine side effects. There is a clear relationship between lack of vaccination and infection.

Pathogenesis of pertussis — *Bordetella pertussis*, a Gram-negative coccobacillus, is a strictly human pathogen that has an affinity for ciliated bronchial epithelium. Once attached, the organism produces a **tracheal toxin** that immobilizes and progressively destroys the ciliated cells. The persistent coughing associated with this infection is caused by the inability to move the mucus that builds up in the respiratory tract up and out.

Pertussis does not invade cells of the respiratory tract or deeper tissues. After an incubation period of seven to ten days, the infection follows three stages:

- The catarrhal stage features persistent perfuse and mucoid **rhinorrhea** (runny nose) for one to two weeks. There may also be sneezing, malaise, and **anorexia**. The infection is most communicable during this stage.
- The paroxysmal stage starts with persistent coughing that can build to 50 times a day for two to four weeks. This is when the characteristic "whooping" sound is heard. It is the result of the patient's trying to catch his or her breath during a coughing episode. In addition, apnea may follow the coughing, especially in infants. There is also a significant increase in lymphocytes during this period, with counts sometimes reaching 40,000 per mm^3 of blood.
- In the convalescent stage, the frequency and severity of coughing and other symptoms gradually decreases.

The most common complications of pertussis are superinfection with *Streptococcus pneumoniae* and convulsions. However, subconjunctival and cerebral bleeding as well as anoxia can occur due to the persistent coughing.

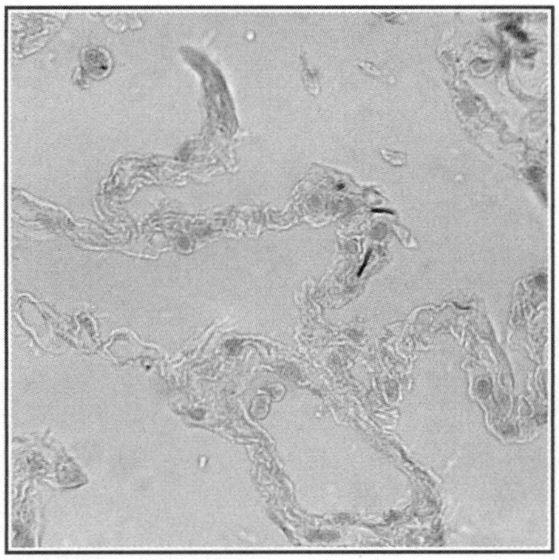

Figure 21.14 A photomicrograph of *Bacillus anthracis* in the lung tissue of a case of fatal inhalation anthrax.

Treatment of pertussis — Antibiotics such as erythromycin and clarithromycin can be used in early stages of the infection, and this treatment limits the spread of the infection to others. However, once the paroxysmal stage is reached, therapy is only supportive. Vaccination against pertussis is the best prevention.

Inhalation Anthrax

This infection produces a fulminate pneumonia (one that comes on suddenly with great severity) leading to respiratory failure and death. Anthrax is primarily a disease of herbivores such as cattle, horses, and sheep and is acquired from spores found in pastures. If these spores are inhaled, anthrax can occur in the respiratory tract. Infection is infrequently seen in healthy individuals and usually affects farmers, veterinarians, and meat handlers; it usually presents as localized lesions. However, it has in recent years become of interest as a biological weapon, a topic we discuss in Chapter 28. In the 1980s there was an unconfirmed report of an explosion at a biological-weapons plant in the Soviet Union that killed 50 workers and released anthrax spores into the environment. More recently, in October 2001 letters contaminated with powdered anthrax spores were mailed to various locations in the United States.

Pathogenesis of inhalation anthrax — *Bacillus anthracis* is a Gram-positive spore-forming rod (**Figure 21.14**) commonly found in the soil of pastures. When they reach the rich environment of human tissues, the spores germinate into the bacilli. The anti-phagocytic properties of the capsule that surrounds the organism aid in its survival and growth in large numbers. Pathology is the result of the exotoxin produced by the organism.

Symptoms of pulmonary anthrax are one to five days of nonspecific malaise, mild fever, and a nonproductive cough that leads to progressive respiratory distress and **cyanosis**. There is rapid and massive spread to the central nervous system and bloodstream, followed by death.

Treatment of inhalation anthrax — Antibiotic therapy can be successful, and *B. anthracis* is susceptible to penicillin. Doxicycline and ciprofloxacin are alternatives recommended for prophylaxis.

Legionella Pneumonia (Legionnaires' Disease)

This infection is caused by *Legionella pneumophila*, a Gram-negative rod that cannot be stained or grown using normal techniques (**Figure 21.15**). Unrecognized before 1976, it was the cause of a widely publicized outbreak at an American Legion convention held in Philadelphia that year. There is evidence that the bacterium may have been around since the 1950s but not identified properly.

In nature, *Legionella* is ubiquitously found in fresh water, particularly in warm weather. In water, *Legionella* is found living within *Acanthamoeba* organisms, and these are the infectious reservoirs. The bacterium is transmitted to humans as a humidified aerosol from contaminated water systems, and person-to-person transmission has never been seen. Most outbreaks occur in large buildings that use cooling towers for their air-conditioning system.

Fast Fact

Legionella can persist in the water systems of air-conditioning cooling towers even in the presence of chlorine.

Healthy people do not seem to be affected very often, and many cases of infection may go undetected. In fact, *Legionella* seems to have a low virulence for humans, with infections occurring in less than 5% of the population; these people are usually immunocompromised in some way. It is estimated that there are about 25,000 cases per year in the United States.

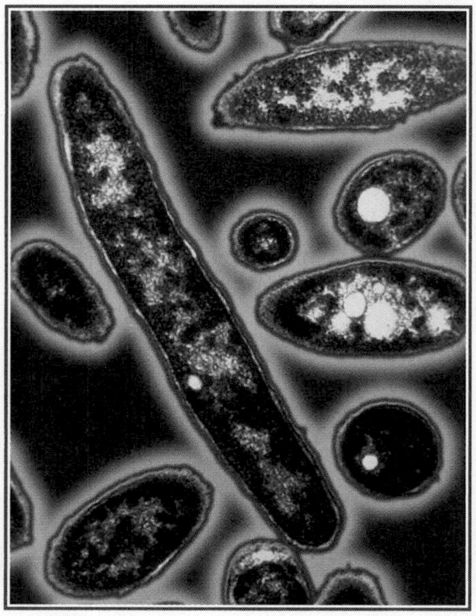

Figure 21.15 A transmission electron micrograph of *Legionella pneumophila*, the etiologic agent of Legionnaires' disease. The reservoir for this organism is amoeboid organisms found in water.

Pathogenesis of *Legionella* pneumonia — *Legionella* is a facultative intracellular parasite that aggressively attacks the lungs, producing a necrotizing multifocal pneumonia involving the alveoli and terminal bronchioles but not so much the larger bronchioles. An inflammatory response occurs during which an exudate is formed containing fibrin, polymorphonuclear leukocytes, and red blood cells. *Legionella* organisms are inhaled and enter the alveoli, where they infect the alveolar macrophages through the production of an endocytic vesicle. Inside this vesicle, the bacteria continue to replicate and prevent the fusion of the vesicle with lysosomes. The infected macrophage takes on a coiled morphology (**Figure 21.16**) and eventually dies, releasing the infectious bacteria.

Legionnaires' disease is a severe toxic pneumonia that begins with myalgia and headache followed by a rapidly rising fever. There are chills, pleuritic chest pain, vomiting, diarrhea, and occasional confusion and delirium. Patchy or interstitial infiltrates in the lung can be seen on X-ray, and there can also be hepatic dysfunction. Serious cases show progressive illness for three to six days and end in shock or respiratory failure, or both. The mortality rate for *Legionella* infection is about 15% but can be as high as 50% in hospital outbreaks because in these cases the infected population is immunocompromised or immunosuppressed.

Treatment of *Legionella* pneumonia — Erythromycin is better than penicillin because *Legionella* produces a β-lactamase enzyme that destroys penicillin. Although tetracycline, rifampin, and quinolones are good, azithromycin and clarithromycin are the agents of choice.

Q Fever

Q fever is a zoonotic infection seen throughout the world, with cattle, sheep, and goats being the primary reservoirs for humans. Q fever is caused by *Coxiella burnetii*, which is the only Gram-negative spore-forming bacterium. This organism grows well in the placenta of animals so that, as birth occurs, large numbers of *Coxiella* are liberated onto the ground. The organism can survive in the soil for very long periods (even years), and workers involved with animal births or in slaughtering animals are the most prone to infection. *Coxiella* can be transmitted by inhalation as well as by the ingestion of unpasteurized milk.

Pathogenesis of Q fever — The pathology of this infection is not clearly understood. It usually begins nine to twenty days after inhalation, with the abrupt onset of fever, chills, and headache. There can also be a mild, dry, hacking cough, and a patchy interstitial pneumonia may develop. In some cases, there can also be abnormal liver function. One or two percent of patients with acute infection die.

Treatment of Q fever — Most cases of Q fever resolve spontaneously without therapy, but tetracycline can be given to shorten the fever and reduce the risk of a rare chronic infection. One or two percent of patients with acute infection die.

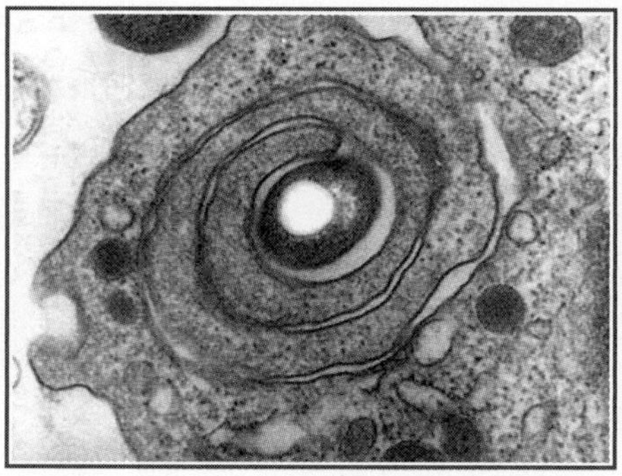

Figure 21.16 The coiled vesicle seen in macrophages that have been infected by *Legionella*.

Psittacosis (Ornithosis)

This infection of the lower respiratory tract is also known as *human psittacosis*. It is a zoonotic pneumonia contracted by the inhalation of bird droppings infected with the bacterium *Chlamydia psittaci*. Infection was originally seen in parrots and parakeets, but it is now found in many birds, including turkeys. The infection is usually latent in birds, but the stress of confinement in cages may cause large numbers of *C. psittaci* bacteria to be shed into the feces. Some strains of *C. psittaci* are extremely contagious.

Pathogenesis of psittacosis — In humans, psittacosis presents as an acute infection of the lower respiratory tract. There is acute onset of fever, headache, malaise, and muscle aches, accompanied by a dry, hacking cough and bilateral pneumonia. There can also be occasional systemic complications, such as myocarditis, endocarditis, and hepatitis. In some cases, there may be **splenomegaly** and **hepatomegaly**.

Treatment of psittacosis — Both tetracycline and erythromycin are effective, but only if given early in the infection.

Keep in Mind

- Bacterial infections of the lower respiratory tract include both nosocomial and community-acquired bacterial pneumonia.
- *Chlamydia* and *Mycoplasma* can also cause pneumonia.
- Tuberculosis is a serious lower respiratory tract infection caused by *Mycobacterium tuberculosis*, an organism that is becoming more resistant to antibiotic treatment.
- Pertussis (whooping cough), inhalation anthrax, legionellosis, Q fever, and psittacosis are serious infections of the lower respiratory tract.

VIRAL INFECTIONS OF THE LOWER RESPIRATORY TRACT

When both upper and lower tracts are considered, respiratory infections account for 75–80% of all the acute infections in the United States, and most of these are of viral origin. In fact, there are three or four viral infections per person per year in this country. The incidence of these infections varies inversely with age and is greatest in young children. Most viral infections are also seasonal, with the lowest rates in summer and the highest in winter.

The two viruses that cause the majority of acute viral infections in the lower respiratory tract are influenza and respiratory syncytial virus. A common characteristic of infections with these viruses is a short incubation period of about one to four days and transmission from person to person. This transmission can be either direct (through droplets) or indirect (through hand transfer of contaminated secretions to conjunctival or nasal epithelium). Influenza and syncytial infections are seen worldwide. We have already mentioned both of them in earlier discussions (Chapters 8, 12, and 13). Please make sure to refer to these sections to broaden your understanding.

Influenza

The influenza virus is a member of the *orthomyxovirus* group, and its virions are surrounded by an envelope (**Figure 21.17**). The virions contain single-stranded RNA in eight segments, allowing the virus to undergo a high rate of mutation, and it is this ability to mutate so easily that makes the virus a repetitive problem for humans. There are three major serotypes — A, B, and C — with the differences based on the antigens associated with the nucleoprotein. Because influenza A is the best-documented of the three serotypes and the most virulent, our discussions will center on infection with this virus.

Influenza is a significant health concern, and there is currently concern about its potential to combine with an avian influenza strain to produce a significant pathogenic influenza virus (Chapter 8). Recall from Chapter 13 that the ability of a virus to mutate gives it the ability to undergo antigenic drift and antigenic shift, both of which have a role in our lack of ability to defend against it. Humans are the hosts for influenza, but the reservoir for this virus is birds. Severe respiratory problems are the primary manifestation of influenza infections, and outbreaks have been described since the sixteenth century. Outbreaks of differing severity occur nearly every year; the most severe pandemics occurred in 1743, 1889, 1918 (Spanish flu), 1957 (Asian flu), 1968 (Hong Kong flu), and 2009 (Swine Flu).

Direct droplet transmission is the most common method by which influenza spreads, with outbreaks occurring more frequently in the winter months. There seems to be an interval of two to three years between major outbreaks, with the typical epidemic lasting three to six weeks and involving as many as 10% of the population. However, illness rates may exceed 30% in school-aged children and in residents of closed institutions such as prisons and convalescent homes. The identification of an influenza outbreak as an epidemic is based on excessive mortality rates (in other words, more deaths from the infection than expected).

Pathogenesis of influenza — The influenza virus has a predilection for the respiratory epithelium, and viremia is not a feature of infection. The virus multiplies in the ciliated cells of the lower respiratory tract, resulting in functional and structural abnormalities in these cells. As part of the infection, cellular synthesis of nucleic acids and proteins is shut down, and both the ciliated and mucus-producing epithelial cells are shed, resulting in substantial interference with the mechanical clearance mechanisms of the lower respiratory tract. In addition, there is localized inflammation associated with the death of these cells. The respiratory epithelium may not be restored to normal for two to ten weeks after the initial infection. There is also viral destruction of tissues accompanied by inflammation and impaired phagocytic and chemotactic responses, which can cause superinfection by bacteria.

Recovery from influenza infection starts with the production of interferon, which limits the spread of the infection (Chapter 15). This is accompanied by the rapid generation of natural killer cells, which reverse the infection. Shortly thereafter, cytotoxic T cells and specific antibodies, which are part of the adaptive immune response, appear in large numbers and control the infection so that tissue repair can begin.

In some cases, patients develop **acute influenzal syndrome**, which differs from the common infection course just described. This acute syndrome has a short incubation time of about two days, and symptoms can develop in a matter of hours. These include fever, myalgia, headache, and occasional shaking chills. The infection reaches maximum severity in just six to twelve hours, and a nonproductive cough develops. These acute symptoms can last three to five days and are usually followed by gradual improvement. However, occasionally these patients develop a progressive infection that involves the tracheobronchial tree and lungs, resulting in lethal pneumonia.

As mentioned above, one of the most common and important complications of an influenza infection are bacterial superinfections. This usually involves the lungs and can develop during the convalescent stage of the viral infection, when the patient is debilitated (recall that these secondary infections can be more dangerous than the primary infection). Superinfection is identified by the development of an abrupt worsening of the

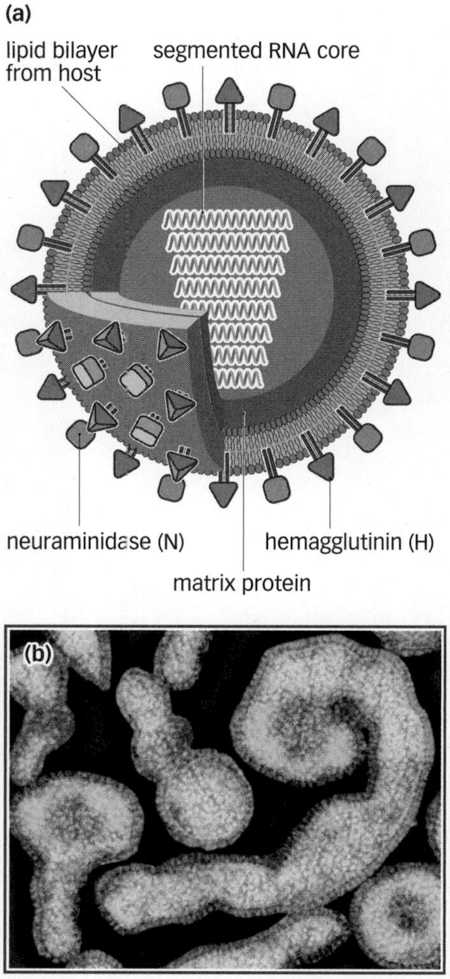

Figure 21.17 The influenza virus Panel **a**: The influenza A virion. Panel **b**: Colorized electron micrograph of influenza virus.

patient's condition after an initial stable period. Most often, the secondary superinfection involves *Streptococcus pneumoniae*, *Haemophilus influenzae*, or *Staphylococcus aureus*.

It is important to note that there are three ways in which an influenza infection can cause death:

- Underlying disease can occur in people with limited cardiovascular activity or pulmonary function. This progression usually occurs in the elderly.
- Superinfection can lead to bacterial pneumonia or in some cases disseminated bacterial disease.
- Direct rapid progression of the infection can lead to overwhelming viral pneumonia and asphyxia.

Treatment of influenza — There are two basic approaches for managing influenza: symptomatic care and anticipation of potential complications. The best treatments for influenza are rest, adequate fluid intake, conservative use of analgesics for myalgia and headache, and cough suppressants.

When influenza is diagnosed, four to five days of administration of amantidine or rimantadine may be considered, but this is useful only if the infection is diagnosed within 12–24 hours of onset (**Table 21.2**).

Respiratory Syncytial Virus (RSV)

This virus is so named because of the syncytia associated with it. (Recall from Chapter 5 that a syncytium is a large multiple-nucleus cell produced during a viral infection.) Community outbreaks of RSV occur annually in the late fall to early spring, with the usual outbreak lasting about 8–12 weeks. These outbreaks can involve 50% of families with small children. It is usually an older sibling that brings the virus into the home, but it infects young children or infants most often. Virus is normally shed for 5–7 days but can be shed for up to 20 days in infants. This virus is a major cause of nosocomial infections, and control of these infections in a hospital setting is difficult. However, attention to hand washing as well as the exclusion of staff and visitors with respiratory symptoms helps to hold down the level of infection.

Fast Fact

Wearing surgical masks is of no use in controlling RSV infection because the virus is small enough to move through the material of the mask easily.

Pathogenesis of RSV infection — RSV spreads to the upper respiratory tract by contact with infectious secretions, and infection is usually confined to the respiratory epithelium with progression to the middle and lower airways. Viremia is rare, and the effect of the virus on the respiratory epithelium is similar to that seen in influenza infections. Cytotoxic T cells seem to have a significant role in controlling acute RSV infection.

Major pathological consequences involve the bronchi, bronchioles, and alveoli, including necrosis of the epithelial cells, interstitial mononuclear cell infiltration, and inflammation that can involve the alveoli and the alveolar ducts. These developments can result in the plugging of the small

Table 21.2 Comparison of Antiviral Drugs for Influenza

Feature	Amantadine	Rimantadine	Zanamivir	Oseltamivir
Susceptible viruses	Influenza A only	Influenza A only	Influenza A and B	Influenza A and B
Emergent resistant strains	Yes	Yes	Not known	Not known
Administration	Oral	Oral	Inhalation	Oral

airways with mucus, necrotic cells, and fibrin. The incubation period is usually two to four days, followed by the onset of rhinitis. The severity of the infection peaks within three days.

Clinical signs can include hyperexpansion of the lungs, **hypoxia** (low oxygenation of the blood), and hypercapnia (retention of CO_2). There can also be pulmonary collapse seen on X-ray, but the normal duration of acute signs is only 10–14 days. The fatality rate among hospitalized infants is about 1%, but it can be as high as 15% in compromised children. RSV infection is usually mild in adults and older children.

Treatment of RSV infection — Treatment is primarily directed at the underlying clinical pathology and includes adequate oxygenation and ventilation. Additionally, close observation to deal with potential bacterial superinfections is indicated. There is no vaccine for this infection.

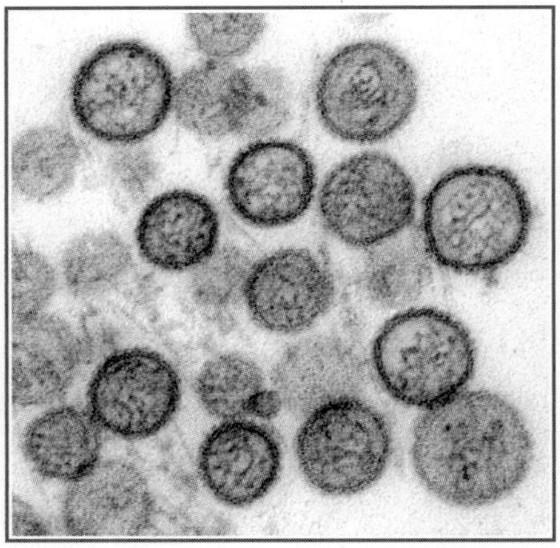

Figure 21.18 An electron micrograph of the hanta (Sin Nombre) virus.

Hantavirus Pulmonary Syndrome (HPS)

It has been known for some time that rodents in the United States are infected with hantavirus (**Figure 21.18**), but there was no recognized infection in humans until an outbreak in 1993. This infection usually occurs in the southwestern United States (where the outbreak in 1993 occurred), but cases have been reported in 21 states. The virus causes a fulminant respiratory infection with high (50–70%) mortality.

Three types of hantavirus cause HPS, with the type known as Sin Nombre being the most common. Hantavirus infections are associated with increases in the rodent population.

Pathogenesis of hantavirus pulmonary syndrome — Transmission is through inhalation of dried rodent excreta, by the conjunctival route, or by direct contact through breaks in the skin. Human-to-human transmission has not been observed with this virus. The early symptoms of this infection include fatigue, fever, and muscle aches especially in the large muscle groups, thighs, hips, and back. There can also be headache, dizziness, and abdominal problems. Four to ten days after the initial phase of the illness, late symptoms appear. These include coughing and shortness of breath as the lungs fill with fluid.

Treatment of hantavirus pulmonary syndrome — The major control of HPS infection involves controlling the rodent population. For the infection, aggressive respiratory support is required, although intravenous ribavirin has been used with some success.

FUNGAL INFECTIONS OF THE RESPIRATORY SYSTEM

As you read this discussion of fungal infections of the respiratory system, it is important to remember that fungal infections are for the most part seen only in patients who are in some way compromised (Chapter 14). There are two major factors governing the incidence and spread of fungal infections: (1) the ubiquity of the infectious organisms, which are found not only in the soil but also as resident flora of the body; and (2) the fact that the adaptive immune response usually keeps these infections under control. Consequently, in patients who are compromised in some way, the risk for fungal infection rises drastically. Although a variety of fungal organisms can cause respiratory infections in compromised hosts, we will focus on only the most common.

Pneumocystis Pneumonia (PCP)

In immunocompromised patients, particularly those with acquired immunodeficiency syndrome (AIDS), the fungus *Pneumocystis (carinii) jiroveci*

(**Figure 21.19**) causes a lethal pneumonia called *Pneumocystis* pneumonia (PCP). This fungus has never been grown in culture, and most of what is known about it comes from clinical information obtained from patients with infected lungs. For a long time, the infection was thought to be caused by a protozoan because of the shape, nucleus, and spores of *Pneumocystis*, which resembled structures seen in protozoans. In addition, this organism has cholesterol in the plasma membrane rather than the ergosterol seen in other fungal cells. Eventually, RNA typing was used to confirm that *P. (carinii) jiroveci* is indeed a fungus.

Pulmonary infection with this fungus occurs in humans and animals throughout the world, and antibodies against it are found in almost all children by the age of 4 years. Although the reservoirs and modes of transmission have yet to be defined, it has been shown in animal models that the fungus can be transmitted by aerosol.

Before the AIDS pandemic, *Pneumocystis* pneumonia was seen only sporadically in infants with congenital immunodeficiencies and in some compromised adults. Now, AIDS is the most common predisposing factor for PCP. In fact, before aggressive therapy for AIDS was introduced, PCP was seen in more than 50% of patients presenting with AIDS, and most patients with AIDS develop this form of pneumonia as the infection progresses.

Pathogenesis of *Pneumocystis* pneumonia — *Pneumocystis (carinii) jiroveci* has a low level of virulence and therefore rarely affects immunocompetent hosts who have normal T cell function. However, in AIDS the risk of PCP increases as CD4 T cells disappear. Little is known about the early stages of the infection, but it is thought that the organism may attach by way of a surface glycoprotein that binds to host cell proteins or fibronectin. This fungal glycoprotein also seems to undergo antigenic variation, which may be the reason for the persistence of the infection.

Pneumocystis pneumonia is characterized by alveoli filled with sloughed-off alveolar cells, monocytes, and fluid, producing a distinct foamy honeycombed appearance. In compromised hosts, this infection presents as progressive diffuse **pneumonitis**. For patients with AIDS and infants with this infection, the onset of the infection is insidious, and the infection can be present for three to four weeks before it is discovered.

The principal manifestation of infection is progressive **dyspnea** and tracheal pneumonia, with eventual cyanosis and hypoxia. A nonproductive cough appears in 50% of patients, with X-rays showing some alveolar infiltrates that spread out from the hili of the lung, eventually affecting the entire lung. Radiographic abnormalities are accompanied by decreased O_2 saturation of arterial blood as well as decreased lung vital capacity. Death occurs through progressive asphyxiation. There can also be lesions in the lymph nodes, bone marrow, spleen, liver, eyes, thyroid, adrenal glands, and kidneys.

Treatment of *Pneumocystis* pneumonia — In patients who do not have AIDS, a combination of trimethoprim and sulfamethoxazole (TMP-SMX) for 14–21 days is the treatment of choice. In patients with AIDS,

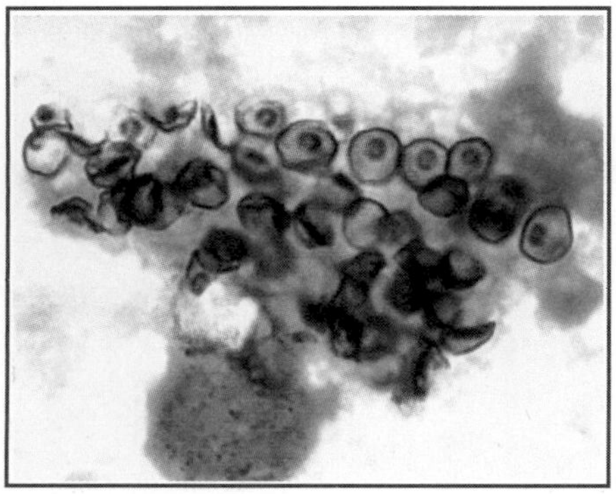

Figure 21.19 A photomicrograph of *Pneumocystis (carinii) jiroveci* in the sputum of a patient with *Pneumocystis* pneumonia. This organism can not yet be grown in a laboratory setting, and consequently all images of it have been made from clinical samples.

because most of them have severe side effects with TMP-SMX therapy, the preferred alternative therapy is with pentamidine and trimetrexate. In addition, patients with AIDS usually receive treatment for longer than 21 days because in most cases (1) they present with more advanced infection and (2) they respond more slowly and relapse more often.

Blastomycosis

Blastomycosis (also known as Gilchrist's disease) is an infection caused by the fungus *Blastomyces dermatitidis*. The spores of the fungi enter the body through the respiratory system and primarily affect the lungs. However, the disease can occasionally spread through the bloodstream and affect other parts of the body, including the skin. Most infections occur in the United States but they have also been seen widely spread in Africa. Men between the ages of 20 and 40 years are the most commonly infected with this fungal disease. Unlike most fungal infections, blastomycosis is not seen more often in people who have AIDS.

Pathogenesis of blastomycosis — Infection of the lungs begins gradually with fever, chills, and drenching sweats. Chest pain, difficulty in breathing, and a cough may also develop. Infection in the lungs develops slowly and can sometimes heal without treatment. When the infection spreads it can affect many areas of the body, but the skin, bones, and genitourinary tract are most often affected. In the skin, the infection begins as papules, which may contain pus. Warty patches then develop and are surrounded by tiny painless abscesses. Painful swelling of the bones can also occur. Men may experience prostatitis or painful swelling of the epididymis.

Treatment of blastomycosis — Blastomycosis can be treated with intravenous amphotericin B or oral itraconazole. With treatment the patient begins to feel better quickly, but the drug therapy must be continued for months. Without treatment, the infection slowly worsens and leads to death.

Histoplasmosis

This fungal infection is caused by *Histoplasma capsulatum*, a fungus commonly found in temperate, subtropical, and tropical zones in soil contaminated with bat or bird droppings. In the United States, *H. capsulatum* is most often found in areas around the Ohio and Mississippi rivers.

Fast Fact

There is no person-to-person transmission of histoplasmosis, and for reasons that are not clear it seems to be more prevalent in men than in women.

Between 50% and 90% of residents in areas where this organism is found test positive for it, indicating that they have been exposed. People who live and work in the vicinity of bat or bird droppings are at increased risk of developing this infection.

Pathogenesis of histoplasmosis — Transmission is through inhalation of conidia of the organism, which are small enough to reach the bronchioles and alveoli of the lung. Because of their minute size, these spores are usually referred to as microconidia (**Figure 21.20**).

Most cases of histoplasmosis are asymptomatic or present with only fever and mild cough. Initial infection is pulmonary, but the lymph nodes, spleen, bone marrow, and other elements of the mononuclear phagocytic system (Chapter 15) can become involved. During the infection, microconidia are inhaled and converted to the yeast forms, which are phagocytosed by macrophages and polymorphonuclear leukocytes. The yeast forms survive the formation of the phagolysosome in phagocytic cells by capturing iron and lowering the pH of the phagolysosome. They then continue to divide in the cytoplasm of the phagocytic cell.

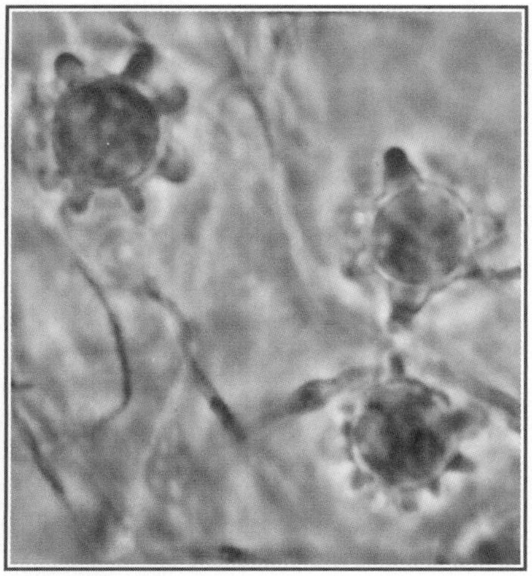

Figure 21.20 The microconidia of *Histoplasma capsulatum*, the causative agent of the fungal infection histoplasmosis. These structures are small enough to bypass some of the defenses of the upper respiratory tract and enter directly into the lung, causing pneumonia-like symptoms. Like most other fungal infections, histoplasmosis is seen only in patients whose defenses are compromised in some way.

As growth continues, there is formation of a tubercle similar to that seen in tuberculosis. The vast majority of cases never go further in the infectious process, although some patients may develop fever and cough for a few days or even weeks. In severe cases there may be chills, malaise, chest pain, and extensive pulmonary infiltration, but even these severe infections usually resolve spontaneously.

Treatment of histoplasmosis — Usually the infection resolves spontaneously, and there is no need for treatment. If necessary, amphotericin B is the treatment of choice, but it is toxic and used only for short times, and then only in severe cases. Itraconazole and ketoconazole have been used in patients with AIDS who have histoplasmosis.

Coccidioidomycosis

This infection is caused by *Coccidioides immitis*. Infection can occur either in an asymptomatic form or in the infectious form known as **valley fever**. Like histoplasmosis, coccidioidomycosis is restricted to certain geographical areas. This fungus grows only in alkaline soil and semiarid climates known as the *lower Sonoran life zone*, usually in places with hot, dry summers, mild winters, and an annual rainfall of 10 inches. These locations are scattered throughout the United States (Arizona, New Mexico, western Texas, and California), and 50–90% of long-term residents in these areas test positive for exposure to *Coccidioides*.

Pathogenesis of coccidioidomycosis — The arthroconidia of the fungus are inhaled and are small enough to bypass defenses of the upper respiratory tract and lodge directly in the bronchioles. The outer wall of *Coccidioides immitis* has antiphagocytic properties that prevent elimination of the organism, and the arthroconidia convert to spherules (**Figure 21.21**), which slowly grow and completely inhibit phagocytosis. It is thought that, in the spherule stage, cell wall proteases may contain virulence factors, but they have not yet been identified.

More than half of infected individuals show no signs of infection, but the remainder progress to valley fever and present with malaise, cough, chest pain, fever, and arthralgia (joint pain), all of which occur one to three weeks after the infection begins. The signs can last for up to six weeks; however, most patients spontaneously resolve the infection, and only 10% of patients ever experience pulmonary symptoms. Disseminated coccidioidomycosis has been seen in patients with AIDS and in individuals on immunosuppressive therapy. In addition, a form of coccidioidal meningitis can occur, which can be fatal if not treated aggressively.

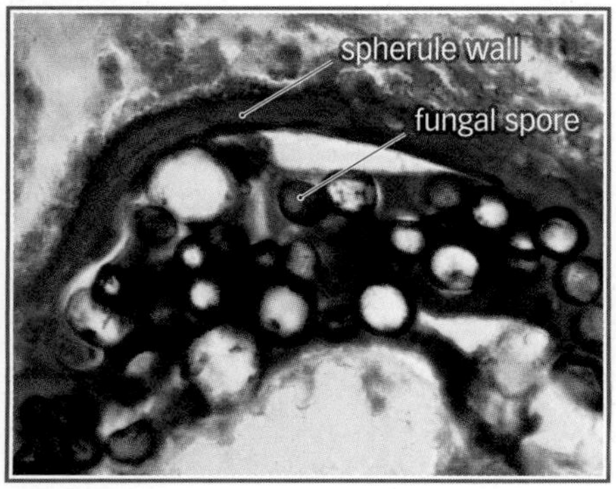

Figure 21.21 The spherule structure seen in coccidioidomycosis. The fungal spores are contained inside the spherule.

Treatment of coccidioidomycosis — Because this infection is in most cases self-limiting, no treatment is indicated. However, in cases of progressive pulmonary infection or infection of the central nervous system, amphotericin B can be used. Fluconazole and itraconazole are also effective.

Aspergillosis

Invasive aspergillosis is distinguished in immunocompromised individuals by rapid progression to death. This infection is typically seen in immunocompromised patients, in particular patients with leukemia or AIDS and those undergoing bone marrow transplantation. The fungus *Aspergillus* (**Figure 21.22**) is widely distributed in nature and found throughout the world.

Dispersal occurs through inhalation of resistant conidia and has been seen more and more in nosocomial infections associated with air-conditioning systems.

Fast Fact

Aspergillus can be present in older homes and become an aerosol when disturbed as the homes are remodeled.

Pathogenesis of aspergillosis — The conidia of *Aspergillus* are small enough to reach the alveoli of the lung when inhaled. Infection is rare when good immune potential is present. The mechanism by which the conidia attach to host cells is not well understood, but it has been shown that, after attachment, the fungus produces extracellular proteases and phospholipases. However, the involvement of these extracellular products in the infection process is not yet understood. Most *Aspergillus* species also produce toxic metabolites, but their involvement in the infection process is also not clear.

Aspergillosis usually occurs in individuals with a preexisting pulmonary disease, such as chronic bronchitis, asthma, or tuberculosis, and also in immunocompromised hosts. Colonization with *Aspergillus* leads to invasion of tissues by the branching septate hyphae. Invasion of lung tissue can lead to penetration of blood vessels, causing hemoptysis (coughing up blood) and/or acute pneumonia in immunocompromised patients. This is accompanied by multifocal pulmonary infiltrates that consolidate and present with high fever. The prognosis of this infection is grave, and the mortality for invasive aspergillosis is 100%.

Treatment of aspergillosis — Amphotericin B and itraconazole for the invasive form of the infection can be used but are usually ineffective.

Keep in Mind

- Viral infections of the lower respiratory tract include influenza, respiratory syncytial virus, and hantavirus pulmonary syndrome.
- The adaptive immune response usually keeps fungal infections under control.
- Fungal infections of the respiratory tract are usually opportunistic infections seen in immunocompromised individuals.
- Fungal infections of the respiratory tract include *Pneumocystis* pneumonia, histoplasmosis, coccidioidomycosis, and aspergillosis.

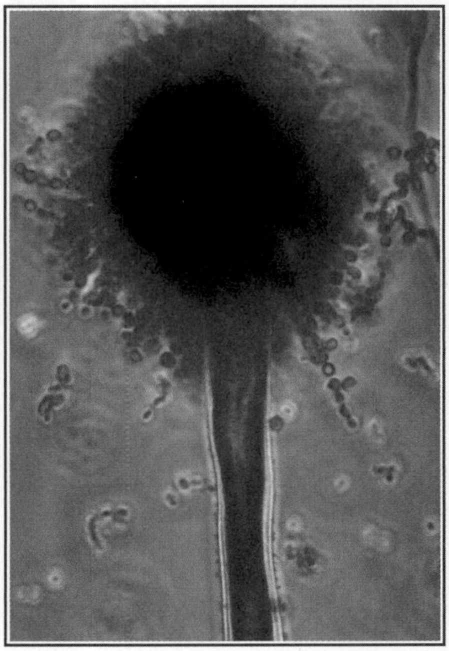

Figure 21.22 This photomicrograph shows the appearance of a conidiophore of the fungus *Aspergillus*. *Aspergillus* is the etiologic agent of aspergillosis, an infection that occurs in immunocompromised patients.

SUMMARY

- The respiratory system is a major portal of entry for pathogens.
- The upper respiratory tract is continuously exposed to pathogens, whereas the lower respiratory tract is essentially sterile.
- The respiratory system is also a good portal of exit for pathogens.
- Bacterial infections of the lower respiratory tract include nosocomial and community-acquired bacterial pneumonia.
- Tuberculosis, inhalation anthrax, pertussis, legionellosis, psittacosis, and Q fever are all serious infections of the lower respiratory tract.
- Viral infections of the lower respiratory tract include influenza, respiratory syncytial virus, and hantavirus pulmonary syndrome.
- Fungal infections of the respiratory tract are usually opportunistic infections seen in immunocompromised individuals.
- *Pneumocystis pneumonia*, histoplasmosis, coccidioidomycosis, and aspergillosis are fungal infections of the respiratory tract.
- The adaptive immune response keeps fungal infections under control.

In this chapter, we have seen that bacteria, viruses, and fungi can all use the respiratory tract as a portal of entry and cause infection. This picture will be repeated in all of the body systems that we explore in the following chapters. It is important to remember that although many pathogens can gain entry to the respiratory system, the impressive defense mechanisms associated with the system work to control the occurrence of infections, especially in the lower respiratory tract.

SELF EVALUATION AND CHAPTER CONFIDENCE

Multiple Choice

Answers are given in the back of the book and help can be found in the student resources at:
www.garlandscience.com/micro

1. Penicillin is used to treat all of the following except

   A. Pneumococcal pneumonia
   B. Mycoplasmal pneumonia
   C. Scarlet fever
   D. Diphtheria
   E. Streptococcal sore throat

2. Pneumonia can be caused by all of the following except

   A. *Haemophilus*
   B. *Mycoplasma*
   C. *Streptococcus*
   D. *Legionella*
   E. All of the above

3. Which of the following does not produce any exotoxin?

   A. *Bordetella pertussis*
   B. *Streptococcus pyogenes*
   C. *Corynebacterium diphtheriae*
   D. *Mycobacterium tuberculosis*
   E. None of the above

4. Infection by which of the following results in the formation of Ghon complexes?

   A. *Corynebacterium diphtheriae*
   B. *Streptococcus pyogenes*
   C. *Bordetella pertussis*
   D. *Mycobacterium tuberculosis*
   E. None of the above

5. Which of the following produces the most potent exotoxin?

   A. *Streptococcus pyogenes*
   B. *Bordetella pertussis*
   C. *Corynebacterium diphtheriae*
   D. *Mycobacterium tuberculosis*

6. The recurrence of influenza epidemics is due to

   A. Antigenic shift
   B. Lack of antiviral drugs
   C. Lack of naturally acquired active immunity
   D. All of the above

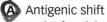

7. Legionella is transmitted by

A. Fomites
B. Airborne transmission
C. Vectors
D. Foodborne transmission
E. Person-to-person contact

8. The patient has a sore throat. Which of the following would be involved?

A. *Mycobacterium*
B. *Haemophilus*
C. *Corynebacterium*
D. *Bordetella*
E. Can't tell

DEPTH OF UNDERSTANDING

Questions listed here require you to bring together the concepts you have learned in this chapter into a discussion format. This helps you to increase your depth of understanding of the material you have learned. Help can be found in the student resources at:
www.garlandscience.com/micro

1. From the perspective of infection, discuss the differences between the upper and the lower respiratory tracts.

2. Discuss the pathogenesis of tuberculosis, including the host defense or lack thereof.

CLINICAL CORNER

Help can be found in the student resources at: www.garlandscience.com/micro

1. You are working at a hospital in the industrial district of Indianapolis. A new office building opened near the hospital approximately eight months ago and has been touted to be a state-of-the-art, energy-efficient structure. On a late Thursday afternoon, the emergency room is overwhelmed with 53 patients complaining of different levels of respiratory distress ranging from mild coughing to severe pneumonia. In fact, two have died of pneumonia shortly after being admitted. As it turns out, 47 of the 53 patients with these complaints work at the new office building but not on the same floor. In fact, the offices of these individuals range from the first floor to the 25th floor.

   A. What would be your initial guess on the cause of this outbreak?
   B. How would you go about proving your hypothesis?
   C. What would be your recommendation for treatment?

2. You work in the hospital nursery where you take care of new born babies. You have been having influenza-like symptoms the last few days but they are nowhere near strong enough to make you stay home from work. In addition, you really need the money and don't want to miss days simply because you don't feel that great. A coworker has noticed that you are ill and has told you to stay home until you feel better. To placate her you have decided to wear a mask while at work but even this step has not made her happy. She has threatened to report you to the nursing supervisor if you do not stay home until you feel better.

   A. Why is she so insistent about your not coming to work?
   B. Why is wearing the mask not enough to overcome her objections?
   C. What can you do to allay her concerns?

Fast Fact

There is a 50% chance of children in underdeveloped countries dying from gastrointestinal diseases before they reach the age of seven years.

Outbreaks of digestive diseases are usually associated with crowding, poor hygiene, and contaminated food or water. In the developed world, the bacterium *Campylobacter* is the most common cause of gastrointestinal infections, with the bacteria *Salmonella* and *Shigella* next in order of most common causes of infections. In the underdeveloped world and the tropics, the most common cause of gastrointestinal infections is the bacterium *Vibrio cholerae*. Sources and transmission routes for several bacterial infections in the digestive tract are shown in **Figure 22.1**.

A variety of pathogens cause infections in the digestive system. For example, *Clostridium difficile* causes serious gastrointestinal infections in patients who are overmedicated with antibiotics. Nosocomial infections with *C. difficile* have become a serious problem in health care. Overuse of antibiotics is also associated with the development of opportunistic infections that are normally prevented by the microbial flora present in our bodies (**Figure 22.2**).

In the United States, gastrointestinal infections are one of the top three clinical problems seen by physicians; however, the majority of those infected do not seek medical attention.

Infections in the digestive system can be classified in two groups: **exogenous infections**, which are caused by pathogens that come into the body from the outside, and **endogenous infections**, which are caused by organisms that are part of the normal microbial flora of the body (**Figure 22.3**). Exogenous organisms are brought into the digestive system through contaminated food or water. This group includes *C. difficile*, which is frequently brought into the body from the hospital environment, as just noted above, and *Helicobacter pylori*, which spreads from human to human through oral–oral or fecal–oral contact. Infection with these exogenous pathogens can cause nausea and vomiting within six hours of ingestion.

Endogenous organisms, such as the bacteria *Streptococcus* and *Enterococcus* and anaerobes of the intestinal tract, are found as part of the normal flora. However, given the right set of circumstances, these organisms can cause dental diseases; infections of the bowel, appendix, and liver; and diverticular abscesses.

Figure 22.1 Examples of the sources and modes of transmission of some common bacterial causes of gastrointestinal infections.

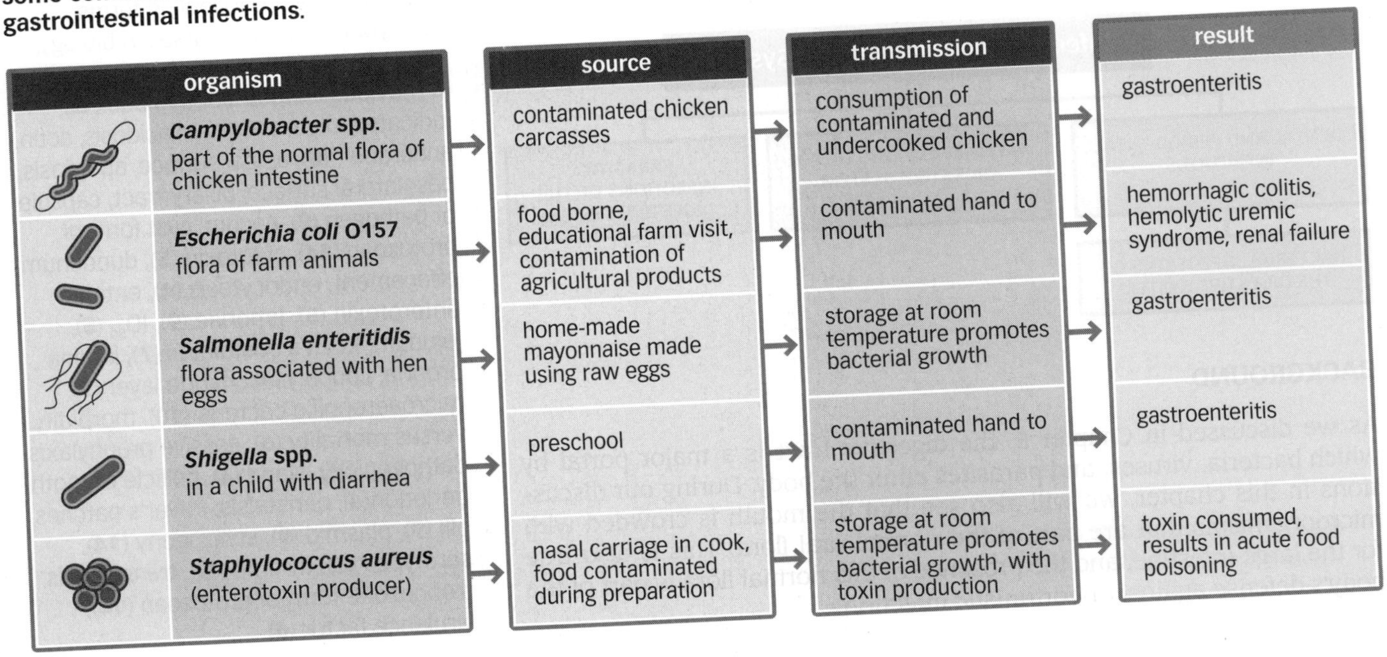

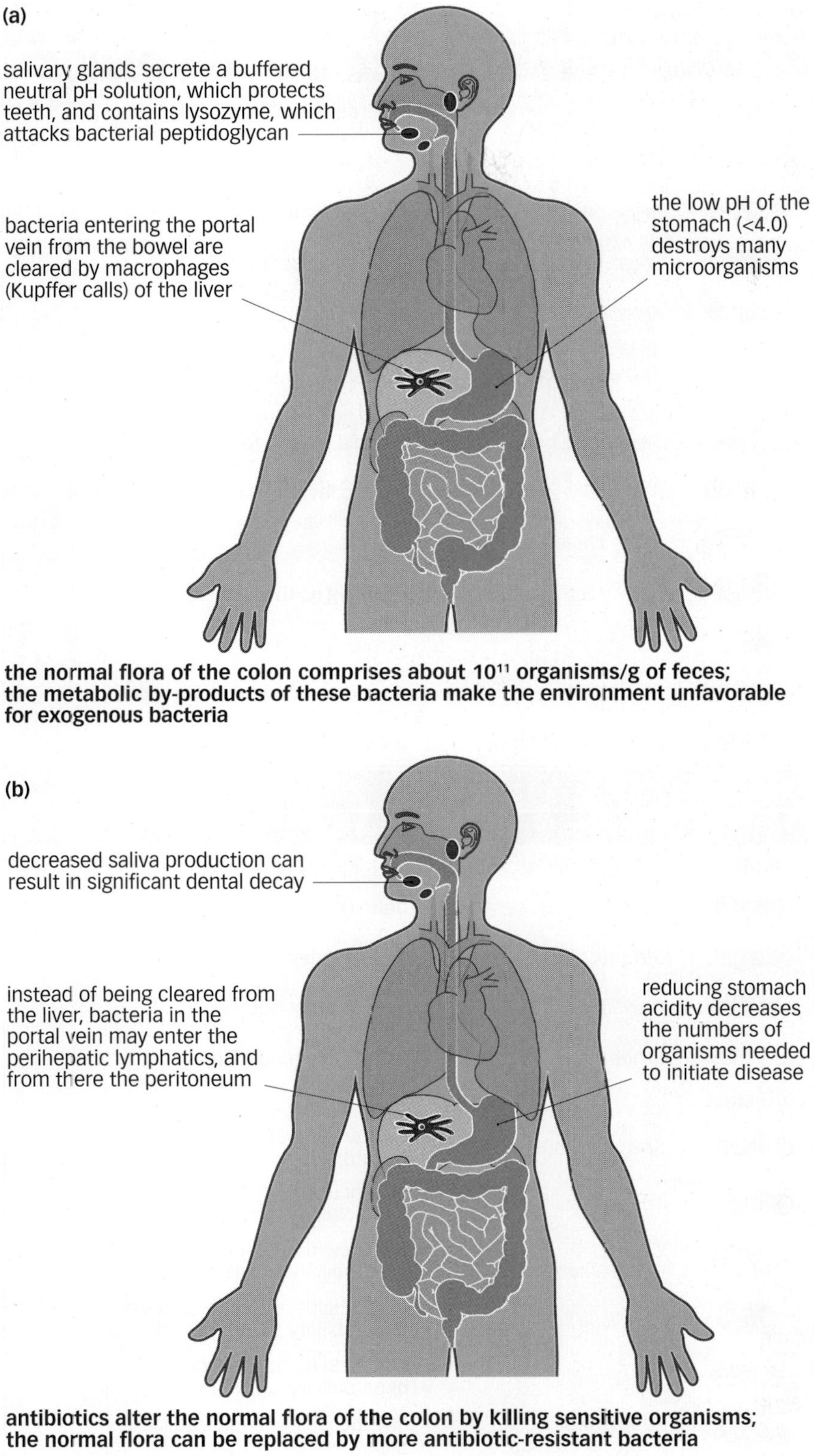

Figure 22.2 The body's defenses against gastrointestinal infection. Panel **a**: Anatomical features and normal bacterial flora that are part of the human body's defense against gastrointestinal infection. Panel **b**: How changes to these anatomical features or defenses can predispose to gastrointestinal disease.

Clinical Symptoms of Gastrointestinal Infections

The most common symptoms of gastrointestinal infections are fever, vomiting, abdominal pain, and diarrhea. These symptoms vary with the specific infection and can also vary with the stages of infection. In all cases, however, the central feature is diarrhea, and the presence and nature of diarrhea can be used to classify gastrointestinal infections into three categories: *watery diarrhea*, *dysentery*, and *enteric fever*.

Figure 22.3 Some bacteria that cause infections of the digestive system. For exogenous organisms the main mechanisms of disease are indicated by In = invasion; Inf = inflammation; P = penetration; T = toxin.

exogenous

organism	disease	mech.
Campylobacter spp.	gastroenteritis	In/Inf/P
Salmonella enteritidis	gastroenteritis	In/Inf
Salmonella typhimurium	gastroenteritis	In/Inf
Shigella dysenteriae	bacillary dysentery	In/Inf
Shigella sonnei	gastroenteritis	In/Inf
Escherichia coli O157	hemorrhagic colitis hemolytic uremic syndrome acute and chronic renal failure	T
Salmonella typhi	typhoid/enteric fever	P
Salmonella paratyphi	enteric fever	P
Vibrio cholerae	cholera	T
Staphylococcus aureus	vomiting (with diarrhea)	T
Bacillus cereus	vomiting (with diarrhea)	T
Helicobacter pylori	peptic ulcer/gastric malignancies	T/Inf
Clostridium difficile	antibiotic-associated diarrhea	T

endogenous

organism	disease
mouth	
Streptococcus sanguis	dental caries, dental abscess
Streptococcus mutans	dental caries, dental abscess
Prevotella spp.	gingivitis, periodontitis
Fusobacterium spp.	gingivitis, periodontitis
intestine	
Streptococcus spp.	diverticulitis, appendix abscess, hepatobiliary sepsis, periodontitis
Enterococcus spp.	diverticulitis, appendix abscess, hepatobiliary sepsis, periodontitis
Coliforms (e.g. *Escherichia coli*)	diverticulitis, appendix abscess, hepatobiliary sepsis, periodontitis
Klebsiella spp.	diverticulitis, appendix abscess, hepatobiliary sepsis, periodontitis
Bacteroides spp.	diverticulitis, appendix abscess, hepatobiliary sepsis, periodontitis
Peptostreptococcus *Peptococcus* spp.	diverticulitis, appendix abscess, hepatobiliary sepsis, periodontitis
Clostridium spp.	diverticulitis, appendix abscess, hepatobiliary sepsis, periodontitis

Watery diarrhea is the most common type of gastrointestinal infection. It develops rapidly and results in frequent voiding. There may also be accompanying vomiting, fever, and abdominal pain. This category of gastrointestinal infection is caused by pathogenic mechanisms that attack the intestines. The purest form of watery diarrhea is caused by enterotoxin-secreting bacteria, such as *Vibrio cholerae* and enterotoxigenic *Escherichia coli*. These pathogens cause the diarrhea that is the main sign of the infection known as watery diarrhea without destroying the cells of the intestines, whereas other pathogens, such as rotavirus, cause diarrhea and in the process destroy intestinal cells. Watery diarrhea infection is usually acute but brief (lasting for one to three days) and self-limiting. However, watery diarrhea caused by either *V. cholerae* or the parasite *Giardia* can last for weeks and be very severe.

Dysentery has a rapid onset with frequent intestinal evacuations. These are smaller in volume than those seen in watery diarrhea, and they contain blood and pus. There can be accompanying cramps and abdominal pain, but there is very little vomiting. The pathologic consequences of dysentery center in the colon. Organisms that cause dysentery damage the colonic mucosa either directly or through the production of toxins. This tissue damage results in the blood and pus seen in the stool. There is not as much fluid loss in dysentery as in watery diarrhea, but dysentery lasts longer. Even so, most cases resolve in three to seven days.

Enteric fever is a systemic infection with a focus in the gastrointestinal tract. The prominent features of this condition are a fever and abdominal pain that take days to develop. In this condition, diarrhea is mild and does not usually appear until late in the infection. The pathogenesis of enteric fever is more complex than that of either watery diarrhea or dysentery and involves the penetration of **enterocytes**, which are epithelial cells of the small and large intestine. Pathogens spread from here to the biliary (bile) tract, liver, and organs of the reticuloendothelial system. The most investigated form of enteric fever is **typhoid fever**, which is caused by *Salmonella enterica* serotype Typhi (see discussion below). Although it is usually self-limiting, enteric fever can be serious and result in significant mortality.

There are five major types of gastrointestinal infection, with each type caused by a different organism: endemic, epidemic, traveler's diarrhea, food poisoning, and nosocomial infections.

Endemic Gastrointestinal Infections

Endemic intestinal infections are those that occur sporadically. There are some that are seen throughout the world and others that are geographically restricted. In addition, some organisms show a seasonal variation and age-related predisposition. In developed countries, the most common causes of endemic gastrointestinal infections are the bacteria *Campylobacter*, *Salmonella*, and *Shigella*. As mentioned above, these infections are more commonly seen in children because of the underdeveloped immunity and prevalence of fecal–oral contact. Some organisms — *V. cholerae* is one example — are geographically restricted because they need a warm climate.

Fast Fact

Vibrio organisms, once found only in Asia, Africa, and the Middle East, are now found along the coast of South America and along the Gulf Coast and southern Atlantic coasts of the United States.

Epidemic Gastrointestinal Infections

Epidemic gastrointestinal infections are defined as those that involve regional, national, and international populations, with the most common epidemic gastrointestinal infections being cholera, typhoid fever, and shigellosis. All of these infections are directly related to failures in public health management: cholera and typhoid fever are associated with contamination of water, and shigellosis is associated with wars, crowding, and poor sanitation. Because the ID_{50} of *Shigella* infection is very low, this infection is easily spread. In the United States, the most frequently seen epidemic gastrointestinal infections are seen with *E. coli* O157:H7, *Cryptosporidium*, and *Giardia*.

Traveler's Diarrhea

Many travelers to underdeveloped countries are concerned about infections. In fact, 20–50% of these travelers will get diarrhea in the first week of their visit to these countries. Usually, the problem is brief and self-limiting, but in some cases it can be serious. The most extensive studies of traveler's diarrhea have involved subjects who travel from the United States to Latin America, particularly Mexico. In 50% of the cases studied, the causative agent was enterotoxigenic strains of *E. coli*, with 10–20% being caused by *Shigella* and the remainder by infection from unknown sources. Ingestion of improperly cooked food was found to be the major source of transmission, with some toxigenic *E. coli* also being found in salads and vegetables.

Food Poisoning

Many gastrointestinal infections involve food; food poisoning is usually connected to one single meal as the source and almost always involves improper food handling. Food poisoning is typically seen with multiple patients connected to a single source of contamination. The spread of this type of food-borne infection has become greater with the popularity of fast food.

Food poisoning can result in two ways: as an *infection*, in which case a pathogen is directly involved in the process, or as an *intoxication*, in which case a toxin produced by a pathogen is involved. In both types, the incubation time and severity of the illness usually depend on the number of pathogens ingested (in infection) or the amount of toxin ingested (in intoxication). In general, the incubation time is shorter in intoxication than in infection because in the former the toxin is already present in the ingested food. Intoxication may also involve organs outside the digestive tract, as in botulism, which affects the central nervous system. The most common features of both types of food poisoning are shown in **Table 22.1**.

Table 22.1 Features of Microbial Food Poisoning

Etiology	Percentage of Cases[a]	Typical Incubation Period	Clinical Findings	Foods
Intoxication[b]				
Clostridium botulinum	5–15	12–72 hours	Vomiting, diarrhea	Improperly preserved vegetables, meat, fish
Staphylococcus aureus	5–25	2–4 hours	Vomiting	Meats, custard, salads
Bacillus cereus	1–2	1–6 hours	Vomiting, diarrhea	Rice, meat, vegetables
Infections[c]				
Clostridium perfringens	5–15	9–15 hours	Watery diarrhea	Meat, poultry
Salmonella	10–30	6–48 hours	Dysentery	Poultry, eggs, meat
Shigella	2–5	12–48 hours	Dysentery	Variable
Vibrio parahaemolyticus	1–2	10–24 hours	Watery diarrhea	Shellfish
Trichinella spiralis	1–3	3–30 days	Fever, myalgia	Meat, especially pork
Hepatitis A	1–3	10–45 days	Hepatitis	Shellfish

[a]Based on documented outbreaks reported to the Center for Disease Control.
[b]Disease caused by toxin in the food at time of ingestion.
[c]Disease caused by infection after ingestion.

Most cases of both types of food poisoning involve failing to cook food adequately and then allowing the undercooked food to sit warm for some length of time. The latter serves as an incubation period for the pathogen, allowing it either to multiply or else to produce a sufficient amount of toxin to sicken a person. In about 80% of investigated cases, an additional contributing factor is improper storage of food. Of the 400–500 reported outbreaks of food poisoning in the United States each year (which involved over 15,000 people), fewer than 200 were "solved." However, *Salmonella*, *Clostridium perfringens*, and *Staphylococcus aureus* account for more than 70% of cases in which a microbial source was identified.

Nosocomial Gastrointestinal Infections

Infections acquired in the hospital can usually be traced either to employees who were ill but came to work anyway or to contaminated food prepared outside the hospital and then brought in by friends or relatives of the patient. Two major pathogens are responsible for nosocomial gastrointestinal infections: *E. coli* in hospitals is usually seen in infants and small children; and *C. difficile* (**Figure 22.4**), which accounts for 90% of the cases. Symptoms typical of nosocomial *Clostridium* infections vary from mild diarrhea to fulminant pseudomembranous colitis, which arises either during or after treatment with antibiotics. In these cases, the *C. difficile* organisms may be part of the microbial flora of the patient or they may have been acquired from other patients in the hospital.

Treatment and Management Options for Gastrointestinal Infections

Treatment for most of the gastrointestinal infections described above involves supportive care with liquid replacement and rest. In some cases, such as with cholera, where liquid loss can be substantial, intravenous replacement of liquids may be required. Infection with *E. coli* O157:H7 can result in renal failure, which requires dialysis or in some cases a transplant. Gloves, gowns, and handwashing in the hospital prevent the spread of *C. difficile*. A summary of treatment options for several pathogens that cause gastrointestinal infections is shown in **Figure 22.5**.

Keep in Mind

- The digestive system is a major portal by which pathogens may enter the body.
- Gastrointestinal infections can be exogenous (coming from the outside) or endogenous (caused by organisms found in a person's normal intestinal flora).
- The common symptoms of infection of the digestive system include diarrhea, fever, vomiting, and abdominal pain.
- Dysentery differs from diarrhea in that in dysentery the stool contains mucus, blood, and pus.
- Digestive diseases can be endemic or epidemic, and many involve food poisoning.
- Treatment for most gastrointestinal infections includes supportive care and the replacement of liquids lost as a result of diarrhea and/or vomiting.

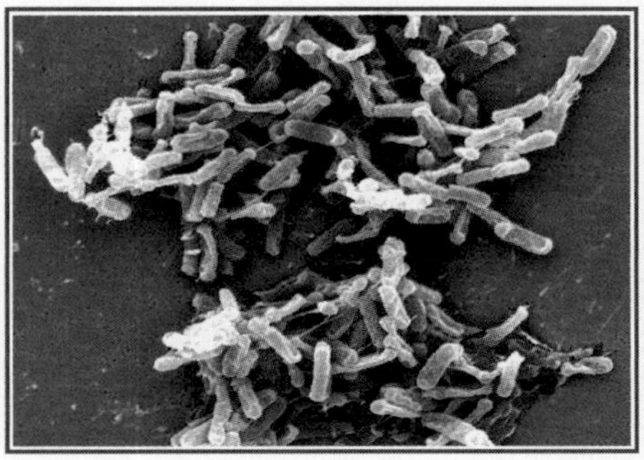

Figure 22.4 This photomicrograph depicts Gram-positive *Clostridium difficile* bacteria isolated from a stool sample. This organism accounts for 90% of hospital-associated gastrointestinal diarrhea.

	organism	treatment
exogenous		
	Campylobacter spp.	supportive, occasionally erythromycin
	Salmonella enteritidis	supportive, occasionally ciprofloxacin
	Salmonella typhimurium	supportive, occasionally ciprofloxacin
	Shigella dysenteriae	supportive, occasionally ciprofloxacin
	Shigella sonnei	supportive
	Escherichia coli O157	supportive, dialysis, renal transplantation, NO antibiotics
	Salmonella typhi	supportive, ciprofloxacin
	Salmonella paratyphi	supportive, ciprofloxacin
	Vibrio cholerae	supportive
	Staphylococcus aureus	supportive
	Helicobacter pylori	amoxicillin, clarithromycin
	Clostridium difficile	metronidazole, vancomycin
endogenous		
	Streptococcus mutans	dental care, amoxicillin, clindamicin as needed
	Prevotella spp.	dental care, amoxicillin, clindamicin as needed
	Streptococcus spp.	amoxicillin, cefuroxime, vancomycin
	Enterococcus spp.	amoxicillin, vancomycin
	Coliforms (e.g. *Escherichia coli*)	ciprofloxacin, gentamycin, cefuroxime,
	Clostridium spp.	metronidazole, vancomycin

Figure 22.5 Examples of treatment options for infections of the digestive system.

DENTAL AND PERIODONTAL INFECTIONS

The mouth is the portal of entry for many pathogens of the digestive tract. In addition, many opportunistic pathogens reside here. The most commonly seen infections in the mouth are dental caries (also called cavities) and infections of the gum tissue (gingivitis). In both cases, the major source of the problem is *plaque*, which is a soft, adherent dental deposit that forms as a result of bacterial colonization of the surface of the teeth. Plaque is insoluble in aqueous media, including saliva, and resists removal by all but the most vigorous brushing and flossing. The formation of caries in a tooth results from the progressive destruction of

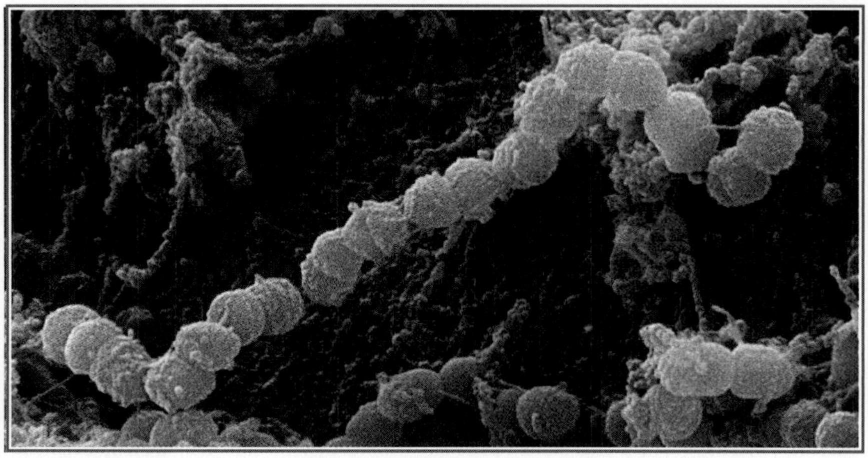

Figure 22.6 Scanning electron micrograph of a biofilm of *Streptococcus mutans*. This organism is one of the first to colonize the pellicle around a tooth.

the mineralized tissue of the tooth, destruction that occurs because of the organic acids produced as a part of the metabolic activity of bacteria located in the plaque coating the tooth.

Formation of Dental Plaque

The tooth surface is normally covered by a thin organic film, the pellicle (Chapter 5), that results from the absorption and binding of specific molecules (mainly proteins and glycoproteins) found in saliva. The pellicle forms around the tooth and because of it bacteria never interact directly with the surface of the tooth but instead adhere to the pellicle.

This adherence is facilitated through the interaction of bacterial adhesion molecules, and the initial adhesion of bacteria (usually *Streptococcus mutans*; **Figure 22.6**) to the pellicle is followed by growth of the bacteria and additional aggregation of other organisms onto the primary layer of bacteria. Primary among these additional organisms are Gram-positive cocci such as streptococcal species and short, Gram-positive rods such as actinomycetes. After two to four days, new layers of organisms have piled on to the growing plaque, and they are followed by Gram-negative motile anaerobic organisms. By this time, there can be as many as 400 species of bacteria affixed to the now "mature" plaque adhering to the pellicle covering the tooth (**Figure 22.7**).

Dental plaque is considered a biofilm because it is permeated with channels that transport nutrients to the bacteria making up the plaque and remove the waste products of bacterial metabolism. It also acts as a mediator for many types of interaction between the organisms making up the plaque. It is important to remember that antiseptic substances, such as biguanides, chlorhexidine, fluorides, triclosan, and quaternary ammonium compounds (discussed in Chapter 18), can inhibit plaque formation and reduce the level of plaque buildup.

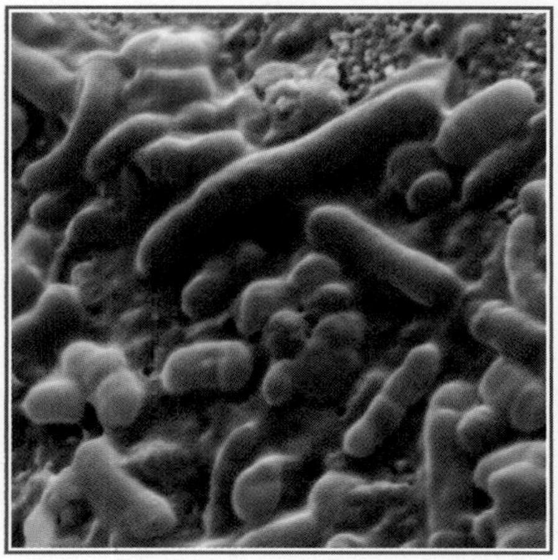

Figure 22.7 Colorized scanning electron micrograph of dental plaque, which can be made up of as many as 400 different species of bacteria.

Dental Caries

Caries are the single greatest cause of tooth loss. There are several factors involved in the development of caries, including tooth structure, types of microflora residing on the tooth, and types of substrate available to the microflora to produce the organic acids that destroy a tooth. Normally, saliva will protect against the establishment of many bacteria because it contains lysozyme, IgA, and other antibacterial products. However, when organisms that produce acid by-products colonize the pellicle, they can cause dental caries. The most common acid-producing organisms are *Streptococcus mutans*, *Streptococcus salivarius*, *Lactobacillus*, *Acidophilus*, and *Actinomyces* species.

Bacteria that cause tooth decay must continue to have an appropriate substrate to metabolize if they are to continually produce the acids that destroy the teeth. The most readily usable substrates are sugars, and when sugars are available in the mouth, they are absorbed and metabolized by the bacteria so quickly that the acid by-products begin to accumulate. The longer this acid accumulation lasts, the more damage is done to the structure of the tooth.

Studies in humans have shown that *S. mutans* is a major cause of dental caries, and this organism initiates the problem, but other organisms also contribute to the destruction of the tooth structure. Carbohydrates easily enter the plaque where all these organisms are located, and are readily metabolized by the bacteria. The frequency of application of substrate is very important in the overall process, and repeated snacking on sugar or carbohydrates keeps the acid level high and continues the demineralization of the tooth.

Gingivitis and Periodontitis

There are two forms of plaque-induced periodontal disease:

- **Gingivitis** is an inflammatory condition limited to the marginal surfaces of the **gingiva**, which is the medical term for gum tissue. This condition does not involve the loss of bone and, depending on the degree of severity, can be corrected.
- **Periodontitis** is an infection of the gingiva that results in the loss of supportive bone and ligaments, and is responsible for most of the tooth loss seen in people over 35 years of age. There are very aggressive forms of periodontitis in which the loss of bone is very rapid.

Both of these problems are thought to be caused by certain bacteria in the dental plaque that lie next to the gingival surfaces. Gingivitis is an inflammation of the connective tissue attached to the tooth and causes a loss of collagen. The tissue destruction is caused by enzymes such as hyaluronidase and collagenase as well as by bacterial toxins such as leukotoxin and endotoxin. In mild forms of gingivitis, there is no bacterial invasion of tissues, but such invasion does occur in aggressive forms of periodontitis.

Gingivitis will continue as long as dental plaque remains on a tooth. If it is removed and the tooth is kept plaque-free, gingivitis can completely resolve, and the tissue will return to normal. However, when the supporting bone begins to be resorbed the condition goes from gingivitis to periodontitis. In this case, the bone is not replaced. Gingivitis can begin in as little as two weeks if teeth are not effectively cleaned.

Necrotizing Periodontal Disease

The condition known as **necrotizing periodontal disease** (NPD), previously referred to as **Vincent's disease** or **trench mouth**, is a spectrum of acute inflammatory diseases, starting with destruction of the mouth's soft tissue as well as the bone and ligaments associated with the teeth. The onset of NPD can be acute, and there is an association of this condition with emotional stress and poor oral hygiene. Necrotizing periodontal disease causes rapid ulceration of the tissues and pronounced bone loss, with spirochete bacteria in direct contact with and invading the tissues. The condition is very painful and causes extremely bad breath. Both systemic and topical administration of antibiotics relieve the symptoms, but resolution depends on professional cleaning of the teeth.

It is important to note that the dental plaque we have discussed in this section must be viewed as a hazard for immunocompromised patients and patients with heart valve malfunctions, in whom there can be an increased incidence of endocarditis. In immunocompromised patients, plaque can give rise to serious systemic infections. In fact, one of the most frequent sources of lethal infections in leukemia patients is the oral cavity.

Keep in Mind

- Many opportunistically pathogenic organisms are found in the mouth.
- Most infections in the oral cavity involve the formation of dental caries, which destroy the teeth, and gum disease, with dental plaque being the source of the infections.
- Dental plaque is made up of hundreds of different organisms.
- Gum diseases include gingivitis, chronic periodontitis, and necrotizing periodontal disease.

BACTERIAL INFECTIONS OF THE DIGESTIVE SYSTEM

Enterobacteriaceae is the name of a diverse family of Gram-negative rod-shaped bacteria, some of which are free-living and some of which are part of the indigenous microflora of the human body. These bacteria grow rapidly under both aerobic and anaerobic conditions, and a small number of them are important etiologic agents of diarrheal diseases. In addition, spread of these organisms to the blood can cause endotoxic shock, which can be fatal. As the family name indicates (*entero-* means "pertaining to the intestines"), all these bacteria do their damage in the human intestines only, never in the stomach.

Members of the family Enterobacteriaceae are some of the largest bacteria, and they have a variety of morphologies, ranging from coccobacilli to elongated filamentous rods. They do not form spores, and the cell wall, plasma membrane, and internal structures are morphologically similar in all species of the family. Some of the cell wall components and other bacterial structures are antigenic, and, as we learned in Chapter 9, these structures have been used to divide species into serotypes. The lipopolysaccharide in the outer membrane is referred to as the **O antigen** (**Figure 22.8**). Cell surface polysaccharides may form a well-defined capsule that is referred to as the **K antigen**. The motile strains use flagella, and the proteins that make up the flagella can be distinguished antigenically and are collectively called the **H antigens**.

Most members of the family Enterobacteriaceae are colonizers of the lower gastrointestinal tract of humans. However, some readily survive in nature and live freely anywhere that water and minimal energy sources are available. In humans, many of these organisms are components of the normal colonic flora. It should be noted that the enterobacteria *Shigella* and *Salmonella* are *not* part of the normal flora, but members of these genera are not free-living; they are strictly animal and human pathogens.

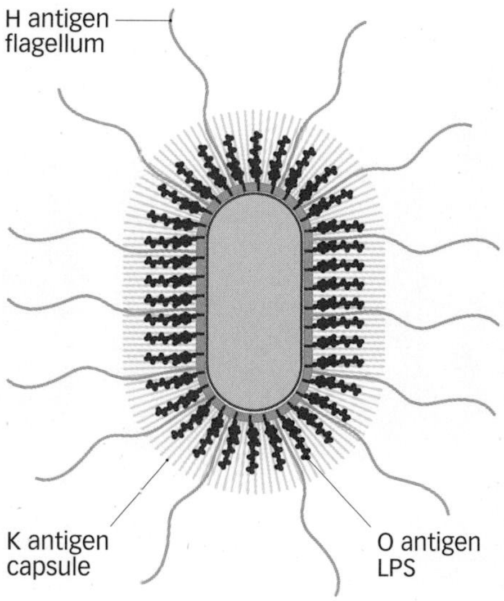

Figure 22.8 Antigens associated with the outer layer of members of the family Enterobacteriaceae. Depending on the organism, there can be a variety of combinations of these antigens. LPS = lipopolysaccharide.

Any enterobacteria that *are* part of the resident flora in the human digestive system always have the potential to cause infection, and it is important to remember that many infections are simply cases of these resident organisms finding their way out of the digestive system to other systems in the body. It is known that surface structures on the bacterial cells, such as fimbriae, aid these organisms in causing infections, but once they get inside the deep tissue of a human host, their ability to persist and cause injury is not really understood except for the action of exotoxins, endotoxins, and capsules.

Salmonella, *Shigella*, *Yersinia enterocolitica* and some strains of *E. coli*, all members of the family Enterobacteriaceae, produce disease in the gastrointestinal tract, and these pathogens have invasive properties. In addition, some produce virulence factors, such as cytotoxins and enterotoxins, that correlate with the type of diarrhea they produce. Enterotoxin-producing bacteria normally cause a watery diarrhea, whereas invasive and cytotoxic strains produce dysentery. For some species, such as *Salmonella*, the gastrointestinal tract is just the portal of entry but the actual disease is systemic.

In addition to adherence by means of fimbriae and the production of endotoxins and exotoxins, enterobacteria can also produce a variety of virulence factors that cause disease. Some of these bacteria use a *contact secretion system* to introduce virulence factors into the cytoplasm of the host cell. The genes for these factors are located either on the chromosome of the bacteria or in a plasmid (sometimes in both) and are organized into pathogenicity islands. Enterobacteria produce the widest variety of infections of any group of microbial organisms. Let's look at several of the most clinically important members of this group in more detail.

Escherichia coli

The first enterobacterium we look at is *E. coli*. These Gram-negative rods ferment lactose and can be distinguished from other organisms that harm the human digestive system by the biochemical reactions they perform. There are over 150 different groups of *E. coli* classified according to what O, K, and H antigens they have. All these distinct groups are defined by numbers (for example, *E. coli* O157:H7). Both fimbriae and pili are often found on these bacteria, and these structures, which are how the bacteria attach to host cells, have an important role in the virulence of the organism.

Escherichia coli can produce every toxin seen in the Enterobacteriaceae family, including a pore-forming cytotoxin, protein inhibition toxins, and toxins that affect cellular signaling pathways in infected cells. Some of the major toxins produced by pathogenic *E. coli* are:

- Pore-forming toxin which is a form of hemolysin that inserts itself into the membrane of the host cell, destroying the integrity of the cell, which eventually dies.

- Shiga toxin, which was originally thought to be exclusive to *Shigella* bacteria but is now known to occur with *E. coli*. It is a two-chain (A and B) toxin (Chapter 5). The B chain attaches to host cells, and this attachment allows the A chain to move into the cell cytoplasm, where it enzymatically modifies the host ribosomal RNA, thereby inhibiting the binding of tRNA to the ribosome. This blocks protein synthesis and causes the death of the host cell.

- Heat-labile toxin is also a two-chain AB toxin and is called heat-labile because of its sensitivity to heat. The B chain binds to the membrane of the host cell and allows entry of the A chain, which causes a breakdown in the signaling mechanisms of the cell. The result is an accumulation of water and other liquids in the lumen of the intestine, an accumulation that becomes a basis for the diarrhea seen in infections involving these bacteria.

- Heat-stable toxin is a small peptide that is more resistant to heat than heat-labile toxin. It binds directly to glycoprotein receptors on target cells. This results in the secretion of liquids and electrolytes into the lumen of the intestine (in a similar manner to that seen with heat-labile toxin).

With virulence as a method of classification, *E. coli* can be divided into the following categories:

- Entero**toxigenic** *E. coli* (ETEC)
- Entero**pathogenic** *E. coli* (EPEC)
- Entero**invasive** *E. coli* (EIEC)
- Entero**hemorrhagic** *E. coli* (EHEC)
- Entero**aggretive** *E. coli* (EAEC)

Each type causes disease by a different mechanism, and the diseases differ both clinically and epidemiologically. A summary of the pathogenesis, clinical syndromes, and epidemiology for these five categories of *E. coli* is shown in **Table 22.2**, but we will look at the enterotoxigenic, enteropathogenic, and enterohemorrhagic forms in more detail here.

Enterotoxigenic E. coli

As the name indicates, these pathogens release a toxin that is the cause of the disease, and they are the most important cause of traveler's diarrhea. These organisms also produce diarrhea in infants and are the leading cause of morbidity and mortality in the first two years of life. In underdeveloped countries, they are a leading cause of mental retardation and malnutrition, but these symptoms are rarely seen in developed countries.

Transmission of enterotoxigenic *E. coli* is by consumption of food or water that is contaminated either by actively infected individuals or by carriers of the organism. Person-to-person transmission is very rare. Uncooked foods are the greatest risk.

Pathogenesis of enterotoxigenic *E. coli* infections — The organism adheres to the cells of the small and large intestine, and the diarrhea seen in this infection is caused by the heat-labile or heat-stable enterotoxins (described above), which are secreted into the small intestine. Genes for these toxins are found on plasmids. Recall our discussions of transfer of genetic information in Chapter 11 and keep in mind that these toxin genes can be readily moved from organism to organism. The bacteria remain attached to the surface of the host cell, and the toxin causes water and electrolytes to flow into the intestine. There is no invasion of the host cells, no damage to those cells, and no inflammatory response in these infections.

Enteropathogenic E. coli

Enteropathogenic *E. coli* were first seen in hospital outbreaks of diarrhea in the 1950s but have essentially disappeared in developed countries. They account for about 20% of diarrhea seen in bottle-fed infants less than one year of age, and transmission is through the fecal–oral route, with infants being the reservoir.

Organism	Toxin	Lesion	Pathogenic Gene Location	Transmission	Disease
Escherichia coli					
Common	Hemolysin	Inflammation	None	Adjacent flora	Opportunistic
ETEC	Heat-labile and heat-stable	Hypersecretion	Plasmid	Fecal–oral	Watery diarrhea
EPEC		Effacement (covering) of the small intestine	PAI	Fecal–oral	Watery diarrhea
EIEC		Invasion and inflammation	Large plasmid and PAI	Fecal–oral	Dysentery
EHEC	Shiga toxin	Effacement (covering) of colon and hemorrhage	PAI	Fecal–oral, cattle	Bloody diarrhea
EAEC		Adherent biofilm			Mucoid watery diarrhea
Shigella					
Shigella dysenteriae	Shiga toxin	Invasion, inflammation	Large plasmid, PAI	Fecal–oral	Severe dysentery
Shigella flexneri	Shiga toxin	Invasion, inflammation	Large plasmid, PAI	Fecal–oral	Dysentery
Shigella boydii	Shiga toxin	Invasion, inflammation	Large plasmid, PAI	Fecal–oral	Dysentery
Salmonella					
Salmonella enterica		Ruffles, invasion, inflammation	PAI	Fecal–oral, animals and humans	Gastroenteritis, sepsis
Salmonella enterica serotype Typhi		Macrophage survival, RES growth	PAI	Fecal–oral	Typhoid fever

Abbreviations: ETEC, enterotoxigenic *E. coli*; EPEC, enteropathogenic *E. coli*; EIEC, enteroinvasive *E. coli*; EHEC, enterohemorrhagic *E. coli*; EAEC, enteroaggregative *E. coli*; PAI, pathogenicity islands.

Table 22.2 Characteristics of *Escherichia coli*, *Shigella*, and *Salmonella*

Pathogenesis of enteropathogenic *E. coli* infections — These pathogens initially attach to cells in the intestine, using fimbriae to form clustered "colonies" on the surface of the cells. The lesions resulting from attachment cause effacement of the microvilli of the intestinal cells and change the overall morphology of the cells. The secretion system of these *E. coli* then delivers at least five different proteins into the target cell cytoplasm, and these proteins then inhibit cell signaling and induce modifications to cytoskeletal proteins. The cause of the diarrhea seen in enteropathogenic *E. coli* infections is not understood but may have something to do with morphological changes to the microvilli.

Enterohemorrhagic E. coli

These *E. coli* organisms are so named because they involve the production of Shiga toxin, which causes capillary thrombosis and blood in the stool. They are associated with diseases in which a host consumes products from animals colonized with these pathogens. However, person-to-person transmission can occur. Interestingly, these infections are seen more in developed industrialized countries than in underdeveloped nations. The most talked about of enterohemorrhagic *E. coli* infections occurred in the 1980s in several fast-food restaurants and was linked to the serotype *E. coli* O157:H7, which causes bloody diarrhea and is associated with ground meat and unpasteurized juices.

Pathogenesis of enterohemorrhagic *E. coli* infections — These pathogens have a very low ID_{50} and a common reservoir (cattle), both factors that increase the possibility of human infection. In the food-processing industry in the United States and Europe, changes that provide "fresher" meat have caused outbreaks of disease, with the worst outbreaks seen in countries with the most modern food processing. Because meat cooked only to the rare stage can harbor live pathogens, most states have mandated that all ground meat must be thoroughly cooked. Fruits and vegetables can also be contaminated with enterohemorrhagic *E. coli* and should be thoroughly washed before eating.

Fast Fact

The *E. coli* infection of California spinach in the summer of 2006 caused several deaths and enormous financial losses for farmers. The form responsible was enterohemorrhagic *E. coli* O157:H7.

The distinguishing clinical factors for enterohemorrhagic *E. coli* are the production of Shiga toxin and the intestinal microvilli effacement seen with enteropathogenic strains. Enterohemorrhagic strains attack the colon by adhering through attachment proteins and using the secretion injection system to deliver proteins into the target cells. The proteins then radically alter the cytoskeletal components of the cells. Attachment and effacement causes the diarrhea, while the Shiga toxin produces capillary thrombosis and inflammation of the colonic mucosa, leading to hemorrhagic colitis. The Shiga toxin can also circulate in the blood and bind to renal tissue, causing glomerular swelling and the deposition of fibrin and platelets in the blood vessels.

Treatment of all five forms of *E. coli* infection — Most *E. coli* diarrhea is mild and self-limiting, and therefore treatment is not required. When diarrhea is severe, liquid replacement may be required. In enterohemorrhagic infections, more heroic measures such as dialysis may be necessary. Treatment with trimethoprim/sulfamethoxazole or quinolones can reduce the duration of the diarrhea, but antibiotics will have no effect if hemorrhagic colitis has occurred.

Shigella

Shigella species are members of the family Enterobacteriaceae and are closely related to *E. coli*, but *Shigella* does not ferment lactose. They also lack flagella and cannot be identified by H antigens. There are four species of *Shigella* — *S. dysenteriae*, *S. flexneri*, *S. boydii*, and *S. sonnei* — all of which are able to invade and multiply inside a wide variety of cells, including enterocytes (see **Table 22.2**). All species also produce the Shiga toxin, with *S. dysenteriae* producing the most.

Shigella causes dysentery, which is spread from person to person under unsanitary conditions and is a strictly human pathogen. There are 8–12 cases of shigellosis per 100,000 people per year in the United States, but shigellosis is one of the most common causes of diarrhea worldwide and is responsible for more than 600,000 deaths each year. Transmission of the infection can occur by the fecal–oral route, by person-to-person transmission, and by the consumption of contaminated food or water. The ID_{50} of *Shigella* is fewer than 200 organisms, making it easily transmissible. In fact, 40% of patients get this infection from a family member. There is also a direct connection between the incidence of *Shigella* infections and community sanitary practices.

The most common *Shigella* species are *S. flexneri* and *S. sonnei*. *Shigella dysenteriae* is found mainly in underdeveloped countries, where it causes the most severe form of infection, **bacillary dysentery**.

Pathogenesis of shigellosis — *Shigella* cells are acid-resistant and survive passage through the stomach and small intestine into the large intestine. Once there, the bacteria invade the cells of the colonic mucosa, triggering an intense acute inflammatory response that causes mucosal ulcerations and abscess formation. This multi-step process is illustrated in **Figure 22.9**.

Shigella cells cross the mucosal membrane by entering the M cells of the intestine. The bacteria selectively adhere to these cells and use transcytosis to move through them and into the underlying phagocytic cells of the host, where the bacteria induce apoptosis and kill both the M cells and the phagocytes. Any *Shigella* cells released from the M cells contact the basolateral side of the neighboring enterocytes and then initiate a sequence of steps to invade these cells. *Shigella* is nonmotile, and during this process each bacterial cell creates an actin tail to be used as a means of transport through the cytoplasm of the infected host cell.

Shigella moves to the membrane of the host cell, and here some rebound back into the cytoplasm while the rest push into the adjacent cell. This pushing causes the formation of a fingerlike projection into the next cell (the fourth step in **Figure 22.9**), and eventually this pinches off, forming a vacuole that contains *Shigella* and is surrounded by a double membrane that protects it from the host immune response. *Shigella* then lyses both membranes and is released into the cytoplasm of the newly invaded host cell, and the process starts all over again. This cell-to-cell extension creates localized ulcers in the mucosa, particularly in the colon, and adds to the infection a hemorrhagic component that allows *Shigella* to enter the lamina propria. An intense inflammatory response results and diarrhea begins. The stool is small and contains white blood cells, red blood cells, and bacteria.

Shigella cells produce the Shiga toxin, which also contributes to the overall severity of the illness, resulting in an acute inflammatory colitis and bloody diarrhea that presents as dysentery with cramps and bloody mucoid discharges. The symptoms of fever, malaise, and anorexia are the initial indications, and these are followed by the dysentery. The majority of shigellosis cases spontaneously resolve after two to five days, but the mortality in shigellosis epidemics in Asia, Latin America, and Africa has been as high as 20%.

Treatment of shigellosis — Several antibiotics have been effective at shortening the period of illness by limiting excretion of the bacteria, with trimethoprim and sulfamethoxazole being the antibiotics of choice. Standard sanitation disposal and water chlorination are important in preventing the spread of *Shigella*.

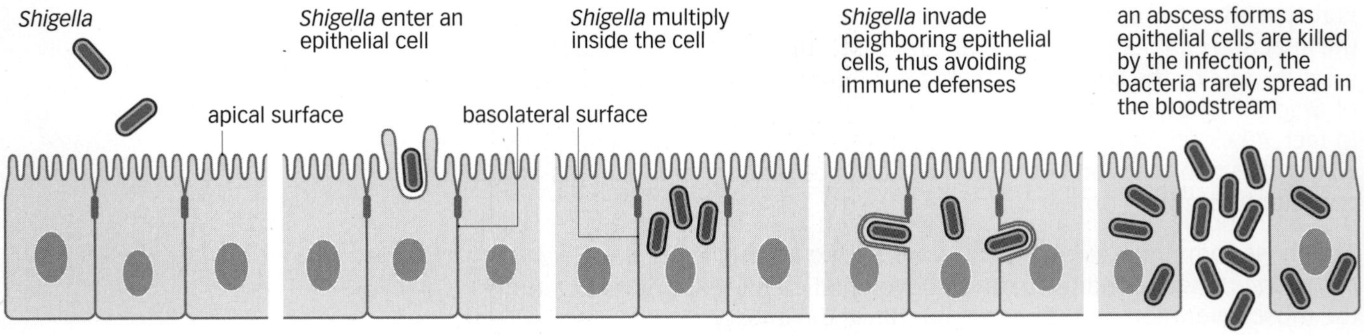

Figure 22.9 Sequence of events in *Shigella* infection. Notice that the initial invasion of the target cell is by way of the apical surface but subsequent movement of the pathogen is by way of the basolateral surfaces. Also note the projection of the fingerlike structure into the neighboring cell.

Salmonella

There are many types of the enterobacterium *Salmonella* and they used to be named in a variety of ways. However, they are all now classified as one species, *enterica*, a name that indicates the target of these organisms: the intestine portion of digestive systems. *Salmonella enterica* can be subdivided into serotypes based on the different lipopolysaccharide O antigens found in the cell wall and on the variety of capsular K antigens. The specific O antigens are organized into serogroups, and the K and H antigens are added to the "formula." You can also distinguish *Salmonella* serotypes by host range, and some, such as *Salmonella enterica* serotype Typhi (sometimes called *S. typhi*), are strictly adapted to humans (see **Table 22.2**).

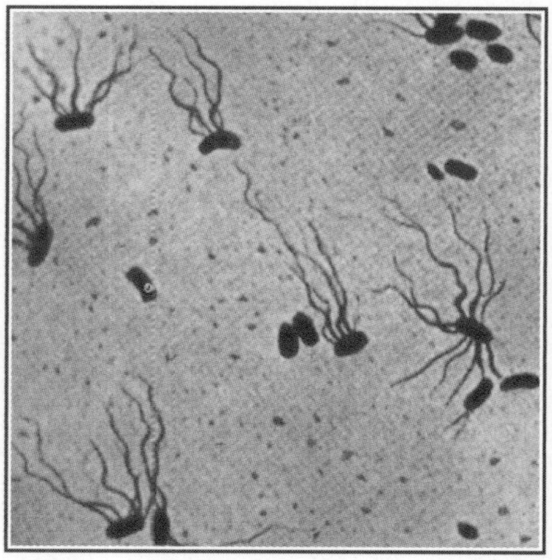

Figure 22.10 A photomicrograph of flagella stained *Salmonella* bacteria. The flagella make *Salmonella* very motile.

Salmonella cells possess multiple pili that bind to mannose receptors on various eukaryotic cells, and most of these bacteria are very motile (**Figure 22.10**). Using common clinical patterns of pathogenesis, we can divide *Salmonella* infections into five groups: gastroenteritis, bacteremia, enteric fever, chronic infections, and typhoid fever.

Gastroenteritis (discussed in detail below) typically begins 24–48 hours after ingestion of the pathogen, with nausea and vomiting being the initial symptoms. This stage of the infection is followed by cramps and diarrhea that persist for three to four days and then spontaneously resolve. Fever occurs in about 50% of patients, and diarrhea varies from loose stool to severe.

Bacteremia is a form of gastroenteritis caused by *Salmonella enterica*. Approximately 70% of AIDS patients get *Salmonella* bacteremia, an infection that can lead to septic shock and death. Bacteremia also leads to the spread of the pathogens to the meninges, the bones, and sites with preexisting abnormalities, such as atherosclerosis plaque, or sites of malignancy.

Enteric fever is a multiorgan *Salmonella* infection with sustained bacteremia and profound infection of organs in the mononuclear phagocytic system (in particular the lymph nodes, liver, and spleen). Incubation takes about 13 days, with the first symptoms being fever and headache. The fever can increase over the first 72 hours and if untreated can last for weeks. In addition, some patients will exhibit constipation whereas others will have diarrhea.

Chronic infection by *Salmonella* is a very serious problem if the bacteria enter the blood of the host continuously, because the continuous release of endotoxin can lead to myocarditis, encephalopathy, intravascular coagulation, or infection of distal sites in the body. This is particularly true of the biliary tract, which can continue to harbor organisms that reinfect the intestines and cause diarrhea late in the disease.

Let's now look in more detail first at gastroenteritis caused by *Salmonella* and then at typhoid fever.

Salmonella *Gastroenteritis*

As the name indicates, *Salmonella* gastroenteritis occurs both in the stomach (gastro-) and in the intestines (entero-). This infection is predominantly a disease of industrialized societies and results from improper food handling. Transmission is from animal or human reservoirs to humans, but the ID_{50} is higher than that seen with *Shigella*, making *Salmonella* a less infectious problem: 1000 or more organisms are required to cause infection.

Salmonella is the leading cause of food-borne gastrointestinal infections, with poultry and infected eggs being the transmission vehicle implicated most often. In addition, poor food handling and preparation can also be implicated in this infection, and the infection can also be transmitted by exotic pets such as turtles. Incidence in the United States is double that of *Shigella* infections: 40,000–50,000 *Salmonella* cases are reported each year. It is important to note that this may represent only 1–4% of the total cases that occur, because many people do not report it. Nearly 30% of *Salmonella* infections occur in nursing homes, hospitals and mental health facilities. Approximately 5% of patients recovering from salmonellosis will shed the organism in their feces for up to 20 weeks, and chronic carriers who are food handlers can be an important reservoir for these bacteria.

Fast Fact

The geographical distribution of salmonellosis has changed with the advent of interstate and international distribution systems that can deliver large amounts of contaminated food to many different places in a relatively short time.

Pathogenesis of *Salmonella* gastroenteritis — Ingested *S. enterica* cells pass through the stomach acid and swim through the intestinal mucous layer. Eventually *Salmonella* organisms reach the enterocytes and M cells of the large and small intestine, and here they use their pili to mediate adherence to M cells, causing the surface of the M cells to "ruffle" (**Figure 22.11**). These ruffles are specialized plasma membrane sites where there has been rearrangement of the filamentous actin cytoskeleton. This rearrangement of the host-cell cytoskeleton is stimulated by 12 or more proteins that are produced by the bacteria and are coded for by genes located on pathogenicity islands in the *Salmonella* chromosome.

The ruffled surface of the host M cells engulf the adherent bacteria into an endocytic vacuole that transcytoses from the apical surface of the cell to the basolateral surface. From here the bacteria enter the lamina propria, initiating a powerful inflammatory response. *Salmonella* can withstand the phagocytic defense of the host cell by inducing apoptosis after being taken up by a host phagocytic cell. It seems that the combination of transcytosis of the pathogen and the vascular permeability associated with the inflammatory response causes the onset of diarrhea. With *Salmonella*, however, this process remains localized in the mucosa and submucosa of the host intestinal cells.

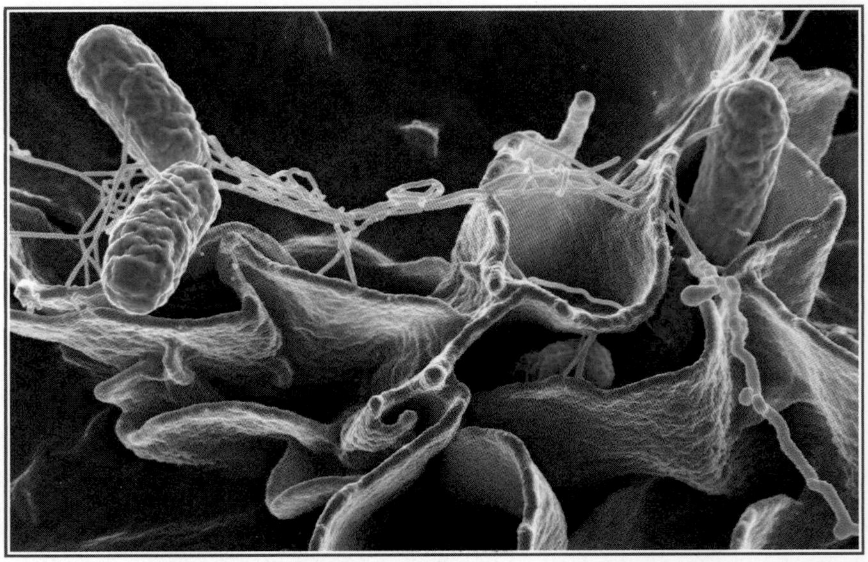

Figure 22.11 Colorized scanning electron micrograph of *Salmonella* (green) invading the epithelial cells of an intestine (purple). Notice the "ruffling" of the intestinal cell membrane.

Typhoid Fever

Typhoid fever, the result of infection by *Salmonella enterica* serotype Typhi, is a strictly human disease, with chronic carriers being the primary reservoirs. Some patients carry the disease for years, one example being the infamous "Typhoid Mary" (Chapter 1). Long-term carrying occurs because the bacteria become sequestered in the gall bladder and biliary tract when stones are present. The bacteria are transmitted to the water supply when sewage contaminates drinking water and are passed from person to person by the fecal–oral route. The ID_{50} of *S. enterica* serotype Typhi is not as low as that of *Shigella* (fewer than 200 organisms) but can become lower if the organism is encapsulated. Although the incidence of typhoid fever is low in the United States, there is still significant morbidity and mortality from this organism in Latin America, Asia, and India.

Pathogenesis of typhoid fever — Because there is no animal model in which we can study this infection, it is difficult to show positively the events that occur. It is presumed that there is a killing of intestinal M cells and macrophages similar to that seen in *Salmonella* gastroenteritis infections. However, unlike other *Salmonella* serotypes, *S. enterica* serotype Typhi can survive for long periods inside viable host macrophages by inhibiting the release of the oxidative poisons used by macrophages to kill invading bacteria. This allows *S. enterica* serotype Typhi to multiply and infect new macrophages, which eventually leads to spilling of these bacteria into the lymphatic circulation, allowing them to migrate to the lymph nodes, spleen, liver, and bone marrow of the host.

This systemic infection is exacerbated by the release of lipopolysaccharide endotoxin, which causes a fever that increases and persists. The bacteria can also spread to the host urinary system and other organs, eventually coming full circle and reinfecting the intestine. The most important complication of typhoid fever is hemorrhaging that causes perforation of the wall of the colon or the ileum at the site of Peyer's patches that have become necrotic. The entire cycle from intestine back to intestine takes only two weeks.

General treatment of *Salmonella* infections — The primary therapy for all forms of *Salmonella* infections is the replacement of fluid and electrolytes and the control of nausea and vomiting. Antibiotic therapy is not appropriate because it increases the duration and frequency of the carrier state. However, therapy with chloramphenicol, ampicillin, trimethoprim sulfamethoxazole, or some of the cephalosporins can be used prophylactically to prevent the spread of the disease. A vaccine for typhoid fever has been available for many years in both the oral and injectable form. It is also essential to provide clean water and treatment for those carrying the disease.

This ends our discussion of bacteria belonging to the family Enterobacteriaceae. Next let's look at some bacteria belonging to the family Spirillaceae.

Vibrio

Members of the genus *Vibrio* are Gram-negative, non spore-forming rod-shaped bacteria commonly found in salt water. They have a unique morphology in that they form S shapes and half-spirals (comma shapes), and they are highly motile by means of a single polar flagella (**Figure 22.12**). *Vibrio* can grow either aerobically or anaerobically and have a cell structure similar to that of other Gram-negative bacteria (**Table 22.3**). They have a low tolerance for acidic conditions but grow well in mildly alkaline environments.

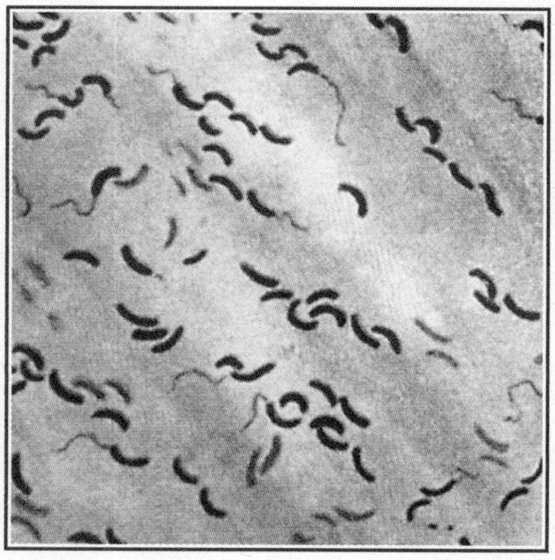

Figure 22.12 A photomicrograph of *Vibrio cholerae*.

Organism	Growth	Urease	Epidemiology	Pathogenesis	Disease
Vibrio cholerae	Facultative	–	Fecal–oral, water	Cholera toxin	Watery diarrhea
Campylobacter jejuni	Microaerophilic	–	Animals, unpasteurized dairy products	Unknown	Dysentery, watery diarrhea
Helicobacter pylori	Microaerophilic	+	Transmission not understood	Vacuolating cytotoxin, urease	Chronic gastritis, ulcers
All of these organisms are Gram-negative rods with a similar morphology.					

Table 22.3 Features of *Vibrio*, *Campylobacter*, and *Helicobacter*

The species responsible for the disease cholera, *Vibrio cholerae*, produces a toxin (discussed in Chapter 5) that causes a devastating intestinal infection. There have been eight cholera pandemics, historically lasting from 5 to 25 years. Today cholera remains endemic on the Indian subcontinent and in Africa but has also spread to other parts of Asia and to Europe. The cholera bacterium first appeared in the United States in the early 1970s, entering Texas and Louisiana at the coastline of each state, and is now found in states along the southern Atlantic coast as well.

Pathogenesis of *V. cholerae* — Epidemic cholera is spread primarily by contaminated water and poor sanitation. Its short incubation period, about two days, ensures a rapid epidemic cycle. To produce disease, *V. cholerae* must reach the small intestine in sufficient numbers to multiply and colonize. The bacteria possess long filamentous pili that form bundles on the bacterial surface, and these are used for colonization. Interestingly, *V. cholerae* will colonize the entire intestinal tract, from the jejunum to the colon, and liquid loss resulting from colonization depends on a balance between bacterial growth, toxin production, and host liquid secretion and absorption. The loss of liquids and electrolytes, which can amount to multiple liters a day, is greatest in the small intestine, and the results of liquid loss are dehydration, hypokalemia (loss of potassium), and metabolic acidosis resulting from loss of bicarbonate.

Vibrio cholerae does not invade or damage the enterocytes of the digestive system but instead uses its toxin plus a variety of virulence factors to cause disease. These factors, which are part of a remarkable controlled, coordinated system involving environmental sensors, are all coded for by genes located on pathogenicity islands in *V. cholerae*.

Fast Fact

The extreme dehydration seen with cholera can lead to hypotension and death within a matter of hours.

Cholera has a rapid onset, beginning with abdominal fullness and discomfort, rushes of peristalsis and loose stools. The stools quickly become watery, voluminous, and almost odorless and can progress to what is referred to as "rice" stool because it contains bits of mucus. There is no fever with cholera and no blood in the stool.

Treatment of cholera — The outcome of cholera depends on balancing liquid and electrolyte loss. This balance can be accomplished by oral or intravenous liquid replacement, and this is all that is required except in the most severe cases. Tetracycline can shorten the duration of diarrhea and the magnitude of liquid loss. It is important to note that epidemic cholera does not exist in areas where waste disposal is adequately dealt with.

Campylobacter Enteritis

The digestive infection *Campylobacter* enteritis is caused by *Campylobacter jejuni* (**Figure 22.13**), which was not recognized as a human pathogen until 1973 but is now one of the most common causes of diarrhea. It is found in 4–30% of diarrheal stools, making it the leading cause of gastrointestinal infection in developed countries, and has an ID_{50} of only a few hundred, making it easily transmissible. In fact, there are more than 2 million cases of *C. jejuni* infection in the United States each year.

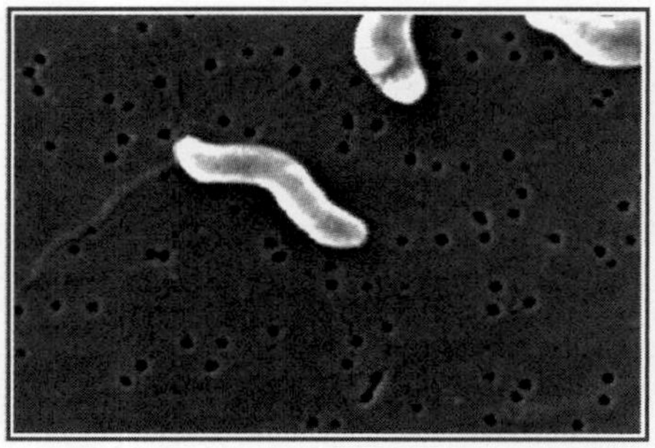

Figure 22.13 A scanning electron micrograph of *Campylobacter jejuni*, which has a corkscrew appearance.

The primary reservoir of *C. jejuni* is animals, and it is transmitted to humans either by ingestion of contaminated food or by direct contact with pets that harbor the organism. The most common source of human infection is undercooked poultry, but transmission can also occur via contaminated water or unpasteurized milk. It should be noted that *Campylobacter* is commonly found as part of the gastrointestinal and genitourinary tract flora of warm-blooded animals, and for this reason domestic pets may have a significant role in transmission to humans.

Fast Fact

It is possible to re-contaminate poultry after cooking by placing the cooked bird on the cutting board used to prepare it for cooking without first thoroughly washing the board and/or by cutting the cooked bird with the same knife used to prepare it for cooking without first thoroughly washing the knife.

Pathogenesis of *Campylobacter* enteritis — Oral ingestion of the pathogen is followed by colonization of the intestinal mucosa, where bacteria adhere to and then enter cells in endocytic vacuoles. Once inside the cells, the bacteria move in association with microtubule structures and produce a lethal **distending cytotoxin** that arrests cell division while the cytoplasm continues to increase. How this leads to the common symptom of diarrhea is not known. Illness with *C. jejuni* begins about seven days after ingestion of the bacteria, with fever and lower abdominal pain that may be severe enough to mimic appendicitis. This is followed within hours by dysenteric stools containing blood, mucus, and pus.

Interestingly, there is an association between *C. jejuni* infection and Guillain–Barré syndrome. Up to 40% of patients with this acute demyelinating disease have evidence of infection with *Campylobacter* at the time that neurologic symptoms arise. This is probably due to an autoimmune reaction in which antibody against the bacteria cross-reacts with the myelin of the neurons.

Treatment of *Campylobacter* enteritis — This infection is self-limiting, and signs and symptoms last for only three to five days (though they can last for up to two weeks in severe cases). Fewer than 50% of patients benefit from antibiotic treatment; however, if the infection is severe, ciprofloxicin is the treatment of choice.

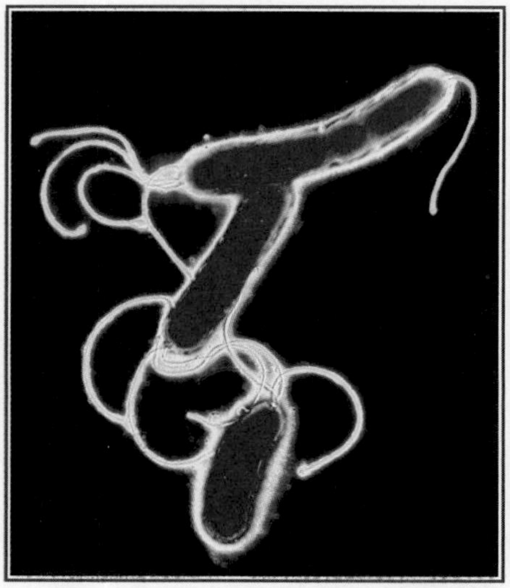

Figure 22.14 A colorized transmission electron micrograph of *Helicobacter pylori*, which is a causative agent of ulcers. Notice the multiple flagella, which help to propel this organism through the acidic stomach.

Helicobacter pylori

This organism is similar to *Campylobacter* in morphology and growth characteristics. It is a slender, microaerophilic, Gram-negative curved rod with polar flagella (**Figure 22.14**) and a cell wall structure similar to that seen in most other Gram-negative organisms. However, the lipopolysaccharides in the outer layer of *Helicobacter pylori* cells may be more toxic than those found in other Gram-negative pathogens.

Fast Fact

Helicobacter pylori was one of the first bacteria to be classified as a carcinogen by the World Heath Organization.

One of the unique features of *H. pylori* is the production of a urease enzyme that allows it to survive in very acidic environments by generating ammonia. Another circulating protein associated with this pathogen, **vacuolating cytotoxin**, causes apoptosis in infected eukaryotic cells. These infected cells are easily identified microscopically because they have large vacuoles throughout their cytoplasm. Like many other intestinal pathogens, *H. pylori* cells possess a contact injection system that introduces proteins that disrupt many of the proteins in the infected cell. As in other bacteria, these virulence factors are coded for by genes located in pathogenicity islands.

Helicobacter pylori infection causes what may be the most prevalent disease in the world, ulcers (**Figure 22.15**). The organism is found in 30–50% of all adults in developed countries and in practically 100% of adults in developing countries. The mode of transmission for the pathogen is not known but is presumed to be person-to-person by the fecal–oral route. It has been shown that colonization increases with a patient's age and persists for decades. *Helicobacter* is the most common cause of gastritis and of gastric and duodenal ulcers and is also the predisposing factor for gastric adenocarcinoma, which is one of the most common causes of cancer deaths in the world.

Pathogenesis of *H. pylori* gastritis — *Helicobacter pylori* uses multiple mechanisms to adhere to a host's gastric mucosa and survive the acidic environment. The bacterium is highly motile and swims to less acidic areas located below the mucous layer surrounding the lining of the stomach. Once in these less acidic areas, the bacterium adheres by using surface proteins, some of which bind to blood group antigens. Colonization is almost always accompanied by cellular infiltration as part of the host's inflammatory response to the infection. The inflammatory response can be extensive and can cause the formation of microabscesses that contribute to the ulceration and are also accompanied by the release of virulence factors that enhance cellular erosion.

The primary infection with *H. pylori* is either without symptoms or may cause some nausea and mild upper abdominal pain that lasts for about two weeks. Years later, however, there can be gastritis, or peptic ulcer disease with nausea, anorexia, vomiting, and pain. Many patients are asymptomatic for decades even up to perforation of the ulcer, which leads to extensive internal bleeding.

Treatment of *H. pylori* gastritis — *Helicobacter pylori* is sensitive to a wide variety of antimicrobial agents. First line therapy is amoxicillin, clarithromycin, and omeprozole. Treatment with bismuth salts and a combination of metronidazole and tetracycline for patients that fail initial treatment.

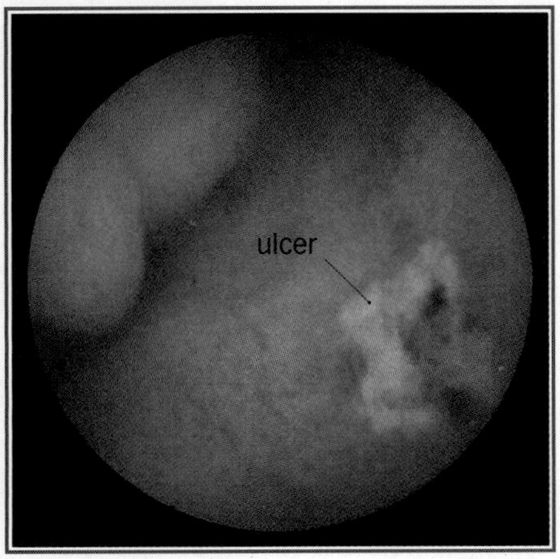

Figure 22.15 An ulcer located in the duodenum of a patient. This lesion was caused by *Helicobacter* infection.

Keep in Mind

- The family Enterobacteriaceae comprises Gram-negative rods and contains a variety of organisms involved in infections of the digestive system.
- Many bacteria of the family Enterobacteriaceae are resident flora of the human digestive system.
- O antigens are part of the lipopolysaccharide in the outer membrane, whereas H antigens are found on the flagella, and K antigens are part of the capsule.
- The enterobacteria *Escherichia coli*, *Salmonella*, *Shigella*, and *Yersinia* cause infections of the digestive system through the production of toxin as well as the invasion of tissue.

- Several toxins are produced by *E. coli*, …h one of the most dangerous being the Shiga toxin.
- Virulence properties can be used t… divide *E. coli* into five groups: enterotoxigenic, enteropathogen…roinvasive, enterohemorrhagic, and enteroaggretive.
- Four species of *Shigella* are … digestive infections, and all produce the Shiga toxin: *S. dysenter*… …*neri*, *S. boydii*, and *S. sonnei*.
- Two prominent infection… …stive system that are caused by *Salmonella* are gastroe… …yphoid fever.
- *Vibrio cholerae*, a me… …amily Spirillaceae, produces a powerful toxin that causes d… …ctions of the digestive system.
- *Campylobacter* e… the most common causes of diarrhea.
- *Helicobacter p*… …ommon cause of gastric and duodenal ulcers.

VIRAL INF… DIGESTIVE SYSTEM

As you … estive system is an important portal of entry for vir… poliovirus is one example — use the digestive syste… oint. These viruses infect intestinal epithelial cel… ed by the virus occurs somewhere else in the b… discuss some of the most common viruses that … gestive tract.

…st common sign associated with these viral infec… …lly the diarrhea has a rapid onset (within hours … he body) but lasts for less than three weeks. There … of virions in the stool, with amounts in excess of …l. These viral particles are easily identified in stool … microscopy.

…e required to implicate viruses in digestive infec… …re similar to the Koch postulates used for bacteria

- … must be detected in ill patients, and viral shedding must … with the onset of symptoms.
- …ere must be a significant antibody response in patients who are …dding virus.
- The disease must be reproduced through experimental inoculation of nonimmune humans or animals with the virus in question. This is a very difficult requirement to satisfy, because many viruses cannot be grown in culture.
- Other causes of the signs and symptoms of the infection (mainly diarrhea) must be excluded.

On the basis of these criteria, several groups of viruses that cause gastrointestinal infections have been established including rotavirus, calicivirus, astrovirus, some serotypes of adenovirus, and enterovirus. The infections caused by all these viruses have similar characteristics, including brief incubation periods, fecal–oral routes of transmission and vomiting that either precedes or accompanies diarrhea. We have listed the characteristics of some of the viruses that cause digestive disease (**Table 22.4**) but because of space limitations we will only discuss rotavirus and enterovirus in detail.

Feature	Rotavirus	Calicivirus	Astro	Adenovirus
Biological features				
Nucleic acid	ds RNA	+ss RNA	+ss RNA	ds DNA
Shape	Naked, double-shelled capsid	Naked, round	Naked, sta	Naked, icosahedral
Number of serotypes	Four that infect humans	More than four	At least five	nknown
Pathogenic features				
Site of infection	Duodenum, jejunum	Jejunum	Small intestine	stine
Epidemiologic features				
Seasonality	Usually winter	None known	None known	
Ages primarily affected	Infants, children <2 years old	Older children and adults	Infants, children	
Transmission	Fecal–oral	Fecal–oral; contaminated water and shellfish	Fecal–oral	Fe
Incubation	1–3 days	12 hours to 2 days	1–2 days	8–1

Table 22.4 Characteristics of Viruses that Cause Ga Infections

Rotavirus

The rotaviruses belong to the family *Reoviridae* and are non-en spherical, double-stranded RNA viruses. Under an electron micr a rotavirus looks like a wheel (*rota* is Latin for "wheel") and has a d capsid structure in which the outer capsid is attached to the inner o short, spokelike structures. Rotaviruses were not discovered until 1 when electron microscopy was used to examine biopsy specimens fr infants with diarrhea. Since then, members of the rotavirus genus ha been found around the world and are believed to account for 40–60% of cases of acute gastroenteritis. Rotavirus causes acute gastrointestinal disease in a variety of species in addition to *Homo sapiens*, and it has the ability to undergo genetic re-assortment, a property that makes this virus difficult to deal with immunologically (recall our discussion of influenza virus in Chapters 12 and 13).

Outbreaks of rotavirus infections are common in infants and children under two years of age, but adults are usually only minimally affected. These infections can affect elderly institutionalized people. Usually, rotavirus infection in infants results in little or no clinical illness, and by the age of four years more than 90% of individuals have developed antibody against the virus. It is estimated that rotavirus infections kill more than 1 million infants worldwide each year, but in the United States there are fewer than 100 deaths attributed to rotavirus each year. However, rotavirus infection is still a major cause of hospitalization early in life.

Pathogenesis of rotavirus infection — The virus localizes primarily in the duodenum and proximal jejunum, where it "blunts" (shortens) the microvilli of the epithelial cells. This change in the microvilli causes a mild infiltration of mononuclear and polymorphonuclear leukocytes into them. The primary effect of this is a decreased absorptive surface on the microvilli coupled with decreased enzymatic function. This results in malabsorption and defective handling of fats and carbohydrates. Interestingly, the gastric and colonic mucosa are not affected. It can take eight weeks to restore normal histology and function to the damaged area after the infection is over.

Incubation is between one and three days and begins with the abrupt onset of vomiting, followed within hours by frequent copious watery brown stools. There can often be low-grade fever as well, and the vomiting and diarrhea can last for several days. As in many other intestinal infections, the major complication of these effects is dehydration.

Treatment of rotavirus infection — There is no specific treatment, but severe cases require vigorous replacement of fluid and electrolytes. Rotavirus is highly infectious and can spread rapidly in institutional settings. Therefore, control measures involving hygienic practices are important to inhibit the spread of the infection. A live attenuated vaccine has been developed.

Enterovirus

Enteroviruses are members of the *Picornaviridae*. They are very resistant to acidic environments, which helps them survive passage through the stomach on their way to the intestines. They also resist common disinfectants and various detergents, especially if embedded in organic material. Humans are the natural hosts for these enteroviruses, and asymptomatic infection is common. However, there are some enteroviruses that do not infect humans. Although whether infection is symptomatic or not depends on the species of enterovirus in question and on the age of the host, all the enteroviruses that infect the digestive system show a seasonal infection pattern and are predisposed to temperate climates.

Direct or indirect fecal–oral transmission from person to person is the most common way these infections are spread, and the virus will normally spend one to four weeks in the oropharynx after infection. The virus can be shed for 18 weeks in the feces.

Pathogenesis of enterovirus infection — About 60% of infections with enteroviruses occur in children aged nine years or younger, and the incubation time is two to ten days. Attachment usually occurs between attachment proteins on the surface of the virion that are specific for receptors on the host cell. The virus is brought into the host cell by envelopment in the host membrane, and viral RNA is released into the cytoplasm. Here it binds directly to ribosomes and begins synthesis of viral proteins. Enteroviruses are lytic, and the end result of the infectious cycle is destruction of the host cell and release of new virions that infect other cells. The primary infection occurs in the digestive system but then spreads to other sites in the body.

Treatment of enterovirus infection — None of the currently available antiviral agents have been shown to be effective against the *Picornaviridae* enteroviruses.

Hepatitis Viruses

The name **hepatitis** describes any disease that affects the hepatocytes of the liver, and these diseases can be caused by a variety of agents, including bacteria, protozoans, viruses, toxins, and drugs. There are at least six different viruses that cause hepatitis, and we will look at each of these in detail. It is important to note that the hepatitis viruses are distinctly different from one another. Three of the most frequently encountered types are summarized in **Table 22.5**.

Feature	Hepatitis A	Hepatitis B	Hepatitis C
Virus type	+ss RNA	ds DNA	+ss RNA
Percentage of viral hepatitis	50	41	5
Incubation period (days)	15–45	7–160	15–160
Onset	Usually sudden	Usually slow	Insidious
Age preference	Children, young adults	All ages	All ages
Transmission			
Fecal–oral	+++	+/–	–
Sexual	+	++	+
Transfusion	–	++	+++
Severity	Usually mild	Moderate	Mild
Chronicity (%)	None	10	>50
Carrier state	None	Yes	Yes
Protection by immune serum globulin	Yes	Yes	Yes

Table 22.5 Comparison of Hepatitis A, B, and C Viruses

Hepatitis A

This is a non-enveloped, single-stranded RNA virus with cubic symmetry. It resists inactivation and is stable at –20°C and low pH. Hepatitis A is classified as a member of the genus *Hepatovirus*, a member of the family *Picornaviridae*. There is only one serotype of this virus, and humans are the most common natural host for it. The major form of transmission is the fecal–oral route, and hepatitis A infections are commonly seen in situations where there is crowding and poor hygiene. The rates of hepatitis A infection are higher in lower socioeconomic groups. The rates of infection in the United States have been declining since 1970. In developing nations, however, as many as 90% of the population show evidence of previous infection with hepatitis A virus.

Pathogenesis of hepatitis A infection — This infection often results from poor personal hygiene in food handlers. Patients are most contagious one to two weeks before the onset of clinical symptoms of infection. The virus is believed to replicate initially in the intestinal mucosa and can be seen in feces by electron microscopy 10–14 days before the onset of symptoms. It is interesting to note that when symptoms begin, there is no longer shedding of the virus in feces.

Multiplication in the intestines is followed by spread to the liver. This causes lymphoid infiltration into the liver, which leads to necrosis of the parenchymal cells and to the proliferation of Kupffer cells. The extent of the necrosis coincides with the severity of the infection. Patients with anti-hepatitis A antibodies cannot be reinfected with this virus, indicating that the immune response to this virus is a protective one.

Incubation times for hepatitis A vary from 10 to 50 days, followed by the onset of fever, anorexia, nausea, pain in the upper right abdominal quadrant, and jaundice. The urine of infected patients with jaundice can become dark, and their stool can become clay-colored one to five days before onset of the jaundice. Many people infected with hepatitis A will be asymptomatic or only mildly affected and do not develop jaundice. This form of the infection, referred to as *anicteric* hepatitis A, is a function of the patient's age. It is seen more in children and less in adults.

Treatment of hepatitis A infection — There is no treatment for hepatitis A infection, and supportive measures such as rest and adequate nutrition are the only recommendation. A passive prophylactic treatment with human immune serum globulin obtained from pools of donors can give protection during the period of incubation and is 80–90% effective. There is also an active immunization protocol that uses either live attenuated virions or killed virions for those who are repeatedly exposed to hepatitis A virus.

Hepatitis B

Unlike some of the other hepatitis viruses, which are RNA viruses, hepatitis B is an enveloped DNA virus. It belongs to the family *Hepadnaviridae* and is unrelated to any other human virus. It has a spherical shape with a surrounding envelope and a viral genome that is unique (**Figure 22.16**). This DNA genome is only partly double-stranded and contains a short stretch that is single-stranded. It also carries with it its own viral DNA polymerase. The envelope of hepatitis B virus contains viral surface antigens called HBsAg, and aggregates of these antigens are often found in abundance during the infection. Hepatitis B DNA can be found in the nucleus of infected hepatocytes, and viral HBsAg can be found in the cytoplasm.

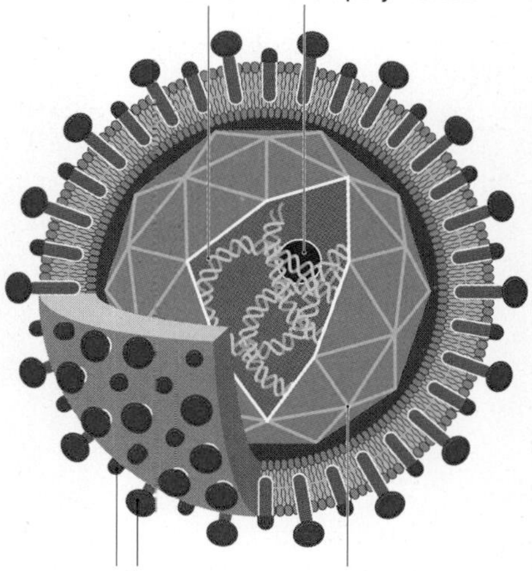

Figure 22.16 A diagrammatic representation of the hepatitis B virus. This DNA virus carries its own viral DNA polymerase.

Hepatitis B virus has a unique replication cycle as a result of its incompletely double-stranded DNA. During viral replication, full-length positive viral RNA transcripts are inserted into the core of the virus, and these RNAs are used to form a template for reverse transcription in which a negative DNA strand is made. Then a positive DNA strand is made from this negative DNA strand, but it never gets finished before the virus is released. Therefore the new virions have a double-stranded DNA plus a stretch of single-stranded DNA.

Hepatitis B is found worldwide, though its prevalence varies from one country to another. Chronic carriers are the main reservoir for the virus. In the United States, it is estimated that 1.5 million people are infected with hepatitis B, and there are 300,000 new cases each year. Five to ten per cent of these people will become chronic carriers of the virus, and 300 others will die of acute viral infection. One striking statistic is that up to 4000 of the 300,000 Americans who contract the infection in a given year will develop hepatitis B cirrhosis, and 1000 will get hepatocarcinoma as a result of the infection. Fifty per cent of infections are sexually transmitted, but screening of blood donors has markedly reduced the incidence of transfusion transmission.

Fast Fact

There is a high risk of hepatitis B infection in health care workers.

Pathogenesis of hepatitis B infection — The major mode of transmission is through close contact with body liquids from infected individuals. Hepatitis B antigens have been found in most body liquids, including saliva, semen, and cervical secretions, and it has been shown that as little as 0.0001 milliliters of infected blood is able to produce an infection. Consequently, inadequately sterilized hypodermic needles or instruments used in tattooing or piercing can easily transmit this viral infection.

Factors that determine the appearance of clinical symptoms are not yet completely understood, but some seem to involve the immune response of the host. For example, there is an arthritis that sometimes precedes the jaundice with the arthritis being mediated by the complement system of the host. Lesions in acute hepatitis B infections resemble those seen with other hepatitis viruses in that chronic active infection causes a continued inflammation resulting in necrosis of hepatocytes in the liver. This destruction of liver cells can lead to collapse of the reticular framework of the liver and progressive fibrosis.

The clinical symptoms are variable, and the incubation time is anywhere between 7 and 160 days. Acute infection is usually manifested by a gradual onset of fatigue, loss of appetite, nausea, and pain in the upper right abdominal quadrant. Early in the infection, painful swollen joints and arthritis may appear, and some patients will develop a rash during this period. Jaundice can also be a complication, and these symptoms can last for months. The ratio of infection to disease varies with age of the host, but fulminant hepatitis is seen in less than 1% of infections. Approximately 10% of infected individuals will develop chronic hepatitis, and this risk is higher in children and immunocompromised persons. Chronic hepatitis B can lead to cirrhosis, liver failure, or hepatocellular carcinoma.

Treatment of hepatitis B infection — There is no effective treatment for acute hepatitis B infection, but it can be prevented by using safe sex and avoiding needle sticks. Hepatitis B serum globulin is available, and administration of this soon after exposure can reduce development of the disease. A vaccine that is recombinant and made in yeast provides excellent protection. Health care workers are required to receive it.

Hepatitis C

This RNA virus is in the *Flaviviridae* family (as are yellow fever and dengue fever) and has a very simple genome consisting of only eight genes. There are six major genotypes and multiple subtypes of hepatitis C virus, and the genotypes are related to the geographic distribution of the virus and the severity of the disease produced. Transmission of hepatitis C via blood transfusions is well known, but it can also be transmitted sexually. Needle sharing accounts for more than 40% of infections, and hemodialysis patients are also at risk. More than 3.5 million people in the United States are infected with hepatitis C.

Pathogenesis of hepatitis C infection — The incubation time for this infection varies between 6 and 12 weeks, and the infection is usually mild or asymptomatic. However, 85% of those infected will become carriers of the infection and progress to chronic hepatitis (this can take 10–18 years). Cirrhosis and hepatocellular carcinoma are late consequences of chronic hepatitis C infection (**Figure 22.17**). This condition is the leading cause of liver transplants in the United States.

Treatment of hepatitis C infection — Combination therapy with interferon-α and ribavirin is the treatment of choice for hepatitis C infection.

In addition to the three main strains of hepatitis — A, B, and C — there are three others worth mentioning here: D, E, and G.

Hepatitis D

Hepatitis D is a small single-stranded RNA virus that is currently referred to as a satellite virus because it requires the presence of hepatitis B antigens to complete its replication cycle and become infective. It is seen only in people infected with hepatitis B. How hepatitis D virions

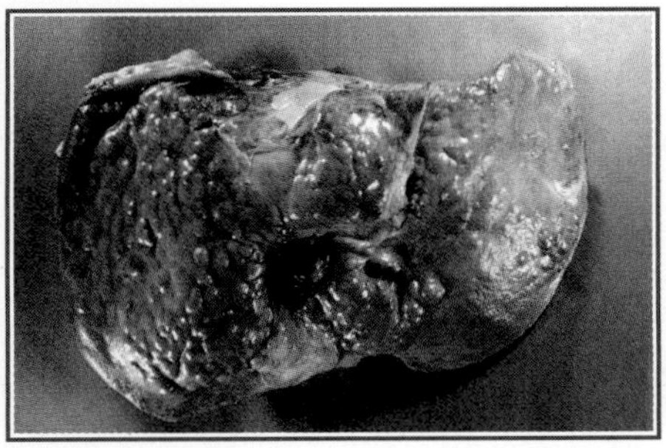

Figure 22.17 A liver damaged by hepatitis C virus infection.

replicate is unknown, but it has been shown that the capsid of hepatitis D is made up of hepatitis B proteins. This infection is seen most often in intravenous drug abusers, a group that also happens to be at high risk for hepatitis B infection.

Pathogenesis of hepatitis D infection — There are two types of infection with this virus. The first involves co-infection with hepatitis B, and the second presents as a superinfection of people already infected with hepatitis B. In simultaneous infection, the clinical picture mirrors that seen with hepatitis B. In superinfection, the hepatitis D infection can cause a relapse, reoccurrence of jaundice, and increased risk of cirrhosis. A rapid progression of liver disease and death is also seen in 20% of these cases.

Treatment of hepatitis D infection — Interferon-α is given to patients doubly infected, but it is not as effective as with hepatitis B alone, and only about 15–25% of patients improve. Prevention methods include safe sex and no sharing of needles.

Hepatitis E

This virus, which is in the family *Caliciviridae*, causes a form of hepatitis that is transmitted by the fecal–oral route. Hepatitis E resembles hepatitis A virus (although the two belong to different families) and is a non-enveloped RNA virus that has pronounced spikes on the capsid. Infection with hepatitis E is frequently subclinical; it causes acute disease only in pregnant women. The highest attack rates for this virus are seen in young adults and are associated with contaminated drinking water. However, the virus does not seem to be transmitted by person-to-person contact. Incubation takes about 40 days, and there is no treatment.

Hepatitis G

This RNA virus was discovered in 1995. It is a member of the family *Flaviviridae* and is similar to hepatitis C virus. Approximately 2% of blood donors are found to be positive for hepatitis G RNA, but the pathogenesis and disease process associated with this virus are not yet understood. There is no treatment for infection with hepatitis G.

Keep in Mind

- The digestive tract is an important portal of entry for viruses.
- The most common symptom of viral infection of the digestive system is diarrhea.
- Viral pathogens of the digestive system include rotavirus, calicivirus, astrovirus, and adenovirus.
- Hepatitis describes any disease that affects hepatocytes of the liver, and there are six different viruses that are classified as hepatitis viruses.

PARASITIC INFECTIONS OF THE DIGESTIVE SYSTEM

There are several protozoan and helminthic infections associated with the human digestive system. In Chapter 14, we looked at amebiasis, pinworms, and tapeworms, and here we discuss other common parasitic diseases that affect the digestive system. The protozoa we look at here are *Giardia* and *Cryptosporidium*, and the helminths are whipworms (*Trichuris trichiura*) and hookworms (*Necator americanus* and *Ancylostoma duodenale*).

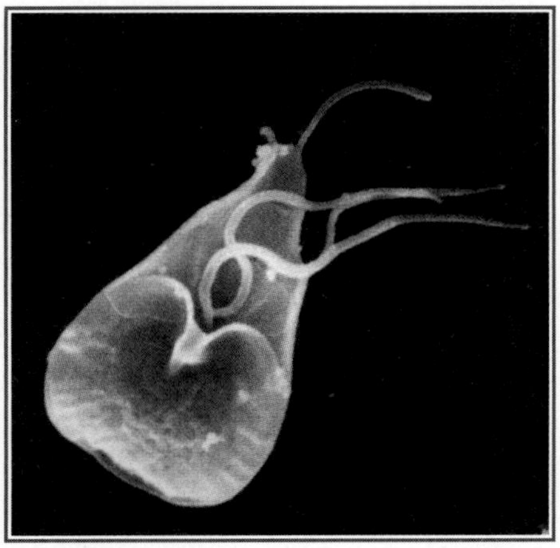

Figure 22.18 A micrograph of one species of *Giardia*, a protozoan parasite that causes diarrhea and is found throughout the world. This image shows the trophozoite stage, with the characteristic four pairs of flagella.

Giardiasis (*Giardia duodenalis*)

Giardia duodenalis is a flagellate protozoan parasite found throughout the world, and the infection of the digestive system that they cause is **giardiasis**. *Giardia* exists in both a trophozoite form and a cyst form, and the organism is quite large. The trophozoite form resembles a stingray and has four pairs of flagella located ventrally, laterally, anteriorly, and posteriorly (**Figure 22.18**). These flagellates reside in the duodenum and jejunum of the human small intestine, where they absorb nutrients from the host digestive tract. They move about and through the mucous layer at the base of the intestinal microvilli in two ways, either by using a tumbling motion or through a large ventral sucker that attaches to the epithelium. Unattached *Giardia* is carried in the fecal stream to the large intestine.

While *Giardia* is in the descending colon, the flagella are retracted into cytoplasmic sheaths and a clear cyst wall is secreted and encloses the organism. While in the cyst, *Giardia* divides, producing a quadri-nucleate organism with two sucking disks. This mature cyst is the infectious form of the parasite. It can survive in cold water for months and is also resistant to the chlorine used in most municipal water supplies. Transmission of the infection is by the fecal–oral route. Once inside a host, the cyst divides into two trophozoites.

Giardia may be one of the most widely distributed intestinal protozoans and is found in fish, amphibians, reptiles, birds, and mammals. There are three morphologically distinct groups of *Giardia*, and the prevalence of the infection in young children and young adults is highest in areas of poor hygiene and sanitation. Direct fecal spread is also high among homosexuals who engage in oral–anal sex. Travelers and hikers can become infected through contaminated water or food, and there have been more than 20 water-borne outbreaks in the United States associated with the contamination of municipal water supplies by sewage.

Fast Fact

Domestic cats and dogs have a high rate of infection with *Giardia* and can act as reservoirs for human infection.

Pathogenesis of giardiasis — The disease is associated with malabsorption by the intestinal tract, in particular fats and carbohydrates. The precise pathogenic mechanisms are unknown, but there can be blockage of the intestine by large numbers of *Giardia*. Destruction of microvilli by the parasite's sucking disk is also possible, as are damage to bile-production pathways and altered intestinal motility. The parasite may also cause an accelerated turnover of the mucosal epithelium and invasion of the mucosal tissue. Because none of these events correlates with the symptoms normally seen, pathogenesis remains a mystery.

In endemic situations, more than 60% of the infected patients are asymptomatic, but in acute outbreaks most of those infected will show symptoms. When these symptoms occur, they begin one to three weeks after exposure to the parasite and include diarrhea that is sudden in onset and explosive. The stool is foul-smelling and greasy in appearance, but there is no blood or mucus. Abdominal cramping is also common, and large amounts of gas are produced, resulting in abdominal distension and abundant flatulence. Nausea and low-grade fever are also possible.

Acute illness usually resolves itself in one to four weeks but may persist in children and lead to significant weight loss and malnutrition. In many adults, the acute phase may give way to a subacute or chronic stage consisting of heartburn, weight loss, and flatulence that can last for weeks to months.

Treatment of giardiasis — There are four drugs currently available for this infection in the United States: quinacrine hydrochloride, metronidazole, furazolidone, and paromomycin. These drugs require five to seven days of therapy and should not be used during pregnancy.

Cryptosporidiosis (*Cryptosporidium*)

Cryptosporidium is a small protozoan parasite that infects the intestinal tract of both humans and other animals, triggering the infection called **cryptosporidiosis**. It is an obligate intracellular parasite that alternates between sexual and asexual reproduction cycles, both of which are completed in the gastrointestinal tract. This parasite was not identified as a human pathogen until 1976, when it was shown that the species *Cryptosporidium parvum* could infect humans. The organism is small and spherical and arranges itself in rows along the microvilli of the intestinal tract (**Figure 22.19**). It is interesting to note that the parasites remain external to the cell cytoplasm but eventually become covered over by the membrane of the host cell, which is why they are referred to as intracellular.

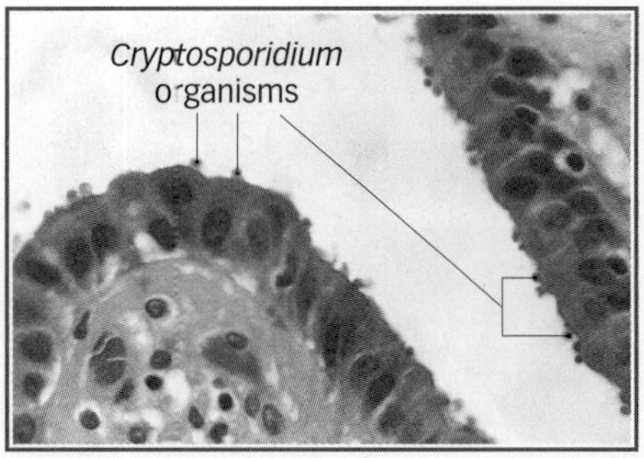

Figure 22.19 A histological section of the gall bladder of an AIDS patient, showing numerous *Cryptosporidium* organisms along the luminal surfaces of the epithelial cells.

Infectious *C. parvum* oocytes are excreted in the stool of infected persons, and the oocytes are fully mature and infectious on passage into the feces. On ingestion of these oocytes by a new host, sporozoites are released from the oocyte and attach to the microvilli of the epithelial cells in the small intestine. Here they transform into trophozoites, which divide by schizogony (fission) to form schizonts containing eight daughter cells. When released from the schizont, each daughter cell attaches to another epithelial cell, where the schizogony cycle is repeated.

After several rounds of schizogony, the trophozoites develop into male and female forms, and the sexual reproductive cycle takes place. The resulting zygote develops into an oocyte that is shed into the lumen of the small intestine. The zygote has a thick protective wall that ensures safe passage both in the fecal stream and in the external environment. Approximately 20% of the zygotes formed will not develop protective walls, and the resulting oocytes will rupture, releasing infectious sporozoites back into the lumen.

Various species of *Cryptosporidium* can infect most vertebrates, with infection usually occurring in the young or immature. Domestic animals are the reservoir for this parasite, and most human infections of *C. parvum* result from person-to-person transmission. In developed countries, one to four per cent of children harbor oocytes, but this number doubles in underdeveloped nations, with the highest rate of infection being seen in children, families of infected children, medical workers, and travelers to countries where the disease is endemic. The principal route of transmission is fecal–oral, but there can also be transmission by contamination of food and water.

Pathogenesis of cryptosporidiosis — The jejunum is the most heavily involved area for these parasites, but in some severe cases they can be seen throughout the entire digestive system. Bowel changes are minimal for this infection, with mild to moderate destruction of microvilli and some mononuclear cell infiltration of the lamina propria. The pathology of the resulting diarrhea is not understood but may be similar to that seen in cholera. Antibodies against *C. parvum* are protective against infection, and it has been shown that CD4 T cells and interferon can also have a role in clearance of the parasites.

In immunocompetent patients, there is onset of profuse, explosive watery diarrhea one to two weeks after exposure to the parasite. The illness lasts for about five days and then rapidly clears, although there may be continued mild weight loss for up to a month. In contrast, for those patients who are immunodeficient, the diarrhea can be more severe with a liquid loss of 25 liters per day. Unless the deficiency is reversed, the disease can last for life. Half of the cryptosporidiosis patients who have AIDS will die within 6 months.

Figure 22.20 The whipworm *Trichuris trichiura*, which infects more than a billion people worldwide. Notice how the first two-thirds of the worm are quite thin, whereas the last part is thick and bulbous. This example is a female, which is able to shed between 3000 and 20,000 eggs per day.

Treatment of cryptosporidiosis — In the immunocompetent patient, the disease is self-limiting, and therefore no treatment is required, although it may be necessary to rehydrate small children. In immunocompromised patients, the diarrhea is so pronounced that rehydration is essential. There is no uniformly effective therapy for this infection.

Whipworm (*Trichuris trichiura*)

The adult whipworm is 30–50 mm in length, with the first two-thirds thin and the last one-third bulbous (**Figure 22.20**). The female worm can produce 3000–10,000 eggs per day, and the eggs have a brown shell with translucent knobs at each end. Whipworms infect about 1 billion people worldwide, and infection is usually concentrated where there is indiscriminate defecation and a warm, humid environment. The infection rate in tropical environments can be 80%. This infection affects more than 2 million people in rural areas of the southeastern United States. The adult worm can live for eight years.

Pathogenesis of whipworm infection — Adult worms live with their anterior end attached to the host colonic mucosa. While attached in the cecum, the female releases eggs into the lumen of the intestine. These eggs are passed out of the body with the feces and deposited in the soil. At this stage, the eggs are immature and not infective, and they must incubate in the soil for 10 days before they embryonate and become infectious. The infectious eggs are picked up on hands, passed into the mouth and swallowed. They move into the duodenum, where the larvae mature for about one month before migrating to the cecum.

Attachment of adult worms to the colonic mucosa produces hemorrhaging and localized ulcerations that can be used as portals of entry for opportunistic bacterial infections. Consequently, concomitant bacteremias can often be seen with this parasitic infection. Although light infections are usually asymptomatic, with more moderate worm loads there can be damage to the intestinal mucosa, accompanied by nausea, diarrhea, and abdominal pain. Some children can have worm loads as high as 800, and in these cases there is significant mucosal damage, blood loss, and anemia. When these children strain to defecate, the sheer force of the fecal stream on the worm bodies can cause prolapse of the colonic or rectal mucosa out through the anus.

Treatment of whipworm infection — There is no need for treatment of asymptomatic infections; however, for more severe disease, mebendazole is very effective and is the treatment of choice. The cure rate is 60–70%, with 90% of the worms being expelled with this treatment. However, even though the patient becomes asymptomatic, worms may still be present. Prevention of the infection involves good sanitation.

Hookworms

There are two species of this parasite that infect humans: *Necator americanus* and *Ancylostoma duodenale*. Adult worms are pinkish and about 10 mm long. Because the direction in which the anterior end curves is often opposite to the direction in which the rest of the body curves, the organism takes on the appearance of a hook. The hookworm is found worldwide, with transmission through deposition of eggs into shady, well-drained soil. Infection with these worms can be significant in densely populated communities. In the United States, there are more than 700,000 cases of hookworm.

The hookworm *Necator americanus* is found in southern Asia, Africa, and the Americas; *Ancylostoma duodenale* is found in the Mediterranean, the Middle East, northern India, China, and Japan.

Fast Fact

It has been estimated that together *Necator americanus* and *Ancylostoma duodenale* extract 7 million liters of blood each day from more than 700 million infected people.

Pathogenesis of hookworm infection — *Ancylostoma duodenale* uses four sharp toothlike structures (**Figure 22.21**), whereas *Necator americanus* has dorsal and ventral cutting plates for attachment to the mucosa of the small intestine. Fertilized females can release 10,000–20,000 eggs per day, which are passed in the feces. These eggs are passed in a four-cell or eight-cell stage of development and on reaching the soil hatch within 48 hours, releasing larvae that feed on soil bacteria. They double in size and molt to become infectious larvae that can survive four to six weeks in the soil.

On contact with human skin, the larvae penetrate the epidermis and move into the blood and lymph. Using the blood system, they move to the heart and eventually the lungs, where they rupture in the alveolar spaces and are coughed up and swallowed. They then move into the small intestine and mature into adult worms. As the worms migrate in the intestine, they leave behind bleeding points at the former attachment site, and because the worms can live up to 14 years, the blood loss can be enormous.

In the overwhelming majority of patients, the worm load is small and the infection is asymptomatic. However, in symptomatic cases there can be rash and swelling where the worm makes initial entry (usually between the toes), and this can persist for several days. Pulmonary problems are infrequent and less severe, but in the gut, the worms can cause abdominal pain and abnormal peristalsis. The major clinical manifestation of this infection is blood loss and concomitant anemia. The severity of the anemia depends on worm load and on the patient's dietary iron intake. Usually the development of severe anemia can take months or even years, but in children there can be earlier problems, including heart failure and retardation of mental and physical development.

Treatment of hookworm infection — The anemia that results from this infection is the primary consideration and must be corrected. The three most widely used anti-helminthic drugs for this infection are pyrantel pamoate, mebendazole, and albendazole, which are all highly effective. Prevention of this infection is tied to proper sanitation.

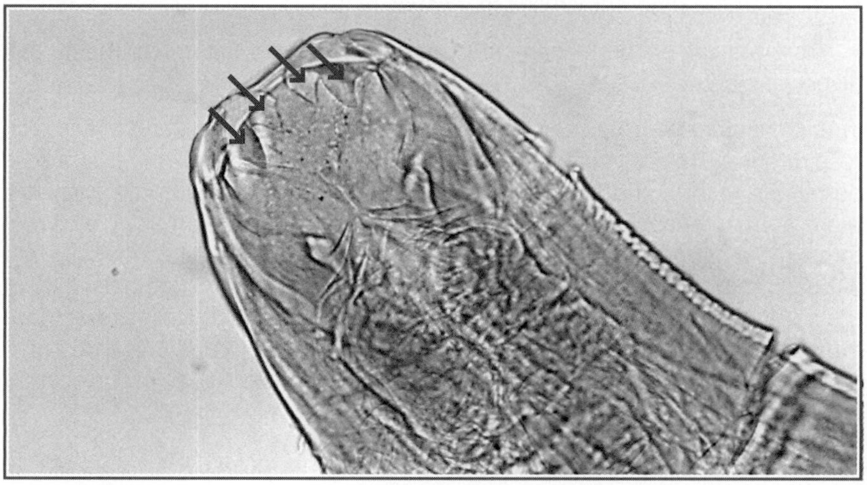

Figure 22.21 This unstained micrograph of the hookworm *Ancylostoma duodenale* shows the four sharp toothlike structures. These structures aid the worm in attaching to the mucous membranes of the host's small intestine.

Keep in Mind

- There are several protozoa and helminths that infect the human digestive tract.
- Parasitic protozoan pathogens of the digestive tract include *Giardia* and *Cryptosporidia*.
- Infection of the digestive tract by helminths can be caused by whipworms or hookworms.

SUMMARY

- The digestive system is a major portal of entry by which pathogens can enter the body.
- Gastrointestinal infections can be exogenous (caused by organisms from the outside) or endogenous (caused by organisms in the normal intestinal flora).
- Common symptoms of infections of the digestive system include bacterial diarrhea, fever, vomiting, and abdominal pain.
- Treatment for most infections of the digestive system includes supportive care and replacement of fluids.
- Most infections in the oral cavity involve dental caries or diseases of the gums.
- Many of the organisms that cause infections in the digestive system are Gram-negative rods.
- The O antigens are part of the lipopolysaccharides in the outer membrane, the H antigens are found on the flagella, and the K antigens are part of the capsule.
- Many of the organisms that produce infection in the digestive system do so through the production of enterotoxins such as the Shiga toxin.
- *Campylobacter* infection is the most common cause of diarrhea.
- *Helicobacter pylori* is the most common cause of gastric and duodenal ulcers.
- The digestive system is a portal of entry for many viruses.
- There are six different viruses that cause hepatitis.
- There are several protozoan parasites and helminths that infect the human digestive system.

In this chapter, we have seen that the digestive system is second only to the respiratory system as a portal of entry for infectious organisms. It is also a portal of exit through the fecal–oral route of infection. As with other systems, there are bacteria, viruses, and parasites that can produce an array of symptoms; the most prominent is diarrhea. It is important to recall continually that although this system seems to be overwhelmed by potential pathogens, there are significant defenses that are located in and around the digestive tract that protect us from many of these diseases (see Chapters 15 and 16). In the next chapter, we look at the third of the systems that provides a good portal of entry for pathogens, the genitourinary tract.

SELF EVALUATION AND CHAPTER CONFIDENCE

Multiple Choice

Answers are given in the back of the book and help can be found in the student resources at:
www.garlandscience.com/micro

1. The most common source of gastrointestinal infection in the developed world is
   A. *Salmonella*
   B. *Shigella*
   C. *Escherichia coli*
   D. *Campylobacter*
   E. *Staphylococcus aureus*

2. The causative agent for 80% of ulcers is
   A. *Clostridium difficile*
   B. *Helicobacter pylori*
   C. *Escherichia coli*
   D. *Campylobacter*
   E. None of the above

3. An infection that is limited to the marginal surfaces of the gum tissue is called
   A. Trench mouth
   B. Chronic periodontitis
   C. Gingivitis
   D. None of the above
   E. All of the above

4. Dental plaque is an example of
   A. A pellicle
   B. A biofilm
   C. Gingivitis
   D. Trench mouth
   E. All of the above

5. Vincent's disease is an example of
   A. Gingivitis
   B. Periodontitis
   C. Necrotizing periodontal disease
   D. All of the above

6. Which of the following is produced by *E. coli*?
   A. Pore-forming toxin
   B. Shiga toxin
   C. Heat-labile toxin
   D. Heat-stable toxin
   E. All of the above

7. Transmission of traveler's diarrhea occurs through
   A. Sexual contact with infected individuals
   B. The respiratory route
   C. The skin
   D. Contaminated food or water

8. Bacillary dysentery is caused by
   A. *Salmonella*
   B. *Shigella*
   C. *Escherichia coli*
   D. *Campylobacter*
   E. All of the above

9. Serotypes of *Salmonella* can be identified by
   A. H antigens
   B. O antigens
   C. Capsules
   D. A and B
   E. All of the above

10. The most important complication of typhoid fever is
    A. High fever
    B. Diarrhea
    C. Perforation of the colon wall
    D. Dementia
    E. None of the above

11. The disease cholera is caused by
    A. Colonization of the entire intestinal tract
    B. An endotoxin
    C. Perforation of the colon wall
    D. Retention of fluid electrolytes
    E. All of the above

12. Which of the following has now been classified as Hepatovirus?
    A. Hepatitis C
    B. Hepatitis E
    C. Echovirus
    D. Hepatitis A
    E. None of the above

13. Chronic carriers are the main reservoirs for
    A. Hepatitis A
    B. Hepatitis C
    C. Hepatitis E
    D. Hepatitis B
    E. None of the above

14. The intestinal parasite *Giardia* is found in
    A. Fish
    B. Mammals
    C. Reptiles
    D. Birds
    E. All of the above

15. Rice stools are characteristic of
    A. Amoebic dysentery
    B. Bacillary dysentery
    C. Shigellosis
    D. Cholera

16. Most of the normal microbial flora of the digestive system are found in the
    A. Stomach
    B. Mouth
    C. Small intestines
    D. Large intestine

17. Poultry products are a likely source of infection by
    A. *Salmonella*
    B. *Shigella*
    C. *Vibrio*
    D. *Streptococcus*

18. Most gastrointestinal infections are treated with
    A. Water and electrolytes
    B. Quinacrine
    C. Penicillin
    D. None of the above

DEPTH OF UNDERSTANDING

Questions listed here require you to bring together the concepts you have learned in this chapter into a discussion format. This helps you to increase your depth of understanding of the material you have learned. Help can be found in the student resources at:
www.garlandscience.com/micro

1. Given the disaster that befell New Orleans, describe how infectious diseases of the digestive system can be a hazard after flooding.

2. Describe the pathogenic features of enterohemorrhagic bacteria.

CLINICAL CORNER

Help can be found in the student resources at:
www.garlandscience.com/micro

1. A patient comes into the clinic showing signs of dehydration and explains that she has had serious diarrhea for several days. The patient reports that she and some friends went out to dinner a couple of days before and ordered a hamburger cooked rare. One of her friends ordered a steak also cooked rare and the other friend had a salad. She suspects that she may have contracted food poisoning. She says her friend who ate salad is also experiencing diarrhea, but the friend who had steak is not ill.
    A. Would you consider this a consequence of food poisoning?
    B. What would you tell the patient about possible causes for her problem?
    C. What tests would you recommend be done to confirm your suspicions?

2. Your patient is dehydrated and has metabolic acidosis. She has had severe diarrhea with intermittent bouts of cramping and a feeling of fullness. Her stools are voluminous but odorless and look to be composed mainly of fluids. These symptoms began about 24 hours after she attended a party at the yacht club, where she consumed several raw oysters that were served as appetizers. She delayed coming to the doctor because she thought that she had a simple case of food poisoning.
    A. Does she have a simple case of food poisoning?
    B. How does the case history help you to formulate possible explanations for her problems?

Infections of the Genitourinary System

Chapter 23

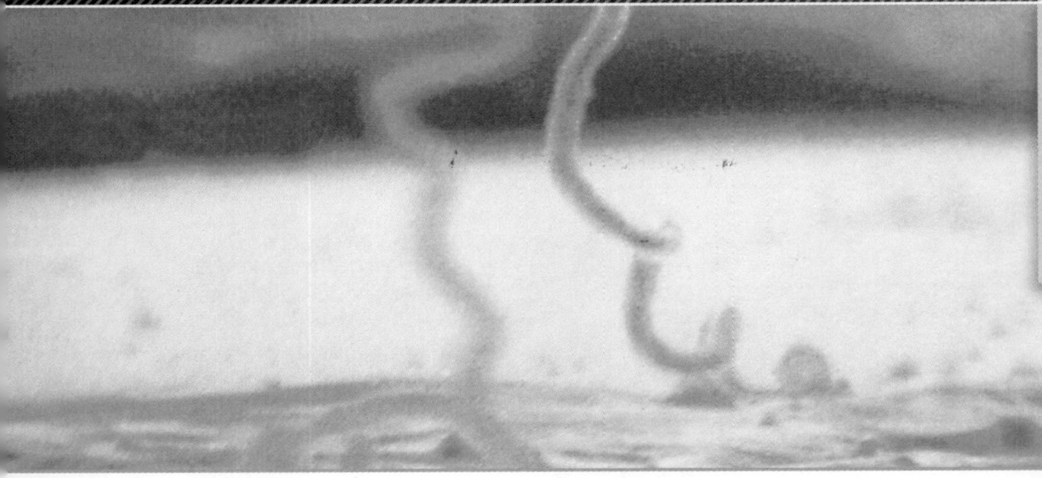

Why Is This Important?

The genitourinary system is the third of the body's systems that are open to the outside world, and many pathogens use this system as a portal of entry into the body.

OVERVIEW

In this chapter, we look at diseases that occur in the urinary and reproductive systems, collectively called the genitourinary system. This system, which is open to the outside environment, is the third of the three major portals by which pathogens enter the body. As we will see, some urinary system infections begin in the urethra and can travel up to the bladder and in severe situations even to the kidneys. The outcome of these infections can be severe and potentially life-threatening. We will look at a variety of reproductive system infections, many of which are designated sexually transmitted diseases (STDs). Infections of the genitourinary system can be caused by bacteria, viruses, yeast, and protozoan organisms. We will divide our discussions into the urinary and reproductive systems, keeping in mind that the anatomy of these systems has a role in the kinds of infection that occur.

We will divide our discussions into the following topics:

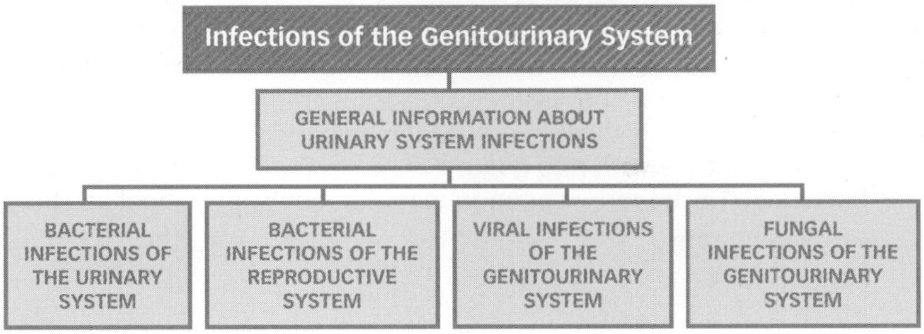

GENERAL INFORMATION ABOUT URINARY SYSTEM INFECTIONS

Because urine is essentially sterile, the presence of pathogens or inflammatory cells in the urine is an indication of a urinary tract infection (UTI). UTIs are more common in women than in men, and the pathogen involved is usually a bacterium or yeast.

In hospitals, UTI is a serious problem that is usually seen associated with indwelling catheters. These catheters compromise the structure and physical barriers of the urethra and the bladder, with the result that bacteria or yeast can ascend either the outside or the lumen of the catheter to reach the bladder. It should be noted that after removal of the catheter a 14 day course of antibiotics should be given to halt the infection.

What Do I Need to Know?

To get the most out of this chapter, please review the following terms from your previous courses in biology, anatomy, physiology, or chemistry or in previous chapters of this book as indicated in parentheses: apoptosis, bacteremia (**5**), blood cell cast, catheter, diplococcus (**4**), endogenous infection (**22**), genotype (**11**), inclusion body (**9**), keratinized, low oxygen tension, neutrophil (**15**), pilus (**9**), prokaryote (**4**), prostate, pyuria, septic shock (**5**), sequela, serotype (**16**), tachycardia, urethra, viral shedding (**12,13**).

Fast Fact

A single catheterization has a 1% risk of causing a UTI, and 10% of people with catheters get a UTI.

If we look at the anatomy of the urinary system (**Figure 23.1**), we see that urine flow is ideally in one direction: from kidney to bladder to urethra and then out of the body. Notice from **Figure 23.1**b, however, that this flow pattern can cause a reflux action, and this is one of the methods that pathogens can use to infect the urinary tract. A UTI is called **urethritis** when it occurs in the urethra, **cystitis** when it occurs in the bladder, and **nephritis** when it occurs in the kidneys. In males, infection of the urinary tract can lead to **prostatitis** (infection of the prostate).

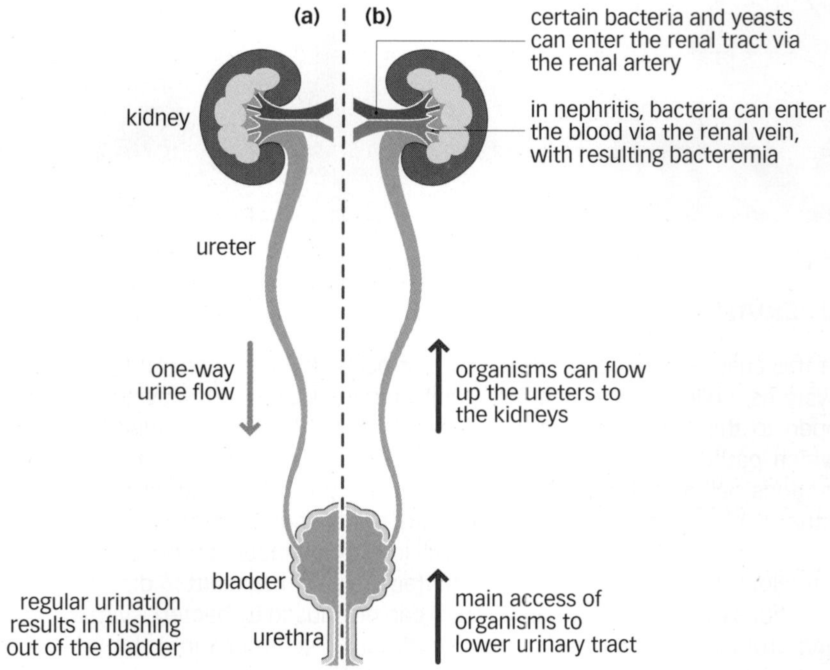

Figure 23.1 Generic version of the human urinary tract. Panel **a**: Normal structure and physiology permit descending flow and a sterile environment for urine. Panel **b**: Pathogens can enter the urinary tract from the external environment and produce infection that may spread all the way up the urinary tract. In an alternative pathway, pathogens in the blood enter the urinary system from the renal arteries, as indicated at the top of the illustration.

BACTERIAL INFECTIONS OF THE URINARY SYSTEM

It is likely that a few bacteria routinely enter the urinary tract, either from the external environment or from blood passing through the renal artery, but these are normally flushed out during urination. The prevalence of bacterial UTIs varies with age (**Figure 23.2**). In the first three months of life, bacterial UTIs are more common in males, but by preschool age they

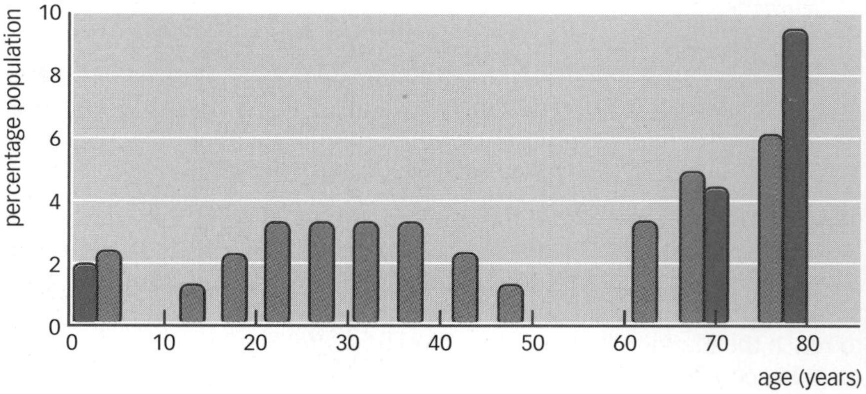

Figure 23.2 The prevalence of urinary tract infections in females (orange) and males (blue), according to age.

are more common in females as a result of refluxing of urine from the bladder into the ureters. In older males and females, anatomical changes associated with aging predispose to chronic bacteria in the urine (bacteriuria), but this condition is often asymptomatic. In males, enlargement of the prostate can increase the incidence of these infections, and in the elderly of both sexes, gynecological or prostatic surgery, incontinence, or chronic catheterization increases the rates of bacterial UTIs to 30–40%.

The problem of nosocomial bacterial UTIs is complicated by the increasing number of antibiotic-resistant bacteria found in hospitals, and in particular resistant bacteria such as *Pseudomonas*, *Streptococcus*, *Enterococcus*, and MRSA (methicillin-resistant *Staphylococcus aureus*).

The bacterium *Escherichia coli* accounts for more than 90% of the 7 million or more cases of cystitis and the 250,000 cases of nephritis estimated to occur in otherwise healthy people in the United States each year. A list of the most common causes of bacterial UTIs is shown in **Figure 23.3**.

Pathogenesis of Urinary System Bacterial Infections

Infection occurs when bacteria are able to get into the urine and remain there. Because all portions of the urinary tract are connected to one another by a liquid medium, infection is spread easily. Much of the pathogenesis seen in these infections is the result of anatomy. Males are protected from bacterial infections to some degree because the male urethra is longer than that in the female, and the shorter female urethra means that bacteria invading from the external environment have a shorter route to the bladder. In addition, the female urethra ends in the vaginal area, which is colonized by a variety of bacteria as part of the normal flora, many of which can initiate a UTI. Furthermore, sexual intercourse and manipulation of the female anatomy are likely to increase the chances of bacteria entering the bladder.

Any process that interferes with the complete emptying of the bladder can allow some bacteria to remain and increase in numbers. For example, uropathogenic *E. coli* have adherent molecules on their surface, and when these organisms are able to climb to the bladder, they can use these molecules to adhere to the walls of the bladder such that the normal flushing associated with urination is unable to remove them. As we have seen throughout our discussions, both the ability to remain in one location (staying in) and increases in pathogen numbers (defeating the host defenses) are requirements for a successful infection.

	organism	treatment
	*Escherichia coli*	trimethoprim, cephalexin, gentamicin
	*Proteus* spp.	trimethoprim, cephalexin, gentamicin
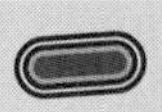	*Klebsiella* spp.	trimethoprim, cephalexin, gentamicin
	*Pseudomonas aeruginosa*	ciprofloxacin, gentamicin
	*Enterococcus* spp.	amoxicillin, vancomycin
	*Staphylococcus aureus*	trimethoprim, cephalexin, gentamicin
	coagulase-negative staphylococci	trimethoprim, cephalexin, gentamicin

Figure 23.3 Some of the bacteria that commonly cause urinary tract infections and the antibiotics that can be used to treat them.

Bladder infections cause an inflammatory response in which neutrophils migrate to the infection site and, along with bacterial toxins and enzymes, irritate the lining of the bladder and urethra. This causes the symptoms of increased frequency of urination, urgency, and dysuria (painful urination), which are commonly seen in these infections. In males the infection can reach the prostate from the urethra, lymphatics, and blood to cause prostatitis. The inflammation caused by prostatitis can lead to compression of the urethra lumen, a compression that can obstruct or retard the flow of urine. Prostatitis can be either acute or chronic.

As mentioned above, *E. coli* is responsible for most bladder infections and is the most potent of all the pathogens that cause UTIs. Yet there are fewer than 10 serotypes of *E. coli* that account for all these infections. The ability of these *E. coli* serotypes to produce UTIs is associated with a variety of virulence factors, such as α hemolysins and specialized pili referred to as **P pili**, which bind to receptors on epithelial cells of the urinary tract. This binding is very avid, allowing successful adherence and subsequent colonization by the bacteria.

The clinical sequelae associated with UTIs can vary, and more than 50% of these infections do not produce recognizable illness. We can divide UTIs into four types: urethritis, cystitis, nephritis, and prostatitis (men only).

- **Urethritis** (urethra infection) and **cystitis** (bladder infection) — with bacterial urethritis and cystitis, the symptoms are dysuria, frequency, and urgency. There can also be low back pain and abdominal pain or tenderness over the bladder area. In addition, the urine may be cloudy (**Figure 23.4**). Cystitis can be distinguished from urethritis by the fact that the former has a more acute onset, more severe symptoms, and the presence of bacteria and blood in the urine.

- **Nephritis** — bacterial kidney infection usually presents with pain in the flanks of the body and a fever above 38.3°C. It may be preceded or accompanied by symptoms of cystitis and in more severe cases can present with diarrhea, vomiting, and tachycardia. Nephritis can occasionally result in septic shock. Usually, the symptoms resolve themselves, and there is no damage to kidney function. In 20–50% of pregnant women, however, the infection causes premature birth. Some people develop chronic nephritis without ever showing any symptoms of a UTI.

- **Prostatitis** — prostate bacterial infection usually presents with lower back pain and pain in the perirectal area and testicles. In acute prostatitis, there can be high fever, chills, and symptoms of bacterial cystitis. Inflammatory swelling during bacterial prostatitis can lead to obstruction of the urethra and the retention of urine. In addition, there can be abscess formation, epididymitis, and seminal vesiculitis. Acute prostatitis is usually seen in young men, whereas the chronic form is associated with the elderly and usually with catheterization.

The diagnosis of a bacterial UTI is based on the examination of urine for evidence of bacteria or accompanying inflammation. This requires the collection of a clean voided midstream urine sample. For about 90% of patients, UTIs are identified as pyuria (more than 10 white blood cells per

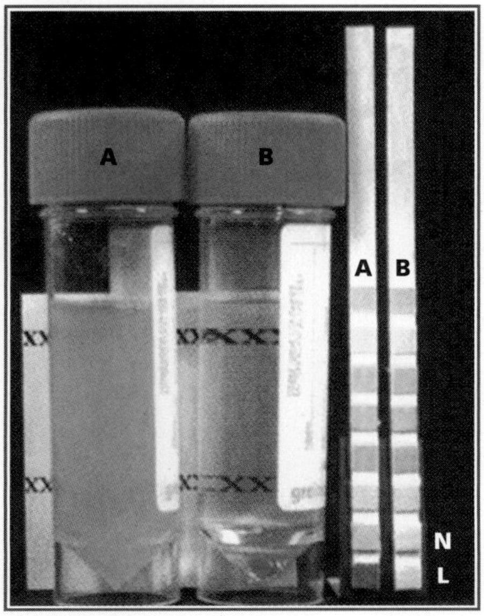

Figure 23.4 Urine specimens and diagnostic dipsticks. Specimen A is cloudy, and both the leukocyte (L, purple) and nitrate (N, deep pink) results shown on the dipstick are positive, indicating a bacterial infection in the genitourinary tract. Specimen B is clear, and the dipstick for it is negative (the colors in the L and N regions are unchanged).

cubic millimeter of urine) or, more specifically, by the presence of white blood cell casts in the urine. The most positive way to confirm a UTI is to Gram-stain a urine sample. The presence of at least one bacterium per microscopic oil-immersion field is an indication of infection.

Treatment of Urinary System Bacterial Infections

The antibiotics used to treat some of the most common organisms involved in bacterial UTIs are shown in **Figure 23.3**. Although sulfonamides and trimethoprim, alone or in combination with sulfamethoxazole and a fluoroquinolone, are commonly used, treatment is best guided by results of cultures and antimicrobial susceptibility tests. The duration of treatment depends on the severity of the bacterial infection and on the risk to the patient. The success of the treatment is determined by a culture of urine two weeks after therapy.

It is important to keep in mind that antibiotic resistance is becoming an increasing problem, especially in hospital and other institutional settings, where coliform bacteria account for 85% of the bacteria isolated from urine specimens. At least 50% of these isolated bacteria are resistant to amoxicillin, and 20% are resistant to trimethoprim.

Keep in Mind

- The presence of pathogens or inflammatory cells in the urine is an indication of a urinary tract infection.
- UTIs are very serious in hospital settings and are usually related to indwelling catheters.
- Infection in the urethra is called urethritis; in the bladder, cystitis; in the prostate, prostatitis; and in the kidney, nephritis.
- Bacterial UTIs are seen more in women than in men because of the difference in the lengths of the urethra.
- *Escherichia coli* is responsible for the majority of bacterial bladder infections and is the most potent of all pathogens that cause UTIs.
- Bacterial UTIs are routinely treated with antibiotics for various periods, depending on the severity of the infection.

BACTERIAL INFECTIONS OF THE REPRODUCTIVE SYSTEM

The major bacterial infections seen in the reproductive system are sexually transmitted and most often affect women. These infections are caused by a wide range of organisms, including group B streptococci, *Neisseria gonorrhoeae*, *Treponema pallidum* (the bacterium that causes syphilis), and *Chlamydia trachomatis* (the bacterium that causes non-gonococcal urethritis). Even though many STDs have been well studied and even though successful treatments are available for them, in many cases the infected individuals will not seek medical help because they are asymptomatic.

Gonorrhea and other STDs can cause urethritis, **cervicitis** (bacterial infection of the cervix), prostatitis, and pharyngitis. Although the pharyngitis is usually asymptomatic, the infected individual is a carrier and can readily infect sexual partners. In addition, **pelvic inflammatory disease** (PID) can result from gonococcal or chlamydial infection and can cause infertility and ectopic pregnancy. These bacterial infections can also infect the fetus and newborn. Consequently, expectant mothers are routinely screened during prenatal visits for exposure to sexually transmitted disease.

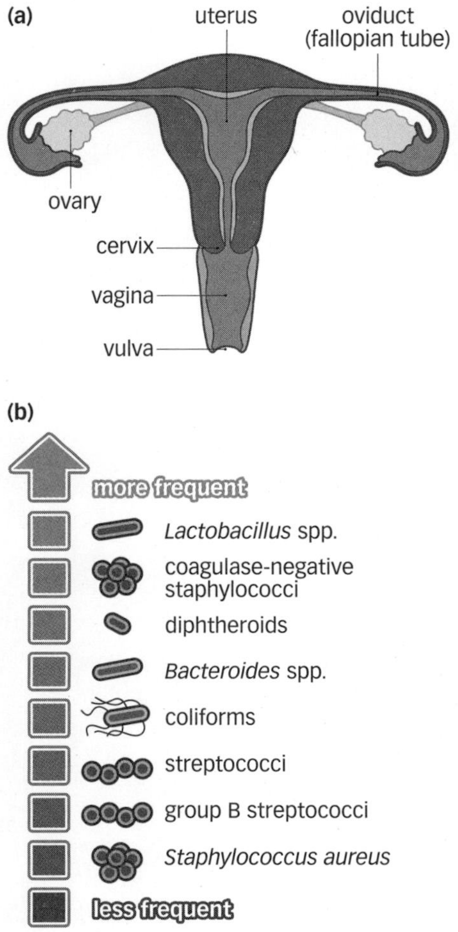

Figure 23.5 The microbial flora of the female reproductive tract. Panel **a**: The female reproductive tract. Panel **b**: Bacteria that make up the normal vaginal flora.

Figure 23.5 shows the bacterial species found in the microflora present in the vagina of the female genital tract. It is easy to see in this illustration how these vaginal organisms can move up to the uterus and fallopian tubes. In addition, the close proximity of the anus to the vagina makes it easy for some of the organisms normally found in the digestive tract to move into and infect the female reproductive tract. Furthermore, the vaginal epithelium in prepubescent females is not yet keratinized and can easily support the growth of gonococcal organisms. Changes in the vaginal flora that occur as part of the menstrual cycle can be involved in potential infection. The bacteria involved in infections of the reproductive tract are shown in **Figure 23.6**.

organism	disease
sexually transmitted disease	
Neisseria gonorrheae	urethritis, cervicitis, proctitis, pharyngitis
Chlamydia trachomatis	urethritis, cervicitis, proctitis
HSV type 2	genital and oral ulcers
Treponema pallidum	primary, secondary, and latent syphilis
HIV	AIDS
vaginitis	
Trichomonas vaginalis	
Candida albicans	thrush
Gardnerella vaginalis	bacterial vaginitis
pelvic inflammatory disease (PID)	
Neisseria gonorrheae	acute and chronic PID
Chlamydia trachomatis	acute and chronic PID
coliforms	chronic PID
congenital and peritoneal infection	
rubella virus	spontaneous abortion, neurological and cardiac defects
Treponema pallidum	osteochondritis, hepatosplenomegaly, mental retardation
hepatitis B virus	chronic liver disease, cirrhosis, carcinoma in later life
HIV	AIDS
group B streptococcus	neonatal sepsis and meningitis
Listeria monocytogenes	neonatal sepsis and meningitis
post partum infection, gynecological infection, and septic abortion	
group A streptococcus and other streptococci	
Staphylococcus aureus	
coliforms	

Figure 23.6 Organisms involved in various infections of the female genitourinary tract.

Sexually transmitted infections have been around for hundreds of years and were first identified in the 1600s. They affect all populations and social strata, and interest in them has been rekindled by the appearance of HIV and AIDS. The most common STDs are caused by the bacteria *C. trachomatis* and *N. gonorrhoeae* and by the papilloma, herpes simplex, and HIV viruses. However, other organisms, including hepatitis B virus, cytomegalovirus, and the bacterium *T. pallidum* can be spread by sexual contact. **Table 23.1** lists the major sexually transmitted pathogens and the infections they cause. The bacterial agents are discussed in this section, and the viral and fungal agents in the following two sections. The one parasitic entry in **Table 23.1**, *Trichomonas vaginalis*, is there for the sake of completeness; this parasite was discussed in Chapter 14.

Depending on the pathogen, a sexually transmitted infection can be either localized or systemic. For localized infection, the most common manifestations are inflammatory (for example, urethritis or cervicitis) and may not be noticed by the patient. In some cases, the infection can involve deeper tissues and structures, causing epididymitis and **salpingitis** (inflammation of the fallopian tubes). It is important to understand that these latter types of infection can become systemic.

Common Clinical Conditions Associated With STDs

In the remainder of this section, we look briefly at some of the common clinical syndromes seen with bacterial STDs and then move on to a more in-depth look at three bacteria responsible for reproductive-system infections: *Treponema*, *Neisseria*, and *Chlamydia*.

Genital ulcers are lesions (either single or multiple) that arise on the genitalia. They begin as either a papule (solid skin bump) or a pustule (skin bump containing pus) and evolve into ulcers. The features of these ulcers are described in **Table 23.2**. Some of these differences can be significant. For example, the syphilis ulcer is single and painless, whereas genital herpes ulcers are often multiple and very painful.

Table 23.1 Sexually Transmitted Pathogens and the Infections and Diseases They Cause

Pathogen	Genitourinary Infections
Bacteria	
Neisseria gonorrhoeae	Urethritis, cervicitis, prostatitis, pharyngitis, pelvic inflammatory disease, and disseminated gonococcal infection
Chlamydia trachomatis	Non-gonococcal urethritis, epididymitis, cervicitis, salpingitis, and lymphogranuloma venereum
Treponema pallidum	Syphilis
Haemophilus ducreyi	Chancroid
Ureaplasma urealyticum	Non-gonococcal urethritis
Viruses	
HIV	AIDS, AIDS-related complex
Herpes simplex virus	Primary and recurrent genital herpes
Papillomavirus	Genital warts and association with cervical carcinoma
Fungi	
Candida albicans	Vulvovaginitis and penile candidiasis
Protozoan parasites	
Trichomonas vaginalis	Trichomonal vaginitis

Infection	Type of Ulcer	Involvement of Inguinal Lymph Nodes
Genital herpes	Multiple grouped vesicles and painful coalesced ulcers	Tender and nonsuppurative
Syphilis	Non-tender indurated ulcer that occurs singly	Rubbery consistency
Chancroid	Tender, shallow, painful ulcer	Suppurative
Lymphogranuloma venereum	Painless, small ulcer or papule usually healed by the time of presentation	Discrete progressing to suppurative

Table 23.2 Causes and Types of Genital Ulcers

The condition known clinically as **sexually transmitted urethritis** usually presents as dysuria or urethral discharge, or both. The discharge may be prominent or may have to be milked from the urethra. The major causes of sexually transmitted urethritis are *N. gonorrhoeae* and *C. trachomatis*, and in many cases the infection will involve both pathogens. Diagnosis of gonococcal urethritis usually requires culture of the organism, but it can also be done by direct microscopic examination of Gram-stained samples as well as by DNA analysis. However, for detection of *C. trachomatis*, DNA amplification analysis is required. Successful treatment of sexually transmitted urethritis depends on the agent causing the infection and also on whether the infection has spread beyond the local area.

Epididymitis, which is unilateral swelling of the epididymis, is commonly seen in sexually active men and is usually quite painful, with fever and swelling of the testicles. Two bacteria are usually implicated in this infection, *N. gonorrhoeae* and *C. trachomatis*. In men over 35 years of age and in homosexual men, Enterobacteriaceae and coagulase-negative staphylococci may also cause this condition.

Fast Fact

The incidence of PID is 5–10 times higher in women with intrauterine devices than in women not using contraception.

The microbial etiology of **cervicitis**, infection of the cervix, can vary, but it is usually caused by *N. gonorrhoeae* and *C. trachomatis* infecting the stratified squamous epithelium of the cervix. This infection usually involves a mucopurulent vaginal discharge as the cervix becomes inflamed, and phagocytic polymorphonuclear leukocytes will be found in the discharge.

Bacterial vaginitis is the most common form of **vaginitis** and is associated with overgrowth of multiple members of the vaginal anaerobic flora and the Gram-negative rod *Gardnerella vaginalis*. In bacterial vaginitis, there can be a yellowish discharge that is homogeneous and stays adhered to the vaginal wall. Vaginal epithelial cells called *clue cells* (**Figure 23.7**) found in the discharge are covered with bacteria. Vaginal discharge can occur alone or in connection with salpingitis, endometritis, or cervicitis, and the clinical findings of this condition vary with the etiological agent.

The clinical manifestations of PID vary, but they generally include abdominal pain. Approximately 50% of PID is caused by *N. gonorrhoeae* infection, but there can be non-gonococcal infections caused by a combination of bacteria including *C. trachomatis*, *Bacteroides* and anaerobic streptococci. The infections caused by these non-gonococcal PID agents are more complex than the gonococcal variety, but they are usually milder than that caused by *Neisseria*.

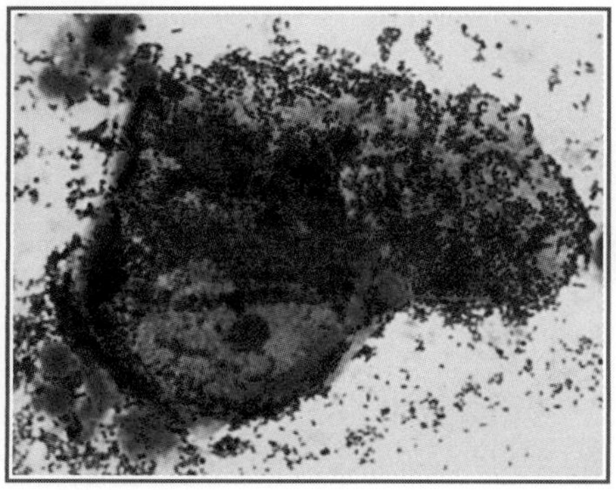

Figure 23.7 A clue cell found in a vaginal smear. Notice the bacteria (the small dark circles) clinging to the surface of the clue cell.

The condition **lymphadenitis** — inflammation of lymph nodes — is seen in several sexually transmitted infections, especially in herpes infections (discussed below) and lymphogranuloma venereum, which is caused by *C. trachomatis*. Lymphadenitis usually begins as a small genital ulcer that is frequently unnoticed. Usually the first evidence is a tender swollen lymph node in the groin (**Figure 23.8**).

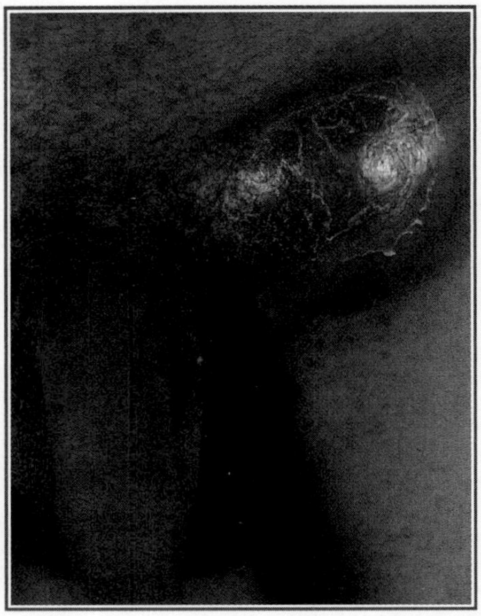

Figure 23.8 Early development of bilateral enlarged lymph nodes called buboes, which are seen in lymphogranuloma venereum caused by *Chlamydia trachomatis*.

Now let us look at three of the most common bacterial sexually transmitted infections.

Syphilis (Treponema pallidum)

Syphilis may be the earliest recorded sexually transmitted infection, first described in the 1600s. It is caused by the bacterium *T. pallidum* (**Figure 23.9**), a slim spirochete with regular spirals that resemble a corkscrew. This pathogen shows a characteristic slow rotating motility with occasional 90° flexion (resembling a gentleman bowing from the waist). *Treponema* cannot be grown on bacterial media but can be grown in mammalian cell cultures at a low oxygen tension. It is extremely susceptible to any changes in its environment and to any deviation in physical conditions. *Treponema* dies rapidly if dehydrated or heated and is very sensitive to detergents and disinfectants. The transmission method for this pathogen is restricted to direct contact.

Treponema is exclusively a human pathogen, and infection is acquired by sexual contact with someone that has active primary or secondary syphilis. However, it has been shown that there is a possibility of transmission through the sharing of contaminated needles and also transplacentally. Since 1990 the number of syphilis cases in the United States has steadily declined and is now below 40,000 per year. Of these 40,000 patients, 20% have primary or secondary disease, and the rest are either latent or in the tertiary stage of the disease. Syphilis is still a major health problem worldwide, with more than 12 million cases reported each year.

Pathogenesis of syphilis — Because there is no animal model for this disease, the pathogenesis has been extrapolated from observations of patients with syphilis. The spirochetes reach the subepithelial tissues either by moving through breaks in the skin or by passing between the epithelial cells of the mucous membranes of the reproductive tract. Because the organism multiplies slowly, there is little or no inflammatory response during the initial stage of the infection. As a lesion develops, small arterioles begin to swell, and the endothelial cells of these vessels proliferate. This increase in the number of endothelial cells reduces the blood flow through the arterioles, leading to the necrotic ulceration seen at the primary infection site. This is followed by an influx of granulocytes, lymphocytes, monocytes, and plasma cells that surround the affected blood vessels. The primary lesion heals spontaneously, but by this time the bacteria have spread to other locations by way of the blood and lymph. For reasons that are not yet understood, syphilis then goes silent for a period before the secondary stage develops. The disease also undergoes this silent period before the onset of the tertiary stage.

Treponema binds to host proteins such as immunoglobulins and complement components, and it has been suggested that this binding is a type of camouflage to protect the bacterium. It does not produce any virulence factors during the progression of the disease, which has several clinically defined stages:

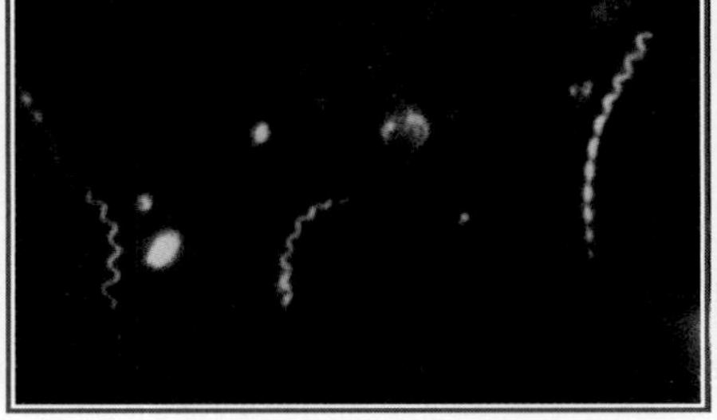

Figure 23.9 A darkfield preparation of the spirochete *Treponema pallidum*, which causes syphilis.

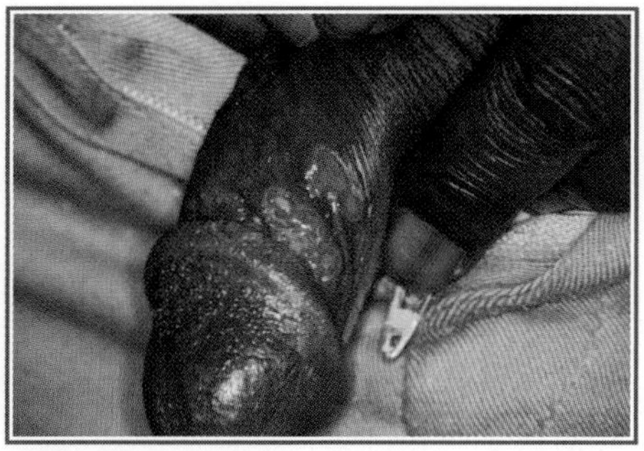

Figure 23.10 The chancre seen in primary syphilis on the genital area of a male.

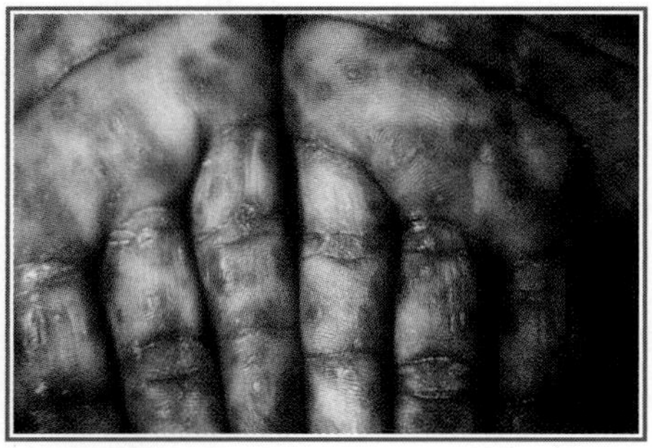

Figure 23.11 Palm lesions in secondary syphilis. As well as on the palms, these lesions can be seen on any other surface of the body.

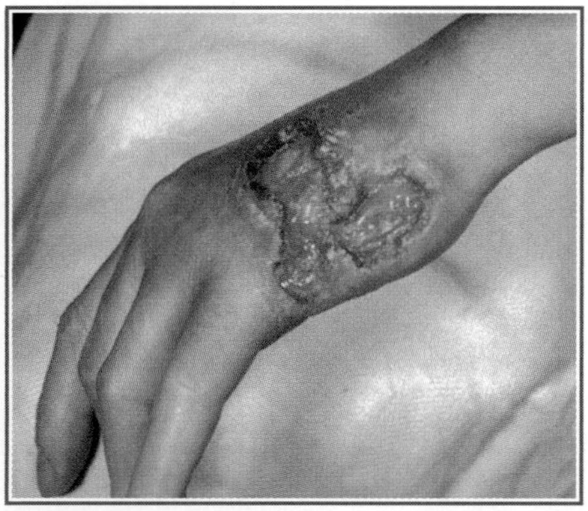

Figure 23.12 Gummas, lesions seen in tertiary syphilis, are rarely seen today because of treatments that prevent the progression of the disease to the tertiary stage.

- *Primary syphilis* is associated with the appearance of the primary syphilitic lesion, which is a papule that evolves into an ulcer. This ulcer is usually located on the external genitalia or on the cervix but can also be found in the oral cavity or anus, depending on the type of sexual interaction. The ulcer remains painless and is referred to as a **chancre** (**Figure 23.10**). The incubation time from contact to chancre is about three weeks, with lymphadenopathy arising within one week of the appearance of the initial lesion. The lymphadenopathy can persist for months, but the chancre will disappear in four to six weeks. Primary syphilis can also be associated with unilateral or bilateral enlargement of the lymph nodes of the groin.

- *Secondary syphilis* is also known as *disseminated syphilis* and develops two to eight weeks after the chancre formed in the primary stage has disappeared. It is characterized by generalized lymphadenopathy and the appearance of a symmetric mucocutaneous maculopapular rash accompanied by fever, malaise, and lymphadenitis. The skin rash is seen on the face, trunk, and extremities, including the palms of the hands (**Figure 23.11**) and soles of the feet. The lesions associated with this rash are teeming with spirochetes and are extremely infectious. The lesions resolve in a few days, but in one-third of cases, resolution takes many weeks. In the two-thirds of cases where the lesions are gone in a few days, the next step is development of latent syphilis.

- *Latent syphilis* is a stage that can last for years. There are no clinical signs or symptoms, but the infection is continuing. During the first few years, latency can be interrupted by progressively less severe bouts of secondary syphilis. When the infection is in the latent stage, transmission of the disease to others is possible only during these relapse periods, but transmission from mother to fetus is possible throughout the latent period.

- *Tertiary syphilis* is the stage reached by about one-third of untreated patients. Signs and symptoms can appear as soon as five years after the initial infection but are usually not seen for 15–20 years. The clinical findings seen in tertiary syphilis depend on whether the infection spreads to the cardiovascular system (*cardiovascular syphilis*) or to the nervous system (*neurosyphilis*). In cardiovascular syphilis, the bacteria move to the vaso vasorum of the aorta, causing necrosis and destruction of elasticity. This can lead to the development of aneurisms and aortic valvular incompetency. During this process **gummas** appear. These are localized granulomatous lesions seen in the skin, bones, joints, and internal organs (**Figure 23.12**). Neurosyphilis is characterized by a mixture of meningovasculitis and degenerative changes in the parenchymal tissue, changes that can occur in any area of the body. The most common symptoms of neurosyphilis are chronic meningitis, fever, headache, and increased numbers of cells and protein in the cerebrospinal fluid. There can also be cortical degeneration of the brain, causing such mental changes as decreased memory, hallucinations, and psychoses. These findings of the central nervous system are classified as PARESIS (**p**ersonality, **a**ffect, **r**eflexes, **e**yes, **s**enses, **i**ntellect, and **s**peech).

- *Congenital syphilis* is the form passed from a mother to her fetus. The fetus is susceptible to infection with *T. pallidum* only after the fourth month of gestation, but the infection can have devastating consequences. If the mother is treated for the disease before the fourth month of pregnancy, the fetus will show no signs of infection. However, if she is not treated she can lose the baby or the baby can be born with congenital syphilis, which is clinically similar to secondary syphilis. The condition can result in changes to the entire skeletal structure of the newborn, anemia, thrombocytopenia, and liver failure.

Treatment of syphilis — *Treponema pallidum* has remained very sensitive to penicillin, which is the treatment of choice. If there are allergies to penicillin, treatment with tetracycline, azithromycin, or cephalosporin is successful. Safe sex is effective for prevention of this infection.

Gonorrhea (Neisseria gonorrhoeae)

The STD gonorrhea is caused by *N. gonorrhoeae*, which is a Gram-negative diplococcus that possesses numerous pili extending through the outer membrane of the cell. This outer membrane also contains phospholipids, lipopolysaccharides, and several distinct outer membrane proteins, including porins and adherence proteins.

Neisseria grows well on chocolate agar and requires supplementation with CO_2. The organism can also change the antigens associated with its surface from generation to generation, and this antigenic variability is also seen in the pili, which undergo multiple changes through recombinant gene exchanges. Because *Neisseria* is easily transformed, there can be extensive genetic changes in this pathogen. These genetic changes are important because they allow the pathogen to escape the host defensive response and also make it possible for the organism to bind to a variety of different receptors, thereby maximizing the potential for infection.

Although reported cases of gonorrhea have been decreasing in the United States for the past 20 years, the reported cases probably represent only 50% of the actual number of cases. Therefore, gonorrhea is still a major public health problem. The overall incidence in the United States is about 130 cases per 100,000 people, but the rate of infection among adolescents is very high and increasing by 10% per year, with the highest rates of infection being seen in women aged between 15 and 19 years and men between 20 and 24 years.

Fast Fact

There is no truly effective means of controlling gonorrhea because it is difficult to detect asymptomatic cases (which are still infectious) and also because of increased resistance to antibiotic treatment.

The major reservoir for *N. gonorrhoeae* is asymptomatic patients, with almost 50% of these individuals being infectious. In fact, the infection rate can be as high as 20–50% if there is sexual intercourse with an asymptomatic individual.

Pathogenesis of gonorrhea — The infectious process can be divided into three parts:

- **Attachment and invasion.** *Neisseria gonorrhoeae* is not part of the normal microbial flora found in the body. This bacterium contains pili and adherence proteins that are used to attach to the urethral and vaginal epithelium, as well as to sperm, and parts of the fallopian tubes. After attachment, the pathogen invades host epithelial cells through a unique process in which the microvilli of the epithelial cells surround the organism and escort it into the cytoplasm. This process has been termed parasite-directed endocytosis because it seems to be initiated by the bacterium and not the host cell. This entry involves nonphagocytic cells. Once inside the target cell, the bacterium transcytoses through the cytoplasm and exits through the basal membrane of the cell to enter the submucosa.

- **Survival in the submucosa.** In the submucosa, the bacterium is immediately exposed to the host defenses that we learned about in Section IV of this book. The evasion of these defenses occurs in a variety of ways. *Neisseria* can block the deposition of the C3 component of complement, effectively shutting down the complement pathway. Additionally, the antibody response is inhibited by *Neisseria* surface proteins that bind to host antibodies in such a way that the bacteriocidal activity of the antibodies is blocked. These blocked antibodies are readily found in patients who have repeated gonococcal infections. *Neisseria* can also survive phagocytosis by interfering with the attachment of phagocytic cells and also by producing excessive amounts of catalase, the enzyme that neutralizes the oxidative killing that is part of phagocytosis.

- **Spread and dissemination.** *Neisseria* organisms tend to stay localized in the genital structures, an immobility that facilitates transmission of the infection and also causes increased inflammation and localized tissue injury. Purulent exudates containing sticky clusters of *Neisseria* held together with bacterial proteins can be found, and these are probably the primary infectious units. Infection may spread to deeper structures by progressive extension to adjacent mucosal glandular epithelial cells of the prostate, cervical glands, and fallopian tubes. In women, this spread can be facilitated by bacteria adhering to sperm. In addition, small numbers of infectious *Neisseria* can reach the blood and produce a systemic infection.

There are several clinical manifestations seen in gonorrhea, including these three:

- Genital gonorrhea in men occurs primarily in the urethra. Symptoms appear two to seven days after infection and consist of purulent urethral discharge (**Figure 23.13**) and dysuria. Infection can spread to the epididymis and the prostate. In women, symptoms include increased vaginal discharge, urinary frequency, dysuria, abdominal pain, and menstrual abnormalities. It should be noted that all these symptoms can be mild or completely absent, particularly in women.

- **Pelvic inflammatory disease** develops in about 10–20% of women infected with *Neisseria*. Symptoms include fever, bilateral abdominal tenderness, and leukocytosis and are caused by the spread of the bacteria along the fallopian tubes, producing salpingitis (**Figure 23.14**), and into the pelvic cavity, causing pelvic peritonitis and abscess formation. The most serious complications of PID resulting from gonorrhea are infertility and ectopic pregnancy.

- Disseminated gonococcal infection is the result of either localized gonorrhea or PID. The clinical features are fever, polyarthralgia, and petechial maculopapular or pustular rash. In fact, some of these signs may be caused by the host response to the bacteremia. The spread of the bacteria can lead to endocarditis or meningitis but it most commonly appears in the form of purulent arthritis.

Treatment of gonorrhea — Both individual patient issues and public health concerns must be considered in treating gonorrhea. Patients who discontinue treatment early risk continuing to transmit the disease and also developing antibiotic-resistant strains of *Neisseria*. In fact, resistance to penicillin has rendered this drug useless for the treatment of gonorrhea. This is due to alterations in the penicillin-binding protein (Chapter 20),

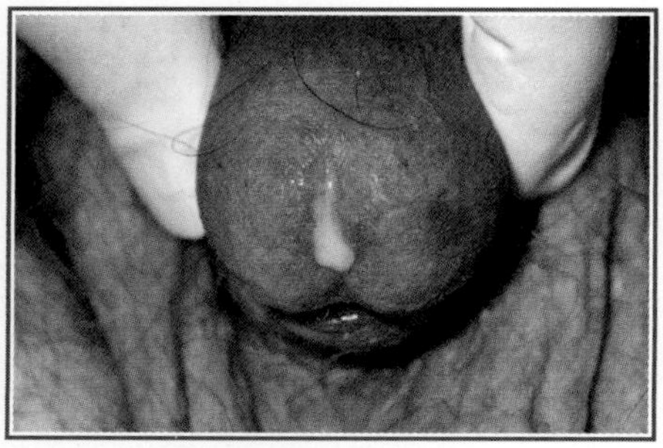

Figure 23.13 Purulent discharge from the male urethra of a patient with gonorrhea.

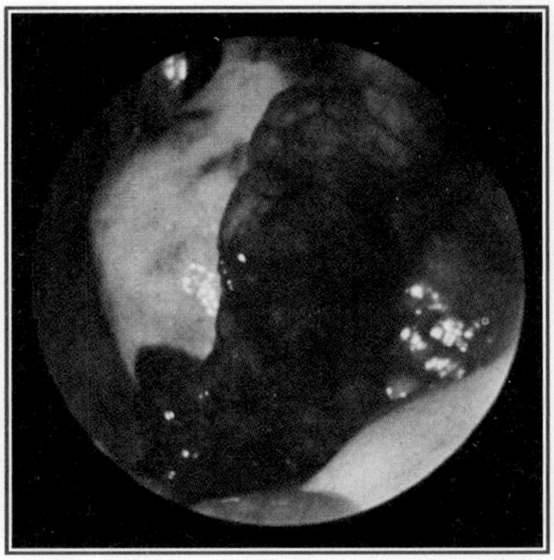

Figure 23.14 Endoscopic view of an infected and inflamed fallopian tube (bright red). Such tubal inflammation is called salpingitis. The ovary (white) can also be seen.

alterations that require administration of such large amounts of penicillin they would be toxic. In addition, *Neisseria* acquired a new β-lactamase (penicillinase) during the Vietnam War that is now found worldwide.

The development of resistant *Neisseria* strains has caused treatment options to shift to third-generation cephalosporins, which are not affected by penicillinase. In addition, these drugs have such high activity that treatment involves only one dose of antibiotic. Antibiotics such as fluoroquinolones, azithromycin, and doxycycline are also effective.

Non-gonococcal urethritis (Chlamydia trachomatis)

Non-gonococcal urethritis (NGU) is the most common form of sexually transmitted disease, with the number of people contracting it being twice as high as the number contracting gonorrhea. NGU is caused by a unique form of bacteria called *Chlamydia*, which are obligate intracellular bacteria. *Chlamydia trachomatis* is the most common cause of NGU. In addition to being an agent of sexually transmitted disease, this organism also causes the eye infection called trachoma (discussed in Chapter 26).

Chlamydia is a round cell surrounded by an envelope. The envelope is a trilaminar outer membrane containing lipopolysaccharides and proteins similar to those seen in Gram-negative bacteria. However, *Chlamydia* does not contain peptidoglycan. These organisms cannot be grown outside host cells, and they contain one of the smallest genomes of all the prokaryotes. Although they contain ribosomes and can produce energy in the same way as other bacteria, they lack the genes necessary to synthesize amino acids. *Chlamydia trachomatis* contains multiple outer-membrane proteins that further divide the members of this species into strains.

Replication of *Chlamydia* — *Chlamydia* has a unique replication cycle that involves two forms of the bacterium. The first is a small, hardy, infectious form, the **elementary body** (EB), and the second is a larger, more fragile, replicative form called the **reticulate body** (RB; sometimes called the *initial body*). The EB form is metabolically inert, and the proteins of this form contain a large number of disulfide bonds, making the EB structurally strong.

The replication cycle (**Figure 23.15**) begins when an EB attaches to unknown receptors on the plasma membrane of susceptible host cells (usually columnar epithelial cells), and enters the cell through endocytosis. While in the endocytic vesicle, the EB converts to an RB. An interesting observation is that endocytic vesicles carrying *C. trachomatis* do not fuse with lysosomes in the way you would expect them to. Instead, they fuse with other endocytic vesicles carrying the pathogen, and as the number of *Chlamydia* in each vacuole increases, the endosome membrane expands by fusing with the lipids of the Golgi apparatus to form a large inclusion body. After 24–72 hours, the process reverses and the RB form re-organizes and condenses to the EB form. The endosome membrane then either disintegrates or fuses with the host cell membrane, releasing the EBs, which go on to infect new targets. *Chlamydia trachomatis* inhibits apoptosis of epithelial cells, an inhibition that allows it to complete its replication cycle.

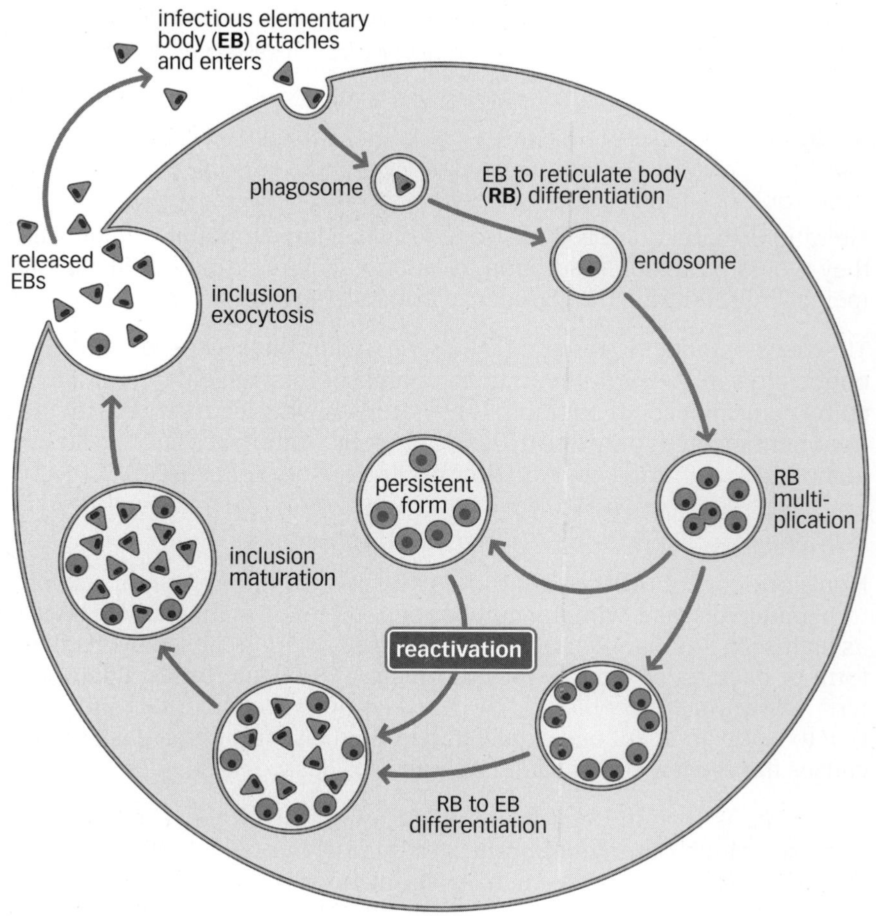

Figure 23.15 The replication cycle of *Chlamydia*. The elementary bodies do not replicate but are involved in the initial infection of the host cells. The elementary bodies transform into replicative reticulate bodies.

Pathogenesis of *C. trachomatis* genital infections — As noted earlier, infection with *C. trachomatis* is the most common STD, with more than 700,000 cases each year in the United States. Humans are the only reservoir for this pathogen, and the prevalence of chlamydial urethritis in men and women ranges from 5% in the general population to as high as 20% in populations receiving medical treatment for STDs. Approximately one-third of men who have sexual contact with women infected with *C. trachomatis* develop NGU after an incubation period of two to six weeks. It is important to note that many of these infected men will show no symptoms of the disease. Re-infection with NGU is a common occurrence.

Chlamydia has an affinity for epithelial cells of the endocervix and upper genital tract of women, and for the urethra and rectum of both men and women. Once an NGU infection is established, there is also a release of pro-inflammatory cytokines by the infected epithelial cells. It is thought that this inflammatory response may be generated by the lipopolysaccharides of the *Chlamydia* outer membrane. The inflammatory response results in tissue infiltration by polymorphonuclear leukocytes followed by lymphocytes, macrophages, plasma cells, and eosinophils. If the infection is not treated or if there is a failure in the immune response, aggregates of lymphocytes and macrophages may form in the submucosa, causing necrosis, fibrosis, and scarring. These symptoms can become chronic.

Fast Fact

More than 50% of all infants born to mothers infected with *Chlamydia trachomatis* who are excreting the organism during labor will be infected, usually with the eye condition known as conjunctivitis (Chapter 26).

The same sorts of clinical sequelae seen in gonorrhea — urethritis and epididymitis in men and cervicitis, salpingitis, and urethritis in women — can also be seen in *C. trachomatis* NGU infections, which usually present with dysuria and a thin urethral discharge. Infection of the cervix is usually asymptomatic but can also manifest as a vaginal discharge. In addition, ascending infections resulting in salpingitis and PID can occur in 5–30% of infected women.

Treatment of *C. trachomatis* genital infections — *Chlamydia* organisms are sensitive to doxycycline, azithromycin, and some fluoroquinolones. Topical administration of erythromycin or silver nitrate has little effect on the conjunctivitis seen in newborn infants of infected mothers, and 15–20% of these infants will develop conjunctivitis. There is no vaccine against NGU.

Keep in Mind

- The majority of bacterial infections of the reproductive system are sexually transmitted and most often affect women.
- STDs can also be caused by viruses, fungi, and protozoan parasites.
- The most common STDs are syphilis (caused by *Treponema pallidum*), gonorrhea (*Neisseria gonorrhoeae*), and non-gonococcal urethritis (*Chlamydia trachomatis*).

VIRAL INFECTIONS OF THE GENITOURINARY SYSTEM

The most important viral infection of the genitourinary system is HIV, which results in acquired immunodeficiency disease (AIDS), which we discussed in detail in Chapter 17. We will use this section to discuss two other prominent viruses, herpes simplex virus type 2 and human papillomavirus (HPV).

Herpes Simplex Virus Type 2

There are two distinct epidemiological and antigenic types of herpes simplex virus (HSV), HSV type 1 and HSV type 2. These are DNA viruses with linear double-stranded DNA molecules, and the two types share many of the same antigens and have 50% homology in their genomes. HSV-1 is described as an *above-the-waist* virus because it causes cold sores in the area of the mouth, and HSV-2 is a *below-the-waist* virus because it causes the infection known as genital herpes. HSV-1 can also cause genital herpes but on a much smaller scale than HSV-2. Thus, what follows here in this discussion applies to type 2 herpes only.

HSV-2 is distributed throughout the world, and humans seem to be the only reservoir for it. Transmission of infection is through direct contact with infected secretions. Interestingly, antibodies against HSV-1 can be found in a large portion of the population, but antibodies against HSV-2 are rarely seen before puberty. In adults, 15–30% of sexually active persons in Western industrialized countries carry antibodies against HSV-2.

Many patients infected with HSV-2 either are asymptomatic or else have small lesions on the penile or vulvar skin that go unnoticed. It is important to note that even though these people are asymptomatic, shedding of the virus can still occur. This accounts for transmission of the virus by individuals who have no active genital lesions and often no history of genital herpes. Although genital herpes is not a reportable disease in the United States, it is estimated that there are 1 million new cases each year.

Pathogenesis of genital herpes — There are two types of infection seen with genital herpes: acute and latent. In the acute infection, the obvious pathology is the appearance of multinucleate giant cells, ballooning degeneration of epithelial cells, focal necrosis, eosinophilic intraneural inclusion bodies, and an inflammatory response. The virus can spread intraneuronally, interneuronally, or through supporting cellular networks of axons or nerves. Spread of virus can also be through cell-to-cell transfer, a pathway that inhibits the effects of circulating antibody.

In the latent infection, the virus has been demonstrated in the sacral region (S2–S3). Although infection does not result in the death of the infected neuron, the effects of the virus on the host cell are not understood. It is known that there can be several copies of the viral genome in each infected host cell, and these are found in a circular form. Because latent infections do not require the synthesis of viral proteins, most antiviral drugs do not eradicate the virus when in its latent state.

Reactivation of latent virus from ganglionic cells with subsequent release of infectious virions accounts for most recurrent genital infections. The mechanisms associated with the reactivation are not yet known, but several precipitating factors, such as exposure to ultraviolet radiation, fever, and trauma, are known to initiate reactivation.

Over 70% of first episodes of genital herpes in the United States are caused by HSV-2. However, 90% of patients who test positive for HSV-2 antibody have never had a clinical genital episode, and in many cases the first episode does not occur for years after the initial exposure and infection.

Herpes infections can be primary, recurrent, or neonatal.

Primary genital herpes. Relatively few people develop clinically evident primary genital herpes. For those that do, the incubation time from sexual contact to the onset of lesions is about five days. The lesions begin as small erythematous papules that form vesicles and then pustules on the mucosal tissues (**Figure 23.16**). Within three to five days, these pustular lesions break to form painful coalesced ulcers. Some of these ulcers may crust over before healing, but they all eventually do so without scarring.

In primary genital herpes, the lesions are usually multiple (about 20), bilateral, and extensive. The urethra and cervix can also be involved, and bilateral enlarged and tender lymph nodes in the groin may persist for weeks or even months. About one-third of patients with primary genital herpes show systemic symptoms such as fever, malaise, myalgia, and, in some cases, aseptic meningitis. First episodes of primary genital herpes last an average of 12 days.

Recurrent genital herpes. In contrast to primary genital herpes, *recurrent genital herpes* has a shorter duration, is usually localized in the genital region and usually has symptoms, such as a burning or prickly sensation in the pelvic area, that occur 12–24 hours before the appearance of grouped vesicular lesions in the external genitalia. There is accompanying pain and itching that can last four to five days, but the lesions usually disappear in two to five days. At least 80% of patients with primary genital HSV-2 infection develop recurrent episodes within 12 months, and in these patients the median number of recurrences is four to five per year. These are not evenly spaced in time, however; sometimes flare-ups repeatedly occur for several months and are followed by long periods with no symptoms. Recurrent viral shedding from the genital tract may occur without clinically evident disease.

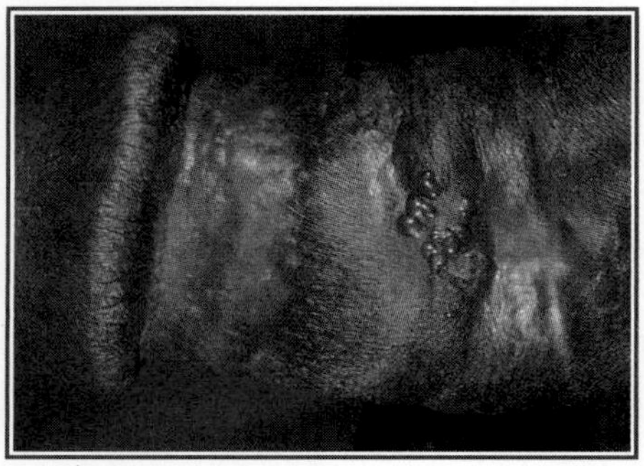

Figure 23.16 Vesicular lesions associated with genital herpes.

Neonatal herpes infection. Herpes infection in newborn infants usually results from transmission of the virus during delivery as the infant is exposed to infected maternal genital secretions. Most cases of severe neonatal herpes infection are associated with a woman who has a primary infection at or near the time of delivery. This results in an intense viral exposure to the infant during the birth. The incidence rate of neonatal herpes in the United States is approximately 1 in every 2500 live births; this infection is very serious, with an overall mortality rate of approximately 60%. Even those infants that survive will experience abnormal nervous system function and can show disseminated vesicular lesions with widespread internal organ involvement, necrosis of the liver and adrenal glands, listlessness, or seizures.

Treatment of genital herpes — Several antiviral drugs inhibit HSV, but the most effective and most commonly used is the nucleoside analog acyclovir. This drug inhibits the function of viral DNA polymerase and significantly decreases the duration of a primary infection. If taken daily, it can also suppress recurrent infections. Resistant HSV virions have been recovered from immunocompromised patients who have persistent lesions, and in these cases the drug foscarnet has been effective. In 1996, the FDA approved valacyclovir and famciclovir for the treatment of recurrent genital herpes.

Prevention can be accomplished by avoiding contact with infected individuals who are expressing lesions. Although this strategy can limit the spread of the disease, it is important to remember that virus is still being shed in asymptomatic individuals and can be transmitted not only via urethral and cervical secretions but also via saliva.

Human Papillomavirus

Papillomaviruses are small, non-enveloped, double-stranded DNA viruses with an icosahedral symmetry. They cause papillomas (which are benign tumors) or warts in a wide range of higher vertebrates, and the infections are species specific. In some cases, tumors caused by these viruses can be malignant. Because papillomavirus has not been grown in culture, most of what we know about them comes from molecular studies. There is wide genetic diversity among papillomaviruses that infect humans — human papillomaviruses (HPVs) —and this diversity is indicated by using numbers to identify different genotypes. Currently more than 70 genotypes of HPV have been identified, and some are associated with specific lesions.

HPVs have been identified in a variety of genital hyperplastic epithelial lesions, including cervical, vulvar, and penile warts. They are also associated with premalignant and malignant cervical cancer. Twelve of the HPV genotypes have been identified in human genital lesions, but many of the other genotypes of HPV may be associated with "silent" infections. It is possible to be infected with more than one genotype of HPV simultaneously.

The incidence of HPV infection is rising, and today between 20% and 60% of women in the United States are infected with one or another of the genotypes. HPV types 6 and 11 are associated with benign genital warts in males and females, and types 16, 18, 31, and 45 cause warty lesions of the vulva, cervix, and penis. Infections with any of the latter four, especially type 16, may progress to malignancy, and the genomes of these four types have been found in cases of dysplastic uterine cervical cells and in malignant lesions. (**Dysplasia** is abnormal tissue development.)

Fast Fact

HPV infection is now considered to be a cause of the majority of carcinomas of the cervix, and HPV DNA has been found in 95% of cervical carcinoma specimens.

Pathogenesis of HPV infection — Papillomaviruses have a predilection for infection at the junction of squamous and columnar epithelium (for instance, in the cervix). The mechanism of malignant transformation is not understood and is difficult to study because HPV is difficult to grow in culture. However, it has been shown that part of the viral genome can be found integrated into the host cell chromosome, and this integration does not seem to be site-specific. Host cells normally produce a protein that inhibits the expression of papillomavirus-transforming genes, but the virus seems to inactivate that protein.

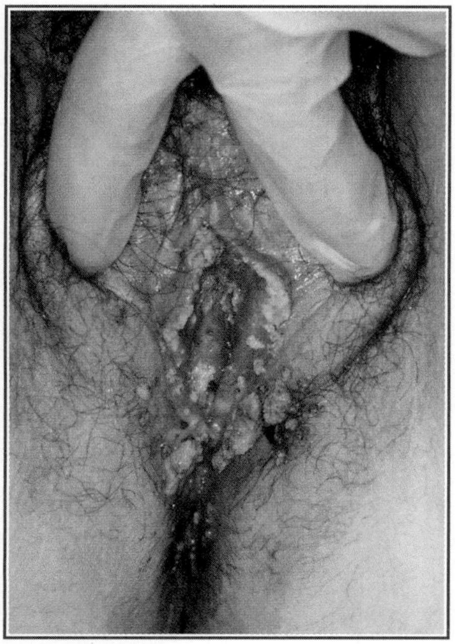

Figure 23.17 Genital warts caused by human papillomavirus occurring on the vulva. It is estimated that there are nearly 1 million new cases of genital warts in the United States each year.

External genital HPV infection presents in the form of genital warts (**Figure 23.17**). This is most often caused by genotypes 6 or 11, and these lesions may grow to a cauliflower-like appearance during pregnancy or immunosuppression. Genital HPV infections are usually benign, and many lesions reverse spontaneously. However, they may become dysplastic and proceed to severe dysplasia or carcinoma. Paradoxically, the most common malignant lesion is caused by HPV type 16, but the lesions of this genotype are also those that reverse most quickly.

Treatment of HPV infection — Currently, the only treatments for HPV infection are surgical, cytotoxic drugs, and cryotherapy with liquid nitrogen. The most commonly used cytotoxic drugs are podophyllin, podophyllotoxin, 5-fluoroacetic acid, and trichloroacetic acid. Recurrence of the infection is common after cessation of the treatment, because the virus is able to survive in the basal layers of the epithelium.

Fast Fact

Recently, a vaccine for HPV has become available, and vaccination of young girls has been recommended in the United States.

FUNGAL INFECTIONS OF THE GENITOURINARY SYSTEM

Because it is so open to the external environment, the genitourinary tract can be exposed to many fungal organisms. However, as we have discussed before, fungal infections are usually opportunistic and caused by part of the normal microbial flora. This is true of one of the most prominent fungal infections, vaginal candidiasis, caused by *Candida albicans*.

Vaginal Candidiasis (*Candida albicans*)

Candida albicans is part of the normal microbial flora of the oropharyngeal and gastrointestinal regions in both males and females, and the genital tract of females. It can grow in multiple morphological forms but is most often seen as yeast. The general name for infections caused by *C. albicans* is **candidiasis**, and the infection can be either local or systemic. In vaginal candidiasis, the main symptoms are itching and a thick white discharge. Infections with *Candida* are normally endogenous except in cases of direct mucosal contact with a person expressing lesions, such as would occur during sexual intercourse. Indwelling catheters and overuse of antibiotics provide additional opportunities for *Candida* to become opportunistically pathogenic.

Pathogenesis of *C. albicans* infection — Because this organism is a part of the normal flora found on mucosal surfaces, it must undergo changes to become pathogenic. One of the changes is the appearance of hyphae, which are seen when *Candida* invades tissues. It is not known what causes the hyphae to form, but their appearance is accompanied by the production of several factors that permit strong attachment of the organism to human epithelial cells. This attachment involves the usurping of host cell enzymes.

The hyphae excrete proteases and phospholipases that digest epithelial cells and further facilitate the invasion of tissues. *Candida* can also protect itself by binding to the C3 fragment of complement in such a way that the complement is not available to opsonize the yeast. In point of fact, any factors that inhibit T cell function or compromise the host's immune response enhance the ability of *Candida* to invade tissue. Infection of the vagina by *C. albicans* produces a thick discharge with the consistency of cottage cheese, and accompanying itching of the vulva. Most women will have at least one episode of vaginal candidiasis in their lifetime. In addition, a small percentage of women may become chronically infected and experience recurrent symptoms. *Candida* can also infect the urinary tract, leading to cystitis, nephritis, abscesses, and expanding fungus ball lesions in the renal pelvis.

Treatment of *C. albicans* infection — This fungus is usually susceptible to azole drugs, amphotericin B, nystatin, and flucytosine. In many cases, the lesions seen in this infection will resolve spontaneously on the elimination of predisposing conditions (mainly the removal of a catheter).

Keep in Mind

- There are several viral STDs, with AIDS being the most important one today.
- Humans seem to be the only reservoir for herpes simplex virus type 2, and many people infected with genital herpes are asymptomatic but still infectious.
- Although drug therapy can inhibit the severity or duration of an infection, there is no cure for genital herpes infection.
- Human papillomavirus (HPV) can cause genital warts and also is a cause of cervical cancer.
- The incidence of HPV infection is rising, and it is estimated that as many as 60% of women in the United States are infected with one or another of the genotypes of HPV.
- A vaccine is now available for the prevention of HPV infection.
- The most prominent form of fungal infection of the genitourinary tract is vaginal candidiasis caused by *Candida albicans*.
- Indwelling catheters and the overuse of antibiotics provide additional opportunities for opportunistic *Candida* infection.

SUMMARY

- The presence of pathogens or inflammatory cells in urine is an indication of a urinary tract infection.
- Urinary tract infections that occur in hospitals are usually related to indwelling catheters.
- Infections of the urethra are called urethritis; infections of the bladder are called cystitis; and infections of the kidneys are referred to as nephritis.
- Urinary tract infections are seen in women more than men because of the difference in the lengths of the urethra.
- The majority of bacterial infections seen in the reproductive system are sexually transmitted diseases (STDs).
- STDs can be caused by bacteria, viruses, fungi, and protozoan parasites.

- The most common STDs are non-gonococcal urethritis, which is caused by *Chlamydia trachomatis*, and gonorrhea, caused by *Neisseria gonorrhoeae*.
- There are several viral STDs, including those caused by HIV and herpes simplex virus type 2.
- Humans are the only reservoir for herpes simplex virus type 2, and many people infected with this virus are asymptomatic.
- The most common form of fungal infection is candidiasis caused by indwelling catheters coupled with the overuse of antibiotics, bringing about a superinfection.

This chapter completes the discussion of the three systems that are the most open portals of entry for microorganisms. As we have seen, the genitourinary tract, like the respiratory and digestive tracts, is susceptible to a variety of infections. We saw that in the urinary system there can be an ascending infection going from the urethra to the bladder, to the prostate in males, and on up to the kidneys. We also saw that most infections of the reproductive system were classified as sexually transmitted diseases. The same ascending sequence seen in urinary tract infection can be seen with STDs, which can go from the urethra to the reproductive organs.

SELF EVALUATION AND CHAPTER CONFIDENCE

Multiple Choice

Answers are given in the back of the book and help can be found in the student resources at:
www.garlandscience.com/micro

1. Cystitis is most often caused by
   A. Gram-positive rods
   B. Gram-negative rods
   C. Gram-negative cocci
   D. Gram-positive cocci
   E. All of the above

2. Nephritis may result from
   A. Systemic infections
   B. Cystitis
   C. Urethritis
   D. All of the above

3. Which of the following is treated with penicillin?
   A. Candidiasis
   B. Syphilis
   C. Genital warts
   D. Lymphogranuloma venereum

4. Lymphogranuloma venereum is caused by
   A. *Candida albicans*
   B. *Neisseria gonorrhoeae*
   C. *Chlamydia trachomatis*
   D. *Treponema pallidum*

5. Which of the following recurs at the site of the infection?
   A. Gonorrhea
   B. Genital herpes
   C. Syphilis
   D. Chancroid

6. Which of the following is caused by an opportunistic infection?
   A. Gonorrhea
   B. Candidiasis
   C. Syphilis
   D. Genital herpes

7. Most nosocomial urinary tract infections are caused by
   A. *E. coli*
   B. *Enterococcus*
   C. *Pseudomonas*
   D. *Staphylococcus*

8. Recurring vesicles are symptoms of
   A. Genital herpes
   B. Candidiasis
   C. Lymphogranuloma venereum
   D. Syphilis

9. The most common STD in the United States is
   A. Gonorrhea
   B. Syphilis
   C. Non-gonococcal urethritis
   D. Herpes

10. Which of the following is caused by *Chlamydia*?
    A. Genital herpes
    B. Candidiasis
    C. Syphilis
    D. Non-gonococcal urethritis

DEPTH OF UNDERSTANDING

Questions listed here require you to bring together the concepts you have learned in this chapter into a discussion format. This helps you to increase your depth of understanding of the material you have learned. Help can be found in the student resources at:
www.garlandscience.com/micro

1. Describe how urinary tract infections can begin as urethritis and develop into nephritis.

2. Discuss the phases of a syphilis infection and the symptoms seen as the infection progresses.

CLINICAL CORNER

Help can be found in the student resources at:
www.garlandscience.com/micro

1. Your best friend comes to you with great news; she has found a new boyfriend. You can see that she is really excited and the romance continues for several dates and then stops. When you ask her what happened, she breaks down and tells you that when she tried to take the relationship to the next level her boyfriend told her that he had contracted genital herpes several months ago from a former girlfriend. Although he assured her that they could still be intimate by his wearing a condom, she is still very fearful that she will be infected if they have intercourse. She really cares for him and wants to continue the relationship and asks your advice.
   A. What do you tell her about the possible risks of becoming infected?
   B. Will his wearing a condom be able to protect her?
   C. Can using the prescription medicines available for genital herpes help them be together?
2. Your patient was admitted to the hospital for elective cosmetic surgery. After the surgery, which went very well, it was discovered that she had a urinary tract infection with *Pseudomonas*. She was placed on broad-spectrum antibiotics for two days, during which she developed vulvovaginal candidiasis. By day four she showed signs of nephritis and was put on dialysis. Obviously her family is distressed and does not understand how she has gone from having a simple "tummy tuck" to dialysis.
   A. How can you explain what has happened?
   B. What is the significance of the candidiasis?
   C. What are the major concerns you have regarding this patient?

Infections of the Central Nervous System

Chapter 24

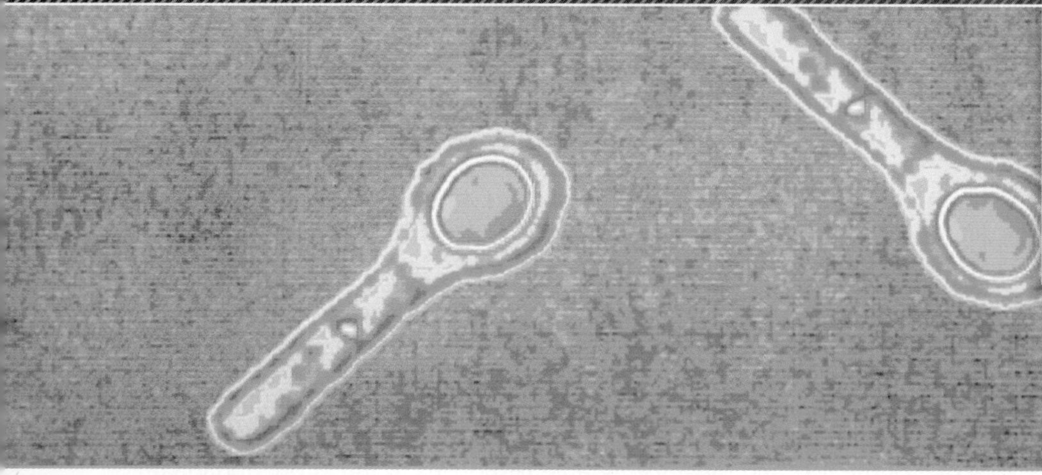

Why Is This Important?

Because the central nervous system controls the function of the body, infections that affect the central nervous system can be catastrophic and potentially lethal for the host.

OVERVIEW

In this chapter, we look at some of the infections seen in the central nervous system. We will find that in some cases this system is the main target of the infection, whereas in others the effect on the central nervous system is secondary. As in all the other systems of the body, infections of the central nervous system can be caused by bacteria, viruses, fungi, and parasites. In addition, in the central nervous system we will see infections caused by infectious proteins called prions. Once again, we must think about the anatomy of the nervous system to put these infection agents into perspective.

We will divide our discussions into the following topics:

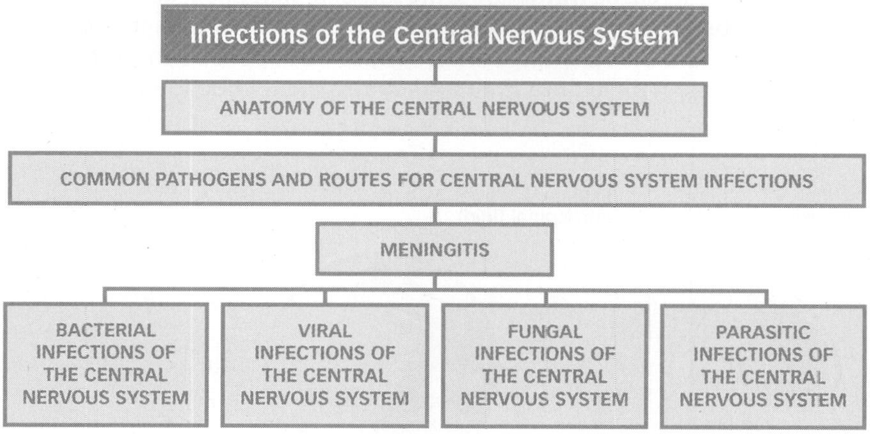

ANATOMY OF THE CENTRAL NERVOUS SYSTEM

The central nervous system (CNS) is divided into two major parts, the brain and the spinal cord. Both of these structures reside in the dorsal body cavity and are protected by the skull and vertebral bones. There is a good reason for this protection because the CNS is perhaps the most critical part of the body. It controls everything else, and when it is compromised in some way, the results can be devastating. Both the brain and the spinal cord are surrounded by three layers of connective tissue called the *meninges*. The *dura mater*, which is the most superficial meningeal layer, is the most durable, with the consistency of wax paper. The middle layer is called the *arachnoid mater* because it has spidery extensions that connect it to the deepest layer, the *pia mater*.

What Do I Need to Know?

To get the most out of this chapter, please review the following terms from your previous courses in biology, anatomy, physiology, or chemistry or in previous chapters of this book as indicated in parentheses: abscess, acute infection (**5**), ameba, anterior horn cell, astrocyte, autonomic nervous system, bacteremia (**5**), complement (**15**), cribriform plate, enterovirus (**12**), etiologic agent, exudate, granulomatous (**21**), gray matter of central nervous system, hematogenous route, mycosis (**14**), Negri body (**5**), opsonization (**15**), persistent infection (**7**), presynaptic/postsynaptic, prion (**8**), shunt, slow infection (**12**), sudden infant death syndrome, titer (**16**), trigeminal nerve, trophozoite (**14**), vacuole (**4**), viremia (**5**).

Between the arachnoid mater and pia mater is a space referred to as the *subarachnoid space*, and it is here that *cerebrospinal fluid* is found (**Figure 24.1**). This fluid, which is produced by the choroid plexus and bathes the brain and spinal chord, is the intermediary for nutrients required by nervous tissue and is also a vehicle for carrying cellular waste away from the CNS. The brain and spinal chord are protected from the rest of the body by the *blood–brain barrier*, a structural arrangement in which the capillary endothelial cells rest on basement membranes that do not allow the passage of certain materials from the blood to the brain. This barrier protects the body against infectious disease by preventing pathogens that might find their way into the blood from infecting nervous tissue of the brain. As we will see, some pathogens can pass through the blood–brain barrier.

Because there is no room for swelling of the nervous tissue enclosed by the skull, any pathology that occurs in the brain and results in increased intracranial pressure can have disastrous and potentially lethal consequences, all the result of compression, herniation, and brain cell death. In addition, as we learned throughout our discussions of infectious disease, inflammation is one of the first and most formidable responses to infection, and inflammation always causes swelling. Therefore, an inflammatory response in the subarachnoid space, with the requisite release of cytokines, results in loosening of the tight junctions between vascular endothelial cells. The loosened junctions allow albumin into the cerebrospinal fluid and the result is the form of swelling known as **vasogenic edema**.

Swelling can also result from the toxic substances produced by bacteria and from neutrophil invasion, both of which result in oxygen and nutrient deficiencies. This produces another kind of swelling, the kind known as **cytotoxic edema** (**Figure 24.2**). Infection can also affect the biochemical CNS reactions required for proper brain function through acidosis, hypoxia, and the destruction of neurons.

The effects of increased intracranial pressure, biochemical abnormalities, and neural tissue necrosis can be profound and sometimes irreversible. Furthermore, the blood–brain barrier, which is supposed to protect the brain, can make it difficult to treat and control CNS infections.

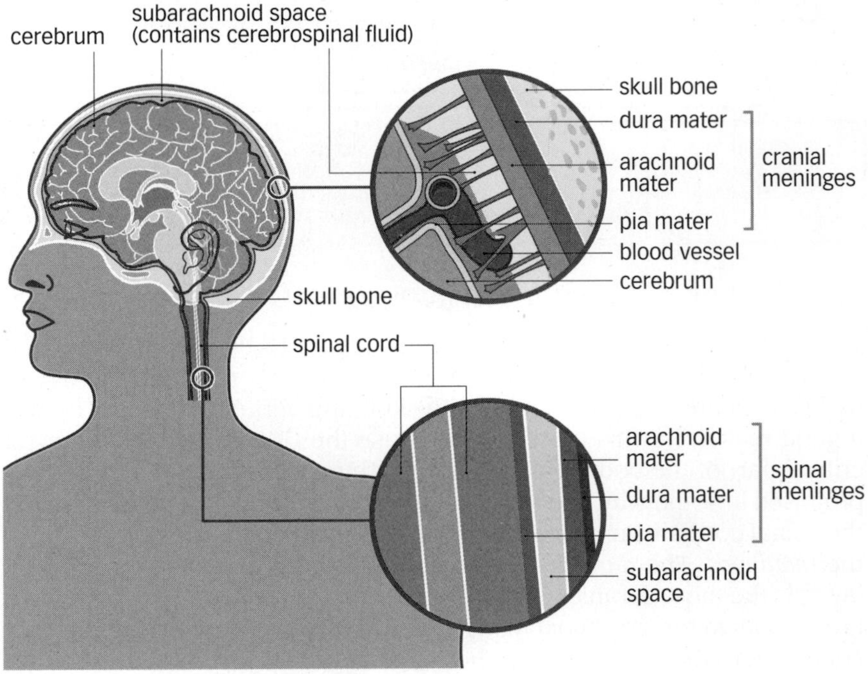

Figure 24.1 The meninges and the cerebrospinal fluid. Three layers of meninges surround the brain and the spinal cord: the dura mater, which is the most superficial; the arachnoid mater, which is the middle layer; and the pia mater, which is the deepest layer. Between the arachnoid and pia maters there is the subarachnoid space, through which the cerebrospinal fluid circulates. The central nervous system is vulnerable to contamination by any bloodborne pathogens that can penetrate the blood–brain barrier.

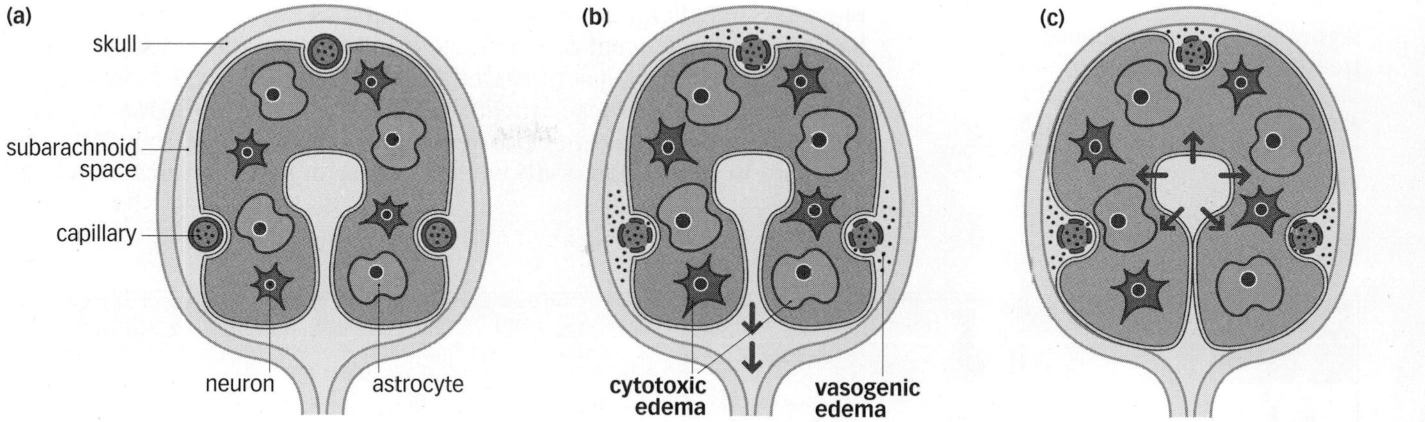

Figure 24.2 The effects of edema (swelling) on the brain Panel **a**: The normal brain. Panel **b**: Vasogenic and cytotoxic edema caused by inflammation in the meninges. Panel **c**: The effects of these types of edema include swelling of the brain, and oxygen and nutrient deficiencies.

COMMON PATHOGENS AND ROUTES FOR CENTRAL NERVOUS SYSTEM INFECTIONS

Organisms reach the brain and spinal cord in a variety of ways, with blood being one of the most obvious. Because there is extensive vascularization in the CNS, any organism that circulates in the blood and can enter the cerebrospinal fluid can cause **meningitis** (infection of the meninges). In addition, the skull contains various *sinuses* and *mastoid air spaces* that are separated from the brain by the bone of the skull. Infections in these areas can eventually cause the erosion of the bone, and then organisms can enter the brain and cause abscesses to form (**Figure 24.3**).

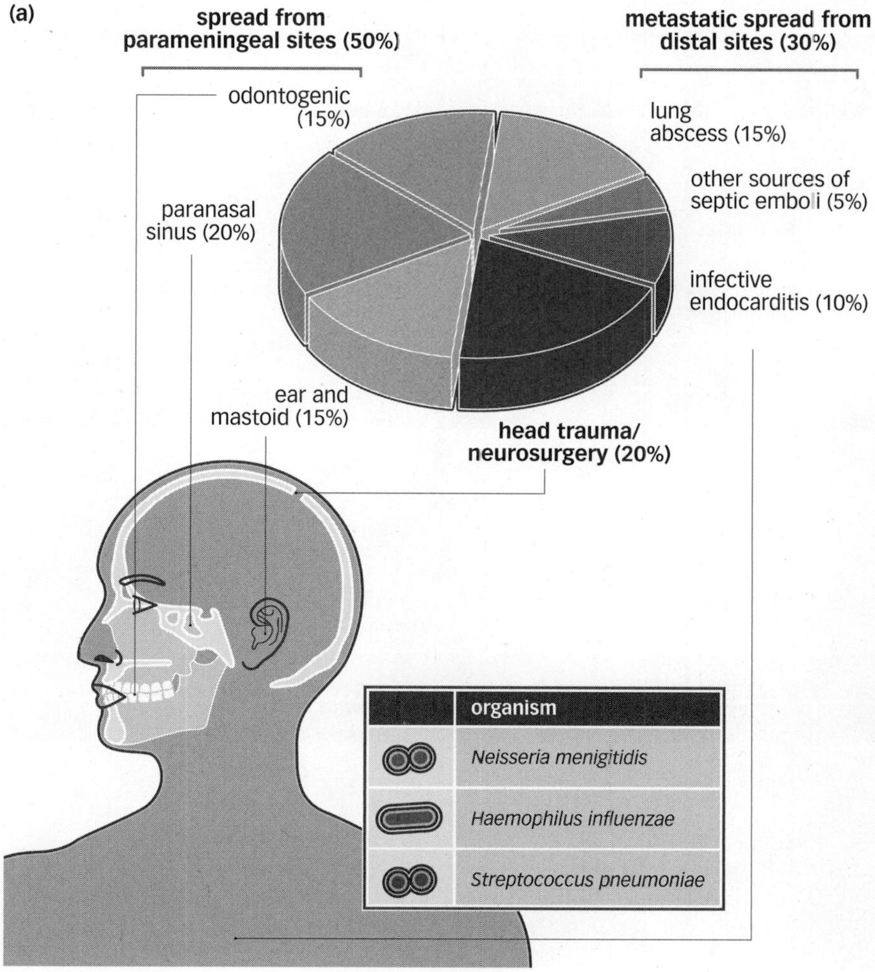

Figure 24.3 Meningitis and brain abscesses. Panel **a**: The main routes by which organisms reach the central nervous system to cause meningitis and brain abscesses. Panel **b**: Purulent basilar meningitis due to infection with *Streptococcus pneumoniae*.

(b)

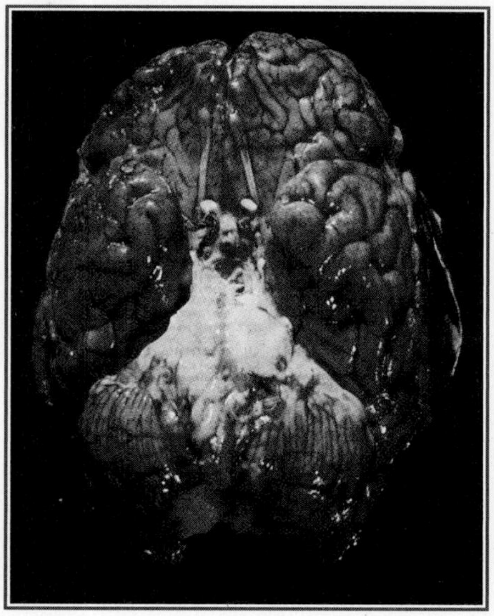

Most CNS infections result from passage across the blood–brain barrier of bacteria or viruses that are circulating in the blood as a result of some non-CNS infection. A list of bacteria that infect the CNS is shown in **Figure 24.4**. Some of these, such as *Neisseria meningitidis*, *Haemophilus influenzae* type b and *Streptococcus pneumoniae*, can be found as normal flora in the human body, whereas other organisms, such as *Listeria*

Figure 24.4 Brain infections and treatments. A description of some of the bacteria that cause meningitis, brain abscesses, shunt infections, and toxin-mediated infections in the central nervous system.

	organism		treatment
bacterial meningitis			
	Streptococcus pneumoniae	acquired member of the normal flora	benzylpenicillin, cefotaxime, chloramphenicol
	Haemophilus influenzae	acquired member of the normal flora	cefotaxime, chloramphenicol
	Neisseria meningitidis	acquired member of the normal flora	benzylpenicillin, cefotaxime, chloramphenicol
	Escherichia coli	neonatal meningitis acquired from maternal vaginal flora	cefotaxime, gentamicin, chloramphenicol, meropenem
	Streptococcus agalactiae (group B)	neonatal meningitis acquired from maternal vaginal flora	benzylpenicillin + gentamicin, cefotaxime
	Listeria monocytogenes	meningitis in newborns and the immunocompromised	benzylpenicillin + gentamicin, cefotaxime
	Staphylococcus aureus	a member of the normal flora	floxacillin + rifampicin
brain abscess			
	streptococci	e.g., *Streptococcus anginosus*	benzylpenicillin, chloramphenicol, clindamycin
	coliforms	fecal flora	cefotaxime, chloramphenicol
	Staphylococcus aureus	a member of the normal flora	floxacillin, chloramphenicol, clindamycin
ventriculo-peritoneal shunt infections			
	coagulase-negative staphylococci	skin commensals	vancomycin + rifampicin
	Staphylococcus aureus	normal flora	floxacillin, vancomycin + rifampicin
	Pseudomonas aeruginosa	a member of the flora of the hospitalized patient	ceftazidime, gentamycin, meropenem
toxin and immune-mediated disorders			
	Clostridium botulinum	contaminated food products	ventilatory support
	Clostridium tetani	contaminated wounds, intravenous drug use	ventilatory support, surgery, wound debridement, benzylpenicillin
	Campylobacter spp.	food poisoning, gastroenteritis	ventilatory support

monocytogenes, are ingested. Group B *Streptococcus* can also be acquired during the birthing process. It is important to note that many CNS infections occur through the contamination of shunts, in particular with coagulase-negative *Staphylococcus* species.

Some of the viruses that can infect the nervous system are shown in **Table 24.1**.

The initial source of a CNS infection can be either occult (infection of mononuclear phagocytic system cells being one example) or overt, through complications of, for example, pneumonia, pharyngitis, skin abscesses, or **infectious endocarditis** (which is an infection in the heart). In some cases the infection site is either close to or in direct contact with the CNS. For example, infections in the middle ear, mastoiditis, or sinusitis can easily make the short trip into the CNS under the proper set of circumstances.

Anatomical defects in the structures that encase the CNS can also permit infection. Surgical, traumatic, or congenital developmental abnormalities can leave openings into the CNS that are used by pathogens. For example, a fracture at the base of the skull is in close proximity to the body's respiratory system, which contains many organisms and presents them with ready access to the CNS. It is interesting to note that the least common route to the CNS is by intraneural pathways. However, there are exceptions to this statement. The rabies virus, for example, uses the peripheral sensory nerves to gain access to the CNS, and herpesvirus uses the trigeminal nerve to gain access to the CNS.

Although brain abscesses are relatively rare, they present a special problem. They can be found in the *subdural space* (between the dura mater and the arachnoid mater), in the *epidural space* (the region superficial to the dura mater), or directly in the brain tissue and are commonly formed by bacteria or fungi from a distant site. Brain abscesses can also result from extensions of pathogens located at the site of mastoiditis or sinusitis or from complications of surgery.

Table 24.1 Primary Acute Viral Infections of the CNS

Agent	Major Age Group Affected	Seasonal Predominance
Enteroviruses	Infants and children	Summer–fall
Mumps	Children	Winter–spring
Herpes simplex		
Type 1	Adults	None
Type 2	Newborns, young children	None
Arboviruses		
Western equine encephalitis	Infants and children	Summer–fall
Eastern equine encephalitis	Infants and children	Summer–fall
West Nile encephalitis	Adults	Summer–fall
Rabies	All ages	Summer–fall
Measles	Infants and children	Spring
Varicella-zoster	Infants and children	Spring
Epstein–Barr virus	Children and young adults	None
HIV and cytomegalovirus	All ages	Variable

Clinical Features of CNS Infections

There are several terms used for infections of the CNS. Some of these infections are forms of meningitis; others affect the nervous tissue directly.

- *Purulent meningitis* is an infection of the meninges marked by acute inflammatory exudate. It is usually caused by a bacterial infection and in most cases has an acute onset with progression characterized by fever, irritability, and various degrees of neurological dysfunction.

- *Chronic meningitis* has a more insidious onset than the purulent form, with signs and symptoms developing over weeks. Chronic meningitis is usually caused by *Mycobacterium tuberculosis*, by fungi, or by protozoan parasites (**Table 24.2**).

- *Aseptic meningitis* is associated with meningeal inflammation evidenced by increased numbers of lymphocytes and mononuclear cells in the cerebrospinal fluid, with an absence of culturable bacteria or fungi. Although aseptic meningitis is most often attributed to viral infection and is usually a self-limiting problem, it can also be seen in syphilis infections and other spirochete infections. The primary site of inflammation is in the meninges, with no clinical involvement of neural tissue. Symptoms include fever, headache, stiff neck and back, nausea, and vomiting.

- **Encephalitis** is a term used to describe patients who have no symptoms of aseptic meningitis but who show obvious signs of CNS dysfunction, such as seizures, paralysis, or defective mental faculties. In this case, the problem is associated not with the meninges but with the actual nervous tissue.

- **Poliomyelitis** targets and destroys the cells associated with the anterior horn of the spinal cord and brain stem. Infection causes weakness or paralysis of muscle groups and occasionally respiratory difficulties. The hallmark of this CNS infection is asymmetrical paralysis.

- Acute **polyneuritis** is an inflammatory infection of the peripheral nervous system characterized by symmetrical paralysis of the muscles. This infection can be associated with diphtheria toxin, enteric pathogens, cytomegalovirus, and Epstein–Barr virus.

Common Pathogens of the CNS

Acute purulent meningitis is usually caused by one of three bacteria — *H. influenzae*, *N. meningitidis*, and *S. pneumoniae* — but in recent years the vaccine used to protect against *H. influenzae* has reduced the number of these infections. In newborns, group B streptococci and *Escherichia coli* are frequently involved in meningitis acquired through the birthing process.

Table 24.2 Causes of Chronic Meningitis

Form of Chronic Meningitis	Agents
Chronic granulomatous infection	*Mycobacterium tuberculosis, Coccidioides immitis, Cryptococcus neoformans, Histoplasma capsulatum*
Protozoan parasitic infection	*Toxoplasma gondii*
	Trypanosoma
	Acanthamoeba species
Nematode parasitic infection	*Trichinella spiralis*
Other parasitic infections	*Leptospira* species, *Treponema pallidum, Borrelia burgdorferi*

The most common viral causes of acute CNS infections are enteroviruses, herpes simplex virus, HIV, Epstein–Barr virus, and several arthropod-borne viruses. Viral infections of the CNS manifest as aseptic meningitis, encephalitis, or poliomyelitis. The viral CNS infections referred to as *subacute sclerosing panencephalitis* usually result from measles or rubella infections. Still other CNS infections are caused by *M. tuberculosis*, as noted in **Table 24.2**, and deep fungal mycoses are caused by *Cryptococcus neoformans* and *Coccidioides immitis*.

General Treatment of CNS Infections

Bacterial and fungal infections of the CNS require prompt and aggressive treatment, and treatment periods vary depending on the type of infection. For uncomplicated bacterial meningitis the treatments can last from 10 days to 12 months, and even longer if the infection is caused by *M. tuberculosis*. For fungal infections, treatment can last for years. In contrast, treatment of viral infections of the CNS is mostly supportive.

Now let's look at some of the more common CNS infections. As in previous chapters, we will separate our discussions into bacterial, viral, fungal, and parasitic infections, and we will also include a review of prion diseases, which we learned about in Chapter 8.

Keep in Mind

- The CNS is well protected structurally and employs a blood–brain barrier to prevent infections.
- Organisms that move from the blood to the cerebrospinal fluid can cause meningitis.
- Most CNS infections result from bacteremia or viremia from distal infections.
- In some cases, infection comes from locations close to or in direct contact with the CNS, such as the middle ear or the sinuses.
- Bacterial and fungal CNS infections require immediate and aggressive treatment. Viral CNS infections are more difficult to treat, and therapy includes mostly supportive care.

MENINGITIS

Before we look at bacterial and viral infections of the CNS, let's take a look at a broad category of infections referred to as meningitis. Meningitis is an infection of the fluid surrounding the spinal chord and brain, usually caused by either a viral or a bacterial infection. It is important to know which type of infection it is, because the severity of the illness as well as the treatment will differ.

Bacterial meningitis can be quite severe and ultimately cause brain damage, hearing loss, or learning disability. It is also important to know what type of bacterium is causing the meningitis, because antibiotic therapy can prevent the spread of the infection. Before 1990, *H. influenzae* type b (Hib) was the leading cause of bacterial meningitis, but vaccines now given to all children as part of their routine immunizations have reduced the occurrence of *H. influenzae* meningitis. Today, *S. pneumoniae* and *N. meningitidis* are the leading cause of bacterial meningitis.

Some forms of bacterial meningitis are contagious, with bacteria being spread through exchange of respiratory and throat secretions. However, meningitis is not as contagious as the common cold or flu. Meningitis caused by *N. meningitidis* can spread to other people who have had close or prolonged contact with infected individuals. This is why cases of meningitis are causes of concern for daycare centers and schools.

Viral meningitis (also known as aseptic meningitis) is generally less severe and usually resolves without treatment. It is caused by infection with one of several types of virus. About 90% of cases are caused by enterovirus, although herpesvirus and mumps virus can also cause the disease.

Although there are 25,000–50,000 hospitalizations due to viral meningitis in the United States each year, viral meningitis is rarely fatal if the patient has a competent immune system. The symptoms last 7–10 days and the patient usually recovers completely.

Symptoms of Meningitis

Symptoms of meningitis include high fever, stiff neck, and headache. These symptoms can develop over several hours or may take one to two days. Other accompanying symptoms can include nausea, vomiting, sensitivity to bright light, confusion, and sleepiness. As the disease progresses, patients may develop seizures. It is important to note that symptoms in newborns and infants may be difficult to detect.

Diagnosis and Treatment of Meningitis

If symptoms of meningitis occur, the patient should see a doctor immediately. The diagnosis is usually made by growing bacteria from a sample of spinal fluid. The identification of the specific type of bacterium is then made so that effective antibiotic therapy can begin. If antibiotic therapy is started early, it can limit the risk of death to 15%, although this may be higher in the elderly.

Viral meningitis is also diagnosed using spinal fluid but no antibiotic therapy is necessary. Patients are given bed rest, plenty of fluids and medicines to relieve fever and headache.

BACTERIAL INFECTIONS OF THE CENTRAL NERVOUS SYSTEM

As we mentioned above, many bacteria cause the family of CNS infections known as meningitis, characterized by either growth of the organisms or a powerful inflammatory response directed against the infection. However, there are two bacterial infections, tetanus and botulism, that affect the CNS in a different way. These infections occur through the production of exotoxins that have an affinity for CNS tissue. Recall from Chapter 5 that exotoxins are extremely toxic proteins that are produced by Gram-positive bacteria and usually have a two-chain configuration, with one chain binding to the target cell and the other chain causing the damage. It is also important to remember that once an exotoxin is produced, it no longer matters whether the organism that produced it survives. Therefore, antibiotic therapy against exotoxin-producing organisms will be ineffective once the exotoxin has been produced. In addition, remember that the binding of exotoxins is irreversible.

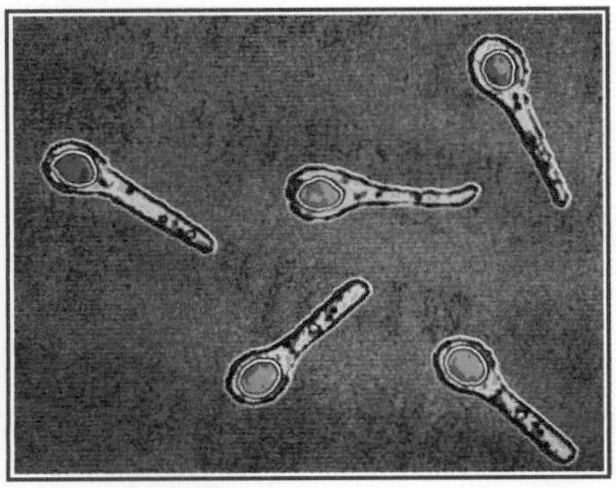

Figure 24.5 False color transmission electron micrograph of *Clostridium tetani* showing the drumstick shape of the spore.

Tetanus (*Clostridium tetani*)

Tetanus is caused by the bacterium *Clostridium tetani*, which is a Gram-positive, anaerobic, spore-forming rod. It yields a terminal spore that gives it a drumstick appearance (**Figure 24.5**). This organism is a strict anaerobe and cannot survive in the presence of oxygen. It is commonly found in the soil, particularly soil that has been contaminated by manure, and its spores can survive in the soil for years. These spores, which are very resistant to disinfectants and can withstand boiling, are introduced into wounds contaminated with soil.

The toxin produced by *C. tetani* is a *neurogenic* toxin, meaning that it has an affinity for and targets nervous tissue. Called either **tetanospasmin** or **tetanus toxin**, it acts by enzymatically degrading proteins required for normal physiology in nervous tissue.

Pathogenesis of tetanus — Tetanus spores require areas of low oxygen content if they are to germinate, and the area of necrosis around a tissue injury is perfect for the initiation of this process. The spores germinate, and *Clostridium* begins to grow in that location. The bacteria do not cause any damage to the tissue where they reside; their role is to produce the neurogenic toxin, which then enters the presynaptic terminals of the lower motor neurons and from there they gain access to the CNS. In the spinal cord, the toxin acts at the level of the anterior horn cells and blocks postsynaptic inhibition of the spinal motor reflexes. This produces spasmodic contraction of the muscles, contractions that occur locally at first but then may extend up and down the spinal cord.

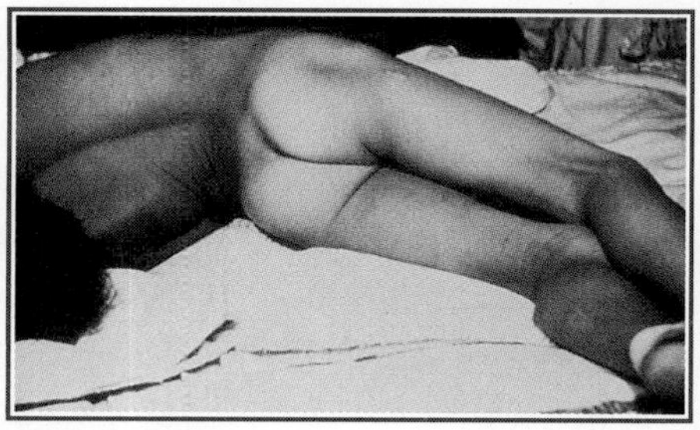

Figure 24.6 A patient with opisthotonos resulting from advanced tetanus. The characteristic sign is a curvature of the body that draws the heels toward the head.

The incubation period for *Clostridium* can vary between four days and several weeks. Generally, the shorter the incubation period, the more severe the infection will be. This neurogenic toxin is systemic for muscles, and the masseter muscle of the jaw is usually the first to be affected, which means the mouth cannot be opened. (This is why this condition is sometimes referred to as **lockjaw**.) Eventually the muscles responsible for respiration and swallowing can be compromised, and in severe cases the patient can also suffer from **opisthotonos**, a condition in which the head and heels move toward each other and the body bows out dramatically (**Figure 24.6**). Death can result from exhaustion and respiratory failure, and the mortality for untreated tetanus can vary from 15% to 60%. Several factors affect the mortality rate, including location of the lesion, incubation period, and the patient's age. Mortality rates are highest in infants and the elderly.

Treatment of tetanus — Because the toxin is the problem, once it is expressed, the death of the *C. tetani* organism becomes secondary. Therefore, neutralization of the toxin with large doses of human tetanus immunoglobulin (derived from volunteers who are hyper-immunized with the toxoid of this toxin) is important for initial therapy. In addition, supportive measures — including maintenance of a dark, quiet environment; sedation; and provision of an adequate airway for breathing — should be provided if necessary. Antibiotics are not effective once the toxin has been produced.

Botulism (*Clostridium botulinum*)

The etiologic agent of botulism is *Clostridium botulinum*, a Gram-positive, anaerobic, spore-forming rod. The spores are found in the soil as well as in the sediments of ponds and lakes worldwide. If they contaminate food under anaerobic conditions, these spores can convert to the vegetative state and begin to produce toxin. This toxin is among the most poisonous in the world, and contamination of food with botulinum toxin can occur without affecting the smell, taste, or color of the contaminated food. Botulism is commonly seen in cases of home canning in which the temperatures used in the process are not high enough to prevent contamination.

Pathogenesis of botulism — Botulism begins with cranial nerve palsy and develops into a descending symmetrical motor paralysis that may involve the respiratory muscles. There is no fever or inflammation and no obvious sign of infection, and the time course of the infection depends on the amount of toxin produced and on whether the toxin was ingested in a preformed state or was produced in the intestinal tract.

Foodborne botulism is classified as an intoxication (Chapter 22), not an infection. The toxin is ingested as a preformed component and is absorbed directly through the intestinal tract, reaching the neuromuscular junction target via the bloodstream. Once bound there, the toxin inhibits the release of acetylcholine, causing muscular paralysis. The symptoms observed depend on which nerves are bound by the toxin, and damage to the nerves after toxin has bound is permanent. Any recovery of function requires the formation of new synapses.

Foodborne botulism usually starts 12–36 hours after ingestion of the toxin. The first symptoms are nausea, dry mouth, and in some cases diarrhea. Symptoms of nervous system dysfunction start later and include blurred vision, pupillary dilation, and rapid eye movements. Symmetrical paralysis begins with ocular, laryngeal, and respiratory muscles and spreads to the trunk and extremities. The most serious complication is complete respiratory paralysis, with the mortality for botulism varying from 10% to 20%.

There are two other forms of botulism, the *infant form* and the *wound form*. Infant botulism occurs in infants between the ages of three weeks and eight months and is the most commonly diagnosed form of botulism. The organism is introduced either on weaning or through dietary supplements, particularly honey. It multiplies in the infant's colon, and the botulinum toxin is then absorbed into the blood. The symptoms of infant botulism include constipation, poor muscle tone, lethargy, and feeding problems. In severe cases, vision problems and paralysis can also occur. One of the difficulties in diagnosing infant botulism is that it mimics sudden infant death syndrome.

Wound botulism occurs very rarely and is usually seen in intravenous drug users. The symptoms are similar to those of food poisoning and usually begin with muscle weakness in the extremities used for injection.

Treatment of botulism — The single most important determinant is the availability of intensive support measures, in particular mechanical ventilation. If proper ventilation support is provided, mortality is less than 10%. Because botulism is caused by an exotoxin, antibiotic therapy is given only to patients with the wound form.

Keep in Mind

- Two of the most serious CNS bacterial infections, tetanus and botulism, can cause devastating infection.
- Both tetanus and botulism are caused by exotoxins produced by the pathogens.
- Tetanus toxin is a neurogenic toxin that blocks postsynaptic inhibition of the spinal motor reflexes, which leads to spasmodic contraction of muscles.
- Botulism toxin moves from the intestinal tract to the CNS via the blood and inhibits the release of acetylcholine, causing muscular paralysis.

VIRAL INFECTIONS OF THE CENTRAL NERVOUS SYSTEM

As with bacteria, viruses with an affinity for the CNS can cause meningitis symptoms, principally increased intracranial pressure and inflammation. We will look at two types of viral CNS infections, acute viral infections and persistent (slow) virus infections.

Rabies

An acute and fatal viral CNS infection, rabies was first reported more than 3000 years ago, and the term *rabies* is derived from the Latin word meaning rage. This is fitting, because part of the infection process is overproduction of saliva and inability to swallow that make patients seem to be foaming at the mouth. Rabies can affect all mammals and is transmitted by infected secretions (usually through a bite). The virus is large and bullet-shaped (**Figure 24.7**), with glycoproteins that cover the entire virion.

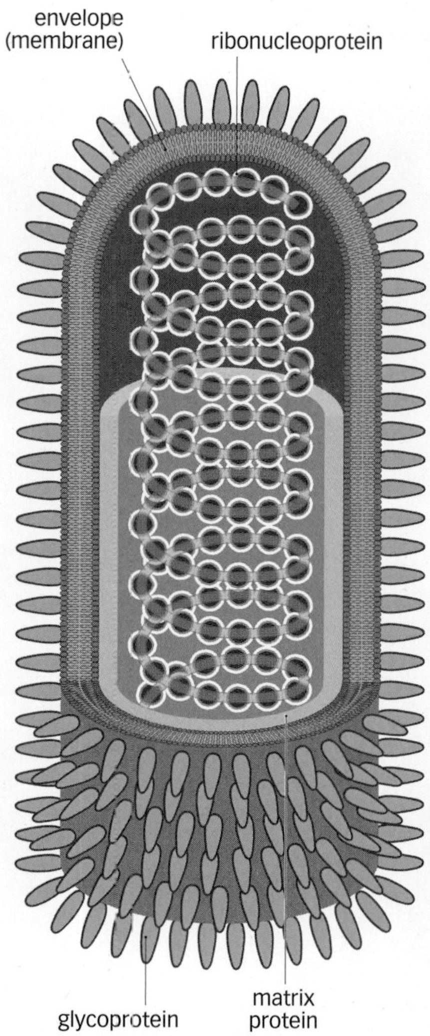

Figure 24.7 A diagrammatic illustration of the *rabies* virus.

Pathogenesis of rabies — Rabies involves severe neurological symptoms and signs in a patient bitten by a rabid (infected) animal. The CNS abnormalities are characteristic and include a relentless progression of excess motor activity, agitation, hallucinations, and excessive salivation. There can also be severe throat contraction when swallowing is attempted.

Rabies exists in two forms, *urban* and *sylvatic*, with the urban form being associated with unimmunized dogs and cats, and the sylvatic form being seen in wild animals. Infection in humans is incidental and does not contribute to maintenance or transmission of the infection. In the United States, 75% of cases occur in wild animals, and the occurrence of human rabies worldwide is about 15,000 cases per year. Approximately two cases of rabies infection in humans are reported in the United States each year.

The first event of rabies infection is introduction of the virus, usually through the epidermis via an animal bite, but inhalation of a heavily contaminated material such as bat droppings can also introduce the virus. Rabies virus first replicates in the striated muscle at the site of the bite (or in the lungs if the virus was inhaled), and immunization at this time will keep the virus from migrating into the nervous tissue. Without intervention, however, the virus moves into the peripheral nervous system at the neuromuscular junction and spreads into the CNS, where it replicates exclusively in the gray matter. After replication, the virus moves into other tissue, such as the adrenal medulla, kidneys, lungs, and salivary glands. At the same time, lymphocytes and plasma cells are infiltrating into the CNS tissue, and these agents destroy nerve cells. The primary lesion seen in the nervous tissue is the Negri body (**Figure 24.8**).

The incubation period for rabies can vary from 10 days to as long as a year, depending on the amount of virus initially deposited with the bite or inhalation, the amount of tissue infected, the host's immune response, innervation at the site, and the distance the virus must travel to reach the CNS.

Rabies presents as acute, fatal encephalitis, and only a handful of infected people have ever recovered without immediate treatment. Once the symptoms have started, the infection is irreversibly fatal. The illness begins with nonspecific fever, headache, malaise, nausea, and vomiting. The onset of encephalitis is marked by periods of excessive motor activity and agitation accompanied by hallucinations, combativeness, muscle spasms, and seizures followed by coma. There can also be excessive salivation, dysfunction of the brain stem and cranial nerves, double vision, facial palsy, and difficulty in swallowing. Involvement of the respiratory centers of the brain causes respiratory paralysis, which is the major cause of death. The median survival after the onset of symptoms is four days, and the maximum is 20 days.

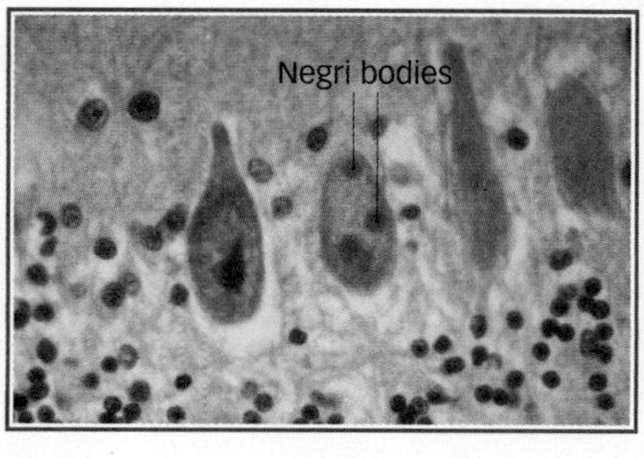

Figure 24.8 Negri bodies are a diagnostic sign of rabies infection.

Treatment of rabies — Prevention is the main way to control this infection. Treatment consists of a course of injections that are beneficial only if administered before the onset of symptoms. Intensive supportive care can result in longer-term survival. However, even with supportive care, the mortality for rabies is 90%.

Polio

Polio virus is one of the most important enteroviruses in the world; the infection it causes first emerged as important during the latter half of the twentieth century. Although this condition was first known as *infantile paralysis*, the risk of paralysis from infection with this virus actually increases with age. In most modern countries, poliomyelitis is essentially nonexistent because an effective vaccine exists, but in underdeveloped countries this infection is still a major problem.

Pathogenesis of polio — The polio virus has an affinity for the CNS and normally reaches it by crossing the blood–brain barrier. The virus can also use the axons or perineural sheath of the peripheral nervous system to gain access. Motor neurons are particularly vulnerable to infection, and various levels of neuronal destruction cause necrosis of the neural tissue and infiltration by mononuclear cells, primarily lymphocytes. About 90% of poliomyelitis infections are very mild and are subclinical, with incubation times varying from 4 to 35 days but averaging about 10 days.

Three types of infection are caused by the polio virus:

- Abortive poliomyelitis is a nonspecific febrile illness lasting two to three days and having no signs or symptoms of CNS involvement.
- Aseptic meningitis (nonparalytic poliomyelitis) is characterized by meningeal irritation, a stiff neck, back pain, and back stiffness. Recovery from this form of poliomyelitis is rapid and complete.
- Paralytic poliomyelitis occurs in approximately two per cent of persons infected by the virus. The hallmark of this form of the infection is asymmetric flaccid paralysis, with the extent of the paralysis varying from case to case. Temporarily damaged neurons can regain their function, but this recovery can take six months. Paralysis persisting after this recovery period is permanent.

Prevention of polio — As mentioned above, the construction of the polio vaccine has essentially wiped out this infection in developed countries. Two types of vaccine are licensed for use in the United States, both developed in the 1950s: the inactive form designed by Jonas Salk, and the live attenuated vaccine designed by Albert Sabin. There are three serotypes of the virus, and both vaccines contain all three serotypes, a design that effectively inhibits infection. In fact, no cases of poliomyelitis attributed to indigenously acquired wild poliovirus have been reported in the United Sates since 1979. (Recall from our discussion of influenza and rhinovirus infections in Chapter 21 that there is no way of producing a completely effective vaccine because of the multiple numbers of serotypes of these viruses.)

Fast Fact

There have been cases of polio that could be attributed to reversion of the live virus back to the infectious form in the Sabin vaccine.

Viral Encephalitis

The family of neurological infections classified as viral encephalitis are caused by mosquito-borne viruses called arboviruses (arthropod-borne), which is not a microbial taxonomic group. These viruses are common in the United States, and there is an increased occurrence of these infections in the summer months as the number of mosquitoes increases. A variety of clinical types of arbovirus cause infections that range in severity from subclinical symptoms to rapid death. These infections are all usually characterized by chills, headache, and a fever that can lead to mental confusion and coma. Survivors of viral encephalitis infections can subsequently develop permanent neurological disease.

Both horses and people are affect y arboviruses, and the names of some of the infections reflect th t. For example, **eastern equine encephalitis** (**EEE**) and **west quine encephalitis** (**WEE**) both cause severe infection in hum act, EEE has a 35% mortality rate in humans, and survivors n ffer brain damage, deafness, and other neurological problem **uis encephalitis** is the most common form of arbovirus e but fewer than 1% of cases show clinical symptoms.

West Nile virus, an halitis infection (discussed in Chapter 8), was first see 1999. It mostly affects birds but can also infect huma ugh most human cases of West Nile disease are sub severe infection and rapid death in the elderly.

Persiste

A varie al diseases in both humans and other anim es or by filterable agents that share the pro ermed *slow viral diseases* but a better te *ctions*, both because of the long period f disease and because of the prolonged

ent viral CNS infections caused by con- et's look at some of them now.

ematu
sure. When

itis is a rare chronic measles infection that progressive neurological disease. It is char- of personality change, progressive intellec- ion of the autonomic nervous system.

e CNS, most
on initially
for a psy-
entation,
changes
an last
monia,
ately

n is seen in patients with congenital or . It is a chronic CNS infection characterized by argy, seizures, and increased numbers of mono- . This infection is caused by both echoviruses and py with human hyperimmune globulin can cause ent, but relapses occur if the therapy is discontinued.

on-
n-

complex is part of the pathology of HIV infection and a persistent CNS infection in asymptomatic AIDS patients. al course of this infection can vary from mild to very severe sive dementia.

conventional Agents

As we learned in Chapter 8, prions have been proved to cause bovine spongiform encephalopathy in cattle, scrapie in sheep, and five fatal CNS infections in humans. Persistent infections can be divided into two groups: those associated with "conventional" viral agents (described above) and those associated with the "unconventional" agents known as prions (discussed in Chapter 8). Prions are transmissible infectious agents, but they do not contain nucleic acids in the way that viruses do. They also do not seem to elicit inflammatory or immune responses by a host. Some prion properties are listed in **Table 24.4**.

Pathogenesis of these infections is not well understood, but the pathological features of each of the human infections are similar, including loss of neurons and proliferation of astrocytes. They are called spongiform because

Infection	Agent
Subacute sclerosing panencephalitis	Measles virus
Progressive panencephalitis following Rubella	Rubella virus
AIDS dementia complex	Human immuno-deficiency virus
Persistent enterovirus infection of the immunodeficient	Enterovirus

Table 24.3 Conventional Viruses That Cause Persistent Viral CNS Infections

Property
Chronic progressive pathology without remission or recovery
No initiation of host inflammatory response
No alteration in pathogenesis by immunosuppression
Diameter 5–100 nm
No virion structure and no nucleic acids
Replication to high titers in susceptible tissue
Transmissible to experimental animals
Do not elicit interferon production
Unusual resistance to ultraviolet irradiation, alcohol, formalin, boiling, proteases, and nucleases
Can be inactivated by prolonged exposure to steam autoclaving or to 1 M or 2 M NaOH

Table 24.4 Properties of Prions

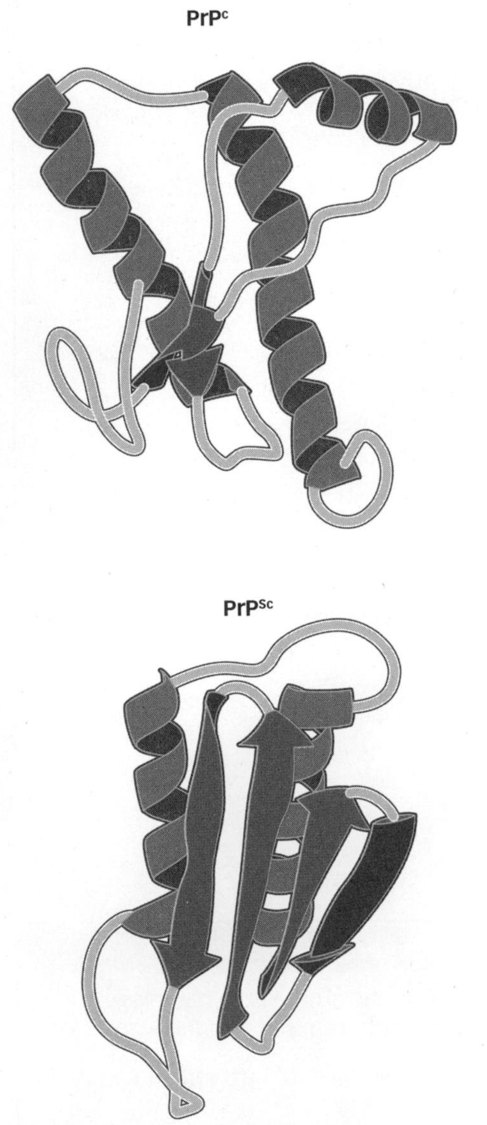

Figure 24.9 Change in conformation that converts the PrP^c protein into the infectious PrP^{sc} prion.

of vacuoles seen in the brain corte[illegible] cerebellum of infected patients. The incubation period can be from m[illegible] years, with the course of the infection being protracted and alway[illegible]ions can remain viable in brain tissue even after years of being [illegible] formalin, and they are very resistant to ionizing radiation and disinfectants.

A prion is a protein coded for by a nor[illegible] protein is designated **PrP^c** and can be converted into the [illegible]lled **PrP^{sc}** (the sc stands for scrapie) by a change in confo[illegible]9). Contact of normal PrP^c with the infectious PrP^{sc} for[illegible] form to reconfigure into the infectious form, and t[illegible]h the infectious form "multiplies." It is the prolifer[illegible]at causes the pathology seen in prion infections [illegible]

There are several prion infections that affect hu[illegible] *kuru* is a subacute progressive neurological dise[illegible] in 1957 in the Fore people of New Guinea. Epide[illegible] at that time showed that kuru usually affected a[illegible] dren of both sexes. Symptoms were a failure of m[illegible] (ataxia), hyperactive reflexes, and muscular spasms [illegible] sive dementia and death. There was diffuse neurona[illegible] spongiform change of the cerebral cortex and basal g[illegible] showed that kuru was transmitted through the ingest[illegible] from human brains. (The Fore people cannibalized their [illegible] a way of celebrating their lives, rather than burying or c[illegible] Incubation could take as long as 20 years after initial expo[illegible] practice of cannibalism stopped, kuru disappeared.

Creutzfeldt–Jacob disease is a progressive fatal infection of th[illegible] often seen in patients who are 60–70 years old. The infect[illegible] presents as a change in cerebral function that can be mistaken [illegible] chiatric disorder. The patient exhibits forgetfulness and disor[illegible] and there is a progression to overt dementia that can involve [illegible] in gait, involuntary movements, and seizures. This progression [illegible] from four to seven months, with eventual paralysis, wasting, pneu[illegible] and death. The infection is seen throughout the world, and approxi[illegible] one case per million people is reported each year.

The mode of transmission is unknown but has been attributed to co[illegible]taminated dura mater grafts, corneal transplants, and contact with co[illegible]taminated electrodes or instruments used in neurosurgical procedures. Transmission has also been linked to contaminated growth hormone, but there is no evidence of transmission by direct contact or airborne spread.

The incubation period for Creutzfeldt–Jacob disease is anywhere from 3 to 20 years, and the pathology is identical to that seen in kuru, with high levels of infectious prions found in the brain. In point of fact, brains of individuals with Creutzfeldt–Jacob disease show the same fibrils as those seen in sheep that have died of scrapie (**Figure 24.11**). Examination of brain tissue is the only way to confirm Creutzfeldt–Jacob disease, and there is no therapy.

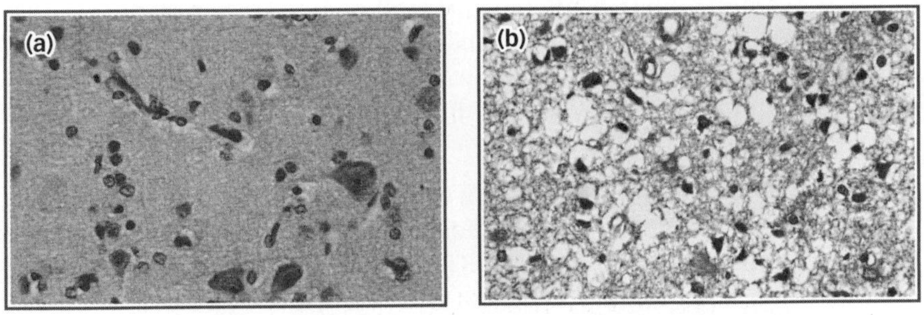

Figure 24.10 Changes in brain tissue after infection with prion disease. Light micrographs of a section of normal brain (panel **a**) and a section of brain infected with prions (panel **b**). Note the large number of vacuoles visible in the latter.

The prion infection called *fatal familial insomnia* presents as a difficulty in sleeping followed by increasingly progressive dementia. It occurs in people between the ages of 35 and 61 years and is fatal, with death occurring between 13 and 25 months after diagnosis.

Bovine spongiform encephalopathy ("mad cow disease") was first identified in the United Kingdom in 1986. The source of the prions that caused infection was traced to cattle feed that contained meat and bone meal from sheep that had scrapie. The cows that ate the feed became infected. Humans who ate beef from the infected cows developed an infection known as *variant Creutzfeldt–Jacob disease*. So far, more than 100 people with variant Creutzfeldt–Jacob disease have died in the United Kingdom.

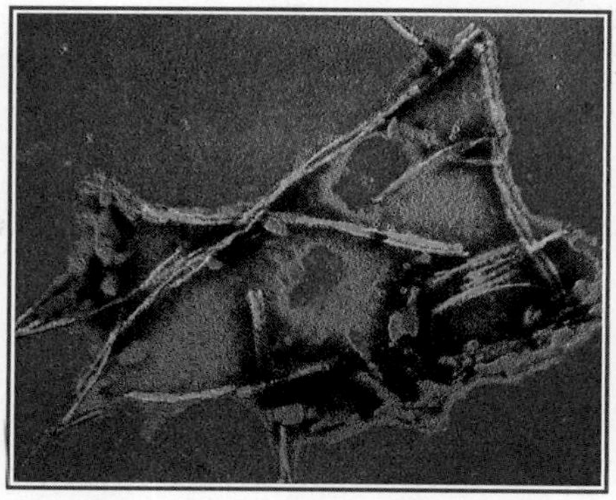

Figure 24.11 Characteristic fibrils seen in brain tissue of patients with prion-caused infections.

Cases frequently present in young adults as psychiatric problems that progress to neurological changes and eventually dementia. The average life expectancy after diagnosis is 14 months.

Keep in Mind

- Many viruses have an affinity for the CNS and cause meningitis symptoms.
- There are two types of viral CNS infection: acute and persistent.
- Acute viral CNS infections include rabies, polio, and viral encephalitis.
- Some persistent CNS infections are caused by conventional agents such as measles, rubella, and enterovirus.
- Other persistent CNS infections are caused by the unconventional agents known as prions.

FUNGAL INFECTIONS OF THE CENTRAL NERVOUS SYSTEM

As we discussed in Chapter 14, fungal infections are primarily opportunistic and are usually seen in immunocompromised patients. In our discussions of CNS infections, the most important fungal infection is cryptococcosis.

Cryptococcosis

The fungal infection called cryptococcosis is caused by *Cryptococcus neoformans*, which is an encapsulated form of yeast. Capsule production varies with the strain and is associated with environmental conditions. This fungus grows best at 35–37°C and in culture produces colonies within two to three days. *C. neoformans* can be found throughout the world, especially in soil contaminated with bird droppings (even though birds are never sick from this fungus).

Fast Fact

Cryptococcosis can occur in individuals in whom there is no evidence of immunocompromise.

Pathogenesis of cryptococcosis — *C. neoformans* causes a chronic form of meningitis, with a slow, insidious onset of infection. Symptoms include low-grade fever and headache, progressing to altered mental status and seizures. This infection is usually seen in patients who are immunocompromised and is common in AIDS patients.

The infection begins with inhalation of the yeast cells. Once inhaled, each yeast cell begins to overproduce its capsule (**Figure 24.12**). As we saw for encapsulated bacteria, this capsule is anti-phagocytic and can bind to

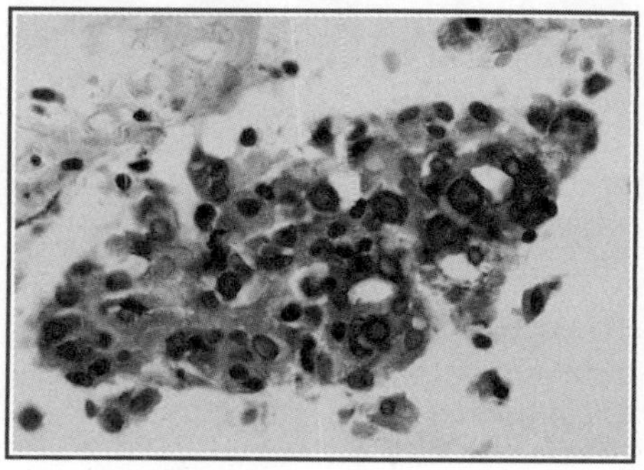

Figure 24.12 Micrograph depicting the histopathology associated with cryptococcosis of the lung using Mucicarmine stain.

complement components, thereby reducing the opsonization defense of the host. The capsule can also interfere with the presentation of antigens to T cells, thereby inhibiting the adaptive immune response. After inhalation, the yeast cells multiply outside the lungs and move into the nervous system. This results in intermittent headache, dizziness, and difficulty with complex cerebral functions, symptoms that continue over a period of weeks or months. Seizures, cranial nerve damage, and papilledema (edema of the optic nerve) appear in the later stages of the infection, accompanied by dementia and decreased levels of consciousness. These symptoms are accelerated in patients with AIDS.

Treatment of cryptococcosis — Amphotericin B or fluconazole is used to treat cryptococcosis infections, and 75% of patients with cryptococcal meningitis initially respond to treatment. However, a significant portion of these patients relapse when the therapy is stopped, and many patients with chronic infection require repeated courses of therapy. There is residual neurological damage in more than half of cured patients.

PARASITIC INFECTIONS OF THE CENTRAL NERVOUS SYSTEM

Parasitic Amebic Meningoencephalitis

The CNS infection known as **parasitic amebic meningoencephalitis** is caused by free-living amebas belonging to the genera *Naegleria* and *Acanthamoeba*. *Naegleria* infections affect children and young adults and are acquired by swimming in fresh water. This infection occurs infrequently but is almost always fatal, and *Naegleria* organisms are found in large numbers in shallow freshwater ponds, particularly during warm weather.

Parasitic meningoencephalitis resulting from infection by *Acanthamoeba* species is a subacute or chronic illness that is almost always fatal. *Acanthamoeba* is found in soil and also in fresh brackish water and has been found in the oropharynx of asymptomatic humans. Most *Acanthamoeba* infections occur in the southeastern United States, and infected patients typically fall ill during the summer months after swimming or water skiing in small, shallow freshwater lakes.

Pathogenesis of amebic meningoencephalitis — Histologic evidence suggests that *Naegleria* organisms enter the CNS by traversing the nasal mucosa and the cribriform plate. Once in the CNS, the organisms produce a purulent, hemorrhagic inflammatory reaction that extends perivascularly from the olfactory bulb to other regions of the brain. The infection is characterized by rapid onset of severe bifrontal headache, seizures, and occasionally abnormal taste and smell, progressing to coma and death within days. Examination of the cerebrospinal fluid shows blood and an intensive neutrophil response, and wet mounts of cerebrospinal fluid reveal trophozoite forms of the parasite.

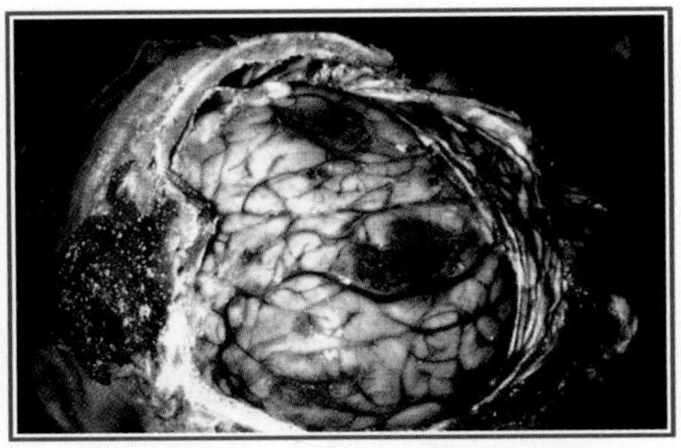

Figure 24.13 Autopsied brain from a patient with *Acanthamoeba* encephalitis. Notice the necrotizing granulocytic lesions.

The epidemiology of *Acanthamoeba* encephalitis has not been clearly defined, but it is known that infection involves the elderly and immunocompromised. It is thought that the ameba reaches the brain by hematogenous dissemination from an unknown site, possibly the respiratory system, eye, or skin. The infection produces diffuse, necrotizing granulomatous encephalitis (**Figure 24.13**), frequently involving the midbrain, with both cysts and trophozoites being found in the lesions. In AIDS patients, there can also be cutaneous ulcers and hard nodules containing amebas; in these patients, amebas are also seen in the cerebrospinal fluid. The clinical course of *Acanthamoeba* infection is more prolonged than that of *Naegleria* infection, and the former can occasionally end in spontaneous recovery. In immunocompromised individuals, however, *Acanthamoeba* infection is invariably fatal.

Treatment of amebic meningoencephalitis — To date, only four patients have ever survived infection with *Naegleria*, and all of these individuals were diagnosed early and treated with high doses of amphotericin B together with rifampin.

Studies on drug therapy for *Acanthamoeba* infections have not yet been completed.

Keep in Mind

- Fungal CNS infections are primarily opportunistic and are usually restricted to patients who are immunocompromised.
- The most important fungal CNS infection is cryptococcosis, caused by *Cryptococcus neoformans*, the only encapsulated fungus.
- Parasitic CNS infections are usually caused by free-living amebas such as *Naegleria* and *Acanthamoeba*.

SUMMARY

- Central nervous system infections can be very dangerous and require immediate and aggressive therapy.
- The central nervous system has a blood–brain barrier to prevent access of blood borne pathogens.
- Pathogens that can move from the blood to the cerebrospinal fluid cause meningitis.
- Most central nervous system infections result from bacteremia or viremia.
- Many viruses have an affinity for the central nervous system.
- Viral infection of the central nervous system can be acute or persistent.
- Acute viral infections include rabies, polio, and encephalitis.
- Persistent viral infections include measles and rubella.
- Prions cause persistent central nervous system infections.
- Fungal infections of the central nervous system are primarily opportunistic and occur in immunocompromised individuals.
- Parasitic infections of the central nervous system are usually caused by free-living amebas.

In this chapter we have looked at infections that involve the CNS. These infections can be the result of bacterial, viral, fungal, or parasitic organisms, all of which can take advantage of the anatomy of the CNS to invade and cause infection. In addition, there are brain infections caused by infectious proteins called prions. When reviewing these infections, remember the anatomy of the CNS and how it is implicated in the infectious process. Also keep in mind that the condition referred to as meningitis, which is an infection of the meninges of the CNS, can be caused by a variety of organisms that all produce the same symptoms.

SELF EVALUATION AND CHAPTER CONFIDENCE

Multiple Choice

Answers are given in the back of the book and help can be found in the student resources at:
www.garlandscience.com/micro

1. One of the most dangerous results of infection of the central nervous system is
   - **A.** Production of cerebrospinal fluid
   - **B.** Vasogenic edema
   - **C.** Disruption of the choroid plexus
   - **D.** None of the above

2. Which of the following can infect the central nervous system?
   - **A.** *Neisseria meningitidis*
   - **B.** *Listeria monocytogenes*
   - **C.** *Streptococcus pneumoniae*
   - **D.** *Haemophilus influenzae*
   - **E.** All of the above

3. Infections of the CNS can result from which of the following?
   - **A.** Infections of the middle ear
   - **B.** Sinusitis
   - **C.** Mastoiditis
   - **D.** Pneumonia
   - **E.** All of the above

4. The clinical features of CNS infections include
   - **A.** Aseptic meningitis
   - **B.** Encephalitis
   - **C.** Poliomyelitis
   - **D.** Polyneuritis
   - **E.** All of the above

5. The clinical effects of botulism are caused by
   - **A.** Bacteria entering the brain
   - **B.** Viruses entering the brain
   - **C.** Exotoxins from *Clostridium tetani*
   - **D.** Exotoxins
   - **E.** None of the above

6. Antibiotic therapy for tetanus would be
   - **A.** Very effective at any time during the infection
   - **B.** Moderately effective late in the infection
   - **C.** Not effective at any time
   - **D.** Effective only if given early in the infection

7. The symptoms of infant botulism include
   - **A.** Constipation
   - **B.** Bleeding from the rectum
   - **C.** Increased activity and restlessness
   - **D.** Diarrhea
   - **E.** None of the above

8. The rabies virus is transmitted by
   - **A.** The fecal–oral route of contamination
   - **B.** The *Anopheles* mosquito
   - **C.** The saliva of an infected animal
   - **D.** As a co-infection with the influenza virus

9. The urban form of rabies is associated with
   - **A.** Wild animals
   - **B.** Domesticated cats
   - **C.** Bats
   - **D.** Unimmunized dogs and cats
   - **E.** All of the above

10. The symptoms of rabies include all of the following except
    - **A.** Hallucinations
    - **B.** Headache
    - **C.** Nausea
    - **D.** Lethargy
    - **E.** Vomiting

11. The polio virus binds to receptors on
    - **A.** Astrocytes
    - **B.** Lymphocytes
    - **C.** Motor neurons
    - **D.** Neutrophils
    - **E.** All of the above

12. The live attenuated polio vaccine was developed by
    - **A.** Pasteur
    - **B.** Zinsser
    - **C.** Sabin
    - **D.** Salk
    - E. None of the above

13. Prions are
    - **A.** Viruses
    - **B.** Parasites
    - **C.** Proteins
    - **D.** Bacteria
    - **E.** None of the above

14. All of the following are prion infections except
    - **A.** Kuru
    - **B.** Bovine spongiform encephalopathy
    - **C.** Creutzfeldt–Jacob disease
    - **D.** Subacute sclerosing panencephalitis
    - **E.** Fatal familial insomnia

15. Cryptococcosis is a
   - A. Bacterial CNS infection
   - B. Viral CNS infection
   - C. Parasitic CNS infection
   - D. Fungal CNS infection
   - E. Prion infection

16. *Naegleria* infection is acquired through
   - A. Contaminated food
   - B. Swimming
   - C. Drinking water containing fungal spores
   - D. Working with infected pigs
   - E. None of the above

DEPTH OF UNDERSTANDING

Questions listed here require you to bring together the concepts you have learned in this chapter into a discussion format. This helps you to increase your depth of understanding of the material you have learned. Help can be found in the student resources at:
www.garlandscience.com/micro

1. Describe the various methods by which pathogens can gain entry to the central nervous system.

2. Describe the pathogenesis of tetanus from initial infection to a fatal outcome.

CLINICAL CORNER

Help can be found in the student resources at:
www.garlandscience.com/micro

1. You are an avid jogger and, living in New York, Central Park is the place you love to run. It has great scenery and you enjoy watching the horse-drawn carriages taking tourists on sightseeing trips through the park. One morning as you enter the park you are told that the park is closed for mosquito spraying and will not be open again for several days because of West Nile virus. Although this ruins your jogging plans, you understand the problem. Several other joggers and drivers of the carriages are not so understanding and demand to know what the problem really is. You want to be helpful.
   - A. What do you tell them about West Nile virus?
   - B. Why are they spraying for mosquitoes?
   - C. Who are the people most at risk of this infection?
   - D. When one of the joggers points out that the Nile River is thousands of miles away, how do you explain to him about West Nile virus being found in New York City's Central Park?

2. Your patient presents complaining of headache and dizziness, which have been going on for more than a month. He has also become forgetful to the point that he can no longer manage the automotive parts store he owns. His medical history shows that he received a corneal transplant 20 years ago but he has never suffered any side effects from it. Tests have shown no indication of bacterial, fungal, or viral infection.
   - A. What are some of the possible explanations for his problems?
   - B. Does the fact that he is a transplant recipient mean anything?

Infections of the Blood

Chapter 25

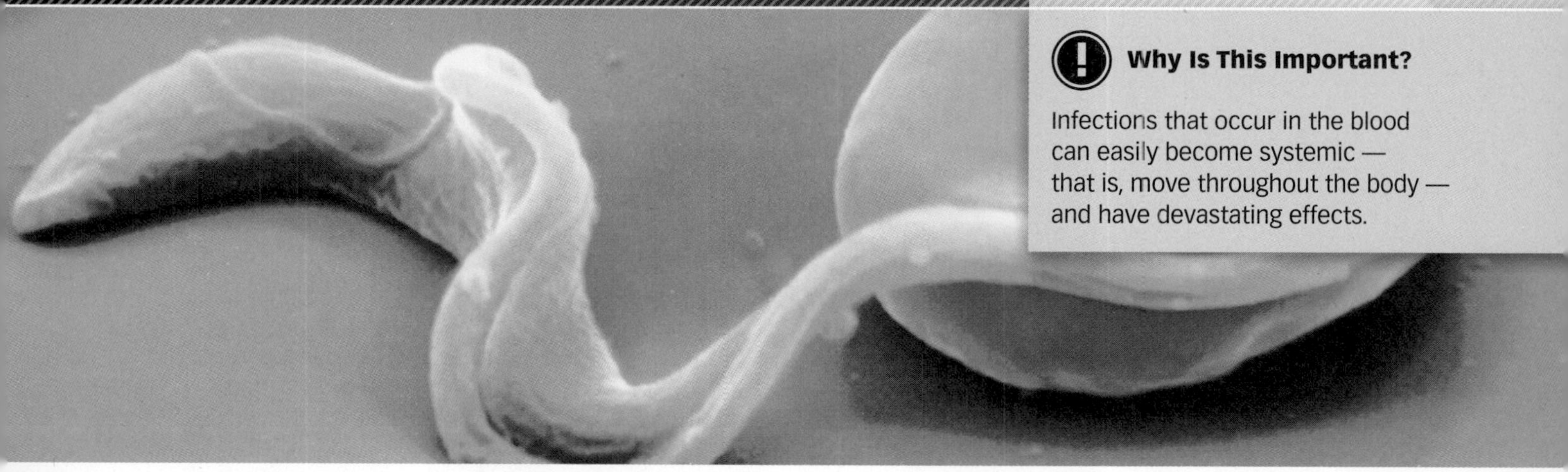

Why Is This Important?

Infections that occur in the blood can easily become systemic — that is, move throughout the body — and have devastating effects.

OVERVIEW

In this chapter, we look at infections that occur in the blood. As we will see, the systemic nature of these infections requires that we deal with them aggressively. We will also see that there are bacterial, viral, and parasitic infections of the blood and that several of these infections involve transmission by vectors.

We will divide our discussions into the following topics:

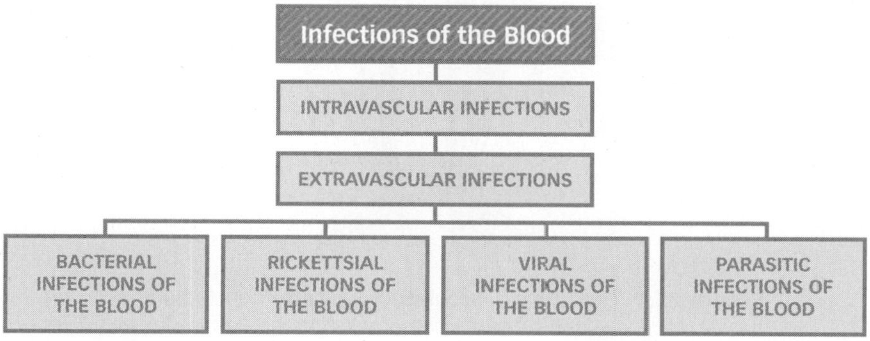

To understand the importance of blood infections, you must remember that blood has access to the entire body. It therefore represents a "highway" that traverses all of the systems of the body (**Figure 25.1**) and can easily carry pathogens from one location to another.

In many ways the presence of circulating pathogens in the blood is part of the natural progression of an infection, and these organisms are quickly removed from the blood by the host defenses. However, in some cases, pathogens in the blood can be a reflection of serious, uncontrollable infection. Depending on the type of organism, this infection is classified as one of the following:

- Bacteremia — bacteria in the blood (**Figure 25.2**)
- Viremia — viruses in the blood
- Fungemia — fungi in the blood
- Parasitemia — parasites in the blood

What Do I Need to Know?

To get the most out of this chapter, please review the following terms from your previous courses in biology, anatomy, physiology, or chemistry or in previous chapters of this book as indicated in parentheses: analog, arbovirus (**12**), arthropod (**6**), complement (**15**), edema (**24**), etiologic (**7**), infectious endocarditis, lymphadenopathy (**17**), lymphatics, mononuclear phagocytic system (**15**), nosocomial infection (**6**), nucleoside (**11**), phagolysosome (**15**), sepsis (**7**), septic shock (**7**), septicemia (**5**), septum, spirochete (**4**), sporozoa (**14**), tachycardia, trypomastigote (**14**).

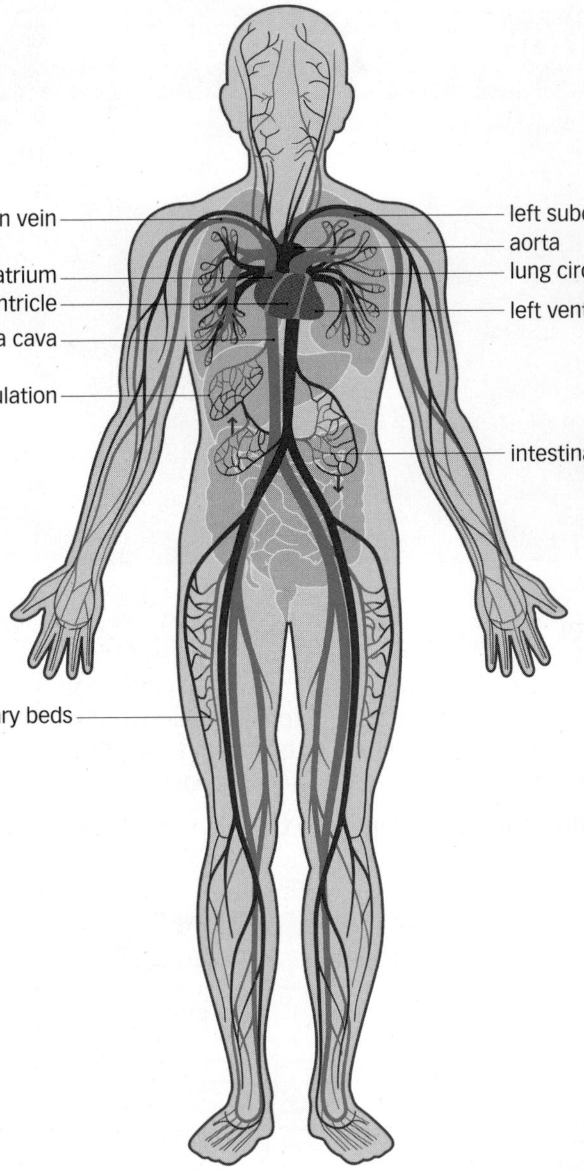

Figure 25.1 The human circulatory system. Notice that this system has access to every area of the body, making it the perfect "highway" for pathogens moving through the body.

Pathogenic organisms in the blood can also lead to bacteremia in which organisms are growing in the blood. This condition may lead to sepsis or septic shock. In either case, the presence of bacteria in the blood can be a life-threatening situation.

Bacteremia and fungemia may also be caused by pathogens growing on the inside or the outside of intravenous devices. The infection may start out as a minor problem but can quickly become serious. This is particularly true in patients who are debilitated from long hospital stays. Transient bacteremia can occur if the normal bacterial flora are exposed to the blood through manipulation or trauma to the body. These organisms are normally of no clinical significance but can lead to bacteremia and sepsis if they enter the blood.

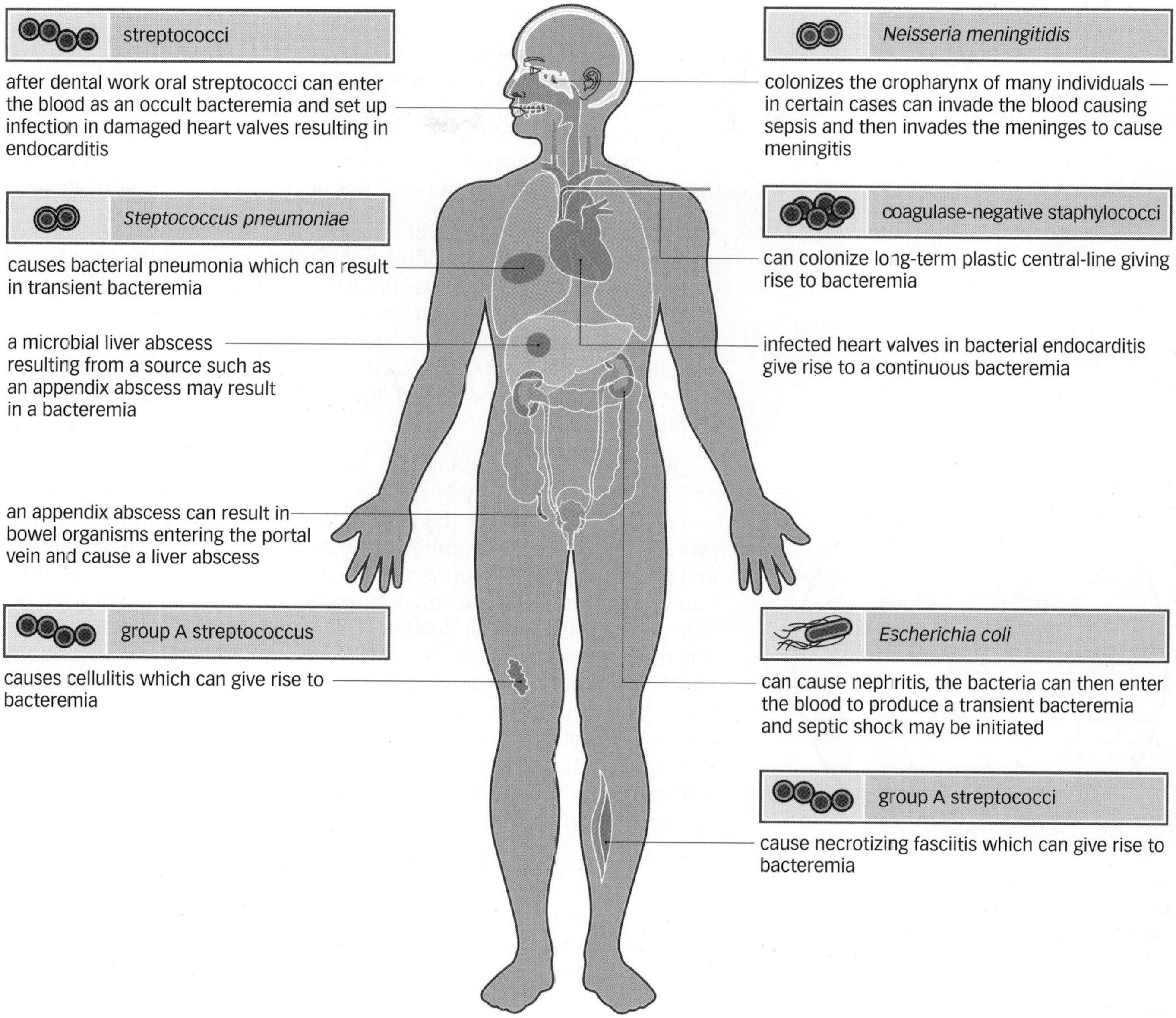

Figure 25.2 Some examples of the sources of transient, intermittent, occult, or continuous bacteremias.

INTRAVASCULAR INFECTIONS

Intravascular infections arise when pathogens gain entrance to the blood and damage the structures of the cardiovascular system. Infectious **endocarditis** (infection of the heart), **thrombophlebitis** (infection in the veins), and **endoarteritis** (infection of the arteries) are most commonly caused by bacteria, but fungi can sometimes be involved. In any case, infections of the cardiovascular system are very dangerous and can be fatal if not promptly and adequately treated. These infections commonly produce constant shedding of organisms into the blood, a condition that causes persistent low-grade fever.

Let's look more closely at two intravascular infections, infectious endocarditis and bacteremia, caused by intravenous lines or catheterization.

Infectious Endocarditis

Infectious endocarditis was once referred to as *bacterial endocarditis*, but the more general name was adopted once it was realized that organisms other than bacteria can cause it. This infection affects the heart valves, either the body's natural valves or synthetic ones in patients who have received replacement valves. The infection can also develop on the septa of the heart and on cardiac shunts. This infection involves several factors:

- The endothelium of the heart is altered by the infecting pathogens, and this alteration facilitates the colonization of the area by the pathogens. This in turn causes the deposition of platelets and fibrin at the site (**Figure 25.3**). The turbulence of the blood flow around these deposits can lead to further irregularities of the endothelial surfaces, irregularities that facilitate the further deposition of platelets and fibrin. Eddies caused by slower blood flow in these areas cause even more pathogen growth here.

- Circulating pathogens adhere to the fibrin and platelets, which causes an inflammatory response that includes both activation of complement and further damage to the endothelial surfaces. As this process continues, a thrombotic "mesh" composed of platelets, fibrin, and inflammatory cells forms and leads to the formation of a structure known as **mature vegetation** (**Figure 25.4**). This structure protects the pathogens from the host defense and also helps keep out antibiotics. The mature vegetation causes alterations in the cardiac endothelium, which obstructs blood flow and increases turbulence. The increased turbulence can cause part of the mature vegetation to fall off and form an **embolus** ("plug"; the plural is ***emboli***) that may move into the smaller vessels and also obstruct blood flow to other parts of the body (**Figure 25.5**). Transport of these emboli to the brain or the coronary arteries can be a lethal event.

- Transient bacteremia is common after some medical procedures and in these cases it is of no importance clinically. This is often seen in dental procedures, bronchoscopy, sigmoidoscopy, and some surgical procedures. The organisms responsible for transient bacteremia are normally common surface flora with low levels of virulence. However, even these normal floras can colonize the valve areas of the heart if there is sufficient change in the cardiac endothelium.

- Another complication of colonization of the heart valves is the formation of immune complexes by the host. These complexes form as antibodies against the infecting organisms bind to them and form large aggregates. These aggregates activate the complement system, which then causes such peripheral problems as nephritis, arthritis, and cutaneous vascular lesions.

Infectious endocarditis can be classified in two ways: *acute*, which presents with high fever and toxicity and can result in death within a few days or weeks, and *subacute*, which presents with low-grade fever, weight loss and night sweats. In this form, death can take weeks to months to occur. **Table 25.1** lists some of the more common bacterial causes of infectious endocarditis.

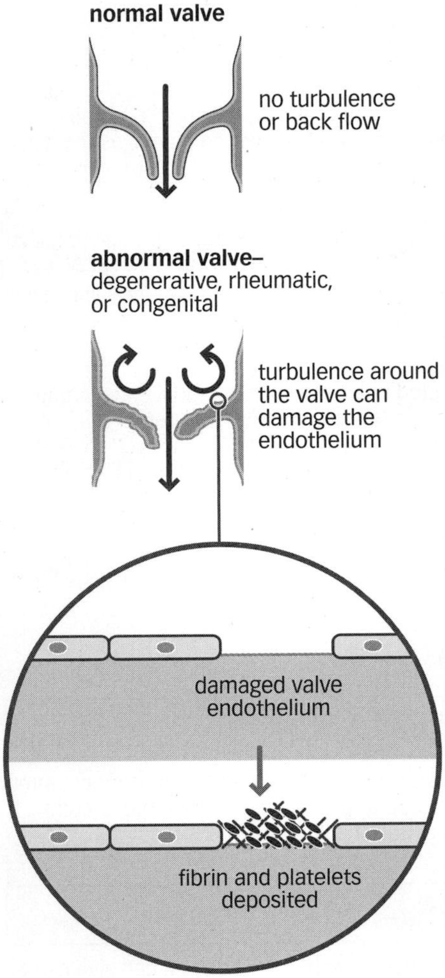

Figure 25.3 Changes that occur in heart valves that predispose to infectious endocarditis. The turbulence that occurs around the heart valves gives the infecting organisms an opportunity for additional growth.

Fast Fact

Before antibiotics, death was inevitable for all forms of endocarditis.

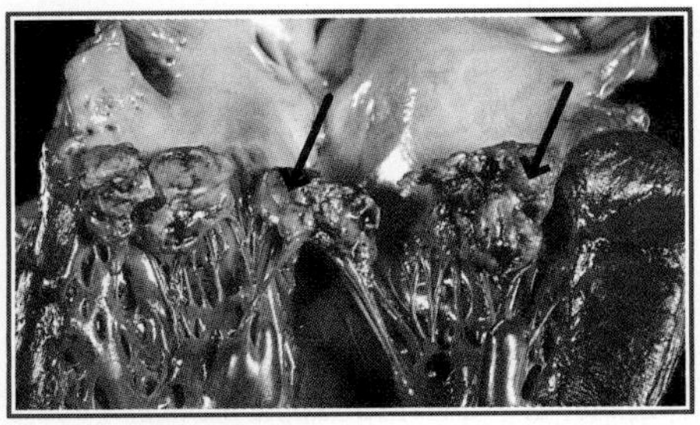

Figure 25.4 A heart that has been subjected to subacute infectious endocarditis. Unlike the acute form of the infection, this form occurs over a period of weeks or months. The endocarditis develops as pathogens attach to the heart surface and multiply. This causes the formation of the fibrin-platelet mature vegetation (arrows), a structure that protects the accumulating pathogens from the host defenses.

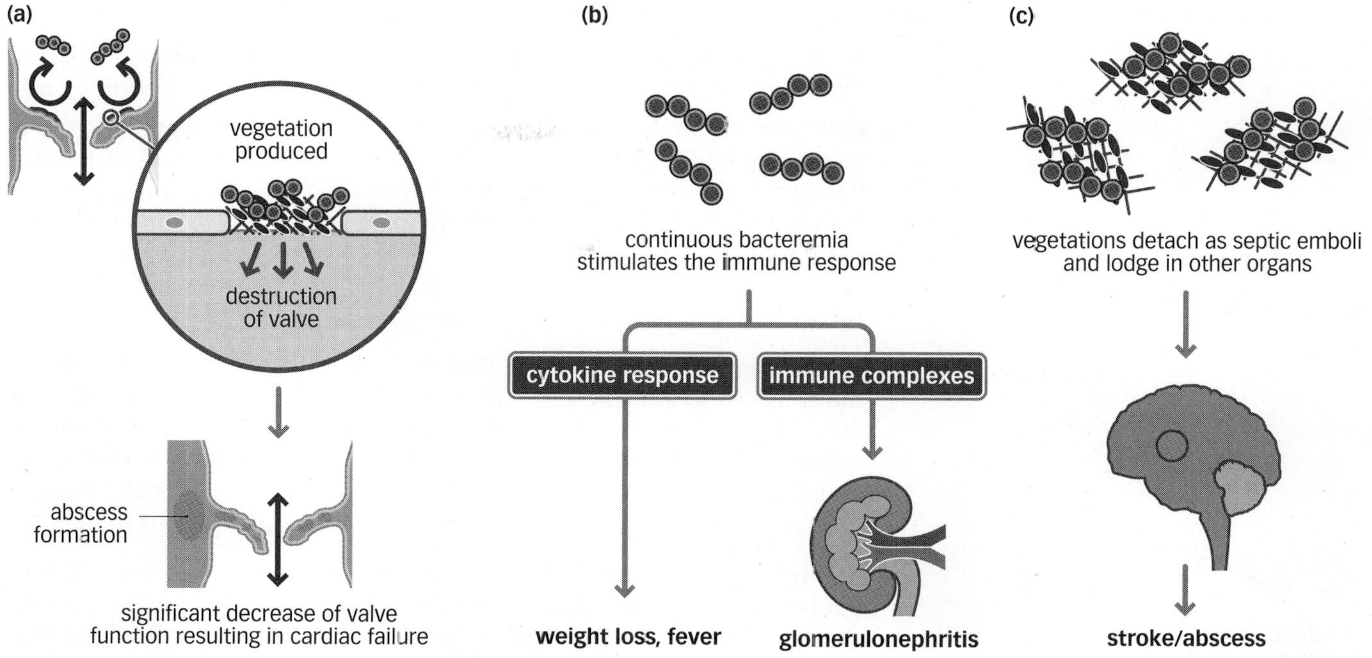

Figure 25.5 Effects of bacterial endocarditis. Bacteria can initiate infectious endocarditis (panel **a**) and stimulate the immune system (panel **b**). Panel **c**: The mature vegetation formed during the infection can detach from an infected heart valve and become an embolus that can then lodge in some distal organ.

Complications from infectious endocarditis include risk of congestive heart failure, rupture of the chordae tendineae of the heart valves, and perforation of the valves. In addition, the kidneys are usually affected, and blood can be found in the urine. If the infection occurs on the left side of the heart, coronary emboli can form. Infection of the right side of the heart can lead to infection of the lung in addition to the formation of emboli.

In addition to treatment, management of infectious endocarditis is important. The nature of the infection causes response to therapy to be slow and cure to be difficult. Therefore, the antibiotic therapy must be aggressive and include bacteriocidal drugs that can be given in high enough concentration to guarantee for levels to remain high in the blood without causing toxicity, a balancing act that can be difficult to achieve. In

Table 25.1 Common Bacterial Sources of Infectious Endocarditis

Organism	Approximate Percentage of Cases Caused by Organism
Streptococcus viridans	30–40
Enterococci	5–18
Other streptococci	15–25
Staphylococcus aureus	15–40
Coagulase-negative staphylococci	4–30
Gram-negative bacilli	2–13

addition, the course of therapy can be prolonged, lasting more than four weeks (**Figure 25.6**). Removal or replacement of the infected valve may be necessary, and this involves its own set of potential consequences.

Dental procedures can be dangerous for people who are predisposed to infectious endocarditis. These are people who already have cardiac valvular lesions, such as those seen in heart disease. They need to take high doses of antibiotics before dental procedures and continue them for 6–12 hours after the procedure.

Intravenous-Line and Catheter Bacteremia

Recall from our discussions in Chapter 6 that nosocomial infections are the cause of serious problems for patients who are hospitalized because these patients are in many cases debilitated or immunocompromised. Additionally, there are always large numbers of antibiotic-resistant bacteria in hospitals. The combination of these two factors makes the implantation of intravenous lines and catheters a potentially serious matter.

In any infection resulting from either an intravenous line or a catheter, the blood can become colonized with organisms normally found on the skin. If this were a transient bacteremia in a healthy host, the infection would quickly be dealt with by the host defense. In debilitated hosts, however, the bacteremia can persist and increase the chances of such secondary complications as infectious endocarditis and distal infections. Pathogens most commonly involved in this type of secondary infection include *Staphylococcus epidermidis*, *Staphylococcus aureus* and *Corynebacterium* species.

In some cases, it is the intravenous fluid that is contaminated rather than the apparatus, and in these cases the organisms involved are usually Gram-negative rods such as *Pseudomonas* species. In cases of catheter bacteremia, removal of the contaminated device and treatment with antibiotics can normally remedy the situation.

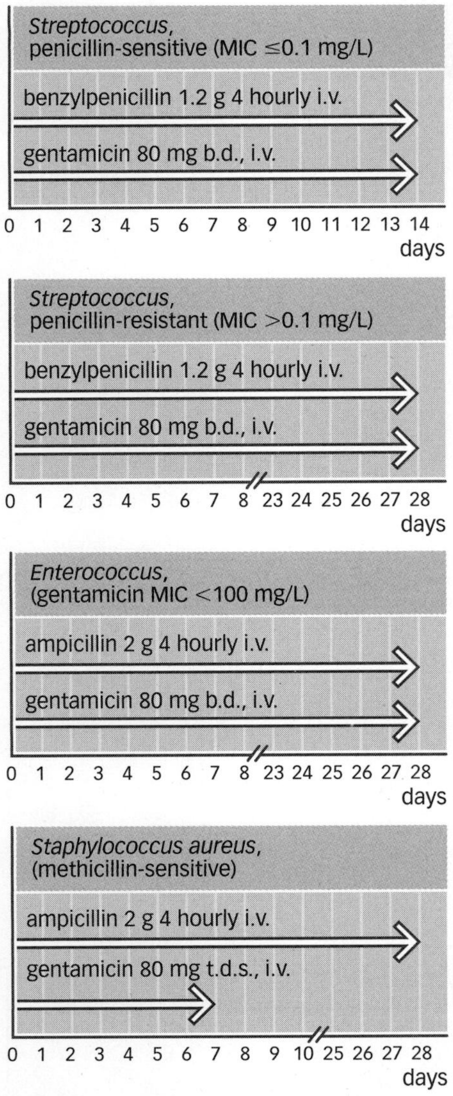

Figure 25.6 Some antibiotics used in the treatment of infectious endocarditis and the times required for treatment. All doses are for adults with normal renal and liver function. MIC, minimum inhibitory concentration.

EXTRAVASCULAR INFECTIONS

Most cases of significant bacteremia result from an overflow of the infecting organisms from an extravascular infection. The organisms move into the lymphatics from infected tissues and from there into the blood. Recall that the lymphatics have to pass through the lymph nodes, and many pathogens are captured in the nodes. However, if there are overwhelming numbers of pathogens present, some of them can make it to the blood. With staphylococcal pneumonia, for example, there can be thousands of pathogens per milliliter of blood. **Table 25.2** shows the frequency with which certain bacteria produce bacteremias. The most common sources of bacteremia are urinary tract infections (Chapter 23), respiratory infections (Chapter 21), and skin infections (Chapter 26). Organisms that produce meningitis (Chapter 24) usually also produce bacteremia at the same time.

Sepsis and Septic Shock

Both sepsis and septic shock result from the progression of bacteremia. Gram-positive and Gram-negative bacteria, as well as some fungi, protozoa, and viruses, can produce these conditions. As noted in Chapter 7, **sepsis** is the suspicion or proof of infection for which the evidence is a host response such as fever, chills, or tachycardia. If sepsis causes altered blood flow to organs, it is termed **sepsis syndrome**, for which the evidence is reduced urine output, systemic acidosis, or changes in mental status. If

Table 25.2 Organisms Most Often Involved In Bacteremia

Frequency and Organism
Often (more than 90% of cases)
Haemophilus influenzae type b
Neisseria meningitidis
Streptococcus pneumoniae
Brucella
Salmonella enterica serovar Typhi
Listeria
Variable frequency, depending on stage and severity of infection (10–90% of cases)
Beta-hemolytic streptococci
Streptococcus pneumoniae
Staphylococcus aureus
Neisseria gonorrhoeae
Leptospira
Borrelia
Acinetobacter
Shigella dysenteriae
Enterobacteriaceae
Pseudomonas species

this syndrome continues, it leads to **septic shock**, which is characterized by the development of hypotension. If not treated, septic shock leads to **refractory septic shock**, in which the hypotension cannot be resolved by standard treatments. Continuation of septic shock leads to multiorgan failure, **disseminating intravascular coagulation** (**DIC**) and death.

The initial events in sepsis are vasodilation with decreased peripheral resistance and increased cardiac output. This is followed by capillary leakage, reduced blood volumes, and multiorgan failure. Most features of septic shock can be caused by the cell walls of Gram-negative bacteria containing endotoxin that stimulates the release of pro-inflammatory cytokines. Early recognition of this problem is critical because resolution depends on more than treatment with antibiotics. Treatment of sepsis conditions also requires adequate maintenance of tissue perfusion, and careful fluid and electrolyte management.

Keep in Mind

- Blood and lymph travel throughout the body, and infection in any part of the body can be spread throughout the body if pathogens gain entry to the blood or lymph.
- Pathogenic organisms in the blood include bacteria (causing bacteremia), viruses (causing viremia), fungi (causing fungemia), and parasites (causing parasitemia).

- Pathogens in the blood can lead to sepsis and septicemia.
- Intravascular infections arise from pathogens that gain entrance to the blood and cause damage to cardiovascular structures.
- Extravascular infections occur when pathogens from infected tissues overflow into the blood.
- Extravascular infections can result in either sepsis or septic shock, which can result in multiorgan failure and death.

BACTERIAL INFECTIONS OF THE BLOOD

Now let us examine some of the circulatory-system infections caused by bacteria, including plague, tularemia, brucellosis, Lyme disease, and relapsing fever.

Plague

Plague is an infection that is vector-transmitted to humans and may be the most explosively virulent bacterial infection ever known. The infection spreads from the lymph nodes to the blood and can spread from there to the lungs. Once spread to the lungs, the infection is referred to as **pneumonic plague** and can be easily spread from person to person. This is the form of plague referred to as the **Black Death** in the Middle Ages.

Fast Fact

In the fourteenth century, the population of Europe was about 105 million. In less than 4 years (1346–1350), the Black Death killed 25 million of them.

All forms of plague can lead to toxic shock and death in days! In fact, no other bacterial infection kills previously healthy people so quickly.

Plague is caused by the bacterium *Yersinia pestis* and is an infection of rodents that can be transferred to humans by the bite of the rat flea (*Xenopsylla cheopis*). Plague exists in two forms: *sylvatic*, which is seen in wild rodents and is the primary reservoir of the organism, and *urban*, which is seen in cities. The urban form is the more infectious form because there are more potential hosts in the city than in the wild.

The plague life cycle begins when fleas become infected after feeding on infected rats. The *Y. pestis* organisms multiply in the intestine of the flea and are regurgitated into a host as the flea bites. The bite of the infected fleas leads to **bubonic plague**, which is not normally contagious. However, some infected individuals will develop the pneumonic form of the infection through bacteremia spreading to the lungs, and this form is highly contagious.

Pathogenesis of plague — Once *Y. pestis* is injected past the barrier of the skin, the difference between flea body temperature and human body temperature and the difference in ionic environment cause the bacterium to produce virulence factors. One of these factors, the **F1 protein**, forms a gel-like capsule that prevents phagocytosis and allows the bacteria to multiply in the submucosa. The bacteria eventually reach the lymph nodes and multiply there very rapidly. This produces a **bubo**, which is a suppurative, hemorrhagic, bulging lymph node (**Figure 25.7**) and is the source of the name *bubonic plague*. From the bubo, the organism spreads rapidly into the blood.

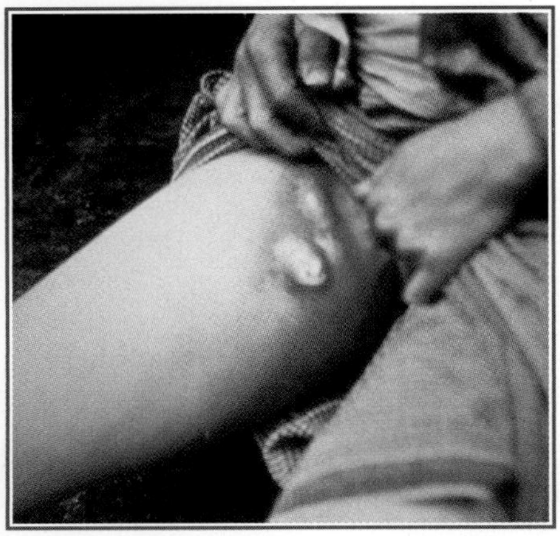

Figure 25.7 A bubo, the salient feature of bubonic plague, on the thigh of a patient. The infecting organism is *Yersinia pestis*.

In the blood, *Yersinia* produces extensive systemic toxicity resulting from the release of lipopolysaccharides and virulence enzymes. The bacteremia spreads to other organs, most notably the lungs, where the result is necrotizing hemorrhagic pneumonia. The incubation period, which is the length of time between when the patient is bitten and when the first symptoms appear, is two to seven days. The main signs are fever and painful buboes, most often in the groin. Without treatment, 50–75% of patients get bacteremia and die of septic shock within days. If plague localizes in the lungs, death can occur within two to three days. Plague is almost always fatal if treatment is delayed more than a day after the onset of symptoms. Approximately 5% of patients develop the pneumonic form of the infection, which is very contagious.

Treatment of plague — Streptomycin is the treatment of choice for both the bubonic and pneumonic forms of plague, but doxycycline, ciprofloxacin, gentamycin, and, chloramphenicol can also be used. If treatment is timely, the mortality from plague is less than 10%.

Tularemia

Tularemia is a zoonotic infection that moves from animals to humans and is caused by *Francisella tularensis*, a small Gram-negative facultative coccobacillus. This organism grows only on a specialized medium that contains sulfhydryl compounds, and even in this supportive environment it takes up to 10 days of incubation before tiny colonies appear. Tularemia is an infection of wild animals, most often ground squirrels and rabbits in the United States. Transmission can be by inhalation, tick bite, ingestion of contaminated meat or water, or directly by contact with an abrasion or cut while skinning an infected animal. Infected animals may not show any signs of this infection. Because the ID_{50} is small (less than 100 organisms) and because humans can become infected from a bite, there are many routes to tularemia infection. It has been shown that tularemia can be acquired through the inhalation of *F. tularensis* organisms as an infected animal is being skinned. Tularemia is distributed throughout the Northern Hemisphere, but distribution patterns vary widely from one region to another. In the United States between 100 and 200 cases are reported each year.

Fast Fact

Tularemia is considered to have potential as a bioterrorist weapon (discussed in Chapter 28).

Pathogenesis of tularemia — The incubation phase for tularemia lasts between two and five days and usually results in an ulcerated lesion at the inoculation site. From this site, the infecting organisms move to the organs of the mononuclear phagocytic system and form granulomas. *Francisella* multiplies in macrophages, using acidification to disrupt the fusion of the phagosome and the lysosome, and also multiplies in hepatocytes and endothelial cells. The presence of the pathogens in the latter cells can result in abscess formation.

Tularemia can follow a number of courses depending on the inoculation site and on how far into the body the pathogens spread. All possible courses, however, begin with the acute onset of fever, chills, and malaise. In **ulceroglandular tularemia**, a localized papule (a small, solid elevation on the skin) forms at the inoculation site and becomes ulcerated and necrotic. This leads to swelling of the regional lymph nodes, which become very painful.

The infection can also be acquired through the eyes, in which case it is referred to as **oculoglandular tularemia** and produces a painful purulent conjunctivitis. A person ingesting a large number of *Francisella* bacilli can develop **typhoidal tularemia**, which presents with symptoms and signs similar to those seen in typhoid fever.

Any form of tularemia can progress to a systemic infection in which lesions are found in multiple organs. The mortality of untreated tularemia can be as high as 30%.

Treatment of tularemia — Streptomycin is the drug of choice for all forms of this infection, but gentamycin, doxycycline, and ciprofloxacin are also effective. Two important factors for prevention are the use of rubber gloves, a mask, and eye protection when in contact with potentially infected animals, and the prompt removal of any ticks found on a person's body.

Brucellosis

The zoonotic infection *brucellosis* involves infection of the reproductive tract and is seen in sheep, cattle, pigs, goats, dogs, and other animals. The infecting organisms are various bacteria of the genus *Brucella*. Humans become infected by occupational contact or by ingesting contaminated animal products. In humans, brucellosis is a chronic illness characterized by fever, night sweats, and weight loss that can last for weeks or months. In some cases, the patient develops a cycling pattern of symptoms referred to as **undulant fever**.

Pathogenesis of brucellosis — *Brucella* organisms gain access through cuts in the skin, contact with mucous membranes, inhalation, or ingestion. After they have penetrated the skin or mucous membranes, the organisms multiply in macrophages in the host's liver, spleen, bone marrow, and other components of the mononuclear phagocytic system. Their ability to keep host phagosomes and lysosomes from fusing allows the organisms to survive and multiply. In addition, *Brucella* impairs the host's ability to produce and release cytokines as part of the immune response. If not controlled locally, a *Brucella* infection causes the formation of small granulomas at sites throughout the mononuclear phagocytic system. At these sites, bacteria multiply and are released back into the circulation, intermittently causing recurrent bouts of chills and fever (thus the name *undulant* fever).

Symptoms start with malaise, chills, and fever, continuing for 7–21 days after initial infection. During this period, drenching night sweats and fever reaching 40°C are common. This nocturnal fever can continue for weeks, months, or even years. Other symptoms include headaches, body aches, and weight loss. Fewer than 25% of infected individuals show any signs in the mononuclear phagocytic system, but those that do exhibit splenomegaly (enlargement of the spleen), hepatomegaly (enlargement of the liver), and lymphadenopathy.

Treatment of brucellosis — Doxycycline and gentamycin are the primary antibiotics used for brucellosis, and doxycycline is preferred because of its pharmacologic characteristics. In seriously ill patients, these drugs can be supplemented with streptomycin, gentamycin, or rifampin.

Lyme Disease

Lyme disease is caused by the spirochete *Borrelia burgdorferi*, which is transmitted to humans by the *Ixodes* tick. This bacterium is a Gram-negative spirochete with properties similar to *Treponema pallidum* (the causative agent of syphilis). It requires a specialized medium for growth, and even in that medium the doubling time is between 8 and 24 hours. Consequently, isolation can be difficult. In addition, there are at least 10 subspecies of *B. burgdorferi*, all of which are geographically localized. *B. burgdorferi* exists as part of a complex life cycle involving ticks, mice, and deer (**Figure 25.8**). Humans are incidentally involved when ticks feed on people who enter their habitat.

Lyme disease is endemic in several parts of the United States, Canada, and temperate areas of Europe and Asia. Approximately 90% of the 10,000–15,000 cases of Lyme disease reported worldwide are reported in the United States.

The primary reservoir for *B. burgdorferi* is mice. Mice serve as the host for the early stages of the *Ixodes* life cycle, and deer serve as the host for the final stages. In spring, fertile adult female ticks living on the deer become engorged on blood, fall from the deer, and lay their eggs in the soil. When hatched, the tick larvae seek out mice for blood meals. Because mice are

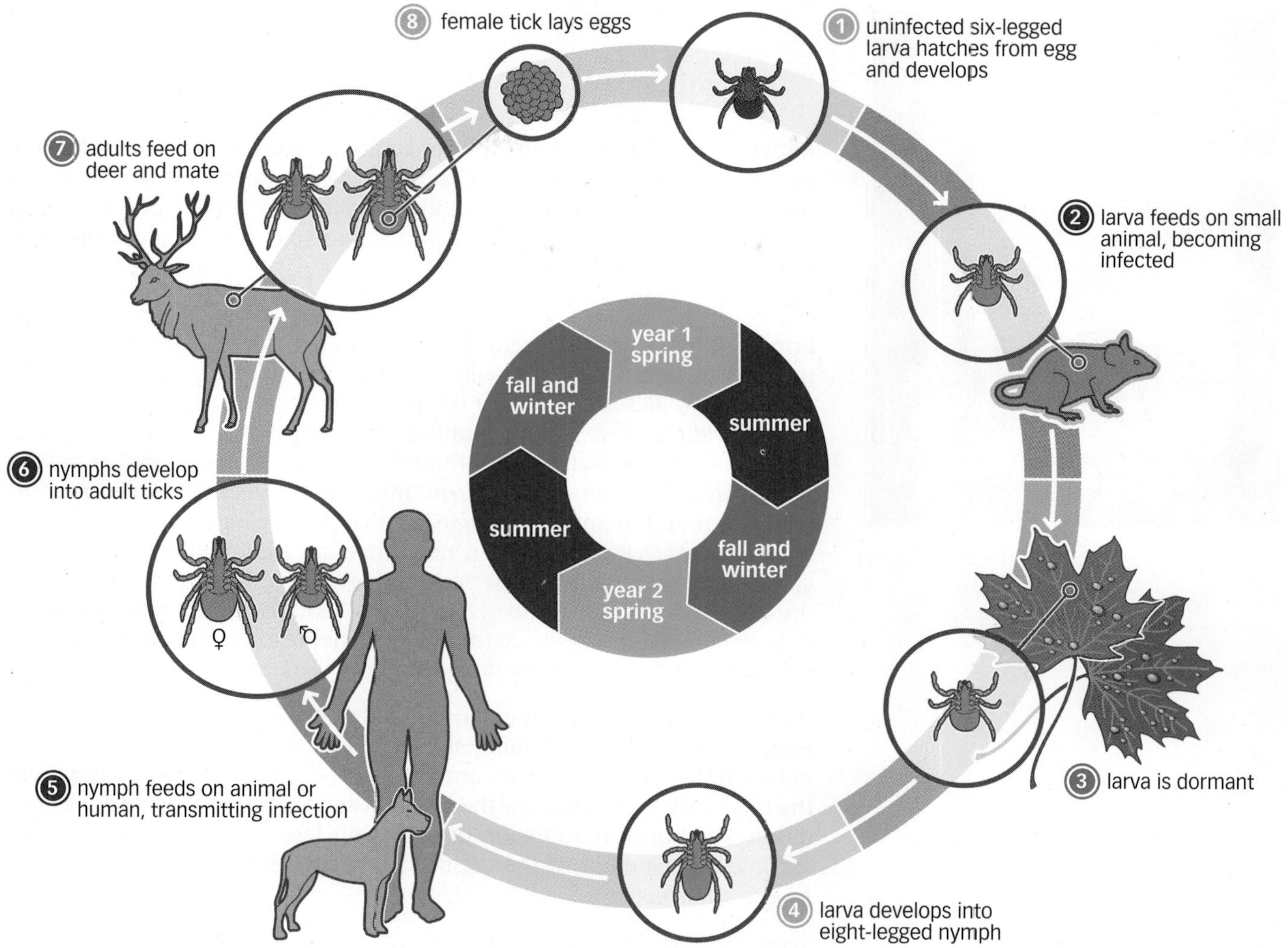

Figure 25.8 The life cycle of *Ixodes*, the tick vector of Lyme disease. Notice that the tick depends on the availability of deer to complete its life cycle.

the host for *B. burgdorferi*, the bacterium is picked up by the tick larvae feeding on the mice and then remains with the ticks throughout their development. During the following spring or summer, the larvae mature to adulthood and parasitize deer. This complex life cycle takes two years, and the deer host is essential to the existence of Lyme disease because this is where *Ixodes* matures and mates.

Humans contract Lyme disease primarily through tick larvae, but the pathogenic mechanisms of acquisition are not yet understood. It is known that the outer membrane of *B. burgdorferi* contains proteins and a toxic form of lipopolysaccharide that differs from the usual Gram-negative endotoxin. These bacteria also contain peptidoglycan, which has inflammatory properties that can survive for long periods in tissues and may cause arthritis if deposited in joint tissues.

Patients with Lyme disease exhibit modulations in their immune response, including inhibition of mononuclear and natural killer cell function, proliferation of lymphocytes, and the production of cytokines. In fact, chronic Lyme disease has aspects very similar to those seen in autoimmune disease.

Pathogenesis of Lyme disease — Acute Lyme disease is characterized by fever, a migratory "bull's-eye" rash, and muscular and joint pain. Additionally, there is often meningeal irritation associated with the acute form. Chronic Lyme disease is characterized by the evolution of meningoencephalitis, myocarditis, and disabling recurrent arthritis.

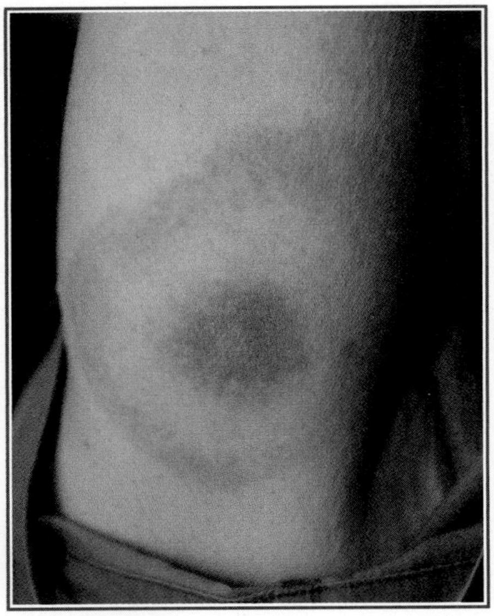

Figure 25.9 The bull's-eye rash associated with Lyme disease.

Both the acute and chronic forms of Lyme disease are highly variable and involve multiple body systems. Signs and symptoms occur in multiple overlapping patterns that come and go at different times, but skin lesions spreading from the site of the tick bite and relapsing arthritis are the most persistent findings. Lyme disease is rarely fatal, but if untreated it can be the source of chronic ill health.

Primary lesions appear in the first month after the tick bite and expand to become annular lesions with a raised red border and central clear area (**Figure 25.9**). The ring of the bull's-eye expands and forms an *erythema migrans lesion* accompanied by fever, myalgia, headache, and joint pain. If the infection is untreated, the skin lesions disappear but the other symptoms can persist for months. Days, weeks, or even months later, a secondary infection stage involving the central nervous system and the cardiovascular system may develop. Neurologic abnormalities can include meningitis, facial nerve palsy, and peripheral nerve destruction, and cardiovascular involvement can lead to acute myocarditis and enlargement of the heart. Normally both neurologic and cardiovascular symptoms resolve spontaneously in a matter of weeks.

Weeks to years after the initial onset of the infection, arthritis can begin, and this marks the continuing stage of Lyme disease. This stage occurs in two-thirds of untreated patients, fluctuates intermittently, and involves the large joints, in particular the knee. In serious cases, this arthritic condition may cause the erosion of bone, but less frequently there can be chronic involvement of the central nervous system, affecting memory, mood, and sleep.

Treatment of Lyme disease — Prevention invloves prompt tick removal. Doxycycline and amoxicillin are the preferred antibiotics for treatment of *early-stage* Lyme disease, but intravenous penicillin G is used for patients with neurologic and cardiovascular involvement. The response to therapy is very slow and requires the continuous use of antibiotics for 30–60 days.

Relapsing Fever

The infection known as **relapsing fever** is caused by *Borrelia* species other than *B. burgdorferi* and is transmitted to humans by the bite of either ticks or body lice. There are two forms of relapsing fever, and which of them a patient contracts depends on whether the vector was a tick or a louse and also on the *Borrelia* species involved. The louse-borne form is usually seen in epidemics, whereas the tick-borne form is not.

Occurrence and distribution of the tick-borne form are determined by the biology of the tick and by the relationship between the tick and the *Borrelia* reservoir, which can be rodents, rabbits, birds, or lizards. Ticks that harbor *Borrelia* can remain infectious for several years. In addition, an infected tick is able to transfer the spirochete it carries to its progeny, making the offspring carriers even if the offspring do not feed on infected rodents or other *Borrelia* vectors.

The cycle is different for the louse-borne form of relapsing fever because lice have only human hosts. In addition, lice live for only about two months and do not pass the pathogenic spirochete to progeny. The louse-borne form is caused by *Borrelia recurrentis*, and lice become infected with the spirochete after biting an infected human.

B. recurrentis multiplies in the endolymph of the louse but never moves to the salivary glands or the feces. Therefore, to infect an uninfected human, the louse must be crushed and scratched into the bite wound.

Pathogenesis of relapsing fever — This illness presents with fever, headache, muscle pain, and weakness. The symptoms last about one week, disappear, and then return a few days later. During the relapse, the spirochetes can be found in the patient's blood. Relapsing fever develops when thousands of spirochetes are circulating per milliliter of blood, and although the exact mechanisms of the infection are not known, it has been shown that these circulating organisms disappear during the periods between relapses. It is thought that they may sequester in the organs, and at each relapse the spirochetes appear with new antigenic markers, causing the synthesis of new antibodies to the new antigens. The periods of relapse correspond to the development of new antibodies.

The incubation period is approximately seven days and is characterized by a huge number of spirochetes in the blood (this condition is called **spirochetemia**). For tick-borne relapsing fever there are only two relapses, but for the louse-borne form there can be as many as ten relapses. Fatalities are rare in the tick-borne form, but mortality in the louse-borne form can reach 40% in untreated individuals, usually from myocarditis, cerebral hemorrhage, and liver failure.

Treatment for relapsing fever — A single dose of doxycycline or erythromycin is sufficient to deal with this infection.

RICKETTSIAL INFECTIONS OF THE BLOOD

Rickettsia species are coccobacilli, but we treat them here in a section separate from the section on bacterial infections because they have characteristics of both bacteria and viruses. These coccobacilli divide by binary fission and are very small (0.3–0.5 μm). Although they are Gram-negative organisms, they stain very poorly and are therefore better resolved with Giemsa stain. They have a peptidoglycan cell wall with a Gram-negative outer layer containing lipopolysaccharides and outer membranes that extend to the cell surface. All of these traits are similar to those of bacteria, but *Rickettsia*, like viruses, are obligate intracellular parasites.

Rickettsia organisms cause spotted fevers and typhus-related illnesses (**Table 25.3**). They use animal reservoirs and are transmitted by arthropod vectors. Most *Rickettsia* species have animal reservoirs and are spread through the vector's life cycle. Infections produced by *Rickettsia* pathogens are typically fevers that are often accompanied by vasculitis. The most common of these is **Rocky Mountain spotted fever** (**RMSF**), which, despite its name, is seen throughout the world.

Rickettsia grows freely in the cytoplasm of infected eukaryotic cells and can be cultured only in cell cultures and fertile eggs. These pathogens grow in cytoplasmic vacuoles of the host and then escape the vacuole and begin growing in the cytoplasm. They also use directional actin polymerization like that seen with some viruses to move through the cell and from cell to cell. Eventually their numbers become so large that they rupture the host cell, which is another similarity to viruses. *Rickettsia* cannot survive outside host cells, and if they do not find a host they cease metabolic activity and begin to leak proteins, nucleic acids, and other essential molecules. This instability leads to a rapid loss of potential infectivity.

The classic example of a rickettsial infection is **epidemic typhus**, but the most prevalent rickettsial infection in the United States is RMSF. Both of these infections are characterized by fever, rash, and muscle aches. Both diseases may be fatal as a result of vascular collapse.

Infection	Pathogen	Distribution	Vector	Reservoir
Spotted fever group — Rocky Mountain spotted fever	*Rickettsia rickettsii*	North and South America	Tick	Rodents and dogs
Rickettsial pox	*Rickettsia akari*	USA, Russia, Korea, and Africa	Mite	Mouse
Typhus group	*Rickettsia prowazekii*	Africa, Asia, and South America	Body louse	Humans

Table 25.3 Some Pathogenic Rickettsiae

Rocky Mountain Spotted Fever

The etiologic agent for Rocky Mountain spotted fever is *Rickettsia rickettsii*. The infection is an acute febrile illness that occurs in association with residential and recreational exposure to wooded areas infested with ticks, which are the vectors for the *Rickettsia* species that cause RMSF.

Different RMSF vectors are found in different geographic locations: the wood tick (*Dermacentor andersoni*) in the western United States, the dog tick (*Dermacentor variabilis*) in the eastern part of the country, and the lone star tick (*Amblyomma americanum*) in the Southwest and Midwest of the country. It is interesting to note that the *Rickettsia* pathogens do not harm the tick but live in the endolymph and are passed from one vector generation to the next. More than two-thirds of RMSF cases occur in children younger than 15 years of age between April and September.

Pathogenesis of Rocky Mountain spotted fever — The incubation period from time of bite to onset of symptoms is usually two to six days. The symptoms include fever, headache, muscle aches, mental confusion, and a rash that appears first on the soles, palms, wrists, and ankles (**Figure 25.10**) and then moves toward the trunk. The rash is the most characteristic feature of the infection. It usually develops on the third day of illness, and its appearance makes it easy to distinguish RMSF from viral infections in which a rash appears. Muscle tenderness, which can become extreme, particularly in the calf muscles, is also a feature of this infection. If untreated and in some cases, even with treatment, RMSF can include complications such as disseminated vascular collapse and renal and heart failure, resulting in death.

In most cases, infection with *Rickettsia* causes vascular lesions, and the pathogens multiply in the endothelial cells lining the patient's small blood vessels. This leads to thrombocytosis (development of clots) and leakage of blood into the surrounding tissue (causing the rash). Although these vascular lesions occur throughout the body, they are most apparent in the skin and most serious in the adrenal glands.

Treatment of Rocky Mountain spotted fever — Antibiotic therapy is highly effective if given in the first week of illness. However, if treatment is delayed, it is much less effective. Doxycycline is the antibiotic of choice. The mortality associated with untreated infection can be as high as 25%, but with treatment the mortality is only five to seven per cent.

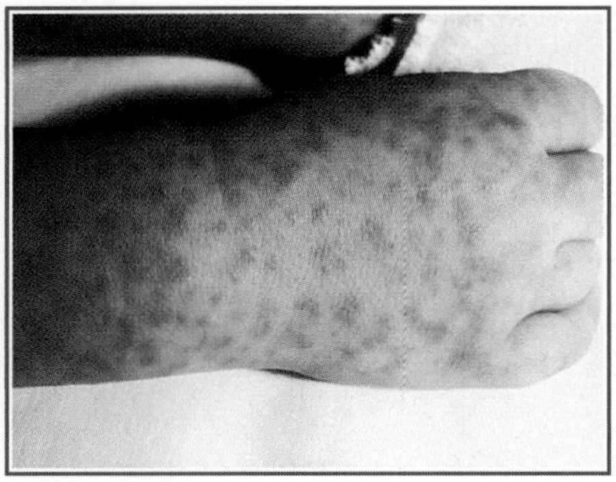

Figure 25.10 The rash caused by Rocky Mountain spotted fever, a rash often mistaken for measles.

It is important to note that *sulfonamides contribute to the infection process* and are therefore strongly contraindicated.

Typhus

There are several types of typhus, but we will restrict our discussion to the epidemic and endemic forms.

Epidemic Typhus

Epidemic typhus is caused by *Rickettsia prowazekii*, which is transmitted by the human louse *Pediculus humanus corporis*. It is historically seen in times of war or famine, in situations — such as crowding and infrequent bathing — that favor body lice.

Pathogenesis of epidemic typhus — *Rickettsia* circulates through the blood during acute febrile illness, and lice feeding on the body of an infected human become infected. After 5–10 days, the number of *Rickettsia* organisms in the infected lice increases and the pathogen is found in the louse feces. The lice defecate while feeding, and the *Rickettsia* organisms are rubbed into the bite wounds when they are scratched. Dried louse feces can also be infectious by entering a human through the eyes, respiratory tract, and mucous membranes.

Within two weeks of being bitten, a patient with epidemic typhus gets fever, headache, and a rash that begins on the trunk of the body and moves to the extremities. Complications of epidemic typhus are myocarditis and central nervous system dysfunction.

As with RMSF, infection causes vascular lesions, and the pathogens multiply in the endothelial cells lining the patient's small blood vessels. This leads to thrombocytosis (development of clots) and leakage of blood into the surrounding tissue (causing the rash). Although these vascular lesions occur throughout the body, they are most apparent in the skin and most serious in the adrenal glands.

Treatment of epidemic typhus — Doxycycline and chloramphenicol are very effective in treating this infection. In addition, control of lice is very important for preventing infection. If untreated, the fatality rate increases with age to as high as 60%.

Endemic Typhus

Endemic typhus is caused by *Rickettsia typhi* and is transmitted to humans by the rat flea (*Xenopsylla cheopis*). Human infection is incidental, and the primary infection is from rodent to rodent. About 30–60 cases of endemic typhus are reported in the United States each year, with more than half being along the Gulf Coast and southern California.

Pathogenesis of endemic typhus — The pathogenesis of endemic typhus is similar to that of epidemic typhus, except for the vector. The rat flea defecates into the bite wound, and the symptoms of infection — headache, muscle aches, and fever — appear one to two weeks later. A **maculopapular rash**, which is one made up of broad lesions that slope away from a central papule, also forms. If untreated, the fever may last 12–14 days, but mortality and clinical complications are rare in this infection even if untreated.

Treatment of endemic typhus — Doxycycline or chloramphenicol can reduce the fever period from two weeks to two or three days. In addition, control of rats helps prevent the development of this infection.

Keep in Mind

- Bacterial infections of the blood include plague, tularemia, brucellosis, Lyme disease, and relapsing fever.
- Plague is the most virulent bacterial infection ever known.
- Plague can present as either pneumonic plague or bubonic plague.
- Tularemia and brucellosis are zoonotic infections.
- Lyme disease and relapsing fever are caused by the spirochete *Borrelia*.
- Rickettsial infections of the blood include Rocky Mountain spotted fever, epidemic typhus, and endemic typhus.

VIRAL INFECTIONS OF THE BLOOD

Recall from Chapters 12 and 13 that (1) the primary goal of viruses is to make more copies of themselves, and (2) viruses bind only to specific receptors on host cells. Like other microbes, viruses can be found in the blood of an infected person, and this condition of viral bodies in blood is referred to as viremia. The viruses use the blood to carry infection throughout the host's body. In this section we discuss viruses whose cellular hosts are found in the blood. Note that we will not include HIV here because it was discussed in detail in Chapter 17.

Cytomegalovirus

Cytomegalovirus (**CMV**), which causes the formation of perinuclear cytoplasmic inclusions and enlargement of the host cell, has the largest viral genome. There are innumerable strains of this ubiquitous virus. In developed countries, more than half of the population have antibodies to it, indicating that they have been exposed, and 10–15% of children are infected in the first five years of life. CMV can be isolated from saliva, cervical secretions, semen, urine, and white blood cells for years after the initial infection.

The rate of congenital infection, which means the fetus is infected *in utero*, is 1% worldwide (about 40,000 a year in the United States). These infants excrete CMV either in urine or in nasopharyngeal secretions. Most of these infections are asymptomatic, but about 20% of infants born infected can have neurological impairment resulting from the infection, either sensory-nerve hearing loss or psychomotor retardation, or both. Infants with systemic infection can develop **hepatosplenomegaly**, jaundice, anemia, low birth weight, microencephaly, and chorioretinitis.

In contrast to congenital infection, *neonatal infection*, which is infection acquired during or shortly after birth, rarely has consequences. CMV can also be efficiently transmitted by breast milk, but infections acquired via this route are usually asymptomatic.

Pathogenesis of cytomegalovirus infection — Cytomegalovirus causes a latent infection in leukocytes and bone marrow cells. It is responsible for the infections associated with blood transfusions and organ transplants. Although CMV is a herpesvirus (human herpesvirus 5), it differs from herpes simplex in that it does not cause skin infections. Instead, it produces a visceral infection that affects the organs and triggers a mononucleosis syndrome. The virus infects both epithelial cells and leukocytes and can cause both tissue damage and immunological damage.

In healthy young adults, CMV can cause a mononucleosis syndrome. In immunosuppressed patients, both primary infections and reactivation of latent infections can be quite severe. For example, for patients receiving bone marrow transplants, CMV causes interstitial pneumonia, which is the leading cause of death in these patients (50–90% mortality). In patients with AIDS, CMV often disseminates to visceral organs, causing gastroenteritis, chorioretinitis, and infection of the central nervous system.

Treatment of cytomegalovirus infection — CMV does not respond well to any antiviral drugs. However, ganciclovir, a nucleoside analog of acyclovir, has been shown to inhibit CMV replication and prevent infection in AIDS and transplant patients and to reduce the retinitis caused by the virus. Combinations of immunoglobulin and ganciclovir have been shown to reduce the high mortality rate seen with CMV pneumonia infections in patients with a bone marrow transplant, but the long-term survival of these patients is not good.

Epstein–Barr Virus

Epstein–Barr (EB) virus is the major etiologic agent of infectious mononucleosis and of Burkitt's lymphoma. This virus is human herpesvirus 4; it has a small genome, which has been completely mapped. The EB virus can be grown in culture with long-term lymphoblastoid cell lines derived from humans (these are cells that are transformed and grow continually) and has an affinity for human B lymphocytes and epithelial cells. The infection is nonproductive in B cells but productive in epithelial cells.

Although EB virus can be cultured from the saliva of 20–25% of healthy adults, the infection is not highly contagious. Approximately 90–95% of adults worldwide are seropositive, indicating that they have been exposed to this virus. It is important to note that if a primary EB virus infection does not occur until the second decade of life or later, it is usually accompanied by the signs and symptoms of infectious mononucleosis.

Pathogenesis of Epstein–Barr virus infection — This virus is not very contagious and is transmitted only after repeated contact with infected individuals. One of the most important findings about EB virus infection is the role of the virus in the development of malignant infections, including Burkitt's lymphoma, nasopharyngeal carcinoma, and lymphoproliferative infections in immunocompromised patients.

EB virus does not produce cytopathic effects or inclusion bodies like those seen with herpes viruses, and the major consequence for infected B cells is transformation. When this occurs, only a small amount of the viral DNA integrates into the host chromosome, with most of the viral DNA staying in a separate circular **episome** form. After infection, viral proteins called EBNAs (Epstein–Barr nuclear antigens) appear in the nucleus of the infected cell just before the initiation of virus-directed protein synthesis.

The virus enters a human B lymphocyte by means of glycoproteins located on its envelope, and the glycoproteins bind to CD21 receptors on the lymphocyte. These receptors are normally used by the B cell as a complement receptor, but recall from Chapter 12 that viruses often usurp normal receptors to gain access to target cells. After about 18 hours, EBNA proteins are detectable in the nucleus of the infected cell, and the infected B cells begin to express these proteins. It is these viral proteins that mark the cell as infected, and they are the targets for cell-mediated killing by the host defense. During the acute phase of mononucleosis, more than 20% of B lymphocytes express EBNA proteins.

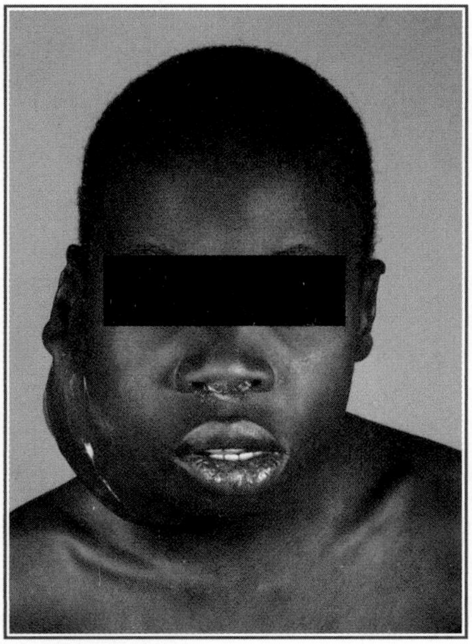

Figure 25.11 A child with Burkitt's lymphoma, which is caused by the Epstein–Barr virus.

Patients with infectious mononucleosis experience fever, malaise, pharyngitis, tender lymphadenitis, and splenomegaly. These symptoms persist for days to weeks and then slowly resolve spontaneously. In one to five per cent of cases, however, there can be complications, such as laryngeal obstruction, meningitis, encephalitis, hemolytic anemia, thrombocytopenia, and splenic rupture.

The Burkitt's lymphoma caused by the EB virus is a common malignancy in children in sub-Saharan Africa (**Figure 25.11**). The highest number of lymphomas seem to cluster in areas where malaria is prevalent, and there is some thought that malaria acts as an infectious co-factor or possibly a predisposing factor for Burkitt's lymphoma.

Treatment of Epstein–Barr virus infection — The treatment for infectious mononucleosis is mostly supportive, and more than 95% of those infected make a full recovery. In a small minority, the spleen becomes at risk of rupture, and the patient is restricted from taking part in contact sports. In laboratory tests, the EB virus has been shown to be sensitive to acyclovir, but systemic administration of this drug has little effect on clinical illness.

Arbovirus Infections

A variety of infections are seen with arboviruses (**Table 25.4**), and some of them affect the blood. In the blood, arbovirus infections can be classified as *fever infections* because this symptom is always present. We will discuss two of these infections: *dengue fever* and *yellow fever*. Recall that these infections were part of our discussion of emerging infectious diseases in Chapter 8.

Arbovirus infections are usually transmitted by mosquitoes, and in some cases the virus can be transferred from one generation of mosquitoes to the next. The vector suffers no ill effects from the virus it harbors. The virus multiplies in the vector during what is called the *extrinsic incubation period*, which increases its numbers and enhances the chance of causing infection when a host is bitten. Transient viremia that can last for a week or more is a feature of dengue and yellow fever infection in humans.

Pathogenesis of Arbovirus infections — There are three major manifestations of arbovirus infection in humans that are associated with the affinity of the virus for various target organs. Some arboviruses target the *central nervous system*, others attack *major organs*, particularly the

Table 25.4 Selected Arboviruses That Cause Infection in Humans

Organism and Infection	Distribution	Vector	Infection Expression
Togaviruses			
Western equine encephalitis	North America	Mosquito	Encephalitis
Eastern equine encephalitis	North America	Mosquito	Encephalitis
Flaviviruses			
St Louis encephalitis	North America	Mosquito	Encephalitis
Dengue fever	All tropical zones	Mosquito	Febrile illness
Yellow fever	Africa, South America, and the Caribbean	Mosquito	Hepatic necrosis
West Nile fever	Africa, Eastern Europe, Middle East, Asia, and North America	Mosquito	Febrile illness and encephalitis

liver (the yellow fever virus works in this way), and the virus that causes hemorrhagic fever damages *small blood vessels* (this is what causes hemorrhaging). In addition, all arboviruses produce a cellular necrosis that instigates inflammation and leads to fever.

Yellow fever — In yellow fever, the arbovirus attacks the liver and causes necrosis of the hepatocytes. The virus can also affect the urinary system by destroying the renal tubules, cause brain hemorrhaging, and destroy the myocardium. Hemorrhage is the major complication in yellow fever, which is distributed throughout the Caribbean, Central and South America, and Africa. Yellow fever is also a potential threat to the southeastern United States because the vector for this virus (the *Aedes aegypti* mosquito) has migrated to this area. The clinical symptoms associated with yellow fever include the abrupt onset of fever, chills, headache, and hemorrhaging that may become severe and cause bradycardia (a slow heart rate) and shock.

Fast Fact

Yellow fever had plagued the southeastern United States and Caribbean for 200 years and had caused the French to stop work on the Panama canal. Walter Reed, an army medical officer, found that the infection was transmitted by the mosquitoes. When steps were taken to control the mosquito population, the Panama Canal was able to be completed. Today, the Walter Reed Army Medical Center in Washington is named after this medical pioneer.

Dengue fever — There are four related serotypes of the virus responsible for dengue fever. The serotypes are spread throughout the world, particularly in the Middle East, Africa, the Far East, and the Caribbean. The vector for dengue fever is the same as for yellow fever, the *Aedes aegypti* mosquito, but the clinical infection seen in dengue is different from that seen in yellow fever. In dengue fever the infection brings about fever, rash, and severe pain in the back, head, muscles, and joints. Some of the more severe forms of the infection are characterized by shock, pleural effusion, hemorrhage, and death.

Treatment of Arbovirus infections — There is no specific treatment for arbovirus infections other than supportive care. Prevention can be enhanced by control of the vector population, but this is not easy. There is a live attenuated vaccine for yellow fever, and many countries where it occurs require travelers to these countries to have been vaccinated.

Filovirus Fevers

Filoviruses are filamentous viruses that occur in branched, fishhook, and circular configurations. They vary in length and are negative single-stranded RNA viruses. Two of the best-known (thanks to television and movies) are Ebola and Marburg, the only two filoviruses that infect humans.

Ebola caused outbreaks of hemorrhagic fever in 1976, with a mortality rate of 88% in Zaire and 50% in Sudan. Then in 1995, an outbreak in Zaire made headlines with more than 200 cases and a 75% mortality rate. As it turns out, about 10% of the population in rural Central Africa carries antibodies against the virus, indicating that they have been exposed. Transmission is person-to-person, and the infection is very contagious.

Marburg was first recognized in Germany when technicians working with monkey kidney cells began dying of hemorrhagic infection. This virus has also been seen in nosocomial settings, with 25% mortality.

Pathogenesis of Ebola and Marburg infection — Both Ebola and Marburg cause hemorrhaging in the skin, mucous membranes, and internal organs. Liver cells, lymphoid tissue, kidneys, and gonads are all destroyed, and there can also be brain edema.

The reasons for such rapid lethal hemorrhaging are still not clear, but there is evidence that Marburg replicates in vascular endothelial cells. This causes necrosis and bleeding. Ebola may destroy cells by secreting a glycoprotein that interacts with neutrophils and inhibits the inflammatory defense reaction of the host. Ebola causes symptoms in as little as four to six days after infection, and the mortality rate is an extremely high 30–80%.

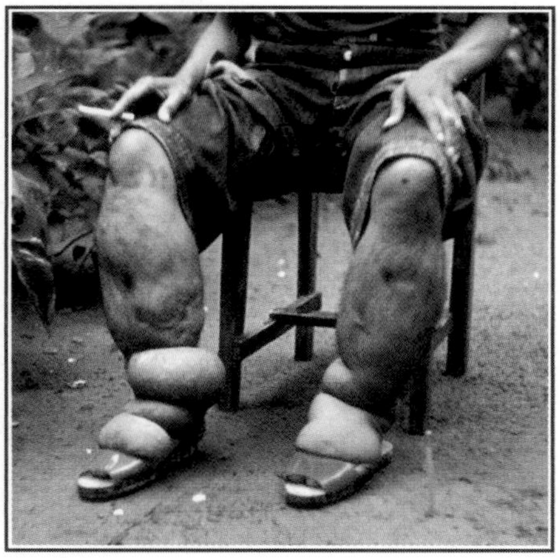

Figure 25.13 Elephantiasis of the leg caused by the parasite *Wuchereria bancrofti*. The swelling is the result of a massive blockade of the lymphatics by the adult worms.

induces an acute inflammatory response, chronic lymphatic blockade, and, in some cases, the swelling of the extremities and genitalia known as **elephantiasis** (**Figure 25.13**).

There are two parasites most commonly involved in filariasis infections: *Wuchereria bancrofti* and *Brugia malayi*. Both are threadlike worms that lie coiled up in the lymphatic vessels of a human host for decades. Gravid females produce large numbers of fertile eggs, and once the eggs are laid the embryos uncoil to their full length of between 200 and 300 μm. At this point they are referred to as **microfilariae**. The shell of each egg elongates and becomes a flexible sheath around its microfilaria.

The microfilariae eventually leave the lymph and enter the host's blood, where most of them accumulate in the pulmonary vessels during the day. However, at night the gas pressure in the host pulmonary system changes and the change in pressure causes the worms to spill into the systemic circulation, where they are found in greatest numbers between 9 p.m. and 2 a.m. This movement between the pulmonary circulation and the systemic circulation is called *periodicity*, and it is important because it determines which type of mosquito will serve as the vector and intermediate host.

A mosquito feeding on an infected human ingests the microfilariae along with the human's blood. The microfilariae are transformed into the larval form in the thoracic muscles of the mosquito, and when the mosquito bites an uninfected human, the microfilariae penetrate the feeding site. In the new host, the microfilariae migrate to the lymphatic vessels and go through a series of molts until they reach adulthood, a process that can take 6–12 months.

Pathogenesis of filariasis — Currently there are about 120 million people infected with either *W. bancrofti* or *B. malayi*, mostly in Africa, Latin America, the Pacific islands, and Asia, with more than 75% of the cases being in Asia. Humans are the only known vertebrate hosts for most strains of *Brugia* and *Wuchereria*.

The pathology of filariasis infections is confined primarily to the lymphatic system, and there are two types of infection, acute and chronic. In the acute form, the presence of molting adolescent and dying adult worms stimulates dilatation of the lymphatics and hyperplastic changes to the vessel endothelium. This brings about infiltration of lymphocytes, plasma cells, and eosinophils, and this infiltration results in the formation of a granuloma, fibrosis, and permanent lymphatic blockade. Repeated infections eventually result in massive lymphatic blockade, which causes the skin and subcutaneous tissues to fill with edematous fluids and fibrous tissue. At this point, there can be bacterial and fungal superinfections of the skin, which contribute to further tissue damage.

The chronic form of filariasis usually develops 10–15 years after the onset of the first acute attack. The incidence and severity of chronic clinical manifestations tend to increase with age. The main characteristic features of chronic filariasis are chronic **lymphangitis**, thickened lymphatic trunk, chronic lymphedema, and elephantiasis.

People indigenous to areas where *W. bancrofti* and *B. malayi* are found usually remain asymptomatic after infection; however, some can experience filarial fevers and lymphadenitis for 8–12 months. These fevers are usually low-grade and are accompanied by chills and muscle aches. Lymphadenitis will normally first be noted in the femoral areas as an enlarged, red, tender lump. Inflammation then spreads down the lymphatic channel of the leg, and vessels become enlarged and tender, with the overlying skin red and edematous.

Acute manifestations last a few days, resolve spontaneously, and then reoccur periodically for weeks or even months. If infection occurs repeatedly, however, there can be permanent lymphatic obstruction involving edema, ascites, and pleural and joint effusion. If the lymphadenopathy persists, lymphatic channels can rupture and cause the formation of abscesses. In patients heavily and repeatedly infected over periods of decades, the infection will result in elephantiasis.

Treatment of filariasis — Diethyl carbimazine eliminates the microfilariae from the blood and kills or injures the adult worms, resulting in long-term suppression of the infection and a potential cure. However, the dying worms can elicit an allergic reaction in the host. This reaction is occasionally severe and requires the use of antihistamines and corticosteroids.

Tissue changes seen in elephantiasis are often irreversible, but enlarging of the extremities may be ameliorated through the use of pressure bandages. As with any vector-transmitted infection, control of the vector (in this case, mosquitoes) will help prevent infection.

Keep in Mind

- The three most important parasitic blood infections are malaria, toxoplasmosis, and schistosomiasis, all examined in Chapter 14.
- Two additional parasitic blood infections are Chagas' disease and filariasis.
- Chagas' disease is caused by the flagellate protozoan *Trypanosoma cruzi*, which is vector-transmitted by reduviids, flying insects known as "kissing bugs."
- Filariasis is caused by the parasites *Wuchereria bancrofti* and *Brugia malayi*, which infect the human lymphatic system and can cause acute inflammatory responses and, in severe cases, elephantiasis.

SUMMARY

- Because blood and lymph travel to all parts of the body, they are good ways to spread infection.
- The presence of bacteria in the blood is referred to as bacteremia, viruses in the blood as viremia, fungi in the blood as fungemia, and parasites in the blood as parasitemia.
- Intravascular infections arise from pathogens gaining entrance to the blood; extravascular infections occur when tissue infections overflow into the blood.
- Bacterial infections of the blood include plague, tularemia, brucellosis, Lyme disease, and relapsing fever.
- Rickettsial infections of the blood include Rocky Mountain spotted fever, and epidemic or endemic typhus.
- Viruses that cause infection of the blood include cytomegalovirus, Epstein–Barr virus, and Ebola and Marburg viruses.
- The three most important parasitic infections of the blood are malaria, toxoplasmosis, and schistosomiasis, but Chagas' disease and filariasis are also parasitic infections of the blood.

In this chapter, we have seen that, like the body's other systems, the blood and lymph can be targets for a variety of infections caused by all of the "usual suspects" (bacteria, viruses, and parasites). Notice that we did not discuss fungal infections in this chapter, but keep in mind that systemic fungal infections make use of the blood to move to distal locations in the body.

SELF EVALUATION AND CHAPTER CONFIDENCE

Multiple Choice

Answers are given in the back of the book and help can be found in the student resources at:
www.garlandscience.com/micro

1. A condition in which bacteria are growing in the blood is called
   A. Bacteremia
   B. Septicemia
   C. Endotoxic
   D. Parasitemia
   E. None of the above
2. Infection of the veins is called
   A. Bacteremia
   B. Endocarditis
   C. Thrombophlebitis
   D. Endoarteritis
   F. Veinitis
3. Endocarditis involves
   A. Colonization of the heart by bacteria
   B. Deposition of excessive amounts of platelets at the site
   C. Deposition of fibrin at the site
   D. Formation of mature vegetation
   E. All of the above
4. Complications of infectious endocarditis include
   A. Rupture of the chordae tendineae
   B. Congestive heart failure
   C. Perforation of the heart valves
   D. Formation of emboli
   E. All of the above
5. In patients who are NOT debilitated it would be unusual to find the following growing in the blood:
   A. *Staphylococcus aureus*
   B. *Candida albicans*
   C. *Staphylococcus epidermidis*
   D. *Corynebacterium*
   E. None of the above
6. Plague is transmitted to humans by
   A. The exchange of bodily fluids
   B. The use of contaminated needles
   C. The bite of the rat flea
   D. The bite of a mosquito
   E. The bite of a tick
7. Plague is caused by
   A. *Streptococcus mutans*
   B. *Yersinia pestis*
   C. *Neisseria plagus*
   D. *Xenopsylla cheopis*
   E. None of the above
8. The most contagious form of plague is
   A. The cutaneous form
   B. The neurologic form
   C. The pneumonic form
   D. The oral form
   E. The genitourinary form
9. Organisms of the genus *Francisella* cause
   A. Plague
   B. Typhoid fever
   C. Tularemia
   D. Relapsing fever
   E. None of the above
10. Lyme disease is caused by
    A. *Staphylococcus*
    B. *Borrelia*
    C. *Candida*
    D. *Chlamydia*
    E. *Francisella*
11. Lyme disease is transmitted from
    A. Human to human
    B. Tick to human
    C. Deer to human
    D. Sand flea to human
12. Lyme disease involves all of the following except
    A. Humans
    B. Deer
    C. Ticks
    D. Rats
13. Epidemic typhus is caused by
    A. Ticks
    B. Rat fleas
    C. Body lice
    D. *Rickettsia typhi*
    E. None of the above
14. The etiologic agent for mononucleosis is
    A. Variola major
    B. Epstein–Barr virus
    C. Cytomegalovirus
    D. Herpes simplex type 1
    E. Herpes simplex type 2

15. The protozoan *Trypanosoma cruzi* causes
   - A. Sleeping sickness
   - B. Malaria
   - C. Relapsing fever
   - D. Chagas' disease
   - E. Dengue fever

DEPTH OF UNDERSTANDING

Questions listed here require you to bring together the concepts you have learned in this chapter into a discussion format. This helps you to increase your depth of understanding of the material you have learned. Help can be found in the student resources at:
www.garlandscience.com/micro

1. Describe the events associated with the development of endocarditis and why this condition is so dangerous.

2. Discuss the different aspects of infection with *Borrelia burgdorferi* and how this infection can be controlled environmentally.

CLINICAL CORNER

Help can be found in the student resources at:
www.garlandscience.com/micro

1. A patient comes into the clinic for follow-up after having had his spleen removed three months earlier as a result of trauma from an automobile accident. After checking to make sure that he is feeling well and healing properly, the doctor has given him a prescription for antibiotics and told him it is important that he take these before and after he visits the dentist, even if he is only going to have his teeth cleaned. He asks you why this has been ordered.
   - A. What do you tell him?
   - B. What are the potential consequences if he neglects to follow the doctor's orders?
2. You are working as a physician's assistant in a Doctors Without Borders clinic in the Congo. Your patient is a 60-year-old female who is a Catholic nun that has been on a religious mission to the interior areas of the Congo. She is complaining of difficulty in breathing, pain, and cramping in the groin, which has been going on for about four days. On examination you discover several bulging inguinal lymph nodes, some of which are necrotic and bleeding.
   - A. What are your initial thoughts about this patient?
   - B. Why would these inguinal nodes be large and necrotic?

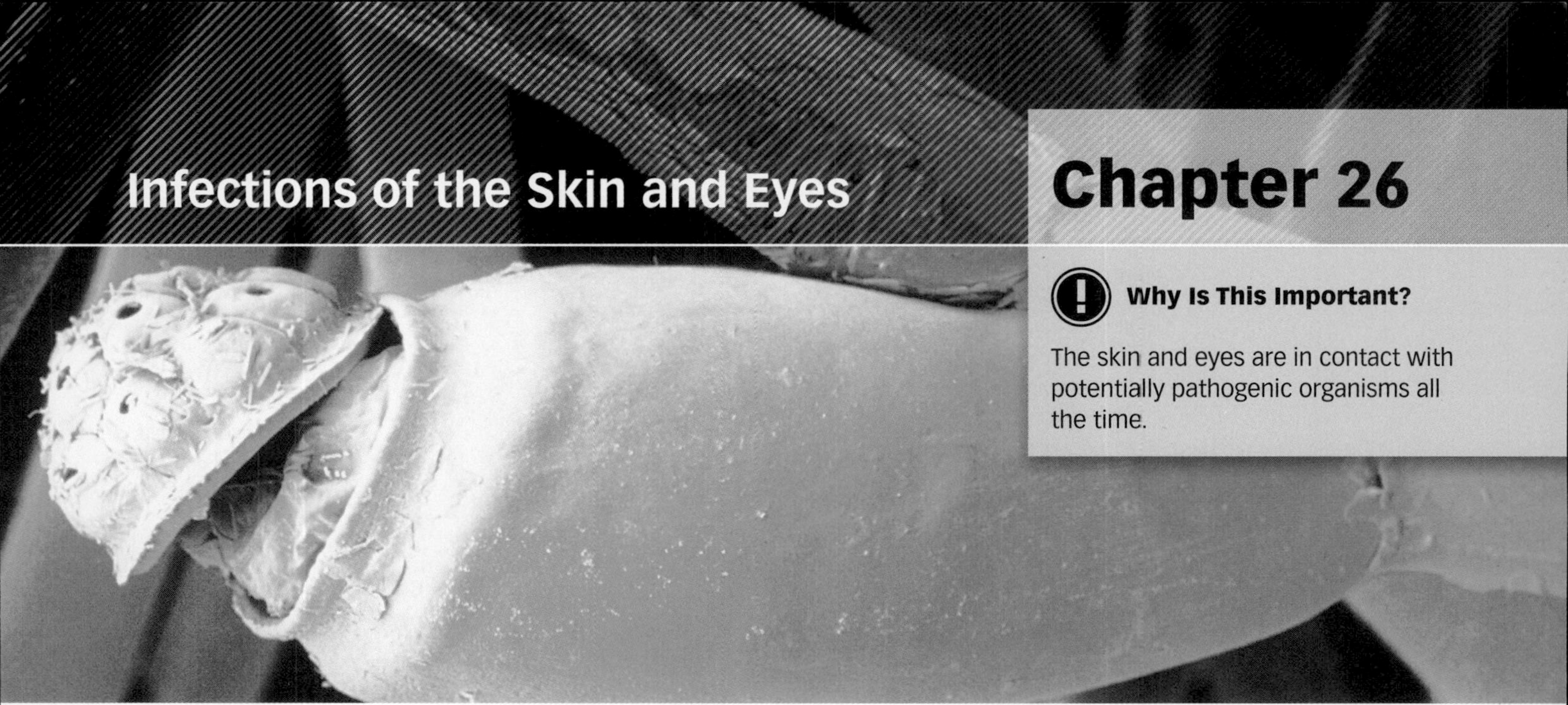

Chapter 26

Infections of the Skin and Eyes

Why Is This Important?

The skin and eyes are in contact with potentially pathogenic organisms all the time.

OVERVIEW

In this chapter, we conclude our discussions of infections that affect different parts of the body with a look at infections of the skin and eyes. For the skin, we will see that many infections can occur here and that there are two main reasons for this: first, the skin is always exposed to pathogenic organisms, and second, the soft tissue just below the skin is an excellent breeding ground for infection. Eyes are also open to the outside world, and infections here can be extremely dangerous because of the proximity to the nervous system and the potential for loss of vision.

We will divide our discussions into the following topics:

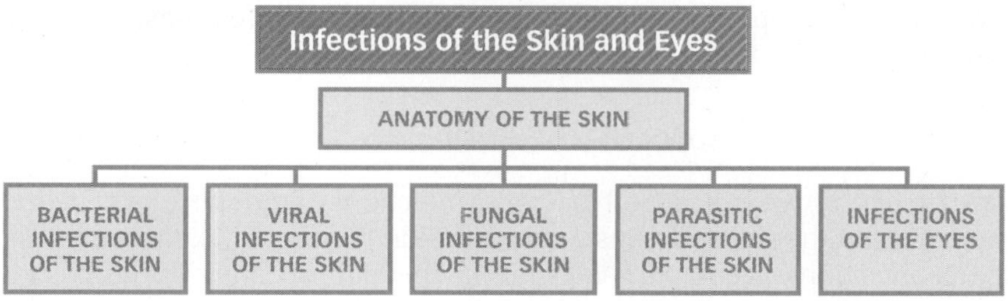

What Do I Need to Know?

To get the most out of this chapter, please review the following terms from your previous courses in biology, anatomy, physiology, or chemistry or in previous chapters of this book as indicated in parentheses:antigenic drift (**8**), conjunctiva, cornea, dorsal root ganglion, edema, epidermis (**5**), exotoxin (**5**), focus of infection (**5**), fomites (**6**), hematoma, hyphae (**14**), lymphadenopathy (**16**), mycosis (**14**), necrotic, nosocomial (**5**), nucleoside analog (**11**), opsonization (**5, 15**), septicemia (**5**), single-stranded RNA virus (**12**), syncytia (**5,12**), synergy, viruria (**13**), yeast (**14**).

ANATOMY OF THE SKIN

Before we look at the kinds of infection that occur in the skin, it would be wise to revisit the anatomy of the skin briefly. The skin is the largest organ in the body and accounts for a significant portion of our weight. It is the barrier between our body and the outside world and, as we learned in Chapter 15, is the first line of defense against invading microorganisms. When you look closely at the skin, it is apparent that this organ is very well engineered to fit the role of protective barrier (**Figure 26.1**). The outer layer (the epidermis), which comes into direct contact with the environment, is essentially contiguous and made up of several layers of dead or dying cells. There is constant shedding of these cells, and this shedding keeps pathogens from successfully attaching to the skin.

In addition, there are several mechanical mechanisms that discourage pathogens from colonizing the skin. For example, the production of perspiration has a flushing action that constantly moves pathogens off the surface of the skin. More importantly, *sebum*, which is produced by the sebaceous glands, is a natural antibacterial substance.

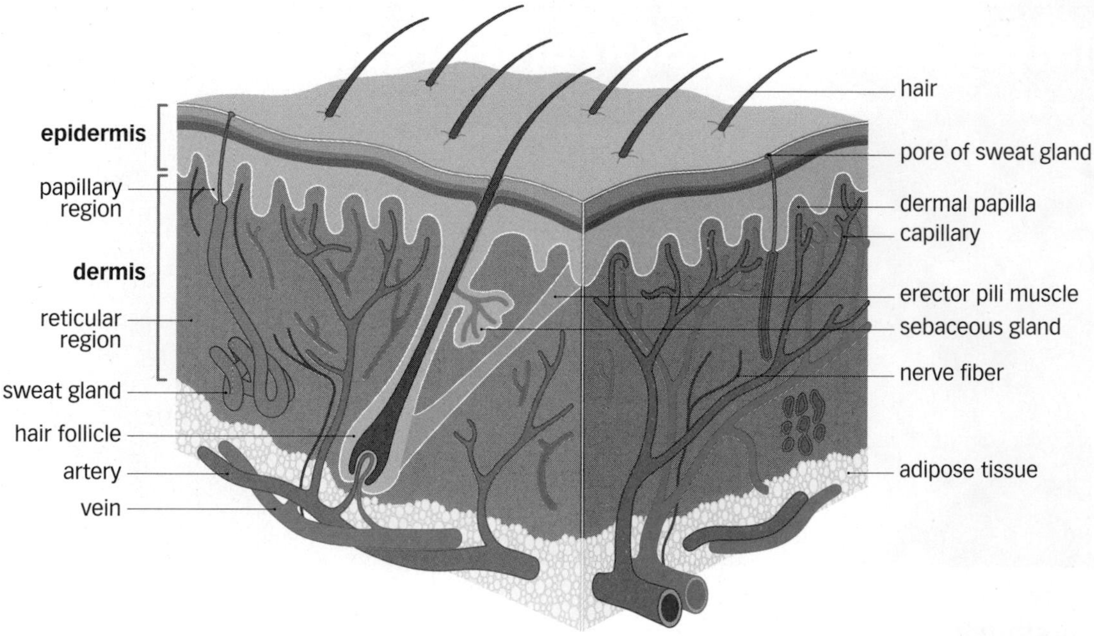

Figure 26.1 A cross section of the skin, a major barrier against infection.

In spite of these protective mechanisms, the skin is often breached by trauma, such as abrasions, wounds, punctures, and even bites from insects or animals. When this occurs, pathogens can get into the underlying tissue. Infections of the skin can be caused by bacteria, viruses, and fungi. As we have done in the other systems' chapters, we will look at each of these types of infection separately.

A summary of some skin infections is shown in **Table 26.1**, and there are four main types of skin lesions: **macules**, **papules**, **vesicles**, and **pustules** (**Figure 26.2**).

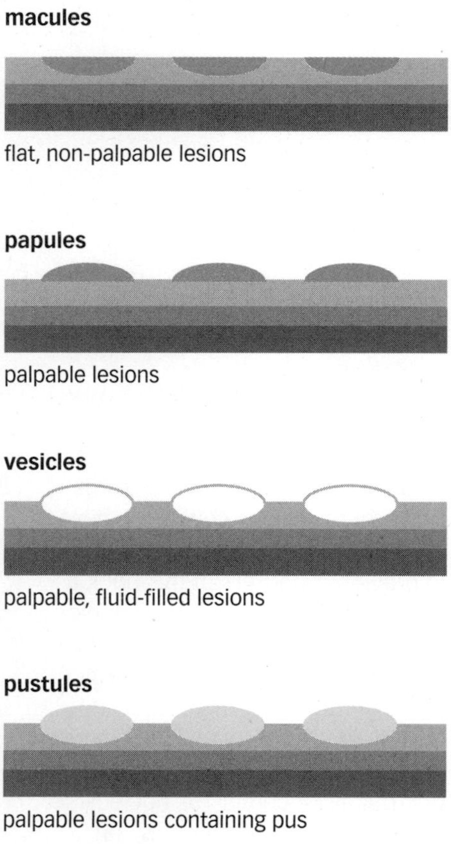

Figure 26.2 Differences between skin macules, papules, vesicles, and pustules, the four types of skin lesion.

BACTERIAL INFECTIONS OF THE SKIN

In this section, we look at both bacterial infections of the skin and bacterial infections of the soft tissue just under the skin (**Figure 26.3**). Most of these latter infections accompany a break in the skin because of the barrier nature of this organ. Once the skin is breached, organisms can enter the soft tissue below, which contains the blood supply and is a rich environment for pathogen growth. In addition, the nearness of the body's blood supply to this soft tissue increases the possibility of systemic infection.

Because any surgical procedure breaches the skin, wound infections are an important problem in hospital settings. In this situation, *Staphylococcus aureus* and group A streptococci are major concerns for nosocomial infections. Similarly, the bite of a dog or cat can introduce organisms from their oral flora into the soft tissues, and this can result in serious infection.

The intact skin is relatively dry, and the cells of the epidermis are constantly being sloughed off. The surface of the skin usually has a pH of 5.0 to 6.0, which is too acidic for many pathogens. The skin also harbors a large number of coagulase-negative staphylococci and other Gram-positive organisms that are part of the skin's normal flora and act as additional barriers to opportunistic pathogens. Sebum can be converted to free fatty acids by the normal flora of the skin, and these fatty acids inhibit the growth of such pathogens as group A streptococci.

Infection	Organism Causing the Infection	Characteristics of the Infection
Bacterial		
Folliculitis	*Staphylococcus aureus*	Skin abscess
Scalded skin syndrome	*Staphylococcus aureus*	Vesicular lesions over the skin. Mostly seen in infants
Erysipelas	*Streptococcus pyogenes*	Skin lesions that can spread to systemic infection
Acne	*Propionibacterium acnes*	Skin lesions caused by excess of hormones, which is common in teens
Impetigo	Staphylococci and streptococci	Skin lesions seen in children spread by hands and fomites
Viral		
Rubella	Rubella virus	Mild maculopapular rash. Dangerous in pregnancy
Measles	Rubeola virus	Severe infection with fever, conjunctivitis, cough, and rash
Chickenpox	Varicella-zoster	Generalized macular skin lesions
Shingles	Varicella-zoster	Pain and skin lesions usually on the trunk
Smallpox	Smallpox virus	This infection has been eradicated through immunization but is a favorite possibility as a biological weapon
Fungal		
Dermatomycoses	Dermatophytes	Dry scaly lesions on various parts of the skin
Candidiasis	*Candida albicans*	Patchy inflammation of the mucous membranes of the mouth or vagina

Table 26.1 Infections of the Skin

Bacteria enter the skin via minor abrasions, hair follicles, and surgical or traumatic wounds. One of the most difficult-to-treat skin infections is fasciitis. Fasciae are sheets of fibrous tissue lying just below the subcutaneous layer of the skin, and **fasciitis** is any infection in a fascia. **Necrotizing fasciitis** (**Figure 26.4**), an infection in which fascia is destroyed, can be caused by an organism such as group A streptococci acting either alone or in combination with other bacteria through a synergistic effect. Usually, this synergy comes from bowel flora, which can cause fasciitis in the abdominal wall after surgery. After trauma or surgery, bacteria enter the fascia. In patients with bacteremia, group A streptococci can reach the fascia and settle in small hematomas or in bruised areas.

Once bacteria enter the fascia, they spread rapidly, and the resulting inflammatory response then affects the neurovascular bundles found in the fascia. Thrombosis of these vessels compromises the blood supply and nerves to that area, and the area rapidly becomes necrotic. In fact, this infection can be so rapid that only surgical removal of the tissue can resolve the spread.

Infections of the Hair Follicles, Sebaceous Glands, and Sweat Glands

The surface of the skin is normally penetrated by ducts, hairs, and sweat glands. These structures serve specific functions that are part of the normal physiology of the body. Microbial invasion can occur along any of these routes, especially through any ducts that are obstructed for some reason.

Figure 26.3 Some of the organisms involved in infections of the skin and underlying soft tissue.

	organism	disease
infections of the skin		
	group A streptococcus	impetigo, ecthyma, erysipelas, cellulitis, paronychia
	Staphylococcus aureus	impetigo, folliculitis, carbuncles, cellulitis, paronychia
	Pseudomonas aeruginosa	folliculitis, paronychia, cellulitis
	Propionibacterium acnes	acne
soft tissue infections		
	group A streptococcus	necrotizing fasciitis, myositis
	coliforms	synergistic necrotizing fasciitis, Fournier's gangrene
	anaerobes of the bowel	synergistic necrotizing fasciitis, Fournier's gangrene
	Clostridium perfringens	gas gangrene

Folliculitis, Acne, Impetigo, and Erysipelas

Pathogenesis of folliculitis — This minor infection of hair follicles, usually caused by *S. aureus*, is associated with sweat gland activity. Folliculitis infections are therefore most often seen in areas where these glands predominate, such as the neck, face, axillae, and buttocks. Blockage of the gland ducts predisposes to folliculitis, and serious infections can result in the formation of **boils** (furuncles), which are a type of abscess. In this case, there is a localized region of pus surrounded by tissue inflammation. Antibiotics cannot penetrate a boil very well, which means that boils are difficult to treat. Draining the abscess is usually the initial step in treatment. If the body defenses do not wall off this localized infection, neighboring tissues become infected and there is usually extensive damage and the onset of fever. This enlarged infected region is referred to as a **carbuncle**.

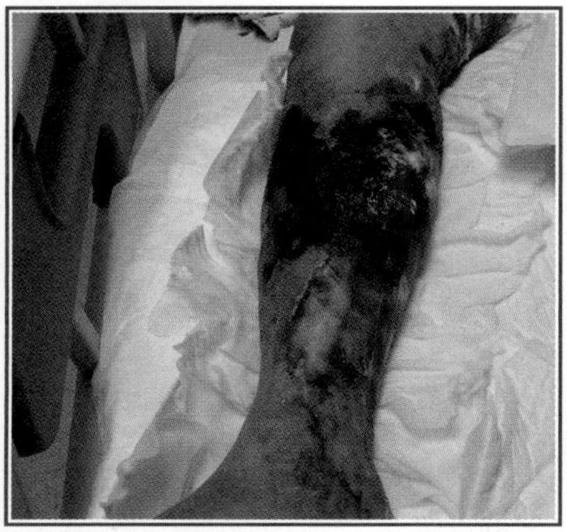

Figure 26.4 Necrotizing fasciitis arising from infection with group A *streptococci*.

Folliculitis can also be caused by *Pseudomonas aeruginosa*. In fact there has been an increase in these infections associated with hot tubs and whirlpools. At the elevated temperatures used in these devices, *Pseudomonas* can grow in large numbers, causing large areas of folliculitis in those parts of the body that are immersed. The symptoms normally subside when the insult is discontinued.

Pathogenesis of acne — Acne is the most common skin infection in humans, affecting 17 million people in the United States alone, with 85% being teenagers. There are three categories of this skin infection: *comedonal acne*, *inflammatory acne*, and *nodular cystic acne*. Comedonal acne results from inflammation of the hair follicles of the face and the associated sebaceous glands, which become plugged by a mixture of shedding skin cells and sebum. As the sebum backs up and accumulates, whiteheads, called *comedos*, appear on the skin. If the blockage protrudes through the skin, blackheads, called *comedones*, appear on the skin.

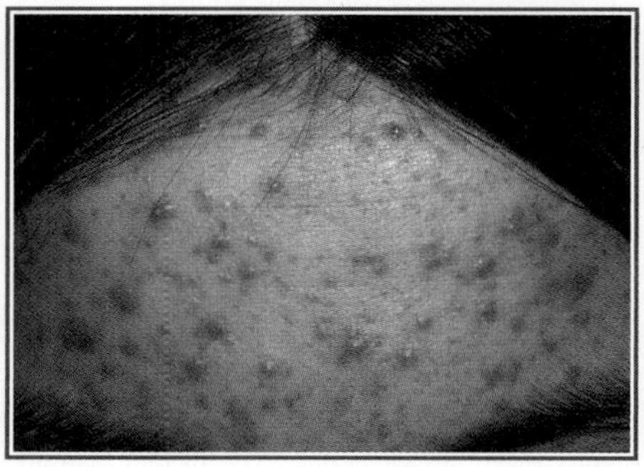

Figure 26.5 Inflammatory acne, a skin infection seen in young people, is caused by the bacterium *Propionibacterium acnes*.

Inflammatory acne (**Figure 26.5**) is caused by the bacterium *Propionibacterium acnes*, which is the predominant anaerobe of the skin and metabolizes the glycerol component of sebum. This glycerol metabolism causes free fatty acids to form, and the presence of the fatty acids initiates an inflammatory response. Neutrophils, which are part of the inflammation process, secrete enzymes that damage the wall of the hair follicles, causing pustules and papules to form. The primary cause of inflammatory acne is hormonal influence on the secretion of sebum, which is increased during puberty. Usually, inflammatory acne resolves spontaneously in adulthood.

The nodular cystic form of acne is characterized by the formation of **cysts**, which are inflamed lesions that lie deep in the skin and are filled with pus. These lesions leave prominent scars on the skin.

Fast Fact

The use of antibiotics for minor comedonal acne can be a problem because the acne will resolve naturally but the use of these drugs contributes to antibiotic resistance.

Treatment of folliculitis and acne — Folliculitis is usually treated locally by drainage without antibiotics. All three types of acne can usually be effectively treated with topical drying agents.

Pathogenesis of impetigo — **Impetigo** is a common, sometimes epidemic, skin lesion that is caused primarily by group A streptococci or staphylococci. The initial lesion is often a small vesicle that develops at the site where the bacteria have entered the skin. The vesicle ruptures and causes superficial spread of the bacteria, with the spread being characterized by skin erosion and serous exudate. This area dries to a honey-colored crust that contains numerous infectious *Streptococcus* or *Staphylococcus* species (**Figure 26.6**). The infection is usually found in children and is associated with heat, humidity, and poor hygiene. It can be spread by fomite transmission when people share contaminated towels or clothing.

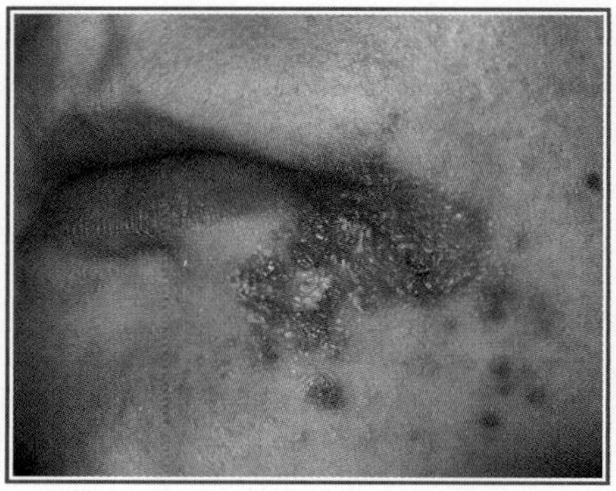

Figure 26.6 Impetigo infection caused by staphylococci. This infection is characterized by isolated pustules that become crusted over.

Treatment of impetigo — The usual treatment involves penicillin or erythromycin taken orally and topical skin antiseptics to limit the spread of the infection.

Pathogenesis of erysipelas — The skin condition known as **erysipelas** is characterized by a rapidly spreading infection of the deeper layers of the dermis. The infection is always caused by group A streptococci. Symptoms include edema of the skin marked by erythema (redness), pain, and systemic infection, including lymphadenopathy and fever. Erysipelas can progress to septicemia and local necrosis of the skin and can be serious, requiring immediate treatment.

Treatment of erysipelas — Either penicillin or streptomycin is effective in dealing with this infection.

Scalded Skin Syndrome

The bacterial infection **scalded skin syndrome** derives its name from the fact that the salient sign is the blistering and peeling off of large sheets of skin (**Figure 26.7**). The infection is caused by two exotoxins secreted by certain strains of *S. aureus*. The exotoxins are referred to as **exfoliatins**.

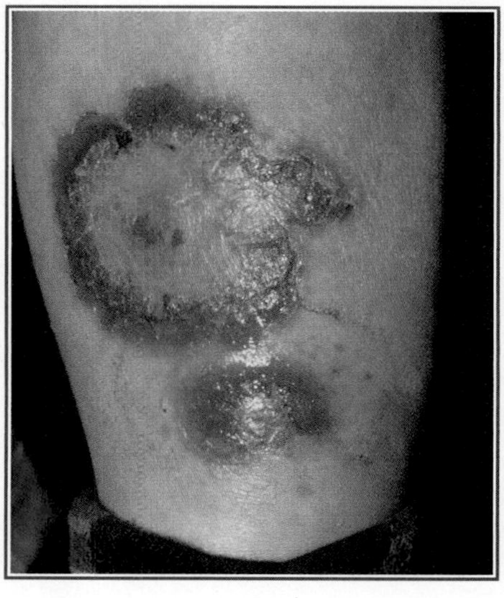

Figure 26.7 Scalded skin syndrome (also known as Lyell's disease), which is an infection caused by *Staphylococcus aureus*.

The gene for one exfoliatin is located on the bacterial chromosome, and the gene for the other exfoliatin is located on a plasmid. It is therefore possible for some strains of *S. aureus* to have only one of the genes and for other strains to have both genes. Scalded skin syndrome is normally restricted to infants but can occur in adults, especially in the late stages of toxic shock syndrome.

Pathogenesis of scalded skin syndrome — Exotoxins are transported through the blood to distal sites, where they cause the upper layers of the skin to separate and peel off. The first sign is a reddened area of the skin, usually around the mouth. The area soon spreads out, forming large, soft vesicles over the whole body. As the top layer of skin peels away, it leaves the exposed dermal layer looking scalded. This condition is only temporary, because the skin will regenerate in 7–10 days. There is an accompanying high fever throughout the infection.

Treatment of scalded skin syndrome — There is a good immune response to this infection, and so recurrence is unlikely. Most of the bacteria responsible for the infection are sensitive to penicillin, and in cases of penicillin-resistant strains, cephalosporins are effective. Alternatives include vancomycin, clindamycin, and erythromycin.

Gas Gangrene (*Clostridium perfringens*)

Gangrene is tissue necrosis resulting from an obstructed blood supply, and the skin infection known as **gas gangrene** gets its name from the fact that the bacteria responsible for the infection release gases as part of their metabolic activity. The infection is caused by *Clostridium perfringens* and is usually associated with deep tissue wounds often seen in battlefield injuries. However, car and motorcycle accidents can also result in this type of infection.

Pathogenesis of gas gangrene — *C. perfringens* is a Gram-positive, anaerobic, spore-forming rod. The spores are introduced into tissue where blood circulation has been impaired and the tissue is dead. Because this environment is perfect for anaerobic growth, the spores germinate and the bacteria multiply. They produce toxins, proteinase, lipase, hyaluronidase, and collagenase, all of which destroy the tissue surrounding the already-dead tissue that is the focus of the infection. The destruction of the surrounding tissue expands the anaerobic environment, and the infection spreads.

The onset of gas gangrene is sudden, appearing anywhere from 12 to 48 hours after the initial injury to the tissue. As the bacteria grow, they ferment and produce hydrogen gas, which causes breaks in the tissue. (This tissue is referred to as crepitant tissue.) Movement of the affected area causes snap, crackle, and popping sounds. There is also a foul smell associated with the tissue destruction, making it obvious that infection has set in. The infection is accompanied by a high fever, massive tissue destruction, shock, and blackened skin (**Figure 26.8**). If not treated quickly, a gas gangrene infection is lethal.

Treatment of gas gangrene — Excision of the affected tissue opens the area to oxygen, which kills the anaerobic pathogens. Treatment in hyperbaric chambers is effective in reducing the spread of infection. Penicillin and clindamycin should also be given.

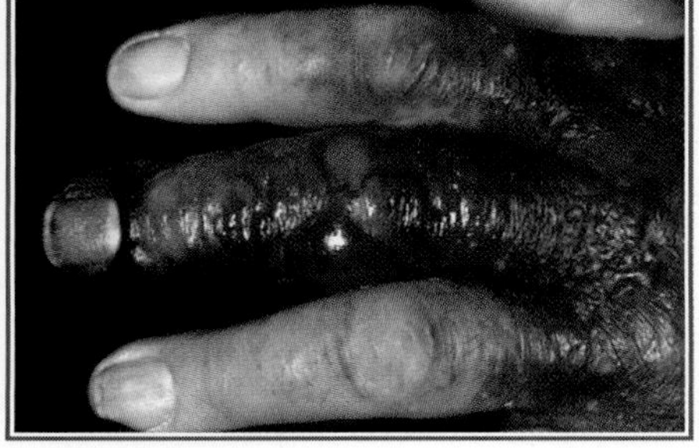

Figure 26.8 Gas gangrene of the hand.

Cutaneous Anthrax

As we saw in Chapter 21, *Bacillus anthracis* can cause serious respiratory infection. Because of that feature, this bacterium has received notoriety as a potential bioweapon (see Chapter 28). This organism can also cause less harmful infections in the skin, and the one we look at here is **cutaneous anthrax**.

Pathogenesis of cutaneous anthrax — The first signs of a cutaneous anthrax infection usually appear two to five days after anthrax spores have been inoculated into an opening in the skin, most often the forearm or hand. The initial lesion is a papule that looks like an insect bite. This papule progresses through vesicular and ulcerative stages in seven to ten days to form a black *eschar* (scab) surrounded by edema. Although the eschar is often referred to as a malignant pustule, it is neither.

The symptoms of cutaneous anthrax are normally mild, and the lesions typically heal slowly after the scab falls off. In some cases, however, the infection can become systemic, progressing to a massive edema and toxemia that can be fatal.

Treatment of cutaneous anthrax — Antibiotics have little effect on this infection except to protect against dissemination. *Bacillus anthracis* is susceptible to ciprofloxacin, which is the treatment of choice.

Keep in Mind

- Skin is a barrier impenetrable to pathogens and must be broken for infection to occur.
- A wide variety of bacteria cause skin infections, with necrotizing fasciitis being one of the worst.
- Bacterial infections of the hair follicles, sebaceous glands, and sweat glands include the following: folliculitis; comedonal, inflammatory, and nodular cystic acne; impetigo; and erysipelas.
- Scalded skin syndrome is caused by *Staphylococcus aureus*.
- Gas gangrene causes necrosis of tissues and can be rapidly fatal.
- Cutaneous anthrax is usually a mild infection unless it becomes systemic, in which case it can be fatal.

VIRAL INFECTIONS OF THE SKIN

The skin is also a barrier to viral pathogens, which, as we have seen throughout our discussions, are constantly surrounding us. The same barrier constraints that apply to bacteria also apply to viruses, which means there needs to be an entry point for the pathogens. There are several common viral infections that manifest their signs on the skin after systemic infection.

Measles

Measles is an extremely contagious infection caused by a single-stranded RNA virus of the paramyxovirus family. It is the leading cause of vaccine-preventable disease worldwide. Common forms of measles include **rubeola** (five-day measles) and **hard measles**, which last 7–18 days. The measles virus can produce severe infection in children, with accompanying high fever, widespread rash, and transient immunosuppression.

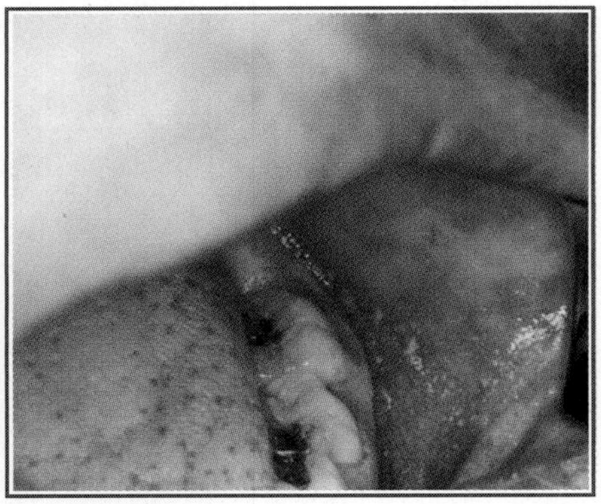

Figure 26.9 A red skin rash plus Koplik's spots on the inner surface of the cheek are early signs of a measles infection.

Measles usually occurs in preschool children who have not yet received the MMR (measles/mumps/rubella) vaccination. Although there is only one serotype of the measles virus and it is restricted to humans, there may be some antigenic drift associated with the virus.

Pathogenesis of measles — Between 9 and 11 days after exposure to the virus, the infection begins in the respiratory tract with cough, runny nose, and fever. These initial signs are followed by viremia and lymphatic spread of the virus throughout the body, including the lymph tissue, bone marrow, and skin. Virus can be found in the blood during the first week of illness, and viruria can persist for up to four days after the rash appears. There is also a depressed immunity and susceptibility to bacterial superinfections during this time. One to three days after the onset of the respiratory signs but before the skin rash breaks out, small red spots known as **Koplik's spots** appear on the mucous membrane of the cheeks (**Figure 26.9**). One day after the Koplik's spots comes the characteristic red skin rash. Significant numbers of virions can be found in the Koplik's spots as well as in the areas around the skin rash once it appears. Lymphadenopathy is common during this infection.

Measles can be very severe in individuals who are immunosuppressed and can result in death. In developing countries, a 15–25% mortality is associated with this infection. There can also be complications in up to 15% of cases, including otitis media, sinusitis, mastoiditis, pneumonia, and sepsis. In addition, 1 in every 1000 cases will develop encephalitis, which can cause permanent nerve damage or death.

Treatment of measles — There is no therapy other than supportive care and close observation for potential complications. There is a very effective vaccine that is routinely given to children as part of their vaccination schedule.

Rubella (German Measles)

Rubella is a very mild or asymptomatic infection that can involve low-grade fever, lymphadenopathy, and faint macular rash. However, this infection is very serious in pregnant women and can cause congenital abnormalities in the developing fetus. Rubella is usually seen in the spring, and 30–60% of susceptible individuals can develop a clinical infection. An infected individual is contagious for seven days before and seven days after the appearance of the rash.

Fast Fact

Many states require a blood test for rubella as part of the procedure to obtain a marriage license.

Pathogenesis of rubella — The virus enters through the respiratory tract and spreads to the blood, lymph, organs, and skin. The viremia is seen up to eight days before the rash appears, and viral shedding can be seen in the oropharynx up to eight days after the onset of infection. As mentioned above, in a pregnant woman there can be transplacental transfer of virions to the fetus.

Treatment of rubella — There is no specific therapy for rubella infection or for the congenital infection that sometimes accompanies it. Since 1969, administration of live attenuated vaccine (MMR) has caused a marked decrease (90%) in the number of rubella cases, and this vaccination is recommended in the first year of life.

Smallpox (Variola)

Smallpox, an infection caused by a DNA poxvirus, has been known since the Middle Ages, when 80% of the population of Europe suffered from this infection. It was introduced to the Americas by European colonists.

There are two forms of smallpox infection: **variola major**, for which the mortality rate can be 20% or higher, and **variola minor**, which has a very low mortality rate (approximately 1%). Technically speaking, smallpox has been eradicated from the entire world, with the last victim being seen in Somalia in 1977. Because the only reservoir for this virus is humans, there should be no more smallpox cases ever. However, there are stocks of the smallpox virus in both Russia and the United States, a situation that leaves open the possibility of further infections. This is further compounded by the decrease in herd immunity to smallpox. The dominant feature of smallpox infection is the appearance of a papulovesicular rash involving pustules that form within the first two weeks of the onset of infection.

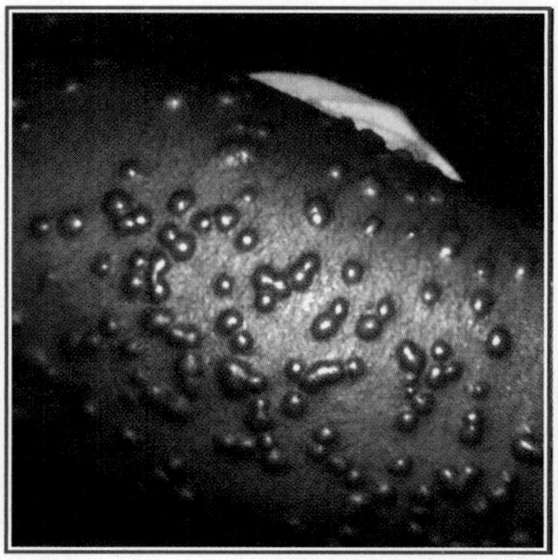

Figure 26.10 Photograph (from 1968) showing the right arm of a person with smallpox in the late pustular stage. The disease manifested itself as the classical maculopapular rash.

Pathogenesis of smallpox — The incubation period is 12–14 days but can be as short as 4–5 days, accompanied by the abrupt onset of fever, chills and muscle aches. A rash appears 3–4 days later and evolves to papulovesicles, which are seen most prominently on the head and extremities and become pustular over 10–12 days (**Figure 26.10**). These pustules appear only once, and they crust over and slowly heal. Death from smallpox results from either overwhelming virus infection or bacterial superinfection.

Treatment of smallpox — Because vaccination for smallpox was so effective, the infection was wiped off the planet. However, the potential use of this virus as a bioweapon has caused many countries to begin stockpiling the vaccine. Because there is no infected population, it is difficult to test potential antibiotics, but several candidate drugs, such as cidofovir, are being evaluated.

Chickenpox and Shingles

Infections caused by the virus varicella-zoster are the most frequently reported infections in the United States, with almost 90% of the US population being infected by the age of 10 years. Approximately 95% of the population in the United States have been infected, but the mortality rate is very low (100 per year). The infection has two clinical manifestations: **chickenpox** in children and **shingles** in adults.

Pathogenesis of chickenpox and shingles — The virus is spread through secretions of the respiratory tract, and the infection occurs in the upper respiratory tract and the lymph nodes. Varicella causes a generalized vesicular rash, usually found initially on the back of the head and ears and then appearing on the face, trunk, and proximal extremities. There is commonly involvement of mucous membranes and fever early in the infection. As few as 10 or as many as several hundred irritating, itchy lesions can appear during the course of the infection. Secondary viremia includes infection of the skin.

In the early days of medicine, varicella and zoster were considered separate entities, the former the agent for chickenpox and the latter the agent for shingles. The possibility that varicella and zoster were clinical manifestations of the same virus was shown as early as 1892 and was confirmed in 1954. A latent form of the virus resides in the dorsal root ganglia of adults who had chickenpox as children. When this latent virus is reactivated, it multiplies in a sensory neuron and then travels down that neuron to the skin. The shingles rash comprises vesicles similar to those seen in chickenpox, but in shingles the vesicles are localized in distinct areas of the body, usually the waist and in some cases the upper chest and back (**Figure 26.11**). These areas are usually the areas of skin innervated by the sensory neuron harboring the latent virus, and it seems that reactivation increases in frequency with advancing age. The shingles

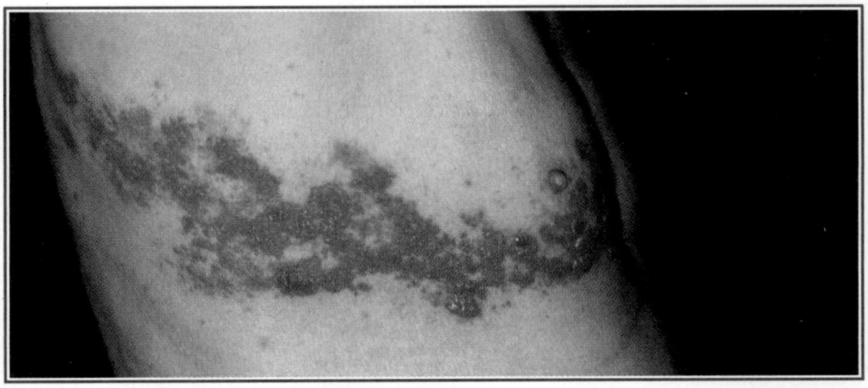

Figure 26.11 The lesions associated with shingles, which occurs when a latent form of the varicella-zoster virus is activated in an adult.

lesions are very painful; they appear several days to several weeks after pain is experienced in the affected area and can persist for months. In immunocompromised patients, there can be multiple organ involvement and a significant mortality rate (about 17%).

Fast Fact

A vaccine to prevent shingles is available and is recommended for people aged 60 and over in the United States.

Treatment of chickenpox and shingles — Acyclovir and famciclovir can reduce fever and skin lesions and are the recommended treatments for patients over 18 years old and patients with immunodeficiency.

Herpes Simplex Type 1

Infections caused by the herpes simplex type 1 virus (HSV-1) are referred to as above-the-waist infections to distinguish them from the genital infections caused by herpes simplex type 2 (Chapter 23). HSV-1 causes a latent infection, with signs appearing only when the virus is reactivated. This virus is found throughout the world, and humans are the only reservoir. In developing countries, 90% of the population has antibodies against HSV-1 by the age of 30 years, whereas in the United States antibodies against this virus are found in 60–70% of adults.

Pathogenesis of herpes simplex type 1 — During this acute infection, syncytia develop, and there is a degeneration of epithelial cells that brings about necrosis at the infection focus. The inflammatory response occurs, with infiltration by neutrophils followed by mononuclear cells. The virus spreads either interneuronally or intraneuronally, and intraneural spread means the virus can hide from the immune response and lie latent, sometimes for years. The latent virus resides on the trigeminal, superior cervical, and vagus nerve ganglia, but the reactivation stimuli are not yet understood.

HSV-1 infection is often asymptomatic, but when that is not the case the principal sign is grouped or single vesicular lesions that become pustular and coalesce to form multiple ulcers. There can be painful ulcerative lesions on the tongue, gums, and pharynx. These lesions usually persist for 5–12 days, and after the primary infection, latent reactivation can occur in the form of *cold sores* (also called *fever blisters*) that appear on or near the lips and can last for as long as 7 days (**Figure 26.12**).

HSV-1 sometimes infects the fingers in the area of the nails. This usually occurs because of a break in the skin and causes the formation of painful pustular lesions on the fingers. This virus can also infect the eye (ophthalmic herpes) and is the most common cause of corneal damage and blindness in the developed world. The damage occurs because of dendritic ulcerations in the conjunctiva and the cornea, which cause scarring.

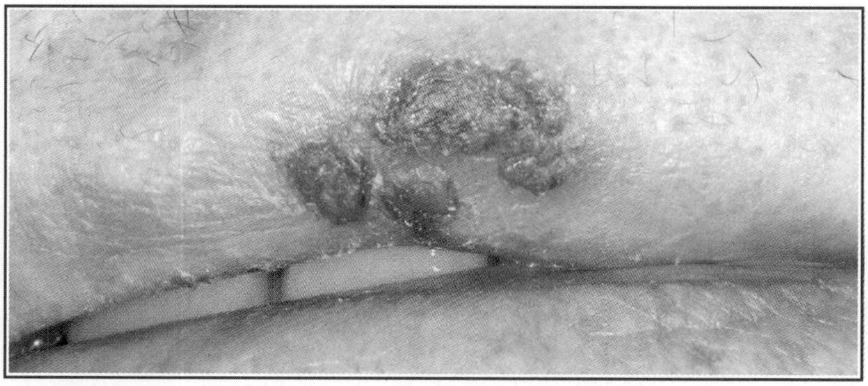

Figure 26.12 Cold sores (fever blisters) occur because of reactivation of latent herpes simplex type 1 virus.

Treatment of herpes simplex type 1 — The most effective treatment for HSV-1 infection is the nucleoside analog acyclovir. This antibiotic significantly reduces the duration of the primary infection and can suppress recurrence. Immunocompromised patients may harbor resistant HSV, in which case foscarnet can be used.

Warts

Warts are small growths that appear either on the skin or on the mucous membrane of the respiratory tract, genital tract, and interior of the mouth. They are caused by human papillomavirus (HPV). HPV infection is a lifelong infection, and warts can return even after removal. This is because the virus is still associated with the tissue adjacent to where a wart was removed.

Warts vary in appearance, location, and pathogenicity. Some are small and self-limiting, others are large but benign, and still others are malignant. Genital warts do not generally become malignant, but some strains of HPV do cause cervical cancer. Warts are larger and occur more frequently in people who are immunodeficient.

Pathogenesis of warts — HPV infection is transmitted by direct contact between humans and by fomites. In dermal warts, which incubate for one to four weeks, the virus gains entry through broken skin. Genital warts (**Figure 26.13**) can be sexually transmitted and require 8–20 months of incubation before signs are expressed.

In dermal warts, the virus infects epithelial cells, and it is a proliferation of these cells that forms the warts. This occurs at the boundary between the dermis and the epidermis (see Figure 26.1), and dermal warts are more common in children than in adults. Usually only one or at most a few warts occur during an outbreak. The removal of one wart in a cluster causes the others in the cluster to regress, and in some cases there is spontaneous regression without a first wart being removed. It is thought that this may be the result of an immune response.

Treatment of warts — There is no satisfactory treatment for warts. However, in many cases the growth can be removed using liquid-nitrogen cryotherapy accompanied by the removal of adjacent infected tissue. It has been shown that antimetabolites, such as 5-fluorouracil and interferon, can block HPV infection.

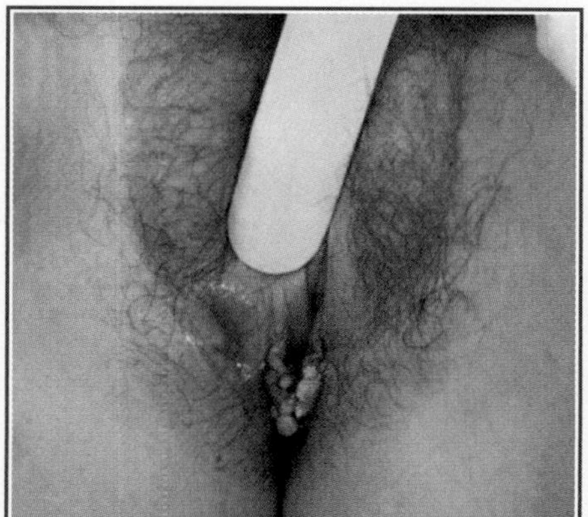

Figure 26.13 Genital warts.

Keep in Mind

- Viral pathogens require a portal of entry to infect the skin.
- Viral infections that cause skin lesions include measles, rubella, smallpox, chickenpox, herpes simplex type 1 (cold sores), and human papillomavirus (warts).

FUNGAL INFECTIONS OF THE SKIN

Remember that fungi are always present on the skin, and as long as we have a competent immune system, they rarely bother us. However, if there is some compromise of an individual's health, fungi can be opportunistically infectious. The unbroken skin is a barrier to fungi, just as it is to bacteria and viruses. Recall that in Chapter 14 we looked at fungal infections. We will revisit some of them here.

Cutaneous Candidiasis

The fungal infection known as **cutaneous candidiasis** is caused by *Candida albicans*, which we discussed in Chapter 14. This organism is a member of the normal flora in the oropharyngeal, gastrointestinal, and genitourinary tracts. *Candida albicans* can grow in multiple morphological forms but is most often seen as a yeast. It does have the capacity to form hyphae, and the change from the single-celled yeast form to the formation of hyphae is strongly associated with pathogenicity.

Pathogenesis of candidiasis — *Candida albicans* organisms use their hyphae to invade deep into tissues. The stimuli for this change in morphology have not yet been identified, but it is known that the change is associated with the appearance of factors that increase adherence to and destruction of tissues.

The hyphae form strong attachments to human epithelial cells and secrete proteinases and phospholipases that digest epithelial cells and further facilitate tissue invasion by the hyphae. *Candida albicans* contains surface proteins that bind to C3 receptors, thereby preventing opsonization. The compromise of T-cell function or the overuse of antibiotics also allows *C. albicans* to increase in numbers, resulting in local and invasive infection. Indwelling catheters and chemotherapy may also advance the invasion.

Candida albicans infections of the skin usually occur in folds of skin and other areas in which two wet skin surfaces are opposed to each other (on the buttocks of infants, for example, causing diaper rash). The initial lesions are erythematous papules or areas of tenderness and fissured skin. Infants, whose normal flora has not yet been established, can get the infection called **thrush**, which is the development of a whitish overgrowth in the oral cavity. This same condition can be seen in patients with immunodeficiencies (**Figure 26.14**).

A *C. albicans* infection usually remains confined to chronically irritated areas. However, in rare cases in which the infected individual is immunocompromised, the condition called **mucocutaneous candidiasis** is seen. In this case, infections of the hair and skin fail to heal and require therapy. This form of the infection can result in considerable discomfort and in some cases disfigurement with extensive areas of lesions.

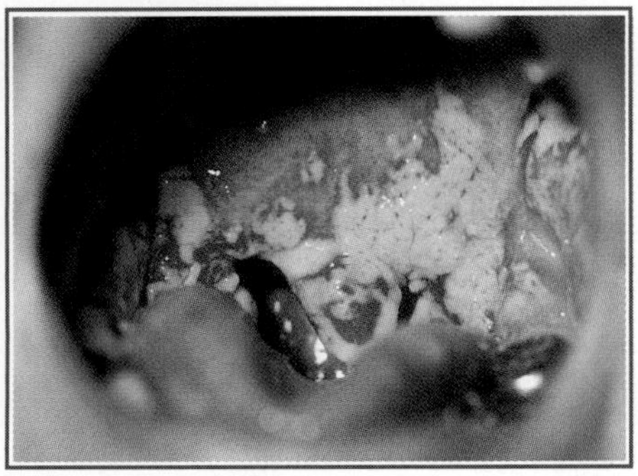

Figure 26.14 **Oral candidiasis, also known as thrush.**

Treatment of candidiasis — *Candida albicans* is usually susceptible to nystatin, flucytosol, the azole antifungals, and in cases of deep tissue involvement, amphotericin B. In addition, measures should be taken to decrease moisture and chronic trauma. Fluconazole is the most effective treatment for mucocutaneous candidiasis.

Dermatophytosis

Dermatophytes are fungi that are pathogenic to the skin, and the cutaneous mycoses they cause result in slow, progressive eruptions of the skin that are unsightly but not painful or life-threatening. There are several forms of dermatophyte-caused infection, classified according to the inflammatory response in the skin, but all forms typically involve erythema, induration, itching, and scaling. The umbrella term for all these conditions is **dermatophytosis**.

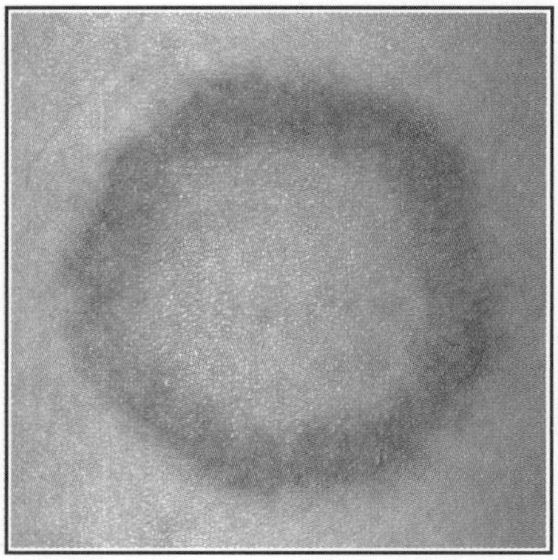

Figure 26.15 **The classic lesions of tinea capitis, also known as ringworm.**

The most familiar dermatophyte infection is *tinea capitis*, more commonly known as scalp ringworm (**Figure 26.15**), which gets its name from the shape of the creeping margin at the edge of the dermatophyte growth. Dermatophytosis in the groin is referred to as *tinea cruris* or jock itch, and dermatophytosis on the feet is *tinea pedis* or athlete's foot. There are ecological and geographic differences in the occurrence of different dermatophyte infections, and many domestic cats and dogs as well as cows and horses are infected and act as reservoirs for these fungi.

Human-to-human transfer requires close contact because dermatophytes are poorly infective and have very low virulence. The infection is usually seen in families, barber shops, or locker rooms.

Pathogenesis of dermatophytosis — All three forms of tinea infection begin when minor traumatic skin lesions come into contact with the dermatophyte hyphae from an ongoing infection. Once the *stratum corneum*, which is the most superficial layer of the epidermis, is penetrated by the hyphae, the dermatophytes proliferate in the keratinized layers of the skin, aided by the production of protease enzymes. The course of the infection depends on anatomical location, moisture, and the rate of skin-cell shedding. The speed and strength of the inflammatory response also have a role in the infection. It is of interest to note that the faster the skin is shed, the less time it takes to get over the infection. Inflammation can increase the shedding rate and therefore affect the time course of the infection. In contrast, immunosuppression increases the length of the infectious period. In any case, invasion of deep tissues during a tinea infection is extremely rare. Most of these infections are self-limiting, but in cases where the inflammatory response is poor, the infection can become chronic. Dermatophyte infections can also affect the nails and hair follicles, in the latter case by plugging them up and causing the hair to become brittle.

Treatment of dermatophyte infections — Most skin dermatophyte infections resolve without therapy. Those that do not can be treated with topical agents like tolnaftate, allylamines, and azoles. More extensive infections in the nail beds require systemic therapy for weeks or months with griseofulvin or itraconazole. Tinea capitis also requires systemic therapy.

PARASITIC INFECTIONS OF THE SKIN

Cutaneous Leishmaniasis

The zoonotic parasitic infection **cutaneous leishmaniasis** is seen in tropical and subtropical rodents. As we discussed in Chapter 14, it is common in central Asia, India, the Middle East, and South and Central America. Humans contract this infection when they enter areas where the rodents live. The infection is vector-transmitted from rodent to human, with sand fleas being the vector. Because domestic dogs can also be a reservoir for *Leishmania*, cutaneous leishmaniasis can be transmitted through these animals; in this case, dog fleas are the vector.

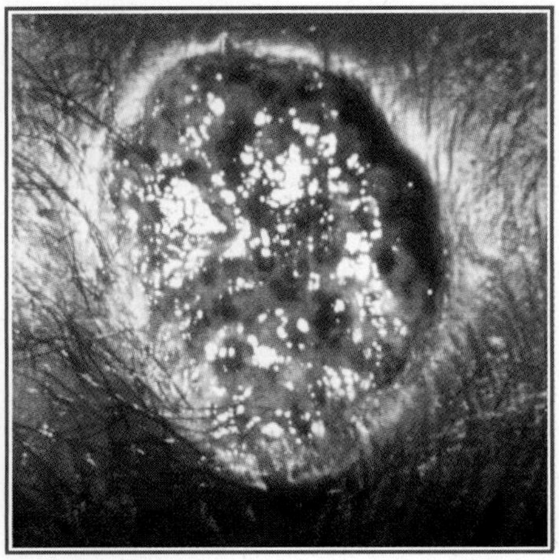

Figure 26.16 A lesion caused by cutaneous leishmaniasis.

Pathogenesis of cutaneous leishmaniasis — The lesions associated with cutaneous leishmaniasis appear on the extremities or the face weeks to months after the bite of the sand flea. They appear as itchy pustules accompanied by lymphadenopathy. Within a few months, the pustules ulcerate (**Figure 26.16**), and there can be several of these lesions on the body. There is spontaneous healing in 5–12 months, leaving depigmented scars on the skin. If the lesions occur on the ear, the infection can cause destruction of the pinna, and in patients with AIDS this infection causes multiple nonhealing lesions.

Treatment of cutaneous leishmaniasis — If there is no involvement of the mucous membranes, no treatment is required. Amphotericin B, ketoconazole, and itraconazole can be used for more severe lesions.

Lice

The parasitic infection caused by lice on the human body is called **pediculosis**. Infection can occur on the head (in which case the parasite is the head louse, *Pediculus humanus capitis*) or on the body (in which case the parasite is the body louse or clothes louse, *Pediculus humanus corporis*). Infection on the body is often in the genital region, causing the condition referred to colloquially as *crabs*. Body lice can transmit infections such as epidemic typhus. Lice have been around for centuries as infecting agents, and outbreaks of lice infections, usually head lice, are frequently seen in schools, where the transfer of lice occurs easily.

Pathogenesis of pediculosis — Lice require blood and feed several times a day. The itching that accompanies a pediculosis infection is due to a reaction against the saliva of the biting louse. Scratching can result in secondary bacterial infection. The head louse has specialized legs with which to grasp the hair shaft, usually at the base (**Figure 26.17**).

Treatment of pediculosis — Treatment usually involves combing out the lice and nits with a fine-toothed comb in combination with nonprescription drugs such as Nix and Rid. However, there has been increasing resistance to these drugs in recent years. Other topical ointments, such as lindane and malathion, can be effective but are very toxic.

Figure 26.17 A louse nit (egg) attached to a scalp hair. The nymphal stage hatches from the cap (operculum) by gulping air and forcing it out of the anus until the nymph pops free.

Keep in Mind

- Fungi are always present on the skin and are rarely infectious.
- The most common fungal infection of the skin is candidiasis (caused by *Candida albicans*).
- Several forms of dermatophyte-caused infection are classified according to the inflammatory response seen in the skin.
- One of the most common parasitic infections of the skin is cutaneous leishmaniasis.

INFECTIONS OF THE EYES

We are including the eyes in this chapter because they are affected by infections in the same manner as the skin. It is important to remember that the eyes are a *privileged site*, which means they are essentially protected from the immune response. This makes sense because it would be

Infection	Causative Organism	Infection Characteristics
Bacterial		
Ophthalmia neonatorum	*Neisseria gonorrhoeae*	An infection acquired during passage through the birth canal; can lead to blindness
Conjunctivitis	Several bacterial species	Highly contagious inflammation of the conjunctiva of the eye
Trachoma	*Chlamydia trachomatis*	Infection and eventual destruction of the cornea and conjunctiva; can cause blindness
Keratitis	Bacteria, viruses, and fungi	Ulceration of the cornea; occurs mainly in immunodeficient and debilitated patients
Parasitic		
Onchocerciasis	*Onchocerca volvulus*	Blackflies transmit the microfilariae through bite, microfilariae infect other tissues; causes blindness
Loaiasis	*Loa loa*	Transmitted by the bite of the deer fly; causes inflammation of the conjunctiva and the cornea

Table 26.2 Infections of the Eye

devastating if a defensive response were triggered every time a foreign particle entered the eye. Recall from our discussions of the host defense in Chapter 15 that mechanical barriers, such as tears, protect the eye from infecting organisms.

A summary of some common eye infections is shown in **Table 26.2**.

Conjunctivitis and Other Eye Infections

Eye infection usually involves pain and the potential for vision loss. **Conjunctivitis** is the main infection seen in the eye (**Figure 26.18**) and can occur in all age groups. Conjunctivitis is inflammation of the inner membranes of the eyelids with an associated discharge. The infection is easily spread if a contaminated hand rubs the eyes.

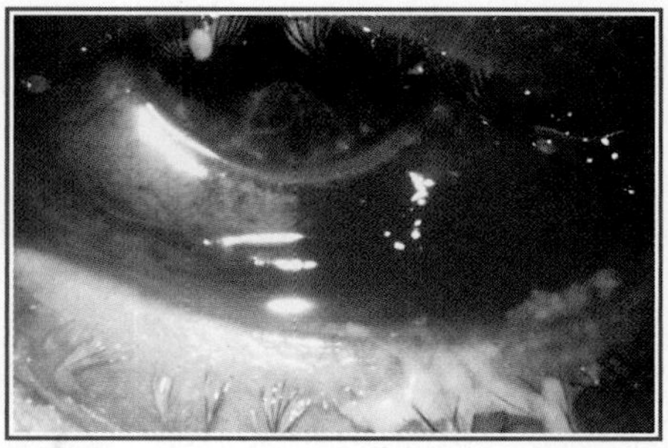

Figure 26.18 A case of gonorrheal conjunctivitis that resulted in partial blindness due to the spread of *Neisseria gonorrhoeae* bacteria.

Parts of the eye other than the conjunctiva can also become infected, including the cornea (in which case the infection is called **keratitis**) and the anterior and posterior chambers. Structures such as the orbital sinuses can also be involved, and because of the close proximity to the central nervous system, these infections can become life-threatening. The use of contact lenses, which can be washed in contaminated water, has led to an increased number of cases of lens keratitis. In the hospital, aggressive keratitis can be caused by *Pseudomonas aeruginosa*. This condition is painful, and loss of vision can occur.

Eyelid abscesses such as *styes* are fairly common problems in the eyes, and infection of the entire eyelid is also possible. In addition, the lacrimal gland and duct can be infected. Infection of the gland is called **dacryocystis**, and infection of the duct is called *canaliculitis*.

In underdeveloped countries, a leading eye infection is **trachoma** (**Figure 26.19**), caused by *Chlamydia trachomatis*. It is estimated that 500 million people are affected by this condition, with 10 million having been blinded by it. The pathogen is spread to the eyes by hands, by such fomites as towels and clothing, and by flies. Trachoma, which is essentially a chronic conjunctivitis, leads to scarring, corneal ulceration, and eventual vision loss.

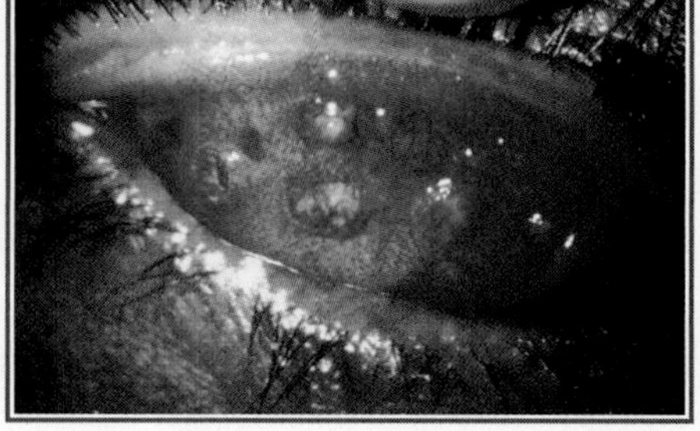

Figure 26.19 Inflammation of the eyelid in trachoma, a chronic and contagious eye disease caused by *Chlamydia trachomatis*.

The parasite *Onchocerca volvulus* causes the eye infection known as **river blindness**, which affects 20 million people in the world. It is spread by the blood-sucking blackfly, which transmits the parasite to the eye. The parasite invades the anterior chamber of the eye, causing corneal ulceration, fibrosis, and blindness.

In most instances topical eye drops or ointments containing erythromycin or gentamycin are used to treat acute bacterial conjunctivitis, whereas fluoroquinolones can be used for eye infections caused by *Pseudomonas*. Quinolones such as ciprofloxacin are also useful for all types of eye infection.

Neonatal Eye Infections

A serious conjunctivitis caused by the bacterium *Neisseria gonorrhoeae* is **neonatal gonorrheal ophthalmia**, which can be contracted as an infant is passing down the birth canal of an infected mother. The infant acquires large numbers of *N. gonorrhoeae* organisms during the passage.

In many cases, mothers infected with *N. gonorrhoeae* are also infected with *Chlamydia trachomatis*, which is involved in non-gonococcal urethritis, and this organism can also make its way into the eyes of newborn infants. Both neonatal gonorrheal ophthalmia and the infection caused by *C. trachomatis* cause large amounts of pus to form in the eyes (**Figure 26.20**), and without treatment it will lead to ulceration and scarring of the cornea.

It is common practice to treat the eyes of newborn infants with erythromycin to prevent the onset of both neonatal gonorrheal ophthalmia and the problems caused by *C. trachomatis* infections. This practice originally involved the application of silver nitrate to the eyes, but this has no effect on *Chlamydia*, so antibiotics are used instead.

Loaiasis

The eye infection **loaiasis** is caused by the parasitic worm *Loa loa*, found in the African rain forests. The worm is transmitted to humans by the deer fly, which feeds during the day and acquires the parasite from infected humans.

Like many other parasites, *Loa loa* matures in its vector (in this case, the deer fly). The worm migrates to the mouth of the fly and is then transmitted to a human when the fly bites. The microfilariae migrate through the infected person's subcutaneous tissue, causing inflammation as they go, and settle in the cornea and conjunctiva. The worms can grow to more than an inch in length and are easily seen on ophthalmic examination (**Figure 26.21**). There is a very elevated eosinophil count throughout this infection, and the drug albendazole can reduce the problem.

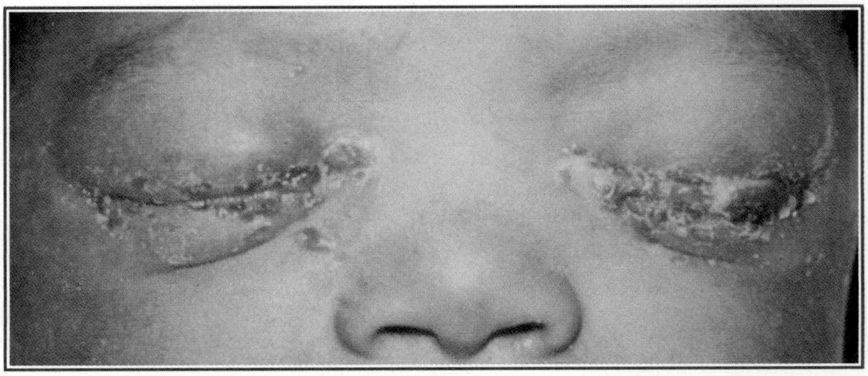

Figure 26.20 Ophthalmia neonatorum is a gonococcal infection of the eyes most often seen in infants.

Keep in Mind

- Like the skin, the eyes are affected by infections through direct exposure to pathogens.
- Trachoma, which is a form of chronic conjunctivitis, is caused by *Chlamydia trachomatis*.
- *Neisseria* pathogens in infected mothers can cause eye infection in babies as they move down the birth canal.
- Loaiasis is a parasitic infection of the eyes, caused by the worm *Loa loa*.

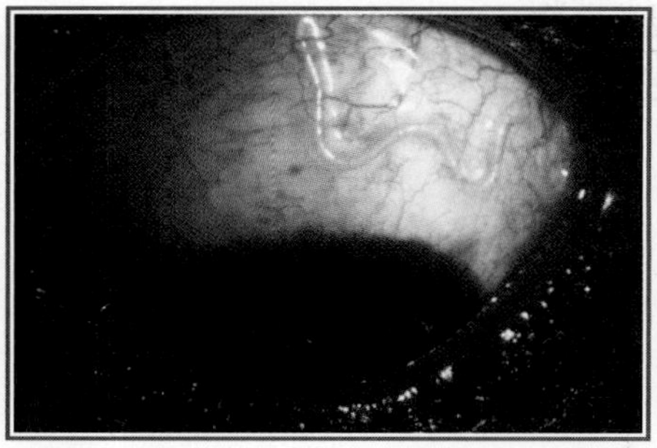

Figure 26.21 The *Loa loa* worm, which infects the eyes and can grow to an inch long.

SUMMARY

- Skin is an impermeable barrier to pathogens.
- A wide variety of bacteria can cause infection of the skin, with necrotizing fasciitis being one of the worst infections.
- Bacteria can infect hair follicles, sebaceous glands, and sweat glands.
- Viral pathogens also require a portal of entry to infect the skin.
- Viral infections that cause lesions on the skin include measles, rubella, smallpox, chickenpox, herpes simplex type 1, and human papillomavirus.
- Fungi are always present on the skin but rarely cause infection.
- The most common fungal infection of the skin is candidiasis.
- Dermatophytosis can be seen as ringworm or athlete's foot.
- One of the most common parasitic infections of the skin is leishmaniasis.
- Eyes are infected through direct exposure to pathogens.
- A common eye infection is trachoma, which is caused by *Chlamydia trachomatis*.
- Parasitic infections of the eye include loaiasis.

In this chapter we have examined the skin and eyes as targets for infectious pathogens. Keep in mind that the barrier nature of the skin is very effective in limiting the number of these types of infection, and only after that barrier has been breached do we see the development of infection. Remember that inflammation, which is a very important defense against infection, can also cause damaging side effects that are clearly seen in infections of the skin and eyes.

SELF EVALUATION AND CHAPTER CONFIDENCE

Multiple Choice

Answers are given in the back of the book and help can be found in the student resources at:
www.garlandscience.com/micro

1. The skin is important in protecting against infectious pathogens because
   - **A.** It is made up of dead cells
   - **B.** It is a barrier to invading organisms
   - **C.** It is easily broken
   - **D.** It is able to transport pathogens into the body
   - **E.** None of the above
2. All of the following act as barriers to invasion of the skin by pathogens except
   - **A.** Sebum
   - **B.** Perspiration
   - **C.** Acidic pH of the skin
   - **D.** The hair follicles
3. Necrotizing fasciitis is caused by
   - **A.** *Neisseria*
   - **B.** *Staphylococcus*
   - **C.** Group A streptococci
   - **D.** *Borrelia* organisms
4. Inflammatory acne is caused by
   - **A.** *Borrelia*
   - **B.** Group A streptococci
   - **C.** *Pseudomonas aeruginosa*
   - **D.** *Propionibacterium*
   - **E.** *E. coli*
5. Scalded skin syndrome is caused by
   - **A.** *Staphylococcus aureus* endotoxin
   - **B.** *Streptococcus pyogenes*
   - **C.** *Streptococcus exotoxin*
   - **D.** *Staphylococcus aureus* exotoxin
6. The etiologic agent of gas gangrene is
   - **A.** Endotoxin of *Clostridium botulinum*
   - **B.** *Clostridium tetani*
   - **C.** *Bacillus anthracis*
   - **D.** *Clostridium perfringens*
   - **E.** None of the above
7. The characteristic signs of smallpox are
   - **A.** Pimples that form only on the face
   - **B.** Papulovesicular rash on the head and extremities
   - **C.** Open sores on the face and neck
   - **D.** None of the above
8. The etiologic agent of chickenpox is
   - **A.** Variola minor
   - **B.** Variola major
   - **C.** Paramyxovirus
   - **D.** Varicella-zoster
   - **E.** None of the above
9. Cold sores are formed as a result of infection with
   - **A.** *Streptococcus pyogenes*
   - **B.** Variola major
   - **C.** Herpes simplex type 1
   - **D.** Herpes simplex type 2
   - **E.** Paramyxovirus
10. Cutaneous candidiasis results from infection with
    - **A.** *Cryptococcus neoformans*
    - **B.** *Streptococcus mutans*
    - **C.** *Candida albicans*
    - **D.** *Staphylococcus aureus*
    - **E.** None of the above
11. Humans acquire cutaneous leishmaniasis through
    - **A.** Contaminated food
    - **B.** Contaminated water
    - **C.** The bite of the tsetse fly
    - **D.** The bite of the *Anopheles* mosquito
    - **E.** The bite of the sand flea
12. Conjunctivitis is
    - **A.** Infection of the epidermis
    - **B.** Infection of the eye
    - **C.** Infection of the dermis
    - **D.** Infection of the hair follicle
13. The treatment for gonococcal ophthalmia is with
    - **A.** Mercury
    - **B.** Silver nitrate
    - **C.** Antibiotics
    - **D.** Eye wash

DEPTH OF UNDERSTANDING

Questions listed here require you to bring together the concepts you have learned in this chapter into a discussion format. This helps you to increase your depth of understanding of the material you have learned. Help can be found in the student resources at:
www.garlandscience.com/micro

1. Using the structure of the skin, describe the mechanisms that control the level of infection through this portal of entry.

2. Alice has a sexually transmitted chlamydial infection and is about to give birth. Describe the precautions that will be taken at delivery and why they are necessary.

CLINICAL CORNER

Help can be found in the student resources at:
www.garlandscience.com/micro

1. A patient is brought into the emergency room with a very swollen arm. He is in a lot of pain, and as you examine him you notice what looks like a surgical scar that has not healed properly and is now surrounded with dark purple skin. He explains that he had surgery to reattach his bicep, which had been hurt lifting weights, and the site of surgery has continued to hurt. You suspect that this might be a case of gangrene but there is no smell associated with wound. After examination, the patient is moved from the emergency room and the next day you hear that he had his arm amputated and he is on long-term heavy antibiotic therapy and will be for some time. In addition, the operating room nurse says he will be hospitalized and closely monitored for some time.
   - **A.** What condition did he have?
   - **B.** Why did you think it was gangrene and how would you have proven it was or wasn't?
   - **C.** Why is the long-term antibiotic therapy needed?

2. Curtis Worthy is a 13-year-old male who has come to the clinic with his mother. After your initial workup, Curtis seems to be in good health and seems to be well developed both physically and emotionally. He has a moderate case of facial acne with several whiteheads and blackheads on his face and neck. His skin also has an oily feel to it. He seems to be handling his skin condition very well but his mother is very concerned. She is adamant that Curtis be given a prescription for antibiotics that will alleviate her son's condition.
   - **A.** What can you tell her about her son's condition?
   - **B.** Would a prescription benefit Curtis?

PART VII

The Best and the Worst; Important Issues in Microbiology

We have reached the end of our journey, and it has been an exciting one. We have learned so many things about microbiology and infection and here in this last part of our studies, we will turn to the best and the worst of microbiology. The best involves the amazing things we have been able to do using microbiology and the burgeoning field of Biotechnology. The worst involves our manipulation of microbiology for the purposes of terrorism. Unlike the other chapters in this book, these two are intended for information. As a health care professional you need to know the basics about biotechnology because you will see the benefits it will provide throughout your career. Sadly, you may have to face the consequences of bioterrorism at some point in your career. In these two chapters we give you a general understanding of these topics and do not go into a great deal of detail.

In **Chapter 27**, we will take a brief look at some of the new and novel ways that scientists are combining microbiology and our ability to engineer organisms at the molecular level. Using these approaches, we have made magnificent strides in perfecting new treatments and new potential therapeutic modalities. What was only a decade ago the stuff that dreams are made of is know not only easily used but in many cases routine. Biotechnology has given us the potential to truly go where no man has gone before and correct nature's mistakes.

Unfortunately, **Chapter 28** takes us back to earth by showing us the dark side of microbiology. The world is currently faced with evil men who are trying to use microbiology to cause great harm to others. Many of the things that you have learned in previous chapters will make understanding what you read here in **Chapter 28** easier to understand. The use of infectious pathogens as weapons is not a new idea, but the ability to use new techniques such as we see in **Chapter 27**, make the possibility of bioterrorism something we cannot ignore. Sadly, the effect of success in the use of these weapons will have dramatic consequences on the health care profession. Therefore, it is important that students entering this field be aware of this potential danger.

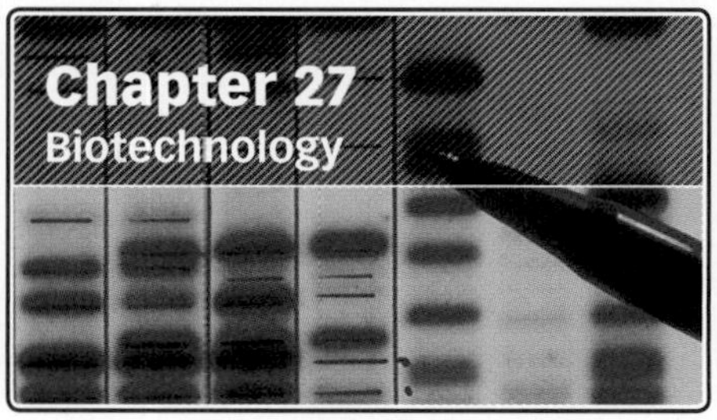

When you have finished with **PART VII** you will:

- Have an understanding of some of the techniques routinely employed by the biotechnology industry.
- Become familiar with the potential use of microorganisms as biological weapons.
- See how the health care community is coming to terms with the possibility of bioterrorism.

This brings to a close our study of Microbiology. Without doubt, there is a great deal of information for you to master. We hope that our presentation of the material and the ancillary materials has made the journey easier for you. We wish you well in your future as a health care professional.

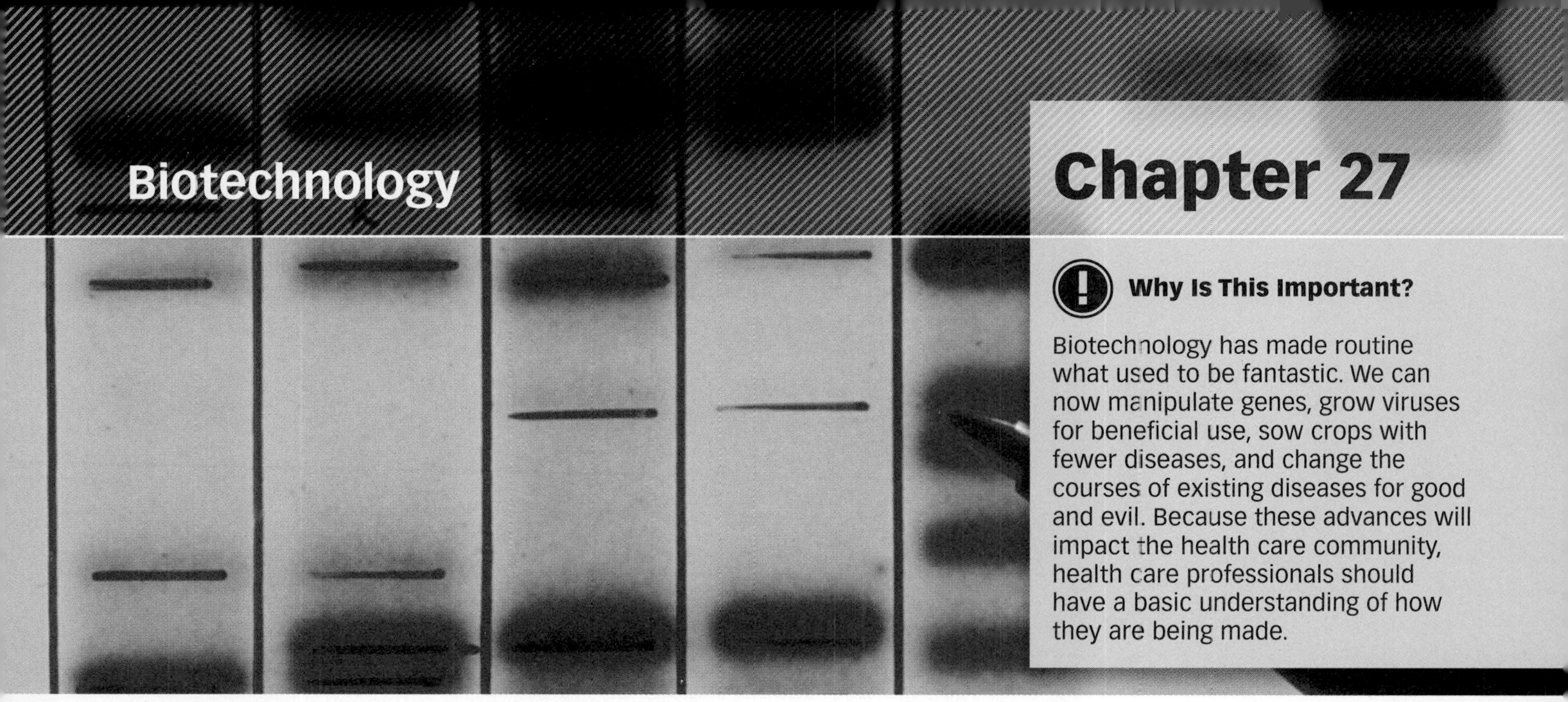

Biotechnology

Chapter 27

Why Is This Important?

Biotechnology has made routine what used to be fantastic. We can now manipulate genes, grow viruses for beneficial use, sow crops with fewer diseases, and change the courses of existing diseases for good and evil. Because these advances will impact the health care community, health care professionals should have a basic understanding of how they are being made.

OVERVIEW

Historically, biotechnology as a subject of study has been relegated to mentions of specific applications as they relate to other topics of interest. Today, however, biotechnology is ubiquitous in our day-to-day existence. It affects every aspect of our lives, from food to reproduction to religion to politics. There is no area that cannot somehow be related to some aspect of new biological technology. Tools of biotechnology are used in everything from manufacturing soaps to genetically engineering crops to turning bacteria into weapons. Because of the increasing importance of biotechnology in our lives, this chapter is designed to introduce and explain some of the tools available to us in this rapidly growing field.

We will divide our discussions into the following topics:

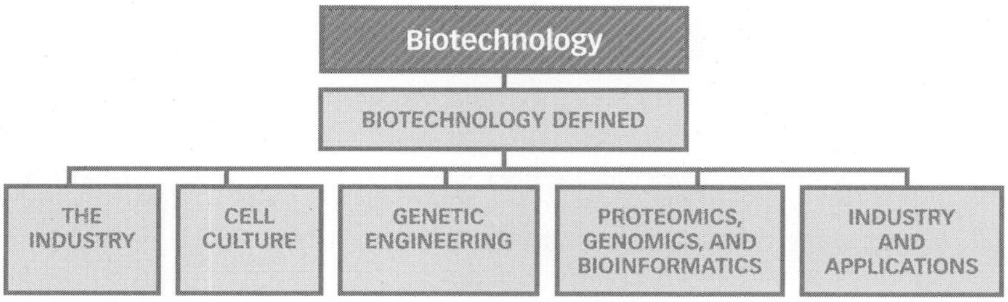

If a group of children come to you and ask you why they should not drink from streams or puddles, after reading Chapters 14 and 22 you can tell them about the presence of *Giardia* organisms and what these pathogens can do to them if ingested. What if they ask you why the water in the house is safe to drink, however? Will you know the reasons? If they ask why Uncle Jimmy, who is in the military, had to be vaccinated against smallpox before going overseas, you can explain how the disease works and about vaccines, but if they then ask why we all don't get those shots, or why the shots are necessary if the disease has been eradicated, can you tell them about the weaponization of bacteria? In the 1700s, the development of the microscope changed the field of science in fundamental and profound ways. Scientists were able to see cells and their functions. The germ theory was developed and disease mechanisms were studied at a molecular level. Now, more than 300 years later with the advent of microscopy, genetic engineering, and molecular biology and chemistry, biotechnology stands to change the very foundations of society.

What Do I Need to Know?

To get the most out of this chapter, please review the following terms from your previous courses in biology, anatomy, physiology, or chemistry or in previous chapters of this book as indicated in parentheses: Antibody (**16**), antigen (**16**), clone, cloning, codon (**11**), genetic transformation (**11**), genome (**11**), Human Genome Project, *in vitro*, nitrogenous base (**2**), nucleotide (**2**), plasmid (**11**), purine (**2**), pyrimidine (**2**), recombinant (**11**), somatically, somatic cell, stromal cell, transgenic, virulence factor (**1**).

BIOTECHNOLOGY DEFINED

The word *biotechnology* is derived from *bio-*, meaning "associated with life," and *technology*, meaning "the study of science and scientific techniques." Thus, a very simplified definition of biotechnology is the study of scientific technology as it affects living organisms. Actually, *biotechnology* as a term has undergone almost as many transformations as the field itself. It was not until 1969, for instance, that this word was used in the context of providing tools for medical research. Today, the definition of **biotechnology** has evolved to

> The field devoted to applying the techniques of biochemistry, cell biology, biophysics, and molecular biology to addressing issues related to human beings and the environment in which they live.

Over time, scientists have developed the ability to use the basic cellular processes carried out by microbes to create a large number of useful biotechnological tools. Despite the diversity of cell types in nature, most have the same basic properties: (1) they all need energy to function, (2) that energy is supplied in chemical form, and (3) they all use some sort of protein synthesis that requires the creation of a *map*, or *design template*. These shared basic characteristics provide opportunities for external manipulation by scientists.

DNA and RNA are the genetic materials that direct and map protein construction and design. They permit a basic sameness in the processes occurring in all cells while creating tremendous diversity and specificity of function. Because all living things use DNA and RNA in some form, information for the coding of a protein can be shared or exchanged by different organisms. The genetic directions encoded in any strand of DNA or RNA are specific for generating a specific protein. This specificity allows us to use nature's diversity to create that specific protein in a myriad of different organisms, taking advantage of their speed or of their environment for our own betterment. This is the heart of biotechnology today.

THE INDUSTRY

The biotechnology industry has grown explosively in the past 20 years. It is an international enterprise with well over $40 *billion* dollars in annual revenues. In the United States alone, there are over 1473 "biotech" companies employing more than 198,000 people. Every state in the nation currently offers economic development initiatives to assist bioscience companies. US revenues went from $8 billion dollars in 1992 to almost $40 billion in 2003. **Table 27.1** gives an overview of the industry statistics.

Table 27.1 US Biotechnology Industry from 1994 to 2003
*US dollars in billions

Year	2003	2002	2001	2000	1999	1998	1997	1996	1995	1994
Sales*	28.4	24.3	21.4	19.3	16.1	14.5	13	10.8	9.3	7.7
Revenues*	39.2	29.6	29.6	26.7	22.3	20.2	17.4	14.6	12.7	11.2
R&D expense*	17.9	20.5	15.7	14.2	10.7	10.6	9.0	7.9	7.7	7.0
Net loss*	5.4	9.4	4.6	5.6	4.4	4.1	4.5	4.6	4.1	3.6
Number of public companies	314	318	342	339	300	316	317	294	260	265
Number of companies	1473	1466	1457	1379	1273	1311	1274	1278	1308	1311
Employees	198,300	194,600	191,000	174,000	162,000	155,000	141,000	118,000	108,000	103,000

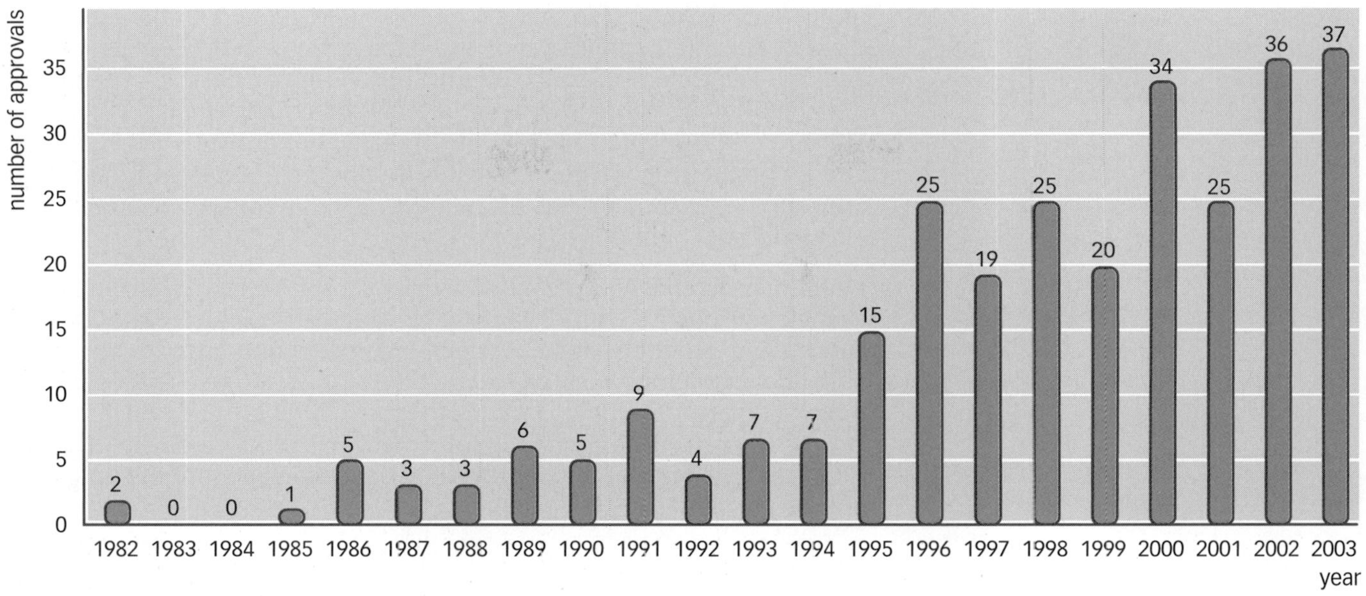

Figure 27.1 Increase in number of new biotechnology-derived drugs and vaccines approved over a 20 year period in the United States.

More than 200 diseases are currently being targeted as possibly being amenable to treatment by drugs produced using biotechnology. These diseases include Alzheimer's, AIDS, cancer, multiple sclerosis, diabetes, and arthritis. So far, more than 190 medicines have been developed through biotechnology and are available to treat diseases, compared with only 45 in 1995 — an increase of almost 150 in less than 10 years! Each year, an average of 30 new drugs and vaccines developed through biotechnological research are approved in the United States (**Figure 27.1**).

The process of discovering and developing a new drug in today's market is both expensive and daunting, with the average time interval for the development of a new drug being up to 16 years (**Figure 27.2**). Finally, and this is one of the most important statistics to you, the student, biotech companies in the United States employ almost 200,000 people and spend up to $100,000 per employee on research and development.

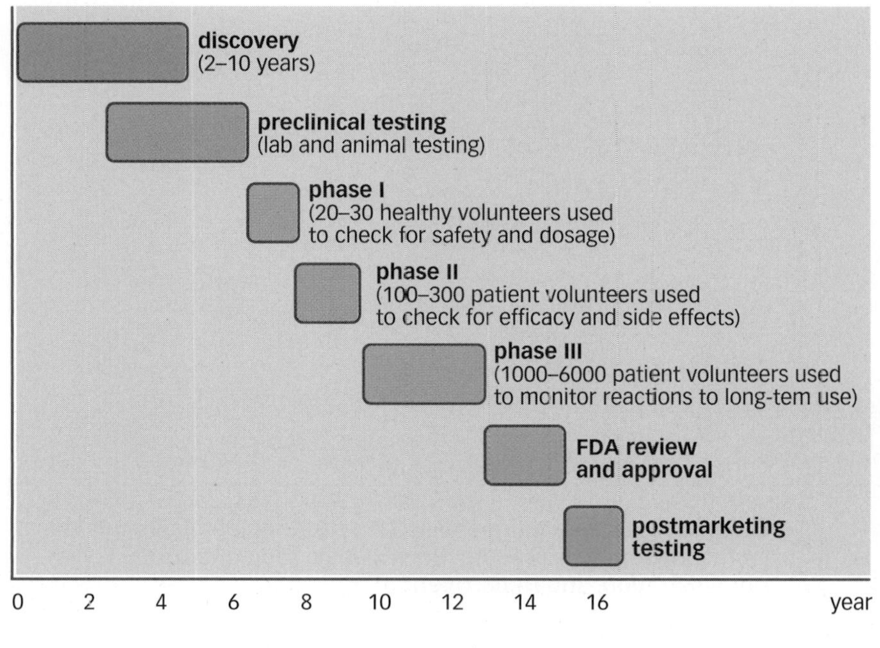

Figure 27.2 Steps involved in developing a new pharmaceutical. The process can take 16 years, with an estimated cost in the hundreds of millions of dollars. (In some cases, drugs that are particularly important can be fast-tracked, resulting in a time to approval shorter than 16 years.)

In the United States, the biotech industry is regulated by the Food and Drug Administration (FDA), the Environmental Protection Agency (EPA), and the Department of Agriculture (USDA) (**Table 27.2**). The level of governmental regulation allows a certain amount of consumer confidence. The process by which each new product is approved for public use is a long and arduous one. Despite the time and expense required to introduce new products to the market, the funding for research and development continues to increase yearly.

Worldwide regulatory bodies do not exist *per se*. There are governmental agencies in each country that regulate biotech products with various degrees of restrictions and enforcement. Most countries try to use existing trade regulations adapted and modified for new biotech products, but there are many developing nations that have little or no regulation. This lack of regulatory clarity and enforcement has led to a decline in consumer confidence and has raised ethical questions about how drugs are tested and sold.

Products already in use around the world include genetically engineered foods, hundreds of medical diagnostic tests and treatments, environmentally sound hazardous-waste disposal techniques, and more efficient processes for the use of water and energy in the production of textiles, metals, and enzymes. So, with these products in mind, let's look at some historical perspectives, modern discoveries, and future uses in biotechnology.

History of Biotechnology

Biotechnology was nominally in use by about 2000–4000 B.C., as humans learned to use plants and plant by-products medicinally and simple yeasts to modify both solid foods (bread) and liquids (wine and beer). If we skip along the timeline to the 1590s, we see the Dutch lens maker Zacharias Janssen inventing the compound microscope and then, in 1663 and 1675, respectively, the English scientist Robert Hooke using one to observe

New trait / organism	Regulatory tests conducted by	Reason for tests
Viral resistance in a food crop	USDA EPA FDA	Is it safe to grow? Is it safe for the environment? Is it safe to eat?
Herbicide tolerance in a food crop	USDA EPA FDA	Is it safe to grow? New use of companion herbicide Is it safe to eat?
Herbicide tolerance in an ornamental crop	USDA EPA	Is it safe to grow? New use of companion herbicide
Modified oil content in a food crop	USDA FDA	Is it safe to grow? Is it safe to eat?
Modified flower color in an ornamental crop	USDA	Is it safe to grow?
Modified soil bacteria that degrades pollutants	EPA	Is it safe for the environment?

Table 27.2 Areas of Biotech and Genetic Engineering Regulated by the Environmental Protection Agency, Food and Drug Administration, and Department of Agriculture

individual cells in plant and animal tissue and the Dutch naturalist Antony van Leeuwenhoek using one to observe individual bacteria. Because of their pioneering work in this area, these men are technically the fathers of modern biotechnology.

It was not until the late 1700s that human health care received its next exponential jump, when Edward Jenner used the cowpox virus to inoculate people against smallpox. He is often credited for the first use of a viral vaccine, but in reality the procedure of introducing tiny amounts of live smallpox virus into the skin so as to produce immunity had been in practice for at least 75 years. In England, Mary Montagu, the wife of an ambassador returning from Turkey in 1718, brought the technique with her and had her son inoculated to set an example; in the United States, a medical doctor named Zabdiel Boyleston used it experimentally and in a very controversial fashion in the Boston outbreak in 1721.

The reason that vaccination was so controversial at the time was simply that people had no knowledge of why and how it worked. Louis Pasteur's "germ theory" was not proposed until 1857, after the discovery and isolation of proteins and enzymes and their functions in 1830.

In the mid-1800s there was an explosion of knowledge relating to microorganisms and molecules at a cellular level. The decades 1870–1890 found the German bacteriologist Robert Koch developing new ways of visualizing and identifying bacteria and their structures. In 1928, the Scottish bacteriologist Alexander Fleming and his discovery of penicillin as an antibiotic heralded the beginning of truly explosive growth in the health care industry. From that point onward, new discoveries were made every few years. In only 18 years we went from the discovery of penicillin to recognizing that genetic material from different viruses could be combined to form new types of virus. Seven years later, in 1953, James Watson and Francis Crick discovered the double-helical structure of DNA, and by 1966 the human genetic code was cracked.

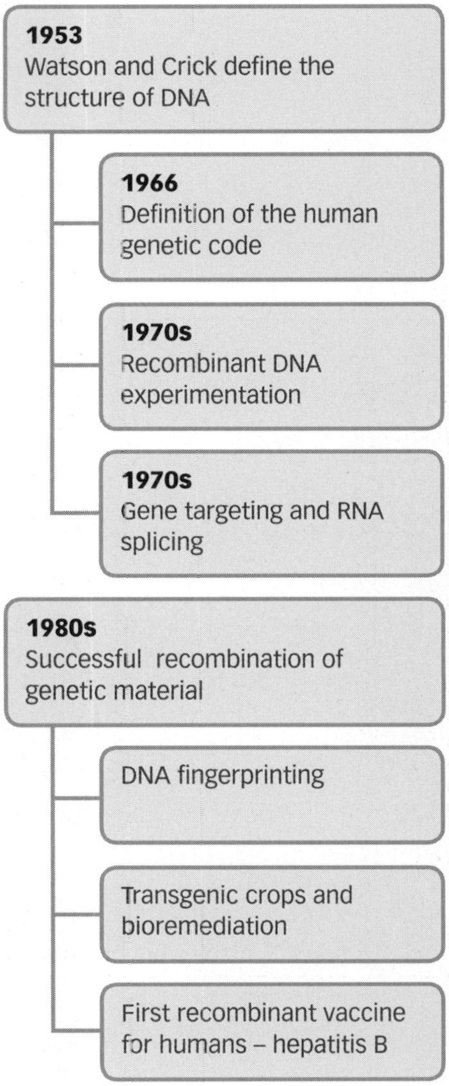

In the outline to the right, we can see an overview of the beginnings of biotechnology. The 1970s gave us recombinant DNA experimentation, gene targeting, and RNA splicing. The 1980s saw human genes successfully introduced into a bacterium. Human insulin was produced by genetically altered bacteria, a development that heralded the production of the first official biotech drug. The 1980s also gave us the first DNA "fingerprinting" used in a court of law as evidence, the EPA-approved release of the first transgenic crop (tobacco), the first recombinant vaccine for humans (for hepatitis B), and the beginnings of **bioremediation**, which is the science of using microorganisms to clean up the environment. The first use of this promising technique was the use of microbes on oil spills.

The 1990s saw the beginning of the Human Genome Project, an international effort to map all of the genes in the human body. We created the first transgenic cow, successfully gave baboon bone marrow to a human, and birthed the first sheep cloned from an adult sheep cell. The FDA declared in 1992 that there was no danger inherent in the release of transgenic crops into the environment. Yet today there is still disagreement about the truth of that statement and about the effects of these crops on the environment and world economy. The Biotechnology Industry Organization, where much of this information is made available to the general public, was formed in 1993 and provides references to information about the continuing evolution of biotechnology. As the 1990s progressed, there were more and more claims of successfully cloned animals around the world and a number of human embryonic cell lines were established. In 2002, the first part of the Human Genome Project was published. Astonishingly, in just 10 years, a rough draft of the entire human genetic code was developed.

All of this fast-moving progress naturally leads us to ask where we are headed in the twenty-first century. In 2004, Korean researchers announced the discovery of a way to clone an embryonic stem cell by nuclear transfer. Are we prepared for where science will take us?

Tools

To understand the limitless applications of biotechnology, you need a basic understanding of the fundamental molecules and techniques used in developing biotech products. The techniques used most often today are *cell culture*, *genetic engineering*, *monoclonal antibodies*, *recombinant DNA*, and the various specialties that fall under the umbrella name *molecular biology* (**Figure 27.3**).

CELL CULTURE

The growing of cells outside living organisms has become a necessary and routine technique of the biotechnology industry. Enzymes are used to isolate some cells from the tissue of an organism, and the cells are then cultured in a medium that supplies them with the nutrients and growth factors needed for their growth. Cell culturing is used in biotechnology to maintain the life of a cell line so as to harvest useful products made by particular cells. Some types of cell are relatively easy to grow, such as certain plant cells and fibroblasts (skin cells). However, some specialized cells are still in experimental stages for growth *in vitro*. Plant cell cultures have been used to harvest products such as flavorings for foods, thickening agents, and emulsifiers. Insect cell cultures are used to identify and improve biocontrol agents for insect pests. This can be in the form of mass production of a pest-specific bacterium that targets a particular insect without harming any other insect species or causing a build-up of environmental toxins specific for an insect pest.

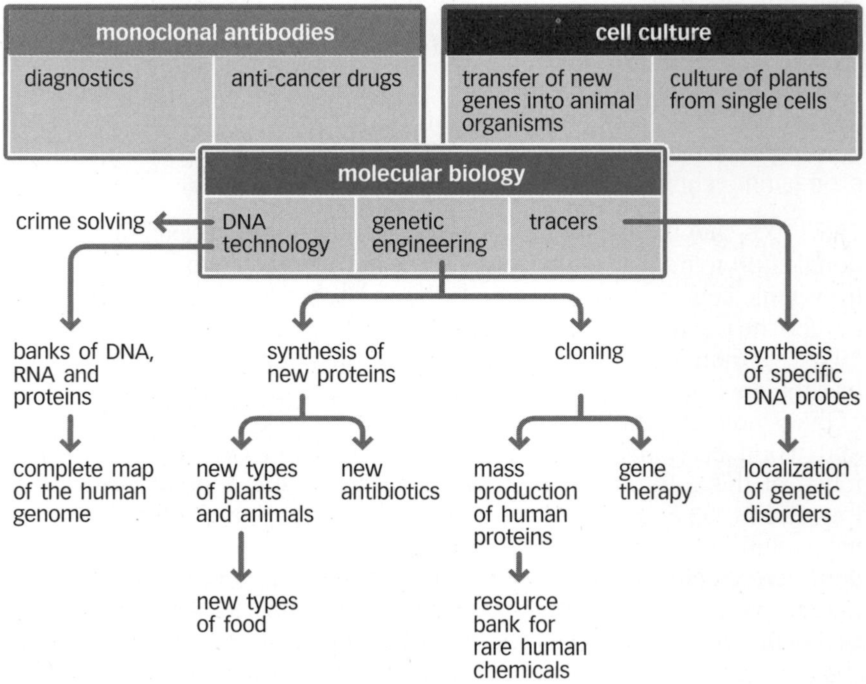

Figure 27.3 An overview of the main areas in which biotechnology is used.

Mammalian Cell Cultures

Mammalian cell cultures are widely used in microbiology. They can be grown as a medium for the growth and/or identification of bacterial and viral organisms, or they can be studied to increase the understanding of the mammalian cell life cycle. Mammalian cell cultures allow us to work with fertility issues and also to make advances in diagnosing and treating human diseases. These cultures may one day remove the need for testing new pharmaceuticals on live animals. In a way, mammalian cell culture is at the heart of all the advances we use currently in biotechnology. It is a useful technique because it provides a renewable source of cells for isolating DNA. In addition, scientists can use mammalian cells grown in culture to study how various chemicals and drugs affect certain cells and, by extrapolation, the whole organism.

Viral Cell Cultures

The study of viruses can provide fundamental information about aspects of cell biology and metabolism. Because viruses contain small, simple packages of DNA or RNA as their genome and have easily reproducible life cycles (Chapter 12), they are ideal models for experiments. In the laboratory, viruses can be grown either in a *primary* cell line or a *continuous* cell line. A *primary cell line* is one derived directly from a tissue sample. This type of culture is not used frequently because it deteriorates and dies after only a few generations.

A *continuous cell line* is one in which the cells have been transformed so that they grow continuously in culture. The cells in a continuous line do not produce a normal monolayer; instead, they clump and grow without any inhibiting factors.

Finally, there is another type of cell culture on which viral cells can be cultured: the *diploid cell culture*. As its name implies, this is a culture developed from human embryos. The cells in a diploid culture can survive about 100 generations in culture. They are used when the virus being cultured requires a human host for growth.

Fast Fact

One example of the use of diploid cell lines in viral culture is the growth of rabies virus to create rabies vaccines.

A type of cell culture that is often discussed in the mainstream media today is the stem cell culture. A stem cell is an unspecialized cell that can renew itself for long periods through cell division and remain unspecialized throughout the process. When offered the proper genetic, chemical, and environmental stimuli, however, stem cells have the ability to specialize into many different types of cell. Stem cells were isolated during a study of early mouse embryos over 20 years ago. Over a period of months, the proliferation seen in stem cell culture can lead to millions of unspecialized cells that can be used therapeutically or in research trials.

When an unspecialized cell gives rise to a specialized cell, the process is called differentiation. It is not known exactly what factors cause stem cells to differentiate, but scientists have done experiments leading to the belief that the process is controlled by genes, chemical signals, and microenvironmental factors, such as physical contact with neighboring cells. The ability of a stem cell to differentiate into multiple different cell types on the basis of the internal and external stimuli applied is called *plasticity* (**Figure 27.4**).

Adult Stem Cells

Adult stem cells are populations of undifferentiated cells that reside in various tissues of an organism, such as the brain, heart, bone marrow, liver, and skin. They were initially thought to be responsible only for repopulating the tissue in which they reside, to compensate for the normal loss of

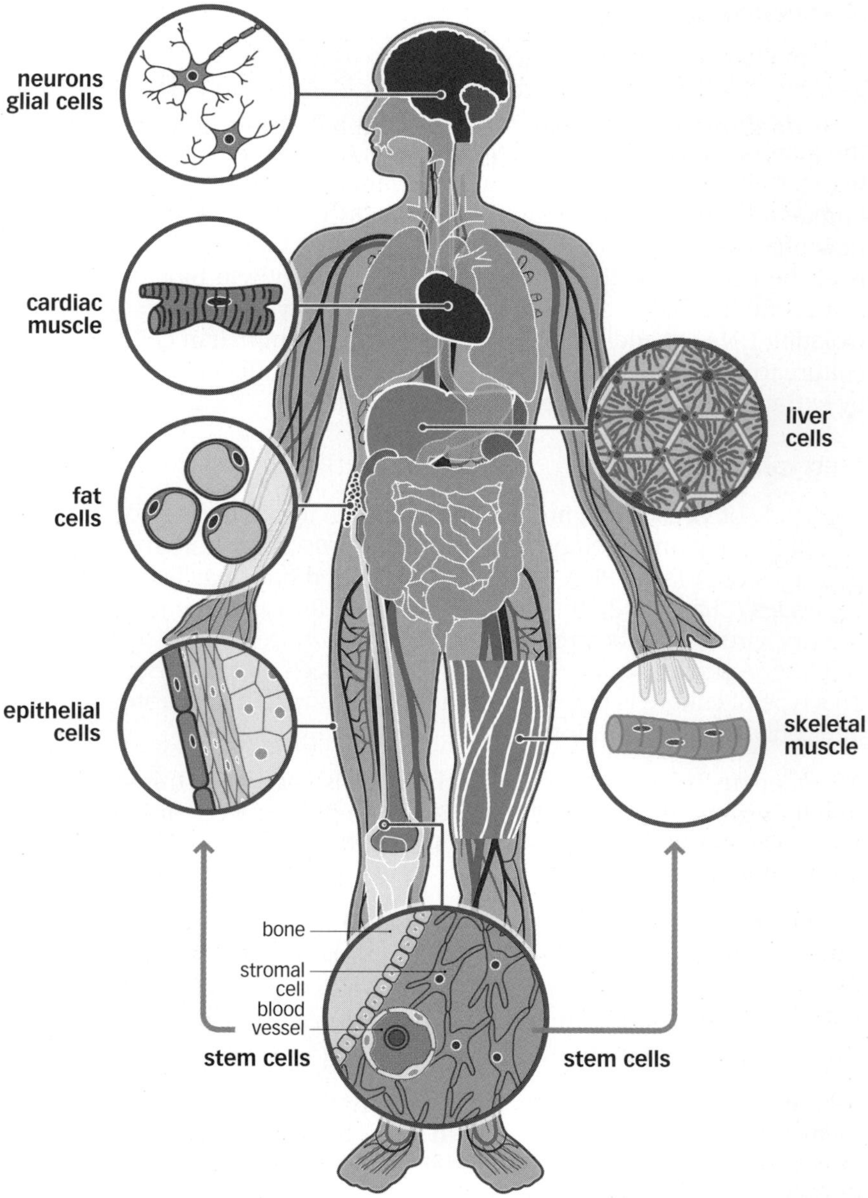

Figure 27.4 Adult stem cell plasticity. In this example, bone marrow stem cells differentiate into a variety of cell types.

cells by wear and tear. Adult stem cells in a person's liver, for instance, were thought capable of producing only liver cells, those in a brain only brain cells, and so on. However, the recent discovery of adult stem cells in tissues not previously believed to contain them has scientists exploring the possibility that these cells might be used for patients requiring new tissue or even new organs. The inherent advantage of using a person's own population of cells to regenerate tissue after injury or disease is that there would be no host rejection of the transplanted tissue or cells. Recent experiments have shown that there is evidence that some adult cell lines are capable of *pluripotency* — that is, they are able to differentiate into tissues other than that from which they were taken. The following list offers examples of adult stem cell plasticity that have been reported over the past several years.

- Hematopoietic stem cells can differentiate into three major types of brain cell — neurons, oligodendrocytes, and astrocytes — as well as skeletal muscle cells, cardiac muscle cells, and liver cells.
- Bone marrow stromal cells can differentiate into cardiac muscle cells and skeletal muscle cells.
- Brain stem cells can differentiate into blood cells and skeletal muscle cells.

Embryonic Stem Cells

The drawback to using adult stem cells in research and development is that most adult stem cell lines are rare in mature tissues and are difficult to proliferate in cell culture.

Embryonic stem cells are completely pluripotent and so can become any cell type in the body. They are easily grown in the laboratory and maintain their undifferentiated state through numerous generations, making them ideal for research on cell-based therapies. Human embryonic stem cells derive from eggs fertilized *in vitro* at clinics serving women being treated for infertility. The embryos not used in the infertility treatment are donated for research with full knowledge of the donors. When a donated embryo is three to five days old (the blastocyst stage), the inner cell mass, comprising about 30 cells, is removed from the blastocoel and cultured on what is called a *feeder layer* of mouse embryonic skin cells. These mouse cells provide a foundation for the human stem cells to grow on and also provide initial nutrition to them. The culture is re-plated when the original plate becomes crowded, and over a period of months the original 30 human embryonic stem cells generate millions of undifferentiated cells.

When conditions remain constant, the cells continue to proliferate in an undifferentiated fashion. However, if the cells are allowed to accumulate and form clumps called embryoid bodies, spontaneous differentiation can occur (**Figure 27.5**)

According to the National Institutes of Health, stem cell research — with both adult stem cells and embryonic ones — is essential to the future of health care. There remain many questions, however. One major issue is regulation. During the administration of President Bill Clinton, guidelines were published about the use and registration of stem cell lines. Then when George W. Bush became president, he announced that stem cell research would be allowed to continue but only with lines that had been registered before August 9, 2001. Since that time, several bills have been presented to Congress about a complete ban on human cloning, a technique that includes somatic cell nuclear-transfer technology for the purpose of deriving stem cells. There has been some effort to redefine *cloning* so that the bans involve only reproductive cloning (for example, creating a lamb from a cell obtained from an adult sheep) and not therapeutic cloning, but so far no definitive legislature has been passed. In January 2009, the FDA gave approval for the world's first study on embryonic stem cell therapy. In March 2009 President Barrack Obama lifted restrictions on federal funding for research on new cell lines. There is public funding of embryonic stem cells in other countries, though the research is tightly regulated.

Fast Fact

Diseases that might be treated by cells generated from human embryonic stem cells include Parkinson's disease, diabetes, traumatic spinal cord injury, Purkinje cell degeneration, Duchenne's muscular dystrophy, heart disease, and vision and hearing loss. However, using embryonic stem cells for this work is a topic of great debate.

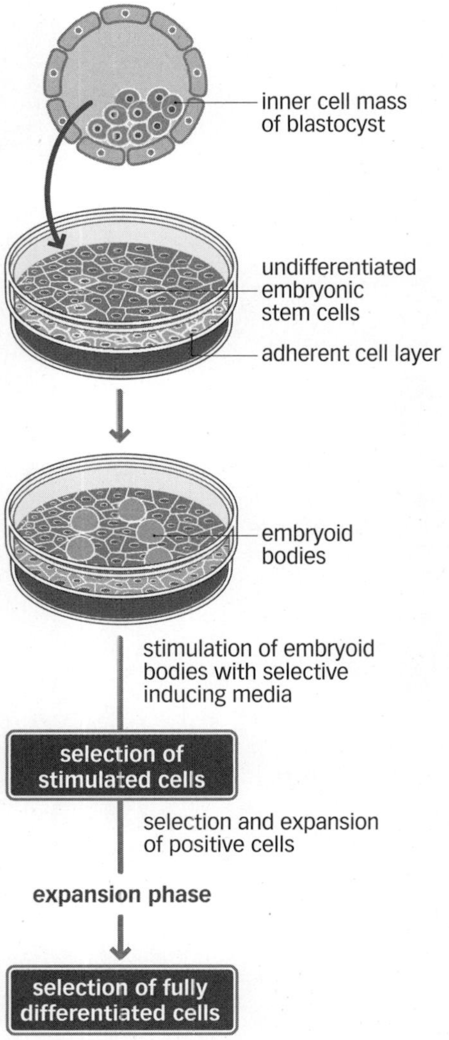

Figure 27.5 Embryonic stem cells can be used to develop fully differentiated cells of different types.

GENETIC ENGINEERING

There are several distinct tools available under this heading. The most basic and widely used are *monoclonal antibodies* and *recombinant DNA technology*.

Monoclonal Antibodies

As we discussed in Chapter 16, when a host's immune system is exposed to an antigen, specific host cells respond by producing antibodies that have different degrees of responsiveness to each available binding site on the antigen. In 1975, Georges Köhler and César Milstein developed an advanced type of cell culture for creating antibodies from a single cell such that all populations of those antibodies are exactly the same in their specificity and binding abilities. The single cell from which the antibodies are produced is called a clone cell, so the antibodies are called monoclonal antibodies. They have exceptional purity and specificity, are capable of binding to very specific antigens, and are also capable of recognizing other *nonself molecules*, such as drugs and viral or bacterial products.

The basic obstacles in the original technology used to produce monoclonal antibodies involved difficulty in growing them in culture and in harvesting them. Originally, antibody was harvested by introducing a foreign particle into a mouse so that the mouse's immune system would produce antibodies against the particle. These antibodies were then harvested from the ascites, abdominal fluid produced by the mouse in response to the particle. Aside from the obvious humane-treatment issues with this technique, there was the additional problem that the harvested product was not always pure and so might contain molecules dangerous to humans. There were also insufficient quantities produced for use in research and therapy. To solve these problems, the goal of biotechnology scientists was a fast, clean, safe, humane procedure that would produce large amounts of good-quality antibodies and do so repeatedly for long periods.

What Köhler and Milstein determined was that it is not productive to culture antibody-producing cells because these cells have a relatively short life span. However, malignant tumor cell lines are virtually immortal but are incapable of producing antibodies. The biotechnology was in the fusing of an antibody-producing cell line with a tumor cell line to create a hybridoma, which is a combination (that is, a *hybrid*) of nuclei from the two cell lines (**Figure 27.6**). The hybridoma cells were then tested for the production of specific antibody, and the successful hybridomas became the tool for producing monoclonal antibodies.

Fast Fact

For their work on the development of hybridoma cells, Köhler and Milstein won a Nobel prize in 1984.

Monoclonal antibodies can be used for everything from diagnostic tests to "magic bullets" in cancer therapy. Scientists are even able to attach, or *conjugate*, toxins and drugs onto these antibodies and then use the conjugated molecules as delivery vehicles serving a particular antigen-binding site. Because of this elegant versatility, there is a very high commercial demand for monoclonal antibodies. Their uses today include:

- Locating environmental pollutants
- Detecting harmful microorganisms in food
- Distinguishing cancer cells from normal cells
- Diagnosing infectious diseases
- Delivering chemotherapeutic drugs to cancer cells while avoiding the normal healthy cells

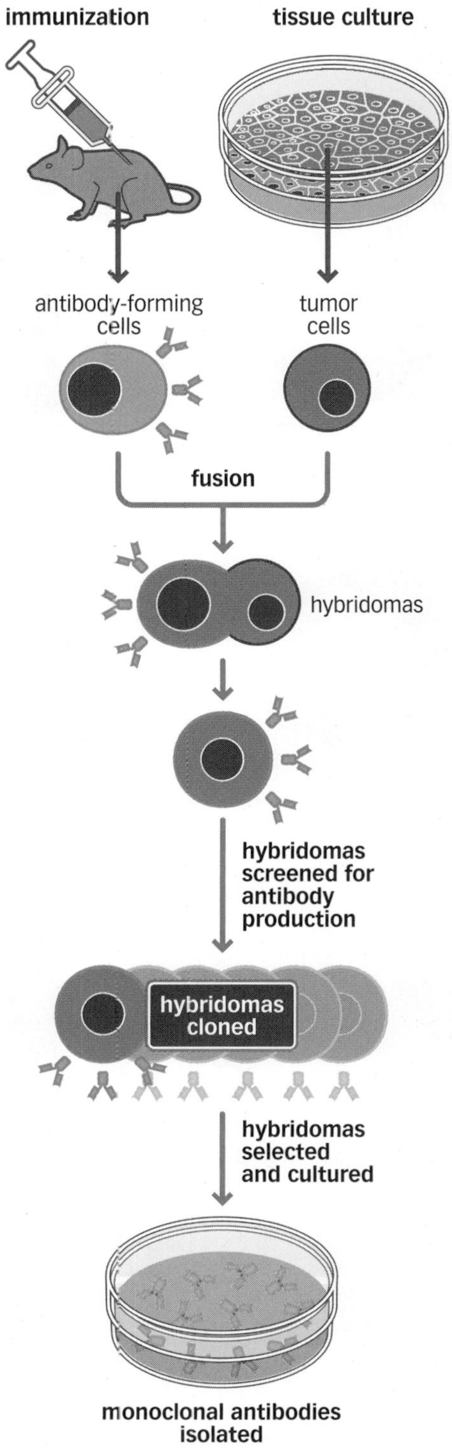

Figure 27.6 Construction of a hybridoma that secretes a specific monoclonal antibody. After the antibody producing cell has been cloned, the cloned cells can be used to produce large quantities of monocloncal antibodies.

Monoclonal antibodies can even be used to attack specific parts of the immune system of an organ-transplant patient to prevent rejection while leaving the rest of the immune system free to fight off the common cold!

One current problem with the production of monoclonal antibodies revolves around the use of mice. The most commonly accepted and simplest method of production involves creating hybridomas by combining mouse myeloma cells with spleen cells from one mouse and then injecting the cloned cells into another mouse. This injection causes a severe inflammatory response in the host mouse, which leads to the development of antibodies that are then harvested repeatedly from the ascites, as noted earlier. Although this technique has been in use for many years and has proved to be an effective, efficient method of production, there are questions regarding the humane treatment of the mice. Because of these questions, many countries now prohibit this method of producing monoclonal antibodies.

There are ways to produce monoclonal antibodies in cell cultures, and the FDA ruled, after a lengthy study, that there are occasions when this technique must be used to achieve a specific research goal. It stressed, however, that other techniques must be proven ineffective before live mice were used for monoclonal antibody production. Two popular alternatives to the use of live mice are bulk tissue cultures in encapsulated or hollow fiber systems, and the expression of cloned antibody genes in high-producing eukaryotes through recombinant DNA techniques, the topic we turn to next.

Recombinant DNA Technology

As you learned in Chapter 11, DNA is the genetic blueprint for every living organism. It is composed of nucleotides and some associated proteins. Recall that there are four nitrogenous bases found in nucleotides: adenine, guanine, cytosine, and thymine. Adenine and guanine are purines, and thymine and cytosine are pyrimidines. Because adenine binds only to thymine and cytosine binds only to guanine, every DNA molecule contains equal numbers of purines and pyrimidines. By now, you are familiar with the image of the A–T, C–G double helix that is the shape of every DNA molecule and should recall that it is the nucleotide sequence in DNA that holds the code for all human traits and characteristics.

The two-strand structure of DNA has an essential role in the ability of scientists either to take apart a DNA molecule and *recombine* it in different configurations or cut the molecule and *recombine* it with different ribonucleic materials. These techniques are the foundation of recombinant DNA technology. In short, scientists are now able to separate strands of DNA, cut them up into individual genes, and then decode each gene to determine which gene codes for the protein the researchers wish to make. The gene that codes for the desired protein is then replicated.

Two frequently used techniques for replicating genes are cloning using vectors and polymerase chain reaction techniques. In the vector method, the gene is inserted into a vector, which is a piece of genetic material capable of replicating itself. The most commonly used vector is the plasmid, free-floating, circular DNA that exists in the cytoplasm of cells (discussed in Chapter 11). The plasmid can replicate in the host cell and is easily transferable from that cell to a recipient cell. For recombinant DNA amplification, these are ideal characteristics.

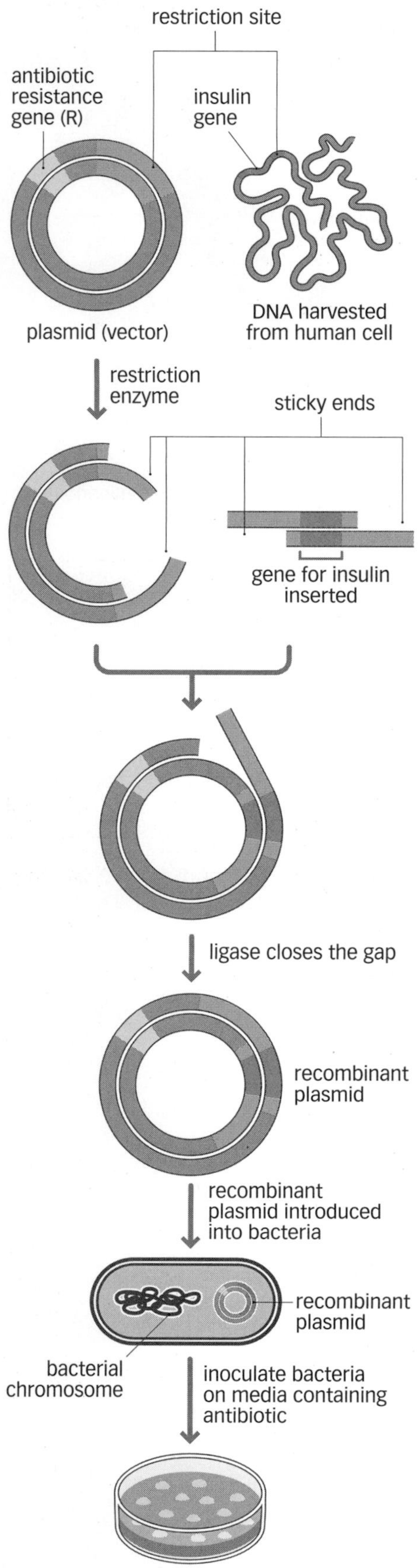

The DNA of the plasmid acting as the vector is cut with restriction enzymes, the gene to be replicated is inserted into the plasmid using DNA ligase enzyme, and then the recombined plasmid is inserted into a common bacterium for replication (**Figure 27.7**). Unfortunately, this method is restricted by the size of the gene that can fit into the plasmid vector. For example, a bacterial plasmid can accept only 15,000 nitrogenous bases for transfer. A bacteriophage (bacterial virus) can accept up to 20,000 bases, and a *cosmid*, which is a recombinant plasmid vector containing bacteriophage components, can accept up to 45,000 bases.

When replication is performed via the *polymerase chain reaction (PCR)* technique, genes can be replicated in large numbers in only a few hours. Here, a length of DNA is incubated with primers, nucleotides, and DNA polymerase (**Figure 27.8**). The goal is to produce large numbers of the segment of DNA labeled "target" in **Figure 27.8**. The first step is to unwind the target segment into its two single strands. One primer is complementary to one end of strand 1, and another primer is complementary to one end of strand 2. The primers attach as shown in the "anneal primers" step of **Figure 27.8**. In the extension step, the presence of primer at the right-hand end of strand 1 triggers production of the complementary bases to the left of the primer site, and the presence of primer at the left-hand end of strand 2 triggers production of the complementary bases to the right of the primer site. The result is two double-stranded DNA segments, each identical to the original target segment.

After each cycle, the newly made DNA segments are heated to separate them into single strands that are then used as templates for making additional replicas of the original target segment in a "chain reaction" that gives this procedure its name. The chain reaction, in which each cycle doubles the number of segments, allows the production to occur at an exponential rate. The drawback to PCR replication is that it can be used to replicate only small sequences of DNA. This technique cannot successfully replicate an entire genome.

To summarize what we have just covered, we saw that it is possible to insert DNA from one organism into a plasmid of another organism to produce a *recombinant* DNA. This recombinant DNA can then be cloned in a plasmid (or some other type of vector), or it can be replicated using PCR technology. After the DNA has been replicated in sufficient quantities, it can be inserted into a living host cell. Once this process is complete, products coded for by the replicated DNA can be harvested.

Any cell can be chemically treated to make the insertion of new DNA possible, or the cell can be exposed to low levels of electricity in a process known as *electroporation*. Both of these techniques make cells more accessible to foreign DNA when they are mixed in solution. The use of a device called a *gene gun* can be used to introduce DNA into plant cells, and direct injection using micropipettes can be used with animal cells (**Figure 27.9**).

Figure 27.7 The construction of a recombinant plasmid vector. Restriction enzymes open the plasmid at a specific location, and the gene to be replicated is inserted. The recombinant plasmid is then introduced into a bacterium that will express the inserted gene.

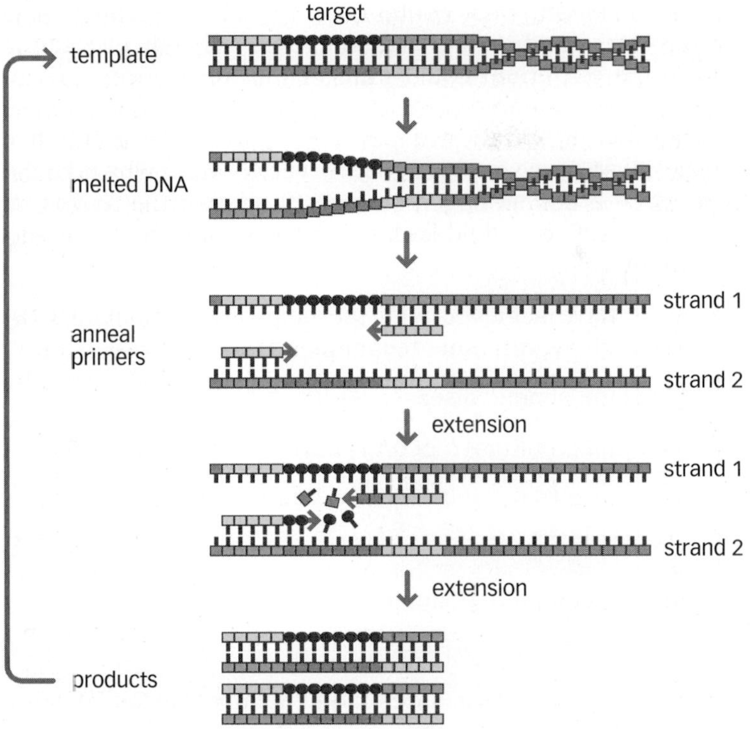

Figure 27.8 The polymerase chain reaction (PCR), which uses heat and complementary base pairing to replicate a target DNA strand.

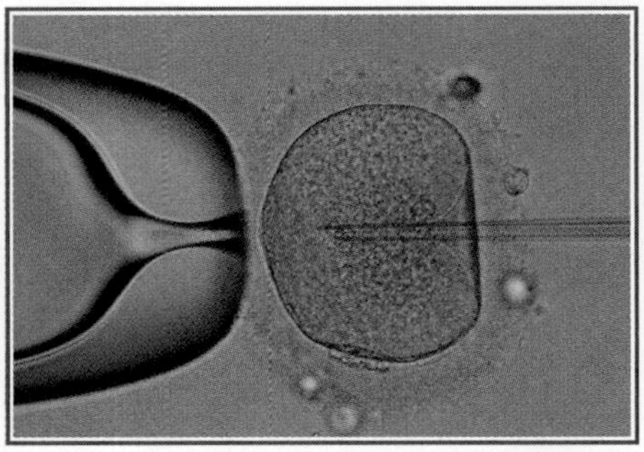

Figure 27.9 Implantation of DNA into a eukaryotic cell.

There are many applications of recombinant DNA technology (**Figure 27.10**). The production of insulin uses many of the techniques described above and was one of the earliest successes of recombinant DNA technology. The technology enables researchers to mass-produce a product — insulin — for the treatment of a complex and deadly disease — diabetes — but there is a downside to this technology. What if the vector cell

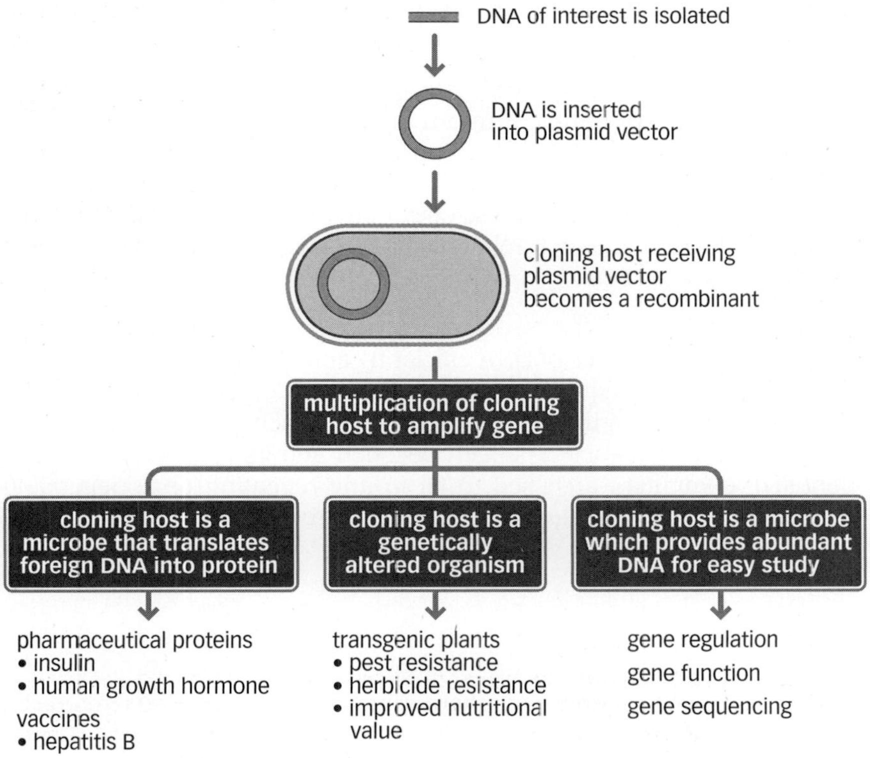

Figure 27.10 A few of the many uses of recombinant DNA technology.

producing the recombinant DNA contains an antibiotic-resistant gene that is easily transferred to the "new" host bacterial cell? What about the ease with which the bacterium could transfer that resistance to other bacteria? In our efforts to harness the production capabilities of these microscopic factories, we have "weaponized" a lot more potentially hazardous bacteria, and we work with them every day. So why take the risk and create such a double-edged sword? To answer that question, think about this: what if your child had juvenile diabetes and there was no insulin to be found?

Other advances we have achieved with the help of recombinant DNA technology in conjunction with other techniques are:

- Treatment of some genetic diseases (**Figure 27.11**)
- Enhancement of agricultural biocontrol agents (**Figure 27.12**)
- Improvement of food's nutritional value (think of orange juice).
- Development of biodegradable plastics
- Decrease in environmental pollutants
- Improved control of viral diseases

A list such as this makes it easy to understand the importance of cloning and genetic manipulation. These technologies provide the foundation for virtually all biotechnological research and development related to pharmaceutical manufacturing, production of transgenic crops, and environmental decontamination.

PROTEOMICS, GENOMICS, AND BIOINFORMATICS

Recombinant DNA, cell culture, and monoclonal antibodies are tools that are very familiar, almost commonplace, in research today. Much like kidney transplantation and vaccine production, these tools require skills and have a degree of difficulty but are now considered "basic" biotechnology. The use of each tool has an established procedure and is used in day-to-day operations.

Now the new biotech fields are *proteomics*, *genomics* and *bioinformatics*. The development of banks of information and the application of computer-based analysis programs to assess what impact each individual piece of information has on a system are what these fields of research are all about. Proteomics is the science of determining all of the proteins expressed in a cell and their relationships to one another and to the host organism.

Recombinant DNA and the associated replication techniques have enabled scientists to clone pieces of DNA efficiently and quickly. Once a piece has been cloned, DNA sequencing allows them to decode the sequence of nitrogenous bases on the piece to determine which proteins it codes for. Once the base sequence has been decoded and its proteins have been identified, computers are used to fill in any repeating or nonencoding segments, and the decoded piece is then stored in electronic databases

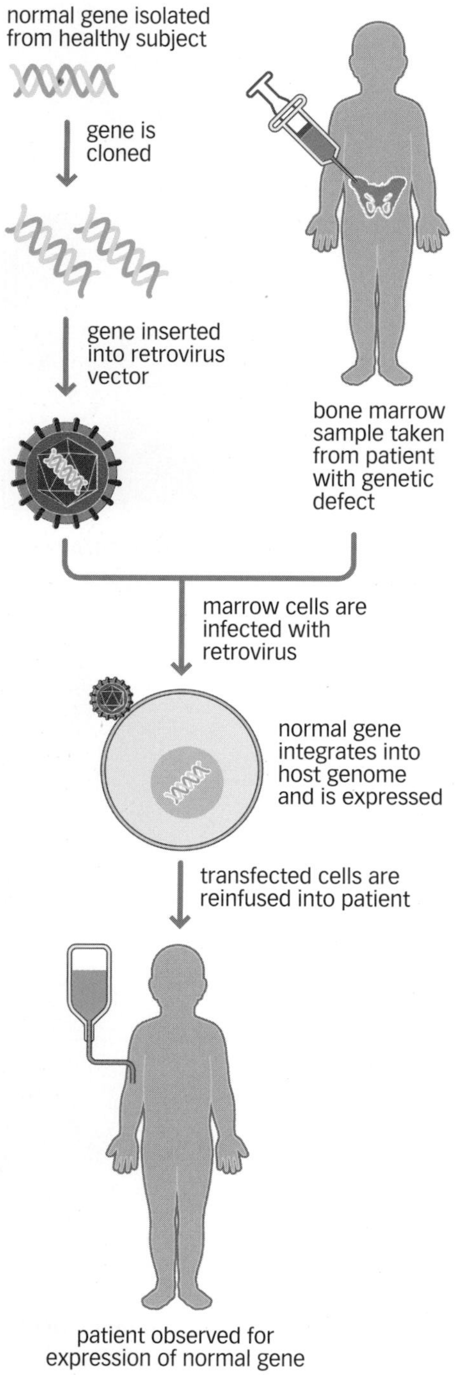
normal gene isolated from healthy subject
gene is cloned
gene inserted into retrovirus vector
bone marrow sample taken from patient with genetic defect
marrow cells are infected with retrovirus
normal gene integrates into host genome and is expressed
transfected cells are reinfused into patient
patient observed for expression of normal gene

Figure 27.11 How gene therapy may someday be used to correct genetic diseases. The new gene will be carried into the host by a viral vector.

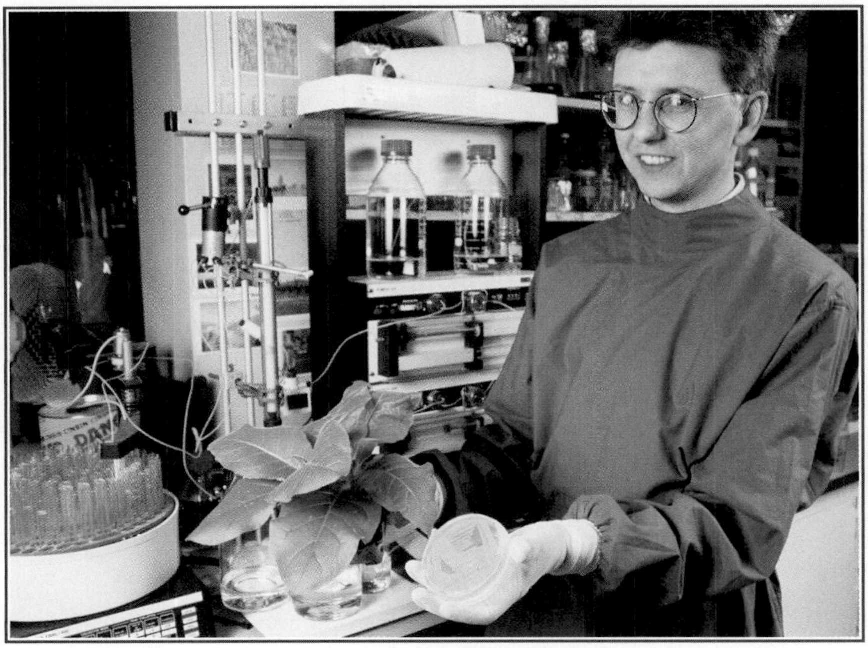

Figure 27.12 Biotechnology used to improve agriculture. The scientist in the photograph is holding a specimen of a transgenic tobacco plant, *Nicotiana* sp., which has been genetically engineered with the soil bacterium *Agrobacterium tumifaciens* (in the culture dish).

available to anyone who has a computer and access to the Internet. From this database, pieces of DNA can be cross-referenced between species, and similar base sequences can be searched for in different organisms. Once any similar base sequences are found, the protein products from the organisms being researched can be assigned to various sequences and cross-referenced between various species as well.

The cross-referencing is where proteomics ends and genomics begins. For example, if a certain piece of DNA from an organism is known to produce a virulence factor, the base sequence in that piece of DNA can be entered into an electronic database, which then searches through all the stored base sequences looking for those that either match or are similar to the sequences that produce virulence factors. When matching or similar sequences are found, researchers can use the techniques of proteomics to examine the proteins these sequences produce, checking those proteins for virulence factor activity. The information gathered in the proteomics step is then analyzed and stored for possible antitoxin or vaccine production against all organisms bearing a similar genetic code sequence.

It may seem as though the field of proteomics will follow the path of genomics closely. For example, genomics was faced with a large-scale mapping project when the Human Genome Project was begun. The differences lie in the inherent differences between genes and proteins. Genetic material is based on a largely static composite of nucleic acids with a complement of four nitrogenous bases. Proteins, however, have 20 amino acids and an almost infinite array of possibilities for expression and combination. Because proteins are expressed with various degrees of folding, the expression of genetic material may contain the same protein subunits but lead to wholly different end products (**Figure 27.13**).

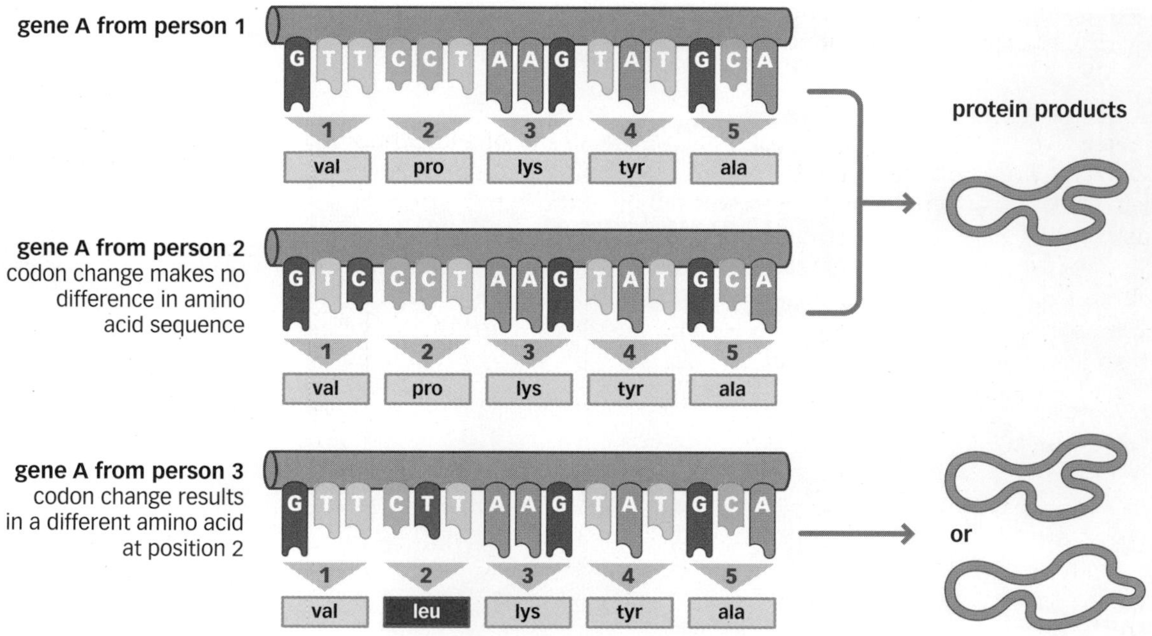

Figure 27.13 Changes in the sequence of nitrogenous bases in a piece of DNA cause changes in the amino acid sequence of the proteins produced.

In addition to mapping the subunits coded for in a functional genome, researchers must address other objectives, such as:

- Determining how age, environmental conditions and disease affect the proteins a cell produces
- Discovering the functions of these proteins
- Charting the progression of a process (such as disease development, the steps in the infection process, or the biochemical response of a plant to insect feeding) by measuring waxing and waning protein production
- Discovering how a given protein interacts both with other proteins in the cell and with proteins that enter the cell from its environment

The field of bioinformatics developed to solve the problem of "information overload" brought on by the ever-expanding pool of research data accumulating from workers in proteomics and genomics. Bioinformatics is a combination of biology, computer science, and information technology, and the main goals of this new discipline are:

- To create a database of information, such as sequences of amino acids in proteins and nucleotides in DNA segments
- To develop interfaces for accessing information in electronic data banks and for submitting new data
- To combine all existing genomics and proteomics information and use the result to study normal cell processes

One example of bioinformatics at work is the existence of the Protein Data Bank (PDB). This was the first database able to store three-dimensional data. The PDB morphed into a database called the Molecular Modeling DataBase (MMDB), which reorganized and validated PDB information and stored it in a way that allowed cross-referencing between the chemical structure of a macromolecule and its three-dimensional structure. Each successive database has made information more readily available and more easily accessible, both nationally and internationally.

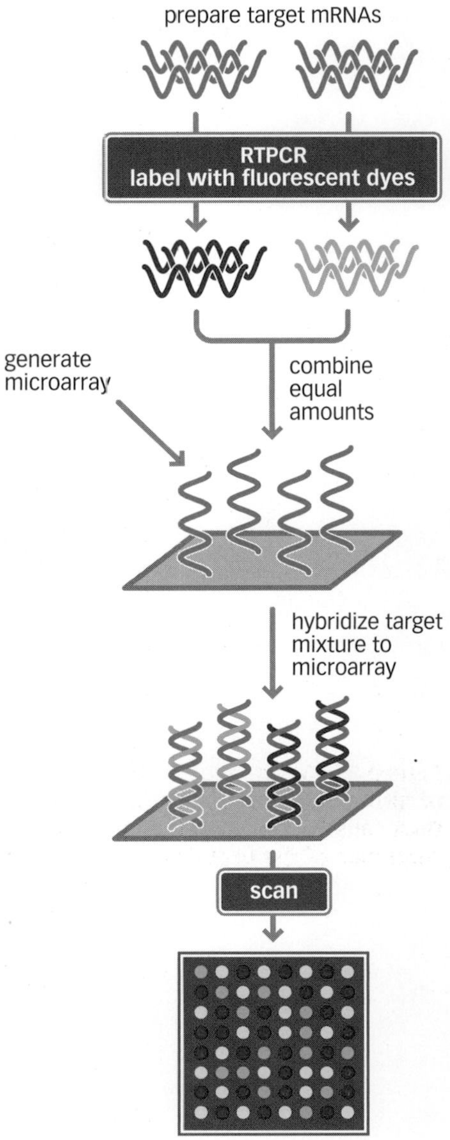

Figure 27.14 The microarray procedure permits rapid and economical review of gene expression.

Finally, we come to the truly futuristic applications of biotechnology, which may not be as far off in the future as we think. Protein engineering has allowed us to create *super enzymes* — which are better, stronger and faster catalysts — for use in industry and environmental applications. The chemical, pharmaceutical, textile, food, and energy industries are all benefiting from cleaner, more energy-efficient production processes using *biocatalysts*. Medical researchers have engineered proteins that can bind to and deactivate viruses and tumor cells.

Microarray Analysis

Microarray analysis is an exciting new technique being used to study genes in DNA samples. The technique is especially useful in situations where the DNA sample is very small or contains a very large number of unknown genes. Microarray technology is also useful in comparing healthy and diseased tissue samples, and the steps for creating and using a microarray are shown in **Figure 27.14**:

- An **array**, or solid support, on which thousands of genes are attached at fixed locations, is established from a known, sequenced segment of DNA. Arrays are usually created on a glass slide, silicone chip or nylon membrane.
- A fluorescently labeled nucleic acid molecule is added to act as a probe to look for specific DNA fragments.
- The fluorescent probe molecule binds with any nucleic acids in the test population of genes that are complementary to the nucleic acids in the probe molecule. The complementary nucleic acids on these two molecules combine to form a hybrid molecule.
- The hybrid molecules are detected via laser techniques, a microscope, and digital photography.
- The data are analyzed by computational methods.

This technique allows scientists to catalog and analyze hundreds of thousands of genes at once. It also significantly improves the ability of researchers to predict which proteins are produced by any given nucleic acid sequence. A third benefit is that microarray analysis yields information that helps researchers determine which nucleic acid sequences are important codons and which are nonsense sequences and so can be discarded.

Forensic Science

DNA technology has become a fundamental part of forensic science, and DNA "fingerprinting" is routinely used in paternity cases (**Figure 27.15**) and criminal prosecutions. This technique has become streamlined and

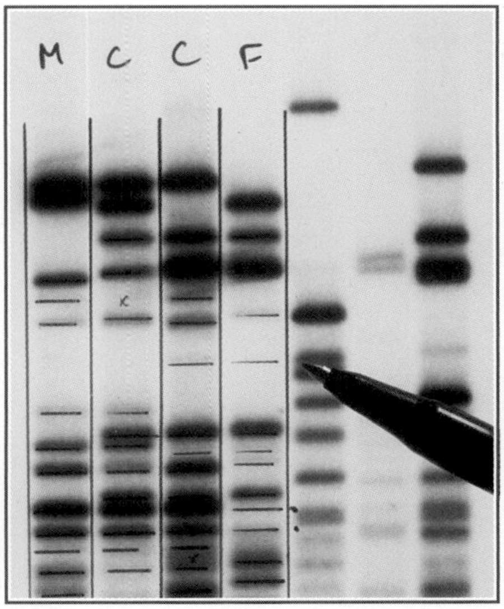

Figure 27.15 DNA fingerprinting used to prove family relationships. The photograph shows part of an autoradiograph of bands of DNA produced by the technique gel electrophoresis in an agarose gel. The pattern of these DNA bands is unique to each individual, but some bands are shared by related people, such as parent and child. DNA fingerprints can be used to prove whether people are related. The bands in these DNA fingerprints are marked **M** for mother, **C** for child, and **F** for father. Both children share some bands with each parent, proving that they are indeed related

automated and is a powerful tool because the genetic make-up of an individual is unique. Several high-profile murder and rape cases have turned on DNA evidence, and many convicted felons who were incarcerated before the advent of this technology have subsequently used it to prove their innocence.

Biosensors and Nanotechnology

Biosensors and nanotechnology are two of the more "far-out" technologies that are growing in popularity and use. They take advantage of our knowledge of biology and biological systems and combine this knowledge with microelectronics to create machinery that can, to name just two examples, run bedside chemical analyses of a patient's blood or detect and quantify a biological warfare agent, either remotely or in minute quantities.

INDUSTRY AND APPLICATIONS

Now that we have seen some of the research being conducted in the field of biotechnology today, we need to see how the discoveries that are the fruit of this research are being implemented in the workforce. Where will these new ideas take us? In another 50 years, will we be able to "grow" new organs to replace damaged ones, making organ transplantation obsolete? Will treating disease by bombarding a patient's whole system with antibiotics be considered a barbaric and crude method once therapeutic intervention at the genetic level becomes standard? And, perhaps the most important question of all, where do you see yourself in this age of discovery?

Industry is making full use of each product coming from biotechnology research and development companies worldwide. The exchange of biotechnological information is rapid and international. There are few borders that remain to either knowledge or diseases. In health care, biotech applications are numerous and relatively obvious, mainly in diagnostics and therapeutics. The advances in molecular technology and genetics allow us to detect diseases more quickly and with greater accuracy. Biotechnology has reached the home with Glucometer instruments for measuring blood sugar levels, and home pregnancy tests that are so accurate they are the same as those used in doctors' offices.

Early detection of genetic diseases has led to the treatment of previously untreatable diseases. You, as a student preparing to be part of the health care profession, have the opportunity to be a part of these advances. Many of the companies involved in the development of biotech products hire specifically from the biological sciences for their employees. Students of microbiology, molecular biology, biochemistry, and pharmacology, to name just a few disciplines, are being actively recruited in ever-greater numbers.

Even if you have no interest in the field of human health care, there are industrial uses of biotechnology everywhere that have little to do with health care. For example, something near and dear to the hearts of pharmaceutical companies everywhere is *product discovery and development.* An advanced knowledge of biotechnological systems is essential to researchers trying to create both new diagnostic tests for diseases and protocols for gene-based therapies. To be able to deconstruct a disease and find ways to treat it, you must first understand the pathways and processes of the disease. This understanding requires a more laboratory-research-oriented profession.

Say, however, that even that is too close to the human health care services for your tastes. If so, perhaps your future is in the environmental sciences, which take advantage of biotech applications that many of us never hear about. For example, industry has long used fossil fuels to power their processes. Now, through biotechnology, there are options for alternative fuels, such as renewable biomass-based fuels. Producing biomass fuels is cleaner than processing crude oil, and less waste is generated. For those processes in which plant biomass products cannot be used as the main fuel source, biotechnology has helped develop ways to clean fossil fuels so that their use generates fewer pollutants.

Companies involved in industry are constantly searching for new enzymes to streamline their production processes. When new enzymes are not being made, they are being searched for. About 30,000 bacterial species are currently represented in an electronic database being used to catalogue the discovery of novel species. However, a recent study suggested that the world's oceans might support as many as 2 million different bacteria and that just one ton of topsoil might support upwards of 4 million bacterial species!

As you can see, we are a long way from knowing what exists right now, not counting the recombinant creations we are making. Companies are especially interested in the discovery of organisms adapted to harsh environments. This is because many industrial processes require a harsh environment — for example, the day-to-day operation of a paper mill. Sampling "wild" microbial populations adapted to living in extreme temperature and pH ranges allows us to create a more efficient and less polluting atmosphere in which these products can be produced.

Finally, one of the most important developments in biotechnology is in the field of renewable energy. Billions of dollars and multiple presidential initiatives have called for the creation of a *bio-based economy* for the United States. This would involve the creation of bio-based fuels and products. We have already seen huge advances with the production of ethanol from wheat straw by using biotech enzymes. Given the state of the world today and the dependence of the economically advanced nations on readily available fuel sources, advances such as this make biotechnology the wave of the future.

Living organisms are being used not only in the production of new fuels but also in the cleanup of the by-products of the old ones. As noted earlier in this chapter, this process is called **bioremediation** (**Figure 27.16**).

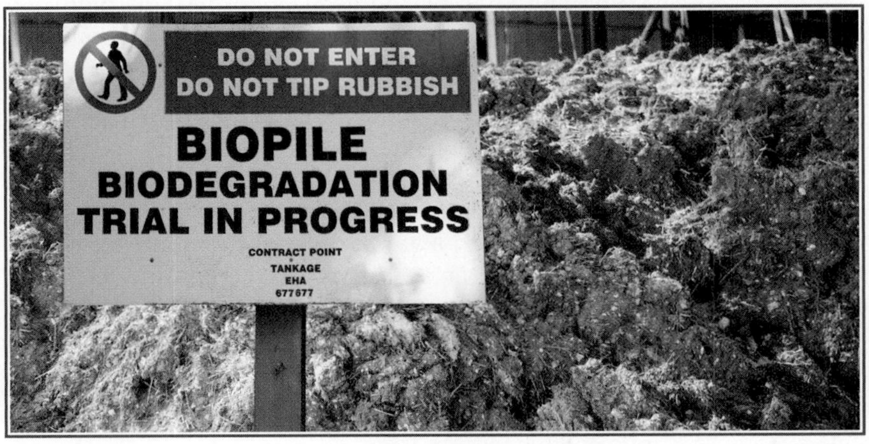

Figure 27.16 A biopile. This is a facility to economically clean up soils that have been contaminated with hydrocarbons such as gasoline, diesel, and jet fuels.

We use bioremediation fungi to clean up toxic wastes produced by the paper industry. We use bioremediation microbes in toxic waste dumps to degrade harmful compounds into harmless ones. Environmental biotechnology can clean hazardous wastes more efficiently than traditional methods such as incineration. In addition, bioremediation methods take much of the impetus away from the NIMBY (Not In My Back Yard) principle of waste storage.

Bioremediation involves either dumping a toxic waste at a site where bacteria that can feed on the waste are already present, or dumping the waste at a site where no bacteria are present initially and then adding bacteria that will degrade the waste. The idea is that when the toxic waste sustaining the population of bacteria is gone, the bacteria will either die or return to their naturally occurring level. The dangers inherent in this method for cleaning up toxic waste are evident. What if the bacteria mutate into some dangerous form? Or what if they maintain an overpopulated state once the waste has been consumed? Something to consider is that in all new technologies there are unexpected consequences that must be dealt with.

WE KNOW THAT WE CAN, BUT SHOULD WE?

Sprinkled throughout this chapter are questions and statements designed to make you think about the scientific advances we have made in the past 100 years and where you stand on these issues as a professional. Biotechnology has the potential to improve the health and wellbeing of people worldwide. However, it also has the potential to create repercussions of unforeseen magnitude both in our own society and in the social and ethical development of underdeveloped nations. We know we can do these things, but should we?

Here, for instance, is an expression of concern dating back to when recombinant DNA technology was in its infancy:

> *"In 1973, a few days after Drs. Herbert Boyer and Stanley Cohen described their successful attempt to recombine DNA from one organism with that of another, a group of scientists ... sent a letter to the National Academy of Sciences (NAS) and the ... journal* Science, *calling for a self-imposed moratorium on certain scientific experiments using recombinant DNA technology. ... They suggested that an international group of scientists from various disciplines meet, share up-to-date information and decide how the global scientific community should proceed."* (Bio 2006–2006 Guide to Technology)

In February 1975, in response to this call to action, 150 scientists met and discussed recombinant DNA. They replaced the voluntary moratorium that the scientific community had placed itself under with a complicated set of rules regarding laboratory experimentation with recombinant DNA.

In the United States, the National Institutes of Health formed a recombinant DNA advisory committee, which adopted the rules set out by the international meeting. In 1988, the NIH also sponsored the development of a National Center for Biotechnology Information. Its purpose is to develop and coordinate access to a variety of recombinant-DNA databases and software for scientific and medical communities, both nationally and internationally.

Now that recombinant DNA technology has become accepted, there are other, newer ethical considerations on the horizon. Cloning and stem cell research are two of the most familiar and contentious issues that biotech researchers must face in the near future.

There has been no voluntary moratorium on cloning techniques, and the backlash from the regulation or attempted regulation of access to embryonic stem cell lines has reached international proportions. The production and purchase of transgenic crops is rapidly becoming an international powder keg as the countries of the European Union and also in Africa line up to deny access to their economic markets. Massive quantities of agricultural products are being denied access to world markets, which causes prices in the United States to spiral downwards, which then causes international repercussions in world markets trading in agricultural companies.

The realities of international trade create an atmosphere in which science is going to have to apply itself to the rigorous standards of public scrutiny. Whether we, as professionals, pass that test will depend on our willingness and ability to understand the issues, defend our positions, and hold on to our view of society as a whole. Without knowledge, we fall prey to ignorance. Nothing breeds misunderstanding, fear, and eventually violence faster than ignorance. With the potential to bring great advances to society, biotechnology also trails with it the seeds of great fear.

SUMMARY

In this chapter, we have reviewed some of the major points associated with the burgeoning field of biotechnology. As a health care professional, you may never have the opportunity to use some of these techniques, but you will undoubtedly be exposed to the benefits provided by this technology on an almost daily basis. It is important that you have an understanding of the potential of this technology and how it will affect health care in the years to come. Perhaps more importantly, as a member of the human race you will be influenced in a number of ways by the "tricks" we can perform using biotechnology to clean up the environment and feed the world. It is important to remember that these advances come with a price. In the next chapter we examine how these microbiological systems can be manipulated to the detriment of humankind through bioterrorism.

NOTE: There are no self-evaluation and chapter confidence questions for this chapter. The information provided here is for your understanding of new techniques that can make health care easier and more precise. This information is not intended to be part of your educational requirements for microbiology. Rather, it is information that can make you a better health care professional.

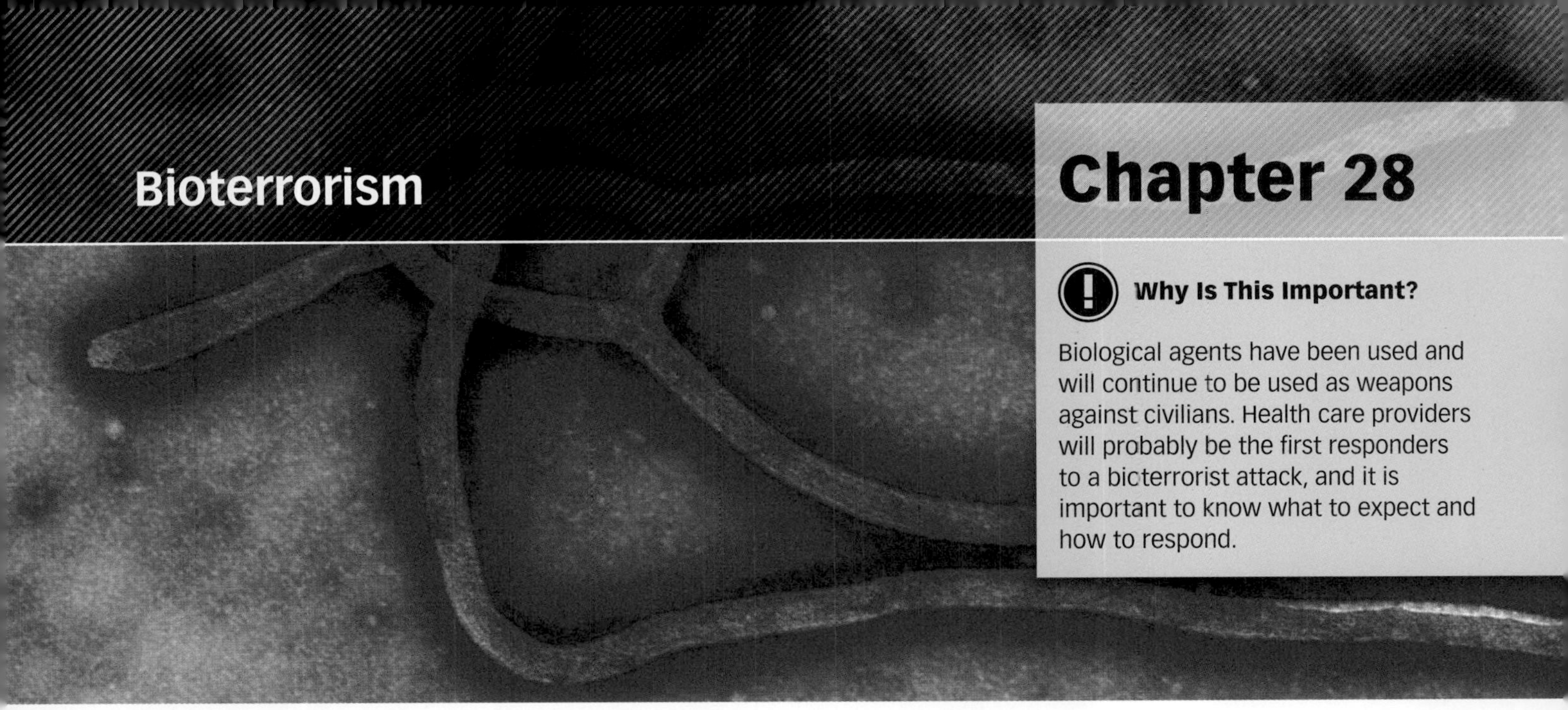

Bioterrorism

Chapter 28

Why Is This Important?

Biological agents have been used and will continue to be used as weapons against civilians. Health care providers will probably be the first responders to a bicterrorist attack, and it is important to know what to expect and how to respond.

OVERVIEW

In this chapter, we look at bioterrorism and the biological agents that could be responsible for attack. Biological weapons have been used throughout history and as recently as the 2001 anthrax attacks through the US postal system. It is important for health care providers to be familiar with the symptoms and presentations of bioterrorism agents because hospitals will probably be the first to respond to a bioterrorist attack. Here we discuss probable biological agents, such as anthrax and smallpox. We also discuss clues and indications that a bioterrorist attack has occurred and how to respond to such an event.

We will divide our discussions into the following topics:

- Bioterrorism
  - WHAT IS BIOTERRORISM?
    - HISTORY OF BIOTERRORISM
    - BIOLOGICAL WEAPONS
    - PROBABILITY AND EFFECTS OF A BIOLOGICAL ATTACK

Suppose that some day you pick up your mail from your mailbox and open an envelope that has a white powder on it. Perhaps you don't see the white powder. You become ill in a couple of days, but how can you know your being ill has anything to do with the mail? The five individuals who died of anthrax in October 2001 didn't know. At least, they didn't know until it was too late.

All indications are that the anthrax powder was mailed to numerous public officials and publications (**Figure 28.1**). The first victim was a photo editor at a Florida newspaper. One morning, he woke up feeling nauseated and began vomiting. The emergency room where he was seen knew something wasn't right. This wasn't a typical case of the flu. He was admitted to the hospital, but it took health providers two more days to make the diagnosis on 4 October. The patient died the next day. Was the medical center incompetent? It's not likely. Inhalation anthrax is so uncommon in an urban environment that there would be no reason to check for it — unless you suspected there had been a bioterrorist attack.

What Do I Need to Know?

To get the most out of this chapter, please review the following terms from your previous courses in biology, anatomy, physiology, or chemistry or in previous chapters of this book as indicated in parentheses: acetylcholine, alphavirus (**12**), bacteremia (**5**), bubo (**25**), septicemia (**5**), cholinergic receptor, eschar (**26**), labile, maculopapular rash (**26**), pandemic (**7**), papule (**26**), phagocytosis (**15**), pustule (**26**), tachycardia, triage, viremia (**12**), zoonotic disease (**6**).

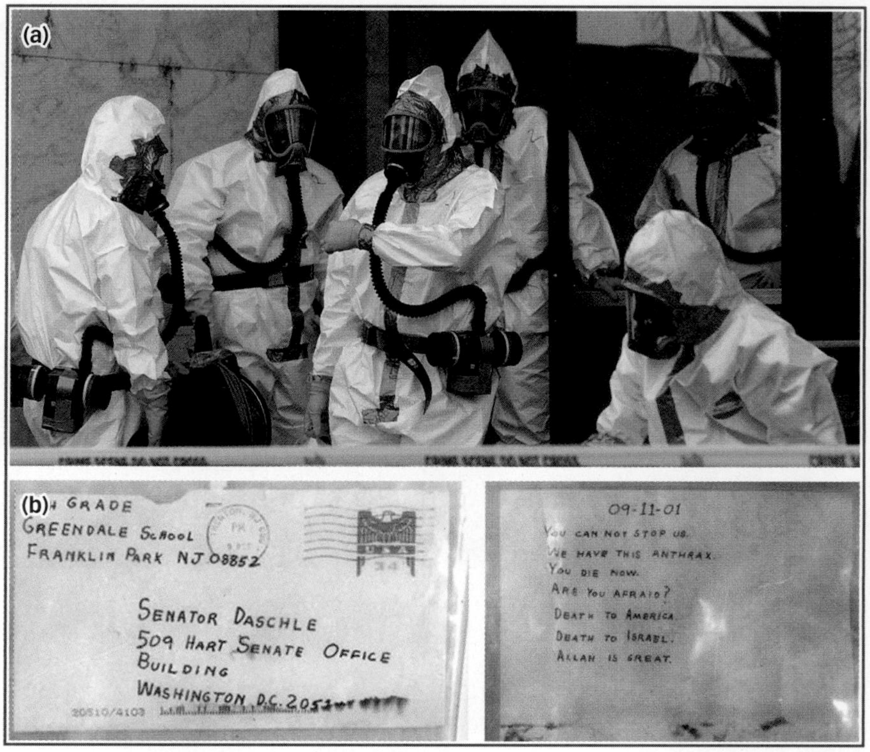

Figure 28.1 Letters laced with anthrax spores were sent to Congress through the US mail in 2001. Panel **a** shows workers in protective garments preparing to enter the Hart Building in Washington DC. Panel **b** shows a contaminated envelope and letter.

Four more deaths occurred in the following weeks as a result of anthrax: two postal workers, a hospital supply worker in New York City, and a 94-year-old woman in rural Connecticut.

WHAT IS BIOTERRORISM?

The Centers for Disease Control (CDC) in Atlanta defines **bioterrorism** as the intentional release of bacteria, viruses, or toxins for the purpose of harming or killing civilians. Although these agents can be found in nature, they may be altered to make them more effective in causing disease, spreading disease, or resisting treatment. Biological agents may appeal to terrorists because they are difficult to detect and may not cause disease for hours or days after initial exposure, making it challenging to trace the illness back to the source. Biological agents used as weapons can be spread in a variety of ways, including from person to person, as occurs with smallpox and plague.

The most effective biological weapons have a high mortality rate, but large numbers of deaths are not necessary to cause widespread panic. Just the suggestion of a bioterrorist attack can quickly disrupt society. During the 2001 anthrax attack, 33,000 individuals were given prophylactic antibiotics for possible exposure. Think of the implications of providing care for thousands of individuals who were exposed to a biological agent and may or may not develop the disease. Think of the difficulty of treating those who are not killed by the disease but remain critically ill for weeks or months.

HISTORY OF BIOTERRORISM

Biological weapons were around centuries before the discovery of bacteria in the nineteenth century and antibiotics in the twentieth century. Armies knew that people who were sick could make other people sick, even if they didn't know how. The intentional contamination of drinking water has occurred throughout history, but the Greeks probably initiated this practice when they threw animal carcasses into wells in about 300 B.C. During the Middle Ages, Tartar forces would catapult plague victims into besieged villages to infect and decimate the enemy population. European conquerors entering the New World did not need to rely on firepower alone to fight the indigenous peoples already there. Smallpox devastated various Indian tribes in both North and South America. Europeans gave Indians blankets infested with smallpox during the French and Indian Wars. (The Europeans didn't need to go through the trouble of passing out blankets because the natural transmission of smallpox through respiratory droplets was far more efficient.)

Wars of the twentieth century were not all that different when it came to biological weapons. During World War I, Germany developed a biological weapons program that involved infecting sheep with *Bacillus anthracis*. Luckily, the cultures were confiscated before they could be used against the enemy. During World War II, a Japanese biological weapons section known as Unit 731 became notorious for its experiments involving biological warfare (**Figure 28.2**). Chinese prisoners and prisoners of war held by this unit were routinely used for medical experiments, which included deliberately infecting the prisoners with plague, anthrax, and typhoid. The Japanese put their experiments into practice when they dropped plague-infected fleas over populated areas, attacks that killed thousands of Chinese during World War II. However, the Japanese were not very adept at protecting themselves from the biological weapons they created. More than 1700 Japanese deaths occurred as a result of improper handling of their own biological agents.

Although the United States and its allies did not use biological weapons during World War II, they did not want to be caught unprepared in the future. After the war, the United States offered protection from criminal charges to some of the Unit 731 leaders in exchange for the data they had collected. The United States also developed its own biological weapons programs, including a controversial study involving dispersal methods for the bacterium *Serratia marcescens* near San Francisco that may have infected American citizens in 1950. In 1969, President Richard Nixon ended the US offensive biological weapons program. Then, in 1972 a Biological and Toxin Weapons Convention was held. Although more than 100 countries signed this agreement, including the United States, biological weapons programs continued in some countries. During the Cold War between the United States and the Soviet Union, there was evidence that the Soviets had several biological weapons facilities, including one that housed smallpox and hemorrhagic viruses. After the fall of the Soviet Union in 1991, several of the scientists who worked on developing biological weapons left the country. Their knowledge could be dangerous if given to terrorist organizations.

Biological weapons have also been used outside wars. The largest mass casualty as a result of anthrax occurred in 1979, when the accidental release of anthrax spores from a Soviet Union bioweapons facility resulted in the infection of 77 individuals, with 66 deaths. In 1984, a religious cult used *Salmonella* to contaminate a restaurant salad bar in Oregon. Although no one died, the incident led to 751 patients and 45

Figure 28.2 Body disposal at Unit 731, a Japanese testing facility during World War II where humans were exposed to biological agents in a program developing biological weapons.

hospitalizations. The Japanese cult Aum Shinrikyo released sarin nerve gas in a Tokyo subway, killing 20 people. It was later discovered that the cult had developed other biological weapons including anthrax and hemorrhagic viruses. In 2001, anthrax released through the US postal system caused 22 infections and 5 deaths.

BIOLOGICAL WEAPONS

The intent of a bioterrorist attack is to cause public panic and social disruption. With this goal in mind, we can ask: What does it take to be a good biological weapon, commonly referred to as a **bioweapon**? First, it must be an agent that can either be easily disseminated over a large population or is highly contagious so that it will spread quickly even when only a few individuals are infected by the initial attack. Second, high mortality rates are preferred, but some survivors are necessary so as to transmit the disease from person to person. Last, a good biological weapon will cause disease that will have a significant impact on the resources of the health care system.

The CDC separates bioterrorism agents into three categories based on how easily they can be spread and on the severity of illness they cause (**Table 28.1**). Category A agents are given the highest priority because they pose the greatest risk to national security. They are easily disseminated,

Table 28.1 CDC Categories of Biological Weapons

Category	Risk	Agent
A	Greatest	Anthrax (*Bacillus anthracis*)
		Botulism (*Clostridium botulinum* toxin)
		Plague (*Yersinia pestis*)
		Smallpox (variola major)
		Tularemia (*Francisella tularensis*)
		Viral hemorrhagic fevers (filovirus and arenaviruses)
B	Lower than with category A agents	Brucellosis (*Brucella* species)
		Epsilon toxin of *Clostridium perfringens*
		Food safety threats (*Salmonella* species, *Escherichia coli* O157:H7, *Shigella*)
		Glanders (*Burkholderia mallei*)
		Melioidosis (*Burkholderia pseudomallei*)
		Psittacosis (*Chlamydia psittaci*)
		Q fever (*Coxiella burnetii*)
		Ricin toxin from *Ricinus communis* (castor beans)
		Staphylococcal enterotoxin B
		Typhus (*Rickettsia prowazekii*)
		Viral encephalitis (alphaviruses)
		Water safety threats (*Vibrio cholerae*, *Cryptosporidium parvum*)
C	Lower than with category B agents	Emerging infectious diseases

have high mortality rates, cause public panic, and require special preparations by public health authorities. Category B agents are moderately easy to disseminate, have low mortality rates, and require increased disease surveillance. Category C agents are emerging infectious diseases that could be engineered in the future as bioweapons because of their availability, ease of dissemination, and high mortality rates.

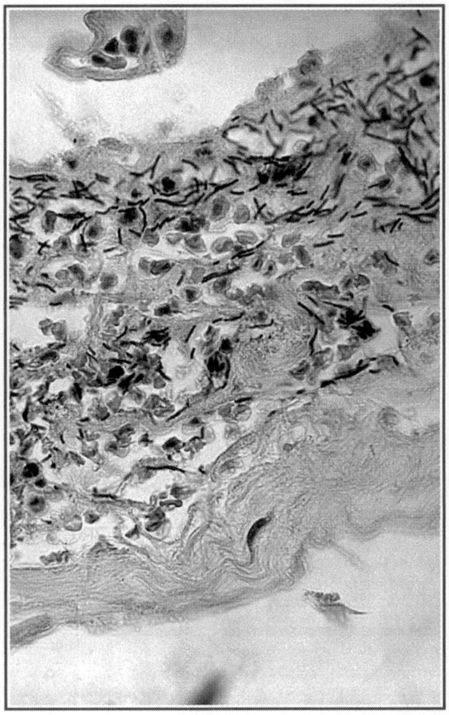

Figure 28.3 A photomicrograph of the meninges showing the presence of *Bacillus anthracis* in a case of fatal inhalation anthrax.

Anthrax

Bacillus anthracis is an encapsulated, aerobic, Gram-positive, rod-shaped bacterium (**Figure 28.3**). It produces spores, which are highly resistant and can survive for decades. It was one of the first bacterial pathogens recognized when the German scientist Robert Koch discovered the life cycle of the organism in the 1870s. From his studies of *B. anthracis* came Koch's postulates, which, as we saw in Chapter 7, were the basis on which was established the theory that a specific pathogen causes a specific disease.

There are three routes by which anthrax can infect the body: through the skin (**cutaneous anthrax**), through the digestive tract (gastrointestinal anthrax), and through inhalation (**inhalation anthrax**, also referred to as *respiratory anthrax*). Cutaneous anthrax is the most common form of the disease and occurs when the skin comes into direct contact with the *Bacillus* spore. The infected individual develops localized itching after an incubation period of one day. A papular lesion forms and turns into a black eschar in seven to ten days. The cutaneous form of anthrax is not considered lethal and can easily be treated with antibiotics.

Gastrointestinal anthrax occurs after eating undercooked meat or dairy products from infected animals. The incubation period can be anywhere from one to seven days, with initial symptoms of nausea, vomiting, anorexia, fever, bloody diarrhea, bloody vomit, and abdominal pain. Shock and death occur two to five days after the onset of symptoms.

It would be difficult to infect a large number of individuals through the cutaneous or digestive routes of infection, and consequently these routes would most probably not be used by terrorists. Instead, anthrax would most probably be released as an aerosol in a terrorist attack. This would result in inhalation anthrax, which is the rarest and most deadly form of the disease. The incubation period is usually one week but can be as long as two months. Infected individuals first develop nonspecific signs and symptoms such as fever, nonproductive cough, malaise, fatigue, muscle aches, and chest discomfort. Because these signs and symptoms mimic those of much milder infections, a diagnosis of inhalation anthrax could easily be delayed. There is frequently a short respite during which the person seems to be getting better, much as one would recover from a cold or the flu. However, in a couple of days the individual begins to feel much worse, developing a high fever and respiratory distress. Without antibiotic or supportive treatment, shock and death ensue in 24–36 hours. The mortality rate used to be 80–90%, but in the 2001 attack in the United States the mortality rate decreased to 45–50%. This is probably due to better antibiotic regimens and improved intensive care capabilities.

All forms of anthrax require treatment with intravenous antibiotics, but the amount of supportive care required is different for the three types. Inhalation anthrax, the most likely to develop from a bioterrorist attack, requires the skills of a critical care team. The patient may be admitted to an intensive care unit that can offer ventilator assistance. If the patient survives the initial complications, several weeks of care may be needed for a full recovery. For this reason, a mass casualty involving large numbers of cases with inhalation anthrax will cause an enormous strain on the health care system, especially in areas where intensive care beds are in short supply.

A vaccine for all forms of anthrax has been developed and is currently given to all US military personnel. It is a series of six inoculations given over a period of 18 months. For any significant increase in survivability rates, however, even vaccinated patients require antibiotic therapy after exposure. The civilian use of the anthrax vaccine is limited to those at a high risk of exposure. New treatments and vaccines are currently being studied.

Botulism

Clostridium botulinum is a Gram-positive, spore-forming, obligate anaerobic bacterium found mainly in soil and in marine and agricultural products. The spores are heat-resistant, surviving at 100°C for several hours, but the toxin is heat labile and easily denatured at 80°C. Because the spore is dormant, the disease cannot be passed from person to person. It is the toxin that *C. botulinum* produces that causes serious physical effects in individuals unlucky enough to come into contact with it. Absorption of the toxin can be through a mucosal surface, such as the gut or lung, or through a wound, but the toxin is unable to penetrate the skin.

There are five ways in which botulism can be acquired. *Food-borne botulism* occurs through the ingestion of food contaminated by the toxin. *Infant botulism* develops when spores are ingested and colonize an infant's gastrointestinal tract. The spores germinate into organisms that produce and release toxin into the gastrointestinal tract. *Adult infectious botulism* is similar to infant botulism but occurs in adults. In *wound botulism*, the wound is infected with *C. botulinum*, which then produces the toxin. The final form, *inhalational botulism*, occurs only if the toxin has been aerosolized and released.

Botulism toxins are distributed by the bloodstream to the body's cholinergic receptors, where the toxins block the release of acetylcholine. This results in a descending paralysis that begins with the face and progresses to the upper and lower extremities and eventually the respiratory muscles. All forms of botulism cause almost identical neurological symptoms, beginning with cranial neuropathies and descending symmetric weakness, but the central nervous system and sensory nerves are not affected. Naturally occurring food-borne botulism may be preceded by nausea, vomiting, and diarrhea. However, these signs may be due to the toxin's interaction with other microbes in the food and may not occur if purified toxin is used for a bioterrorist attack.

The incubation period for botulism infection ranges from 2 hours to 8 days but is most commonly between 12 and 36 hours. The initial symptoms — dry mouth, double vision, difficulty in speaking, difficulty in swallowing, and facial weakness — commonly proceed to shortness of breath, loss of muscle tone, and eventually complete paralysis (**Figure 28.4**). Intubation and mechanical ventilation are often required to combat respiratory difficulties. Other symptoms include decreased salivation, ileus (intestinal obstruction), and urine retention. Clinical manifestations can vary widely from one person to the next, with some patients mildly affected and others, at the opposite extreme, dying within 24 hours of developing symptoms. Because of these variables, botulism is frequently misdiagnosed as Guillain–Barré syndrome or some other neurological disorder.

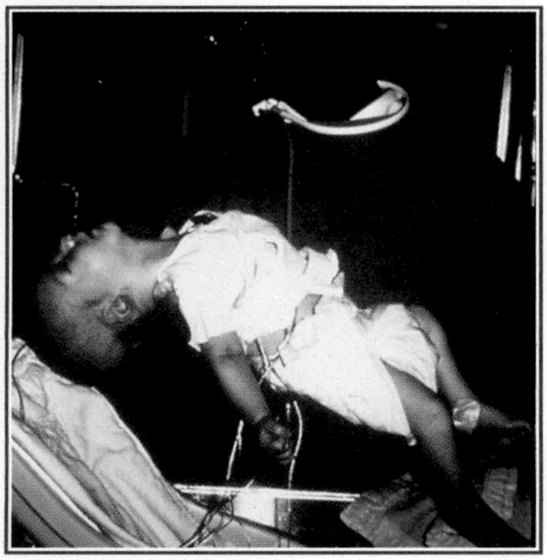

Figure 28.4 A limp infant showing the poor muscle tone that is an early symptom of botulism infections.

The mortality rate for botulism has decreased from 25% to 5–10%, an improvement that is most probably due to modern medical care. However, paralysis can last for weeks or months and require extensive medical support, including assisted ventilation, tube-feeding, and treatment of secondary infections. The only treatment for botulism is supportive care and administration of antitoxin after exposure. Antitoxin can reduce the severity of the disease but will not necessarily decrease the duration of the paralysis.

Inhalational botulism, the most likely form to be used in a bioterrorist attack, would not easily be detected initially. Symptoms would begin with a mild irritation of the upper airway followed by different degrees of paralysis. The first patients who were maximally affected would not present to a medical facility until a few days after exposure. This delay in diagnosis and the limited supply of antitoxin would result in a significant number of patients requiring ventilator support, which would immobilize intensive care units for months. Despite the low mortality rate, botulism toxin makes a good biological agent because of its significant morbidity and the toll it will take on the health care system.

Plague

Yersinia pestis is a nonmotile, Gram-negative, bipolar coccobacillus that causes plague. In naturally occurring plague, there is usually an inoculation of *Y. pestis* under the patient's skin after a bite by a plague-infected flea. The bacteria migrate to regional lymph nodes and are phagocytosed but not destroyed. They quickly cause destruction and necrosis of the lymph nodes, followed by bacteremia and septicemia that can lead to shock, disseminated intravascular coagulation, and coma.

Plague is rapidly fatal, highly contagious, and not easily contained, properties that make *Y. pestis* an excellent bioweapon candidate. The bacterium is endemic to the southwestern United States, with approximately 10 cases being documented each year. Reservoirs for this organism are usually rodents, and the vector for transmission is the oriental rat flea. Humans can become infected when bitten by an infected flea, inhaling respiratory secretions from infected animals, or handling infected animal tissues.

There are three main types of clinical syndrome associated with plague: bubonic, septicemic, and pneumonic. Bubonic plague, the most common type, occurs when a person is bitten by a *Y. pestis*-infected flea or when the skin comes into direct contact with an infected animal. Rarely, bacteremia develops without a preceding bubo, which has been termed septicemic plague. Primary pneumonic plague is the result of inhaling the microorganism in an aerosolized form. Pneumonic plague can also develop secondarily to the hematogenous spread of bubonic plague.

The incubation period for bubonic plague is two to eight days. Patients present with a sudden onset of fever, chills, weakness, and headache. Tender lymphadenopathy usually occurs in the groin, axilla, and neck. It can take hours to days for a bubo to develop (**Figure 28.5**), which is characterized by surrounding edema (swelling) and extreme pain limiting the range of motion of the affected limb.

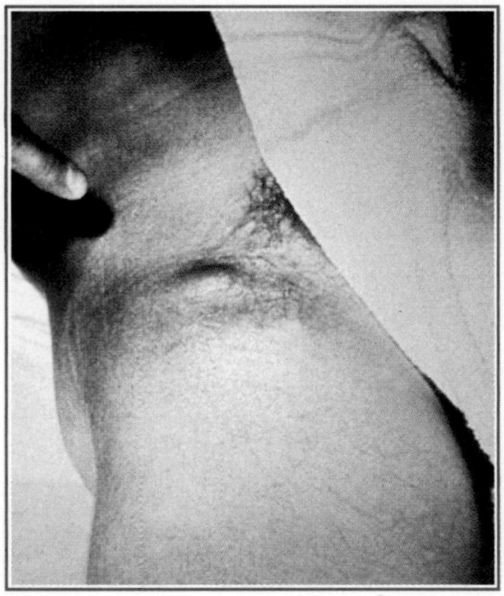

Figure 28.5 This patient shows symptoms that include a swollen inguinal lymph node referred to as a bubo.

The buboes in bubonic plague are 1–10 cm in diameter, firm, and tender. In the two to four days after buboes form, the patient deteriorates rapidly, with a high fever, tachycardia, malaise, headache, vomiting, chills, and altered mental status as bacteremia develops. In some patients the *Y. pestis* organisms are contained in the lymphatic system and bacteremia never develops, but without treatment the estimated mortality rate is 50%. Treatment consists of antibiotics and supportive treatment. The cyanosis (dark blue color of the skin as a result of lack of oxygen in the blood) and gangrene of the peripheral tissues that are characteristic of bubonic plague led to the term "Black Death" that was used during the European epidemics (**Figure 28.6**).

Figure 28.6 An illustration depicting the bubonic plague in seventeenth-century Italy.

Septicemic plague can be difficult to diagnose because there may be no detectable lymph node involvement and therefore no buboes to alert medical personnel. Patients have a high concentration of bacteria in their blood, and apparently the bacteremia overwhelms patients before lymphadenitis occurs. In septicemic plague, infected individuals are febrile without localized signs or symptoms. Without treatment, hypotension and multiorgan failure leads rapidly to death. Standard treatment includes antibiotics and supportive treatments that will often require the resources of an intensive care unit.

Pneumonic plague has an incubation period of one to three days and is rapidly fatal. Patients present with fever, chills, headache, body pains, weakness, and chest discomfort. As the disease progresses, sputum production, chest pain, hemoptysis (bloody sputum), and respiratory failure develop. Buboes do not present in pneumonic plague. This form of plague is 100% fatal without antibiotic treatment, with death occurring 18–24 hours after the onset of symptoms. Patients are highly contagious from the onset of disease, and their cough is highly infectious. Therefore strict respiratory isolation is critical in those patients in whom pneumonic plague is suspected.

Fast Fact

If 50 kg of *Yersinia pestis* were aerosolized and released over a city of 5 million inhabitants, as many as 150,000 persons could be infected with pneumonic plague and 30% would die.

Antibiotics are very effective in treating pneumonic plague, but they should be given in the first 24 hours after symptoms appear because delay in treatment greatly increases the mortality rate. In a bioterrorist attack, anyone with a temperature greater than 38.5°C or a new cough should be treated with prophylactic antibiotics. Because patients with pneumonic plague are highly infectious, precautions should be taken to isolate respiratory droplets, which usually mean having patients wear facemasks.

The World Health Organization studied the potential consequences of using plague as a biological weapon. Plague would most probably be disseminated in an aerosolized form in the event of a bioterrorist attack, which would probably result in an outbreak of pneumonic plague. Symptoms would initially resemble those of other respiratory illnesses. Therefore a patient waiting in a crowded emergency department could potentially infect dozens of individuals. Providers would not initially think of pneumonic plague because its natural occurrence is so rare, especially in the United States. The patient would most probably be treated for a simple respiratory infection and would continue to infect others until properly treated. Clues that a bioterrorism attack had taken place would include the occurrence of disease in nonendemic locations, disease in persons without risk factors, and the absence of previous rodent deaths.

Smallpox

The last documented case of naturally occurring smallpox was in the African nation of Somalia in 1977, and in 1980 vaccinations ceased worldwide. However, stocks of the virus are kept in laboratories in Moscow, Russia, and at the CDC in Atlanta.

Smallpox, a DNA virus, is one of the largest and most complex of viruses. It takes only a few smallpox virions to infect a human. The virus migrates from the respiratory tract to nearby lymph nodes. An asymptomatic viremia develops as virions move to the spleen, bone marrow, and lymph nodes. As the virions replicate in these organs, a secondary viremia develops, and it is at this time that signs and symptoms appear. The incubation period is typically 12–14 days with initial signs/symptoms of fever, malaise, and headache.

About 12–14 days after exposure, a maculopapular rash begins with lesions on the tongue and palate and small red spots on the face. The rash spreads to the proximal extremities, trunk, and distal extremities (**Figure 28.7**). After a couple of days, the rash becomes vesicular and pustular. The pustules are round, tense, and deeply implanted in the dermis. Approximately 8–9 days later, crusts form over the pustules, followed by scabs and scarring. As the oral lesions form vesicles and break down, a large number of virions are released. This is when infectivity is at its greatest, which occurs in the days when the maculopapular rash first appears. Death is secondary to hypotension and multiorgan failure, and usually occurs during the second week.

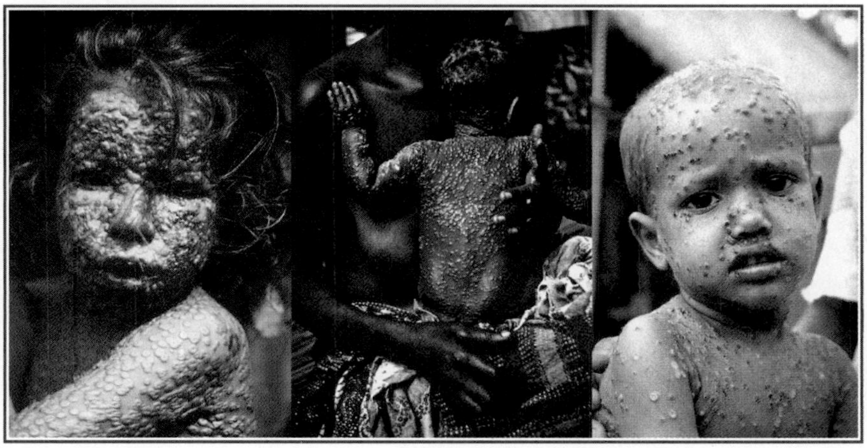

Figure 28.7 Images of smallpox lesions.

Before 1972, every US child received a smallpox vaccination at one year of age. The duration of the immunity is unclear, but antibodies seem to decline over a 5–10-year period, which means that the vaccine does not confer lifelong immunity. Therefore, adults who were vaccinated as children are probably no longer protected. The vaccine is administered with a bifurcated needle (**Figure 28.8**), and the skin is broken with 10–15 strokes, leaving a characteristic scar, usually on the arm. The policy in the United States today is that only military personnel serving in combat environments are vaccinated.

The vaccine can be administered after exposure to smallpox. If given within four days of exposure, the vaccine may prevent or decrease the effects of disease. The only other treatment for smallpox is supportive care and antibiotics for secondary bacterial infections, which are rare. Because of the threat of a bioterrorist attack, new treatments are being studied.

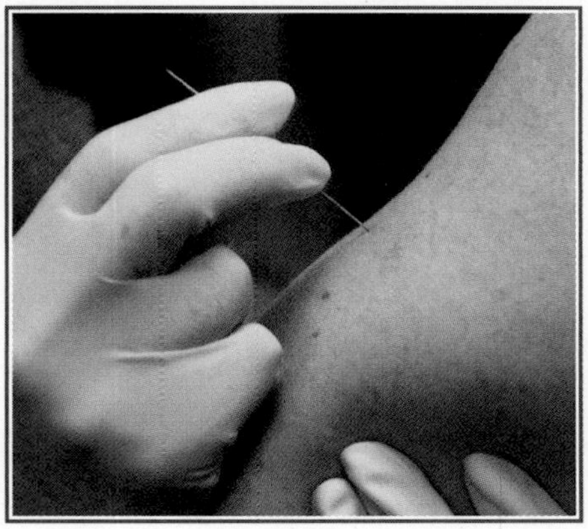

Figure 28.8 A CDC clinician demonstrates the use of a bifurcated needle during the 2002 Smallpox Vaccinator Workshop.

Fast Fact

Just one case of smallpox is considered an international health emergency.

If used as a bioterrorist agent, smallpox virus would most probably be aerosolized. One feature of smallpox infection making this virus an attractive bioweapon is that, because of the long incubation period, many humans could be infected before detection of the virus. Because very few individuals would still have immunity against smallpox, a bioterrorism attack would require administering the limited amount of vaccine we have on hand to thousands of exposed individuals, a task that would prove difficult after an attack. This is why the initial diagnosis of smallpox is so critical.

Tularemia

Tularemia is a zoonotic disease with a large range of hosts covering many different environments in several countries. The causative agent, *Francisella tularensis*, is a Gram-negative, facultatively anaerobic, intracellular bacterium. It is highly resistant and can survive for long periods in soil, water, and animal carcasses. The most significant sources of human infection are rodents, hares, and rabbits, especially the cotton-tailed rabbit. Humans can be infected by the bite of an infected animal or through vectors such as ticks and deerflies. Often outbreaks of human infection parallel outbreaks in other animals. Humans can also be infected by ingestion or inhalation of the bacterium.

There are several clinical syndromes associated with tularemia. The most common is ulceroglandular, which accounts for 60–80% of cases. Patients present with an abrupt onset of fever, chills, malaise, sore throat, and headache after an incubation period of two to five days. If the route of infection is through the skin or mucous membranes, a primary ulcer will develop at that site (**Figure 28.9**). The ulcer begins as a solitary papule that changes to a pustule with surrounding inflammation. Sometimes the lesion will go unnoticed and will heal in a week. If antibiotics are not given within seven to ten days of the original infection, lymph node enlargement can occur with suppuration (formation of pus). This occurs in 30–40% of cases.

If the route of infection is via inhalation, the signs and symptoms of respiratory tularemia are variable and can resemble those of several less threatening infections. Patients present with a mild respiratory infection or high fever, chills, malaise, and cough, with the severity of symptoms being dependent on the subspecies of the infecting *F. tularensis*.

Francisella tularensis, subspecies *tularensis*, is the most likely subspecies to be used as a bioweapon through aerosolization. It causes fever, lymph node enlargement, dry cough, and retrosternal pain and carries a 30% mortality rate if left untreated. It takes just 10–50 bacteria to cause disease in humans, which makes *Francisella* much more infectious than the causative agent of anthrax, *Bacillus anthracis*. The World Health Organization predicted that a release of 50 kg of virulent *F. tularensis* in a population of 5 million would result in 250,000 cases. Ten per cent of the population would require hospitalization, and there would be 2,500 deaths. One million people would require preventive antibiotic treatment for 10 days. This is assuming that the form of the bacterium had not been altered and was sensitive to antibiotics. Even with no alteration of the biological agent, the spread of tularemia would quickly overwhelm health care communities.

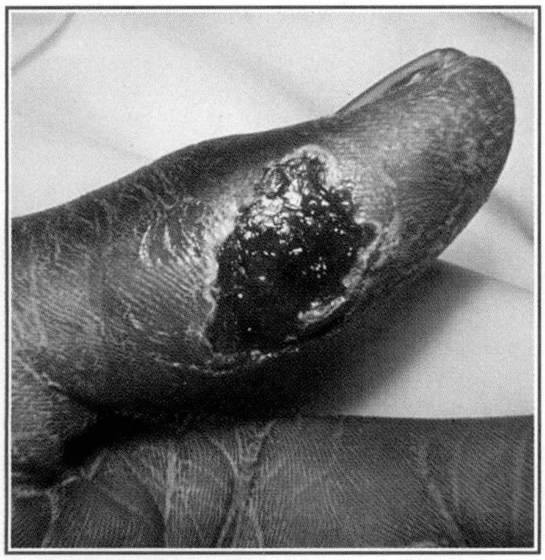

Figure 28.9 Skin lesion in a patient diagnosed with tularemia.

Tularemia can be treated effectively with antibiotics, and most patients recover in five to seven days. The advent of antibiotics has resulted in a significant decrease in the mortality rate, from 33% to 4%. Until recently, laboratory workers were vaccinated with a live attenuated vaccine. This vaccine was administered through scarification, in much the same way as the smallpox vaccine. It was used as an investigational new drug in the 1960s but is no longer approved by the Food and Drug Administration under this status. There is the possibility that reversal of the attenuation could occur, resulting in partial or full virulence. In addition, it is unclear how the vaccine provides immunity against infection.

Scientists continue to research options for a viable tularemia vaccine. Isolation is not necessary for tularemia patients because human-to-human transmission does not occur.

Viral Hemorrhagic Fevers

A hemorrhagic fever is one that has as a principal sign hemorrhaging of the capillaries of the patient's circulatory system, and there are at least 18 viruses that cause human hemorrhagic fevers. Several of these viruses are listed as category A agents (**Table 28.1**), including the filoviruses Ebola (**Figure 28.10**) and Marburg and the arenaviruses Lassa and Machupo. The United States and Russia have experimented with weaponizing these viruses and have found that aerosolizing them produces a high infectivity rate in nonhuman primates (*Clin. Lab. Med.* 26(2), 345–386 (1 June 2006)). In the 1990s, the Aum Shinrikyo cult in Japan attempted to weaponize the Ebola virus (*Clin. Lab. Med.* 26(2), 345–386 (1 June 2006); *Infect. Dis. Clin. N. Am.* 20, 179–211 (2006)). There is no known incident in which these viruses have been used in a biological attack, but the interest exists.

Viral hemorrhagic fevers are caused by RNA viruses not naturally found in humans. Rodents and insects are the reservoirs, and humans become infected through contact with either an infected animal or an infected insect. Once an individual has been infected, the disease is transmissible from person to person. The virus can cause vascular damage and permeability by targeting the vascular endothelium. Patients present with fever and aching muscles after an incubation period of several days. Depending on the virus, patients may have evidence of capillary leak, hypotension, flushing, and petechiae (minute hemorrhages seen on the skin). Death is usually the result of vascular permeability, loss of blood, and multiple organ failure.

Although the various viruses that cause hemorrhagic fever have similar characteristics, each of them has unique features. The virus responsible for Lassa fever is found in West Africa. This infection has a high mortality rate in children and pregnant women, and deafness is a frequent complication in survivors. The Rift Valley fever virus is found in sub-Saharan Africa and rarely causes serious hemorrhagic disease. However, 10% of patients develop eye pain and blindness from retinitis (inflammation of the retina). Marburg and Ebola viruses produce distinct rashes, and severe bleeding is common. However, extensive bleeding is not an indicator of a patient's chance of surviving; deaths are seen in cases of mild bleeding, and recovery is seen in cases of severe bleeding. The fatality rate for Ebola hemorrhagic fever, for instance, is 80%.

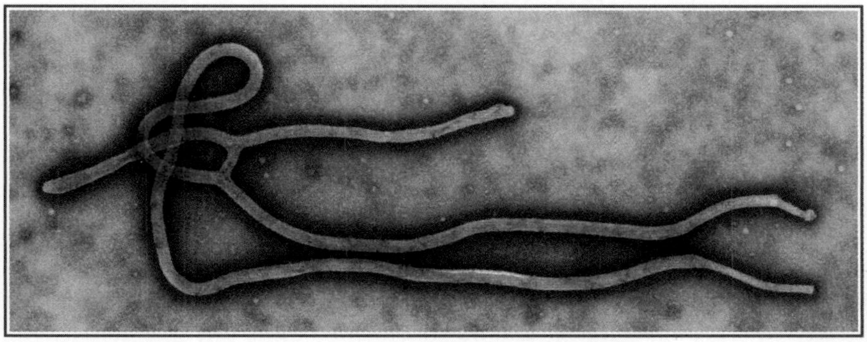

Figure 28.10 A transmission electron micrograph of the Ebola virus.

Treatment of viral hemorrhagic fevers is mainly supportive. Patients should be handled with great care to prevent bleeding, and aspirin should be avoided. Blood transfusions may be indicated for hemorrhage and clotting disorders. There is only one specific antiviral therapy, ribavirin, which has shown potential for a few of the viruses. The only vaccine for viral hemorrhagic fevers is indicated for yellow fever, and all travelers to areas where yellow fever occurs must receive this vaccine. Infected individuals should be isolated because the virus is transmissible through close contact. Respiratory isolation is ideal but may not be possible in a large outbreak. Contact-barrier precautions are essential in these situations to prevent spread of the disease.

PROBABILITY AND EFFECTS OF A BIOLOGICAL ATTACK

Biological attacks have occurred throughout history and are likely to continue in the future. Bioweapons are cheaper to produce than chemical weapons and can cause mass destruction. Several countries have established bioweapons programs for experimentation, and countries that sponsor international terrorism are suspected of possessing harmful biological agents. Additionally, individuals who possess knowledge of genetic engineering could alter simple biological agents to make them more virulent and resistant to antibiotics.

The biggest consequence of a bioterrorist attack may be not the physical casualties but the psychological impact. Just a few casualties, as seen in the 2001 anthrax attacks, could cause alarm and serious economic impact. Even the suspicion of a biological weapon being released could instigate mass panic and disruption of communities, health care systems, and governments. Global disorder is a serious threat as people travel routinely and quickly around the world, potentially spreading infectious diseases far from where they originated.

Perhaps more importantly, the initial symptoms may not lead health care providers to suspect bioterrorism. As a result, proper precautions may not be used at first, potentially increasing the number of people exposed and infected. The majority of any population is susceptible to infection with a category A agent. A few initial exposures could quickly turn into mass casualties, especially when the infection is one that can be transferred through human contact.

Fast Fact

It is estimated that it will cost the United States $26.2 billion dollars for every 100,000 persons exposed to anthrax in a bioterrorism situation.

The economic impact of a bioterrorism attack will most probably be devastating. During the 2001 anthrax attacks, it took $23 million to decontaminate areas of a government building in Washington DC. Although studies have shown that early intervention can significantly decrease the costs resulting from a bioterrorist attack, it will be expensive to provide prophylactic antibiotics to a large number of individuals. However, the economic outcome of fewer infections and hospital admissions will greatly outweigh the initial costs.

Warning Signs

In any location hit by a bioterrorism act, the public health system will probably be the first to detect the attack and respond. It may not be realistic to wait for confirmation of a diagnosis because the potential for spread of the disease increases. An emergency response may need to be activated on the basis of patterns, timing, and patient presentation. There are important clues that can help alert hospitals to a bioterrorist attack. This information can be passed on to the appropriate authorities, who can coordinate health care facilities, local and state departments, and emergency response teams. Some clues are listed in **Table 28.2**, but every health care professional should be suspicious of any unusual activity.

Clues
Increased incidence of a specific disease in an otherwise healthy population
Presence in a given geographical area of a disease not typically found in that area
Disease outbreak at an atypical time of year
Large numbers of people presenting to medical facilities with similar symptoms
Increase in number of rapidly fatal cases
Infectious diseases not responsive to antibiotics that were formerly successful in treating the disease
Single case of a relatively uncommon disease (smallpox, pneumonic plague, tularemia)
Increased numbers of emergency calls
Discovery of a package that has been tampered with

Table 28.2 Clues to a Bioterrorist Attack

It Takes a Village...

It will take many people in a variety of fields to control the impact of a biological attack. Many infectious diseases are zoonotic, and veterinarians will have invaluable expertise in this field. Scientists, epidemiologists, doctors, and nurses will all need to work together to battle the impact of a biological disaster. Law enforcement will be particularly important when it comes to reporting disease and controlling public reaction. Bioterrorism is a matter of national and international security that will require the coordination of local, state, and federal agencies.

What can you, as a health care professional, do to prepare for a bioterrorist attack? The first step is to become familiar with your medical facility's disaster and emergency response plans. Because there is little real-world experience with bioterrorist attacks, protocol is often based on the protocol for dealing with a natural disaster. However, a bioterrorist attack will present several unique challenges not encountered in a natural disaster, such as mass panic resulting from the unknown consequences of releasing a biological agent. Depending on the location and size of your particular facility, needs and requirements will differ. A small rural hospital may be responsible for alerting local emergency services and coordinating the subsequent transfer of patients to a higher level of care. At the other end of the spectrum, a large urban hospital may be the receiving point for numerous casualties, the command center for disseminating information, and the focal point for communication.

Once you have reviewed the emergency response plan, it is imperative that you understand your role. The time to determine your particular duties is not during a disaster or mass casualty incident. Prepare ahead of time to become familiar with the location of important telephone numbers and resources so that you will be ready to assist at a moment's notice. Your day-to-day responsibilities may be much different during the response to a bioterrorist attack. The first step after identifying that a bioterrorist attack has occurred is notifying the proper officials. For example, a nurse may be asked to call police and fire departments. Knowing how to contact these agencies in advance may save crucial minutes during a time of chaos.

Although several actions may be occurring at once, it is important to know what tasks may be required of you. An essential part of preventing the spread of disease will be infection control. Review — often and thoroughly — all you have learned about likely biological agents, about how the infections they cause are spread, and about how to prevent spread of the infections.

All patients exhibiting symptoms that are suspected of being related to a bioterrorist agent should be managed using standard precautions. These precautions are designed to decrease the transmission rate in medical facilities. Standard precautions include no contact with any body liquids from infected patients, frequent hand-washing, gloves, eye protection, masks, and gowns. Additional precautions, such as respiratory isolation, may need to be implemented for certain diseases such as smallpox and pneumonic plague.

In the event of mass casualties, the placement of patients will be a critical part of treatment. Patients presenting with similar symptoms should be grouped in a designated section or ward in the hospital. These sections and wards should be identified before the incident to ensure that proper ventilation, plumbing, and resources are available. A triage area may be set up to facilitate the movement of patients. In addition, transporting patients who are victims of a bioterrorism attack may need certain requirements. Medical equipment may require special cleaning, disinfection, and sterilization, although soap and water will work most of the time.

Finally, education and training for bioterrorist attacks should be routine at all medical facilities. Disaster and mass casualty drills will improve coordination between departments and allow health care professionals to practice their skills. In 1996, the US Department of Defense was given $36 million for first-responder training in confronting terrorist attacks, which is just one indication of how important it is for health care professionals to know their role in responding to bioterrorism. You can take the first step by learning more about bioterrorism through the Centers of Disease Control (http://www.bt.cdc.gov/).

SUMMARY

Bioterrorism continues to be a real threat, and most experts believe it is simply a matter of time until the next attack occurs in the United States. The public health community will be heavily involved in any response to a bioterrorist attack. Being familiar with organisms likely to be used as bioweapons will prepare you to act efficiently and effectively during an emergency. It is important for you to review your medical facility's emergency response plan before an actual event. Preparation is the key to controlling the effects of a bioterrorist attack.

NOTE: There are no self-evaluation and chapter confidence questions for this chapter. The information provided here is for your understanding of the potential for the development of bioweapons. This information is not intended to be part of your educational requirements for microbiology. Rather, it is information that can make you a better health care professional.

Multiple Choice Answers

Chapter 1

1. E
2. C
3. B
4. D
5. D
6. E
7. C
8. D
9. C
10. E

Chapter 2

1. B
2. D
3. D
4. B
5. B
6. B
7. B
8. E
9. C
10. D

Chapter 3

1. B
2. A
3. C
4. A
5. D
6. D
7. C
8. C
9. E
10. D
11. C
12. D
13. C
14. E
15. C
16. D
17. C
18. A
19. C
20. C
21. B
22. D
23. D
24. B
25. A
26. D
27. C
28. D

Chapter 4

1. A
2. B
3. D
4. C
5. D
6. B
7. C
8. D
9. D
10. D
11. C

Chapter 5

1. B
2. C
3. E
4. C
5. B
6. E
7. C
8. D
9. C
10. D
11. A
12. B
13. A
14. C

Chapter 6

1. E
2. A
3. C
4. D
5. C
6. B
7. E
8. A
9. E
10. D
11. D
12. C
13. C
14. D
15. B
16. B
17. D
18. C

Chapter 7

1. C
2. D
3. B
4. E
5. C
6. C
7. A
8. D
9. D
10. D

Chapter 8

1. C
2. D
3. C
4. C
5. E
6. D
7. D
8. E
9. C
10. B
11. D
12. B
13. D
14. E
15. D
16. C
17. C
18. D
19. D
20. C
21. C
22. D

Chapter 9

1. B
2. E
3. E
4. E
5. C
6. E
7. C
8. B
9. D
10. D
11. D
12. C
13. D
14. D
15. C
16. C
17. B
18. C
19. B
20. B
21. B
22. D
23. A
24. A
25. C
26. B
27. B
28. D
29. C
30. B
31. E
32. B
33. C
34. A

Chapter 10

1. A
2. E
3. B
4. C
5. B
6. B
7. C
8. A
9. A
10. C
11. D
12. C
13. B
14. A
15. C
16. E
17. C
18. C
19. D
20. D

Chapter 11

1. E
2. E
3. D
4. D
5. C
6. C
7. B
8. D
9. D
10. E
11. C
12. C
13. C
14. C
15. C
16. D
17. B
18. E
19. B
20. E
21. D
22. D
23. C
24. C
25. C
26. C
27. E
28. D
29. C
30. E
31. D
32. A
33. A
34. C
35. D
36. E
37. E
38. A
39. D
40. D

Chapter 12

1. A
2. B
3. C
4. D
5. A
6. D
7. C
8. A
9. D
10. C
11. D
12. A
13. C
14. B
15. C
16. B
17. A
18. C
19. D
20. D

Chapter 13

1. D
2. C
3. C
4. C
5. C
6. D
7. C
8. E
9. B
10. C
11. C
12. D
13. D
14. B
15. D
16. B
17. E
18. D
19. B
20. A
21. B
22. C

Chapter 14

1. D
2. D
3. B
4. C
5. D
6. C
7. C
8. B
9. D
10. C
11. C
12. D
13. D
14. D
15. D
16. D
17. D
18. C
19. B
20. E
21. C
22. D
23. D
24. C
25. C
26. D
27. C
28. E
29. E

Chapter 15

1. A
2. C
3. E
4. D
5. D
6. E
7. D
8. E
9. C
10. D
11. E
12. B
13. D
14. C
15. D
16. E
17. A
18. C
19. E
20. C
21. D
22. C
23. C
24. D
25. D
26. D
27. D
28. B

Chapter 16

1. A
2. C
3. D
4. D
5. D
6. C
7. D
8. A
9. C
10. B
11. D
12. C
13. C
14. D
15. C
16. C
17. B
18. D
19. D
20. D
21. C
22. C
23. C
24. D
25. B
26. E
27. E
28. C

Chapter 17

1. D
2. C
3. C
4. B
5. B
6. C
7. C
8. B
9. C
10 E
11 A
12 D
13. B
14. D
16. A

Chapter 18

1. B
2. E
3. A
4. E
5. E
6. D
7. E
8. C
9. D
10. B
11. E
12. C
13. D
14. C
15. B
16. C
17. B
18. B

Chapter 19

1. D
2. C
3. D
4. A
5. E
6. A
7. B
8. A
9. E
10. A
11. E
12 D
13. D
14. C
15 D
16. C
17. D
18. D
19. C
20. C
21. A
22 D
23. E
24. B
25. E
26. C
27. C

Chapter 20

1. C
2. C
3. A
4. B
5. D
6. E
7. C
8. D
9. C
10. C

Chapter 21

1. B
2. E
3. D
4. D
5. C
6. A
7. B
8. E

Chapter 22

1. A
2. B
3. C
4. B
5. C
6. E
7. D
8. B
9. B
10. C
11. A
12. D
13. D
14. E
15. D
16. D
17 A
18. A

Chapter 23

1. B
2. D
3. B
4. C
5. B
6. B
7. A
8. A
9. C
10. D

Chapter 24

1. B
2. E
3. E
4. E
5. D
6. D
7. A
8. C
9. D
10. A
11. C
12. C
13. C
14. D
15. D
16. B

Chapter 25

1. B
2. C
3. E
4. E
5. E
6. C
7. B
8. C
9. C
10. B
11. B
12. D
13. C
14. B
15. D

Chapter 26

1. B
2. D
3. C
4. D
5. D
6. D
7. B
8. D
9. C
10. C
11. E
12. B
13. C

ABC transport systems A type of active transport including several proteins and several steps in which the molecule being transported is handed off from one protein to the next.

Abscess A lesion that contains pus in a cavity hollowed out by tissue damage.

Acidophiles Organisms that grow best in an environment with a pH of 4.0 to 5.4.

Active immunization An immunization in which antigen representing the infectious agent is administered and confers immunity. This is used to increase herd immunity.

Active transport The mechanism used to transport things from one side of a membrane to the other in a process that requires ATP.

Active viremia Viruses replicating in the blood.

Acute congestion A phenomenon that occurs where local capillaries become engorged with neutrophils.

Acute croup Also known as laryngotracheitis. A severe infection of the larynx and trachea.

Acute disease The opposite of chronic disease; one in which infection occurs quickly and lasts a short time.

Acute inflammatory colitis Rapid inflammation of the colon.

Acute influenzal syndrome An acute influenza in which symptoms can develop in a matter of hours.

Acute pancreatitis The rapid onset of serious inflammation of the pancreas.

Acute phase of HIV infection This phase occurs in the first few days after initial infection: the virus is reproducing in the lymphocytes of the lymph nodes, resulting in lymphadenopathy and flu like symptoms.

Acute phase protein A protein such as C-reactive protein or mannose-binding protein that is related to the development of the inflammatory response but is seen only in acutely ill patients.

Acute septic shock The rapid onset of hypotension associated with overwhelming infection. It is thought to result from the action of endotoxins.

Acute urticaria Sudden onset of vascular reactions in the upper dermis marked by transient appearance of slightly elevated patches that are redder than the surrounding tissue.

Acyclovir An antibiotic used for viral infection. Acyclovir is a synthetic purine nucleoside with activity against herpes simplex virus.

Adaptive immune response A specific host defense composed of the humoral (antibody) and cellular immune responses. It has memory and takes several days to get started.

Adhesin A protein or glycoprotein found on attachment pili or in capsules that helps microorganisms attach to host cells.

Aerobe An organism that uses oxygen.

Aerobic respiration Metabolism that uses oxygen as the final electron acceptor.

Aerotolerant bacteria Bacteria that grow in the presence of oxygen but do not use oxygen for metabolism.

African sleeping sickness *see* Trypanosomiasis.

Agar A low-melting-point complex polysaccharide derived from marine algae and used to solidify media for the growth of microorganisms.

Agranulocytes Leukocytes (monocytes or lymphocytes) that lack granules in the cytoplasm.

AIDS (acquired immune deficiency syndrome) A viral infection caused by the human immunodeficiency virus (HIV), which destroys a patient's helper T cells and thereby his or her adaptive immune response.

AIDS dementia A metabolic encephalopathy induced by HIV infection manifested by cognitive, behavioral, and motor abnormalities.

Alcohol fermentation Fermentation in which pyruvic acid is reduced to ethyl alcohol.

Allergen An ordinarily innocuous foreign substance that can elicit an adverse immunological response in a sensitized individual.

Allergic conjunctivitis Inflammation of the conjunctiva caused by allergic reaction and characterized by itching, tearing, and redness.

Allergic rhinitis Allergic reaction of the nasal mucosa.

Allergy *see* Hypersensitivity.

Allolactose A fragment of the sugar lactose that acts as an inducer molecule and allows the lactose operon to be turned on.

Allosteric inhibition Also known as the allosteric effect. It involves the binding of a non-competitive inhibitor to a site on the enzyme molecule that causes a change in the shape of the active site and inhibits the binding of the substrate in the active site.

Alpha hemolysis Partial lysis of red blood cells, leaving a greenish ring in the blood agar medium around the colonies.

Alternative pathway of complement activation One of the sequences of reactions seen in the complement system of innate immune response initiated by factors B, D, or P.

Alveolar macrophage A macrophage that is found in the alveoli of the lungs.

Amantadine An antiviral agent that prevents the penetration of host cells by the influenza virus.

Amastigote The intracellular morphologic stage in the development of certain hemoflagellates.

Amebiasis Infection with ameboid protozoan parasites, in particular *Entamoeba histolytica*.

American sleeping sickness *see* Chagas' disease.

Aminoacyl-tRNA synthetase The enzyme required for binding the amino acid to the transfer RNA molecule.

Aminoglycoside antibiotics Chemical agents that block bacterial protein synthesis.

Amphitrichous flagella One of the four forms of flagellar arrangement. In this case there are flagella at both ends of the bacterial cell.

Anabolism Chemical reactions in which energy is used to synthesize large molecules by using smaller molecules (this process involves synthesis reactions).

Anaerobe An organism that does not use oxygen, including some organisms that are killed by oxygen.

Anaerobic respiration Metabolism that takes place without oxygen. In this case the final electron acceptor is an inorganic molecule other than oxygen.

Analytical epidemiology A form of bioinformatics that focuses on establishing the cause-and-effect relationship and can use prospective or retrospective data.

Anaphylactic shock Condition resulting from a sudden extreme decrease in blood pressure due to an allergic reaction.

Anaphylaxis An immediate exaggerated allergic response to antigens.

Anergy Automatic inactivation of a lymphocyte due to lack of co-stimulation.

Angiogenesis Development of blood vessels.

Animal virus A virus that infects animal cells. Human cells are considered animal for this definition.

Anion A negatively charged ion.

Anorexia Loss of appetite.

Anthrax A fatal disease of ruminants contracted by humans through contact with contaminated wool or animal products, or inhalation of spores. *See* Inhalation Anthrax and Cutaneous Anthrax.

Anthrax toxin A powerful cytotoxin produced by *Bacillus anthracis*, which increases vascular permeability. It has three components: edema factor, protective antigen, and lethal factor.

Antibiotics Chemicals that control the growth of microorganisms.

Antibody *see* Immunoglobulin.

Antibody affinity A term used to describe how well the antigen fits within the antibody-binding site of the antibody molecule.

Antibody avidity A term used to describe how tightly the antigen is bound by the antibody molecule.

Antibody titer *see* Titer of antibody in serum.

Antibody-dependent cellular cytotoxicity A process involving cytotoxic cells that can destroy targets which are covered with antibody molecules.

Anticodon loop The portion of the transfer RNA molecule that associates with the codon of the messenger RNA molecule at the ribosome.

Anticodon region The whole cloverleaf portion of a transfer RNA molecule.

Antigen Also known as an immunogen. Any molecule that will elicit an immune response.

Antigen-binding site The site on the antibody molecule that interacts with epitopes on the antigen. It is made up of a portion of the variable region of the heavy chain and a portion of the variable region of the light chain.

Antigenic determinant *see* Epitope.

Antigenic drift The process of antigenic variation that results from mutations.

Antigenic shift The process of antigenic variation caused by the re-assortment of genes.

Antigenic variation Mutations seen in the influenza virus that occur by antigenic drift and antigenic shift.

Antigen-presenting cell A cell that processes antigens and places them on their surface for presentation to helper T cells. The predominant ones are macrophages or dendritic cells, but the B cell can also present antigen.

Antimicrobial An agent that will inhibit the growth of a microbial organisms.

Antiseptic A chemical substance that can be used on tissues to control the growth of microorganisms.

Antiviral proteins Proteins produced through stimulation by interferon that become active in the presence of double-stranded RNA and can protect cells from being infected with virus.

Apoptosis Genetically programmed cell death; a process in which the cell commits a kind of suicide. This occurs when cells are worn out or if specific signals are given to the cell, as in cytotoxic reactions.

Arbovirus A group of viruses that are transmitted to humans by mosquitoes or ticks, including agents of yellow fever and viral encephalitis.

Armed effector T cells Primed effector T lymphocytes that can be triggered to perform effector functions immediately on contact with cells bearing the peptide–MHC complex for which they are specific.

Arthroconidia A specialized stalk structure found in some fungi that holds conidia.

Arthropod An organism that is characterized by a jointed chitinous exoskeleton, segmented body, and jointed appendages associated with some or all of the segments.

Arthus reaction A localized inflammatory response with increased vascular permeability at the site.

Ascospore One of the eight sexual spores produced in each ascus of certain fungi.

Aseptic The use of techniques that minimize the chances of cultures becoming contaminated.

Asian influenza The term used to describe an influenza pandemic that occurred in 1957 and caused more than 70,000 deaths in the United States.

Aspergillosis Disease caused by *Aspergillus* fungi which causes the development of inflammatory granulomatous lesions in the skin, nasal passages, lungs, bones, and meninges.

Asthma Recurrent attacks of difficulty in breathing and wheezing due to spasmodic contractions of the bronchi.

Ataxia telangiectasia Primary immunodeficiency syndrome involving reduced numbers of T cells.

Athlete's foot *see* Tinea pedis.

Atomic number The number of protons found in the nucleus of the atom.

Atopy Hypersensitivity to environmental allergens.

ATP The biological energy molecule consisting of adenosine coupled with three phosphate atoms. Breaking the third phosphate off releases energy and ADP, that can be converted back to ATP with the addition of a phosphate and energy.

Attachment pilus *see* Fimbria.

Atypical pneumonia A lower respiratory infection caused by atypical organisms such as *Mycoplasma*, rickettsial organisms, and *Chlamydia*. It causes fever, shortness of breath, labored breathing, joint pain, malaise, confusion, and diarrhea.

Autoclave An instrument for sterilizing by means of moist heat and pressure.

Autoimmune hemolytic anemia Autoimmune disease in which antibodies are generated against red blood cell antigens.

Autotrophy Obtaining carbon atoms from CO_2.

Avian influenza A type of influenza seen in avian species that has been able to infect humans. It is similar to the influenza virus that caused the Spanish flu pandemic of 1918.

Axial filament Also known as an endoflagellum. It is a sub-surface filament found in spirochete organisms that causes the spirochete to rotate in a corkscrew fashion.

Axostyle A microtubule used for attachment and tissue damage found on the parasite *Trichomonas vaginalis*.

β-lactamase An enzyme produced by penicillin-resistant bacteria that destroys the β-lactam ring of penicillin.

B cell lymphoma Non-Hodgkin's form of neoplasm seen in lymphoid tissue.

B lymphocyte A lymphocyte that is part of the adaptive immune response. This cell is born in the bone marrow and matures there; it is responsible for the humoral (antibody) response of the adaptive response and differentiates into a plasma cell which produces antibody.

Bacillary dysentery The most serious form of gastrointestinal infection caused by *Shigella* organisms.

Bacillus (plural bacilli) A rod-shaped organism.

Bacitracin An antibacterial polypeptide produced by *Bacillus subtilis* that prevents cell wall synthesis.

Bacteremia Bacteria in the bloodstream.

Bacterial hemorrhagic disease Loss of blood from vessels caused by certain bacterial species.

Bacteriocide Any agent that kills bacteria.

Bacteriocins Chemicals produced by bacteria that inhibit the growth of other microorganisms but have no effect on the organism producing them.

Bacteriophage Viruses that infect bacteria.

Bacteriostatic Term used for the activity of chemicals that inhibit microbial growth but do not kill the organisms.

Bacteriuria Bacteria in the urine.

BALT Bronchial-associated lymphoid tissue.

Basidiospore The sexual spore of club fungi.

Basophil A type of white blood cell containing large numbers of granules. This cell releases histamine and other molecules as part of the inflammatory response.

Bed sore A decubitus ulcer on the skin.

Beta hemolysis Complete destruction of red blood cells by bacterial enzymes.

Betadine An iodophor routinely used to prepare human skin for surgery and injection.

Binary fission The process in which a bacterial cell duplicates its components and divides into two identical daughter cells.

Biofilm A structure composed of a variety of bacterial species that coexist. It is often seen as the beginning layer of plaque, which builds up on teeth.

Bioinformatics Science that uses many techniques such as applied mathematics, statistics, informatics, and artificial intelligence to solve biological problems.

Bioremediation A process that uses naturally occurring or genetically engineered microorganisms to transform harmful substances into less toxic or nontoxic compounds.

Biotechnology An industry that uses molecular biology and microorganisms for specialized purposes, including genetic engineering.

Bioterrorism The use of microorganisms as weapons.

Bird flu *see* Avian influenza.

Bisphenolics Chemical disinfectants consisting of two covalently linked phenolic compounds.

Black plague *see* Pneumonic plague.

Blastoconidia The products of the budding form of reproduction seen in fungi.

Blood agar Bacterial growth medium that contains blood; used to identify organisms that produce hemolysins, which break down red blood cells.

Boils *see* Carbuncle.

Botulinum toxin An extremely poisonous neurotoxin associated with botulism, a digestive intoxication resulting from food poisoning.

Broad-spectrum antibiotics Chemicals that attack a wide variety of microorganisms.

Bronchiolitis Inflammation of the bronchioles.

Bronchitis Acute or chronic inflammation of one or more of the bronchioles.

Bronchopneumonia Inflammation of the lungs beginning at the terminal bronchioles.

Broth dilution test A test used to determine the minimal microbicidal concentration.

Bubo Enlargement of infected lymph nodes, especially in the groin and armpit, due to the accumulation of pus. This is seen in bubonic plague and other diseases.

Bubonic plague A noncontagious form of plague caused by *Yersinia pestis* and transmitted to humans by the bite of a flea. It is a systemic disease that spreads through the blood and lymph fluid.

Budding The process that occurs in yeast and a few bacteria in which a small new cell develops from the surface of an existing cell.

Burkitt's lymphoma A tumor caused by the Epstein–Barr virus and seen mainly in African children.

Cadherin A transmembrane protein that has a role in cell adhesion.

Candidiasis A yeast infection caused by *Candida albicans*. It can appear as thrush in the oral cavity or as vulvovaginitis.

Capillary thrombosis Formation of clots in the capillaries.

Capsid A protein outer coat of a virus. It is made up of repeating protein subunits known as capsomeres.

Capsomere A protein subunit that, along with other identical subunits, makes up the capsid (protein coat) of a virus.

Capsule A protective structure found around the outside of a bacterial cell. It can be made up of polysaccharides, polypeptides, or a combination of both.

Capsule stain Also known as a negative stain; the background is stained, making the capsule visible.

Carbapenem antibiotics Antibiotics that attack the cell wall of bacteria.

Carbuncle Also known as a boil. A massive pus-filled lesion resulting from an infection in which the invading pathogens have been walled off. It is usually seen in the neck and back.

Carrier molecule Usually a co-factor or coenzyme used in metabolic pathways.

Catabolism The chemical breakdown of organic molecules in which there is a release of energy.

Catalase An enzyme that converts hydrogen peroxide to water and oxygen.

Cation A positively charged ion.

CBC (complete blood count) A laboratory procedure in which the formed elements (cells) of the blood are counted. This information can be of great value in determining the stage of infection seen in a patient.

CD40 A B-cell surface receptor that binds the CD40 ligand found on activated T cells.

CD40 ligand Molecule found on activated helper T cells that binds to the CD40 receptor on B cells.

Cellular adaptive immune response Part of the adaptive immune response. It comprises the activity of T cells in response to antigens.

Cephalosporin family A group of antibiotics derived from the fungus *Cephalosporium*. These chemicals attack the cell wall of bacteria.

Cercariae A tail-bearing larval form of trematodes.

Cervicitis Infection of the cervix.

Cestode A tapeworm.

Cestodiasis Infection with parasitic flatworms (cestodes).

Chagas' disease The American form of trypanosomiasis.

Chagoma A local chancre associated with Chagas' disease, filled with neutrophils, lymphocytes, and tissue fluids.

Chancre An ulcer located on the external genitalia or cervix seen in cases of syphilis.

Chediak – Higashi syndrome Primary immunodeficiency disease leading to lethal progressive systemic disorders and recurrent or chronic bacterial infections.

Chemical agent A general term that includes both antiseptics and disinfectants.

Chemically defined growth medium Growth medium in which each of the ingredients is precisely defined.

Chemoautotroph An autotroph that obtains energy by oxidizing inorganic substances such as nitrites and sulfides.

Chemoheterotroph Organism that obtains energy from breaking down already-formed organic molecules.

Chemokine A class of cytokines that attract additional phagocytes to the site of infection.

Chemosmosis The process of capturing energy in which a proton gradient is created in the electron transport chain and is used to provide energy for the formation of ATP.

Chemostat A device used to maintain the logarithmic growth of bacteria by the continuous addition of fresh growth medium.

Chemotaxis A nonrandom movement of an organism toward or away from a chemical.

Chickenpox A highly contagious disease characterized by skin lesions and caused by the varicella-zoster virus. It is usually seen in children.

Chloramines Combinations of chlorine and ammonia used in wound dressings.

Chocolate agar Type of medium made with heated blood and used for growing organisms that require heme for growth.

Cholera An acute infectious disease caused by *Vibrio cholerae*. It causes severe diarrhea with extreme fluid and electrolyte depletion.

Cholera toxin An exotoxin produced by *Vibrio cholerae* pathogens. It is an enterotoxin that causes increased permeability in the intestinal tract and a consequent loss of fluids.

Chorioretinitis Infection of the central nervous system

Chromatin The form of DNA seen in cells that are not ready to divide. It has the appearance of threads.

Chromosome The structure containing DNA that is seen in cells that are ready to divide. It is derived from the condensation of chromatin.

Chronic disease The opposite of acute disease. Such a disease starts slowly and lasts a long time.

Chronic gastritis Chronic infection of the stomach.

Chronic granulomatous disease Primary immunodeficiency disease causing frequent severe infections of the skin, oral mucosa, intestinal tract, bones, lungs, and genitourinary tract.

Cilia (singular cilium) Short cellular projections seen in eukaryotic cells used for movement that results from the beating of these projections in coordinated waves.

Ciliates Organisms that move by the use of cilia; they are rarely parasitic to humans.

Circulatory shock The profound hemodynamic disturbance caused by the failure of the circulatory system to maintain adequate perfusion of organs.

Cirrhosis A general term for a group of liver diseases marked by interstitial inflammation and fibrosis of the liver.

Clade A taxonomic term used to divide closely related organisms such as HIV-1 and HIV-2.

Class I MHC A group of cell-surface proteins that are essential to immune recognition reactions. Class I MHC molecules are involved in natural killing functions and are found on all cells *except* those involved in the adaptive immune response.

Class II MHC A group of proteins found only on the surface of cells, involved in the adaptive immune response. These proteins are absent until an immune response is generated and they have an important role in antigen presentation.

Classical pathway of complement One of the ways in which the innate immune response works to protect the body. This pathway responds to antigens that have been seen previously and involves antibody against those antigens.

Clinical latency The effect seen in HIV infections in which virus replication is very low and is mainly in the lymph nodes.

Clonal selection A theory that explains how exposure to an antigen selects and stimulates a specific lymphocyte to proliferate, giving rise to a clone of identical responsive cells.

Clonorchiasis Infection of the biliary passages of the liver by the liver fluke *Clonorchis sinensis*.

Coccidioidomycosis Also called valley fever. This is a fungal respiratory disease caused by *Coccidioides immitis*, which is commonly found in soil.

Coccus (plural cocci) A spherical bacterium.

Codon A sequence of three bases in messenger RNA that codes for a particular amino acid.

Coenzyme An organic molecule bound to or associated with an enzyme.

Co-factor (1) An inorganic ion required for the function of an enzyme. (2) A component necessary for viral attachment to the host cell.

Cold sore A lesion caused by herpes simplex type 1 virus that is usually found on the lips or gingiva of the oral cavity.

Commensalism A symbiotic relationship in which one organism benefits and the other is unaffected.

Common-source outbreak An outbreak arising from contact with contaminated substances such as water.

Communicable disease An infection that can be spread from person to person.

Competitive inhibition A reaction in which a molecule that is similar in structure to a substance competes with that substance by binding to the active site of an enzyme.

Complement system A set of more than 20 proteins found in the blood that when activated can destroy bacteria by making holes in the bacterial cell wall. This system also amplifies the inflammatory response to infection.

Complex growth medium A growth medium that contains ingredients such as beef extract or blood, whose exact composition is not known.

Complex viruses Viruses without either helical or icosahedral symmetry.

Concerted assembly The process in which the virion is assembled while the viral genome is being synthesized.

Condyloma *see* Genital wart.

Congenital rubella syndrome Transplacental infection of the fetus with rubella that is usually seen in the first trimester of pregnancy.

Conidia The asexual reproductive elements found in fungi.

Conjugation The transfer of genetic information from one bacterial cell to another. In Gram-negative bacteria this occurs through the use of a pilus, whereas in Gram-positive cells it occurs when two cells stick together.

Conjunctivitis An infection of the conjunctiva of the eye.

Constant region of the antibody molecule The part of the immunoglobulin molecule with a relatively constant amino acid sequence. It determines the effector function of the immunoglobulin molecule.

Constitutive gene A gene that is always turned on, in contrast with inducible or repressible genes.

Contact transmission A mode of disease transmission; it can be direct, indirect, or through droplets.

Contagious A disease that is communicable on contact and spreads quickly.

Continuous strand *see* Leading strand.

Coronavirus A virus with clublike spikes that causes colds and acute respiratory infections.

Co-stimulatory signal A signal generated from, and regulated by, cells of the innate immune system.

Covalent bond A bond between atoms in which electrons are shared.

C-reactive protein An acute-phase protein that binds to phospholipids and is seen in acute infections.

Creutzfeldt-Jacob disease A form of transmissible spongiform encephalopathy of the human brain, caused by prions.

Croup Acute obstruction of the larynx that causes a hoarseness and barking cough.

Cryptosporidiosis Disease caused by protozoans of the genus *Cryptosporidium*, commonly seen in patients who are immunocompromised.

Cutaneous anthrax Infection with *Bacillus anthracis* that appears on the skin two to five days after endospores enter the epithelial layers of the skin.

Cutaneous candidiasis A fungal infection of the skin caused by *Candida albicans*.

Cutaneous leishmaniasis An infection of the skin caused by the parasitic protozoan *Leishmania* and transmitted to humans by the bite of a sand flea.

Cyanosis Bluish discoloration of the skin and mucous membranes caused by a lack of oxygen in the blood.

Cyst A spherical thick-walled structure that resembles an endospore; formed by certain parasites.

Cysticercus An embedded form of tapeworm found in meat, giving the meat a "measly" appearance.

Cystitis Infection of the bladder

Cytocidal effect Part of the pathology seen when host defenses kill virally infected cells.

Cytokines Low-molecular-weight proteins that are released by a variety of cells in the body. There are two types: the hematopoietin family, which includes growth hormones and interleukins, and the tumor necrosis factor family.

Cytomegalovirus (CMV) One of a widespread and diverse group of herpesviruses that often produce severe effects in immunodeficient patients.

Cytopathic effect (CPE) The destruction of host cells through the lytic cycle of viruses.

Cytopathic virus Viruses that use the lytic infection cycle to kill host cells.

Cytoplasm The semifluid substance inside cells, excluding the nucleus of eukaryotic cells.

Cytoskeletal structures Structures found inside cells made up of protein fibers that give rigidity and support to eukaryotic cells and permit cell movement.

Cytotoxic chemotherapy The use of cytotoxic drugs to kill malignant cells or depress the immune system to allow the survival of transplants. These drugs cause leucopenia, which is a decrease in the number of white blood cells.

Cytotoxic edema Swelling in the brain caused by the production of toxic substances by bacteria and from neutrophil invasion.

Cytotoxic T cell A specific thymus-derived T lymphocyte that kills other cells and has memory.

Cytotoxins Toxins produced by cytotoxic cells that kill infected host cells.

Dacryocystis Infection of the lachrymal gland.

Death phase Also known as logarithmic decline phase. The fourth phase of bacterial growth, in which bacteria die faster than they divide.

Defensins Antibacterial peptides produced by humans.

Definitive host An organism that harbors the adult sexually reproducing form of a parasite.

Degenerate genetic code The genetic code is said to be degenerate because more than one codon can code for the same amino acid. This allows for mistakes that can take place in the DNA sequence: the appropriate amino acid can still be placed in the primary protein sequence.

De-germing The method of moving organisms away from a place on the body; an example is the application of alchohol before injections.

Dehydration synthesis A chemical reaction in which water is removed so as to build a complex organic molecule.

Demyelinating panencephalitis Infection of the central nervous system, usually viral in origin, which causes demyelination of neurons.

Dendritic cell A phagocytic cell found in the dermis; responsible for antigen presentation to helper T cells in the adaptive immune response.

Dental plaque Adherent deposits that form as a result of bacterial colonization of the surface of the teeth.

Deoxyribose The sugar found in DNA.

Dermatophytosis The general term for a variety of dermatophyte infections involving erythremia, induration, itching, and scaling.

Descriptive epidemiology The study of the physical aspects of patients and the spread of disease.

Detergents Positively charged organic surfactants that are more soluble in water than soap.

Diapedesis The process in which white blood cells make their way to the site of infection. It includes margination, in which cells adhere to vessel walls, and emigration, in which cells leave the blood vessels.

Diarrhea Abnormally frequent evacuation of watery feces.

Differential blood analysis A routine lab test in which the percentage of each of the white blood cell populations is determined.

Differential medium A growth medium that includes components that cause an observable change in color or pH when a particular chemical reaction occurs, making it possible to distinguish between organisms.

Differential stain The use of two or more dyes to differentiate between bacterial species or distinguish different structures of an organism.

DiGeorge syndrome Primary immune deficiency disease in which the thymus does not develop properly, causing a deficiency of T-cell functions.

Dimorphism The term used to describe some fungi that can grow in either the mold or the yeast form.

Diphtheria An acute bacterial infection of the nose, throat, or larynx caused by *Corynebacterium diphtheriae*. It is marked by formation of a grey/white pseudomembrane.

Diphtheria toxin An exotoxin produced by *Corynebacterium diphtheriae* that affects the membranes of the nose, throat, and larynx. It can also affect the heart and the central nervous system.

Diplococcus The arrangement of bacteria that grow in pairs. *Streptococcus pneumoniae*, the leading cause of pneumonia, is an important example.

Discontinuous strand *see* Lagging strand.

Disease A negative disturbance in the state of health during which the body does not function properly.

Disinfectants Chemical agents used on inanimate objects to destroy microorganisms.

Disk-diffusion test Also known as the Kirby–Bauer method. It is used to determine microbial sensitivity to antimicrobial agents through the use of a zone of inhibition around filter paper disks placed on plates with growing bacteria.

Disseminated intravascular coagulation (DIC) Widespread coagulation of the blood in different areas of the body.

Dissimilation plasmid An extrachromosomal piece of genetic material that contains genes that enable organisms to be resistant to disinfectants and environmental pressure.

Distending cytotoxin A toxin produced by *Campylobacter* that arrests cell division while the cytoplasm continues to increase.

Disulfide bridge A bond that forms between sulfur-containing amino acids. It is one of the primary ways in which the three-dimensional folded shapes of proteins are maintained.

Diverticular abscess A pus-containing lesion occurring in the diverticulum of the intestinal tract.

DNA (deoxyribonucleic acid) The molecule that carries the hereditary genetic information from one generation to the next.

DNA helicase An enzyme that unwinds the helical structure of the DNA molecule.

DNA ligase An enzyme that is used to connect sections (fragments) of DNA together by filling in gaps. It is seen on the lagging strand of DNA during replication.

DNA polymerase The enzyme used to match and bond complementary base pairs during the process of DNA replication.

Dysentery A severe diarrhea in which the fecal material contains mucus and blood.

Dyspnea Labored or difficult breathing.

Dysuria Pain and burning sensations during urination.

E site A site on the ribosome where the tRNA exits after the removal of its amino acid.

E test A diffusion method used to determine the MIC (minimal inhibitory concentration) of a drug.

Eastern equine encephalitis (EEE) Type of viral encephalitis seen most often in the eastern United States; it primarily infects horses but can also infect humans.

Ebola hemorrhagic fever Viral infection of the endothelial cells lining blood vessels and platelets by a filovirus. It is characterized by loss of coagulation, hemorrhage, and shock.

Ecchymosis A small hemorrhagic spot in the skin or mucous membranes forming a nonelevated irregular blue or purplish patch.

Efflux pumping A mechanism used by bacteria to resist attack by antibiotics. It takes advantage of pumps in the plasma membrane that pump out the antibiotics.

Elementary body The infectious form of *Chlamydia*.

Elephantiasis Swelling of the extremities and genitalia as a result of an intense inflammatory response and lymphatic blockade by the parasite *Wuchereria bancrofti*.

Embolus A mass of clotted blood or other material that is brought by the blood and forced into a smaller vessel, obstructing the circulation.

Emerging infectious disease Diseases that have not been seen before. They arise from movement into areas where humans have not ventured before. They can also be diseases that move from certain areas of the world to places where the diseases have not been before.

Encephalitis Inflammation of the brain.

Endemic disease Refers to a disease that is constantly present in the population.

Endemic typhus A flea-borne typhus caused by *Rickettsia typhi.*

Endoarteritis Inflammatory lesions that occur in the arteries.

Endocarditis Exudative and proliferative inflammatory lesions found on the endocardium and valves of the heart.

Endocytosis The process in which vesicles form by invagination of the plasma membrane of the cell and move substances into the cell.

Endoenzyme An enzyme that works within the cell.

Endogenous infection An infection caused by organisms that are part of the normal microbial flora.

Endometritis Infection of the endometrium.

Endoplasmic reticulum An extensive network of membranes that form tubes and plates in the cytoplasm of eukaryotic cells. They are involved in the synthesis and transport of proteins and lipids.

Endosome A membrane enclosure seen during the virus penetration step of the infection process.

Endospore A highly resistant dormant structure that is formed by certain bacteria.

Endosymbiotic theory Evolutionary theory holding that the organelles of eukaryotic cells arose from bacteria that came to live in a symbiotic relationship inside eucaryotic cells.

Endotoxin A toxin incorporated into the outer layer of Gram-negative cells that is released when the cell dies.

Enteric Having to do with the intestine.

Enteric fever A systemic infection with a focus on the intestinal tract.

Enterobacteriaceae A family of bacteria, many of which are intestinal Gram-negative facultative anaerobes that have flagella.

Enterocytes Epithelial cells of the large and small intestine.

Enterohemorrhagic *E. coli* A type of *Escherichia coli* that causes destruction of the blood vessels; the causative agent in infections seen with contaminated ground beef.

Enterotoxicosis *see* Food poisoning.

Enterotoxin An exotoxin that attacks the tissues of the intestinal tract.

Enterovirus One of three major groups of picornaviruses that infect nerves, muscle cells, and the lining of the respiratory tract.

Enveloped virus A virus with a lipid bilayer surrounding its capsid.

Enzyme active site The location on the enzyme where reactants are made into products.

Enzyme–substrate complex The association of the enzyme with its substrate.

Eosin methylene blue A blue dye used for a variety of bacterial stains, both simple and differential.

Eosinophil A white blood cell normally found in very low numbers in the blood but in very high numbers during a parasitic infection.

Eosinophilia An abnormal increase of eosinophils in the blood.

Epidemic disease A disease that has a higher than normal incidence in the population over a short period.

Epidemic typhus A louse-borne rickettsial disease caused by *Rickettsia prowazekii*; seen most often in conditions of overcrowding and poor sanitation.

Epidemiology The study of factors and mechanisms involved in the spread of disease within a population.

Epididymitis Infection of the epididymis.

Epithelial hyperplasia An oral viral infection causing many pink or whitish flat mucosal masses which are painless but very contagious.

Epitope Area of an antigen molecule to which specific antibodies bind; also called an antigenic determinant.

Epstein–Barr virus A virus that causes Burkitt's lymphoma and mononucleosis.

Erysipelas A rapidly spreading infection of the deeper layers of the dermis.

Erythema Redness of the skin due to congestion of the capillaries.

Erythema infectiosum A mildly contagious disease of children between 4 and 12 years of age marked by a rose colored macular rash.

Eschar The thick crust or scab that forms over a severe burn.

Etiology The cause of a disease.

Eukaryote A cell that has a distinct nucleus and other membrane-enclosed organelles.

Excision repair A repair process in which thymine dimers are removed from DNA.

Exfoliatin An exotoxin produced by *Staphylococcus aureus* and seen in scalded skin syndrome.

Exocytosis Process in which vesicles inside a cell fuse with the plasma membrane and release their contents to the outside of the cell.

Exogenous infection An infection caused by organisms that enter the body from the outside.

Exotoxin A soluble protein toxin produced by living bacteria. It is seen in many types of systemic infection.

Extreme halophile A bacterium that requires extremely high concentration of salt in order to grow.

Extreme thermophile A bacterium that grows at temperatures above 80°C.

Exudate Fluid containing high levels of protein and cellular debris that is deposited in tissues or on tissue surfaces. It is usually the result of inflammation.

F1 protein A factor seen in plague that forms a gel-like capsule preventing phagocytosis and allowing the bacteria to multiply in the submucosa.

Facilitated diffusion Diffusion across a membrane that is carried out by a nonspecific carrier molecule and does not require ATP.

Factors B, D, and P (properdin) Chemical molecules found in the blood that respond to carbohydrates found on bacteria and initiate the alternative pathway of complement.

Facultative anaerobe A bacterium that uses oxygen for metabolism but shifts to anaerobic metabolism when oxygen is no longer available.

Facultative halophile A bacterium that can grow in high concentrations of salt as well as normal concentrations of salt.

FAD (flavin adenine dinucleotide) A coenzyme that carries hydrogen atoms and electrons.

Fas Membrane receptor on the surface of cytotoxic T cells that is involved in killing of the cytotoxic cell.

Fasciitis Infection of the fascia of the body.

Fastidious bacteria Bacteria that do not grow well without specific supplements added to the medium. Even then, these bacteria can take longer to grow than other organisms.

Fatal familial insomnia A very rare prion disease of the brain causing complete sleeplessness. It is untreatable and fatal.

Fatty acid A long chain of carbon and hydrogen atoms with a carboxyl group at one end.

Fc fragment of immunoglobulin The tail region of an antibody molecule that may contain sites for complement and macrophage binding.

Fecal coliform count A test done to determine the level of contamination in water.

Fecal–oral route of contamination A major route of infection associated with poor sanitation and hygiene.

Fever A body temperature that is abnormally high.

Fibrinolysis The enzymatic breakdown of fibrin.

Filariasis Disease of the blood and lymph caused by any of several different roundworms.

Fimbria (plural fimbriae) Also known as an attachment pilus. A short hairlike appendage found exterior to the cell wall. It is used as a mechanism for staying in the host during infection.

Flagella (singular flagellum) Long thin helical appendages found on certain cells. They provide a means of locomotion.

Flagella stain A technique used for observing flagella by coating the surface of the flagella with multiple layers of dye or metal such as silver.

Flagellates Parasitic protozoan organisms.

Flagellin protein Globular protein that forms the flagella.

Flucytosine An antifungal antibiotic used in the treatment of severe candidal and cryptococcal infections.

Fluid mosaic model of plasma membrane Model incorporating all of the structures found in the membrane and based on the phospholipid bilayer structure of the membrane.

Flukes Adult trematodes that can live for decades in human tissue and blood vessels.

Flushing Transient redness of the face and neck.

Focus of infection An area where the infection is localized and from which it can spread to other parts of the body.

Follicles Areas of the lymph node where different cells are found.

Folliculitis Inflammation of the follicles.

Fomites Nonliving substances (such as clothing, dishes, or paper money) that are capable of transmitting disease.

Food poisoning Also called enterotoxicosis. It is a gastrointestinal disease caused by ingesting foods contaminated with preformed toxins or other toxic substances.

Forespore The structure formed during the development of a spore in which the plasma membrane wraps around the developing spore.

Formalin A 37% solution of formaldehyde used in clinical settings for disinfection.

Foscarnet A virustatic antibiotic used to treat cytomegalovirus and herpesvirus infections in immunocompromised individuals.

Fosfomycin Antibiotic that targets the *mur*A gene and prevents peptidoglycan subunits from being produced.

Frameshift mutation Mutation resulting from an insertion or deletion of one or more bases in DNA.

Freeze-drying *see* Lyophilization.

Fungal pneumonia Inflammation of the lungs, caused by a fungal infection, with the formation of exudates..

Fungemia A systemic fungal infection.

Fungicide A substance that kills fungi.

Fungistatic Term used for the activity of chemicals that prevent the growth of fungi but does not kill them.

GALT (gut-associated lymphoid tissue) A collective name for tissues of lymphoid nodules, especially those in the digestive tract.

Gametocyte A male or female sex cell.

Gametogony The development of merozoites into male and female gametes, that later fuse to form a zygote.

Gamma hemolysis A type of hemolysis in which there is no destruction of red blood cells.

Ganciclovir A derivative of acyclovir used to treat retinitis caused by cytomegalovirus infection.

Gas gangrene A deep wound infection in which tissue is destroyed, often caused by species of the genus *Clostridium*.

GasPak™ jar A container used to incubate obligate anaerobic organisms.

Gastritis Inflammation of the stomach.

Gene constellations The term used to describe clusters of genes that determine virulence.

Gene expression The process of transcription and translation.

Generalized transduction Type of transduction in which a fragment of DNA from the degraded chromosome of an infected bacterial cell is accidentally incorporated into a new bacterial virus particle, which can then be transferred to another cell.

Generation time The time required for a population of organisms to double in number.

Genetic code The one-to-one relationship between each codon and a specific amino acid.

Genital wart Also known as a condyloma. It is an often malignant lesion that arises on the genitalia and is usually associated with sexually transmitted viral diseases.

Genital herpes *see* Herpes genitalis.

Genome The genetic information in an organism or a virus.

Genus A taxonomic term consisting of one or more species.

German measles *see* Rubella.

Germination The mechanism by which organisms develop from the endospore state to the vegetative (growing) state.

Gerstmann-Sträussler-Schenker syndrome A group of rare prion diseases having the common characteristic of cognitive and motor disturbances.

Ghon complex Calcified necroses seen in the tubercles, associated with tuberculosis.

Giardiasis A gastrointestinal disorder caused by the flagellated protozoan *Giardia intestinalis*.

Gingiva The soft tissue (gums) in the mouth.

Gingivitis The mildest form of periodontal disease characterized by inflammation of the gingiva.

Glycocalyx A term used to refer to all substances containing polysaccharides that are found external to the cell wall.

Glycogen A storage molecule for carbohydrates.

Glycolipid A lipid molecule that contains carbohydrate.

Glycolysis An anaerobic metabolic pathway used to break down glucose into pyruvate; it produces some ATP.

Golgi apparatus An organelle found in eukaryotic cells that receives, modifies, and transports substances coming from the endoplasmic reticulum.

Gonorrhea A sexually transmitted infection caused by *Neisseria gonorrhoaea*, commonly asymptomatic in females but marked by a painful purulent discharge from the urethra in males.

Goodpasture's syndrome Autoimmune disease affecting the kidneys and causing glomerulonephritis with pulmonary hemorrhaging, leading to renal failure.

gp41 A protein found on HIV.

gp120 A protein found on HIV.

Gram stain A differential stain that differentiates bacteria into either Gram-positive or Gram-negative groups.

Granulocyte A leukocyte (white blood cell) with granular cytoplasm and an irregularly shaped nucleus.

Granulocyte-macrophage colony-stimulating factor A substance that recruits large numbers of phagocytic cells to the site of infection.

Granuloma A collection of epithelial cells, macrophages, lymphocytes, and fibers seen in chronic inflammation.

Granulysin A peptide secreted by killer cells that causes the onset of apoptosis.

Granzyme A protease enzyme secreted by killer cells that causes the death of a target cell by triggering apoptosis.

Graves's disease Autoimmune disease in which antibodies attack thyroid stimulating hormone causing it to be overproduced, leading to hyperthyroidism.

Griseofulvin An chemical agent that inhibits fungal growth.

Group translocation One of three types of active transport seen in bacteria, in which chemical substances brought into the cell are chemically modified so they cannot diffuse back out of the cell.

Guillain–Barré syndrome Acute autoimmune disease of the peripheral nervous system, usually triggered by acute infection.

Gummas Localized granulomatous lesions seen in the skin, bones, joints, and internal organs in tertiary syphilis.

H antigen Refers to the proteins that make up the flagella of bacteria; they are antigenic and can initiate an immune response.

Hairy Leukoplakia White patch on the tongue or buccal mucosa caused by Epstein-Barr viral infection and associated with HIV infection.

Halophile A salt-loving organism that requires a moderate to extreme level of salt so as to grow. There are three categories: obligate, facultative, and extreme.

Hanta pulmonary syndrome The severe respiratory syndrome caused by the "Sin Nombre" hantavirus.

Hard measles *see* Rubeola.

Hashimoto's thyroiditis Autoimmune disease causing the destruction of thyroid tissue.

Helical symmetry The "spiral staircase" arrangement of the DNA molecule, which has a precise symmetry.

Helicase An enzyme that unwinds the helical structure of the DNA during replication.

Helicobacter pylori An organism found in the stomach and duodenum of the small intestine; it accounts for up to 80% of ulcers found in these regions of the digestive tract.

Helminth A worm with bilateral symmetry; includes roundworms and flatworms.

Helper T cell The T lymphocyte that stimulates the function of other immune cells such as B cells, cytotoxic T cells, and macrophages.

Hematogenous dissemination Movement through the blood; the most efficient way for systemic viruses to disseminate throughout the body.

Hemolysin An enzyme that destroys red blood cells.

Hemoptysis The spitting of blood or bloodstained sputum.

Hemorrhagic colitis Infection of the colon with bloody discharge.

Hemorrhagic cystitis Inflammation of the bladder accompanied by severe hemorrhage.

HEPA filter A filter containing holes that are too small for microorganisms to pass through. It is used to keep rooms sterile.

Hepatitis An inflammation of the liver usually caused by viral infection but can also be caused by an ameba or by toxic chemical damage to the liver.

Hepatocellular carcinoma Malignancy involving cells of the liver.

Hepatomegaly Enlargement of the liver.

Hepatosplenomegaly Enlargement of the spleen and liver

Herd immunity The proportion of individuals in a population who are immune to a particular disease.

Hermaphrodites Term used to describe the reproductive system of some flukes that have both male and female reproductive systems.

Herpes genitalis Primarily sexually transmitted disease of the genital region which can give rise to confluent ulcerations.

Heterotrophy The use of carbon atoms from preexisting organic molecules to produce new biomolecules.

Hfr (high frequency of recombination) cell An F^+ bacterial cell that harbors a plasmid incorporated in its chromosome.

Histoplasmosis A fungal respiratory disease endemic to the central and eastern United States, caused by *Histoplasma capsulatum*.

Hong Kong influenza One of the three major influenza epidemics in the United States. It occurred in 1968 and caused more than 30,000 deaths.

Human papilloma virus (HPV) Virus that attacks skin and mucous membranes, causing papillomas or warts. It has been proved to cause cervical carcinoma.

Humoral immune response A response to infection involving the production of antibodies.

Hydrogen bond A relatively weak bond between a hydrogen carrying a partial positive charge and an oxygen or nitrogen molecule carrying a partial negative charge.

Hydrolysis A chemical reaction that produces smaller products by using water to break down larger molecules.

Hygiene hypothesis The theory that a less hygienic environment can help protect against certain allergens.

Hypersensitivity Also called allergy. It is a disorder in which the immune system reacts to antigens that it would normally ignore.

Hypertonic A solution containing a concentration of dissolved material greater than that within a cell.

Hypervariable region of the antibody molecule The regions of the heavy and light chains found in the antigen-binding site of the immunoglobulin molecule that have a highly variable protein structure.

Hyphae Long threadlike extensions of the cytoplasm seen in molds.

Hypokalemia Abnormally low levels of potassium in the blood.

Hypotonic A solution containing a concentration of dissolved materials lower than that within the cell.

Hypoxia Condition resulting from reduced levels of oxygen being supplied to the tissues.

Iatrogenic transmission Transmission of infection because of poor techniques employed by health care workers.

ICAM-1 A cell receptor used for rhinovirus infection.

Icosahedral symmetry Capsid arrangement seen in some viruses in which there are 20 geometric sides that make up the virus protein coat.

ID_{50} Infectious dose 50%. The number of organisms required to infect 50% of the subject population.

Immune escape Refers to the ability of some pathogens to evade the host immune response.

Immunocompromised Refers to an individual whose immune defenses are weakened. It can result from a variety of causes such as HIV infection, drug therapy for malignancy or transplantation as well as old age or debilitation.

Immunogen Also known as an antigen. Anything that can elicit an immune response.

Immunoglobulin Also known as an antibody. A class of proteins that are produced by the humoral immune response and respond to a specific antigen.

Immunological memory The ability of the adaptive immune response to immediately recognize and respond against antigens it has previously been exposed to.

Impetigo A highly contagious pyoderma caused by staphylococci or streptococci, or both.

Inactive vaccine Also known as a killed vaccine. A vaccine composed of virus that is either dead or inactivated.

Incidence of disease The number of new cases of a particular diseases seen in a population over a specific period of time.

Inclusion bodies (1) Form of cytopathic effect in which viral particles, viral components, or remnants of virus aggregate in the cytoplasm of infected cells. (2) The aggregation of reticulate bodies seen in *Chlamydia*.

Incubation period The time between the infection and the first appearance of signs and symptoms of the infection.

Index case The first person to be identified as having the disease.

Inducible gene A gene that is off and can be turned on.

Infectious endocarditis Infection of the heart.

Infectious mononucleosis An acute disease that affects many systems, caused by the Epstein–Barr virus.

Inflammation The body's defensive response to any trauma or infection of the body.

Influenza An acute viral infection of the respiratory tract marked by inflammation of the nasal mucosa, pharynx, and conjunctiva.

Inhalation anthrax One of several types of anthrax infection caused by *Bacillus anthracis*, resulting in highly fatal pneumonia.

Initiation codon The sequence of messenger RNA that sets the reading frame.

Initiator protein The protein that recognizes the initiator sequence and begins transcription.

Innate immune response The nonspecific immune response that is present at birth.

Integral protein Membrane protein found attached to one side of the plasma membrane.

Interferon A group of small protein molecules often released in response to a viral infection that bind to noninfected cells, causing them to produce antiviral proteins that protect against viral infection of the cell.

Interleukin 1 (IL-1) The best-known endogenous pyrogen, which causes the onset of fever during an infection.

Interleukin 6 (IL-6) A cytokine produced during an infection that causes the liver to produce acute-phase proteins.

Interleukins A class of cytokine produced by leukocytes.

Intermediate host An organism that harbors the sexually immature stage of a parasite.

Interstitial macrophages Cells found in the stroma of the lung. They are smaller and less phagocytic than alveolar macrophages.

Interstitial pneumonia Pneumonia characterized by thickening of the interstitial tissue.

Invasin A virulence factor.

Iodophor An organic compound that incorporates iodine in such a way that the iodine is released slowly.

Ionic bond A bond formed by the attraction between the opposite charges of a cation and anion.

Ionizing radiation Form of radiation that causes disruption of the electron clouds surrounding atoms, causing the ionization of the atoms

Ions Electrically charged atoms produced when atoms either gain or lose electrons.

Isotonic Fluid containing the same concentration of dissolved materials as that inside the cell.

J chain The protein found on the pentameric form of IgM and the dimeric form of IgA that helps to hold the monomers together.

Jaundice Yellowness of the skin, sclera, mucous membranes, and excretions caused by hyperbilirubinemia.

K antigen Refers to the polysaccharide associated with the capsule found around certain bacteria.

Kaposi's sarcoma A malignancy associated with immunodeficiency in which blood vessels grow in tangled masses that are filled with blood and easily ruptured.

Keratitis Infection of the cornea of the eye.

Killed vaccine *see* Inactive vaccine.

Kirby–Bauer method *see* Disk-diffusion test.

Koch's postulates Four postulates proposed by Robert Koch in the nineteenth century, used to prove that a particular organism causes a particular disease.

Koplik's spots Red spots with centralized bluish specks that appear on the mucous membranes in the early stages of measles.

Kupffer cell A phagocytic cell that is stationed in the sinusoids of the liver.

Kuru Transmissible spongiform encephalopathy of the human brain. The disease is caused by prions and is associated with cannibalism and with tissue or organ transplantation.

***Lac* operon** A sequence of genes that controls the production of enzymes required to break down the sugar lactose.

Lacrimal apparatus The structures associated with the production and recycling of tears.

Lacrimal gland The gland that produces tears. It is part of the lacrimal apparatus.

Lactic acid A fermentation product produced through the anaerobic fermentation of glucose.

Lag phase of bacterial growth The first phase of the bacterial growth curve, in which organisms acclimate to their surrounding; they grow in size but do not increase in number.

Lagging strand Also known as the discontinuous strand. The strand of DNA that is replicated in pieces called Okazaki fragments.

LAL assay The *Limulus* amebocyte lysate assay, which is used to determine whether there is endotoxin contamination.

Langerhans cells Dendritic cells found in the layers of the skin. They are phagocytes that are one type of antigen-presenting cell.

Lantibiotics Antibacterial peptides produced by some Gram-positive bacteria.

Laryngitis Inflammation of the larynx.

Laryngotracheitis *see* Acute croup.

Lassa fever Acute viral hemorrhagic disease spread by exposure to rodent feces. It infects almost all human tissues and progresses to systemic vascular disease and fulminant viremia.

Latency The strategy adopted by some pathogens to avoid triggering an immune response in which the pathogen stops replicating and becomes inactive. Latent pathogens can be reactivated by a number of factors.

Latent infection A disease in which there are periods of inactivity either before the onset of symptoms or between attacks.

LD_{50} Lethal dose 50%. The number of organisms required to kill 50% of the subject population.

Leading strand Also known as the continuous strand. The strand of DNA that is replicated continuously.

Lectin-binding pathway The complement pathway that is activated by the carbohydrate mannose.

Leishmaniasis Infection with the parasite *Leishmania*.

Leprosy A chronic disease caused by *Mycobacterium leprae* which is characterized by granulomatous lesions of the skin, mucous membranes, and central nervous system.

Leukocidin An exotoxin produced by many bacteria that kills white blood cells, including phagocytic cells.

Leukocyte A white blood cell.

Leukocyte endogenous mediator A factor that lowers plasma iron concentration, which limits the availability of iron and thereby inhibits the growth of some pathogens.

Leukotrienes A chemical substance released from mast cells that causes prolonged airway constriction, increased dilation and permeability of capillaries, increased secretion of mucus, and stimulation of nerve endings that causes pain and itching.

Light repair *see* Photoreactivation.

Lipid raft A portion of the plasma membrane of host cells that contains increased numbers of receptors required for virus attachment.

Lipocalin A substance that inhibits the scavenging of iron by pathogens, thereby inhibiting their growth.

Lipopolysaccharide layer An outer layer of the cell wall found around Gram-negative cells. It contains endotoxin that is released when the organism dies and this layer falls apart.

Lipoteichoic acid A molecule that is found only in Gram-positive bacteria that penetrates the entire cell wall and attaches to the plasma membrane of the cell.

Live attenuated vaccine Live virus that has been chemically modified to reduce infectivity, used for some forms of vaccination.

Loaiasis A disease caused by the parasitic *Loa Loa* worm and transmitted to humans by the deer fly.

Local infection An infection confined to a specific area of the body.

Lockjaw The early symptom of tetanus, an infection of the nervous system that initially affects the small muscles of the face, preventing them from relaxing.

Log (exponential) phase of bacterial growth The second of the four phases of bacterial growth, in which cells divide at an exponential rate.

Logarithmic decline phase *see* Death phase.

Lophotrichous flagella The arrangement in which there are two or more flagella at one or both ends of a bacterial cell.

Lyme disease Recurrent multisystemic disorder caused by *Borrelia burgdorferi*.

Lymph nodes Encapsulated globular structures found along the lymphatic vessels throughout the body that help clean the lymph fluid before it is put back in the blood. These structures are filled with lymphocytes that react against anything that is nonself in the lymph fluid.

Lymphadenitis Inflammation of one or more lymph nodes.

Lymphadenopathy A swelling of the lymph nodes due to increased numbers of cells in the node. It is usually seen as a signal of ongoing infection.

Lymphangitis Inflammation of the lymphatic vessels.

Lymphocyte A form of white blood cell that is involved with the adaptive immune response of the body.

Lymphogranuloma venereum Venereal infection caused by *Chlamydia trachomatis* marked by a primary transient ulcerative lesion in the genital region.

Lymphokine A cytokine secreted by T cells after encountering an antigen.

Lymphoma Any neoplastic disorder of the lymphoid tissues.

Lymphopenia A reduced number of lymphocytes resulting from certain forms of infection and also from some kinds of therapy for malignancy or transplantation.

Lyophilization Also referred to as freeze-drying. It is a method of extracting water from a frozen state.

Lysogenic infection A viral infection in which the viral nucleic acid is incorporated into the chromosome of the infected host cell and can be carried there for long periods as the host cell continues to divide.

Lysogenic virus A virus that can cause a lysogenic infection.

Lysosome A small membrane-enclosed organelle seen in eukaryotic cells that contains digestive enzymes.

Lysozyme A antibacterial enzyme found in secretions such as tears and saliva.

Lytic infection A viral infection in which the production and release of large numbers of new virions leads to the death of the host cell.

Lytic virus A virus that causes a lytic infection.

M cells Specialized cells found in the digestive tract. They are used by pathogens to enter the tissues of the body.

M protein A chemical virulence factor found on the bacterial cell surface and also on fimbriae that helps pathogens stay in the host.

Macrophage A highly phagocytic white blood cell.

Macrophage activating factor A cytokine produced by the Th1 helper T cell population.

Macrophage activation A process in which macrophages are stimulated by Th1 helper cells to eliminate pathogens that proliferate in macrophages.

Macule A discolored spot on the skin that is not raised above the surface.

Maculopapular rash Broad lesions that slope away from a centrally located papule.

Mad cow disease The name given to the spongiform encephalopathy seen in cattle.

Major histocompatability complex (MHC) A group of cell surface proteins that are required for the development of an immune response.

Malaise A vague feeling of discomfort.

Malaria An infectious febrile disease caused by protozoan parasites of the genus *Plasmodium*. It is transmitted by the *Anopheles* mosquito.

Malarial paroxysm The term used to describe the acute phase of infection seen in malaria.

MALT Mucosal-associated lymphoid tissue.

Mannose-binding protein (MBP) Acute-phase protein that binds to mannose sugars found on many bacterial and fungal cell membranes.

Margination The process in which white blood cells traveling in the blood are able to slow down and stop adjacent to the area where the tissue injury has occurred. It is caused by the localized secretion of selectin, which is a sticky molecule on the inner side of the vessels.

Maginins Antibacterial peptides produced by frogs.

Mass number The total number of protons and neutrons found in the nucleus of the atom.

Mast cell A white blood cell that releases histamines during an allergic response.

Mastoiditis Inflammation of the mastoid cavity and cells.

Matrix protein The second layer of viral protein found just inside the envelope of the virus.

Mature vegetation A "mesh" of platelets, fibrin, and inflammatory cells seen in infectious endocarditis.

Maximum growth temperature The highest temperature at which an organism will grow.

Measles A highly contagious viral infection usually seen in children. It involves the respiratory tract and is marked by discrete red papules, which become confluent. *See* Rubeloa.

Mechanical vector transmission The passive movement of organisms from a vector to a human or fomite.

Membrane attack complex The complex formed during the final stages of the complement pathway. This complex produces a hole in the cell wall of bacteria, leading to their death.

Memory cell A long-lived T or B lymphocyte that is derived as part of the adaptive immune response. It retains a memory of the antigen that it was sensitized to.

Meningitis Infection of the meninges (layers) that surround the brain and spinal cord.

Meningoencephalitis Inflammation of the brain and meninges.

Merozoite A trophozoite form of the malaria parasite derived from sporozoites. It is found in red blood cells and hepatocytes during the malaria infection cycle.

Mesophiles Organisms that grow best at temperatures between 25°C and 40°C.

Messenger RNA (mRNA) A type of RNA that results from the transcription of DNA and carries information about the arrangement of amino acids in a protein.

Metabolic acidosis The shift in acid base balance due to loss of bicarbonate.

Metabolism The sum of catabolism (the breakdown of organic molecules) and anabolism (the building up of organic molecules).

Metacercariae An encysted form of cercariae seen in the development of a fluke.

Metachromatic granules Polyphosphate granules seen in the cytoplasm of certain bacteria.

MHC restriction Major histocompatability complex restriction associated with the interaction of antigen-processing cells and helper T cells.

Microaerophilic bacteria Organisms that grow best in the presence of small amounts of oxygen.

Microarray analysis A type of biological analysis method that is used in a variety of biotechnology processes.

Microbial antagonism The ability of normal microbial flora to inhibit the growth of pathogens by competing for resources.

Microbial flora Microorganisms that are normally found living in and providing important benefits to the host.

Microconidia Smaller of two types of conidia produced by fungi.

Microfilament A protein fiber that makes up part of the cytoskeleton in eukaryotic cells.

Microfilariae Live offspring of tissue nematodes that circulate in the blood and subcutaneous tissues until they are ingested by specific blood-sucking insects.

Microglial cells Resident macrophages found in the central nervous system. There are two forms, ameboid (which travel through developing brain tissue and are also found in damaged brain tissue) and ramified (found in normal brain tissue).

Microtubule A protein tubule that forms the structure of cilia, flagella, and part of the cytoskeleton in eukaryotic cells.

Minimal inhibitory concentration (MIC) The lowest concentration of an antimicrobial agent that prevents growth in the dilution method of determining antibiotic sensitivity.

Minimal microbicidal concentration (MMC) The lowest concentration of an antimicrobial agent that kills microorganisms, as indicated by the absence of growth when the antimicrobial agent is removed.

Minimum growth temperature The lowest temperature at which an organism will grow.

Miracidia The ciliated free-swimming first-stage fluke larvae that have emerged from eggs.

Missense mutation Also known as a point mutation. A change in a single base of the DNA sequence.

Mitochondria A membrane-enclosed organelle found in eukaryotic cells that is responsible for the production of ATP.

Mold The filamentous multicellular form of fungi.

Monocyte A nonphagocytic white blood cell found in the blood that will differentiate into a phagocytic macrophage in response to an infection.

Mononuclear phagocytic system Formerly called the reticuloendothelial system. It is a collection of phagocytic cells and tissues that contain phagocytic cells, located throughout the body.

Mononucleosis An excess of monocytes in the blood which can be seen in chronic or infectious forms.

Monotrichous flagellum One of the four types of flagellar arrangement in which there is one flagellum seen on the cell.

Morbidity rate The number of individuals affected by a disease during a set period.

Mortality rate The number of deaths caused by a disease during a set period.

MRSA Methicillin-resistant *Staphylococcus aureus*.

Mucociliary escalator Mechanism involving ciliated cells that allows materials in the bronchi, trapped in mucus, to be lifted up into the pharynx and subsequently swallowed or spat out.

Mucocutaneous candidiasis A fungal infection of the mucosal tissues caused by the fungus *Candida albicans*.

Mucoid watery diarrhea Watery stool containing mucus.

Mucopurulent discharge Discharge containing mucus and pus.

Multiple sclerosis Neurological disease in which autoimmunity causes production of antibody against antigens on the myelin sheath of neurons.

Mumps An acute contagious disease usually seen in children. It is caused by a paramyxovirus and chiefly affects the parotid salivary glands.

Mupirocin An antibiotic that prevents colonization of the nasal passages with *S. aureus*.

Mutagen An agent that causes mutations in DNA.

Mutualism A form of symbiosis in which two organisms of different species live in a relationship in which both benefit.

Myalgia Muscular pain.

Myasthenia gravis Genetic disorder of infants characterized by apnea, weakness, and fatigue.

Mycelium A mass of long threadlike intertwining structures called hyphae.

Mycolic acid A waxy substance found in the cell wall of certain bacteria, such as the genus *Mycobacterium*.

Mycology The study of fungi.

Mycosis A disease caused by fungi.

Myocarditis Inflammation of the muscular walls of the heart.

***N*-acetyl glucosamine (NAG)** One of the repeating disaccharides that make up the cell wall of bacteria.

***N*-acetyl muramic acid (NAM)** One of the repeating disaccharides that make up the cell wall of bacteria.

NAD^+ (nicotinamide adenine dinucleotide) A coenzyme that carries hydrogen atoms and electrons.

Naive T cell A T cell that has not found the antigen that binds to its receptor.

Naked virus A virus that does not have an envelope.

Narrow-spectrum antibiotics The range of activity of an antimicrobial agent that attacks only a few kinds of microorganism.

Nationally notifiable disease A disease that must be reported to the Centers for Disease Control and Prevention (CDC).

Natural killer cells Large granular cells found in the peripheral tissue and blood that kill tumor cells, virus-infected cells, bacteria, fungi, and parasites.

Necrotizing fasciitis An infection in which the fascia is destroyed by organisms such as group A streptococci either alone or in a synergistic way with other bacteria.

Necrotizing periodontal disease (NPD) Also known as Vincent's disease or trench mouth. A spectrum of acute inflammatory diseases resulting in the destruction of the soft tissue of the oral cavity.

Negative stain The technique of staining the background around a specimen but leaving the specimen clear and unstained.

Negri body A characteristic cytopathology seen in rabies virus infections. These are areas in the cytoplasm that contain masses of viral particles.

Nematode Roundworm.

Neonatal gonorrheal ophthalmia Infection of the eyes of newborn babies from women with gonorrhea. It is caused by infection as the newborn travels down the birth canal.

Nephritis A serious infection of the kidneys.

Neurotoxin A toxin that acts on the tissues of the nervous system.

Neutropenia A lower than normal number of neutrophil white blood cells.

Neutrophil A phagocytic white blood cell.

Noncytocidal effect Viral cytopathology that causes a shutdown of host cell function.

Noncytopathic virus A virus that does not cause destruction of the host cell.

Non-gonococcal urethritis The most prevalent sexually transmitted disease. Caused by *Chlamydia trachomatis*.

Non-ionizing radiation A type of radiation, such as ultraviolet radiation, that causes the formation of thymine dimers in DNA.

Nonpolar covalent bond The equal sharing of electrons between two atoms.

Nonsense mutation A mutation in the DNA sequence that codes for a stop codon.

Nonseptate fungus A fungus in which the hyphae do not have septa.

Nosocomial infection An infection that occurs during a hospital stay.

Nuclear membrane Also called the nuclear envelope. It is the membrane surrounding the nucleus seen in eukaryotic cells.

Nuclear region of bacterial cell Also called a nucleoid. It is the central location where DNA, RNA, and some proteins are found in bacterial cells.

Nucleocapsid The area of the virus in which the capsid and the nucleic acid are found.

Nucleoli (singular nucleolus) Areas in the nucleus of eucaryotic cells where ribosomal RNA is made and ribosomal assembly takes place.

Nucleoplasm The semifluid portion of the cell nucleus of eukaryotic cells that is surrounded by the nuclear membrane.

Nucleoproteins Viral proteins attached to the inner side of the capsid that holds viral nucleic acid in place.

Nucleotide excision The repair mechanism in which enzymes look for distortions in the helical structure of DNA and excise those regions.

Nutrient agar A formulation of media solidified by agar, used for the growth of many types of bacteria.

Nutrient broth A formulation of media in fluid form, used for the growth of many types of bacteria.

O antigen Lipopolysaccharide found on the outer layer of Gram-negative bacteria.

O polysaccharide A type of polysaccharide found on the cell wall of bacteria.

Obligate aerobe A microorganism that cannot grow without oxygen.

Obligate anaerobe A microorganism that cannot grow in the presence of oxygen.

Obligate halophile A microorganism that requires higher than normal concentrations of salt for its growth.

Obligate intracellular parasite The definition of a virus that requires entry into a host cell to reproduce.

Oculoglandular tularemia Purulent conjunctivitis caused by *Francisella tularensis*.

Okazaki fragments Pieces of DNA that are made on the lagging strand of DNA during replication.

Oncogenic virus A virus that causes transformation of the host cell.

Oocyst The encysted or encapsulated stage in the development of any sporozoan.

Open reading frame The starting point for protein synthesis on messenger RNA.

Operator site on DNA A gene in an operon that can bind repressor proteins and inhibit the transcription of the structural genes of the operon.

Operon A sequence of genes that includes both structural and regulatory genes controlling transcription.

Opisthotonos The last stages of tetanus in which the body bends backwards as a result of the relentless contraction of the muscles without relaxation.

Opportunistic pathogen Resident or transient microorganisms that do not ordinarily cause disease but can do so under certain circumstances.

Opportunistic infection An infection caused by an opportunistic pathogen.

Opsonization The process by which microorganisms are rendered more attractive to phagocytes by being coated with antibodies (opsonins) and/or C3b complement proteins.

Optimal growth temperature The temperature at which microorganisms grow best.

Organ specific autoimmunity Autoimmune reactions that are confined to the organs.

Organelle A structure found in the cytoplasm of eukaryotic cells.

Origin of replication The point on the DNA where replication begins.

Osmosis A special type of diffusion in which water "chases" (moves toward) a higher concentration across the plasma membrane.

Osmotic lysis The rupture of a cell after excess water moves across the membrane through osmosis.

Otitis media Infection of the middle ear.

P pilus A structure that projects from the exterior of a bacterial cell wall and binds to receptors on the epithelial cells of the urinary tract.

P site The area of the ribosome that holds the growing chain of amino acids.

Pancreatitis Inflammation of the pancreas.

Pandemic disease An epidemic disease that has become worldwide.

Papilledema Edema of the optic nerve.

Papilloma virus A virus that can cause warts; it is associated with human cervical cancer.

Papule Small elevated lesion of the skin.

Paracortical area of the lymph node An area of the lymph node where T cells are found.

Paragonimiasis Infection with the lung fluke *Paragonimus*.

Parasite-directed endocytosis A unique process in which the microvilli of the epithelial cells surround the organism and escort it into the cell cytoplasm. This process is seen in gonorrhea infection.

Parasitemia Parasites in the blood.

Parasitic amebic meningoencephalitis Infection of tissues of the nervous system caused by parasites that have ameboid characteristics.

Parasitism A symbiotic relationship in which one organism (the parasite) benefits at the expense of the host.

Parenchyma The essential or functional elements of an organ.

Parenteral route A portal of entry in which the barrier of the skin is broken, as in a cut, puncture, or surgical procedure.

Passive immunization A procedure in which an already formed immune product such as antibody is administered to a patient.

Passive transport Movement of materials across the membrane without the expenditure of ATP.

Pasteurization Mild heating to destroy pathogens and other organisms that cause spoilage.

Pathogen An organism capable of causing disease.

Pathogenesis The process involved in the development of a disease.

Pathogenicity islands Sections of the genome that include groups of genes coding for virulence factors that increase the pathogenicity of a microorganism.

PD_{50} Paralytic dose 50%. The number of organisms required to cause paralysis in 50% of a subject population.

Pediculosis A form of lice infection.

Pediculus humanus capitis The head louse.

Pediculus humanus corporis Louse that infests portions of the body or clothes..

Pellicle (1) A thin protein film over the tooth that is the base for the development of a biofilm leading to plaque formation and tooth decay. (2) A strengthened plasma membrane seen in some protozoa.

Pelvic inflammatory disease (PID) An infection of the pelvic cavity in females, caused by any of several organisms.

Penicillin G A naturally occurring form of penicillin that is taken by injection.

Penicillin V A naturally occurring form of penicillin that is taken by mouth.

Penicillin-binding proteins Proteins found in the cell walls of bacteria that function in the building of the wall structure. The β-lactam ring of penicillin binds to these proteins.

Pentamidine An antibiotic used in the treatment of *Pneumocystis* infections, leishmaniasis, and African sleeping sickness.

Peptide bond The bond that forms between amino acids in constructing the primary sequence of proteins.

Peptidoglycan The major component of bacterial cell walls.

Peptidyl transferase reaction The enzymatic reaction that links amino acids together by forming the peptide bond.

Perforin An enzyme released by cytotoxic cells that leads to the destruction of target cells.

Period of convalescence One of the five periods that characterize disease, in which the patient is recuperating.

Period of decline The fourth period used to characterize the disease process, in which the patient is getting better and symptoms are declining. This is the period when secondary infections can occur.

Period of illness The third of the periods that are used to characterize disease, in which the symptoms are greatest. During this period the immune response is functioning at its maximum. This is also the period during which a patient may die.

Periodontitis Inflammation of the gingiva of the oral cavity.

Periplasmic space The space between the cell membrane and the outer membrane in Gram-negative bacteria.

Peritrichous flagella This is the arrangement in which flagella are distributed all over the cell.

Permease An enzyme complex involved in the active transport of materials through the cell membrane.

Peroxidase The enzyme used by bacteria to convert hydrogen peroxide to oxygen and water.

Peroxisome Vesicles in eukaryotic cells that contain the enzymes peroxidase, catalase, and oxidase.

Persistent infection A viral infection in which virus is continuously produced over long periods.

Pertussis *see* Whooping cough.

Petechiae Pinpoint-sized hemorrhages most commonly found in skin folds. They are often seen in rickettsial diseases.

Peyer's patches A collection of lymphoid nodules found at the junction between the small and large intestine.

Phagocyte A cell that can carry out phagocytosis.

Phagocytosis Ingestion of materials into cells by means of vacuole formation.

Phagolysosome A structure resulting from the fusion of a phagosome with a lysosome.

Phagosome A vacuole that forms around an organism within the phagocyte that engulfed it.

Pharyngitis Inflammation of the pharynx commonly called a sore throat.

Phase variation A mechanism used by some microorganisms in which the number of pili decreases to prevent the binding of antibody.

Phenol A powerful disinfectant compound used as the standard by which other disinfectants are measured.

Phenol coefficient A numerical expression of the effectiveness of a disinfectant relative to that of phenol.

Phosphoenol pyruvate (PEP) A high-energy molecule used by some bacteria for translocation.

Phospholipid A lipid composed of glycerol, two fatty acids, and a polar head group. It is the basic unit seen in membrane structures.

Phosphorylation The addition of a phosphate group to a molecule, often from ATP, which generally increases the molecule's energy.

Photoautotroph An autotroph that obtains energy from light.

Photoheterotroph A heterotroph that obtains energy from light.

Photoreactivation Also known as light repair. The process of using the enzyme photolyase to unlink thymine dimers in DNA.

Pili (singular pilus) Tiny hollow projections used to attach bacteria to surfaces (called attachment pili) or for the transfer of genetic material during conjugation.

Pilin protein Globular proteins that make up the pilus structure.

Pinocytosis The taking in of small molecules by invagination of the cell membrane.

Plasmid An extrachromosomal piece of DNA that is small and circular and replicates independently. It can be transferred to another cell.

Plasmolysis The shrinking of a cell as a result of changes in the osmotic concentration resulting from loss of water in a hypertonic solution.

Pneumonia Inflammation of the lungs.

Pneumonic plague Also known as the black plague. An infection of the lungs with *Yersinia pestis*, causing a highly contagious form of plague.

Pneumonitis Inflammation of the lungs.

Point mutation *see* Missense mutation.

Polar covalent bond A covalent bond in which there is unequal sharing of electrons.

Poliomyelitis An acute viral disease marked by fever, sore throat, headache, and often stiffness in the neck and back during minor disease. Major disease can involve the central nervous system with the potential for paralysis.

Polyenes Antifungal agents that increase membrane permeability.

Polymorphonuclear leukocyte Another name for a neutrophil.

Polyneuritis Infection of the peripheral nervous system with systemic paralysis of muscles.

Polyribosome A long chain of ribosomes attached at different points along a strand of messenger RNA.

Porin proteins Proteins in the outer layer of Gram-negative bacteria that nonselectively transport polar molecules into the periplasmic space.

Portal of entry A site at which microorganisms enter the body.

Portal of exit A site at which microorganisms can leave the body.

Postherpetic neuralgia Persistent burning pain in the skin following an attack of varicella-zoster virus.

Post-translational modification Changing the structure of proteins so that antibodies do not recognize them.

Prevalence The number of people infected with a particular disease at any given time.

Primary immune response The initial adaptive immune response to an antigen.

Primary immunodeficiency disease A genetic or developmental defect in which T cells or B cells are lacking or nonfunctional.

Primary infection An initial infection in a previously healthy individual.

Primase An enzyme that puts the RNA primer on the lagging strand of DNA during replication.

Primer:template junction The area where the RNA primer is located on the strand of DNA. It is required for the replication of DNA.

Prion An infectious protein.

Prion protein scrapie (PrP^{sc}) The abnormally folded, infectious, form of the prion protein.

Processivity A term used to define how many bases DNA polymerase can add at a time.

Prodromal period of disease The second phase of the disease process, in which nonspecific symptoms such as headache and malaise appear.

Prodrug The inactive form of a drug that must be activated enzymatically once in the patient's body.

Proglottids Reproductive segments seen in cestodes, which contain both male and female gonads.

Progressive multifocal leukoencephalopathy An opportunistic demyelinating infection of the central nervous system.

Prokaryote Microorganism that lacks a cell nucleus and membrane-enclosed organelles. All bacteria are prokaryotes.

Promoter site on DNA The site where RNA polymerase binds to the DNA strand to begin transcription.

Propagated epidemic Disease that involves people-to-people contact and stays in the population for a long time.

Properdin pathway The alternative complement pathway that is activated by contact between lipopolysaccharides and endotoxins on the surface of pathogens and three factors found in the blood (factor B, factor D, and factor P).

Prophage A sequence of DNA from a bacterial virus that is incorporated into the bacterial chromosome.

Prospective analytical study A type of epidemiology study in which analysis is ongoing while the disease is occurring.

Prostaglandins Chemical mediators that act as cell regulators and are produced during the inflammatory response. They can stimulate pain and fever responses.

Prostatitis Inflammation of the prostate gland.

Proteasome Large proteins found in both prokaryotes and eukaryotes that degrade unneeded or damaged proteins.

Proton motive force A concentration gradient of protons seen during the chemiosmosis step during electron transport.

Protozoa Single-celled, microscopic, animal-like organisms.

PrP^c protein The normally folded form of the prion protein.

PrP^{sc} protein The abnormally folded, infectious, form of the prion protein.

Pseudocyst An aggregate of trypanosome protozoa that forms in lymph nodes in Chagas' disease.

Psychotrophs Bacteria that grow best at temperatures between 20°C and 30°C.

Psychrophiles Bacteria that grow best at very cold temperatures, between 0°C and 15°C.

Purines The nucleotide bases adenine and guanine.

Pustule A small elevated pus-containing lesion of the skin.

Pyogenic bacteria Pus-forming bacteria.

Pyrimidines The nucleotides thymine and cytosine.

Pyrogens Chemicals that can induce a fever response.

Pyuria Pus in the urine.

Quaternary ammonium compounds (QUATS) Popular detergents containing ammonium cations. They are low-level disinfectants/antiseptics but are odorless, tasteless, and harmless to humans (except at high concentrations).

Quorum sensing Type of decision-making process seen in bacteria that is based on the density of the population of bacteria.

Rabies A viral disease that affects the brain and nervous system with symptoms such as hydrophobia and aerophobia. It is transmitted by animal bites.

Radiation The energy that is emitted from atomic activities and dispersed at high velocity through matter or space.

Reading frame The mechanism used to read the DNA-coded sequence for transcription.

Re-assortment Term used to describe changes in the DNA sequence.

Receptor-mediated endocytosis The process of uptake of materials into the cell through binding to specific receptors on the cell membrane.

Recombination The combining of DNA from two different cells, resulting in a recombinant DNA molecule.

Redox reaction The oxidation and reduction reactions that move electrons from donor to acceptor molecules.

Reduviid A large winged insect that feeds on sleeping hosts in the evening hours.

Re-emerging infectious disease A disease that was thought to be under control through the administration of antibiotics but reappears in the form of drug-resistant disease. A good example is tuberculosis.

Refractory septic shock An irreversible, fatal hypotension caused by septic shock.

Relapsing fever Disease caused by *Borrelia* species. It is vector transmitted through tick or louse bites and has a poor prognosis, including severe jaundice bleeding and changes in mental status.

Replicase *see* RNA-dependent RNA polymerase.

Replication Process by which an organism or structure duplicates itself.

Replication fork Location along the double-stranded DNA helix where replication is ongoing.

Replicator sequence An easily opened sequence of A-T pairs in the DNA.

Repressible gene A gene that is on and can be turned off.

Reservoir of infection Site where microorganisms can persist and maintain their ability to infect.

Resident (fixed) macrophages Phagocytic cells that are stationed in specific tissues throughout the body.

Residual body The exocytotic vesicle containing the elements of the destroyed organism. It is seen at the end of the phagocytic process.

Resistance islands Areas of the chromosome in which there is an accumulation of genes associated with resistance to antimicrobial agents.

Respiratory syncytial virus A virus that causes lower respiratory infections affecting children below the age of one year.

Reticular fibers Structural fibers made of collagen that form a meshwork supporting soft tissue.

Reticulate body The larger, more fragile, replicative form of *Chlamydia*.

Reticuloendothelial system Old term for what is now known as the mononuclear phagocytic system.

Retrospective analytical study A type of epidemiological study involving the analysis of data after the episode is over.

Retrovirus An enveloped RNA virus that uses its own reverse transcriptase to transcribe its RNA into DNA in the cytoplasm of the host cell.

Reverse transcriptase An enzyme found in retroviruses that can convert RNA into DNA.

Rheumatic fever Febrile illness caused by infection with Group A hemolytic streptococci.

Rhinitis Inflammation of the nasal mucous membranes.

Rhinorrhea Discharge of thin nasal mucous.

Rhizopoda A group of free-living protozoans, many of which form commensal relationships in the human intestines.

Ribavirin A broad-spectrum antiviral antibiotic used to treat severe viral pneumonia caused by respiratory syncytial virus, particularly in high-risk infants. Also used in conjunction with interferon for therapy of hepatitis C infection.

Ribose The form of sugar found in RNA molecules.

Ribosomal RNA The form of RNA that is part of the structure of the ribosomal subunits.

Ribosome The structure in which translation of mRNA into proteins occurs.

Rice stool A symptom of late stages of cholera infection.

Rickettsia Small, nonmotile Gram-negative organisms. They are obligate intracellular parasites of human cells.

Ring stage of malaria The appearance given to red blood cells that are harboring merozoites.

Ringworm A skin lesion that can be found all over the body and is characterized by red margins, numerous scales and reddish itching skin. *See* Tinea capitis, Tinea cruris, and Tinea pedis.

River blindness Parasitic infection of the eye spread by the sucking blackfly, which causes corneal ulceration, fibrosis, and blindness.

RNA polymerase The enzyme that is involved in transcription of DNA into RNA.

RNAase H The enzyme that removes the RNA primer from the fragments being made on the lagging strand of DNA during replication.

RNA-dependent RNA polymerase Also known as replicase. An enzyme seen in RNA viruses that catalyzes the replication of RNA from an RNA template.

Rocky Mountain spotted fever Disease caused by *Rickettsia rickettsii*, which is transmitted to humans by ticks.

Rostellum Retractable chitinous hooks that are found on the scolex of tapeworms and are used for attachment.

Rubella Also known as German measles. A viral infection characterized by skin rash that can cause severe congenital damage.

Rubeola Common form of measles; also called hard measles.

S pili Fragments of pili intended to bind to and inactivate antibody molecules.

Saint Louis encephalitis Type of viral encephalitis most often seen in humans in the central United States.

Salpingitis Inflammation of the fallopian tubes.

SARS Severe acute respiratory syndrome.

Scabies Highly contagious skin disease caused by the mite *Sarcoptes scabiei*.

Scalded skin syndrome An infection caused by staphylococci that produces large soft vesicles over the whole body.

Scarlet fever Infection caused by *Streptococcus pyogenes* that produces erythrogenic toxin.

Schistosomes A term used to describe the reproductive system of some flukes.

Schistosomiasis Infection with the protozoan parasite *Schistosoma*.

Schizogony A reproductive cycle of simple fission followed by sexual reproduction (gametogony).

Scleroderma Hardening and thickening of the skin occurring in generalized or localized forms.

Sclerosing panencephalitis A progressive debilitating and fatal brain disorder caused by infection with a mutated measles virus.

Scolex The head of the tapeworm, used for attachment and for absorbing nutrients.

Scrapie The name given to the spongiform encephalopathy seen in sheep.

Sebaceous gland Epidermal structure associated with hair follicles that secretes an oily substance called sebum.

Sebum Oily substance secreted by the sebaceous glands.

Secondary immune response The adaptive immune response that occurs when an antigen that has been previously seen is encountered again. This response is quicker and more powerful than the primary adaptive immune response.

Secondary infection Infection that can occur in patients recovering from a primary infection. It can be worse than the primary infection because of the weakened immune response resulting from fighting off the primary infection.

Secretory piece A protein found on the secretory IgA molecule that attaches it to mucins of the tissues.

Secretion The process of moving substances from the inside to the outside of a cell, across the plasma membrane.

Selectin A molecule that is secreted from the epithelial cells of blood vessels that causes the margination of white blood cells at the site of the tissue damage.

Selective medium Medium that encourages the growth of some organisms while inhibiting the growth of others.

Selective permeability The ability to prevent the free passage of certain molecules and ions across the membrane while allowing others to pass through.

Selective toxicity The ability of an antimicrobial agent to kill microbes without causing significant damage to the host.

Semi-synthetic penicillin Natural penicillin that has been chemically modified in a laboratory.

Sense strand of RNA (+) An RNA strand that encodes information for making proteins needed by virus.

Sepsis Presence in the blood of pathogenic microorganisms or their toxins.

Sepsis syndrome Sepsis that causes altered blood flow to organs.

Septa (singular septum) Crosswalls seen in the hyphae of some molds.

Septate fungus Fungi that have septated hyphae.

Septic shock A life-threatening hypotensive event caused by endotoxins, in which blood vessels collapse.

Septicemia An infection caused by the rapid multiplication of pathogens in the blood.

Sequential assembly The mechanism in which the viral genome is inserted into already formed capsids.

Severe combined immunodeficiency syndrome (SCID) A primary immunodeficiency congenital condition in which there is no T-cell function and no B-cell function.

Severe sepsis Systemic inflammation due to infection.

Sexually transmitted urethritis An infection that presents as dysuria, urethral discharge, or both.

Shiga toxin A dangerous enterotoxin produced by *Shigella* organisms but also found in other Enterobacteriaceae that have acquired the genes for the production of the toxin.

Shigellosis Gastrointestinal disease caused by several strains of *Shigella* that invade the intestinal tract.

Shingles Sporadic disease caused by reactivation of varicella-zoster virus. It appears mostly in elderly and immunocompromised individuals.

Simple diffusion The net movement of particles from a region of high concentration to a region of lower concentration. It requires no ATP.

Simple stain A single dye used to reveal cell shapes and arrangements.

Sinusitis Inflammation of the sinuses.

Sjögren's syndrome Autoimmune disease which attacks the exocrine glands that produce tears and saliva.

SLE Autoimmune disease in which autoantibodies are produced against DNA, RNA, and proteins associated with nucleic acids. It causes damage to small blood vessels, especially in the kidneys.

Sleeping sickness Disease caused by the parasite *Trypanosoma*. It is marked by intermittent and progressive loss of consciousness.

Slime layer A thin protective structure loosely bound to the cell wall that protects some cells against drying and is sometimes used to bind cells together.

Slow viral disease A persistent viral infection.

Smallpox A formerly worldwide and serious viral infection that is reputed to have been eradicated.

Sodium thioglycolate A reducing medium used for growing anaerobic bacteria.

Spanish influenza An epidemic of influenza that occurred in 1918 and is believed to have killed 50 million people.

Specialized transduction Type of transduction (transfer of genetic material) in which the DNA being transduced is limited to one or a few genes lying adjacent to the viral insert in the host chromosome that are accidentally included with the viral insert when it is excised from the host chromosome.

Spectrophotometry A technique in which light scattering can be used to give an indication of the number of organisms in a solution.

Sphere of hydration The surrounding of an ion with water molecules. It occurs because of the nonpolar covalent bonding seen in water molecules.

Spirillum (plural spirilla) A spiral-shaped rigid bacterium.

Spirochete A corkscrew-shaped motile bacterium.

Spirochetemia Spirochetes growing in the blood.

Splenomegaly Enlargement of the spleen.

Spontaneous mutation A mutation that occurs in the absence of any agent known to cause changes in DNA.

Sporadic disease A disease that is limited to a small number of isolated cases, posing no threat to a large population.

Sporogenesis The development of a spore.

Sporotrichosis Fungal skin infection caused by *Sporothrix schenckii* and often transmitted to the body from plants.

Sporozoa A group of protozoan parasites that use both sexual and asexual reproduction during infection.

Sporozoite A malaria trophozoite present in the salivary glands of infected mosquitoes.

Sporulation The formation of a spore.

Stationary phase of bacterial growth The third of the four phases of the bacterial growth curve, in which new cells are produced at the same rate as the old cells die, leaving the number of live cells constant.

Sterilization The killing or removal of all microorganisms in a material or on an object.

Stop codon The last codon of mRNA to be translated at the ribosome, causing the release of the mRNA molecule.

Strep throat Serious infection of the pharynx by streptococcal bacteria.

Subacute disease A disease that is intermediate between acute and chronic.

Subacute sclerosing panencephalitis Viral infection of the central nervous system, usually caused by measles or rubella. It is universally fatal.

Subclinical infection An infection in which there are no apparent symptoms, either because there were insufficient organisms to produce them or because the immune response disposed of the infection before symptoms could appear.

Subunit vaccine A vaccine composed of immunogenic parts of virus derived from genetic engineering and recombinant DNA techniques.

Superantigen Powerful antigens such as bacterial toxins that activate large numbers of T cells, causing a powerful immune response that can lead to toxic shock.

Supercoiling of DNA The characteristic seen in DNA in which coils of the helix are themselves coiled.

Superficial mycosis Fungal infection that does not involve a tissue response.

Superinfection A secondary infection caused by the loss of normal microbial flora, which permits colonization of the body by pathogenic and often antibiotic-resistant microbes.

Superoxide dismutase An enzyme that converts superoxide to molecular oxygen and hydrogen peroxide.

Suppressor mutations Mutations reversing other previous mutations.

Surfactants Chemicals that reduce the surface tension of solvents. They are very effective for disrupting the plasma membrane of cells.

Symptomatic phase of HIV infection The end stage of the infection in which a variety of infections occur as a result of the loss of the patient's immune defenses.

Symptoms of disease Characteristics such as headache and nausea that can be observed or felt only by the patient. These are nonmeasurable.

Syncytia Multinucleate masses formed by the fusion of many virally infected cells into one gigantic cell.

Syndrome A combination of signs and symptoms that occur together.

Synercid antibiotics Antibiotics containing a pair of antibiotics, quinupristin and dalfopristin.

Syphilis Venereal disease caused by *Treponema pallidum* leading to many structural and cutaneous lesions.

Systemic autoimmunity Autoimmune disease in which the immune response attacks multiple organs.

Systemic infection An infection that occurs throughout the body.

Systemic lupus erythematosus *See* SLE.

T lymphocyte A thymus-derived lymphocyte involved in the adaptive immune response.

Tachycardia Rapid heartbeat.

Tapeworms Long ribbonlike helminths (worms). They are the largest of the intestinal parasites.

Tegument The viral protein layer located between the capsid and the envelope in complex icosahedral viruses.

Teichoic acid A polymer attached to peptidoglycan in Gram-positive cells.

Tetanospasmin The neurotoxic component of tetanus toxin.

Tetanus An acute and often fatal infection caused by the neurotoxin produced by *Clostridium tetani*.

Tetanus toxin A neurotoxin that inhibits the relaxation phase of muscle function.

Tetracycline An antibacterial agent that inhibits protein synthesis.

Th1 cells A subset of helper T cells involved with intracellular pathogens and dendritic cells.

Th2 cells A subset of helper T cells involved with extracellular pathogens.

Thermal death point (TDP) The lowest temperature required to kill all the organisms in a sample in 10 minutes.

Thermal death time (TDT) The shortest length of time needed to kill all organisms at a specific temperature.

Thermophiles Bacteria that grow at high temperatures above 45°C.

Thrombocytopenia A decrease in the number of platelets found in the circulating blood.

Thrombophlebitis Inflammation of veins.

Thrush Milky patches of inflammation on oral mucous membranes. It is a symptom of oral candidiasis caused by *Candida albicans*.

Tincture A solution used to carry other antimicrobial chemicals.

T-independent antigen An antigen not requiring Th2 helper T cell activity to activate B cells.

Tinea capitis Also called scalp ringworm. It is a form of ringworm in which hyphae grow in hair follicles, often leaving circular patterns of baldness.

Tinea corporis A form of ringworm found between the fingers and in the palm.

Tinea cruris Also called jock itch. It is groin ringworm that occurs in skin folds in the pubic area of the body.

Tinea pedis Also called athlete's foot. It is a foot infection in which hyphae invade the skin between the toes, causing dry scaly lesions.

Titer of antibody in serum The quantity of antibody found in serum.

TNF *see* Tumor necrosis factor.

Tolerance to self antigens A state in which the adaptive immune response no longer reacts to host antigens.

Toll-like receptors Molecules located on the surface of cells that defend the body. These receptors bind to antigens found on pathogens.

Tonsillitis Inflammation of the tonsils in particular the palatine tonsils.

Topoisomerase The enzyme that separates the strands of DNA during replication.

Toxemia The presence and spread of toxins in the blood.

Toxic shock syndrome Condition caused by certain toxigenic types of *Staphylococcus aureus*. It is often associated with the use of certain superabsorbent but abrasive tampons.

Toxoid An exotoxin that has been chemically inactivated but remains antigenic and can therefore be used to immunize against the toxin.

Tracheal toxin The toxin produced by *Bordetella pertussis*. It immobilizes and progressively destroys ciliated cells.

Tracheitis Inflammation of the trachea.

Tracheobronchitis Inflammation of the trachea and the bronchi.

Trachoma A chronic conjunctivitis of the eye that leads to scarring, corneal ulceration, and eventual blindness.

Transcription The process of producing RNA from DNA.

Transcytosis The movement of IgA across the epithelial cell barrier.

Transduction A form of transfer of genetic material from one cell to another by a virus.

Transfer RNA Type of RNA that transfers amino acids from the cytoplasm to the ribosome for placement into newly developing peptides.

Transferrin A substance that binds iron.

Transformation A form of transfer of genetic information in which naked DNA is taken up by recipient cells.

Transitional mutation A kind of point mutation in which one pyrimidine base is exchanged for the other.

Translation The synthesis of protein using the information carried by RNA.

Translational apparatus The large and small subunits of ribosomal RNA and proteins, which come together to form the intact ribosome along with transfer and messenger RNA.

Translocation protein systems Proteins in the cell that move substances out of the cell.

Transmissible spongiform encephalitis The neurological disease caused by prions.

Transposition The process in which genetic sequences can move from one location to another in the DNA.

Transposon A mobile genetic sequence that contains the genes for transposition as well as one or more genes not related to transposition.

Transverse mutations A pyrimidine base exchanged for a purine.

Treg cell A T lymphocyte that regulates the cellular and humoral immune response.

Trematode A fluke.

Trench mouth *see* Necrotizing periodontal disease.

Trichinosis Disease caused by eating inadequately prepared meat infected with *Trichinella spiralis*. It is marked by diarrhea, nausea, fever, stiffness, and pain.

Trophozoite The vegetative form of a protozoan.

Trypanosomiasis A neurological disease caused by the parasite *Trypanosoma*. It is commonly referred to as African sleeping sickness.

Trypomastigote The form of the parasite *Trypanosoma* that is found in the blood.

Tuberculosis Severe infection of the lower respiratory tract, primarily caused by *Mycobacterium tuberculosis*.

Tularemia A zoonotic disease caused by *Francisella tularensis*.

Tumor necrosis factor (TNF) A chemical cytokine seen in the inflammatory and immune response to infection.

Typhoid fever A form of enteric fever caused by *Salmonella enterica* serotype Typhi.

Typhoidal tularemia Tularemia infection with symptoms similar to those of typhoid fever.

Typhus A group of closely related rickettsial diseases all characterized by headache, chills, fever, stupor, and papulovesicular eruptions.

Ubiquitin Molecular carrier tags for materials to be taken to the proteasome for recycling.

Ulceroglandular tularemia A localized papule that forms at the inoculation site and becomes ulcerated and necrotized.

Ulcer A defect or excavation of the surface of an organ or tissue produced by sloughing off of necrotic inflammatory tissue.

Undulant fever A cycling pattern of symptoms seen in *Brucellosis* infections.

Urban rabies Rabies infections associated with domesticated animals such as dogs and cats.

Urethritis Inflammation of the urethra.

Use dilution method A method used to determine whether a chemical substance is bacteriostatic or bacteriocidal.

Vacuolating cytotoxin A circulating protein associated with *Helicobacter pylori* infections.

Vaginal candidiasis A fungal infection of the vaginal area caused by the yeast *Candida albicans*.

Vaginitis An infection of the vaginal area.

Valley fever The disease caused by the infectious form of coccidioidomycosis.

Vancomycin-resistant enterococci Also known as VREs. Enterococcal organisms that are resistant to vancomycin.

Variable region of the antibody molecule The most variable regions of the immunoglobulin molecule made up of variable regions of the heavy and light chain. It contains the antigen-binding site.

Variant Creutzfeldt–Jacob disease One of the forms of prion disease seen in humans.

Varicella-zoster virus A type of herpesvirus that causes chickenpox.

Variola major A form of smallpox with a high mortality rate (20% or higher).

Variola minor A form of smallpox with a low mortality rate.

Vasogenic edema A swelling of the brain caused by albumin entering the cerebrospinal fluid.

Vector An organism that transmits a disease-causing organism from one host to another.

Vector transmission Movement of an organism from one host to another. This can be through biological mechanisms such as bites, or through mechanical mechanisms such as shedding from the vector's body.

Vehicle transmission of disease The movement of a disease-causing organism through the use of a nonliving carrier.

Vesicles Membrane-enclosed inclusions in the cytoplasm of cells.

Vibrio toxin An enterotoxin that damages the intestinal lining.

Vincent's disease *see* Necrotizing periodontal disease.

Viral envelope The structure seen around the capsid of some viruses. It is made up of the membrane components obtained when the mature virion leaves the host cell during the process of budding release.

Viremia An infection in which viruses are transported in the blood but do not replicate there.

Virion A mature fully developed viral particle.

Virucide A chemical agent that can kill viruses.

Virulence The degree of intensity of the disease produced by a pathogen.

Virulence factors A structural or physiological characteristic that helps pathogens cause infection and disease.

Viruria Virus in the urine.

VREs *see* Vancomycin-resistant enterococci.

VRSA Vancomycin-resistant *Staphylococcus aureus*.

Wall teichoic acid The form of teichoic acid that extends through only a portion of the Gram-positive cell wall of bacteria as opposed to all the way through.

Wandering macrophages Phagocytic cells that circulate in the blood or move into tissues when microbes or other foreign materials are present.

Watery diarrhea The most common type of gastrointestinal infection symptom. It develops rapidly and results in frequent voiding.

West Nile encephalitis (meningitis) The most damaging form of West Nile virus infection with mild early symptoms similar to West Nile fever. It progresses to loss of consciousness, near coma, and hyperactive deep tendon reflexes that later diminish.

West Nile fever A less damaging form of infection seen with West Nile virus. Symptoms include headache, chills, diaphoresis (excessive sweating), lymphadenopathy, joint pain, and cold like symptoms.

West Nile virus A virus that causes an emerging infectious disease transmitted by mosquitoes. It causes problems of the central nervous system, including seizures and encephalitis.

Western equine encephalitis Type of encephalitis seen most often in the western United States. It primarily affects horses but can also infect humans.

Whooping cough A highly contagious respiratory infection caused by *Bordetella pertussis*.

Wiskott–Aldrich syndrome A disease in which there are defects in the cytoskeleton of eucaryotic cells, causing a predisposition to infection with pyrogenic bacteria.

Wobble hypothesis The theory developed by Francis Crick which states that errors in the third base of a codon can be acceptable because of the degeneracy of the genetic code.

X linked gammaglobulinemia Primary immunodeficiency disease in which there is an absence of B cells leading to low levels of antibody production, which results in increased bacterial infection.

Yeast The single cell form of fungi.

Yellow fever Disease caused by the flavivirus known as the yellow fever virus and transmitted by mosquitoes. It causes fever, chills, bleeding into the skin, headache, back pain, jaundice, and prostration.

Zone of inhibition A clear area that appears on agar in the disk-diffusion method, indicating where the agent being tested has inhibited the growth of the organism.

Zoonotic disease A disease that can be transmitted from animals to humans.

Pathogen List

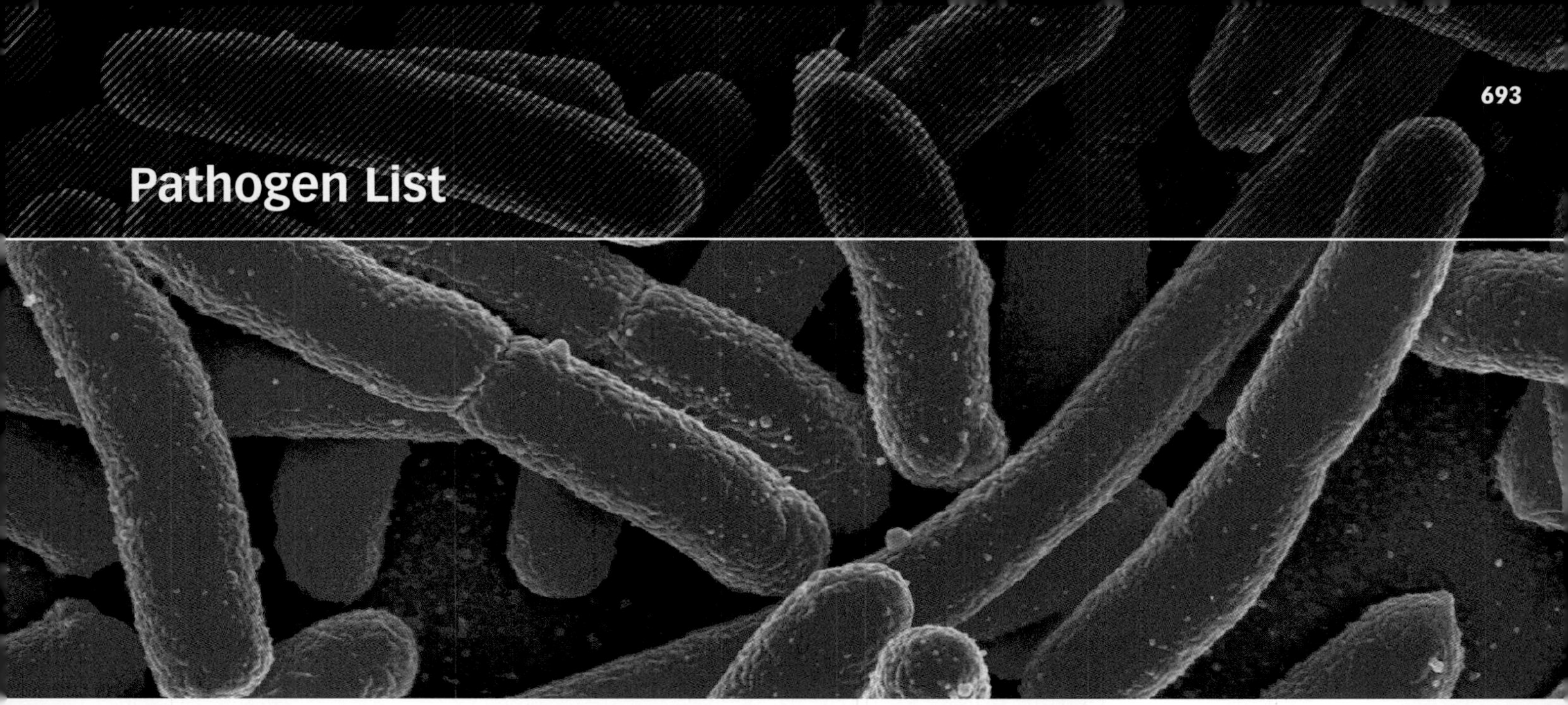

Bacteria				
Latin name	**Gram stain**	**Morphology**	**Disease**	**Chapter**
Acinetobacter	Gram –		bacteremia	25
Actinomyces species	Gram +		tooth decay	22
Bacillus anthracis	Gram +		cutaneous anthrax, inhalation anthrax	1, 5, 9, 19, 21, 26, 28
Bacillus cereus	Gram +		food poisoning	5, 22
Bacteroides	Gram –		pelvic inflammatory disease	23
Bartonella henselae	Gram –		bacillary angiomatosis, cat-scratch fever	20, 25
Bordetella pertussis	Gram –		pertussis (whooping cough)	5, 7, 21
Brucella species	Gram –		bacteremia, brucellosis	28, 25
Campylobacter jejuni	Gram –		acute intestinal disease, Campylobacter enteritis, dysentery, watery diarrhea,	18, 19, 22
Clostridium difficile	Gram +		gastrointestinal infections, superinfections,	19, 22
Clostridium botulinum	Gram +		botulism, food poisoning	4, 5, 9, 10, 18, 22, 24, 28
Clostridium perfringens	Gram +		gas gangrene, food poisoning, watery diarrhea	5, 10, 22, 26, 28
Clostridium tetani	Gram +		tetanus (lockjaw)	4, 5, 6, 18, 24
Corynebacterium species	Gram +		catheter bacteremia, intravenous-line bacteremia	7, 9, 25
Corynebacterium diphtheriae	Gram +		diphtheria	5, 8, 11, 21

The Bug Parade, with additional pronunciation guide, can be found at www.garlandscience.com/micro

Latin name	Gram stain	Morphology	Disease	Chapter
Coxiella burnetii	Gram –		atypical pneumonia, Q fever	9, 18, 21, 25, 28
Enterobacter species	Gram –	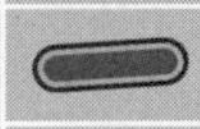	bacteremia, nosocomial pneumonia	7, 21, 25
Enterococcus species	Gram +		dental diseases, diverticular abscesses, endocarditis, infections of the bowel, appendix, and liver	6, 7, 22, 23, 25
Enterococcus faecalis	Gram +		endocarditis	20
Escherichia coli	Gram –	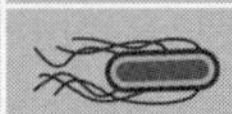	cystitis, meningitis, opportunistic infections, nephritis, urethritis, watery diarrhea	1, 7, 9, 11, 18, 19, 20, 22, 23, 24
Escherichia coli O157:H7	Gram –	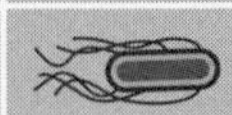	gastrointestinal infections, hemolytic uremic syndrome, renal failure	1, 5, 8, 20, 22, 28
Francisella tularensis	Gram –	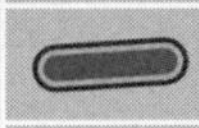	tularaemia	25, 28
Gardnerella vaginalis	Gram variable	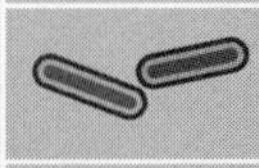	bacterial vaginitis	23
Haemophilus ducreyi	Gram –	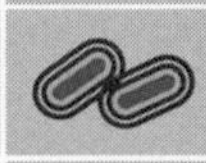	chancroid	23
Haemophilus influenzae	Gram –		acute purulent meningitis	5, 6, 10, 21, 24
Haemophilus influenzae type B	Gram –		bacteremia, meningitis, pneumonia	7, 24, 25
Helicobacter pylori	Gram –	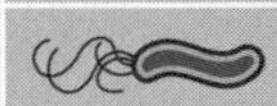	chronic gastritis, gastric cancer, gastroduodenal ulcer, peptic ulcers	5, 7, 8, 9, 10, 15, 20, 22
Klebsiella pneumoniae	Gram –	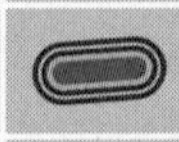	pneumonia	5, 7, 8
Legionella pneumophila	Gram –		atypical pneumonia, legionellosis (Legionnaires' disease)	8, 20, 21
Leptospira species	Gram –		bacteremia, chronic meningitis	24, 25
Leptospira interrogans	Gram –		leptospirosis	9
Listeria species	Gram +	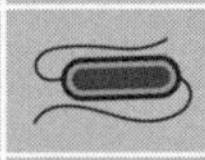	bacteremia	19, 25
Listeria monocytogenes	Gram +	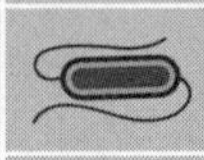	listeriosis	17, 18, 24
Mycobacterium leprae	Gram +	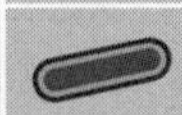	leprosy	4, 7, 9, 10, 15, 17, 19
Mycobacterium tuberculosis	Gram +	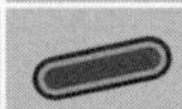	chronic meningitis, infections of the central nervous system, tuberculosis	1, 4, 5, 7, 8, 9, 10, 15, 16, 17, 18, 19, 20, 21, 24
Mycoplasma	no Gram reaction		mild pneumonia	2, 4, 8, 19
Mycoplasma pneumoniae	no Gram reaction		*Mycoplasma* pneumonia	5, 21
Neisseria species	Gram –		bacteremia	7, 11, 15, 16, 17

The Bug Parade, with additional pronunciation guide, can be found at www.garlandscience.com/micro

Latin name	Gram stain	Morphology	Disease	Chapter
Neisseria gonorrhoeae	Gram –		bacteremia, gonorrhea, neonatorum (neonatal gonorrheal ophthalmia), ophthalmia	4, 5, 7, 9, 18, 19, 20, 23, 25, 26
Neisseria meningitides	Gram –		acute purulent meningitis, bacteremia, meningitis	5, 24, 25
Nocardia	Gram +	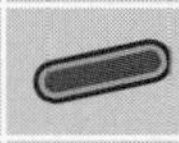	nocardiasis	4
Proteus species	Gram –	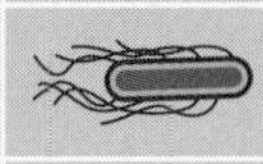	nosocomial urinary tract infections	7
Pseudomonas species	Gram –	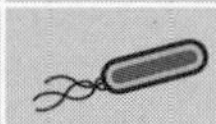	catheter bacteremia, intravenous-line bacteremia	11, 18, 19, 20, 23, 25
Pseudomonas aeruginosa	Gram –		aggressive keratitis, folliculitis, nosocomial infections	4, 6, 10, 19, 21, 26
Salmonella species	Gram –	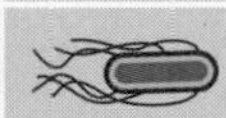	dysentery, food poisoning, gastrointestinal infections,	4, 8, 9, 11, 15, 16, 17, 18, 19, 22, 28
Salmonella enterica	Gram –	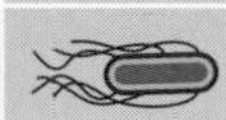	bacteremia, gastroenteritis, salmonellosis, sepsis	5, 22
Salmonella enterica serovar Typhi	Gram –		bacteremia, typhoid fever	5, 7, 10, 18, 22, 25
Serratia marcescens	Gram –	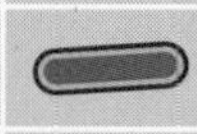	respiratory disease	28
Shigella species	Gram –	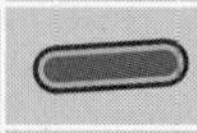	shigellosis (bacillary dysentery)	4, 5, 7, 9, 11, 28
Shigella boydii	Gram –	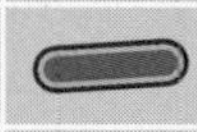	dysentery	22
Shigella dysenteriae	Gram –		bacillary dysentery, bacteremia, food poisoning, gastrointestinal infections,	22, 25
Shigella flexneri	Gram –	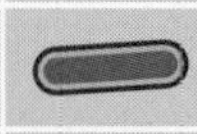	dysentery	22
Shigella sonnei	Gram –	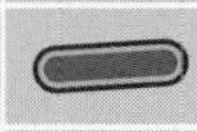	dysentery	22
Staphylococcus species	Gram +		impetigo	11, 15, 26
coagulase-negative staphylococci	Gram +		endocarditis	25
Staphylococcus aureus	Gram +		boils (furuncles), carbuncles, catheter bacteremia, endocarditis, folliculitis, food poisoning, intravenous-line bacteremia, nosocomial pneumonia, scalded skin syndrome, skin infections, staphylococcal pneumonia, toxic shock	1, 4, 5, 6, 7, 9, 10, 18, 19, 20, 21, 22, 25, 26
methicillin-resistant *Staphylococcus aureus* (MRSA)	Gram +		antibiotic resistant nosocomial infections	1, 20, 21, 23
vancomycin-resistant *Staphylococcus aureus* (VRSA)	Gram +		antibiotic resistant nosocomial infections	1, 20

The Bug Parade, with additional pronunciation guide, can be found at www.garlandscience.com/micro

Latin name	Gram stain	Morphology	Disease	Chapter
Staphylococcus epidermidis	Gram +		catheter bacteremia, intravenous-line bacteremia	7, 10, 25
Streptococcus species	Gram +		dental diseases, diverticular abscesses, infections of the bowel, appendix, and liver	15, 19, 22, 23
group A streptococci	Gram +		erysipelas, impetigo, necrotizing fasciitis	26
group B streptococci	Gram +		neonatal sepsis, post partum infections	24
β-hemolytic streptococci	Gram +		bacteremia	25
Streptococcus mutans	Gram +		tooth decay	5, 22
Streptococcus pneumoniae	Gram +		acute otitis media, acute purulent meningitis, bacteremia, meningitis, middle ear infections, pneumonia, sinusitis	4, 5, 6, 7, 9, 17, 20, 21, 24, 25
Streptococcus pyogenes	Gram +		abscesses on the tonsils, acute tonsillitis, endocarditis, erysipelas, necrotizing fasciitis, pharyngitis, pneumonia, scarlet fever, strep throat, toxic shock	4, 5, 9, 19, 21, 26
Streptococcus salivarius	Gram +		tooth decay	
Streptococcus viridans	Gram +		endocarditis	25
Treponema pallidum	Gram –		chronic meningitis, syphilis	4, 5, 7, 9, 10, 17, 23, 24
Ureaplasma urealyticum	No Gram reaction		non-gonococcal urethritis	23
Vibrio cholerae	Gram –	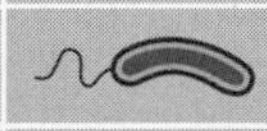	cholera	5, 9, 20, 22, 28
Vibrio cholerae 0139	Gram –		cholera	8, 20
Vibrio parahaemolyticus	Gram –	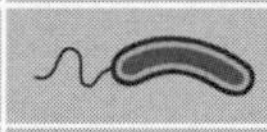	food poisoning, watery diarrhea	22
Yersinia enterocolitica	Gram –	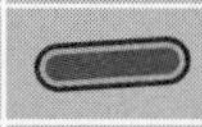	gastrointestinal infections	11, 19, 22
Yersinia pestis	Gram –		plague (bubonic plague, pneumonic plague)	1, 5, 6, 7, 15, 25, 28

Viruses				
Latin name	**Nucleic Acid Type**	**Morphology**	**Disease**	**Chapter**
adenovirus	dsDNA	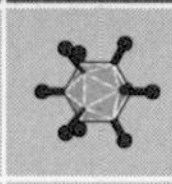	acute respiratory disease, diarrhea, gastrointestinal infections	12, 13, 22
alphaviruses	+ ssRNA		encephalitis	28
arboviruses	+ ssRNA + ssRNA – ssRNA	 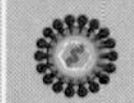 	hemorrhagic fever, hepatic necrosis, West Nile fever (fever and encephalitis)	8, 24, 25
arena virus	– ssRNA	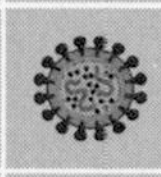	zoonotic central nervous system infections	13
Arenaviridae	– ssRNA	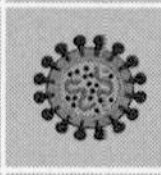	Lassa fever	8
avian influenza virus	– ssRNA		avian influenza (bird flu)	8
Bunyaviridae	– ssRNA	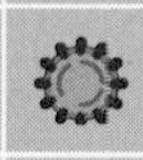	hantavirus pulmonary syndrome	8
calicivirus	+ ssRNA		diarrhea, gastrointestinal infections	22
coronavirus	+ ssRNA	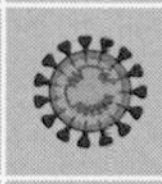	acute mild respiratory infections	13
cowpox virus	dsDNA	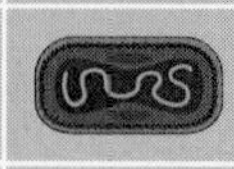	cowpox	13, 16
coxsackievirus A16	+ ssRNA		hand–foot–and–mouth disease	13
cytomegalovirus	dsDNA		acute polyneuritis, mononucleosis syndrome	13, 17, 19, 23, 24, 25
eastern equine encephalitis virus	+ ssRNA		encephalitis	24, 25
Ebola virus	– ssRNA	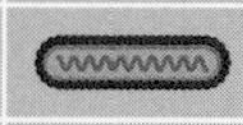	hemorrhagic fever	4, 8, 12, 25, 28
enterovirus	+ ssRNA		diarrhea, gastrointestinal infections, meningitis, persistent enterovirus infection of the immunodeficient	13, 22, 24
Epstein–Barr virus	dsDNA		acute polyneuritis, Burkitt's lymphoma, infectious mononucleosis, lymphoproliferative infections in immunocompromised patients, nasopharyngeal carcinoma	12, 13, 16, 25
flavivirus	+ ssRNA		hemorrhagic fever, yellow fever	6, 8
hantavirus	– ssRNA		hanta fever, hantavirus pulmonary syndrome	8, 19, 21

Latin name	Nucleic Acid Type	Morphology	Disease	Chapter
hepatitis A virus	dsDNA		food poisoning, hepatitis A	5, 22
hepatitis B virus	dsDNA		cirrhosis, hepatitis B, hepatocellular carcinoma	5, 6, 13, 22, 23
hepatitis C virus	+ ssRNA		cirrhosis, hepatitis C, hepatocellular carcinoma	5, 13, 15, 22
hepatitis G virus	+ ssRNA	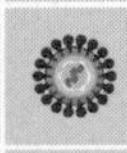	hepatitis	22
herpesvirus	dsDNA		meningitis	12, 19, 24
herpes simplex virus type 1	dsDNA		cold sores (fever blisters), ophthalmic herpes	13, 15, 17, 19, 23, 24, 26
herpes simplex virus type 2	dsDNA		genital herpes	5, 13, 15, 17, 19, 23, 24
human immunodeficiency virus (HIV)	+ ssRNA		acquired immunodeficiency syndrome (AIDS)	1, 4, 5, 6, 8, 12, 13, 15, 17, 19, 23, 24
human papillomavirus	dsDNA	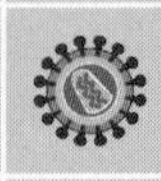	cervical cancer, warts (dermal, genital)	16, 23, 26
human lymphotropic virus	+ ssRNA	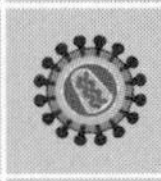	lymphoma, T cell leukemia	13
influenza virus (influenza A and B)	– ssRNA	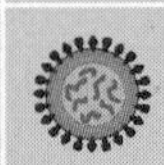	influenza	1, 5, 7, 8, 12, 13, 15, 16, 19,21
Junin virus	– ssRNA	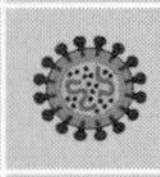	hemorrhagic fever	8
Lassa virus	– ssRNA	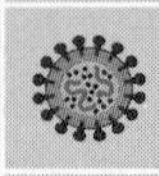	Lassa fever	28
Marburg virus	– ssRNA	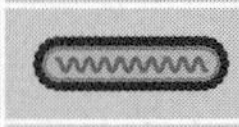	hemorrhagic fever	8, 25, 28
measles virus	– ssRNA		measles, subacute sclerosing panencephalitis	5, 13, 24
mumps virus	– ssRNA		meningitis, mumps	5, 13, 24
Nipah virus	– ssRNA		central nervous system infections	8
Norwalk viruses	+ ssRNA		intestinal infections	13
papillomavirus	dsDNA	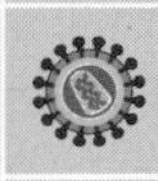	carcinomas, papillomas, skin warts	12, 13
parainfluenza virus	– ssRNA		parainfluenza	21

Latin name	Nucleic Acid Type	Morphology	Disease	Chapter
parvovirus	ssDNA		erythema infectiosum	13
poliovirus	+ ssRNA		poliomyelitis (abortive, aseptic, paralytic)	5, 12, 13, 24
polyoma virus BK	dsDNA		hemorrhagic cystitis	13
polyoma virus JC	dsDNA		progressive multifocal leukoencephalopathy	13
poxviruses	dsDNA		smallpox	12
rabies virus	– ssRNA		rabies	5, 6, 24
respiratory syncytial virus	– ssRNA		respiratory infections	19, 21
rhabdovirus	– ssRNA		rabies	8
rhinovirus	+ ssRNA		common cold	12, 13, 21
Rift Valley fever virus	– ssRNA		Rift Valley fever	8, 28
rotavirus	dsRNA	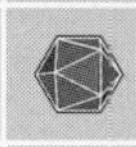	diarrhea, gastrointestinal infections	22
rubella virus	+ ssRNA		progressive rubella panencephalitis, rubella (German measles)	5, 13, 24, 26
rubeola virus	– ssRNA		measles (five–day measles, hard measles)	26
SARS virus (coronavirus)	+ ssRNA	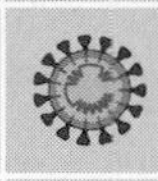	severe acute respiratory syndrome (SARS)	8, 20
St Louis encephalitis virus	+ ssRNA	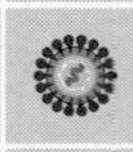	encephalitis	25
varicella–zoster virus (VZV)	dsDNA		chickenpox, postherpetic neuralgia, shingles	5, 13, 15, 17, 19, 24, 26
variola minor; variola major virus	dsDNA	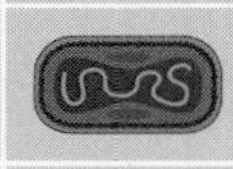	smallpox	12, 16, 26, 28
Venezuelan encephalitis virus	+ ssRNA		central nervous system infections	13
western equine encephalitis virus	+ ssRNA		encephalitis	24, 25
West Nile virus	+ ssRNA	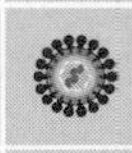	West Nile fever (fever and encephalitis)	6, 8, 24
yellow fever virus	+ ssRNA	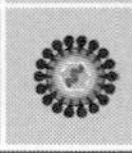	hemorrhagic fever, hepatic necrosis	8, 13

The Bug Parade, with additional pronunciation guide, can be found at www.garlandscience.com/micro

Protozoan Parasites And Worms		
Latin name	**Disease**	**Chapter**
Acanthamoeba species	chronic meningitis, parasitic amebic meningoencephalitis	24
Ancylostoma duodenale	hookworm infection	22
Ascaris lumbricoides	ascariasis	14, 19
Brugia malayi	filariasis (lymphadenitis, lymphadenopathy, elephantiasis)	25
Burkholderia mallei	glanders	28
Burkholderia pseudomallei	melioidosis	28
Clonorchis sinensis (Asian liver fluke)	clonorchiasis	14
Echinococcus species	hydatidosis	19
Entamoeba histolytica	amebiasis	14
Enterobius vermicularis (pin worm)	enterobiasis	14, 19
Giardia duodenalis	gastrointestinal infections, giardiasis, watery diarrhea	22
Giardia intestinales (lamblia)	giardiasis	5, 14
Histoplasma capsulatum	chronic meningitis, histoplasmosis	14, 17, 21, 24
Hortaea werneckii	tinea nigra	14
Leishmania species	cutaneous leishmaniasis (skin lesions)	14, 15, 17, 19, 26
Loa loa	Loaiasis	26
Naegleria species	parasitic amebic meningoencephalitis	24
Necator americanus (hookworm)	hookworm disease	14, 22
Onchocerca volvulus	onchocerciasis (river blindness)	26
Paragonimus species (lung fluke)	paragonimiasis	14
Pediculus humanus capitis (head louse)	head lice	26
Pediculus humanus corporis (body louse, clothes louse)	body lice, genital lice (crabs)	26
Plasmodium species	malaria	5, 6, 8, 15, 17, 19
Plasmodium falciparum	malaria	14
Plasmodium vivax	malaria	14
Schistosoma species	schistosomiasis	14, 25
Sporothrix species	sporotrichosis	14
Strongyloides species	strogyloidiasis	19
Strongyloides stercoralis (roundworm)	gastrointestinal parasitic infection	14
Taenia saginata (beef tapeworm)	beef tapeworm	14
Taenia solium	pork tapeworm	19
Toxoplasma species	toxoplasmosis	19

Latin name	Disease	Chapter
Toxoplasma gondii	chronic meningitis, toxoplasmosis,	6, 14, 17, 24
Trichinella species	trichinosis	14
Trichinella spiralis	chronic meningitis, food poisoning, myalgia	14, 18, 22, 24
Trichomonas vaginalis	trichomoniasis, vaginitis	14, 19, 23
Trichophyton rubrum	athlete's foot, ringworm	14
Trichuris trichiura (whipworm)	whipworm infection	14, 19, 22
Trypanosoma brucei brucei, Trypanosoma brucei gambiense, Trypanosoma brucei rhodesiense	trypanosomiasis (sleeping sickness)	6, 14, 19, 24, 25
Trypanosoma cruzi	Chagas' disease, heart and gastrointestinal lesions	6, 14, 25
Wuchereria bancrofti	filariasis (lymphadenitis, lymphadenopathy, elephantiasis)	25

Chlamydia		
Latin name	**Disease**	**Chapter**
Chlamydia species	genital tract infections, lymphogranuloma venereum, respiratory and cardiovascular disease, trachoma,	4, 7, 19
Chlamydia pneumoniae	atypical pneumonia, chlamydial pneumonia	21
Chlamydia psittaci	atypical pneumonia, psittacosis (ornithosis)	21, 28
Chlamydia trachomatis	cervicitis, epididymitis, lymphogranuloma venereum, non-gonococcal urethritis, salpingitis, trachoma	5, 6, 18, 23, 26

Rickettsia		
Latin name	**Disease**	**Chapter**
Borrelia species	bacteremia, relapsing fever	6, 25
Borrelia burgdorferi	chronic meningitis, Lyme disease	6, 7, 8, 9, 20, 24, 25
Borrelia recurrentis	relapsing fever	25
Rickettsia species	vector transmitted diseases	4, 7, 19
Rickettsia akari	rickettsial pox	25
Rickettsia prowazekii	typhus	6, 25, 28
Rickettsia rickettsii	Rocky Mountain spotted fever	5, 6, 25
Rickettsia typhi	typhus	25

Figure Acknowledgments

Cover Images

Front: © Science Source / Science Photo Library.

Back: Top Row Center Left: © Biology Media / Science Photo Library

Top Row Center Right: Courtesy of the National Museum of Health and Medicine, Armed Forces Institute of Pathology, Washington, D.C., United States.

Top Row Far Right: © Andrew Crump / Science Photo Library

Middle Row Far Left: © Dennis Kunkel.

Middle Row Center Left: From Centers for Disease Control Public Health Image Library / James Gathany.

Middle Row Center Right: © NIAID / CDC / Science Photo Library.

Middle Row Far Right: From Centers for Disease Control DPDx Parasite Image Library.

Bottom Row Center Left: From Centers for Disease Control Public Health Image Library.

Bottom Row Center Right: Reprinted with kind permission from *Diseases of the Skin*, 2006. © Elsevier.

Learning Skills: Using Your Brain Effectively

Opening Image: © 2009 Courtesy of Elsa Fuster-Mears and Matthew McClements.

Chapter 1

Chapter Opening Image: Reprinted with permission given by the Special Programme for Research & Training in Tropical Diseases (TDR), TDR Image Library, World Health Organization / Andy Crump.

Figure 1.1: Reprinted with permission given by the Special Programme for Research & Training in Tropical Diseases (TDR), TDR Image Library, World Health Organization / Andy Crump.

Figure 1.4: Reprinted with permission given by the Special Programme for Research & Training in Tropical Diseases (TDR), TDR Image Library, World Health Organization / Andy Crump.

Figure 1.5 Panel A and Panel B: Courtesy of the National Museum of Health and Medicine, Armed Forces Institute of Pathology, Washington, D.C., United States.

Chapter 2

Chapter Opening Image: © Charles D. Winters / Science Photo Library.

Chapter 3

Chapter Opening Image: Courtesy of Brian Oates.

Chapter 4

Figure 4.4: Panel B: From Centers for Disease Control Public Health Image Library / Courtesy of Dr. Richard Facklam; Panel C: From Centers for Disease Control Public Health Image Library / Courtesy of Dr. Mike Miller.

Figure 4.5: Panel B: Reprinted with permission from *Nature Reviews Microbiology*, **2.6.** Colonization control. © 2004 Macmillan Publishers Ltd. ; Panel C © Centers for Disease Control Public Health Image Library.

Figure 4.6: Panel B: From Centers for Disease Control Public Health Image Library / Courtesy of Janice Haney Carr and Jeff Hageman, M. H. S; Panel C: From Centers for Disease Control Public Health Image Library / Courtesy of Thomas F. Sellers and Emory University.

Figure 4.8: From Centers for Disease Control Public Health Image Library / Courtesy of Larry Stauffer of Oregon State Public Health Laboratory.

Figure 4.9: From Centers for Disease Control Public Health Image Library / Courtesy of Dr. William A. Clark.

Figure 4.10: From Centers for Disease Control Public Health Image Library / Courtesy of Dr. George P. Kubica.

Figure 4.11: From Centers for Disease Control Public Health Image Library / Courtesy of Larry Stauffer of Oregon State Public Health Laboratory.

Figure 4.17: Reprinted with permission from *Trends in Microbiology*, **6.6.** Bacterial entry into epithelial cells: the paradigm of *Shigella*. © 1996 Elsevier.

Figure 4.18: Reprinted with permission from *Cell Research*, **17.6.** Cilia containing 9 2 structures grown from immortalized cells, © 2007 Nature Publishing Group, Macmillan Publishers, Ltd.

Figure 4.19: © NIBSC / Science Photo Library.

Figure 4.24: Panel A: Reprinted with permission from *Cell*, **68.4.** Caveolin, a protein component of caveolae membrane coats. © 1992 Elsevier; Panel B: Reprinted with permission *Phagocytic Mechanisms in Health and Disease* © 1972 Intercontinental Medical Book Corp, digitized by Thieme Publishers.

Figure 4.25: © Eye of Science / Science Photo Library.

Figure 4.27: Courtesy of Lelio Orci, University of Geneva.

Chapter 5

Chapter Opening Image: © Professor P. M. Motta et al / Science Photo Library.

Figure 5.5: © Gilette Corporation / Science Photo Library.

Figure 5.7: © Professors P. Motta and F. Caprino / University "La Sapienze", Rome / Science Photo Library.

Figure 5.8: © Professor P. M. Motta et al / Science Photo Library.

Figure 5.9: © Professor P. M. Motta et al / Science Photo Library.

Figure 5.10: © David Scharf / Science Photo Library.

Figure 5.11: From Centers for Disease Control Public Health Image Library / Courtesy of Dr. David Cox.

Figure 5.15: Reprinted with kind permission from *Diseases of the Skin*, 2006. © Elsevier.

Figure 5.18: From National Center for immunization and Respiratory Diseases, Division of Viral Diseases, Centers for Disease Control.

Figure 5.19: From Centers for Disease Control Public Health Image Library / Courtesy of Dr. Daniel P. Perl.

Figure 5.20: From Centers for Disease Control Public Health Image Library/ Courtesy of Dr. Edwin P. Ewing, Jr.

Chapter 6

Chapter Opening Image: From Centers for Disease Control Public Health Image Library.

Figure 6.2: © Dr. Gary Settles / Science Photo Library Public Health Image Library.

Figure 6.5: Centers for Disease Control Public Health Image Library.

Chapter 7

Chapter Opening Image: © Andy Crump / Science Photo Library.

Figure 7.4: Reprinted with permission from *Nature Reviews Microbiology*, **3.1.** Tuberculosis – metabolism and respiration in the absence of growth. © 2005 Nature Publishing Group, Macmillan Publishers, Ltd.

Chapter 8

Chapter Opening Image: Courtesy of the National Museum of Health and Medicine, Armed Forces Institute of Pathology, Washington, D.C., United States.

Figure 8.3: From Centers for Disease Control Public Health Image Library / Courtesy of Dr. Amanda Loftis, Dr. William Nicholson, Dr. Will Reeves, Dr. Chris Paddock.

Figure 8.4: Reprinted with kind permission from *Nature Medicine*, **10.** Molecular constraints to interspecies transmission of viral pathogens. © 2004 Nature Publishing Group, Macmillan Publishers, Ltd.

Figure 8.5: From Centers for Disease Control Public Health Image Library/ Courtesy of Dr. Thomas Hooten.

Figure 8.8: Reprinted with permission from *Nature Medicine,* **10.12s.** Exotic emerging viral diseases: progress and challenges. © 2004 Nature Publishing Group, Macmillan Publishers, Ltd.

Figure 8.10 Panel A: © Science Photo Library; Panel B: Courtesy of the National Museum of Health and Medicine, Armed Forces Institute of Pathology, Washington, D.C., United States.

Chapter 9

Chapter Opening Image: © Thomas Deerinck, NCMIR / Science Photo Library.

Figure 9.5 Panel A: © Dr. Immo Rantala / Science Photo Library; Panel B: From Centers for Disease Control Public Health Image Library; Panel C: Courtesy of Professor Bill Costerton.

Figure 9.6 Panel A: © Thomas Deerinck, NCMIR / Science Photo Library; Panel B: © Dennis Kunkel Microscopy, Inc.

Figure 9.7 Panel A: From Centers for Disease Control Public Health Image Library/ NCID / Rob Weyant; Panel C: Reprinted with permission from *Journal of Bacteriology*, **88:** 17. Electron Microscopy of Axial Fibrils,Outer Envelope, and Cell Division of Certain Oral Spirochetes. © 1964 American Society of Microbiology.

Figure 9.8 Panel A: From Centers for Disease Control Public Health Image Library / Courtesy of Dr. David Cox; Panel B: From Centers for Disease Control Public Health Image Library.

Figure 9.9: © Kwangshin Kim / Science Photo Library.

Figure 9.12 Panel A: From Centers for Disease Control Public Health Image Library; Panel B: From Centers for Disease Control Public Health Image Library / Courtesy of Janice Carr; Panel C: From Centers for Disease Control Public Health Image Library / Courtesy of Dr. William A. Clark; Panel D: From Centers for Disease Control Public Health Image Library.

Figure 9.22: © Biophoto Associates / Science Photo Library.

Figure 9.23: ©Dr. Gopal Murti / Science Photo Library.

Figure 9.24 Panel B: © CNRI / Science Photo Library

Chapter 10

Chapter Opening Image: © Eye of Science / Science Photo Library.

Figure 10.2 Panel A: From Centers for Disease Control Public Health Image Library / Courtesy of Dr. Gavin Hart and Dr. N. J. Fiumara; Panel B: From Centers for Disease Control Public Health Image Library.

Figure 10.3: Reprinted with permission given by the Special Programme for Research & Training in Tropical Diseases (TDR), TDR Image Library, World Health Organization / Dr. Colin McDougall.

Figure 10.4: © Eye of Science / Science Photo Library.

Figure 10.8: Reprinted with permission given by American Society of Microbiology, Microbe Library Image Website (http://www.microbelibrary.org/).

Figure 10.9 Panel A: Reprinted with permission given by American Society of Microbiology, Microbe Library Image Website (http://www.microbelibrary.org/); Panel B: Reprinted with permission given by American Society of Microbiology, Microbe Library Image Website (http://www.microbelibrary.org/).

Figure 10.10: From Centers for Disease Control Public Health Image Library / Courtesy of Dr. Richard Facklam.

Figure 10.11 Panel B: © Dr Kari Lounatmaa / Science Photo Library.

Figure 10.14 Panel A: ©Simon Fraser / Science Photo Library; Panel B: Appalachian Farming Systems Research Center, Beaver, West Virgina, Agricultural Research Service, United States Department of Agriculture, United States Government.

Figure 10.16: From Centers for Disease Control Public Health Image Library / Courtesy.of Larry Stauffer, Oregon State Public Health Laboratory.

Chapter 11

Chapter Opening Image: © Dennis Kunkel Microscopy, Inc.

Figure 11.31: © Dennis Kunkel Microscopy, Inc.

Chapter 12

Chapter Opening Image: © Eye of Science / Science Photo Library.

Figure 12.3 Panel A: © Dr. Klaus Boller / Science Photo Library.

Figure 12.5 Panel A: © Eye of Science / Science Photo Library.

Figure 12.17: © CNRI / Science Photo Library.

Figure 12. 18: From *Science* Sept 12, 1997. STATs and Gene Regulation. © 1997 Reprinted with permission from AAAS (The American Association for the Advancement of Science).

Figure 12. 19: Reprinted with permission from *Journal of Virology*, **73.11:** 8. Observation of measles virus cell-to-cell spread in astrocytoma cells by using a green fluorescent protein-expressing recombinant virus. ©1999 American Society for Mirobiology.

Chapter 13

Chapter Opening Image: © CNRI / Science Photo Library.

Figure 13.3: St. Bartholomew's Hospital / Science Photo Library.

Figure 13.7 Top Panel: Reprinted with permission from *Trends in Microbiology* **9.2.** Emerging Infections, Volume 4. © 2001 Elsevier.

Figure 13.9: From Centers for Disease Control Public Health Image Library / Courtesy of Dr. K. L. Hermann.

Figure 13.10 Panel A: © CNRI / Science Photo Library; Panel B: From Centers for Disease Control Public Health Image Library / Courtesy of Dr. Daniel P. Perl.

Figure 13.11: From Centers for Disease Control Public Health Image Library / Courtesy of Dr. Hermann.

Figure 13.13: From Centers for Disease Control Public Health Image Library / Courtesy of Dr. N. J. Flumara and Dr. Gavin Hart.

Figure 13.14: © Kent Wood / Science Photo Library

Figure 13.15: From Centers for Disease Control Public Health Image Library.

Chapter 14

Chapter Opening Image: © D. Phillips / Science Photo Library

Figure 14.1: © D. Phillips / Science Photo Library

Figure 14. 2: From Centers for Disease Control DPDx Parasite Image Library.

Figure 14.3: From Centers for Disease Control Public Health Image Library.

Figure 14.4: © E. R. Degginger / Science Photo Library.

Figure 14.5: © London School of Hygiene & Tropical Medicine / Science Photo Library.

Figure 14.6: © Eye of Science / Science Photo Library.

Figure 14.8: Reprinted with permission from *Nature Medicine* © 2004. Differentiating the pathologies of cerebral malaria by postmortem parasite counts. Nature Publishing Group, Macmillan Publishers Ltd.

Figure 14.9: © Martin Dohrn / Science Photo Library.

Figure 14.10: © Dr. M. A. Ansary / Science Photo Library.

Figure 14.11: Reprinted with permission from the *New England Journal of Medicine*, **328:** 927. Photo by Martin Weber. Images in Clinical Medicine – Pinworms. © 1993 Massachusetts Medical Society.

Figure 14.12: From Centers for Disease Control Public Health Image Library.

Figure 14.13: From Centers for Disease Control Public Health Image Library.

Figure 14.14: From Centers for Disease Control Public Health Image Library / Courtesy Dr. Sulzer.

Figure 14.15: Courtesy of the Aspergillus Trust: www.aspergillus.ork.uk.

Figure 14.16: From Centers for Disease Control Public Health Image Library.

Figure 14.17: From Centers for Disease Control Public Health Image Library / Courtesy of Dr. Libero Ajello.

Figure 14.18: From Centers for Disease Control Public Health Image Library / Courtesy of Dr. Lucille K. Georg.

Figure 14.19: From Centers for Disease Control Public Health Image Library.

Figure 14.20: © Reprinted with permission from www.doctorfungus.org.

Figure 14.21: © Reprinted with permission from www.doctorfungus.org.

Figure 14.22: From Centers for Disease Control Public Health Image Library / Susan Lindsley, VD.

Chapter 15

Chapter Opening Image: © Photo Insolite Realite / Science Photo Library.

Figure 15.1: © Andrew Syred / Science Photo Library.

Figure 15.2: © D. Phillips / Science Photo Library.

Figure 15.3: © Photo Insolite Realite / Science Photo Library.

Figure 15.6: © David Becker / Science Photo Library.

Figure 15.7: Courtesy of Anthony Butterworth.

Figure 15.8 Panel A: ©Biophoto Associates / Science Photo Library; Panel B: © Science Photo Library.

Figure 15.10: © Professor P. Motta / Department of Anatomy / University "La Sapienza", Rome / Science Photo Library.

Figure 15.16: Courtesy of Dr. Dorothy F. Bainton, M.D.

Figure 15.17 Panel A: © Dr. Kari Lounatmaa / Science Photo Library; Panel B: Biology Media / Science Photo Library.

Figure 15.19 Panel A: Courtesy of Professor Sucharit Bhakdi, University Medical Center, Mainz, Germany; Panel B: © Schreiber et al, 1979. Originally published in *The Journal of Experimental Medicine*, 149: 870-882.

Chapter 16

Chapter Opening Image: Courtesy of Professor Ann Dvorak.

Figure 16.3: © David Scharf / Science Photo Library.

Figure 16.4: © Dr. Olivier Schwartz, Institute Pasteur / Science Photo Library

Figure 16.6: Reprinted with permission from *Immunology Today* **20.3:** 11. Regional specialization in the mucosal immune system: what happens in the microcompartments? © 1999 Elsevier.

Figure 16.13: Courtesy of Professor Ann Dvorak.

Chapter 17

Chapter Opening Image: © Authors: Dominika Rudnicka; Nathalie Sol-Foulon; Olivier Schwartz / Institut Pasteur.

Figure 17.1 Panel A: Courtesy Hans Gelderblom, Berlin.

Figure 17.3: © Authors: Dominika Rudnicka; Nathalie Sol-Foulon; Olivier Schwartz / Institut Pasteur.

Figure 17.7: © Department of Medical Photography, St Stephen's Hospital, London / Science Photo Library.

Figure 17.9: Dr. P Marazzi / Science Photo Library.

Figure 17.11: Courtesy of Irene Visintin, Dr. Gil Mor Laboratory, Department of Obstetrics and Gynecology, Yale University.

Figure 17.12: Courtesy of Euan Tovey, Woolcock Institute of Medical Research, Australia.

Figure 17.14: Courtesy of Professor Barry Kay, Imperial College London, England.

Chapter 18

Chapter Opening Image: © Health Protection Agency / Science Photo Library

Figure 18.5: © Visuals Unlimited.

Figure 18.9: © Phototake.

Chapter 19

Chapter Opening Image: From Centers for Disease Control Public Health Image Library / Courtesy of Gilda L. Jones.

Figure 19.7: From Centers for Disease Control Public Health Image Library / Courtesy of Gilda L. Jones.

Figure 19.8: From Centers for Disease Control Public Health Image Library / Dr. Richard Facklam.

Chapter 20

Chapter Opening Image: Courtesy of http://www.morguefile.com.

Chapter 21

Chapter Opening Image: © Dennis Kunkel Microscopy, Inc.

Figure 21.4: © Steve Gschmeissner / Science Photo Library.

Figure 21.6: From Centers for Disease Control Public Health Image Library / Courtesy of Dr. Heinz F. Eichenwald.

Figure 21.8: © Visuals Unlimited.

Figure 21.9: From Centers for Disease Control Public Health Image Library.

Figure 21.10: © A. D. Dowsett / Science Photo Library.

Figure 21.12: From Centers for Disease Control Public Health Image Library / Courtesy of Dr. George P. Kubica.

Figure 21.13: From Centers for Disease Control Public Health Image Library.

Figure 21.14: From Centers for Disease Control Public Health Image Library / Dr. LaForce.

Figure 21.15: CDC / Science Photo Library.

Figure 21.16 Reprint with permission from *Cell*, **36.1**:27-33. Phagocytosis of the Legionnaires' disease bacterium (Legionella pneumophila) occurs by a novel mechanism: Engulfment within a pseudopod coil. © 1984 Elsevier. Kind permission also provided by the author, M. A. Horwitz.

Figure 21.17 Panel B: © Dennis Kunkel Microscopy, Inc.

Figure 21.18: From Centers for Disease Control Public Health Image Library / Courtesy of Cynthia Goldsmith and Luanne Elliott.

Figure 21.19: From Centers for Disease Control Public Health Image Library / Courtesy of Dr. Russell K. Byrnes.

Figure 21.20: From Centers for Disease Control Public Health Image Library / Courtesy of Dr. Libero Ajello.

Figure 21.21: © Michael Abbey / Science Photo Library.

Figure 21.22: From Centers for Disease Control Public Health Image Library / Courtesy of Dr. Libero Ajello.

Chapter 22

Chapter Opening Image: From Centers for Disease Control Public Health Image Library / Courtesy of Janice Carr.

Figure 22.1: Reprinted with permission from *Clinical Bacteriology* by J. Keith Struthers and Roger P. Westran, © 2003. Manson Publishing, Ltd.

Figure 22.3: Reprinted with permission from *Clinical Bacteriology* by J. Keith Struthers and Roger P. Westran, © 2003. Manson Publishing, Ltd.

Figure 22.4: From Centers for Disease Control Public Health Image Library / Courtesy of Lois S. Wiggs.

Figure 22.6: © David Scharf / Science Photo Library.

Figure 22.7: © SCIMAT / Science Photo Library.

Figure 22.10: From Centers for Disease Control Public Health Image Library.

Figure 22.11: © NIAID / CDC / Science Photo Library.

Figure 22.12: From Centers for Disease Control Public Health Image Library.

Figure 22.13: From Centers for Disease Control Public Health Image Library / Dr. Patricia Fields and Dr. Collette Fitzgerald.

Figure 22.14: © A. B. Dowsett / Science Photo Library.

Figure 22.15: © David M. Martin, MD / Science Photo Library.

Figure 22.17: © martin M. Rotker / Science Photo Library.

Figure 22.18: From Centers for Disease

Control Public Health Image Library / Courtesy of Janice Carr.

Figure 22.19: From Centers for Disease Control Public Health Image Library / Courtesy of Dr. Edwin P. Ewing, Jr.

Figure 22.20: From Centers for Disease Control Public Health Image Library / Courtesy of Dr. Mae Melvin.

Figure 22.21: From Centers for Disease Control Public Health Image Library / Courtesy of Dr. Mae Melvin.

Chapter 23

Chapter Opening Image: From Centers for Disease Control Public Health Image Library / Dr. David Cox.

Figure 23.4: Reprinted with permission from *Clinical Bacteriology* by J. Keith Struthers and Roger P. Westran, © 2003 ASM Press.

Figure 23.7: © Dr. Y. Boussougan / CNRI / Science Photo Library.

Figure 23.8: © Dr. M. A. Ansary / Science Photo Library.

Figure 23.9: From Centers for Disease Control Public Health Image Library / Courtesy of Susan Lindsley.

Figure 23.10: From Centers for Disease Control Public Health Image Library / Courtesy of M. Rein, MD.

Figure 23.11: From Centers for Disease Control Public Health Image Library.

Figure 23.12: From Centers for Disease Control Public Health Image Library / Courtesy of Susan Lindsley.

Figure 23.13: © Dr. P. Marazzi / Science Photo Library.

Figure 23.14: © CNRI / Science Photo Library.

Figure 23.16: © Dr. M. A. Ansary / Science Photo Library.

Figrue 23.17: From Centers for Disease Control Public Health Image Library / Courtesy of Joe Millar.

Chapter 24

Chapter Opening Image: © PASIEKA / Science Photo Library.

Figure 24.2: Reprinted with permission from *Clinical Bacteriology* by J. Keith Struthers and Roger P. Westran, © 2003. Manson Publishing, Ltd.

Figure 24.3 Panel B: From Centers for Disease Control Public Health Image Library.

Figrue 24.5: © PASIEKA / Science Photo Library.

Figure 24.6: From Centers for Disease Control Public Health Image Library.

Figure 24.8: © Visuals Unlimited.

Figure 24.10 Panel A: © Visuals Unlimited; Panel B: © Visuals Unlimited.

Figure 24.11: © EM Unit, Veterinary Laboratory Agency, UK / Science Photo Library.

Figure 24.12: From Centers for Disease Control Public Health Image Library / Courtesy of Dr. Leanor Haley.

Figure 24.13: Courtesy of Dr. Andrew Bollen and Dr. Walter Finkbeiner.

Chapter 25

Chapter Opening Image: © Eye of Science / Science Photo Library.

Figure 25.4: From Centers for Disease Control Public Health Image Library / Courtesy of Dr. Edwin P. Ewing, Jr.

Figure 25.7: From Centers for Disease Control Public Health Image Library.

Figure 25.9: From Centers for Disease Control Public Health Image Library / James Gathany.

Figure 25.10: From Centers for Disease Control Public Health Image Library.

Figure 25.11: © Dr. M. A. Ansary / Science Photo Library.

Figure 25.12: © Eye of Science / Science Photo Library.

Figure 25.13: From Centers for Disease Control Public Health Image Library.

Chapter 26

Chapter Opening Image: © Eye of Science / Science Photo Library.

Figure 26.4: © Necrotizing faciitis, Steve Burdette. In www.antimicrobe.org; Empiric, 2008. ESun Technologies, Pittsburgh, PA.

Figure 26.5: © Dr. P. Marazzi / Science Photo Library.

Figure 26.6: © Dr. P. Marazzi / Science Photo Library.

Figure 26.7: © Science Photo Library.

Figure 26.8: © Stevie Grand / Science Photo Library.

Figure 26.9: From Centers for Disease Control Public Health Image Library.

Figure 26.10: From Centers for Disease Control Public Health Image Library / Courtesy of Dr. John Noble, Jr.

Figure 26.11: © Dr. P. Marazzi / Science Photo Library.

Figure 26.12: © Dr. P. Marazzi / Science Photo Library.

Figure 26.13: From Centers for Disease Control Public Health Image Library / Courtesy of Susan Lindsley.

Figure 26.14: From Centers for Disease Control Public Health Image Library.

Figure 26.15: © Science Photo Library.

Figure 26.16: From Centers for Disease Control Public Health Image Library.

Figure 26.17: © Eye of Science / Science Photo Library.

Figure 26.18: From Centers for Disease Control Public Health Image Library.

Figure 26.19: © Western Ophthalmic Hospital / Science Photo Library.

Figure 26.20: From Centers for Disease Control Public Health Image Library / Courtesy of J. Pledger.

Figure 26.21: © Sue Ford / Science Photo Library.

Chapter 27

Chapter Opening Image: © David Parker / Science Photo Library.

Figure 27.2: © Ernst & Young LLP, *Biotechnology Industry Report: Convergence*, 2000.

Figure 27.9: © CC Studio / Science Photo Library.

Figure 27.12: © Sinclair Stammers / Science Photo Library.

Figure 27.15: © David Parker / Science Photo Library.

Figure 27.16: © Paul Rapson / Science Photo Library.

Chapter 28

Chapter Opening Image: From Centers for Disease Control Public Health Image Library / Courtesy of Cynthia Goldsmith.

Figure 28.1 Panel A: © Getty Images.

Figure 28.3: From Centers for Disease Control Public Health Image Library / Courtesy of Dr. La Force.

Figure 28.4: From Centers for Disease Control Public Health Image Library.

Figure 28.5: From Centers for Disease Control Public Health Image Library.

Figure 28.6: Oil Painting by Josse Lieferinxe, 1497 – 1499.

Figure 28.7 Panel A: From Centers for Disease Control Public Health Image Library / Courtesy of James Hicks; Panel B: From Centers for Disease Control Public Health Image Library / Courtesy of Dr. Lyle Conrad; Panel C: From Centers for Disease Control Public Health Image Library / Courtesy of Dr. Stan Foster.

Figure 28.8: From Centers for Disease Control Public Health Image Library.

Figure 28.9: From Centers for Disease Control Public Health Image Library / Courtesy of Emory University and Dr. Sellers.

Figure 28.10: From Centers for Disease Control Public Health Image Library / Courtesy of Cynthia Goldsmith.

Index

Page numbers in **boldface** refer to major discussion of a topic; page numbers followed by F refer to figures, and those followed by T refer to tables.

M